UDOCUS HONDIUS NATUS IN
AGO FLANDRIÆ DICTO WACKENE XVI
ALEND. NOVEMBRIS ANNO CIƆIƆLXIII:
IXIT ANN. XLVII. M. VII. D. XXIX: DENAT
S XIV KAL. MARTII ANNO CIƆIƆCXII.

TOOLEY'S
DICTIONARY
OF
MAPMAKERS

TOOLEY'S
DICTIONARY
OF
MAPMAKERS

Compiled by
RONALD VERE TOOLEY

With a Preface by
HELEN WALLIS

Alan R. Liss, Inc., New York, New York
Meridian Publishing Company, Amsterdam, The Netherlands

Tooley's Dictionary of Mapmakers *is based on an unfinished part-work series, published by Map Collector's Circle between 1965 and 1975, which is now completed, revised and expanded.*

Illustration on the dust jacket is a detail from the map 'Novam Hanc Territorii Francofurtensis Tabulam' by J. and C. Blaeu, first published in 1640.

End papers show the engraved portraits (reduced in size) of Mercator and Hondius from the 1613 edition of Gerardi Mercatoris Atlas Sive Cosmographicae Meditationes de Fabrica Mundi Et Fabricati Figura.

The ship on the binding is from a woodcut by E. Reuwich illustrating Von Breydenbach's Travels to the Holy Land, printed in Mayence in 1486.

Address all Inquiries to the Publishers:

Alan R. Liss, Inc.
150 Fifth Avenue,
New York, New York, 10011

and

Meridian Publishing Company
P.O. Box 4061
Amsterdam, The Netherlands

Library of Congress Cataloging in Publication Data

Tooley, Ronald Vere
Tooley's Dictionary of Mapmakers

History
1. Title
ISBN 0-8451-1701-7   LC 79-1936

Printed in the United States of America

Foreword

The aim of this Dictionary is to give information in the most compact form possible on persons associated with the production of maps from the earliest times to the year 1900.

The information is based on my observation and study of maps over a period of fifty years, and it has been amplified and supplemented by consulting a number of reference works and the catalogues of antiquarian book- and mapsellers. A list of the main works which I have consulted will be found on page ix.

I have tried to be as comprehensive in the compilation of this Dictionary as possible, but the available data on many obscure cartographers, engravers, and publishers are scant and in many cases such persons are known from a single map or work only.

Essentially, I have provided an alphabetical presentation, giving name, dates of birth and death (whenever known), titles of honour (if any), working addresses and changes of addresses, (which in many cases enable the user to establish the approximate date of publication of a map), main output of maps or atlases, with dates and editions wherever known.

In order to present this information in a compact form, I have used many abbreviations. A list of the most commonly used abbreviations will be found on page vii.

When a certain map, atlas or group of maps has a special historical significance (e.g., the first time a certain area has been mapped) I have added this information.

Originally, the first half of the work appeared in parts in *Map Collectors' Circle*, which has now been discontinued. Those portions have been revised and considerably amplified, in some cases more than doubling the number of entries. For the first half of the alphabet the present volume is thus a new edition; the second half is here published for the first time.

Inevitably, in so extensive a subject there will be many omissions, but I hope that the users will nevertheless find the work useful. Any additional information will be appreciated.

My thanks are due to previous writers on the subject, to the staffs of the map rooms of the British Library (formerly The British Museum) and the Royal Geographical Society, and to a number of individuals who have placed their specialised knowledge at my disposal. Among these are Mr. Eran Laor, Mrs. Helena Malinowski, Professor Klaus Stopp, and the late R. A. Skelton. For assembling the autograph signatures illustrating the work my thanks are due to Miss R. Slifer and Mr. Nico Israel. Finally, I wish to thank the copy editor, compositors, and proofreaders for the way they performed a somewhat unusual task.

November 1978
R. V. Tooley

List of Abbreviations

acc.	accurate; account	*E.I. Co.*	East India Company
anc.	ancien(ne)(s); ancient	*Eng.*	England
Amer.	America	*eng.*	engraved engineer
ant.	antique	*emb.*	embouchure (estuary)
Arch.	Archipelago	*fl.*	flourished
Aust.	Australia	*ft.*	foot, feet
B.	British	*Frhr.*	Freiherr (title)
b.	born	*front.*	frontispiece
B.C.	British Columbia	*geogr.*	geography
Beschr.	Beschreibung (description)	*géogr.*	géographe
B.I.	British Isles	*geolog.*	geological
B.M.	British Museum	*Gr.B.*	Great Britain
B.N.	Bibliothèque Nationale (Paris)	*gdn.*	garden(s)
C.	Cape	*gen.*	general
ca.	circa	*Glo(c).*	Gloucester
Cal.	California	*gov.*	government
celest.	celestial	*hist.*	history, historical
Cdr.	Commander (naval)	*hydr.*	hydrography
choro.	chorographical	*ich.*	ichnographia
col.	coloured	*i.e.*	id est (this is)
comm.	commentary(ies) commercial	*ills.*	illustrations
co.	country company	*inf.*	inferior
Ct.	Court	*int(ro)*	introduction
ded.	dedication, dedicated	*I.*	Isle
dept.	département, department	*Is.*	Island(s)
del.	delineated, delineatio	*it.*	itinerary (ies)
desc.	description, described	*jnl.*	journal
dict.	dictionary	*K*	Karte, (map)
Div.	Division	*La.*	Louisiana
E.	East	*L & W.*	Laurie and Whittle
eccl.	ecclesiastical, ecclésiastique	*Lt.*	Lieutenant
ed.	edited, editor, edition(s)	*Lt. de V.*	Lieutenant de Vaisseau = Naval Lieutenant
		Md.	Maryland
		mag.	magazine
		Med(it).	Mediterranean
		mérid.	méridionale
		Middx.	Middlesex

min.	. . . .	miniature
mil.	. . . .	military
mod.	. . . .	modern
miss.	. . . .	missionary
MS(S)	. . . .	manuscript(s)
mt.	. . . .	mount
mts.	. . . .	mountains
N.	. . . .	North
nav.	. . . .	navigation
N.B.	. . . .	New Brunswick
n.d.	. . . .	no date
N.E.	. . . .	New England
Nept.	. . . .	Neptune
NSW	. . . .	New South Wales
N.Z.	. . . .	New Zealand
Pa.	. . . .	Pennsylvania
phil.	. . . .	philosophical
P.O.	. . . .	Post Office
pop.	. . . .	population
poss.	. . . .	possession
prop.	. . . .	proposed
pt.	. . . .	point
pub.	. . . .	published
Q.M.G.	. . . .	Quarter Master General
q.v.	. . . .	quem vide (see under)
r.	. . . .	river
Reg.	. . . .	Regierung (government)
reg.	. . . .	region(s)
Rly.	. . . .	Railway
R.N.	. . . .	Royal Navy
R.P.	. . . .	Révérend Père (Reverend Father)
r.r.	. . . .	rail roads
S.	. . . .	South
sh.	. . . .	sheet(s)
S.D.U.K.	. . . .	Society for the Diffusion of Useful Knowledge
sect.	. . . .	secretary
sept.	. . . .	septentrional(e)
Sh.	. . . .	Shire
S.J.	. . . .	Society of Jesus
Sn.	. . . .	San(to)
soc.	. . . .	society
Staffs.	. . . .	Staffordshire
stat.	. . . .	statistical
tab.	. . . .	tabula(e)
terrest.	. . . .	terrestial
trans.	. . . .	transaction(s)
trav.	. . . .	traveller (-ing)
thro.	. . . .	through
topo.	. . . .	topographical
trig.	. . . .	trigonometrical
univ.	. . . .	universal, university
v. (ol.)	. . . .	volume(s)
V.D.L.	. . . .	Van Diemens Land
voy.	. . . .	voyage(s)
Vte.	. . . .	Vicomte (Viscount)
Vve.	. . . .	veuve (widow)
W.	. . . .	West
Wash.	. . . .	Washington
W.I.	. . . .	West Indies
W.O.	. . . .	War Office

Works Consulted

American Geogr. Society, Catalogue of Maps of Hispanic America. 4 vols. New York, 1932.

Arrigoni, P. & Bertarelli, A. Le Carte Geografice dell'Italia. Milano, 1930.

Bagrow, L., A. Ortelii Catalogus cartographorum. Gotha, 1928.

——, History of Cartography (revised Skelton). London, 1964.

Baltimore Museum of Art, Catalogue of Exhibition. 1952.

Barwick, G. F., Pocket Remembrances. Eyre & Spottiswoode. n. d.

Beazley, C. R., Dawn of Modern Geography. 3 vols. Reprint New York, 1949.

Bénézit, E., Dictionnaire des Peintres, Sculpteurs, Dessinateurs et Graveurs. 8 vols
    Paris, 1948—55.

Berthaut, (Col.), Les Ingénieurs-géographes militaires. 2 vols. Paris, 1902.

Boase, F., Modern English Biography. 6 vols. London, 1965.

Bonacker, W., Kartenmacher aller Länder und Zeiten. Stuttgart, 1966.

British Museum Catalogue of Printed Maps. 15 vols. London, 1967.

Brown, L. A., Story of Maps. Boston, 1949.

Bryant, Dictionary of Painters & Engravers. 5 vols. Washington, 1964.

Canada Maps Dominion Archives. Ottawa, 1912.

—— Map Collection Public Reference Library. Toronto, 1923.

Catalogues Booksellers, Cappelen, Eberstadt, Francis Edwards, N. Israel, Kraus, Maggs,
    Muller, Orion, Rosenthal, Stevens, Son & Stiles, etc.

Cortesão, A., History of Portuguese Cartography. Lisboa, 1969—71.

Cohen, Cartes de la Sibérie. Paris, 1911.

Cumming, W. P., South-East in Early Maps. Princeton, 1958.

Chubb, T., Atlases of Great Britain & Ireland. London, 1927.

Denucé, J., Oud-Nederlandsche Kaartmakers. 2 vols. Reprint Amsterdam, 1964.

Dictionary of National Biography. 22 vols. London, 1908—09.

Engelmann, W., Bibliotheca Geographica. 2 vols. Leipzig, 1857—58.

Fite & Freeman, A Book of Old Maps. Cambridge, 1936.

Fockema Andreae, S. J., Kartografie van Nederland, The Hague, 1947.

Guthorn, P. J., American Maps & Mapmakers of the Revolution. New York, 1966.

Harms, H., Künstler des Kartenbildes. Oldenburg, 1962.

Harrisse, H., Découverte et évolution cartographique de Terre-Neuve. Paris, 1900.

Hoefer, Nouvelle Biographie Générale. 46 vols. Paris, 1855—56.

Hollstein, F. W. H. Dutch & Flemish Engravings & Woodcuts. 19 vols. Amsterdam, 1949—1976.

Humphreys, A. L., Old Decorative Maps, London, 1926.

Hyamson, A. M., Dictionary of Universal Biography. London, 1916.

Imago Mundi. 28 vols. Stockholm & Amsterdam, 1935—78.

Karpinski, L. C., Printed Maps of Michigan. Reprint Amsterdam, 1977.

Koeman, C., Atlantes Neerlandici. Vols. 1—V. Amsterdam, 1967—70.

Lang, A. W., Seekarten d. Südlichen Nord- u. Ostsee. Hamburg, 1968.

Leithauser, J. G., Mappae Mundi. Berlin, 1958.

Lemosof, P., Livre d'Or de la Géographie. Paris, 1902.

Library of Congress. List of Geographical Atlases. 4 vols. Washington, 1909–20.
    Continuations. Vols. V, VI, VII, and VIII.

Lister, R., How to Identify Old Maps and Globes. London, 1965.

Luany, R. M., Early Maps of North America.

Lynam, E., Mapmakers' Art. London, 1953.

Map Collectors' Circle. 11 vols. London, 1963–75.

Marshall, D. W., Catalog of Maps of America in the W. I. Clements Library. 4 vols.
    Boston, 1972.

Matsutaro, N., Old Maps in Japan. Osaka, 1973.

Michigan University Maps, William L. Clements Library. 2 vols. Boston, 1972.

Nagler, G. K., Neues Allgemeines Künstler Lexikon. 22 vols. Munich, 1835–52.

Nederlands Historisch Scheepvaart Museum Catalogue. 2 vols. Amsterdam, 1960.

Nordenskiöld, A. Facsimile Atlas. Stockholm, 1889.

——, Periplus. Stockholm, 1897.

Oehme, R., Geschichte d. Kartographie des Deutschen Südwestens. Konstanz, 1961.

Phelps Stokes, I. N. & Haskell, D. C. American Historical Prints. New York, 1932.

Phillips, P. L., Maps of America in the Library of Congress. Washington, 1901.

Raisz, E., General Cartography. New York, 1948.

Robinson, A. H. W., Marine Cartography in Britain. Leicester, 1962.

Royal Geographical Society, Card Index.

Société Royale de Géographie d'Anvers. Exposition Anvers, 1876–1926.

Taylor, E. G. R. Tudor Geography. London, 1930.

——, Late Tudor and Early Stuart Geography. London, 1934.

Thieme and Becker, Allgemeines Lexikon der bildenden Künstler. 37 vols. Leipzig, 1908–47.

Thomas, B. L., Biographical List of Cartographers, Engravers and Publishers. Lawrence, 1961.

Tiele, P. A., Nederlandsche Bibliographie van Land- en Volkenkunde. Amsterdam, 1884.

Tooley, R. V., Maps and Mapmakers. London, 1968.

Wagner, H. R., Cartography of NW Coast America. Berkeley, 1937.

Waller, F. G., Biogr. Woordenboek van Noord-Nederlandsche Graveurs. The Hague, 1938.

Wheat, C. I., Mapping of Transmississipi West. San Francisco, 1957.

Wieder, F. C., Monumenta Cartographica. 5 vols. The Hague, 1925–33.

Winsor, J., Narrative & Critical History of America. 8 vols. New York, 1889.

Wurzbach, A. von., Niederländisches Künstler Lexikon. 3 vols. Amsterdam, 1963.

Preface

The Dutch cartographer Abraham Ortelius was the first to compile and publish a list of mapmakers. In his *Theatrum Orbis Terrarum* (1570) he included the names of some 90 cartographers who had sent him maps for his atlas. This "Catalogus auctorum tabularum geographicarum" provides valuable evidence on the leading mapmakers of the day, some of whom would otherwise have remained unknown. Ortelius also cited the names of geographers and cartographers ancient and modern, whose work he had used as a source for his own maps, from Ptolemy, Pliny and Strabo onwards. By 1603 the *Theatrum's* "Catalogue" had been enlarged to 183 names. The list of 1570 formed the basis of Leo Bagrow's *Catalogus Cartographorum* (1928), comprising a modern study of what may be called the "Ortelian cartographers."

The Minorite Friar Vincenzo Coronelli, Cosmographer of Venice, with typical encyclopaedic zeal, was the next to turn his hand to the compilation of this kind of record. In his *Cronologia Universale* (1707), designed as an introduction to his massive *Biblioteca Universale,* pp. 522–24, he provided a "Cronologia de'geografi antichi, e moderni," from Homer to Ponza, naming 96 geographers and mapmakers and supplying biographical notes for 69 of them. Scholars active in the early years of the 18th century were now pursuing a lively antiquarian interest in the historic past. Johann Gottfried Gregorii was the next to pioneer in this field. In his *Curieuse Gedancken von den vornehmsten und accuratesten Alt- und Neuen Land-Charten* (1713) he cast his eye back to very remote times. In Chapter VIII "Von den Vornehmsten Geographis," he named Moses as the first geographer (p. 120, Moses Omnium Geographorum & Historicorum Facile Princeps), and proceeded to list in chronological order the leading geographers up to his own time. In the next chapters he turned to mapmakers, starting with Johann Baptista Homann (p. 511), his contemporary and fellow-countryman.

The growing interest in the history of geography in recent years has added a variety of catalogues, bio-bibliographies and carto-bibliographies as source materials for a general directory of cartographers. Regional directories have been compiled, for example, O. Regele's for Austrian military cartographers, R. Oehme's for south-west Germany, B. Olszewicz for Polish cartographers. Author entries are included in catalogues of major collections, such as the *British Museum Catalogue of Printed Maps* (1967), and Philip Lee Phillips' invaluable work *A List of Geographical Atlases in the Library of Congress* (1909–20) with its several volumes arranged under authors and its *Author List Index.* Of the histories of cartography which provide lists of cartographers the most notable is Leo Bagrow's *Die Geschichte der Kartographie* (1951). Its "Verzeichnis der Kartographen" names 1210 persons. The revised and enlarged *History of Cartography* edited by R. A. Skelton, published in 1964, claims to have 1500 names in its "List of Cartographers (to 1750)." More than 2000 names (providing a fuller record of British mapmakers) appear in Raymond Lister's *How to identify old maps and globes, with a list of cartographers, engravers, publishers and printers concerned with printed maps and globes from c. 1500 to c. 1850.* (1965).

In the mid-sixties two other major projects quite independently reached the stage of publication. Wilhelm Bonacker's *Kartenmacher aller Länder und Zeiten* appeared in 1966. Prepared as a sequel to the author's list of globemakers, this comprised the first published directory of the most important mapmakers of all tongues and periods, and is particularly strong for German cartography. Unknown to Dr. Bonacker, R. V. Tooley had also been preparing over the years a dictionary of mapmakers. Part 1, A to Callan, appeared as no. 16 in the *Map Collectors' Series* in 1965, and was in time to be noted in Bonacker's Introduction. When the *Map Collectors' Circle* discontinued publication in 1974, ten parts of Tooley's *Dictionary* (covering A to P) had been printed. These parts are now enlarged and revised for the complete *Dictionary of Mapmakers.*

The *Dictionary* thus ranks as the first comprehensive work of this kind in English. It contains some 21,450 entries recording every person associated with the production of maps from the earliest times to the year 1900.

For over 50 years Ronald Vere Tooley has been one of the leading 'map-men' in London, developing a specialized knowledge from the constant handling of maps and from many days of research in libraries. During this time he has been amassing the information for this Dictionary. Philosophers and cosmographers, geographers and astronomers, explorers and surveyors, engravers and lithographers are among the diverse persons featuring in the list, and some had other callings, trades or professions, as princes, governors of provinces, and glove-makers to name a few of them. What all have in common is the fact that in one way or another they were associated with map-making. Only the later ones would or could call themselves cartographers — if the title was appropriate — for the word 'cartography' ('cartografia') was not coined until 1839. Some names such as Mercator's and Ptolemy's are household words, others emerge from obscurity to be recorded here. The emphasis in the entries is on the work or works produced.

Finally, there is of course no Moses or Joshua on this list. Antiquity nevertheless has a place in the record. A collective term indicates the activities of astronomers and mapmakers in ancient Babylon. This is a reminder that in all ages many practitioners of the art and craft of mapmaking have remained hidden from the pages of history.

Helen Wallis,
February, 1979

A

A., A. *Rep. Argentina* 1868

A., G. L. = George Lilius Anglus. See Lily, G.

A., J. *Plan Montreal* 1756

A., J. L. de. *Atlas Geogr . . . de Espana* 1848

A., M. J. See Armstrong, Mostyn, John

A., See Aelst, Nicholas van

A., [T] , *Carolina,* 1682

Aa, Baudouin, Printer. Worked for his brother Pieter (below)

Aa, Cornelis van der (1749—1816) Bookseller at Haarlem and Utrecht. *Atlas van de Zeehavens der Batav. Rep.* 1805

Aa, Hildebrand. Engraver. Brother of following.

Aa, Pieter van der (1659—1733) of Leiden, publisher, editor and bookseller, *Atlas Nouveau* 1700, *East and West Indies* 28 vols. in 12, 1707, *Galérie Agréable du Monde* 66 vols. 1729, *Forces de l'Europe &c.* 20 pts. in 2 vols. 1726

Aägaard, Ulrik Frederick (1694—1727) Danish cartographer, painter of Stavanger

Aanum, Ole M. *Reykjavik* 1801

Abadie, Philibert. *Atlas Historique* 1846

Abancourt, Charles Frerot d' (b. Paris, d. Munich 1801) *Map of Switzerland*

Abbadie, Antoine-Thomson d'(1810—1897) *Géodesie d'Ethiopie* 1890

Abbela y Casanego. Spanish geologist. *Phillippines* 1886 *I. de Panay* 1890

Abbéville, d'. See Sanson d'Abbéville; also see Du Val, Pierre

Abbatt, Richard. *Orthographic Projection of World* 1857, *World Antipodes* 1857

Abbe, Cleveland (1838—1916) Mathematician, astronomer, globemaker of New York

Abbot, K. E. (d. 1873) *Account of Persia* 1849—50

Abbot, Charles, Lord Colchester. *Route Brit. Embassy along Yang-tse-kiang Admiralty Chart* 1841

Abbott, E. *Dallas Co. Iowa, N.Y.* (1874)

Abbott, Capt. F. *Kabul & Environs* 1842, *War Office* 1879

Abbot, Lieut. (later Capt.) Henry L. American cartographer. *Routes proposed for Pacific Rail Road* 1855, *Basins Mississippi & Trib.* 1861, *Army Potomac* 1864

Abbott, Lieut. (later Capt.) J. *District Bareilly* 1833—7, pub. 1858. *Shajehanpor* 1838—9, pub. 1866.

Abbott, Lieut. S.A. Revenue Surveyor. *Bandah* 1840—1, pub. 1865 *Cawnpore* 1845, *Benares* 1846

Abbott, Moses. *Plan Town Tewkesbury* 1794 MS

Abbott, William. English cartographer. *Submarine cables of World* 1872–76. *Canada* 1884

Abbs, Lieut. R. N. *British Honduras.* London, Weller 1867 (after J.H. Faber E.L. Rhys)

Abd-er-Rahman. Arabian globemaker, 14th century

Abdu-r-Rahmani. Prof. *General Atlas* in Turkish, Cairo, 1804

Abe, Yoshito. *Japanese World map*(1840)

Abeken, Heinrich A. *Sinai-Halbinsel* 1859

Abel, Gottlieb, Friedrich. Royal engraver court Stuttgart. *Swabia for Bohnenberger,* for Jaeger's *Atlas d'Allemagne* 1789 and 1796. *Plan Stuttgart* 1811

Abel, Caspar (1676–1763) Theologian and historian of Hindenburg. *Preuss. Staats Geographie* 1735

Abela, Giovanni Franceso. *Descrittione di Malta,* map 1647·

Abelin, Hans. Engraver. *Kempten* 1569

Abelin, Johann Philipp [pseud. for Gottfried, J.L.] b. Strassburg. *Newe Archontologia Cosmica* 1638. *Topo. Helvetiae* 1642, *New Welt,* Merian 1655

Abendroth, E. *Elb-Mündungen* 1846

Abent, Leonard. *Plan of Passau* for Braun & Hogenberg's *Civitates* 1581

Abercrombie, James. *Country round Ticonderoga (New York)* 1758 MS

Abercrombie, Capt. Thos. *Plans of Forts & Towns USA* 1756–9 MSS

Abercromby, Sir Ralph (1734–1801). *Trinidad* 1797

Aberdeen, A. *Orkney* (1770) with William Aberdeen

Aberdeen, William. *Orkney* with A. Aberdeen (above). *Lerwick*1776

Abergh, Oscar. *Sverige och Norge* (1871)

Aberl, L. *Island of St. John* 1806

Abert, Col. John James (fl. 1824–1851) American cartographer. Chief Topo. Bureau 1851. *E. Florida* 1837, *St. Mary's River, Md.* 1824, *Mississippi River* 1843, *Texas* 1844, *Arkansas* 1847

Abert, Lieut. J.W. *Territory of New Mexico, Washington* 1845–7 (with Lt. Peck)

Abernethie. Engraver & Surveyor. *Charleston Harbour* 1785. *Project for Map of the Roads of State of S. Carolina* 1787

Abernethy, George. Engineer. *Outer Bay, Port Adelaide* 1858

Abich, Otto Wilhelm Hermann (1806–1886). Mineralogist and geologist of Berlin. *Atlases of Volcanoes from* 1839. *Caucasian Atlas* (1882), *Geologische Fragmente* 1887

Abilgaard, Søren (1718–91) Drew maps for Pontopiddans *Danske Atlas*

Abington, William. Bookseller at the Three Silk Worms in Ludgate Street . *Blome's Cosmography* 1683

Ablancourt, N.F. d'. See Frémont d'Ablaincourt

Abollado, Don Yldefonso de Aragon y. *Map Philippines* 1821

Aboulfeda. See Abulfeda

Abou Obaid al Bekri (1040–1094) Arab geographer and historian. *Description of World*

Abou Rihan. See **Albyrouny**

Abrahams. *Eiland Walcheren* 1865

Abreu Gorgan Joao. *Brazil* 1747

Abrial. *Dépt. du Tarn* 1873

Abrille, F. V. See Villa Abrille

Abu Bakr Ibn Bahram. Syrian geographer of Damascus, 17th century

Abu Jafar of Khiva. *Map World* ca. 800

Abulfeda [Aboulfeda] Edmadeddin
Ismael (1273–1331) of Damascus.
Chronicle and a *Geography* pub. in
Europe 1754, in Paris 1840

Abul, Hassan. Moroccan astronomer
in Cairo 13th century

Abu Rihan. See Albyrouny

Aburrow, Charles. *Diamond Fields
Griqualand W.* 1886

Acacius. See Hoheneck

Academia Cosmografica Degli Argonauti,
Venice. First Geographical Society.
Founded by V.M. Coronelli in 17th
century.

Acad. Reg. Scient. el Eleg. Litt. Boruss
ca. 1755 (Prussian Royal Academy)
Atlas, Berlin 1760

Accerlebout, J. *Hatfield Chase* (1662)

Accinelli. *Contorni di Genova* 1746

Acerbi, Joseph (1773–1846) *Collected
geographical documents on Scandinavia
and Egypt* 1826–36

Achephol. Dortmund 1872

Achin. *Plan of Paris* 1825, *Environs
of Paris* 1829

Ackerman. *Orange & Alexandria
RR (USA)* 1851

Ackerman[n] Lithographer of New
York City, 379 Broadway and 120
Fulton Street. *Colton's U.S.* 1849,
Surveys W. Utah (1852)

Ackerman, G. *Konstanz u. Umgebung*
(1891)

Ackermann, Rudolph & Co. London
Publishers. 101 Strand, and 96 Strand
and 191 Regent Street. *Poirson's
Mexico* 1826, *General Atlas* 1838

Ackers. *Gold Reefs N. Queensland* 1887

Acosta, F. de F. *Venezuela* 1870

Acosta, Joaquin (1799–1852) Col. of
Artillery. American historian and
geographer. *Nueva Granada* 1847

Acosta, José (1539–1600) Spanish
Jesuit missionary, & historian, and
cosmographer. *Hist. Natural de las
Indias* 1590 (including map of New
World) German edns. 1598–1605
(30 maps)

Acosta, Juan Luis de, Capitan y Pilato
mayor. *Rio de Cagayan* (1720) MS

Acqua, Giovanni Domenico. *Icon.
Vicenza* 1711

Acrelius. See **Akrel,** Fridrick

Acton, Oliver. *Plan Bridewell Estate
Thames* 1737

Acuna [Acugnal] *Christ. d'* (1597–1655)
Born Burgos. *Amazon*

Acuna, *P.P. d' Carte de la Terre Firme*
2 sh. Paris 1703

Adair, James. *American Indian
Nations* (1775)

Adair, John (ca. 1655–1719) Scottish
surveyor, F.R.S. 1688. MS *Maps of
Scot. Shires* 1682–88. *Hydro.
Descript. Sea Coast Scotland* 1688–1703,
Strathern 1690, *Montrose* 1693,
posthumously, *Clyde* 1731, *Lothians*
3 sh. 1745

Adair, Robert Alex. Shafto. *Military
sketch map England & Wales,* 1853

Adam. Compiled maps (with Giraldon)
for Lapie's *Atlas Classique* 1812,
Océanie 1816

Adam, Alexandre. *Plan de Boulogne sur
Mer* 1863

Adam, H. *Würzburg* (1860)

Adam, J. Engraver. *Ungarn* (1873)

Adam of Bremen (fl. 1069–1076)
Historian and geographer. First writer
to mention Norse discoveries in
America (Vinland). In his *History* he
devoted a section to geography,
particularly of Northern Latitudes

Adami, Karl Christian Ludwig (1802–
1874) Berlin cartographer and globe
maker. *Globe* 1852, *Atlas* 1855,
School Atlas 1868

Adams. *World Atlas,* Vienna (1800)

Adams, Clement (1519–1587) Schoolmaster and author. Cited by Hakluyt as engraver of *Cabot's map* 1549 English edition

Adams, Cyrus Cornelius (1849–1928) Geographer, born Napierville, Illinois

Adams, Daniel (1773–1864) *Atlas,* Boston 1814 and 1823

Adams, Dudley (1797–1825) Globe and mathematical instrument maker to George III, 60 Fleet Street, London. *Globe* 1797

Adams, D. P. *Chesapeake Bay* 1830

Adams, George Elder (ca. 1704–1773) Instrument maker. *Celestial Globe* 1769 *Terrest. Globe* 1760, and 1769

Adams, George Younger (1750–1795) Instrument maker to King. Son of George above. Fleet Street, London. *Terrest. Globe* 1782, 1790

Adams, Henry Martin. *Island of Antigua* 1891

Adams, John (fl. 1670–1796). Topographer and barrister. Attempted *Trigonometrical survey England and Wales* 1681–4. *Index Villaris* 1680, 1690 and 1710

Adams, John. *Map of the Roads & Inland Naviagation of Pennsylvania* (with John Wallis) 1791

Adams, J. *Present Seat of War in Turkey* 1774

Adams, J. Surveyor. *Estate plan Tenterden Kent* 1806 *Hocklebury Est. Kent* 1809. *Dover & Environs* (1828)

Adams, John. Translator of *Michelot's Mediterranean Pilot* 1795

Adams, John. Teacher of mathematics in Edmonton. *School Atlases* 1805–1843

Adams, John. Military surveyor. *Quebec* 1822, pub. 1826

Adams, John Junior. Land surveyor and artist. *Finchcocks Estate [Kent & Sussex]* 1829

Adams, John Couch (1819–1892) President Royal Astronomical Society

Adams, M. *Plan Hounslow* 1836

Adams, Robert (ca. 1540–1595) Surveyor of The Queen's Buildings. Several MS maps extant. Compiled charts for Ubaldini's *Exped. Hispaniorum in Angliam vera descriptio* 1588. *Flushing* 1585, *Portsmouth* 1585, *Thames* 1588, *Poole Harbour* 1590 *Plymouth* 1593

Adams, Robert. Surveyor. *Chart Southampton Water* (1810)

Adams, William. *Deluge of the Fens,* May 1862

Adams, Wm. James [& Sons] Publisher, 59 Fleet Street. *Guide to Environs of London* ca. 1850. *Railway map London* 1879

Adams (with Asher) American publishers *Township map Indiana, State of Iowa* (1868), *Atlas & Gazetteer N.Y.* 1871, *Gazetteer U.S.* 1873

Adams-Reilly. A. *Surveys in the Alps* 1863–1868

Adamy, Heinrich. *Umgegend von Breslau* (1865), *Schlesien* 1867, *Glatz* 1869

Adan, P.J. & H. *Bergen op Zoom.* Large scale map, mid 18th century

Adanson, Jean Baptiste (1732–1804) *Maps Egypt* MSS (1763–84)

Adcock, Benjamin. Surveyor. *Little Pollicott Farm* 1837 MS

Addison, John (fl. 1825–1845) Globemaker. Drew & engraved *Greenwood's Kent* 1839, *Maltby's Celestial Globe* 1845

Addison, Thomas of Rufford, Lancashire. Surveyor *Speke Estate* 1781

Adelbauer, Johann Ernst. Printer, of Nürnberg. *Kohler's Schul-Atlas* 1719, *Homann's Grosser Atlas* 1731

Adelhauser, Hans. *Plans of Gotha & Grodno* 1568. Used by Braun & Hogenburg

Ademos. *Dépt. de la Creuse* 1867

Aderholz, George Philip. *Russische Provincien Lithauen, Wolhynien, Podolien &c.* 1831

Adger [Aedgerus] Cornelis (fl. 1580–1583) Dutch mathematician and geographer. *Map of Cologne* 1583

Adie, A. J. Contributed to *Johnston's plan of Edinburgh* 1850

Adies. *Borneo coast* 1789

Adkins, I.F. *Jefferson Co., Georgia* 1879

Adkinson (D.) Estate surveyor, mid 18th century

Adlard, Alfred. Engraver. *Vignette Modern Atlas* 1849–1850 *10 miles round Ludlow* (1850)

Adlard, J. Printer. Duke St. West Smithfield *Wilkes Co. Atlas* 1810

Adlard, J. & C. Printers. Bartholomew Close. *Harding's map London* (1834)

Adlard & Browne. Printers, Fleet Street, London. *Russell & Price, England Displayed* 1769

Adlard & Jones [Adlard & Co.] Publishers. Ave Maria Lane. *Maps in Wilkes Co. Atlas* 1811

Adler, C. *Handatlas* 1870. *Schul Atlas* (1889)

Adlum, J. & J. Wallis. *Roads & Inland Navig. Pennsylvania* ca. 1795

Admiralty Charts. Hydrographic Office founded 1795

Adnitt & Naunton. Lithographers. *Tisdale's Plan Shrewsbury* 1875

Adrichem [Adrichom] Christian van (1533–1585) b. Delft, d. Cologne. Theologian & surveyor. *Jerusalem,* Cologne 1584, reprinted in Dutch, English, French, Spanish, Italian and German. *Theatrum Terrae Sanctae* (12 maps) Cologne, Birkmann 1590

Adrichomius. See **Adrichem**

Adshead, Joseph. *24 Illust. maps of Manchester,* 23 sh. & Index 1851

Aedgerus, C. See Adger, C.

Aedone, A. d'. *Sicily* (1757)

Aefferden [Afferden, Alferden] Don Francisco de (1653–1709) Belgian cartographer. *El Atlas Abbreviado* Amberes (Antwerp) Juan Duren 1696. Other editions 1697, 1709, 1711 and 1721

Aegidius Bulionius Belga. See Boileau de Bouillon

Aegido, M. *Limburg* 1603

Aelst, Guilliam van. *Leo Belgicus* (1645)

Aelst [Aalst] Nicolas van (ca. 1527–1613) Flemish printseller and engraver resident in Rome, 5 Maire della Pace. Acquired some of Lafreri & Salamanca's plates. *Pozzuoli* from 1550, *Abruzzo* 1587, *Burgundy* 1596, *Rome* 1590, *Catania* 1590

Aeneas Sylvius Picolomini [later Pope Pius II] . Author of a geographical treatise

Aerlebout, [Acerlebout, Arelebout] Josias. Surveyor. *Maps of Hatfield Chase* (1639). Used *Dugdale* in *History Embarking* 1662

Aertsen. See Arsenius family

Aeschler, Jacobus. Editor *Strassburg Ptolemy* 1513

Aeuer, H. German globemaker at Eberfeld, mid 19th century

Afferden. See Aefferden

Africanus, Giovanni Leo. See Leo Africanus

Agas, John. Estate Surveyor. *Gt. Coggeshall* 1619 MS

Agas [Aggas] Ranulph or Ralph (ca. 1540–1621) English land surveyor of Stoke by Nayland in Suffolk. Various *estate plans* in *Essex & Suffolk*

1575–93. *Birds eye plan Oxford* 1578, *Dunwich* 1587, *Cambridge* 1592 Incorrectly credited with *London* ca. 1560

Agassiz, A.R. *Map of Tong-Kin & c.* 5 sh. (1890)

Agassiz, Louis J. R. (1807–1873) Studied hydrography of S. Atlantic and Pacific

Agathadaimon (ca. 250) Greek of Alexandria. Geographer and mechanic. Said to have constructed the first world map for Ptolemy's Geography

Agatharcides (fl. 148–105 B.C.) Greek historian and geographer. Tutor to Ptolemy Soter. *Description of Erythraean Sea*

Agathemere Greek geographer, (fl. 150 A.D.) made an abridgement of *Ptolemy* pub. Amsterdam 1671

Agde, Jean Cavalier d'. See **Cavalier**

Ageminius, P. See **Paulus Ageminius**

Aggas. See **Agas**, Ralph

Aggere, P. See **Petrus ab Aggere**

Aginelli, Nicolo. *Plan of Raab* 1566. Used by Braun & Hogenberg

Aglio, Pietro Martire. *River Po* 1758

Agnelli, Frederico. Engraver for *Sesti's Pianta della Citta, Milan* 1707

Agnese, Battista (1514–1564) Genoese cartographer, worked in Venice. Portolan maker. Over 70 *manuscript atlases,* some with 10 or more maps

Agnew, Daniel. *Fort McIntosh, Pennsylvania* 1896

Agnew, John Vans. *Corea* 1894

Agniet, J. See **Lagniet**

Agnis, Benjamin. *Althorne* 1728 MS, *W. Mersea* 1756 MS.

Agnis, John. *Ardleigh* 1773 MS.

Agnolo, Jacopo d'. See **Angelus Jacopo d'**

Agostini, Tomaso. *Pianta di Firenze* [1877?]

Agostino, Gio. Bat. See **Codazzi**

Agote, Manuel de. *Plan-Macao Admiralty Chart* 1805

Agrigonio, Fra Bono. Venetian cartographer, 16th century

Agrippa (30 B.C. – A.D. 12) Roman General. *Map of the World under Augustus.* Possible source of Peutinger Table

Agueros, Fr. Pedro González. *Chile* 1787 MS

Aguirre, Domingo de. *Topo. de del Real Sitio de Aranjuez* 1775, 16 sh. & 7 views

Agutter, Benjamin. *Borough of St. Albans* 1634

Aher, David. Irish surveyor *Queens Co. and Tipperary. Maps of Bogs* 1811–1812, pub. 1814

Ahmad al Tusi Persian Cosmographer, 12th century. Composed *Atlas of Islam,* ca. 1160

Ahmed, H. See **Hhaggy Ahmed**

Ahmer, V. Engraver. *Eisenbahn-Karte-N. Amerikas* [1865?]

Ahrenfeld, Alfred. *Southampton Row to Strand, Streets of London* 1893

Ahrens, H. *Oesterreich-Ungarn* [1874]

Aigenler, Adam (1635–1673) of Ingoldstadt. Mathematician and cosmographer

Aigner, Hans von. *Plans Breslau* 1857 & 1866

Aikin, Arthur (1773–1854) Chemist, pioneer of Geological Society

Aikin, James. Engraver. *Crisp's Plan of Charles Town* 1794 pub. 1809. *Santa Canal* 1802

Aikin, John (1747–1822) Man of Letters. *England Delineated* (with series of county maps) 1790 and later editions

Aikman, George. Engraver. *Plan Newcastle upon Tyne* 1833. Worked for A. & C. Black *Canals & Railways U.S.* 1838

Ailly, Pierre d' [Petrus de Alliaco y Auliacus] (1350–1420[2] French cardinal, philosopher and teacher. Wrote about 1410 *Ymago Mundi with Schematic World map,* first printed in Louvain 1483

Ainslie, Lieut. C. *Chin Hills* Rangoon 1892

Ainslie, James. Publisher, St. Andrews St. Newtown Edinburgh *John Ainslie's Scotland* 1792

Ainslie, J. Ordnance Engraver, Survey Office Dublin. *Waterford* 1842

Ainslie, John (1745–1828) Scottish surveyor and engraver. St. Andrews St. Newtown, Edinburgh. *Plan Jedburgh 1770 Selkirk* 1772 & 1801, *Kinross* 1775, *Edinburgh* 1780, *Wigton* 1782, *Forfar* 1794, *Kirkcudbright* 1797, *Roads Gt. Britain* 1797, *Renfrew* 1800

Ainslie, Mrs. Publisher. *Forfar* 1794, *Kirkcudbright* 1797, *Berwick* 1797, *Renfrew* 1800

Ainslie, J. & E. Donald. Published large scale county maps. *Cumberland* 1760, *Beds.* 1765, *Westmoreland* 1768–70, *Bucks.* 1770, and *Yorks* 1771–2

Ainsworth. *Country between Mombassa & Machakos* 1892

Ainsworth, A.B. *Allotments Parish of Caramut* 1858

Aird, Lieut. D., R.N. Hydrographer. *E. Australia* 1845–7

Airey, John (fl. 1867–1900) *Railway maps of England and Wales. District maps London* 1869, *Manchester & District* 1869, *Scotland* 1875, *S. Wales* 1880, *Lancs. & Dist.* 1884, *England* 1887, *S. Wales* 1893

Airy, Sir George Biddell (1801–1892) English astronomer and geographer. Director of Greenwich Observatory

Aislabie, G. Surveyor mid 18th century

Aitken, Alexander. d. 1800. Canadian mathematician and land surveyor. *York Harbour* 1793

Aitken, Robert (1734–1802) Bookseller, printer and engraver of Philadelphia. Born Dalkeith in Scotland, moved Phila. 1769 opposite the London Coffee House Front Street Phila. *Maps for Penny Magazine* 1775, *Nova Scotia* 1776, *Virginia* 1776

Aitken, William. *Lothians from Adair's observations by J. Elphinstone* 1744

Aitoff, D. *Plan de Paris* 1886

Aiton, Wm. *Ayrshire* 1811, *Bute* 1816

Aitzing [Aitzinger, Eyzinger, Eitzinger, Eitzing], Michael Freiherr von. (ca. 1530–1598) Historian and cartographer. Born Obereitzing upper Austria, invented the *Leo Belgicus. De Leone Belgico* 1583, *Terra Promiss.* 1581

Ajello, Michele & Doyen. *Liguria Marittima* 1834

Akademia Nauk St. Petersburg [Académie des Sciences] founded 1726 by J. N. De Lisle. *Atlas Russicus Petropoli* 1745, *Plan St. Petersburg* 1753, *Russian Discoveries* 1775

Akamidzu [Akamatsu] de Mito. Japanese cartographer 18th century. *Hist. Atlas of China* 1789, *World* 1796

Akbar Ali. *Map of Punjab,* 1882

Akeleye, Jens Werner. *Chart Kattegat* 1770

Akelund, Erik (1754–1832) Swedish engraver. *Gottland* 1805, *Abo Finland* 1808, *Reskarta ofver Svea och Gotha Riken, Stockholm* 1828, *Rinsed* 1836

Akerman, Andrew (1721–1778) Glovemaker, cartographer and engraver of Uppsala. *Globes* 1759, 1766, 1780, 1792. *Sinus Finnicus* 1768, *Atlas Hydrografica* 1768

Akersloot, W.O. (ca. 1600–1640) *Haarlem maps & plans* 1628

Akhmatov, Ivan. Russian cartographer. *Russia* 1845, *Hist. atlas of Russian Empire* 1831

Akkeringa, A. *Kaart van Europa* (1874)

Akkeringa, J. E. *Map Dutch E. Indies* 1872–3

Akrel, Carl Fredrik (1779–1868) Cartographer of Stockholm. Son of following. *Plan Stockholm* 1805, *Karta ofver Sverige* 1811, *Res Karta ofver Sverige* 1857

Akrel [Akrelius] Fridrick (1748–1804) Swedish engraver. *Plan Stockholm* 1771; *Sweden* 1778, *Sveriges Sjö Atlas* 1795, *Nya Canal* 1800

Akron Map & Atlas Co. *Summit County, Ohio* 1891

Alabern, Pablo. Engraver of Barcelona. *North America* 1830 For *Diccionario geog. univ.* 1830–4

Alabern, Ramon. Engraver of Barcelona. *Africa* 1845, *Océania* 1845, *Atlas Univ.* 1857, *Deposito Hydrografico* 1865

Alagna, J. Giacomo. Hydrographer of Messina. *Charts Portugal & Med.* (1760) and (1767)

Alard. See Allard

Albach, Julius. Publisher. *Maps Austria* 1875–79

Al Baladhuri. See Albeladory

Alban, E. *Mecklenburg-Schwerin* (1891)

Albateny [Al Bateni or Batani] Arabian Astronomer 9th–10th century. Born Batan in Chaldea

Albe, E. E. F. d'. see Fournier D'Albe

Albe. See Bacler d'Albe

Albeca, A. L. d' *Dahomey*, Paris 1892

Albeladory. Historian and geographer 9th century. *Book of the Conquests* ca. 880

Albernas, João Teixeira. See Teixeira

Alberouny. See Albyrouny

Albert I, Prince of Monaco. b. 1848, patron of geography and exploration

Albert, J. J., Col. U.S. (fl. 1824–27) *Texas by W.H. Emory,* U.S. topog. engineer, pub. Washington 1844

Albert, R. *Der Harz Strassen Karte* [1898]

Albert, Th. *Rügen* (1874)

Alberti. *Carta topo. Ducato Piacenzo* 1816

Alberti, A. *Map Saginaw Valley Michigan* 1860

Alberti, F. Leandro. *Bressan, Padouan, Verona* 1701; *Mantoue, Modena* 1702

Alberti, Ignaz d. 1801. Austrian engraver for Reilly, Schraembl

Alberti, Lod. Landrost at Cape. *Alagoa Bay* 1810

Albertin, Johann Heinrich (1713–1790) born Zürich. Cartographer

Albertin de Virga. See Virga

Albertis, L. M. d' *Voy. Beccari & Albertis to West New Guinea* 1873, *Fly River, New Guinea,* Sydney 1876

Alberts, Rutger Christopher. Publisher of The Hague. *Plan Turin* 1706, *Geneva ancient & mod.* The Hague 1725, *Blaeu's Savoy & Piedmont* 1726, *Cadix* 1727

Albertsen, D. Engraver. *Friesland* 1618

Albertus, Gaspar [or Caspar] Publisher. Successor to Palumbi in Rome. *Fortress St. Elmo, Malta* 1566

Albertus, Leander (1479–1553) Dominican of Bologna, *Islands of Italy* 1567, *Corsica* 1567

Albi World map (8th century). The oldest known geographical monument of Western Europe [Cotton MS]

Albin, John. *Isle of Wight, published Newport by L. Albin* 1795, 1802, 1805, 1807, 1823

Albini, Rear Admiral Giuseppe. *Portolano della Sardegna* 1842, *Liguria* 1854

Albino de Canepa. See Canepa

Al Biruni. See Albyrouny

Albizio, Antonio. *Princip. Christ. Stemmata,* Augsburg 1610

Albrecht. *Gen. Karte Ungarn* 17 sh., Vienna 1858

Albrecht, David. Bookseller and publisher of Wroclaw. Helwig's *Silesia* 1605

Albrecht, Ignaz. Swiss engraver *Switzerland pub. Reilly* 1789, *World* 1795

Albrizzi. Publisher. *Stato presente de tutti paese* 1742

Albrizzi, Girolamo. Venetian cartographer and printer (Fl. late 17th century)

Albu, Siegfried. *Flag Maps* (1870)

Albyrouny [Al Birani or Abu Rihan] (973–1048) Arab mathematician, astronomer and geographer. *Treatise on mathematical & astronomical geography*

Alcedo y Herrera, Colonel Antonio (1735–1812). Spanish geographer. *Dict. Geografico Hist.* 1786, English edition 1812–15

Alcuin of York (ca. 735–804)

Alden, E. C. *Plan Oxford* 1888

Alden, Lieut. James, U.S.N. (fl. 1837–1860) *Charts coasts America*

Alden, John Berry (1847–1924) *Home Atlas* 1887

Alden, Ogle & Co. *Plat Books Counties Illinois* 1890–1

Aldenhoven, Ferdinand. *La Grèce* 8 sh., Athens 1838

Aldersey, William. Surveyor. *Ditton Park (Bucks.)* 1695 MS

Alderwerelt, R. P. Cartographer. *Ysselmonde* 1812

Aldger, Cornelius. *Cologne* 1583

Aldjayhany. 10th cent. Statesman and geographer. *Book of Ways*

Aldous, M. *Manitoba & N.W. Territory,* Winnipeg (1895)

Aldrich, Lieut. Edward. *Jerusalem* 1841, *Gaza* 1843

Aldridge, T. Script engraver. *Battle Bridge* for Rennell (1810)

Aldring, fils. Engraver. *Cartes de France* 1744–60. *Grasset's Suisse* 1769, for Beaurain 1782

Aldwell, Thomas. *Cranborne Chace* 1618 MS

Alean, Estienne. Bookseller of Beauvais, Rue S. Pierre. De l'Isle's *Diocèse de Beauvais* 1710

Alefounder, Thomas, elder. *Estate surveys Essex* 1762

Alefounder, Thomas, junior. *Estate Surveys* 1776–1781

Alegre, Francisco. Spanish pilot. *Binan* 1745, *de Cuyo Paragua* 1750 MS

Aleman, N. *Evéché de Strasbourg* 1659

Alemand, P. *Costes de L'Amérique Sept.* 1687

Aleni, Giulio (1582–1642) Italian Jesuit in China. Mathematician and astronomer. *World map* (1630)

Aleotti, Giovanni Battista (1546–1636) of Argenta Ferrara. Engineer N. Italy 1603. *Duchy Ferrara* used by Ortelius 1608 and Vrients 1612

Aleph (pseud.) Compiler of *Cartoon maps Geographical Fun H. and S.* (1869)

Alès, Auguste François. Rue des Mathurins St. Jacques No. 1, Engraver for Monin 1840, for Faynard's *Atlas National* 1870, and for Vuillemin 1873–7

Alessandri, Innocente. Engraver. *Globe* 1784

Alessandro, A. Italian engraver. Worked for Coronelli end 17th century

Alestakhry [Al Istakhri] . See **Istakhri**

Alexander, Andrew. *Headquarters of James River. . . Virginia,* Philadelphia 1814

Alexander, Daniel (fl. 1794–1803) *Plans of London Docks*

Alexander, G. Publisher and engraver 130 Strand. *Plan London* 1817

Alexander, G. H. M. *Etawah District Agra* 1858

Alexander, John. Engraver in Italy early 18th century

Alexander, J. Lyon. Surveyor. 7 Castle Street, Holborn. *Halstead* ca. 1855 MS

Alexander, Capt. Sir J. E. *Niagara Gore and Talbot Districts,* London 1842, *British Guiana* 1832, MS, *Missions Orinoco* 1835

Alexander, William (1726–1783) American general, Surveyor General of New York

Alexander, W. D. Supt. Hawaiian Government. *Survey Hawaiian Is.* 1876

Alfani [Alphane] Dominico (ca. 1483–1553) of Perugia. *Lago di Geneva* (1590)

Alfani [Alfiano] Epiphanus de. Engraver *Gastaldi's Italia* 1597

Alfaraby [Al Faribi] 10th century Turkish geographer. *Book of Latitutdes and Longitudes*

Alferes, Joaquin Cardozo Xavier. *Campanha de Pilvens* 1801 MS

Alfergany [Al Farghani] See Fargany

Afliano, E. See Alfani

Alfonse, Jean. Navigator. *Cosmographie* 1544 MS

Alfonso X the Wise (1252–1284) King of Castille. *Astronomical Tables*

Alfonsus. See Peter Alfonsus

Alford. Engraver. *Plan Clink Liberty,* Southwark 1827

Alfred the Great (849–901) Collected material for geography, particularly of northern lands. The so-called Anglo Saxon or Cotton MS 10th century may have had its origin in Alfred's acquired knowledge

Algarra, Augustin de. *Aduanas de Espana* 1855

Algermissen, Johann Ludwig (b. 1847) Cartographer of Cologne. *Maps of Alsace Lorraine and Metz*

Algoet [Goethals, Panagathus], Lieven [Leuinus] (d. 1547) Poet, calligrapher and cartographer of Ghent. *Map of Northern Lands,* original lost, published

in Antwerp 6 sh. in 1562 and on smaller scale by De Jode 1570

Al Heravi [Herawi] Arab traveler and geographer, early 13th century.

Alibrandi, Gio. d'. Engraver. For *Atlante geografico* (1788–1800)

Alison. *History of Europe. Atlas* 1848 & 1850

Al Istakrhi. See Istakrhi

Alix, Thierry (1530–1594) Historian and geographer of Lorraine. Worked with Mercator. *Lorraine* 1564, *Vosges* 1579

Aljahedh (9th century) One of the earliest Arab geographers

Al Kazwini. See Kazwini

Alkendy, Abu Yusef Yakub. 9th century scholar and geographer. *Book of Routes & Principalities*

Alkharizimi, [Al Kharizmi, Khwarizimi] Mohammed (9th century). Studied trigonometry in India, made *additions to Ptolemy's Geography*

Alladi, P. Engraver. *Vallardi's Oceania* 1881

Allam, E. C. Assisted Admiralty Survey. *Falkland Is.* 1860–63

Allan, A. (fl. 1794–1820) *Mysore* 1794, *Ceylon* 1803, *India* 1818

Allan, A. C. Surveyor. *N.S.W. plans* 1875

Allan (C.) *Map Ceylon,* Faden 1803

Allan, D. Lithographer. *Chart of Tamar River, Van Diemen's Land* 1833

Allan, F. *Nederlandsch Indië* (1876)

Allan, Gordon. *Plan Belize, Honduras* 14 sh. 1885–86 pub. 1887

Allans, Lieut. (Queen's Rangers) *Skirmish at Richmond* 1781

Allard Family. Founder Huych [Hugo] Allard (d. 1691) Amsterdam, Calverstraet in de Werelt Caart. Engraver and publisher, father of Carel. *Guinea* engraved Doetecum 1602, *E. Indies*

1640, *World* 1640, *England* 1657,
N.Y. City 1660, *Leo Belgicus* 1665,
Germany 1666, *New England* 1673

Allard [Allardt] Carel [Carolus, Karel]
(1648–1709) Engraver and publisher
of Amsterdam op den Dam in de
Kaertwinkel. Son of Hugo the first.
Atlas Minor 1696, *Atlas Major* ca. 1705,
Theatre of War 1702, *Orbis habitabilis
oppida* (100 plates) 1698. Transferred
stock to his son Abraham in 1706

Allard, Abraham (1676–1725) Engraver
and printseller of Leiden. *Sedes Belli
Polonia* 1680, *Spanje en Portugaal*
(1700), *Hollands* 1714

Allard, Hugo the younger (1673–91).
Son of Carel. Republished some of
his maps

Allardice, S. Engraver. *Plan Philadelphia*
1793

Allart, Johannes. *Buitenstreeken v.
Parys* (1794)

Allason, Thomas. Architect. *Plans of
Elis and Olympia* 1814–19, *Course
of Alpheus* 1824

Allday, S. L. *Dennis's Official Cab
Radius Map. . . Birmingham* 1886

Alldridge, Capt. George M. Admiralty
Surveyor. *West coast England* 1850–
64, *River Usk* 1867 [70]

Allen, A. M. Surveyor. *Macamish* (1860)

Allen, C. R. *Harrison Co. Iowa* 1884,
Pottawattame Co. Iowa 1885

Allen, D. B. *Peoria County, Illinois*
1861

Allen, Fordyce A. *Comprehensive
Geography* 1866

Allen, F. D. *British Possessions in N.
America* 1813

Allen, George (1832–1907) English
engraver and publisher

Allen, George S. (fl 1790–1821)
Engraver at No. 4 Sadlers Wells Row
Islington 1791–2, later 19 Shoe Lane
Fleet Street. *N. Wales for Harrison*

1791, *Barbados for Bryan Edwards*
1794, *Pacific for Butler* 1799, *Whale
Fishery* 1796, *I. of Wight* 1800,
Arrowsmith's Asia & India 1801–04,
Faden's Channel 1804, *East India Pilot*
1804, *Laurie & Whittle's Caribbes*
1810, *Baker's Cambridge* 1821

Allen, Ira. U.S. Surveyor General.
Vermont 1798

Allen, James, *St. Croix & Misacoda*
1834, *Sources Mississippi R.* 1834

Allen, Joel. Engraver for Blodget's
Connecticut 1792

Allen, John (1796–1851) of Poplar
Row, New Kent Rd, Surrey. Engraver
Rowe's London 1811, *Baker's
Cambridge* 1821, *Ann Arbor,
Michigan* 1825 MS

Allen, John of Oxford. Surveyor.
Estate at Medmenham 1822 MS

Allen, Joseph. *Survey Wexford
Harbour* pub. Scalé & Richards 1764

Allen, J. B. Engraver for Tallis 1851

Allen, Mark. *Roads of Gt. Britain,*
Dublin 1832

Allen, Michael. Estate Surveyor.
Thorneborough Bucks. 1637 MS

Allen, Patrick. *Estate plans Ireland*
1657

Allen, Thomas. Publisher. *Plans London*
in *Hist. of London* 1828. *Panorama of
London* 1830

Allen, W. American engraver. Assisted
Tanner *USA* 1829, *Philadelphia* 1830

Allen, William [& Son] Publishers.
Dane St. Dublin. *Ireland* 1778 &
1783, *Carlow* 1798, *Ireland* 1815,
Meath 4 sh. 1817, *Wicklow* 2 sh. 1834

Allen, Capt. William, R. N. *Charts W.
Coast Africa* 1830–40

Allen, W. H. *Man River San Juan,
Nicaragua* 1851

Allen, William H. & Co. Publishers 7 Leadenhall St. London. *India* 1820 & 1841, *Punjab* 1846, *Java* 1852 *Route Map World* 1856

Allen, Parbury & Co. Publishers, later W. H. Allen & Co. *China* 1833

Allen. See Lackington, Allen & Co.

Allen & West. Publishers Paternoster Row. Maps in Malham's *Naval Gazette* 1795

Alley, William. *Journal of Voyages* 1665–78 (charts of *Mauritius, Maldives, & Ceylon*)

Alleyne, Capt. I. *Zululand Q.M.G. Dept.* 1850

Allezard, Jean Joseph. Hydrographer of Leghorn. *Ports de la Mediterranée* 1795, *Nouveau Recueil* 1800, for *Roux* 1804

Allfree, Edward. Surveyor 1864

Allhusen, E. L. *Geological Map Coolgardie* 1898

Alliaco, Petrus de. See d' Ailly, Pierre

Allies, M. Publisher. *Fisher's County Atlas* (1842–5)

Allin, Henry. Surveyor of Tonbridge. *Estate plans Kent* 1603

Allin, Richard of Robertsbridge (late 16th century) Estate Surveyor. *Manor Eastdean, Sussex* 1597

Allman, Thos. Joseph. *Penny School Series Atlas* [1871]

Allnut, Zachary. Estate Surveyor. *Cookham, Berks.* 1812 MS

Allodi, Pietro (fl. 1859–1878) Maps of Europe & Italy. *Atlante Universale* 1867, *Carta Nautica* 1878

Allom, Thomas. *Scotland* 1846

Alloy, Luigi Seb. *Diocesi di Milano 1821*

Almada, Andreas d'. Professor of Theology, Revised *Hessel Gerritz's Spain,* Jansson 1631. *Maps in Tassin's France* 1640–3

Almazan. Maps *Mexico* 1855–62

Almeda, Romao Eloi de. Portugese cartographer. *Carte militar Portugal* 1808

Almeida, Candido Mendes d' (1818–81) *Atlas do Brasil* 1868

Almeida, Manoel d'. *Ethiopia* 1602 MS, printed 1640. The first European map of Ethiopia. Reissued by Van der Aa ca. 1710

Almeida, R. E. De. See Eloy de Almeida

Almeida, Xavier d'. *Guarapuaia (Brazil)* 1770 MS

Almodovar, Pedro de Gonzaga y Luzon, Duque de. *Plan de Manila* 1787–[90] *Islas Filipinas* (1790)

Almon, John. Publisher opposite Burlington House in Piccadilly. *Camden's Britannia* 1722, *Environs of Boston* (in Remembrancer) 1775, *Middle Brit. Colonies* (in *Pownall's America*) 1776, *Hutchin's Virginia* 1778, *Situation Fleet Sandy Hook* 1779. Succeeded by Debrett 1781

Almonte y Muriel, Enrique d'. *Isla de Luzon* 4 sh., Madrid 1883, *Distrito Leyte* 1898

Aloja, Giuseppe. Engraver for *Carafa's Naples* 1775, and *Citta da Napoli* 1788

Aloja, Vincenzo. Engraver. *Regno di Napoli* 1807

Alonzo de Santa Cruz. See Santa Cruz

Alph, Johann. *View Munster* (1628)

Alphand, Jean Charles Adolphe. *Atlas ancient plans of Paris* 1880

Alphane, D. See Alfani

Alphen, Pieter van. Publisher of Rotterdam by de Roobrug inde Vyerige Colom. *Zee Atlas* Rotterdam 1660 & 1682. *Netherland Prov.* 1691 (47 maps)

Alphonse, Jean de Saintonge, of Rouen. Pilot to Roberval & Cartier. *Sailing directions and Charts of St. Lawrence* 1582, *Cosmographie* 1544 MS

Alphonsus. See Petrus Alfonsus

Al-Qazuini, [Zakariya ibn Muhammad] (1203–1283) Cosmographer, geographer and encyclopedist. Called the Moslem Pliny. Several MS

Alsop, George. *Landskip Prov. Maryland* 1666

Alsop, John. *Map Friends Meeting Rhode Is.* 1782

Alt, August Theodor Heinrich von (1730–1803) b. Kassel. Military cartographer

Alt, Franz. *Alps* 1873–7

Alt, H. Engraver. For Stieler's *Hand-Atlas* 1863

Alt, W. Engraver. For Spruner & Merz 1855 and Stieler [1850]

Alth, Alojzy (1819–86) Czech geologist

Altena, M. van. Engraver, *Plan Muiden* 1787

Altenburger, Peter. *Austria* 1843

Altheer, Publisher of Utrecht. *Nederlandsche ontdekkingen* 1827

Alting, Menso (Elder) 17th century Dutch mapmaker

Alting, Menso (Younger) 1637–1713) Burgomaster of Groningen. Geographer, *Notitia Germaniae Inferioris* Amstd. 1697. *Descriptio Frisiae* 1701

Altischofen, P. von. See Pfyffer von Altischofen

Altmann, J. *Railway & Canals, St. Petersburg & Moscow* 1844, *Neuer Handatlas* 1847

Altmueller, H. W. *Plan v. Jerusalem* 1859. *Sinai & Golgotha* 1861

Alvaredo, Pedro de (1486–1541) Spanish military Commander. Companion of Columbus.

Alvares, Cristovao. Portuguese cartographer 17th century. *Cidade de San Salvador* 1638, *Forte Real* 1629

Alvares [Alvarez] Seco [Seccus] Fernando *Portugal* 1560. Used by De Jode 1563, Ortelius 1570, Blaeu 1640

Alvares 'da Cunha, Fernando. Portuguese cartographer 16th century

Alvarez-Malgorey, Carlos. *Atlas geografico* 1886

Alvarez, Manuel. *Plano Ciudad de Mexico* 1867

Alves, A. F. *Plan Victoria, Hong Kong* 1862

Alves, Capt. Walter, R. N. *Charts of Malaya* 1763–82

Alvord, C. A. Printer. Colby's *Diamond Atlas* 1857

Alwin, Richard. Surveyor. *Quarrendon (Bucks.)* 1621 MS

Alypius of Antioch (fl. 360–371) Geographer

Alzate y Ramirez, José Antonio de (1739–1790) Mexican astronomer and geographer *N. America* 4 sh. 1768, *Laguna de Tesaco* 1786

Amabile, V. M. *Carte Gen. de Cuba* Paris 1896

Amari, W. *Carte de la Sicile* Paris, Lemercier 1859

Amaseo, Grigorio (1464–1541) Cartographer of Udine. *Map of Friuli*, used later by Ortelius

Amat de Tortosa, Andrés. *Yslas de Canaria* 1779

Amat Di San Filippo, Conte Pietro (1822–95) *Studi bibliografici* 1875

Amati, Amato *Dizionario Corografico* 1867–97

Amati, C. Engraver. *Plan de Malte* 1798, *Citta di Torino* 1800

Amato. See Bellamato

Ambassadors' Globe ca. 1525. So called because it is shown in Holbein's painting, *The Ambassadors* 1533

Amberger, Christoph (ca. 1505–1560) Map maker in Munich

Ambile, N. N. *Colorado* MS 1757

Ambridge, J. *Fifteen miles round London* 1855

Ambrosini, Floriano *Corografia. . . Contado di Bologna* 1605

Ambrosius, Marcus (fl. 1568–74) *Livonia* 1570 MS

Amendale, Sr. d'. See Damme, Jan van

à Merica. See Myrica

American Atlas Co. *Plat Books Counties, Illinois, Michigan, Indiana;* 1893–1907

American Express Co. *Tourist's Pocket Altas* 1881, *Note Book & Atlas* 1889

American Photo-Lithographic Co. N.Y. [Osborne's Process] *Railway routes to Pacific* 1869

American Tract Society. *Bible Atlas* [1862]

Amerigo Vespucci. See Vespucci

Amerigius, P. van der. See Myrica

Amery. *Antilles,* pub. by Mondhare 1782.

Ames, I. See Amys

Ames, I. *Co. Monmouth* 1785

Ames, Joseph. *Town & Pier of Ramsgate* 1736

Ames, N. S. Canadian Surveyor. *Westmoreland County* 1857, *Bradford Co.* 1858, *Lawrence & Beaver Counties* 1860

Ameti, Giacomo Filippo (end 17th century) *Map of Papal Territories* 4 sh. Rome, de Rossi 1696, *Il Lazio* 4 sh. Rome, de Rossi 1693, *Tuscia* 1696

Amettier, B. Engraver. *Plano. . . Havana* 1798

Amherst, Edward. *Peninsula in Bay of Fundy* (with George Mitchell) 1735

Amicus, J. B. *De Motibus corporum coelest. Venice* 1536

Amirutzes, Georgius (fl. 1465–1470) of Trapezuntios. Cartographer in Constantinople. *Maps for Ptolemy's Geography*

Amman, Ignaz Ambros von (1753–1830) of Mülheim. *Maps of Swabia*

Amman, Johannes (1695–1751) born Schaffhausen. Engraver.

Amman, Jost (1539–1591) of Zürich. Painter and wood engraver in Nürnberg. *Apian's map of Bavaria* 1568

Amon, Ant. *Poland Reilly* 1796

Amouroux, L. *Plan de Nantes* 1852

Amrien, Caspar Constantin von (1845–1898) of Neudorf. Swiss geographer in St. Gallen

Am Stein, Johann Rudolf (1777–1862) Military cartographer

Amthor, Eduard. *Volkes Atlas* 1869, *Repetitions Atlas* 1876

Amundson, Albert. *Post & Telegraph Map Sweden* 1873

Amyce, I. See Amys

A. Myrica. See Myrica

Amyrilius. See Myrica

Amys, [Amyce, Ames] Israel (fl. 1576–1598) Estate surveyor. *Belchamp St. Pauls* 1576 MS, *Manor of Plomborowe* 1579 MS, *Castle Hedingham* 1592

Anania, Giovanni Lorenzo d' (ca. 1525–1602) of Calabria. Cartographer *L'Univ. Fab. del Mondo* 1576. First edn. with maps 1582 and 1596 (maps of continents)

Ananias of Schirag. 5th century Armenian geographer

Anaximander of Miletus (610–546 B.C.) Called founder of map making. Constructed a *world map* and credited with invention of gnomon

Ambile. *Rios Colorado y Gila* 1757 MS

Anbury, Thomas. *Travels through America* (map) 1789

Ancelin, Pierre (1653–1720) Surveyor and cartographer of Rotterdam. Invented the indication of height in maps by isohypsic lines. *River maps with depths contoured*

Ancelin & Le Grand, P. *Atlas de toutes les Russies,* pub. Moscow & St. Petersburg 1795

Ancessi, Victor-Antoine (1844–78)
Atlas Géog. 1876 *Atlas de l'Ancien et Nouveau Testament* 1885

Anders, Friedrich Gottlieb Eduard.
Atlas d. Evangelischen Kirchen in Schlesien 1856

Andelfinger, J. J. *Constantinople,* pub. by Seutter 1735

Andersen, Karl Gustaf (1738–1808) Swedish land surveyor

Andersen, S. Swedish engraver. *Charts* 1837

Anderson, Alexander (1775–1870) Engraver *Kentucky & U.S. for American Atlas* 1796. *Countries surrounding Garden of Eden* 1796 & 1799. *Maryland & Delaware* 1799

Anderson, A. Engraver *U.S.* 1783, *Kentucky* 1795

Anderson, A. Engraver for *d'Anville's Ancient Geog.* 1814

Anderson, Alexander C. *Frasers River, San Francisco* 1858

Anderson, Capt. Allen. *New Mexico* 1864

Anderson, Anders. See Andersson

Anderson, C. J. *Maps of Damaraland S. Africa* 1866 MS

Anderson, Capt. F. C. *District of Shajehanpoor* 1862–4 pub. 1866

Anderson, Hugh (fl. 1811–1853) Engraver. *State Kentucky* 1818, *Robinson's Mexico* 1819, *Warners USA* 1820, *McCarty & Davis' USA* 1824

Anderson, James. *Hebrides* 1785, *Aberdeen* 1794

Anderson, James. *Plan Perish Ratho* 1818, *Glen Kingle* 1825

Anderson, Lieut. Surveyor and draftsman, *Plan Sebastopol* 1858

Anderson, Lieut. J. *Punjab & Kashmir* 1846

Anderson (Johann) Burgomaster of Hamburg. *Description Iceland, Greenland* 1746

Anderson (John) Map & Bookseller. 62 Holborn Hill, London. *Demerary* 1795.

Anderson, John, Jr. Engraver for *Payne's System Univ. Geog.* 1798

Anderson, Major (later Lt. Col.) John, U.S. Topo. *Engineers Sketch map Crown Point* (MS) 1818 (with I. Roberdeau) and *Portland Harbour* MS 1833

Anderson, J. A. *Railway maps Pa., N.J. & N.Y.* 1871–3

Anderson, John Corbet. *Croydon* 1889, *Great North Wood, Norwood* (1897)

Anderson, J.W. Military Surveyor 1800–1812

Anderson, P. Civil Engineer. *New York –New Haven Railroad* 1845, *Canal Map Hartford, Conn.* 1847

Anderson, Rev. Dr. J. *Geology of Fifeshire* 1841

Anderson, Samuel. Engraver for Hermelin

Anderson, Thomas. *Barnwell District, South Carolina* 1818, for Mills 1825, *Edgefield District, South Carolina* 1817

Anderson, William. Attributed *Chart S. Hemisphere* (1775?)

Andersson [Anderson] Anders (d. 1695). Surveyor of Sodermanland. *Hölbo District* 1879

Anderton, G. Engraver for *Taylor's Oxford* 1750

Andrade, B. M. F. de. *Oceano Atlantico Norte* 1870–73, pub. 1885

Andrau, Lieut. K.F.R., Dutch Navy *Atlantische Oceaan* 1856

André, Aimé. Publisher, son-in-law of Poirson. Quai des Grands Augustins 59 and Rue Christine, Paris. For Fournier 1820, for Malte Brun 1837

André. Engraver. Worked for Rizzi Zannoni 1771, for Raynal 1780, for Bonne et Desmaret 1787, for Mentelle's *Atlas Univ.* 1797–1801

André, Major John. (1751—1780).
Several *MS maps of towns, forts &
counties in U.S.*

Andre, Karl. *N. America* 1851

André, Louis. *Plan Biarritz* (1860)

André, Peter. Surveyor 7 Noel St.
Berwick St. *Essex Surveyed* with John
Chapman 1772—74, 25 sh. pub. 1777
and 1785.

André de Gy. See **Chrysologue**

Andrea[e], Johann Ludwig (fl. 1717—
26). Cartographer of Nuremberg.
Terrest. and *Celest. Globes*

Andrea, Johann Philipp (1720—1757).
Mathematician of Nuremberg

Andrea, Lambert (fl. 1590—98) publisher
of Cologne, Botero's *Tab Geog.* 1596,
Theatrum Principatium 1596

Andrea, P. Publisher of *Muller's
Kleiner Atlas, Frankfort* 1702

Andrea, T. *Plano do Porto e Cidade de
Dilly* 1870

Andreas, Alfred T. American compiler
and publisher of County maps. *Knox*
1870, *Fulton* 1871, *Pike* 1872, *Des
Moines* 1873, *Lee* 1874, *Iowa* 1875
Atlas Dakota 1884

Andreas Lyter & Co. *Co. Atlases
Illinois* 1872—3

Andrée, Otto. *Sachsen* 1862, *Sächsisch-
Röhmischen Schweiz* 1865

Andrée, Dr. Richard (1835—1912).
Born Braunschweig. Geographer and
cartographer of Leipzig. *Volksschul
Atlas* 1876, *Allgemeiner Handatlas*
1880, *Physik. statist. Atlas deutschen
Reichs* 1878 (with O. Peschel).
Africa 1885

Andreossy, François (1633—1688).
Water Engineer. *Canal Royal de
Languedoc* (1669)

Andrews, C. A. *Survey River Pungue,
Mosambique* 1891

Andrews, C. W. *Christmas Island* 1899

Andrews, E. B. *Gallia Co., Ohio* 1874,
Meigs Co. 1875

Andrews, H. of Modbury, Surveyor.
Farms in Plymstock Devon 1850 MS

Andrews, Israel D. *Straits of Florida*
1852, *Railroads in USA* (1853)

Andrews, James. *Chart Skerryvore
Rocks* 1836

Andrews, John (1736—1809). Geog-
rapher, surveyor, engraver, mapseller
at No. 5 Fish Market Westminister
Bridge; at Mr. Branches 40 Corner of
Buckingham St. Strand (1776—7) 29
Long Acre at Mr. Blissets opp. Mercer
St. (1777—82) and 211 facing Air St.
Piccadilly (1781 onwards) *Plans
Capital cities Europe* 1771, *Plan
Canterbury* (with Wren) 1768,
Kent 1769, *Wilts.* 1773, *I. of Wight*
1775, *Rivers England* 1796, *Hist. Atlas
England* 1797, *Geog. Atlas England* 1809

Andrews (J.) & A. Dury. Surveyors.
25 miles round Windsor 1777, *Herts.*
1766, 1777, & 1782, *Kent* 25 sh. 1769,
1775, 1779, & 1794, *65 miles round
London* 1774—93, *15 miles round
Richmond* 1777, *Wilts.* 18 sh. 1755,
& 1810, *I. of Wight* 4 sh. 1769.

Andrews, Norman S. Surveyor. *Map
of Constantine* 1874

Andrews, P. Engraver for *Mary Ann
Rocque's Plans and Forts America*
1763, *Holland's Isle St. John* 1765,
Montresor's N.Y. 1766, *Rocque's
Surrey* 1754, *Marmora* 1770, and
for *American Pilot* 1772

Andrews, Samuel (fl. mid 18th century)
Chart Cork Harbour 1792

Andrews, Thomas C. *World* 1895

Andrews, Thomas C. Publisher of Boston.
H. Harris's Rhode Is., Conn. 1796

Andriveau, Gilbert Gabriel Benjamin
(ca. 1805—1880) Bookseller, publisher
and geographer. Married the daughter of
M. Goujon and founded the firm of
Andriveau Goujon in Paris, Rue du Bac
No. 6. Later moved to 21 Rue du Bac

prés le Pont Royale. *Océanie* 1832 and 1856. *Atlas Classique et Univ.* 1856. In 1858 he passed the business to his son Eugene.

Andriveau-Goujon, Eugene (1832–97). Publisher, editor, geographer and globemaker. Paris, Rue du Bac 21 (1875–6), Rue du Bac 4 (1876). *Atlas de Choix* 1862, *Carte d'Allemagne* 1865, *Planisphère* 1872, *Atlas Classique* 1875

Andriveau-Goujon, J. Publisher, mapseller, and geographer. Rue du Bac 6 (1836) Rue du Bac 17 (1841–50) Rue du Bac 21 (1854). *Suisse* 1831, *Atlas Classique et Univ.* 1835, *Routes de France* 1838, *Atlas A-M* 1856, *Atlas Elément* (1868)

Androuet Du Cerceau, Jacques. *Plan Paris* 1555 (attrib.)

Andruzzi, E. *Sardegna* 1869

Anexarius or Anexerius. See **Buondelmonte**

Angeli, Vittorio. *Lombardia* 1837, *Alessandria* (1860)

Angelieri, Giorgio. Printer of Venice. *Porcacchi's Isole* 1567 and 1575

Angelino, Angel. *Texas* 1788

Angelis, F. Antonio de. Minorite monk, Naples. *Plan Jerusalem* 1578

Angelis, Girolamo (1567–1623) Jesuit missionary in Japan. *Japan* 1615, *Hokkaido* 1621 MS

Angell, Samuel. *Cornhill Ward* 1833, Wm. Standridge lithog.

Angelo, G. N. *Schleswig-Holstein* 1776–1806

Angelo, M. de S. *Carte de Corse* Paris, Lattré 1768 Maps for *Julien's Théatre du Monde* 1768

Angelo, Theodore Gottfried Nicolai (1767–1816) Danish engraver and publisher. B. Schleswig d. Copenhagen. *Aalborghuus* 1793, *Anstrup* 1795

Angelloti, P. *Pianta della valli di Conacchio* 1658

Angelucci, Gio. Batt. *Carta. Topo. Scandoglia* 1788

Angelus [Angiolo, Agnolo] Jacopo d' of Scarperia, Tuscany. Translated Ptolemy into Latin 1406. His text is used in editions 1482, 1486, 1490 and 1525

Angelius, Vadius. See **Vadius Angelus**

D'Anghiera. See **Martyr d'Anghiera**, P.

Angiolini, Giorgio. *Pianto topo. . .citta di Firenze* 1826

Angiolo, Jacopo d'. See **Angelus** Jacopo

Anglois, N. See Langlois

Angory, W. H. *W. Australia* 1887

Anguissola, Leander (1652–1720) Austrian Military engineer and cartographer. *Danube* 1681, Plan *Vienna* 1683, and in conjunction with J. Marinoni *Vienna* 1704 & 1706

Anich, Peter (1723–1766) Austrian cartographer and instrument maker born Oberperfuss, Tyrol. *Globes* 1756–9, and in conjunction with Blasio Hueber *Atlas Tyrolensis* Vienna 1774, reduction in Paris on 9 sh. 1808

Anito, Nicolo. *Pianta. . . citta di Palermo* (1760)

Ankarkkrona. Plan *Baie de Coquimbo* 1853–8

Anker, M. U. *Madagaskar* 1875

Annin, W. B. Engraver for Cummings' *School Atlas* Boston 1818, *State of Maine* 1822

Annin & Smith. Engravers for D. Adams. *Plan Salem* 1820, *School Atlas,* Boston 1823, *Newton* 1831, *Boston School Atlas* 1832

Anoutchine, D. Russian geographer. Directed Geolog. review *Zemliebiedmie*

Anquetil-Duperron, Abr. Hyacinthe (1731–1805) Orientalist and traveller. Wrote on history and geography of India

Ansart, Felix. *Atlas Hist.* 1834, 1836. *Amérique* 1838

Anse, Luggert van (fl. 1690–1716).
Engraver for Visscher & van Keulen,
Amsterdam. Also pub. as Luggardas
van Anse 1715. *Husson's France* 1708

Ansel. *Rochester, Mass.* 1856

Anselin. Printer of Paris. For *Le Boucher's
Atlas* 1830 for *Saint Cyr's Atlas des
Mémoires* 1831, *Jomini's Atlas portatif*
(1840)

Anselm, C. *Telegraphenkarte von
Baden* 1870

Anselmier, Claude Henri Jules (1815–
1895) Topographer of Geneva

Anselmier, G. *Stadt Bern* 1868

Ansfield, J. C. Engraver. *Stieler's Nord-
Amerika* 1825

Ansoleaga, Florencio. *Pacific Ocean*
(1743)

Anson. Engraver for Weiland 1848

Anson, George (1697–1762). Circumnavigator. *Voyage round the World*
1739–44 pub. 1748

Anstatt. *Environs Berlin* 10 sh. 1616–19

Ansted, David Thomas (1814–1880)
Geologist. *Atlas of Physical & Hist.
Geog.* 1852–9

Anstey, Capt. Thos. Henry. *Battlefield
Isandhlwana* 1881

Anthoine, Edouard (1848–1919) *Atlas
de Géog. Mod.* 1890

Anthoine, J. See **Jacobs**

Anthonisz [Anthoniszoon, Antoniades]
Cornelis (1499–1557) Dutch seaman,
cartographer and engraver. Established
himself in Amsterdam. *Book of Sailing
Directions* 1532, *Caerte van Oestlant*
(Baltic Northern Regions) 1543,
Italian version pub. Rome 1558; *Plan
Amsterdam* 12 sh. 1544

Antillon y Marzo, Isidoro de (1778–
1820) Spanish geographer. *N. America*
Madrid 1803, *Atlas* 1804

Antipov. *Stratigraph map Donetz
Coalfield,* St. Petersburgh 1872

Antoine, Louis. Printer of Paris, 70 Rue
des Noyers. For Andriveau-Goujon
1854–74, *Rue de Cluny* for Dufour
(1864), *Grand Atlas Univ.* (1870)

Antoine, Pierre. *Arrond. St. Dié* 1872

Anton, Manuel. Spanish Navy. *Mindanao
Harbour* 1875

Antonelli, Guiseppe. Publisher of Venice.
Dizionario Geografico 1833, *America*
1834, *Oceania* 1839

Antonelli, Juan Battista. Map *Mexico*
1590

Antoniades. See **Anthonisz,** C.

Antonius, Johannes. Publisher. *Map
Rome* 1600

Antony, Carl. *Böhmen* 1881

Anville, H. F. B. d'. See **Gravelot**

Anville, Jean Baptiste Bourguignon d'.
(1697–1782) Royal geographer and
cartographer. Paris, Aux galéries du
Louvre. Elected Académie des Sciences
1773. Compiled over 200 maps.
France 1719, *Atlas de la Chine* 1737,
N. America 1746, *S. America* 1748,
Africa 1749, *Asia* 1751, *World* 1761,
General Atlas from 1737

d'Anville

Anxerinus, C. See **Buondelmonte**

Apeus [Apens, Appeus] Cornelius
(1634–ca. 1688) Publisher and mapmaker. Engraved *Conders ab Helpens
Groeningen* (1640)

Appé, *Dépt. de l'Indre* 1862

Appeus. See **Apeus**

Apian [Apianus] Peter or Peter Bienewitz
(1495–1552) Geographer and
Astronomer Royal, b. Leisnig, Saxony;
d. Ingoldstadt. *World map* in Solinus,
pub. Vienna 1520 and in Pomponius Mela,
pub. Basle 1522. *Cosmographicus
Liber* 1524, *restit. per Gemma Phrysium, ,*
Antwerp 1533, 1540, 1545, 1574 & later

P. Apianus [or Bienewitz] (1495–1552); engraving by Th. de Bry

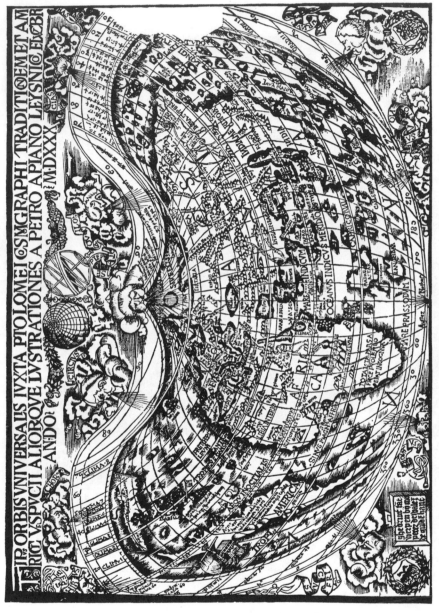

Woodcut cordiform world map by P. Apian (published 1530 at Antwerp) (reduced)

Apian [Apianus] Philip (1531–1589) German cartographer and mathematician, son of Peter Apian [Bienewitz] b. Ingoldstadt, d. Tübingen. *Maps of Bavaria* 1563 and 1568, engraved by Jost Amman. Another edition engr. by Weinherr, Munich 1579. *Terrestial Globe* 1576

Applegate, H. S. & J. Publishers in Cincinnati, Ohio. *Conclin's River Guide* 1850

Appleton, A. *Orfordness* 1568 MS

Appleton, Daniel & Co. Printers of New York. *Railway map U.S.* 1865, *Atlas of U.S.*, New York 1888 *Library Atlas* 1892, 1895

Appleton, J. W. *Bay Bengal* 1876, *N. Amer.* 1880, *Baltic* 6 sh. 1882, *Adriatic* 1884

Après de Mannevillette, Jean Baptiste Nicolas Denis (1707–1780) Hydrographer. Chev. de l'ordre du Roi, Capt. des Vaisseaux de la compag. des Indes, Correspondent de l'Acad. Royale des Sciences, Assoc. de l'Académie Royale de la Marine. *Neptune Oriental* Paris Dezauche 1775, *Supplément* Paris Demonville 1781

Arago, Domenique François-Jean. (1786–1853). Astronomer

Aragon, Anne Alexandrine (b. 1798) *Atlas* 1843

Aragón, Idelfonso. Col. of Engineers. *Campo Santo de Merida* 1812–[23] *Plans Manila* 1814–19, *Prov. Ylocos* 1821

Aragon y Abollado. See **Abollado**

Aragona, S. di. See **Sebastiano di Aragona**

Arana, Barros (b. 1830) Chilean geographer and historian

Arandia, Pedro Manuel de. *Plano Puerto Sisiran* 1757

Araos, Juan de. *Cuba,* pub. Havana 1788

Arason, Magnus (ca. 1680–1728) Map & chart maker of Hagi Bardaströnd . *District maps of Iceland* 1728

Aratus [Aratos] Greek astronomer ca. 272 BC

Araujo, A. J. *Carta da Bahia* 1837, *Rio de Janeiro* 1858

Arazon y Bollardo, Y. de. *Philippines* 1821

Arc, A. d' See **Gautier D'Arc**

Arce, Armando da. *Mapa del Rio Mississipi* 1699

Arch, J. Publisher and bookseller. Cole & Roper's *British Atlas* 1810, Walker's *Univ. Atlas* 1820

Archdale, John (d. 1717) Governor of Carolina 1694–5. *Papers relating to ye Province of Carolina with a Draught of ye Town [Charles Town] Mapps of ye Forts, Rivers, Coasts &c.*

Archer, J. Bookseller of Dublin. *Plan Dublin* 1797

Archer, J. Engraver for R. Wilkinson's *Gen. Atlas* 1802

Archer, Capt. James. Engineer. *Plan York* 1643 MS

Archer, Capt. James. Engraver. *Plan Scarborough* 17th century

Archer, James. *Portsea Island* 1773

Archer, James. Engraver for *Bradford's Illustrated Atlas* 1838

Archer, John Lee. Civil engineer. 1843, *Tasmania Stanley & Forrest Farms, V.D.L.* 1843–4

Archer, Joshua. Draughtsman and engraver. Pentonville, London. *County maps* 1833, *Plan London* 1833–7, *Ecclesiastical maps,* Brit. Mag. 1841, *Colonial Church Atlases* 1842–50, *Maps London* 1848–61, for Hughes' *Atlas* 1866

Archer, Mark. *Collieries of Durham* (1890)

Archer, Thomas. Estate surveyor. *Leyton* 1721 MS, *Walthamstow* 1725 MS

Archer, Thos. *Egypt* 1885

Archimedes of Syracuse. 287–212 B.C. Greek mathematician and physician.

Arciszewski, Krzystof (1592–1656) Polish general in service Dutch West India Co. in Brazil

Arctowski, Dr. H. *Charts Falkland Is.* 1898 MS

Arculf. Saxon monk, visited *Jerusalem.* Drew plan of city A.D. 670

Ardagh, Maj. Gen. Sir John. *Egypt* 1883, *Atlas boundary Guiana-Venezuela* 1898

Ardeiser, Johann. *Vallistelina* 1620, 1625

Arden, Edward. Surveyor. *Surveys in Huntingdon* 1809–11 MS

Arderne, R. *Beveridge Is.* 1875

Arebault, J. See **Aerlebout**

Arenales, J. *Valdivia* 1795, *Carta Gen. Patagonica* 1845 MS

Arendts, Prof. Dr. Carl (1815–1891) Austrian geographer, founded Geogr. Soc. Munich. *Prussia* 1862, *Mittel Europa* 1869, *School Atlas* 1871

Arendt, Charles. *Plan of Peking* (1890)

Arendts, W. *Elsass & Deutsch-Lothringen* 1871

Arenhold, Gerard Justus. *Hildesheim* 1727, *Maps for Homann* 1753

Aretin[us] of Ehrenfeld, Paulus (17th century). Czech surveyor, in Prague 1608–27. *Map of Bohemia* 1619 and later; *N. Moravia* 1623

Arfe[r]ville, Sr. de See **Nicolay**

Arfwedson (C. D.) *De Colonia Nova Suecia* Uppsala 1825 (map of Delaware Bay)

Argaria, Gasparo (fl. 1538–67) Italian cartographer *Naples* 1538 *Messina* 1567

Argelander, Friedrich Wilhelm August (1799–1875) Astronomer. *Celestial Atlas* (18 pl.) 1843, *Atlas des Himmels* 1863

Argensola, Bartolomé Leonardo de. Spanish poet and historian. *Conquista de*

las islas Molucas Madrid 1609 (Draught of Moluccan Is.)

Argoli, Andrea. 17th century Italian astronomer

Argonauti, Academia Cosmografia degli. See **Academia Cosm. degli Argonauti**

Arias, Montano [Montanus] Benito [Benedictus] (1527–1598) Spanish philosopher and theologian. Edited Polyglot Bible *(maps of World & Palestine)* 1572

Arigoni[o], Fra Bono. 16th century Venetian cartographer

Arista, Gen. Manuelo. *Mouvements US Army in Mexico* 1846

Aristogoras of Miletus (495 B.C.) Greek geographer

Aristarchios (264 B.C.) of Samos. Mathematician and astronomer

Aristotle, Nicolo d'. 16th century Venetian printer and publisher. 1st Edition of *Bordone's Isolario* 1528

Arlett, Lieut. William (fl. 1844–1851) Hydrographer. *Survey W. Coast of Africa* 1531–4, *Gran Canaria* 1834–62.

Arluc. *Plan de Cannes* [1895]

Armani, Domin. *Territ. Vicenza* 1775

Armann. Lithographer and publisher. *Plans Cassel* [1869?]

Armato. See **Bell' Armato**

Armendale, Sieur d'. See **Damme**

Armentaria, San J. N. *Bolivia,* pub. La Paz 1888

Armitage, Benjamin. Estate Surveyor. *Latton* 1778 MS

Armour & Ramsey. Publishers in Montreal. *Staveley's Canada* 1846

Arms, C. J. Senior & Junior. Engravers. *Caldwell's Atlas Harrison Co. Ohio* 1875

Arms, Walter F. *Hist. Atlas Adams Co. Ohio* (1880)

Armstrong, Lieut. (later Capt.) Andrew. (fl. 1768–1781) Surveyor. Father of Mostyn John. *Durham* 4 sh. 1768, *Pocket*

Map Scotland 1775, *Lincs.* 8 sh. 1778–9, *Rutland* 1781, and jointly with his son Mostyn *Northumb.* 9 sh. 1769, *Berwick* 4 sh. 1771, *Ayrshire* 6 sh. 1775, *Lothians* 6 sh. 1773

Armstrong, George. *30 miles round Boston* 1775

Armstrong, George. *Maps of Palestine* 1890–1921

Armstrong, J. Estate Surveyor. *Marks Tey* 1809 MS

Armstrong, Col. John. *Navigation Kings Lynn–Cambridge* 1725 (7 maps)

Armstrong, John. Surveyor. Gen. of Ordinance. *Minorca* 1756

Armstrong, John (1755–1816) American Senator and military cartographer. *Illinois River* MS, *Wabash* (1790)

Armstrong, J. B. Civil Engineer. *Grand Island City* [1857]

Armstrong, Marcus. *Scotland* 2 sh. 1782

Armstrong, Mostyn John. (fl. 1770–91) Surveyor and publisher. Son of Andrew above. *Peebleshire* 2 sh. 1775, *Great Roads London–Edinburgh* 1776, *London–Dover* 1777, *Scotch Atlas* 1777, 1787; corrected Kinderley's *Fens* 1781

Armstrong & CIA. *Plano de Mexico* 1890

Arnaud, d'. Engineer. *Bahrel-Abiad* 1843

Arnaud, C. Drew and engraved *California* 1849

Arnd, F. H. *Europ. Russland* [1877?] *Frankreich &c.* [1880?]

Arnetti, Giacomo. *Parte Merid. del Latio,* pub. by Rossi 1693

Arngrimur-Jonsson (1568–1648) *Liber de Gronlandia* MS

Arnold, Brig. Gen. Benedict (1741–1801). *Sleepy Hole to Portsmouth* 1781 MS. *Attack Fort Trumbull* 1781, *Passes into Canada* 1778, *Route Gen. Simcol* 1781 MS

Arnold, E. G. *Topo. Map District Columbia* 1862

Arnold, E. J. *School Atlases* 1889

Arnold, Richard. Estate Surveyor. *Wanstead* 1740 MS

Arnold, Capt. Stuart Amos. Hydrographer. *Thames Estuary* 1777, *Irish Sea* 1796, *Mull & Kintyre to C. Wrath* 1800

Arnold, Dr. Thomas. *Revised Teesdale's General Atlas* 1856

Arnoldi, Fiamengo Aroldo di. (d. 1602) Flemish cartographer, worked in Bologna and Siena. *Maps of the World and the Four Continents* 1600, *World Siena* 1602. Worked for Magini in Bologna 1595–1600 and Floriani 1600–02

Arnoldo, Joh. Gerhardo *Tab. Geog. Frankfort* 1698

Arnoul. Engraver. Rue du Petit Pont No. 26 Paris. For Buchon 1825, Perrot 1827, Andriveau-Goujon (1841–62)

Arnoullet, Balthazar d'. Publisher of Lyons. *Plan Paris* 1530, *Chorogr. Europe* 1553

Arnswaldt, B. von. *Umg. v. Eisenach* 1857

Arnz, A. & Co. (ca. 1825–1850) Lithographers of Dusseldorf & Leiden. *School Atlas* 1829

Arnz (J.) Publisher and cartographer of Dusseldorf. *Atlas der Alten Welt* (1829)

Arosander, Peter (d. 1696) of Upland. Surveyor

Aron & Abraham. Greenlanders. *Maps & diagrams of Godthab District* 1860

Arppe, Wilhelm. *Finland* 1872 with M. Wijkberg

Arquer, Sigismund. *Sardinia* for Munster 1544

Arquiza, Jacob de. *Maps Provinces of Philippines* 1832, *Las Almas* 1845

Arredondo, Antonio de. Spanish Royal Engraver. *Plan S. Augustin Florida* 1737 MS, *San Simon* 1737 MS, *Florida* 1742 MS

Arrianus Nicomadensis (2nd century B.C.) *Periplus of Black Sea*

Arrigoni, Capt. Ferdinando. *Carta postale della Italia* 1846, *Contorni di Milano* 1843, *Lombardia* 1860

Arrivet, J. (fl. 1764–1786) French draughtsman and engraver. For *Hydrographie Française* 1765, for *Vaugondy* 1760–78, for Bonne 1771

Arros, C. H. O. H. d'. See **Hallez d'Arros**

Arrott, W. *Miami Co. Ohio* (1855)

Arrowsmith, Aaron (1750–1833) English cartographer, engraver and publisher. Hydrographer to His Majesty. Born County Durham, died Soho. Established in London ca. 1770 as land surveyor. Worked for Faden and Cary. In 1790 set up his own establishment Castle Street Long Acre. First production *Chart World 1st April* 1790. Moved to 5 Charles St. Soho Sq. 1794, to 24 Rathbone Place 1802, to 10 Soho Sq. 1808. He died aged 73 years. He became Hydrographer to Prince of Wales ca. 1810 and to His Majesty 1820. In all he issued about 200 maps, mostly large scale, and was easily the foremost cartographer of his time. His principal works include *Pacific* 9 sh. 1798, *Marquesas & Oregon* 1798, *Chart Bass's Strait* 1800, Pilot *England to Canton* 1807, *Africa* 1802, *Asia* 1801, *America* Thomson's *Alcedo* 1812–15, *England* 1816

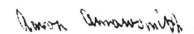

Signature of A. Arrowsmith (1750–1833)

Arrowsmith, Aaron, Junior. Sometimes worked with his brother Samuel, sometimes solo. Son of Aaron above. *Denmark* for Constable 1823, *Ceylon* 1826 (with Kershaw), *Orbis Terrarum Vet. desc.* 1828, *Outlines World* 1828

Arrowsmith, John (1790–1873) Nephew of Aaron. *London Atlas* 1834, continually revised and extended. Later editions are the more valuable, particularly for new surveys in *Australia* and *N. America, Texas,* &c.

A
PILOT

FROM

England to Canton,

IN SEVEN CHARTS,

BY

A. ARROWSMITH.

Strait of Dover	1 Sheet
North Atlantic Ocean	4 Sheets
South Atlantic Ocean	4 Sheets
Cape of Good Hope	4 Sheets
Indian Ocean	4 Sheets
Mouths of the Ganges	2 Sheets
Eastern Passages	4 Sheets

SOLD BY
A. ARROWSMITH, Rathbone-Place,
AND
BLACKS and PARRY, Leadenhall-Street,
London.
1806.

Title-page to A. Arrowsmith's Pilot from England to Canton (London, 1806)

Arrowsmith, Samuel (d. 1839) Son of Aaron Senior. *Bible Atlas* 1835. Also worked with his brother Aaron Junior.

Arsenius [Aertssens] Family of instrument makers and engravers of Antwerp and Louvain. Ambrose and Ferdinand brothers; engraved maps for Plantin 1589–1628, used by Ortelius from 1595. Also worked for Coignet's *Epitome* 1601 and for Vrients 1601–1612. Gauthier, Remy and Regnier (nephews of Gemma Frisius) had a workshop in Louvain for astronomical and geographical instruments, globes, astrolables, quadrants, etc. Gualterius Arsenius was the finest craftsman in the family between 1555 and 1588

Artaria, Carlo (1747–1808) Publisher, art dealer, and founder of firm Artaria & Co. in Vienna.

Artaria, Domenico (1775–1842) Vienna publisher and art dealer

Artaria, Francesco (1744—1808)
Turned family firm into a company

Artaria & Company Publishers & mapsellers of Vienna. Kohlmarkt No. 1151
from 1800. *Hungary* 1801, *Venice* 1805,
Bohemia 4 sh. 1808, *Poland* 1813,
Lombardy 1815

Artemidorus (2nd century B.C.) Greek
geographer

Artero y Gonzalez, Juan de la Gloria.
Atlas de la Espana 1879, *Atlas de Geogr.*
1895

Arthaud, Guy. *Duché d'Angers* 1652

Arthur, Pierse. *Weymouth & Melcombe
Regis* 1857

Arthur, Thomas. Estate Surveyor.
Dagenham 1747 MS

Arthur, T. W. Plan *Carlisle* 1880

Artopaeus, Peter [or Becker] (1491—
1563). Map *Pomerania* for S. Munster.
Used by Ortelius

Artus. Engraver for Brué 1833

Arvinger, Hans. *Plan Kiobenhafn* 1764[6]

A. S. See **Salamanca, A.**

Asaki, Gheorghe (1788—1869) Cartographer

Asaph, Judaeus. 11th century *World Map*,
in Bibliothèque Nationale, Paris

Ascencio, José. *Plan Madrid* 1800 (with
Martinez de la Torre)

Ascension, Father Antoine. *California* MS,
originated conception of California as
an island

Aschbach, G. A. Surveyor. *City Allentown
(Pa)*. 1871, *Lehigh Valley Rail Road,
Penn.* 1871, *Lookout Park* 1872

Asche, A. *Hannover u Linden* (1894)

Aschke, O. *Ems-Mündung* 1895

Aschland, Arent. *Coast of Iceland* 1820—24

Asea, Claudio. *Prov. di Venezia* 32 sh.
1876

Ash. Engraver. *Adams Place Borough*.
For Horwood's *London* 1792—9

Ash, Joshua W. *Delaware Co., Pa.* 1848

Ashbridge, Capt. John. *Plan Quedah Road*
1781

Ashby, H. (1744—1815) Engraver for
Noorthucks London 1772, for *Armstrong's
Scotch Atlas* 1777, *Fannin's Isle of Man*
1789

Ashby, Robert. Cartographer. *Lough
Foyle* 1601

Ashcombe. See **Eisdell & Ashcombe**

Ashcroft, T. H. *Nigeria* 1880

Asher & Adams. *Township map Indiana*
1865, *State Iowa* (1868) *Atlas N.Y.*
1871, *U.S.* (1873)

Ashley, Sir Anthony (1551—1625)
Translated and edited Waghenaer's
Mariners Mirrour 1588

Ashman, Rev. J. *Liberia* (1833)

Asmead, George C. Revised *Plumley's
plan of Bristol* 1829 & 1833

Ashmore, Capt. Samuel. *W. Coast
Sumatra* 1822, *Tracks thro. Barrier
Reef* (1840)

Ashpital, W. H. Surveyor. *Estate Joh
Clark Powell Whitechapel* 1805, *St. John
Hackney* 1831

Ashton, Henry G. G. Surveyor. *Liverpool
Bay* 1896—7

Ashwell, Samuel. Estate surveyor.
Havering 1662 MS

Asiatic Lithographic Co's Press, Calcutta
Road Maps Bengal, Madras and *Bombay*
1828

Aspher, P. *Celestial Globe* 1586

Aspin, Jehoshophat (fl. 1814—1840)
Chart New South Wales 1816, *Maps for
Lavoisne's Atlas* 1807—1820, *Arctic
Regions* (1822), *Planisphere* 1844

Aspley, William. Bookseller and publisher.
Camden's Britannia 1637

Asquith, Richard. Estate surveyor.
Estates in Cumberland 1862

Assa, Jose Frederico d'. *Carta do Goa*
1878

Assaf. 11th century Jewish mathematician and geographer. *Planisphere*

Asselen, C. F. von (fl. 1720–1726) *Ambt Apen* 1721, *Gegend v. Sandfurth* MSS

Assensio, Josef. Engraver for Torfino de San Miguel. *Atlas of Spain* 1788, *Plan Madrid* 1800 (with Torre)

Asser, James. Estate surveyor. *W. Thurrock* 1798 MS, *Bulphen* 1803 MS

Assheton, J. T. *Geogr. Companion England & Wales* 1832, *Historical Palestine* 1825

Assiotti, Francis Aegidius. *Plan Minorca* 2 sh. 1780, *Siege of Castile* 1782

Assum, G. van. See **Gerritsz**

Aster, Friedrich Gottlieb (1788–1813) Military cartographer b. Dresden

Aster, Friedrich Ludwig (1732–1804) Military cartographer of Dresden

Astley, J. Publisher. *London Magazine* 1732–46

Astley, Philip senior. *New map France* 1800, *Europe* 1808

Astley, Thomas. Bookseller. Paternoster Row. *Collection of Voyages* 1745–7.

Aston, D. W. *N. Lines. Railway* 1855

Aston, Publisher of Coventry. *Warwickshire* 1816

Aston, Joseph. Publisher, printer and map-seller No. 84 Deansgate Manchester and Exchange Herald Office St. Anns St. *Manchester & Salford* 1804, 1816

Astor, Diego de. Maps in *Barros' Asia* 1615

Asturlabi (13th century) Instrument maker

Aszalay József (1798–1874) Map draughts-man

Athbridge, Capt. John. *Quedah Road* (1763)

Atherstone, Dr. *Geolog. Sketch map Cape Colony* 1875

Atienza, Jose de. Spanish navigator. *Charts* 1830 MS

Atienza y Cabos, Augusto. *Espana y Portugal* 1888 *Railways Spain*

Atkinson, Alfred. *Levels Marley, Lincs.* 1875

Atkinson, George. Surveyor General. *Ceylon, Faden* 1813

Atkinson, G. W. E. *Turkestan* 1879, *Route map Tibet & Mongolia* pub. by Perthes 1885

Atkinson, James. Mathematical instru-ment maker on ye E. side of S. Saveris Dock over against the Griffen and at his shop at Cheney Garden Stairs on Rotherhithe Wall. Joined with Seller, Thornton, Colson and Fisher in publishing an *Atlas Maritimus* 1675–7

Atkinson, John. *Forest of Dean* 1845

Atkinson, Joseph, of Spanish Town. Chart *Virgin Islands, London, Mount & Page* 1739

Atkinson, Tho. Surveyor. *Plan River Crake (Lancs.)* 18th cent. MS

Atlas, Laureano. Engraver. For Arandia 1757, for *Spanish Empire in form of Woman* 1761

Atlas Publishing Co. *Universal Atlas* 1893, *Erie Co., Ohio* 1896

Atlasov, Vladimir (d. 1711) Russian Cossack. First map of *Kamtschatka*, used by Homann

Atthalin, Baron (1784–1836) Military cartographer

Atwood, (fl. 1840–1865) New York engraver. Worked for Woodbridge 1845; for Colton, *U.S.* 1849, *Ohio* 1852, *U.S.* 1865, and for H. Phelps *National Map U.S.* 1847

Atwood, Holmes & Read. *Pocket Atlas U.S.* 1884

Au, Herman. *Mapa de Guatemala* 1868 and 1876

Aub. See **Friend & Aub**

Aubant, A. d'. See **D'Aubant**

Aubel, O. Compiled *Geographical Games* 1883

Auber, H. P. East India Co. Service. *Survey Canton River* 1815–6 pub. 1818

Auber, L. Engraver for Mentelle 1797–1808

Aubert, Major Benoni (1768–1832) Born Copenhagen moved Norway 1790 and was employed in coastal surveys

Aubert, F. *Geolog. map Tunis* 1892

Aubert, L. Père. Script engraver. Port St. Jacques No. 87 Paris. For La Perouse 1797, for French Admiralty 1799, for Mentelle 1797–1801, for Buache (1798) for Paultre 1803, Humboldt 1811

Aubert, Fils. Engraver. For Chamouin 1812

Aubin. Script engraver. For Bretez. *Plan Paris* 1739

Aubonne, F. von Jenner von. *Carte des Cantons,* Bern (1820)

Aubrée, Armand. Publisher. Rue de Vaugiraud, 17 and Rue Tavanne No. 4, Paris. For Monin (1840)

Aubrey, John. *Map Surrey* 1673

Aubry, Nicolas. Jesuit. *Canada* 1713, *Limitte entre France et Angleterre* 1715

Au Capitaine, Baron Henri (1832–1867) French geographer

Aucher, Anthony (fl. 1538–1558) b. Appledore, d. Calais. *Chart Dover Harbour* ca. 1541

Aucuparius [Vogler] Thomas Heinrich (d. 1532). Printer of Strassburg. Wrote preface for 1522 edn. of Ptolemy

Audebeau. *Basse-Egypte* 1897

Audebert, D. (18th century) Dutch clock and globe maker

Audibert, Jean Pierre (1680–1763) b. Montpellier. Military cartographer

Audiffret, J. B. (1657–1733) French traveller and geographer. *Géographie* 2 v. 1691, *Géogr. Ancienne* 1689–94 and 1702

Audin. Publisher, Paris. *Carte routière d'Allemagne* 1836, *Cours du Rhin* 1840

Audry, C. *Carte des communications du Globe,* Paris 1891

Augouard, Father. *Carte du Congo* (1882)

Augspourger, Samuel. *St. Simons & Jekyll Is.* 1739

Augustin, Caspar. *23 roads leading from Augsburg,* 1631

Augustin, Vincenz. Freiherr von (1780–1859) Hungarian military cartographer

Augustine (St.) Bishop of Hippo. (395–430). His geographical views influenced mediaeval mapmakers

Augustodunensis. See **Honorius**

Augustus. Roman Emperior. (63 B.C.–14 A.D.) Had *World* map constructed

Auhagen, E. *Harzgebirge* 1867

Auld & Litchfield. *Eye sketch of Adelaide River* 1865

Auliacus. See **d'Ailly, Pierre**

Aulick, R. *River Jordan and Dead Sea* 1849

Aumale, Sr. d'. See **Damme**

Aunpekh, George (1423–1461) Viennese mathematician and astronomer

Auracher, Klemens (1748–1790) b. Netherlands, worked in Austria as a military cartographer

Aurelius, Cornelius, of Gouda (ca. 1460–1523) Poet and historian. *World map* woodcut Leiden, Severszoon 1514, *Cronycke van Hollandt,* Leyden 1517

Aurelius. See **Veen Adrian**

Aurelle de Paladines, Louis Jean Baptiste (1804–1877) *Operations army of Loire* 1870–71

Aurigarius. See **Waghenaer, L.J.**

Auriogalus [Aurigallus] Matthaeus (ca. 1490–1543) *Map of Saxony* for Munster

Auroux, Nico. *World* Du Val 1676

Ausfeld, Johann Carl. Engraver for Gaspari 1821, Stieler 1825, Schlieben 1830, Stulpnagel 1832

Austen, E. *Road Book London to Naples* 1835

Austen, Capt. F. W. *Orfordness–Bawdsey* 1812

Austen, Henry. *Liverpool with part of River Mersey* 1854

Austen (John). *Part of Is. Enisherkin Co. Cork* 1709 MS

Austen, J. T. Survey of *Fowey Harbour* 1813

Austen, Robert Goodwin. *Map England and Wales* 1865

Austen, Stephen T. Publisher. Newgate Street. *Map British Empire* 1733, *Salmon's Modern Hist.* 1744–6. Popple's *Brit. Empire America* 1746, *Roads in Highlands* 1746

Austin, Capt. (later Admiral) Horatio T. *Arctic* 1850–1, *Charts N.W. Passage* 1853 *S. Atlantic* 1854

Austin, Lieut. Loring. *E. Canada & New England* 1814 MS

Austin, Robert. Asst. Surveyor. *Interior W. Australia* 1854

Austin, Stephen F. *Texas* 1836, 1839

Autenrieth, E. L. *Isthmus Panama* 1851

Auteroche, Chappé d'. See **Chappé d' Auteroche**

Autolykos (ca. 315–240 B.C.) Greek mathematician and astronomer

Autrechy. Drew *plan Cazenovia, N.Y.* 1794

Auvergne, P. d'. *Topo. plan Isle Trinidada* 1787

Auvray, F. *Plan Caen* 1718

Auvray, P. L. Engraver. *Canton Basel* 1766

Auvray, Th. *L'Europe* (1630)

Auxerinus, C. See **Buondelmonte**

Auzolles, Sieur de la Peyre, Jacques d' (1571–1642) Geographer

Auzzani, G. *Contorni di Firenze* (1870)

Avanzo, Dominique. Lithographer. *Plan Bruges* (1855)

Avaulez. Mapseller Paris, Rue St. Jacques à ville de Rouen. *Brion's America* 1775

Aveele [Avelen] Jan van den (1655–1722) of Leiden. Cartographer, engraver, and mapmaker. Worked Amsterdam and Stockholm. *Zuid Holland* 1682, *Oldenburg* 1683, and for Dahlberg 1700–06

Aveline, A. Engraver for Cassini de Thury 1744–60; for Rocque 1750, for Le Rouge 1762, for Desnos (1769)

Aveline, Antonio. *Views of cities Copenhagen, London, Malta* 1665, *Basle & St. Cloud* 1680, *Rome & Tangier* 1700

Aveline, F. A. Engraver for Après de Mannevillette 1745, for Bellin 1751

Avellana, Miguel. *Maps of Spain* 1858–61

Avellar, Andrea d' (1546–1622) Portuguese cosmographer. *Sphaerae utr. tabella,* Coimbra 1593

Aventinus [Turmair] Johannes (1477–1534) Historian and cartographer. *Bavaria* 1523, used later by Ortelius

Averill, H. K. Jr. *Map of Decorah, Iowa* 1856

Averill & Hagar. *Clinton Co., N.Y.* 1879

Avery, David. Publisher. *Mitchell's Chart of Thames* 1733

Avery, Egbert. Surveyor. *South Nebraska City* (1855)

Avery, J. C. *Isthmus of Tehuantepec* 1851

Avery, Joseph. Hydrographer. *Chart south coast of England* 3 sh. (Arundel to St. Albans. Dorset) 1721, later editions 1731 (1775), revised by Burn 1794

Avet. *Generale Carte Prov. Napolitane* 1874

Avezac de Caster Macaya, Marie Armand Pascal d' (1799–1875) Geographer

Avico, Joseph. *Area covered by Treaty of Turin* 1754 MS

Avinea, Antonio. See **Wyngaerde**, Ant.

Avity, Sieur Pierre de Montmartin (1573–1635) b. Tournon d. Paris. *Estats et Empires du Monde* 1626 and later editions, *Amérique* 1640, *Neuwe Archontologia Cosmica* 1646

Avivas, A. *Plan d'Angers* 1875

Avril, Charles. Engraver for *Fonton's Atlas Asie-Mineure* 1840

Awnsham, Abel Swale. Publisher for R. Morden 1695

Ayala, Juan Manuel de (1745–1797) *Surveys coast of California* 1775–8

Ayling, Nick. *Survey Manor of Langley in Eling* 1692 MS

Ayloffe, Benjamin. Estate surveyor. *Dedham* 1573 MSS

Aymonier, E. Fr. (b. 1844) *Géographie du Cambodge* 1876, 1899

Ayot, D'. See **Dayot**

Ayrouard [Ayrouart] Jacques. 18th century. French hydrographer. *Recueil de plusiers plans des ports Méditerran.* 1732–46

Azambuja, Carlos Augusto Nascientes d'. *Chart Colonia (Brasil)* 1839 MS

Azara, Felix de (1746–1821) French traveller and cartographer. *Voy. dans l'Amérique mérid.* 1809

Azevedo, Jose da Costa. *Charts Brazil* 1860, *Rio Japura, Rio de Janeiro* 1871

Azevedo, P. de. Corrected *Vidal's Madeira* (1855)

Aznar, Pantaleon. Spanish publisher *Vazquez's Atlas Elementar* 1786

Azurara, Gomes Eannes d. Portuguese historian and geographer

Azuz, Abd el. *Map Damascus* 1879 MS.

Azzi, E. V. *Emisfero Occid. & Orient.* 1838

B

B. *Holstem* 1801

B., A. *Chales* (1565) *Genevra* (1565)

B., A. *Chart Tibee Inlet in Georgia* 1776

B., B. de. *Atlas Geographico* 1848

B., D. See **Bertelli**, Donato

B., F. See **Bertelli**, Fernando

B., G. A. *Wanderings apostle Paul* (1655)

B., H. F. *Schlesien* 1742

B., J. *Piscataqua River* (1700)

B., J. B. D. *Postes des Royaumes de Pays Bas* (1827)

B., J. G. *Carte Topo. Cherbourg* 1787

B., N. See **Béatrizet**, Nicolas

B., N. See **Bonifacio**, Natale

B., N. See **Bellin**, N.

B., W. *Hudson River* 1757

B., W. *Set Geographical Playing Cards* 1590

Baade, Heinrich. *Map of Mecklenburg* 1894–9

Baagoe, P. H. *Kart over Siaelland og Moen* 1813

Baalde, S. J. 18th century Dutch publisher. *Zac & Reis Atlas* 1788

Baarle, C, von. See **Barlaeus**

Baars, W. T. *Eilanden Java* 1842

Baarsal, Cornelis van. (fl. 1761–1826)

Dutch engraver for Ollefen. *Douay* 1794, *Amsterdam* 1795, *Haarlem* 1807, *Pays Bas* 1816

Baarsel, William Cornelius van (1791–1854) Son of Cornelis. Engraver *Holland* (1828?) *Amsterdam* (1865)

Babbage, B. H. Explorer in Australia. *Rough sketch Adelaide* (1853)

Babel, Florian (1771–1811) Austrian cartographer

Babel, Père. Traveller in Canada. *Routes Suivies de Babel* 9 sh. MS

Baber, Edward Colborne (1843–1890), *Chart Min River (China)* 1880, MS *Grosvenors Mission* 4 sh. 1878, *Exploration Western China* 4 sh. 1877–9

Babinet, Jacques (1794–1872) Invented *'un nouveau système de projection homolographique'. Atlas Universel* (1861) *Planisphère* 1865

Babinski, A. *Seas of N. America & Europe* 1869

Babylonian Map *Clay tables* ca 2000 B.C.

Babylonians. Introduced sexagesimal system, dividing sky into 360 degrees, degrees into minutes, and minutes into seconds. Also invented the gnomon.

Bach. *Plan Stadt Warschau* 1808 & 1809

Bach, C. P. H. (1812–1870) Geologist and topographer

Bach, Heinrich. Geological charts *Württemberg* 1850, *Germany* 1856, *Europe* 1859, *School Atlas* 1862

Bach, M. *Hist. Plan Nuremberg* 1882

Bachacek, Martin (1539–1612) Czech astronomer and cartographer

Bache, Alex. Dallas Supt. U.S. coast survey. *U.S. coastal charts* 1854, *Albemarle Sound* 1860, *Iso-magnetic lines Penn.* 1862

Bache, Hartman (fl. 1823–36) American cartographer *Charleston S.C.* 1823, *Kennebec River* (Maine) 1834, *Cape May roads* 1836

Bachelder, John B. *Map Battlefield Gettysberg* 1876

Bachelier, Jean. Géographe du Roi 1620

Bachelot, C. M. *Carte statist. de France* 1846

Bachiene, Willem Albert (1712–1783) Preacher and geographer in Maastricht. *Canaan* 4 sh. 1757, *Netherlands* 1768, *Kuilenberg Atlas (after Bowen),* Amsterdam, Schalekamp 1785, *Reis Atlas Duitschland* 1791

Bachiller, Doroteo (d. 1866) *Atlas de España* 1852

Bachmann, Fr. *Post u Reisekarte v. Deutschland* 1865

Bachmann, Isidor. *Carte géolog. Suisse* 26 sh. 1859–87

Bachmeier [Bachmayer] Wolfgang (1597–1685) *Map Ulm* 1653

Bachofen, Johann Heinrich, of Zürich. *Canton of Geneva* 1845

Bachot, Jérome. Géographe du Roi *Pais d'Aunis,* 1627, *Ville et Gouvern. de La Rochelle,* 1627

Bäck, Elias. See **Baeck**

Back, Sir George Capt. (Later Admiral) R.N. (1796–1873) Arctic Explorer. *Charts* in *Franklin's Narrative* 1823. *Expedition in H.M.S. Terror* 1858

Back, Jean Conr. Engraver *America after Popple. Plan chateau de Harburg* (1700)

Back, Joseph. Engraver. *Post u. Reise Karte Deutschland, Frankfort* 1843

Backenberg. *Carte d'une partie de la Saxe* 10 sh. (for Seven Years War)

Backer, Georges de. Mapseller in Brussels. *Plan Bruxelles* (1697)

Backer, Rommet Teunisse. Map for *Danckert's Atlas* (1710)

Backhoff, Edward. *Maps Sweden* 1875–7

Backhouse, James. Publisher in Darlington. *Map of Meetings [Quaker] Darlington* 1773

Backhouse, Thomas. *New Pilot S.E. Coast Nova Scotia* 1798

Backwell, T. Publisher. *Asia* 1742

Bacler D'Albe, Louis Albert Ghislain (1761–1824) Rue des Moulins au coin de St. Thérèse 542. Engineer, geographer, engraver. Director Topographical Bureau of War in Italy. *Theatre of War in Italy* 55 sh. 1798–1802

Bacon, Austin. English surveyor 19th century

Bacon, George Washington [& Co.] (1830–present day) 127 Strand (1874–1890's) succeeded to J. Wyld III ca 1893, *Seat War in Va.* 1862, *Comic war maps* 1877, *Ordnance Map British Isles* 1883, *Commercial & Library Atlas British Isles* 1895, *War Philippines* 1898. Firm survives today with W. & A.K. Johnston

Bacon, J. Barnitz. Surveyor. *Pier Map of City of New York* 1864

Bacon, Roger (1214–1294) Philosopher, geographer, and scientist. *Map of travels of Rubruck to Tartary* 1253–55 and *World Map*

Bacqueville de la Potherie. Maps in *Hist de l'Amérique Septentrionale* 1722

Badder, John. Surveyor. *Plan of Nottingham (with T. Peat)* 1744

Badeslade, Thomas (fl. 1719—45) Surveyor and mapmaker concerned with river improvements, fens reclamation etc. *Great Level of ye Fens* 1723, *River Ouse* 1723, *Thames* 1725, *History Navigation Port of Kings Lyn* 1725, *Wisbech River* 1735, *Chorographia Britannia (county maps)* 1741, editions to 1747

Badge, T. Engraver. *Lough Erne, Lough Neagh* (1815)

Badger, G. P. *Map Abyssinia* (1869), *Oman* 1871

Badger, S. C. *City Concord, New Hampshire* 1855

Badlidge, A. *Middlesex and Essex* (1768)

Badré, L. *Carte de l'Europe* 1838

Baeck [Bäck] Elias (1679—1747) Painter and engraver of Augsburg. *Atlas Geographicus* 1710, *Map Switzerland* 1710, *Amsterdam* 1720, *Augsburg* 1729, *Cologne* 1730, *Moscow* (1738)

Baeda, Venerabilis. See **Bede**, Venerable

Baedeker, Karl (1801—1859) German editor, geographer, and bookseller

Baehr, Johann Leopold von (1793—1893) Engraver of Leipzig. *Neuer Atlas* 1850, *Elbe* 1858, *Weser and Elbe* 1859

Baer, Johann Christian (fl. 1834—71) Engraver for Stieler 1834, *Post Roads Europe, Railway Atlas Germany* 1856

Baer [Bär] Joseph Christoph (1789—1848) Cartographer for Perthes' *Deutschland (with F.U. Stülpnagel)* 1846

Baerentzen, E. & Co. Publishers of Copenhagen. *Nörrejylland* 9 sh. 1841—8, *Sleswig* 1849

Baerle, Caspar Jan. See **Barlaeus**

Baermann. *Königreich Bayern* 5 sh. 1892

Baertls, Jos. Anton. *Plan de Mannheim* 1758

Baeyer, Johans Jacob (1794—1885) Numerous works on mathematical geography

Baffin, William (1584—1622) English navigator. *Charts Hudsons Bay* 1615 MS., *North of Davis Straits* 1616. *Mogol Empire* 1619, first English map of this area

Bagay, Nic. de la Cruz. Engraver and geographer of Manila. *Philippines* 4 sh. 1734, on 1 sheet 1744. *Aspecto geog. del Mundo Hispanico* 1761. *(World in form of a woman,) Velarde's Historia* 1749

Bagetti, Giuseppe Pietro. *Napoleonic Battle plans Italy* 68 pl. 1835

Bagg, Egbert. *Forest Hill cemetery Utica* (1870)

Bagge, Arthur H. *Map of Tenasserim* 4 sh. 1868

Bagge, T. *Map of The Fens* 1796

Baggesen, A. von. *Holstein & Lauenburg* 1827

Baggi, J. M. *Prov. da Bahia* 4 sh. 1888

Bagnoni, Jacopo. Publisher of *Dudley's Arcano* 1661

Bagot. *Carte topo. Dept. Rhone* 1868

Bahadur, Iman Sharif Khan. *Lake Darbat, Dhofar* 1894, *Dhofar* 1895, *Hadramaut* 1894 MS.

Bahr, Dr. *Heilquellen Deutschland und Schweiz* 1844

Bahre, Hans Georg. Engraver. *Map Austria* 1628

Bailey, A. M. *Pikes Peak Gold Region* 1859

Bailey, John (1750—1819) Surveyor, agriculturist, and engraver

Bailey, J. & E. Ramsey. Engravers for *Gill's Wexford* 1811

Bailey, W. L. *Chart Islands Kingsmill Group* 1833—41

Bailleul, [Baillieu, Baillard] F. l'ainé
Engraver. *Canal St. Denis* (1726), *Riv.
Senegal* (1750), *Prov. of France* 1752–4

Bailleul le Jeune. Engraver. *Paris* 1722
and 1745, *Europe* Lyons 1778

Bailleul [Baillieu, Bailleux] Gaspard
(18th century) Engineer and geographer.
Paris, au bout du pont au change vis a
vis l'orloge du Palais au Neptune
François. *Turin* (1700), *La Provence*
1707, *Chastellerie de Lille* 1707, *l'Alsace*
1708, *Portugal* 1725 *Foret de
Compiègne* 1728, *Rouen* 1731, *Brabant
and Limburg* 2 sh. 1744

Bailleul, N. *Guerre en Savoy* 1747,
Gouvern. Gén. Languedoc 1753

Bailley, F. & R. American publishers.
Scott's Atlas of U.S. 1796

Baillie, Alexander. Engraver *Great
Britain* 1746, *Lanark* 1775, *Loch Lomond*
1777, *Laurie's Midlothian* 1763, *Ayr* 1774

Baillie, Charles William (1844–1899)
Supt. Meteorological Office 1888–99.
Charts temperature Atlantic 1884, *Mean
Barometrical Pressure Atlantic &c.*
1887

Baillie, Thomas. *New Brunswick,* London,
Wyld 1832

Bailliere, F. F. *Gazetteers and Road Books:
S. Australia* 1866, *N.S.W.* 1870, *Victoria*
1865

Bailliere, H. Ethnograph. *Maps of Africa,
Europe, America, Polynesia* 1843

Baillieu. See **Bailleul**

Bailly, J. S. *Hist. de l'Astronomie ancienne*
1781

Bailly, Robert [Robertus de] (fl. 1525–
30) *Terrestrial Globe* 1530

Baily, Francis (1774–1844) Astronomer,
vice president Geographical Society.
Writer on longitude

Baily, John. *Map Central America* 1850

Bain, Nisbet. Contributed some maps to
Poole's Historical Atlas 1896

Bain, T. *Geological map S. Africa* 1875–6

Bainbridge. *Seat War Florida* 1839

Bainbridge, John (1582–1643) Astronomer.

Bainbridge, Thomas. Estate surveyor. *Plans
in Oxford, Berks, Worcs., Cheshire &
Shropshire* 1789–92

Baines, Engraver. *Kentucky,* London 1792

Baines, T. *Map Zambesi* 3 sh. 1859 MS.
Route to Transvaal Goldfields 1874, MS.
Goldfields of S.E. Africa 1876

Bains, Jean de (ca. 1577–1631) Royal
geographer. *Dauphiné* pub. Hondius
and Blaeu

Baird Smith, Lt. Col. R. Engineer.
Irrigation Canals Lombardy 1855, *Siege
of Delhi* 1876

Baist, George William. American
publisher. *Town Estate and Property Atlas*
1886–1909, *Atlas of Camden* 1886, *Atlas
of Philadelphia* 1888

Bakalowitz, Jan. Geographer to Stanislaus
Augustus, King of Poland, credited with
Regni Poloniae Magni Ducatus pub. later
in 1770 by Jacob Kanter in Königsberg.
Also plan of *Cracow* pub. 1772 by
Charles Pertees

Baker. Lithographer in New York.
Juliana Wis. 1824, *Brouckville N.Y.* 1836

Baker, Lt. A. *Survey Khyber Pass* 1842

Baker, A. E. *Plano Habana* 1837

Baker, Benjamin (fl. 1766–1824) Engraver
and publisher in the High Street Islington
1791 and Lower Street Islington 1798, for
Faden 1779, for *Univ. Mag.* 1791–7, *Surrey*
1793, *Wilkinson's Atlas Classica* 1796,
Kitchin's Univ. Atlas 1799, *Owen's
Spithead* 1801, *Plan London* ca. 1810. In
his later years Baker was employed by the
Ordnance Office and engraved the sheets
for first edition of the one inch ordnance
survey maps

Baker, Benjamin (the younger) (ca. 1780–1840) Engraver

Baker, B. R. Lithographer. *Course Nile* 1845

Baker, Charles. *County of Worcester* (with J. Pierce) 1793

Baker, E. *France* 1802, *S. America* 1806

Baker, Edward (1780–1821) Maker of Chronometers

Baker, Edward and Benjamin. *Plan Islington* 1793

Baker, F. Surveyor. *Plan Rochester* 1772

Baker, G. Bookseller near Royal Exchange. *Two large hemispheres by F. Lamb* 1679

Baker, Capt. George *Persaim River* 1754, *Coromandel* 1759–84, *Macao* 1780, *Sumatra* 1786, *S. Coast of Borneo* 1786

Baker, George H. *White Pine Mining District Nebraska* 1869

Baker, J. *Estate plans* 1792–1810

Baker, James T. *Map of Chicago* 1854

Baker, John. *Chart Harbour Rio de Janeiro* 1838

Baker, Lieut. Joseph (d. 1853) brother of Benjamin Baker, elder. Prepared *charts for Vancouver's Atlas* under his supervision 1798

Baker, Mathias. Surveyor. *Uffington Berks.* 1785 MS.

Baker, Richard Grey. Surveyor and publisher *Cambridge* 4 sh. 1816–20 pub. 1821

Baker, Robert. Surveyor General of Island of Antigua. *Antigua* 4 sh. 1748–9, *Antigua* 1775, in *Jefferys' West India Atlas* 1794, and later edition Laurie & Whittle 1810

Baker, R. W. *City Lowell, Mass.* 1868

Baker, Lt. Samuel, R.N. *St. Christopher* 1753, another edn. 1780

Baker, Sidney. *Cumberland Co. Maine* 1857, *Sagadahoc Co.* 1858

Baker, Sir S.W. *Route Khartoum to Albert Nyanza* 1862–5 MS.

Baker, T. Bookseller. *Twelve miles round Southampton* 1801, *Plan Southampton* 1802, *Isle of Wight* 1785, and later

Baker, T.W. *Ohio* 1861

Baker, W. *Engraved map N.S.W.* 1841

Baker, William. Surveyor. *Estate at Brill* 1768, *Snells Estate* 1768

Baker, William. *Railways England* (1867)

Baker, W. B. *Kickapoo City Kansas* 1855

Baker, W. E. Bengal Engineers. *Plan Battle Sabraon* 1846

Baker, W. E. *Ashtabula County* 1856

Baker, William S. G. *Atchison, Kansas Terr.* (1854)

Bakewell, Robert (1768–1843) Geologist. *Introduction to Geology* 1813

Bakewell, Thomas (1716–1764) Printer; book, map and print seller Next ye Horn Tavern in Fleet Street and Against Birchin Lane in Cornhill. *England & Wales* 1731, *Europe* 1739, *America* 1740, *Asia* 1742, *Africa* 1742, *View New York* 1746, maps in *Bowen & Kitchin's Royal English Atlas* 1762

Bakker, Frans de (fl. 1736–1767) Map maker and engraver of Amsterdam. *Namur* 1746, *Ville de l'Ecluse* 1748, for Dutch edition of *Anson* 1849

Bakker, L. E. *Ostfriesland* 1876

Bakonyi, Carl *Umgebung von Peterwardein* (with A. Brosch) 1872

Balbi, Adriano, elder (1782–1848) Geographer of Venice & Vienna. *Compendio di geografia* 1819, *L'Empire Russe* 1829, *Abrégé de Géogr.* 1832

Balbi, Eugenio (1812–1884) Geographer, son of Adriano.

Balbi, Johann Fried. (1700–1779) Engineer and cartographer in Berlin.

Balch, George Thatcher (1828–1894) American military cartographer. *Dakota Country* (1856)

Balch, S. W. *Central portion Yonkers* 1892

Balch, V. and S. Stile. Engravers of New

York. *Maps for Farmer's Michigan* 1831, and *Williams N.Y.* 1827.

Bald, William. Irish cartographer. *S. Uist and I. of Benbecula* 1805–36, *County Mayo* 25 sh. 1809–17, pub. 1830. *Western Isles* for *Thomson's Atlas* 1832, *Dublin & Limerick Railway* 1845

Baldaeus, Philippus. *Malabar* 1672

Baldauf, F. *Territoire d'Erfurt* 1813

Balde, G. *Piano Topo. citta Trieste* (ca. 1820)

Baldelli, M. Francesco. Translator of vol. 2 Lilio, *Breve desc. del mondo,* Venice 1551

Balding, R. *Map Town of Geelong* 1864

Baldrick, J. *Plan Cadiz* 1739

Baldwein, Eberhard (d. 1593) Instrument maker

Baldwin, Charles Candee (1834–1895) American geographer.

Baldwin, C. & R. Publishers. Bridge St. Blackfriars. *Honduras* 1809

Baldwin, Cradock & Joy. *Wyld's General Atlas* 4 to 1822, *Baldwin & Cradock Terrestrial Globe in 6 maps* S.D.U.K. 1831

Baldwin, E. Publisher. Hoxton's *Chesapeake Bay* 1735

Baldwin, George R. American surveyor and engraver. *Plans of Charlestown* and *Massachusetts* 19th century.

Baldwin, James Fowle (1782–1862) American engineer and surveyor

Baldwin, Loammi (1780–1838) American engineer and surveyor

Baldwin, Oran. Civil engineer. *Map town Montrose, Iowa* 1853

Baldwin, Richard (fl. 1681–1698) Bookseller, Ball Court near the Black Bull, Great Old Bailey 1681, Near the Oxford Arms Inn Warwick Lane 1690. Plan *Mons* 1692

Baldwin, Richard junior (d. 1770) Publisher and bookseller at the Rose in Paternoster

Row. *London Magazine* 1746 &c. *Virginia* 1755.

Baldwin, Robert. Bookseller. Continued the business of Richard Baldwin 1770 to 1810

Baldwin and Cradock. Publishers 47 Paternoster Row. Later Baldwin Cradock & Joy. *Society for Diffusion of Useful Knowledge* 1833.

Bale and Woodward. *New Celestial Globe* 1845

Baleato, Dr. Andres. *Plano Intend. de Truxillo* 1792, *Cuzco* 1792, *Montanas Peru* 1795

Balestrier, Louis de. *Cartes Commerciales Mexique* 1889

Balfe, James. *Map Mount Wellington* 1860

Balfour, Prof. Isaac Bayley. *Map of Socotra* 1880 published Edinburgh 1888

Ball, Allan Edward. *Trig. Survey N.E. Coast Africa* 1848, *S.E. Coast Arabia* 1839

Ball, John (1818–1889) English geographer and naturalist. Treatises on physical and geographical science

Ball, Richard. *Plan Algiers* 1776

Ball, Sir Robert Stawell (1840–1913) *Atlas of Astronomy* 1892

Ball [Balle] William (d. 1690) Astronomer

Balla, Antonius. *Nova Tab. Geogr.* 1765, *Mappa Spec. Region. Coeli* 1793

Balla, (1739–1815) Hungarian cartographer

Ballanti, G. Publisher. *Carta del Gran Chaco,* Faenza 1763

Ballantine. Lithographer. *Plan Dundee* 1827, *Plan Paisley* 1828

Ballard, Christophe. Publisher of Paris. *Buache's Consid. Géogr.* 1753, *Du Caille's Etrennes Géogr.* 1760

Ballard, Samuel (fl. 1704–09) Publisher at the Blue Ball in Little Britain, London. *Wars of Europe* 1704

Ballard, Capt. V. V. *Gorontalo River* 1798, *Kema Roads* 1801, *Chart Moluccas* (1811)

Ballast Office, Dublin. *Lights on Coast of Ireland* 1830

Balleis. *Grundriss Stuttgart* (1780)

Ballendine, John. *Potomack and James Rivers* 1773 MS.

Ballino, Giulio (fl. 1560–69) Italian geographer. *Illustri città et fortezze del Mondo,* Venice, Zaltieri 1569

Ballivian, José. *Mapa Coro. de la Repub. Bolivia* 1845

Ballivian, Manuel Vicente (b. 1849) Bolivian geographer. *Various works on South American geography*

Ballot, Victor. *Etab. Français Bénin* 1889

Balluseck, Lt. F. V. *Topo. Kaart Krawang (Indonesia)* 1877

Balthasar [Baltesar] van Berckenrode, Floris (1562–1616) Surveyor, goldsmith and engraver of Delft. *Ostend* 1604 (used by Braun and Hogenberg) *Delfland* 1606 (used later by Hondius) *Rijnland* 1610–15

Balthasar, Floriszoon van Berckenrode (1591–1644) of Delft. Surveyor and engraver in Amsterdam and Hague. Son of above

Balthasar, Cornelis Floriszoon van Berkenrode (1607–1635) of Delft. Cartographer in Amsterdam. *Dieren on Visschers Ysel-stroom* (1630), *Siege Bolduc* 1629

Baltimore, Cecil Calvert, 2nd Baron. Founder of Maryland. *Relation of Maryland* 1636 contains first separate map of the colony

Baltzer, A. *Carte géolog. de la Suisse* 26 sh. 1859–87, *Geolog. K. Kanton Bern* 1889

Baltzer, C. H. *Celestial Wall Map* 1854

Balugoli, Alberto (d. 1579) Cartographer of Modena. *Map Modena* 1571

Bamberg, Karl. *Europa* (1877) *Süd America* (1889) *Schulwand Karte* 1894–5

Bampton, Capt. William. Marine surveyor. *Strait of Sunda* 1787, *Chart of Torres Strait from surveys of Cook & Bampton* pub. Arrowsmith. Inset *chart of Bampton Shoal* 1798

Bancker, Evert. *New Jersey* 1754

Bancroft, A. L. & Co. Publishers. *Mineral Map Utah* 1871, *Central California* 1871

Bancroft, Col. H. H. & Co. Booksellers 609 Montgomery St. San Francisco. *Oregon, Washington, Idaho, Montana & British Columbia,* San Francisco 1868

Band, Otto. *Wandkarte des Kreises Colmar* (1874)

Banduri, Anselmo. *Tabula Geografia (Asia Minor and Greece)* (1718), *Constantinople* 1711

Banfield, J. A. Surveyor and geologist. *Mineral map region Lake Superior, Mich.* 1864

Banister, J. Engraver for *Throsby's map of Leicester* 1777

Banister, T. C. Cartographer. *Warwickshire* 1825

Banker, Gerard. City surveyor of *New York* 1772, *Sir Peter Warden Estate* 1773

Bankes, Henry. Surveyor. *North Meols* 1736

Bankes, Richard. Surveyor. *Manor of Hardmead (Bucks.)* 1638 MS.

Bankes, Thomas. *Universal Geography* (together with E. W. Blake and A. Cook) 1787

Bankoku, Sozu. *World Map* 1645 MS.

Banks, Sir Joseph (1743–1820) Accompanied Cook in his voyage round world 1768–71

Banks, T. B. *Map Guernsey* (1897)

Banks, William. Engraver. *Lakes of Cumberland* 1860

Banning, Emile (1836–1898) Belgian geographer. *Studies on Congo Afrique* 1878

Bannister, John. Surveyor. *Town Two Rivers* 1831

Bansemer, J. M. *Atlas of Poland* (with P. Zaleski Falkenhayen) Wyld 1837

Baptist Missionary Society. *Central Africa* 1884, *Cameroons* 1884

Baptista, Pedro Joao. *Lands of Cazembe* 1873

Baptiste, J. *Mayo* 1585, *Mayo and Sligo* 1587 MS.

Baquol, Jacques (1813–1856) of Strassburg. *Atlas Historique* 2 vols. 1860

Baquoy, Ch. *Engraved Front Title Atlas Universel M. Robert* 1757

Bär. See **Baer**

Baraga, Fr. X. *Istria* 1778

Baralle, Alphonse. French cartographer *Atlas Universel* 1877

Baranovski, Stepan Ivanovic (1817–ca. 1890) *Physical map World,* St. Petersburg 1847

Barateri, Marc. Ant. *Stato di Milano* 4 sh. 1636, *Lombardia* 1701

Baratta, Allessandro. *Urbis Neapolitanae delin.* 1627

Baratti, A. L. J. *Map battle of Stockach* 1800

Barbakstini, J. *Lucca* 4 sh. 1804

Barbalan. See **Barbolan[o]**

Barbantini, Tommaso. *Topo. prov. Ferrarese* 1825

Barbari, Jacopo de (ca. 1440–1515) Engraver. *Venice* 1500 woodcut 6 sh.

Barbeau de la Bruyère, Jean Louis. *Mappemonde* 1749

Barber, E. *Geogr. England and Wales* 1877

Barber, John. (fl. 1770–1800) Engraver. *Taylor and Skinner, Roads of Scotland* 1776, *Chandler's Greece* 1776, *Ireland* 1777

Barberi, Rafael. *Plano de la Ciudad de Mexico* 1867 (with Espinosa)

Barbey, Antonio. Engraver. *Dalmatia, Rossi, Roma,* 1689, *Piedmont* 4 sh. 1691,

Hibernia 1692, *Italy* 1693, *Liguria* 1697, *Roma* 1697, *Romagna* 1699

Barbié du Bocage, A. F. *Gallia Ulterior vel Transalpina* 1818

Barbié du Bocage. *Carte de la Province d'Alger* 1830

Barbié du Bocage, Jean Denis. (1760–1825) French geographer and cosmographer. Pupil of d'Anville, one of the founders Geogr. Soc. of Paris. *Voyage Anacharsis* 1781–88. Corrected d'Anville's *Hémisphères* 1786, *Ancient Greece* 1791, *Recueil de cartes géogr.* 1791, *Mer des Indes* 1811, *Plan Constaninople* 1822

Barbie du Bocage, Jean-Guillaume. Junior. (1795–1848) Geographer. Son of preceding. School maps, *Maps for Guigoni's Atlante Geogr. Univ.* 1854

Barberini, Cardinal Francesco (1597–1679) Founder Biblioteca Barberiana

Barbier, J. V. (1840–1898) Author of several maps and geographical memoirs. *Colonies* 1883

Barbier de Menard, Casimir (b. 1827) *Dictionnaire Géogr. et Hist. de la Perse* 1861

Barbieri (18th century) Italian engineer and geographer

Barbin, Alexandre Fred. (1798–1837) *Traité de Géogr. Gén.* 1832

Barbin, Claude. Publisher. *Tavernier's Voyage* 1676, *Relation de la Chine* 1688

Barbinais, L. G. de la. *Charts* in *Voyage autour du Monde* 1728

Barbolan[o] Fra Michelli. *Universale Orbe della Terra,* Venezia 1514 MS.

Barbolano, Hieronymo MS *World* map Venice 1525

Barbot, R. *Atlas Nat., Plan de Bordeaux* (1874) *Brest* (1875) *Nantes* 1875, *Mappemonde* (1878)

Barbou, J. Publisher. *Brion de la Tour's Atlas et Tables Elémentaires* 1777

Barbuda, Luis Jorge de [Ludovicus Georgius] Portugese Jesuit cartographer. *China* for Ortelius 1584

Barcaiztergui, V. *Costa da Cuba* 1793

Barcheston, Warwick. *Tapestry maps.* See **Sheldon**, William

Barclay, G. Engraver. *I. of Wight* 1832

Barclay, Henry. *Exploration between Alice Springs and Eastern Boundary Adelaide* 1878

Barclay, Rev. James. *Complete Universal Dictionary* Virtue (1840—52). Some editions have Moule's county maps, others have maps engraved by E.P. Becker's Omnigraph

Bard, I. L. *Pianta . . . della Citta di Firenze* 1739

Bardin, A. G. *Binghampton N.Y.* 6 sh. Philadelphia 1872

Bardin, J. M. (fl. 1782—1800) Globe-maker of Paris. *Terrestrial Globe* 1799, *Celestial Globe* 1800

Bardin, Wm. and J. M. (fl. 1780—1827) Globe makers. *Celestial Globe* 1785, *Terrestrial Globe* 1807

Barengi, Jean (17th century) Italian astronomer. *Consid. . . sistemi Ptolemaico e Copernicano*, Pisa 1638

Baren, Jodocus van der. *Louvain* (1620?)

Barents [Barendsz, Barentszoon, Bernardus or Bernardi] Willem (1550—1597). Dutch navigator and cartographer, born Amsterdam. Attempted a N.E. passage to China 1594—96. *Caertboeck van de Midlandtsche Zee* 1593—95, the earliest printed portolan atlas of the whole of the Mediterranean. French edition 1598, later editions to 1629. Some *charts* in later editions of *Waghenaer and Linschoten* 1599, post-humously

Barentsen, Dirk. *Il vero disegno dell' Isola di Malta* Venice (1550)

Bareri, Alfonse di. *Teatro delle Citta del Mondo* Venice, Rasciotti 1600

Barfield, J. Publisher, No. 91 Wardour St. London. *Genealogical Chron. Hist. and Geogr. Atlas* 1818, 1834—36

Barkeley, Sir Francis. Surveyor. *Ulster, Tyrone, and Sligo* 1601

Bar Kepha, Moses (d. 982) *Map of South Syria*

Barker, Elihu. *Map Kentucky* Philadelphia, Carey 1794. Reissued smaller scale 1795 and 1797

Barker, James K. Surveyor. *Map City of Lawrence, Mass.* 1853

Barker (Capt. Jno.) Marine surveyor. *Draught Coekle and St. Nicholas Gatts* 1753

Barker, J. N., Surveyor. *Route to Shoa (Abyssinia)* 1841

Barker, Robert (1570—1645) Northumberland House Aldersgate St. King's Printer, licensed to print charts and maps. *Sea chart English coast* 1604

Barker, William. Engraved maps for *Carey's American Atlas* 1794—95, for Payne 1798

Barker, William Charles. *Trig. Survey Africa Coast* 1844

Barker, William J. Surveyor. *Clark's map of Fairfield Co. Connecticut* 1858, *Washington Co. Penn.* 1856

Barker, W. P. Surveyor. *Montgomery, Alabama* (1870)

Barlaeus [van Baarle, Baerle] Caspar (1584—1648) Historian, philosopher and theologian of Amsterdam. *Rerum in Brasilia Gestarum Historia* 1647, with large scale map used by Blaeu. Oversaw the maps in *Herrera* 1622

Barlow. Engraver. *Plan of Canterbury* (1815)

Barlow, Francis (1626—1702) Engraver for title page *Ogilby's Britannia* 1675

Barlow, Fred. *Plan Foundling Estate* 1844

Barlow, Henry William. *St. Helens (Ordnance Survey)* 1851

Barlow, I. Engraver. *St. Albans Abbey* (1790)

Barlow, James S. Engraver. *Chart E. Scotland, West Indies* Dublin, Grierson (1749) *Celest. Hemispheres* (1790) *Telford's Road map Scotland* 1803

Barlow, Peter, elder & younger. *Magnetic Curves* 1833

Barlow, Peter. *Chart of Magnetic Curves Edinburgh,* Black (1842)

Barlow, Peter William Surveyor *South Eastern Railway* 1835

Barlow, Robert. Draughtsman. *Canada* (1855)

Barlow, Roger (d. 1554) Cartographer and navigator. Sailed with Sebastian Cabot

Barlow, T. Engraver. *S. Part of Great Britain* in *Moll's Atlas* Dublin, Grierson 1732

Barlow, William. *Solar System* (1750)

Barmston, A. S. *Vancouver Is.* 1864 MS.

Barnard, Capt. Charles H. *Chart Falkland Is.* 1829

Barnard, John (Fl. 1685–1693) Corrected and enlarged *Bohun's Geographical Dict.*

Barnwell, Col. Engineer to the Province of Carolina. *Mouth of the Altamaha Region* 1722 MS, used later by Popple and Mitchell

Barnard, Col. (later Brig. Gen.) John Gross. *Isthmus of Tehuantepec* 1851. Inset on *Kiepert's Central America* Berlin 1858, *Plan siege Yorktown* (1862)

Barnard, Thomas. *Coast Coromandel* 1778

Barnes, John. *St. Helena* Admiralty 1846, French edn. 1851

Barnes, M. C. *Topo. map of Danville Penn.* 1870

Barnes, R. L. *Railway maps Pennsylvania* 1868–74

Barney, Joshua. *Railway routes in U.S.* 1831–45

Barney, W. J. *North E. Iowa and S.E. Minnesota* 1863

Barnikel [Barnicelius] John [Johannes] Chr. (d. 1746) *Le Duché de Courlande* 1747 for *Homann's Grosser Atlas*

Barnsley, Lt. (later Capt.) Henry (1742–1794) *Draught of Rattan* 1742 and 1775, *Chart New England* (1760)

Barnston, Brevet Major. Surveyor. *S.W. Crimea* 28 sh. 1856

Barnum, Wallace. *Surveys in Nebraska* 1856–60

Barocci, Giovanni Maria (fl. 1560–70) Italian globemaker

Baron, Bernard (1696–1766) Engraver for Popple 1733

Baron, W. H. *Map City of Sydney*, Woolcott & Clarke 1854

Baroncelli, Bernardo. *Plan Vélo des Environs de Paris* (1892)

Barotte, J. *Carte Géolog. Haute Marne* 1859–65

Barraga, Fr. X. *Istria* 1778

Barral, Jean Augustin (1819–1884) *Atlas du Cosmos* 1861–67

Barralier, Capt. F. de (1773–1853) Marine surveyor to expedition to Bass's Strait. *Trig. survey Barbados* 2 sh. 1825 [7]

Barrallier, Ensign. Surveyor. *Coal Harbour & Hunter River N.S.W.* 1801

Barras de la Penne. *Détroit de Gibraltar* 1756

Barraud, Charles D. *New Zealand graphic & descriptive* 1877

Barre, Le Febure de la. *France Equinoctiale* 1666

Barreiro, Fran. Alvarez. *Nueva Mexico* 1727 MS, *Extremadura* 1729, *Nueva Mexico* 1770

Barren, A. *Collectorate of Trichinopoly* 1835

Barreo, Maps for *Wolfgang's Atlas Minor* 1689

Barrera, Daniel. *Peru* 1871

Barrère, Henry, *Afrique* 1893

Barrère, Pierre. *Coast French Guiana & Brasil* 1743

Barres, J. F. W. Des. See **Des Barres**

Barreto, Don Manuel. *Carta nautica Yndias occid.* (1680)

Barrett, Lucas (1837–1862) Geologist. Director of geological survey of Jamaica 1859, *Geolog. map neighbourhood of Cambridge* 1859

Barretté, Thomas George Leonard, 23rd Royal Welch Fusiliers. *Battle plans War of Independence America* 1780

Barrière. Engraver. *Campagne en Italie* (1799). *Répub. Italienne* (1803)

Barrière, Freres. Engravers for *Delamarche Atlas* 1820

Barrière, Marcellin. *Système Planétaire-Barrière Cosmographie* 1872

Barrington, Hon. Daines (1727–1800). *Miscellanies* 1781, *Map North Pole* 1818

Barrois, l'Aîné. Publisher of Paris, Rue de Seine. *Soult's Atlas Militaire* 1809

Barrois, Charles. *Carte géolog. de la France* 4 sh. 1889

Barron. Australian surveyor. *S.W. Lake Eyre* 1877–8, pub. Adelaide 1879

Barron, Edward. *Plans of Forts in Nova Scotia* 1779

Barros Gomes, B. *Cartas Elém. de Portugal* 1878

Barros, João de (1496–1570) Portuguese historian and geographer. *Asie Portugaise,* Lisbon 1552

Barrow, John. Surveyor. *Lady Bay Warrnambool* 1853, *Town of Mount Gambier* 1869

Barrow, John. *Navigatio Britannica* 1750, *History of Discoveries* 1756

Barrow, Sir John (1764–1848) English explorer, private secretary to Lord Macartney at Cape of Good Hope. *Shantung* 1793, *Cape of Good Hope* 1800, *Voyage to Arctic Regions* 1818, pub. 1846

Barrow, T. T. *Plan Tunbridge Wells* 1808

Barruffaldi, A. *Corografia del Ducato di Ferrara* 1758

Barry, Fred. Engineer *Grand Junction Railway Ireland* 1853

Bartelot [Bartlet] John (fl. 1530–1541) Seaman and navigator. *Chart Dover* 1541

Barth, B. *Siege of Vicksburg* (1863)

Barth, F. Plans of *Wien, Gratz, Komorn* (1855)

Barth, Dr. Heinrich (1821–1865) of Hamburg. German explorer in the service of Great Britain. *Travels in North and Central Africa* 1850–55

Barthelemier, J. Publisher of Paris, Quai des Grands-Augustins 39, ci-devant Quai Conti. Successor to Ch. Picquet. Proprietor of the maps and stock of Brué. *Atlas Universel de Géographie, A. Brué, nouvelle édition revue et augmentée . . . par Ch. Piquet,* Paris 1865

Barthélemy, Abbé Jean Jacques (1716–1795) *Recueil de Cartes Géogr.* 1789, *Voyage du Jeune Anacharsis en Gréce* 1789, *Atlas Oeuvres Complètes* 1822

Barthelet [Bartlett] Richard, of Norfolk, d. Donegal 1603. Cartographer *Ulster* 1600–3, *Blackwater* 1602 MS.

Bartholomaeus, Fr. *Atlas von dem Preussischen Staate* 1859

Bartholomew, Firm of Publishers. George 1784–1, John 1805–1861, John II 1831–1893, John George 1860–1920. George the founder apprenticed to Lizars; John II worked under Petermann; John George introduced layer coloring 1888

Bartholomew, J.G.

Bartle, H. (b. 1840) Works on mathematical geography

Bartlett, F. A. *Survey Borough of St. Marylebone* 1832, 1834, 600 feet to inch

Bartlett, G. *Index Geologicus* (1841)

Bartlett, Richard. See **Barthelet**

Bartlett, S. M. *County Atlas Monroe Michigan* 1876

Bartoli, Papirus and Simone. *Civitas Nova in Piceno* 1613

Bartolomeo dalli Sonetti [Zamberti] (fl. 1477–85) Venetian sailor and cartographer. *Isolario* 1485 and 1532 – a description of the Aegean Sea in sonnet form with maps. The first printed atlas of the Greek Islands, printed portolan, the first founded on actual measurements.

Bartolozzi, F. Artist and engraver. Designed and engraved vignettes on title pages of some atlases and contributed to *Wilkinson's Atlas* 1802

Barton, James. *N.W. Borneo* (1780), *Palawen Is.* 1781

Barton, Rev. John. *India* 1873

Barton, T. A. and A. R. Flint. *Owlshead Harbour* 1836

Barton, W. *City Troy New York* 1858, 1869

Bartram, John. *Town of Oswego* 1751

Bartram, J.

Bartruff, Fried. Johann Jakob (1754–1833) Military cartographer

Bartsch, Jakob. Professor Mathematics at Strassburg. Son-in-law of Kepler. *Planisphaerium Stellatum* 1661

Bartsch, Johann Gottfried (d. 1690) Engraver in Berlin

Baruffaldi, A. *Ferrara* 4 sh. 1758, corrected edition 1782

Baruffi, Abbé Guis. (b. 1800) Italian savant. *Peregrinazioni auturnali* 1841–43

Barwise, J. (19th century) American geographer

Bary, H. (1640–1707) *Map of Palestine in Bible* 1672

Bas, F. de. *Waterwegen in Nederland* 1880

Basanta, Joseph. Surveyor General. *Island of Trinidad,* Arrowsmith 1838

Bascarini, Nicolo (fl. 1547–8) Venetian publisher of *Gastaldi's edition of Ptolemy* 1548

Basch, A. & Co. Publishers of Sydney. *Atlas of N.S.W.* 1874

Basch, L. *New Victorian Counties Atlas* 1874 *Settled Co. N.S.W.* 1874

Baschin, Adolph Carl Otto (b. 1865) of Berlin. *Bibliotheca Geographica*

Baseggio, Giacomo. *Carta del Mar Negro* 1793

Basil [Basyll, Basill, Basilius, Simon]. Surveyor of H.M. Buildings 1608, *Plans of Hertford Castle, Hatfield House, Plan Ostend* 1590, *Sherborne Lodge* (1600)

Basire, Isaac (1704–1768) Engraver for *Kirby's Suffolk* 4 sh. 1736, *Drake's Yorks.* 1736, *Lemprière's Cartagena* 1741, for *Rocque* 1749, *Tindal and Rapin* (1750), *London* 1763

Basire, James (first of the name) Engraver. *Gibraltar & Mahon Harbour,* in Rapin 1745. *Siege Pontefract* 1750, *Encampment Portsmouth* 1788, *Siege Boulogne* 1788

Basire, James (second of the name) (1769–1835) Engraver. *Coal Districts Northumberland* 1800, *Parish of Walley* 1801, *River Thames* 1803, *Bogs of Ireland* 66 sh. 1810–14, *Balds Mayo* 1812–14, *Mary-bone Park Estate now called Regents Park* 1824, *Swan River* 1827

Basire, James (third of the name). Engraver. *Plan Roman Villa* 1832, *Exeter Canal* 1839, *Magnetic Curves* 1833

Baskin, O. L. with Everts and Stewart. *Ogle Co. Illinois* 1872, *McHenry Co.* 1872

Bass, George (d. 1812) Explored coast of N.S.W. and circumnavigated Tasmania. *Twofold Bay* 1798, *Pacific Isles* 1802, *Strait Malacca* 1805

Bassac, Edmond. *Carte Golfe de Morbihan* 1870

Bassac, N. *Ville et environs de Vannes* 1898

Bassani [Bassanus] Bernardino. Engraver. *Milan* 4 sh. 1636, *Disegno della Lombardia* 1701

Bassantin, James (d. 1568) Scottish astronomer. *Astronomique Discours,* Lyons 1557

Basset le jeune (fl. 1755–1839) Parisian publisher. Rue St. Jacques 64. *Lorraine & Bas* 1762, *Théatre Guerre en Amérique* 1779, *Golphe du Mexique* 1780, *Asie* 1809, for Lapie 1817–19, for Hérisson 1829

Basset, Thomas (fl. 1659–93) Bookseller at the George in Fleet St., later Westminster Hall. pub. *Heylin's Cosmography* 1669, *Blome's World* 1670, *Speed's Atlas* 1676 (with Chiswell)

Basset & Christian. Surveyors. *New North Road London to Highbury* 1812

Basset, Christian & Hodskinson. Surveyors. *Parish St. Mary Kensington* 1822

Bassett, A. Surveyor. *Boundaries parish of Ealing* 1777

Bassett, N. Surveyor. *Woodgatesen Farm* 1705

Basso, [Bassus] Francesco (fl. 1560–70) Cartographer and instrument maker of Milan. *Globe* 1570

Bassot, General L.A. Leon (b. 1840) *mathematical geography Meridian France* 1870–1892

Bassus, Franciscus. See **Basso**

Bast, Pieter Jacobs (1570–1605) Publisher, engraver and geographer in Emden and Leiden. *Birds eye plan Amsterdam* 1595, reprinted *Allard* 1599, *Visscher* 1611

Bastard, Ben. Surveyor. *Manor Osborne* 1766

Bastide, Capt. Henry. *Plan Harbour & fortifications of Louisbourg* 1745, *Annapolis Royal* 1751, *Harbours of Nova Scotia* 1761

Bastius, Petrus. Author, engraver and publisher. *Plan Leiden,* Rome 1600

Basyll, Simon. See **Basil**

Batacha, Caetano Miria (fl. 1855–78). Portuguese cartographer. *Carta de Belenga* (1855), *Leiria* 1859, *Lisboa* 1857, *Barro do Porto* 1871, *Lisbon* 1878

Batchelor, Thomas. Publisher in conjunction with Christopher Browne ca. 1690

Bate, H. C. *Underground survey prinicipal mines at Stawell (Victoria)* 6 sh. 1879

Bate, R. B. 21 Poultry, London. Agent for *Admiralty Charts,* 1814–49. Sold *Smith's Syria* 1840

Bate, William Thornton. *Mer de Chine* 1863

Batelli & Fanfani. Publishers in Milan. *Mappemonde* 1820

Bateman, Henry William. *District round Shanghai* 1864

Bateman, John. Engraver for Horsburgh *E. India Pilot* 1813–31. *Survey N. & S. Sands Malacca* 1820

Bateman, J. F. *Maps of Wales for Water Supply* 1870

Bateman, Richard. Engraver. *Singapore Harbour* 1829, 1830. *Kyoak Phyoo Harbour* 1833

Bateman, Thomas. Estate surveyor. *Anderson Estate St. Neots* 1757, *Old Priory and Priory Farm* 1757

Batenham, George, Junior. Engraver. *Plan Chester* 1821

Bates, Elias. *Plan Merjee* 1725, *Malabar Coast* 1775

Bates, Henry B. *Walks about Buxton* (1850)

Bates, H. S. *New Plymouth N.Z.* 1863

Bath, Neville. Irish cartographer. *Co. Cork* 6 sh. 1811

Bathalos, Manoel. Portuguese cartographer 17th century

Bathurst, J. *Part Lower Egypt* 1801

Batley, R. Engraver. *River & Sound of Dawfoskie* 1773, *East Coast England by J. Chandler* (1780)

Battatzi [Vatace] Basil, of Constantinople. Greek cartographer of Central Asia 1732

Battersby, William. *Travelling map England* 1680

Battey, Jer. Bookseller at the Dove in Paternoster Row. *De Lisle's S. America* 1725

Bauchope, Robert. Land surveyor. Oversaw maps for *Bute, Linlithgow,* and *Lanarkshire,* for *Thomson's Atlas* 1832

Baudet, Pierre Joseph Henry (1824– 1878) Mathematician and map historian

Baudin, Nicolas (1750–1805) French mariner, leader of expedition to explore Australia 1802–04,

Baudoin. *Maps for Tassin's Provinces de France* 1640–43, *Tavernier's Théatre Géogr. de France* 1634

Baudouin, Gaspar. *Charte de la Suisse* 1625

Baudouin, H. *Plan Stettin* 1828

Baudous, Robert Willemszoon de (1574– 1659) of Brussels. Engraver & publisher in Amsterdam

Baudrand, Abbé Michel Antonio (1633– 1700) Geographer to the King (of France) *La Grèce* (1688) *Postes d'Italie* (1695), *Stato della Chiesa,* Rome, Rossi 1669, *Geografia* 1682

Baudwin, Gabriel S. *St. Sebastian* 1740, *Draught coast Biscay* 1742

Bauer, Friedrich Wilhelm von (d. 1783) Military cartographer in Prussian and Russian service. *Théatre de la guerre* 4 sh. 1759, *Bataille de Crefeld* 1758, pub. 1765, *Vellinghausen* 1762, *Affaire de Meer* 1766

Bauer [Bacher] J. *Carte Hydro. et de poste,* 4 sh., Vienna, Mollo 1808; *Globe* ca. 1800

Bauer, Robert. *Reise u Gebirgs Karte von Salzburg* (1890)

Bauernkeller, George (fl. 1830–70) Relief cartographer and globe maker. *Deutschland* 1832, *Relief maps* 1839–69, *Embossed plan London* 1841, *Hand atlas* 1846

Bauernfeind, Karl Marx. *Railway map Bavaria* 1845

Baugh, Robert. Engraver. *Shropshire* 9 sh. 1808, *Evan's Six Counties of North Wales* 1795

Bauland. *Special Karte von dem Odenwald* 1808

Baum, Heinrich S. S. (fl. 1573–75). Astronomer of Munich

Baum, Johan Christoph. *Map of Mansfeld* 1732

Bauman, Major Sebastian (1739–1803) German, emigrated to New York to serve in the French and Indian Wars. Capt. in New York Regiment of Artillery. *Plan Yorktown in Virginia* 1781, pub. Philadelphia, Tanner 1825. *Investment of York and Gloucester* 1781

Baumann, Joseph Martin (1767–1837) Relief cartographer

Baumann, Dr. Oscar. *Insel Fernando Poo* 1887, *Massai Expedition* 1894, *Mittlere Congo* 1888–90, *Deutsch Ost Afrika* 4 sh. 1894, *Sansibar Archipel.* 1897

Baumann, Victor. *Elsass-Lothringen* 1873, *Rench-Bader* (1877)

Baumann, William. *Pulaski County* 1871

Baumeister, I. Maps for Homann Heirs. *Stadt Atlas* 1762

Baumeister, Johann Jacob von. *Ducatus Mantuani* (1720), *Plan de Carlsruhe* 1739

Baumgartner, A. *Atlas von Sachsen* 1816, *Hypsometric maps of Austria* 1827

Baumgartner, J. J. (18th century) Cartographer of Baden

Baum, Johann (1617–1697) Historian and geographer

Baum, Konrad (ca. 1616–1671) Engraver and bookseller

Baur, Carl Friedrich. German cartographer. *Cairo* 1841, *World maps* 1871, *America* (1872) *Oesterreich* (1874), *Böhmen* (1876) *World maps* 1893

Baur, F. W. See **Bauer**, F. W. von

Bauser, G. W. *Württemberg & Hohenzollern* 1868

Bauza [Bausa] Felipe. Mapmaker to Malaspina Expedition, 1798, *Atlas Maritimo de España* 1789, *Manila Bay* 1792, *Antillas* 1802, *Seño Mejicana* 1829, *Colombia* 1830, *Central America* 1835, *Madrid* 1841

Bavier, S. *Die Strassen der Schweiz* 1878

Bawr, F. G. de. *Carte de la Moldavie* 6 sh. Amsterdam (1770)

Bawr, F. W. See **Bauer**, F. W. von

Baxter, F. *Plan Parish Elton* 1784

Bayer, Fritz. *Neuester Hand Atlas* 1894

Bayer, [Bayr] Johann (1572–1625) Astronomer, first to use Greek letters to denote magnitude of stars. *Uranometria* 1603

Bayfield, Capt. (later Admiral) Henry Wolsey, R.N. (1795–1885) Admiralty surveyor. *Gulf and River of St. Lawrence* 1828–65, 110 charts and plans. *Nova Scotia Pilot* 1856

Bayley, Lieut. T. *Environs Batavia* 1812

Baylor, H. B. and E. B. Latham. *Atlas of Atlanta, Georgia* 1893

Bayly, John (fl. 1755–1794) Engraver. St. Thomas Apostle, London. *Collet's N. Carolina* 1770, *Byres' Bequia & Dominica* 1776, *Stobie's Roxburgh* 1770, *Sheffield's Jamaica* (1770), *Speer's West Indies* 1771, *St. Vincent & Tobago* 1776, *Dorset* 1773, *Pennant's Scotland* 1777, *London & Westminister* 1761, *Martin's Geogr. Magazine* 1782, *Pitcairn* 1773

Bayly, William (1737–1810) Astronomer. Accompanied Capt. James Cook 1772 and 1776

Baylye, Jeremie. *Plott manor of Shingle Hall* 1618 MS.

Baynes, Lieut. *Plan Sevastopol* (with Capt. Anthony Charles Cooke) 1857

Baz, Gustavo. *Hist. Ferro carril Mexicano* 1874

Bazalgette, J. W. Engineer, Surveyor. *Central Kent Co. Railway* 1836

Bazendorff, J. F. von (18th century) Bavarian cartographer

Bazilevski, Ivan & Fedor. *Orenburg Province* 1880

Bazin. *Carte de Champagne et de Brie* 2 sh. 1790

Bazin, Francois. *Atlas de Géogr . . . de la France* (1855) *L'Europe Economique* 1887

Bazley, Thomas Sebastian. *Stars in their courses* 1878

Beadle, B. A. *Lands of Cazembe* 1873

Beal, John. *Hunting and Tourists' map of Sussex* 1885

Beale, E. F. *Surveys in California & Nevada* 1858–1863

Beall, George W. *Ann Arundel Co., Md.* 1860, *Prince George's Co.* 1861

Bean, Charles. *Elementary Atlas* (1854) *School Atlas* 1854

Bean, James Philip. *Classical Geography* 1835

Bean, John. *Charts of Coasts & Harbours Britain* 1781–1802

Beardsley, Charles. Map of *Elkhart* 1871

Beare, James. *World map* in *Beste's True Discourse* 1578 attributed to Beare

Bearhop, And. Map of *Baronry of Stobbo* [Peebles 1740]

Beato [Beatus] San [Saint] Liebanensis (d. 798) Benedictine monk born in Valcorado Spain. Biblical commentator. *World maps* to illustrate his text ca. 776, several variants.

Béatrizet [Beatrice, Beatricetto, Beatricius] Nicolas (ca. 1520–1570) Publisher, painter, and engraver in Rome. *Tramezinni's Germany* 1553, *Malta* (1566), *Rome* (1557), *Thionville* 1558

Beatson, Capt. A. *India campaigns* 1790–91

Beatson, R. *Tor Bay* ca. 1785

Beatty, John. *Survey Manor of Livingston* 1714

Beatty, W. Surveyor. *Township of Armour* (Ontario) 1877, *Township of Strong* 1878

Beatus. See Beato

Beauchamp, Adolphe [Alphonse] de. *Nouvelle Carte du Brésil* 1815

Beauchesne de Gauin. *Carte du Paraguay, Chili, Magellan* (1710)

Beaudouin, Major. *Maroc,* Paris 2 sh. 1848

Beauford, William. *Plan City and Suburbs of Cork* 1801, *Irish Atlas* pub. Grierson (1825)

Beaufort, Rev. Daniel Augustus (1739–1821) Founder Royal Irish Academy *Map Ireland* 1792 and later editions. *Meath* 1797

Beaufort, Capt. (later Admiral) Sir Francis. Hydrographer for Navy 1829–55, son of preceding. *Charts N. Atlantic* 1863, *N. Pacific* 1861. Prepared atlas used by Society for Diffusion of Useful Knowledge

Beaulieu, Sébastien de Pontault, Sieur de (1613–1674) French military engineer. *Conquêtes de Louis le Grand* (1643–84), *Villes de France* (1694)

Beaumont, Henri Bouthillier de (1819–1889) Swiss geographer. Founder and president of Geographical Society of Geneva.

Beaumont, Jean Baptiste Armand Louis Léonce Elie de (1794–1874). French geographer and geologist. *Deutschland* 1838, *Geological map France* 1840, *Europe* 1875

Beauplan. See Le Vasseur de Beauplan

Beaupré. See Beautemps-Beaupré

Beaupré, B. de. Engraver. Rue de Vaugirard Paris. *Territoire de Michigan* 1825

Beaurain, Jean Chev. de (1696–1772) Engineer and Royal Geographer. Quai des Augustins au coin de la rue Gilles Coeur quartier St. André aux Arts. Pupil of Moulart Sanson. *Mantoue* 1734, *Cartagena* 1741, *Flandre* 1755, *Brabant* 1756, *Jersey* 1757, *Cours du Rhin* 1760, *Allemagne* 1765

Beausard [Beausardus] Pierre [Petrus] (fl. 1535–1577) Physician and mathematician of Louvain. *Anneli Astron. Instrumenti usus* 1553, *Universalis Cosmographici* 1556

Beautemps-Beaupré, Charles François (1766–1854) Engineer hydrographer. Sous-chef du Dépôt Général de la Marine contre Amiral de France. *Neptune de la Baltique* 1785, *Atlas de la Mer Baltique* 1796, Compiled charts for Bruny-Dentrecasteaux 1792–3 pub. 1807, and Freycinet 1811. *Coasts of France* 1824–9 *Manuscript atlas of Ports of Gulf of Venice* compiled for Napoleon, 1806

Beauvais, Henri Baron de. *Relation journalière du Levant* 1615

Beauvilliers [Bauvillier] Sieur de. *Partie ouest Louisiane* 1720

Beavan, Capt. R. *Plan Cantonments Kandehar* 1879, *Marri Hills* 1881

Beavan, T. *Jubilee and Philip Rivers, New Guinea* 1887

Beaver, Capt. Philip. *Nautical map of W. Coast Africa* 1805

Beccaria, Giovanni Battista (1716–1781) Astronomer of Turin

Beccario [Becharius] Battista. (fl. 1426–35) Genoese Portolan maker MS. *Sea charts*

Beccaro, Francesco (15th century) Genoese portolan maker *Mediterranean and Black Sea* 1403

Bèche, Sir Henry Thomas de la (1796–1855) Geologist. Founder Geological Magazine. *Map England, 1827, Geological maps Ordnance Survey*

Bechein, J. L. Surveyor. *Saginaw River* 1856

Becher, Arthur. *Plan Simla* 1864

Becher, Comm. A.B., R.N. Hydrographer. *S. America* 1824, *Orkney* 1847–8

Béchet, Bernadin. *Nouvelle carte d'Europe* 1878

Bechler, Gustavus R. Map of *Allegany Co. N.Y.* 1856, *Lenawee Co. Mich.* 1857, *Campaigns Potomac* 1864, *World* 1862

Bechstatt, I. C. *Darmstadt u. umliegenden Gegend* (1820)

Bechtold, Christian von. *Worms und Umgegend* (1860)

Beck, A. A. Engraver. *Plan de la ville de Bronsvic* 1759, *Stadt Inverness* 1760

Beck, Edward (1820–1895) Topographer and relief cartographer of Bern. *Geological map of Switzerland* 1855, *Bern,* 1858, *Terre Sainte* (Relief) 1856

Beck, Hugo. *Central de Bolivia.* London, Wyld 1862

Beck, Lewis Caleb. (1798–1853) *Gazetteer of Illinois and Missouri* 1823 (maps & plans)

Beck, Thomas George. Map *Parish of St. Paul, Covent Garden* 1851

Becker, Capt. *Umgegend von Dresden* 1821

Becker, F. Q.M.G. Staff Darmstadt. *Hessen* (1874)

Becker, Fr. Lithographer. *Carte du District Indust. de Liège* (1875)

Becker, Francis Paul. *China* 1842, *Omnigraph Atlas Modern Geography* 1843, *Africa* 1853, *Post Office map Lincs.* 1861

Becker, George Ferdinand (1847–1919) Geologist. *Geology Comstock Lode, Wash.* 1882, *Quicksilver Deposits of Pacific* 1877–88

Becker, Johann (d. 1756) Engraver of Linz

Becker, Engraver for *Taylor's map of Barbados* 1859

Becker, Moritz Alois R. von (1810–1887) Austrian geographer. General works on geography and topography

Becker, Peter. See **Artopaeus**

Becker, E. P. & Co. Omnigraph (litho.) Engravers. Maps for some editions of *Barclay's Dictionary* (1840–52)

Beckeringh, T. Dutch lawyer and cartographer. *Wall map of Groningen* 4 sh. 1768, 1774

Becket, Robert. Engraver for English edition of *Linschoten* 1598–99.

Beckman[n] Capt. (Later Sir) Martin (17th century) Swedish engineer in the service of Charles II of England. *Fortifications of Hull* 1681–82. *Guernsey & Jersey* 1694, *Plans of Tweed, Humber, and Portsmouth*

Bede [Beda, Baeda] Venerable (637–735) Theologian, historian, and cosmographer. Climate and zonal maps. *Tract. de situ. al de locis sanctis, plans Holy Sepulchre and Mount Sion*

Bedford, E. I. *Approaches to Liverpool* 1834, *Survey N. Wales* 1837

Bedford, George Augustus. *Ascension Island* 1838

Bedford, T. W. *Nemaha Land District Nebraska* 1858

Bedigis. *Carte Topo. l'Ile de Corse* 1824

Bedingfeld, Norman B. *River Ogun* (W. Africa) Admiralty 1861

Beechey, Capt. (Later Rear Admiral) Frederick William (1796–1856) *Voyage of discovery towards N. Pole* 1818, *Survey coasts N. Africa* 1821–3, *Voyage to Pacific & Behring Straits* 1831, *S. America* 1835, *Ireland* 1837. President of Royal Geographic Society 1855

Beechey, Comm. R.B. Admiralty surveyor 1838–49. *Holyhead, Irish Channel, River Severn*

Beeck [Beek], Anna (1678–1710) Engraver, publisher, and colorist of the Hague. *Plans of fortifications &c. Doncquerque* (1730)

Beek, James. Surveyor. *Plan Tower Hamlets* 1843

Beekman, Barnard (1721–1797) American Surveyor

Beeldsnyder [Beeldsnijder] Joost Jansz de [also called Bilhamer] (1541–1590) Architect and cartographer. *Maps of Amsterdam, N. Holland* 1575, *S. Dutch Provinces*

Been, M. Jansz. Dutch cartographer. *Delfland* 1606, *engraved by F. Balthazar*

Beer, G. J. *Mappa selenographica* 1834

Beer, Johann Christoph (1673–1753) *Gross Britannien* 1690

Beer, Josef. *Railway maps Austria-Hungary* 1884–97

Beer, Julio. *Division Buenos Aires* 1859

Beers, D.G., F.W., J.H., and S.N. American publishers and cartographers of *U.S. county atlases* 1860–90

Beetjies, Lieut. P.I. Engineer. *Chart Island Banda Neira* 1801

Beeton, Samuel Orchart. *Maps of War of 1870*

Bégat, Pierre (1800–1882) French hydrographer. *Géogr. mathématique* 1834, *Traité de géodesie 1839, Atlas des côtes mérid. de France* 1850

Begbie, P. Engraver for *Day & Master's Somerset* 1782, *Garden's Kincardineshire* 1776, *Plan Edinburgh* 1768

Begule, L. G. Engraver for De Fer 1705

Behaim, Martin of Bohemia (1459–1507) Geographer and globemaker in Nuremberg. Pupil of Regiomontanus. Settled in Portugal 1482, *Terrestial Globe* 1492, the oldest now extant.

Behm, Ernst. (1830–1884) of Gotha. Contribution to *Petermann's Geogr. Mittheilungen* and founder of *Geographisches Jahrbuch*

Behrens, C. A. *Chart of the Jade & Mouth of the Weser* (1799), *Heiligeland* 1801

Behrens, G. *Maps Lübeck* 1827–43

Behrens, Herman (1854–1902) Engraver for Perthes

Behrens, H. C. *Oldenburg* 1804

Behring, Vitus. See **Bering**

Behrisch, D. *Plan von Dresden* 1865

Beich, D. See **Beuch**

Beiger, Hans Conrad. Engraver. *Helvetia* 1637

Beighton, Henry (1687–1743). Surveyor and engineer. *Survey Bedworth Field* 1707 MS. *Enclosure Surveys* 1729–31, *Warwick* 4 sh. 1728 published E. Beighton 1750

Beighton, Elizabeth. Widow of above. *Warwick* 1750

Beijerinck, Fred. *Rhyn, Waal-Pannerdensche Canaal* 1771

Beijerinck, Martinus. *Nieuwe Kanaal, Rivier de Waal* 1778

Beiling, Carl. *Karte von Palästina* (1846)

Beilby, R. *Plan Newcastle & Gateshead* 1788

Beilly. Engraver for *Murray's Impartial History* 1780

Beins [Bains] Jean de (1577–1651) of Paris. Royal Geographer *Dauphiné* 1611 used by Hondius 1633 and by Blaeu 1631. *Ville de Suze* 1629

Beiton, Athanasius von (17th century) Maps, compiled in service of Russia

Beke, Charles T. *The lands of Cazembe* 1873

Beke, Pieter van der (d. 1567) Flemish cartographer from Ghent. *Flanders* 4 sh. 1538

Bel, Pierre (1742–1813) Engineer and topographer

Belam, Henry. Surveyor. *Liverpool Bay* 1892–93

Bel'Armato, Girolamo. See **Bell'Armato**

Belch. See **Langley & Belch**

Belcher, Comm. (later Admiral) Sir
Edward (1799–1877) *Africa* 1830–33,
Irish Channel 1834, *Pacific* 1835–42,
China Seas 1842–7, *Charts coast America*
1851–3, *Nicaragua* (1860)

Belcher, G. R. *Relief map Yellowstone
National Park U.S.A.* (1876)

Belcher, R. Surveyed *Lake Nyassa* 1889,
Admiralty Chart published 1893

Beleke. *Environs of Berlin* 10 sh. 1816–19

Belga, Aegidius Bulionius. See **Boileau de
Bouillon**

Belga, Jacobus Bossius. See **Bussius**

Belga, Gulielmus Nicolai. *Noua et integ.
Univ. orbis descrip.* 1603

Belgrand, Marie François Eugène (1810–
1878) Hydrographer and Geologist.
Environs Paris 1865

Belgrano, Giovanni Maria. Italian engraver.
For *Borgonio's Savoy* 6 sh. 1680

Belgrano, Luigi Tommaso. *Atlante
Idrografico* 1867

Belidar. *Maps for Maritime Ports of
France,* Faden and Jefferys 1774

Belin De Launay, J. H. Robert (1814–
1883) French historian and geographer

Beleski, J. L. *Dantzig* 1822

Belknap, Jeremy. *New Hampshire* 1791

Bell, Allan & Co. Publishers, Pancras Lane
Warwick Square, London *General Atlas of
the World* 1837–9, *Scotland New General
Atlas* 1837

Bell, Andrew (fl. 1694–1715) Publisher
New Descrip. and State of England 1704,
Fifty new and accurate maps of England
1708

Bell, Andrew (1726–1809) Engraver.
Northumberland 1753, *World* 1759,
Martineco 1759, *River of Northesk* (1760),
Berwick 1771, *Thirty miles round Boston*
1775, *Scotland* 1797

Bell, Clarence. Surveyor. *Stony Lake
District* 1897

Bell, C. G. H. *Basutoland (Berea)* 1881

Bell, Henry. *Ground plat Kings Lyn* 1561,
engraved 1670

Bell, J. Publisher, Oxford St. *Monte Video*
1807, *La Plata* 1807

Bell, James (1769–1833) Geographical
author. Edited *Rollin's Ancient History*
1828, *Gazetteer of England & Wales*
4 vols. 1833–34, *System of Geography,*
Glasgow 1830

Bell, Jared W. *Statistical Sheet Atlas US*
1833

Bell J. T. W. Surveyor. *Hartlepool Railway*
1831, *Great Northern Coal Field* 1843–61

Bell, John. Surveyor. *Fife and Kinross* in
Thomson's Atlas 1832

Bell, Peter. Geographer and engraver. *Lynn*
1762, *N. America* pub. Bowles 1768, *Black
Sea* with Dury 1765, *Environs of London*
Dury 1770, *Andrews' I. of Wight* 1775,
*War Prov. Massachusetts Bay Connecticut
&c.* 1776, *British Isles* 1787

Bell, R. B. *Harbour Belfast* 1862

Bell, Dr. Robert. *Maps of Canada* 1875–
97

Bell, Thomas (1785–1860) Antiquary
and surveyor

Bell, W. Lithographer. *Venango Co.* 1865

Bell, Dr. W. A. Map *S.W. portion U.S.*
1869

Bella, Stefano della (1610–1664) Artist
and engraver of Florence, teacher of
A. F. Lucini. *Schlacht bei Freiburg* 1644,
Siege of Posto de Longone (1650),
published ca. 1680

Bellairs, E. Australian surveyor. *Allotments
and Estates in Victoria* 1855–63

Bellanger. Engraver for *Denis Golphe du
Mexique* 1780

Bell'Armato [Bell Amato, Bellarmatus,
Bel Armato] Girolamo [Hieronimo]
(1493–1560) Born Siena. *Tuscany* 4 sh.

1536 and 1554, used later by Ortelius and
De Jode

Belleau, Richard de. See **Richard of
Haldingham**

Bellefonds, A. Linant de. See **Linant de
Bellefords**

Belleforest, Francois de. (1530–1583)
of Gascony. *Hist. Univ. du Monde* 1575
(a French edition of Munster)

Bellère [Bellerus] Jean (fl. 1550–95)
Antwerp. Printer, publisher, and print-
seller. *Totius novi orbis* 1554 used by
Gomara, Darinel, Cieca de Leon and
Eden 1555, Apian's Cosmographie 1584.
Edited French Edition of Waghenaer 1590

Bellero, Joannis. See **Bellère**, Jean

Belleyme, Pierre de (1747–1819)
Geographer to the King of France. Rue
du Paon No. 1. *Guyenne* (1785)
Aquitaine 4 sh. 1787, 1791, *Pays Bas*
2 sh. (1812)

Bellin, Jacques Nicolas, elder (1703–72)
French Royal Hydrographer. Rue
Dauphine près la Rue Christine.
Charlevoix's Nouvelle France 1744,
Hydrographique Français 2 vols. folio
1756–65, *Petit Atlas Français* 5 vols 4to.
1763, reissued as *Petit Atlas Maritime*
5 vols. 4to. 1764. *Atlas* to accompany
Prévost's Hist. des Voyages 1738–1775,
Atlas de Corse 1769

Bellin, Jacques Nicolas, younger (1745–
1785) Engraver of Paris

Bellinato, F. *Discorso di Cosmographia*
Venice 1573

Bellingero, Giorgio. Genoese portolan-
maker 1687

Bellingshausen, Faddej Faddeevic
(1779–1852) Russian admiral and
polar explorer. *Atlas to his Antarctic
Voyages* 1872 [31], *Karta Yuzhnago
Polushariya,* St. Petersburg 1821

Bellmond, F. B. *Swiss regional maps*
1819–20

Bellot, Joseph René. *Chart Northwest
Passage* 1852–53

Bellue, Pierre. *Atlas ou Neptune
Mediterranée* 1830 [34]

Bellus, J. *Laurea Austriaca (map of
British Isles)* 1610

Belly, Felix. *Canal Interocéanique de
Nicaragua* (1858)

Belmas, Jacques Vital (1792–1864)
Maps Peninsula War 1807–14, *Atlas*
24 plates 1836

Belpaire, Alphonse. *Mouvements de
Transports en Belgique* 1834

Belt, Thomas (1832–1878) Geologist.
Australian Gold Diggings 1852–62

Beltran, Francisco de P. *Plano. Topo.
de Merida* 1864–65

Belville, Alfred. *Gold Fields S.E. Africa*
by T. Baines, assisted A. Belville 1876

Belwe, C. F. G. *Grundriss der Stadt
Aschersleben* 1798

Bembo, Pietro (1470–1547) Venetian
Cardinal. *Terrestrial Globe* 1547

Bemrose, William & Sons. *Map Borough
of Derby* 1854,–67,–71,–76

Bénard, Jacques François. Publisher, Paris
sur le Quai de l'Orloge à la Sphère Royale.
Son-in-law and successor to De Fer.
World 1705, 1717, with Danet 1720

Bénard, Robert (fl. 1750–85) *Amérique
Mérid. Hist. Gen. Voyages* 1754 and
French edition of Cook 1774–85

Benci, Carlo (1616–1676) *Pair Globes*
1671 MS.

Bendorp, K.F. *Charts Rhine* 1785

Benecke, Wilhelm. *Elbe* 1868, *Hamburg
&c.* 1872, *Heidelberg* 1874–77

Benedetti, Giuseppe. Engraver. *Contorni di
Genova* Genoa, ca. 1760

Benedetti, Ignazio. *Topo. di Roma* 1773

Benedetto, Giovanni. Italian cosmographer
in service of France 1535–40, *Portolan*
1543 (attrib.)

Benedict, Lorenz [Laurentz] (d. ca. 1604) Printer and publisher of Copenhagen. *Sea charts of Baltic and North Sea,* Copenhagen 1568

Benedicti, Hieronymus. Engraver for Schraembl 1787–88, *Crimea roads 1787– 89, Lichtenstein 1801–05, West Gallizien* 12 sh. Vienna 1808, *Schönbrunn* (1810)

Beneke, Wilhelm. *Plan der vier Städte Hamburg, Altona, Ottensen und Wandsbeek* 1872

Beneventanus, Marco. Neapolitan cosmographer. Editor 1507 edition of Ptolemy

Benham, Charles E. *Colchester* 1893

Benicken, F. W. *Orbis Terrarum Antiq.* 1829

Benincasca, Andrea (1476–1508) of Ancona. Geographer and Portolan maker

Benincasca, Grazioso (ca. 1400–1482) of Ancona. Portolan maker. *British Isles* 1467, *Black Sea* 1474, *Europe* 1482

Benincasca of Tudela (12th century) Jewish Rabbi and Traveller. Work pub. 1543

Benjamin, Lieut., USA. *Monterey & its Approaches* (1846)

Benitez y Parode, Manuel (1845–1904) Spanish military cartographer.

Bennefeld, Lieut. *Hanover* 1770

Bennet, Roelof Gabriel (1774–1829) *Arctic* 1823, *Australia* (1825), *Atlas with van Wijk* 1827

Bennett, I. R. *Post Office map of Bengal &c.* 1853

Bennett, James, Engraver for *Sellers Atlas Maritimus* 1675, *English Pilot* 1677, and later editions.

Bennett, John (d. 1787) Publisher in partnership with Robert Sayer 1770–84, *American Atlas* 1775, 1776 and later. *West India Atlas* 1775, *American Military Atlas* 1776

Bennett, L. G. *County maps of Illinois and Minnesota* 1861–67

Bennett, Richard. Engineer, draughtsman and engraver. *Africa* 1758, *America* 1759, *Asia* 1759

Bennett, T. Downes. *Geology simplified* 1845

Bennett & Walton. Publishers for Robinson 1807, for Tanner 1832

Benning, Robert. Engraver for Rocque 1746–61. *Berks.* 1761, *London & West* 1755, *Rome* 1750, *Madras* 1751, *Brest* 1757, *Taylor's Hereford* 1757, *Hants.* 1759, *Loup's Oberland* 1766

Bennison, Jonathan. *Liverpool* 1835

Benoist. Engraver. *Plan Bruxelles* 1825

Benoist, Father Michel. Engraved *Rochas & Espinha's Asia and Europe, Peking* 1775

Benoit, Francois Chrétien de (1729– 1812) Military cartographer *Osnabrugensis Episcopatus delin.* (1775)

Benson, Charles. *Atlas of Madras Presidency* 1895

Bent, J. T. *Dhofar* (Arabia) 1895, *Hadramaut* 1894, *Fadhli Country* 1897

Bent, Theodore. *Travels in Arabia and E. Africa* 1890–92

Bent, W. Publisher at the Kings Arms Paternoster Row. *Universal Magazine* 1784–1803, *Barker's City Washington* 1793

Benthuysen, C. van Lithographer. *Map of the Cities of New York and Brooklyn* 1855

Bentinck, William (1764–1813) *Liscomb Harbour* 1784 MS.

Bentze, Hans. *Oost Zee, Copenhagen* 1700

Bentzen, Lauritz Christian von (1803– 1881) Danish military cartographer

Benzinger, J. *Plans Jerusalem* 1896–97

Benzoni [Benzo] Girolamo [Jerome] *Occid. Americae partis* 1594

Ber., Fer. See **Bertelli**, F.

Beranger, Jean. Plan *Baya de St. Joseph* 1718

Beranek, Karl (1779–1842) Austrian military cartographer

Berard, Antoine. *Côte Sept. d'Afrique* 1836, *Mer Méditerranée* 1859–60

Beraud, Major. *Sahara* (central) 1862

Bercham, J. P. van. *Carte pétrographique du Saint Gothard* 1791

Berchem, Nicolas Pietersz (1620–1683) Printer of Haarlem and Amsterdam. *World* engraved Visscher ca. 1660, *Flight Israel* 1669, *Voyage St. Paul* 1669

Bercheyck, Laurens Lodewyck van. *River Demarary* 1759, English edn. 1781

Berckenrode. See **Balthasar** van

Berecz, Antal (1836–1905) Hungarian school cartographer

Berendt, C. H. *Yucatan* 1879, *Grao Para* 1888

Berendts, F. *MS map Campen* 1625

Berey, A. Engraver for *Frézier's Voyage* 1716

Berey, Charles Amadeus de. Engraver in Paris, Rue St. Jacques. For De Lisle 1700, *World* & *Source du Po* 1702, *Cours du Danube* 1703, *Diocèse de Rouen* 1715, *Malte* 1720

Berey, Charles Amadeus de, junior. *Villes Forts et Chateaux de Malte* 1724

Berey, Nicolas (1606–1665) Editor and Enlumineur de la Reine, au bout du Pont Neuf proche les Augustins aux Deux Globes. Father-in-law of A.H. Jaillot. *Plans et Profils de la Ville de Paris* 1650, *Provinces de France et Espagne* 1655, *Carte Générale de l'Isle de France* 1648, *Plan Paris* 1645

Berg, L. W. C. van den. *Hadhramaut* 1886

Berg, Petrus van der. See **Montanus**, Petrus

Bergano, Georgius Jodocus. Italian printseller. *Lago di Garda* 1546

Bergen, Ernst Gottlieb (1649–1722) of Berlin. *Ukraine* MS 1680

Bergen, J. H. *Great Circle Chart N. Atlantic* (1880)

Berger, C. *Heilquellen im Nordwesten Böhmen* (1840)

Berger, Daniel. *Stadt Hirschberg in Schlesien* 1810

Berger, Franz. *Plan de Dôle* 1874, *Wien* 1873

Berger, Friederich Gottlieb (1713–ca. 1800) Engraver, of Berlin. *Prussia* (1765) *Plan Berlin* 1772, *Plan London* 1770

Berger, George. *Military Map India* (1854)

Berger, J. G. *War Chart* (1854), *War Map Northwest and Central Europe* 1870

Berger, Levrault. Widow. *Dépt. Bas Rhin Vve Bl et fils* 1854, *Haut Rhin* 1857, *Strassburg* 1857

Berggreen, V. F. A. *Maps of Denmark* 1870–86

Berghaus, Dr. H. (1828–1890) *Austrian Atlas* 1855

Berghaus, Heinrich Carl Wilhelm (1797–1884) *Africa* 1826, *Atlas Asia* 1832, *National Atlas* 1843, *Physical Atlas* 1845, *Stieler Handatlas* 1847, *Weltkarte* 1859

Berghes, Carl de. *Distritos Minerales . . . de Mexico* 1827

Bergman, Carl Johan. *Gotland och Wisby* (1875)

Bergman, P. C. *Plankarta Jököping* 1868, 1876

Bergman and McGonigale. American surveyors. *Plots land Vicinity Louisville* 1860, *Preston* 1865

Bergquist, Carl. Swedish engraver. *Stockholm Stad* 1771

Bergstrand, P. E. *Bohus hän* (1867)

Bering [Behring] Vitus Jonassen (1680–1741) Danish navigator in Russian service. *Northern Siberia* 1729 MS, first printed in D'Anville's *Atlas China* 1737

Berjon, Jean. Publisher *Champlain's Voyages* 1613

P. Bertius (1565–1629)

Besley, H. *Route maps Roads Devon* (1850) *Plan Ilfracombe* (1892)

Besoet, Iven van (1720–1769) Engraver and publisher of the Hague

Bessel, Friedrich Wilhelm (1784–1846) German astronomer

Bessels, Emile (1847–1888) German geographer and explorer

Besserer, Wilhelm (16th century) Painter and map maker

Besson, J. Publisher in Paris, L'Isle du Palais. *Placide's Siam* 1686, *Baudrand's British Isles* 1701, *Portugal* 4 sh. 1704, *Savoy* 6 sh. 1704

Best, Edward. Geologist. *British Isles* 1874, 1887

Best[e], George (d. 1584) Navigator. *Beste's True Discourse* 1578 *(Map of Frobisher's Straits & World)*

Best, Thomas (ca. 1570–1638) Capt. in Navy, perhaps son of George above

Beste, J. R. Digby. *Pianta di Roma* 1858

Bestehorn, C. B. *Map for Homann Heirs Stadt Atlas* 1762

Bétemps, Adolphe Marie Francis (1813–1888) Military cartographer in Geneva

Betgen, I. F. *Gumbinnen, Lithuania* Homann Heirs 1733

Bettencourt, Emiliano Augusto de. *Atlas of Portugal* 1870

Bettison, Samuel. *Country 25 miles round Cheltenham* (1830) *Plan Cheltenham* (1830)

Betts, A. C. *Beveridge Is. (Australia)* 1875

Betts, Henry D. Civil engineer and surveyor. *Educational maps* 1852–1872, *Nebraska* 1857

Betts, John (fl. 1844–63) Publisher London, 7 Compton St. Brunswick Sq. and 115 Strand. *Itinerant and commercial map England* 1839, *Family Atlas* 1848, 1863 (with Carson), *Sixpenny Maps* 1846, *Portable Globe* (1850), *Educational Maps* 1852–1861

Betts, W. Publisher. *Hoxton's Chesapeake Bay* 1735 (with E. Baldwin)

Beuch [Beich, Beuech] Daniel (d. 1670) Painter in Ravenstein and Munich. *Baden* [1635]

Beudant, Fr. S. (1787–1850) French physician & geologist. *Voyage Minéral et Géologique en Hongrie* 1822

Beudeker, Christofel. Famous collector, formed a personal atlas in 24 vols. The maps from 1600–1750

Beughem, Cornelis van. *Poliometria Britannica* (1695)

Beugo, John. *Scotland with the roads* 1789

Beusch, Th. *Plan von Stettin* (1879)

Beust, F. *Hist. Atlas Kantons,* Zürich 1873

Beuvelot. *Paris* 1817

Bevan, B. *Grand Union Canal* 1809

Bevan, George Philips (1829–1889) *Statistical Atlas Great Britain* 1881–82

Bevan, John. Surveyor, publisher, lithographer. *County Maps USA* 1850–70

Bevan, W. L. Commentary on *Hereford Mappa Mundi* (1873)

Bever, Th. von. *Ville d'Anvers* 1865, 1872

Bevers, Fendol. American surveyor. *Wake County* 1871

Bevis [Bevans] John (1693–1771) Astronomer. *Uranographia Britannica* (1750) *Atlas Céleste* 1786

Bew, John. (d. 1793) Bookseller and publisher London 29 Paternoster Row. *Political Mag.* 1780–85 (with maps)

Bewsher, Lieut. J. B. Surveyor in Iraq 1862–68. *Mesopotamia* 1868

Beyer, Carlos. *Paraguay* 1886, *Atlas Repub. Argentina* 1889

Beyer, Herman. *Schul Atlas* (1868)

Beyer, I. C. *Nederlanden,* Luxemburg 1866

Beyer, Johannes (1673–1751) Instrument maker of Hamburg. *Globes* 1718, 1720

Beyer, L. Engraver for Wieland 1848

Beyer, Th. *Liegnitz* 1877, 1891

Beyerinck, Fred. (ca. 1720–1783) Dutch surveyor *Boven Rhijn* 1772

Béze, Theodore de. *Mappemonde Nouvelle Papistique* (1566)

Biancani, Giuseppe S. J. (1566–1624) Italian mathematician, astronomer and cosmographer

Bianchini, Francesco (1662–1729) of Verona and Rome. Astronomer and archaeologist

Bianco, Andrea (fl. 1436–58) Venetian cartographer. *Portolan Atlas, Chart W. Africa* 1448

Bianconi, F. *Bulgarie* 1887, *Guatemale* 1890, *Venezuela* 1888, Various *Cartes Commerciales* 1885–96

Biart, Alph (1825–1898) Belgian geologist

Bibb, Capt. *Plan Harbour Messaua* 1763, pub. 1784

Bibbs, Thomas Franklin. Australian cartographer. *Co. Mornington Victoria* 1838, *Co. Bourke* 1857, 1866

Biberger, Ulrich Johann. (fl. 1700). Austrian engraver

Bickham, George, senior (1684–1758) James Street Bunhill Fields (1743), later in Featherstone Street Bunhill Fields. Author, engraver, writing master. *Universal Penman* 1733–41 (in parts), *British Monarchy* 1743, reissued 1748 and 1749

Bickham, George, Junior (fl. 1735–67) Ye corner of Bedford Bury New St. Covent Garden and later Mays Buildings Bedford Court Covent Garden. Engraver, publisher, drawing master. *Counties of S. Britain* 1750, *American Colonies* (1743), *Musical Entertainer* 1737–8, *St. George's Parish Hanover Sq.* 1761, *Topo. Survey Parish of Kensington* 1766

Bickham, John. Engraver and publisher at Seven Stars in King St. Covent Garden.

Assisted the George Bickhams with *American Colonies* 1743 and *British Monarchy*

Bidder, George Parker. *Holyhead Junction Railway* 1845

Biddulph, Michael Antony Shrapnel. *Battle Alma* 1854, *Telegraph Routes India* 1860, *Inkermann* 1855

Bidwell, Oliver Beckworth. *Missionary maps Africa* 1860, *China* 1850, *Asia* 1865

Bieling, Johann Heinrich Gottfried. Printer *Homann's Atlas* 1737

Biedermann, H. A. Surveyor. *Elham Park Norfolk* 1778, *Bettswells & Hollybush Hall* 1772

Bielitz, Robert. *Walachia* 2 sh. 1854

Bien, Joseph R. *Atlas State of New York* pub. Julius Bien 1895.

Bien, Julius (1829–1909) Publisher, engraver and cartographer. Worked for U.S. Government *Stat. Atlas U.S.* 1874

Bien & Co. Lithographers Fulton St. New York

Bienewitz. See **Apian**, P.

Biens, Cornelis. *Topographia Enchusae delin.* (1670)

Bierens de Haan, Dr. D. *Overyssel* 1889

Biermann, J. P. *Plan Luxembourg* 1878

Bigges, Walter. *Drake's West India Voyage* 1589, maps by Boazio

Bigrel. *Cochin Chine Français* 1872–73

Bigsby, John Jeremiah (1792–1881) Geologist

Bilhamer. See **Beeldsnyder**

Bill, John (fl. 1604–30) London. King's Printer, and bookseller Northumberland House, St. Martins Lane, Aldersgate St. and Hunsdon House, Blackfriars. Apprenticed to John Norden 1592. *Abridgement of Camden's Britannia* 1626 *(small county maps)*

Billaine, Louis (fl. 1660–79) Publisher, St. Augustin rue St. Jacques Paris. *Davity* 1660, *Voyage de Laval* 1679

Biller, Bernard, elder (1790–1838)
Viennese engraver and globemaker

Biller, Bernard, younger (fl. 1802–40)
Engraver and globemaker

Biller, Dominick. Engraver. *Plan Wien*
1848. Engraved for Scheda 1879

Billig, G. A. *Plans of Berlin* 1872–76

Billinge, Thomas. Engraver. *Liverpool
Harbour* 1771, *Burdett's Cheshire* 1777,
Yates' Lancs. 1786

Billings, J. English sailor, accompanied
Cook on his 3rd Voyage. Later entered
service of Russia. Oceanography *N.E.
coasts of Russia*

Billings, T. *Lincolnshire* (with W. Yates)
(1790)

Billington, William. Engineer. Surveyor.
Plan Bertrand 1859, *Masonic Map of
Illinois* 1862, *Illinois* 1867

Binder. Engraver. *Prov. de Moxos*
(S. America) 1791

Binder, H. *Carte du Kurdistan* (1888)

Bindon, John. Surveyor. *Cullompton
Court* 1820 MS.

Bindoni, Augustino and Francesco di
(fl. 1528–40) Venetian booksellers and
printers. *Coppo's Portolans* 1528, *Istria*
1540

Binetau, P. Geographer. Printer, engraver.
Cochin Chine 1860–63, *Europe* 1876,
Wall maps France &c. 1877, *Kurdistan*
1888

Binfield, R. Surveyor. *Great Greenford
Middlesex* 1775

Binger, L. G. *Maps Niger & Senegal*
1886–98, *Haut Niger* 4 sh. 1889

Bingham, Lieut. H. M. *Crimea* Wyld 1855

Bingley, James. Engraver of Goswell Road
(1823) later Sidney St. Islington. For
Teesdale's Yorks 1828, *Hennet's Lancs.*
1830, *Moule's English Counties* 1837

Binkerd, J. S. *Map City of Dayton (Ohio)*
1862

Binneman, Walter. Engraver. *Ogilby's
Middlesex* (1673) and *London* (1677),
R. Daniel's English Empire in America
pub. Morden 1679, and *Berry's New
England and Annapolis,* reissued
Christopher Browne (1690)

Binney, Edward William (1812–1881)
Geologist. Founded Manchester
Geographical Society

Binns, Jonathan. Surveyor and publisher.
Lancashire 1820–21, 1825. *Lancaster
& Preston Railway* 1836

Bion, Nicolas (1652–1733) Royal
Geographer and instrument maker to
the French King, died Paris. *Use of
Globes* 1700, *Pair Globes* 1708, 1721,
1728

Biot, Jean B. (1774–1862) French
geographer and astronomer

Birago, Karl Frhr von (1792–1845) Military
cartographer

Birch, E. *Southampton and Salisbury
Railway* (1845) *East and West of England
Railway* (1845)

Birch, J. B. *Railway maps* 1844–45 (with
E. Birch)

Birch, J. W. Part of *Malay Peninsula* 1876

Birch, William Russell (1755–1834) *Atlas
of Philadelphia* (Pennsylvania) 1800

Birckman. Published Adrichem's Palestine
1593

Bird, C. *Geological map Yorkshire* 1881

Bird, Caspar. *Plans of Berlin* 1848–68,
Sans Souci 1848, *Gradnetz Atlas* 1868

Bird, J. S. Civil Engineer. *Albion* 1871,
Ionia 1872, *Jackson* 1871, *Monroe* 1871

Birk, Caspar. (fl. 1848–68). Lithographer
and publisher in Berlin

Birken, Sigismond van (1628–1680)
Danube 1664

Birmingham, John (1816–1884)
Astronomer

Bironi. *Plan Jerusalem* 1582

Bironius, Gallus. See **Brion**, Martin de

Birt, William Radcliff. *Lunar maps* 1870

Biruni, Al. See **Albyrouny**

Bischof, C. Aug. L. (1762–1814) German geographer

Bischoff, Max. *Flaggen Karte* 1869

Biscoe, John. *Ice Chart Southern Hemisphere* 1866

Bishop, George (ca. 1562–1611) Printer and bookseller at the Bell, St. Paul's Churchyard *Camden's Britannia* 1590, 1594, 1600. *Hakluyt's Principal Navigations* 1589. Succeeded by his cousin Thomas Adams

Bishop, G. *Northumberland N.S.W.* 1864

Bishop, John. Printer Denmark Street, Camberwell, Surrey. *Cottage Maps* for *Society for Promoting Christian Knowledge* ca. 1850

Bishop, Capt. John S. *War map Southern States* 1863

Bishop, Capt. Robert. Hydrographer. *Windward Passages from Jamaica* 1761, *Old Straits Bahamas* 6 charts 1794–96

Bishop, S. E. *Hawaiian Islands* 1886

Bissuel, H. *French West Africa* 1888

Biston, V. J. *Lille* 1822

Bittner, A. *Geolog. Karte Bosnien-Herzegovina* 1880

Biurman, George. *Norike* 1715, *Stockholm* 1720, *Finland* 1740, *Svea* 1747

Bixet, A. Letter engraver for Migeon 1891

Björk, Nils. (1695–1726) Swedish surveyor

Bjorkbom, Karl Niklas (1742–1809) of Abo. Swedish surveyor

Blachford, Robert (fl. 1801–40) Chart-seller. Son-in-law of J.H. Moore. Succeeded Moore in 1801, partner with Imray 1836. *Baltic* 1803, *Cattegat* 1808, *North Sea* 1809, *Sleeve* 1810, *America* 1818

Blachford, William. *Baltic* 1817, *Cattegat* 3 sh. 1817, *Finland* 1819, *E. coast*

England 1830, *Firth of Clyde* 1830–35, *W. coast Scotland* 1835

Black, Adam (1784–1874) and Charles. Publishers and Booksellers to His Majesty 27 North Bridge, Edinburgh. *World Atlases* 1840 to 1898, *County Atlas Scotland* 1848, *Atlas North America* 1856, *Australia & New Zealand* 1844 to 1896

Black, George. *Contour Map Melbourne* 1888

Black, **Kingsbury**, **Parbury** and **Allen**. Publishers and booksellers. *Horsburgh's E.I. Pilot* 1806–36, *Arrowsmith's Delhi to Constantinople* 1814–16, *Mahratta War* 1817–19, *India* 1820

Black, **Parry & Co**. Publishers. Leadenhall St. *East India Pilot* 1885

Blackadder, John. Surveyor. *Berwickshire* 2 sh. 1797, *Part of Skye* Thomson 1832

Blackburn, George. *S. Carolina* 1822

Blackburn, William (1750–1790) Surveyor and architect

Blacker, Lt. Col. Valentine. Surveyor General for India. *Mahratta War* 1821, *Hindostan* 1824

Blackie & Son. Glasgow, Edinburgh, and London 49 & 50 Old Bailey, E.C. *Imperial Atlas* 1860, by Walter Graham Blackie. *Comprehensive Atlas* 1880–03, *Descriptive Atlas of World* 1893

Blackmore, Nathaniel. *Bay of Fundy* (1720) *Annapolis Royal* 1729

Blackwell, Thomas E. *Railway maps London & Newbury* 1845, *Severn and Wye* 1852

Blackwood, Capt. Francis P., R. N. Hydrographer. *S.W. Pacific* 1841–46, *E. Prov. Australia* 1850

Blackwood, James. *Seat of War in Europe* 1871, *Shilling Atlas* 1858, 1870

Blackwood [W.] and Son. Publishers in Edinburgh and London. *Atlas Scotland* 1838 engraved Lizars, *British North America* 1833, *Atlas to Alison's History*

W. Jansz. Blaeu (1597–1638); engraving by J. Falck

*J. Blaeu (1596–1673). Engraving by J. van Rossum when
Blaeu was 65 years old*

1848, *Berwick* 1850, *Atlas of Physical Geography* 1850, *Keith Johnson's Royal Atlas* 1861

Blades, Robson. *Surrey. New Roads* 1828

Bladie, John. *Castricoms Bay on Celebes* 1666

Blaettner, Hans Samuel. *Territ. Episcopatus Spirensis* 1753

Blaeu Family. [Jans Zoon, Janssonius, Johnson, Alcmarianus, Caesius.] Cartographers, instrument makers, publishers and booksellers of Amsterdam. Willem (William) Jans Zoon (1571—1638) Founder of the firm, born Alkmaar, studied under Tycho Brahe, set up on his own account 1596 in Amsterdam. *Pair Globes* 1599, 1602, 1603, 1616 and 1640. *Maps Holland* 1604, *Spain* 1605, *World* 1605, *Continents* 1608. *Licht der Zeevaert* 1608 (his first sea atlas). In 1629 he bought 37 plates from Hondius and issued his first land atlas 1630.

Blaeu, Joan [John] (1596—1673) born and died in Amsterdam. Collaborated with his father Willem Janszoon and with his brother Cornelius (1610—48) continued to expand the atlas from Vol. 3. *Atlas Blauiane* 6 vols. 1649, *Atlas Major* 9 or 11 vols. (according to language) 1662, *Town Plans* 1649—53, *Wall maps* 1647—49. Great Fire of 1672 destroyed his printing house. Blaeu was appointed official cartographer to the Dutch East India Company 1638—73

Signature of J. Blaeu

Blaeu, Cornelius. Has his imprint on editions of 1638, 1642—44

Blagrave, John (ca. 1550—1611) Mathematician. *World* engraved Benj. Wright 1596, in *Astrolabium Uranicum*

Generale. Manor of Fecknam 1591, *Abbey of Marksby* 1608

Blagrave, Lieut. T. C. *Maps of Punjaub* 1852—53

Blair, Lieut. Archibald. *Manhora River* 1777, *Andamon Is.* 4 sh. 1793, 1795. *Dewgar Harbour* 1784

Blair, Blair. *Classical Atlas* 1853

Blair, Rev. John (d. 1782) Chronologist. *Ancient & Modern Geography* 1768, *N. America* 1768, *E. Indies* 1773, *World* 1779, *Chronology & History of World* 1768

Blake, E. J. *N. Warwick, S. Staffordshire, and E. Worcester Mineral Districts* 1888

Blake, E. W. *New System of Universal Geography* (with T. Bankes) 1787

Blake, George S. *Harbour Bridgport* (1835)

Blake, J. *Camp Coxheath* 1778

Blake, J. Engraver. *N.W. Harbour Mauritius* 1738 MS.

Blake, Lieut. J. E. *Seat of War in Florida* 1839, *Boundary U.S. & Texas* 1838 and 1841

Blake, J. L. *Geographical Atlas* 1816

Blake, W. P. *Geological Map U.S.* 1873, *Stikine River* 1863

Blakely, W. R. *Track through Gillolo Passage* 1823, *Passages to Lymoon* 1830

Blakeston, Capt. *Yang-tse-Kiang,* Arrowsmith 1862

Blamey, Jacob. *North America Pilot Pt. II* 1795

Blancardus. See **Blankaart**

Blanchard, Col. Joseph. *Province of New Hampshire* with Rev. Langdon, engraved Jefferys (ca. 1761). Another edition with additions by Abel Sawyer, Jr., 1784

Blanchard, Rufus. Chicago, 146 Lake St. *Chicago* 1862—92. *Minn.* 1867, *Miss.* 1868, *Kansas* 1870, *Mich.* 1875

Blanchin, Jean. Engraver. *Map Garonne* 1628, 1650. *Hondius' Palestine* 1630

W. Jansz. Blaeu's Map of Europe (published Amsterdam, 1630)

Blanco, Manuel. *Philippines,* pub. Manilla 1845

Blanco, Placido. Lithographer. *Maps of Mexico* 1851–52

Blanco y Crespo, Miguel. *Missiones de Moxos y Chiquitos* 1770

Bland, A. Surveyor. *Plan of Clapham* 1849

Bland, Edward. *Map of Virginia,* engraved Goddard 1651, in *Discovery of New Brittaine* London 1651

Bland, Joseph. Joint surveyor with Warburton & Smyth. *York* 1720, *Essex* 1726, *Little Ilford* 1737 MS, *Middlesex* 1745

Blandford, W. E. *Geolog. Karte von Indien* 1879

Blandowski. *Plan Eckernförde* (1849)

Blaney, George A. *Middlesex Canal in Charleston* (1804) MS.

Blanford, Henry Francis (1834–1893) *Geolog. Survey India* 1855, *Rainfall chart India* 1883

Blankaart [Blanckaert, Blancardus], Nicolas (1629–1703) Professor of Leyden. *Ancient Europe, Asia, Africa,* for Jansson 1652

Blanken, Jan Antonius (1790–1837) Surveyor of Utrecht

Blanken, J. W. *Prov. Groningen* 1860

Blankensteiner. *Denmark* 1796

Blanmont de Thierville. *Comté de Lippe* 1761

Blaskowitz, Charles (fl. 1760–1823) *Chart Narragansett Bay,* Faden 1777, *Newport, Rhode Is.* 1777, *New Hampshire* 1784

Blasquez, Antonio. *Terrestrial Globe* 1530

Blatchford, Torrington. *Geolog. map Coolgardie* 4 sh. 1898

Blau, Otto. *Herzegowina im Jahre* 1861

Blaxland, George. *E. Coast China* 1838

Blegen, John. *Veileder og verdens atlas* (1892)

Bleiswyk, F. Designed & engraved front title to *Van de Aa's Le Nouveau Théatre du Monde* 1713

Bleuler, Johann Ludwig (1792–1850) German surveyor

Bligh, Capt. (later Admiral) William (1754–1817) Governor of New South Wales 1806–08. Sailed with Cook. *Chart Van Diemens Land* 1777 MS, *Humber,* Faden 1797, *Dublin Bay* 1800 and 1803, *Dungeness* 1803 MS, *Voyage South Sea* 1792, *North Sea* 1811

Blink, H. *De Bewoners der vreemde werelddeelen,* Amsterdam 1898–99

Bliss, Henry I. *Maps of U.S. Counties* 1857–72

Bliss, Nathaniel (1700–1764) British Astronomer Royal

Bloch, Maurice (1816–1901) *Französischen Kaiserreichs* 1861, *Europ. Staaten* 1862

Block, Adrien. *Map of New England* 1614 MS (Chesapeake Bay to Penobscot)

Blodget, Lorin. *Climatological maps of U.S.* 1856–86

Blodget, Samuel (1724–1806) *Plan Battle near Lake George* 1755, engraved Johnston

Blodget, William. *Topographical map of Vermont* 1789, *Connecticut* 1792

Blödtner, Cyriak (1672–1733) Gen. Staff Austrian Q.M.G. Army. *Maps of Flanders & Rhine*

Bloemswaerdt, C. C. van *Kaert van Loeven,* Covens & Mortier (1730)

Blome, Richard (fl. 1660–1705) Cartographer and publisher. *Description of World* 1670, *Jamaica* 1672, *Britannia* (county maps) 1673, *Speed's maps Epitomised* 1681, *Cosmography* 1682, *Isles & Territories America* 1687

Blomefield, Francis. *Plan Norwich* 2 sh. 1746

Blommendal, A. R. Hydrographer. *Eems* 1859, *Goeree* and *Maas* 1858, 1869; *Zuiderzee* 1870

PRÆSTAT

INDEFESSVS AGENDO

Woodcut printer's marks used by W.J. Blaeu and J. Blaeu

Blondeau, Alexandre (1799–1828)
Royal engraver in Paris, Rue des Francs
Bourgeois No. 131 près la Place Michel,
later Rue des Noyers No. 24. First
engraver to Dépôt de la Guerre, then for
Dépôt de la Marine. Engraved for *Poirson*
1817, *Lapie* 1817, Delamarche 1820. For
the Dépôt de la Marine *Douarnez* 1817,
Port du Croisic 1827, *St. Nazaire* 1828

Blondel, Col. *Atlas de la Guerre d'Orient*
1858

Blondel, Francois & Bullet, Pierre. *Plan
of Paris* 1675

Bloodworth, Charles James. Suveyor.
Estate plans, Huntingdonshire 1778–1859

Blotelingh [Blooteling] A. (1640–1690)
*Six views Amsterdam after Ruysdael.
Situation of Paradise & Canaan* 1669

Blottière. See **Roussel**

Bluhme, I. P. *Grönland* 1866, 1871

Blume, Lieut. J. *Königreich Preussen*
27 sh. 1831, *Bernberg* &c. 1839

Blumenauer. *Residenzstadt Cassel*
9 sh. (1896)

Blunck, Otto. Lithographer. *Kiel und
Umgegend* 1876

Blundell [G.G.] & Co. Publishers in
Melbourne, mid 19th century.

Blundeville, Thomas (ca. 1560–1603)
*Briefe Description of Universal Mappes
& Cardes* 1589

Blunt, Charles John. *Railway Maps*
1843–45

Blunt, Edmund (1770–1862) American
hydrographer. *American Coastal Pilot*
Newburyport, Mass. 1769. 23 editions
up to 1867

Blunt, Edmund and George William
(1802–1878) Chart publishers *N. & S.
Atlantic* 1827–30, *Florida Bank* 1846,
U.S. Coast 1860–64

Board, Stephen. Engraver. For *Seller's
Chart Russia* in *English Pilot* 1671

Board, W. *Chart Exmouth Harbour* 1830

Boardman, Harvey. *White Mountains,
New Hampshire* 1858

Boas, Franz. *Baffin Land* 1885,
British Columbia 1889–96

Boase, Henry Samuel. *Geological Map
of Cornwall* 4 sh. 1828

Boazio, Baptista (fl. 1588–1606) Italian
cartographer, worked in England. Compiled
5 charts for *Drake's West Indian Voyage*
1585–56, for *Bigges' Summarie* 1589,
Isle of Wight 1591, *Raid on Cadiz* 1596,
Chart Azores 1597, *Ireland* 1599 (used
reduced in 1606 edition of Ortelius) and
MS plans of *Calais* 1596, *Northern Ireland*
1602, and *Ostend* 1604 MS.

Bo de Venecia, Gione. *World* 1520

Bobillier, Charles. *Genève et ses environs*
1899

Bobrik, Hermann (1814–1845) Geographer. *Atlas Königsberg* 1838

Bocage. See **Dubocage**

Boccard, Ludwig. *Coblenz* 1854, *Karte
des Rheins* 1858, 1865

Bochart G. *Geogr. Sacrae* 1646

Bock, A. *Plan Streets within Limits of
Wynberg Municipality* 6 sh. 1888

Bockel [Boeckel] Peter. *Ditmar* 1559,
used by Ortelius 1595

Bockelberg, F. von. *Wildbad-Gastein*
(1860)

Bode, Johann Ehlert (1747–1826)
Astronomer of Hamburg. *Astres* 1782,
Celestial Globe 1790, 1804. *Atlas
Céleste* 1801

Bodega y Quadra, Juan Francisco de la.
Charts west coast America 1766–91

Bodenehr, Gabriel, the elder (1634–
1727) Engraver and publisher of Augsburg.
Atlas Curieux (1704) *Curioser Staats und
Kriegs Theatrum* (in Bavaria, Italy, Rhine,
Spain) 1715

Bodenehr, Gabriel, the younger (1664–
1758) Engraver and publisher of Augsburg.
Atlas Curieux &1704), *Europens Pracht
und Macht* (1730)

Bodenehr, George Conrad (1673–1716) Engraver and publisher of Augsburg. *Reichs-Statt Augsburg* (1685)

Bodenehr, Johann Georg (1631–1704) Engraver and publisher of Augsburg. *Totius Germaniae Itinerarum* (1660) *Teutschland* 1677, *Geogr. Prov. Sueviae descrip.* (1670)

Bodenehr, Moritz (d. 1748) Engraver and publisher. *Delin. Norwegiae novissima* Dresden (1719) *Neue Chur Saechsische Post Charte* 1753

Bodener, Jakob. *Bregenz und Umgegend* (1876)

Bodger, John. Land surveyor, publisher and mapseller at Stilton. *Plan Newmarket Heath* 1787, *Whittlesea Mere* 1786

Bodley, Sir Josiah (ca. 1550–1618) English military surveyor. *North Ireland* 1609

Bodmer, Samuel (1652–1724) of Bern. Mathematician and surveyor

Boeck, Caesar Laesar (1766–1832) Danish Norse cartographer born Frederica, died Christiania

Boecke, H. *City of St. Louis and Vicinity* 1859

Boeckel [Bokel, Boekel] Pieter (ca. 1530–1599) of Antwerp. *Dithmar* and *Holstein* in Ortelius 1570

Boeckel, Peter August Gottlieb (1719–1797) Member of Homann Firm 1744–46

Boeck, Richard. *Hist. Karte von Elsass* 1871, *Deutschen in Europa* 8 sh. (1891)

Boeckler [Böckler] George Andreas. *Maps of Palestine* 1655–1665

Boeddicker, Otto. *The Milky Way* 1892

Boehler, F. L. *Geometrischer Plan von Frankfurt* 1874

Boehm, A. *Cours-Karte von Bayern* (1866)

Boehm, F. *Plans of Berlin* 1850–65

Boehm [Böhm] Josef Georg (1807–1868). Astronomer Innsbruck

Boehme [Böhme] August Gottlieb (1719–(1797) Mathematician and military cartographer of Dresden. *Homann's America* 1746, *Asia* 1744, *Scandinavia* 1776

Boekel. See **Boeckel**

Boelen, J. J. *Straat Lagoendi* 1855, *Lampong Baai* 1857

Boelhouwer, F. K. *Map of Gooiland* 1709, engraved by Ottens 1740

Boell, William. Lithographer. *North Adams, Mass.* U.S.A. (1870) *Baltic* 1862, *Licking Co.,* Ohio 1854, *Bergen Heights,* N.J. 1867

Boem, J. *Recueil de diverses histoires* Paris 1543

Boener, Johann A. *Norimberg Territ.* (1650)

Boerio, G. Maps of Italian departments *Mincio, Olona, Po* 1802

Boermsa, H. L. *Kaart van Europa* 1874

Boese, K. G. *Oldenburg* 1865, 1872

Boetius, Christian Friedrich. Engraver. *Prospect . . . Elb Brücken zwischen Neustadt und Dresden* 1762

Boettcher, Carl. *Mitteleuropa* (1889)

Boettger, Hermann. *Railway map Germany and France* 1864

Boetticher, Wilhelm. *Carthage* 1835

Bofinger, G. *North Swabia* 1880

Bogaerts, Adrian J. van (1813–1891) Printer, lithographer, publisher and artist. *Maps of Germany, Holland and Italy* 1858–73

Bogar, Francisco. *Epis. Dioc. Jaurinensis* 1821

Bogardus, Joannes. Printer of Louvain. *Wytfliet* 1597

Bogert, M. L. *Fond du lac County, Wisconsin* (1862)

Bogerts, Cornelis (1745–1817) Dutch

engraver of Amsterdam. *Bachienes Atlas* 1785

Boguslawski, George (1827—1884) Son of following. Oceanographer

Boguslawski, Ph. L. (1789—1851) German astronomer. Discovered a comet.

Bohemio, A. G. See **Boehme**, A. G.

Bohemus, Martinus. See **Behaim**

Bohn, Carl Ernst. Publisher in Hamburg of maps by Sotzman *New Hampshire* 1796; *Connecticut* 1796; *New Jersey* 1797; *Maine* 1798; *New York* 1799

Bohn (François) Publisher in Haarlem. *Asia* 1807, *Netherlands* 1840

Bohn, Henry George (1796—1884) *Seat War North Italy* 1859, *Atlas of Classical Geography* 1861

Bohn, Oscar. *Geolog. Map Wrangel Mountains* 1899

Bohn, P. D. *Sea of Marmora* 1770

Bohnenberger, Johann Gottlieb Friedrich (1765—1831) Astronomer, mathematician, surveyor. *Schwaben u. Württemberg* 41 sh. Tübingen 1798—1813

Bohnert, Friedrich. *Tourist maps of Germany* 1870—93

Bohun, Edmund. *Geographical Dictionary* 1688, *Thea. Geogr.* 1695

Boileau de Buillon [Aegidius Bulionius Belga, Darinel] Gilles (1525—1563). Flemish diplomat and printer. *Map Germany* Munster 1551, *Sphère des deux mondes* 1555, in Honter 1555, Map *Savoy* 1556, 1557, *Campagna di Roma* 1563

Boim. See **Boym**

Boisaye, Bocage. See **Du Bocage**, Georges-Boissaye

Boisgelin, Louis de. *Sea and land chart of Malta* 1804

Boisquenay, Capt. M. de. *Revised plan of Port Louis* 1775 *(Neptune Oriental)*

Boisseau, Jean (fl. 1637—1658) French geographer, topographer, and genealogist. Enlumineur du Roi pour les cartes géographiques. Paris, L'Isle du Palais sur le quay qui regarde la Mégisserie à la fontaine Jouvence. *World* 1637, *Canada* 1641, *Topographie François* 532 plates 1641, and 687 plates 1655. Reissued *Bouguereau's maps Théatre des Gaules* 1642, *Nouvelle France* 1643, *Trésor des cartes Géogr.* 1643, *Clef Géogr.* 1645, *Ville d'Anvers* 1650, *Paris* 1652

Boisse, Ad. *Carte géolog. Dept. Aveyron* 1858

Boissevin, Louis (fl. 1652—1658) Marchand et Enlumineur. Paris. rue St. Jacques près Séverin à l'image Sainte Geneviève. *Trésor des Cartes* 1653, *Topo. Françoise* 1655, *Tableau Hist.* 1658

Boissière, Gilles de la (fl. 1660—1671) *Tables Géographiques* 1669, *Jeu de France* 1671

Boiste, Pierre Claude Victoire (1765—1824) French cartographer. *Carte de l'Amérique* 1795, *Dictionnaire de géographie Universelle* 1806

Boitet, Reiner. *Het Eiland van Kaap Breton* 1746, *Noord America* 1746

Bokushin, Eami (early 18th century) Japanese cartographer. *Plan Arima* 1737, after Shunboki

Bolam, Robert George. Map *Norham* 1844, after Robert Rule

Bolcher, General André (b. 1828) Russian War Office, Topograph. Section

Bold, Edward H. Surveyor. *Township of Ross, New Zealand* 1867

Boldoduc, Ed. *Arrondissement de Béthune* (1867)

Bolger, F. *Township of Whitman*, Toronto 1878, *Bridgeland* 1878

Bolia, C. Lithographer. *Baden* 1850, *Freiburg* 1860

Boling, C. Cartographer. *Map of Texel* 1795

Bollin, J. R. F. *West Sumatra* 1852

Bollin, R. J. *Plan Bern* 1808, 1811

Bollmann, L. von. *Reise Karte der Schweiz* 1839

Bulotov, Aleksei Paulovic (1803–1853) *Asie Mineure* 1853

Bollo, Luis Cincinato. *Atlas Geografico . . . del Uruguay* 1896

Bolstra, Melchior (1704–1779) Dutch surveyor and cartographer of Leiden. *Maps of the Meuse and Merwede Rivers. Rhynland* 15 sh. 1740–55

Bolswert, B. A. (1580–1633) *Views in Amsterdam* 1609

Bolton, Sir Francis. *Maps to show districts supplied by the New River* 1884

Bolton, John. *Parish of Aldingham in Co. of Lancaster* 1848

Bolton, Solomon. Revised d'Anville maps 1750, 1752, 1775, 1772 and 1800 for Sayer, and Laurie & Whittle. *N. America* 4 sh. 1752, *Gold Coast* 1800

Bolts, William. *Bengal* 1773, 1775. *Bengal, Bahar and Orissa* 1773

Boltze, Julius. *Wand Karte der Stadt und Landkreises Metz* (1877) *Saarburg* 1877

Bolzini, Giovanni Andrea. *Disegno del territorio Cremasco* 2 sh. 1741

Bolzoni, Andrea. Engraver. *Stati del Duca di Modena* 1746, *Citta di Ferrara* 1747

Bompario [Bompare, Bomparius] Pierre Jean. *Map of Provence* used by Ortelius 1590, Mercator 1606, Blaeu 1631

Bomsdorff, Oscar von. Maps of German states *Hanover, Saxony, Vogtland* &c. 1870–98

Bomsdorf, R. von. *Mecklenburg-Schwerin* 2 sh. (1895) *Mecklenburg-Strelitz* (1895)

Bomsdorff, Theodor von (d. 1898/9) *Magdeburg* 1853, *Austria* 1867, *Nord Amerika* 1882, *Saxony* 4 sh. 1889, Railway and telegraph maps &c.

Bonacina, Cesare. *Disegno di Pavia* 1654

Bonaldo, Dolfin. (16th century) Venetian *Marine chart, World* 1541 MS.

Bonaparte, Prince Louis-Lucien. *Carte des Sept Provinces Basques* 1863

Bonaparte, Prince Roland (b. 1858) Great patron of geography and exploration

Bonar, William. *Draught of the Creek Nation* 1757

Bonardo, G. M. *La Grandezza . . . di tutte le sfere*, Venetia 1584

Bonatti, C. Engraver. *Atlante Geog. Univ.* Torino 1854

Bonatti, Domenico. *Plan de Gènes* 1829

Bonatti, Enrico. *Genova* 1875, *Messina* 1877, *Torino* 1869

Bonatti, G. Engraver of Turin. For Barbié du Bocage 1854, *Duchy Genoa* 1855, *Julian Alps* 1864, *County of Nice* 1855, *Savoy* 1854

Boncompagni, Hieronymo (16th century) Italian cartographer

Bond, Frank and Frederick. *Wyoming* 1885

Bond, G. H. *Plan Gravesend and Milton* 1890

Bond, Isaak. *Frederick County* 1858

Bondeau. Engraver for *Nouvel Atlas* 1811

Bondelmonte. See **Buondelmonte**

Bone, J. Agar. Surveyor. *Maidstone* (1860)

Bonetti, A. *Linee Ferroviarie e di Navigazione in Italia* 1878

Bonfadini, Guiseppe Tomasso. *Geografia . . . per dimonstrare l'Inondazione del Ferrarese . . .* 1705

Bonhomme. *Environs d'Alger* 1830

Bonière, Hilarion. *Europe* 1785, to show houses of Order of Chartreux

Bonifac, P. *Mappa geog. and hist. Nigrae Silvae* 1788

Bonifacio [Bonifacius] Natale, of Sebenico. Painter and engraver. *Germany*

1553, *Cyprus* 1570, *Leslie's Scotland*
1578, *Maestricht* 1579, *Abruzzo
Ulteriore* 1587, *Palestine* 1590,
Pigafetta's Congo 1591, *Gallipoli* 1591,
Parigi 1591, *Calabria* 1592

Boninsigna, Domenico di Leonardo
(1384–1465). Drew maps for Latin
version of *Ptolemy's Geograpy* (1415)

Bonisel, J. *Plan Paris* 1815

Bonn, Otto. *Plan Würzburg* (1894)

Bonnange, Ferdinand. *Atlas graphique . . .
du commerce de la France* 1878

Bonne, Rigobert (1727–1795) Engineer
and cartographer. *Atlas Maritime* 1762,
Neptune Américo Septent., *Atlas
Moderne* 1762, *Globe* 1771, *Raynal's Hist.
Philos. du Commerce des Indes* 1774,
Abbé Grenet's *Atlas Portatif* 1781, and
Atlas Encyclopédique 2 vols. 1787–88

Bonnecamp. *Voyage fait dans la belle
rivière en la Nouvelle France* 1749

Bonnefont, Louis. *Mural maps France*
1873, *Europe,* 1875, *France Géologique*
1875, *Planisphère* 1875, *Atlas de Géogr.
Ancienne* (1868)

Bonner, Francis. *The Town of Boston in
New England* 1722

Bonner, John (1643–1726) English
Seaman and cartographer. *Chart Canada*
1710, *Plan Boston* 1714, 1722

Bonner. Draughtsman and engraver.
Yorkshire for *Topographical Dictionary*
1822

Bonner, William G. Civil engineer. *Map
of the State of Georgia* 1857

Bonnet, Charles. *Provincia do Alemtejo e
do Reino do Algarve* 1851

Bonnet, Édouard. *Bouche du Rhône*
1864

Bonnet, V. Engraver. *Atlas de Paris* 1836

Bonncy, James. *Map of St. Joseph Co.*
1863

Bonnier, Adolf. Publisher of Stockholm.
Skol Atlas 1852

Bonnisel, J. *Plan de la Ville et faubourgs
de Paris* 1814

Bonnisselle, J. G. Designer & engraver.
Plan London & Westminister 1772

Bonniver, J. F. H. Cartographer. *Map of
Limburg* 4 sh. 1849

Bonnor, F. Engraver for *Long's Jamaica*
1774

Bonomi, Joseph. *Ancient Egypt* 1848,
Jerusalem 1858

Bonomini, Bartolomeo, of Ancona.
Portolan Mediterranean 1570

Bonpland, Aimé. *Voyages de MM
Humbold et Bonpland, Atlas Géographique
et Physique du Nouveau Continent* 1814,
Géographie des Plantes Équinoxiales
1805

Bonsall, Joseph H. Civil engineer.
Philadelphia 1862, *Map of Denver,
Colorado* 1871, *Map of Colorado* 1872

Bonson, E. W. *Western Province Gold
Coast* 1884

Bonstetten, Albrecht von (ca. 1443–
1510) Swiss cartographer. *Description
of Switzerland* 1478

Bonstetten, Wilhelm von. *Archaeological
maps. Bern* 1876, *Fribourg* 1878

Bontein, Archibald. Chief enginner in
Jamaica. *Map of the Caribee Islands*
1751, *Plan of Carthagena* 2 sh. 1744, *Map
of the Island of Jamaica* 1753

Bontein, William. *Annapolis-Royal* 1754,
Hudson River Albany to Fort Edward
1757, *Survey Louisbourg*

Bontems [Bontemps] August François
(1782–1864) of Geneva. Military
cartographer

Bonvalet, H. Engraver for Mallet de
Bassilan 1846

Bonwick, James. *Victoria with parts of
New South Wales* 1855

Bonwicke, J. Publisher of *E. Wells'
Atlas* [1738]

Boog, F. A. *Chart Moulmein Harbour* (Burma) 1809 MS, *Salween River* 1869 MS

Boom, E. H. *Reede van Samarang* 1855, *W. Kust Java* 1855

Boom, Henry and T. *Atlas François* Amsterdam 1667

Boonacker, Sijmon Willems, Purser. *Buyck-slooter Meer,* Visscher 1625

Boosen, C. B. *Nord deel Norge* Kristiania 1862, *Sydlige Deel* 1862

Booth, Charles. Used maps with 7 colors to show variation of poverty in *London Life & Labour*

Booth, George, 1 Paragon Pl., New Kent Road. *Plans of London* 1845—46

Booth, John. Publisher Duke St., Portland Place, *Settlements in Australia* 1810

Booth, John C. *Map of the Mineral District of Lake Superior, Michigan* 1855

Booth, John. Surveyor. *Manor of Ditton (Bucks.)* 1718 MS.

Booth, W. *Chichester to Rye* 1764

Booth, Lieut. William. *Plan of Harbour, Forts, Town & Environs of Fort Royal in Martinique* 1793, 1795

Booth, W. J. *Plans of Newington Green, London* 1826

Boothby, F. *Drew and lithographed Map of Sussex* 1893

Boottats. See **Bouttats**

Booty, C. *Hunting map Environs Brighton* (1860)

Borch. See **Terborch**

Borchers, E. N. W. *Harzgebirge* (1866)

Borcht [Borchius, Verborcht] Pieter van der (1545—1608) of Antwerp. Painter and engraver for *Ortellius' Theatrum* 1563—1593

Borck. *Belgrade,* Berlin 1726

Borckeloo [Borcouloo, Borculo] Harmen van (ca. 1505—1578). Printer, publisher,

wood engraver. *Plan Jerusalem* (woodcut) 8 sh. 1538

Borda, Jean Charles, Chev. de (1739—1799) Mathematician and engineer. *Atlantique* Paris 1775, *Carte des Iles Canaries* 1780, *Coast of Africa* 1794, *Laurie & Whittle's New Chart Coast of Greece,* 1812

Borde. See **La Borde**

Bordee. Maps for *Laurie & Whittle's East India Pilot* &c.

Borden, H. D. Surveyor and engineer. *City of Syracuse* 1868

Borden. S. *Topographical map Massachusetts* 6 sh. 1844

Borden, Will. G. *Map of Town of Tiverton* 1854

Bordiga Bros. Engravers for Bacler d'Albe *Italy* 30 sh. (1798)

Bordiga, Benedetto (fl. 1800—34) Engraver of Milan

Bordiga, Gaudenzio (d. 1837) Cartographer and engraver of Milan

Bordiga, Giovanni Baptista. Engraver and draughtsman. *Military map Etruria* 1806, *Italy,* 1813, *Ancient Greece* 1828, *Europe* 1832

Bordone [Bordonius, Burdonius] Benedetto (1460—1531) of Padua. Miniaturist, cartographer, and engraver. Worked in Venice. *Globe* and *Italy* 1508 (now lost) *Isolario,* Venice 1528, reprinted 1532, 1534, 1537, 1547

Bordone, Girolamo. *Map Corsica* 1598

Borel, Petrus (17th century) Dutch astronomer

Boresfield, D. L. *Plan Ngatapa Pa,* (New Zealand) 1869

Borgen, Dr. *Süd Georgien,* Berlin 1886

Borghi, Bartolomeo (1750—1821) Italian cartographer. *Citta Firenze* 1817, *Atlante Generale* 1819

Borgian, *World Map* on metal plates, early 15th century

Borgonio, Giovanni Tommaso (1620–1683) Military engineer and Piedmontese cartographer. *Savoy* 1670, *Piedmont* 16 sh. 1680, *Sardinia* 25 sh. 1683, *Turin* 1672

Bormeister, J. Publisher in Amsterdam. *World* (pre-1700)

Born, C. L. von. *Kort over Faeroerne* 1789, English edition 1806

Born, Ignaz Edler von (1742–1791) Geologist and mineralogist of Prague and Vienna.

Born, Moritz L. and A. Aschland. *Surveyed parts Iceland* 1815–19

Bornhaupt, C. *Carte von Livonia, Estonia* and *Kurland* 1875

Borodovsky, L. *Map of Manchuria,* St. Petersburg 1897

Borough, Stephen (1526–1584). Navigator, served under Chancellor. *Narratives of the Muscovy Voyages,* the first English *Voyage to Russia* 1553, pub. in Hakluyt

Borough, William (1536–1599) younger brother of Stephen above. Navigator, hydrographer and writer on nautical subjects. Took service under the Moscovy Company. Comptroller of the Navy. *Chart of the Norther Navigation* (1560) *Drew chart of Frobishers Voyage* 1576 *Inward Parts of Russia* 1578 MS *Thames Estuary* 1588

Bors, Johann Jacob von. *Postkarte Deutschland* 16 sh. Homann Heirs 1764

Borven. Maps for *Sayer & Bennett's General Atlas* 1757–94

Bory de Saint-Vincent, Jean Baptiste Marie George. *Carte des Isles Canaries* 1801–03, *Atlas pour servir à l'histoire des Iles Ioniennes* 1823, *Espagne et Portugal 1824, Atlas Encyclopédique* 1827, *Atlas Géographique . . . de la France* (1845)

Bos. See **Busius**

Bos, Pieter Roelof (1847–1902) *School-* *atlas der geheele aarde.* Reprinted many times

Bosari, Bonifacius (18th century) Italian instrument maker

Bosatsu. See **Gyogi**

Bosch, Hendrik. Bookseller of Amsterdam. issued Dutch edition of Moll's maps 1727

Bosch, Baron Jan van den (1780–1844) of the Hague and Batavia. *Atlas of Dutch Colonies* 1817

Bosche, Philip van den. *Prague* 1606

Boschini, Marco (1613–1678) Venetian printer, engraver and mapseller. *Regno tutto di Candia* 1651, *Dalmatia* 1646, *Greek Archipelago,* Venice, Nicolini 1658

Boschke, A. *Topographical map of the District of Columbia* 1861, *Map of Washington City* 1857

Boscovich, Ruggiero Giuseppe, S.J. (1711–1787) *Stato Ecclesiastico* 3 sh. 1745, 1776, *Stato della Chiesa* 1770

Bose, Hugo von (d. 1856) Geographer of Dresden. *Atlas Austria* 1856, *Saxony* (1845)

Bosma, B. *De Geographische Onderwyzer* Amsterdam 1781

Bosse, Paul. *Grossherzogthum Hessen* (1877)

Bosset, Lt. Col. de. *Harbour of Cephalonia* 1814

Bossi, Giacomo. *Carte deux Voyages de Mungo Park* (1851), *Route du Clapperton* 1851, *Cours du Quorra* 1851

Bossi, Luigo (1758–1835) Italian cartographer. *Nuovo Atlante Universale,* Milan 1824

Bossius. See **Busius**

Bossler, Frederick. Engraver. *Map of Virginia by James Madison* 1807

Boswell, Henry. *Hist. descriptions of Antiquities of England and Wales,* Hogg 1786 (maps by Kitchin)

Bostwick, Henry (1787–ca. 1836) *Historical & Classical Atlas,* New York, 1828

Botella y de Hornos, Frederico de. *Prov. Marcia y Albacete* 1868, *España y Portugal* 1879

Botero, Giovanni (1540–1617) Italian priest and geographer born Piedmont, died Turin. *Relationi Universali,* 1591, reissued in Rome, Bergamo and Venice 1594, 1595, 1605. *Tab. Geogr.* 1596

Both, L. Scandinavian cartographer. *Maps of Denmark, France, and Flanders* 1861–77

Both, Major von. *Battle plans in Germany* 1807–10. *Plan von Berlin* (1811)

Botham, W. R. *Railway maps Great Britain* 1852, *London* 1853

Bothams, J. C. *Plan Salisbury* 1860

Botley, Thomas. *Surrey* 8 sh. 1765–69

Botta, C. *Carta Militaria d'Italia* 1800

Bottius, Adrianus. Publisher of *Mercator's Atlas Minor* 1610

Bottle, Alexander (fl. 1723–56) Surveyor of Harrietsham, Kent. *Manor of Tremworth* 1730, *Sutton Valence* 1733, *Kingsworth* 1729 MS.

Bottle, Robert Thomas. *Estate plans of Hoo, Kent* 1808, *Farm in Smarden* 1810 MS.

Botz, H. *Plan Jena* (1858)

Bouchard, E. *Teatro della Guerra in Italia* 7 sh. 1799

Bouchard, Giuseppe. *Pianta della citta di Firenze* 1755, *Pianta di Gibralterra* 1762

Bouchenvoeder, Major Friedrich von. Surveyor. *Essequibo* 1795, *Dutch Guiana* surveyed 1798–1802, pub. 1804

Boucher, M. Engineer. *Plan du Port Dauphin* 1733, *Neptune Français* 1778, *Ile Royale* 1780

Bouchet, J. V. Jesuit. *S. India* in Stocklein, Vienna 1728

Bouchette, Joseph (1744–1841) Lt. Col. Surveyor General of Lower Canada. *Upper and Lower Canada, and District of Gaspé* 1815. *Province of Lower Canada,* Faden 1818, *Town of Three Rivers* 1815, *Districts of Montreal and Quebec,* Wyld 1831

Bouchette, Joseph, Junior. *Upper and Lower Canada* 1831

Bouchette, R. S. M. *City of Quebec* (1850)

Bouchier, M. *Plan de la Ville de Troyes* 1839

Bouclet, (fl. 1793–1820) Engraver for Bruny-D'Entrecasteaux 1807, for Freycinet 1811, for Dépôt de la Marine. *Environs de Versailles* 12 sh. (1808)

Boudart, A. *Chemins de Fer Belges* 1876

Boudet, Antoine (fl. 1745–89) Publisher, of Paris. Imprimeur du Roi, rue St. Jacques. *World* 1752, *Vaugondy's Atlas* 1757, *North America* 1785, *Indes Orientales* 1786

Boudin, J. Ch. M. *Carte du Globe* 1851

Boudistcheff, Lieut. *Atlas of Black Sea* 1807

Boué, Dr. Ami (1794–1881) of Hamburg. Naturalist and physician in Paris and Vienna. *Geological map Scotland* 1820, *Ethnological map of European Turkey* (1840), *Carte Géologique du Globe* 1845, *Physical Atlas* London 1848

Boue, H. *Map Baltic* 1815

Bouffard, L. Engraver for *Sagra's Cuba and Algiers* 1838, *Canaries* 1838, *America* 1842, *Russia* 1844

Bougainville, Hyacinthe-Yves-Philippe Florentin, Baron de (1781–1846). French marine officer from Brest. *Journal de la navigation autour du globe* with accompanying *Atlas* 1837

Bougainville, Louis Antoine, Comte de (1729–1811) French navigator, explorer and hydrographer. *Voyage autour du Monde* 1766–69, pub. 1771–2. *Malouine* (Falkland Is.) 1763–4, pub. Jefferys 1771

Bougard, Réné (17th century) French hydrographer of Le Havre. *Petit Flambeau* *de la Mer,* Havre du Grace, Hubault 1684 Reprinted 1742. English edition *Little Sea Torch* 1801

Bouge, Jean Baptiste de (1757–1833) Surveyor, engraver, geographer and publisher, worked in The Hague. *Pays Bas* 4 sh. 1786, *Tournay* 4 sh. 1789, *Flanders* 1793, *Sardinia* 1799, *Naples* 1801, *Pas Bas* 20 sh. 1823, *Europe* 45 sh. (1829)

Bouguer, Pierre (1689–1758) Mathematician of Paris. *Figure de la Terre* (with Condamine) 1749

Bouguereau, Maurice (d. ca. 1591) Master Printer and map publisher à la Petite Fontaine du Carroy de Beaulire, at Tours. *Théatre François* 1594, the first printed atlas of the provinces of France

Bouillet. *Plan de la Conquête de l'Isle de la Dominique* 1778

Bouillon. See **Boileau de Bouillon**

Bouillet, Marie Nicolas (1798–1864) *Dictionnaire Universel d'Histoire et de Géographie* 1864, *Atlas Universel* 1877

Bouinais, Alb. M. Aristide (1851–1895) French mariner and geographer. *Works on Guadeloupe and Indo Chine, Tonquin* 1892

Boulangier, Edgar (1850–1899) *Études sur le relief terrestre* 1886

Boullanger. Hydrographer on Freycinet's Expedition. *Charts of coast of Australia* 1801–03, pub. 1811–12.

Boulengier [Bolengier, Boulanger] Louis [Ludovicus] French astronomer and geographer. *Universelle Cosmographie* (1514–18) *Globe* ca. 1517 (attributed)

Boulton, S. *Africa* 1787, 1794, based on d'Anville

Boulze, Auguste. *Carte routière industrielle et Minière de l'arrondissement d'Alais* 1867

Bouquet, Col. Henry (1719–1765) *Military maps and charts Ohio* 1764–66

Bourbon, Petit. See **Petit Bourbon**, Pierre

Bourcet, Pierre Joseph de. (1700–1780) *Alpes Françaises* 1754, *Dauphiné* 9 sh. 1758, *Carte des Alpes* (1800)

Bourdon, Jehan (1602–1668) *Plans of Quebec, Montreal, and St. Lawrence* MS 1635–42

Bourg, Marie François Cromot, Baron du (1756–1836) *Plans of French Army at Philipsburg* 1781, *Siège d'York* 1781

Bourgoin, A. *Plan de Chartres* 1889

Bourgoin, Jean François (1750–1811) *Atlas d'Espagne Moderne* 1807, 1808

Bourgoin, P. Engraver and publisher. Paris, Rue de la Harpe vis à vis le passage des Jacobins. *Asie* 1741, *Postes de France* 1748–63. Engraved for *d'Anville Italie* 1743, *S. America* 1774

Bourguignon d'Anville. See **Anville**

Bourjolly. *St. Domingue* 1803

Bourne, A. *State Ohio* 1820

Bourne, A. Engraver. *Street Map Reading* 1889

Bourne, Ebenezer, of City Road, Islington. Writing engraver for Ordnance Survey. Engraved for *Fairburn's Twelve miles round London* 1798, and for Wilkinson 1797 & 1800, *N. America* 1823

Bourne, Nicolas (fl. 1601–57) Publisher and bookseller. South Entrance Royal Exchange, Cornhill. *World Encompassed* 1628

Bourne, Richard. Surveyor. *The Manor Great Linford* (Bucks.) 1678

Bourne, Simon A. G. *Mexico and Guatemala,* Sidney Hall 1823

Bourquin, F. Lithographer. *Plan Land N.W. Auburn, N.Y.* (1866) *People's map of Michigan* 1864

Bourset, M. *Plan Town Cape of Good Hope* 1770, pub. Faden 1795

Bourtruche, A. French cartographer. *Atlas chronologique,* Paris 1837

Bourz von Seethal, Ferdinand (d. 1827). Surveyor

Bousquel, Auguste. *Plan de Castries* 1867, *États Pontificaux* (1868)

Boutats. See **Bouttats**

Bouthillier de Beaumont, Henry (1819–1898) French mathematician and geographer

Boutigny, H. Surveyor and architect. *Rouen* 1817 (with J. Heliot)

Boutin. *Algeri e Contorni* 1808, 1830

Boutin, F. *Maps of Colombia* 1825

Bouttats [Bottats, Bouttate] Gaspar. (b. 1634) Engraver and artist in Vienna. *Ostend* 1676, *Esquemelin's Panama* 1687, *Atlas of Hungary and Dalmatia* 1690, *Nouveau Monde* 1698 (in Hennepin), *Nijmegen* 1718

Bouttats, Philibert (17th century) of Antwerp. Publisher and engraver in Amsterdam and The Hague

Bouvé, E. W. Lithographer and publisher of Boston. *Nantucket by Mitchell* 1834, *Alton & Springfield Railroad* (1860)

Bouvet, Joachim, S.J. (1660–1732) Jesuit missionary and mathematician in China

Bouvet, L. *Atlas de la Confédération Argentina* 1873

Bouvet de Lozier, J. B. Ch. (1705–1786) Discovered Bouvet's Island 1739 *l'Hémisphère Merid.* de L'Isle (1740)

Bovinet, Giraldon. *Royaume de Perse* (with A. Vivien) 1825

Bouville, J. B. *Chemin de Fer Indo-Européen* 1881

Bowden, James. *Map of the Meetings of Friends in England and Wales* (1850) and (1855)

Bowden, Thomas A. *Wall map New Zealand* 1870, *Oceania* 1874, *New Zealand* 1879

Bowdich, Thomas Edward. *Mozambique et Benguela* 1822

Bowditch, Jonathan Ingersoll. Son of following

Bowditch, Nathaniel (1773–1838) born Salem, Massachusetts. American mathematician, astronomer and hydrographer. *New American Practical Navigator* 1799 (40 editions up to 1867). *Chart of Salem Harbour* 1804–06

Bowdoin, James (1752–1811) *Plan Halifax Harbour* 1780 MS.

Bowen, Abel. Engraver and publisher. *Plan Boston* 1828

Bowen, Emanuel (ca. 1720–1767) Engraver, printseller and publisher in London Next ye King of Spain in St. Katherines; Near ye Stairs in St. Katherines; Opposite to the Bolt and Tunn Inn in Fleet Street. Appointed Engraver of Maps to George II of England and Louis XV of France, but died poor and nearly blind. *Asia* 1714, *World* 1717, *Cusco* 1720, *Britannia Depicta* 1720, *Complete Atlas* 1752, *Royal English Atlas* 1762, *America* 1763, *Atlas Anglicanus* 1767

Bowen, M. Engraver. *China,* for Sayer and Bennett (1780) *Thirty-five to forty miles round London* 1787

Bowen, Thomas (fl. 1749–90). Son of Emanuel (above) 9, Charterhouse Lane, Clerkenwell. Engraved for his father's *British American Plantations* 1749, for Cook 1773, *Royal English Atlas* 1767, *Atlas Anglicanus* 1765, for *Morant's Essex* 1768, *Taylor & Skinner's Roads* 1776, *Middleton's Geogr.* 1777–78, for *Carver's New Univ. Traveller* 1779, for Hogg 1785, for *Forster's History of Voyages* 1786. He died in Clerkenwell Workhouse.

Bowen, William. Revised *Lloyd's map of Lower Mississippi* 1862

Bowen & Co. Lithographers of Philadelphia. *Plan Detroit* 1830

Bower A. *Plans Kingston upon Hull* 1787 and 1791, *River Witham* 1803

Bower, Capt. H. *Tibet and W. China* 1892–3

Bower, J. Maps in *Carey's American Pocket Atlas* 1814, *U.S., North Carolina, Missouri Territory*

Bower, Robert A. *New Handy Atlas of U.S. and Canada* 1884

Bowler, L. P. *Northern Goldfields Matabele & Mashonaland*, Pretoria 1889

Bowles, Carington (1724–1793) Printer and publisher in London, St. Pauls Churchyard. *Britannia Depicta* 1764, *Evans' Middle British Colonies* 1765 (pirated), *The Large English Atlas, New Medium English Atlas* 1785, *America* 4 sh. 1790, *Universal Atlas* 1792. Succeeded by Bowles & Carver

Bowles, Charles. *Maps Europe* 1871–2

Bowles, John. Surveyor. *Huntsmoor, Thurney* (Iver Bucks). 1736 MS.

Bowles, John (1701–1779) Map and printseller, publisher, at the Black Horse in Cornhill: over against the Stocks Market or Mercer's Hall Cheapside. Brother of Thomas Bowles. Published some of the later editions of Moll. *Large Atlas of World, County Maps* 1724, *Ireland* 1728, *English Gentlemens Guide* 1738, *Royal English Atlas* 1762. Took his son Carington into partnership 1754–ca. 1764 and then continued alone. He issued two *Catalogues,* one in 1728, another 1753. Succeeded by Robert Wilkinson

Bowles, Thomas. (fl. 1700–1763). Probably the father of John and Thomas. Map and printseller, opened his shop in 1712 next the Chapter House in St. Pauls Churchyard. Published Moll's works 1721–53. Acquired stocks of Morden, Lea and Seller. Pub. *Catalogue* of his works 1720. Collaborated with others: Homann 1739, with Brandreth, Bowles, Taylor and Gouge for *Price's 30 miles round London.* He retired in 1764 and was succeeded by Carington Bowles

Bowles, Thomas II (b. 1712) Engraver of maps and plans, son of John. Engravings in

Théatre de la Grande Bretagne 1724, *Map London* 1725

Bowles & Carver (fl. 1794–1832) Publishers in London, 69 St. Pauls Churchyard. *Universal Atlas* 1784–98, printed *Catalogue* 1795. Successors to Carington Bowles 1793.

Bowman, Alexander. *Maps of Illinois* 1855–67

Bowman, Amos. *Dawson's Explorations Columbia* 1888, *Cariboo Mining District* 1889, *Geological Survey Canada* 9 sh. 1895

Bowra, John. Surveyor of Groombridge. *River Medway* 1739, *Tunbridge Wells* 1738, *Estate plan Oakes Park, Sevenoaks* 1780

Bowyer, William. *Map British Isles* 1567 MS.

Boycot, William. Surveyor. *Buckland,* Kent 1623, *Marden* 1636, *Manor of Sturry* 1643 MSS.

Boyd, Charles, Surveyor. *Chester District, S. Carolina* 1818, for Mills 1825

Boyd, Hugh. *Plan of Ballycastle Harbour* 1743

Boyd, J. Engraver for *Carey's American Atlas* 1822; *Maryland, Ohio, Windward Islands*

Boyd. See **Oliver & Boyd**

Boydell, John. Engraver and publisher (with Josiah). *Basque Road* (1758) *Hill's plan Philadelphia* 1798

Böye, Herman. *Map Virginia,* reduced from the 9 sheet *map of the state* 1827

Boyer, Abel (1667–1729) *Fortified Towns Europe* 44 plates 1701, *Map & Plan of the Town of Preston* (1715)

Boyer, M. French Marine. *Plan du Havre Peraki* (New Zealand) 1840

Boylan, J. D. *Gulf of Mexico* 1861–64

Boyle, George. *Thames Guide* (1839)

Boyle, John Roberts. *Survey of Newcastle upon Tyne* 1723

Boym [Boim] Michal Piotr. (1612—1659) Polish Jesuit. *MS Atlas China* 1653—5, *Printed map China* 1661

Boynton, George W. Engraver. Worked for Edwards 1832, for Goodrich 1830, Bradford 1838, *Gordon's map of New Jersey* 1838, *Boston* 1846, *Lowell Mass.* 1845

Braakman, Adrian. Dutch publisher and cartographer of Amsterdam. *Atlas Minor,* Amsterdam 1706

Braakensiek, Albert (1811—1883) Dutch cartographer, engraver, printer and publisher. *Scandinavia* 1835, *Noordzee Kanaal* (1872)

Braam, Jan van. Bookseller of Dordrecht. Published (with Onder de Linden) *Valentyn's Travels* 1724, *Kaart Borneo* (1750) *Mauritius* (1750) *India Orientalis* 1770

Brabazon, Capt. Luke. *Sketch map Military route Sebastopol* 1856, *Route China* 1860

Brabo, Francisco Xavier. *Cartas Geogr. America Merid.* Madrid 1872

Brachelli, Prof. H. F. *Neuer Atlas der ganzen Erde* 1863

Brackebusch, Dr. Ludwig (b. 1849) Argentine geographer. *Maps of the Republic* 1881, 1885, 1891

Brackenridge, C. S. *Map Fort Wayne* (Indiana) 1871

Bradbury, John. *Map of USA* in *Bradbury's Travels* 1819

Braddock, Edward. *General plan of his campaigns in America* 1755

Braddock, J. H. *Co. Eyre, S. Australia* 1860

Bradford, H. *Plans Rivers Trent and Tame* 1759

Bradford, J. L. Surveyor. *Olathe, Kansas* (1860)

Bradford, J. R. *Post & Telegraph map Queensland,* Brisbane 1892

Bradford, Samuel. Surveyor. Plans *Coventry* 1750, *Birmingham* 1751

Bradford, Thomas Gamaliel (1802—1887) Publisher of Boston and New York. *Comprehensive Atlas* 1835, *Atlas, N.Y.* 1836, *Illustrated Atlas Boston* 1838, *Univ. Atlas* 1842 (with Goodrich)

Bradford, William (1663—1752) *Map of Country of Five Nations* 1724 and 1735. *Plan city New York* 1728

Bradley, Abraham. Junior. *US Post Roads* (1796) 1819, *Northern Part of USA* (1797) *Southern Parts of USA* (1797) *United States* 1804, *Indiana,* Melish 1817

Bradley, J. *Universal Atlas use Schools* 1819

Bradley, J. H. *Plan Isthmus Tehuantepec* 1851

Bradley, T. Engraver. *Naples* 1835

Bradley, William. *Chart Norfolk Island* 1794

Bradley, Wm. M & Bro. *Atlas of the World,* Phila. 1886, 1893

Bradshaw, George (1801—1853) Engraver, draughtsman and publisher, of St. Mary's Gate and Market Street, Manchester. Innovator of *Railway Guides. Canals Manchester* 1830, *Railway Time Tables* 1839, *Railway Companion* 1844, *Plans Important Cities Europe* (1870)

Bradshaw, Richard. *Geological map Terre Haute and adjacent coalfields* 1875

Brahe, Tycho [Tyge] (1546—1601) Danish astronomer. *Pair Globes* 1584, *Astronomiae Instauratae Mechanica* Nürnberg, Hulsius 1602, *Sphaera Stellifera* 1606, Observatory on Hven Island. Teacher, Blaeu among his pupils.

Brahé, Tycho

AN
ILLUSTRATED
ATLAS,
GEOGRAPHICAL, STATISTICAL, AND HISTORICAL,

OF THE

UNITED STATES,

AND THE

ADJACENT COUNTRIES.

By T. G. BRADFORD.

PHILADELPHIA:
E. S. GRANT, AND COMPANY.

Title-page to T.G. Bradford's Atlas of the United States

Effigies Tychonis Brahe,
Dani Domini. Ætatis suæ, Anno 40.
Anno Domini 1586.

Tycho Brahé (1546—1601)

Brahm, Wm. Gerard de (1717–1799)
Surveyor General of S. District of North
America, Capt. of Engineers under
Emperor Charles VI. Went to America
1751 and founded Bethany. *South
Carolina & Georgia* 1757 and 1780.
Ticonderoga 1759, *Amelia Is.* 1770,
Atlantic Pilot 1771, contributed to
American Military Pocket Atlas 1776

Brambilla, Ambrosio. (16th century)
Engraver and cartographer of Rome.
Genoa 1581, *Rome* 1582, 1590;
Augsburg 1602

Brambilla, Valerio. *North Italy* 6 sh.
1840, *Contorni di Torino* 1854

Bramham, James. *Plan St. Johns Harbour
in Newfoundland* 1751

Bramston, W. *Plan Canton* 1840, *City &
Harbour of Macao* 1840

Brancas-Villeneuve, Abbé André (d. 1758)
French cosmographer. *Système de
Cosmographie* La Haye 1747

Branch, William. Surveyor of London.
MS Plan Kent 1786

Branco, Alves. *Plano Hydrographico da
Enseada do Quicembo* 1888

Brandard, E. P. Steel engraver for *Bible
Atlas* 1878

Brandenburg, Lorens Wilhelm (1794–
1850) Swedish engineer, worked for
Official Survey *Sverige och Norge* 1843

Brander-Dunbar, Lt. *Maps Somaliland*
1896–7

Brandes, H. H. (1798–1874) German
geographer. *Geographical Manuals Europa*
1872, 1885; *Wien* 1863 and 1873

Brandis, G. B. à. See **Brender à Brandis**

Brandreth, T. (18th century) Publisher
At the Archimedes & Globe in Ludgate
Stairs: Over against ye Royal Exchange in
Cornhill. Joint ventures with other
publishers: with Price 1711, with Willdey
1712, with T. Bowles, T. Taylor and
J. Gouge's *30 miles round London*

Brandt, T. J. Cartographer. *Map Zeeland*
1860

Brannon, George. Engraver and publisher.
Isle of Wight 1831

Brannon, John. Publisher. *Plan of
Washington* 1828

Bransby, John (1762–1837) Surveyor,
astronomer & globeseller. *Liberties of
Ipswich* 1812, *River Orwell* 1814

Bransfield, Edward. Master, R. N. *Plan St.
George's Bay, S. Shetlands* 1820

Branth, Lieut. T. *Kaart over Kjobenhavn*
1854

Brantz, Lewis. *Plan city Baltimore* 1822

Brasch, B. M. *Plan der Stadt Neu Ruppin*
1789

Braselmann, J. E. *Bibel Atlas* 1868

Brasiel, William. Surveyor. *Leaseholds in
    Paddington* 1742

Brasier, William Furness (1745–1772)
Plan of Oswego 1759, *Fort Ticonderoga*
1759 MS, *English Harbour Antigua* 1756,
Survey of Lake Champlain 1762, pub.
1776

Brask, John (d. 1538) Bishop of Linkoping.
Supposed author of lost map of
Scandinavia

Brassier, William. See **Brasier**, William

Brath, James. *Balaklava Railway* 1855

Brauchitsch, von. *Umgegend von
Braunschweig* (1895)

Braud, E. *Canton de la Chataigneraie*
1877

Brauer, Dr. A. *Die Seychellen* 1896

Brault, Louis Dexie Leon (1839–1885)
Meteorologist and marine cartographer

Braun, Adolphe. *Théatre de la Guerre*
(in Alsace) 1871, photographic views.

Braun, Ernst Eberhard. Engineer. *Carte
von der Gegend um der Stadt Hannover*
1750–57

Braun, Friedrich *Himmels Atlas in
Transparenten Karten* (1855)

Braun [Bruin] George [Joris] (1541–
1622) Topographer, geographer and
publisher. *Civitates Orbis Terrarum* 6

vols. 1572–1618 (with F. Hogenberg)
French edition Cologne 1574. The first
atlas of town plans and views embracing
the whole world.

Braun, W. *Eisenbahn und Strassen
Atlas Kurhessen* (1855)

Brauns, J. C. Cartographer. *Map Drenthe*
1840

Braunschweig, L. *Telegraphen Karte
von Deutschland* 1869

Bravo, Francisco Javier. *Atlas . . . de la
América Meridional* 1872

Bray, I. Master of Missionary Vessel
Morning Star. Corrected Admiralty
Chart of *Truk or Hogolu Is.* 1880

Bray, J. & A. *Hastings & St. Leonards*
(1898)

Bray, Salomon. *Stadt Haerlem* 1644

Brayer, Lucien, Comte de. *République de
Paraguay* 1863

Brayley, Edward Wedlake. Assisted with
Vernor, Hood & Sharp's *British Atlas* 1810

Brazier, E. J. R. B. *Survey Coast Sindh
& Catch* (with Lieut. Grieve) 1853

Brdiczka, Leopold. *Bohemia* 1861

Bream, Lieut. Engineer Assisted William
Brasier. *Plan of the fort at Ticonderoga*
1759

Brecher, Adolph. *Brandenburg & Prussia*
1868–9, editions to 1880

Bredda, Bartolomeo. *Agri Patavini
Chorographia* 1650

Bredsdorff, H. J. and O. N. Olsen. Danish
cartographers. Produced hypsometric
maps. *Esquisse orographique de l'Europe*
1830, *Skolekaart over Europa* 1863

Breen [Brenius] Daniel van (1599–1665)
Engraver in Amsterdam, working
1630–50

Breen, James. *Stanford's maps of Paths
of Comets* 1858

Breese, Samuel. *Cerographic Atlas of U.S.*
1842, *Morse's North America Atlas*
1842–5

Breidenbech. See **Breydenbach**

Breislak, Scipione (1748–1826) *Topografia della Campania* 1797

Breitholtz, Didrik Julius (1748–1812)
Swedish surveyor

Breitinger, D. *Plan Zürich* 1804, 1814

Breitkopf, Johann Gottlob Immanuel
(1719–1794) Publisher in Leipzig

Bremden, D. V. Engraved title for
Janson's Atlas Major 1657, *Novus
Atlas* 1648, and *Nouvel Atlas* 1658

Brémond, Laurent. *Ports et Rades de la
Mer Méditerranée* (with Michelot) 1718,
1727–30. English edition 1802

Brender à Brandis, Gerrit. (1751–1802)
Zak-en Reis-Atlas Amsterdam, S. J.
Baalde 1788, *Handels Kundige Atlas*
(1800)

Brenna, Giovanni. Engineer and
geographer. *Regno Lombardi-Veneto*
1831 and 1837, *Milan* 1858

Brentel, Frederic. *Urbis Nancei [Nancy]
delin.* 1611

Brereton, Capt. W. *Batangas Bay* 1763

Bressani. *Novae Franciae Acc. Delineatio*
1657. Only one copy known to date

Bressanini, Rinaldo (1803–1864) Engraver

Bresson, Jean Antoine. *Marseille et Environs*
1772, 1773

Bresson, Jean Pierre, fils. *Plan Géometral
de la ville de Marseille* 1773

Bretez, Louis. *Plan de Paris commencé
1734 . . . achevé de graver* 1739, index &
20 sh.

Bretonnière, Coudre la. *Charts for Dépôt
de Marine. Calais and Somme Areas*
1792–97

Bretschneider, Emil W. (1833–1901)
China 4 sh. 1896

Brett, Sir Percy. *Ramsgate* 1755, *Coast
Kent and Sussex* 1759, *Map of the
Downs* (1775)

Brettel, Joseph. Mining engineer in service
of Mohammed Ali. *Plan of Acre* 1840

Brettingham, Mathew. *Plans Holkham in Norfolk* 1773

Breugnot. *Pays Bas* 60 sh. Paris, Piquet 1822

Breusing, Arthur (1818–1892) German geographer. Hydrographical works and *History of Geography*

Breval, J. Historian and geographer. *Remarks on Several Parts of Europe* 1726

Brewer, Rev. John Sheeren. *Elementary Atlas* (1854)

Brewster, Sir David. Collaborated in *Johnston's Physical Atlas* 1849

Brewster, Edw. Publisher in London at the Crane in St. Paul's Churchyard. *Blome's Description of the World* 1670 (with Brooks and Basset)

Breydenbach [Breidenbach] Bernard von (ca. 1440–1497) Traveller. *Peregrinatio in Terram Sanctam,* Mainz, Reuwich 1486, several editions. First book with folding plates first printed book of town plan views. The publisher Reuwich also illustrated the work. A German edition by Speier ca. 1500

Brialmont, Alexis Henri. *Military atlases* 1890

Briatico, Cola de. *MS Atlas* of 3 charts: *Atlantic Coast, Adriatic, Black Sea* 1430

Bridge, William. City Surveyor of New York. *Map of the City of New York and Island of Manhattan* 1807, pub 1811

Bridgens, H. F. *Map Elmira* 1852, *Cumberland Co., Penn.* 1858, *Atlas Chester Co., Penn.* 1873

Bridges, Lt. Col. *Military plan of the Cape of Good Hope* for *Barrow's Travels* (1806)

Briet [Brietus] Philip, S. J. (1601–1668] born Abbéville. *Palestine* 1641, *Parallela Geog. Vet. et Novae* 1648, *Japan* ca. 1650, *World* Paris 1653, *Atlas* Paris 1653

Briffaut, Etienne. Publisher in Vienna. *L'État de la guerre en Italie* Vienna 1734, *Théatre de la Guerre sur le Rhein* 1735

Briggs, Major. *Tea Countries of Assam & Cachar* (1865)

Briggs, Henry. Mathematician. *N. Part of America,* engraved R. Elstracke in *Purchas His Pilgrimes* 1625

Briggs, Joseph. *Plan Montevideo* 1823

Brigham, H. G. *Map of the Counties of Clinton and Gratiot, Michigan* 1864

Brigham, J. C. *Nuevo Systema de geografia* New York 1827–8

Brigham, W. T. Surveyor. *Crater of Kelamea, Hawaii* 1865

Bright, Frederick John. *Map Bournemouth* (1885)

Brightly, C. & E. Kinnersley. Printers of Bungay, Suffolk. *Barclay's Dict. N. Amer.* 1806

Brinck, Ernest. Translator of the first Dutch edition of Mercator 1634

Brindley, James. Canal Engineer. *Plan Bridgewater Canal* (1759) MS, *Navigable Canals* 1769, *River Thames* 1771

Brine, Lieut. Frederic R.E. *Plans Crimea* 1857–9, *Sevastapol and surrounding country* 1857, *Valentia* 1859

Brink, C. F. *Map S. Africa* MS 1761

Brink, Ten. Surveyor. *Part West Coast Afrika* 1761 MS

Brinkley, John (1763–1835) Bishop and astronomer

Brinkman, Bernardus. *Sterren Kaart* 1873

Brion, Benjamin. *Maps for Gourné's Atlas* 1763

Brion, Hypacio de. *Plano Hydrographico da Enseada do Quicembo* 1888

Brion, John & Sons. *Relief maps Brighton, England & Wales, Jerusalem, Isle of Wight* 1858–78

Brion, L. *Forts et Ports Brittann. et d'Hanovre* Paris 1756

Brion de la Tour [Bironius Gallus] Louis (1756–1823) Engineer, Geographer to the King. Rue du Petit Pont près la

fontaine St. Séverin maison de M. Landes.
Libraire au l^er (1793) *Côtes Maritimes
de France* 1757, *Atlas Général, Civil et
Ecclésiastique* 1766, *Théatre de la
Guerre en Amérique* 1777, *Atlas
Itineraire Portatif 1766, Atlas et Tables
Elémentaires* 1777, *Atlas Général* 1790—
98, *U.S.* 1783, *N. America* 1783, *World*
1799, *Atlas Géogr. et Statist. de la
France* 1802

Brion, L.

Brion de la Tour, Fils. *Départments de
la France* (1789)

Briot, B. J. *Composite Atlas* ca. 1660

Brisbane, Sir Thomas Macdougall (1773—
1860) Astronomer, erected Observatory
at Parramatta

Briscoe, J. C. *Plan of the Battle of
Gettysburg* 1863

Bristow, Capt. Abraham. Additions to
Butler's map of Pacific 1822

Brito, Pedro Torquado Xavier de. *Imperio
do Brasil* 1867

Britto, Capello H. C. de. *Maps of
Portuguese Africa* 1883—7

Britton, James. *The British Atlas* 1810

Britton, John. Publisher Burton Street,
London. *Survey of St. Marylebone* 1832

Brix, C. C. *Skole Atlas* 1867

Brjus. See **Bruce**, Count Jacob

Broc, E. *Cartes Commerciales* 1887—88

Brocas, S. Engraver. *Map Killarney* 1815

Brocci, G. *Carta Fisica del Suolo di Roma*
1821

Broch, Ole Jacob (1818—1889) Norwegian
geographer. *Royaume de Norvège* 1878

Brochant de Villiers, André Jean Marie.
Carte géologique de la France 1840

Brochard, Bonaventura. *Palestine* 1544
(lost)

Brochet des Roches, P. *Provincia de
Corrientes* 6 sh. 1877, *America del Sur*
1880

Brock, Major Gen. Sir Isaac. *Sketch map
of the American Lakes and adjoining
country* 1815—16

Brockes, C. *Mappa do Imperio do Brasil*
1878

Brockhaus, F. A. *Grundriss v. Hamburg*
1855, *Russland* (1870)

Brocklesby, John. *Elements Physical
Geography* 1869

Brodie [Broadie] Lt. Joseph. *Part of
North Sea (Texel to Naze)* 1798, *General
Chart of the North Sea* 1812

Brodin, Rudolf. *Karta öfver Stockholm*
1870

Broditzky, Joseph. *Post and railway maps
Austria and Hungary* 1874—6

Broeck [Brock] Abraham van den (ca.
1616—1688) Engraver of Amsterdam.
Zürich, Jansson 1641, *Transilvania*
Jansson 1645, *Valois* Jansson 1658,
Germany Visscher (1680)

Broeckhuyzen, J. van. Cartographer.
Collaborated with Broel Houwer in
Large scale Dutch maps. Gooiland
1709

Broedelet, G. Publisher of Utrecht.
Relandus' Imperium Japonicum (1715).
Hennepin's North America 1697

Brooks, J. *Louisville Canal, Ohio
River* 1825—43

Broen, Joannes de (1659—1730) Engaver
of Amsterdam. *Borgonio's Lac Leman,
Piedmont* 1682, *Savoy* 1700

Broenner, Heinrich Ludwig. Bookseller
and publisher of Frankfurt. *War in
Germany* 1760—1, *Hesse Cassel* 1760,
Holland's N.Y. & N.J. 1777

Brognolio [Brognolus] Bernardo (1539—
1583) of Verona and Mantua. *Verona,
Forlani* 1574, used by Ortelius 1579

Broichmann, J. *Volksschul Atlas* 1880

Brolan, C. Publisher. *Reskarta öfver Skåne*
1868

Brolin, Jonas. *Stockholmsstad* 1771, *Uppsala Stad* 1770

Bromley, George W. and W. S. *Atlas Baltimore* 1898, *Atlas Boston* 1888, *Atlas of New York* 1891, *Atlas Philadelphia* 1889

Bromme, Traugott (1802–1866) Bookseller of Stuttgart. *Deutschland* (1867), *Atlas to Humboldt's Kosmos* 1851–3, *Volks Atlas* 1875

Brongniart, Alexandre (1770–1847) French geologist and mineralogist. *Carte géognostique des environs de Paris* 1810, *Carte Géologique des Environs de Paris* 1811 and 1821

Bronn, Heinrich G. *Mineralogical map neighbourhood of Heidelberg* 1830

Bronner, H. Ludwig. *Plan Mainz* 1735

Brooker, Lieut. Edward W., R. N. *Battle Plans Crimea* 1854–5, *Port Derwent*, Hobart 1861(3)

Brookes, P. *Travelling Companion England & Wales* 1812

Brookes, Richard. *General Gazetteer* 1795 and later editions

Brooking, Charles. *Map of the City & Suburbs of Dublin* 1728, 1730, 1740

Brooks, J. *Rapids of Ohio River*. Frankfort, Kentucky 1806

Brooks, Nath. Publisher at the Angel in Cornhill. *Blome's Descript. World* 1670 [jointly with Brewster & Basset]

Brooks, W. A. Civil engineer. *Charts of River Tees* 1833–40

Broom, Jacob (1752–1810) American Surveyor. *Map of the Brandywine Campaign* 1777 MS.

Brosch, August. *Umgebung von Peterwardein* 1872

Brose, H. *Alpine maps* 1822–50

Brose, Heinrich elder (1805–1868) Engraver for Kiepert 1860. *Battle plans Europe* 1821–54

Brose, W. Engraver for Tanner 1833 and later

Brossard de Corbigny. *Senegal-Gambia* 1861

Brosses, Charles. de. of Dijon. *Histoire Navigations aux Terres Australes* 1756

Brostenhuysen, Jan van [Johann] (ca. 1596–1650) Engraver and printer of Leiden. Worked for Allard 1646

Brothen, Alfred. *Star Atlas* 1870

Brouckner [Bruckner] Isaac, elder (1686–1762) Cartographer and engraver. Geographer to Louis XV France. *Globe* for Empress of Russia 1735, *Nouvel Atlas de Marine* 1749, *Nieuwe Atlas* 1759, *Africa* 1749

Broughton, Hugh (1549–1612) *Concent of Scripture* 1588–90, *Map of Earth*. First English map engraved on copper

Broughton, S. H. *Geological Map Rockland and Pewabic Township, Michigan* 1863

Broughton, W. R. *Chosen Harbour, Corea* 1797, Admiralty Chart 1840, *Japan Kuril Islands* 1855

Brouwer, Hendrick (1521–1643) *Maps Brazil and Chile* (1640)

Brown & Sons. Surveyors. *Plan of Maidstone* 1821

Brown. Publisher of Melbourne. *Tulloch & Brown's Map of the Colony of Victoria* 1856

Brown, A. L. *Dubuque County, Iowa* 1866

Brown, Charles. *Plan of Man of War Bay* (1830)

Brown, Charles Barrington. *Geological map Guiana* 1873, *Jamaica* 1865

Brown, Charles F. Engineer. *Rock Island & Alton Railroad* 1857

Brown, George. Surveyor and engineer of Elgin. *Road surveys* 1790–99, *Map Moray and Nairn* 1810, (1813)

Brown, Henry Yorke Lyell. *Part West Australia* (1873) *Geological map South Australia* 4 sh. 1899

Brown, I. Engraver. *Phoenix Park, Dublin* 1789

Brown, Publisher and cartographer. Seymour Place Bryanston Square *Ireland* (roads) 1842, *New General Atlas* (1852)

Brown, Jos. C. Surveyor. *Missouri, Illinois* 1841

Brown, J. W. Surveyor. *Street map of the Manor of Aston* 1883

Brown, John. *Northumberland and Lancashire Junction Railway* 1846, *Wear Dock Railway* 1846

Brown, Milton R. *Continental Atlas* 1889

Brown, N. Surveyor. *Canal Huddersfield-Ashton under Lyne* (1793)

Brown, R. Architect *Plan Plymouth Stonehouse and Devonport* 1830

Brown, Thomas. Engraver and surveyor. *Settlements N. America* (1770) *Edinburgh* 1793. *Plan of the Manchester, Ashton under Lyne and Oldham Canal* 1793

Brown, Thomas. Bookseller and publisher, No. 1 North Bridge, Edinburgh. *Travelling map Scotland* 1791, *Plan Edinburgh* 1793, *Atlas Scotland* 1800(2) *General Atlas* 1801

Brown, T. S. *Map Dunkirk Harbour Chautauque County . . . New York* 1835, *Presqu'ile Bay* 1837

Brown, William of Tring (Herts.) Surveyor. *Farm Great Missenden* (1840) MS.

Brown, Capt. W. Revenue Surveyor. *Districts of India* surveyed 1827–40

Brown, W. S. and C. M. Foote. *Plat Books of Brown, Dunn and Wanpaca counties, Wisconsin* 1889–9

Browne, Capt. B. Revenue Surveyor. *Districts of India* 1846–63

Browne, Christopher (fl. 1684–1712) English cartographer, globe maker and publisher at the Globe near the west end of St. Pauls Church, London. Bought Speed plates from Bassett & Chiswell; sold to J. Overton. Successor to R. Morden. *Moore's Fens* 1684, *Virginia & Maryland* 1685 and 1700, *Flanders* 1700, *England* 15 sh. 1700, *Geogr. Classica* 1712

Browne, Daniel. Publisher and mapseller without Temple Bar. For *Senex Atlas* 1721, *Geogr. Magna Brit.* 1748 and 1750, *Camden's Britannia* 1753

Browne, I. *Tamerein* (Socotra) 1615, pub. by Dalrymple 1774

Browne, James. *Country of Sikhs* 1787, *Scotland* 1845

Browne, J. O. *Map of Municipal Borough of Thetford* 1837

Browne, John (d. 1599) Irish surveyor. *Parte of the County of Mayo* 1584 MS, *Galway* 1583, *Connaught* 1592

Browne, Lord John Thomas. *Milo &c.* 1851, *Cyprus Admiralty* 1878

Browne, Joseph. Engraver. *Staffordshire* 1682, for *Plot's Staffordshire* 1686

Browne, J. L. with Reuss, *Map Sydney* (1856) & (1862)

Browne, Patrick M.D. *New Map of Jamaica* 1755 [1756]

Browne, P. J. *Monroe Co., N.Y.* 1852

Browne, Robert. *Passage of Moon over Sun* 1724

Browne, Thomas (1708–1780) Land surveyor and herald. (Blanch Lyon pursuivant of arms) *Essex plans* 1726–33, for *Michael Dever's Hants.* 1730

Browne, T. Lithographer in Hobart Town, Tasmania. *Wabbs Harbour* 1860

Browne, William George. *Route Soudan Caravan* 1799 (Travels in Africa)

Browning, J. S. *Province of Canterbury, New Zealand* 1863

Brownlee, J. H. *Railway & Guide Map Manitoba* 1887

Brownrigg, John. *Grand Canal, Dublin* 1788(9)

Brownrigg, W. Meadows. Land surveyor. *Gold Country N.S.W.* (1851)

Bruce, Alexander. Surveyor. *Plan Loch Sunart* 1733

Bruce, C. A. *Map Upper Assam* 1839

Bruce [Brjus, Graf Jakov Villimovic] Count Jacob (1670–1737) Russian Field Marshal of Scots descent. Maps (with Mengden) *South Russia* 1699, first map printed in Russian

Bruce, James. *Charts and Plans to Bruce's Travels to the Sources of the Nile* (1805)

Bruce, John, of Symbister. *Nieuwe Paskaard van Hitland* ca. 1711, published by Ottens 1745

Bruce, William Downing. *Kennet Clackmanshire* 1855

Bruckner, G. *Terrestrial and Celestial Globes* 1842

Bruckner, Isaac. See **Brouckner**

Brucks, Lt. (later Capt.) George Barnes. *Charts of the coasts of Arabia, Bushire, Bahrein and Persian Gulf, Admiralty Charts* 1826–72

Bruder, C. Lithographer. *Canton Schaffhausen* (1855)

Brué, Adrien Hubert (1786–1832) French publisher, Rue des Macons Sorbonne No. 9 Géographe du Roy and Géographe de son Altesse Royale Monsieur Comte d'Artois. *Altas de France* 1820–8, *Atlas Classique* 1830, *Atlas Universel* 1816 &c., *L'Empire Français* 1813. Succeeded by Picquet.

Brué, A. H.

Bruechman. *Plan Gand* (1709), *Lille* 1709, *Bataille Malplaquet* 1709

Brueck, P. A. *Plan ville d'Arlon* 1874

Brueckner, Daniel. *Canton Basel* 1766

Brueckner, Lieut. Johann. *Umgegend v. Dresden* 1868, *Dresden-Freiberg-Meissen* 1868

Bruegner, C. *Plan v. Alger* 1830, *Schlacht bei La Belle Alliance* 1815, *Ligny* 1815, *Schlacht v. Laon* (1821), *Oporto* 1833

Bruegnor, C. Engraver. *Hafens von Navarin* 1827

Bruehl, H. *Ancien Mogontiacum* (1800), *Wiesbaden* 1819

Bruehl, Johann Benjamin. Engraver. *Halle* (1710)

Bruff, J. Goldsborough. *State of Florida* 1846, *Mexican War* 1847, *Tehuantepec* 1851, *Seat War Virginia* 1861

Bruggen, Johann van der (1695–1740) Engraver in Prague and Vienna. *Plan Vienna* ca. 1720, *Autriche* 11 sh. 1737

Brugsch, Heinrich. *Map ancient Egypt* 1857–75

Brugsma, F. C. *Atlas van der Nederlanden* 1856, *Nederland* (1867)

Brugueirolle, Ate. *Dépt. du Gard* (1874)

Bruhms, Dr. Carl Christian (1830–1881) *Hand Atlas* 1864, 1871 and 1879, *Atlas der Astronomie* 1872

Bruin. See **Braun**

Bruletout de Préfontaine. *Carte géographique de l'isle de Cayenne* 1762 (with Louis Charles Buache)

Brummer, V. R. *Karta öfver Åland* 1838

Brun, Giovanni. Engraver. *Roma (antica)* 1801

Brun, Capt. Jayne. *Mapa Nueva Granada* 1843–47, *Port of Sabanilla* 1843–56

Brun, V. A. M. See **Malthe-Brun**

Brunacci [Brunacius] Baldo. Of Pisa. *World* 1516 MS.

Bruneau, Capitaine. *Carte du Sud Oranais* 4 sh. 1884

Bruneau, Robert. Engraver and printer of Antwerp. *Additamenta Ortelius* 1603, *Pisani's Map World* 1613

Brunel, Isambard Kingdom. Civil engineer and inventor. *Great Western Railways* 1834–45, *Northumberland Railway* 1844, *Worcester & Oxford Railway* 1845

Brunet, R. Engraver for *Cassini's Carte de France* (12 sh.) 1744—60

Brunner, Christopher. *Territorium Basileense* 1729

Brunskow, Berthold. *Grundriss Berlin* 1858, *Preussischen Staates* (1867)

Brunton, J. *Railway maps India* 1863

Brunton, R. Henry. *Japan* 4 sh. 1876 and 1880

Bruny-D'Entrecasteaux, Joseph Antoine (1739—1793) Governor of Mauritius and Bourbon. In charge of an expedition in search of La Pérouse. Made surveys coast of Australia. *Atlas du Voyage de Bruny D'Entrecasteaux* 1791—3, Paris 1807

Bruschi, Giacomo. *Topografia del Porto e citta di Genova* 1789

Brush[ius] Caspar (1518—ca. 1550) *Fichtelgebirge* MS.

Brussel, I. K. Composed some *surveys for railways, Vauxhall-Reading* 1833

Bruun, Conrad Malthe. See **Malthe-Brun**

Bruun, Ph. J. (1804—1880) Finnish geographer. Works on *Eastern Mediterranean*

Bruyn, D. L. J. de. *Omstreken van Utrecht* 1871

Bruyn, Marinus Diderius de. *Palestine* 1874

Bruyneel, Jac. *Map Hungary*

Bruzen de la Martinière, Antoine Augustine (1662—1746) *Asia, Africa, America* 1738, *Hungary and Danube, The Hague* 1741, *Atlas* Leipzig 1744—50

Bry, Joannes Théodore de (1561—1623) son of the following. *Pigafetta's Africa* 1598, 1624

Bry [Brij] Théodore de (1528—1598) born in Liege, worked in England 1586(7)—88. Goldsmith, engraver and publisher in Frankfort on Main. *Collection of Voyages, Grands et Petits Voyages.* Engraved charts in the *English edition of Waghenaer* 1588. *Vopel's Rhine* map 1594

Bryan, Capt. Corrected *Thornton's Chart of Antigua* 1701

Bryan, Francis T. *Kansas Gold Regions* (1860)

Bryan, Hugh. Surveyor. *Carolina and Georgia* in conjunction with de Brahm 1757 and Le Rouge 1777

Bryant, A. (fl. 1822—35) English surveyor and publisher, 27 Great Ormond St. Series of *Large scale county maps: Herts.* 1822, *Suffolk, Surrey* 1823, *Norfolk, Oxford, Gloucs.* 1824, *Bucks.* 1825, *Beds.* 1826, *Northants.* 1827, *Lincs.* 1828, *E. Riding* 1829, *Cheshire* 1831, *Hereford* 1835

Bryce, Major. *Harbour of Alexandria* 1802—04, *Lower Egypt* for Arrowsmith 1807

Bryce, Rev. Alexander, of Kirknewton. *N. coast Scotland* 1740—44

Bryce, David. *Pearl Atlas of World* 1893, *Vest Pocket map Glasgow* (1895)

Bryce, J. Annan. *Burma* 1866, *Burma—Siam—China Railway* (1888)

Bryce, James. *Cyclopedia of Geography* 1856, *Family Gazeteer* 1862, *Library Atlas* 1875—6, *Comprehensive Atlas* (1886)

Bryce, James (1806—1877) Geologist. *Papers on North Ireland and Scotland*

Brydone, J. *Plan Edinburgh* and *map Scotland* (1858)

Bryer, F. Engraver. Post Office *Map Surrey* 1874 and 1878

Buache, Louis Charles. *Carte Géographique de l'Isle de Cayenne* 1762

Buache, Philippe (1700—1773) French geographer and publisher, married daughter of De L'Isle 1720. Entered Dépôt des cartes et plans de la Marine (1721). Premier Géographe du Roi (1729). Member Académie des Sciences (1730). Addresses: Paris Sur le Quay de la Mégisserie près le Pont Neuf (1737—41), Quai de l'Horloge (1745—63). *Isles de l'Amérique* 1724 MS, *Martinique*

Th. de Bry (1528–1598)

1732, *Nouvelles Découvertes au Nord*
1753, *Atlas Physique* 1754, *Atlas
Géographique et Universel* 1762. Noted
for his theoretical cartography. Reissued
maps of De L'Isle and Jaillot

Buache de la Neuville, Jean Nicholas
(1741–1825) Conservateur du Dépôt des
cartes de la Marine. Premier Géographe
du Roy (1782) Nephew of the preceding.
Dominica 1778, maps for *Vie de
Washington* 1807, *Charts for La Pérouse.*
Succeeded by Dezauche

Buache, N.

Buasso, V. *Pianta di Torino* 1800

Buch, Leopold Freiherr, von (1774–1853)
German geographer and geologist. *Atlas
physical. Beschreibung der Canarischen
Inseln* 1825, *S. Tyrol* 1823, *Deutshland*
1838

Bucha, Zacharias. *Armillary sphere* 1593

Buchanan, George. *S. Hemisphere for
Thomson's Atlas* 1816, *Plan Estate
Corehouse* 1841

Buchanan, J. *Cork Harbour* 1808

Buchanan, Thomas. *Chart Northern
Ocean* 1820

Buchard. *Fleuve Sénégal* 13 sh. 1893

Bucher, A. *Plan Fort de Gironne* 1812

Bucholtz, Leopold von. *State of Virginia*
1859

Buchon, Jean Alexandre C. (1789–1846)
Atlas Géographique des deux Amériques,
Paris 1825. *Notice d'un atlas en langue
Catalane* 1843

Buchwalder, A. J. *L'ancien Évesché de
Bâle* (1820)

Buck, James. *Estate in St. Marylebone*
1780

Buck, L. von. See **Buch**, Leopold von

Buck, Rufus & A. M. Bouillon. *Mining
maps British Columbia* 1897

Buck, Walter M. *Map New Brunswick*
(1874)

Buckert, Fred. *Chart Cattegat* 1810–11,
published 1825, 27, &c.

Buckhout, Leavitt. *Obey River Oil Region*
(1863)

Buckinck, Arnold [us] Printer of Cologne,
partner of Sweynheym 1497. The *Rome
edition of Ptolemy* 1478

Buckland, J. W. *Map of the Hop-growing
District of Worcestershire and
Herefordshire* 1890

Buckland, William (1781–1856) Geologist.
President Geological Society 1824, 1840.
Wrote geological papers.

Buckland, W. T. Surveyor. *Parish Datchet
(Bucks.)* 1839 MS, *Dorney* 1844 MS,
Parish Horton 1838 MS, *Parish Upton*
1848 MS

Buckle, Comm. C. H. M. *Charts of
Brazil,* Admiralty 1844

Buckley, Samuel. Publisher at the Dolphin
in Fleet Street, London. *Draught of the
City of Carthagena* 1699

Buckman, James (1816–1854) Geologist

Bucknall, Thomas. Surveyor. *Estate plans
Chillerton Farm, Isle of Wight* 1722.
Mannour of Chillerton 1724, *Parnells
Farm* 1724

Budd, T. A. *Harbour of Bridgeport,
Connecticut* (U.S. coast Survey) (1835)

Budgen, Richard (ca. 1824–1879) Surveyor.
Sussex 6 sh. engraved Senex 1724,
also various estate plans

Budgen, T. *Plan Brighthelmstone* 1788

Budker, A. *Plan de Macon* 1893

Buechel, Eman. *Canton Basel* 1766

Buechel, J. *Military plans* 1855–6

Buecking, H. *Geologische Karte von
Attika* 1891

Buehler [Bühler] Adolph. Various *maps
of Switzerland* 1867–97

Buehler [Bühler] Edouard. *Curland* 1848

Buehler, James A. (fl. 1740–1800) Globemaker. *Globe* 1790, 1795.

Buell, Abel (1742–1822) American silversmith and engraver. From 1770 he engraved maps, in 1784 a *Wall map of United States,* New London, Printed & Sold by T. Green 1786. The first map of America by an American publisher.

Bueno, F. A. P. *Mappa de Amazonas* 1865

Buesser, E. *Plan v. Strassburg* 1873

Bufalino, Leonardo. *Plan Rome* 1551, used later by Nolin 1748

Buffa, Frans. *Noord Zee Kanaal* (1876)

Bufford, J. H. Lithographer of Boston. *Barker's City of Charlestown* 1848, *Plan Portland* 1858

Bugarsky, Jean. *Serbie* Belgrad 1845

Bugge, Thomas (1740–1815) Danish geographer and astronomer. Supervised the first triangulation of Denmark from 1762. *Faroe Is.* 1770

Buisson, Burin du. *Dépt. Haute Saône* 1874

Bulifon [Bolifon] Antonio (b. 1649) French bookseller, cartographer and engraver in Naples. *Naples* 21 sh. 1692, *Pianta di Pozzuoli* 1696

Bulionius, A. See **Boileau de Bouillon**

Bull, Adolf (fl. 1850–80) Danish cartographer. *Denmark* 1851, *Skoleatlas* 1870, *Større Skoleatlas* 1876

Bull, G. Engraver for Blachford 1827

Bull, J. (19th century) English surveyor

Bull, William (d. 1755) Surveyor, Lieut. Governor of South Carolina. *South Carolina and Georgia* (with William Brahm) 1757 and Le Rouge 1777

Buller, W. *Anchorage at Blewfield Jamaica* (with Leard) 1793

Bullet, Jean Baptiste. French cartographer. *Plan of Paris* 12 sh. 1710

Bullet, Pierre and François. *Blondel's Plan Paris* 1675

Bullock, Lt. (later Commdr.) Frank. Admiralty Surveyor (1820–54) *Newfoundland* 1821–6, *River Medway* 1840, *River Thames* 1827–47, *Approaches to Harwich* 1847, *Dungenes* 1847–50, *Gravesend Reach* 1852, *Downs* 1857

Bullock, Henry. *MS plans of Scots border* ca. 1552

Bullock, William. *Atlas Historique pour servir au Mexique* 1823

Bulow, Baron A. von. *Nicaragua* and *Panama* 1847, published 1851

Bumble, Otto. *Karte v. Schlesien* (1876)

Buna, Wilhelm C. *Bavaria* (1760), *Partie des cercles du Haut et du Bas Rhin* 6 sh. (1760)

Bünau, Henry. (fl. 1507) German cartographer. *Atlas Historique, Géographique, Topographique et Militaire* (1760)

Bunjiemon, Toshimayo. Map and print publisher of Nagasaki (18th century). *Map of Kyushu province* (1783), *Plan of Nagasaki*

Bunning, J. B. Architect. *Improvements to Smithfield* 1851

Buno, Johannes. *Orbis Terrarum Veteribus Cogniti . . . correcta a J. B.* 1644

Bunou, Philippe (1680–1759) French Jesuit. *Abrégé de Géographie* 1716

Bünting [Buenting] Heinrich (1545–1606) Professor of Theology in Hanover. *Itinerarium Sacrae Scripturae* 1581 and later; *World in shape of a clover leaf, Europe as a virgin,* and *Asia as a winged horse*

Bunyan, C. Stephen Holloway. *South Leaga Bay* (Sumatra) pub. Dalrymple 1774

Buondelmonte [Anexerius, Anxerinus, Auxerinus] Christoforo (15th century) Florentine priest. *Liber Insularum Archipelagi* 1420, used later by Bartolomeo da li Sonetti, Bordone and Boschini

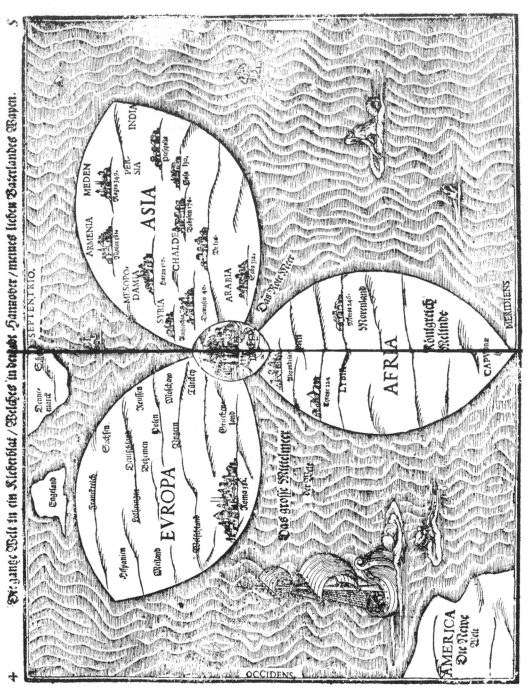

Woodcut world map in the shape of a clover leaf (the device of Hannover) by H. Bünting, published in his "Itinerarium Sacrae Scripturae," 1581

Buono, Florino dal. *Citta di Bologna* 1636

Buonsignori, Stefano [Stephanous Monarchus Montis Oliveti] (d. 1589) Cartographer of Florence, monk of Mont Oliveto. Cosmographer to the Duke of Tuscany. *Map Siena* 1584, *Tuscany* 1589. Maps for the Palazzo Vecchio, Florence

Buraeus. See **Bure**, Anders

Burbank, C. W. *Map of the City of Worcester* 1872

Burchardi, Lieut. Engineer. *Colberg & Environs* 1857–60 16 sh. MS.

Burchell, William John (1782–1863) Explorer and naturalist. *Travels in South Africa* 1822

Burcht, W. van den. Cartographer. *Bois le Duc* pub. Hondius 1630

Burck, A. *Montana* and *Wyoming* (1870)

Burckhardt, John Lewis. *Carte de l'Arabie Pétrée* 1834, *Karte der Sinai-Halbinsel* 1859

Burdett, Peter P. (d. 1793) Surveyor and artist, reported Inventor of aquatinting. *Liverpool Harbour* 1771, *Cheshire* 4 sh. 1777, 2nd edition 1794. *Derbyshire* 6 sh. 1767, 2nd edition 1791

Bure, [Buraeus, Bureus de Boo] Anders (1571–1646) Swedish cartographer, secretary to King of Sweden. *Map of the Northern Provinces of Sweden* 1611, *Scandinavia (Orbis Arctic nova)* 1626 on 6 sh, reissued Mortier ca 1710, used by Hondius and Blaeu

Bureau, E. *Atlas de Géographie Militaire* 1871

Bureau, Jacques. *Carte du Mississippi* 1700

Burgartz, F. *Vorarlberg* (1864)

Burgdorfer, J. J. Publisher. *Plan Bern* 1811, *Berner Oberland* 1816, *Canton de Berne* 1830

Burges, Bartholomew. *Solar System displayed*, Boston 1789

Burgess, F. I. *Topo. survey Dihlee district* 8 sh. 1852

Burgess, J. of Stanton St. John (Oxon) Surveyor. *The Manor Boarstall* (Bucks.) 1697 MS

Burgess, R. W. *Lake Erie* 1852, *Chart of head of navigation of the Potomac River* 1857

Burgess [Burgiss, Burgis] William. Surveyor. *Plan of Boston in New England* 1729, South Prospect City of New York (1720) and 1746

Burgh, Thomas. *Dublin* 1728, with J. Perry

Burghers [Burg, Burgh van de, Burght] Michael (fl. 1670–1720) Born Amsterdam, came to England 1672. Engraver to Oxford University. *Map in Plot's Oxford* 1677, *Wells Atlas* 1700, *Devizes* 1714, *Magna Britannia* (1720)

Burghley, William Cecil, Lord (1520–1598) Patron of map makers and collector. *Lancashire* 1590

Burgh[t] Willebord van der. Flemish cartographer. *Brabant* (Sylvaducis) used by Blaeu 1634 and later.

Bürgi, Jost (1552–1633) Mathematician, astronomer, globemaker, clockmaker. *Globes* 1582 and 1592

Burgkmair, Hans the elder (1475–1531) Engraver of Augsburg. Collaborated with Cusanus.

Burglechner the elder, Matthias (d. 1603) Cartographer of Innsbruck

Burglechner the younger, Matthias (1573–1642) *Map Tyrol* 12 sh. 1608 and 1611. *Tirol in shape of an eagle* 1620

Burhams, S. H. *Maps of Illinois* (1860–1864)

Burkhardt, E. F. *Railway map Germany* (1845)

Burleigh. *Plan of River Wear from Newbridge to Sunderland Boro.* 4 sh. 1737 (with Thompson)

Burlingame, E. H. *Map City Williamsport* 1872

Burman [Burmen] Gerhard von. *Skäne S. Sweden* Stockholm 1756

Burmeister, Herman K. (1807–1892) Argentine geographer

Burmester, Johan (1664–1713) Swedish surveyor of Gotland

Burischek, Franz (18th century) Surveyor

Burn, George. *Channel Pilot of Great Britain* (with J. Stephenson) 1786 and 1791, *Coasting Pilot for G.B. and Ireland* 27 sh. 1791–1803, *New and Correct chart Thames Mouth* 1809

Burnes [Burns], Lt. Alexander. *Cutch* 1829, *Central Asia* pub. Arrowsmith 1834

Burnet, Thomas (1635–1715) *Telluris theoria*, London 1681, *Maps of Western Hemisphere, Europe &c.*

Burnett, Gregory. Surveyor. *Sutherland* 6 sh. 1833 (with W. Scott)

Burnett, J. C. Assisted *Wickham's Moreton Bay* 1846

Burnett, Lieut. *Approaches to Hobart Town* (1868)

Burney, James. *Chart Coast of China* 1811

Burnford, Thomas. Engraver. *French Conquests* 1674, *Seventeen Provinces* 1674, *Blome's Cosmography* 1682, for Admas (1692)

Burr, David H. (1803–1875) Topographer to the Post Office and Geographer to the House of Representatives of the United States. *Universal Atlas* (1835) *Atlas New York* (1829), *American Atlas* 1839, *United States* 1842, *Ohio* pub. by Colton 1845, *World* 1850

Burr, Henry A. *Disturnell's new map U.S.* 1854

Burr, J. *Allotments in N.S.W.* 1814

Bursian, Conrad. *Geographie von Griechenland* (1872)

Burston, John. Chart maker. Dwelling in Radcliff Highway neare London. *Portolan chart of the Mediterranean* (1666)

Burrell, Peter. Surveyor General. Directed *Richardson's Royal Manour of Richmond* 15 sh. 1771

Burriel, Father Andres Marcos. *N. America* 1756, *S. Sea* 1757, *California* 1757, all in *Venega's Noticia*, Madrid 1757

Burritt, Eljah H. *Atlas Geogr. of Heavens* 1835

Burroughs [Burrows] Samuel (fl. 1677–1707] Publisher at Crown in Cornhill, Bible and Three Legs in Poultry Little Britain. *Morden's State of England* 1704

Burslem, F. H. Surveyor and lithographer. *South Australia* 1842, *Barossa Special Survey* 1843, *Green Hills* 1843, *Rapid Bay* 1845

Burstall, Lieut. (later Commander) Edward. Hydrographer. *Newhaven Harbour* 1839, *River Crouch* 1847, *Folkestone* 1848, *Beachy Head* 1852, *N. W. Mull* 1860

Burston, John (fl. 1640–66) Chartmaker Dwelling at the Signe of the Platt in Ratcliff Highway neare London. *Mediterranean* 1658, 1666 partner with Comberford. *East Atlantic* 1660 MS.

Burton, M. Surveyor. *Proposed Roads Buckingham* 1853 MS.

Burton, Charles Edward (1846–1882) Astronomer

Burton, Decimus. Architect. *Tunbridge Wells* 1828

Burton, Giles. (17th century) Estate Surveyor

Burton, James. Surveyor. *Intended Improvements Radford Estate* engraved Tuppin

Burton, Sir Richard Francis (1821–1890) Traveller, orientalist, translator

Burty, Rue de la Veille Estrapade 3 Paris. Engraver for Levasseur 1851

Burucker, Johann Michael (1763–1808) Engraver and geographer, born Nürnberg. *Behring Straits* 1781

Burwell, Lewis. *County of Brant, Canada West* 1859

Burwood, R. Engraver. *County Pembroke* 1827, *St. Pancras* 1850

Busby, B. Surveyor. *Orangeburgh District South Carolina* 1820, for Mills 1825

Busby, Richard. Surveyor. *Ly Farm, Wilts* 1750

Busch, Andreas (1656–1674) Instrument maker of Limberg. *Gottorp Globe* 1664, under direction Olearius

Busch, Edward. Engraver. *Atlas Borough Germantown* 1787

Busch, George (16th century) German astronomer

Busch, Georg Paul. Engraver. *Strasse von Gibraltar* (1720), *Schlesien* (1742), *Bayern* (1745)

Busch, J. G. *Deutschen Reiches* 1872, *Plan v. Leipzig* (1890)

Bush, B. F. *Bay County* 1869, *Saginaw Valley* (1870)

Bush, W. Civil Engineer. *Plan proposed London Steam Docks Deptford* 1841

Bushman, John. Midshipman, assistant surveyor. *Chart N.E. Coast of America* (under Capt. Parry) 1824

Busius, Albertus. Publisher and painter of Düsseldorf. *Atlas Mercator* 1595, 1602

Busk, Capt. H. *Isle of Man* 1861

Bussche, Georg Wilhelm von. *Osnabrug. Episcopatus* (1775)

Bussemacher [Bussmacherus, Buxemacher] Janus [Johannes] (fl. 1580–1613) Cartographer, engraver, printer of Cologne, Off. S. Maximini Strasse. *Westphalia* 1590, *Quad's Atlas* 1592, 1600 and later.

Bussius Belga [Bos, Busius, Bossius] Jacobus (fl. 1551–1577) Engraver for Tramezini and Lafreri. *Rome, Switzerland, and Flanders* 1555. *Brabant, Holland, Guelderland* 1556, *Friesland, Northern Regions* 1558, *Asia* 1561

Butler, B. F. Lithographer in San Francisco. *N. Portion San Francisco County* 1853, *Map San Francisco* 1852

Butler, E. W. *Plan Parish of Gruyere* 1875

Butler, Rev. George. Edited *Public Schools Atlas* 1872–93

Butler, H. *Almanac Map for 1853 showing the principal Theatrical Towns* 1853

Butler, J. Junior. *Plan Paddington* 1824

Butler, James. *Peekskill* (New York) 1852, *Staten Island* (New York) 1853

Butler, J. P. *Manila Bay* 1830

Butler, M. Engraver for Cummings 1818

Butler, Philip and Richard. *Carlow* 9 sh. 1789

Butler, Samuel (1774–1839) Bishop of Lichfield and Coventry. *Ancient Geography* 1822 *Atlas of Modern Geography* 1825 &c.

Butler, Thomas. *W. Part Pacific Ocean* 1799

Butlin (with Noble) *Plan Northampton* 1747

Butterfield, G. W. Surveyor. *Plan Union Village in Medway* 1855

Butterfield, Michael (ca. 1635–1724) English instrument and globe maker.

Butters, R. Publisher & Bookseller, 79 Fleet Street London. Later 22 Fetter Lane, Fleet Street. *Political Magazine* with maps by Lodge 1789–90, *Atlas of England* 1803

Butterworth, E. Engraver for *Thomson's Scotland* 1824–7, *Aberdeen* 1826

Butterworth, J. Engraved for Jos. Priestly 1769, *Proposed Canals* 1770 and 1790, *Whitecloth Hall* 1775

Butts, A. G. *Map of the State of Georgia* revised by A. G. Betts 1870

Butts, J. R. *Map of the State of Georgia* compiled by J. R. Betts 1870

Buvignier, Armand. *Carte Géologique des Environs de Paris* (1850)

Buxemacher. See **Bussemacher**, J.

Buy de Mornas, Claud (d. 1783) Historian and Royal Geographer *Atlas Méthodique* 1761, *Atlas Historique et Géographique* 1762

Engraved world map for M. Quad's "Geographisch Handsbuch"
by J. Bussemacher

Buzin, J. J. von. *Croatien u. Slavonien* 1877

Buzzo, Francesco. *Val Telina e Val Chiavena con suoi confini* (1640)

Bye, Joseph P. Engraver. *Enouy's Scotland* 1803, *Coasts of Sussex* &c. 1804, *Doncaster* 1805, *Bristol Channel* 1806, *Azores* 1807. *Asia* 1809, for Smith 1808, for Playfair 1814

Byers, W. N. City plans. *Omaha City* 1857, *Excelsior, Nebraska* (1857), *Saint Vrain, Colorado* (1860)

Byles, A. D. Publisher in Philadelphia. *Kent Co. Delaware* 1859

Bylica, Marcia. Astronomer and mathematician. Studied in Cracow (1451–59) and Bologna (1463–64) Professor in Bratislava 1467 and Buda 1471. *Globes* 1480

Byres, Chief Surveyor Tobago. *St. Vincent Surveyed* 1765–73(76) *Dominica* 1776, *Tobago* 1776, *Bequia* 1778, all published by S. Hooper, engraved by Bayly

Byrne, A. T. *Republica Honduras* 1886

Byrne, George. Engraver. *Rocque's Kilkenny* 1758, *Nevill's Wicklow* 1760, *Drury's Harbour of Valencia* 1788

Byrne, Joseph James. *Dublin* 1819, *Estate plans of Galway* 1852

Byron, Commodore J. (1723–1786) Circumnavigator. Grandfather of the poet. Governor of Newfoundland 1769–72. *MS charts* 1764–66, *Voyage* 1766

Byron, Samuel. *New Plan Dublin* 1791, *Plan City of Kilkenny* (1795)

C

C***. **See Chatelain**, Henry Abraham

C., F. I. *Atlas de Géographie Physique* 1878

C., H. Courtier, Henri. *Drayning of Fens* 1629

C., J. *Plan of Canals ND. New Kind of Fire* 1628

C., L. S. *Grundriss der Festung. Phillipsburghe* (1734)

C., M. *Carta Topographique del Territoire Friuli* 1793

C. N. *The principal passages of Germany* 1636

C., P. V. See Call, Pieter van

C., R. See Crawford, Robert

C., T. *Short discourse of Newfoundland* 1623

Caamano, Jacinto. *MS maps of N.W. coast America* ca. 1761–2

Cabello. Engraver. *Plan del curso los Rios Huallaga y Ucayali* 1830

Caballero, (1800–1876). Spanish writer and geographer. *Geographical Dictionary of Spain*

Cabeza de Vaca, Alvar Nuñez (1490–1561) Spanish explorer. *Relation* 1542

Cabillet, E. Reduced and engraved *Plan de la Ville de Bourdeaux* 1840

Cabot, [John] (1420–1501) Venetian pilot, Bristol merchant and navigator

Cabot [Gabato, Caboto, Cabotus]

Sebastian (1474–1557) Son of John. Venetian cosmographer and cartographer. Pilot-Major Casa de la contretacion (1518–48). Various MS maps, and *Planisfere* engraved Antwerp 1554 (lost)

Cabral, Francisco Augusto Monteiro. *Ilha Fortificada de Angediva* 1847, *Carta Hydrographica do Porto de Goa* 1842

Cabral, Pedro Alvarez (1468–1520) Portuguese navigator

Cabrera, Fran. Engraver for *Maestre's Guatemala* 1832

Cacaut, François. *Plan de Nantes* 4 sh. 1757–9

Cacciatore, Leonardo (1775–1830) Italian cartographer. *Nuovo atlante istorico* 1831

Cadamosto [Cd da Mosto] Alysius Alvise or Luigi da (1432–1480) Venetian navigator. Voyages 1455–6 *to Gambia and Cape Verde Islands. Nuovo Portolano. . . Levante*

Cadell, Capt. F. (1820–1879) Australian explorer

Cadell, Robert (1788–1849) Publisher

Cadell, Thomas (1742–1802) Bookseller and publisher London, in the Strand. *Cook's Voyages* 1773–7 with Strahan

Cadet, Felix. *Atlas de Géographie physique. . .de France* (1855)

Cadet de Metz, Jean Marcel (1751—1835) French mineralogist. *Various works on the history of geography*

Cadle, J. C. *Railway maps New England* 1831

Cadwalader, A. *White Pine Mining District* 1869

Cady & Burgess. Publishers, 60 John Street New York. *United States* 1848, *Michigan and Wisconsin* 1850 (in *R. C. Smith's Atlas*)

Caesius. See **Blaeu**, W. J.

Cafferty, E. M. *Map of Norwich* 1872

Cagniard de la Tour, Baron Charles (1777—1859) Engineer geographer

Cagnoli, Antonio (1743—1816) of Zante. Mathematician and astronomer. *Works on mathematical geography*

Cagnoni. Engraver for Mauro Fornarı 1786—9

Cahill, D. Irish cartographer. *Queens County* 4 sh. 1806

Caignart de Saulcy, Louis Félicien Joseph. *Imperial map Palestine* (1864)

Caillat, Ph. *Tunisie Ancienne* 1876

Caillaux, Alfred. *Carte Minière de la France* (1880)

Caille, Nicolas Louis. Abbé de la. See **La Caille**

Cailliaud, Frédéric. *Carte générale de l'Egypte* 1827, *Carte géologique de la Loire Inférieure* 1851, *Aethiopien* 1859

Caimox, C. See **Caymox**, C.

Cairncross, T. W. *Map of Cape Town* 1891

Caiz, Jan Las. *Description del seno Mexicano Islas de Barvolento* 1708

Cajamarca. *Atlas Géographique del Peru* 1865

Cal. See **Call**

Calamaeus [Chameau] Joannes [Jean]

Berry 1556. Used by Ortelius 1570, Bouguereau 1594

Calapoda, Giorgio. See **Callapoda**

Caldas, F. J. de. *Rio Grande de la Magdalene (Humboldt's Voyage)* 1805—34

Calder, Admiralty Surveyor. *Maps of Tasmania* 1849—51 (with Sprent), *Seat of War in New Zealand* 1863

Calderinus, Domitius. Publisher of *Ptolemy* 1478

Calderon, Augusto. *Mapa de los Caminos de Hiero de España y Portugal* 1878

Caldwal. Engraver for *Bell's Environs of London* 1769

Caldwell, Joseph A. *Atlas of Franklin County Ohio* 1872, *Knox County* 1871, Ohio 1874, *Beaver Co. Penn.* 1876, *Oil Belt* 1878, *Monroe County* 1898

Calef, John (1725—1812) Chart of *Penobscot* 1781 *(position of troops & fleet Siege of Penobscot)*

Calheiros da Graca, Francisco. *Charts Brazil* 1890—4

Calhoun, J. Surveyor General. *Plan of public surveys of Kansas & Nebraska* 1856

Calkoen, Jean Frederik van Beck [Beek-Calkoen] (1772—1811) Dutch astronomer. *Théorie de construction pour les mappe-mondes,* Utrecht 1810

Call, Jan van (1655—1703) *Maps and Views of Dutch, German* and *Swiss Cities.*

Call, Jan van, II (1698—1748) *Donauwerth* 1704, *Ulm* 1705

Call, John. *Arcot Fort* 1778, *Carangooly* 1778. *History Military Transactions Indostan* 1778, *Geometrical Survey of Alice Holt and Woolmer Forest, Hants by Wyburd* 1790 MS, John Call Commissioner

Call, Pieter van. *Plans Ath* (1698), *Gand* 1709, *Béthune* (1710), *Aire* (1711), *Douay* (1718)

Callahan, Denis. *Chart of head of navigation of the Potomac River* 1857, *Survey of the N. & N.W. Lakes* 1858

Callan, Bernard. American surveyor *Cleveland, Ohio* 1852, *Rochester, N.Y.* 1851

Callan, James. *New Map of Arkansas* 1836

Callander, James. Surveyor. *Chart of the Knysna* 1806

Callapoda [Sideri] Giorgio (fl. 1537–65) Portolan maker of Crete. *World* 1550, 1552. *Atlas of 10 maps* 1563 MSS.

Callaway, Thomas W. *Wilkes Co.* 1877, *Putnam Co., Georgia* 1878

Calléja[s] Félix del Rey. (b. 1750) Spanish general. *Nuevo Santander* (Texas) (1795) MS.

Callendar, B. Engraver for J. Morse *America* 1797, *N. Holland* 1797, and for *Malham's Naval Atlas,* Philadelphia 1804

Callender [Callendar] George. Master, R.N. *Boston Harbour* surveyed 1769, published 1775

Callender, John. *Terra Australis Cognita,* Edinburgh 1766–68

Callender, Joseph. Engraver for *Osgood Carleton's District of Maine* 1790, *Vermont* 1793, *Massachusetts Proper* 1801 (with Samuel Hill)

Callet, J. François (1744–1798) French hydrographer

Callewaert, C. Frères. *Atlas Général* 1858, *Atlas Elémentaire* 1861, *Atlas Classique* 1876

Callier, Camile. Syrie Méridionale 1835(40)

Callipus. Greek astronomer 4th century B.C.

Callot, Freiherr Eduard von. *Frankreich* 1859

Callot, Jacques (1592–1635) Painter and engraver of Nancy. *Siege of Breda* (1628), *Obsidio Rupellae* (1628)

Calmar, Yvez. *Plan of Brava on the East Coast of Africa* 1755, English edition 1780

Calmet, Augustin (1672–1757) Historian and theologian

Calmet-Beauvoisin, Maria Ant. *España y Portugal* 63 sh. Paris 1821

Caloiro, Placido. See **Oliva**

Caloiro, e Oliva, Giovanni Battista. See **Oliva**

Calver, Staff Commander Edward Killwick, R.N. Admiralty Surveyor. *Coasts of England* 1847–1856, *Downs* 1865

Calver, W. B. Master, R. N. *Tobago* 1864–5, published 1867

Calvert, Alexander. *River Swale* 1792

Calvert, A. F. *Expedition N.W. Australia* 1891–3, *Coolgardie* 1896

Calvert, Cecil. See **Baltimore**, Baron

Calvert, Charles. *Parts of Pennsylvania, Maryland and Virginia* 1763

Calvert Lithographing Co. of Detroit. *Maps of Michigan Railway areas &c.* 1860–80, *City Detroit* 1871 and 1879

Calvete, Juan Christoval. *Plan Mahdia* 1551, used by Braun & Hogenberg

Calvi, Ulisse. Engraver. *Plan of Alexandria* 1882

Calvus, M. Fabius. *Antiq. Urbis Romae* 1532 (20 plans)

Calwagen, Erich. Geogr. *Charta Medelpad* 1769

Cam [Canus, Caõ] Diego (15th century) Portuguese navigator, discovered the Congo

Camacho, Josef. Prima Pilot de la Real Madrid. *Costa Septent. de la California* 1779 MS

Camacho, Lucio. *Republica de Bolivia* 1842–59

Cambier, Ernest (b. 1844) Belgian traveller, explorer of Congo

Cambrensis, Giraldus [Gerald de Barri] (1146–ca. 1220) Welsh topographer. *Map of Western Europe* ca. 1200 *Itinerarium Cambriae*

Cambridge, Richard Owen (1717–1802) Poet. *War in India* 1761

Camden, William (1551–1623) Antiquary and historian. *Britannia* 8vo edition 1586, 1600, *maps after Mercator and Kaerius.*

W. Camden (1551–1623)

Britannia 1607 small folio Latin text 57 maps 1610 and 1637, plain backs, *maps after Saxton.* 1695 edition of *Britannia, maps by Robert Morden.* 1789 *Britannia, maps by Cary*

Camerarius [Kammermeister] Elias (d. 1581) Mathematician

Camerarius [Kammermeister] Joachim (1673–1734) German Professor of Mathematics. *Brandenburg* used by

Cameron, George. Engraver. *Breadalbane* (1770) *Ross' Lanarkshire* 1773, *Ross' Dumbartonshire* 1777

Cameron, David Nelson (1820–1916) Geographer. *Higher Geography* 1865, *Intermediate Geography* 1866

Cameron, H. C. *Wall Maps Italy* 1865, *Greece* 1866, *Roman Empire* 1867 (all with A. Guyot)

Cameron, Henry Lovett (1844–1894) African explorer. *Tanganyika* 1876, *Tropical S. Africa* 1876

Cameron, Capt. John R. E. Worked for Ordnance Survey. *Aldershot* 1856, *London* 1857

Cameron, R. M. *Penny Atlas* 1873, *Pupils Atlas* 1873

Cameron, Comr. Verncy Lovett. *Tropical S. Africa* 1876

Cameron, William (1833–1886) Explorer and geologist

Camesina, Albert. Reproductions of early plans of *Vienna* 1863–76

Cammen, Edmond Paridant van der. *Carte Géologique Chypre* 1874

Cammermeyer, A. *Reisekart N. Norge* 4 sh. 1882, *Sydlige Norge* 2 sh. 1887

Camocio [Camocius, Camotio, Cametti] Giovanni Francesco [Joan Francisco] Venetian cartographer, publisher, and mapseller ad Signum Pyramidis (fl. 1558–72) *World, Lombardy Atlantic* (Forlani) 1560, *N. Regions* 1562, *Africa, Dalmatia, France, Friuli, Holland* 1563, *Isole Famose* 1563 and 1564, *Asia Minor*, (Gastaldi)

Brabant; Cyprus, Friesland, Greece, Italy, Naples, Piedmont, Sicily 1566, *Morea* 1569 & 1570, *Venice* 1571, *Ireland* 1572, *Asia* 1575, *America* 9 sh. 1576, *Europe* 1579, Plates reissued by Donalto Bertelli

Camotto. See **Camocio**

Camp, John. *Bommerlerwaard* 1765

Camp, John de la. *United States Mexican Boundary* 1857, *Region between Gettysburg and Appomattox* 1869

Camp, Capt. W. *Ostfries and Harlingerlande,* Berlin 1804

Campana, Pietro. Engraver. *Rome* for Nolli 1748, *Naples* for Carafa 1775

Campana von Splügenberg, Anton R. (1776–1841) Military cartographer

Campanius Holm, Tomas (d. 1702) *Nova Suecia* 1654–5, published Stockholm 1696, *Totius Americae descriptio,* Stockholm 1702, *Virginia* 1702

Campano da Novara (13th century) Physician, mathematician and astronomer

Campbell, Admiral. *Brazil* 1807

Campbell, Alexander F. *Proposed Railway England to India* 1851, *Northern boundary of United States* 1878

Campbell, Lt. (later General) Sir Archibald (1739–1791). Governor of Jamaica. *Guadeloupe* 1760, *Dominica* MS ca. 1770, *N. Frontiers Georgia* 1780, *S. Coast of Jamaica* 1785

Campbell, Charles. *Plan Long Leate* (1730)

Campbell, Charles Dugald. *Chart Kurrachee* 1853

Campbell, Colin. *Belton, Lincs.* (1720) *Vitrivius Britannicus* 1767–71

Campbell, D. Surveyor. *Verplanks Point* 1779, Plan *Stoney Point* 1784

Campbell, Dugald. *Fort Cumberland to Fredericton* (1795)

Campbell, I. C. *Pembrokeshire* 1827

Campbell, Capt. James. *Directions for navigating up Delaware Bay,* Faden 1776

Campbell, John. County Surveyor. *Charleston and the British attack of June 1776. MS Survey on Red River* 1800

Campbell, Major J. L. *Plan Ghuznee* 1839, *Idaho* (1860)

Campbell, Lieut. Marius H. *Map S. Appalachians,* Admiralty 1825

Campbell, Robert. Surveyor. Capt. 4th Company of St. John's Loyalists. *Map of the Great River St. John & Waters from the Bay of Fundy* 1788

Campbell, Lieut. Robert. Admiralty Surveyor. *Scotland* 1790, *St. Kilda* 1809, *African Islands* 1811, *Ascension* 1825

Campbell, Robert A. *State of Illinois* 1870 (with H.F. Walling) *Missouri* 1873, 1880

Campbell, Thomas. *City of Dublin* 1811

Campbell, Ulrich (1510–1583) Engraver and theologian. *First printed map of Rhaetia* 1573

Campbell, W. *Plan proposed harbour of Wick* 1814 MS, *Caithness-Shire* 1822

Campbell, W. M. *Country round Kandahar* 1879

Campe, Friedrich. Publisher of Nuremberg. *Nuremberg* 1813, *Danzig* 1813, *Eastern Hemisphere* 1816, *Australien* 1817, *Bayern* 1838

Campe, Joachim Heinrich (1746–1818) Theologian and publisher

Campen, Gerhart von. Publisher of Cologne. *Netherland Provinces (Leo Belgicus)* 1584

Campen, H. A. Van. *Stad Leiden* 1870

Campen, W. J. van. *Nieuwe Kaart der Stad Leyden*

Campenhausen the Elder, Balthasar (1746–1808) Military cartographer

Camperio, Capt. Manfredo (1826–1899) Italian geographer and explorer. *Tripolitania e Cirenaica* 1884

Camphuysen, J. (1760–1840) *Plan Waterloo* (1816)

Campi [Campus] Antonio (1525–1587) Artist and cartographer of Cremona. *Agri Cremonensis Typus* 1579

Campion, C. W. *Map Collieries Iron Works S. Wales* 1877 and 1894

Campion, Edward. *Description of Ireland* 1571, printed 1633 (with R. Stonyhurst)

Campion Frères. Parisian publishers 18th century

Campo, Josef del. *Plano dela Bahia de Laware y entrada de Filadelphie* 1784

Campos, F. Carneiro de. *Planta Cidade Rio de Janeiro* 1858

Campos, Rafael Torres (19th century) Spanish geographer

Camps, D. J. *Philippines Plans MS* 1833

Campus. See **Campi**

Canaan, William. Publisher of Edinburgh. 1 North Bridge Street (with Ritchie 1818–22), 60 and 61 Princes Street (with William Swinton 1823), 63 Princes Street (Canaan Swinton & Grove 1825). *Wood's Scottish Town Plans* 1818–26

Cancrin, J. A. Engraver *Libya et Aethiopia* 1830

Candish, Candys. See **Cavendish**, Sir Richard

Caneiro, Januensis [Caveri] Nicolo de. Portuguese cartographer. *Marine chart of the World* 11 sh. ca. 1502

Canepa, Albino de (fl. 1480) Portolan maker of Genoa. *Mediterranean* 1489

Canerin, J. A. Engraver of Amsterdam. For Weygand 1827

Canfield, Capt. A. American surveyor. *Clinton River, Michigan* 1853

Caniani, G. Engraver for Rossi 1820

Canina, Luigi. *La Campagna Romana* 6 sh. 1845, *Roma Antica* 1832, *Topografica di Roma Antica* 1833

Canius [Cagno] Paolo. Of Genoa. *Naples* 1582

Canizares, José de. *Plano del Puerto de San Francisco* (1775) MS.

Cannabich, J. G. F. (1777–1859) German geographer. *Geographical Manual* 1816, *Euro Russia* 1833

Cannaci, G. Engraver for Lesage 1811

Cannon, John. Surveyor for Van Diemens Land Company. *Circular Head* 1844

Cannon, Lieut. James, R.N. *Admiralty Chart Trincomalee* 1832 (with H. Loring) published 1834

Cano [Elcano, Delcano] Juan Sebastian del (fl. 1520–26) Spanish navigator and portolan maker. Served under Magellan. *East Indies* (ca. 1523) MS, now lost.

Cano y Olmedilla. See **Cruz Cano y Olmedilla**

Canot, Pierre Charles (d. 1777) French engraver, worked in England. *Milton's Plan of Portsmouth Dockyard* 1754

Cantel, Le Sr. *L'Italie* 1701, *Catalogne* 1762

Cantelli [da Vignola] Giacomo [Jacobo] (1643–1695) Cartographer of Modena. *Alta Lombardia* 1680, *Venice and China* 1682, *Brandenburg* 1687, *Brunswick* 1691, *France* 1691, *Italy* 1695, *Mercurio Geografico* 1688 *(with maps)*. Pair of globes for the Duke of Modena

Cantemir, Prince Constantin Dimitri (1673–1723) Cartographer in Russian Service. *Moldavia* 1716, *Constantinople* 1720. A contemporary unsigned English mezzotint portrait exists

Canter, Johann Jacob (18th century) Dutch cartographer

Cantino, Alberto. Italian cartographer and engraver. *Carta da Navigar (made for Duke of Ferrara)* MS 1502, *Iceland* 1502

Cantino, Francisco. *Ichnographia Villa Tiburtina* 1751

Cantoni, G. *Dept. de Reno* 2 sh. 1810

Cantova, Juan Antonio. *Islas de los Dolores [Carolinas]* 1731

Cantzler, Bernard (17th century) Mathematician. *Comitatus Wertheim* 1603

Canus, D. See **Cam**

Canzler [Canzier, Cantler] Fridrich Gottlieb (1764–1811) Professor at Goettingen. *Polynesia* 1795, Worked for Homann. *West Indies* 1796

Caõ. See **Cam**

Capella, Minneo Felice Marziano (3rd century A.D.) of Rome. Encyclopaedist. His geography a standard in Mediaeval period

Capellaris, Jean Ant. Engineer. *Istria* 1803, *Dalmatia* 1806

Capelle, J. V. D. (1624–1679) *View of Leyden and surrounding country*, engraved Beauvarlet

Capellis, Comte de. *Sierra Leone River* 1779 (1802)

Caper, Johannes. *Plan of Prague* 1562, used by Braun & Hogenberg

Capitaine, H. See **Au Capitaine**

Capitaine, Charles (d. 1789) Engraver

Capitaine, Louis (ca. 1749–1797) Son of Michel. Worked with Belleyme and Chanlaire on *Topographical maps of France* and *Carte Chorographique de la Belgique* 69 sh. 1797

Capitaine, Michel (1746–1804) Military cartographer. *Harbour and City of Annapolis* (1781)

Capitaine, Pierre Charles (d. 1778). Worked with Cassini in Paris Observatory. Military cartographer

Capitaine, Pierre Nicolas (1773–1838) son of Pierre Charles. Abridged edition of Cassini's *France* 1789

Caplin (fl. 1822–29) Engraver for Dépôt de la Marine. *Côtes de Brésil* 1822, *du Pérou* 1824, *Golfe du Mexique* 1826, *Rade de Rio* 1829

Capocacia, Gioseppe. *Ferrara* 1602

Cappelier, Bernard. *Diocèse de Tournay* 1694

Capper, Benjamin Pitts. *Topographical Dictionary of United Kingdom* 1802 (46 maps), 1812 (47 maps) and editions to 1829

Capper, Robert. Revised *Wyld's Plan of Swansea* 1888

Caprarola Palace. *Sala del Mappamondo* some of the earliest surviving mural maps, painted in 1574 by G. A. de Varese

Capriolo [Capreolus, Cavriolo] Elia. (d. 1519) Italian cartographer. *Regional map of Brescia ca. 1502.*

Carabelli, Ferdinando. Drew and engraved *Atlante Geografia Storico* 1876

Carafa [Caraffa] Giovanni, Duca di Noia (1715–1768) *Mappa Topografia citta di Napoli* 1775

Caraffa, Giovanni (16th century) Neapolitan globemaker. *Terrestrial Globe* 1575

Carbonel, Esteban. *MS map of North West America* ca. 1632

Carbonel, Juan José. *Kingdom Valencia* 1812

Carcia de Cespedes. Andreas. See **Cespedes**

Cardano, J. Engraver. *Costas de Yucatan* 1808, *Mexico* 1814

Carden, Hans (fl. 1773–1782) *Plan of Black River on the Mosquito Shore*

Cardenas, Manuel Joseph de. *Lago de San Bernardo* MS 1691

Cardim, Père Antonio Francesco (1596–1649) Portuguese Jesuit, died Macao. *Japan* ca. 1614 (1646)

Cardinael, Sybrandt Hanszoon (1578–1647) Cartographer and mathematician of Amsterdam

Cardon, A. A. J. (1739–1822) Belgian painter and map maker. *Maps of Soignies and Marimont.*

Cardona, Nicolas. *Western Hemisphere* ca. 1632 MS.

Cardono. Engraved for Antillon y Marzo (1802)

Carel, D. W. *Zélande* 5 sh. 1753 (with A. Hattinga)

Carette, E. *Carte de l'Algérie,* Paris 1846, *Kabilie* 1847

Carew, Sir George (1555–1629) *Survey of Kerry & Desmond* (6 maps) 1617 MS, *Wars of Ireland* (maps) 1632

Carew, Richard (1555–1620) Poet and antiquary. *Survey of Cornwall* 1602

Carew, T. See **Carve**

Carey, A. D. *Chinese Turkestan* 1887

Carey, Henry Charles (1793–1879) and Isaac Lea. Publishers Chestnut Street, Philadelphia. *Historical, Chronological, and Geographical American Atlas,* Philadelphia 1822, London edition 1823, *Family Cabinet Atlas* 1832–6

A COMPLETE

HISTORICAL, CHRONOLOGICAL, AND GEOGRAPHICAL

AMERICAN ATLAS,

BEING

A GUIDE TO THE HISTORY

OF

NORTH AND SOUTH AMERICA,

AND THE

WEST INDIES:

EXHIBITING

AN ACCURATE ACCOUNT

OF THE

DISCOVERY, SETTLEMENT, AND PROGRESS OF THEIR VARIOUS KINGDOMS,

STATES, PROVINCES, &c.

TOGETHER WITH THE

WARS, CELEBRATED BATTLES, AND REMARKABLE EVENTS,

TO THE YEAR 1822,

ACCORDING TO THE PLAN OF LE SAGE'S ATLAS,

AND INTENDED AS A COMPANION TO

LAVOISNE'S IMPROVEMENT OF THAT CELEBRATED WORK.

PHILADELPHIA:

H. C. CAREY AND I. LEA,—CHESNUT STREET.

1822.

Title-page to the American Atlas published by Carey and Lea (Philadelphia, 1822)

Carey, Mathew [& Son] (1760–1839) American publisher of Irish descent. No. 122 Market Street Philadelphia. *American Atlas,* Philadelphia 1795 (the earliest atlas of the United States), reissued 1796. *General Atlas* 1794, 1804 and later. *American Pocket Atlas* 1796, 1801, 1805 &c. *Scripture Atlas* 1817, *U.S.A.* 1820, for *Lavoisne* 1820, 1821

Carey, Thomas. *Isles de Guernsey, Aureny* &c. 1746

Carey & Hart. Philadelphia publishers of *Tanner's Atlas* 1842 &c., *New Mexico* 1847

Carez, J. French publisher. *Atlas Géographique by J. A. Buchon* 1825

Carignano, Giovanni Maura da. Of Genoa (14th century) Chartmaker. *World* ca. 1310

Carkas, John. *Barony of Naas* 1655, *Barony of Ballibrit* 1657

Carl. See **Carolus Flandro**

Carl, F. *Maehren und Oestereichisch Schlesien,* Vienna 1810

Carl, Ferd. *Bier Productions Karte von Mittel Europa* 1876

Carl, J. *Hopfenbau Karte von Mittel Europa* 1875

Carles, W. R. *Korea* 1886, *Central China* 1898

Carless, Lieut. (later Capt.) Thomas Grere, Indian Navy. Hydrographer. *Survey East Coast Africa* 1838–1848, *Survey Red Sea* 1830–33, *Gulf Aden* 1854

Carlet, Louis François, Marquis de la Rozière. *Hesse Cassel* 1759, *Paderborn* 1760, *Théatre de la Guerre en Allemagne* 4 sh. (1761)

Carleton, I. D. Land agent Port Huron. *Map of the Great Oil Regions of Michigan & Canada* (1880)

Carleton, Osgood. Surveyor, publisher and teacher of mathematics in Boston. Maps for *Norman's American Pilot* 1792, 1794. *Maine* 4 sh. 1795, *Massachussetts* 4 sh. 1795, *Plan of Boston* 1796 and 1800, *United States* 1806

Carletti, Niccolo. Corrected and engraved *Mappa Topografia di Citta di Napoli* 1775

Carli, Pazzini. Publisher, of Siena. *Sweden* 1794

Carling, J. *Territoire Portuguez de Goa* 1893

Carlo, Sieur du. *Plan City of Rochelle* 1628

Carlsten, Osgood (1748–1816) American military cartographer.

Carmona, Manuel Salvador. Engraver. *Title to Torfino's Atlas* 1789

Carmichael, John. Engraver of Sydney. *Mitchell's New South Wales* (1834) *Entrance Moreton Bay* 1846, *City of Sydney* 1854

Carnall, R. von. *Oberschlesien* 1844, *Geolog. K. Niederschlesischen Gebirge* (1862–68)

Carnan, Thomas (fl. 1751–1788) Publisher in London, 65 St. Pauls Churchyard. Partner with step-brother Francis Newbury. *Paterson's Roads* 1771–86. *County maps* 1770 and 1779

Carneiro, Antonio de Mariz (d. 1642) Portuguese cartographer. *Roteiro de Mariz* (1640)

Caroc, Frederic Carl Wilhelm (1811–1882) Danish military cartographer

Carolus, Frans of Amsterdam, *Holland.* Published Schenk (1705)

Carolus Flandro [Carl] Joris [Georgius] (1601–1625) Pilot and map maker. Compiled *various MS maps* and *map of Iceland,* printed by Hondius 1631, and Blaeu. Worked to improve *Waghenaer's Het niewe Vermeerde Licht* 1634

Caroly [Caroli] Francesco de. *Stati di S.M. il Re di Sardegna* 1779, English edition 1794, *Torino* 1785

Caron, E. *Carte du Niger* 1891, *Tombouctou* 1895, *Atlas du Cours du Niger* 1898

Caron, François (d. 1674) Official of Dutch East India Company *Japan* 1636 [1661]

Carondelet, Baron de. *Map of the Mississippi and Ohio Rivers* (1794) *Fortificaciones Nueva Orleans* 1792

Carpelan, W. M. (1780–1830) Swedish cartographer. *Various maps of Sweden and Norway*

Carpenter, Benjamin D. *Washington D.C.* 1881

Carpenter, Nathaniel. *Geography delineated* 1625

Carpenter, Samuel (19th century) American surveyor

Carpinetti, Joâo Silverio. Of Lisbon. *Maypas das provincies de Portugal* (1762)

Carpini [Pian del Carpine] G. da Piano (1190–1282) Franciscan Traveller. *Embassy to Mongolia* 1254–7

Carr, L. *Lake Nyassa* 9 sh. MS.

Carr, Richard. *A Description of al [sic] the postroads in England from London to Edinburgh drawn and perfected by R. Carr* 1668

Carr, Richardo. Dutch engineer of Amsterdam. *Plano de Cavite* 1663

Carr, Capt. Robert. *Chart of China Seas and Philippines* 1734 and 1778

Carranza, Domingo Gonzalez. Principal Pilot to the Flota of New Spain 1718. *Coasts, Harbours & Sea Ports, Spanish West Indies* 1740

Carraro, Giuseppe (d. 1886) Italian geographer. Wrote on physical and statistical geography.

Carrasco, Ed. (1779–1865) Peruvian hydrographer. Director Marine Academy, Lima. *Province de Lambeyegue*, Wyld 1819

Carré, V. Engraver for Dépôt des Cartes et Plans de la Marine 1843–61

Carres, Louis. (d. 1833) *Atlas of Jesuit Missionary Provinces*

Carrigan, Philip. *New Hampshire* (1816)

Carrington, Frederick A. *Scotland, engraved Dower,* 1842 and later editions. *Plan of the Town and settlement of New Plymouth New Zealand* 1842, *New Zealand* 1846, *Kings Forest* 1849

Carrington, Henry Beebee (1824–1912) *Battles of American Revolution* 1775–1781, published 1876

Carrington, Octavius. Provincial surveyor. *New Plymouth New Zealand* (1844), *Plan Pekapeka* 1860, *Taranaki* 1863

Carrington, Robert Christopher. Hydrographer for Admiralty. *Tonga* 1866, *Charts Coast of England* 1894

Carson, C. W. Engraver and publisher 68 State Street. *Albany* 1845

Carson, W. Publisher with John Betts of *Family Atlas* 1863

Carta Marina. *Title of World Map* by Waldseemüller 1516, Fries 1525, Olaus Magnus 1539

Cartaro [Cartarius, Kartaro, Karterus.] Mario (fl. 1560–1612) Born Viterbo, died Naples. Cartographer and engraver of Rome and latterly Naples. Worked for Fernando Bertelli. *Globes* 1577, *Italy* 1577, *Kingdom of Naples* 1579, *Perugia* (1580), *Atlas of Provinces of Naples* (13 maps)

Cartee, Cornelius S. *School Atlas* 1856

Carte Pisane (1300) Oldest known portolan

Carter, Lieut. *Abyssinia* 1869

Carter, E. O. *Crescent City, California* (1855)

Carter, Henry John. *Geological Atlas Western India* 1857

Carter. James G. *Essex County* (1845)

Carter, S. B. *Marseilles* 1868

Carteret, Lieut. (later Rear Admiral) Philip (d. 1796) Served under Byron and Wallis; discovered Island named after him. *Discoveries New Guinea* 1769–72

Cartero. See **Cartaro**

Cartier, Jacques (1491–1557) French navigator and explorer. *Voyage de découvertes au Canada* 1534–42, *Nova Francia & Hochelaga* 1535 in *Ramusio's Voyages* 1556

Cartilia, Carmelo (fl. 1720) Italian Instrument maker. *Globes*

Cartwright, R. *Bridport* 1832, *Bedford Level* 6 sh. 1842

Cartwright, Samuel. Publisher with M. Sparke of the English edition of *Mercator's Historia Mundi* 1635

Carvalho, I. J. M. *Imperio do Brazil*, Rio de Janeiro 1856

Carvalho da Costa, Antonio Nuñes de (1650–1715) Portuguese cartographer. *Chorographica portugueza* 1706–1712

Carve [Carew, O'Corrain] Thomas (ca. 1590–1672) Irish traveller. *Itinerarium* 1639

Carver, Capt. Jonathan (1732–1780) Traveller, surveyor, cartographer. *Travels interior of North America* 1767–1768, *Map Quebec* (for Jefferys) 1776, *Laconia North America* 1777 MS, *New Universal Traveller* 1779

Carver. See **Bowles & Carver**

Carwitham, J. Engraver for *Gordon's Bedfordshire* 1736, *Draught Bay of Biscay* 1742

Cary, Francis (1756–1836) Engraver. Brother of John

Cary, George, elder (d. 1830) Brother and partner of John, 86 St. James Street, London, 1820

Cary, George, younger (d. 1859) and John, sons of the following. Continued the business which passed to G. F. Cruchley 1844, and from him to Gall & Inglis. *Catalogue of maps* by G. & J. Cary 1800

Cary, John, elder (ca. 1754–1835) English cartographer, engraver, globe maker and publisher 188 Strand (1783–7), New Norfolk Street 181 Strand (1791–1820), 86 St. James Street (1820–32) with George. Apprenticed to William Palmer 1770. Issued many county and foreign atlases. *New and correct English Atlas* 1787, editions to 1831, *Travellers Companion* 1790–1828, *Globes* 1799, *Universal Atlas* 1808, *England and*

Wales scale half inch to mile for Post Office 1832. Commissioned to survey the roads of England 1794

Cary, John

Cary, William (1759–1825) Brother of John, with whom he collaborated. Map publisher and globe maker. *Celestial Globe* 1799, *maps of Africa* 1805, 1811 and 1828

Casalis, Godefroi (d. 1856) Italian geographer. *Dictionnaire géographique et statistique des États Sardes.*

Casanes, F. de. See **Cesanis**

Case, I. Estate surveyor. *Estate Carew Mildmay* (1767)

Case, N. Publisher. *Bible Atlas* 1832

Case, O. D. & Co. *Bible Atlas* 1877

Case, Zophar. *Clinton Co.* (1860)

Casebourne, T. *Maps in Telford's Atlas* 1833

Casembrot, A. Dutch engraver at Messina. *Maps and views of Messina*, some published by C. Allard.

Casement, R. *Maps of West Africa* published by War Office 1894

Cashee, Nathaniel. *Plans of Manila and Tondo* 1819

Casilini, O. de. *Topografia dello Stato d'Ascoli* (1680)

Casket Mappamundi, engraved Paulus Ageminius. *Cordiform World map* engraved on lid of a casket in the Trivulziana collection at Milan. A copy of *Sylvanus' World map* of 1511

Caspari, C. E. (b. 1840) French hydrographer

Cass, Lewis (1782–1866) Governor. *Map of North West Territory of United States* 1821, *Map of the surveyed part of Michigan* 1822

Cassas, Louis François. *Istria & Dalmatia* (1802)

Cassell, John. Publisher of atlases from 1864

Cassell, Peter & Galpin. Publishers La Belle Sauvage Yard, Ludgate Hill. *Dispatch* 1863–7, *British Atlas* 1864–7.

Cassella, Andrea. *North America,* Torino 1849

Cassianus Da Silva, D. F. Engraver for *Bulifon's Naples* 1692

Cassini, Giovanni Maria. Painter and engraver. *L'America* Rome, Calcographia 1788, *Nuovo Atlante Geografico Universale* 1792–1801, *Globes* 1790–2, *Carta generale dell Italia* 15 sh. 1793 *Stato Ecclesiastico* 7 sh. 1805

Cassini, G.M.

Cassini, Jean Dominique [Giovanni Domenico] (1625–1712) Astronomer and geodesist, professor of Astronomy at Bologna. Settled in France 1669 as Director of the Observatory. Supervised the Académie des Sciences' measurement of a meridian for France. Contributed to *Neptune François. Planisphére Terrestre* Paris, Nolin 1696 (a reduction of the Great Planisphere on the floor of Paris Observatory). *Hydrographia Galliae* (1685), London Edition Morden 1699

Cassini, G. D. Astronomer.

Cassini de Thury

Cassini, Jacques [Jean] Dominique, Comte de (1748–1845) Astronomer son of Cesar François with whom he collaborated. Portrait by Cless, engraved Westermayr

Cassini de Thury, César François (1714–1784) born Paris, died Thury. Son of Jacques below. Succeeded him as Director of Observatory and of the Triangulation of France. *Carte Géometrique de la France* 182 sh. 1744–60, *Atlas National de France* 1791, *Atlas topographique, minéralogique et statistique de la France* 1818

Cassini de Thury, César. Portrait by Cless, engraved by Westermayr

Cassini de Thury, Jacques [Giovanni] (1677–1756) Son of Jean Dominique Cassini, whom he succeeded as Director of Paris Observatory. Conducted Triangulation of France with his son César François

Casson, R. and J. **Berry.** *Plan of Manchester & Salford* 2 sh. 1741

Castaldo [Castaldi] See **Gastaldi**

Castlemain, Roger Palmer, Earl of (1634–1705) Cartographer, diplomatist and writer.

Castañera, Ignacio. *Division militar de España y Portugal* 4 sh. 1889

Castell, William. *Short discourse of the coasts and continent of America* 1644

Castell, Prosper von. *District of Kastell* 1792

Castelli, D. *Carte . . . della Romagna* 1621

Castello, Pablo. *Plano de S^n Augustin de la Florida* 1763

Castera, Ignacio de. *Plano geometrico . . . ciudad de Mexico* 1776–(85)

Castiglione [Castillioneus] Bonaventura (1480–1555) Inquisitor General of Milan. *MS map of Alps*

Castiglioni Map [Planisphere of Mantova] *MS World* (1525) in possession of Castiglioni family of Mantua

Castillo, Domingo del. *California* 1541, published 1770

Castillo, Rafael de. *Filipinas* 1890

Castle, H. J. Lithographer on York Street, Toronto. *Great Lakes* 1836

Castonnet Des Fosses (1846–1898) Amateur French geographer. Wrote on French explorers.

Castorius. Roman grammarian and surveyor, credited with authorship of *Peutinger Table* ca. 340

Castro, João de (1500–1548) Portuguese admiral and explorer. Viceroy of the Indies, born Lisbon, died Goa. Compiled 3 celebrated coastal pilots: *Lisbon to Goa* 1538, *Goa to Diu* 1538–9, *Red Sea* 1541–2. Contemporary painted portrait National Museum in Lisbon.

Castro, Joseph de. *Inner Oesterreich* Vienna, Mollo 1812

Castro y Andrade, Don Tomas de. *Plan Manila* 1733 MS, *Balabac* 1753, published by Dalrymple 1774, *Ypoloté* 1774, *Plans Cavite Manila* 1762

Catalan Atlas (1375) Prepared for Charles V of France, attributed to A. Cresques. Reproduced on 12 sh. Paris 1883

Cataneo, A. Engraver for Zannoni 1792

Catesby, Mark. (ca. 1679–1749) Naturalist. *Carolina, Florida and Bahamas* 1731–43

Catherwood, Frederick. *Plan Jerusalem* 1833, *New map of Palestine* 1862

Catlin, George (1796–1872) American explorer, geographer, and ethnologist. *Chart of the River Niagara* 1831. *Indian localties in 1833* (1841), *Indian frontier in 1840*

Cattlin, F. Town plans of Canada *Montreal* 1830, *Quebec* 1830, *Bytown* 1832, *Goderich* 1835, *Gulf* (1835)

Cau, Bonifacius. *Paskaart van Oostende tot den Hoek van Schouwen* 1799

Cauchy, Fr. Th. (1795–1842) Belgian engineer and geologist.

Caucia, M. (18th century) German cartographer

Caucigh, Michael S. J. (18th century) Mathematician and globemaker

Caulin, Antonio. Spanish geographer. *Orénoque* 1759, *Historia corografia . . . Nueva Andalucia* 1779

Caundish. See **Cavendish**

Causton, Isaac. Surveyor. *Briset Hall* 1725

Cauvanilles, A. J. *Mapa del Reyno de Valencia* 1795

Cauvin, Thomas (1762–1846) French geographer and archaeologist

Cavada y Mendez De Vigo, Augustin de la. *Manilla, Philippines* 1876

Cavalier, Jean. Engineer and cartographer of Agde. *Carte de Languedoc* 1643

Cavalier de La Salle, Robert. See **La Salle**

Cavalleri[s] Giovanni Batista de (1530–1597) Italian engraver in Rome. *Terra y Castello di Dieppa* 1589, *Ducatus Polocencis* 1580.

Cavalleri[s] Piero de. Engraver *Vera disegno del Lago di Geneva* 1589

Cavallini, Giovanni Battista. Portolan maker of Livorno (fl. 1639–69) *Portolan atlas of 6 charts* 1636 (with Jaume Oliva)

Cavassi, J. A. *Sicile* 2 sh. 1714

Cave, Edward (1691–1754) Publisher and editor of *Gentleman's Magazine*. *River Forth* engraved Jefferys 1746, *Geography of the Great Solar Eclipse* 1748

Cavendish, Capt. Alfred Edward John. *Korea* 1894

Cavendish [Candish, Caundish, Candys] Sir Richard (fl. 1513–1549) Survey engineer and Master of Ordnance Berwick. *Essex and Suffolk coasts* MS (1525), *Thames Estuary* (1533) MS, *Dover Harbour* 1541. Father of Thomas below.

Cavendish, Sir Thomas (1560–1592)
Circumnavigation 1586–88. *World* 1588,
*Voyage of the Delight to the Straits of
Magellan* 1591 MS.

Caveri[o]. See **Caneiro**

Cavriolo. See **Capriolo**

Cawston, George. *Nyasaland,* Stanford
1891, *Matabili, Mashona countries* 1888,
Africa south of the Equator 1889

Cawthorne, J. Publisher in London,
Catherine Street, Strand. *Jamaica* 1807

Cay, Jakob (1566–1590) Cartographer

Cay, Robert. *Completed Horsleys
Northumberland* 1753

Caylus, de. *Town & Fort Grenada,
Fort Royal, Martinique in Jefferys'
French Dominions* 1760

Caymox [Caimox] Cornelius. Nuremberg
merchant. *Plan of Nuremberg in Braun &
Hogenberg's Civitates* 1575

Cazzul, Matteo and Carlo. *Il Piemonte
con il Monferrato* 1615

Ceard. *Route du Simplon* 1820

Cecil, Sir William. See **Burleigh**, William
Cecil

Cecill, Thomas (fl. 1627–35) Engraver
of London. *Relation of Maryland* (map)
1635

Celi, Francis Mathew. *Maps for Jefferys'
Topography of North America* 1762, 1768

Cella, C. See **Zell**, C.

Cellarius, Andreas (fl. 1656–1702) Dutch
mathematician and geographer. Rector of
Hoorn. *Atlas Coelestis* 1660, *Harmonia
Macrocosmica* 1660, 1661 and 1708,
Description of Poland 1659, 1660

Cellarius [Keller] Christophorus (1638–
1707) Geographer, Professor at Halle,
born Schmalkalden, died Halle. *Geographia
Antiqua,* many editions 1686–1812,
Notitia Orbis Antiquae 1703

Cellarius, Ferimontanus [i.e. of Wildberg]
German geographer and physician born
Wildberg, settled Flushing. *Compiled
text for De Jode's Speculum* 1578

Celle, B. Of Genoa, Publisher. *Levanto's
Specchio del Mare* 1664

Cellius. See **Zell**

Celsius, André (1701–1744) Swedish
astronomer

Celtes, Protucius [Celtis, Pickel, Meissel]
Konrad (1459–1508) Geographer.
Professor in Vienna, born Wipfeld. Discovered
Roman road map and left it to Peutinger.
Published as *Tabula Peutingeriana,* Engraved
1591. See also **Castorius**

Centeno y Garcia, José. *Philippines* 1876

Center, Lieut. A. T., US Army. *Survey
Hawe Bay, Michigan* 1836

Centurion, Manuel. *Guiana* 1770

Cenus, N. See Zeno, N.

Cernoti, Leonardo. *Translated Latin edition
of Ptolemy into Italian* 1597

Cerri, Carlo. *Carta postale Italia* 1849,
Italia carta generale 1859

Cerutti, Agostino. *Corso del Po* 1703

Cerruti, G. E. *Carta dello Stretto di
Galero,* Turin 1873

Certes, Jean. Publisher for Du Val 1676

Cervino, P. *Plano Topo . . . de Buenos
Aires* 1814, pub. Paris, Darmet 1817

Cesanes. See **Cesanis**

Cesani [Alysius Cesanis Ydrontius] Luigi
World 1574, *MS atlas,* 4 maps 1581

Cesanis [Cesanes, Casanes] Francesco de.
Venetian chartmaker. *Mediterranean*
1421 MS.

Cesaris, Giovanni August (1749–1832)
Italian astronomer and topographer

Cespedes, Andreas Garcia [Carcia] de
(16th century) Spanish geographer and
chronicler of the Indies. *General
Geografia e Historia* 1598 *Regimiento
de Navegacion* 1606

Ceulen. See **Keulen**

Cexano, Alvise and Francesco (15th
century) Italian Portolan makers.
Mediterranean (1490)

Cexano, Francesco (15th century) Chart-maker

Ceylat (18th century) Surveyor.

Chabannes, de Viscount, Admiral. *Atlantic Ocean Pilot Chart* 1868

Chabert, Joseph Bernard Marquis de (1723–1805) French Admiral and hydrographer. *Acadie* 1751, *Voyage en l'Amérique Septentrionale* 1750–1, published Paris 1753. *Atlas Général Méditerranée* 1791

Chace, J. American Surveyor. *Fulton Co., N.Y.* 1856, *Rockingham Co., New Hampshire* 1857, *Suffolk Co., N.Y.* 1858

Chadwick, J. Surveyor. *Liverpool* 1725

Chadwick, Joseph. Surveyor. *North America* 1764–1772

Chaffat, Antoine du (fl. 1734–50) French military cartographer and engraver in Bavarian service

Chaffers, E. M. Surveyor. *Port Nicholson, New Zealand* 1840, *Tory Channel* 1841

Chaffrion [Chafrion] José (Gioseppe) (1653–1698) Mathematician and cartographer of Valencia. *Liguria* 1685, *Rivera de Genova* 1685

Chaillé-Long, Charles. *Victoria Nyanza* 1874–5

Chaillou, F. *Brie* 1762

Chaillu, Paul Beloni du (1835–1903) African explorer. *West Africa* 1864–5

Chaix, *Carta volcanologia e topografica dell' Etna* 1892

Chaix, Paul (1808–1901) Swiss geographer and publisher. *Atlas Elémèntaire* 1842, *Savoie* 1832, *Valley Beaufort* 1856, *Gell's Environs of Rome* 1861

Challamel aîné. Publisher of Paris. *Colonies Françaises* 1866

Challoner, Sir Thomas. *De Republica Anglorum* 1579

Chalmandrier, Nicolas. Engraver and publisher of Paris, for Cassini 1744–60.

Plan Madrid 1761, *Plan of Warsaw* 1772, *Carte du Canal Royale . . . du Languedoc* 16 sh. *Toulouse* 1774, maps in *Chappé d'Auteroche's Sibérie* 1768, for *Bonne* 1762, 1779

Chalmers, Cathcart. *British Guiana* 1875

Chalmers, Rev. P. *Dunfermline Coalfields* 1841

Chalonnois, Jean Jubrien. See **Jubrien**, J.

Chamberlain, E. L. Surveyor. *Atlas map of St. Charles County, Mo.* 1875

Chamberlaine, Thomas. *Ireland* 1776

Chamberlin, Thomas Crowder (1843–1928) Geologist. *Survey Wisconsin* 1877

Chambers, Ephraim (1680–1740) Apprenticed to Senex

Chambers, Francis. Surveyor. *Ward of Farringdon Within* 1851 (with R. Tress), *St. Michael le Querne* 1853

Chambers, Frederick C. Surveyor. *Town of Coastland, Westchester County N.Y.* 1858, *Town of Greenburgh* 1857

Chambers, R. *Map of Canada, copied from Mitchell's map* 1839

Chambers, William and Robert. Publishers of Edinburgh (1800–1883) *School Atlas* 1845, *Atlas Ancient and Modern Geography* 1845, *Atlas for the People* 1846, *Atlas of Encyclopedia* 1869, *Handy Atlas* 1877. *Atlas to Chambers' Gazetteer* 1895

Chambertin, G. L. Surveyor. *Map cities of Marietta and Harmer, Ohio* 1869

Chambon. Engraver for *Longchamps & Janvier's Amérique* 1754. *Longchamps' Louisiana* 1756

Chambray, Marquis de. *Atlas for campaign of 1812.* 5 sh. Paris 1823

Chameau, Jean. See **Calameus**

Chamouin, Jean-Baptiste-Marie (b. 1768) Engraver in Paris, Rue de la Harpe No. 35. *Plan de Paris* 1803, *Tyrol* 1808. Engraved for Lapie 1808 and 1812

Champain, I. U. Surveyor. *Plan of Lucknow* 1858

Champeaux. *Vavao for Dumont d'Urville* 1847 (with Duranty)

Champion. *Plan of Bristol Channel from the Key to the Hot Wells* 1767

Champion, J. N. Engineer, geographer, engraver. *Nouvelle carte de l'Allemagne* 30 sh. Leipzig 1806. *Mappemonde* 1816. Reduced *Arrowsmith's America* for Frémin 1828

Championnet, Jean Etienne. *Kingdom of Naples* 1779

Champlain, Sieur Samuel de (1567–1635) Explorer and Governor of Quebec, born Brouage, Saintonge. *St. Lawrence* 1612, *Nouvelle France* 1613

Chancellor, Richard (d. 1556). Navigator under Willoughby, opened up trade with Russia

Chancourtois, A. E. Béguyer de (1820–1886) French engineer and geologist

Chancy, General Antoine Eugène Alfred (1823–1883) *Campagne de 1870–71* (operations of The Army of the Loire) 1871

Chandia, Antonio de. *Philippines* 1727

Chandler, John. Hydrographer. *British Channel* 1782, *North Sea* 1784 and 1795, *North East Britain* 1782

Chandler, Lieut. *Salt Lake City to Cajon Pass* 1859

Chandler, R. W. *United States* 1829

Chandler, William. (18th century) Surveyor in America. Worked with James Helme 1743

Chandler, W. *Maps of Brasil* 1864–69

Chanlaire

Chang, Heng (ca. 70–140 A.D.) Chinese astronomer

Changuion, D. J. Publisher *Raynal's Atlas Portatif* 1773

Chanlaire, Pierre Grégoire (1758–1817) Geographer and publisher, 7, Rue Geoffroy Langevin, later No. 328 (after 1795) *Atlas National de la France* 1790–1811 (with Mentelle). *Atlas Portatif de la France* 1792 (with Dumez) *Europe* 1801 (40 sh.) *Nouvel Atlas de la France* 1802, *Description topographique et statistique de France* (1801–1811) *Atlas Universel* 1807 (with Mentelle) *Carte Chorographique de la Belgique* 69 sh. Paris 1797 (with Capitaine) *Cappaduce,* 1800, *Arménie* 1801 *British Isles* 1818

Chantovoine. *Carte de la Virginie* 1781

Chantry, John (fl. 1660–69) Engraver for *Voyage of Mandelslo, Course of the Great River Wolga* 1682

Chantry, T. Surveyor. *Plan City Bath* 1793 and 1805

Chanzy. See **Chancy**

Chapeaurouge, C. de *Plano Topografico Prov. de Santa Fe* 1872, *Buenos Aires* 1883, *Republica Argentina* 1897

Chapin, A. M. *Map City of Manchester, Hillsborough Co. New Hampshire* 1850

Chapin, William. Engraver of New York. *Lay's map of United States* 1832 and 1839

Chapman, Capt. Charles. *Azores* 1782

Chapman, C. F. *Plan Township Papineau, Toronto* 1878, *Township of Proudfoot* 1878

Chapman, George T. *North part of the province of Auckland* 1865, *Waikato* 1866, *Bay of Plenty* 1873

Chapman, Isaac A. *Plan position Gen. Burgoyne 10 Oct. 1777,* Philadelphia 1818

Chapman, James. Surveyor. *Island of Lewis* 1807–9

Chapman, James. *Gold Fields of S.E. Africa* 1876

Chapman, John (fl. 1772–77) Cartographer and publisher, Royal Arcade, Pall Mall, *Essex*

25 sh. surveyed 1772–76 pub. 1777 and 1785 (with André) *Notts.* 4 sh. 1776, 1785 and 1792. Engraved *Staffs.* for W. Yates 1775

Chapman, Leander (19th century) Surveyor General of Michigan

Chapman, R. M. *Lands of Saco Water Power Co.* 1848

Chapman, Silas. *Sectional map Minnesota* (1856) *Township map the North West* 1858, *Complete map Colorado* 1861, *Sectional map part Dakota* 1869

Chapman, William (1749–1832) Engineer. *Canal Newcastle to Maryport* 1795, *Newcastle to Haydon Bridge*, 1796, *Plymouth* 1809, *New Shoreham Harbour* 1815

Chapman, W. Master, R. N. Admiralty Surveyor. *Entrance Tagus* 1806, *Plymouth Sound* 1809, *Coast of Portugal* for Faden 1810, *Plan part of the rivers of Derwent and Hartford* 1795, *Plan Scarborough Harbour* 1800

Chapman & Hall. Publishers 186 Strand, later 193 Piccadilly (1842) *New British Atlas* 1833 *S.D.U.K.* 1833–42, *Penny Maps* 1852, *Lowry's Universal Atlas* 1853

Chappé D.'Auteroche, Abbé Jean (1722–1769) French astronomer and geographer, born Mauriac, Auvergne; died California. *Voyage en Sibérie* 1761 published 1768, *Carte Géographique* 9 sh.

Chapuy, Jean Baptiste (1760–1813) Engraver.

Charcornac, M. (19th century) French astronomer

Chardin, Jean (1643–1713) French traveller in Persia 1686. In English editions called Sir John Chardin

Charland, Louis. *Land survey maps Lower Canada after Samuel Holland, published Vondenvelden* 1803

Charle, J. B. L. French geographer. Paris, Rue du Conde 14. *Pyrénées* 1831, *Nouvel Atlas National de la France* 1833, *Atlas élémentaire* 1845, *Carte du Royaume de Siam* 1854, *Amérique* 1865 and 1868

Charles, Archduke of Austria. *Atlas to his campaigns* 1796, 1799, pub. 1820

Charles, F. Engraver for Vandermaelen (1840–46)

Charles, Henry (19th century) Publisher and printer of Philadelphia. *United States, published, printed and colored* 1819

Charlemagne (742–814) It is recorded that he had three silver tables showing *maps of Rome, Constantinople,* and *World*

Charlesworth, L. C. *Map of Eagle Lake* 1897

Charlevoix, P. F. Xavier de (1682–1761) French Jesuit, visited Canada 1705–20. *Histoire de la Nouvelle France* 1744, *America, published Le Rouge* 1777, *St. Domingo* 1730–1, *Histoire de Paraguay* 1751, *Histoire du Japon* 1736

Charlton, John. *Ilchester Meadow, Somerset* 1792 MS.

Charnières, Charles F. P. de (1740–80) French sailor and astronomer. *Théorie et pratique des Longitudes en mer* 1772

Charpentier, Charles. *Plan de la ville de Lunéville* 1875

Charpentier, F. *Nouvelle carte de France* 1869

Charpentier, H. *Ile Réunion* 1867

Charpentier, Johann Friedrich Wilhelm von (1728–1805) Mineralogist

Charpentier, J. *Chemins de Fer Belges* (1875)

Charpentier, Le. See **Le Charpentier**

Charpentier, W. H. *New map of Portsmouth* 1892

Charras, Jean Baptiste Adolphe (1810–1865) *Campagne de 1815,* 1858

Charrier, C. *Ville des Sables d'Olonne* (1875)

Charrier, F. *Magnae Britanniae auster Iknographicus* 1637

Chartier, A. T. *Chaine des Alpes Françaises* (1868) *Europe* 1867, *European Railways* 1873, *maps for Migeon* 1874

Chase, G. E. *Plan of the new city of Pensacola* 1836

Chase, J. Cent livres. *Eastern Frontier Colony, Cape of Good Hope* Arrowsmith 1836, *Emigrants Guide to Cape of Good Hope* 1845

Chase, J. G. *Street rail and road Routes in and leading from Boston* 1865, *York County, Maine* 1856, *Plan Cambridge, Mass.* 1674

Chase, W. Bookseller in Norwich, Cockey Lane. *Plan Norwich* (1720), *Map Norfolk* (with J. Goddard) 1731, 1740 and 1745

Chassant. Engraver for Dépôt de la Marine. *Mappemonde* (1823), *Martinique* 1826, for Dumont d'Urville 1833–42

Chassegros de Léry. See **Chaussegros de Léry**

Chassereau, Pierre. *Havana*, Homann 1739–50 *Malta* 1740, *Carthagena*, Homann Heirs 1740, *Survey Parish St. Leonards, Shoreditch* 1745, *Plan Ipswich* 1745, *York* 1750

Chassignet, Daniel. Engraver. *Celestial globe* Rome 1616

Chassun, Sieur de. *Carte du Diocèse d'Aire* 1635

Chasteler-Courcelles, Jean Gabriel Joseph, Marquis du. (1763–1825) Austrian military cartographer

Chastenet-Puiségur, Jacques, Comte de. *Pilote de l'Ile de Saint Dominique* 1787

Chastillon [Chatillon] Claude de (1549–1616) Topographer to Henry IV 1589, *Siege plans &c.* 1612, *Topographie Françoise* 1641

Chatelain, Henry Abraham (1684–1743) Geographer and publisher. *Atlas Historique* 7 vols. Amsterdam 1705–20, 1721 and 1732–39

Chatelain, Zacherie. Worked with Henri Abraham above, on the *Atlas Historique*

Chatelaine, Anatole. *Voies de communication dans le Monde Entier* 1862

Chatillon, C. See **Chastillon**

Chatoff, General. *Black Sea and Caspian Sea* 1819, *Map of Walachia and Bulgaria* 4 sh. 1854

Chauchard, Capt. French military cartographer. *Carte générale d'Allemagne* 9 sh. Paris, Dezauche (1790) *Holland, Italy &c.* Stockdale 1800

Chaucheprat. *Baie de la Concepcion (Chile)* 1827 *Port de l'Ancon* 1828

Chaudière, Guillaume. Publisher of Paris. For Thevet 1575, 1581

Chaulmier. Engraver. *Plan Gibraltar* 1782

Chaumeau, Jean. See **Calamaeus**

Chaumette Des Fosses, Amadée. *Corso de los Rios Huallaga y Ucayali* (1830), *Pampa del Sacramento* 1836

Chaumier, C. J. *Théâtre de la Guerre Flandre*, Brabant 1806, *Tableau général du Royaume de France* 1817

Chauncy, H. S. *Auriferous region of Mount Alexander* 1852

Chauncy, Philip. *Local maps of Victoria, Australia* 1856–59

Chaurand, Enrico de. *Carta . . . della Ethiopia* 8 sh. Florence, 1894

Chaussegros De Lery. *Quebec* 1722, *Lake Champlain* 1748

Chavanne, Dr. Josef (1846–1902) Austrian geographer. *Africa* (1878), *Central Asia* 1880, *Central America* 1882, *Afrique Equatoriale*, Brussels (1883), *Congo* 1885, *Ferro-Carriles Argentina* 1889

Chaves, Alonso de [1492–ca. 1590] Spanish cosmographer royal. *Carta Moderna general* 1536 (lost)

Chaves, Gabriel de (16th century) Spanish topographer in Mexico

Chaves, [Chiaves, Chavez] Geronimo [Hieronymo] (1523–1574) son of preceding, successor to Cabot as Piloto Mayor (1552) Translated *Sacrobosco* (1545) *with map. MS maps of Andalusia and Florida* used later by Ortelius. *Florida* is the first regional map of America

Chaves (18th century) *Malvinas Islands* MS.

Chavignaud, I.. *Plan de la ville de Lille* 1873

Chaymox, Corneille. Engraver of Antwerp. Worked for Braun and Hogenberg

Chazelles, Jean Matthieu de (1657–1710) French hydrographer and astronomer in Marseilles. *Côtes de Provence* (1689), *Maps in Neptune François* 1693

Chedeville, A. *Plan des chutes de Niagara* (1851)

Cheevers F. Engraver. *Rapids in the River Ohio* 1778

Cheevers, John. Engraver. *Otaheite* 1769, *New South Wales* 1770, for Cook 1773, *Roy's Scotland* 1774 [93], *Carnatic* 1778, *Rapids Ohio River* 1778, *Virginia, Penn., Md.* 4 sh. 1778

Cheffins, Charles [& Son] Surveyors, publishers, lithographers. *English & Scotch Railways* 1841–56. For Kelly 1860

Cheffins, C. F. *Republic of Texas,* London 1841

Chelezoff. *Hydrographical chart of Russia* 1801

Chelsham, Robert. Surveyor. *Parish of Bray* 1672

Chemitte, François. *Plan de la ville de Candie* 1692

Ch'en, Lun chiung (1684–1747) Chinese hydrographer

Chenevière, I. I. *Geographical game* (1720)

Cheng Shum Kung (16th century) Chinese cartographer. *Map Japan* 1557

Cherbuliez, A. & Co. (19th century) Publishers in Paris

Chermont, Jean Gabriel, Comte de. *Atlas général* in conjunction with Desnos, Paris 1790

Chermside, Lieut. H. *Survey defensive position near Bulair* 1877, *Sketch Varna & Environs* 1878

Cheron, L. Designed *cartouche title for Moll's World* 1719

Chéron, Louis. *Golfo Dulce (Costa Rica)* 1850

Cherubini, Claudio (19th century) Italian relief cartographer *Alpi occidentali e Appennino ligure* 1882

Chesbrough, E. S. City engineer. *Boston Harbour* (1852), *Boston Water Works* 1852

Chéseaux, J. P. Loys de (1718–1751) Swiss astronomer. *Éléments de Cosmographie* 1747

Chesney, Col. Charles Cornwallis. *Euphrates and Tigris* 12 sh. 1850. *Outline maps for military students* 1868

Chesnoy, Capt. du. *Théâtre de la Guerre en Amérique* 1778

Chesser, William. *Nepean Bay, South Australia* Arrowsmith 1839

Chesteven, Robert. *Relation present state of Summer Islands* MS 1629

Chetwind, Philip. Publisher of *Heylin's Cosmography* (4 maps) 1670, 1674, 1677

Chevalier, François (d. 1738) French mathematician and cartographer

Chevalier, Michel. (1806–1879) French statesman, professor of political economy, cartographer. Advocate of free trade. *Voies de communication aux États Unis* 1841

Chevallier. *L'Évesché de Meaux* 2 sh. 1717

Chevers See **Cheevers**

Chevillot. *Plan de la ville et fort Louis* (La.) 1711

Chevreau, H. *Carte Topographique du Dépt. du Rhône* 1868

Chewitt [Chewett] James G. *Province of Upper Canada* 3 sh. 1826

Chewitt, William. Senior Surveyor. *Sketch from Fort Erie* 1793, *Upper Canada*, Faden 1813

Chewitt, W. C. and Co. Lithographer of Toronto. *Part of the North Shore of Lake Huron* 1863

Cheylat (17th century) French military cartographer

Cheyne, N. B. *Travelling Directory through Scotland* 1792, *Roads of Great Britain* 1797

Chia Tan. (730–805) Chinese cartographer. *Map China* 33 by 30 feet

Chiara, Metheo de. Portolan maker of Rimini. *World* 1519 MS.

Chiaves, Hieronimo. See **Chaves**, Geronimo

Chiesa, Andrea. Cartographer. *Tiber* 1744 (with B. Gambarini) *Territorio Bolognese* 1762

Chiesa, Innocento (18th century) Italian engraver

Chieze, Jacques de, of Orange. *Orange* 1627 used by Blaeu 1630 &c.

Chikhachev, P. *Asie Mineure* 1853, *Carte générale de l'Altai* 2 sh. 1854, *Carte géologique de l'Asie Mineure* (1877)

Chilcott, J. *Eleven miles round Bristol* (1842)

Child, G. Engraver. *Universal Traveller* 2 vols. 1752–3

Child, J. American surveyor (fl. 1849–50) *Sectional View Copper Falls Mine* 1849

Child, Nathaniel. *North half of Township Woodstock* 1772

Child [Childe] Timothy. Publisher London, At the sign of the Unicorn at the West end of St. Pauls Churchyard, and later White Hart west end of St. Pauls Churchyard from 1701. *New body of Geography* 1695 (maps by Moll) *System of Geography* 1701, 1709 (worked in conjunction with A. & J. Churchill and A. Swall)

Childs, C. G. Engraver for Cumings 1829

Childs, O. W. Surveyor. *Canal from Pacific to Atlantic Oceans* 1850–1, 1852. *Part of City of New York* (1862)

Chill, D. H. *Map Indus River* 4 sh. 1835 (with E. Winston)

Chillas, D. Lithographer of Philadelphia *Huron Harbour* 1854

Chilmead, J. (17th century) English globemaker. *Hues' Use of Globes* 1638

Chimmo, Lieut. (later Commander) William, R. N. (d. 1891) English hydrographer. *Survey South West Pacific* 1852–56, *Trinidad* 1866–68, *Labrador* 1867, *Borneo* 1871, *Sulu* 1872

Ch'in Hun-Keung (18th century) *Navigational map* 1744

Chiquet, E. L. M. *World* 1729

Chiquet, Jacques. French cartographer. *Nouveau atlas Français* 1719, *Nouveau et Curieux Atlas* 1719

Chirikov, Aleksei Il'ich. *Observations Astronomiques* 1741, *Découvertes des Vaisseaux Russiens aux Côtes de l'Amérique Septentrionale,* St. Petersburg 1758

Chisholm, G. G. *Atlases* for Longman 1859–95

Chiswell, Richard (1639–1711) Publisher and mapseller at Two Angels and Crown Little Britain, later Rose and Crown St. Pauls Churchyard. *Plan Londonderry* (1689), *Speed's Atlas* 1676 (with Bassett), *Plan London* 1707, also maps for the Royal Society

Chittenden, G. B. *South West Colorado* (1876)

Chitty, M. *Map for Van Diemens Land Company* 1833

Chitty, Simon Casie. *Ceylon Gazetter* 1884

Chkatoff. Engraver for Piadischeff 1823–6

Chlebowski (18th century) Prussian general and cartographer

Chocarne (fl. 1822—29) Engraver for
Pilote Français. Golfo Porto Vecchio
1828

Chodźko, Josef (1800—1881) Polish
geographer and historian, born Oborek,
died Poitiers. Paris, A La librarie
Polonaise 20 Rue de Seine. *Table de
la Pologne* 1830, *Carte Géographique
de la Pologne* 1830, *Atlas des sept
partages* 1831 1846, *Poland* 1863

Choffard, Pierre Philippe (1730—1809)
Artist and engraver. *Plan London* 1756,
Titlepage for Bellin 1758, *Frontispiece
for Étrennes Géographiques* 1761, *West
Indies* 1762, *Boullanger's World* 1781

Choffat, Léon Paul (1849—1919) Geologist

Choiseul-Gouffier, Marie Gabriel Auguste
Florens, Comte de. *Chart Mediterranean
Sea* 1776, English Edition 1794. *Island
of Paros* 1808

Chollet, Louis. *Essequibo colony* (1791),
Pomeroon Coast 1794

Choo Sze Pun (14th century) Compiled
maps for the Kwang-yu-too

Chopy, Antoine (1674—1760) *Carte du
lac de Genève* 1730

Chou Jen-Chi (fl. 1755) *Atlas of CheKiang
1755*

Choudiakoff. Drew and engraved map of
Russia 1795

Chovin, I. A. Engraver for *Engel's Extraits
Raisonnés des Voyages* 1779, *Amérique*
1764

Christensson, Marten (17th century)
Swedish cartographer

Christiaenszoon, Leenhart (17th century)
Amsterdam engraver

Christiani, F. *Maps of Denmark* 1859—78

Christie, Charles. *Canada* 1759, *New
Jersey* 1759, *Sumbawa,* published
Dalrymple 1780

Christie, Major Charles. *New York, New
Jersey* 1759 MS.

Christman, F. *Maps for Ewald* 1861

Christoforo [Cristofano] Girolamo da
(16th century) Florentine cartographer,
worked in Lyons

Christoph [Christoffel, Christopheri]
Joannes (fl. 1598—1619) Painter and
cartographer. *Insularium,* Cologne 1601,
Bremen & Verden for Blaeu

Christopher, Lieut. Indian Navy Hydrographer. *Survey East Coast Africa* 1838—
1848

Christophers, Joseph S. *Emigrants Guide
to Cape of Good Hope* 1845

Christy, William. Surveyor. *Directory
Map Huntingdon Co. Pennsylvania*
1856

Chrysolas [Crysolas] Emanuel (1335—
1415) born Constantinople. Greek
Scholar. Translated *Ptolemy* into Latin

Chrysologue, Noel André. *La Mappemonde*
1774, *Abrégé de l'Astronomie* 1778,
Planisphères célestes 1778

Chrzanowski, Wojciech. (1793—1861) of
Warsaw and Paris. Cartographer.

Chu, Yu. *Atlas of China* 1895

Chudaekow, Jefim. *Plan Stadt St.
Petersburg* 1790

Chukei, Ino Tadataka (fl. 1800—1816)
Japanese cartographer. *Japan,* Tokyo
(1807)

Chun Ching (1328—1392) Chinese
geographer.

Chün Hsii Chao. *North and South
Hemispheres* 1807

Church, Ambrose F. American surveyor.
*Township maps of American counties:
Colchester* 1864, *Kings County* 1864,
Cumberland 1870, *Hampshire* 1871

Church, Commander W. H. Admiralty
Surveyor. *Ireland* 1849, 1855

Churchill, Awnsham (fl. 1681—1728)
Publisher London at Black Swan in
Paternoster Row M.P. 1705—10 partner
with John Churchill below. *Camden's*

Britannia 1695, *Moll's Compleat Geographer* 1709, *Atlas Manuale* 1709, and for Wells 1706

Churchill, Henry A. *Fortress & Field Defences of Kan* 1855, *Map Turco Persian Frontier* 1855

Churchill, John (fl. 1690–1714) Brother and partner of Awnsham above, in atlases mentioned above

Churchman, John. *State of Delaware* (1770), *Peninsula between Delaware & Chesapeak Bay* (1770), *Magnetic Atlas* 1794

Churruca, Don Cosmo de. *Las Caribes* 1793, *Puerto Rico* 1794, *Islas Antillas* 1802, *Islas Caribes* 1804. *West Indies,* Laurie & Whittle *1813*

Churruca, Evaristo de. *Port San Juan,* Admiralty 1874

Churton, William (d. 1767) Appointed Commissioner with Weldan for South Carolina (boundary line). Surveyor of Edenton, N. Carolina. Supplied materials for Fry & Jefferson 1751, Collet 1770, and Mouzon 1775

Chu Shu-Hsüeh (16th century) Chinese astronomer and cartographer

Chu Ssu-Pen (1273–1335) Chinese cartographer. *Terrestrial Atlas* 1579, 1588

Ciera, Michele Antonio (d. 1815) Coimbra astronomer. *Maps of Portugal rectified* 1804 for Dépôt de la Marine

Cierto, Andres. *Antilles* 1755 MS.

Cieza, Peter de. *Travels in Peru,* London 1709

Cigerlus. See **Ziegler**

Cigni, Domenico. Engraver. *Provincia Quitensis Secietatis Jesu in America* (1751)

Cigni, Giulio Cesare. Jesuit. *Quito* (1751) MS.

Cimerlini [Cimerlino, Cimerlinus, Cunerlinus] Giovanni [Joannes] Paolo. Engraver from Verona, worked in

Venice. *Cosmographia universalis ab Orontio olim descripta* 1566

Cingolini, Giovanni Battista. Cartographer of Rome. *Agro Romano* 1692, *Campagna di Roma* 6 sh. 1704

Cintra, C. & C. Riviere. *Prov. do Espirito Santo,* Rio de Janeiro 1878

Cipri, G. *Chemin de Fer Indo Européen* 1881

Cipriani, G. B. *Pianta topografica della Citta di Roma e dintorni* 1851

Cirkoff, Axence. Lieut. in the Corps of Imperial Topographers. *Carte de la Principauté de Serbie* 1848

Ciroldis, J. See **Giroldi**

Cisneros, Carlos B. Peruvian geographer. *Bolivia,* Lima 1897

Civelli, Giuseppe. *Gran carta d'Italia* 28 sh. 1853

Claesz [Claeszoon, Nicolay, Nicolas Cornelis, Corneille] (1560–1609) Dutch printer, publisher and bookseller opt Water imt Schryf-boeck — à l'enseigne du livre à escrire, Amsterdam. *Gerritsz's Zeevaert* 1588, *Waghenaer's Spieghel* 1588, *Plancius' World* 1592, for Linschoten 1595–6, *America* 1605, *Caert Thresoor* 1598–1609. Joint publisher *Mercator Hondius Atlas* 1606–8 and *Atlas Minor* 1607

Clairaut, Alex. Cl. (1713–1765) French Cosmographer. *Théorie de la figure de la Terre* 1743, *Théorie de la lune* 1752

Clamorgan, Jean de (1480–1562) Chartmaker of Dieppe. *MS charts of Atlantic*

Clancy, Jonathon. Master, H. M. S. *Centurion. Draught of Lisbon* 1763, many *Marine Surveys in Mediterranean*

Claparède, Arthur de (1852–1911) Swiss geographer and diplomat. President Geographical Society of Geneva. *Japan* 1889, *A travers le monde* 1894

Clapiès, Jean de (1670–1740) from Montpellier. Astronomer, cartographer and mathematician

Clapp, Lieut. Edward S. *Harbours and Anchorages in the Sandwich Islands* 1878

Clapp, William W., Junior. *Plan of Boston corrected* 1864

Clapperton, Lieut. (later Capt.) Hugh (1788–1827) African explorer. *Narrative of Travels North and Central Africa* 1826 (includes translation of an Arab MS on geography of the interior of Africa)

Claret de Fleurieu [D'Eveau de Fleurieu] Charles Pierre, Comte de (1738–1810) Captain in French Marine, born Lyons. Director of ports and arsenals 1776, Minister 1790. *Voyages* 1768–9, *Découvertes sud est de la Nouvelle Guinée* 1768–9 [90] *Neptune du Cattegat* 1786–90, published 1809, only 100 copies printed. 65 maps, plans and views. Edited by Buache and prepared by Beautemps-Beaupré. *Azores* 1772, *Atlantic* 1777, *W. Africa* 1797

Clarici, G. B. *Ducato di Urbino* 1572

Clarici, Paolo Bartolommeo (1664–1725) Painter and cartographer of Padua, born Ancona. *Diocesi Padovana* 1720, *Il Polesine di Rovigo* 1721, *Carte de Trévisan* 1776

Claridge, John (1791–1828) Estate surveyor

Claringbull, W. *Chart Plymouth Sound and maps of the towns of Plymouth, Devonport and Stonehouse* 1841

Clark, Capt. *Plan of the city of Cadis* 1761

Clark, Lieut. Engraver for *Thornton's World* 1700

Clark, Benjamin A. Surveyor. *Map of the Town of Deerfield, Massachusetts* 1855, *Map of Hamilton, Madison County, New York* 1858

Clark, F. A. *White Pine Mining District* 1870–80

Clark, F. C. *New Sectional map of Missouri* (1860)

Clark, G. T. American surveyor. *City of Flint, Michigan* 1873

Clark, J. Surveyor. *Wenona, Doriphan County, Kansas* (1860)

Clark [e] [Clerk] James. Engraver and printseller in Grays Inn. For *Seller's New Jersey* (1664) and *English Pilot* 1671, for *Hollar's Plan London* (1675), *Star charts for Halley* 1678, *Collins' Coasting Pilot* 1693 for Adair 1703

Clark[e] James. *Survey of the Lakes* (1787)

Clark, John. Land surveyor. *Roxburgh* in *Thomson's Scotland* 1832

Clark, Matthew. American cartographer. *Charts coast America,* Boston 1789–90

Clark, Richard. Publisher of Philadelphia. *Plan Wilmington* 1850. *County maps Connecticut.* Partner with Takabury (1861)

Clark, Rev. Samuel. *British Empire* 1848–51, *Bible Atlas* 1868

Clark, Thomas. *Five miles round the Town of Bridgewater* (with R. Down) 1853

Clark, Capt. William (1770–1838) American explorer, Governor of Missouri Territory. *Mississippi & Missouri Rivers* 1805 MS, *Route maps of Expedition with M. Lewis* 1803–6

Clark, W. C. *Pocket map Sydenham* 1875

Clarke, Capt. Alexander Ross. *Ordnance Trigonometrical Survey of Great Britain and Ireland* 1858

Clarke, Benjamin (fl. 1836) *British Gazetteer* 1852 (43 maps with H. G. Collins)

Clarke, H. Wadsworth. *Location of Salt Wells at Syracuse,* 1868, *Highlands of Syracuse, New York* 1872

Clarke, J. *Map of the Vales of Clwyd & Langollen* (1862)

Clarke, James. English cartographer. *Isle of Wight* 1812

Clarke, J. R. Publisher of Sydney. *North East Australian Gold Fields* 1860, *New South Wales and Queensland* 1860

Clarke, Samuel. *Geographical Description of the known World* 1646

Clarke, Samuel. *Geological map of the United Kingdom of Great Britain and Ireland* 21 sh. *Railway and Steam Navigation map of the United Kingdom* 21 sh. 1860

Clarke, Stephen Reynolds. *New Yorkshire Gazetteer* 1828

Clarke, Thomas. Engraver. *Plan of the Town of Alexandria in the District of Colombia* 1798

Clarke, Capt. Thomas. *Bonny River (African Pilot)* ca. 1785

Clarke, Capt. T. J. *Naval Operation before Canton* 1840 MS.

Clarke, W. B. Drew *town plans for Society for Diffusion of Useful Knowledge* 1831–6, London 1836

Clarke, William Branwhite. *Geological sketch map of New South Wales* 1880, *Geological map of New South Wales* 2 sh. 1893

Clas [Glass] , Capt. G. *Canary Islands for Jefferys* 1775

Classen [Classun] de. *Maps* used by Tavernier 1634, *Carte du Diocèse d'Aire* 1635, *Siège Dax &c.* 1638

Claudianus [Klaudian, Kulhanek] Nicolas [Mikulas] (d. 1522) Physician and printer of Jungbunzlau, Bohemia. *Bohème* 1518, the first independent map of Bohemia

Claudius, Hendrik. *Cap de Bonne Espérance* published 1686

Claus, Jacob. *Plan Philadelphia.* Dutch edition of *Pennsylvania Missive* 1684

Clausen, I. C. G. *France,* published Euler 1753

Clauser, Jacob. Engraver. Worked for *Münster's Cosmographia* 1550

Clausner, Jakob Joseph (1744–97) Swiss surveyor and engraver. *Suisse* 1799

Clauson, C. *Mappa topografia Citta di Napoli* 1775

Claussen, [Claussøn] See **Clavius**

Clavero, Gregorio. *Plans Manilla* 1792–1800 MSS.

Clavigero, Francisco Javier (1731–1787) *Laghi di Messico* (1780–1)

Clav[i]us [Claussøn, Cymbricus] [Svart, Niger] Claudius. Danish geographer (b. 1388) *Map of Scandinavia* (1427) used in early editions of *Ptolemy*

Clay, A. M. *Bergen County, New Jersey* 1874

Claye, Jacques de Vaulx de. *Brazil* (1579)

Clayton, *Henry (fl. 1805–45) English surveyor*

Clayton, P. Surveyor. *Map of Weston Platte County, Missouri* 1858

Clegg, John. *Elements of Geography* 1796

Cleghorn, Dr. A. *Mountain regions of Punjaub* 1864

Cleland, James (1770–1840) Statistician. *Plan Glasgow* (1822)

Clemens, M. *Map of the State of Ohio* 1861 (with T. W. Baker)

Clement, T. *Plan de la Ville d'Anvers* (1814)

Clement, Timothy (1706–1766) Surveyor. *Plan of Hudson River from Albany to Fort Edward . . . to Lake George* 1756

Clements, Capt. John. *Plan Harbour of Chusan* 1756–84, *Natunas Island* 1773

Clements, S. *Chart of the Entrance to the River Thames* 1791

Cleomedes. (1st century) Greek cosmographer

Clerc, le. See **Le Clerc**

Clerc, Pierre Antoine (1774–1843) French surveyor

Clerck [Klerc] Nicolas de (1599–1621) Engraver and cartographer of Delft. Worked for Hondius and Londerseel

Clericus, J. See **Le Clerc**, Jean

Clerk, Ja. See **Clark[e]**, James

Clerk, Thomas (fl. 1810–1835) Engraver of Edinburgh, 265 High Street (1813–23)

22 St. James Square (1831—35). *Thomson's New General Atlas* 1814, *Wood's Scottish Town Plans* 1818—25

Clerke, Capt. Charles (1741—1779) Surveyor. *North West Coast of America* (with Cook) 1778—9, published Carey 1795

Clerke, Thomas. *Lordshippe or Mannor of Holkeham* (Norfolk) 1590 MS.

Clermont, Jacques (1752–1827) Engineer geographer from Savoy

Clérot, Victor. *Auvergne et Lyonnais* (1866), *Carte routière de la Bretagne* (1867) *Chemins de Fer de l'Europe* 1869, *Carte de l'Amérique* (1877), *l'Afrique* (1878)

Clerq, J. H. W. le. See **LeClerq**

Clerville, Louis Nicolas de (d. 1677) French cartographer. *Montagnes de la Haute Auvergne* 1642, *Carte de la Haute Auvergne* 1670

Clesmeur. See **Du Clesmeur**

Cletcher, D. Engraver and map maker. *Maestricht* (1633)

Clevelly, A. *River Dart* 1772

Cleyhens, Bernard (fl. 1700—42) Dutch publisher. *Pocket Atlas West Europe* (1700)

Cliffe, J. *Waterford* 1725

Clinton, George (1686—1761) Surveyor in America. *New Jersey* 1781

Clinton, Sir Henry (1730—1795) MS *Battle plans and Sieges operations in the Jerseys, War Independence in America*

Cload, Charles, *Florida* 1739 MS.

Clobucciarich [Klobucaric, Clobucarius] Johann (1550—1609) Monk, cartographer and engraver *Large scale map of Austria* 109 sh. MS.

Clod [Clodiensis] Sebastianus. See **Ré** Sebastiano di

Clodi[an]us, Marcus [Marcellus] Publisher of Rome. *British Isles* 1589, *Milan* 1589, *Parma* (n.d.)

Cloete, A. J. *South East Africa . . . Cape . . . Natal.* pub. Arrowsmith 1850

Clogher, William. *Map of Chicago and Vicinity* 1849

Cloppenburg, H. Jan Evertsz. [Johannes Everhardus] of Hamersveldt. Publisher in Amsterdam op't Water in den Vergulden Bijbel. *Bible* (with maps) 1609, for Linschoten 1619—44, *Regni Bohemiae desc.* (1620), *Germany* 1630, *Mercator's small atlas* 1630—6

Clopton, J. W. *Map of the City of Helena, Arkansas* 1859

Cloué, G. Ch. (1817—1889) French admiral and hydrographer

Clouet, Jean Baptiste Louis, L'Abbé (b. 1730) Académie Royale des Sciences de Rouen. *Géographie Moderne,* Paris, Mondhare 1767, *America* 1776, *World* 1785, *Wall maps of the Four Continents* 1788, *America* 1793

Clouzier [Clousier] Gervais (fl. 1656—79) Published *Tavernier's Voyages* 1676

Clusius [Ecluse] Carolus [Charles de l'Escluse] (1525—1609) Botanist, born Arras, died Leyden. *Narbonne* 1565, *Spain* 6 sh. 1570

Clutton, John (1809—1896) Surveyor of London. President of the Institution of Surveyors 1868

Cluver[io] Philip[pe] (1580—1622) Geographer of Dantzig, specialised in Near East and Ancient Geography, settled in Leyden. *Foederatae Rhaetiae descriptio* (1600) *Geographicus Academicus,* Leyden 1616. *Intro. in Universam Geographicam* 1629, *Rhaetia* published Visscher 1630, *Germaniae antiquae libri* 1631

Cnobbari, Joannis. Publisher of Antwerp for *Strada's De Bello Belgico* 1635, 1636

Cnobbari, Widow and heirs of Joannis above, published later editions 1640—49

Cnopf, Matthaeus Ferdinand (1715—1771) *Brandenburg* (1740), *Comitatus Oettingensis* (1744)

Coate, Marmaduke. Surveyor. *Lexington, Newberry & Richland Districts, South Carolina* 1820, for Mills 1825

Cobb, Francis B. *Chart of Harbour of Vera Cruz* 1843

Cobbé, Jacque André. *Ganges* pub. Friex 1726

Cobbett, William (1762–1835) London publisher and author, 11, Bolt Court, Fleet Street. *Plan of the blockade of Cadiz* 1797, *Geographical Dictionary England* 1832, 1854

Coburn, W. A. *Plan of the Isthmus of Tehuantepec* 1851

Cocazzi, Agostino [Gio. Batt. Agostino] (1793–1859) *Atlas Venezuela,* Caracas 1840, first survey of Venezuela *Colombie* 1865, *Atlas de Colombia* 1889

Coccejo, I. H. *Jerusalem* for Faber (1715)

Cocchi, I. *Carta Geologica della parte orientale dell' Isola d'Elba* (1872)

Cocchini, Giuseppe. Printer of Florence, all insegna della stella. *Dudley's Arcano del Mare* 1661

Cochado, Antonio Vicente. *Rios Para, Curupae Amazonas* (1623)

Cochelet, Adrien-Louis (1788–1858) French geographer

Cochin, Nicolas (1610–1686) Engraver and artist of Paris. Town plans: *Gravelines* 1645, *Ager* 1647, *Bethune* 1680, for Beaulieu (1694) and de Fer 1705

Cochran, A. B. *Schuyl kill, Pennsylvania* 1875 (with Beers)

Cochran, Lieut. Alex., R. N. *Route map Se Chuan* 1892

Cochrane, A. S. *Athabasca River* 1885, *Geological and Natural History Survey of Canada* 1885

Cochrane, Joseph. *Plan de la Ville de Boulogne sur Mer* 1844

Cock, Hendrick. Flemish cartographer. *Spain* 1581

Cock [Coc, Coquis, Coccius] Hieronymous [Girolamo] (1510–1570) Publisher, engraver and printseller of Antwerp. *Guelders* (1545), *Malta* 1551, *Parma* 1551, *Piedmont* 1552, *Siena* 1555, *Plans Antwerp* 1557, *Milan* 1560, *Ypres* 1562, *America,* 1562, *Sgrooten's Germany* 1565, *Asia* 1566, *Terra Sancta* 1570

Cockburn, William. *Barony of Balquhidder* 1756 MS.

Cockeril[1] Thomas (fl. 1674–1702) Publisher at Atlas in Cornhill (with Morden 1677), Three Bears in Poultry over against the Stock Market (1678), Corner of Warwick Lane Paternoster Row (1697), Three Legs and Bible aginst Grocers Hall in the Poultry (1699). *Morden's Geography Rectified* 1680, *New Description and State of England* 1701

Cockson, Thomas. Engraver of map by Baptista Boazio 1596

Codde, Capt. Pieter, of Enchuysen. *Seehaven ende stadt Van Duynkercken* 1631. Used by Blaeu 1634 and Jansson

Codine, J. *Mer des Indes* 1867

Codogno, O. *Itinerario delle poste per tutto il mondo* 1620

Coeck, Gerard [Geevaerd] Engraver *Jansson's Holland* 1647, *Leeghwater's Haerlemmer Meer* (1651), for Pieter Goos 1659. & for Blaeu, *Ambacht & Poznam* 1662

Coeler, [Koller] George (fl. 1640) Engraver and publisher of Amsterdam, worked for Hondius

Coelho, Gomes. *Costa de Loanda* 1883

Coelho, João. *Estados Unidos do Brazil* 2 sh. 1891

Coello y Quesada, Francisco (1820–1898) Spanish explorer and geographer, Lt. Colonel Spanish corps of Engineers. *Atlas de España* (1848–68), *Canarias* 1649, *Filipinas* 1849–52, *Isla de Cuba* 1853, *Geological map España y Portugal* 1879

Coello y Quesada, Joao. *Brazil* 1891

Coenders van Helpen. See **Conders**
ab Helpen

Coentgen [Cöntgen] Heinrich Hugo.
Engraver in Mainz to the Court and the
University. Maps for Therbu (after 1763),
for Broenner 1777

Coentgen, George Joseph (1752–1799)
Engraver, son of above, moved to Frankfort am Main 1777

Coffey, James. *County of Limerick*
1825

Coffin, J. A. *City of New York* 1847

Coffin, R. Engraver for *Donn's
Bristol* 1769, *Plan of the City of Bath*
1773

Cogan, Lieut. Robert, Indian Navy.
West coast of India 1828, Trigono-
metrical *Survey Bombay Harbour,*
Horsburgh 1833

Cogger, Anbrose. *Estate plans Sussex
and Kent* 1627–8 MS.

Coghlan, David. *City Scranton,
Penn.* 1870

Coghlan, Francis (fl. 1831) Publisher
West Strand. *Strangers Guide to
London* 1831, *Guide to Rhine* 1841,
Davies' London 1843

Coghlan, John. *Buenos Aires Railway*
1860

Coghlan, William C. *Pittsburgh*
1868, *Map City of Louisville,
Kentucky* 1873

Cohen, Dr. E. *Umgegend von
Heidelberg* 1874–77, *Transvaal*
1877

Coignet, Michel (1549–1623)
Mathematician and cosmographer
of Antwerp. *Miniature editions of
Ortelius* 1601–12, *Navigation
Guide* 1581

Coimbra, Heitov de. Portuguese
cartographer. *East Indies* 1524

Coindé, Madame. *Maps for Le Sage*
1813

Col, Juan. *Provincia de Corrientes* 1891

Colart, L. S. *Histories de France et
d'Angleterre* 1841

Colbeck, Charles. Editor *Public
Schools Historical Atlas* 1885

Colberg [Kolberg] Chr. Heinrich
Julius de (1776–1831) *Poland* 8 sh.,
Warsaw 1810, *Regional maps of
Poland* 1826–7

Colbert, Edmond, Surveyor. *Land
allotment in Victoria* 1854–59

Colbert, Edouard Charles. Victurnien,
Comte de Maulevrier (1758–1820)
*MS maps of war in Virginia and
New York* 1781–83

Colbert, Jean Baptiste (1619–1683)
Statesman and supporter of cartography in France

Colbert. *Carolina* 1770

Colbourne, Thomas. *Navigation
Shannon* 1808, *Bogs in County
Clare* 1811–(14)

Colbran, J. *Map Tunbridge Wells* (1850)

Colbran, Thomas (fl. 1768–96) Surveyor.
Estate plans in Sussex

Colbrant, Jules. *Plan de la ville de St.
Omer* 1865

Colburn, Zerah (1832–1870) American
engineer. Railway advocate in New
York 1854

Colby, Charles G. (1830–1866) *Morse's
General Atlas of the World* 1856, *Diamond
Atlas* 1857, *World in Miniature* 1857

Colby, C. J. *Union County, Iowa* 1876

Colby, George N. *Rockland, Me.* 1873,
Lancaster, Penn. 1874, *Aroostook
County, Me.* 1877 (with Roe). Colby
& Co. *Washington County, Me.* 1881,
Somerset County 1883, *Maine* 1884

Colby, Capt. (later Major General) Thomas
Frederick (1784–1852). Director General,
Ordnance Survey 1820–46. *Ordnance
Survey of England and Wales* 1799–1811
(with Mudge) *Ireland 6 inch survey*
1820–47

Colden, Cadwallader (1688–1776) *Map country of the Five Indian Nations* (1724)

Coldewey, Ehrenreich Gerhard (d. 1773) *Tabula Frisiae Orientalis* 1730, Maps in *Homann's Atlas Germaniae* 1753

Cole, Benjamin. Engraver and mapseller. New Wadham College, Oxon (1695), At the Sun and Key near Snow Hill, Conduit Street London; At London Horse Yard St. Pauls Churchyard, At the corner of Kings Head Court near Fetter Lane. For *Wells' Atlas* 1700, *Twenty miles round Oxford* 1706, *Twenty miles round Cambridge* 1710, *Ward maps for Maitland's Survey of London* 1754, *Plan London* 1756, *Jamaica* (1760)

Cole, Charles Nelson (1723–1804) Lawyer. *Bedford Level* 1789

Cole, G. and J. Roper. *British Atlas* 1804–10

Cole, Humphrey (ca. 1530–1591) Engraver, goldsmith, and the instrument maker. *Canaan* 1572, published in Bishop's Bible, the first map known to have been engraved by an Englishman. *Armillary Sphere* 1582

Cole, James. Engraver and mapseller at the Crown in Great Kirby Street, Hatton Garden. *Gibraltar* (1725), *Romney Marsh* (1737), *Actual Survey of the City of Bath* 1743, *Chassereau's St. Leonards* (1745)

Cole, William. *Geographical Companion through England* 1824

Colebrook, R. H. *Course of Ganges* 1806

Coleman, G. D. Surveyor. *Town and Environs of Singapore* 1836

Coleman, Rev. Lynam. *Landscape map Egypt* 1874

Colepepyr, Robert. *Map Rye Harbour* (1698)

Coles, Edward. *Map Village of Peoria* (1820)

Coles, Joseph. Engraver. *True plan of the city of Excester* 1709

Colijn. See **Colin**, M.

Colin. *Philippines* 1659

Colin, Emanuel. Publisher of *Herrera's Indes Occidentales* 1601

Colin [Colijn] Michael (17th century) Dutch publisher of Amsterdam, sur l'Eau au livre domestique. *Novus Orbis* 1622

Colines [Colinaeus] Simon de (1480–1546) Engraver and publisher in Paris

Colinet, Charles. *Forêt de Fontainebleau* 1870

Colins. See **Colius**

Colins, Charles Hubbard. See **Collins**, Charles Hubbard

Colius [Cool, Cools, Colins, Ortelianus] Jacobus [Jacques] (1563–1628) Nephew of Ortelius. Prepared *Tabula Peutingeriana* for publication by his uncle.

Collard, R. Suveyor and publisher of Broadstairs. *Town and Royal Harbour of Ramsgate* (1822 with Hurst), *Broadstairs* 1824

Collard, Thomas W. *Plan City of Canterbury* 1843, *Reduced plan* 1860

Collard, W. Engravers. *Country round Newcastle* 1850

Collegno, H. de. *Carte géologique d'Italie*, Paris 1846

Collen, George William. *Britannia Saxonica* (1833)

Colles, Christopher (1738–1816) Born Dublin, went to America in 1771. Engineer and cartographer. *Survey of the Roads of the United States of America* 1789 (the first American Road Book), *New England Coast Barnstable to Long Island* 1794, *Northern Vermont and New Hampshire &c.* in *Geographical Ledger*, New York 1794, *Part of the State of New Jersey* 1808

Collet, J. *Les Cartes Topographiques*, France 1887

Collet, Capt. John Abraham (1767–1775) Swiss origin. Engineer, surveyor, Governor of Fort Johnston. *Plan of Fort Johnston*

Engraved title to Captain G. Collins' "Coasting Pilot"

1767 MS. *Back country of North Carolina*
1768 MS, *Compleat map of North
Carolina* 1770, *Southern British Colonies
in America* 1776 (Collet & others)

Collet, V. See **Collot**, G. H. V.

Colleton, Capt. James Roupill. *Battle of
Salamanca* 1814

Colley, Col. G. Pomeroy. *Military Sketch
Transkei* 1875, *South African Republic*
(Transvaal) 1877

Collie, J. *Plan Glasgow* 1839 (with D. Smith)

Collier, Sir George, R. N. *Survey Princes
Island, Guinea* 1819 MS.

Collier, George Ralph. *Survey Texel* 1799–
[1803]

Collier, Richard. *River Clyde* 1776, *Plan
Glasgow* 1776

Collier, William. *Plan Town and Castle of
Windsor* 1742

Collier, William Francis (1807–1890) and
Leonard Schmitz. *Crown Atlas* 1871,
International Atlas 1873, *Library Atlas*
1875

Collignon [Colignon] Francesco. Engraver.
Thionville 1644, *Valenza* 1656, *Villes de
l'Europe* 1672 (93 plates) *Brisach* 1677,
Candia 1699

Collignon, Romain Charles Édouard (1831–
1897) *Chemins de Fer Russes* 1857 and
1862, *Atlas* 1868

Collimitius [Tanstetter] George (1482–
1535) Bavarian cartographer. Corrected
Cuspianus' Austria 1526 and *Lazarus'
Hungary*, published Apian 1528

Collin, Etienne (fl. 1798–1829) Engraver
for Dépôt de la Marine. *Golfe de Suez*
1798, *Côte Amérique Méridionale* 1800,
For Bruny d'Entrecasteaux 1807, *Cattegat*
1813, *Mer Baltique* 1815, *Rade du
Brest* 1822, *Laborde's Arabie Pétrée*
1834

Collin, L. *Louisiane et Pays Voisins* 1802

Collin, Richard. Engraver. *Status Belgicus*
(1630)

Collin, T. I. *Carte topographique de Spa*
1788

Collin fils. Engraver. *Afrique Occidentale*
1828, *Mimizan* 1829

Collins, Benjamin. Printer of Salisbury,
on the New Canal. *Naish's plan of Salisbury,
County maps by Tunnicliffe* 1791

Collins, Charles. Engineer. *Lake Shore &
Michigan Southern Railway* 1875

Collins, Charles Hubbard. *Map of
Chignectou Bason* 1754 MS (plans of
English and French forts in Nova Scotia)

Collins, Christopher. Master of H.M.S.
Cumberland. Charts Coast Great Britain
1781–1802, *Survey Varne and Ridge*
1793

Collins, Freeman. Printer of *G. Collins'
Coasting Pilot* 1693, *Camden's Britannia*
1695

Collins, F. Howard. *Tidal Streams North
Sea* 1894

Collins, Capt. Greenville (fl. 1669–98)
Hydrographer to the King. *Great Britain
Coasting Pilot* 1693, the first English Sea
Atlas based on personal survey. Reissued
many times to 1792 with occasional
revisions and additions

Collins, Henry George. Publisher, 22
Paternoster Row London. Editions of
Teesdale's Atlas (1848), *Travelling Atlas
of England & Wales* 1850, *British Gazetteer*
1852 (with B. Clarke), *Indestructible
Atlas* (1858), *One Shilling Atlas* (1858),
Junior Classic Atlas (1859)

Collins, James. Engraved maps for G.
Collins

Collins, R. Engraver. *Warren's plan St.
Edmunds Bury* 1747

Collins, William (Sons & Co.) Publishers
in London, Glasgow and Edinburgh.
Imperial Outline Atlas 1860, *New Series
School Room maps* (1872), *County
Geographies* (1872), *Atlas Australian
Colonies* 1875, *Complete Atlas* 1897, *New
Crown Atlas* 1898

Collinson, Lieut. (later Capt.) R., R.N.
Hydrographer. *Durian & Rhio, Singapore*
1822, *Survey China* 1840—45, *Ordnance
map Hong Kong* 4 sh. 1846, *Survey North
of Behring Straits* 1849—54

Collinson, Richard. *Plan of Eccles* (1887)

Collinson, William S. *Survey Straits of
Durian* 1823, *Straits of Singapore* 1835

Collis, William. Lithographer. *Bourke
County, Victoria* 1865, 1866, *Hampden
County* 1867

Collomb, Édouard. *Geological map
Environs of Paris* 1865, *Carte Géologique
de l'Espagne et du Portugal* 1868

Collon. *Plan communal de la ville de
Liège* (1820)

Collon, Jacques. *Carte Topographique
de l'Ile de Corse* 1827

Collot, George Henry Victor (1752—1805)
Map of Country of the Illinois (1826),
Plan Topographique du Détroit (ca. 1796)

Collot, Victor. Georges Henri (1750—
1805) Military cartographer. *Voyage dans
l'Amérique Septentrionale* 1826

Collyer, Joseph (1748—1827) Engraver
for *Taylor's Ireland* 1793, and for Faden
1797

Colm, Morgan. *Angliae et Hiberniae nova
descriptio* 1643

Colnett, Capt. James. *Japan* 1791, *Voyage
to the South Atlantic* 1798, *Plan Island of
Cocos, Galapagos,* pub. Arrowsmith 1798

Colom, Arnold [Aernout, Arent] (1624—
1668) Chart publisher, engraver and bookseller. Op de Texelsche Kay inde Lichtende
Colom. Son of Jacob Aertz. *Zee Atlas*
1654 (Latin text) and 1658 (Dutch text),
Lighting Colom of the Midland Sea
1660

Colom, Jacob Aertz, elder (1600—1673)
Printer, publisher and bookseller in the
Fiery Colom on the Corn Market,
Amsterdam. Born in Dordrecht. Founded
Navigation School in Amsterdam.
Vyerighe Colom 1632—33, French

edition *L'Ardente ou Flamboyante
Colomne* 1633, English editions *Fiery
Sea Colomne* 1647, *Upright Fiery
Colomne* 1648 and later.

Colombo, Barthélemi [Bartolomeo]
*Les provinces de Veronese, du Vicentin,
de Polésine* 1736

Colombo, Bartolomeo (1460—1514)
Cartographer, brother of Christoph.
African coasts 1506 MS (attributed)

Colombo, Christoph. (1451—1506) born
Genoa, died Valladolid. *De Insulis
inuentis,* Basle 1494 with *map of West
Indies,* the first printed representation of
the New World. *Sketch of North East
Hispaniola* 1492—3

Colombo [Colon] Fernan[do] (1488—
1539) Son of above, priest of Cordoba.
Began *survey of Spain* (produced only
notes), *World* 1517 (attributed)

Colon, Hernando. See **Columbo,** F.

Colquhoun, A. R. *Indo China, South
China* 1882, *South West China* 1887

Colson, John. Mathematician, teacher of
Navigation. Joint publisher with Seller of
Atlas Maritimus 1675

Coltman, Nathaniel. Surveyor, employed
by Post Office. Publisher in Greenwalk,
Blackfriars Road. *South Wales* 1797,
Laurie & Whittle's Welsh Atlas 1805,
*Coach and crossroads of England and
Wales* 1809, *Inland Navigation* 1808,
Plan London 1831, *York* 1833

Colton, Charles B. (1832—1916)
Publisher with G. W. Colton. *Cuba* 1855,
Louisiana 1870

Colton, George Woolworth (1827—1901)
Publisher, geographer and engraver. 172
William Street, New York. *Missouri* 1851,
Atlas of America 1855, *General Atlases*
1855—1888, *Series of Railway maps*
1858, *Illustrated Cabinet Atlas* 1862,
Central America 1889

Colton, G. W. and Co. *Complete Ward
Atlas of New York City* (1892)

J. Aertsz. Colom (1599–1673); engraving by J. Matham

Colton, Joseph Hutchins (1800–1893) Publisher, 86 Cedar Street, New York, and 172 William Street (1854–64). *South America* 1845, *Atlas World* 1855, *General Atlases* 1857–84, *Family Atlas* 1862 (with A. J. Johnson), *American School Geography* 4to (1870)

Columbine, Lieut. (later Capt.) E. H., R. N. Hydrographer. *St. John, Antigua* 1789, published 1793; *Trinidad* 1803, published 1816, *Chart Virgin Islands* 1808

Columbine, G. A. *The Mole at Naples* (1790)

Columbus. See **Colombo**, Christoph

Colvin, Verplank. *Adirondack Survey* 1872

Colyer, J. (1748–1828) London engraver, worked for Faden

Colyns, Alexander. *Plan Innsbruck* used by Braun and Hogenberg 1598

Combatti, Bernardo. Publisher *Lagune di Venezia* 1815

Combe, Querenet de la. *Environs d'York en Virginie . . . la position des armées* 1781

Combe, T. *Map Leicestershire and the surrounding country* 1834

Comberford, Nicholas (fl. 1641–1670) English portolan maker, Dwelling at the signe of the Platt neare the west end of the School House in Ratcliffe. *British Isles and North France* 1641, *South Virginia* 1657, *English Channel, North Europe* 1655, all MSS

Combette, A. Publisher of Paris, Rue de Parcheminerie 15. *Levasseur's Atlas National* 1847, 1854

Combis D'Augustine, Lieut. M. de. French Marine. *Plans Newfoundland* in *Neptune Français* 1786–92

Combitis, Nicolaus de (15th century) Venetian artist. *Mediterranean Sea*

Combletee, D'. *Rivière et Forts de Choueguen* 1756

Comenius [Comenio, Komensky] Johan Amos (1592–1670) Czech cartographer and priest. *Moravia* 1727 used by Hondius, Visscher, & Blaeu

Cometti, Giovanni Francesco. See **Camocio**, G. F.

Cominotti, G. *Pianta Cagliari* 1832 (with Marchesi)

Comotti, G. F. See **Camocio**, G. F.

Compton, Thomas. Surveyor. *Lundy Island* 1804 MS

Comrie, A. *Hull & Selby Railway* 1834

Comstock, Capt. Cyrus Ballou (1831–1910) New York cartographer (Corps of Engineers USA). *Siege Vicksberg* (1863), *Mouth Detroit River* 1874

Comyn, Tomas de. *Islas Filipinas* 1821

Conaway. See **Conway**, M.

Concepcion, Juan de la. *Yslas Philipinas* 1744–[88]

Concha Miera, Francisco de la. *Plan Oviedo* 1777

Conclin, George. *New River Guide,* Cincinnati 1849, 1850, 1855

Condamine, Charles Marie de la (1701–1774) Mathematician and geographer of Paris. *Amazon River* 1745

Conde, Pedro Garcia. *Nouvelle Espagne* 1807 (with Costanzo), *Plan Mexico* 1811, *Republica Mexicana* 1845[8]

Conder, Lieut. (later Major) Claude Regnier (1848–1910) Archaeologist and cartographer, born Cheltenham. *West Palestine* 26 sh. 1872–77 (with Kitchener), reduced edition 1881, *Palestine* 1890

Conder, Thomas (fl. 1775–1801) Engraver and cartographer. *Maritime survey of Ireland* 1775, *Moore's Voyages* 1778, *Stobie's Perth* 1783, *Walpole's British Traveller* 1784, *Virginia* 1788, *Plan Liverpool* 1790, *Wilkinson's Atlas* 1794, *Barker's Kentucky* 1795, *Ohio* 1797, *Well's Geography Old and New Testament* 1801

Conders ab Helpen [Coenders van Helpen] Frederick (1541–1610) Cartographer. *Groninga* (1640)

Conders ab Helpen, William. Brother of above

Condet [Kondet] Family of Engravers. Claas (1685–1725), Gerardus (1731–1764), Harmanus (1725–57) and Johannes (1711–1781). C. and G. worked for D'Anville, the last for Mortier, Popple and Delisle. *Carte de l'Égypte* 1746

Condie, Thomas. Bookseller of Philadelphia. *Plan of the Jail at Philadelphia* (1797–8)

Condima, Sebastiano. Of Messina. *Chart of the Mediterranean* (1615) MS

Cone, J. Engraver for Lucas 1823, and Cumings 1829

Coneys, W. T. *Fond du Lac County, Wisconsin* (1862)

Coninx, Arnold. Printer of Antwerp. *De Jode's Speculum* 1593, *Ortelius' Epitome* 1595

Conklin, E. W. *Map of the Village of Jamaica* 1868

Conklin, George W. *World Atlas* 1889

Connor, J. *City and suburbs of Cork* 1774

Conrad, Frederik Willem, the elder (1769–1808) Surveyor of Delft, *Lower Rhine* 1793, *Heusden en Altena* 1798 (with Engelman)

Conrad, George, Junior. *Franconia* 1638–41

Conrad, J. F. W. *Kaart van het Eiland Zuid Beveland* 1858, *Nederland* 1865

Conradi, F. L. *Plan of Fort St. George, Madras* 1755 MS

Conradi, Karl Theodor (18th century) *Bretten* 1779

Consag, Father Fernando. Sailed round the Gulf of California, showing it to be a peninsula. *California* 1746 MS, printed in *Venegas' Noticia* 1757

Consitt and Goodwill. *New plan of the Town of Kingston upon Hull* 1817

Consoni, Lieut. Joseph. *Plan München* 1806

Constable, Archibald. Publishers of London and Edinburgh. For Mackenzie 1811, for Arrowsmith 1817, *Map Edinburgh District* 1822

Constable, A. & Co. *Hand Atlas of India* 1873–93

Constable, Lieut. (later Commander) C. G. Admiralty Surveyor. *Charts of the Persian Gulf surveyed* 1857–60

Constantin de Ramière. See **Ramière**

Constantinus, von Antiochien. See **Cosmo Indicopleustes**

Contarini [Contareni] Giovanni Matteo (d. 1506) Italian cartographer. *Map of the World* 1506 engraved by Roselli, the oldest known printed map to show America

Contarini [Contareni] Giovanni Pietro (1549–1603) Painter and cartographer of Venice. *Europe* 16 sh., Venice (1564), *Candia* 1564

Conte Hoctomanno. See **Freducci**, Conte di Ottomano

Cöntgen. See **Coentgen**

Conti, Filippo. Engraver for *Atlante Geografico* (1788–1800)

Conti, Niccolo dei (ca. 1395–1469) of Chioggia. Venetian merchant and traveller

Conty, Henri A de. *Carte pratique et officielle des excursions de Normandie* (1867)

Converse, M. S. *City plan of Elmira, New York* 1876

Conway, Sir William Martin. *The Karakoram Himalayas* 1894

Conway [Conaway] Michael, R. N. Master of Trinity House 1762–75. *Approaches to the Thames* (1780)

Conynenberg, Jacob. *Charts for* Theunisz 1707

Coode, John. *Victorian Harbours Gippsland, Lakes Entrance* (1880)

Cook, Alexander. *Universal Geography* 1787 (with T. Bankes & E. W. Blake)

Cook, Franklin. Drew and published *Map of Minneapolis, Minnesota* 1876

Cook, George H. (1818–1889) American geologist

Cook, J. Engraver. *Plan Thames* 1847

Cook, Capt. James, R. N. (1728–79) Explorer, circumnavigator, hydrographer. *Draught of the Bay and Harbour of Gaspee, J. Mynde sculp.* (1759) the first engraved map by Capt. Cook. First Voyage 1768–71, Second Voyage 1772–75, Third Voyage 1776–79, Accounts published 1773–84. *Newfoundland surveys* 1758–1765, published 1766–70, in atlas form as *Newfoundland Pilot* 1770

Cook, James. Surveyor for boundary between North and South Carolina. *Draught of Halifax* 1755, *Draught of Port Royal* 1766, *West Florida* 1766, *South Carolina* 1770. One of the editors of *North American Pilot,* Sayer and Bennett 1777–8

Cook, James B. *Memphis, Tennessee and its Environs* 1860

Cook, John. Publisher. *New Royal . . . system of Universal geography* (1787–1810) (with Bankes and Blake)

Cook, Robert J. *England and Wales* 1866, *Railway maps of United States* 1872, *South America* (1890)

Cook, Thomas. *Circular Tours* 1873, *Literary and Historical map of London* 1899

Cooke. *Royal Map of Dublin &c.* 1831

Cooke, Capt. (later Major) Anthony Charles, R. E. Director Topographical Department of War Office, New Zealand. *Plan Sevastopol* 1857, *Verona* 1859, *Pekin* 1859, *Waikato River, New Zealand* 1864

Cooke, Charles (d. pre-1822) Publisher of London. *Modern British Traveller* (1809)

Cooke, Edward. *Voyage to the South Sea* 1708–11, published 1712, *World Map*

Cooke, Francis. *Principles of Geometry, Astronomy and Geography* 1591

Cooke, George Alexander *Modern British Traveller* 1802–1810, *Universal Geography* 1802, *Topography of Great Britain* 26 vols. 1822

Cooke, H. T. *Map Warwickshire* (1861)

Cooke, John. Engraver, draughtsman, and publisher. Clare Court Drury Lane (1790), 50 Howland Street (1799), 11 Platt Place (1805). For Steel 1790, Faden 1796, Arrowsmith 1807, *Edwards' West Indies* 1794, *Smith's Universal Atlas* 1802, *Chatfield's Historical Review* 1808, *Cyclopedia* 1820

Cooke, John. Engraver and geographer of Plymouth. *Plymouth Breakwater* 1823, *Environs of Plymouth* 1830, *Plan Devonport and Stonehouse* 1834

Cooke, O. D. & Co. Publishers of Hartford, Connecticut. *Willard's Atlas* 1826, *Modern Atlas* 1831, *School Atlas* 1835

Cooke, Lieut. Colonel Philip St. G. (1809–1895) *Santa Fe to Pacific* 1847 MS

Cooke, S. R. Engraver. *Bumpus' Plan London* 1827

Cooke, S. W. Engraver. *Laurie & Whittle's Plan London* 1801

Cooke, Thomas. *City of Westminster* 1847

Cooke, W. D. *Virginia Springs* 1858, *North Carolina* (1860)

Cooke, W. Engraver. Cartouche for *Arrowsmith's America* 1804

Cookingham, E. R. *Sanilac County, Michigan* 1894

Cookson, Arthur Augustus. Surveyor. *Chart Islands Westward of Socotra* 1856

Cooley, William Desborough (d. 1883) *Africa south of the Equator* (1864)

Cooling, Lieut. William John, R. N. Admiralty Surveyor. *Mediterranean* 1830, *Gulf Arta* 1834

Cools, Jacques (1563–1628) of Antwerp and London. Publisher and antiquary, nephew of Ortelius

Cools, M. Engraver of Antwerp. Worked for van den Ende, maps for *Blaeu's Atlas*

Cooper, Capt. Allen. *Track of the Ship Atlas* 1789

Cooper, H. Engraver. 23 Chancery Lane, London. *Rees' Cyclopaedia* 1806–10, for Playfair 1809, for Atkinson 1813 *(Ceylon),* for Philipps 1823

Cooper, J. 36 Fetter Lane, Fleet Street, London. *British Isles* (1835) etc., *England, Wales and Southern Part of Scotland* 1849

Cooper, Mrs. Mary [widow of Thomas] (fl. 1740–1761) Bookseller and publisher at the Globe in Paternoster Row. *Plan Dunkirk* 1743, *Kitchen and Jefferys' Small English Atlas* 1749, *Cowley's New set of Pocket Maps of all the Counties of England* 1745, *London* 1753

Cooper, R. Surveyor. *Map Forest of Mamlome* 1737

Cooper, Richard, the elder (d. 1764) Engraver and publisher of London and Edinburgh. Engraved maps for Adair 1730–45, for Bruce 1733, for Jeffray 1739, for Moir 1739, for Edgar 1741, for Bryce 1744, *Plan Leith* 1759

Cooper, Richard, the younger (ca. 1740–1814) Painter and engraver of Edinburgh. Engraved for Capper 1802

Cooper, Robert (fl. 1800–36) Surveyor, engraver and publisher of York. Maps for Dawson, *County and City of York* 1832

Cooper, T. Publisher of *Wells' Atlas* 1700

Cooper, T. W. Assistant Surveyor. *Plans of Land Grants in Victoria* 1857–9

Coordes, G. (1839–1890) German geographer

Coore, John Hamilton. *Chart Spithead* 1791

Copeland, Commander Richard. Admiralty Surveyor. *Mediterranean* 1826–36, *Charts* 1852–1866

Copernicus [Kopernicus] Nicolas (1473–1543) *Maps of Prussia* 1529, *Lithuania,* now lost

Copland, Robert. Translator of the *Rutter of the Sea* 1550, 1555

Copley, Charles. *Tierra del Fuego* 1849, *Charts Coast of United States* 1863–4

Copmayer (17th century). Engraver and publisher of Augsburg. Maps and Views after Merian

Coppens, P. *Atlas de la Belgique* (1852)

Coppo [Coppus, Copus] Pietro (1470–1565/6) Venetian cartographer domiciled in Istria. Woodcut *World Maps* 1524–6, *British Isles* (1525) *Portolano* 1528 MS, *Istria* 1528. Used by Bertelli, Camocio and Ortelius

Coquand, H. *Geological map of Charente* 1859

Coquart, A. Engraver. *Camps Ortnaw* 1690. Engraved for De Fer 1696, *Plans of Turin* 1705, *Gibraltar* 1706, *Anvers* 1708, *Barcelona* 1711. Engraved for Blottière and Roussel 1730

Coquus. See **Cock,** Hieronymous

Cora, Guido (b. 1851) Geographer of Turin. Founder of Cosmos 1873, *Baia d'Assab* 1883, *Italia* 1891

Coran, T. *South Carolina,* Charleston 1802

Corbelletti, Francesco. Publisher of Rome. Strada *De Bello Belgico* 1632 and 1647. Continued by Heirs. *Rome* 1658, *Santa Theresa's Brazil* 1698

Corbet, C. *Plan Culloden* 1746

Corbet, W. *New Plan of the City of Dublin* 1803

Corbett, D. Engraver. *Plan Cork* 1750

Corbett, John A. Surveyor. *Agricultural Lots, Oxley, Victoria* 1857, *Country Lands on the Reedy Creek* 1856

Corbett, V. P. *Seat of War showing battles of July 18th, 21st and Oct. 21st.* 1861

Corbigny. See **Brossard de Corbigny**

Corbridge, James, Surveyor. *Newcastle on Tyne* (1723), *Norwich* 1727, *Norfolk* 1730 and 1755, *Suffolk* (1765)

Cordeiro, Luciano (1844–1900) Portuguese geographer. Founder of Societa de Geografia

Cordier, Henri (1849–1925) French orientalist of New Orleans. *Atlas Sino-Coréen* 1896

Cordier, L. Engraver. *Chorographica Regni Poloniae* 1729

Cordier, Louis [Ludovicus] (d. 1711) Engraver for Sanson D'Abbeville elder, 1661–75

Cordier, Robert (d. ca. 1673) Engraver for Sanson D'Abbeville elder ca. 1650, for Jaillot 1669 *(Brest)*, for Du Val 1670

Cordier, Robert. Engraver. *Cercle d'Autriche* 1782

Cordova, J. de. *Map State of Texas* 1854

Corenwinder, N. *Ville et Port de Dunkerque* 1841

Corfield, M. Surveyor. *Plan Manor of Staunton St. Bernard Wilts.* 1784

Corne, Louis, Engraver for Ayrouard (1732–1746)

Corneille, Jean-Baptiste (1649–1695) Painter and engraver of Paris

Cornelis, Lambert (1546–1601) Engraver. *Zaltbommel* for Linschoten and for Mercator *Magellan* (1606)

Corneliszoon, Cornelis (1493–1544) Painter of Leiden. *Friesland* 1531, *South Holland* 1537

Corneliussen, O. A. Geologist. *Nordlige Norge* 1879

Cornell, Sarah Sophia. *Companion Atlas to Cornell's High School Geography* 1859, editions to 1879, *Physical Geography* 1870

Cornell, S. *Map City of Rochester* 1838

Cornell, Thomas C. *Town of Yonkers, N.Y.* 1892

Corner, G. F. (17th century) Cartographer

Cornet, F. L. (d. 1887) Belgian geologist

Cornetus, Petrus, of Rotterdam. *Portolan* 1618

Cornput, J. van den. Military engineer with Prince Maurice. *Steenwyck* 1581

Cornwall, Capt. Henry. *Aden* 1703, published Dalrymple. *Voyages to India* 1720

Cornwall, James. *School Atlases* 1852–1862

Coronelli [Corneille] Vicenzo Maria [Marc Vicent] (1650–1718) Cosmographer to the Republic of Venice 1685, globemaker, Franciscan Friar and General of the Order 1701. Founded first Geographical Society 1680, the Academia Cosmografica degli Argonauti. *Morea* 1687 (maps from 1684), *Atlante Veneto* 1690–6, *Corso Geografico Universale* 1692–94, *Epitome Cosmografica* 1693, *Isolario* 1696–7, *Miniature globes and large globes* 1683–1704, *Viaggi* 1687, *Specchio del Mare* 1698, *Singalente di Venezia* (1716)

Corputius [Corput Broedanas] Johannes de. *Plan Duisburg* 1566, used by Braun and Hogenberg

Correa, Manuel. *Coast of Luzon* 1759

Corréard, Joseph (1792–1870) *Théâtre de la guerre en orient* 1853, *Atlas de la Mer Noire* 1854

V.M. Coronelli (1650–1718). Founder of the
"Academia degli Argonauti"

Corry, Capt. Armar L., R. N. *Observations used by Arrowsmith on map of Syria* 1823

Corsulensis, Vicentius (16th century) Italian cartographer, author first known printed *map of Vavassore*, Venice 1532

Cortambert, Eugène. *Cochinchine Française* (1860), *Atlas Général de la France* 1868, *Géographie Élémentaire* (1874)

Cortambert, Pierre François (1805–1881) French geographer (with son) Revised *Malte-Brun's Nouvel Atlas* 1865, *Petit Atlas de Géographie Anciene* 1865, *Globe Illustré* 1875

Cortazar, Daniel de. *Mapa Géologico Provincia de Almeria* 1873–4

Corte, Carlo da, of Genoa. *Chart* (1592)

Cortés, Hernán Fernando (1485–1547) Conquistador of Mexico. *Praeclara Ferdinandi*, Nuremberg 1524, Woodcut map of *Mexico and Gulf of Mexico* marking *La Florida.* First use on a printed map

Cortés, Martin. Spanish geographer and mathematician. *Compendio de la Esfèra y de la carte de naveguar,* Cadiz 1551. English translation by Eden 1584

Cortès, Phelipe Feringan. *Isla de Santa Rosa* 1775 MS, *Pensacola* 1755 MS

Cortés, Tomás. *Plan of Manila* 1823

Cortese, Emilio. *Carta geologica della Calabria* 1895

Cortiro, Nicola dal (16th century) Venetian cartographer. *Friuli* 24 sh. 1535 MS

Cortona, Giovanni Antonio di (d. ca. 1560) Painter of Udine. *Udine* ca. 1540 MS

Corvinus, F. *Grundriss der Stadt Posen* 1856

Corvinus, Johann August (1683–1738) of Leipzig. Engraver

Cosa, Juan de la (d. 1509) Pilot to Columbus, circumnavigator, chartmaker *Voyage to America* 1493–1507. *Chart of World* 1500 (MS), the earliest map to show the discoveries of Columbus

Cosel, E. von. *Provinz Brandenburg* 1861

Cosin, S. English surveyor (17th century)

Cosmo [Cosmas, Kosmos], **Indicopleustes** (6th century) Theory of a flat world in *Christian Topography* 535–547 A.D.

Cosmo de Churruca. See **Churruca**

Cossin, Johan. Marine cartographer. *Carte géografique Universelle,* Dieppe 1570

Cossinet, Francis (fl. 1659) Book, map and instrument seller

Cossins, John. *Plan City of York* 1748

Costa, Antonio Carvalho da. See **Carvalho da Costa**

Costa, Euzebio da. Geographer. *Planisfere,* Lisboa 1720 MS.

Costa, Giovanni Francesco (d. 1773) Venetian engineer and engraver.

Costa, J. da. Publisher. Red Cross Street, Southwark. *Plan Town and Harbour of Boston* July 29th 1775, earliest battle plan of War of Independence

Costa e Miranda, Joseph da. *West Europe, Africa and America* 1688 MS.

Costanzo [Constanso] Miguel (fl. 1769–1811) Spanish cosmographer, accompanied Portola to California 1769–70. *Oceano Asiatico o Mar del Sur* 1770, *Nueva España* 1777, *Nouvelle Espagne* 1807 (with P.G. Conde)

Costaz, Baron Louis (1767–1842) French mathematician and geographer

Coste, Pascual. *Basse Égypte* 1818–27, published 1830

Costi, Ago. Engraver for *Atlante Geografico* (1788–1800)

Costo, Gieronimo. Genoese chartmaker working in Barcelona late 16th century

Costo, Giovanni of Genoa. *Portolan World* 1602 MS

Cotilla, Juan. *Isla de San Carlos . . . Misisipi* 1769 MS.

Cotsford, Edward. *Itchapour District* 1778

Cotta, Bernard von (1804–1879) German geologist. *Koenigreich Sachsen* 1830–46, *Thuringia* 1836, *Umgegend von Dresden* 1868

Cotta, J. G. *Die Bayerische Pfalz* (1867)

Cottam, Arthur. *Charts of the Constellations* 1889

Cotterell, H. F. *Country round Bath and Bristol* (1840)

Cotterell, J. H. *City and Borough of Bath* 1852

Cotterell, Thomas Sturge. *Historic map of Bath* 1898[9]

Cotton, George H. *Plat of St. John, Harrison County, Iowa* 1857

Cotton, Sir Robert Bruce (1571–1631) Antiquary, collector of books, maps and manuscripts. Friend of Speed, assisting him in his survey. His collection now in British Museum

Couceiro, L. *East Africa* 1892

Couché fils. Engraver for Perrot 1845

Couchman, Thomas. *Plan Pental Island, N.S.W.* 1870

Coucke, J. (1783–1853) Painter and engraver. Town plans and views

Coudet, Johannes (1711–1781) Engraver of Amsterdam

Coulier, Philip J. *Atlas Général,* Paris 1850

Coulon. *Plan de la Ville de Nantes* (1795)

Coulon, A. von (1779–1855) *Post Karte von Baiern* 1810, *Süd Deutschland (1820)*

Counradi, M. *Grundriss der Stadt Braunschweig* (1735)

Coupvent-Desbois, Aimé Augustus Elie. Maps for Dumont d'Urville 1837–40

Courbé, Augustin (fl. 1635–49) Publisher for d'Avity 1635

Courbet, A. *Mouillage d'Alexandrette* 1854–(6)

Court, Capt. Charles, R.N. Admiralty Surveyor. *Red Sea* 1804–6, *Palmiras*

1817, *Moscos Islands* 1826, *Mootapilly Bay* 1830

Courtalon, l'Albe. *Atlas Élémentaire* 1774

Courtier, L. *Plan de Rheims* 1899

Courtry, Alexandre. *Congo Français* 1897–8

Cousin, Hugues. See **Cusinus**, Hugo

Cousin, Paul. *Topographical Plan City of Quebec* 1875

Coussin [Cousin] Honoré. Engraver for Ayroud (1732–46), *Provence* 1758, *Ville d'Aix* (1765)

Coutans, Guillaume (b. 1724) *Atlas topographique environs de Paris* 16 sh. (1775), *Route de Paris à Rheims* 1825

Coutaut, E. *Carte routière et vinicole de la Gironde* 1877

Coutgon. Engraver. *Course Rhein, Neuwied to Mainz* (1780)

Couts, Lieut. Cave Johnson. *Colorado and Gila Rivers, California* 1849 MS.

Couzens, M. K. Civil Engineer. *Key to East Virginia* 1861, *Key to the South* 1861

Covarrubias, Fran. Diaz. *Carta Hydrografica del Valle de Mexico* 1862

Covel, Thomas. Surveyor. *Ickworth Survey Booke* 1665

Covens, Cornelis. *Bataafsche Republiek* 1799

Covens, Jean [Jan, Johannes] & Corneille Mortier. Map printers and publishers of Amsterdam. Succeeded to business of Pierre Mortier 1711, and continued by heirs to 1866. Numerous atlases mostly using maps after De Lisle, Sanson and Jaillot. *Nieuwe Atlas* 1730 &c., *North America* 1757, *Louisiane* 1758, *Zak Atlas* 1794, Covens Mortier and Covens & fils 1774, Coven J. & Son 1792, Mortier Covens et fils 1795, Mortier & Zoon 1810

Covert, Walter. *Coast of Sussex* 1587 (with Sir T. Palmer) published 1870

Cowan, Frank. *South West Pennsylvania* 1874

Cowan, J. *Chart Shannon* 1771

Cowe, Richard. *Plan of Plymouth* 1778
2 sh. published 1780

Cowley, Capt. Circumnavigator. *New map of the World* (1699)

Cowley, H. *Plan Annapolis Royal* 1752
MS.

Cowley, John. Geographer to His
Majesty and engineer. *North Britain* 1734,
Coasts of Florida and Carolina published
Vaughan (1740), *Mediterranean* 1744,
Dodsley's Geography of England 1744,
Counties of England 1745

Cowley, R. *Charlestown (South Carolina)*
(1780)

Cowper, Thomas. *Almanack for 1637 . . .
with principal highways in England &
Wales* 1637

Cowperthwait, Desilver & Butler (formerly
Thomas, Cowperthwait & Co.) Publishers
253 Market St. Philadelphia, later
Cowperthwait & Co. (1872). *Mitchell's
Atlases* 1839–59, *New Universal Atlas*
1855, *Warren's Common School
Geography* 1872, editions to 1887

Cowperthwait, H. Publisher. Philadelphia.
Mitchell's School Atlas 1856 & 1858

Cowperthwait, J. *Warren's Physical
Geography* (1845)

Cox, Professor E. T. *Coalfields of South
West Indiana* 1874

Cox, George. Publisher in London, King
Street, Covent Garden. Issued *Atlas* for
Society for Diffusion of Useful Knowledge
1852–3

Cox, Gustavo. *Distrito Minerales de
San Antonio* 1865

Cox, Commander H. L. Admiralty
Surveyor. *England* 1853–60, *Salcombe
River* 1863

Cox, Capt. J. H. *Oyster Bay Port St.
Francisco* 1789, published Dalrymple

Cox, Rev. Thomas (ca. 1660–1734)
Bookseller and topographer. Corner of St.

Swithin's Alley, Cornhill. Edited *Camden's
Britannia* 1720–31, *Magna Britannia* 1724

Cox, W. J. *Schuylkill County, Pennsylvania,*
1864 (with W. Scott)

Cox, Zachariah. *Part of Mississipi Territory*
(1815) MS.

Coxe, Dr. Daniel (1640–1730) Physician to
Charles II and Queen Anne. Governor of New
Jersey (1687–92) *Description of Carolina*
(with map) published by his son Daniel
(1673–1739), *Map Carolina,* London 1722,
1726, and 1741

Coxe, Rev. William (1747–1828) Historian.
*Russian Discoveries between Asia and
America* 1780, *Switzerland* 1789, *Travels
in Poland* 1784

Coxen, J. U. *City of Nagpoor* 1865(–70)

Cozens, William H. *Map City of Pilot Knob,
Iron County, Missouri* 1858

Cozuela, Vasquez de. *Royaume de
Portugal et d'Algarve* (1762)

Craalinge. See **Cralingen**

Craan, W. B. *Waterloo* 1816, *Plan
Bruxelles* (1836)

Cracalescu, D. M. *Romania* 1888 and
1892

Cracklow, Charles Thomas. *Dock Plans*
1796, 1803

Cracroft, John. *Map Greenland, Iceland
. . . Nova Zembla* 1740

Crad[d]ock. See **Baldwin and Cradock**
1758–9

Craddock and Joy. Publishers of *Lambert's
Travels through Canada and United
States* 1813 *(map of British Settle-
ment and plan of Quebec)*

Craft, N. Henry. *Plans Boston* 1864, *Plan
Boston and Roxbury* 1867

Craig, Francis S. Surveyor for *Spalding's
Plan of Cambridge* 1875

Craig, George A. *Spencer Town,
Massachusetts* 1884

Craig, James (d. 1795) Architect, *Plan Bridge over North Loch* 1763, *Plan Edinburgh* 1768, 1773

Craig, John. Surveyor. *Plans Edinburgh and Leith* 1804, *Ross and Cromarty* for Thomson 1826

Cralingen [Craalinge] Jan. Publisher of Amsterdam. *Leo Belgicus* 1640, *Plan London* 1666

Cram, D. H. *Alabama and West Florida* 1875

Cram, George [& Co.] (1841–1928) Engravers and publishers, 66 Lake St., Chicago. *Wisconsin* 1872, *Colorado, Indiana, Iowa, Kansas, Missouri* 1873, *Unrivalled Family Atlas* 1883, *Illustrated Atlas* 1855, *Unrivalled Atlas* 1887, *American Railway system Atlas* 1891

Cram, Capt. T. J. *Boundary between Michigan and Wisconsin* 1840–1842

Cramer, John Anthony. *Ancient Italy* 1825, *Ancient Greece* 1827, *Asia* 1832, *Italy* 1836

Cramer, Louis H. *Saratoga & Ballston* 1876 (with F. W. Beers)

Cramer, Zadok (1773–1813) *Navigator* 1806, editions to 1818, earliest navigation guide for inland waters of North America

Cramm, Gustav. *Plan Hamburg* 1873–6, *Karte von Kephalonia* (1874)

Cramoisy, Sebastian. Publisher of *Tassin's Plans* 1634–6, *Du Creux's Nova Francia* 1660

Campton, Prof. R. C. *City of Jacksonville, Illinois* 1868

Crane, Capt. Gilbert. Surveyor. *Charts in Seller's English Pilot* 1671

Crane, John. *Butler County, Ohio* 1855

Cranston, George. Road surveyor. *Roxburgh* for Thomson 1832

Crantz, David. *History Greenland* 1767

Craskell, Thomas. Engineer. *General map and county maps Jamaica* 1756–61, published 1763 (with J. Simpson)

Crassus, Valerius. *Valle di Piedmont* 1640

Crates of Mallus (fl. 180–145 B.C.) *Globe.*

Crato [Krafft] Johann (d. ca. 1578) Publisher and printer of Wittenberg. *Stella's Palestine* 1552, *Itinerary of the Israelites* 1557, *Itinerary of St. Paul* 1562, *Moravia* 1570 MS (sent to Ortelius)

Craven, A. *Geological survey of California* 1874

Craven, T. A. M. (1813–1864) U.S. Navy. *Harbour of Bridgeport, Connecticut* (1835)

Crawford, D. C. *Ionia County, Michigan* 1891

Crawford, J. W. *Richmond, Ohio* 1875

Crawford, Lieut. John. Hydrographer. *Macassar* 1814, *Straits of Bally* 1816, *China Sea* 1821, *Coast of Ava* 1849

Crawford, Joseph T. *Galveston Bay (Texas)* 1837

Crawford, Robert, Draughtsman. *Farm Plans East Grinstead* (1840) MS (attributed)

Crawford, William [later & Son] Surveyors *Dundee* 1777, *Dumfrieshire* 4 sh. 1804

Crawford & Son. *Northern & Southern part of Dumfrieshire* 1828

Crawley, Richard. Estate Surveyor *Bedford* 1634

Crealock, H. Hope. *Plan of action of Palechiao* (in China) 1861

Crease, Major Anthony R. V. *Plan Kertch* 1856

Creassy, James. *River Kyme Eau and New Sleaford* 1773

Credner, Heinrich. *Thüringen* 1855, *Hanover* 1865, 1874

Credner, Hermann. *Königreich Sachsen* 1875

Creighton, R. Drew English Counties for Greenwood 1818–1830, London 1831, *County Atlas* 1842, *England and Wales* 9 sh. (1855)

Creitz. See **Cruys**

Cremer [Kramer] See **Mercator**

Crenne, Verdun de la. See **Verdun de la Crenne**.

Crépy. Publisher Paris, Rue St. Jacques à St. Pierre près la rue de la Parcheminerie *Asia* 1735, reissued *Le Rouge* in *Atlas Nouveau* 1767–73, *America after Popple, Moithey's Poland* 1769, *Possessions Angloises en Amérique* 1777

Crépy, Paul (1835–1899) Founder, Geological Society of Lille

Crespi, Fr. Juan (1721–1782) *San Francisco* 1772 MS.

Cresques, Abraham. Catalan chart maker. Credited with *Catalan Atlas* 1375

Cresques, Jaf[f]uda, [Jacme]. Son of above. *Portolan of the Mediterranean* 1416 (attributed)

Crest, Jacques Bartholome Micheli du. (1690–1766) of Geneva *Trigonometrical survey of Switzerland*

Creuse. Worked for Dépôt de la Guerre on *map of France* 1832

Creux [ius] F. du. See **Du Creux**

Creuzbaur, Robert. Communication maps of *United States* 1849, Draughtsman for *Cordova's Texas* 1854

Crevaux, Dr. Jules Nicolas (1847–1882) *Fleuves de l'Amérique du Sud* 1877–79

Crevier, Jean Baptiste Louis (1693–1765) *Atlas de géographie ancienne.* 1819

Crew[e] Randolph (1631–1657) *Cheshire* in *King's Vale-Royall of England* 1656

Crichton, Capt. C. David. *Patta* 1751, *Kissen Bay* (Arabia) published Dalrymple 1775

Crichton & Bell. Engravers. *District of Kyntyre* 1793

Crickenburg, Johan van. Painter of Ghent. *Boundaries of France* (1506)

Criddle, Lieut. J. S. *Island of Thwart-the-Way* 1820

Crighton, George. *Ooloogan Bay, Palawan* 1836 MS.

Criginger [Grigingerus, Kriegengerus, Kringerius, Krüginger, Grisviger, Krüger, Krieger, Kriegner, Crispingius] Johann (1521–1571) Theologian and cartographer of Prague. *Meissen, Thuringia, Saxony* 1567, *Bohemia* 1568, used by Ortelius 1570 and de Jode 1578

Cring, Henry. *Holmes County, Ohio* 1875

Crippen, John. Estate Surveyor of Canterbury. *Crablice and Pairs Farms, Essex* 1731 MS.

Crisp, Edward. Surveyor and publisher at the Carolina Coffee House in Birchin Lane London. *Plan of Charles Town* 1704, *Carolina* 1711

Crisp, May. *Plan of Reception Bay* (Tristan d'Acunha) 1814

Crispi, Giambattista. *Plan of Gallipoli* 1591 (used by Braun & Hogenberg)

Crispingius. See **Criginer**

Crispinianus. See **Cuspinianus**

Cristofano. See **Christoforo**

Crivellari, F. *Citta di Vicenza* 1821

Crivelli, Taddeo (ca. 1425–1479) Painter of miniatures in Ferrara 1452–71. Accredited with maps for 1477 edition of Ptolemy

Croc, H. de. See **Croock**

Croce, Enrico. *Italy* 1875, *Stato di S. Paulo* 1893

Crocker, H. S. & Co. *Sacramento County* 1871

Crocker, P. *Hundred of Mere* 1822

Crocker, William P. *Alton & Terrehaute Rail Road* 1850, *Stansteed–Montreal Rail Road* 1845

Croes, J. J. R. (1834–1906) *West Side New York* 1873 (with Van Winkle)

Croghan, William. American surveyor 1800–04

Croiset. Engraver for Durand 1802

Croisey. Engraver for Bellin 1763. *Dépôt de la Marine* 1768–98, Desnos 1790–2, Bruny d'Entrecasteaux 1807

Croissant, J. Engraver of Antwerp. Map for *Nicolay's Voyages en Turquie*

Croix, F. de la. Engraver. *North America* (1687)

Croix, L. A. N. Abbé de la. See **Delacroix**

Croll, Dr. J. (1821–1890) English geologist

Cromack, Irwin C. *Map City of Boston* 1896

Cromberger, Jacobus. Printer in Seville. *Peter Martyr's Opera* 1511

Crome, August Frederick Wilhelm. *Europe* 1782, *Netherlands* 1785

Cronenberg, F. *Sumatra* 1852, *Boni* 1860

Cronk, Henry H. *Tunbridge Wells* for Brackett 1863

Cronk, W. *Chart Sandwich* 1775

Cronstedt, F. A. U. *Hellsingland* 1796, *Herjeaedalen* 1797

Croock, Hubert de (1490–1554) of Bruges. Painter, engraver, publisher

Croock [de Croc] Willem Hendricksz. Engraver. *North Holland* 1529

Crophius, Martin Gottfried. Published *Fridericks Thall* (1680?) Engraved for *Seutter's Atlas Minor* 1744 and *Lotter's Novus Atlas* (1770)

Cropsey, I. E. *Plan for Panama Canal* 1852

Crosley. See **Crossley**

Cross, Edward (1798–1887) Surveyor. *Arkansas* 1837

Cross, Joseph [from 1844 **and Son**] Engravers and publishers, 18 Holborn Hill opposite Furnival's Inn. *Edgeworth's Roscommon* 1817, *Lambeth* 1824, *Charts of Van Diemens Land* 1826–42, *Oxley's New South Wales* 1827, *New Plan of London* 1828, editions to 1854; *London Guide* 1837, editions to 1851; *Dixon's New South Wales* 1837

Cross, Thomas (fl. 1632–82) Engraver. *Fens* 1642, for Fuller 1650, *Faroe Is.* 1676

Cros[s]ley, William (d. 1797) Surveyor. *Surrey* 1790, editions to 1874 (with J. Lindley). *Rochdale Canal* (proposed) 1791

Cros[s]thwaite, Peter. *Maps of the Lake District* "surveyed & planned by P. Crosthwaite. Admiral at Keswick Regatta, who keeps the Museum at Keswick & is Guide, Pilot, Geographer & Hydrographer to the Nobility & Gentry who make the Tour of the Lakes." Many reissues

Crousaz, Abraham de (1619–1710) of Lausanne. Military cartographer. *Grundriss Pässen des Bern Gebiets* 1670

Crow, James. Estate surveyor. *Hendon* 1754, *Estate plans Essex* 1771 MSS (with J. Marsh)

Crozet, M. *Côtes des Terres de Diemen* (in *Marion's Voyage*) 1783. The first French mapping of part of the coast of Tasmania

Crozet, C. *Internal Improvements of Virginia* 1848

Crozier, Commander (later Capt.) F. R. M. Admiralty Surveyor *Antarctic* exploration 1839–43, *North West Passage* 1845 (lost with Franklin)

Cruchley, George Frederick (fl. 1823–76) Mapseller, engraver and globemaker 349 Oxford St. (1823–25) 38 Ludgate St. (1825–33), 81 Fleet St. (1833–75). Trained by A. Arrowsmith, reissued some of his maps. Bought Cary's plates 1844. *Environs of London* 1824, *Lancashire* 1836, *General Atlas* 1843, *Cary's County Atlas* 1862, *Guide Books London* 1867, *Railway and Telegraph County Atlas* 1855–8

Cruijs. See **Cruys**

Cruise, Richard J. *Geological Survey of Ireland, Leitrum* 1876–7.

Cruquius, Jacob. Surveyor. Brother of Nicolas. *Delftland* 1712 (with brother)

Cruquius [Kruikius] Nicolas Samuelsz (1678–1754) of Delft. Hydrographer, Supervisor of Sluices and Dams. *Isobath map of Merwede River* 1729, earliest printed map with contour lines.

Cruttenden, R. Publisher. *Neal's History of New England* (with map) 1720

Cruttwell[s] Clement (1743–1808) Publisher and surveyor of Bath. *Atlas to Cruttwells Gazetteer* 1799–1808

Cruttwell, Lieut. James. Additions to *Roads of Portugal* 1832

Cruys [Cruis, Crys, Cruijs, Creitz] (1657–1727) Dutch Admiral in Russian Service. *Atlas of Don and Sea of Azof* (1680–1700), *Volga–Don Canal* (1703–4)

Cruz, Alonso de Santa. See **Santa Cruz**

Cruz Cano y Olmedilla, Juan de la (1734–1790) Spanish cartographer and engraver. *Plano de Charlestown* (1750), *Golfo de Mexico* 1755, *Florida and Louisiana* 1755, *Magellan* for Ortego 1769. Worked for Lopez and Torfino de san Miguel 1789. *South America* 8 sh. 1775

Cruz, Lieut. Manuel de la. *Plano puerto de Braga* 1838 MS.

Cruz, Manuel. *Argentina, Chili etc.* 1896

Cruz Bagay, N. de la. See **Bagay**

Cryftz, N. See **Cusa**, N. de

Cuaranta, B. Maps for Bachiller 1852 and Grilo 1876

Cubitt, Thomas. Surveyor. *Scilly Is.* 1791, *Jersey* 4 sh. 1795 (with Gardner)

Cubitt, Sir William (1785–1861) Civil Engineer. *Plans Belfast Harbour* 1835

Cuco, P. Engraver for *Atlante Español* 1778–95

Cudlip, Lieut. (later Commander) F. A. Admiralty Surveyor. *Singapore, Durian & Rio* 1831 (1840), *Heligoland* 1855

Cuipen, P. *Republicas de America Central* 1860

Culbertson, Alexander. *Yellowstone and Platte Rivers* 1857 MS.

Culemann, Friedrich. *Grundriss der Stadt Braunschweig* 1789

Culen-Burgh, Jacob. *Rhynberck,* published Visscher 1633

Cullingworth, J. Publisher and Stationer, Natal Star Office, Durban. *Colony of Port Natal* 1862 (with Masser)

Cullingworth, William. *Plan Stonely, Hunts.* 1769

Culver, John P. Civil Engineer and surveyor. *Covington, Nebraska* 1857 (with Betts), *St. Helena, Nebraska* (1858)

Cumberland, Capt. Charles Edward. *Sebastopol* 1854–5 (1858)

Cumings, Samuel. *Western Navigator* 1822, *Western Pilot* 1825 editions to 1854

Cumming, G. W. *Banda Island* 1818

Cumming, John. Publisher. 16 Lower Ormond Quay, Dublin. *Travellers Guide through Ireland* 1815, *New General Atlas* 1817 (with Thomson), *Atlas of Scotland* 1832

Cumming, William. *Plan Inverness* 1802

Cummings, Benjamin F. *Salmon River* 1856 MS.

Cummings, Jacob Abbot (1773–1820) American geographer. *Ancient and Modern Geography* 1813, *School Atlas* (1817)

Cummings, J. A. J. *Colombia and Montour Counties, Pennsylvania* 1860

Cummings, J. B. *Portland, Maine* 1854

Cundall, H. J. *Prince Edward Island* 1851, 1861

Cundee, James (later J. and J.) Publisher, Albion Press. *Plan London* 1808, Maps for Cole and Roper 1810, Maps for *Robins' Atlas of England and Wales* 1812–14, for Dugdale 1912–15. Succeeded by J. Robbins & Co.

Cunerlinus, J. See **Cimerlini**, G.

Cuningham, W. See **Cunningham**

Cuninghame, William. *Plan of Port Royal Ratan* 1743

Cunningham, Allan (1791–1839) Botanist

& explorer. *Geographical Memoirs on New South Wales* 1825

Cunningham, George Godfrey. *Gazetteer of World* 1850–56, *Bell's System of Geography* 1860

Cunningham, P. A. *Auburn, New York* 1871, *Howard County, Indiana* 1875

Cunningham, R. J. Hay. Geological surveyor. *Lothians* 1838, *Sutherland* 1841, *Stewarty of Kirkcudbright* 1843

Cun[n]ingham, William. *Cosmographical Glasse* (with *Plan Norwich*) 1559, oldest printed British town plan

Cupet, Pierre Paul. *Carte de l'Indo-Chine* 4 sh. 1893

Curado, Paiva. *Bahia do Turrafaf* (N. Atlantic) 1890

Currier, Nathaniel (1813–1888) Lithographer 148 Nassau, corner of Spruce Street. *Northern Part Illinois* 1836, *Ann Arbour,* (Michigan) 1836, *Surveyed part of Wisconsin* 1838

Currey, Edward. *Witwatersrand Gold Fields* 1892

Curtice & Stateler. *Map of St. Paul, West St. Paul and Minneapolis* (Minnesota) 1876

Curtis, Cockburn. *Navigation Yarmouth to Norwich* 1846

Curtis, H. K. American surveyor 1834–39

Curtis, Samuel R. *Des Moines River* 1849

Curtis, W. J. Engineer. *Map Island of Trinidad* 1845, *London Railway* (1861)

Curtius, Ernst (1814–1896) *Athens* 1868, *Atlas von Athens* 1878, *Attika* 1881–94

Curtius, Ludovicus. Plan *Pavia* (1600)

Cusa[nus] [Kusa, Koebs, Cryftz, Khrypffs] Cardinal Nicolas (1401–1464) Cartographer. First modern *map of Germany* 1491, used in Nuremberg Chronicle

Cushee, E. (fl. 1729–60) Globemaker and publisher in conjunction with Senex and

Bowles. Worked with Thomas Wright. *Surrey* 1729, *Globe* (1750)

Cushee, L. (18th century) London Globemaker

Cushee, Richard. Surveyor and globemaker of London. *Manor of Ruckhott, Essex* 1728 MS, *Berden* 1732

Cushing, S. B. *Map City of Providence* 1849 (with H. F. Walling)

Cusi, Giuseppe. *Provincia di Sondrio* 1825

Cusinus [Cousin] Hugo [Hugues] *Burgundy* ca. 1580. Engraved for Ortelius 1589

Cuspinianus [Crispinianus, Spiesshaimer, Spiesshamer] Johannes (1473–1529) Diplomat, physician and geographer. Born Schweinfurt, died Vienna. Corrected *Apian's Hungary* 1528, *Danube, Plan of Vienna* (both lost), *Austria* engraved 1506

Cust, Robert Needham. *Language map of Africa* 1884, *Missionary Map of Africa* 1891

Custance, William. Surveyor and publisher. *Plan Cambridge* 1798, *Estate plans* 1808

Custo[[di]s, David. Engraver. *Hungary, Austria* (1610), *Prussia* 1627

Custo[di]s, Raphael (ca. 1590–1651) Engraver and publisher, brother of David above. *Hunter's Swabia* 1619, *Apian's Bavaria* 1632

Custos, Dominikus (1560–1612) Engraver in Augsburg

Custos, Jacob. *Posthauss zu Augspurg* 1628

Cutfield, Commander W. Admiralty Surveyor. *Africa* 1821–6 (with Commander Vidal)

Cuthbertson, Walter R. *New Guinea* 1887

Cutler, Manasseh (1742–1823) Geologist and botanist in Massachusetts. *Western Pennsylvania to Scioto River* 1787, the first map of *Ohio*

Cutler, Nathaniel. *Coasting Pilot* 1728, *Atlas Maritimus et Commercialis* 1728

Cuvier, G. L. C. F. D. Baron De. *Carte géognostique des Environs de Paris* 1810, *Carte géologique des Environs de Paris* 1865

Cuyk, P. (1720–1787) Engraver of the Hague. *Texel* 1780

Cuylenburg, John van. *Island of Singapore* 1898

Cuynat, C. S. (1774–1853) French geographer. *Topographie de Barcelone* 1841, *Topographie des Asturias* 1849

Cylov, Nikolay Ivanovic (1801–1879) Military cartographer, of St. Petersburg.

Cysat[us] Johann Leopold (1601–1663) Historian of Lucerne. *Lake Lucerne* 1645 (1661)

Cysat, Renwart. *Japan,* Freyburg by Gemperlin 1586

Czajkowski, Franciszek (1742–1821) of Warsaw. Historian and cartographer. *Palatinate of Sandomierz* 1786 MS.

Czaki, Ferensz Florian (d. 1772) Military cartographer. *Vistula* 1760, *Spisz* (Hungary) 1762

Czeel. See **Zeel**

Czermak, E. *Schlesien* (1869)

Czerny, Joseph (early 19th century) Cartographer and globemaker of Vienna

Czerwenka, J. *Plan der Stadt Mittweida* (1880)

Czetter, Samuel. Engraver for Korabinsky 1817

Czjzek, J. Geologist. *Manhard* 1849, *Wien* 1860

Czoernig van Czernhausen, Karl Joseph Freiherr von. Austrian geographer. *Oesterreichischen Monarchie* 1855

D

For majority of names with prefixes D',
Da, or De, see under terminal name.

D., S. *Plan de Pensacole* 1719

D. B. See **Bertelli, Donato**

D., C. See **Doedes, C**

D., S. *Plan de Pensacole* 1719

D., J. See **Davies, John**

D.V.T.Y. See **Avity**

D'Abbeville, S. See **Sanson d'Abbeville**

Dabenzara, J. See **Debenzara**

Dabey, J. See **Deventer**

Da Bisticci, V. See **Vespasiano da Bisticci**

D'Abla[i]ncourt, N. P. See **Perot
d'Abla[i]ncourt**

Dablon, Claude (1618/19–1697) Born
Dieppe, Jesuit missionary in Canada.
*Relation . . . en Nouvelle France 1671 &
1672*, Paris 1673

Da Cadamosto. See **Cadamosto**

Da Carignano, G. See **Carignano**

Da Corte, C. & D. See **Corte**

Da Costa, A. Carvalho. See **Carvalho
da Costa**

Da Costa, E. & J. See **Costa**

D'Acosta, José. See **Acosta**

Da Costa e Miranda, J. See **Costa e
Miranda**

D'Acuna. See **Acuna**

Daddow, S. Harries. *Anthracite fields of
Pennsylvania* (1868)

Dade, Thomas. Land surveyor.
Rattesden (Suffolk) 1704

Daetz, C. See **Doedes**

Dagelet, M. *La Côte du Nord-Ouest de
l'Amérique* 1797 (in Lapérouse)

Dagen, Hendrik Rochusz. van (b. 1634).
Engraver, born Amsterdam. Worked
for Blaeu

Dagg, G. A. de M. E. *Devia Hibernia* 1893

Daggett and Ely. *Connecticut* 1858

D'Agnolo, J. See **Agnolo**

Dagustine, M. de Combis. See **Combis
Daugustine**

Dahl, Peter M. *Plat Book Hennepin &
Ramsey Counties, Minnesota* 1898

Dahlberg[h] [Dalberck], Count Erik
Jönsson (1625–1703). Military engineer
and cartographer of Stockholm. *Funen
1657, Bogense 1658, Denmark 1660,
Suecia Antiqua 1667–1716, Atlas of
Sweden 1698* (presented to Charles XII)

Dahll, Tellef (b. 1825) Norwegian
geologist. *Geolog. maps Norway*
1858–79

Dahlman, Carl Edward (1828–1900)
Swedish cartographer. *Frövi-Fala
Jernbanan 1867 Stockholm 1870,
Sverige 1876*

Dahlstierna [Eurelius] , Gunnar af (1661–1709). Swedish mathematician and surveyor of Stralsund

Dahse, Paul. *Goldküste* 1881

D'Ailly, P. See **Ailly**

Daines, Isaac C. Surveyor, 20 St. Swithin's Lane, London. *County Courts Districts* (ca. 1846)

Daintree, Richard (1831 1878). English geologist, specialized in mines of Australia

Dajot. *Carte Routière de Seine et Marne* 1872

D'Albe, L. A. G. B. See **Bacler d'Albe**

D'Albeca, A. L. See **Albeca**

Dalberck, E. J. See **Dahlbergh**

D'Albertis, L. M. See **Albertis**

Dal Buono, F. See **Buono**

Dalby, Isaac (1744–1824) Prof. of Mathematics at Woolwich; assisted Roy & Mudge in triangulation of England 1787–96

Dal Cortivo, N. See **Cortivo**

Dale, Lieut. R. *Jordan & Dead Sea* (1849)

Dalekovic, Feliks. *Political map Servia* 1878

Dalen, Cornelis van (1602–1665) Engraver of Amsterdam. Roberts' *Merchants Map* 1638, and for *Vegetius' De re militari* (1645)

Dall, Nicholas Thomas (d. 1777) Danish landscape painter; settled London ca. 1760. Designed cartouches for *Jefferys' York* 1771–2

Dall, P. Master, R. N. *Leith Harbour* 1831

Dall, William Healey (1845–1927) Geographer, naturalist and explorer of Boston. *Alaska* 1869, 1884, *Indian tribes of Washington County* 1876, *Pacific coast Pilot* 1879, 1883

Dallas, W. L. *Bay of Bengal* 1886

Dalla Vedova, Giuseppe (b. 1834) Geographer, born Padua

Dallinger, J. Engraver. *Millard & Manning's Plan Norwich* 1830

Dallinger, Robert. Estate surveyor in Essex. *Rivenhall* 1773 MS, *Wickham Bishops* 1775

Dalli Sonetti, B. See **Bartolomeo dalli Sonetti**

Dallington, Sir Robert. Survey *State Tuscany* 1596 MS, printed 1605

Dallman, Capt. Ed (d. 1896) German seaman and explorer. *Greenland* 1873–4, *Sea of Kara* 1878, *New Guinea* 1884–5

Dall'Olmo, R. See **Olmo**

Dally, A. *Régions milit. de la France* (1874)

Dally, N. *Amérique Centrale et les Antilles* Brussels, 1840

D'Almada, A. See **Almada**

Dalmas, J. B. *Geolog. map Ardèche* 1859

Dalmatino, Natale. See **Bonifacio**

D'Almeida, C. M. & M. See **Almeida**

Dalmont, Isidore. Letter engraver. *Paris Historique* (1865), & for Migeon (1874)

Dalorto [**Dell'Orto**] , Angel [lin] o (fl. 1339). Catalan or Italian portolan maker. *La carta nautica* 1325, *World,* Genoa 1330. Probably identical with Dulcert, Angelino, q.v.

Dal Re, M. A. See **Re**

Dalrymple, Alexander (1737–1808) Writer to E. India Co. at Madras 1752–62, 1st Hydrographer to E. India Co. 1779, 1st Hydrographer to Admiralty 1795–1808. Established Soho Square 1771, later High Street, Marylebone (1806). MS. surveys *E. Indies* 1758–61. *Part Borneo & Sooloo* 1761–4, *Discoveries in S. Pacific Ocean before 1764,* 1767. *Bay of Bengal* 1772, *Collection of Charts* 1774, *Formosa* 1792 &c., *Town Cheltenham* 1806

Dalrymple, G. Elphinston (d. 1876) Australian explorer. *Narrative & Reports Queensland N.E. Exped*

Dalrymple, Capt. John (d. 1766) Scottish officer. Organized publication of revised *Fry & Jefferson's Virginia* 1755

D'Altischofen, P. See **Pfyffer von Altischofen**

Dalton, I. *Level of Ancholme, Lincs.* 1791(with E. Johnson)

Dalton, J. *Coti River, Borneo* 1828–(37)

Dalton, William Hugh. Improved *Walpoole's English Traveller* 1794

Dal Trozzo, A. Publisher. *Poland,* Warsaw 1810

Daly, D. D. *River Linghy* 1875, *Part of Malay Peninsula* 1876, *N. Perak* 1882, *Brit. N. Borneo* 1888

Dalyell, Sir R. *Vilayet of Danube* 1869

Dalzel, Capt. Archibald. *Gato or Regio Creek* 1785, *Coast of Africa* 1797 & '99, *Plan Loanda St. Paul* 1797, *Atlantic* 1802

Da Massaio, P. See **Massaio**

Dameld, J. See **Daniell**

Damell, Giovanni. *Charts Atlantic & W. Europe* 1637–9 MSS.

D'Amendale. See **Damme**, J. van

Damerum, William. *Southern New York* 1815

Damianus. See **Schoonebeck**, P. D.

Damien de Templeux. See **Templeux**

Damilano, J. *Plan Alexandria* 1887

Damme, Alfred (d. 1892). German cartographer

Damme, Jan van, Sr. d'Armendale [Aumale]. Dutch cartographer. Géographe du Roy de France 1634. *Charolois,* Hondius 1633, used also by Blaeu, Tavernier &c.

Dammert, I. C. *Plan Pirmont* 1790

Damoreau. See **Smith & Damoreau**

Da Mosto, Alvise & Luigi Ca'. See **Cadamosto**

Dampier, Capt. William (1652–1715). Seaman, hydrographer and circum-

navigator. *New Voyage around the World* 1697–1709, *Sharks Bay* 1701

Damrong, Prince Rajanubhad. *Siam,* Calcutta 1897

Dana, James Dwight (1813–1895) American naturalist and explorer. Accompanied Wilkes as geologist 1838–42 & published manuals of geology and geography

D'Anania, G. L. See **Anania**

Danby, Thomas. Engraver. Drake's *Plan Cincinnati* 1815

Dance, C. Architect. *Plan Excise Office* 1768, *Plan Newgate* 1780

Dance, George, the elder (1700–1768). Architect and surveyor to Corporation of London. *Plans drains, conduits Marylebone* 1746

Dance, George, the younger (1741–1825). Architect and City Surveyor in London. *Port of London* 1799–1803, *River Thames* 1803

Dance, N. *Plan Fleet Market, London* 1737

Dancker[t]s [Danckaerts, Danquerts]. Family of Amsterdam publishers and engravers with extensive ramifications. To be distinguished from Danckerts de Rij

—— Cornelis (1561–1631). Printseller in Antwerp, born Amsterdam

—— Cornelis, the elder (1603–1656). Engraver and publisher in Amsterdam at the Sign of the Atlas (1) Kalverstraet, (2) Nieuwendijk. First cousin of C. Danckerts de Rij. Succeeded father above 1631. Engraved for Speed 1627. *Charte Universelle* 1628 (with Tavernier), Ryther's *Plan London* (1633), Bertius' *Asia* (1640), *Europe* 1643, *Danube* 1647, *View London* 1647, *7 United Provs.* 1651, *Holland* 1656

—— Cornelis, the younger (1664–1717). Engraver and publisher in Amsterdam; son of Justus the elder. *Fortresse van Napels* (1680), *World* (1690). Business continued by widow until her death 1719

—— Danker [Danckert] (1634—1666) Publisher in Amsterdam; son of Cornelis the elder. *N. Holland* 1657, *Denmark* 1657, *Germany* 1661, *Nova Totius Terrarum Orbis Tabula* (1660)

—— Hendrik. *West Indische Paskaert* 1659, *E. Indies* 1660, *Nieuwe Stuurmans Zee Spiegel,* 1704

—— Just[in]us (1635—1701). Engraver, mapseller and publisher in the Kalverstraet, Amsterdam; son of Cornelis the elder. *Hungary* 1662, published atlases ca. 1670—1700 with his sons; also made globes

—— Justus, the younger (d. 1692). Engraver and publisher in Amsterdam; son of Justus the elder

—— Theodore, the elder (1640—1690). Engraver and publisher in Amsterdam

Dancker[t]s, Theodore, the younger (1663—ca. 1727). Son of Justus the elder

Danckerts de Rij, Cornelis (1596—1662) First cousin of C. Danckerts, the elder. Surveyor of Amsterdam. *Plan Grol,* Visscher 1627. *Kaertboeck . . . stadt Amsterdam* 1642—3

Danckwerth, Caspar (d. 1672). *Schleswig-Holstein* 1652 (with J. Mejer of Husum)

Danckwerth, F. See **Dankworth**

Danckwerth, Joachim. Collaborated with brother Caspar above

Dancy, Francis L. Surveyor General. *Portsmouth, Ohio, to Linville, N. Carolina* 1836, *State of Florida* 1860

Dandalle. *Map France* 22 sh. Paris, Picquet 1822

Dandeleux, H. Engraver. *Plan Dantzig* (1813), for Saint-Cyr (1821), Dufour 1822, Brué 1822, Lorrain (1836)

Danes, Richard. Estate surveyor. *Epping* 1634 MS.

Danet, Guillaume. Publisher sur le Pont N. Dame a la Sphère Royale, Paris. Son-in-law and part successor of De Fer (with Bernard) 1760, reissued his maps. *India* 1721, *Santo Domingo* 1723, *Paris* 1724, *Gibraltar* 1727, *London* 1727, *Roussillon* 1730, *L'Amérique* 1731, 1740, *Africa* 1732

Danet, Thomas. Translated Guicciardini into English 1591

Dangeau, Louis de Courcillon, Abbé de (1643—1723). Cartographer of Paris

D'Anghiera [D'Angleria]. See **Martyr,** P.

Daniel, H. Ad. (1812—1871). German geographer. Manuals of geography 1845—59

Daniel, J. See **Daniell**

Daniel, Richard. *Compendium of useful hands* 1664, *English Empire in America* 1679

Daniel, R. W. T. *Mississippi* 1857

Daniel[l] [Dameld]. John (fl. 1642) Chartmaker, in St. Katherines near unto the iron gate by the Tower of London. *Chart Spitsbergen* 1612 for Muscovy Co., spirited to Holland & published by Hessel Gerritsz. *Cart W. coast Europe . . . E. coast S. America* 1614, *W. America* 1637 MSS.

Danielow, Ivan. *Empire de Russie* 12 sh., Vienna 1812, *Empire Ottoman* 1813

Danielson, J. *Mellersta* 1880

Danit. Printer Amsterdam, inde Werreltcaert. *Barents' Polar Regions* (1599) in Linschoten

Danizy, A. See **Danysi**

Dankerts. See **Dancker[t]s**

Dankworth, F. American engraver, for Tanner 1833, 1844, *World* 1838, & for Mitchell 1834—49

Danquerts. See **Dancker[t]s**

Danti [Dante], Egnazio [Ignazio, Pellegrino Danti de Rinaldi] (1536—1586) b. Perugia, d. Rome. Astronomer, monk, Papal Cosmographer 1580, professor Univ. of Bologna, Bishop of Alatri. Designed *fresco maps* in *Palazzo Vecchio,*

Florence (1563–75) & *Galleria del Belvedere,* Vatican (1580–3), *globes* 1567, first work on astrolabes in Italian 1568, *Brit, Isles* 1570, *Use of Globes* 1573, *Arabia* 1575, *Perugia* 1580 on wall of *Palazzo del Governato,* Bologna. *Orvieto* 1583

Danti, Pietro Vincenzo di Bartolomeo Rainaldi (ca. 1430–1513). Mathematician and globe-maker of Perugia

D'Anville, J. B. B. See **Anville, D'**

Danysi [Danizy], Aug. H. (1698–1777). French astronomer and cartographer, Avignon, d. Montpellier

Danz, C. F. *Schmalkalden* 1848 (with Fuchs, C. F.)

Danziger, F. E. dei. Assisted L. H. Mitchell's expedition Abyssinia 1876–7

Dapper, Dr. Olivier [Olfert] (1636–1689). Doctor and geographer of Amsterdam. Translated works of Montanus into German. *Mesopotamia* 1661, *Africa* 1668, *China* 1670, *Persia* 1672, *Asia* 1672, *America* 1673, *African Is.* 1676, *Syria & Palestine* 1677, *Archipelago* 1688, *Morea* 1688

D'Après de Mannevillette, J. B. N. D. See **Après de Mannevillette**

Darbamont (fl. 1700). French cartographer. *Toul & surrounding country*

Darbishire, B. V. *Asia Minor* 1895, *Explorations of MacGregor* 1896

Darby, Henry William. *London Guide* 1842, *Plan Cheltenham* 1843

Darby, John. Publisher and printer in Bartholomew Close, London. For Seller 1669. Seller's *English Pilot* 1671–7, *Atlas Maritimus* 1675, Senex's *Atlas* 1720, *Atlas Maritimus et Commerc.* 1728, Camden's *Britannia* (1730)

Darby, M. Engraver. *Grundy's Plan E. Fen* (1744)

Darby, William (1775–1854). Geographer of Hanover, Pa. *Louisiana,* Phila. 1816, *Plan Pittsburgh* 1817, chart *Mobile* 1818, *U.S.A.* 1818, *Florida* 1821

Darbyshire, George C. *Railway Geelong to Ballaarat* 1857

Darbyshire, John. *Township of Gooramadda* 1859

Dare, Nathaniel B. *Cleveland* 1868

Daret, [Darret], Pierre (1610–1675). Engraver in ordinary to the King, sur le Quai des Gesures, Paris. Worked for Boissevin

Darinel. See **Boileau de Bouillon**

D'Aristotile. See **Aristotile**

Darley, J. M. (d. 1898). Cartographer, in American Geog. Society

Darling, Orlando. Surveyor. *Fort Scott. Bourbon County* (1860)

Darly, P. *Peregrinatio Israelitarum* 1621

Darme. Engraver. *Oakley's Plan Metz* 1754

D'Armendale. See **Damme**, J. van

Darmentier, L. I. Maps for de Fer

Darmet. Publisher in Paris. *Plano Topo. ciudad Buenos Ayres* 1817, *L'Amérique Méridionale* 1825

Darmet, J. M. *Napoleonic Battle Plans* 1820–4, *S. America* 1825, *Plan Lyon* 1830, *Alger* 1836, *Colombie* (1840)

Darmstadt, Johann Adolph. Engraver. *Plan Ratzeburg* 1799

D'Arnoullet, B. See **Arnoullet**

Darondeau, Benoit-Henri (1805–1869). French Hydrographer

Darré, C. *Plan d'Elbeuf* 1866

Darre, Nils Stockfleth (1765–1809). Scandinavian geographer. Surveys 1799 onwards & Director Topo. Survey from 1806

Darret, P. See **Daret**

Darsy, Eugène. *Atlas de Géographie Physique, Pol. et Hist.,* Paris (1890)

Dartois, Jean. *Mappemonde* 1750

Darton, John Maw. See under **Wm. Darton Sr.**

Darton, L. Engraver. *View of York* on *Tuke's York* 1787

Darton, N. H. *Dakota artesian basin*

Darton, Thomas. Printseller and publisher, 25 Great Surrey St., near Blackfriars Bridge, London. With William Jr. (1807–10), then on own. *Strangers' Guide* 1808, Rowe's *Plan London* 1811

Darton, William. Engraver in Tottenham, for Dilly & Robinson 1785, *Yorks,* 1787, *New Miniature Atlas* 1820

Darton, William Senior [& Successors] Engravers and publishers, map, print, and chart warehouse, 40 Holborn Hill, London (to 1806), then 58 Holborn Hill. Succession of business: (1) Wm. Sr. & Jr. together, (2) Wm. Jr. alone [1806–7], (3) Wm. Jr. & Thomas [1807–10], (4) Wm. Jr. on own again [1810–34], (5) Wm. & Son [1835–7], (6) Darton & Clark [1838], (7) John Maw Darton [1841], (8) Darton & Co. [1851], Dix's *Juvenile Atlas* 1811, *Union Atlas* 1812, *London* 1814, *S. America* 1820, *New Miniature Atlas Eng. & Wales* 1820, *Atlas Eng. Counties* 1822 (with Dix), republished *Wilkinson's General Atlas* 1826, *Darby's London Guide* 1841, *12 miles round London* 1843, *Atlas* 1871

Darton, Wm. & Clark. See **Darton, Wm.**, above

Darton, Wm., & Harvey, Josiah [later & Darton; later Darton, Harvey & Co.] (fl. 1843). Printers and printsellers, Gracechurch St., London. *Plans London* 1792, [1814], *Atlas to Walker's Geog.* 1802, *Geog. present* 1817, *Walker's Universal Atlas* 1820

Das, N. C. *Ancient India* [1899]

Das, Sarat Chandra. *Routes Tibet* 1882–(3)

Dasauville, W. See **Dassauville**

Daser, P. *Plan Tellichery,* pub. Dalrymple 1780

Dash, T. A. *London Parks* 1862–3, *Brompton Cemetery* 1884

Dashiell, S. L. *N. America* ca. 1800, *N. part State of Maine,* Washington 1830

Dasori, Giovanni Batt. Engraver. *Pontine Marshes* 1795

Dassauville [Dasauville] William. Engraver of Edinburgh. *Scotland* 1805, engd. for Thomson 1815–32, *Johnson's Peebles* 1821, *Western Isles* 1822–3

Dassie, F. *Routier des Indes,* Paris 1677. *Description générale des costes de l'Amérique* Rouen 1677, *Le pilote expert* 1683

Dasypodius [Hasenfratz], Conrad (1529/30–1600) Swiss cartographer, Professor of Mathematics. Strasbourg. *Globes* 1574 in Strasbourg clock

Datan, E. *Boras* 1728

Dati, Gregorio di Anastasio [Gorio di Stagio] (1363–1436). Brother of Leonardo. Author of *La Sfera* ca. 1400

Dati, Leonardo (ca. 1360–ca. 1425). Cosmographer and poet of Florence. *La Sfera* sometimes attributed to him–see Gregorio above

Dau, Friedrich von (1796–1867) German cartographer

D'Aubant, A. *Jersey* 1737, *Goat Island* (1778), *Fort Brown* 1799, *Environs of Newport, Rhode Island* 1779

Daubenton, P. P. Jesuit. *Islas Carolinas* 1757

D'Aubert, B. See **Aubert**

Daublebsky, Moritz Freiherr von (1884–1917) Austrian general and cartographer

Daubree, Auguste (1814–1896) French geologist. *Eaux souterraines* 1887, *Régions invisibles du globe* 1888

Daudet. Map seller, Rue Mercière, Lyons. For Bailleul le Jeune from 1748, *Africa* 1752

—— Veuve. Widow and successor of above. *Plan Lyons* 1767 (with Joubert)

Daudet, Chev. Louis Pierre. Royal Geographer and engraver (1722–44) b. Nîmes *Plan Rheims* 1722, *Chemins de France* 1724–68, *Route map Paris-Compiègne-Soissons* 1728, Chopy's *Lac de Genève* 1730

Daugustine, M. de Combis. See **Combis Daugustine**

Daulphinois, N. See **Nicolay**

Daumas, Lt.-Col. *N. Africa* 1845

Daumas, Rev. F. *Voyage Basutoland* 1836, *Missionary stations S. Africa* 1840, *Sketch Basutoland* 1868 MS.

Daumont. Publisher and mapseller, Rue de la Ferronnerie a l'aigle d'or and Rue St. Martin, Paris successor to Nolin. *Nolin's Amérique,* 1754, 1759. For Mayer 1757, *Nolin's British Isles* 1773

Daussy, Pierre (1792–1860). French engineer and astronomer, Directeur d'Hydrographie, Dépôt de la Marine. *Côte Occid. d'Afrique* 1833, *Océan Atlantique Méridionale et Septentrionale* 1834–66

D'Auteroche, Chappé. See **Chappé d'Auteroche**

Dauthendey [Dauthendeij], Caspar. Architect and mathematician. *Brunswick, Blaeu* 1640

Dauty & Rovet. Publishers, *Charle's Atlas* 1833

D'Auvergne, Philip, Prince de Bouillon (1754–1816) *Plan Fairhaven* 1744

D'Aux, Chevalier. *Plan Boston* 1693

Davenet, Capt. *Communications in Crimea* 1855

Davenport. See **Dewitt & Davenport**

Davenport, S. Engraver. *Quebec* 1835

Davenport & Strippelman. *Underground Chicago* 1874

Daventer[iu]s, [Daventria]. See **Deventer**

Da Verrazano. See **Verrazano**

Davet. Engraver. *St. Honoré de Lérin* 1635

Davey. Surveyor. *Estates at Weston* 1774

Davey, J. M. Land surveyor of Canterbury. *Staple & Ash* ca. 1830

Davey, N. T. Surveyor. *Cachar* 1870, 1877

Davey, P. Bookseller. *Osborne's Geog. Magnae Brit.* 1748, *Kitchen's Geog. Scotiae* 1756, *Salmon's Geog. Grammar* 1758

David, Johann, elder (1796–1846) Cartographer and lithographer of Vienna. *Tyrol* [1823], for Meyer ca. 1835

David, M. Cartographer. *Hannover* (1803)

David, Martin Alois (1757–1836) Mathematician and astronomer of Prague. *Bohemia* 1800 and 1834

David, Pierre. Publisher, sur le Pont Neuf devant la Samaritaine, Paris. *Voyage aux Indes Orient.* 1646

David, R. *English Empire in America,* Morden (1679)

David, T. *Wien* 1823

David, Sir Tannatt William Edgeworth (1858–1934). Geologist, b. Cardiff. *Geolog. map Vegetable Creek, N.S.W.,* Sydney 1885, *Tingha & Elsmore Districts* 1895

Davidos, C. F. *Spanien und Portugal,* Vienna 1820

Davidson, A. F. *Lower Platte River* 1845 MS.

Davidson, C. Wright. *Atlas of Minneapolis* 1887

Davidson, George (1825–1911). American astronomer and hydrographer of San Francisco, b. Nottingham, England. *Coast Pilot of Calif. & Oregon* 1857, *Coast Pilot Alaska* 1868

Davidson, James. Partner in 1792 with Mount [Mount & Davidson], on his own from 1800. *Collins' Coasting Pilot* 1792, *Mediterranean Pilot* 1795. See also **Mount & Page**

Davidson, Capt. John Wynn, of Dragoons (1823–1881). *Ft. Tejon to Utah* 1859 MS.

Davidson, R. *Leamington Spa* 1876

Davidson, Thomas (b. 1817). English geologist. *Illust. Silurian life* 1868

Davidson, Walker Rennie. Surveyor General N.S.W. *N.S.W.* in *Gazetteer* 1866

Davidson, William. Engineer. *Watts River Scheme, Melbourne* 1885

Davidszoon, Dirck [Davidsz Dirck] Charts for Goos 1666 and Colom 1669

Davie. Publisher. *Atlas* 1863

Davies, B. *River Loddon* (1769)

Davies, Benjamin. *City and suburbs of Philadelphia* 1794

Davies, Benjamin Rees. London engraver and publisher (1) 30 Goodge St. [1811], (2) 10 Compton St., Brunswick Sq. [1813], (3) 34 Compton St., (4) 16 George St., Euston Sq. 1845. *Cadiz* 1811, *Arrowsmith's Bible Atlas* 1835, *London & environs* 1841, 1875, 1879; drew for Weekly Dispatch 1858–60, for Kelly 1860

Davies, Griffith. Estate surveyor. *Harwich* 1745 MS.

Davies, Henry. *Cheltenham* 1834, *Plan Luton* 1842

Davies, John, of Kidwelly (ca. 1627–1693). Translator, Mandelslo's *voyages,* maps of *India, Livonia & Volga* 1662

Davies, J. *Country round Bath* (1865)

Davies, Richard. *Birmingham* 1858 See also **Davis**

Davies, Thomas (ca. 1712–1785). Actor and bookseller fl. 1736–46 & from 1762, 8 Russell St., Covent Gdn., London. *Camden's Britannia* 1772

Davies, Thomas, Lieut. General (ca. 1737–1812) *Draught of the River St. Lawrence* (1760)

Davies, Thomas. *Atlas* 1859 (with T. Nelson)

Davies, W. Augustus. *Aston Manor* (1895)

Davies, Wm. Henry (fl. 1858–60) Land surveyor of Abingdon, Berks.

Davies & Bryer [Davies & Co.] Engravers. *Admiralty charts W. Indies* 1867, *Suez* 1873

Davies & Eldridge. Booksellers, Exeter. *Wallis' Pocket County Atlas* 1810

Davies & Powell. Engravers for Admiralty 1864

Davies & Tebb. Surveyors. *Ellesmere Canal* (1796)

Da Vignola, C. See **Cantelli da Vignola**

Da Vinchio, P. M. See **Vinchio**

Da Vinci, L. See **Vinci**

Davis, B. Printer of Philadelphia. *Scott's Atlas of U.S.* 1796

Davis, Charles (d. 1755). London bookseller (1) Fleet St., (2) Paternoster Row, (3) Holborn opp. Gray's Inn Gate. Succeeded by nephew Lockyer Davis, q.v. *Camden's Britannia* 1753

Davis, Admiral Charles Henry (1807–1877). American hydrographer, b. Boston, d. Washington. *American Nautical Almanac* 1849–62, *Interoceanic Railroads & Canals* 1867

Davis, Cornelius Butter (fl. 1845–55) Land surveyor E. Woodhay, Hants.

Davis, F. A. & Co. *Hist. atlases Berks County, Pa. & Rockland County, N.Y.* 1876

Davis, Hon. Jefferson (1808–1889) President Confederate States U.S. *Route Pacific Railroad* (to accompany Report) 1855

Davis [Davys], John (ca. 1550–1605) b. Standridge. Voyages to discover N.W. passage 1585–87. Assisted with Molyneux *globe.* Invented backstaff *The World's Hydrographical Description* 1595

Davis, John. Printer of Philadelphia. *Delaware Bay* 1756 (with J. Fisher), *Scull's Improved part of Pennsylvania* 1759

Davis, J. Publisher, 36 Essex St., Strand, London. *Ancient Egypt* 1814

Davis, John. Estate surveyor. *Bloxham, Oxon.* 1817

Davis, John. *Eastbourne* (1897)

Davis, Sir J. F. *Is. Chusan* 1853

Davis, John Lambe. Estate surveyor. *White end farm Chesham* 1760 MS., *Plans Chesham* 1762, estate maps *Herts.* 1770–6 MSS.

Davis, Lockyer (1719–1791). Bookseller and publisher, Holborn, opp. Grays Inn Gate, succeeded uncle Charles Davis, q.v., 1755. *Camden's Britannia* 1772

Davis, O. F. *Kearny City, Neb.* (1857), *Dakota County, Neb.* (1860)

Davis [Davies], Richard (fl. 1708) Publisher at The Three Ink Bottles in Castle Alley near the Royal Exchange, London. *Theat. Hist.* 1709, *Seller's Middx.* 1710, *Moll's N. America* (1720)

Davis [Davies], Richard (fl. 1784–1814) of Lewknor, Oxfordsh. Farmer and agricultural writer. Topographer to His Majesty. *Oxfordshire,* 16 sh. 1797

Davis, Thomas W. *Plans Districts of Boston* 1868–73

Davis, U. J. *Gloucestershire* (1858)

Davis, William J. *Geography of Kentucky* (1881)

Davis, Wm. Morris (1850–1934) Geographer and geologist of Cambridge, Mass. *New England states* 1895

Davis, W. T. Engraver for Wilkinson 1797

Davis. See **McCarty & Davis**

Davison, Robert. *Oristano Bay* 1803

D'Avity. See **Avity**

Davy. *Atlas Mitchell County, Kansas* 1884 (with Gillen)

Davy, J. Chart *Tor Bay* 1761

Davy, Marie. *World Communications* 1868

Davys, J. See **Davis**

Daw, Edmund. Publisher, 114 Fetter Lane, London. *St. Pancras* 1849, *Parish St. Mary Abbotts* 1852, 1863

Daw, M. E. & Co. *British Isles* 1875

Dawes, William (1762–1836). Scientist, officer of marines and surveyor. *Port*

Jackson, Stockdale 1789, *Territory New South Wales,* Stockdale 1791–(2)

Dawes, W. Governor Sierra Leone. *Sierra Leone* 1794 on *Wilkinson's Africa* 1800

Dawes. Rev. Wm. Rutter (1799–1868). Astronomer. Star maps for Society for Diffusion of Useful Knowledge 1844

Dawkins, Henry (fl. 1753–80). Engraver *Coast of Louisiana* 1763, *Scull's Pennsylvania* 1770

Dawson. Schoolmaster. *Plan Stonehaven* 1818

Dawson, E. B. American engraver. Assisted and engd. for Tanner 1829–36

Dawson, Edward J. *Yukon Gold Fields* 1897

Dawson, George Mercier. Geologist. *Queen Charlotte Is.,* Gotha 1881. Maps *Canada* 1884–98

Dawson, John. Printer for 1627 & 1631 editions of Speed

Dawson, Joseph (fl. 1787). Estate surveyor in Essex. *Good Easter* 1754, *Danbury* 1756–8 MSS.

Dawson, J. *Nova Scotia & Pr. Edward Is.,* Pictou, N.S. 1847

Dawson, Sir John Wm. (1820–1899) *Geolog. map Canada* 1865

Dawson, Lieut. L. S., R.N. Hydrographer. *Fiji* 1876

Dawson, Robert. *Survey Flintshire* 1821–6 MS.

Dawson, Lieut. (later Lt.-Col.) Robert Kearsley, R.E. (1798–1861). *County maps and City plans Eng. & Wales,* Hansard 1832

Dawson, S. J. Explor. surveys *Lake Superior* 1859–70

Dawson, William and John Prior, Publishers No. 7 Paternoster Row, London. *Whyman's Leicestershire* 1779

Dawson Bros. Publishers, Montreal. *J. W. Dawson's Geolog. survey Canada* 1865, 1879

Dax, Paul (1503–1561). Cartographer of Innsbruck. *Region of Acherthal* 1545 (lost)

Day, Charles. Land surveyor. *Keyston, Hunts.* 1842

Day, Lieut. G. F. *River Paraguay* 1853, for Admiralty 1856

Day, [John] **& Sons.** Lithographers to the Queen. *Europe* for *Dispatch Atlas* 1857–63, *King's School Atlas* (1868)

Day, William. Surveyor and publisher of Blagdon. *Somerset,* 9 sh. 1782, 1800 (with Masters, C. H.)

Day, Wm. Printer and lithographer. *Plan Hull* (1830), for Royal Geographic Society from 1833, *Pocock's Globe* (1835)

Day & Haghe. Lithographers to the King/Queen, 17 Gate St., Lincoln Inn Fields, London. *Plan Abbey Cum Hir Estate* 1837, for Greenhow 1846

Daye, John (fl. 1577). Printer of London. Cunningham's *Cosmographical Glasse* 1559

Dayles, Matthew. Estate surveyor of Buttsbury. *Mayland* 1739, *Prittlewell* 1746 MSS.

Dayman, Charles Orchard. Revised Lubbock's star maps for Society for Diffusion Useful Knowledge (1860)

Dayman, Lieut. Joseph, R. N. Admiralty surveyor. *Gt. Sandy Is. Strait* 1847, *Cape Colony* 1852–6, *Port Natal* 1856, *Algoa Bay* 1857

Daynes, John. *Plan Manor Bloomsbury* 1664–5 MS.

Dayot [D'Ayot], Capt. Jean Marie. *Port Quinhou* 1793 pub. Dalrymple, *Gulf Siam* 1798, *Cochin China* 1798, *Saigon* 1799, *Amoy* 1805, pub. 1840

Dayton, A. W. *Frederich Co., Va.* 1855 (with J. M. Lathrop)

Dayton. See under Wentworth, H.

For majority of Names with prefix "De" see under terminal name

Deacon, Charles Wm. [later & Co.]. Publisher, Charing Cross Chambers, W.C. *Survey* 1878, *Yorks.* (1893)

Deacon, Thomas. Publisher, 15 Furnivals Inn, Holborn, London. *Environs London* 1831

De Aefferden, F. See **Aefferden**

Dean. *National Atlas* 1852 (with Gilmour), *Seat of war in Italy* (1859)

Dean. Chas. Youle. *Kimberley District, W. Australia* 1886

Dean, James (1777–1849). American publisher and cartographer. *Gazetteer of Vermont* 1808

Dean, James. Surveyor. *Manor Doddiscombe Sleigh, Devon* 1814

Dean, R. & W. & Co. Printers, publishers, and mapsellers, 33 Market street lane, Manchester (1811), later 80 Market St. Pigot's *Plans Manchester & Salford* 1804–1824

Dean & Munday. Publishers, Threadneedle St., London. *Smith's Syria* 1840

Dearborn, Benjamin (b. 1755) *Parts of Boston* 1814, *Plan Boston and Roxbury Mill Dam* 1834

Dearborn, W. L. Civil engineer. *Railroad St. Louis–San Francisco* 1849

Dearsley, W. Estate surveyor. *Estate plan Essex* 1819 MS.

Deas, James. *Firth of Clyde* used Reid 1890

Dease, Peter W. *Discoveries N. coast America* 1840 (with Simpson)

Deasy, Capt. H. H. P. *Tibet* 1896, *W. China & Tibet* 1897

Debar, J. H. D. *Oil region W. Virginia* [1864]

Debarsy. *Plan Namur* (1840) (with Leroy)

Debbeig, Lieut. (later General) Hugh (1731–1810). Engineer. Assisted survey Scotland 1749–51, *Plan St. Johns, Newfoundland,* 1762 MS.

De Beins, J. See **Beins**

Debembourg, Georges. *Atlas hist. du Rhône* 1862

Debenzara [Dabenzara], Jehu, of Alexandria (fl. 1480—1505) Charts of *Mediterranean* 1497, 1505 MSS.

Debes, Ernst (1840—1923). Publisher and cartographer of Leipzig. *Grossherzogthum Hessen* (1875), *Kleiner Schul Atlas* 1897 *Schul Atlas* 1893, *Neuer Handatlas* 1895 (with Wagner, H.)

Debes, Lucas Jacobson (1623—1676) Danish cartographer and naturalist, Pastor of Thorshaven, Faroes. *Faroer* engraved 1673

De Bhuyne, Komm. A. *Swedish & Dutch Arctic Expeditions* 1878, Stanford 1879

de Blosseville, Jules. *Carte de l'Ile Ika-na-Mauwi (New Zealand)* 1824

De Brahm, John Gerar William [later signed himself William Gerard] (1717—1799). Surveyor and engineer, Capt. of Engineers under Emp. Charles VI, born Holland, came to America 1751 and founded Bethany. Surveyor for Georgia 1754, Surveyor General 1757, Surv. Gen. of S. District of N. America 1764. Surveyed and fortified towns in S. Carolina 1755—7, *S. Car.* 1757, 1780, *Ticonderoga* 1759, *Georgia* (1760) MS., *Amelia I.* 1770, *Atlantic Pilot* 1771, *Report of the Gen. survey in the S. Dist. of N. Am.* (1772—8), many MS. maps, contributed to *Amer. Military Pocket Atlas* 1776, *S. Car. & Georgia* pub. Faden 1780

Debray, *Rep. Mexicana* 1867

Debrett, John (d. 1822). Publisher opp. Burlington House, Piccadilly, London. Succeeded J. Almon 1781, retired 1814. For Imray 1797, *Egypt* 1799, *Chart Indian Ocean* 1800, *Serres Little Sea Torch* 1801

Debrie, G. F. L. Engraver, 1730 edition of *Ptolemy*

De Bruls, Godhart. *Plan Niagara with adjacent country* 1762

De Bruyn, Henry. Land surveyor. *N. Fambridge* 1807 MS.

De Bry. See **Bry**

De Bure Globe. See **Paris Gilt Globe**

Decan. Printer, Rue Richer 19, Paris. *Arnaud's Calif.* 1849

Decembrio, Pietro Candido (1392—1477) b. Pavia, d. Milan. Cosmographer and litterateur

De Cespedes, A. G. See **Cespedes**

Dechen, Dr. Ernest Heinrich von (1800—1889). German geologist. *Rheinlaender* 1825, *Skye* 1829, *Deutschland* 1838, 1869, *Rhein u. Westfalen* 1855—65, *Wiesbaden* (1876)

Deckar, Capt. *I. of Barbuda*, Admiralty 1814

Decken, Baron C. von der. Maps of *Tanganyika* 1862

Decker, Carl. Engraver. *Bantam* (1700), for *Aa's Galérie Agréable* 1729, *Coylang* (1730)

Decker, Major Carl D. von (1784—1844). Prussian milit. cartographer. *Environs Berlin* 1816—19, *Batailles de Guerre de Sept Ans* 1839

Decker, Coenraet (fl. 1650—85). Engraver and draughtsman, Amsterdam. *Kennemerlandt ende W. Vrieslandt* (1650), *Delft* (1676), *Prague* (1679). Also worked for Jansson's heirs

Decker, J. A. *Ballon Géographique* 1889

Decker, P. *Attack on Gibralter* (1730)

Decluy. See **Le Gendre-Decluy**

Decomberousse [Comberousse], Joséphine. *Plan de Lyon* 1813, 1816

Decombes, Victor. *Atlas milit.* (1876—80)

De Costa, J. See **Costa**

Decouange. *Carte du Canada* (1711), *Cours de la Rivière d'Orange* (1750)

Dedenroth, H. von. *Lombardei & Venedig* (1859)

Dedier. *Poolo Condor,* pub. Dalrymple (1807)

Dedovich. Ingenieurs Oberstleutenant. *Austria* &c. 14 sh. 1803 MS.

Dee, Dr. John (1527–1608). Mathematician, friend of navigators & geographers; strong advocate for N.W. Passage. *N. America* (1580), *Artic Regions* ca. 1582 MSS.

Deeley, Wm. *Survey Edgbaston* 1701

Deering, Charles. *Plan Nottingham* 1751

Defehrt. Danish publisher. *Siaeland* 1768

Defehrt. Engraver. *Jersey* [1780]

De Fer, A. & N. See **Fer**

De Fer de la Nouerre. *Carte élém. navig. du royaume,* Paris 1787

Deferes, L. Lithographer of Ghent. *Amérique Septentrionale* 1844

Deforrest, J. H. Jr. *Gold Region* (N. America) 1859

Defremery, Ch. (1822–1883). *Fragments de géog. et d'hist. arabes et persanes* 1849

Deftedar-Bey (d. 1831) *Upper Nubia,* for expeditions of Ismael Pasha

Degaulle, J. B. (1732–1810). French hydrographer and mathematician. *Usage d'un nouveau calendrier perpetuel* 1768, *Usage compas azimuthal* 1779, *Neptune François* 1788

Degen, J. V. *Plan de la Ville de Vienne* (1830)

Degenhardt, O. *Oberschlesisch-Polnische Berg-district* 1871

Degli Uberti, L. See **Uberti**

De Goeje, M. J. (b. 1836). Dutch geographer and orientalist, translator of works of Arab geographers

De Graaff, I. See **Graaff**

Degrees. Sexagesimal system introduced by Babylonians

De Groot. *Washoe Mines, San Francisco*

1860, *Calif. mine surveys* 1863 MSS., *Nevada Territory* 1863

Deharmé. *Plan Paris* 1763

Deharme, L. F. Engraver for *Rocque's Berks* 1761

De Haven, Lieut. Edwin Jesse (1816–1865) American naval officer. *Wind chart Atlantic* 1851, *Amer. Arctic expedition* 1851, *Wellington Channel* 1852

De Hooghe, R. See **Hooghe**

Deichmann, Louis (fl. 1881–99) of Kassel. *Relief maps*

De Jode. See **Jode**

De Jonghe, C. See **Jonghe**

Dekinder, K. Engraver for Kiepert 1860

De Krafft, F. C. Surveyor of Washington, D. C. *Plan City Washington* 1828

De la Beche, H. T. See **Beche**

De la Bella. See **Domenico Machaneus**

De la Caille, N. L. See **Caille**

De la Cases, Comte. See **Las Cases**

Delachaux, Enrique. *Catamarca* 1893, *Prov. Buenos Aires* 1893, *Sud-Amérique* 1895, *Central Argentina* 1899

De la Cosa, J. See **Cosa**

Delacre, Louis. *Charleroi* 1875

De la Croix, A. P. Geographer to the King of France. *Wereld Beschryving* 1705

De la Croix, F. See **Croix**

Delacroix, Abbé Louis Antoine Nicolle. French geographer. Information used by De l'Isle, Maclot, Brion de la Tour. Engraver for Robert de Vaugondy 1750. *Nouvel Atlas Portatif* 1778

De la Croyère, D. See **L'Isle de la Croyère**

De Lacy, Walter Washington (1819–1892). Military engineer and surveyor of Helena, Mont. *Territory of Montana* 1865, 1870, 1879, *Idaho Territ.* 1866 MS.

De Laet, J. See **Laet**

Delafage, N. *Plan du passage de Ségre* (1646)

De La Feuille. See **La Feuille**

Delafield & Haven. American surveyors and engineers. *Eddytha Farm, Madison Co., Ill.* [1856]

Delafosse. Engraved cartouches for d'Anville 1751–5

Delafosse, Chatry. See **La Fosse**

Delafosse, Sr. Geographer. *L'Amérique,* Lyon 1771

De la Garde, J. *Terrest. globe* 1552

Delagrave, Charles (b. 1842). Publisher Geogr. Soc. Paris, 15 Rue Soufflot and 58 Rue des Ecoles. *Atlas de Géog. Mod. Phys. et Politique* (1870), *Brué's Atlas Univ. de Géog.* 1876, *Darsy's Atlas* (1890)

Delagrive, Abbé Jean (1689–1757). Mathematician, geographer and engraver of Paris. *Plan Paris* 1737, *Plan Versailles* 1746, reduced *Plan Berlin* 1749

Delaguette, Veuve. Printer, rue S. Jacques à l'Olivier Duschesne, Paris. *Isles Jersey et Guernesey* 1757

Delahaye, Guillaume Nicolas (1727–1802). Engraver and geographer. *l'Amérique méridionale* 1748, *Canada* 1755. Engd. for Robert de Vaugondy 1750–6, Dépôt de la Marine 1752–85, Buache 1753, Barbié du Bocage 1786–8.

Delahaye, Jean Baptiste (L'Ainé). Engraver for De l'Isle 1724, B. Jaillot 1721–5, d'Anville from 1727, & Robert de Vaugondy 1752–3

De la Haye, Mark W. Surveyor General, U.S. *Public Surveys Kansas and Nebraska* 1863

De la Hire, Philippe (1640–1718) French geodesist, determined astronomical points in France 1667–81 (with Picard). *France* 1682 (ascribed)

De la Houoe. See **Houoe**

Delaistre, H. *Empire Français* 1812

Delalain. Publisher Paris. *Atlas moderne* 1771 (with Lattré)

Delamain, Richard. *Making of horizontal quadrant* 1632

Delamarche, Alexandre (1815–1884). *Atlas de Géographie ancienne et moderne* 1850

Delamarche [Lamarche], Charles François (1740–1817) Geographer, publisher, globemaker. Successor to Robert de Vaugondy (1792) whose atlases he re-issued and to Fortin (1795). Rue du Foin St. Jacques au Collège de Mtre. Gervais, Paris (1778–1804) and Rue du Jardinet No. 13 vis à vis celle de l'Eperon (or Quartier St. André des Arts) (1805–7) *USA* 1785, *Globes* 1785–1800, *Les usages de la sphère et des globes* 1791, *Tableaux géographiques* 1794, *Institutions géographiques* 1795, *Atlas d'Etude* 1797, *World* and *Continents,* each 4 sh. 1805, *Atlas Elémentaire* 1820.

De la Marche fils, Felix. Ingenieur-Mécanicien pour les Globes et Sphères Rue du Jardinet No. 13. Son of above. *N. America* 1811, *America* 1818, partner with C. Dien 1819–20, *Atlas de géog.* 1829, *Atlas de géog. anc* 1839

Delamarche, François Alexandre. *France après la Guerre de 1870–1*

Delamarche. Engraver, St. André des Arts 45, Paris. For Escudero 1861

Delamare, F. Engraver, Rue de la Harpe 26, Paris. For Andriveau-Goujon 1850, *Genève* 1856, for Colton 1855

Delamarre, Casimir. *Carte linguistique, ethno. Europe Orient.* 1868

Delambre, Jean Baptiste Joseph (1749–1822). Parisian astronomer and geodesist, determined meridian of France. *Base du système métrique* 1810, *Hist. de l'astron.* 1817

Delamotte, Wm. Draughtsman. Wren's *plan City of London* 1800

Delapointe, F. Engraver. *Malte* (1650), Vivier's *Environs de Paris* 1674–(78)

Delaporte, Lieut. Louis, French navy. Charts *Indo-Chine* 1873

Delaram, Francis (fl. 1615–24) English surveyor.

Delaroche, Charles F. *Plan Alger* (1850) *Algérie* 1851, *Sahara Algérien* 1853

Delarochette, Louis Stanislas d'Arcy (1731–1802). Cartographer and engraver, associated with Faden. *N. America,* Bowles 1765, *Caribee Is.* 1768, *North America* 1781, *C. Good Hope* 1782, *Lower Egypt* 1802

Delarue, François. Publisher Paris for Vuillemin 1864

De la Rue, Philippe. *Terre Saincte* 1646, *Regnum Judaeorum* 1651, *Assyria Vetus* 1651

De Lasaux, P. *Diocese Canterbury* 1782

De Lat, J. See **Lat**

De la Tour, B. *Eng. possessions in N. America* 1778, 1780

De la Tour, C. See **Cagniard de la Tour**

De la Tour, L. See **Brion de la Tour**

De la Tour, Le Blond. See **Le Blond de la Tour**

De la Tour, Veuve (fl. 1764). Printer, Rue de la Harpe, Paris

Delaune. Engraver for Vivien St. Martin 1883–1900

Delaval, J. B. Publisher and map seller, Rue Geoffroy-Langevin No. 7, Paris. *Dépt. de Drome, Ardèche, Loire, Gard,* 1792, 1818; *Asie* 1819

De Lavaux, Alexander. Engineer. *Colonie of Provintie van Suriname* (1700), *Plan Chester* (1745)

Delavigne, N. *Chantilly* 1725

Delbalat, Jean Baptiste James. *Plan Bouches de Tigre* 1848, *Plan Port Gènes* 1853 pub. 1857

Delbos, Joseph. *Carte géolog. Haut Rhein* 1865

Del Cano, J. S. See **Cano**

Delcroz, François Joseph (1777–1865) French military cartographer in Florence

Delebecque, André (b. 1861). *Atlas des Lacs Français* (1895)

Delerue, Louis. *Plan Ville St. Omer* 1847

De l'Escluse, C. See **Clusius**

Delesquellen. *Fernando Noronha* 1735 MS., pub. Dalrymple

Delesse, Achille E. O. J. (1817–1881). French geologist and mineralogist, Director *Lithologie des fonds des Mers,* Rue de Géologie 1860–78. *Carte Hydrologique de Paris* 1858, hydro. charts depts. France 1862–73, *carte lith. des mers de l'Amérique N.* (1870)

De Leth, H. See **Leth**

Deletre. Engraver for Dufour 1857

Delfos, Abraham (1731–1820). Engraver of Leyden. Maps for *Van Mieris' Beschryving v. Leyden* 1762

Delgado, J. F. N. *Carta geog. de Portugal* 1876, 1899

Delight, E. *Boxley Farm, Maidstone* ca. 1760 MS.

De l'Isle, G. See **L'Isle**, Guillaume de

Delitsch, Prof. Otto (1821–1882) Engraver and geographer of Leipzig. *Europa* (1860), *Neuer Atlas* 1863, *Grosser Hand Atlas* 1871, *Schul Wandkarte* 1874

Delius, C. *Plan Berlin* 1868, 1874

Delius, Hermann. Lithographer for Kiepert. *Bromberger Kreis* 1843, *Plans Berlin* (1859), 1867, *Schweiz* 1866

Delius, Th. *Plans Berlin* 1874, 1875, *Grünewald* 1875, *Potsdam* 1876, *Brandenburg* [1890]

Delkeskamp, Friedrich Wilhelm (1794–1872). Artist, cartographer and engraver. Panoramas of *Rhine, Mainz, Baden Baden* &c. from 1820, *Altas Pittoresque du Rhin* 1847

Della Bella, Stefano (1610–1664). Florentine painter and engraver, teacher of A. F. Lucini

Della Gatta, Giovanni Francesco. Publisher in Rome. *Palestine* 1557

Delleker, George. Engraver of Philadelphia; worked with Young, G. H. For Macpherson 1806, Carey 1817, Carey & Lea 1822, Fielding Lucas 1823

Dellinger, A. Engraver for Kiepert (1869)

Dell'Orto, A. See **Dalorto**

Delmas, Jacques. French geographer. *L'Aude* 1867, *Provence* 1878

Delmenhorst. *Münster & Osnabrug* (1730), *L'Evêché de Münster* 1757, *Bremen* 1767

Deloche, J. Ed. Maximin (1817–1900) French historian. *Géog. hist. de la Gaule* 1864

Deloncle, François. *Indo Chine* 1889

Delpech, Ernest. *Plan ville Bordeaux* 1869, 1876

Delph[inas], O. F. See **Finé**

Delpino. Spanish cartographer. *Pyrenées* 1768 (unfinished)

Delsenbach, Johann Adam (1687–1765). Artist and engraver of Nuremberg

Delsol, Th. Engraver for Lemaitre (1840), *Canton de Fribourg* 1855

Deluc, J. A. (1727–1817). Swiss geologist

Delure, J. B. *Globe,* Paris 1707, *Gibraltar* 1727

Delves-Broughton, W. E. *Map of country in dispute with U.S.* [Highlands Conn.], Wyld 1842 (with Featherstonhaugh, J.D.)

Demanne. Mapseller, Rue de l'Ortie, vis-à-vis le Logement du Premier Géog. du Roi, Paris. For d'Anville, Barbié du Bocage 1786

De Mannevillette, A. See **Après de Mannevillette**

Demarest, P. *Plan Marseille* 1808

Demarque-Geoffroy. *Plan Forêt Compiègne* 1874

De Mayne, Anthony. Admiralty surveyor. *Surveys W. Indies* 1811–28, *Chesapeake* 1820, *Cape Palmas* 1831

Demersay, L. M. Alfred. French geographer and traveller. *Hist. physique Paraguay* 1860–4 (16 sh.)

Demezynshi. *Pianta di Milano* 1844

Demidoff, Anatole de. *Russie Mérid.* 1837, pub. 1857

Demiege. *Maconnais* 4 sh. 1776

De Mole, G. E. *Lacipede Bay* (S. Australia) 1859

De Mongenet, François (d. 1592). Cosmographer and globemaker. *World* 1552

Demonville. Imprimeur-libraire de l'Académie Franc., Rue S. Sévérin, aux Armes de Dombes, Paris. For Après de Mannevillette 1775–81

Demotier. Lithographer. *Plan Calais* (1845)

De Munk, F. See **Franciscus Monarchus**

De Musis, Julius. Engraver *World map in two hemispheres,* Venice, Tramezini 1554

Denaix [Denais], Maxime Auguste (d. 1844). Chef de Bataillon au Corps Royal d'Etat Major. *Atlas Physique, Polit. et Hist. de l'Europe* 1829, *Carte de l'Analyse Géog.* 1831, *Atlas de la France* 1836–7

Denarowski, Carl. Polish cartographer. *Sanitäts Karte der Bukowina* (1880)

Dencède, A. *France kilométrique* 1893

Denecourt, C. F. *Forêt de Fontainebleau* (1844), 1870, 1878

Dengelsted. Maps for *Schenk's Sächsischer Atlas* 1752–8

Denham, Lieut. (later Capt.) Henry
Mangles. Admiralty surveyor. *Fiji* 1822,
Bristol Channel 1827–35, *W. coast Eng.*
1842–4, *Niger & Guinea coast* 1845–6,
River Dee 1847, *S.W. Pacific* 1852–60

Denhardt, C. *Äquatoriales Ost. Afrika,*
Perthes 1881, *Unt. Tana Gebiet*
Berlin 1884

Denis. *Umgebungen von Metz* 1872

Denis, Ferdinand (1736–1805)
Mannheim 1780, *Mannheim u. Umgebung*
1782

Denis, J. Ferdinand (1798–1890) French
traveller and geographic writer

Denis, Louis (1725–1794). Engraver,
geographer to Duc de Berry. *Plan Paris*
1758 (with Pasquier, J. J.), *Mappemonde*
1764, *Atlas géog.* (1764), *Atlas de
Normandie* 1770; *Routes de Paris* 1774,
Théâtre de la Guerre en Amérique 1779,
Corrected *Nolin's World* 1785, *Amérique*
1788

Denis, Nicolas. *Europe* (1480–5)

Denison, Charles. *Pocket atlas Climatic
maps U.S.* 1885

Denison, J. *Maryland & Delaware,*
engraved A. Doolittle, Boston, Thomas &
Andrews, n.d. (early 19th century. *Maps
for Morse's American Geography* 1796

Dennis, A. (1779–1817) *Monmouth
County* (with Williams, I., reduced from
Hills) 1781 MS.

Dennis, Wm. Charts *Gt. Britain* 1781–
1802

Dennis, W. Luke. *Transport map
Birmingham* 1886

De Nobilibus, P. See **Nobilibus**

De Nova, P. *Straits Malacca,* Sayer 1779

Dent, Lieut. Albert. Admiralty surveyor.
Sound of Harris 1859, *Hebrides* 1865

Dent, Capt. Frederick Tracy (1821–
1892). American soldier. Assisted
Simpson surveys *New Mexico* 1849

Dent, R. *Plan Islington Parish* 1805–6,
1822

Denton, Guillermo. See **Denton,** Wm.

Denton, Thomas (fl. 1803–18). Estate
surveyor of Staines. *Plan Kirtling* (1815),
Plan E. Bedfont 1816 MS. (with Wm.
Denton)

Denton, Wm. *Plan E. Bedfont* 1816 MS.
(with Thos. Denton)

Denton, Wm. [Guillermo] (1823–1863).
American geologist. *Distritos Minerales
de San Antonio* 1865

D'Entrecasteaux. See **Bruny D'Entrecasteaux**

Dentu. Printer and bookseller, Palais du
Tribunat, galéries de bois No. 240, Paris.
Recueil de Cartes 1802

Denys, Jehan. Norman pilot of Honfleur.
Newfoundland (1506), *Canada* (1506),
St. Lawrence (1508)

Denys, Wm. *Mounts Bay* 1794 pub.
Laurie & Whittle

Denza, Francesco. *Correnti Marine*
(1875?)

Denzler, Johann Heinrich (1814–1876)
Topographer of Bern

De Palmeus, G. See **Palmeus**

Depéret, Charles. *Carte géolog. de la
France* (1889)

Depósito Hydrográfico, Madrid. Founded
1797 by Espinosa, q.v.

Dépôt de la Guerre, Paris. Founded by
Louvois 1688. Published *Roussel and La
Blottière's Pyrenées* 1730, *Atlas des
campagnes de Napoléon* 1844, *Atlas de
l'expéd. de Chine* 1861–2

**Dépôt des Cartes et Plans de la Marine,
Paris.** Founded 1720

Depping, Georges Bernard (1784–1853).
Collaborated with Malte-Brun on *Annales
des Voyages* &c. *L'Angleterre* 1828

De Puy, Wm. Harrison (1821–1901)
People's Atlas (1885)

De Ram, J. See **Ram**

Derbishire & Desbaratas. Publishers
Toronto. *Devine's N.W. Canada* 1857

Derby, Lieut. George Horatio (1823–1861). U.S. Topographical engineer. *Colorado River* 1850

Derendinger, Joh. Stephan. (fl. mid 18th century) Swiss cartographer and surveyor

Derfelden van Hinderstein, Baron Gijsbert Franco (1783–1857) Dutch geographer. *Kaart van Ned. Oostindië* 1842, *Sumatra* (1872?)

Derksen, J. See **Diricks**

Dermott, James R. *Plan Washington, D.C.* 1796

De Rogier, Johan (1600–1684) Swedish cartographer

Derond, J. (fl. 1770). Engraver and cartographer in Amsterdam

De Roos, Comm. the Hon. John Fred. Fitzgerald. Admiralty charts. *Port Maceio* 1834, *River Parahyba* 1834

Derosaire, Michel. Lithographer. *Allahabad* 1862, *Bareilly* 1886

Derozier. See **Des Rosiers**

Derrien, Capt. I. *Galilée* 1870, *Oran* 1874, *Algérie* 1876, *Haut Sénégal* 1882

Desaguiliers, Henri. *Prov. des Pais Bas* (1710?)

Desaint, J. Ch. Publisher. *Cassini's France* 1783

Desandrouins. *Carte des environs de Williamsburg Armée de Rochambeau* 1782. *Entrée de la Baye de Chesapeack* 1782. *Partie du Nord de l'Isle de New-York* 1781.

De S'Angelo. See **Angelo**

Desbaratas. See **Derbishire & Desbaratas**

Des Barres, Joseph Frederick Wallet (1722–1824). Hydrographer, born Basle, educ. Woolwich, Lieut. Royal Amer. Regt. 1756, engineer under Wolfe at Quebec, Governor Pr. Edward Is. *Rhode Is. Harbour* 1776, *Atlantic Neptune* 1777–81, (about 800 charts & views in various states)

Desbois. Publisher, Rue St. Jacques à la Sphère Royale, Paris. *Danet's Amérique*

Desbordes, Charles. *Plan Turin* 1705, *Plan Menin,* Visscher (1706)

Desbrosiers. Engraver *French provinces* 1707–15

Desbruslins, F. [Père] . Engraver for B. Jaillot 1719, Buache 1741–57, Charlevoix 1744, Gendron 1754

Desbruslins, Fils. Engraver. *Hesse Cassel* 1760–1, maps for *Courtalon's Atlas Elémentaire* 1774

Desbuissons. Engraver, Rue des Noyers No. 8, Paris. For Perrot 1827, for Migeon 1882

Desbuissons, E. *Opérations Mexique* 1862, *Basse Egypte* 1869, *Guinée* 1886

Desbuissons, L. E. (b. 1827). French geographer. Maps for Migeon 1874, *Nouvel Atlas* 1891

Descalvi, Dr. N. *Plan del Rio Bermejo* (Argentina) 1831

Descelliers [Desceliers, Descalier] , Pierre (1487–1553). Cartographer of Dieppe. MS. *world maps,* 1546 (for Dauphin), 1550, 1553

Deschmann, K. (d. 1889). Austrian geographer and naturalist.

Descubes, A. *Mauritius* 1880–(1)

Deseine, François Jacques (d. 1715) Parisian geographer and bookseller working in Rome

Desenne. See **Menard & Desenne**

Desessards, C. *Carte géolog. Dépt. Deux Sèvres* 1870

Desfontaines, B. Engraver. *France* 1744–60

Desfontaines, J. B. *California* 1849

Desfossés, Romain Joseph. *Baie de l'Ouest* 1832–(45), *Plan Isles du Salut* 1834–(63)

Desgodins, Abbé C. II. *Frontière du Thibet oriental* 1885

Desgranges. Géographe du Roy. *Palatinat du Rhin* 1688, *Angleterre* &c. 1689

Desgranges. *Plan du Mans* 1862

Deshayes [Deshaies, Deshaise] Sieur de. *Sept Isles* 1686, *Riv. de S. Laurens* (1695), *Parish map Canada* (1700), *Coste du Canada* 1704, *Grande Rivière de Canada* 1715

Des Hayes, Louis, Baron de Courmenin (1592–1632). *Voyage de Levant* 1624, *Voyage en Danemarc* 1664, *Itinéraire à Constantinople* 1665

Desilver. See **Cowperthwait, Desilver & Butler**

Desimoni, Cornelio (1813–1899) Italian geographer

Desire Roblet, Père. *Mission cath. Emirne, Madagascar* 1881

Desjardins. Insets on *St. Lawrence* chart, Sayer & Bennett 1777

Desjardins, Prof. Constant. *Hauteurs des montagnes* 1830, *Länge der Ströme* (1855)

Desjardins, Ernest E. A. (1823–1886) Geographer of Paris. *Atlas géog. de l'Italie* 1852, *Table de Peutinger* 1869, *Atlas Universel* 1877

Desjeans, Jean Bernard Louis, Sieur de Pointis (1645–1707). *Cartagena* 1697

Desliens, Nicolas. Chartmaker of Dieppe. MS. *world maps* 1541, 1566, 1567

Desmadryl, Frères. Draughtsmen for Schneider's *Iles Ioniennes* 1823

Desmadryl, N. *Plan Ville Mans* (1855?)

Desmarest, Nicolas (1725–1815) French physician and geologist. *Dict. de géog. physique, Atlas encyclopédique* 1787 (with Bonne), *Puy de Dôme* 1823

Desmarets, Jean Bapt. Franç., Marquis de Maillebois. *Cartes géog. topo . . . Italie* 1775

Desmaret[z], Capt. John. Engineer. *River Medway* 1724, *Harwich Harbour* 1732, *Portsmouth* 1750, *Shoreham* 1753, *Ramsgate* 1755 (with Brett)

De Smet, Pierre-Jean (1801–1873) Belgian Jesuit. *Rockies* 1844–7

Desmond, Hugo A. Engineer. *Railway Mejillones to La Paz* 1880

Desmond, John T. *Atlas Haverhill & Bradford, Mass.* 1892

Desmoulins, Léandre Eugène. *Côtes Syrie* 1861–(2)

Desnos, Louis Charles (fl. 1750–70) Publisher of Paris. Ingénieur Géographe pour les Globes et Sphères, Ingénieur Géographe du Roi de Danemark. 1 Rue St. Julien le pauvre (1753), Rue St. Jacques au Globe No. 254 (1766). *Globes* ca. 1750 –82. *Le Rouge's Atlas Nouveau Portatif* 1756. Reissued *Danet's Asie* 1760, *Atlas Méthodique* 1761, *Atlas chorographique de la France* 1763, *Nouvelle Histoire Générale* 1766, *Isles Britanniques* 4 sh. 1766, *Nouvel Atlas d'Angleterre* 1767, *Atlas Général* 1768, *Almanach Géographique* 1770, *Amérique* 1795

Desor, Ed. (1811–1882). Swiss geologist. *Orographie des Alpes* 1862

De Soto, H. See **Soto**

Desparées, Sieur. See **Erault**

Desprez, Germain. *Guide chemins Angleterre* 1571 MS.

Despriez. *Plan de Caen* (1840)

Desray. Publisher, Rue Haute-feuille, No. 4, près celle St. André des Arts, Paris. Hérisson's *Atlas* 1807, for Barbié du Bocage 1811, & Brué 1815, 1821

Desrosiers. Engraver for Blaeu's *Novus Atlas* (1640)

Des Rosiers [Derozier] . Engraver for de L'Isle 1705–15

Dessalines d'Orbigny, Alcide. *Voyage Amér. Mérid.* 1835–47, *Rép. Argentine* 1835, *Rép. de Bolivia* 1839

Dessans. *Entrée de Charles Town* 1777

Dessel, Camille van. *Carte archéolog. de la Belgigue* (1877)

Dessel, F. van. *Frieslant* &c. (1720)

Dessingy. *Côte de la Guyane* 1765 for Bellin

Dessiou, Joseph Foss, Master, R.N., Hydrographer. *River Dart* 1782,

Adriatic 1806, *W. Indies* 1808, revised
Lane's charts 1809, *Bury River & Rosilly
Bay* 1821

Destar, John. *Navigation London to
Lowestoft,* 5 sh. [advt. 1778]

Desterbecq, François (fl. 1807–49) Artist
and lithographer. Director of "Etablissement
géog.", The Hague. *Atlas der Nederlanden*
(1841)

Destouches, E. *Côte occidentale d'Afrique*
1887

Destrée, Jacques [or Maillard] . *Plan
Ypres* 1564

Detaille. *St. Lucia* 1849

D'Etaples, L. See **Leferre d'Etaples**

Deterville. Bookseller, Rue Hautefeuille
No. 8, Paris. *Coutans' Environs de Paris*
1800

De Thury, Cassini. See **Cassini de Thury**

Detterline, T. *Battle of Gettysburg* (1863)

Dettmers, A. Printer for Kiepert (1857)

Detz, L. B. J. S. *City Bangkok,* Singapore
1871

Deur [Deuer] , Abraham Jansz. (fl.
1666–1714). Engraver of Amsterdam
for de Wit, Visscher, Goos' *Atlantic Ocean*
1669, *Canada* 1670, *Gelderland* 1684

Deur, Jacob (fl. 1709–14). Engraver of
Amsterdam, son of Abraham,
collaborated with brother Jan. *Delflant*
(Polder map), *Allard's Italy*

Deur, Jan. Engraver of Amsterdam,
collaborated with brother Jacob. *Italy*
1706, *Globes* 1720

Deutecum. See **Doetichum**

Deutsch, H. *Ballarat Municipalities*
(1865)

Deutsch, Hans Rudolf Manuel. Wood
engraver. *Würzburg* 1548, *Nördlingen*
1549, both used by Münster

Deuvez, Arnold. Artist of Paris. Painted
Nolin's globe 1693

De Vaugondy, R. See **Robert de Vaugondy**

Devaux, Pierre. *Portolan Atlantic,* Le
Havre 1613

Devaux, V. *Is. Maurice* 1848

Devel, Pieter. Engraver of Brussels for
Fricx 1712

Deventer [Dabey, Daventria, Daventers] ,
Jacob van/a [Jacob Roelofs, Roelafsz]
(1500–1575) Geographer to Charles V
& Philip II, b. Deventer, worked Malines,
d. Cologne. *Brabant* 1536–(46), 6 sh.
1558, *Holland* 1540, 9 sh. 1558, *Gelderland*
9 sh. 1542–(56), *Frisia* 9 sh. 1545, *Zeeland* 4 sh. 1560. Over 300 Dutch town
plans of which 220 survive, *plan Basle*
pub. Münster, plans 1558–72 for Braun
& Hogenberg

De Vere, Lieut. *Plan fortification
Sebastopol* 1855

Deverell, F. H. *Vall. de Andorra* 1890

Deverell, W. *Interior Gold Lakes Otago,
N.Z.,* Wellington 1888

Devereux. Engraver. *N. America* (1849)

Devert, B. A. H. *Plan Paris* (1830)

Deville, E. (1849–1924) French Canadian
Surveyor. *N.W. Territ.* 1881, *Prov.
Manitoba,* Ottawa 1884

Devilliers, P. Engraver for Vuillemin
1873

Devin, *Louisiana* 1719–20 MS. *Baye de
St. Louis* 1720

Devine, Thomas. Deputy Surv. Gen. for
Ontario, draughtsman. *N.W. Canada,*
Toronto 1857, *Gov't map Canada* 1859,
Huron & Ottawa Territ. 1861, *Canada*
1873, *N. Amer.* 1878

Devi Prasada, Munshi. *Hindi atlas for
schools* 1882

Devos, Rev. *Ortos Country & parts
Mongolia,* Calcutta 1876

De Vou, Johannes. *Plan Rotterdam* 1694

Devoux, Esprit. *Provence* 1758, *Plan
Ville d'Aix* (1765?)

De Vries, Abel. Dutch publisher. *Geheele
Alblasserwaert* 1716

De Vries, J. Fr. *Ostfriesland* 1880

De Vries, Jacob Pieter. *Inkoomen van de O. en W. Eemze* 1797

De Vries, Nikolas. Maps for Keulen 1696, for Loots 1698, *Middland Zee* (1717), *Oost-Zee* 1718, *Falmouth* 1720

De Vries, Simon. *Curieuse Aenmerckingen . . . O. & W. Ind.,* Utrecht 1682, edited *Wereld-Beschryvingh* 1683

Dew, J. Estate surveyor. *S. Hanningfield* 1799 MS.

Dewald, G. A. Stephan. *Halbkugel* 1860, *Deutsches Reich* 1876

De Wale, P. See **Wale**

Dewalque, Gustave (b. 1826). Belgian geologist. *Carte géolog. de la Belgique* 1879

Dewar, J. Engraver. *Teesdale's World* 1844

Dewar, Capt. James. *Soundings on the Chagos [Solomon Is.]* 1763, pub. Dalrymple

Dewarat, Peter (d. 1800). Surveyor of Mannheim. War maps, *Rhine* area 1794–9, *Course Neckar* 1798

Deward, John. *Survey Franefield* 1618 MS.

De Wasme, Pletinckx. Lithographer. *Malta,* Brussels 1835

Dewe, I. & R. *Pictorial plan Oxford* (1850?)

Dewe, John. *Canada West* 1860

Dewe, R. See **Dewe**, I.

Dewey, William. *Des Moines River Improvement* 1849

Dewhirst & Nichols. *Plan Bedford* 1836

Dewick & Clarke. Printers for Wilson's *plan London* 1791

Dewing, Francis (fl. 1716, d. after 1745). Engraver and printer, London, established in Providence, then Boston, later returned England. *Southack's Eng. Empire in N. America,* Providence 1717 (first chart engraved on copper in N. Amer.), *Bonner's town Boston,* Boston 1722

De Wit. See **Wit**

De Witt, Simeon (1756–1834) Pioneer American cartographer, geographer, and Surveyor General to Continental Army 1780. Surveyor General New York state 1784, *Winter Cantonment of the American army* 1778, *Part of Rockland Modern County* 1780, *State map New York* 1792–1802, *plan Albany N.Y.* 1794

Dewitt & Davenport. Booksellers & publishers, Tribune Buildings, Nassau St., N.Y. *Lawson's Upper Calif.* 1849

Dexmier de Saint Simon, Etienne Jules Adolphe, Visc. d'Archiac. *Carte géolog. Dépt. de l'Aisne* 1845, *Carte géol. Environs de Paris* 1865

Dezauche, J. A. (fl. 1831). Editeur, marchand de cartes, Géographe et Graveur, Successeur et Possesseur du Fond Géog. des Srs. De l'Isle et Buache et chargés de l'Entrepôt Gén. de Cartes de la Marine, Rue des Noyers [later No. 40] près celle des Anglois, Paris 1820–27. Reissued Après de Mannevillette's works from 1775, pub. atlases using the plates of de l'Isle and Buache. Succeeded Buache de la Neuville ca. 1780. *Topog. de la Zélande* 1790, *Carte itinéraire de la France* 4 sh. 1796

Dezauche J.A.

Dezendorf, J. F. Surveyor. *Cities Norfolk & Portsmouth* 1876

Dezoteux. Staff officer, French army. *Chart for Marquis de Chastellux Journal* 1786, English edition 1787

Dheulland [Dheulan], Guillaume (1700–1770). Publisher, engraver and draughtsman of Paris. *Plan d'Angers,* 4 sh. 1736. Engraved for Charlevoix 1736, Bellin 1744, Cassini 1744—60, Prévost 1752. *Atlas of Brabant* 1747, *Plan Paris* 1756

D'Houdan [Doudan]. Engraver for *Atlas Nat. de la France* (1793), Barbié du Bocage 1800, Bruny D'Entrecasteaux 1807, *Environs de Versailles,* 12 sh. (1808), and Dépôt de la Marine

Diamanti, E. *Plan Alexandrie* 1877

Dias, Antonio. Portuguese cartographer mid 18th century

Dias, Manuel, the younger [Yang ma-no] (1574—1659). Portuguese Jesuit. *Chinese globe* 1623 MS. (with Longobardi)

Dias, Lieut. V. P. *Rio Amazones* 1865

Diaz Romero, F. See **Romero**

Dibbets, G. J. *Omstreken van Arnhem* 1821

Dibdin, Thomas. *River of Nagore* 1779—(85)

Dicaearchus (350—290 B.C.) of Messina. Pupil of Aristotle. *Map of Alexander the Great's campaigns* 320 B.C.

Dicey, Cluer & Co. London publishers and printers Aldermary Church yard. In partnership with Marshall, R. (1764). *Plan London* 1765, edn. Speed *c.* 1770. See also under Wm. below

Dicey [Wm.] & Cluer. Printers and print-sellers at the printing office in Bow Church yard. *Seale's Europe* (1740), *Scotland* 1746

Dickens, Lt.-Col. C. H. *Canals from R. Soane, S. Behar* 1861

Dickert, Lorenz. *Relief karte von Central Europa* 1882

Dickert, Thomas (1801—1883). German cartographer. Numerous *relief maps*

Dickinson, Bis. Printer & map seller with T. Millward at Inigo Jones Head, London. *Wit's Edinburgh* (1690?), *Ground plot Hamburg* (1706), *Eng. & Wales* 1734

Dickinson, John. *Survey Sutton Super*

Darwent, East Riding Yorks 1725 MS., *County of York* 1740

Dickinson, Joseph. *Plan of the Level of Axholme,* Lincs. 1767—8, pub. 1791

Dickinson, Joseph. *Keswick Estate, Cumb.* (1805), *Lowbyer Demesne* (1805)

Dickinson, Rudolphus. *Elements of Geography* 1813

Dickinson, Lieut. S., Royal Eng's. *Chart Heligoland* 1809, *Plan Porto Praya* 1812, *Cape Verd Is.* 1813

Dickinson, Samuel. Surveyor. *Course of Rivers Baine & Waring* 1792 (with Stickney, R.)

Dickinson, S. N. *Plan City Boston* 1844

Dickinson, Capt. T. Surveyor. *Is. of Bombay* 1843

Dickmann, Aegidius. *Danzig,* 7 sh. 1617

Dickson, George K. *Atlas Madison Co., Ill.* 1892 (with Riniker & Hagnauer)

Dickson, John. *Seaman's Guide round Gt. Brit.,* Liverpool 1700

Dickson, J. Bookseller of Edinburgh. *Ainslie's Selkirk* 1772

Dickson, Samuel. Estate surveyor of Writtle. *Essex* 1844, *Chelmsford* 1846 MS., *Hadleigh* 1852 MS., *Paslow Wood* 1859 MS.

Dickson, Lieut. W. B., R.N. *Gulf of Bengal* 1861, *S. Andaman I.* 1863

Di[c]quemare, L'Abbé Jacques-François (1733—1789). *Index géog.* 1769. Revised plates for the *Neptune Oriental* 1775, *British Channel & Bay of Biscay* 1778

Dicuil (b. ca. 780) Irish astronomer and geographer. *De mensura orbis terrae*

Diderich, H. *Region du Mayumbe, Congo* 1899

Diderot. *Encyclopaedie* 1770—79 with atlas (maps by Vaugondy)

Didier, C. *Ost Afrika* 1861

Didier & Tebbett. Publishers 75 St. James St., London. *Plan London* 1807

Didient. *Côte d'Or* 1867

Didot. Publisher, Quai des Augustins à la Bible d'or, Paris. *Charlevoix's Nouvelle France* 1744

Didot, Firmin (1736–1827). Publisher, Paris. For Bellin 1758, *Mappemonde* 1808

Didot, Firmin, Frères. Publishers, Rue Jacob 56, Paris. *France* 1840

Didot, J. *Europe Politique* 1823

Didot, P., l'ainé. Publisher and printer at the Louvre, Paris. *World* 1802.

Didot, P. N., le jeune. *Recueil de cartes géog.* 1799

Didrichsen, Ferd. (1814–1887) Danish geographer and botanist

Diebel, Elias. *Plan Lübeck* (1550) used by Braun & Hogenberg

Diebison, C. Lithographer. *Plan Breslau* 1858, *Neuer Plan Breslau* 1867

Diebitch, Marshall. *Sketch ground near Varna* 1847 MS.

Diebitsch-Narten, Karl Fred. Wilhelm, Freiherr von (1738–1822) Military cartographer of Wittenberg, in Russian service

Diedrich, Rudolf. *Nederlanden* 1850

Diefenbach, C. *Schulwandkarte Wiesbaden* (1872?)

Diéguez, Juan. *Bolinao* (Philippines) 1800

Diehl, Johann Philip. *Panorames de Francfort* 1834, *Schul Atlas* [1866]

Dien, Charles (1809–1870). Engraver, publisher, and globemaker of Paris, Rue du Foin St. Jacques, au Collège de Mtre. Gervais and Rue de Jardinet No. 13. Engraved for Griwtoon 1801, *Jefferys' St. Lucie* 1802, partner with Delamarche (1819), *Raynal's Atlas du Globe Terrestre* 1820.

Dien, Charles (1809–1870). French cosmographer, son of above. *Atlas des phénomènes célestes* 1843, *Atlas céleste* 1855, *Uranographie* (1870)

Diener, Carl. *Mittel Syrien* 1885

Dienz, *Saar & Mosel Weinbau-Karte* 1868

Diercke, Karl (1842–1913) b. Kyritz, compiled school maps *Karte d. Harzes* 1887

Dieskau, General. French army. *Carte du Lac St. Sacrement* (1755)

Diesner, L. *Hessen-Nassau* (1877)

Diespecker, L. *Environs Cognac* (1864)

Diest, W. von (1851–1932). *Itin. Kleinasien* 1895, *Anatolische Eisenbahn Linien* 1898

Diestel. *Plan Travemünde* 1891

Diesth, Aegidius Coppenius [Gillis Coppens]. Printer of Antwerp for Apian 1534–44, Honter 1552, *Plan Antwerp* 1565, for Ortelius 1570–8

Diet, Leo. *Egypte* 1884

Dietell, Chris. Engraver of Vienna. Maps in Stöcklein's *Travels* 1728–32

Dietlein, Woldemar. *Hist. map Weltgeschichte* (1874?)

Dietrich, Lieut. *Moscow* 1812, *Posen* (1863), *Plan Cilli* (1873)

Dietrich, A. *Kohlenreviere von Lugau* 1859

Dietrich, G. Engraver for Meyer 1867

Dietrici, K. F. W. (1790–1859). German geographer and statistician

Dietz, Ad. *Schweinfurt* 1892

Dietz, C. W. *Resid. Stad Kiobenhavn* 1769

Dieu. Designed frontispiece for *Jaillot's Atlas François* 1695

Diewald, Johann Nepomuk (1774–1830). b. Salzburg, cartographer active in Nuremberg. *Holstein* 1815, *Niederlande* 1817, *Weltcharte* 1829

Diez, F. M. *Post Karte von Teutschland* Berlin 1795, *Deutschland, Niederlande &c.* 1838, *Post und Reise Karte Deutschland* 1833, 1855

Digges, Sir Dudley (1583–1639). *Circumference of earth* 1612

Digges, Leonard (d. 1571?). Mathematician *A Book named Tectonicon* 1556 (first important English book on surveying), *Pantometria* 1571 (includes description of theodolite, invented by Digges)

Digges, Thomas (ca. 1543–1595). Mathematician, hydrographer and geographical theorist, son of above, published father's *Pantometria*. *Winchelsea Harbour* (1577), *Platte of Dover Haven* 1581, 1595 MSS.

Dignoscyo, L. de. *Plan ville Lyon* 1847, 1874

Diguja, D. J. *Cumana* 1898

Dijck, H. See **Dyck**

Dijk, P. van. Maps *Java* 1872–3

Dikaiarchos (350–290 B.C.) Geographer of Messina

Diko, W. *Schul Atlas* (1868)

Dilbergerin, Lucas. Cartographer latter 17th century

Dilcher [Dilger], Heinrich (1824–1885). Map lithographer in Amsterdam

Dilich [-Schäffer], Wilhelm Schaeffer ca. 1571–1655. Geographer, map colorist and mining surveyor, b. Cassel. *Hessen* (1591–1630) MS.

Dilleben, Christian (fl. 1780). German cartographer

Dillon, J. P. Master, R.N. *Admiralty chart Tobago* 1864–5

Dillon, C. H. Admiralty charts *Beirout Bay* 1842–(4), *Montevideo Bay* 1850

Dillon, Theobald. *Mouvements Armées du Rhin* 1796

Dillon, Wm. 2nd Master H.M.S. *Wellesley. S. Juan de Nicaragua* 1833 MS, *Mia-Tao* 1840

Dilly, Charles (1739–1807). Publisher and bookseller, partner with brother Edward at 22 Poultry, London; on own from 1779, collaborated with G.G. & J. Robinson 1783–1801, succeeded by J. Mawman ca. 1801. *U.S* 1783, *Atlas to Guthrie's System of Geog.* 1785 &c.

Dilworth, Thomas (d. 1780). English globeseller and schoolteacher

Dimat ad Din Muhammad. 16th century Arab mathematician

Dimes, W. Piercy. *Ecclesiastical Div. England & Wales* 1874

Dinesen, Jorgen. Danish cartographer. *Øresund* 1688

Dingelsted, Christopher Adolph. *Tab. geog. Halberstadtiensis* (1710)

Dinglinger, Georg Friedrich. *Plan Stadt Hannover* 1748

Dinomé, Abbé S. E. Achille (1787–1871). French geographer, translated many geog. works, edited *Annales des voyages*

Dinsmore. *R'roads & canals of U.S. & Canada* 1856

Diodymus. See **Aucuparius**

Diognetus. Surveyor to Alexander the Great

Diogo, Mestre. 16th century Portuguese cartographer

Dion, Philippe. *Carte vinicole de l'Algérie* 1891

Dionysius Periegetes [Alexandrinas] (fl. 2nd century A.D.) of Alexandria. Manual of geography translated 6th century by Priscianus of Caesarea. *Cosmographia De Situ Orbis,* Venice 1478

Diquemare, J. F. See **Dicquemare**

Dircks, Henry. *Plan Castle & Citadel Raglan* 1862

Diricksen, Jan. Engraver. *Copenhagen* 1611

Dirckx, Jan. Portolan maker "in't Vergulde Compas tot Edam." *Atlantique* 1599 MS.

Diricks [Van Campen, Derksen], J. (b. ca. 1590) of Campen. *Plan Hamburg* 1611

Diriks, C. *Norske Kyst fra Arnedal til Christiansand* 1856

Dirom, Alexander. *Dumfrieshire* 1812

Diron. *Baye de la Mobile* 1725

Diruf, Arthur. *Stromkarte der Elbe* 1885

Dirwalt [Dirwald, Dirwaldt] , Joseph.
Schwaben 1809, *Bavaria* 9 sh. 1813,
Allgemeiner Hand Atlas, Vienna 1816,
North America 1823

Dispatch. See **Weekly Dispatch**

Diston, John. Chart maker. *Lowestoft-
London* 1760, *River Thames* 1767

Disturnell, John (1801–1877). Publisher,
102 Broadway, N.Y. *Guide to City N.Y.*
1836, *30 miles round N.Y.* 1839, *Méjico*
1846, 1847, *N. America* 1850, 1854,
U.S.–Mexico boundary 1853 (Disturnell
& Schroeter)

Dittli[n]ger, Albrecht Anton (1704–
1780). Surveyor of Berne

Dittmar, Heinrich (1792–1866).
Historischer Atlas [1856]

Ditzinger, Ludwig (fl. 1600). Goldsmith
and engraver of Tübingen

Divinus Eustachius de. 17th century
Italian astronomer

Dix, I. C. & W. Lithographers. *Map illust.
Ancient Hist.* (1847)

Dix, Thomas (fl. 1799). Surveyor and
schoolmaster of N. Walsham, Norfolk,
Juvenile Atlas 1811, *Beds.* 1818, *County
of York* 1820, 1835, *County Atlas* 1822
(with Darton), *Chester* 1830

Dix, W. See **Dix**, I. C.

Dixon, Capt., from Whitby. *Scott's V.D.L.*
1824

Dixon, Arthur. *Wangamatta, Owen's
River* 1855

Dixon, Capt. George (d. 1800?). Sailed
with Cook on 3rd voyage. Voyages round
world 1785–8 published 1789, *Chart
N.W. coast America* 1788

Dixon, G. G. *British Guiana* 1894

Dixon, Jeremiah. *Boundary Maryland &
Penn.* 1768 (with Mason)

Dixon, John (1740–1811). Engraver,
b. Dublin, moved London ca 1765.
Rocque's Dublin, 4 sh. 1760

Dixon, Lieut. Joseph, U.S. Top. Engs.
Fort Dallas to Gt. Salt Lake 1859

Dixon, Robert (1800–1858). Surveyor
and explorer. *Aracati River* 1831, *N.S.W.*
1837, *Moreton Bay* 1840

Dixon, R. Surveyor of Godalming. *Tithe
map Warlingham & Chelsham* 1844

Dixon, Thomas. *Borough Bradford* 1856

Dixon. See **Griffing, Dixon & Co.**; **Stannard
& Dixon**

Dixson & Kasson. *Overland mail routes,*
San Francisco 1859

Dixwell, James. Printer St. Martin's-Lane,
London. *Rocque's Berks.* 1761 ·

Djurberg, Daniel (1744–1834). Member
Cosmog. Soc. of Uppsala. *Polynesia,*
Stockholm 1780 (first Swedish map to
show Cook's discoveries, introduced name
Ulimaroa for Australia)

Doane, Thomas (1821–1897). American
engineer. *Stanstead & Montreal R.R.* 1845

Dobbie, W. H. *Plan Bay Kossier* 1799 pub.
Dalrymple 1802

Dobbs, Arthur (1689–1765) Governor
N. Carolina 1754. *N. America* 1744, used
Wigate 1746

Dobert, W. *Wandkarte Sachsen* 1876

Doblado, J. Published *Ascensio's Madrid*
1800

Dobler, Alfred. *Eisenbahn u. Postkarte
Deutschland* (1864), . . . *des Rheins*
(1873)

Dobner von Dobenau, Johann (1816–1889).
Viennese general and cartographer

Dobrée, Capt. Nicholas. Agent to the
Admiralty. *Alderney, Guernsey, chart
Channel Is.,* all 1746

Dobson, James. *Chart globe* 1794 (based
on Halley)

Dobson, Joseph. Pilot. *Chart entrance R.
Tees* 1762

Dobson, Thomas. *Map Paradise* in Bible,
Philadelphia 1799

Dobson & Rhoades. *Atlas town Calumet*
[1875]

Dodd, George (1783–1827). Civil engineer,

son of Ralph below. *Strand Bridge* (London) 1808

Dodd, James Solus. *Travellers' Directory through Ireland* 1801

Dodd, Moses. 10 *miles round Reading* 1846

Dodd, Ralph (1756—1822). Civil engineer and surveyor. *Newcastle—Carlisle canal* 1795, *Plans Port of London* 1798—1800, *Hartlepool* 1802

Dodd, Mead & Co. *Universal Atlas* 1892

Dodge, Robert Perly (1817—1887) *Plats Squares W. Washington* 1833

Dodsley, James (1724—1797). Publisher and bookseller, partner with brother Robert at Tully's Head in Pall Mall, London, from 1755, continued business alone 1764. Kitchin's *English Atlas* (1770). See also Dodsley, R. & J.

Dodsley, Robert (1703—1764). Publisher, writer, poet, dramatist, founded *Annual Register* 1758. At Tully's Head in Pall Mall, London, from 1735, in partnership with brother James from 1755. *Cowley's Geography of Eng.* 1744, and his *Pocket maps Eng. & Wales* 1745. Evans' *Middle British Colonies* 1755. See also under Dodsley, R. & J.

Dodsley, R. & J. Publishers 1755—64. *Roads of Eng. & Wales* 1756, *Plan London* 1761, *England Illustrated* 1764. See also Dodsley, James and Robert separately

Dodson, James, F.R.S. (d. 1757). Teacher of mathematics. Revised Halley's *magnetic variation chart* (1745)

Dodson, Colonel Wm. *Mapp Great Levell of Fenns* 1665

Dodsworth. *Siege Delhi* 1876

Dodt, H. *Prussia* 1862

Dodwell. *Andaman & Nicobar Is.* 1778

Doebeli, Samuel (1858—1919). Relief maps

Doedes [Daetz, Doetz, Doedtszoon], Cornelis. Chart maker of Edam "inde vier heems Kinderen." *Chart Baltic* 1589,

reissued Visscher 1610, *Coast Asia* 1598, *Coast Europe* 1600, *Ganges* 1600

Doergens. *Imperial map Palestine* (1864)

Doering [Döring], Ferdinand von (1820—1889). maps *Prussia* 1827, 1836, *Admin. Statistischer Atlas v. Preussischen* 1845

Doerrbecker [Dörrbecker], P. A. *Beira Harbour* 1891

Doesborgh [Doesburgh], Thomas (fl. 1677—1714). Engraver. *C. Good Hope,* Allard (1667), maps for Allard's *Orbis habit. oppida* (1698), *Delfland* 12 sh., maps of *Utrecht*

Doetichum [Deutecum, Doet, Doetecum, Doetechum, Doetecomius, Dotecum, Duetechum]. Family of engravers and map publishers working in Deventer and Haarlem
—— Baptista van. *Belgium* 1588, Doedes' *Baltic* 1589, helped his father Jan elder in engr. for Linschoten 1595, for Plancius, Waghenaer 1602, Bible 1609, *Hannonia,* Jansson, 1633

Doetichum, Jan van, elder (fl. 1558—1601) b. Deventer, worked Haarlem. Brother of Lucas. Engraved for Plantin 1559 &c., Jode's *Speculum* 1566—71, for Waghenaer 1584—92, Ortelius, Linschoten
—— Jan van, younger (fl. 1592, d. 1630). Engraved for Plancius 1592—94, *Africa* 1610, Visscher's *Hungary* 1624
—— Lucas van (fl. 1558—93). Worked with his brother, Jan elder. Gastaldo's *Africa* (1560), for Jode 1566—71, *America* 1578, *Bohemia* (1590)

Doetsch, Johann Anton von. *Imp. Russicum* 1785

Doetszoon, C. See **Doedes**

Doetz, C. See **Doedes**

D'Ogerolles, I. See **Ogerolles**

Doggett. *U.S.* 1851

Doharty [Dougharty], John the Elder (1677—1755). Estate surveyor and mathematician of Worcester. Estate map *Lindridge* 1721 MS, *Plan Worcester* 1741—(2)

Doharty [Dougharty] , John the Younger. Estate surveyor of Worcester. Estate plans *Droitwich* 1731—3, *Plan Worcester* 1742, *Ruislip* 1750, *Kidderminster* 1753

Doidge, W. & H. Surveyors. *Plan Ancient City Canterbury* 1752

Doidge & Co. *Cycling & Touring Map Plymouth* 1890

Doin, Ochikochi. Pseudonym, see **Hanchi, Fujii**

Doino, Catharino. *Vicenza* 1611

Dolben, William Digby Mackworth. Admiralty chart *R. Volga* 1862

Dolce, Lodovico (1508—1568) b. Venice. *World* 1555 in *Le Transformationi*

Dolcetta, Giuseppe. Engraver for Coronelli's *Atlante Veneto* 1690—6

Dolendo, Bartholomew (1571—1629). Engraver of Leyden. Maps for *Wahrhaftige Beschreybung aller Züge und Victorien* (1612)

Doležal, A. *Oesterreich-Ung. Mon.* 1870, *Galicya* (1875)

Doležal, Franz. *Plan Brünn* (1858)

Dolfinado, N. del. See **Nicolay**

Dolfinatto, N. del. See **Nicolay**

Dolier, F. See **Dollier de Casson**

Dolivar[t] , Juan. *Angleterre &c.* 1689

Doll, G. *Valdivia* 1852, Route map *Atacama* 1854

Doll, J. *Zuid America,* Amsterdam 1783

Doll, M. *Umgebung von Karlsruhe* 1895

Dolland. *Miniature globe,* London 1839

Dollfus, Aug. (1810—1869). French traveller and geologist. *Voyage géolog. dans Guatemala* 1868

Dollier de Casson [Dolier] , François. *Lac Ontario* 1670

Dolling, James. *Pocket map London* (1863), *Paddington* 1863

Dolter, Cornelio (b. 1850). Austrian geologist. *Reiseskizze aus Sardinien* 1878, *Vulkane Kapverden* 1882

Dom, S. Engraver for Homann Erben 1746

Domann, Bruno. Engraver. For Blumentritt 1882, *Africa* 1885, worked for Stieler; *World,* Perthes 1897, *Hand Atlas* 1907

Domenech[e] [Domeneth] , Arnald. Catalan portolan maker of Naples, pupil of Roselli. *Mediterranean* 1486

Domenego. See **Venetiano**, D.

Domeneth, A. See **Domenech**

Domenico Machaneus [De la Bella, De Belli, Bellius] (d. 1530) b. Maccagno. *Lago Maggiore* (1546)

Domergue, Eug. (b. 1849). French geog. publisher. *Géog. pittoresque* 1875

Domeyko, Ignacy (1802—1889). Chilean geologist and cartographer

Domingues, Fr. Francisco Atanacĭo. *Moqui province* 1777 MS, *New Mexico* 1777 MS.

Dominguez, Francisco. *Pacific ca.* 1584 MS.

Dominus, N.G. See **Nicolaus Germanus**

Domvill, Silas. *Harwich* 1667

Domville, W. Bookseller Royal Exchange, London. *England Displayed* 1769

Donald, Thomas (d. 1802). Surveyor. *Beds.* 1765, *Bucks.* 1770, *Yorks.* 1771—2 (all with Ainslie), *Cumb.* 6 sh. 1770—1 pub. (1774), *Environs Keswick* 1789, *Norfolk* 1790—4 pub. (1797) (with Milne)

Donat. *Chart Tristan da Cunha* 1767— (81)

Donck, Adriaen van der (1620—1655). Lawyer and Dutch colonial administrator. *Besch. van Nieuw Nederlant* 1655, 1656 (with map of *New England)*

Doncker, Hendrik (1626—1699). Bookseller, instrument maker, hydrographer and publisher "Inde Nieuwbrugh St. in't Stuurmans gereetschap," a l'enseigne des utensils de Pilots. *British Isles* 1658, *Zee atlas* 1659 &c. to 1697, *Atlas del Mundo* (1665), *Grand & nouveau miroir ou*

flambeau de la Mer 1667–84, *Nieuw Groot Stuurmans Zee-spieghel* 1676, *Portulan pour la Mer Baltique* 1694, *Nieuwe Groote Vermeerderde Zee Atlas* 1696

Dondey, Dupré. *Atlas Portatif de France* 1823

Dondi, Jacopo (1298–1359) of Chioggia Astrologer

Donelly, J. P. *Mining District Lake Superior* 1872

Donelson, John. *Virginia and Cherokee boundaries* (1771)

Donia, Franciscus. Engraver for *Mercurio Geografico* 1688

Donis. See **Nicolaus Germanus**

Donn, B. See **Donne**

Donn, J. W. *Battlefield of Chattanooga* (1864)

Donne, [Donn], Benjamin (1729–1798). Surveyor and mathematician, b. Bideford, *Plan Plymouth* 1759, *Devon* 12 sh. 1765, *11 miles round Bristol* 1769, *Glos.* 1769, *Plans Bristol* ca. 1775, 1784, 1800 *11 miles round Bath* 1790, *City of Bath* 1810

Donne, Samuel. Surveyor of Melbury Osmond. *Manor of Leigh upon Mendip, Dorset* 1764, *Manor Little Winsor, Dorset*, 1775 MS.

Donnelley, Reuben H. Publisher Chicago. *Sectional atlas Chicago* 1891, *Atlas St. Paul, Minn.* 1892

Donnelly, Capt. *Chart Rio de la Plata* 1809

Donnelly, Dominick. Estate surveyor. *Epping* 1719 MS.

Donnelly, John Fretcheville Dykes, *Plan Sevastopol* 1857, *Plan attacks on Sevastopol* 1858

Donnelly, Ross. *Skye to Aberdeen* 1797

Donnet, Alexis. Ingenieur-Géographe. *France* 1818 (25 sh. reduction of Cassini), *Arrond. Corbeil* 1834, *Atlas de la France* (1844) (with Frémin), *Environs de Paris* (1867)

Donnet, Hendrik (fl. 1766) in Amsterdam. *Dantzig* 1785

Donnet, Johannes (1754–1779). Engraver in Dantzig

Donnet, Samuel (1699–1734). Artist, publisher and engraver of Amsterdam

Donnus. See **Nicolaus Germanus**

Donoso, Capt. de Ings. *Costa de Chile* 1855

Donovan, W. *Lands of Jackson, Lansing & Saginaw Rail Road Co.* 1873

Donus. See **Nicolaus Germanus**

Donzel, Anton. French lettering engraver. *Santa Teresa's Brazil* 1700, for Petrini 1700

Doolittle, Amos (1754–1832). Engraver of Philadelphia. For Morse 1784, *Purcell's Virginia, Carolina, Florida, &c.* 1788, *Carey's Amer. Atlas* 1795, *Whitelaw's Vermont*, 3 sh. 1796, *Connecticut* 1797, *American Pocket Atlas* 1813

Doolittle, A. J. *California* 1863 (with Ransom)

Doolittle & Munson. Draughtsmen and engravers. *U.S.*, Cincinnati 1847

Doornick. Engraver for *Voyage Astrolabe* 1838

Doornick, Marcus Willemsz. Publisher of Amsterdam. *Plan London* 1666

Doppelmayr [Doppelmayer, Doppelmaier], Johann Gabriel (1677–1750). Astronomer and cartographer of Nuremberg. *Star chart* 1709, *globes* 1728–40, *Hemisphaerium Coeli* 1730, completed Homann's *Atlas Coelestis* 1742, *Tab. Selenographica* (1750)

Dopter. Printer of Paris. *France* 1837

Doran, Edmond (fl. 1586). Irish portolan maker

Doran, Hercules. Portolan maker, son of Edmond above. *Mediterranean* (1586)

D'Orbigny, A. See **Dessalines**

Dore, Edward. Surveyor. *Plan Borough Devizes* 2 sh. (1759?), *Estate Baseldon, Berks* 1769 MS.

Dore, Capt. Hastings. *Route chart Sumatra* 1805 MS.

Döring, F. van. See **Doering**

Dormier. Engraver. *Plans Madras & Pondicherry* for *Langlès Monuments de l'Hindostan* 1821

Dorn, Capt. *Santo Tomas de Guatemala* 1850

Dorn, Hans (1430–1509). Dominican monk, globemaker

Dorn, Johann Martin. Engraver of Nuremberg, 18th century

Dorn, Sebastian. (d. ca. 1778) Engraver for Homann Heirs 1746–68

Dornius, Conde. *Princip. de Cataluña* 1776

Dornseiffen, I. *Atlas van Nederland* 1865, *Java* 1866, *Oost Indie* 1871, *Atlas Ned. O. & W. Indie* 1894

Dorph, E. *Tasermint, Greenland* 1874

Dorr, Capt. F. W. *Battlefield of Chattanooga* (1864)

Dörrbecker, P. A. See **Doerrbecker**

Dorret, James. Land surveyor to the Duke of Argyll and engraver. *Scotland* 1750, *Correct map Scotland* 1751, used by Bowles 1794, Schraembl and Von Reilly

Dorrington, George. Engraver. *Chart Baltic Sea* 1854, *River Danube* 1854

Dorta, Bento Sanches (1739–1795). Astonomer

Dortet de Tessan, A. *Plan Rade de Cherbourg* 1853

Dorward, Lt.-Col. A. R. F. *Trig. survey country round Kingston, Jamaica* 1879

Dossaiga, Jaimes. Spanish portolan maker. *Atlas* 1590

Dosseray, J. *Plan Bruxelles* 1876

Dosseville. Printer, Coutans' *Environs de Paris* 1800

Dossow, R. von. *Fredericia und Umgegend* (1864?)

D'Ostening, H. von. *Teutschlandt* (1650)

Doswell, George. *Plan Southampton* 1842

Dotecum. See **Doetichum**

Dou. See **Douw**

Doublebsky, Robert (1839–1910). Czech astonomer and military geographer

Doudan. See **D'Houdan**

Doué, Martin (1572–1638). Artist of Lille. *Gallo Flandria,* Blaeu 1635

Dougal, W. H. American engraver. *Lake Michigan* 1852, *New Mexico* 1859

Dougharty, J. See **Doharty**

Doughty, C. M. *Reise* (Arabien) 1875–8, *W. Arabia* 1884

Doughty, W. E. *Lapeer County, Mich.* 1863

Douglas. *Plans Port of London* 1800 (with Telford)

Douglas, Bloomfield. Admiralty charts *S. Australia* 1856–60

Douglas, C. E. *Plan Waiho Country, Wellington* 1893

Douglas, D. B. *Map Niagara River* 1814, *Siege and Defence of Fort Erie* 1816

Douglas, J. *Sketch map N.W. district Otago* 1863

Douglas, N. Engraver of Edinburgh. *Wood's Plan St. Andrews* 1820, *Plan Brechin* 1822–(27)

Douglas, Dr. Robert. *Roxburgh & Selkirk* 1798

Douglas, William (1692–1752). Surveyor. *Brit. Dominions of New England* (1753), *Glasgow* 1777, *Environs of Greenock* 1778 MS.

Douglass, C. C. *Michigan* 1850

Douglass, L. F. Surveyor. *Jersey City* 1841–(76)

Doumerc, J. *Province of d'Oran* 1889

Doupe, J. *N. Saskatchewan,* Ottawa 1885

Dourado, F. Vaz. See **Vaz Dourado**

Douw [Dou], Jan Jansz. Cartographer and mathematician of Rijnland and notary of Leyden, son of following. *Alckmaer & Bergen,* Blaeu 1640, *Rijnland* 1646, first modern map on trigonometrical basis, *Kennemerland ende West-Vrieslandt,* 12 sh. 1647, reprinted 1687, 1746, (with Balthasar Florisz)

Douw [Dou], Jan Pietersz. Surveyor and cartographer at Leyden; author of first Dutch book on surveying 1625

Dove, Capt. Hastings. *Chart march detachment of H.E.I. Co. troops in Sumatra* 1805 MS.

Dove, H. Percy. *Plans Sydney* (1880)

Dove, Heinrich Wilhelm. *Nördlichen Hem.* 1855, *Nordpolar-Länder* 1868, *Atlantic Ocean* 1868

Dovers, Capt. William. *Admiralty Bay, Bequia* 1811

Dowcett, Ralph. Estate surveyor. *Danbury* 1667 MS, *Little Baddow* 1677 MS.

Dowde, J. *Lake of Killarney* (1800)

Dower, John. Engraver, draughtsman and publisher Pentonville, London. Engd. for Sharp 1825–8, *Swire & Hutchings' Cheshire* 1830, for Greenwood 1830, Teesdale 1831–56. *New Gen. Atlas* 1831, *V.D.L.* 1831, for Moule 1834–7, Orr & Co. 1850–60, Weekly Dispatch 1858–63, revised *Bacon's Thames* 1869, *Bacon's Illust. Complete Atlas* 1871

Dowie, David. *Forrest of Mamlorne* 1735–(7)

Dowling, P. H. *Maps of Townships for Everts and Stewart's Combination atlas of Washington County* 1874

Down, Robert. *5 miles round Bridgewater* 1853 (with T. Clark)

Downer, W. *Berbice,* 3 sh. Arrowsmith 1844

Downes, C. (fl. 1778). Engraver of 112 Fetter Lane, London. For *Taylor's Kildare* 1783, *Whitworth's Canal Ashby de la Zouche-Coventry* 1792

Downes, Wm. (fl. 1829). Estate surveyor of Colchester. *Layer Marney* 1817 MS.

Downie, James. *Pt. Romania to Anambas Is.,* Dalrymple 1805, *Achen Road,* &c., Dalrymple 1806

Downie, J. *Plan Aberdeen* 1811

Downie, James. *Sketch River Thames in New Zealand* 1820

Downie, Murdo, Master R.N. (fl. 1788). Surveys *F. of Forth, R., Tay & Cromarty Firth* from 1790, *New Pilot for E. coast Scotland* 1792, *New Anchorage Bermuda* 1798, Collaborated with Telford, *Road map Scotland* 1803, *Intended Caledonian Canal* (1800)

Downing, Alex. *Plan Bury St. Edmunds* 1740

Downsborough, J. *Bexhill-on-Sea* 1888

Dowthwaite, A. *Tottenham (River Lea)* 1834

Doyle, D. *Gaza Land* 1891

Doyle, Wm. *Waterford* 1738, *Harbours Rineshark to Waterford* 1738, *Brit. Dominions* 1770

Doyley, John (fl. 1847). Land surveyor 10 Greville St., Holborn. *Farms Felsted & Gt. Waltham* 1796, *Fyfield* 1798 MSS, *Leyton, Wanstead & Woodford* (90 maps) 1811–12 MSS, *Aldersbrook* 1818

Doyley, John, younger. Surveyor of Grays Inn. London *Railway survey* 1835

D'Oyley, Wm. *Epping Forest* 1874

Dozy, G. C. J. *School Atlas* 1877

D'Perac. See **Du Perac**

Draak, F. Engraver for Aa (1715)

Drach, F. See **Drake**

Draeck, F. See **Drake**

Draeck, [Dragk] Pieter. 16th century Dutch cartographer

Drage, Theodore Swaine. *N.W. parts Hudson Bay* 1746–(68)

Drake. *Plan Cincinnati* 1815

Drake, C. F. Tyrwhitt. *Route map Negeb* (1871)

Drake [Drach, Draeck], Sir Francis (ca. 1540–1596). Admiral; circumnavigator (1577–80). *La Heroike enterprinse faict par le Signeur Draeck* (1581?), *Expeditio . . . in Indias Occident.* 1588 (contains view of St. Augustine; the earliest engd. view of any city in U.S.A.), *Vera totius expeditionis nauticae desc.,* engd. Hondius (1590)

Drake, Sir Francis . . . Nephew of above. *World encompassed* 1628, *(Mappe of the World)*

Drake, Francis. *Plan City York* 1736, *Roman roads County York* (1745)

Draper, Henry (1837–1882). American geographer and meteorologist

Draper, John. Engraver. *Moore & Jones' Travellers' Directory* 1802

Draper, W. A. Surveyor. *Colorado* 1861

Drapeyron, Ludovic (1839–1901). Geographer and historian of Paris

Drasche, Richard von. *Geolog. map Luzon* 1881

Drayton, F. Surveyor. *Plan City of Philadelphia* Carey & Lea 1824

Drayton, John. Engraver. *South Carolina* Carey & Lea 1822, *Alabama,* &c. for *Cabinet Atlas* 1830

Drayton, Michael (1563–1631). Poet. *Polyolbion* 1612, 2nd part 1622 (maps of rivers of Eng. & Wales)

Drayton, T. F. *Surveys for ship canals round Niagara, & to connect Lakes Erie & Ontario* 1835

Drebbe[r] [Drebbell], Cornelis Jacobsz. (1572–1634). Mathematician, engraver and publisher, b. Alkmaar, d. London. *Plan Alcmaer* 1597, *globe,* presented James I 1621

Drechsel, Wolf (d. 1644). Artist and surveyor of Nuremberg

Drenckan, C. A. (fl. 1750). Cartographer of Kiel

Drentwet[t], Abraham, Junior. Cartographer. Maps for Seutter 1725–41

Dresel. See **Kuchel & Dresel**

Dreselly, A. *Schliersee* (1896)

Dreuer, Domenicus. *Lower Elbe* (1569) MS.

Drew, Frederick. *Geology country between Folkestone & Rye* 1864

Drew, John. Publisher, No. 31 Fetter Lane, Fleet St., London. *Plan London* 1799, *Chron. charts Anc. Geog.* 1835

Drew, Lieut. (later Capt.) John Clarke. Admiralty surveyor. *Chart Petaldi Bay (Greece)* 1865 MS.

Drew, J. W. *Atlas Clarke County, Ohio* (with Lake, &c.) 1870

Drewry, W. S. *Slocan Mining Camp, B.C.* 1897, *Topog. survey Canada* 1900

Drewyer, G. See **Drouillard**

Drexel. *Forest County Penn.* (1868)

Dreyskorn, C. R. Engraver for Streit 1817, Schlieben 1828–30

Driat. *Plan Lyon* (1874)

Driesemann, C. *Plan Stadt Halle* 1876, 1892

Drigalski, Guido de. *Mapa ferro-carriles vapores Argentina* 1896

Drikerhoff. *Environs of Manheim* 1814

Dring, Thomas (fl. 1655–95). Publisher. *Olearius' Voyage* 1662

Drinkwater, Capt. *Labouan Is. & Bruni River* 1845 for Admiralty

Drinkwater, John. *I. of Man* 1826, *Regents Canal* 1830

Drioux, Claude Joseph (1820–1898). *Atlas d'Hist. et de Géog.* 1867, *Atlas univ. et class.* 1867 &c., *Nouvel Atlas de Géog. Moderne* 1876, *Atlas Universel* 1890

Dripps, Matthew. Publisher, 103 Fulton Street. *Plan N.Y. city* 1854, *Atlas of New Utrecht, N.Y.* (1887)

Driver, Abraham P. Estate surveyor. *New Forest* 1789 (with W. below, Richardson & King), *Romford* 1797 MS, *E. & W. Ham* 1799 MS, *Stapleford Abbots* 1803 MS (with W. below), *Manor Lambeth*

POLY-OLBION

GREAT BRITAINE

By
Michael Drayton
Esqr:

London printed for { M. Lownes. I. Browne. I. Hanaue.
I. Helme. I. Bushie. W. Hole.

Engraved title to M. Drayton's Poly-Olbion

1812, *Leigh Swatch (Essex)* 1814 (with E., W. & S.), *Leyton* 1820 MS, (with E. Driver)

Driver, Edward (1800–1853). Estate surveyor London, collaborated with G. N. Driver 1824–46. *Leigh Swatch (Essex)* 1814 (with A., W. & S.), *Hornchurch* 1814 MS, *Leyton* 1820 MS (with A. Driver), *Scilly Is.* 1833

Driver, George Neale. Estate surveyor, New Bridge St., London, later Richmond Terrace. Collaborated with E. Driver 1824–46. *Stapleford Abbots* 1824 MS, *Scilly Is.* 1833

Driver, Robert C. Surveyor. *Plan waste lands Epping Forest* 1877

Driver, Samuel. Surveyor. *Dane-End Farm, Weston (Herts.)* 1768

Driver, S. Land surveyor. *Manor Lambeth* 1812, *Leigh Swatch (Essex)* 1814 (with A., W., & E. Driver)

Driver, William. Estate surveyor. *Plan New Forest* 1789 (with A. above, Richardson & King), *Stapleford Abbots* 1803 MS, (with A. above), *Leigh Swatch (Essex)* 1814 (with A., E., & S.)

Drivet, F. Relief map *Environs Paris* 1875, *Afrique, Amérique* 1878

Droeshout [Drowshot], Michiel (b. 1570) d. London. *Allegory on London & Gunpowder Plot* 1605, *Plan Breitenstein* 1631

Drogenham, G. See **Droogenham**

Dromeslawer, Robert. *Plan Hartlepool Harbour* ca. 1585 MS.

Dron, Henri. *France Militaire* 1867, *L'Europe des Points Noirs* (1869)

Droogenham [Drogenham], Gerrit [Gerard]. Land surveyor and engraver of Amsterdam. *Plan Amsterdam,* Visscher (1700), *Nova tab. Imp. Russici* 1704, for Du Sauzet 1734

Drouaillet. Lithographer, Corner Washington & Kearney, San Fran., for Milleson 1864

Drouet, Santiago. Engraver for Torfiño de San Miguel 1786

Drouillard [Drewyer], George. *Yellowstone & Big Horn* 1808 MSS.

Drouville, Col. G. *Contrées entre l'Indus et l'Euphrate* 1827

Drowshot, M. See **Droeshout**

Droysen, Prof. Gustav (1838–1908). *Handatlas* 1886

Droysen, W. *Goldminen Gebiete . . . W. Australien* 1896

Dru, Léon. (b. 1837). French geographer and engineer. *Mission géolog. au Caucasie* 1882

Drude, Dr. Oscar (1852–1933). Botanist and geographer in Dresden. *Physikalischer Atlas* 1886–92

Drugeon. *Petit Atlas Commercial* 1824, *Carte Routière France* 1868

Drummond, A. *Cyprus* (1753)

Drummond, Willis. *U.S.* 1871

Drury, Comm. Byron. Admiralty surveyor. *Lowang Channel* 1840 (with Woolcombe), *Survey New Zealand* 1850–6, *Ahuriri Road & Port Napier* 1857

Drury, Luke (d. 1845) *Geog. for Schools* 1822

Drury, Capt. William O'Brien, R.N. *Charts Ireland, Bantry Bay &c.* 1788, *surveys Harbours Rutland . . . Blacksod . . . Corke & Waterford,* 1789, *Cork* 1790

Druisiani, S. *Linie Telegrafiche d'Italia* 1869

Druten, J. van. *Guinea,* Utrecht 1895

Dryander [Eichmann], Johannes (1500–1560) b. Hesse d. Marburg. Mathematician and astronomer, Professor at Marburg. *De Globulo Terestri* 1537, *De usu Instrumenti* 1538, *Hesse,* used Münster 1540 & Ortelius (1579), *Cosmographiae Introductio* 1543

Drysdale, Robert. Surveyor of Dunfermline. Corrected *Fife* for Thomson 1832

Du Bac. *Viconté de Turenne* (1670)

Dubail, Lieut. Augustin von Edmond. (b. 1851) *Europe centrale* 1875, *Atlas Classique de Géog. Univ.* 1877, *Atlas de l'Europe militaire* 1880

Dubar. *Belgique* 1832

Dubercelle. Engraver. *Ville et Abbaye de Tournes* (1753), *Besançon* 1735

Duberger, John B. *Plan part Prov. Lower Canada* 1795 MS, *Country west of L. Ontario* 1800 MS, *Plan Quebec* 1800 MS.

Du Bocage, Georges-Boissaye, elder (1626–1696) Prof. Roial en la Navigation au Havre de Grâce. *Carte Ronde* 1679

Du Bocage, Georges-Boissaye, younger (ca. 1661–1717). Son of above, hydrographer of Le Havre. Assisted father in observations on tides

Dubocage, J. D. Barbié. See **Barbié du Bocage**

Dubois. Engraver for Lapie 1816

Dubois, A. *Carte routière et hydro. Seine et Oise* 1876

Dubois, Edmond-Marcel (1856–1916). Geographer of Paris.

Dubois, François. Imprimeur du Roy. *Toul* 1608

Dubois, J. *Cours de Weser* (1758)

Dubois, Jean. *Mont Blanc* 1825

Du Bois, Lieut. J. V. *Camp Floyd to Fort Union* 1860 MS.

Du Bois, N. Draughtsman for U.S. Topographical Engineers 1868

Dubois, T. *Plan Battle Culloden* 1746 MS.

Dubois de Montpereux, Frédéric (1798–1850) Swiss geologist. *Voyage antour du Caucase* 1839–43

Dubon. *Atlas* 1878

Dubosc, Claude. *Bethune* 1731, *Arras* 1736, *Plan Gibraltar* (1750)

Du Bouchet, Michel. *Pais d'Auvergne* 1645

Dubrena, V. *Tab. Géog. de la Navig. de l'Emp. Franç.* 1811, *Carte Hydro. de France* 1828, *Navig. de France* 1838

Dubreuil, Pierre Justin Charles. *Plan Baie Coquimbo, Chile* and *Iquique* 1826–(36)

Dubrovin, Mark. Russian seaman and cartographer. *Bokhara* 1731 for Kirilov

Dubuisson. Engraver for Tardieu ca. 1765, Mentelle's *Atlas Univ* 1797

Dubuisson, Fr. R. André (1763–1836) French chemist and mineralogist. *Geolog. studies in Loire Inf., Essai d'une méthode géolog.,* Nantes 1819

Duby, Père. *Mission de Sénégambie* 1877

Ducarla-Bonifas, Marcellin (1738–1816). *Cosmogonie* 1779, *Hist. Nat. du Monde* 1782

Du Carlo. Geographer to French King. *Plan de La Rochelle* 1628

Du Chaffat, Antoine (fl. 1750). French engraver and cartographer in Bavarian service. *Philippsburg* (1734), *Plan Alt-Brisach* 1735, *Plan Memmingen* 1737, *Crimea* (1738)

Du Chaillu, P. B. See **Chaillu**

Duchaxel, Marie Edme Félix. *Mer médit., Côte d'Afrique* 1857

Duchemin, N. L. *Tableaux des villes de France* 1777, *Royaume de France* 1815

Duchesne. Bookseller, Rue St. Jacques, Paris. *Hist. Jersey et Guernsey* 1757

Duchetti [Duchet, Ducheto, Ducheti], Claudio (1554–1597). Publisher of Rome, nephew and part successor to Lafreri 1577, *World* 1570, *Europe* 1571, *Africa* 1579, *Naples* 1585. Business continued by heirs. *Pozzuoli* 1586. *Calais* 1602

Duchier, E. Engraver for Vuillemin 1873

Ducker, E. Estate surveyor. *Fryerning* 1770 MS, *E. Tilbury* 1773 MS.

Dücker, F. See **Duecker**

Du Clesmeur, Chevalier de. Commander of *Castries* in Marion's expedition to New Zealand. *Terres de Diemen* (1772), *Mouth of Dwina* 1785

Duclout, J. *Pampa Central* 1887, *Rep. Argentina* 1888

Ducrest, Ch. Louis, Marquis de (1747– 1824). *Machines hydrauliques* 1777, *Courants d'eau* 1800, *Nouveau système de navig.* 1809

Du Creast, J. B. M. See **Crest**

Du Creux [Creuxius], François (1596– 1666) b. Saintes, d. Bordeaux. Jesuit. *Historia Canadensis* 1656, 1664, *Tabula Novae Franciae* 1664

Ducros de Saint Germain, A. M. P. *Dépt. Puy-De-Dôme* 1881

Dudith, Andreas. 16th century Hungarian astronomer

Dudley, J. *Fort William & Mary (Piscataqua River)* 1699 MS

Dudley, Sir Robert [Styled Duke of Northumberland, Earl of Warwick] (1574– 1649). Engineer and geographer, living in Italy, d. Florence. *Arcano del Mare*, Florence 1646–7, 1661 (First sea atlas by an Englishman)

Duetechum. See **Doetichum**

Dufart, P. *Atlas Géog. Anc. Hist. et Comparée* 1839

Du Fayen, J. See **Fayen**

Duff, Patrick. *Elgin & Nairnshire* 1842

Diffee, Wm. Printer for *Tanner's N. America* 1822

Duffin, Capt. Robert. *Pescadore Is.* 1792, pub. Dalrymple

Duffossat, G. See **Soniat du Fossat**

Du Fief, Jean Bap. Ant. Jos. *Atlas de Belgique* 1894, *Congo* 4 sh. 1895, *Atlas Général* 1895

Duflot de Mofras, Eugène (1810–1884). Cartographer. *Exploration du territoire de l'Orégon, des Calif. et de la Mer Vermeille*, Paris 1844 (contains the first plan of Sitka)

Du Fossé, Nicolas. Publisher, Rue Sainct Jacques au Vase d'Or, Paris. *Avity's Estats du Monde* 1619

Dufour, Adolphe Hippolyte (1798–1865). Geographer. *Atlas Classique* 1830, *Atlas Historique* 1840, *Océanie* 1851, 1863 *Atlas historico* 1852 (with Duvotenay) *Atlas universel* 1860, *Atlas Géog.* 1861

Dudley, Sir R.

Duecker [Dücker, Dükher] **von Hasslau zu Urstein u. Winkel**, Franz (1609–1671) b. Innsbruck. *Salzburg* 1666

Duems, W. *Mittel Europa* 1889

Duentzfeld [Düntzfeld], Johann Friedrich. *Nieder-Isenburgischen Länder* 1772

Duerer [Dürer], Albrecht See **Dürer** (1471–1528).

Duerrich, Major von. *Höhen-Karte v. Würtemberg* 1865

Duerrich [Dürrich], Ferdinand von. *Atlas der Schlachten* 1857

Dufour, Alphonse. *Carte routière de France* 1878

Dufour, Auguste Henri (1798–1865). Cartographer. Maps for Letronne 1827, *Atlas classique* (1835), *Atlas de géographie numismatique* 1838, *Amérique du Nord* 1860, *Atlas hist.* (1864)

DELL'ARCANO
DEL MARE,
DI D. RVBERTO DVDLEO
DVCA DI NORTVMBRIA.
E CONTE DI WARVICH.

PARTE SECONDA DEL TOMO TERZO
CONTENENTE IL LIBRO SESTO,

Nel quale si tratta delle Carte sue Corografiche , e Particolari.

AL SERENISSIMO
FERDINANDO SECONDO
GRAN DVCA DI TOSCANA
SVO SIGNORE.

Orsa Minore.

A F°L Fe

IN FIRENZE,
Nella Stamperia di Francesco Onofri. 1647. Con licenza de' Superiori.

Title-page to the first edition of Sir Robert Dudley's
Arcano del Mare (published Florence, 1647)

Dufour, Guillaume Henri (1787—1875). Swiss general. Survey of Switzerland from 1830. *Grande carte de la Suisse* 1833—65. *Eisenbahn-Karte Schweiz* 4 sh. 1874

Dufour, G.H.

Dufour, H. *Atlas Universel,* Paris 1860

Dufour, Piotr [Pierre]. French painter in Warsaw. *Geog. Dict.* (Polish) 1782

Dufrénoy. Pierre Armand (1792—1850+). Geologist. *Deutschland* 1838, *Carte géol. de la France* 1840, 1865

Dufresne, A. *Possessions et colonies françaises,* for Levasseur 1879

Dufresne, M. See **Marion-Dufresne**

Du Fresnoy, L. See **Lenglet du Fresnoy**

Dugan, Frances L. S. *Kentucky* 1818

Dugay, Capt. A. *L'entrée du Mississippi* 1803

Du Guay-Trouin, René. *Expédition de Rio de Janeiro* 1712

Dugdale, James. *New Brit. Traveller* 1819

Dugdale, Thomas (fl. 1860) *Curiosities of Gt. Brit.* (maps by Cole) 1835 &c.

Dugdale, Sir William (1605—1686). Garter King of Arms, antiquary and geographer. *Warwick* 4 sh. 1656, *Romney Marsh* 1662, *Fens* 1662

Du Halde, Jean Baptiste (1674—1743). French Jesuit, geographer of Paris. *Description de la Chine* 1735

Duhamel, A. *Carte géolog. Haute-Marne* 1857—60

Duka von Kadar, Friedrich Peter, Frhr. von (1756—1822). Military cartographer of Vienna

Düker, F. See **Duecker**

Duker, Jonas Pedersson. Surveyor late 17th century, son of following

Duker, Peter Jönsson. Surveyor mid 17th century

Duke's Map (anon.) *Description of the Towne of Mannados or New Amsterdam (New York)* 1664 MS. in British Museum. Information probably supplied by Jacques Courtelyon and possibly drawn by W. Hacke

Dukes, Thomas. *Planisphere* (1820)

Du Lac, P. See **Perrin du Lac**

Dulague, V. Fr. J. N. (1729—1805). French hydrographer b. Dieppe, d. Rouen. *Leçons de Navig.* 1768, *Principe de Navig.* 1787

Dulaure, Jacques Antoine (1755—1835). *Nouvelle Topog.* 1784, *Hist. plans Paris* 1823, 1829

Dulcert [Dulceti], Angellino. Chart-maker of Majorca or Genoa. Probably identical with Dalorto q.v. *Mappemonde* 1339

Duller, Eduard. *Rhenish Atlas* 1845

Dumaresq, J. Surveyor, mid 17th century

Dumaresq, Philip (?1650—1690). Seigneur of Saumarez. *Jersey* in *Falle's Account of Jersey* 1694, also used in *Camden's Britannia* 1789

Dumas, Emilien. *Cartes géolog. Dépt. Gard* 1844—75

Dumas, Comte Guillaume Mathieu (1753—1837). *Evénements militaires* (1816—26) (maps by Tardieu)

Dumas de Fores, Isaac. *Clef de la Géographie Générale* 1645 (with Boisseau)

Dumbleton, John. *Plan Parish St. Neots, Hunts.* 1770

Du Meé, L. *Baay van Vigos, & Cadix,* used Beek 1702

Duménil. Printer for Dumont d'Urville 1833

Dumeresq, Jean (1749—1819) Surveyor

Dumez [La Veuve]. Publisher, Rue de la Harpe No. 26, Paris. *Atlas National Portatif de France* 1790 (with Chanlaire)

Duminy, Chevalier. *Coast Africa* 1781—5 pub. 1787, *Plan Beschermer's Harbour* 1793 pub. Dalrymple

Dummer, Edmund. Surveyor of the Navy. *Plymouth Sound* 1682, *Portsmouth* 1698, *Ports of S. Coast* 1698. MSS, (17 plans)

Dumont, André H. (1809—1857). *Carte géolog. de la Belgique* 1856, *carte géolog. de l'Europe* 1875

Dumont D'Urville, Jules Sébastian César (1790—1842) French aristocrat, navigator and geographer. Great contribution to mapping of Australia. *Voyage of Corvette Astrolabe* 1833, *Voyage autour du Monde* 1834, *Voyage au Pole Sud* 1841—54

Dumortier. Engraver for Gouvion Saint-Cyr 1821

Dumoulin. See **Vincendon-Dumoulin**

Dun, S. See **Dunn**, Samuel

Dunbar. *Chart Portland* (Dorset) 1790

Dunbar, Colonel David. *Sketch map New Hampshire* 1730

Dunbar, Samuel. *Map Town of Taunton* 1836

Dunbar, Thomas. Surveyor. *Jamaica Estates* 1730 MS.

Dunbar, Wm. (1749—1810). U.S. surveyor General. *District Natchez, La.* 1798 *Washita River, La* 1804

Dunbibin[e], Daniel. Pilot of North Carolina. *Chart of Cape Hateras to Cape Roman* [1755?], pub. in *Norman's American Pilot* 1792

Duncan. *Tramways Birmingham & S. Staffs.* 1855, *Tramways London* (1855)

Duncan, Capt. Charles, R.N. *Charts W. coast America* 1788, pub. Dalrymple 1789, *Strait Juan de Fuca* 1790

Duncan, D. See **Duncan**, W. & D.

Duncan, James. Publisher, Paternoster Row, London. *Scotch Itinerary* 1805, *Lothian's Atlas of Scotland* 1826, *County Atlas of England and Wales* 1833 and editions to 1845

Duncan, Robert. *Survey Nargen Is.* 1809, pub. Admiralty 1810

Duncan, William. Principal Draughtsman Q.M.G. Department. *Dublin to Wexford* 1806, *County Dublin* 8 sh. 1821, *Ireland* 1823, *Dublin Castle* 1847

Duncan, W. & D. Engravers for *Macfarlane's Plan Greenock* 1842

Duncan. See **Ogles, Duncan & Co.**

Duncker, Johann Heinrich. Printer. *Orbis Terrarum* 1677

Dundas, A. *Triangulation Prov. of Wellington (N.Z.)* 1872

Dundas, Lieut., Colin M. *Easter Is.* 1868 MS, pub. Admiralty 1869

Dundas, Lieut. [later General Sir], David (1735—1820). *Military survey Scotland* 1752—5

Dundas, D. *Portland to Southampton* 1757

Dundas, Comm. F. G. *Caravan route Hameye to Mt. Kenia* 1892, *River Juba* 1893 MS, *Delta River Niger* 1895

Dundee. Engraved for Seller 1675

Duner, A. *Spitsbergen* (1865), *Orig. Karte von Spitsbergen* 1865

Dunham, A. *Middlesex* 1781 MS, (with Rue)

Dunham, C. T. *County Stephenson, Ill.* 1859

Dunham, F. H. *Atlas Plainfield, N.J.* 1894

Dunham, J. R. *Atlas Hamilton County, Nebr.* 1888

Dunker. *Stettin* (1897)

Dunker, Balthasar Anton (1746—1807). Artist and engraver in Berne

Dunker, Philipp Heinrich (ca. 1780—1836). Artist and engraver, b. Berne, d. Nuremberg. Son of B. A. Dunker

Dunker, W. *Specialkarte Grafschaft Schaumburg* 1867

Dunkin, Edwin. 19th century English astronomer

Dunlap, Lauren. *Territory Dakota* 1885

Dunlap, W. *Plan Louisbourg* Philadelphia 1758

Dunn, E. J. *Geolog. maps; Beechworth, Victoria* 1871, *Cape Colony* 1874, *S. Africa* 1876, 1887

Dunn [Dun] Samuel (d. 1794). London mathematician and publisher, teacher of mathematics from 1751, "Boards young gentlemen and teacheth Penmanship, Merchants Accounts, Navigation, Fortification, Astronomy &c." Chelsea. *Directory for East Indies* 1767, *General Atlas* Sayer & Jefferys (1768), *New Atlas of Mundane System* 1774, *American Military Pocket Atlas* (1776). Drew maps for Gregory 1779, *World* 2 sh. 1780–1, *England and Wales* 1788

Dunnavant. See **Ritchie & Dunnavant**

Dunnett. Lithographer for *Gould's W. Tasmania* 1860

Dunnica, W. F. *City Glasgow, Howard Co., Missouri* (1860)

Dunning, A. G. *Ancient Classical & Scriptural Geography,* N.Y. 1850

Dunning. See **Prior & Dunning**

Dunnington. See **Gittings & Dunnington**

Dunod, Pierre Joseph, S.J. (1657–1725). Historian, revised maps

Du Noyer, George V. Geologist. *Ireland* 1854, *Geolog. survey Ireland* 1863

Dunoyer, Peter. Book and mapseller at the sign of Erasmus's Head, London. *Map & Plan of Preston* 1715

Dunstable, John. *Latit. & longit. ali. urbium* (1450) MS.

Dunthorne, J. Draughtsman. Cole's *Copford* (Essex) 1810 MS.

Dunton, John. *Journal Sallee Fleet* 1637

Duntze, A. *Hansestadt Bremen* 1851

Düntzfeld, J. F. See **Duentzfeld**

Dunwoody, Col. H. H. C. *Military telegraph lines in Cuba,* Washington 1899

Dupain de Montesson (1720–1790+). Geographer of Paris

Dupain-Triel, Jean Louis (1722–1805). Engraver and geographer, Clôitre Notre Dame, Paris. Engraved *Cassini's France* 1744–60, *Plan Toulouse* 1780, *Carte de la France* 1799

Duparc, L. *Carte géolog. du Massif du Mont Blanc,* Geneva 1898

Dupard, A. M. *Plan Lima* 1859

Duparquet, Père. *Côte du Loango* 1875, *Ovampo* 1881, *Okavango River (Bechuanaland)* (1880)

Du Perac [Dupeyrac], Stefano [Etienne] (ca. 1525–1604) Engraver, artist, architect and mapseller of Paris; in Rome 1559–82, Architect of the Papal Conclave 1572, *Naples* 1566, *Romae Descr.* 1577, *Jerusalem* n.d.

Duperrey, Louis Isidore (1786–1865). Charts for Freycinet 1818–20, *Opérations géographiques de la corvette La Coquille, Voyage autour du monde* 1822–25, *Atlas Hydrographie et Physique* 1827

Duperrier. *Prov. de Normandie* 1767

Duperron, Jean. *Eyland Ceylon* 1789, MS.

Du Petit-Thouars, Abel Aubert (1793–1864). French admiral and hydrographer. Circumnavigation in the *Venus* 1837–39. *Voyage autour du monde* 1836–39, published Paris 1841 *Plan de la Rade . . . des Isles St. Pierre et Miquelon* 1818–19, *Plan St. Pierre* 1824, *Nouvelle Zélande,* Dépôt 1845

Dupeyrac. See **Du Perac**

Du Pinet, Antoine, Sieur de Norroy (ca. 1510–ca. 1566). *Plantz villes et forteresses de l'Europe, Asie & Afrique,* Lyon 1564, used later by Braun and Hogenberg

Du Plat, Anton Heinrich (1738–1781). Cartographer of Hanover

Du Plat, Georg Josua (1722–1795). Cartographer of Hanover

Du Plat, Johann Wilhelm (1735–1806). Cartographer of Hameln

Du Plat, P. J. (1737–1776). Saxon military cartographer

Du Plessis, D. M. See **Martineau du Plessis**

Du Plouich, V. See **Plouich**

Dupon, Jean. See **Dupont,** Jean

Duponchel, Adolphe (b. 1821). French engineer and geographer. *Géog. du Dépt. de l'Herault*

Dupond, Guillaume. *Chart coast of Africa* 1755, pub. Dalrymple

Dupont, Edouard (b. 1841). Belgian geologist. *Lettres Congo* 1889, *Cartes géolog.*

Dupont, G. *Road Calais–Paris* 1814

Dupont, Jean. Chartmaker of Dieppe. *Chart W. coast Africa* 1625

Dupont, Paul. *Chemins de Fer d'Europe Centrale* 1857

Duportail, Louis le Beque de Presle (1743–1802). French military engineer. MSS, military plans, contributed to design for fortifications West Point, *Plan Position Valley Forge* 1778

Du Pratz, Le P. See **Le Page du Pratz**

Du Pré, Galliot. Publisher of Paris, *Finé's World* 1531

Dupré, Julius V. *Atlas Milwaukee* 1881 &c.

Dupuis. *Plan de St. Jean de Luz* (1790?)

Dupuis, André. *Plan de Jérusalem,* Nantes 1841

Du Puis, C. *Sonora* 1794

Dupuis, L. A. Belgian engraver. *Switzerland* 1769, for Ferraris 1777, *Cours du Danube* 1785

Dupuy, August Myionnet. *Union des deux Océans, Atlant. et Pacifique* 1855

Durado, F. Vaz. See **Vaz Dourado**

Durand. Engraver and mapseller of Paris. *Mantouë* 1734, for Cassini 1744–60, sold *Robert de Vaugondy's Atlas Portatif*

1748, engraved *Bataille de Chotemitz* 1758

Durand, Alfred. *Diego Suarez* 1890

Durand, F. *Chemin de Fer Dunkerque à Furnes* (1866)

Durand, G. *Plan de Niort* 1868

Durand, Maj. Gen. Sir Henry Marion, R.E. (1812–1871). Surveys used for *Walker's N. Punjab* 1846

Durand, Jean Baptiste Léonard (1742–1812). Diplomat and traveller. *Atlas au voyage du Sénégal* 1802

Durant, J. Engraver for *Tavernier's Japan* 1679

Duranty. *Plan de Vavao* used by Dumont d'Urville (1842–8)

D'Urban, Lt. Gen. Sir Benjamin (1777–1849). First Governor Brit. Guiana (1831), Governor Cape Good Hope (1833–7). *Settlements in Brit. Guiana* 1828

Durban, Wm. *Channel between Sardinia Sicily & Africa* 1810

Durbin, John Price (1800–1876) Methodist clergyman b. Kentucky. *Jerusalem after Catherwood,* Philadelphia 1850

Dureau de la Malle, Ad. J. C. Aug. (1777–1857). French geographer and historian. *Géog. Mer Noire* 1807

Durell, Lieut. (later Vice-Adm.) Philip. *Plan Guantanimo* 1740, *Porto Bello* 1740, *Cumberland Harbour* (Cuba), *Carthagena* 1741, *Louisbourg Harbour* 1745, *Dunkirk* 1764 MS.

Durell, Capt. Thos. *Harbour & Is. Canso* (1732), *Sea coast New Eng., Nova Scotia, New'f'd* n.d.

Dürer [Duerer], Albrecht (1471–1528). Artist and engraver of Nuremberg. *Imagines coeli septent.* 1515 (woodcut map of eastern hemisphere), supplied decoration for *J. Stabius' globe* 1515, *Underweysung* 1525 (contains instructions for drawing globe gores), *Hemisphaerium Australe* 1527

Duret-Noel (ca. 1599–1650). French cosmographer and mathematician

Durham, Thomas. Cartographer. *Isle of Man* 1595 used by Speed 1611

Durheim, C. *Malta* [1840]

Durieu, Antoine. *Traité de Turin* 1754 MS.

Durnford, Lieut. Elias. Engineer. *Plan Pensacola* 1765, *River Mobile* (1770), *River Mississippi* 1771, *Rivers Iberville &c.* (1771), all MSS, *Battle plan Walmstock* 1777–(80)

Durnford, E. P. *Survey St. Nicholas, Cape Verde Is.* 1823, *Plan Is. St. Mary Madagascar* 1827

Duroch. Charts for Dumont d'Urville (1842–8)

Durocher, J. (1817–1860). French geologist

Duroi, B. See **DuRoy**

Duroslan. *Baie d'Antongil* (Madagascar). 1770

Du Roy, Bernard. *Utrecht*, pub. Visscher (1700)

Duruy, Victor (1811–1894). *Atlas Hist. de France* (1857), *Carte comp. des Dépts.* 1865

D'Urville, J. S. D. See **Dumont d'Urville**

Dury, Andrew (fl. 1742–78). Publisher, printer, engraver, mapseller and surveyor 1768 at The Map & Printshop No. 92 under the Royal Exchange, Cornhill, London; 1742 at the Indian Queen in Duke's Ct., St. Martin's Lane. Rocque's *Dublin* 1757, *New Gen. & Univ. Atlas* 1761, *Plans Principal Cities Great Britain* 1764, *Porte-feuille . . . d'Italie* 1774, *Easburn's Plan Philadelphia* 1776, pub. *Kitchin's Atlas* (with Sayer), Collaborated with J. Andrews in county surveys &c. *Herts.* 9 sh. (1766), 1777, 1782, *Kent* 25 sh. 1769, 1775, 1779, 1794 (with Herbert), *Wilts.* 18 sh. 1773, *65 miles round London* 1774–93, *Roads of Italy* 1777

—— Mrs. Publisher. *Taylor's Louth* 1778

Dusault, Sr. *L'Isle Dauphine* 1717

Du Sauzet, A. G. *Geography epitomized* 1727

Du Sauzet, Henri. Publisher in Amsterdam. *Atlas portatif* 1734–8, *European city plans* 1739

Dusent, C. C. See **Duysend**

Du Simitière, P. G. See **Simitière**

Dussieux, Louis Etienne (1815–1894). French geographer. *Géog. hist. de France* 1844, *Cours de géog., Atlas Gén.* 1848, *Cartes du Canada* 1851, *Atlas Gén de géog. physique* 1854, *Atlas de Géog.* (1882)

Dussy, E. Engraver and draughtsman. For Robert de Vaugondy 1754–78, Caille 1760, Desnos 1767, *Mexico* 1784

Du Temple, V. *Plan de Saint Malo* (1865)

Du Temporal, J. See **Temporal**

Du Temps, J. See **Temporius**

Dutens, Louis (1730–1812). *Itinéraire des routes . . . de l'Europe* 1768–71

Du Tertre, Jean Baptiste. *Histoire générale des Indes* 1667, maps of *St. Christophe, Guadeloupe, Martinique, St. Croix* 1667

Duthie, James. Land surveyor. *Kincardine* for Thomson's *Scotland* 1832

Du Tral[l]age, Jean Nicolas, Sieur de Tillemon[t] (d. 1699). Revised Coronelli's maps for Nolin 1688–9, *Partie occid. de Canada* 1688, *partie orient.* 1689, *globe terrestre* (1690), *Duché de Bretagne* 1703

Dutreuil de Rhins, Jules Léon (1846–1894). French geographer. *Indo-Chine Orientale* 1881, *L'Asie Centrale* 1889, *Haute Asie* 1898

Dutton, Clarence Edward (1841–1912). American soldier and geologist. *Atlas high plateaus Utah* 1879, *Grand Canyon, Colorado* 1882, *Mt. Taylor & Zuni Plateau* 1886

Dutton, H. R. Surveyor. *Brooklyn* (1857)

Dutton, Walter. Surveyor. Estates *Bucks.* 1769 MS.

Petrus Du Val Abbavillæus, Geographus Regius, Ingenio perspicaci, laborare impigro, Eruditione non vulgari clarus; Ab Anno 1645. ad extremam usque ætatem, quam plurima opera Geographica, Mappas ex quisitissime delineatas, Chronologiam ordine commodo dispositam, Genealogias, itineraria tum antiqua tum recentia, Artem denique Heraldicam, in lucem dedit. Obiit. 29ª. Septembris An. D. 1683. Ætatis 65.

P. Du Val (1618–1683); engraving by Langlois

Duval. *Eske-Fiord (Iceland)* 1854

Duval, Ferdinand. *Atlas municipal,* Paris 1878

Duval, Henri Louis Nicolas (1783–1854). Cartographer, printseller. *Atlas univ. des sciences* 1837–1844

Duval, Jules (1813–1870). French geographer and publisher

Duval, Marie Françoise. Engraver for Dezauche's *Histoire Sainte* 1783

Du Val [Du Vall], Pierre (1618–1683) b. Abbeville d. Paris Géog. ordinaire du Roy, nephew and pupil of N. Sanson, Secretary to Bishop of Aire. (1) En l'Isle du palais sur le grand cours de l'eau (1653) (2) Rue Barillerie près le Palais (1655) (3) Proche le For l'Evesque en l'isle du Palais (1664) (4) En l'Isle du Palais sur le Quay de l'Orloge proche le coin de la Rue de Harley à l'ancien Buis (from 1665). *Aiguillon, Aire* for Jansson 1645, maps in *Jansson's Ancient Atlas* 1653, *Canada* 1653, *Amérique* 1655, *Cartes de géog.* 1662, *Le Monde ou la Géog. Univ.* 1662, *L'A.B.C. du Monde* 1670, *Provinces Unies* 1672, *Géog. Franç* 1677, *Diverses Cartes* 1677, *Géog. Univ.* 1682

—— Widow of above. Reissued his 4 sh. *Amérique* 1684, reissued Placide's maps 1692–3

—— Mlle. daughter of above, inherited business at father's last address, later at Rue St. Jacques au Dauphin. Reissued father's *World Amérique* (4 sh.) still dated 1684, *Placide's Siam* 1686, *Danube* 1703

Duval, P. S. Printer and lithographer of Philadelphia, Steam Lith. Press (1849), at 22 & 24 S. 5th St. (1861), Napier's *Peninsula War* 1842, *Plan Wilmington* 1850, *Milit. map U.S.* 1861

Du Vall, P. See **Du Val**

Du Verdier (1544–1600). *Le Voyage de France* 1655, 1673

Duveyrier, Henri (1842–1892). *Explor. du Sahara* 1859–61, Paris 1864, *Carte murale de l'Afrique* 4 sh. 1879

Du Villard, Jean. *Carte du Léman [Lake Geneva]* 1588

Du Voisin, L. See **Popellinière**

Duvotenay, Thunot (1796–1875). Geographer and publisher *Atlas hist.* 1840, *Atlas des campagnes de la Rév. Franç* (1846), *Atlas Géog. Hist. de la Suisse* 1851, *Atlas histórico* 1852, *Atlas de France et ses colonies* 1860 (with Dufour, A.H.); *United States* 1875

Duxmoulin, C. A. V. See **Vincendon-Dumoulin**

Duyckinck, Gerardus. Publisher. *Maerschalck's Plan New York* 1755

Duysend [Dusent], Cornelis Claes. Engraver and publisher of Leyden. Title to *Laet's W. Indies* 1630, *Hondius' Picardy* 1644

Dvořák, S. See **Dworzak**

Dwelling. Daniel. 17th century hydrographer

Dwight, Theodore Jr. (1796–1866). *Gazetteer of U.S.,* Hartford 1833

Dworžak [Dvořák], Samuel (d. 1689). Engraver of Prague. *Moravia* 1677

Dyck [Dijck], Heyman van den. Dutch land surveyor late 17th century. Maps of *Voorne* 1701

Dÿckerhoff, Jacob Friedrich. *Gegend von Mannheim* 1814

Dyer, R. Estate surveyor. *Little Yeldham* 1766 MS, *Layer Breton* 1767 MS.

Dyer, Thomas. Estate plans *Westmeon* 1809 MS.

Dyer. See **Longmans**

Dymock [Dymork], Abraham. Estate surveyor. *Gt. Dunmow* (Essex) 1801, *Manor Fifield & E. Overton (Wilts.)* 1811 MSS.

Dymond, C. W. *Worlebury nr. Weston-s-Mare* 1872

Dymork, A. See **Dymock**

Dyonnet. Charles. Engraver for Dufour 1838–63, Duvotenay (1846), Dépôt de la Marine 1850–61, *Plan Paris* 1850

E

E., A. W. *Survey of Woodford Row* 1863

E. V. See **Vico**, Enea

Eachard [Echard], Laurence (1670–1730) Historian and geographer. *Compleat compendium of geog.* 1691, *Gazetteer* 1692 (16th edition, 1744)

Eagleton, Thomas. Surveyor. Estate map *Bucks.* 1777 MS.

Eandavus. See **Ziegler**

Earhart, J. S. *City of Hamilton, Ohio* 1864

Earl, George Windsor. Admiralty chart *Arafura Sea* 1837 reduced 1840

Earl, Thomas. Admiralty chart *Lagos River* 1857

Earle, Augustine. *Survey Cawston Warren* 1732

Early, Eleazer. Publisher of Savannah. *Map of the State of Georgia* 1818

Earnshaw, J. *Plan Bridlington Quay* 1891

Easburn, Benjamin. Surveyor General of Philadelphia 1732–41. *Plan City Philadelphia,* A. Dury 1776, *Parts of Pennsylvania and Maryland* 1740

East, Thomas (1540–1608). Printer. *The Post of the World* 1576

Eastgate, John. Engraver, *chart coast Northumberland and Durham* 1819

Eastin, Lucien Johnston. *Emigrant's Guide to Pike's Peak* Leavenworth City 1859

Eastman, Capt. S. *Indian tribes U.S.A.* 1852–(3), *Nebraska & Kansas Territ.* 1854

Eastoe. Engraver. *Texel, Vlieter Roads &c.* 1812–(3)

Easton, Alex. *Mailroads Milford to Caermarthen* 1824–(7), *N. Pennsylvania R'road* 1857

Easton, Capt. L. C. *Ft. Laramie to Ft. Leavenworth* 1849

Eastwood [Estwood, Eschuid], John (d. 1380) b. Ashingdon, Essex. *Summa Astrologiae,* Venice 1489 (with *World* map)

Eaton, Amos (1776–1842). Geologist. *Econ. geology of N.Y.* 1830, *survey Erie Canal* (1825)

Eaton, Charles Y. Surveyor. Assisted Colby *Atlas Somerset County Maine* 1883

Eaton, W. C. *Map & Town of Enfield, New Hampshire* 1855, *Tolland Co., Conn.,* 1857

Eaton. See **Hunt & Eaton**

Eayre, T. See **Eyre**, Thomas

Ebano, Elevdoro. *Bay of Paranagua* 1653

Ebden, William. *Atlas of English Counties* 1828, *Spanish America* 1820, *Twelve miles round London* 1848, *Railway and Telegraphic map of Yorkshire* 1856

Ebdy, T. C. Surveyor. *Plan City Durham* 1865

Ebel, J. A. *Elbe-Mündungen* 1846

Ebel, Johann Gottfried (1764—1830).
German geologist. *Traveller's guide through
Switzerland* 1793—(1818)

Ebeling, Chr. D. (1741—1817). German
geographer. *Geschichte von Amer.* 1799—
1816, *Portugies u amerik Landkarten*
1800

Ebeling, E. *Plan Stadt Braunschweig*
(1860)

Eberhard, Dupré. *Madagascar* 1667

Eberhard, H. See **Eberhardt**

Eberhardt. Surveyor. *Plan Tübingen*
1876

Eberhar[d]t, H. Engraver for Stein 1852,
Berghaus 1863

Eberhart, G. A. *Hist. Atlas Carroll County,
Ohio* 1874, *Hancock County, Ohio* 1875

Eberle, J. M. Swiss surveyor. *Appenzell*
1846

Ebersperger [Ebersberger], Johann
Georg (1695—1760). Engraver and
publisher. Son-in-law of Homann, continued the business with J. M. Franz
as Homännische Erben (1730)

Ebert, Frederick J. *Colorado Territory*
1862

Ebert von Ehrentreu, Karl (1754—1813).
Austrian cartographer

Ebhard. Engraver, 42 Rue Bonaparte,
Paris. For Bineteau 1860

Ebner, G. Bookseller of Stuttgart. *König.
Würtemberg* 1831

Ebray. *Carte géolog. du Dépt. de la Nièvre*
1864 (with Bertera)

Ebstorf Map. Circular *world map* 3.58 ×
3.56 metres, perhaps by Gervase of Tilbury
q.v. ca. 1284. Destroyed in World War II

Ecckebrecht, P. See **Eckebrecht**

Echard, L. See **Eachard**

Echegaray, Martin de. Pilot. *Golfo de
Mexico & Amer. Sept.* 1686 MS.

Echenique, Santiago. *Prov. de Cordoba*
1866

Echeverrai, José de. *New Mexico* 1747 MS.

Eck, C. W. A. von. *Gegend von Magdeburg*
1825

Eck, Dr. Heinrich. *Karte geolog.
Rüdersdorf* 1872, *Schwarzwaldbahn* 1884

Eckard. Engraver. Wiebeking's *Holland
u Utrecht,* Darnstadt 1796

Eckebrecht [Ecckebrecht] Philip (1594—
1667) Mathematician and astronomer of
Nuremberg. Friend of Kepler for whom
he drew a *world* map, embraced by a double
headed eagle, for the *Tab. Rudolphinae,
Ulm* 1630. The first printed map to show
the Dutch discoveries in Australia

Eckener, F. Lithographer and printer.
Schleswig-Holstein (1867)

Ecker, Johann Anton (1755—1820?).
b. Graz, artist and cartographer in
Vienna. *Nördliche and Südliche
Halbkugel der Erde* 1800

Ecker, John Alexander (1766—1829).
Bohemian physician. *World* 1794, *S.
Hemis.* 1800

Eckersberg, Chr. W. (1783—1853). German painter and engraver. Town plans

Eckhardt, Dr. Christian Leonhard [or
Christoph Ludwig] Philipp (1784—1866].
Land surveyor. *Hessen u Nassau* (1823),
Stern-Karte 1853; various globes

Eckstein, C. A. (1840—1925). Cartographer and Director Topog. Service of
Dutch Army H.Q. *Java* 1883

Ecluse, C. de l'. See **Clusius**

Eddison, J. Surveyor. *Roundhay Park
Estate* 1871

Eddy, James. Lithographer. *Lake Huron*
1820, *Lake St. Clair* 1828, *Detroit* 1828

Eddy, John H. 30 *miles round New York*
1812, *State of N.Y.* 1818

Eddy, Wm. M. Surveyor of the Town of
San Francisco. *Official map San Fran.*
1849, *Official map State of Calif.* 1854

Edelgestein, R. See **Gemma Frisius,** R.

Edeling, A. C. J. *Eilanden Java* 1842

Edelmuller, Friedrich (1840–1905). Austrian military cartographer

Eden, P. Surveyor of Iver, Bucks. *Part of Iver* 1850 MS.

Eden, Richard (ca. 1521–1576). Translator. Münster's *Cosmography* 1553, *Decades of the newe worlde* 1555 (map by Bellère)

Eden, William. *Windsor Park & part forest* 1800

Eden, W. & Co. Printers and publishers 14 Basinghall St., London. *Smith's Syria* 1840

Eder, Franciscus Xavier, S.J. *Prov. Moxos* (Peru) 1791

Eder, W. *Hand Atlas der allgem. Erdkunde* 1861

Edgar, John. Surveyor. *Manor Strickland, Dorset* 1735

Edgar, Lieut. Thomas. Marine surveyor. *W. Falklands* 1786–7, pub. Arrowsmith 1797

Edgar, Thomas. Master of the *Discovery. Plan of Ship Cove in Charlotte Sound*

Edgar, William. Surveyor. *Peebles* 1741, *Plans Edinburgh* 1742, 1765, *Surveys Scotland* 1743–6, MSS., *Stirling* 1745, *Kinlock Rannock* 1756

Edgcome, W. H. *British Burma, Pegu Div.*, 4 sh. 1885 (with Fitzroy, F.)

Edgell, Capt. Harry Fowkes. *Paquet Harbour, Newfoundland* 1810–(15)

Edgerton, H. H. *Atlin Gold Fields* (Canada) 1899

Edgeworth, Richard Lovell. *Bogs County Longford & Westmeath* 4 sh. 1811, *Bogs River Shannon* 2 sh. 1814

Edgeworth, Wm. Surveyor. *County Longford* 4 sh. 1814, *Co. Roscommon* (with Griffith, R.), 1825

Edkins, Joseph. *E. & W. Hemispheres* 1864

Edkins, S. S. Globe maker and seller, son-in-law of W. Bardin. Pair *globes* 1799–1800 (terrestial globe ded. to Sir J. Banks)

Edler, Anton. *Würm-oder Starnberger-See's* (1866)

Edler, E. G. Engraver for Stieler (1816–50), for Spruner von Merz 1855

Edmands, Benjamin Franklin. Cartographer. *Boston School Atlas* 1832 (5th edition)

Edmonds, A. M. *Railways of Canada* 3 sh. 1891

Edmonds, Christopher. *Plan Parishes St. Mary Lambeth, St. Mary Newington &c.* 1833

Edmonds, Capt. Joseph. *Marmorice Harbour* 1802, Laurie & Whittle's *New Chart Greece* 1812

Edmonston, Dr. A. *Shetland* 1809

Edmunds, William. *Plan Margate* 1821

Edrisi. See **Idrisi**

Edward, Charles. *Plan Dundee* 1846

Edwardes. Major. *Country around Shanghai* 1865

Edwards. Lieut. Col. *Shire Highlands* 1897

Edwards, A. F. *Sacketts Harbor* 1835

Edwards, Bryan (1743–1800) W. India merchant. *History of the British West Indies* 1794 &c. to 1818, *Atlas of W. India Is.* 1810, 1818

Edwards, C. H. *Atlas Lorain Co.*, Ohio 1874 (with Lake, J. P. Edwards &c.)

Edwards, D. Publisher. Archer's *plan Manchester* 1834

Edwards, D. C. *Atlas Illinois* 1879

Edwards, James E. of Dorking. Publisher and surveyor No. 23 Belvidere Place, Southwark, London. *Steyning, Sussex*, 1791–(3), *Plan Lewes* 1799, *Trig. land chart* 1800, *Companion London to Brighthelmston* 1801, *View of Brighton* 1817

Edwards, John P. *Atlas Lorain County Ohio* 1874 (with Lake, C. H. Edwards

&c.), *Labette Co., Kansas* 1880, *Sumner Co.* (1833), *Cloud Co.* 1885

Edwards, Rev. Joseph. *Diocese Lichfield* 1873

Edwards, Langley. *Fens district* 1761

Edwards, Leicester. *Chart Boom Kittam (Sierra Leone)* 1878–[9]

Edwards [Edward] , Rev. Robert. Minister of Murroes. *Angus* (1680)

Edwards, Robert. Engraver. *Orkney Is.* 1711

Edwards, Robert. Hydrographer. *Harbour Kittie, Caroline Is.* 1839

Edwards, Talbot. Military engineer. *Portsmouth Harbour* 1716

Edwards, W. Printer. *Rocque's Environs of London* 1748

Edwards Brothers. Publishers Philadelphia. County atlases *Kansas & Missouri* 1876–82

Edwardson, Capt. *Charts of New Zealand, Foveaux Strait* 1822, *Rouabouki Road* 1823, *Admiralty Chart* 1840

Eek, P. van. Dutch cartographer. *Polder near Dordrecht* 1811

Eekhoff, Wopke (1809–1880). *Atlas van Friesland* 1849–59

Egan, F. W. Contributed to geolog. maps *Ireland* 1871–88

Egede, Chr. Th. (1761–1803). Grandson of following. Travels in E. Greenland

Egede, Hans (1686–1758). Norwegian missionary. *Relation grønlandske Missions* (1741) (2 maps Greenland)

Egerton, F. W. *Gettysburg Bank,* Admiralty 1877

Egerton, I. F. *Districts Midnapur & Hijellee* 1849

Eggenberger, Kiadja. Hungarian cartographer. *Australia* (1870)

Eggers, Aug. *Münz-Weltkarte* 1873

Eggers, Christian Ulrich Detlev. *Iceland* 1786

Eggler, Franz Anton (1779–1854). Draughtsman of Baden

Egglofstein [Egloffstein] , Baron F. W. von. Topographer, Geog. Institute No. 164 Broadway, N.Y. *Mississippi–Pacific Railroad* 1855, *W. coast surveys* 1858

Eglesefeild, Francis. Printer. *Hollar's Mappe of England* 1664

Egleston, Thomas (1832–1900). American geologist

Egli, J. J. (1825–1896). Swiss geographer, b. Zürich. *Nomina Geographica* 1872

Eglin, G. *Kanton Luzern* 1838

Egloffstein, F. W. von. See **Egglofstein**

Egmond, A. van. *Haarlemmermeer Polder* 1867

Egmont, *Madagascar* 1773 (with Cordé)

Egon, F. Buddhist priest and mapmaker. *Map of Jambudvipa* woodcut 1845

Ehnlich, I. F. C. Engraver for Gaspari 1808

Ehrenberg, Christian Gottfried. *Egypt* 1826, *Arabia Petrée* 1828–(34)

Ehrenberg, Herman. Civil engineer. *Klamath Gold Region* 1850, *Gadsden Purchase* 1854, 1858, *Silver regions Tubac* 1857

Ehrenreich, Dr. P. *Ethno. Karte v. Brasilien* 1891

Ehrenstein, Heinrich Wilhelm von (1811–1874). Cartographer of Dresden. *Königreich Sachsen* 1856

Ehrentreu, K. E. von. See **Ebert von Ehrentreu**

Ehrgott & Forbriger & Co. Lithographers of Cincinnatti. *Union Pacific Railroad* 1866

Ehrhardt. *Umgebung von Ruppin* 1888, *Umg. v. Marburg* (1896)

Ehrhardt, Ludwig. *Palestine* 1834

Ehricht, C. Engraver for Meyer 1830–49

Ehrmann, Theodor Friedrich (1762–1811). Geographer and printseller in Weimar. *Guinea* 1793, *Kafferland* 1797, *Senegal* 1804

Eian, Yamazaki. Japanese surveyor, 19th century. *Plan Hakodate* 1860

Eiberger, F. *Höhenkarte Württemberg* 1893

Eichbaum, C. M. Drew Bacon's *hist. picture map Eng. & Wales* (1875)

Eichler, Johann Gottfried, Jun. (ca. 1720–1770). Engraver of Augsburg. *Atlas Geogr. Portatilis,* Lotter (1760)

Eichler, Matthias Gottfried (1748–1818+). Engraver in Berne & Augsburg. *Weiss' Atlas Suisse* 1786–1802, *Grundriss Berne* 1790

Eichmann, J. See **Dryander**

Eichoff [Eichov, Eichovius], Cyprian. German geographer, publisher and print-seller. *Itineraries Italy, Spain, Germany &c.* 1603–6, *Liber Insignium* 1606

Eicholz, D. *Plan Königsberg* (1863?)

Eichovius, C. See **Eichoff**

Eichwald, Ed. (1795–1876). Russian naturalist and traveller. *Kaspischen Meers* 1834, *Géog. ancienne de la mer Caspienne* 1838

Eick, Albert. *Reise-Karte Mittel-Deutschland* 1869

Eicken, Fr. Lithographer. *Vienna* 1848

Eigenbrodt, G. Geographer, engraver and publisher. *Voies de Communic. de la Belgique* 1870, *Bruxelles* 1874

Eikelenberg, S. *Westfriesland, Kennemerland en Waterland* (1720?)

Eillarts, Johannes (1568–1612?) Engraver

Eimmart [Eimmerto], Georg Christoph, younger (1638–1705). b. Regensburg, d. Nuremberg. Astronomer, geographer, painter and engraver, fl. 1660 in Nuremberg. *Planis. caeleste* (1690), pair *globes* 1705, worked for Homann

Eisdell & Ashcombe. Surveyors. Navigation plans 1854 MSS.

Eisen, C. F. *Chemins de Fer de Prusse* (1846)

Eisenschmidt, Johann Caspar (1656–1712). German cartographer. *Tab. Germaniae,* Homann (1710)

Eitel, Edward E. *County atlas California*

(1894), *County atlas Oregon & Washington* (1894)

Eitoku Kano. (1543–1590). Japanese Artist. *Folding screen map of Japan* and *World map.*

Eitzing, M. See **Aitzing**

Ekeberg [Elkberg], Capt. Carl Gustav. *S. E. coast Hainan* 1760, *Dirck Vries & Maurice Bays* 1762, (both pub. Dalrymple), *Plan Cocos Is.* 1787

Ekel, F. C. *Plan ville Reinsberg* 1773

Ekholm, Nils (1730–1778). Swedish land surveyor of Vasa

Ekholm, N. *Karta öfver Amsterdamön* 1896–7

Eklund, A. W. (1796–1885) *Fürstendömet Finland* 1857

Ekman, F. L. (1831–1890). Swedish naturalist. Studies in oceanography

Eland[t], Hendrik (d. 1705) of Amsterdam. Designed and engraved cartouche for Halma's *Anc. Geog.* 1704

Elandt[s] [Elands], Cornelis (fl. 1660–70). Artist, surveyor and engraver of The Hague. *s'Graven-Hage* 1666

Elberts, W. A. *Overijssel* (1866?)

Elcaño, J. S. See **Cano.**

Eldad, the Danite. Hebrew traveller, 11th century

Eldberg, Carl. Swedish chartmaker, late 17th century

Elder, John of Caithness. *Scotland* ca. 1545, MS. for Henry VIII

Eldred, Edward. Estate surveyor. *Shalford* 1603 MS.

Eldridge, George. Surveyor. *Charts coast N. America* 1861–5

Eldridge. See **Davies & Eldridge**

Elekes, Ferenc. Hungarian globemaker in Vienna mid-19th century

Elekes, Franz von. *Grundriss Residenzstadt Wien* 1857

Elena, Félix. Prof. of drawing. *Railroads South America* 1893

Elephant, Gabriel (d. 1722). Swedish land surveyor of Gotland

Elfert, Paul (1861–1898). Geographer of Leipzig for Wagner and Debes

Elford, I. M. *S. Carolina* 1822 (with Blackburn)

Elfwing, Nere A. Schönberg's *World* 1861

Elias, José Antonio. *Atlas geog. de España* 1848

Elias, Ney. *Plan Tsien-tang river* 1867 MS, *Ortos Country* 1876 MS, *Yellow River* 1871 MS, *Sketch journey in Pamir & Upper Oxus regions* 1886

Elie de Beaumont, J. B. See **Beaumont**

Eliot, J. B. Aide-de-Camp to Gen. Washington. *Théâtre de la Guerre* (North America), Paris 1781

Eliseo, P. *Calabria Ult.* (1780)

Eliza, Don Francisco. Spanish explorer. *Vancouver Is.* 1791 MS.

Elkan, G. Bookseller. *Grundriss Harburg* (1859?)

Elkberg, C. G. See **Ekeberg**

Ellacott, C. H. *Rossland & its mines* 1897

Ellery, Robert Louis John (1827–1908). Astronomer, President Royal Soc. of Victoria. Melbourne Dept. Lands & Survey. *Rainfall maps Victoria* (1883), *S.E. Australia* 1884

Ellet, Charles Jr. (1810–1862). American civil engineer. *Pennsylvania Railroad* 1851

Ellicott, Andrew (1754–1820). Surveyor. b. Solebury, Pa. *Plan City of Washington* 1792 (with S. Hill; first official engraved plan) *Journal* 1803

Elliger, Ottmar (1666–1735). German painter and engraver, b. Hamburg, moved to Amsterdam 1679. Appointed court artist Mainz 1716. Moved to St. Petersburg, engraver to the Academy 1726. *Map France* pub. Visscher (1704)

Elliot, Capt. See **Elliott,** Capt. Wm.

Elliott, Charles (1776–1856). Engineer and geographer in E. India Co.

Elliott, Chas. L. *Atlas Newport, R.I.* 1893 (with T. Flynn)

Elliott, Lieut. Chas. P. *Mt. St. Helens* 1897

Elliott, George. *Entrance to S. Harbour Balembangan* 1845

Elliott, Henry C. *Valley Amazon,* to accompany Lt. Herndon's Report (1854)

Elliott, Henry W, Surveyor. *Yellowstone Lake* 1871, *St. Paul, Bering Sea* 1880, *Bering Sea* 1881

Elliott, S. G. Civil engineer. *Central Calif.* 1860, *Battle plans,* 1864

Elliott [Elliot], Capt. Wm. *Bay & Town Kingston, St. Vincent* 1817 MS. (with Langley) Admiralty 1820, *Courland Bay, Tobago,* Admiralty 1820

Ellis, A. G. *Town of Astor, Wisconsin* 1835, *Plat of Navarino, Wisconsin* 1836

Ellis, Charles H. *Detroit & environs*

Ellis, G. Publisher. Smith's Square, Westminister. *Atlas Eng. & Wales* (1819), *Gen. Atlas* (1823)

Ellis, George. Engraver for *T. J. Ellis' Hunts.* 1824

Ellis, Henry (1721–1806). Traveller and hydrographer. *Voyage to Hudson's Bay* 1746–7

Ellis, Henry T. *Manilla* 1859

Ellis, James. Estate surveyor. *Havering-atte-Bower* 1755 MS.

Ellis, John (1750–1800). Engraver and publisher of Clerkenwell. *New Eng. Atlas,* 1765, *Ellis' Eng. Atlas* 1766, *Scalé's Hibernian Atlas* 1776, *Bowles' Pocket Plan London* 1780, *Diocese Canterbury* 1782, *Paterson's Plan environs London* 1800

Ellis, J. *Bay S.W. of Suez* 1801, pub. Dalrymple

Ellis, Robert. *Java* 1811 MS, *Dist. of Boondelkhund (India)* 1813 MS.

Ellis, R. G. *Part New Brunswick Geology* 3 sh. 1880

Ellis, Thomas Joseph. Surveyor Vauxhall Bridge Rd., Westminster. *Hunts.* 1823–4, *Notts.* 1824–5, pub. 1827, *England* 1832

Ellis, William. *Chart Hawaii Archipelago* 1840

Ellison. *Survey Borough Bradford* used by T. Dixon 1856

Ellison, James. *Soondurbuns* 1891

Ellison, Richard. *Survey Rivers Swale & Ouze* 1735

Ellobet, Francisco. *Seville,* Homann Heirs 1781

Ellung, I. *Kort over Jylland* 1820

Elmes, James. *Survey Harbour London* (20 plans) 1838

Elmore, Capt. *Straits Singapore* 1799

Elmpt, Philipp, Frhr. von (1724—1795) Austrian general and cartographer, b. Damerschied

Elmesl[e]y, Peter (1736—1802). Bookseller. *Camden's Britannia* 1772

Elorriaga, Miguel. *Palaos (Philippines)* 1709

Eloy de Almeida, Romão. Portuguese engraver. *Carta militar Portugal* 1808

Elphinston, Mountstuart (1779—1859). Governor of Bombay. *Afghanistan* 1808—9

Elphinstone, Capt. J. *Saldanha Bay,* Admiralty 1813

Elphinstone, John. (d. 1753). Surveyor and engineer. *Lothians* 1744, *North-Britain* 1745, (used by both sides Jacobite Rising) *Scotland* 1745, *Fort William* 1748

Elsenwanger, Anton. Bookseller and publisher of Prague. District maps *Bohemia* (1771), 1794

Elsheimer [Elshaimer, Elzheimer], Adam (ca. 1574—1620) Artist and engraver, b. Frankfurt, d. Rome. *World,* Frankfurt 1598

Elson, Capt. Thomas. Admiralty charts. *Port Mandri* 1829—(43), *Graham Shoal* 1851

Elstobb, William. (1737—1793). Engineer, teacher of mathematics, of Lynn and Cambridge. *Sutton & Mepall Levels, I. of Ely* 1750, *Wisbech Channel* 1773—5 MS,

Bottom River Ouse 4 sh. 1776, *Great Level of Fens* 1793

Elstrack[e], Reynold [Renier] (1571—1630) English engraver, b. London. For Linschoten 1598, *Boazio's Ireland* (1599), *Baffin's E. India* 1619, signed *Speed's Norfolk* (1623 edn.), *Purchas' N. part of America* 1625

Elsworth, Richard. *Ulti. Aethiopum* 1739

Elton, Andrew. *Calabria & Naples* (1780)

Elton, Capt. James Frederick. *Route Tati . . . Delagoa Bay* 1870 MS, *Slave Caravan Route* 1873—4, *Gold Fields S.E. Africa* 1876, *Transvaal* 1877

Elton, John. English naval captain in Russian service *Caspian Sea* 1753

Eltzner, Adolf. *Plan Leipzig* 1847, *Saechsische Schweiz* (1867), *Plan Breslau* (1870), *Plan Bremen* [1871]

Elwe, Jan Barend [Baret]. Publisher of Amsterdam. *Zak atlas, Zeventien Neder. Prov.* 1786 (with Langeveld), *Reis Atlas . . . Duitschland* 1791, *Atlas* 1792

Elwon, Capt. T. Indian Navy. *Red Sea* 1830—4, *Jiddah* 1858

Elwood, S. D. *Upper Peninsula Michigan* (1870)

Ely, A. E. M. *City Palmyra, Marion Co., Missouri* (1860?)

Ely, W. W. *New York Wilderness* 1868, 1874

Elzevier, Isaac (1596—1651). Publisher in Leyden. Bertius' *Theatr. Geogr.* 1618

Elzheimer, A. See **Elsheimer**

Emery, Henry. Estate surveyor. MSS surveys Essex estates. *Tolleshunt d'Arcy* 1805, *Gt. Coggeshall* 1806, *Cressing* 1830

Emery, Louis. *Suisse* 1798

Emeterius of Valcavado. Copies of *Beatus' World map* made 968—978

Emmerich, N. Geographer and publisher. *Regierungs—Bezirks Arnsberg* 1845, 1860

Emmert. *Umgebung Münchens* (1866?)

Emmett, S. B. *N.W. Tasmania* (1865)

Emminger, Eberhard (1808–1885).
Artist and lithographer of Munich.
Erinnerung an den Rhein (1870?)

Emmius [Emmio], Ubbo[ne] (1547–
1625). Cartographer and historian, d.
Groningen. *Frisia Orientalis* 1590–(5),
maps of *Groningen & Friesland* 1616
used Hondius, Blaeu &c.

Emmons, Samuel Franklin (1841–1911).
American geologist and mining engineer.
Geolog. atlas Leadville, Colorado 1882,
1883, *Washington Geolog. Survey* 1883

Emmoser, Gerhard (fl. 1573). Astronomer,
clockmaker to the Emperor. *Celest. globe,*
Vienna 1579

Emory, Lieut. (later Lt. Col.) William
Hemsley (1811–1887). Topog. engineer,
U.S. army. Reduced Nicollet's
Mississippi River 1843, *Texas* 1844, *Ft.
Leavenworth to San Diego* 1847, *U.S.
1854*–(7), *U.S.A.–Mexico boundary*
1857

Emperger, Josef Edler von (d. 1818)
Cartographer and lawyer of Klagenfurt

Emphinger. *Geognostische Charte
Sachsen Schlesien* 1836

Emslie, (1813–1875). Draughtsman
and engraver of London. Reynolds'
Travelling Atlas 1848

Emslie, J. P. & W. R. Engravers. *Eng. &
Wales* 2 sh. 1877, 4 sh. 1877, 4 sh. 1896,
Railway maps Cumb. & Westm. 1897,
Edinburgh & Glasgow 1898

Enackel von Hoheneck, G. A. See
Enenckel von Hoheneck

Enagrins, C. E. *Sweden* 1797

Enaro, I. de. Publisher of Madrid. *Tour's
Cataluna* 1643

Enciso, Martin Fernandez de. Spanish
navigator and geographer. *Suma de
geografia,* Seville 1519

Encke, Johann Franz. *Carte céleste* 1835,
Paths of Comets 1858

Ende, Ferdinand Ad. (1760–1817).
German astronomer

Ende [Endenus], Joos van den (1576–
1618+). Cartographer and engraver of
Zierikzee. *Zirizea,* Blaeu 1649

Ende [Eynde], Josua van den (ca. 1584–
1634+). Engraver of Amsterdam. *Plancius'
World* 1604, *Blaeu's World* 1606, *France*
1607, *Hondius' Europe* 1617, *Jansson's
Gallia Vetus* (1636) & *Dombes* (1638).
Also worked for Waghenaer

Endenus, I. See **Ende**, Joos van den

Enderlin. Surveyor of Baden, late 18th
century

Enders[ch], Johann Friedrich (1705–
1769). Engraver. *Prussia,* Homann Heirs
1753, for Schraembl and Reilly, *Episcopatum
Warmiensern in Prussia* 1755

Endicott & Co. Lithographers of New
York. *Atlantic to Mississippi* (1854)

Endlicher, Stephan Ladislaus. *Atlas von
China* 1843

Endner, Gustav Georg (1754–1824).
Grundriss Stadt Cadix (1810), *Heligoland,
plan Flensburg*

Endter, Wolfgang Moritz. Bookseller of
Nuremberg. *Carinthia* 1688

Enenckel [Enackel] **von Hoheneck,**
Baron George Acacius. Austrian historian
and cartographer. *Greece* in 1614 edn. of
Thucydides

Enfant, P. C. See **L'Enfant**

Engall, J. Sherwin. *Belgian Congo* 1890

Engel, B. F. *Mecklenburg-Schwerin* (1846),
Mecklenburg Strelitz (1850)

Engel, Ernst (1821–1896). German
geographer and statistician

Engle, Samuel (1702–1784). b. Berne.
Mémoires et observ. géog. 1765 (maps
N.W. America)

Engelbrecht, Christian (1672–1735). Engraver, publisher and printseller in
Augsburg, pupil of Sandrart. *Vienna* 1706

Engelbrecht, Martin (1684–1756).
Publisher, worked with brother C. *Basel*
(1700), *Strassbourg* (1720), *Berlin* (1730)

Engeldue, J. Master, R.N. Additions to
Plan of Maldonado Bay 1820

Engelhardt, Lieut. Friedrich Bernhard
(1768–1854). Topographer and surveyor.
Preussen 1811, *Plan Danzig* 1813,
Environs Berlin 10 sh. 1816–19,
Deutschland 1824, *Preussischen Staate*
1827

Engelhardt, Maurice (1779–1842).
Russian geologist

Engelhardt, Paul. *Central Ost Afrika*
1886, *Eisenbahn-Karte Mittel Europa*
1887–(8)

Engelman, J. (fl. 1784–1802). *Rhynstroom*
1793, *Heusden en Altena* 1798 (with F. W.
Conrad)

Engelmann, Julius Bernhard. *Post und Reise
karte von Deutschland,* Frankfurt 1835

Engelmann, L. Dutch draughtsman, late
18th century

Engelmann, Wilhelm (1808–1878).
Publisher of Leipzig

Engelvaart, P. Dutch cartographer.
Voorne 1675

Enouy, Joseph. Geographer. *Invasions of
England* 1797, *Egypt* 1801, *Eng. &
Wales* 1805, 1857, *Scotland* 1803, *Ireland*
1808, *Europe* 1809

Enouy, J. C. Engraver. *Guernsey* 1832

Enrile, Nicolás. *Mindanao* 1826 MS,
Burias 1832 MS.

Enriques, Luis. *Rio Grande de la Magdalene*
1601

Ens, Gaspar. Editor. *Magnae Britanniae
Delicae . . . descriptio* 1613, 1617 edition
Ptolemy

Enschede, Isaac (fl. 1729). Engraver and
publisher of Haarlem with Jan Enschede.
Buache's *North West America* 1754

Ensenius, C. See **Buondelmonte**

Ensign, D. W. [**& Co.**]. *Black Hawk*

County, Iowa 1869, *Kane Co., Ill., De
Kalb. Co., Ill.* 1892, *Cass, Grand Forks
&c. Counties, N. Dak.* 1893

Ensign, Bridgman [& Fanning]. Publishers
156 William St., N.Y.; successors to
Thayer, Bridgman & Fanning q.v. 1854.
Reissued *Travellers Guide to U.S.* 1854.
T.B. & F.'s *U.S.A.* &c. 1854, *map Western
States* 1858, *R'road map U.S.* 1859,
U.S.A. 1865

Ensign, Thayer & Co. [Ensigns & Thayer].
Engravers and publishers, (1) 25 Park Row,
N.Y. & 10 Main St., Buffalo; and (2)
1849–51 at 50 Ann St., N.Y. Probably
successors to Phelps, Ensigns & Thayer
q.v. *Ornamental map U.S.* 1848, Phelps'
National map U.S. 1851

Ensign. See **Everts, Ensign & Everts**

Ensigns. See **Phelps, Ensigns & Thayer**

Ensigns & Thayer. See **Ensign, Thayer &
Co.**

Ensinck, Francis Jan (1806–1856+).
Lithographer of Breda. *Antwerpen* 1856

Enthofer, Joseph (1818–1901). Austrian
geographer and engraver in Washington
for U.S. coast survey

Entick, John (ca. 1703–1773). Schoolmaster. *N. America* 1763 in *Gen. Hist.
of late war*

Entrecasteaux. See **Bruny D'Entrecasteaux**

Entresz, K. A. W. *Prov. Posen* 1842

Eosander, N. German cartographer, 18th
century

Epler. Engineer. *Humboldt County* 1866

Epner, Gustavus. *Gold Regions in Brit.
Columbia* 1862

Epworth, Christopher. Surveyor. *Chart
part River Humber* 1820

Eratosthenes. (ca. 276–196 B.C.). Alexandrian
geographer and mathematician. Founded
scientific geography ca. 200 B.C.,
measured circumference of earth

Erault, Sieur Desparées. *Carte Marine de
l'Isle de Ré* (1680?)

Erbe, Ludwig. Relief maps *Europe* 1844–50

Erben, Josef (1830–1910). Geographer and globemaker of Prague. *Böhmen* 1873

Erdeswicke, Sampson. *A survey of Staffordshire* 1593 MS, printed 1717

Erdmann, A. J. (1814–1865). Swedish geologist. *Carte géolog. de la Suède* 1860–5

Eredia, M. G. de. See **Godinho de Herédia**

Erhard, Georges Erhard Schièble (1821–1880). French engraver of German origin. *Plan Paris* 1875, *Alpes Maritimes* 1880, *Cannes & environs* (1895)

Erhard[t]. Engraver, Rue Bonaparte 42, Paris, for Andriveau-Goujon 1841, Bonange 1878, Soc. de Géog. 1879

Erich[ius], Adolar[ius]. Cartographer Gaudersleben. *Thuringia* 24 sh. 1625, used Blaeu 1634

Erichsen, Hr. C. R. *S. Norge* 1785

Erichsen [Erichson], J. See **Eriksson**

Ericsson, Emil. *Sverige* 1876

Ericsson, H. See **Eriksson**

Eriksson [Ericsson], Harald. *Frövi-Falu Jernbanan* 1867

Eriksson [Erichsen, Erichson], Prof. Jon (1728–1787) *Iceland* 1771–(2), 1780 (with Schönning), *Sydlige Norge* 1785

Erimeln. Engraver for Moreau 1772

Erizzo, Count F. M. *Terre Polare* 1853

Erkel, M. van. Architect. *Plattegrond stad Arnhem* 1853

Erkert, R. d'. *Atlas ethnog. des Polonais* 1863

Erlinger, Georg (ca. 1485–1541). b. Augsburg, printer in Bamberg. *Teutscher Lanndt* 1524

Erman, Georg Adolf. *Ural Gebirge* 1837, *Kamtchatka* 1838

Ermirio, Girolamo. *Contorni di Firenze* (1840)

Erns, E. *Duchy of Würtenberg* 1842, railway maps 1844 *Karte Württemberg* (1853)

Ernst, Adolph (1832–1899). German geographer, worked in Venezuela

Ernst, Hans. Globemaker, late 16th century

Ernst, K. German cartographer. *World* (wall map) 1830, *Atlas of Biblical geog.* 1852

Eroedi [Erödi], Kálmán. Hungarian geographer, 19th century. *School atlases*

Eroedi-Harrach, Bela (1846–1936). Hungarian geographer

Erp, Theodore van (b. 1874). Worked Indonesia as officer in army Topog. service

Erskine, J. F. *Clackmannan* 1795

Erskine, R. Cartographer. *Plans Iberian ports* 1727–34, *Gibraltar Bay* 1744

Erskine, Robert (1735–1780) b. Dunfermline. Land surveyor and engineer next to the Crown, Scotland Yard, London. F.R.S. 1771, emigrated America 1771, geographer to Washington's army, over 150 *maps and manuscript sketches of the roads* 1777, *Part of States of N.Y. & N.J., Map Highlands N.Y.* 1779 MS, *Roads in U.S.* 1778–9, maps *Yorktown Campaign* (1781) MSS.

Erskine, Saint Vincent. *Gold Fields S.E. Africa* 1876, *Transvaal* 1877, *Kaap Gold Fields* 1887

Erslev, Ed. (1824–1892). Danish geographer and naturalist, founder Geog. Soc. of Copenhagen. *Skoleatlas* 1870; pub. various manuals of geog. 1873–6

Ertl, Anton Wilhelm (1654–1715+). Geographer of Munich

Erwin, John W. *Elklast County, Indiana* 1871

Erzey, Jan (1780–1842+). Cartographer of Amsterdam

Escalante, V. de. See **Velez de Escalante**

Escandon, José de, Conde de la Sierra

Gorda. *Texas* 1747, *Mexico* 1747, *Nuevo Santander* 1748 MSS.

Eschauzier, Brand. *Encyklopedische Atlas* 1838

Eschenard, F. See **Eschinardi**

Escher, A. See **Escher von der Linth**

Escher, Lieut. B. G. *Vaarwaters naar de Reede van B.* 1840–(57).

Escher von der Linth, Arnold (1807–1872). Swiss geologist. *Geolog. Karte des Sentis* 1837–72, pub. 1873, *carte géolog. de la Suisse,* Berlin 1853, 4 sh. 1867

Eschinardi [Eschenard] , Francisco (1623–ca. 1700). Jesuit, mathematician of Rome. *Imp. Abassini tab. Geog.* 1674, & 4 sh. 1684

Eschmann, Johannes (1808–1852). Astronomer and cartographer of Zürich. *Canton St. Gallen* (1856?). Assisted Dufour in *Carte de la Suisse*

Eschuid, J. See **Eastwood**

Eschwege, W. Cartographer of Weimar. *America* (1818), *Rio de Janeiro* (1819), *Brazil* (1821)

Escluse, C. del'. See **Clusius**

Escorche-Messes, Frangidelphe [pseudonym] . Italian cartographer. *Mappemonde papistique* 1566

Escudero, Ramon. *Plan Yquique, Chile* 1861

Eseitz, D. F. van. See **Fabricus**, David

Esenius, C. See **Buondelmonte**

Esents, D. F. van. See **Fabricius,** David

Eskes, H. P. *Platte Grond Amsterdam* 1842

Esmarck, J. (1763–1839). Norwegian geologist

Esnault. (fl. 1780–1807) Marchand d'Estampes, Boulevard Montmartre, Terrase Frascaty No. 7 près la Rue de Richelieu, Paris

Esnault, le Jeune. Marchand Estampes et

de Géog., Boulevard Montmartre, Paris. *Plan Paris* 1814

Esnauts & Rapilly. Publishers, Rue St. Jacques à la Ville de Coutances, Paris. *Amérique* 1777, *N. Amer.* (1779), *Theatre War in N. Amer.* 1782, *Brion de la Tour's Amér.* 1783, *Paris* 1784

España, Casildo. Guatemalian engraver for Maestre 1832

Espedic, J. *Carte des Vignobles du Médoc* 1868

Espin, W. M. Surveyor. *Matabele-land* 4 sh. 1896, *Bulawayo Township* 1898 (both with Fletcher)

Espina[u]lt y Garcia, Bernardo. *Atlante Español* 1778–95, *Postas de España* 1794

Espinha, Joseph da. (1722–1788). Jesuit. *Map Asia & Europe* in Chinese 1756–9, pub. Peking 1775

Espinosa, Ant. Vasquez (d. 1630). Spanish geographer, b. near Seville

Espinosa [y Tello] , José de (1763–1815). Traveller and hydrographer of Madrid, founded *Depósito Hydrográfico* 1797, *Voyage 1792 N.W. Amer., Atlas* 1802–6, *Carta esferica Int. Amer. Merid.* 1810, *Antillas Mayores* 1811, *Navegaciones a le Indie Orien.* 6 sh. 1812 (pub. by him, 17 Dean St., Soho, London)

Espinosa, Luis & Manuel. *Plano Cuidad México* 1867

Espinosa de los Monteros, Antonio. *Plano Top. Villa Madrid* 1769

Espinosa y Tello, J. See **Espinosa**, J.

Espinoza, Enrique (b. 1848). *Atlas de Chile* (1897)

Esquemeling [Exquemelin. English forms of Oexmelin] , Alexandre Oliver (ca. 1645–1707+) *Isthme de Panama* 1686, *Honduras, Costa Rica* &c. 1688

Esquirol, Pedro de. Spanish cartographer, late 16th century

Essen, Derk van (1829–1880). Lithographer of Rotterdam. *Plans Rotterdam*

Essen, John van. *Charts E. Indies* ca. 1700 MSS.

Estancelin, Lieut. Gen. *Woods and forest Eu.* 40 sh. 1768

Este World Map. [Catalan] . Circular map ca. 1450 in *Biblioteca Estence,* Modena

Esteller, Eduardo Morena. *Theatro de la Guerra de Oriente* (1876?)

Estévanez, Nicolas. *Atlas de América* (1896)

Esteve, R. Engraver *Buenos Ayres* 1812– 28

Estienne, Charles (1504–1564). French traveller, physician and printer. *Guide des chemins de France* 1552 (attributed); first French road book

Estorff-Veerssen, A. von. *Kreises Uelzen* 1890

Estorgo y Gallegos, Francisco Xavier. *Quiapo (Philippines)* 1746, *Plan Manila* 1770

Estrabou. *État-Major du Soudan Français* (1890?)

Estrada, Bartolome Ruiz de. Pilot to Pizarro. *Coast Panama & Gorgona* MS.

Estrémont de Maucroix d'. *Banc au S.O. des Îles Farrë* 1846

Estridge, A. W. Survey *Wanstead Flats* 1862 MS.

Estruc, Domingo (1796–1851) Engraver. *Diccionario geog. univ.* 1830–2

Estwood, J. See **Eastwood**

Esveldt-Holtrop, J. S. van. Publisher of Amsterdam. *Leo Belgicus* 1806, *Atlas van Saxen* 1810

Eszler, Jacobus. Editor and publisher of 1513 Strassburg edition *Ptolemy*

Etallon, Cl. Ang. (1828–1862). French geologist

Etanduère, l'. See **L'Etanduère**

Ethersey, Lieut. Richard, Indian Navy. Admiralty surveyor. *India* 1836, *Chart Paumben Pass* 1838, *Gulf Cambay* 1845, *Malacca Banks* 1845

Ethicus. Roman geographer, 6th century *Cosmographia,* first printed Venice 1513

Étienne, Jean d'(1725–1798) b. Cernay, cartographer in Bückeburg

Ettling, Theodor (b. 1823). Engraver, lithographer and draughtsman of Amsterdam, later 3 Red Lion Sq., Holborn, London. *Opper Californie* 1849, *Drawing Room Atlas of Europe* 1855, drew & engd. maps for *Dispatch Atlas* 1858–63, maps of *U.S.* for *Illust. London News* 1861, *Times Map North America* 1861

Etzel, F. A. von. See **Oetzel**

Etzlaub, Erhard (ca. 1460–1532) Physician, compass maker and cartographer of Nuremberg. *Der Rom Weg* ca. 1492 (first road map), *Germany* 1492, *Environs of Nuremberg* 1492, *world map* on compass lids 1511–13 (attributed)

Eudoxos (408–355 B.C.) Greek philosopher and astronomer

Eufredutius. See **Freducci**

'Eugene [Eugenius] Atlas'. Compiled by by Blaeu for Laurens van der Hem, later presented to Prince Eugene of Savoy, 46 vols., now in Vienna State Library

Eugene, Carl Ludwig. *Hypsographic map Sweden* 1858

Eulenstein, F. Engraver for Kiepert 1857

Euler, Leonhard (1707–1783) b. Basle, d. St. Petersburg. Cartographer, physician and mathematician. *Atlas geog.,* Berlin 1753, 1756, 1760. Supervised *Atlas of Russia,* St. Petersburg 1745

Eulzofen, T. L. *Guyane Française* (1881)

Eunson, George. Surveyor. *Chart Orkneys* 1795

Eurelius. See **Dahlstierna**

Eussen, Jacob (fl. 1745). Engraver in Amsterdam

Eustachius. See **Divinus Eustachius**

Eustis, Henry Lawrence (1819–1885). American engineer and professor. *Plan ravages of Tornado Aug. 1851,* Boston 1853

Eutling, Julius. *Odilienberg u. Umgebung* 1874

Evans, Lieut, See **Evans,** T.

Evans, A. E. & Sons. Publishers, 403 Strand, London. *Plan London* 1857

Evans, Albert S. *White Pine, Nevada* 1869

Evans, Cadwalader (1762–1841). Surveyor. *Draught of a Tract of Land in Bucks County, Pennsylvania* 1786

Evans, Charles. *Plan Freehold, Mulberry Garden (Buckingham House, London)* 1760 MS.

Evans, Edward. *Rivers of Ireland* 1887

Evans, E. J. *N.E. coast Australia,* Admiralty 1855

Evans, Master (later Capt. Sir) Frederick John Owen (1815–1885). Hydrographer of the Navy 1874–84. *Survey Central Amer.* 1833–6, *N.E. Australia* 1841–6, *New Zealand* 1847–51, *Tasmania* 1860, *Atlantic Ocean Pilot* 1868

Evans, Comm. G., R.N. *Port Louis, Mauritius,* 1819

Evans, G. *Topo. map Madison County, N.Y.* 1853

Evans, Capt. George. *Plan Birkenhead Docks* 1847

Evans, George Wm. (1780–1852). Deputy Surveyor General, Port Dalrymple (Tasmania) 1809; Hobart Town 1812. Exploration in *New South Wales* 1813–18 (with Oxley &c.). *Chart V.D.L.* 1821–(2), *Macquarie Harbour* 1822–(45)

Evans, Henry Smith. *World* 1847, 1851; *Geology made easy* 1851

Evans, John. *Missouri River from St. Charles, Mo., to North Dakota* 1795 MS.

Evans, Rev. John (1767–1827) Publisher, No. 41 & 42 Long Lane, W. Smithfield, London. *Eng. & Wales* 1794, *N. Wales* 9 sh. Lwynygoes 1794–5, 1797, *Plan London* 1796

Evans, John of Shrewsbury. Son of Rev. John. *New Royal Atlas* (1810), Aberystwyth 1824

Evans, Lewis (ca. 1700–1756) Surveyor. b. Pennsylvania. *Bucks County, Penn.* (1738) MS, possibly designed *Cape Hatteras to Boston* for J. Turner 1747; *Middle British Colonies* Philadelphia 1749, numerous later editions in London. *Penn. &c.* 1750 MS. *Pennsylvania, New Jersey, New York and 3 Delaware counties* 1769, *Atlas Amer. Septen.* 1778

Evans, R. M. Surveyor. *Washoe Mines, Cal.* 1860, *Gold Hill, American Flat & the Divide (Nevada Terr.)* 1864

Evans, Smith. *World . . . Coal fields & Steam Navigation* (1851)

Evans, Thomas (1739–1803). Bookseller (1) 54 Paternoster Row, London, (2) No. 41 Long Lane, West Smithfield. *England Displayed* 1769, *London & Westminster* 1796

Evans, Lieut. Thomas, R.N. from Fishguard. *Straits of Sincapore* 1804, survey *Liverpool & Chester* 1812–28, *Bay & Harbour Fishguard* 1817, *Barrow* 1832

Evelyn, Sir John. MS map *Deptford,* with additions by J.E. 1750, *Plan for rebuilding London* (1755)

Everaert, Martin. Translated *Waghenaer's Spieghel* into Latin 1586

Everest, Sir George (1790–1866). English military engineer, directed triangulation of India. *Merid. arc of India*

Everett, Arthur. *Victoria* 1869, drew and coloured *geolog. map Australia* 1875, 1879, 1887, *Continental Australia* 6 sh. 1887

Everett, Edward. *City of Quincy, Adams County, Ill.* 1857

Everett, J. Printer, bookseller and publisher, Market St., Manchester. *Plan Manchester & Salford* 1832–4

Everett, Joseph David (1831–1904) b. Ipswich, globeseller and geographer in Belfast

Evert, E. *Plan Stadt Posen* 1896

Everts, L. H. [& Co.]. *Combination Atlases Ohio* 1874–5, *State Atlas Kansas* 1887

Everts, [Baskin] & **Stewart.** *McHenry County, Ill.* 1872, *Rock Co., Wisc.* 1873, *County atlases Mich., Pa., N.J.* 1874–6

Everts, Ensign & Everts. *Genesee Co., & Yates Co., N.Y.* 1876

Everts & Kirk. *State atlas Nebraska* 1885

Everts & Richards. *Bristol County, Mass.* 1895, *Providence Co., R.I.* 1895

Eveux de Fleurieu. See **Claret de Fleurieu**

Evia, Jose de. *Costa de la Florida* 1783

Evia, Simon de. *Louisiana* 1736 MS.

Evreinov, Ivan Michajlovič (d. 1724). Russian cartographer and geodesist. First map of *Kurile Is.* 1719–23

Ewald, Julius W. (1801–1891). German geologist. Geolog. maps Germany; *Sachsen von Magdeburg* 1865

Ewald, Ludwig Wilhelm (1813–1881). German geographer and statistician. *Erdkarte* 1853, *Europaeischen Staaten* 1858, *Hand atlas der allgem. Erdkunde* 1860

Eward, J. Mapseller at the Beehive against Northumberland St., Strand, London. *Edgar's Plan Edinburgh* 1742

Ewart, Lieut. John Spencer. *Kassassin & Tel El-Kebir (Egypt)* 1882–(7)

Ewich, Hermann. Historian of Wesel. *Ancient Gulik,* pub. Jansson (1652)

Ewing, Thomas. Cartographer. *New General Atlas* pub. Edinburgh 1817, editions to 1862. *Gold Region W. Kansas* (1860)

Ewing, W. C. *Columbia River* 1837, *St. Louis Harbour* 1837

Ewoutsz[oon], Jan. Woodcut engraver *Sailing directions* 1560–1

Ewyk, Nicholas van. Capt. in Dutch East India Co. *Australia* 1750. his own *S. Hemis.* (after De l'Isle), Amsterdam 1752

Exchaquet, Charles-François (1746–1792). *Carte pétrographique du St. Gothard* 1791

Exiles, A. F. P. d'. See **Prévost d' Exiles**

Expilly, Jean Jos. Georges, Abbé d' (1719–1793). French diplomat, traveller and geographer. *Géog. manuel* 1757 &c. *Topog. de l'Univers* 1757–8, *Dict. géog. des Gaules* 1762–70, *Mappemonde* 1765

Exquemelin, A. O. See **Esquemeling**

Exshaw, John (fl. 1798). Publisher at the Bible opposite Castle Lane in Dame Street, Dublin. *Eng. & French Territ. in N. America* 1756

Eyb, Otto Franz von (1858–82) Engraver for Kiepert 1860

Eyes, Charles. Surveyor. *Plan Liverpool* 1785

Eyes, John. Land surveyor and chart publisher of Liverpool, collaborated with Fearon 1736–8. *Liverpool Bay* 1725, *survey Lancs. coast* 1736–7, *chart Furness & Anglesea* 1738, *Plan Liverpool Docks* 1742, *Survey of Dee* 1755 (with Sumner), resurveyed *Liverpool Bay* 1764, *Plan Liverpool* 1769

Eyes, John. *Plan Liverpool* 1865

Eyler. *Topo. map Humboldt Co., Nevada* 1865

Eynde, J. van. See **Ende,** Josua van den

Eynon, I. Publisher and mapseller, Corner Castle Alley, Threadneedle St., Royal Exchange, London Withy's *Plan London* 1760

Eynon, R. Publisher and mapseller behind the Royal Exchange, London. *Staten Is.* 1776

Eyre, Edward John. Estate surveyor. *New Park Richmond* 1764, Essex Estate plans 1758–67, *Plans Minterne* 1767 MS.

Eyre, Edward John (1815–1901). Explorer Lieut.-Governor New Zealand, Governor Jamaica. *Mt. Remarkable (S. Australia)* 1857, *Australasia* 1863

Eyre, Capt. John. *Plans fortifications London* 1643 MS.

Eyre, Mary. *Nottingham described* 1632

Eyre [Eayre], Thomas (1691–1758). Surveyor of Kettering, bell founder and clockmaker. *Plan Kettering, Northants.* n.d. *Co. of Northampton,* revised Jefferys 1779–1791

Eyre, William (d. 1765). Engineer. *Fort Edward* 1755, *Plan Fort William-Henry* ca. 1755 and ca. 1756

Eyries, J. B. Benoit (1767–1846). French geographer

Eyring, Hans. Publisher at Wroclaw. *Helwig's Silesia* 1627

Eysbroek, H. Pilot. *Wind-Kaart van den N. Atlant. Oceaan* 1856

Eyzinger, M. See **Aitzing**

F

F., A. *Milano* (1580?)

F., B. *Celestial globe* 1600

F., D. N. *Plan des Armées Autrich. et Franç. aux deux Côtés du Rhin* (1744?)

F., F. *Vitenbergo* (1565?)

F., G. W. *Indian country to E. & W. of Mississippi* (1836?)

F., H. *Chorog. Bavaria* 1568

F., I. G. *Town of Appanoose, Hancock Co., Ill.* 1857

F., M. See **Florimi**, M.

F., T. *Chart coasts England* ca. 1603 MS.

F., W. See **Fairbank**, Wm.

Faber, Christopher (1800–1869). Painter and cartographer

Faber, Conrad (ca. 1500–1552/3) b. Creuznach, artist in Frankfurt. *Plan Frankfurt* 1552 used by Braun and Hogenberg

Faber, Ernst (fl. 1865–99). Missionary in China, documents on geography and natural history

Faber, Johannes (1650–1721). Engraver and publisher, The Hague, Oxford & Bristol

Faber, J. H. *British Honduras* 1891

Faber, Martin (1587–1648). Architect, painter, engraver and cartographer of Emden. *Pascaerte Riviere O. ende W. Eems* 1642

Faber, Samuel (1657–1716). Theologian and geographer b. Altdorf, d. Nuremberg. *Globes* 1705, *Atlas scholastichodoeporicus* (171–), *Atlas scholasticus* (1720)

Fabert, Abraham (1560–1638). French printer, b. Metz. *Pays Messin* 1605, used by Blaeu, Hondius &c.

Fabert, Abraham de (1599–1662). Cartographer of Metz. *Territorium Metense,* Blaeu 1660

Fabius, Nicasius. *Ancient Flanders* 1641

Fabre, Jean. French cartographer. *Sevenes* 1629

Fabré, Lieut. M. *Chart Anchorage St. Pierre* 1824, charts for Bougainville 1828

Fabregat, J. Joaquin (1748–1807). Engraver, for Aguirre 1775, and Torfiño de San Miguel 1786

Fabretto [Fabretio], Raffaelle. *Rome* (1695?)

Fabri, François. Printer of Douai for Wytfliet 1603–11

Fabri, J. E. E. (1755–1825). German geographer. *Handbuch der neuesten Geog.*

Fabricius, Antonius Bleyrianus. Cartographer. *France* 1624

Fabricius [Van Esents, Eseitz], David (1564–1617). Cartographer, astronomer and pastor of Essen. *E. Friesland* 1598, *Oldenburg* 1592, *O. en W. Vrieslant* (1610), *Plan Emden* 1619

Fabricius, P. See **Fabritius**

Fabricius, Dr. Wilhelm. *Atlas der Rheinprovinz* 1894

Fabris, Dominik Tomiotti de (1725–1789). Austrian general and cartographer

Fabritius [Fabricius], Paulus (1519–1589). Astronomer, physician and cartographer, Prof. Mathematics, Vienna. *Moravia* 6 sh. 1568, reduced 1575, *Austria (lost)*

Facius, Georg Siegmund (b. 1750). Artist and engraver, in London from 1776. For Jäger 1789

Facius, Johann Gottlieb. Worked with brother Georg

Fackenhofen, Georg Carl Adam Joseph, Frhr. von (1745–1804). Cartographer of Würzburg. *Survey Würzburg* 1788–90, engraved 1804

Faden, William (1750–1836). Publisher and cartographer, Geographer to His Majesty & the Prince of Wales, the Corner of St. Martins Lane, No. 5 Charing Cross, London [1802]. Succeeded T. Jefferys 1771, firm styled Jefferys & Faden [1773–83], succeeded by J. Wyld elder 1823. *Mitchell's N. American* (still dated 1755). *World* 1775, *Ratzer's plan New York* 1776, *N. Amer. Atlas* 1776, *General Atlas* 1778 &c. pub. surveys of Knight & Dessiou 1790–1816, *Battles of American Revolution* 1793, *Petit Neptune Français* 1793, *Atlas Minimus* 1798, *Ottoman Dominions* 1822. Pub. early Ordnance Survey maps

Faden, Wm.

Faehtz, E. F. M. *Real estate directory City Washington* 1874 (with E.W. Pratt)

Fagan, L. *Town New Hartford, Conn.* 1852, *Berks Co., Penn* 1861. Many town & county maps Conn., N.Y., Mass., N.H.

Fage. Lieut. (later Lt.-Colonel) Edward, Royal Artillery. (d. 1809). *Rhode Island* 1777, *Newport County* (1779) MS, *Plan of the Posts of York and Gloucester, Virginia* 1782, *Woolwich* 1797 MS.

Fage, Robt. *Cosmography* 1667

Fa Hien (ca. 400–414 A.D.) Chinese explorer

Fahlberg, Samuel. *Chart channels St. Barth, St. Martin & Anguilla* 1814

Faiano, J. See **Fayen**

Faidherbe, Louis L.C. (1818–1889). Soldier, geographer and explorer in Senegal & Soudan. *La Guerre de 1870–71*

Faille, P. de. See **La Feuille**

Faindrick, Sieur. *Haute et basse Allemagne* 1669

Fairbairn, James. *Chambers Educ. course, school room maps* (1861)

Fairbairn, Wm. *Plan improvements Manchester* 1836

Fairbank, Wm. (1730–1801). Schoolmaster and surveyor. *Plan Sheffield* 1771, 1808 (with son), *Navig. canal Stainworth cut to R. Trent* 1793, *Plan Leeds & Liverpool Canal* 1793

Fairbank, Wm. Estate surveyor. *Brexted (Essex)* 1779 MS.

Fairburn, John. Publisher, geographer and map seller, 146 Minories, London. *Plans London* 1795–1806, *Proposed Wet Docks Millwall* 1796, *Environs London* 1798, 1831, *North America* 1798, *Spain & Portugal* 1808

Fairchild, John F. *Atlas Mount Vernon & Pelham, N.Y.* 1899

Fairchild, W. Surveyor. *Shalford (Essex)* 1760 MS, *Manors Hillesden & Cowley (Bucks.)* 1763 MS.

Fairfax, Capt. H., R.N. *Kerguelen Is.* 1875

Fairfax, W. M. C. *Boundary U.S. & Brit. Provs.* 1843

Fairfowl, George. Surgeon, *HMS Dromedary.* *Sketch of the coast of New Zealand* (1820)

Fairlove, Ichabod. Surveyor. *Plan Excester* 1709

Fairman, D. Engraver of Philadelphia for Pinkerton (1804)

Fairman, Francis. Estate plan *Wrothman, Kent* 1705 MS.

Fairman, Gideon. Engraver of Albany, N.Y. *Oneida Reservation* (1796), *Ontario and Steuben Counties* (1798), *State of New York* 1804

Fairman, John. Estate plan *Wrotham, Layborne, & W. Peckham* 1726 MS.

Fairweather, Capt. Pat. *Chart Entrance Old Calebar* 1790

Faithorne, William, the elder (1616–1691). Engraver and publisher of London, at the sign of the Drake against Palsgrave's Head Tavern (1650–ca. 80), later Printing House Yard, Blackfriars. *The art of Graveing and Etching* 1662, *Newcourt's London and Westminster* 12 sh. 1658, *River System between Mediterranean and Toulouse* 1670, *Herrman's Virginia and Maryland* 1673. Also engraved titlepages, bookplates, and portraits.

Fakymolano. Brother of Sultan of Magendanao. *Magendanao* ca. 1774 MS.

Falb, Rudolf (1838–1873). Austrian geologist

Falbe, Lieut. *Côtes de Sicile et Tunis* 1840–56, *Régence de Tunis* 1842, 1857, *Gronland Vestkyst* 1863–(6)

Falckenstein, Albert de. Ingeniero del Estado. *Atlas géog. de la Rép. de Pérou* 1862–(5)

Falckenstein, E. V. von. See **Vogel von Falckenstein**

Falckenstein [Falkenstein], Johann Heinrich von/de (1682–1760). *Del. Nordgoviae Veteris* 1733, maps for Homann Heirs 1753

Falda, Giovanni Battista [da Valduggia] (1648–1678?). Italian engraver working in Rome. *Provincie Unite* 1672, *Citta di Roma*, Rossi 1676

Faleiro [Falero], Francisco. Portuguese mathematician and cartographer, early 16th century

Faleiro [Falero], Rui (d. 1523). Portuguese seaman and cartographer

Faleleeff. Engraver for Piadischeff 1823–26

Falero, F. & R. See **Faleiro**

Faleti, Bartolomeo. Publisher in Rome. *Rome* 1564

Falger, Anton (b. 1791, fl. 1818). Tyrolese lithographer, engraver and artist in Munich and Weimar. For Gaspari 1821, *Tyrol* 1826

Falk. *Königsberger Land-Kreise* 1841

Falk, R. Printer and publisher. *Stadt von Paris* 1870, *Frankreich* (1871)

Falkenstein, E. V. von. See **Vogel von Falckenstein**

Falkenstein, J. H. von. See **Falckenstein**

Falkner, Thomas (1708–1784). Jesuit missionary. *Description Patagonia*, 1774, engraved Kitchen

Falle, P. J. Surveyed *Madras Roadstead* 1876

Falledo y Rivera, Vicente. *Colombia* ca. 1805 MS.

Fallen, O. *Oesterreich* 3 sh. Vienna 1822

Fallon, Ludwig August, Frhr. von (1776–1828). Austrian general and cartographer, chief of Austrian Topog. Bureau

Fallow[e]s, Benjamin. Estate surveyor of Maldon, Essex. MSS, estate plans; *Dengie* 1714 & *Rivenhall Place* 1716–17 *(Essex), Castle Abergavenny* 1718–26, *Newberry, Worcs.* 1718

Falmar. *Road from Chagres to Panama* 1851

Fan, Cho. Chinese geographer 9th century

Faneua [Faveau] **Quesada**, Don Antonio.

California 1768 MS, *Chart Balabac & E. coast Palawan* 1775

Fangel, H. Lithographer. *Plan Hadersleben* 1862

Fannin, Peter, Master, R.N. *Isle of Man* 1789

Fanning. *New York City* 1857

Fanning, J. F. *City Manchester, New Hampshire* 1873

Fanning. See **Ensign, Bridgman & Fanning**. Also **Thayer, Bridgman & Fanning**

Faqih, Ahmad B. Muhammed. Arab cartographer. *Atlas of Islam* 903

Farey, John. Estate surveyor. *Great Clacton* 1809 MS.

Farey, John. Senior (1766–1826). Mineral surveyor and geologist. *Section of the principal strata of Eng.* (earliest geolog. section across country), 1808, *Survey Derby* 1811–13

Farey, John, Junior (1791–1851). Drew *Derbyshire* for his father's *Survey Derby* 1811–13 and for *Gray's New Book of Roads* 1824

Farey, Wm. *S.W. District St. Pancras* 1820

Fargany, Ahmed Ibn Muhammed. Arabian astronomer, 9th century. First Arab to write on the astrolabe

Faria, J. C. de Sá e. See **Sá e Faria**

Faribault, Eugene Rudolphe. Surveyor. *Forest Hill Gold District* 1898, many gold region maps to (1908)

Farini, G. A. *Routes in Kalahari Desert* 1886

Farler, Rev. J. P. Maps of *Tanganyika* 1877–82

Farley, Minard H. *Tramontane silver mines S. Calif. & N. Mexico* 1861

Farley, Wm. *Chart Rabbit Is.* 1820–(30)

Farmer, Capt. George (1732–1779) Information for *chart Bay of Bengal* 1794

Farmer, John (1798–1859). Geographer and engraver of Detroit. *Michigan & Wisconsin* 1826, 1830, (1835), 1858,

City Detroit (1835?), *State Mich.* 1844, *R'road & township map Mich.* 1862, 1865, 1867

Farmer, Silas. *Lake Superior* 1872, *Railroad & township map Wisconsin,* 1872

Farnau, J. (fl. 1680–98). Cartographer. *Heligoland*

Farnham, Thomas. Publisher and geographer of N. York. *California* for Morse & Breese [1842], *Oregon* 1844, *Mexico &c.* 1846

Farquhar. *Plan Huntly* (1845) (with Ogg)

Farquharson, John of Invercald. *Forest of Mar* 1703 MS.

Farquharson, John. *Loch Tay* 1769 MS. (with M'Arthur)

Farrand, William. *Donnabrooke & Tannee (Down survey)* 1655–7, *Rathdown,* Dublin 1656–7 (with Storck)

Farrant, R. G. *Gosha (Somalia)* 1896 MS.

Farrer [Farrar, Ferrar, Ferrer], John. Cartographer, Deputy Treasurer of Virginia pre-1654. Born Little Gidding, d. 1687. *Discovery of New Britain,* London 1651, containing map of *Virginia*

Farrer [Farrar, Ferrar, Ferrer], Virginia. Daughter of above. Issued later states of map of *Virginia* under her own name

Farrington, J. *Plan Makasar* 1815 MS.

Farsky. Lithographer. *Umgebung v. Teplitz* (1875?)

Fassbinder, P. *Palästina zur Zeit Jesu Christi* (1899)

Fatio de Duillier, Jean Christophe (1656–1720). Astronomer.

Fatio, Nicholas (1664–1753). Worked with J. C. Fatio

Fatout. Publisher, Bvd. Poisonnière 17, Paris. For Vuillemin 1859

Fatout, Henry B. *Atlas Indianapolis & Marion County, Ind.* 1889

Fauche, Samuel (1732–1802). Printer, publisher and bookseller of Neuenburg

Faucit, Thomas. Publisher *Vermuiden's Fenns* 1642

Fauck, A. *Idaho Territory* 1865

Faucou, Lucien. *Hist. plans Paris* (1889)

Faujas de Saint-Fond, B. (1741–1819). French geologist, b. Montelimar. *Volcans du Vivarois*

Faul, August. *Orange & Alexandria Railroad* (1850), *Swann Lake & Aqueduct* (Baltimore) 1862

Faulhaber, Johann Matthias, younger (1670–1742). Cartographer of Ulm

Faulkner, George (ca. 1699–1775). Publisher (1) Christchurch Lane, Skinner's Row, Dublin (with Js. Hooey) (1728), (2) Essex St., Dublin (1730). Kitchen & Jefferys' *Small English Atlas* 1751

Faulkner & Smith. Mapsellers, Dublin. Jefferys' *Ireland* 1759

Faunthorpe, Rev. John Pincher. *Element. Physical Atlas* 1867, *Outline Atlas* 1874, *Projection Atlas* 1874

Faure. Marine surveyor. Charts of *Australia* 1802–3 for Freycinet 1812

Faure, A. le. See **Le Faure**

Faure, Guillaume Stanislas (1765–1826). Hydrographer of Le Havre. *Nouveau Flambeau de la Mer* 1822–4

Faure, P. Surveyor. *Plan Paroisse Royale de St. Germain l'Auxerrois* 1739

Fauvel, A. A. *Shantung* 1876

Fauvel, Maurice (1851–1908). French geographer

Fauvet. *Nouvelle mappemonde* 1805

Fava, Duarte Jozé. Capt of Engineers. *Plan Lisbon* 1807–(26)

Favarger, H. F. *Golfo Adriatico* 1854

Faveau Quesada, A. See **Faneau Quesada**

Favoli[us], Hugo [Ugo] (1523–1585). Dutch poet, doctor and traveller, b. Middelburg, d. Antwerp. Latin translation of Ortelius-Galle miniature atlas 1585

Favre, Alphonse. *Savoie, Piedmont, Suisse* 1861, *Carte géolog. de la Savoie* 1862, . . . *du Canton de Genève* 1878, *Glaciers des Alpes Suisses* 1884

Favre, Maximin. *Chart Mediterranean Sea* 1772–[4]

Favrot. French cartographer. *Cours du Rhin* 1696–7

Fawcett, John. Estate surveyor New Ormond St., London. *Corringham* 1818 MS.

Fawcett, T. *Gold Region, Yukon* (1898)

Fawekner, W. II. 2nd Master H.M.S. Phoenix. Drew *N.W. Passage,* Admiralty 1853

Fay, General Charles Alexander. *Théatre des opérations en Bohême* 1867, *Campagne de* 1870–(89)

Fay, Theodore Sedgwick (1807–1898). Diplomat and printseller of New York. *Outline of geography* 1867, *Atlas Univ. Geog.* 1869–71, *First steps in Geog.* 1873

Fayano, J. See **Fayen**

Fayard de la Brugère, Jean Arthême (b. 1836). *Atlas Universel* 1877 (with A. Baralle)

Faye, H. Aug. E. A (b. 1814). French astronomer. *Leçons de cosmographie sur l'origine du monde*

Fayen [Fayano, Faiano], Jean du (ca. 1530–ca. 1612). French physician and geographer. *Limousin* for Bouguereau 1594, used by Ortelius, Mercator, Blaeu &c. (first known map of Limousin)

Fayolle. *Plan de la rade de Dantzig* 1815

Fayram, Francis. Bookseller. *Camden's Britannia* (1730)

Fayram, J. *Stockholm ab Occ.* 1725

Fazari. Arabian astronomer 10th century

Fazello [Fazelli], Padre Tomaso (1490–1570). Sicilian historian, b. Sacca, d. Palermo. *Isola di Sicilia* 1682

Fea, Capt. Peter. *Chart part W. coast Madagascar* 1784 MS.

Fearnside, T. *Hull Citadel* 1745 MS.

Fearon, Samuel (fl. 1766). Shipwright, hydrographer and publisher of Liverpool, collaborated with Eyes. *Survey N.W. coast England* 1736–7 pub. 1738 (first printed charts to show Greenwich as Prime Meridian), *Furness & Anglesea* 1738, corrected *chart Liverpool Bay* 1757

Featherstone, Joseph. *Deeping Fen* 1763

Featherstonhaugh, George Wm. American geologist. *St. Peters River* 1835, *New Brunswick & Lower Canada* 1839

Featherstonhaugh, J. D. Secretary and draughtsman to Commissioners. *Map of country in dispute with U.S. Highlands Conn.)* (1840) (with Delves-Broughton)

Febiger, Rear Admiral John Carson, U.S. Navy. *Yellow Sea* 1868

Febure, Le. See **Le Febure**

Febure de la Barre. See **Barre**

Fedchenko A. P. Russian cartographer. Maps *Asiatic Russia* 1870–2

Feddes [Harliensis, Harlingensis], Pieter (1586–1634). Poet, artist, engraver and publisher of Harlingen

Feer, Johannes (1762–1825). Swiss Cartographer. *Canton Zürich; Rheinthal* 1797

Fegraeus, Ludv. *Karta öfver Wisby* 1879

Fehlbaum, Ed. Lithographer. *Plan Stadt Bern* (1876?)

Fehr, Johannes (1763–1825) Topographer of Zürich

Fehse, C. *Atlas von Groningen* 1862

Feichtmayr [Feuchtmayr], Joseph Anton 1696–1770). Engraver and sculptor b. Linz, d. Mimmenhausen. *Plan von Madach*

Feige, Lieut. *Environs Berlin* 10 sh. 1816–19

Feignet, Johan Carl de (1740–1816). Cartographer and surveyor

Feild [Field], John. Surveyor. Estate plans in Bucks. *Dorney* 1781–2, *Boveney* 1790, *Burnham* 1796 MSS.

Feilden, H. W. *Routes Spitzbergen* 1898

Feistmantel, O. (1848–1891). Czech geologist, secretary to Geol. survey Calcutta

Felberthann, E. Engraver for Kiepert 1860

Felbiger, Johann Ignatz, Sieur de. *Breslau* 1751

Felbinger, Kaspar (fl. 1564–76). Artist and printer of Königsberg

Felden van Hinderstein, G. F. D. van de. See **Derfelden van Hinderstein**

Felgate, Robert. Estate surveyor of Gravesend. *Aldham &c.* 1675 MS.

Felkl, Jan (1817–1887). Globemaker of Prague. *Terrestrial and celestial globes* 1856

Felkin, R. W. *Reiseroute Ladó bis Dara* (1881)

Fell, Thomas M. *Australian Gold Fields* (1854?)

Fell, Lieut. (later Comm.) Wm., Indian Navy. *St. Bees Head to Duddon* 1824, *Survey coasts India* 1841–8, *Pegu* 1852, *N. coast Sumatra* 1853

Fellows, John. Surveyor. *Estate Radclive, Bucks* 1773 MS.

Fellows, W. Surveyor. *Public Foreign Sufferance Wharfs* 1803

Fellweck, Johann Georg. (fl. 1772). Globemaker of Würzburg

Felsecker, Adam Jonathan (d. before 1742). Printer, publisher and engraver of Nuremberg

Felsing, Johann Conrad (1766–1819). Engraver of Darmstadt. *Trier u. Saarburg* 1793–4, *Frankfurt* 1811, *Situations-carte Darmstadt* 24 sh. (1820?)

Felsing, Johann Heinrich (1800–1875). Engraver of Darmstadt, son of J.C. Felsing. *Situations Karte v. Rhein* 1822

Felter[o]us, Kietell [Kettil] Classon (d. 1690). Swedish surveyor. *Västergötland* 1689 MS.

Feltham, Mrs. Mapseller in Westminister Hall, London. *C. Browne's England* (1693)

Felton, John, of Oswestry. Printed, coloured and sold Williams' *Denbigh & Flint* 1720

Felton, S. M. American surveyor. *Beckford Estate in Charles Town* 1837

Fembo, Christoph (d. 1848). Printer, publisher and geographer of Nuremberg. Took over from Homann Heirs 1813 and gave firm his own name. *Atlas Silesiae* 1813, *Grossbrit. u. Ireland* 1818, *Australien* 1820, *Kur Hessen* 1836

Fembo, Christoph Melchior (1805–1876). Son of above, publisher and mapseller of Nuremberg. *Charte von Australien* 1830

Fennema, R. *Geol. Kaart van Java en Madoera*, 26 sh. 1886–94

Fenner, R. Publisher and engraver. *Pocket Atlas of Modern & Anc. Geog.* (1830?)

Fenner, Walter. *Is. of Carriacou* 1784

Fenner, W. A. *Trig. survey of tidal islands in Indus* 1852

Fenner, Sears & Co. Engravers and printers. Hinton's *Atlas* 1832

Fenning, Daniel. Engraver, geographer and globeseller. *New System of geog..* (with J. Collyer) 1764–5

Fenton, J. Land surveyor of Chelmsford. *Great Burstead, Essex* 1831 MS.

Fenwick, T. H. Maps and plans *Chatham* 1821 MS.

Fenwick, Lt.-Col. W. *Plans Cork* 1810–12 MSS.

Fenwick, W. *Plan Peninsula & Harbour Halifax* (1850?)

Fenyes, Elek [Alex.) (1870–1876). Hungarian cartographer from Csokaly

Fen Yeh Yu T'u. (17th century) *Atlas China*

Feodoseyev, V. Russian scribe and draughtsman 17th century. *Map Sevsk* (1650)

Fer, Antoine de (fl. 1644). Mapseller, 'demeurant sur le Quay qui regarde la Mégisserie près l'Orloge du Palais à l'Age de Fer', Paris. *Plans de Villes de France* 1652, *geographical games* 1670–2

Fer, F. Cartographer. *Galizia,* Ortelius (1606)

Fer, Nicolas de (1646–1720). Géographe de sa Majesté Cath., publisher "dans l'Isle du Palais sur le Quay de l'Orloge à la Sphère Royale"; Succeeded by sons-in-law J. F. Benard and G. Danet. *France ecclés.* 1674, *Haute Lombardie* 1682, *Costes de France* 1690, *Liège &c.* 1693, *Introd. à la fortification* 1693, *Atlas Royal* 1695, 1699–1702, *Petit Atlas* 1697, *Atlas curieux* 1700–5, *Atlas ou Recueil de Cartes Géog.* 1709–22, *Introd. à la Géog.* 1717, *Californie* 1720

Ferber, J. J. (1743–1790). Swedish geologist. *Mineralgeschichte von Böhmen*

Fer de la Nouerre, de. See **De Fer de la Nouerre**

Ferdinando, S. M. *Contorni di Napoli* 1819

Ferdinandus, Filips. Artist of Antwerp. *Plan Papa* 1566 for Braun & Hogenberg

Ferguson. Engraver. *Africa* 1792

Ferguson, A. *Lake Lacroix* 1850 MS.

Ferguson, A. M. *Ceylon, Colombo* 1875

Ferguson, Charles. *Entrance Gipps Land Lakes* 1866

Ferguson, George E. *Divisions of the Gold Coast Protectorate* 1884, *Hinterland Gold Coast Colony* 1893, 1897

Ferguson, G. R. *Isthmus of Tehuantepec* 1851

Ferguson, James (1710–1776). Scottish philosopher, and astronomer. *Globes* ca. 1750 MS. *Terrestrial globe* 1773

Ferguson, J. H. D. *Rossland, B.C.* 1898

Ferguson, Patrick (1744–1780). American surveyor. *Long Island* 1779, *Hudson River* 1779, *Part New Jersey* 1780, *Fortifications Savannah* 1780 MSS.

Ferguson, W. (1820–1887). Geologist and botanist in Ceylon

Fergusson, J. *Muirkirk Coalfield, Ayr* 1841

N. de Fer (1646–1720)

Fergusson & Mitchell. Lithographers and publishers. *Provs. Otago & Southland, New Zealand* 1866

Ferimontanus, D. C. See **Cellarius**

Ferlettig, A. *Pianta di Trieste* 1871, 1892

Ferlin. *Plans Toulon* [1873]

Fern, Edmund W. *Railway map India* (1878)

Fernandes, Marcos. Portuguese cartographer. *Chart* 1592 MS.

Fernandes [Fernandez], Pero (fl. 1558). Portuguese pilot and cartographer. *Chart Atlantic* 1528 MS

Fernandes, R. A. *Navig. distances on Amazon,* Manaos 1869

Fernandes, S. See **Fernando**

Fernandes, Valentim. *Ilhas Cabo Verde* 1506–8

Fernandez, Antonio (fl. 1781). Printer in Madrid. Espinault y Garcia's *Atlante Español* 1778

Fernandez, Francisco. *S. America* ca. 1800

Fernandex, P. See **Fernandes**

Fernandez, Pascual. *Aspecio geographico del mondo Hispanico* 1761

Fernandez, R. G. See **Gonzales Fernández**

Fernandez, S. See **Fernando**

Fernandez de Cordoba, Fernando. *Rio Mindanao* 1754 MS.

Fernandez de Enciso, Martin. See **Enciso**

Fernandez de Medrano, Sebastian. *Geog. o moderna descr. de el Mundo,* Brussels 1701, Antwerp 1709

Fernandez de Roxas, Antonio. *Topographia ciudad de Manila* 1717

Fernando [Fernandes, Fernandez], Simon [Fernando Simon]. Portuguese pilot and cartographer in English service, b. Azores. Son of Pero Fernandes. *America* [1580] MS.

Fernel, Jean. French doctor. Measured the meridian, Paris—Amiens 1625

Ferra, J. *Carta Esferica . . . Isla de Sto. Domingo* 1802

Ferrande, Pierre Gracie. *Le Grand Routier et Pilotage* 1520 MS.

Ferrang, P. J. du. 19th century lithographer

Ferrari [Galateus Leccensis], Antonio (1444–1516). Physician, poet and geographer, b. Galatina. Maps used by Albertus

Ferrari, Filippo (1551–1626). Mathematician and geographer of Pavia

Ferrari, G. G. See **Giolito**

Ferraris [Francis], Joseph Johann, Grâf von (1726–1814). Austrian General and cartographer, b. Luneville, d. Vienna. Director General of Artillery in Neths. Survey *Austrian Netherlands* 1770–4, (275 sh.) MSS., pub. 1773–7 (25 sh.), *Frontiers of Emperor and the Dutch* 1789

Ferraz, Guilherme Ivens. *Costa de Moçambique* (1894)

Ferre, de. "Capt. du Regt. de la Marine et ing. ord. du Roy" *Basse Alsace* 1692 MS.

Ferreira de Loureiro, Adolpho, *Atlas* 1886

Ferreiro, Martin (1830–1896). Geographer of Madrid. *Atlas de España* 1864, *España y Port.* (1867?) *Cartagena* 1873, *Prov. de Zambales* [1880]

Ferrel, Will. (1817–1891). American meteorologist. Coast & geodetic survey

Ferrer, Fermin. *Nicaragua* 1855, *Colton's Central America* 1859

Ferrer, Jaume [Jayme]. Catalan cosmographer. *World* (in hemispheres, N. & S.) 1495 MS. (lost)

Ferrer, John. See **Farrer**

Ferrer, J. S. y. See **Seyra y Ferrer**

Ferrer, Prof. Dr. K. *Karte von Palästina* 1891

Ferrer, V. See **Farrer**

Ferreri, Giovanni Paolo. Worked in Paris. *World* ca. 1554 MS.

Ferreri, Giovanni Paolo (fl. 1600–24). Globe and instrument maker in Rome

Ferrerios, Rainaldo Bartolomeo de. Majorcan chart maker. *Chart* 1592 (with Prunes)

Ferrero, Annibale (1839–1902). Mathematician and geographer in Florence, Director Inst. Geog. Milit.

Ferrero della Mamora, Alberto. *Sardegna* 1839, *carte géolog. d'Italie* 1846

Ferret, Pierre Victor Adolphe. *Carte d'Acir – Hedjaz* 1840, *Abyssinie* 1843, *Carte géologique du Tigre* 1848

Ferretti, Capt. Francesco. *Diporti Notturni* Ancona 1580 (28 maps)

Ferri, Fr. F. M. Engraver for Coronelli 1691–6

Ferris, George Titus (b. 1840) Appleton's *Atlas U.S.* 1888

Ferris, Warren Angus. *Rocky Mt. fur region* 1836 MS.

Ferron. *Carte des Chemins de Fer Français* (1874?)

Ferry, A. C. *Atlas Lorrain County, Ohio* 1874 (with Lake &c.)

Ferry, Hypolite. *Nouv. Californie* 1850

Ferslew, Martin William (1801–1852) Lithographer in Copenhagen. Maps *Sleswic*

Ferslew, W. Eugene. Geographer and publisher. *City Richmond, Va.* 1859

Fery, J. H. *Cariboo Gold Region* 1861

Fesca, A. (fl. 1850). Engraver in Kiel

Fest, Matthias (1736–1788). Austrian cartographer of Eisenstadt

Fester, Diderich Christian (1732–1811). Mathematician and cartographer of Drontheim. *Sleswig* 1766, *Zealand n.d.*

Festing, Major A. M. *Caravan routes Sierre Leone* 1885, *Plan Quiah Territory* 1866

Feterus, Olasson. Scandinavian cartographer, late 17th century

Feuchtmayr, J. A. See **Feichtmayr**

Feuersfeld, Franz Schönfelder von. See **Schoenfelder von Feuersfeld**

Feuerstein, W. Engraver for Meyer 1867

Feuille, de la. See **La Feuille**

Feuillée, Louis. Measured longitude between Fero & Paris Observatory 1724

Feunema, R. *Geol. Karte v. Java* 1898

Févre de Montigny, G. F. Dutch Col. of Engineers, late 18th–early 19th century

Fēz. See **Fernandez**

Fiala, John T. *Sect. map Missouri* 1860, 1865 (with Haren)

Fiamengo, A. di A. See **Arnoldi**

Fickler [Wylensis] , Johann Baptist. *Scandinavia* (after Olaus Magnus), Basle 1567

Fidalgo, Joaquin Francisco. *Carta esférica Islas Antillas* 1802, *Plan Porto Cabello* 1804, *Costa del Darien* 1817, *New chart Caribee Is.* 1835

Fiddes, Lieut. James. Engineer. *Plan James Is., River Gambia* 1783 MS.

Fidler, Peter. Explorer N.W. America. *Upper Missouri* 1801 MS. (copy of Blackfoot original)

Fidling, T. Ch. *Plan Babic bei Warschau* (ca. 1750)

Fiedler, F. *Kreises Westprignitz* (1899)

Fiedler, Ferdinand Ambrosius (d. 1756). Surveyor of Magdeburg

Fiedler, Karl Gustav. *Königreich Griechenland* 1840

Fief, J. B. A. J. du. See **Du Fief**

Field, Capt. A. Mostyn, R. N. *Vavau Group (Tonga)*, Admiralty 1898

Field, Barnum. *Atlas,* Boston 1832

Field, John. See **Feild**

Field, John. *Scarborough, Plans* 1812, *Ireland* 1816

Field, Richard. Publisher of London *Boazio's map Drake's Voyage* 1589

Field, Samuel. *Draught Windward Coast* (1740)

Field Publishing Co. *Sangamon Co., Illinois* 1894

Fielding, John. Publisher 23 Paternoster Row, London. Fielding's *London Guide with Plans London, & Environs* 1782. *N. America* 1782, *St. Kitts & Nevis* 1782, *Andrews' History of the War [in America]* 1785

Fielitz, Friedrich August von (fl. 1821–5). Engraver in Berlin

Figari, Antonio. *Études géog.* 1866

Figg, William. Land surveyor of Lewes. *Lands in E. Hoathly* 1822 MS., corrected *Sussex* 6 sh. 1861

Figueiredo, Manoel de (1558–ca. 1630). Portuguese cartographer. *Chronographia* 1603, *Hydrographia* 1608, *Roteiro e navegaçâo das Indias*

Figuera, F. de. *Barra de Ocos* 1802 MS.

Figueroa, G. de la R. y. See **Rocha y Figueroa**

Figueroa, José Vazquez. 'Secret. de Estado y del Despacho Univ. de Marina'. *Estrecho de S. Bernardino (Philippines),* Madrid 1816

Figueroa, Rodrigo de. *Porto Rico* 1519 MS.

Figurative Map. Anon. map of *New Netherland* 1616 MS

Fijnje, H. F. See **Fÿnje**

Filaeus, Petrus. Swedish cartographer, mid 18th century

Filastre, G. See **Fillastre**

Filiberto, Emanuele. Globemaker in Rome. *Globe* ca. 1570 (attributed)

Filla[s]tre [Filastre], Guillaume (1344–1428). French prelate, geographer and classicist, b. La Suze, d. Rome. Text to Clavus' map *N. Regions* which he in-

corporated in MS. *Ptolemy's Geog.* 1427, (1st modern map in version of Ptolemy)

Fillian, John. Engraver. Title for Heylin's *Cosmographie* 1682

Fillingham, Wm. Surveyor. *Beckingham* (1770)

Fillot, H. *Plan de Mojanga (Madagascar)* 1895

Filosi, Giuseppe. *Pianta di Verona* 1737

Fils, Lieut. August Wilhelm von (1799–1878). Artillery officer, cartographer in Berlin and Schleusingen. *Glatz* 1832, *Sachsen* 1836, *Rudolfstadt* 1848, *Thüringer Wald* 1860–7, *Gegend von Ilmenau* (1876?)

Filson, John (ca. 1747–1788). Surveyor & historian of Pennsylvania. *Kentucky,* Phila. 1784

Finaeus, Orontius. See **Finé**

Finaghenof. Engraver for Piadischeff (1823–6)

Finckh, G. P. See **Finkh**

Finden, Edward J. (1792–1857). Engraver, worked with brother Wm. *Road Book London–Naples* 1835

Finden, William (1787–1852). Engraver. Worked with brother Edward.

Findlater, Rev. Charles (1754–1838). *Peebles* 1802

Findlay, Alexander (1790–1870). Engraver and draughtsman of Pentonville, London. Engraved charts 1816–65 (many by A. G. Findlay), *Rees Cyclopaedia* 1820, *Laurie's Environs London* 1829, *Barclay's Eng. Dict.* 1835–43

Findlay, Alexander George (1812–1875). Geographer and publisher, son of above, successor to Laurie & Whittle in 1858. *Chart Med.* 1839, *Modern Atlas* 1843, 1850, *Classical Atlas* 1847, *Comparative Atlas* 1853, *Six Nautical Directors of Great Oceans* 1858–78, *Route between Zanzibar and Great Lakes* 1860

Findlay, John. *Tahiti* 1877 MS.

Findlay, Lieut. J. K. *Survey Kennebeck River* (1826)

Findorf, Dietrich (1722–1772). Painter and engraver. *Lauenburg* n.d.

Findorf, Friedrich C. (fl. 1784–1805). *Gen. vharte H. Bremen* 1795

Fine, E. E. Surveyor. *Humboldt Mining Region* 1864

Finé, [Finaeus, Finnaeus], Oronce [Orontius Delphinas] (1494–1555). French astronomer and cartographer, b. Briançon, d. Paris. *World* (single cordiform) ca. 1521, pub. 1536, reissued by Cimerlinus 1566, *France* 1525, 1536, 1557, *World* (double cordiform) 1531, *Le Sphère du Monde* 1549 MS, 1551 *Cosmographia Universalis,* Cimerlinus 1556

Finger, Fr. Aug. (1808–1888). German geographer

Finiels, Don Nicolas de. *Cours de Mississippi* 1797–8

Fink, Major. *Plan von Lübeck* 1872

Fink, Albert (1827–1897). American engineer. *Baltimore & Ohio Rail Road* 1850

Finkh [Finckh], Georg Philipp, (d. 1679). Bishop. Improved *Apian's Bavaria* 1671, reissued by his son 1684

Finlay, Lieut. John, 83rd Regt. *Guernsey* 1782

Finlay, Capt. John, R.E. *Plans Waltham Abbey* 1800–1, *Faversham* 1801 MSS.

Finlayson, James. Drew for Carey & Lea's *American Atlas* 1822–3

Finlayson, John. *Plan battle Culloden* 1746 MS.

Finley, Anthony (ca. 1790–1840). Cartographer and publisher in Philadelphia. *New General Atlas* 1824, edns. to 1831; *New Amer. Atlas* 1826, *Atlas classica* 1829

Finley, Jos. Estate surveyor of Billericay. *Ramsden Bellhouse &c. (Essex)* 1747 MS.

Finnaeus, O. See **Finé**

Finne, Capt. E. *Omegnen of Mjösen* 1845, *Palaestina til Skolebrug* (1876?)

Finnie, James. Printer Dept. Mines, Melbourne. *Smyth's geolog. map Australia* 1875

Finsch, Dr. O. *Brandenburg Küste* 1894, *Krakau* 1896

Finsler, Hans Conrad (1765–1839). Cartographer of Zürich

Fiorin, N. See **Florino**

Fiorini, Matteo (1827–1901). Italian geographer

Fiorini, M. See **Florimi**, Matteo

Firminger, R. E. Sketch surveys *Kricor* 1886

First Printed Map. A "TO" map in the 1472 edition of *Isidore of Seville's Origines*

Firth, C. Lithographer. *Plans Bermondsey & Southwark* 1828–34

Firth, C. M. Zincographer, 7 St. Michael's Alley, Cornhill, London. *Plan London* 1841

Fisch, J. C. *Plan Lyon* 1875

Fischbein, George. Engraver. *Lasius' Harz Gebirge* 1789

Fischer, A. *Wand-Karte v. H. Anhalt* (1875)

Fischer, Albert. *Umgebungen v. Wildbad* (1846?), *Würtemburg* 1849

Fischer, Carl. *Saxony* 1836–50

Fischer, Eduard. *Bukowina* 1898, 1899

Fischer, Friedrich von (1826–1907). Austrian general and cartographer

Fischer, Dr. G. A. Maps of *E. Africa* 1891–5

Fischer, Georg Peter, of Munich. *Plans German towns* for Merian 1644

Fischer & Co., H. Publishers. *Africa* 1825, *N.W. Germany* 1829

Fischer, Hans. *Hauran* 1889, *Palästina* 1890

Fischer, Hermann von (b. 1851). *Atlas zur Geog. der Schwäbischen Mundart* 1895

O. Finé [Orontius Finaeus] (1494–1555)

Fischer, J. Engineer. *Plan v. Constantinopel* 1877

Fischer, J. B. *Nassau* 1858

Fischer, Johann Christoph (1729—1801). Cartographer in Darmstadt

Fischer, Johann Jakob (1709—1753). Surveyor of Bern

Fischer, Max. *Landeshauptstadt Brünn* (1871?)

Fischer, Major. N. *Klein Asien* 1844, *Plan-Atlas v. Klein Asien* 1854

Fischer, Theobald (1846—1910). Geographer and map historian in Marburg. *Raccolta di mappamondi* 1871—81

Fischer, Wilhelm. Cartographer. *Hist. u. geog. Atlas,* Berlin 1834—7 (with Streit)

Fishbourne. Lithographers, 529 Clay St., San Francisco. For Baker & Barber 1855, for Fine 1864. See also **Nagel, Fishbourne & Kuchel**

Fisher, Arthur A'Court. *Plan Sevastopol* 1857

Fisher, Charles H. *Lake Ontario Shore R'Road* 1868

Fisher, I. Engraver. *Faroe Is.* 1781

Fisher, J. Engraver. Dewe's *Pict. Plan Oxford* (1850)

Fisher, James. Surveyor. *Plan Hurstpierpoint* 1841

Fisher, John. Surveyor. *Whitehall Palace* 1670, *Survey Plot Whitehall* 1680—(1740), *Oak End, Gerrard Cross* 1680 MS, *Bulstrode Estate (Bucks.)* 1686 MS.

Fisher, Jonathan. *Waterford* 1772

Fisher, Joshua (1707—1783). Cartographer. *Chart Delaware Bay,* Phila. 1756 (with J. Davis), contributed to *N. American Pilot Pt. II* 1795

Fisher, Ralph. *Plan River Sherbro* 1773—(94)

Fisher, Richard Swainson. *S. America* 1851, *R'roads U.S. & Canada* 1856, Colton's *Atlas America* 1859, Colton's *Illust. Cabinet Atlas* 1862, *American Atlas* 1865

Fisher, Samuel B. *Coal Fields Pennsylvania* 1849

Fisher, Thomas (1741—1810) *Part of Pennsylvania and Maryland* 1771 & 1789

Fisher, T. *Hull* 1813, *Plymouth* 1814 MSS.

Fisher, William (fl. 1669—1691). Publisher at the Postern-Gate on Tower Hill, London. In 1671 joined John Seller in publication of the *Coasting Pilot.* Later entered into partnership with John Thornton, Fisher supplying the finance and Thornton the charts. They issued jointly the *English Pilot the 4th Book* in 1689. Richard Mount, apprenticed to William Fisher in 1669, married Wm. Fisher's daughter Sarah in 1682, was taken into partnership and succeeded him.

Fisher, Son & Co. Publishers Caxton Press, Angel St., St. Martin-le-Grand, London. *County atlas* 1842—(5)

Fisherman, A. *Chart Armegon Shoals* 1762 pub. Dalrymple

Fisk & Russell. Engravers of New York. For Watson 1867, *Chicago* 1869, *Hannibal & St. Joseph R'Road* 1869

Fisquet, Honoré Jean Pierre (1818—1883). *Grand Atlas Départmentale de la France* (1878)

Fisquet, Théodore Auguste. *Chart coast California* 1851

Fisscher, J. W. van Overmeer. MS. plans *W. African forts* 1786

Fitch, Charles H. *Indian Territory* 1895—9

Fitch, George W. Colton & Fitch's *School Geog.* (1868), 1870; *Primer of Geog.* 1870

Fitch, John (1743—1798). American metal craftsman and inventor. *North West [America]* 1785

Fitch, John. *Golden Square, Church Passage, Piccadilly* 1834 MS.

Fitchatt, J. Estate surveyor. *Cheese Cross & Noak Hill* 1768 MS, *South Weald* 1772 MS. (both in Essex)

Fitton, H. G. Surveys *Sudan* 1894–6, *Routes to Kuror* 1897

Fitzer, Wilhelm (fl. 1650). Born London, worked in Frankfurt for Merian

Fitzgerald, E. A. *Sketch map S. Alps* 1896

Fitzgerald, G. *Projected canal Nicaragua* 1850–1

Fitzgerald, J. W. Surveyor. *Pringle Parry Sound* 1878

Fitzgerald, Lieut. R. *Plan Meanee* (1843?)

FitzHenry, Capt. C. B. *County round Pietermaritzburg* 1897

Fitzhugh, Augustine (fl. 1683–1694). Chartmaker, "living next doare to the Shippe in Virgine Street" and "at the corner of the Minnories neare little Tower Hill". Drew Southack's *Draught Boston Harbour* 1694, *Coasts of Newfoundland* 1693

Fitzmaurice, L. R. Hydrographer. *Bird is.* 1816, *R. Zaire* 1817, *North Sea* 1817–18, *England* 1819–24, *Port Talbot* 1846

FitzOsborne, James. Surveyor. *Manor of Hawards Hoth & Trubweeke* 1638

Fitzpatrick. Engraver. *McCrea's Monaghan* (1795), *Donegal* 1801

Is. 1834, *Galapagos Is.* 1836, *Cocos Is.* 1856

Fitzwilliam, Brian. *Plan Portsmouth* ca. 1585 MS.

Fitzwilliams, E. C. L. *Aberporth Bay* (1860)

Fitzwilliams, Michael. *Belfast Lough* 1569

Fix, W. *Wandkarte Preussischen Staats* 1855, *Rheinland u. Westfalen* 1864

Flachat, Eugène. *Chemin de Fer de Paris à Meaux* 1839

Flacourt, Étienne de (1607–1660). French traveller and administrator in Madagascar. *Madagascar* 1656

Flahaut. Engraver, for Dépôt de la Marine 1818, Letronne 1827, Dufour 1836, Andriveau-Goujon

Flahaut, Mlle. Engraver for Andriveau-Goujon 1854–6

Flahaut, P. Engraver. *Europe* 1841

Flammand, G. B. M. (b. 1861). French explorer and geologist. *Voyage Algérie* 1899–1900

Flammarion, Camille. *Atlas Astron.* (1875), *Atlas Céleste* 1877

Flamsteed [Flamstead], John (1646–1719). First Astronomer Royal (1675). *Cat. of Observations of Stars* 1707, *Atlas Coelestis* 1729 &c.

Flamsteed, J.

Fitzroy, Alexander. *Kentuckie* 1786

Fitzroy, F. *British Burma, Pegu Div.* 4 sh. 1885 (with W.H. Edgcome)

Fitzroy, Comm. (later Vice-Admiral) Robert (1805–1865). Navigator, commanded H.M.S. *Beagle* 1828–36, Governor New Zealand 1843–5. Surveyed voyages of *Adventure* (1826) & *Beagle* (1836), *S. America* 1828–30, *Falkland*

Flandrus, G. C. See **Carolus**

Flatters, Col. Paul François Xavier (1832–1881). French explorer. *R.way survey Algeria & Soudan, Route Sahara* MSS.

Flattman, Surveyor of Canterbury. *Map of Margate* 1681 MS. (lost)

Flavio [Flavius], Gioja [Gisia] of Amalfi. Alleged invention of compass ca. 1320

Flaxman, Charles. *Surveys sources Rhine* 1839–(43)

Flèche, J. le L. de la. See **Le Loyer de la Flèche**

Fleck, Dr. E. *German S.W. Africa* 1899

Fleck, G. *Agric. lands Kyneton, Victoria* 1854

Fleck, Matthew. Surveyor. *Ryton Inclosures* 1829

Fleetwood, St. William. *Survey crown lands* 1610–29 MSS.

Flegel, E. Robert (1852–1886). German explorer. Maps *Nigeria* 1880–5

Fleischmann, Andreas (1811–1878). Engraver in Munich

Fleischmann, Friedrich (1791–1834). Engraver in Munich. *Heligoland* 1814–15, Battle plans 1818

Fleischmann, Johannes Joseph. Printer in Nuremberg. Homann Heirs' *Nouvel Atlas* 1748

Fleming, J. Surveyed portion *Brit. Bechuanaland* 7 sh. 1894, *trade route Transkei* 1899

Fleming, John (1785–1857). Scottish naturalist and mineralogist. *Mineralogy of Orkney & Shetland Is.* 1808

Fleming, Peter. *Plan Glasgow* 6 sh. 1807

Fleming, Sandford A. Engineer-in-chief. *Newcastle & Colborne Districts* 1848, *Canadian Pacific Railway* 1874, 1876

Flemming, Carl (fl. 1842). Publisher of Glogau. *Mexico &c.* 1852, *Austral* 1855, *Texas* (1860), *Graudenz* (1893?)

Flemming, M. *Grosser Atlas der Eisenbahnen v. Mittel-Europa* 1892

Flender, R. *Silesia* 1854, *Special Karte der Krim* 1855

Fletcher. *I. of Wight* 1806 (with T. Baker)

Fletcher. Surveyor. *Plan Birkenhead & Claughton-cum-Grange* 1858 (with Mills)

Fletcher, E. G. *Plan Warley Camp (Essex)* 1779, *Intended canal Cromford to Langley-Bridge* 1789

Fletcher, F. E. & W. *Kootenay* 1897

Fletcher, Hugh. *Geolog. map Cape Breton* 1879

Fletcher, James. Bookseller, the Turle, Oxford. Cole's *20 miles round Oxford* (1705)

Fletcher, Miles. Printer. Miniature edition Speed's *Prospect* 1646

Fletcher, P. *Matabeleland* 1896, 1897, *Plan Bulawayo* 1898 (all with W.M. Espin)

Fleuriais, George Ernest. *Plan Mouillage de Campèche* 1865

Fleurieu, C. P. C. de See **Claret de Fleurieu**

Fleurieu, D'Eveux de. See **Claret de Fleurieu**

Fleurieu de Bellevue, L. B. (1761–1852). French geologist, b. La Rochelle. Works on geology, mineralogy & meteorology 1790–1847

Fleuriot de Langle, Alphonse Jean René (1809–1881). French sailor, b. Finistère. *Spitsberg* 1838–(40). *Cap de Monte* 1842–(52), *Bumbalda* 1843–(5), *Baie de Santa Isabel* 1846–(51)

Fleury, E. de. *Nueva Mapa de Sonora &c.* 1864

Fleury [Flury], François [sometimes misnamed Henri] de (fl. 1792). Served in America under Rochambeau. *Sketch Siege Fort Schuyler* 1777, *Mud Island, Fort Mifflin* 1777, *View Fleet before Philadelphia* 1778

Fleury. See **Flury**

Flexney, W. Bookseller. *Camden's Britannia* 1772

Fliegner, Ferdinand. Cartographer and lithographer in Breslau. *Poland* 1848, *plan Vienna* [1848]

Fligely, August von. (1810–1879). Austrian general and cartographer

Flinders, Lieut. (later Capt.) Matthew (1774–1814). Navigator and hydrographer, b. Donnington, Lincs. Circumnavigated Tasmania with Bass 1798–9 and was the first to circumnavigate Australia 1801–3.

Charts *V.D.L.* 1798–(01), *Twofold Bay, N.S.W.* 1798–(01), *Ram Head to Northumberland Is., N.S.W.* 1800, *Port Philip* 1802–(3), *Gulfs St. Vincent & Spencer* 1807–*(55), Voyage to Terra Australis* 1801–3, *Atlas* 1814

Flint, A. R. *Chart Owlshead Harbour* (1836) (with T. A. Barton)

Flint, Ole Nielsen (1739–1808). *Grund Tegning Kiobenhavn* 1784

Flint, W. Printer for Guthrie 1808

Flintoff, J. Surveyor. *Sutton Grange* 1751 MS.

Flinzer, Theodor (fl. 1850). Lithographer

Flitcroft, H. Surveyor of Westminister 1730–44

Flocquet, J. A. *Course du canal de Province* 3 sh., n.d

Floder, A. *Strassen-Karte Oesterr. Monarchie* 1835

Flore, Capt. *Faroe Is.* 1781

Floreda, Pascual. *Zifur (Philippines)* 1768 MS.

Florence – Palazzo Vecchio. See **Palazzo Vecchio**

Florentinus, S. See **Buonsignori**

Florentius, J. See **Langren**

Flores, Manuel Antonio. *Spanish Guyana* 1777 MS.

Florez, Don I. F. *Ferrol Harbour* 1846, *C. Finisterre to Vigo Bay* 1846

Floriani, A. See **Florianus**

Floriano, A. See **Florianus**

Floriantschitsch de Grienfeld. See **Grienfeld**

Florianus [Floriani, Floriano], Antonius. Painter, architect and cartographer, b. Udine, fl. Venice 1545–55. *World* in form of globe gores (1555)

Florianus [Blommaerts], Joannes [Jean] (1522–1585). Teacher of Bergen op Zoom, b. Antwerp. *E. Frisia* 1579 pub. Ortelius

Florimi [Florini], Matteo (fl. 1580–1612). Publisher in Siena. *Tuscany, Italy, Orvieto,* all dated 1600; many other undated maps in "Lafreri" atlases

Florin, Johan (1739–1796). Swedish land surveyor

Florino [Fiorin], Nicolo. Portolan maker of Venice. *Charts Medit.* 1462, 1489

Floris. See **Balthazar**

Flosie, Michael (1724–1794). Engraver. *Flensburg*

Flotte de Roquevaire, R. de. *Maroc* 1897

Flower, W. L. *Cook Co., Ill.* 1862, *Chicago* 1863

Floyd, Charles. *Siège de Maestricht* 1632–3

Floyd, Wm. H. Jr., & Co. *Atlas St. Joseph, Mo.* 1884

Floyer, Ernest A. *S. Persia* 1882, *Egypt, Eastern Desert,* 3 sh. 1891

Fludd, Robert. *Disert. Cosmographicam* 1621

Fluddes. See **Lhuyd**

Fluddus. See **Lhyud**

Flury [Fleury]. *Charts of coast of Peru* 1824 (with Lartigue)

Flury, F. or H. See **Fleury**, François

Flury, L. *Sketch siege Ft. Schuyler,* N.Y. (1850?)

Flyn, J. Engraver. *Plan London* 1770, Taylor & Skinner's *Roads of Scotland* 1776

Flynn, John. *Florence, Missouri* (1857)

Flynn, Thomas. *Atlas Newport, R.I.* 1893 (with C.L. Elliott), *Atlas suburbs Cleveland, Ohio* 1898

Foa, E. (b. 1862). French explorer. *Zambezi to Congo* 1894–7, pub. 1898

Fock, H. (1766–1822). Dutch engraver. *Haarlem* (1801)

Focken, Hendrik. Publisher "inde

Molsteegh. Amsterdam." *Geneva* (1650), *Batavia* ca. 1680

Foeh [Föh]. *Hafen von Marca (E. Africa)* 1875

Foeltz-Eberle [Föltz-Eberle], E. Drew & Lithographed *Karte von Texas* (1839), *Grundriss v. Frankfurt* (1868?)

Foerster [Förster], Dr. Fr. *Neuester Plan Wien* 1872

Foerster [Förster], H. G. *Plan v. Leipzig* 1878

Foerster [Förster], L. *Plan Trieste* (1860?)

Foetterle [Fötterle], Franz. *Geological maps S. America, Austria & Bohemia* 1854–68

Foge, Edward. *Plan Rhode Is.* 1778

Föh. See **Foeh**

Foisse, Jacques de. Engraver in Hamburg, for *Hess' Topographie* (1785)

Folderbach, A. (d. 1656). Land surveyor of Friesland

Folger, Capt. O. Nantucket whaleman. *Gulf Stream* 1787, *Ship Is., Argentina,* Faden 1816

Folie, A. P. French geographer. *Plan Baltimore* 1792, *Philadelphia* 1794

Foligne, Lieut. *Plan Cap François* 1781

Folin[o], Bartolomeo (1730–1808+). Venetian engraver in Warsaw. *Pologne,* 4 sh. 1770, reissued Schraembl 1788, 1801

Folkard, Thomas. Surveyor. *Mendham (Suffolk)* 1721, *Stokesby Common (Norfolk)* 1721 MSS.

Folkema, T. Engraver for A. Allard, and Sterringa 1717

Folkingham, Wm. *Feudigraphia: the synopsis or epitome of surveying* 1610

Follenweider, R. (1779–1847). German engraver. Plans Swiss and German towns

Follin, O. W. *Plan Isthmus Tehuantepec* 1851

Folque, General Filippe. *Portugal* (1865?); *Cidade de Lisboa* 1871, *Plano hydro. Lisboa* 1878

Folsom, Charles J. Publisher, No. 40 Fulton St., New York. *Mexico & Texas* 1842

Föltz-Eberle, E. See **Foeltz-Eberle**

Fomine, Alex. (1713–1802). Russian historian and geographer. *Description of White Sea* 1797

Fonbonne, Quirin (b. ca. 1680). Engraver, worked Amsterdam ca. 1705, in Paris 1714–34. *Environs de Reims* 1722

Foncin, Pierre (1841–1916). French historian and geographer, b. Limoges. *Géog. hist.* 1888

Fonseca, J. G. da. See **Gonsalves da Fonseca**

Font, Fray Pedro (d. 1781). Franciscan missionary and cartographer. *Monterey– San Francisco* 1776–7 MS.

Fontaine, La. See **La Fontaine**

Fontaine, Mme. Engraver for Andriveau-Goujon 1854–73

Fontaine, Rev. Peter. Chaplain and surveyor to Virginian Commissioners 1749. *Va. & N. Carolina* 1752 MS.

Fontaine. *Yazoo County (Missouri)* with Mercer 1874

Fontaine, P. F. L. *Plans de plusieurs chateaux* 1831

Fontan, Dr. D. D. *Carta Geomet. de Galicia,* Santiago 1845

Fontana, Giovanni Baptista (1525–1587). Artist and engraver of Verona, worked in Innsbruck, court painter to Archduke Ferdinand. *Palestine,* Zaltieri 1569

Fontana, Jacopo. Cartographer of Ancona. *Ancona* 1569 (used by Braun & Hogenberg)

Fonten, G. G. Swedish surveyor, early 19th century

Fonton, Feliks Petrovich (1801–1862). *Atlas Russie dans l'Asie-Mineure* 1840

Fonville, de. *New France,* Quebec 1699 MS.

Fooks, C. E. Surveyor. *Christchurch, Canterbury, N.Z.,* 1862

Foord, Wm. Surveyor. *Caldicott; Shire Newton* 1771 MS.

Foot, Peter. *Middlesex* 1794

Foot, Thomas. Principal engraver to Board of Ordnance, Wester Place, St. Pancras, and Weston Place, Battle Bridge. Engraved for Meares's *Voyage* 1790, for Arrowsmith 1794–8, Faden's *St. Eustatius* 1795, *Gream's Sussex* 1795, *Yates's Staffs.* 1798, *Mudge's Kent* 1801, *Ellis's Hunts.* 1824, *& Notts.* 1827

Foote, Charles M. [& Co.]. Publisher in Philadelphia 1877–9, & in Minneapolis 1884–96, with various partners: Warner, G. E. 1877–86; Hood, E. C. 1887–99; Brown, W. S. 1888–9; Henion, J. W. 1890–5. *Plat Books* of Counties in Iowa, Michigan, Minnesota, & Wisconsin

Foppens, J. F. *Bib. Belgica* 1739

Foppiani, Celestino Luigi. *Genova pianta topo.,* Genoa, 1846, (1860?)

Fora,de la. See **Lafora**

Forbes, A. Estate surveyor. *Walthamstow* 1669 MS.

Forbes, Alexander. Topographical draughtsman and surveyor. *Galloway* 1690, *Ground Plott Conventre* 1691, *Newcastle under Lyne* 1691 MSS., *Plan Barcelona,* Beek (1705)

Forbes, David (1828–1878). English philologist and geologist, b. Douglas, I. of Man. Explor. S. America

Forbes, Edward. Cooperated with A. K. Johnston in *Physical Atlas* 1848

Forbes, H. O. *Timor Laut* 1882, *S.E. New Guinea,* Brisbane 1889

Forbes, James (1749–1819). Explorer and geographer. *Oriental Memoirs* (1813)

Forbes, James D. (1809–1868). Doctor and geologist. b. Edinburgh. Assisted Johnston in *Physical Atlas* 1848

Forbes, W. H. Maps *Richmond (Virginia)* 1863–4

Forbiger, Albert. *Orbis Terr. Antiq.* (1853), (1864), 1865

Forbriger. See **Ehrgott, Forbriger & Co.**

Forchhammer, J. G. (1794–1865). Danish geologist, b. Husum, d. Copenhagen. Works on geology and prehistoric Scandinavia

Forchhammer, P. W. (1803–1894). Brother of above, topographer. *Topo. v. Athens* 1841, *Tenedos u. Festlande* 1856

Ford, Augustus. *Chart Lake Ontario* 1836

Ford, James (d. 1812+). Engraver of Dublin. *Lake of Killarney* [1786], *Cahill's Queens Co.* 1806

Ford, R. Publisher with Clark and Crittenden of *Neal's New England* 1720 and 1747

Ford, Reuben W. *City Austin, Texas* 1872

Ford, Richard. Surveyor. *Barbados* 1675–(81)

Ford, Richard (1796–1858). *Travelling map Spain* 1845

Ford, Major W. H. *Dover* 1811, 1817 MSS.

Ford & West. Lithographers, 54 Hatton Gdn., London. *Oregon* (1855)

Forde, Sir Edward. *River Colne* 1641

Fordyce, General Charles Francis. *Plan Heights Inkermann* 1854–(5), *Battlefield Alma* [1855]

Fordyce, John. Surveyor Gen. of Land Revenue. *Crown Leases London* 1804 MS., *Improvements Hamilton Place & Piccadilly* 1805–(12)

Fordyce, Capt. (later Lt. Gen. Sir) John. *District Agra, Calcutta* 1837–(46)

Forel, Fr. Alph. (b. 1841). Swiss geologist. *Lac Leman* 1892, 1895

Fores, D. See **Dumas de Fores**

Fores, Samuel W. Publisher, Piccadilly, London – at various No's.: 3 (1789), 50 (1807), 41 (1836). *Plans London* 1789–1836

Fores, W. *Trav. Comp. London to I. of Wight* 1819

Forest, F. de Belle. See **Belle Forest**

Foret, Aug. *Sénégal* 1888

Forlani [Furlani], Paolo de (fl. 1576). Engraver and publisher of Verona working in Venice "al segno della Colona in merzaria" and also "in Merzaria alla libreria della Nave". Worked with F. Bertelli, Zaltieri & Zenoi. Prolific output includes: *World* 1560, *Africa* 1562, *Greece* 1562, *Tuscany* 1563, *Venice* 1565, *France* 1566, *Poland* 1568, *Cyprus* 1570, *America* 1574. MS. copies of sea charts 1562–9

Forling, Gustaf Johann. Swedish land surveyor, mid 19th century

Formaleoni, Ab. V. Script engraver. *Zatta's Africa* 1776

Forman, J. E. Civil Engineer, Dubuque, Iowa. *Nebraska railraods* 1857

Formentus. Engraver. *Ivrea,* J. Blaeu late 17th century

Fornari, Mauro. *Prov. di Lodi* 1789

Fornaseri, Jacques de. Italian engraver. *Siege of Bricherasio* (1594)

Förnebohn, A. E. *Geolog. oversigs-karta Sveriges* 1873

Fornel, L. *Carte des Esquimaux* 1748 MS.

Forrest, Lieut. C. R. *Chart Lake Borgne* 1815

Forrest, John (later Sir John; Baron Forrest of Banbury) (1847–1918). Explorer and statesman, b. Bunbury, W. Australia. Deputy Surveyor General of W. Aust. 1876, Premier W. Aust. 1890, Ministerial posts in Federal Govt. Travels in W. Aust. 1869–74

Forrest, Capt. Thomas (ca. 1729–ca. 1802). *Voyage to New Guinea 1774–6,*

pub. 1779, *Journal . . . Bengal to Quedah* 1783–(9). Charts of E. Indies, pub. Dalrymple: *Keysers Bay* 1774, *Track of Tartar* 1789

Forrest, Wm. Surveyor. *Drumpellier & Coats* 1801 MS., *Haddington* 4 sh., 1799, *E. Lothian* 1802, *Lanarks.* 1813, *W. Lothian* 1818, *Linlithgow* 1818

Forrester, A. Printer of Edinburgh. *Wood's Plan Inverary* 1825

Forrester, Joseph James. *Wine Dist. Alto Douro* 1843, *Douro Portuguez,* 3 sh. 1848

Forresters & Co. Lithographer. *Plan estate Strichen* (1860?)

Forsell, C. G. See **Forssell**

Forsell, Jacob. *Gota Canal* 1823 (with Carl Forssell)

Forsell, Lars. Finnish surveyor. *Plan Helsinki* 1696

Forshaw, W. Lithographer of Liverpool. *Rivers of Parana & Paraguay* 1842

Forssell [Forsell], Carl Gustaf af (1783–1848). Mathematician and cartographer of Stockholm. Maps of *Sweden* 1805–7 for Hermelin, *Sweden & Norway,* 9 sh., 1815–26, *Gota Canal* 1823 (with Jacob above)

Forsmann, M. *Transvaal* 1868

Forster. In some cases see **Foerster**

Forster, A. Engraver. *Grèce,* Athens 1838

Forster, D. I. R. *Cape of Good Hope* 1797

Forster, Johann George (1754–1794). F.R.S. Son of J. R. Forster, q.v., sailed with Cook on 2nd Voyage as father's assistant. *Chart S. Hemisphere* 1777 (in *Cook's Voyage*)

Forster, J. H. American surveyor. *W. end Lake Erie* 1849, *Parts Michigan* 1856–8

Forster, Johann Reinhold (1729–1798). German naturalist and explorer, accompanied Cook on 2nd Voyage 1772–5. *Middle part Asia* 1783

Forster, R. P. *E. India Is.* 1818

Forster, T. *Plan City of Durham* 1754

Forster, Thos. *S. shore Lake Erie* 1830 (with J. Maurice)

Forster, Westgarth. *Alston* 1823 MS.

Forster. See **Baskin, Forster & Co.**

Forster-Heddle, M. *Geolog. map Shetland Is.* 1879, *Sutherland* 1881

Forstman, G. See **Fosman**

Forstman, G. A. (1773–1830). Engraver in Hamburg. Map for *Dittmann's Geog. Lehrbuch*

Forsyth, Charles. *Geolog. map W. Lothian* 1847

Forsyth, Lieut. (later Comm.) Charles C. *Waterloo Bay* 1846, *Buffalo R.* 1847, *Port Michael Seymour* 1856

Forsyth, Sir D. *E. Turkestan,* Calcutta 1875, *Central Asia* 1878

Forsyth, Wm. *Plats city of Washington* 1856

Fort Dearborn Publishing Co. *Nat. Standard family & business atlas* (1896) &c., *Internat. office & family atlas* 1897

Forten, M. Master H.M.S. *Herione.* Improvements to Après de Mannevillette's *Junkseylon I.* 1798

Fortier, C. F. (1775–1835). Engraver. Maps to *Voyage en Espagne* 1803

Fortin. Engineer. *St. Pierre* 1763, *globe* 1769, *chart Miquelon* 1782

Fortin, Jean (1750–1831). Publisher & instrument-maker, Ing. Mécanicien du Roy, Rue de la Harpe, près la Rue du Foin, Paris. Succeeded by Delamarche. Pair *globes* 1770, reissued Robert de Vaugondy's *Nouvel Atlas Portatif* 1778, *terr. globe* 1780, sold Messier's *cel. globe* 1780, French edn. *Flamsteed* 1795

Fortún, J. M. See **Martinez Fortún**

Fortune, Wm. (d. 1804?). Canadian surveyor

Fosman [Forstman], Gregorio (fl. 1653). Spanish engraver, worked for Aefferden 1696

Fossati, Giorgio Domenico (1706–1778). Venetian architect and engraver

Fosse, Chatry de la. See **La Fosse**

Fossé, J. B. de la. *Provinces Unies* 1779, *Brabant* 1780, *Flandre,* Paris, Mondhare 1780, *Denmark* 1807

Fossé, N. du. See **Du Fossé**

Fosses, Castonnet des. See **Castonnet des Fosses**

Fosses, Chaumette des. See **Chaumette des Fosses**

Fosset. Printer, R. du Faubourg St. Jacques No. 19, Paris. *Oregon* 1854

Foster. *Bay of Colonia* 1819

Foster, Elizabeth. Publisher, White Horse, Ludgate Hill. Reissued G. Foster's *Plan London* 1752

Foster, George. Publisher, printer & mapseller at the White Horse, St. Paul's Churchyard. *Plans London* 1738–9, *Norfolk* 1739, *Seat War West Indies* 1739–40, *Jamaica* 1740

Foster, G. E. W. *Ontonagon group of mines, Lake Superior* 1864

Foster, Comm. Henry (1796–1831). Admiralty surveyor and astronomer, F.R.S. (1824), Copley Medal of R.S. (1827). *Survey La Plata* 1819, *E. coast Greenland* 1823, sailed as astronomer with Parry 1824–5 & 1827, led expedition to study ellipticity of earth 1827, *chart Atlantic* 1827–31, *River Para* 1831

Foster, John (1648–1681). Printer and engraver. b. Dorchester, Mass., d. Boston. First printer in Boston, "over against the sign of the Dove", 1675. *New England* in Hubbard's *New England,* Boston 1677: first woodcut map produced in English America

Foster, John. Estate surveyor. *Wix Essex* 1712 MS.

Foster, John, Junior. *N.W. part Liverpool* 1793 MS., *Plan site in Liverpool* 1803

Foster, J. G. *Cyclists' & pocket maps Ontario & Toronto* 1895–1900

Foster, J. W. *Geolog. map Lake Superior land Dist.* 1847, *Route Peninsula R'road* (1868)

Foster, N. G. *Cranford, New Jersey* 1870

Foster & Marion. *Denver, Auraria & Highland* (1860?)

Fotherby, John. Surveyor. *Level lying upon the River Ankholme* 1640

Fothergill, J. Engraver. *Tinker's plan Manchester & Salford* 1772

Fothergill, J. Engraver, Market St., Manchester. *Everett's plan Manch. & Salford* 1832–4

Fotheringhame, T. *Survey Yarmouth*, Steel 1804

Fötterle, F. See **Foetterle**

Fouché. *Seine Inférieure* 1857, *Arrond. Dieppe* 1858, *St. Valéry* 1867

Foucherot. *Mer de Marmara* 1784

Foucquet, H. Surveyor. *Sheerness* 1732, *Deal Castle* 1733, *Dover Castle* 1736 MSS., *Plan Dover* 1737

Foudrinier, P. See **Fourdrinier**

Fougasse, Thomas de. Map in *History of Venice* 1612

Fouilliand, Francisco. *Prov. Corrientes* 1891

Foulerton, Capt. John. *Royal Sovereign Shoals* 1813, *Dartmouth to Start Point* 1813, *Breakwater at Portland* (1825)

Foulkes. A. *Llandudno & environs* (1886)

Foulkes, C. H. *Accra* 1898

Foullon-Norbeck, H. de (1850–1896). Austrian geologist. Scientific mission to Australia 1893

Fouqué, P. A. (b. 1828). French geologist. Collaborated in production of *Carte géolog. de la France*

Fouquet, P., younger. *Amsterdam* 1783

Fourcault, Jean Baptiste. *Dépt. Maine et Loire* 1860

Fourcy, Eugène de. *Carte géolog. Côtes du Nord* 1843, *Atlas Souterrain de Paris* 1855

Fourdrinier [Fourdrinière], Peter [Pierre] (fl. 1720–60) French engraver, publisher & mapseller, corner of Craig's Ct., Charing Cross; came to London 1720. *Gordon's View of Savannah* 1734, *Plan Carthagena* 1741, *Plans Edinburgh* 1742, 1573, *Plan anc. city Westminister* (1761)

Foureau, Fernand. *Itin.* 1886

Fourier, Baron, J. B. Jos (1768–1830). French physician & surveyor. Wrote on the geography of Egypt

Fournel, Henri (1799–1876). French engineer & explorer. *Carte géolog. du Bocage Vendeen* 1835, *Explor. géolog. Algérie* 1843–7, *Richesse Minéral. Algérie* 1850

Fournet, J. J. B. X (1801–1869). French geologist and meteorologist

Fournier, Daniel (ca. 1710–1766). "A la mode Beefseller, shoemaker & engraver. Drawing master & teacher of perspective." Published Craskell & Simpson's *Jamaica* 1763

Fournier, H. Printer, Rue de Seine, Paris. *Atlas Ceran*, Lemonnier 1837

Fournier, Joseph Marie Martial. *Plan Iles Chatham* 1838–(40), *Plan Baies de Tokolabo et de Koko-Rarata (N.Z.)* 1838–(44).

Fournier d'Albe, E. E. Statistical maps *Ireland* 1891–(4?)

Fournier de Saint-Martin, Désiré. *Tab. géog. Pays Bas* (1825?)

Fournier des Ormes, C. (1777–1850). French engraver. *Plan Battle of Waterloo*

Fouzy, Lieut. H. *Route de E. Obeiyad à El Facher*, Cairo 1876

Fowkes, Francis. Convict. *Settlement at Sydney Cove* 1789 (earliest printed plan)

Fowler, A. G. *Prov. Bahia* 1886, *Routes Lagos to Niger* 1893–4

Fowler, Charles. Surveyor of Leeds. *Plan Leeds* 1821, *E. Lothian* 1825, *Fife, Midlothian* 1828, *Hungerford Market* 1829, *Yorks.* 1836

Fowler, Capt. Charles J., R.E. *I. of Wight* 1859 MS.

Fowler, H. *Plan River Hondo* 1886–7 MS.

Fowler, John. *Plan London* 1860, *Wady Halfa-Shendy* 1871–2, *Egypt & Equat. Africa* 1876, *Plan R. Tees* 1879, *Ambukol-Shendy* 1884

Fowler, L. D. *Assessment map Jersey City* 1870–(83)

Fowler, W. Engraver. Gibson's *Oxfordshire* 1765, for Ellis' *English Atlas* 1766

Fowler, Wm. Publisher from Wakefield with Greenwood, q.v. [Fowler, Greenwood & Co., Sharp, Greenwood & Fowler], *Lancs.* 1818, *Cheshire* 1819, *Staffs.* 1820, *Berwick* 1825–6, *Fife & Kinross* 1828, *County of Edinburgh* 1845

Fowles, Arthur, W. *I. of Wight* (1896?)

Fownes. Improvements to Huddart's *False Bay* 1798

Fox, Sir D. *Proposed Channel Tunnel* 1883

Fox, G. Maps for Pinkerton's *Mod. Geog.*, Philadelphia (1804)

Fox, H. I. *Plan estates Cambridgeshire* 1804

Fox, H. W. Drew and lithographed *Queensland* 1878

Fox [Foxe], Capt. Luke (1586–1635). English navigator and explorer, b. Hull. Explored *Hudsons Bay* 1631–2. Map in *North-West Fox, or Fox from the North West Passage* 1635

Fox, Samuel. *Derbyshire* 1760

Fox, W. R. *Nile Provinces* 1884

Fox. See **Martin & Fox**

Fox & Otley. *Sect. map Lee County, Iowa* 1861

Foxe, L. See **Fox**, Capt. Luke

Fraas, Oscar (1824–1897). German geologist and explorer, b. Lorch, d. Stuttgart. *Geogn. Karte v. Württemberg, Baden u. Hohenzollern* 1870, *Drei Monate in Libanon* 1876

Fracastoro. Italian cosmographer, d. 1553. Friend of Ramusio and supposedly responsible for the map of America in Ramusio's *Navigationi* 1556

Fracanzano da Montalboddo. See **Montalboddo**

Frachus, G. or I. See **Franco**

Fraitteur, W. *Plan Stadt Mannheim* 1813

Frambottus, Paulus. *Verona fidelis* (1660)

Frame, Edward H. *Flushing* 1871

Frampton, John. *Description of portes, creekes bayes and havens of the Weast India* 1578

Franc, Junctinus. *America N. & S.,* in *Comment.,* Lyons 1577

France, J. la. See **La France**

Franceschi, Domenico de. Edited *Venice,* 6 sh. 1565

Franceschi, Girolamo. Publisher in Florence. Buonsignori's *Florence* 1594

Franceschini, F. *Citta de Bologna* 1822

Francesco, Jean. See **Della Gatta**

Francesco Padovani, Patrizio. MS. *Atlas* 1540 (27 maps)

Francese, Stefano. See **Tabourot**, E.

Francheville, de. *Côte occid. d'Afrique; Baie de Santo Antonio; Cote d'Or* – all 1828

Franchus, G. or J. See **Franco**

Francia y Ponce de Leon, Benito. *Is. of Luzon* 1889

Francis, Jacopo, of Bologna. *World* n.d. attributed; used by Hajji Ahmed 1560

Francini, Alessandro (d. 1648). Florentine sculptor working in France. *Maison Royale de Fontaine Belleau* 1614, *Chasteaux Royaux de Sainct Germain en Laye* 1614

Francis, Bernardoni. *Chorog. desc. prov. Fratrum Minorum Capucinorum* 1649

Francis, J. 1873 copy of Agas' *map London*

Francis, Richard. *Plot of Killbeg (co. Wicklow)* 1656 MS.

Francis, Rev. Wm. F. Diocesan maps 1864, *Lichfield* (1867?)

Francisco [Francoso, Froncoso] , Diego. Engraver of Mexico for Palau's *California antigua e nueva* 1787

Francisco [Fernandes Pereira] , Fr. of Prazeres Maranhão. *Diccion. Geog. de Portugal* 1862

Franciscus Mona[r]chus Mechliniensis [Francois de Malines; François le Moyne, Frans Munnicks, F. de Munk] (fl. 1524, d. 1555). Flemish cartographer of Malines. *De Orbis Situ Epistola,* Antwerp 1524, 1526 (woodcut *world* map), *Globe* ca 1525

Francke, R. *Carlshafen u. Umgegend* 1896

Franco [Franchus, Frachus, Francus] , Giacomo [Jacomo] (1550?–1620). Engraver of Venice. Frontispiece to Sanuto's *Geographia* 1588, *Rome* 1589, *Hungary &c.* n.d. *Venice* n.d.

Franco [François, Frank, Francus] , Isaac (1566–1649). Architect of Tours, Royal overseer Touraine. *Touraine* 1592 used by Bouguereau, Ortelius, Blaeu &c.

Francois, Lieut. C. von. *Belgian Congo* 1885–6

Francois, Isaac. See **Franco**, I.

Francoso, D. See **Francisco**, Diego

Francq, B. le. See **Lefrancq**

Francus, Jacobus. See **Franco**, G.

Francus, Ysaacus. See **Franco**, I.

Frank, I. See **Franco**, I.

Franke, A. *Geogn. Karte Grafschaft Schaumburg* 1867

Franke, Alcuin Rud. *Environs Leipzig* 1874, *Sächsisches Vogtland* (1891?)

Franke, C. *Schul-Atlas* (1866?)

Franke, Julius. *Königl. Preussichen Prov. Sachsen* 1858, *Planiglob in zwei Wandkarten* 1861

Frankendaal, Nicolaas van (fl. 1765). Engraver of Amsterdam. *Naumburg* 1749

Frankland, George (1800–1838). Asst. surveyor V.D.L. 1826, Surveyor General & Commissioner of Crown Lands, Tasmania. Explorations along Derwent, Gordon, Huon & Nive Rivers 1828–35. *Military operations against Aboriginals of V.D.L.* 1831, *V.D.L.* 1837 & 1839

Franklin, Capt. *Plan Brit. Settlement Singapore* 1828

Franklin, Dr. Benjamin (1706–1790). Printer, author, scientist, diplomat, statesman, Dept. Postmaster Gen. for N. Amer. Colonies. Map in *Articles of Agreement* (between Penn. & Baltimore), Philadelphia 1733, authorized publication of the first map of the *Gulf Stream,* Mount & Page 1770, note on Pownall's *Atlantic Ocean* 1787

Franklin, Jane, Lady (1792–1875). Wife of Sir John, below. Travelled with husband in V.D.L., Australia & N.Z. Received Founders' Medal of Geog. Soc. 1860

Franklin, John. Estate surveyor. *Melksham, Wilts.* 1724, *Hawns, Beds.* 1767 MS.

Franklin, Lieut. (later Rear-Admiral Sir) John (1786–1847). English explorer and hydrographer, b. Spilsbury, Lincs., assisted Flinders' explorations S. Pacific. Lieut-Gov. of Tasmania 1837–43. Expeditions to N. coast America, 1818–22, 1825–27, died on third expedition 1845–7. *Journey Polar Sea* 1819–22, pub. 1823

Franklin, J. J., R.N. *Chart Palks Strait* 1838–45, *Tuticorin* 1842, *Tinnevelly* 1846

Franklin, Robert. Land surveyor, Thaxted. *Langley* ca. 1851 MS.

Franklin, Wm. Buel (1823–1903). Corps Topo. Engs. *Rocky Mts.* 1845

Franks, J. H. Engraver, Commutation Row, Liverpool, *Canada* 1820, *Plan Liverpool* 1821, for Sheriff's *Liverpool* 1823

Franks, Theodore. *Public Land States & Terr.* 1865, *U.S.* 1866

Franks & Johnson. Engravers. Swire's *Plan Manchester* 1824

Franquelin, Jean Baptiste Louis (1653– ca. 1725). French cartographer, "hydrographe du roi". *Nouvelle France* 1681, *Quebec* 1683, *Louisiane* 1684, *N. America* 1688, *Environs Boston* 1693, *Mississippi* 1697, *E. Coast* 1699–all MSS.

Franseckij, Ernest von. *Umgegend v. Düsseldorf* 1842

Frantzius, A. von. *Costa Rica*, pub. in Gotha, by Perthes

Frantzl, Augusto. *Plan Trieste* 1843, 1844

Franz. Engraver. *Prussia* 1802–9, for Schmidt 1820

Franz, H. *Schlesien* (1850?)

Franz, J. *Railway maps Europe* 1867–91

Franz, Johann Georg. Globemaker, antiquary and printseller of Nuremberg. Constructed *celestial and terrestrial globes* 1790–1810

Franz, J. Heinrich. Succeeded brother J. M. as Director of Homann Heirs 1761

Franz, Johann Michael (1700–1761). Publisher Nuremberg. Founded Homann Heirs 1730 (with Ebersperger)

Franz, Leopold. *Dist. Kennedy, Queensland* 1860

Franza, Peter (1767–1830). Mapseller and publisher of Prague. Succeeded by son of same name (1797–1834)

Franzini, Major Marino Miguel, Corps of Engineers. Charts *Portugal* 1811–16

Frary, E. *Atlas Herkimer County, N.Y.* 1868

Fraser, A. Military plans 1785

Fraser, Col. A. *Central Ceylon* 1845

Fraser, Capt. A., Bengal Engineers.

Isthmus Kraw 1861, *Route Bengal–Siam* 1862 MSS.

Fraser, Lieut. Alexander, S. Carolina Regt. *Battle Savannah* 1778 MS.

Fraser, F. A. Mackenzie. *Mauritius* 1835

Fraser, G. A. See **Frazer**

Fraser, Lieut. G. J. *Dist. Mozufurnugur* 1827–32, pub. 1858; *Dist. Bareilly* 1833–7, pub. 1858; *Dist. Budaon Agra* 1857

Fraser, J. *Plymouth Sound* 1788

Fraser, J. Engraver. *Behring Straits* 1790, *Palestine* (1790)

Fraser, J. *Fife & Kinross*, 4 sh. 1841

Fraser, James (d. 1841). Publisher, 215 Regent St., London. *Panoramic Plan London* 1831, ca. 1835

Fraser, James. *Guide through Ireland*, Dublin 1838, editions to 1854, *Trav. maps Ireland* 1852, 1884, Dublin 1860

Fraser, Major-Gen. John. *Ceylon* 1862

Fraser, Malcolm. *Route W. Australian Expl. Exped.* 1875

Fraser, Robert. *Soil of Devonshire* 1794

Fraser, R. H. Lithographer. *Mersey R.* 1832

Fraser, Simon. *Plan of attack on Fort Mifflin* 1777 MS.

Fraser, Capt. Wm. *Lands around Dampiers Strait* 1782

Fraser, Wm. *Proposed Tay & Loch Earn Canal* 1807

Fraslin, Pedro. *Chart Philippines to Acapulco* ca. 1765 MS.

Frattino, Giulio Carlo. *Stato di Milano* 1703

Frauenberger, George. *Oil Territories, Penn.* (1860?)

Frauendorff, Carl von, of the Acad., St. Petersburg. *Theatrum Belli Crimea* 1737– 8 pub. (1740?), *Crimeae Conspectus* (1740?)

Frazar, A. *Firth of Forth* 1785 MS.

Frazer, Capt. *Chart Holy I.* (1800)

Frazer, Lieut. Dan, Staff Corps. Military plans 1800

Frazer [Fraser] , Lieut. (later Capt.) George Alex., R.N. Admiralty surveyor. *Carlingford Lough* 1831 MS., *surveys Ireland & St. Georges Channel 1837–52, Scotland* 1841–4.

Frazer, Robert. Map for Lewis & Clark Expeditions 1807 MS.

Frazer, Wm., Shadwell Water Works, London. *Plan fire Ratcliffe* 1794

Frear, Thomas. *City Philadelphia* 1869

Frederic, J. A. *Insul Anticosti* 1758–(60)

Frederici, M. Dutch surveyor. Collaborated with Schotanus à Sterringa in *Atlas of Frisia* 1718

Frederick, J. L. *State Capitol Harrisburg, Penn.* 1820

Frederick, W. Publisher and mapseller. *5 miles round Bath,* Bath 1773, Hibbart's *Plan Bath* (1780)

Fredonyer. *N.E. Calif. & N.W. Nevada* 1865 MS.

Fredricci, J. C. Lieut. in Artillery. *C. of Good Hope* 1789–90

Freducci [Eufredutius] . Italian family of portolan makers in Ancona

Freducci, Angelo. *Atlas* 1555, *Atlas of Asia, Eur. & part America* 1556 MS.

Freducci, Ugo Conte Hoctomanno [Conte di Ottomano] (fl. 1539). 14 charts & atlases known. *Portolan chart Europe* 1497, *Atlas* 1537, *Chart Medit.* 1538

Freebairn. Engraver. Siborne's *War of 1815*

Freed. *Blair County, Penn.* 1859, *Michigan* 1860

Freeling. *Railway Companion London & Southampton* 1839

Freeling. Surveyor Gen. Australia. *S. Australia* 1857

Freeling, Sir Francis (1764–1836). Secretary to the Post Office. Revised Paterson's *Roads* 1811

Freeman. Engraver, for F. Lucas 1823

Freeman, Edw. Aug. (1823–1892). English historian & geographer. *Hist. Geog. of Europe* 1881

Freeman, G. L. B. *Hist. map Anglo Saxon & Roman Britain* 1838

Freeman, John. Estate surveyor. *Gt. Dunmow* 1768, *Aveley* 1782 (copied) – both in Essex

Freeman, W. *King County, N.S.W.* 1868, *Georgiana, N.S.W.* 1877

Freeman, W. J. Steam-Lithographer, 2 Old Swan Lane, Upper Thames St., E.C. Dix's *Warwick* (1878)

Freher, Marquand. *Arch. Karte des Nördlichen Oberrheingebietes* 1618

Freire [Freiral] , João. Portuguese cartographer. *Atlas* 1546 MS., *Chart W. Africa,* n.d.

Freisauff, F. von. *Extypographischer Schul-Atlas* (1848?)

Freitag [Freytag] , Gerard. *W. Frisia,* in Winsemius' *Chronicle of Frisia* 1622

Frellon, Johann. *E. Medit.,* Lyons 1568

Freloff. Engraver for Piadischeff 1823–6

Frémin, A. R. Geographer and publisher, Rue des Fossés St. Jacques No. 34, Paris; pupil of Poirson, attached to the Dépôt Gén. de la Guerre. *Etats Unis* 1820. *Atlas de la France* 1844 (with Donnet), corrected Hérisson's *Océanie* 1854, *Carte physique et routière de la France 1859,* editions to 1878, *Mappemonde* 1868

Fremont. Cartographer of Dieppe. *Diocèse de Rouen* 6 sh.

Frémont, Lieut. (later Gen.) John Charles (1813–1890), Corps of Topo. Engineers, American explorer of French origin, b. Savannah, d. New York. *Missouri to Rockies* 1842, 1843, *Report explor. expd. Rocky Mts.* 1845, *Missouri to Oregon* (1846), *Oregon & Upper Calif.* 1848, *Gold Region W. Kansas* (1860)

Frémont d'Abla[i]ncourt, [d'Ablancourt] Nicolas (1625–1693). Nephew Perrot d'Ablancourt, went to Spain for Turenne, acquired maps which were later published in *Suite du Neptune François,* Mortier 1700

French, E. R. Printer of Brattleboro. Vermont. Greenleaf's *Atlas* 1840, *Richland, Oswego County, N.Y.* 1860, *Sandy Creek* 1861

French, Frank. *Town Milo, N.Y.* 1856, *Albion, Orleans County* 1857, *Ovid, N.Y.* 1858

French, Frederick. American surveyor. *Plan town Dracut* 1791

French, Capt. George. *Coast of Sumatra* 1784–5, pub. 1786

French, John H. Gillette's *Wayne County, N. Y.* 1858

French, J. O. *Prov. of La Rioja* 1839

French & Bryant. *Atlas Brookline, Mass.* 1897

Frend, A. B. *Approaches proposed Ranelagh suspension bridge* 1843

Frentzel, George Friedrich Jonas (1754–1799). Engraver of Leipzig, maps and town plans of Saxony. *Saxony & Bohemia,* 20 sh. (1780)

Frenzel, Johann (1788–1858). Engraver of Dresden. *Dresden* 1808

Frere, Israel. *Liberties Oswestrie* 1602 (with J. Norden)

Frese, D. See **Friese**

Frese, G. W. de. *Carlstads Stift* 1871

Fresnoy, N. L. du. See **Lenglet Du Fresnoy**

Frestoy, Sr. du D. de T. See **Templeux**

Freud, Alexander. *Railway & Post map Austria-Hungary* 1899

Freudenfeldt, H. *Preussische Staat* (1867?) *Preussens u. Deutschlands* 1872

Freudenham[m]er, Georgius [Jerzy]. *Posnan* (1645), used by Blaeu & Jansson

Freusberg, Marquard Rudolph von.

Cartographer of Wurttemburg, early 18th century

Frey & Nell, L. Publishers, 79 Nassau St., New York. *Railroad map California* 1868

Freycinet, Henri-Louis (1777–1840). French navigator, sailed with his brother L.C.D. de Freycinet.

Freycinet, Louis Claude Desaules de (1779–1842). French navigator and hydrographer, b. Montélimar, d. Loriel Sailed with Baudin to Australia 1800–04; expedition round world 1817–19. *Baie des Chiens Marins* 1803, *Détroit de Bas* (1804?), *Voyage découv. Terres Aust.* 1807–17, *Voyage autour du Monde* 1824–44

Freycinet, Louis de. *Plan du Havre de Balade* 1854–(6)

Freyer, H. *Herzogthums Krain,* Vienna 1844–6

Freyhold, Alex. von (1813–1871). Revised maps of Berlin & Stettin. *Method Netz-Atlas* 1846, *Vollständiger Atlas* 1850

Freyhold, Edward. Drew maps for G.K. Warren 1855. *Ft. Lorraine to Great Salt Lake* 1858, *Milit. map U.S.* 1869, *Western U.S.A.* 1879

Freytag, Adolf. *Mitteleuropa* (1889?)

Freytag, Gerard. See **Freitag**

Freytag, Gustav (1852–1938). Cartographer of Vienna, founded Freytag & Berndt 1879. *Eisenbahnen Russlands* 1884, *Afghanistan* [1885]

Frézier, Amédée François (1682–1773). French navigator, "ingénieur Ordinaire du Roy". *Voyage au Mer du Sud* 1712–14, pub. 1716, *St. Domingue* 1722, *Carte du Pérou* 1739

Fricx [Friex], Eugène Henri (d. 1733). Publisher, bookseller, "imprimeur du Roi". Rue de la Madeleine, Brussels. *Théâtre Guerre Pays Bas* 1703, *Pays Bas* 1712, *Partie du Ganges* 1726, De l'Isle's *Amérique* 1730 (copy), *Atlas Militäire* 1746

Frid, O. See **Friedrichs**

Friderich, Johannes. *Ausburg (Germany)* (1624) MS.

Friderici, J. Ch. Surveyor. *S. coast Africa, Cape Aquilles to Swart Kop's River Bay* 1789–90 MS.

Fridrich. Engraver. Pontoppidan's *N. Norway* 1795

Fridrich, I. G. Engraver of Copenhagen. Erichsen & Schönning's *Iceland* 1780, Godich's *Sleswig* 1781

Fried, Franx. Austrian geographer. *Moldau* 1811, *Australien,* 1839, *Amérique* 1841, *Austria* 1857, *Deutschland* 1868

Friedberg, Emanuel von. *Gen. Karte Serbien* 1853

Friedemann, Hugo. *Schul-Wandkarte Sachsen* (1877?)

Friedenreich, P. C. *Danmark . . . Slesvig . . . Faeröerne* 1861

Friederichs, J. Publisher, 5 Nassau St., Soho Sq., London. *Distance map London* (1847?), 1850

Friederichsen, Ludwig F. [& Co.] (1841–1915). Cartographer and publisher of Hamburg, founder Carto. Inst. Hamburg. Worked for Stieler 1863; *Costa Rica* 1876, *Concessionsgebietes S. W. Africa Co.* 1897

Friederichsen, Peter (fl. 1830–1865). Cartographer for Perthes

Friedlein, D. E. Bookseller of Cracow. *Plan Krakow* (1830)

Friedrich, Dr. Ernst. *Kleinasien,* Halle 1898

Friedrich, L. *Post maps Middle Europe* 1859–66

Friedrichs [Frid], Otto[ne]. Engraver for *Ubbo Emmius' Frisia* 1595

Friend, John. Estate surveyor. *Laindon, Essex* 1705 MS.

Friend, John. Chartmaker, "East Lain, Redderif", Rotherhithe. *Chart Whitby-Flamborough Head* 1707, *Chart New-foundland* (1713), *Land van Eendracht* 1739

Friend, John. *Gulf Persia* 1787 MS.

Friend, N. Lithographer of Philadelphia. *Plan Wilmington* 1850, *Jackson Co., Michigan* 1874

Friend, N. M. *Sheboygan, Wisconsin* 1830

Friend, Robert. Chartmaker "to be heard of at the Jerusalem Coffee House in Change Alley." *E. Indies* 1719 MS., *Chart part Malaya, Tanasam &c.* 1738 MS.

Friend, Wm. *Straits of Jubal* 1802, *Tor Harbour* 1804

Friend & Aub. Lithographers and engravers of Philadelphia. McIntyre's *Boston* 1852, Bailey's *Barbados* 1855

Fries [Friess, Frisius, Phrisius, Phryes, Phrijsen], Lorenz [Laurentius] (ca. 1490–ca. 1532). Physician, astrologer and geographer in Metz and Strassburg. *World* in Apian's *Cosmography* 1520, edited & drew *World* for 1522 edn. *Ptolemy,* revised Waldseemüller's maps: *World* 12 sh., 1522, *Yslegung de Mer Carthen,* Strassburg 1525, 1527, 1530; *Hydrographia* 1530

Friesack, Ch. (1821–1891). Austrian geographer and astonomer

Friese [Frese], Daniel (1540–1611). Artist and cartographer, b. Dittmarch, d. Lüneburg; worked Hamburg and Lüneburg, fl. 1568. *Plans Bardowieck* 1588, *Heide* 1596 & *Meldorf* 1596, all used by Braun & Hogenberg; *Schauenburg* 1602

Frieseman, H. Publisher of Amsterdam. *Crimée* 1787, *Mer de Marmora,* 4 sh. 1791

Friesen, Karl Friedrich (1785–1814). Mathematician, architect and cartographer of Magdeburg

Friess, L. See **Fries**

Friex, E. H. See **Fricx**

Friis, Prof. Jens Andreas (1821–1896). *Finmark,* 6 sh. 1861, *Russian Lapland* 1870, *Tromsø Christiana* 1890

Frijlink [Frulink] , Hendrik (1800–1886). Publisher of Amsterdam. *Kleine School-atlas* 1843, 1870, *Australie* 1858, *Nieuwe Hand Atlas* 1854

Frijman, Jonas Andersson (d. 1704+), Swedish surveyor of Gotland

Frintzel. Engraver for Funk 1781

Friquegnon, Capt. Nicolas. *Annam,* 6 sh. 1890, *Indo Chine* 1893, *Chine Mérid.* 1899

Frisak, Capt. H. von. *Skagestrands Bugt i Iisland* 1808–10, pub. 1818, *S. Iisland* 1823

Frisch, F. (fl. 1800). Worked in Augsburg. Town plans after Bacler d'Albe

Frisius, G. See **Gemma Frisius**

Frisius, L. See **Fries**

Fritsch, Andräa Erik. Cartographer of Pressburg, mid 18th century

Fritsch, Carl von (b. 1838). German geologist and explorer, b. Weimar. Visited Canaries and Morocco. *Kanar Inseln* 1867, *Teneriffe* 1867, *Sanct Gotthard* 1873

Fritsch, G. Th. (b. 1838). German naturalist and explorer. *Drei Jahre S. Africa* 1868, *Eingeborenen S. Afrikas* 1872

Fritsch, Dr. I. H. *Charte vom Harz. Magdeburg* 1833

Fritsch. Johann Theobald. *Zweybrücken* 1774, 1794

Fritsch, K. U. *Tenerife,* Winterthur 1867

Fritsche, Herm. P. H. (b. 1839). Russian explorer of German origin. Geog. studies Siberia

Fritsche, W. H. (1859–1894). German cartographer; one of founders Carto. Inst., Rome

Fritschi, J. N. *Umgebungen v. Baden* (1855?), *Umg. v. Heidelberg* (1877?)

Fritsen, P. E. *Noord Braband* 1841

Fritz, A. *Plan v. Carlsruhe* (1872?)

Fritz, H. *Nordlichtes (Arctic),* Gotha 1874

Fritz, Father Samuel (1656–1725). German Jesuit. *Rio Maranon o Amazonas,* 2 sh. 1691, *Amazon* 1695–7, pub. in Quito 1707, English edition by Moll 1717

Fritzsche, G. E. *Sudan* 1885, *Nuovo Atlante Geografico,* Rome 1887, *Karawanstrasse,* Gotha 1890, *Regno d'Italia,* 20 sh. 1893

Frizell, Richard. Surveyor. *Estate Stoke Hammond* 1774 MS.

Frizon, G. See **Gemma Frisius**

Fröbel, J. See **Froebel**

Froben, Hieronymus (1501–1563). Printer and publisher of Basle, son of J. Froben. edition of *Ptolemy* 1533

Froben, Johannes (1460–1527). Publisher and printer of Basle

Frobisher, Sir Martin (ca. 1535–1594). English navigator and hydrographer, b. Doncaster, d. Portsmouth. First English attempt to find N.W. Passage, 1576, 1577 & third voyage to Greenland 1578; Vice-Admiral on Drake's expedition to W. Indies 1586. *World & N. Atlantic* in Beste's *True Discourse* 1578, *Plot of Croyzon* 1594 MS.

Froebel [Fröbel] , J. (1805–1893). German mineralogist. *Aus Amerika* 1857–8

Froger, François (1676–1715+). French engineer and traveller. *Relation d'un voyage . . . Afrique &c.* 1698, *Cayenne* 1698

Froggatt, Walter Wilson (1858–1937), b. Blackwood, Victoria. *Maps of Australian coast & New Guinea* 1885–92

Froggett, J. Engraver for Wilkinson 1809–21

Froggett, John Walter. Engraver and publisher, No. 3 West Sq., London. *Rape of Arundel* 1819, engraved *World* 1825, *30 miles round London* 1831, *15 miles round London* ca. 1842

Frogley, Arthur. Estate surveyor. *Ramsey, Essex* ca. 1750 MS.

Froiseth, B. A. M. *Utah* 1870, 1871, 1875, *Little Cottonwoods, Utah* 1873

Fromann, Gustav. *Special-Karte des Odenwaldes* 1867

Frome, Capt. Edward Charles (1802–1890), Royal Engineers; b. Gibraltar, Surveyor Gen. of S. Australia 1839–49. *S. Australia*, Arrowsmith 1843, *Country E. of Flinders' Range* 1843, Arrowsmith 1844

Fromann, Maximilien. Worked for Glaser (1840), *Hessen* 1867

Fromman, A. B. *Coburg* 1784

Frommel, Gustaf F. Cartographer of Baden, late 18th century

Frommel, Karl Ludwig (1789–1863). Engraver and artist in Karlsruhe, b. Birkenfeld, d. Ispringen. In 1824 founded an engraving workshop in England with H. Winkles

Froncoso, D. See **Francisco**

Frontpertius, Ad. Front de (b. 1825). French publisher and geographer, b. Rennes. *Canada* 1867, *Etats Unis* 1873, *Etats Latins de l'Amér.* 1883

Frosch[auer], Christoffel (ca. 1490–1564). Printer and publisher of Zürich Pub. Vadianus' *Epitome* 1534, Stumpff's *Schwyzer Chronik* 1548

Frost, H. Engraver. Greenwood's *Sussex* 1829, & *Dorset* 1829

Frugoni, Juan. *Estancia de San Jorge Uruguay*, London 1880

Frulink, H. See **Frijlink**

Fry & Son. Surveyors of Grays Inn. *Estate plan Calverley* 1852 MSS.

Fry, Col. Joshua (ca. 1700–1754). American surveyor and mathematician, Commander of the Virginian Regt. Surveyed boundary between Virginia and N. Carolina 1749–51 (with P. Jefferson). *Inhabited Parts of Virginia* 1751 (also with Jefferson), French edition Le Rouge 1777

Fry, Ludwig (fl. 1570–95). Engraver of Zürich. *Plan Zürich*, used by Münster.

Fry, W. Ellerton. *Victoria Falls* 1892 MS.

Frye, Alexis Everett (1859–1936). *Home & School Atlas* 1895, 1896, *Complete geog.* 1895

Fryer, C. E. *Fishery Districts England & Wales* 1884–5

Fryer, George. *Plan Borough Kingston upon Hull* 1885, *Plan City Hull* 1898

Fryer, John. Surveyor. *R. Tyne* 1773, *Castle Garth, Newcastle* 1777, *Alston Moor* 1797, *Hexamshire* 1800, *Lanercost* 1804, & 1806; all MSS.

Fryer, John & Sons. *Northumberland* 1820 (Index 1822)

Fryer, J. H. Surveyor. *Park Ennerdale* 1805, *Ulverston* 1805

Fuca [Valerianos], Juan de la (d. 1602). Greek navigator, in the service of Spain, b. Cephalonia. Originated the strait bearing his name and the misconception of California as an island 1592

Fuchs, C. Friedrich (1803–1874). Engraver in Hamburg. *Schmalkalden* 1848 (with Danz), *Map Heligoland*

Fuchs, C. W. C. *Carta geolog. Ischia* (1872)

Fuchs, Johann Conrad. *Rhine,* Augsburg 1707

Fuchs, Ph. J. Edmond (1837–1889). French geologist and explorer

Fuchs, Dr. W. *Die Venetianer Alpen,* Vienna 1844

Fuehrer [Führer], Lieut. Carl. *Plan attack Fort Washington* 1776

Fuentes, Bartolomeo (17th century). Discoveries of Admiral De Fonte 1752

Fuerst [Fürst], Paul (ca. 1605–1666). Publisher and printseller of Nuremberg. *Globe* for Otterschaden ca. 1606, *Koppenhagen* [1640?], *Nürnberg* 1664, *Regensburg* 1666, *Cölln* 1667, *Hungary* n.d.

Fuerstaller [Fürstaller], Josef (1730–1775). Globemaker and cartographer in Salzburg and Bamberg

Fuerstenhoff [Fürstenhoff], Johann Georg Maximilian von (1686–1753). Cartographer of Dresden

Fuessle & Co. Publishers of Zürich. Brué's *N. Amér.* 1826

Führer, C. See **Fuehrer**

Fuhrmann, Matthias (ca. 1690–1773). Engraver of Vienna

Fuko, Mineta. Japanese author and cartographer. *Map of China* 1849

Fulger, Antony. Engraver, 18th century

Fulgosio, Fernando. *Islas Filipinas* 1871

Fullarton, Archibald & Co. Publishers and engravers of Glasgow (1834–70); London, Edinburgh & Glasgow (1840–3); London, Edin. & Dublin (1845). *New & Comprehensive* [later *Parliamentary*] *Gazetteer of Eng. & Wales* 1834, editions to 1849, Wilson's *Imp. Gaz. Scotland* 1854–7, *Gaz. of World* 1856, *Royal Illus. Atlas* (1864) (pub. in 27 parts 1854–62), *Hand Atlas of World* 1870–2

Fuller, C. A. Engineer, U.S. Coprs. *Buffalo* (1850), *Island of Montreal* (1850), *Memphis* (1850), *Sketch of Red River* 1858

Fuller, Edward Bostock, *Survey Manor Cholsey, Berks.* 1695–6

Fuller, Edward. *Exmoor Forrest* 1817, *Delamere Forrest* 1817

Fuller, Francis. Surveyor. *Plan Muswell Hill Park* 1860

Fuller, George. *Lewes* 1877, *Eastbourne* (1879)

Fuller, John F. *Whitman Co., Wash.* 1895 (with W. J. Roberts)

Fuller, S. P. *Boston* 1838

Fuller, Dr. Thomas (1608–1661). *Bird's-eye view Cambridge* 1634, map to *History of Holy War* 1639, *Pisgah Sight of Palestine* 1650

Fuller, Capt. Wm. *Insets* 1769 on Jefferys' *Plan Amelia I. (E. Florida)* 1770

Fullerton, Lieut. J. D. *Plan City of Harar,* 4 sh. 1885, *Somaliland* 1885

Fulljames [Fulliames], Thomas. Surveyor of Gloucester. *Stead Quarter Farm* 1792 MS, *Desborough Hundred* 1796, *Dean Forest* 1804

Fulmer, F. S. *Rutland Co., Vt.* 1869 (with F. W. Beers)

Fulneck Academy. *Moravian Atlas* 1853

Fulton, David. *Arkansas* 1839

Fulton, H. Surveyor. *Plan Stamford Junction Navig.* 1810

Fulton, Hamilton H. *Kings Lynn* 1846

Fulton, Henry. *Farm line map City Brooklyn* 1874

Fulton, J. A. Surveyor. *Boundary line Ohio* 1818

Funck, Carl Oscar. *Karta öfver Stockholm* 1846

Funck[e] [Funk, Funke], David. (fl. 1680–1705). Publisher of Nuremberg, commissioned maps from Homann. *Saxonia Inf.* (1690), *Cracau* 1695, *Crete* (1700), *Circ. Suevicus* (1725)

Fundanus, Nicander Philippinus. *Totius Ungariae descript.* 1595

Funes de Pavia, Juan Batista. *Derrotero General, Guaiaquil* 1700

Funk, Christlieb. Benedict (1734–1814). Prof. of Nat. Hist. Leipzig. *Southern Hemisphere* 1781 (based on Cook)

Funk, D. *Umg. v. Ingoldstadt* 1865

Funk, David. See **Funck**

Funk, Capt. James de. *Chart Malabar coast,* Dalrymple 1755

Funk & Wagnalls Co. *Standard Atlas* (1896)

Funke, Carl Philipp (1752–1807). *Atlas der alten Welt,* Weimar 1819 (with Vieth)

Funke, D. See **Funcke**

Funnell, Wm. *Plan Bays of Le Grand, Brazil* 1703, *Voyage round the World* (Dampier's) 1707

Funter, Capt. Robert. *Raft Cove,* Dalrymple 1789, charts and *Port Cox* in Meares' *Voyage* 1790

Furck, Sebastian. *S. Salvador* (1650)

Furlanetto, Lodovico. *Laguna Veneta* (1780?) *Dalmazia* 1787, *Terr. di Friul* 1793

Furlani, P. de. See **Forlani**

Furlong, Lawrence. *Amer. Coast Pilot,* Newburyport 1809

Furnas, Boyd Edwin (1848–1897). *Atlas-directory Miami Co., Ohio* 1883

Furne & Cie. Publishers and Booksellers, Rue St. André des Arts No. 55, Paris. Segur's *Atlas pour l'Hist. Univ.* 1842

Furneaux, Capt. Tobias (1735–1781). Circumnaviagtor, sailed with Wallis 1766–8, with Cook 1772–5. *V.D.L.* 1773–(7), first chart of Tasmania

Furnival, E. Engraver. Baugh's *Shropshire* 1811

Fürst, P. See **Fuerst**

Fürstaller, J. See **Fuerstaller**

Fürstenhoff, J. G. M. von. See **Fuerstenhoff**

Furtado, L. C. C. P. See **Pinheiro Furtado**

Fuser [Fusier], Lewis V., Lieut. Royal Amer. Regt. (d. 1780). Maps *St. Lawrence* area 1761–3 MSS.

Fustinoni, A. *Prov. di Como* 1884

Fyers, Lieut. Wm. (fl. 1773–1812) Engineer *Plan of Post at Portsmouth (U.S.A.)* 1781 MS.

Fÿnje [Fijnje], H. F. Dutch surveyor, 19th century. Ordnance maps Dutch provinces

G

G., C. F. *Plan Milford Haven* 1785–(90)

G., D. E. *Swabia* (1680?)

G., E. *Europa für Schulen* (1877?)

G., F. Signs dedication of *World* map in Hakluyt's edition of P. Martyr's *Decades,* Paris 1587. (Filips Galle?)

G., F. *Schleswig* 1863

G., F. L. See **Guessefeld**

G., G. B. V. *Mare Adriatico* 1815

G., G. W. Drew Higgie's *I. of Bute* (1886)

G., H. M. C. D. See **Mallet**, H.

G., J. F. *Guipuzcoa* 1875

G., L. Engraver *Map Sachetts Harbour* 1815

Gabato. See **Cabot**

Gabauer, Johann Jacob. *Naturgraenzenkarte v. Europe, Asien u. Afrika,* Halle 1787

Gabb, William More (1839–1878). Paleontologist, b. Philadelphia. *Californische Halbinsel,* Gotha 1868, geol. map *St. Domingo* 1872, *Costa Rica* 1877

Gäbler, F. E. See **Gaebler**

Gaboriaud. *Sahara Algérien* 1845, *Partie sept. de l'Afrique* 1845

Gabote, Gaboto. See **Cabot**

Gadea, Joseph. *Sclavonia,* Vienna (1718)

Gädertz, A. See **Gaedertz**

Gadner[us] [Gadmer], Georg (1522–1605) Swabian surveyor. *Würtenberg* used by

Ortelius 1579, completed version, 20 sh. 1596; *Stuttgarter Amt* 1587

Gadolin, Jacob (1719–1802). Swedish Bishop, mathematician, astronomer and surveyor

Gaebler [Gäbler], Friedrich Eduard (1842–1911). Cartographer and publisher in Leipzig. *Taschen Atlas* 1886, *Umg. v. Chemnitz* (1898)

Gaedertz [Gädertz], A. *Schan-Tung* 1898

Gaffarel, Paul J. L. (1843–1920) French historian and geographer b. Moulins. *Hist. colon. Français, Hist. de la Floride Françoise* 1875

Gage, Michael Alex. (fl. 1852). *Plan Liverpool* 1836

Gage, Thomas (ca. 1600–1656). English traveller and Dominican, lived in Central America, returned to Europe 1637, died Jamaica. *Survey West-Indias* 1655 (2nd edition with 4 maps)

Gage, Wm. Leonard (1832–1889). *Modern hist. atlas* 1869

Gagen, G. R. Surveyor, 14 Stacey St., Stepney. *Ground plan Hampton Court* 1835 MS.

Gagenhart, Peter (fl. 1490–2). Globe-maker of Nuremberg

Gagnières des Granges, Abbé Claude François (1722–1792). French globeseller

Gail, Jean Baptiste (1755–1829). French hellenist. *Atlas à la géog. d'Hérodote* 1823

Gaillard, Louis, S. J. *Plan de Nanking* 1898

Gaillard, Tacitus. *S. Carolina* 1770 MS (with J. Cook)

Gaillardot, C. (1814–1883). French explorer and geographer, b. Luneville, d. Beyrouth

Gailloüe, Pierre. Mapseller, "dans la Cour du Palais, Rouen". Le Vasseur de Beauplan's *Normandie* 1667

Gaio, Matteo. *World* 1516 MS.

Gaishi [or Gensui] Ebi (17th century). Japanese surveyor. *Coastal navigation chart* pub. 1680

Gaitte. Engraver, pupil of Gardette. For Le Gentil de la Galaisière 1781

Gajetius, L. See **Guyet**

Galabert, Louis. *Carte minéralogique des Pyrénées* 1831

Galaisière, G. J. H. J. B. Le G. de la. See **Le Gentil de la Galaisière**

Galateus Leccensis. See **Ferrari**

Galaup, J. F. de. See **La Pérouse**

Galauti, Abbé Luigi (1765–1830). Italian geographer, b. Naples. *Istit. di geogr. fisica e politica* 1807

Gale, A. C. Land agent, Winchester. *Glenbervie Alice Holt Forest* (1850)

Gale, J. Engraver. *Hull* 1791

Gale, Richard C. Land surveyor, Winchester. *Compton* 1833, *Plan Winchester* 1836, *Survey Durrington* 1843 MS.

Gale, Samuel. *Part of Prov. of Lower Canada* 1795 MS. (with Duberger)

Gale, Thomas (1635/6–1702). English Hellenist, Dean of York, b. Scruton, Yorks. *Britannia Romana*, Weigel (1720)

Gale & Butler. Engravers, Crooked Lane, London. *Plan part of London* 1800, Janvrin's *Guernsey* (1810)

Galeatius, Carolus. *Stato di Milano* 1777

Galego, João Goncalves. Portuguese chartmaker, latter 16th century

Galiano, Lieut. (later Brigadier de Marina)

Dionisio Alcala. *Straits Magellan* 1785–6 MSS (5 charts), *Carta esferica N. O. America* 1795, *Arch. de Graecia* 1806, *Mer de Marmora* 1812

Galignani, A. *Plan Paris* (1850)

Galignani, A. & W. *Plan Paris &c.* (1867?)

Galignani Fratelli. Imprint used: (1) G. Battista & Giorgio. Publishers of Venice. *Ptolemy* 1598 edition. (2) Paolo & Francesco. Publishers of Padua. Porcacchi's *Isole* 1620, *Ptolemy* 1621 edition

Galignani de Karera, Simon (fl. 1576). Publisher in Venice. Porcacchi's *Isole* 1572 (with Porro). His Heirs published the 1590 edition of the above and the 1596 edition of *Ptolemy*

Galindo, Col. D. Juan. *Usumasinta R.* 1833, *Costa Rica* 1836

Galinier, Joseph Germain. *Aoir* 1840, *Abyssinie* 1843, *Voyage en Abyssinie* 1847–8 (with Ferret), *Carte géolog. de Tigre* 1848

Galissoniere, Marquis de la. *Nouvelles découvertes dans l'ouest de Canada* (1750)

Gall, Carl. Publisher of Berlin. *Plan Gegend Danzig* 1813

Gall, Rev. James, Junior. *People's Atlas of the Stars,* Gall & Inglis 1862

Gall & Inglis. Publishers, 38 North Bridge, Edinburgh, later 25 Paternoster Sq., London and 6 George St., Edinburgh [later Bernard Terrace]. Took over Cary's plates from Cruchley 1876, still using them in 1910. *Edinburgh Imperial Atlas* 1850. *Danubian Provs.* 1854, *One Shilling Atlas* (1871), *School Atlas* (1871), *Sixpenny Atlas* (1872), *Imperial Globe Atlas* (1887?), *5 inch map London* (1898)

Gallaeus. See **Galle**

Gallaham, Wilhelm von (1751–1788) Austrian military cartographer

Gallaher. See **White, Gallaher & White**

Gallardo, Ygnacio P. *Plano de Mexico* 1867

Gallatin, Hon. [Abraham Alphonse] Albert. (1761–1849). Secretary of Treasury, diplomat. *Indian tribes of N. America* 1836

Galle, F. See **Galle**, P.

Galle, Jean [Giovanni]. French cartographer. *Picardie* 1540

Galle [Gallaeus], Philippe [Filips] (1537–1612). Editor, engraver and printseller, b. Haarlem, active in Antwerp. *Spieghel der Werelt* 1577 (miniature atlas after Ortelius; he published numerous editions up to 1595), engraved *Portrait of Ortelius* in *Theatrum Orbis Terr.* 1579, *World* in P. Martyr's *Decades,* Paris 1587 (signed F. G. – Filips Galle?)

Galle, Theodor (1571–1633). Engraver and printseller of Antwerp, son of Philippe. *Artois* (1600), *Palatinatus Rheni* (1620?)

Gallego, Hernán (b. 1508/17). Pilot to Mendaña. *Chart Solomon Is.* 1568 (lost)

Gallegos, F. J. E. y. See **Estorgo y Gallegos**

Galleri, Hieronymi. Printer of Oppenheim, for De Bry 1619

Gallet. Engraver. *Plan de Paris* 1824

Gallet, George. Publisher of Amsterdam. *Suite du Neptune François* 1700, Jaillot's *Intro. à la Géographie* 1695

Galletti, Johann Georg August (1750–1828). Cartographer and historian of Gotha, b. Altenburg. *Allgemeine Weltkunde,* Leipzig 1807–10

Galliard. *Plan Geneva* 1760

Gallois. *Plan v. Hamburg u. Altona* 1868

Galluci, Giovanni Paolo. (fl. 1569–97). Italian astronomer, b. Salo. *Theatrum Mundi* 1586 (map *America),* De Fabrica *et usu Hemisphaerii* 1596

Gallus Pugnans. See **Vadianus**

Gall von Gallenstein, Josef Freiherr (19th century). Cartographer of Graz

Gallwey, Capt. H. L. *Sketch map Benin River* 1892, *Gwato Creek* 1893 MSS.

Galton, Francis. *Damar Land,* London 1854, *Isochronic Passage Chart for Travellers* 1881

Galvão, Antonio. Portuguese captain and geographer born in India ca. 1500 died Lisbon 1557. Governor of Moluccas. *Tract. dos descobrièmentos antiguos e modernos* 1731

Galves, Don Manuel. *Lampon Bay* 1754, *Capa Luan (Luzon)* 1774, *Part Salomagues* 1781 – all pub. by Dalrymple

Galvez, J. *Prov. de Santa Fe.* Rosario 1888

Galvez, Dr. Mariano. *Atlas Guatemalteco* 1832

Gama, Luiz Philippe de Saldanha da. *Plano da Guerra do Paraguay* 1869

Gambarini, Bern. *Tiber* 1744 (with Chiesa)

Gambdene, G. See **Camden**, W.

Gambillo, Enrico. *Strade Ferrate Italiane,* 4 sh. Bologna 1886; 2 sh. 1888

Gambino, D. *Pianta Topo. città di Palermo* 1862

Gambino, Prof. Giuseppe (1841–1913). *Carta Murale della Sicilia,* 6 sh., Palermo 1886, *Atlante scolastico muto* 1888, *La Sicilia Itin.* 1888

Gamble, Col. D.Q.M.G. *Part North Is. of New Zealand* 1863–4, pub. 1867

Gamble, Wm. H. Draughtsman and engraver of Philadelphia. Worked for Mitchell 1861–7, for White's *Atlas W. Virginia* 1873, for Bradley 1884

Gamidge, S. Mapseller and publisher, Prior's Head, Worcester. *Plan Worcester* (1780?)

Gamond, A. Thomé de. *Carte d'étude pour servir à l'avant project du Canal Interocéanique de Nicaragua* 1858, *France (bassins hydrog.)* (1873?)

Ganahl, Johann Ritter von (1817–1879). Austrian military cartographer

Gandini, G. Engraver and mapmaker. *Atlas of the Wolga* (8 maps) 1767

Ganeau, Veuve. Bookseller, Rue S. Jacques, près la rue du Plâtre, Paris. Charlevoix's *Nouvelle France* 1744

Ganeparo, Prof. Salvino. *Città di Torino* 1870

Gangoiti, Pedro Manuel (1779–1830). Spanish artist, b. Bilbao, d. Madrid. *Charts Chile & Peru* 1798–9, worked for Antillon 1801–4

Ganière, Pierre (1663–1721). Script engraver and geographer of Paris. Jesuit map *Mojos,* Peru (1713)

Gannett, Henry (1846–1914). Geographer, Chief Geog. of U.S. Geological Survey 1882. *Montana & Wyoming Territ.* (1870), *Hypsometric map U.S.,* Washington 1877, *Scribner's statistical atlas U.S.* (1883), *U.S.* 1891, 1899

Gannon, P. Assistant surveyor. *Allotments Ballarat East* 1859

Ganocchi, Gio. *Canal du lac de Trasimene* 2 sh. 1788

Gantrel[l], Estienne (1646–1706) "Graveur ordinaire du Roy" and globe-maker, b. Metz.

Ganxales, T. Engraver. *Ocean Atlantico* 1813

Gapp, John. Surveyor. *Crown lands Walsoken* 1835

Garandolet, Giovanni. Globe maker at Palermo 1527. Also pub. *De Orbis witu ac descriptione*

Garay, José de. *Boca del Rio Coatzacoalcos* 1843

Garbs, F. A. *Palaestina* (1862?)

Garci-Aguirro, Pedro. *Baia del Salvador* 1798

García, A. C. y. See **Centeno y Garcia**

García, B. E. y. See **Espina[u]lt y Garcia**

García, Nuno. See **Toreno**

García, R. E. *Mapa de Bolivia* 1897

García, Conde, D. & P. See **Conde**

García de Cespedes. See **Cespedes**

García de Leon y Pizarro, Ramon. *Plan Oran, Argentine* MS. 1794

García, de Toreno, N. See **Toreno**

García Martinez, José. Capitán é Ingeniero. *Isla de Iviza* 1765–(78)

García y Cubas, Antonio (1832–1912). *Atlas de la rep. Mexicano* 1856–8, *Arzobispado de Mexico* 1872, *Estados Unidos Mexicanos* 1885

García y García, Capt. Aurelio. *Port Mollendo* 1871

García y González, Lieut. Emilio. *Port of Valencia* 1867–(9)

Garcie [Gracie], Pierre [sometimes called Ferrande] (b. 1430, d. early 1500s). Considered the first French Hydrographer. *Routier de la Mer,* Rouen 1502, *Grant Routtier* 1520 (editions to 1630s)

Garcin, Étienne. *Celto Lygie ou la Provence* 1847

Garde, J. de la. See **De la Garde**

Garden, Major. *Country round Kandahar* 1878, *Country round Cabul,* Calcutta 1879

Garden, Francis. Engraver, b. London, worked in Boston. Maps in Salmon's *Universal Traveller* 1752–3

Garden, William. Surveyor. *Kincardine-shire,* 2 sh. 1776

Gardener, T. & J. Engravers. *Plan Westminster* 1847

Gardette. Engraver for Le Gentil de la Galaisière 1779

Gardette, Veuve de la [widow of], Print-seller in Paris. *Chesapeak Baie* 1781

Gardiner, Capt. *Country of Natal* 1835

Gardiner, Alfonso. Arnold's *Pupil Teacher's Year Book of Memory Maps* 1889

Gardiner, C. K. Surveyor General. *Oregon Territory* 1855

Gardiner [Gardner], James Terry (1842–1912). Surveyor and engineer. *Sierra Nevada* 1863–7, *Yosemite Valley* 1865,

Geolog. map Washoe Mining District (1870–80), *Central Colorado* 1873, *State New York,* 2 sh. 1879, *Colorado* 1881

Gard[i]ner, Ralph. *Chart Tyne* 1650

Gardiner, Samuel Rawson (1829–1902). *School atlas English hist.* 1891 &c.

Gardiner, Wm. Land surveyor. *Newport & Widdington, Essex* 1727 MS.

Gardner. *Atlas Amériquain Sept.* 1778

Gardner. Engraver. *Huntingdonshire* (1822)

Gardner, A. W. & Co. *Plan City Kingston,* London 1889

Gardner, B. H. *Plan Parish St. Sepulchre* 1823

Gardner, C. J. *Maps Chinese Empire* 1880 MSS

Gardner, H. *Plan Nunnery, Bishopgate St.* 1817, *Christ Church, Surrey* 1821

Gardner, H. *Colchester-Harwich Railway* 1836 MS.

Gardner, James, Senior. Publisher and engraver, 163 Regent St. (later No. 129), London. Worked for Smith 1822, *Plans London* 1827, 1845, Bradshaw's *Canals* 1830, for Dawson 1832, engraved 6 sh. *Ireland* 1837, Walsh's *Barbados* 1847; also agent for the sale of Ordnance maps 1828–ca. 1840

Gardner, James, Junior. Mapseller, 33 Brewer St., Golden Sq., London. Agent for the sale of Ordnance maps from ca. 1840. *Geol. map Eng. & Wales* 1846

Gardner, J. T. See **Gardiner**

Gardner, Ralph. See **Gardiner**

Gardner, Thomas (1690–1769). Engraver *Pocket guide to the English Traveller* 1719, copy of Agas' *Dunwich* 1754

Gardner, Wm. Surveyor, worked with Yeakell. *Plan Chichester* 1769, *Sussex* 1778–83, *Guernsey* 1787, *Jersey* 1795 (with Cubitt), drew Ordnance Survey *Kent* 1801

Gardner, W. R. (fl. 1816). Engraver and geographer, 17 Oxford St., London. Smith's

12 miles round London 1822, *Brighton,* 4 sh. 1826, Smith's *Pocket Companion* 1827, *London & Westminster* 1828, *Scotland* (1830?), C. Smith's *New Holland* 1837

Gardner, Wm. Wells. *Sixpenny Elementary Atlas* (1872)

Garella, N. *Central America* 1850, *Road Chagres to Panama* 1851

Garfield, Aquila. Estate surveyor. *Roxwell; W. Ham* (both Essex) 1682 MSS.

Garfurth, W. *Plan fortifications Carlisle* (1580?)

Gariboldi, Hauptmann. *Umg. v. Cilli* (1873)

Garipuy, François (1711–1782). Cartographer of Languedoc. *Canal Royal de Languedoc* 3 sh. 1771, 15 sh. 1774

Garlato, G. B. Engraver, for Miniscalchi 1855

Garner, Georg. *Würtemberg* in de Jode's *Speculum* 1578

Garnett, John. *Travelling maps Lake District* (1860?)

Garnier, l'Abbé. *Recueil de cartes* Paris, Nyon 1787

Garnier, Frères. Publishers, 6 Rue des Saints-Pères, Paris. *Liais' Hydrog. du Haut San Francisco* (Brazil) 1865

Garnier, Adolphe. *Bassins des Vosges* 1872, *Carte routière des Vosges* 1874

Garnier, B. L. Bookseller, Rua do Ouvidor 69, Rio de Janeiro. *Liais' Hydrog. du Haut San Francisco* (Brazil) 1865

Garnier, F. A. (1803–1863). Geographer *Australie* 1860, *Atlas Sphéroïdal et Universel* 1860

Garnier, Francis Marie Joseph (1839–1873). *Atlas d'Explorations en Indo-Chine* 1873

Garnier, Hippolyte. *Baie de Diego-Saurez* 1833–(7), *Port de Santa Barbara* 1847– (56)

Garnot, Eugène Germain. *Atlas expédition Française de Formose* 1884–5, pub. 1894

Garny. *Plan Paris* 1818

Garofalo. Engraver. Martinon's *Sicilia* 1812

Garran, Andrew (1825—1901) Journalist and politician, b. London. Edited *Picturesque atlas Australasia,* Sydney & Melbourne 1886, *Australasia illustrated,* London 1892

Garrard, H. M. *Geelong* 1848—(60?)

Garrard, J. Jervis. *Zululand* 1895

Garretson, Cox & Co. *Columbian Atlas* 1891 &c.

Garrett, G. H. *Sierra Leone,* London 1892

Garrett, Henry A. *Bournemouth* 1883, (1889)

Garrett, John (fl. 1697). Bookseller and publisher "near the stairs of the Royal Exchange in Cornhill"; brother-in-law of J. Overton, successor to T. Jenner after 1672. *New Booke of Maps* 1676, *Book of the Names of all Parishes [Direction for the Eng. Traveller]* 1677 *Quartermaster's map* (1688), *New and exact map America* Hollar, pub. by Garnett 1676

Garrett, Wm. Publisher. Bought Speed plates from W. Humble 1658—9, sold them soon after to Roger Rea

Garrigou, T. E. A. (1802—1893) French Geographer b. Tarascon. *Géog. Aquitaine Sous Caesar* 1863, *Iberes* 1884

Garstin, Symon. *Castletowne, Co. Lowth* 1655 MS.

Gartman, Ewen H. *Atlas der Oost Indien* (1832)

Gartman, V. A. *Telegraph map Russian Empire,* 4 sh. St. Petersburg 1870

Gartner, Amandus. *Wolffenbüttel* 1627

Gartze, Johannes (1530—1574). German astronomer

Garvie, A. *Prov. Otago, New Zealand* (1860?)

Gascoign[e], Joel. See **Gascoyne**

Gascoigne, John. *Draught of Tower Liberties* 1597 MS, pub. 1742

Gascoigne, [Gascoyne], Capt. John. *Plans Port Royal & D'Awfoskee Sound* 1729,

used by De Brahm 1757, *S. Carolina & Part of Georgia,* Jefferys 1768. Charts in *N. American Pilot Part II*

Gascoyne [Gascoin, Gascoigne] , Joel. Surveyor, map & chart-maker and publisher "at ye Signe of the Platt at Wapping Old Stayres, 3 doares belowe ye Chappell". Charts in Sellers' *English Pilot* 1677, *America* 1678 MS (for Capt. J. Smith), *Carolina* (1682), *estates Deptford* 1692, *Cornwall,* 9 sh. 1700 & 14 sh. ca. 1730, *Falmouth* (1700), *Enfield Chase* 1701 MS, *Channel* 1702; *Bethnal Green, Hamlet Limehouse, Mile End, St. Dunstan Stepney,* all 1703; engraved *Straits of Gibraltar* in *English Pilot* (1715), *Greenstead* ca. 1725 MS.

Gascoyne, John. See **Gascoigne**

Gaskell. *Family Atlas* 1887

Gasp, Raphaelo Fabretto. Engraver of Urbino, for Aa 1729

Gaspari, Adam Christian (1752—1830). Geographer of Königsberg. *Schul-atlas* 1793, *Allgemeiner Hand-atlas* 1804, 1821

Gaspary, Nicolas. *Plan v. Metz* 1872

Gasser [Gassarus] , Dr. Achilles Pirum (1505—1577). Physician and topographer, b. Linden, d. Augsburg; active in Feldkirch & Augsburg. *Allgäu, Bodensee &c.* 1534 MS.

Gasson, John. *Hampshire & Berks.* (1646?)

Gast, A. *Plastischer Schul-Atlas* (1876?)

Gast. L. & Bro. Lithographers of St. Louis. *Gold region W. Kansas* 1859

Gastaldi, [Gastaldo, Castaldi, Castallo] , Giacomo [Jacopo] (ca. 1500—ca. 1565). Venetian cartographer, Cosmographer to Rep. of Venice, b. Villa Franca, 109 of his maps traced. *Spain* 1544, *Sicily* 1545, *Europe* 1546, *World* 1546, 1561 (lost), 1564, new edition *Ptolemy* 1548, *Germany* 1552, *Piedmont* 1556, *Asia,* 8 sh. 1559—61, *Europe* 1559, *Italy* 1561, *Poland* 1562, *Africa* 8 sh. 1564, *Madagascar* 1567, *Padua* 1568, *Lombardy* 1570, *America,* 12 sh., n.d., *Europe,* 12 sh., n.d.

Gastambide, Pedro. *Town & fort Pachira (Dumaran),* Dalrymple 1762

Gaston, Samuel N. [Gaston & Johnson]. Publisher of New York. *U.S.A.* 1854, Colby's *Diamond Atlas* 1857, *Campaign atlas for 1861.* See also **Morse & Gaston**

Gastrell, Capt. (later Col.) J. E. In charge of Indian cadastral Surveys, *Jessore* 1854–6, pub. 1870, *Nagpoor* 1865–(70)

Gates, B. C. *Sullivan Co., N.Y.* 1856, *Otsego Co., N.Y.* 1856 (with C. Gates)

Gates, C. *Otsego Co., N.Y.* 1856 (with B. C. Gates)

Gates, G. H. Drew Butler's *Dramatic Almanac map for 1853*

Gatliff, Charles. *Plan London* 1875

Gatonbe, John. *Iceland & Greenland* 1612

Gatta, G. F. della. See **Della Gatta**

Gatti, Aletinus. Publisher of Rome, 16th century. *Geneva,* n.d.

Gaubil, Père Antoine, S. J. (1689–1759). Mathematician, astronomer and missionary, b. Gaillac, d. Peking: went to China 1722. *Lieou-Kieou* 1752, pub. Buache 1754

Gaudi, J. See **Gaudy**

Gaudry, Albert. *Agric. map Cyprus* 1855, *geol. map Cyprus* 1860

Gaudy [Gaudi], John. Marine surveyor. *Barcelona* 1705, *Coast of Newfoundland* 1715, *Levant* [1740?]

Gauld [Gould], George (1732–1782). "Surveyor of the Coast" (1767) *Coast W. Florida & Louisiana* 1764–71. *Plan Kingston* 1772 MS, *W. Florida* 1773 MS., *Plan Manchac* 1774 MS, *Cuba & Colorados* 1773–(90), *Tortugas & Florida Keys* 1773–5 pub. 1790, *Pensacola,* used by Des Barres 1780

Gaule, de. *Embouchure de la Seine* 1788

Gaulle, F. See **Galle,** P.

Gaultherot, V. Printer of Paris. *Angliae descript.* 1545. Apian's *Cosmography* 1551

Gaultier, Abbé Aloisius Édouard Camille (ca. 1745–1818). Cartographer, b. Piedmont, d. Paris; in Eng. during French Revolution. *Complete course of geog.* 1792 &c. *Atlas de géog.* [1810]

Gaultier, J *Palestine* [1875?], *Cordillère des Andes* 1880

Gauss, C. F. (1777–1855). German astronomer, born Brunswick

Gaussin Pièrre Louis Jean Baptiste. *Îles Marquises* 1844–(9)

Gauthey. *Montagnes de France* 1782

Gauthey, Charles. Ordnance Surveyor, ca. 1780

Gauthiot, C. French geographer, born Dijon 1832

Gautier, Hubert (1660–1737). French engineer, "Ingénieur du Roi dans la Marine" b. Nimes, d. Paris. *Nismes,* Nolin 1698, *Uzes* (1698?), *Montagnes des Sevennes* 1703

Gautier de Châtillon. TO map. 13th century

Gautier de Metz. French poet, first half 13th century *Image du Monde Bruges*— several MSS, earliest 1250 in Paris

Gauttier, Pierre. Enseigne de Vaisseau. *I. des Saintes* 1803–(13), *Mer Noire* 1820, *Archipel* 1854–64

Gauttier d'Arc, L. Ed. (1799–1843) French historian and geographer

Gauvreau, N. B. *Surveys in Northern British Columbia* 1891

Gavard. Engraver. *Dufour's plan London* (1838?)

Gavarrete, Juan. *America-Central* (1880?)

Gavin, H. Engraver. *Virginia & Md.* (1750?) *Scotland* 1767, for Armstong's *Berwick* 1772, *Sketch water of Druie* 1776

Gavin & Son. Engravers for Brown's *Atlas of Scotland* 1800, *& General Atlas* 1801

Gavit, John (18th century) Engraver and printer of Albany

Gawler, Col. George (1795–1869). Second Governor of S. Australia 1838–41. *S. Australia* 1840 MS.

Gawthorpe, J. W. Master H.M.S. *Lion.* Charts *S. Africa* 1832–60

Gayangos, A. M. de. See **Martel de Gayangos**

Gazü, Archimandrite. *World map* 4 sh. Vienna

Gazulić [Ghazulus Ragusinus] , Ivan (1430– 1476). Mathematician, astronomer and globemaker of Dubrovnik

Gebert, Martin. *Schwarzbildes* 1783

Gebhard, Franz Xaver (b. 1775). Painter and engraver of Munich

Gebhard, H. *Plan v. Nürnberg* (1868?)

Gedda, Johann Persson (d. 1680). Surveyor of Uppland

Gedda, Petter (1661–1697). Swedish pilot and surveyor in government service, brother of J.P. Gedda. *Chart Kattegat* 1695, *Atlas of Baltic* (1694)

Gedde, Christian. *Plans Kiöbenhafn* 1757

Geddes, George. *Salt Wells, Syracuse, N. Y.* 1868

Geddes, James. *Map and profile Champlain Canal* 1825, *Harbor of Oswego* 1825

Gedney, Joseph F. Lithographer, engraver and plate printer of Washington (fl. 1869). *Ft. Abercrombie to Ft. Benton* 1863, worked for Keeler 1867

Gedney, Thomas R. *Baie de New York* 1880

Gee, Richard. Surveyor. *Rev. Hogg's Estate, Emberton, Bucks.* 1799 MS.

Gee, Thomas. Surveyor. *Crown Allot. Featherstone* 1811

Geelkerken [Geelkerck, Geerkerken, Geylekerck, Geylkercke, Geilenkerlken, Geilkeckiol] Arnold van (d. 1619). Dutch surveyor, d. Italy; brother of Nicolaas. *Palestine* 1619 (engraved by Nicolaas)

Geelkerken, Isaak van (d. 1672). Traveller and surveyor of Arnheim; son of Nicolaas

Geelkerken, Ivan. *Plan Nijmegen,* Wit (1680)

Geelkerken, Jacob van (fl. 1625–50). Dutch surveyor.

Geelkerken, Nicolaas van (d. 1657?). Publisher, engraver and surveyor of Amsterdam (1615), Leiden (1624) & Arnhem. *World* 1610 &c., maps in Emmius' *Frisia* 1616, for Cluver's *Germania Antiqua* 1616, *America* 1617, *O. & W. Indische Spiegel,* Leiden 1619, engraved for Lubinus' *Pomerania* 1628, *Belegeringe Bruijnswijck* 1635

Geelmuyden, B. *Lomme-Atlas over Norge* 1892

Geer. *City Hartford, Conn.* 1864

Geerkerken. See **Geelkerken**

Geerling, W. J. Dutch cartographer. *Nieuwe Atlas* 1859, *Nederlanden* 1886

Geernaert, P. T. *Atlas ecclés. de la Belgique* (1850?)

Geerz, Franz Heinrich Julius (1817–1888). Military cartographer of Berlin. *Holstein u. Lauenburg* 1838–45, *Schleswig, Holstein &c.* 1859, *Nordfriesische Inseln* 1888

Geest, E. de. *Nederlandsch Oost-Indië* 1871, *Nederlanden* 1888–97, *Sumatra,* 12 sh. 1892

Geevards, M. See **Gerards**

Gefferys, T. See **Jefferys**

Geiger [Geyger, Gÿger, Beiger] , Hans [Johann] Conrad (1599–1674). Cartographer of Zürich; many maps of Cantons, 56 of Zürich. *Meÿenfeldt* (1623?), *Switzerland* 1634, (1635), 1637, *Zürich* 1667, (1685)

Geiger, Hans [Johann] Georg. *Tiguiria* 1685 (with H.C. Geiger)

Geiger, J. G. See **Geiger**, H. G.

Geiger, Ph. *Plan Kaiserslautern* (1883)

Geiger, W. *Wien* 1848

Geikie, Sir Archibald (b. 1835), b. Edinburgh. Director Geological survey Scotland 1868–72. *Geolog. map Scotland* 1861, *Edinburghshire* 1861–4, *Ayrshire* 1868–72, *Linlithgow* 1878, *geolog. map Eng. & Wales,* 15 sh. 1896

Geikie, Prof. James (b. 1839). Geologist, b. Edinburgh, worked with brother A. Geikie. *Ayrshire* 1868–72, edited R. S. G. S. *Atlas Scotland* 1895

Geil, John F. *Lorain C., Ohio* 1857, *Laporte Co., Ind.* 1862

Geil, Samuel. Publisher, 602 Chestnut St., Philadelphia. *County maps, Ind., Mich., N. Y., Ohio, Pa.* 1852–64, *geolog. maps* 1865

Geilenkerlken, Geilkeckio. See **Geelkerken**

Geinitz, F. E. *Geol. map Mecklenburg* 1883

Geinitz, H. B. (1814–1899), b. Altenburg. *Atlas zur Geologie der Steinkohlen Deutschlands* 1865

Geisendörfer, J. *Tonquin* 1883

Geisler, A. D. Publisher of Bremen. For Uhlenhuth 1849

Geispitz von Geispitzheim, Baron Carl Heinrich (d. 1787). Austrian military cartographer

Geissel, Wilhelm. *Regierungs-Bezirk Wiesbaden* (1872?)

Geissendörfer, L. Lithographer and printer. *Plan v. Mannheim* (1866?)

Geistbeck, Dr. A. *Deutschen Alpen*, Leipzig 1885

Gelais, Io. Vincentius. Engraver. *Maps of Padua* 1699

Gelb[c]ke, C. H. von (ca. 1783–1840). Military cartographer and historian of Weimar. *Herrschaft Schmalkalden* 1807, *Königreiche Württemberg*, 4 sh. 1813, *Rothenburg* 1816

Gelbrecht, H. *Plan v. Bremerhaven* 1878

Gelder, Jacob de (d. 1848). Mathematician and cartographer of Rotterdam. *Netherlands* 1809, *Holland* 1811

Gelder, W. van. *Schoolatlas Neder. Oost Indië* 1893

Gelio, Capt. d'Etat Major. *Nivellement de Jérusalem*, Paris 1863

Gelis, Capt. *Dead Sea*, Jerusalem 1863

Gell, Sir, Wm. (1777–1836) *Topo. of Troy* 1804, *Geog. and antiquities of Ithaca* 1807, *Roma* 1827, *Peloponnesus* 1830

Gellatly, J. Engraver, draughtsman and publisher of Edinburgh. *Johnson's 12 miles round Edinburgh* 1840, pub. *Lothian's Netherlands* (1844), for George Philip & Son 1851–2

Gelle, Johannes. Engraver (fl. 1625) at Cologne. *Voorhout of the Hague* n.d. *Middleborough* 1622

Gellibrand, Henry (1597–1636). Mathematician, Gresham professor of astronomy, b. London. Works on nautical geography 1633–5

Gemeinhardt, Ch. Geographer of Speyer mid 18th century

Gemellis, John Francis. *Hydrographical draught of Mexico* 1732

Gemini [Geminus, Geminy, Lambrit, Lambert] , Thomas (1500–1570). Publisher, engraver, surgeon, instrument maker; born near Lille, active in London 1524–63. *Astrolabe* 1552, *Spain* 1555, *Gt. Britain* 1555

Gemma Frisius [Frizon, Edelgestein] , Regnier [Reinerus, Reinserszoon] (1508–1555). Astronomer, Prof. of Maths. Louvain, Mathematician to Charles V, b. Dockum, d. Louvain. Founder of Belgian school of geography. *World* 1529 (in Apian's *Cosmographus Liber,* 4th edition), corrected editions of Apian's works from 1529, *globes* 1530, 1537 (with Mercator), *De Principiis Astronomiae et Cosmog.* Antwerp 1530, *World* 1540, 1544

Gemmellaro, Giuseppe (1787–1866). Italian geologist, b. Catane, founder Academia Gionia 1824, *Map eruptions of Etna*, Wyld 1828

Gemperlin, Publisher of Freyburg. *Cysat's Japan* 1586

Gendebien, Albert. *Coupe Géolog. du Bassin du Centre* 1876

Gendre. See **Le Gendre-Decluy**

Gendron, Pedro. Cartographer. *Atlas*, Madrid 1756–8

Genest, P. M. A. *Nouvelle France* 1875

Genist, F. X. *Dominion of Canada* 1883

Genoese World Map 1457. Attributed to Toscanelli, q.v.

Genoi, D. See **Zenoi**

Genovese, Battista. *Portolan* 1514

Genshu Nagakubo. See **Sekisui**

Gensoul, Adrien. Publisher "Pacific Map Depot", 511 Montgomery St., San Francisco. Fine's *Humboldt Mining Region* 1864, for Gird 1865, for Owens 1866

Gent, Thos. (1693–1778). Printer and topographer, settled York 1724. *Hist. York* 1730, *Ripon* 1734, *Hull* 1735. *Plan city New York* (1771)

Gent, Wm. Surveyor. *Manor of Belsize & St. John's Wood* 1679

Gentet, Jaspar. *Golfe de Siam* 1739

Genthe, Herman. *Etruskische Tauschhandel* (1873?)

Gentil, L. Le. See **Le Gentil**

Gentil de la. Galaisière, G. J. H. J. B. Le. See **Le Gentil de la Galaisière**

Gentleman's Magazine. Periodical containing numerous maps 1731–1858

Genton, Jean Louis Ambroise de, Chevalier de Villefranche. *Fortifications River Delaware* 1779

Gentot. Mapseller, Rue Mercière, Lyon. Delafosse's *Amérique* 1771

Gentot. *Rade de Cherbourg* (1787?)

Genus. See **Zeno**, N.

Geny-Gros. Printer, R. de la Montagne, Ste. Geneviève 34, Paris. For Dufour (1870)

Geoffroy, August. *Navigation de Grand Cercle* 1867

Geoffroy, Lislet. See **Lislet-Geoffroy**

Geographische Anstalt (Munich), *Georgien u. Hochland Armenien* 1829

Geographisches Inst., Weimar. See **Weimar**

George of Cyprus. Byzantine geographer, 7th century

George & Co. Lithographers, 54 Huston Gdn., London. *Columbia* 1849

George, C. *Changes Kennedy Channel (Arctic)* 1858 MS.

George, F. E. Engraver, for Levasseur 1849

George, S. A. Electrotype, 607 Sansome St., Philadelphia. For Rulison 1860

George, Thomas. Surveyor. *Part Iver, Bucks.* 1856

Georget, Adj. Drew maps for Suchet 1828

Georgio, Giovanni. See **Georjio**

Georgio, Johannes. See **Giorgi**, G.

Georgia, Ludovicus. See **Barbuda**

Georgius [Schinbain, Tibianus], Johannes [Johann Georg] (1541–1611). Latinist and mapmaker. *Schwartzwald,* Constance 1603 (woodcut, first map of Black Forest)

Georgius, Ludovicus. See **Barbuda**

Georjio [Georgio], Giovanni [Georgius Johannis]. Portolan maker of Venice, *Chart Black Sea, Med. &c.* 1494 MS.

Gephart, Christian. *Plan Oran,* Homann Heirs 1732

Geppert, Georg von (1774–1835). Austrian general and cartographer

Geppert, Ludwig von (1777–1836). Austrian military cartographer

Geraghty, J. P. Surveyor of Ross. *Iron Mines, Dean Forest* 1849 MS

Geraghty, T. R. *Etheridge Goldfield, Queensland* 1898

Gerald, John. *Draught Bellese (Central Amer.)* ca. 1750 MS.

Gerald, M. F. *Costes de Terre Ferme* 1737

Gerald de Barri. See **Cambrensis**, G.

Gérard, Capt. Alexander. *Hohen Himalaja* 1832, *Koonawur* 1841

Gerard, H. *Plan Battle Waterloo,* Brussels 1840

Gerard, H. See **Gerritsz**

Gerard, Col. M. J. *Sketch Gandamak to Tezin Valley,* 3 sh. 1880, *Mesop. & Persia* 1882–6

Gérard, Paul (1796–1866). Geographer in Geneva & Brussels, worked for Vandermaelen. *Hohen Himalaja* (with A. Gérard) 1832, *Belgique* 1847–59

Gerard, Thomas, of Trent. *Survey of Dorset* 1633 MS.

Gerardo, Paulo. *Portulano del Mare,* Venice 1612

Gerards [Gerardus, Geevards], Marcus (1530–1590). Artist and engraver, b. Bruges, d. England. *Plan Brugge* 1562

Gerardus, H. See **Gerritsz**

Gerardus, M. See **Gerards**

Gerasimov, Dmitrij [Demetrius] (b. 1465). Muscovite Ambassador to Rome 1525; his information on Muscovy used by Agnese & Gastaldi

Geray, Richard. *Manor Desford* 1644

Gerbaud, Georg. *Atlas der Völkerkunde* 1886–92

Gerbel, Nicolaus (ca. 1485–1550) Historical geographer

Gerber, Henrique. *Minas Geraes (Brazil)* 1862

Gerber, Johann Gustav (ca. 1690–1734). From Brandenburg; in Russian service from 1710. *E. Caucasus, St. Petersburg* 1735

Gerber, W. Russian colonel. *Chart of the Caspian West coast* 1735

Gerbert, Martin. *Schwarzwald* 1783

Gerbier, Sir Balthasar (1591?–1667). Painter and architect, b. Middelburg, came to England 1616. *Works on cosmography and geography* 1649

Gerbig, C. *Kreis Höxter* (1892)

Gerbillon, Père Jean François (1634–1707), b. Verdun, d. Peking. *Voyages en Tartarie* 1688–98

Gergens, A. *Plan Stadt Mainz* 1795

Gerhard, Carl August (d. 1817). Cartographer of Baden

Gericke, F. E. Engraver. Euler's *Africa* 1753

Gérin. Engraver, for Andriveau-Goujon (1841–73) Vuillemin 1857

Gerini, Giovanni. *Carta corog. Littorale Austro-Illirico,* 6 sh. 1847

Geringius, Erik (1707–1747). Engraver of Stockholm. *Södermanland* (1745?)

Gerlach, C. W. *Plan Leipzig* 1814 and 1845

Gerlach, H. American lithographer. *Lake Superior* 1858

Gerlach, H. D. *Plan Siége de Cassel* 1762–(3)

Gerlach, J. W. R. *Chemins de Fer de l'Europe Centrale* 1873

Gerlach [Gurlach], P. Depty. Q.M.G. *Plan action at Huberton (Vermont)* 1777, 1780, *Prince Ann, Norfolk, Nansemond Counties, Va.* 1781 MS

Gerland, D. G. *Ethno. Weltkarte,* Gotha 1871, *Polynesien* 1872

Germain, Adrien (1837–1895). French hydrographer. *Traité d'hydrog.* 1882

Germain [Germaini], Jean. Geographer, mid 15th century

Germaine. *Mascate et Khulbos* 1862–(3). *Chart Madagascar* 1863–(4)

Germain, Louis (1733–1770). French engraver, for Chappé d'Auteroche 1768

Germaini, J. See **Germain**

German & Br. *Navarro Co., Texas* (1868)

Germanus, H. M. See **Metellus,** J.

Germanus, N. See **Nicolaus Germanus**

Germontius. See **Gourmont,** J.

Gerner, *Deutscher Zollverein* 1869

Gerold, Carl. *Situations plan der Weltausstellung* 1873

Gerolt, Federico de. *Distritos minerales de Mexico* 1827, *Central Mexico,* used by Kiepert 1858

Gerrich, J. C. Engraver, for Weiland 1848

Gerri[j]tsz, Adriaen (ca. 1525–1579). Seaman of Haarlem. *De Zeevaert* 1588, *Generale Pascaerte* 1591

Gerritsz, Cornelis of Zuidland. *Sea Journal into Java,* Wolfe 1598 *(map of Bali)*

Gerritsz, Dirck. *Charts Ireland,* pub. H. Gerritsz. 1612

Gerritsz. [Gerard, Gerardus, Gherritszoon van Assum] , Hessel (1581–1632). Engraver, cartographer, publisher and bookseller, b. Assum, apprenticed as engraver to Blaeu, Cartographer to Dutch E. India Co. 1617, fl. 1607; from 1612 using sign "in de Paskaert" or "sub signo Tabulae Nauticae". Addresses: (1) opt Water bij die oude Brug [1609] , (2) by die Lienbaens Brugh [1616] , (3) Nieuwe Zijds Voorburgwal [1624] , (4) Doelestraat [1627–32] . Engraved 17 *Netherlands* (1608?, lost) *Gulick Cleve* 1610, *Spain* 1612–(15), *Besch. van de Samoyeden Landt* 1612, *Besch. van de Zeecusten van Ierlandt* 1612, engd. Blaeu's *Lithuania* 1613, *Russia* 1613, *Italy* 1617, MS charts *E. Indies, Pacific* &c. 1617–22, maps in Laet's *Novus Orbis* 1625, *Eendrachts' Land (W. Australia)* 1627, *Rotario W. Indies & S. America* 1628–31 MSS

Gerritsz, Martin. Sailed with Vries. *Japan* (1643) MS.

Gerritsz. Pieter. Dutch geographer, early 16th century

Gersdorff, A. *Grundriss . . . Danzig* 1822

Gerster, B. *Canal de Corinthe* 1882

Gerster, Prof. J. S. *Kanton Schaffhausen* (1870), *Suisse Atlas,* Neuchatel 1871, *Vorarlberg u. Liechtenstein* (1895)

Gervais. Publisher of Paris. Delamarche's *Usages de la Sph*ère 1799

Gervais, Henry. *France physique* 1862, *Java* 1872, *France* (1874?)

Gervais[e] de Palmeus, A. F. See **Palmeus**

Gervaize. *Ile Tsis,* used by Dumont d'Urville 1842–8

Gervase of Tilbury [Belmotus] (ca. 1160– ca. 1211). Chronicler. Author of *Mappaemundi* – and possibly of the Ebstorf *map* ca. 1284, q.v.

Gessi, R. *Cours du Nil* 1876

Gessner, Abraham (1552–1613). Artist and goldsmith of Zürich. *Globes* in form of drinking vessels (1595) and 1600

Getkant, Friedrich (d. 1666). Cartographer of the Rhineland in Polish service. *Baltic* 1637, *Ukraine* 1638

Gette, O. *Stadt Bremen* 1875

Geus, G. A. de. *Haarlemmermeer* 1857, *Atlas à l'usage des voyageurs* 1858

Geus, G. K. de. *Atlas à l'usage des voyageurs* (1858)

Geusau, Lieut.-Gen Levin von (1725– 1808). Cartographer of Berlin. *Topo. Milit. Karte v. Neu Altpreussen,* Berlin 1807

Geyer, Lieut. *Regensburg* (1866?)

Geyer, F. *Zuckerfabriken Deutschlands* 1879

Geyer, G. *Hesse u. Nassau* (1868)

Geyer, V. *Schweiz* 1864, *Karolinen, Palau u. Marianen* 1899

Geyger, H. C. & H. J. See **Geiger**

Geylekerck, Geylkercke. See **Geelkerken**

Gfug, Sek. Lt. von. *Umg. v. Graudenz* 1874

Ghandia, Antonio de. Sargento Mayor. *Islas Philipinas* 1727

Ghazulus Ragusinus. See **Gazulić**

Ghebellinus, Stephanus. *Venuxini comitatus,* used by Ortelius 1584

Ghelen, Johann Peter von (1673–1754). Printer, publisher and bookseller in Vienna

Gherri[j]tszoon, H. See **Gerritsz**.

Gheyn, Jacob de [Jaque II] (1565–1629). Artist, engraver and publisher, b. Antwerp, worked in Amsterdam

Ghillany, F. W. (1807–1878). Published a copy of Schöner's *Nuremberg globe* 1853

Ghisi[us] , Giovanni Baptista. Painter, sculptor, editor and engraver. Edition of

Cellarius' *Geog. Antiqua* 1774, *Sicilia,* 4 sh., Rome 1779

Ghisolfi [Gisolfo] , Francesco. Genoese cartographer, copied Agnese's and Gastaldi's maps. *MS atlases* 1546–53 (7 recorded)

Ghisulfus, J. *L'Italie* 1757

Ghymmius, Gualterus. Life of Mercator in Mercator's *Atlas* 1595

Giambullari, Pier Francesco (ca. 1494–1564). Italian philosopher and priest, b. Florence. *Terra Incognita* in his *Descrizione,* Florence 1544

Gianelli [Janellus] , Giovanni. Glove and clock-maker in Cremona, mid 16th century. *Globe,* Milan 1549

Giarré, Gaetano. Engraver, for Le Sage 1809–13, *Contorni di Firenze* (1840?)

Gibb, A. *Plans Aberdeen* 1861–71 (some with Keith)

Gibbard, Wm. Revised *Simcoe Co.* 1853, *Plan Townley Property (Canada West)* 1857

Gibbes, Charles Drayton. *Gold Region, Calif.* 1851, *Alpine Co.* 1866 MS *Calif. & Nevada* 1873, 1879

Gibbings, Lieut. R. *Malawa & adjoining countries,* Calcutta 1845 ,

Gibbins, H. de B. *Atlas of Commercial Geog.* (1893)

Gibbon, John. *Middlesbrough* 1619 MS

Gibbon, Lieut. Lardner, U.S. Navy. *Andes* 1854, *Bolivia* 1854, *Purus* 1856 MS.

Gibbons, Capt. (later Major) A. St. H. *Kingdom Marutse (Rhodesia)* 1897, *Zambesi R.* 1898 MS.

Gibbons, Edward. Parish maps *Cambridgeshire & Hunts.* 1800–16

Gibbons, H. A. *Cape of Good Hope* 1822 for Admiralty

Gibbons, Sir John, Bart. (1717–1776). *Plan Manor Stanwell* (1760)

Gibbs, George (1815–1873). Ethnologist. *N.W. California; Wallamette Valley*

1851, *W. coast* 1855 MSS, assisted *N.W. Boundary Survey* 1857

Gibbs, Joseph. Engineer, London. *Intended Railway Paddington–Clewer* 1835

Gibbs, L. *Drainage of Fenland* 1888

Gibbs, S. *Plan City of Bath* (1870?)

Gibbs, Shallard & Co. Letterpress printers and lithographers, 108 Pitt St., Sydney. Jones' *N.S.W.* 1871

Gibson, Arthur. *Plat book Cottonwood Co., Minn.* 1896

Gibson, A. F. Estate surveyor. *Hatfield Peverel* 1786 MS

Gibson, Edmund (1669–1748). Bishop of Lincoln, later London. Revised and published Camden's *Britannia* 1695 & 1722

Gibson, Lieut. Francis (1753–1805). Seaman and writer of Whitby, Yorks. *Plans parts of Whitby* 1782–94 MSS.

Gibson, George E. *Omaha & S. Omaha* (1887) (with W. Gibson)

Gibson, John (fl. 1750–92). Geographer, engraver and draughtsman, No. 18 George's Court, Clerkenwell. Salmon's *Univ. Traveller* 1752–3, *Middle Brit. Colonies* 1756, *Gents. Mag.* 1758–63, *Atlas Minimus* 1758 (with Bowen), *Counties of Eng. & Wales* (1759), maps in *Amer. Gazetteer* 1762, *Univ. Mag.* 1763–9, engd. for Speer 1771, & for Cook (1773), *Collieries on R. Tyne & Wear* 1787–(8)

Gibson, T. *Channel I.,* Bowen (1750)

Gibson, Wm. *Omaha & S. Omaha* (1887) (with G. E. Gibson)

Gibson, Wm. T. *Seneca Co., N.Y.* 1852

Gidmore, J. *Bath* 1694

Giegler, J. P. *Carte routière d'Italie,* Milan 1818

Gier, H. *Plan Bremen* 1870, *Plan Hannover* (1895)

Giesecke, K. L. [Sir Charles Lewis] (1761–

1833). German professor geology. *Mineralogy of Faeroes & Greenland* 1805–7, *Greenland* 1832

Giesemann. See **Keller & Giesemann**

Gifford, Franklin. *Niagara Co., N.Y.* 1852, *Broome Co., N.Y.* 1855

Gifford, John. *France* 1798, *Environs Toulon* 1798

Gigas [Gigante, Gigus], Joannes Michael (1580–1650+). Doctor, mathematician and geographer. *Atlas of Archbishopric of Cologne* 1620 (maps used later by Blaeu & Hondius), *Paderborn*, J. Hondius (pre-1629), *Osnabrug, Westphalia, Munster, Hildesheim* – all n.d.

Gignilliat, T. Heyward. *Valley Orinoco River* 1896

Gigus, J. M. See **Gigas**

Gilbert, A. W. *Hamilton Co., Ohio* 1856, *Cincinnati & Vicinity (Ohio)* 1861

Gilbert, Frank Theodore (b. 1846). *Atlas Yolo Co., Calif.* 1879

Gilbert, Giles. *Barony of Clonmoghan* 1675 MS

Gilbert, George (fl. 1828). Estate surveyor of Colchester & Furnivals Inn, London. *Gt. Warley, Essex*, ca. 1843 MS.

Gilbert, H. & Co. Publishers, 16 Beekman St., New York. *Pennsylvania Railway Company Atlas* 1875

Gilbert [Gylbert], Sir Humphrey (ca. 1539–1583). Navigator, planted first colony in N. America in Newfoundland 1583. *Petition concerning North [West] Passage* 1566, *Discourse of A Discoverie*, London 1576 (*World Map* to show N.W. Passage)

Gilbert, James. Publisher, 49 Paternoster Row, London. *Modern Atlas of World* (1841), maps in Fisher's *County Atlas* 1842–5, *College Atlas* 1844, *World* 1850

Gilbert, Capt. John. *Part of Cuba* 1755

Gilbert, John. Missionary map of World (1815)

Gilbert, Joseph, Master of the *Pearl*, & Master of *Resolution* on Cook's 2nd Voyage. *Coast Labradore* 1767, *S. coast Newfoundland* 1768–(9), charts in *N. American Pilot* 1775, *Dusky Bay, N.Z.* 1775

Gilbert, J. L. *Regent Park* 1824 MS.

Gilbert, Samuel A. *Plan Charleston* 1849, (on map of *S. Carolina* 1854)

Gilbert & Tayspill. Land surveyors of Colchester. *Little Bromley, Essex* 1839 MS.

Gilchrist, General Charles A. *Atlas Hancock Co., Ill.* 1874

Gilchrist, James. *City of Wheeling &c. Ohio Co., Va.* 1871

Giles, Ernest (1835?–1897). Australian explorer, b. Bristol, d. Coolgardie. *Map of discoveries* 1872–6

Giles, Francis (1787–1847). Civil engineer and surveyor, Salisbury St., London. Surveys rivers & harbours, worked on railways. *Bishop's Stortford to Cambridge Canal* 1810 MS. (with N. Giles). *Plans Thames Haven, Portsmouth, Liverpool & Liffey* 1818–40, MSS, *Whitechapel to Shell Haven* 1839

Giles, George. Master, R.N. *Malooda Bay (Borneo)* 1845–(7)

Giles, H. *Hawaiian Islands* 1876

Giles, Joel. *Inner Harbour, Boston* (1850?), *Plan Back Bay, Boston* 1852

Giles, Netlam (d. 1817). Surveyor of London. *Town & Harbour Dover* 1805 MS, *Bishop's Stortford to Cambridge Canal* 1810 MS (with Francis Giles), *Intended Arundel–Portsmouth Canal* 1815

Gilkison, W. S. *City Fort Wayne, Ind.* 1866

Gilks, Edward. Lithographer. *New England, Natal* (1848), *Town Lots Ballarat, Melbourne* 1859, *S. Australia* 1867

Gill. *Penny Atlas* (1871)

Gill, C. B. *Roadstead of Salina Cruz* 1871

Gill, Michael. Surveyed and published *Plaquemine, La* 1854

Gill, Valentine, Surveyor. *Wexford,* 4 sh. 1811

Gill, Lieut. (later Capt.) Wm. John. R.E. (1843—1882). *N.E. Persia* 1873, *W. China & E. Tibet* 1877

Gilleland, J. C. Cartographer. *Ohio & Miss. Pilot,* Pittsburgh 1820, *Plan battle of Braddock's defeat* 1838

Gillen. *Atlas Mitchell Co., Kansas* 1884 (with Davy)

Gilleron. *Carte géolog. de la Suisse* (1867?)

Gillespie. *Forest Co., Pa.* (1868?)

Gillet, Lithographer. *Plano di Valparaiso* 1835

Gillet, George (1771—1853). *Connecticut,* Hartford 1811—(13) (with Warren), *Northern part of New England* 1821

Gillet, N. J. *Prince Albert Gold District* 1891

Gillet, T. Printer, Salisbury Square, London. Chauchard's *Germany* 1800

Gillette. *Wayne Co., N.Y.* 1858

Gilley, Lieut. F. W., R.E. *Military plan Devonport* 1865 MS.

Gillier, J. *Down,* 2 sh. 1755

Gillig, Charles Alvin. *London* (1889)

Gillingham, E. Engraver for Tanner 1833, for *Hale's Boston* 1819, *New Universal Atlas* 1844

Gillion, D. J. *Mining Locations Seine River* 1899

Gillman, Henry (1833—1915.) Assistant Geodetic Survey Great Lakes, b. Kinsale, Ireland. Topo. Engineer. *Charts harbours on L. Superior* 1855—64

Gillone, John. County Surveyor of Galloway. *Kirkcudbright* 1794, *Plan vicinity Newton Stewart* 1807, *Dee River* 1808

Gillot. Engraver. Thiers' *Atlas de l'Empire* 1875

Gilly, David (1748—1808). Architect of Berlin. *Prussia* 1789, *S. Prussia,* Berlin 1802—3

Gilman, E. Draughtsman. *U.S.A.* (1848)

Gilman, Henry. American surveyor. *Survey N. & N.W. Lakes* 1864

Gilmore, Joseph. *City of Bath* 1697

Gilmour. *National Atlas* 1852 (with Dean)

Gilpin, Col. George. *Plan town of Alexandria, Columbia* 1798

Gilpin, Robert. *Kagosima Harbour,* Admiralty 1863

Gilpin, Thomas. *Proposed Chesapeake and Delaware Canal* 1769, pub. 1821

Gilpin, Wm. (1813—1894). First territ. governor Colorado (1861). *Hydrog. maps N. America in Central Gold Region* 1860

Gilseman, Isaac. Tasman's Pilot. *Charts of Tasman's Voyage* 1642 (lost)

Gilson, Robert. Surveyor. *River Foss* 1792 MS

Gineste, Capt. de. *Insel Thera oder Santorin* 1848

Gingell, Engraver of Bath. *Country round Cheltenham* 1810. *Plan city Bath* (1830)

Ginji, K. A. *Yokohama* 1871

Ginn Bros. Publishers of Boston. *Lessons in Geog.* 1872—5

Ginn & Co. *Classical atlas* (1882), 1895

Ginver, N. *Scilly Is.* 1779

Ginville, Vincent de. Engraver for N. de Fer 1705

Gioane de Bo da Venecia. See **Bo da Venecia**

Giöding, O. I. *Kungsholmen eller Stockholms* 1754

Giolito, Gabriele (1510+/—1578). Publisher, engraver, print and bookseller of Ferrara, worked in Venice "al segno della Fenice", fl. 1536. *Germany* 1552, *Piedmont* 1556

Giordana, F. *Carta geologica del S. Gottardo* (1872?)

Giorgi [Georgio] , Giovanni [Johannes] . *Carta Nautica* 1494, *Milan* used by Ortelius *1570*

Giorgi, de. *Nieuwe Atlas van den Nederlande* 1845

Giorgius, L. See **Barbuda**

Giovanni Battista da Cassini (17th century). Cartographer and globemaker

Giovanni da Carignano. See **Carignano**

Giovio [Jovius] , Paolo (1483–1552). Bishop of Nocera, historian and cartographer, B. Como, d. Florence. Papal embassy to Russia, supplied information for map of *Russia* used by Agnese & in 1548 edition *Ptolemy. Descriptio Britanniae* 1548

Giraldes, M. J. *Reino de Portugal* (1820)

Giraldi[s] , G. See **Giroldi**

Giraldon. Engraver. French edition Rennell's *Hindoostan* 1800, for Freycinet 1807–16, Tardieu 1809, Lapie 1809–16, *globe* 1815

Giraldon-Bovinet. Engraver. *Hist. plans of Paris* 1823, for Vivien de St. Martin 1823–34, *Etâts Unis* 1825, *L'Empire Ottoman* 1827

Giraldus Cambrensis. See **Cambrensis**

Girard. *Morondava* (1725)

Girard. *Seine Inférieure* 1830 (with Carbonnie)

Girard. Geologist. *Rheinprovinz u. Westfalen* [1855–65]

Girard, Capt. *Carte milit. et hist. de la France* (1860)

Girard. *Espagne et Portugal* (1868?) (with Charle)

Girard, Xavier. *Plan Paris* 1820, 1840, *Atlas Portatif de France* 1823

Girardet, A. L. Engraver. Bollin's *Plan v. Bern* 1808

Giraud, Stefano [Etienne] . Engraver. *Plan Naples* 1767

Giraud, V. *Itin. Dar es Salaam—Moero (Tanganyika)* 1885

Girault, P. A. (1791–1855). French geographer, b. Saint Forgeau. *Dict. de Géog. phys. et polit.* 1826

Girava [Giriva] , Jeronimo. Cosmographer to Charles V, b. Tarragona, d. Milan. *Portolan* 1552 (attributed), *Cosmographia,* Milan 1556 (contains *World* map, based on Vopel), & Venice 1570, *Sea atlas*1567 MS

Gird, Richard. *Territ. Arizona* 1865

Girelli, Pietro Paolo (fl. 1690). Engraver of Rome. Cingolani's *Campagna di Roma* 1704

Giriva, H. See **Girava, J.**

Giroldi [Ciroldis, Giraldis, Ziraldis, Zeroldis, Ziredis, Ziroldis] , Giacomo [Jacobus de] (fl. 1422). Portolan maker of Venice. *Portolan* 1425 (attrib.), *Atlas* 1426 (6 charts), *Atlas* 1443 (7 charts), *Atlas* 1446 (6 charts)

Giron, Vicomte de. See **Grenier**

Gironière, P. de la. See **La Gironière**

Girtin, J. Engraver, 8 Broad St., Golden Square, London. *Vicinity of Naples* 1815

Gisborne, L. *Parte del Darien,* Cartagena 1854

Gisolfo. See **Ghisolfi**

Gisolfo, Giacomo. *Atlas* ca. 1550 MS.

Gissing, Capt. C. E. *Teita (Kenya)* 1884

Gist, Christopher (ca. 1706–1759). Explorer and soldier, b. Maryland. *Ohio River* survey 1750, survey *Virginia & Md.* 1751–4

Gittings & Dunnington. *Clarksburg, Harrison Co., W. Va.* 1867

Gittins's Pupils. *Twickenham* 1849

Giunti, Lucantonio [Heirs] . Publisher of Venice. Ramusio's *Navigationi* 1550–65

Giustiniani [Justinianus] , Agostino (1470/9–1536). Dominican Bishop and orientalist, b. Genoa. *Corsica* 1567

Giustiniani, Francisco. Italian cartographer. *Atlas abreviado* 1739

Givry, A. P. Ingénieur hydrographe. *Baie de Cadiz* 1807, *Côte du Brésil* 1822–6 for *Pilote du Brésil* 1826, *Afrique Occid.* 1828.

Gjessing, S. C. (1812–1897). Norwegian artillery officer, official surveyor for government. *Christians Amt.* 1845, *Stavanger Amt* 1866.

Glade, Carlos. *Plano de Buenos Ayres* 1867

Gladwin, George. (fl. 1838). Engraver. *London* 1828

Glaeser [Gläser]. Geographer. *Thiergarten (Berlin)* 1822, *Geg. zwischen Berlin u. Potsdam* 1840

Glaeser [Gläser], A. *Plan Mainz* 1852

Glaeser [Gläser], Dr. Carl. *Deutschland* 1835, *Vollständiger Atlas* (1840), *Schul-Atlas* 1846, *Geog. Hand Atlas* 1863, *Kreis Waldenburg* (1892)

Glaeser [Gläser], Friedr. Gottlob. *Graffschaft Henneberg* 1774

Glareanus [Loritz, Loritus], Henricus [Heinrich] of Mollis [de Gloria] (1448–1563). Swiss humanist and geographer, b. Glarus. d. Freiburg; Prof. of Mathematics in Basle, Paris and Freiburg. *World* 1510; *E. Asia, America &c.* 1510 MSS, *America* 1520, *Geographia,* Venice 1527 (with directions for making globe gores)

Glas, Gustav *Deutsch-Tyrol* 1855, (1890?) *Bayerischer Wald* (1889)

Glas, J. See **Glass**, G.

Glasbach. See **Glassbach**

Glascott, Lieut. A. G. *Route through Armenia* 1835 MS, *Asia Minor* (1836), *S. Shore Black Sea* (1838) MS, *Port of Batoom* (1840?), *Plan of Skutari* 1856

Glasener. *Diocese Trier* (1869?), *Trier u. Cublenz* (1871?)

Gläser. See **Glaeser**

Glaser, C. German cartographer *World Atlas,* Mayence 1851

Glaser, G. F. *Principauté d'Henneberg* 1774

Glaser, Johann Heinrich. Publisher. *Vallistelina* 1625

Glass, Charles E. *Road guide to Gold Fields* (1855?)

Glass [Glas], Capt. George [Jorge]. *Islas Canarias* 1772, *El Rio Harbour* 1794, *Puerto de Naos* 1797

Glass. See **Judd & Glass**

Glassbach [Glasbach], Benjamin. Engraver. Sotzmann's *Prussia* 1802

Glassbach, Carl Christian (b. 1751). Engraver, b. Berlin, son of C.B. Glassbach. *Military maps* 1782–1800

Glassbach, Christian Benjamin (1724–1779). Engraver, b. Magdeburg, d. Berlin. *Poland, Lithuania* 1770

Glavač, Stjepan, S. J. (1627–1680). Philosopher and geographer of Warasdin

Glazier, Capt. *The Father of Waters (Mississippi)* 1881, *Lake Itasca* 1881, *Lake Glazier* 1894

Glegg, John. Surveyor. *Parish Longham, Norham* 1816 MS.

Glegg, Capt. J. B. *Plan Quebec* 1759, reduced 1813

Gliemann, Johann Georg Theodor (1793–1828). Topographer in Copenhagen. *Post-Kort over Danmark* 1820, *Amptskortatlas over Danmark* 1824–9

Glietsch, G. *Communications Télégraphiques* 1871

Glimmerveen, D. J. Dutch cartographer, 19th century. *Alblasserwaard* 1840

Glindemann, C. *Eisenbahnen Deutschlands* 1861

Globe Doré. See **Paris Gilt Globe**.

Globe Vert. 1515, in Bib. Nat., Paris

Globic z Bŭćina, Samuel (ca. 1618–1693). Czech surveyor

Globuciarich, J. See **Clobucciarich**

Glockendon [Glogkendon, Glockenthon], Georg [Jorg] (d. 1514/15). Miniaturist, painter and publisher of Nuremberg. Assisted Behaim with *globe* 1492, pub.

Etzlaub's *Nuremberg* 1492, *& Road map* 1500, *Highways through Roman Empire* 1501

Glocker, Adolf. *Korea* (1887?)

Glocksperger, Jan (1678–1771). Bohemian surveyor and cartographer

Glogkendon, G. See **Glockendon**

Gloria, H. L. de. See **Glareanus**

Glot. Engraver, for Yeakell & Gardner (1776)

Glot, C. B. Engraver. Herisson's *Plan Génève* 1777

Glotsch, Ludwig Christoph (d. 1719). Engraver of Nuremberg. *Tirol* n.d.

Glover, Edward. *Varty's Educ. Series of School Wall Maps* (1853)

Glover, George. Architect and surveyor. *Plans St. Johns Hospital, Huntingdon* 1840 MSS.

Glover, Lieut. (later Sir) John Hawley (1829–1885). Served in Navy 1841–77, later Governor of Newfoundland, & Leeward Is. *R. Kwara* 1857–9, *Bight of Benin* 1858–62, *R. Niger* 1863–(99)

Glover, Moses (fl. 1620–40). Painter and architect of Isleworth, Middx. *Survey Syon House, Isleworth* 1635 MS. *(Plan Petworth House* 1615, attributed to him)

Glover, Robert (1544–1588). Somerset Herald (1571), b. Ashford, Kent. *Kent* 1571 MS, *Survey Herewood Castle, Yorks* 1584

Glover, Stephen (d. 1869). Compiled and published *Derby Co.* (1850?)

Gluemer [Glümer], Bodo von. *Estados Unidos Mexicanos* 1882

Gluss, I. R. *Chart Entrances Goerhee & Quex Deep* 1801

Glynn, Lieut. James (1801–1871). American Naval Officer. *Chart Cape Fear River* 1839, *Beaufort harbour* 1839

Glynne, Richard. Publisher with A. Lea at "the Atlas & Hercules in Cheapside". *London, Westminister and Southwark,*

8 sh. (1690?), Morden & Lea's *Plan London* 1716, (editions to ca. 1725)

Gmelin, J. G. (1709–1755), b. Tübingen. *Reise durch Sibirien* 1751–2

Gmunden, J. S. von. See **Sartorius von Gmunden**

Gnoli, Bartolomeo. Artist and architect of Ferrara. *Ducato di Ferrara* 1645, *Valli di Comacchio* 1650

Goad, Charles E. (fl. 1914). Compiled insurance plans of towns. *Atlas Toronto* 1884

Goad, Thomas W. *District New Mexico, Fort Leavenworth* 1875, *Costilla Estate* 1887

Goalen, W. N. Navig. Lieut. *Port Adelaide* 1875–(6), *Murray River* 1876–(9)

Gobanz, Dr. Josef. *Steiermark* [1864]

Gobert, Martin. Publisher, 'au Palais en la Gallerie des Prisonniers,' Paris. Tassin's *Cartes générales . . . Allemagne* 1633

Gobert, Th. Engraver, Rue St. Jacques 171, Paris, *Océanie* 1858, for Vuillemin 1873

Gobin, Engraver. Cartouche for Robert de Vaugondy's *Afrique* 1749

Goche, Barnabe. *Plan Galway* 1583

Gochet, A. M. (b. 1835). Belgian geographer

Gôczel, S. *Gold Region W. Australia* 1896

Godbid, **Arthur & J. Playford**. Publishers. Symonson's *Kent* 1659, Adams' *Index Villaris* 1680, Seller's *Atlas Maritimus* 1682

Goddard, George H. *California* 1860

Goddard, John (fl. 1645–71). Engraver, for Fuller 1650, Farrer's *Virginia* 1651, *Asia* in Heylin's *Cosmographie* 1652

Goddard, John, Junior. Engraver. *Norfolk* in Speed (1665?)

Goddard, T. Bookseller in Norwich. *Norfolk* 1731, (with Chase), & 1740 (with Goodman)

Goddard, W. Surveyor. *Bonham Estate, Hants.* 1819 MS

Göde, N. See **Goede**

Godefin. *Plan Ville St. Etienne* 1866

Godefroy, (fl. 1784). Publisher, Rue des Francs Bourgeois, Paris

Godet, le Jeune. Designed and engraved *Ville et Faubourgs de Paris* 1825

Godfray, Hugh. *Jersey,* 2 sh. 1849, *I. of Wight* 1849, *South latitudes* 1858

Godfrey, Jonathan. *Dinington Park, Berks.* 1654 MS.

Godiche, A. H. Maps *Denmark* 1763–8, *Kiöbenhafn* 1764–(6)

Godin, H. J. Engraver. *Plan de Spa*

Godin, Louis. *Carte du Pérou* pub. Buache 1739

Godinho, Manoel (1632–1712). Portuguese Jesuit. *Babylon* (1660?)

Godinho de [H]erédia, Manoel (1563–1623). Portuguese, mathematician and cosmographer, b. Malacca. *Goa* 1610, 1616

Godman, T. *Plan St. Albans* 1815

Godrecio [Godreccius], W. See **Grodecki**

Godrich, A. H. [Heirs]. Publishers. *Sleswig* 1781

Godschalk, S. K. *Ashtabula Co., Ohio* 1856, *Cattaraugus Co., N.Y.* 1856

Godson, Richard. *Prop. Canal Warwick– Bramston* 1795

Godson, W. *World,* Willdey ca. 1715

Godson, Wm. Surveyor and publisher of Basingstoke. *Mannour of King-Sumborne* 1734, *Odiam Park* 1739, *Plan Winchester* 1750

Godunov, Fëdor Broisovič (1589–1605). Cartographer. *Russia* used by Gerritsz 1612

Godunov, Peter Ivanovič (d. 1670). Governor of Siberia. *Tab. Russiae* 1614, *Siberia* 1667 (first map of the country)

Godunov, Simon (17th century) Russian mapmaker. Engraved *map of Siberia* published in a Bible 1663, revised edition 1672; *Siberia* 1667 MS. (with Remezov)

Godwin, J. *Passage of the Douro* 1809

Godwin, M. J. & Co. Publishers, 41 Skinner St., London; later 195 Strand [1822]. *Guide through London* 1821, 1823, *Pedestrians' Companion 15 miles round London* 1822

Godwin, R. W. *West Khasi Hills* 1868

Goede [Göde], Nikolaus (1561–1633). Pastor, geographer

Goedecke, John Frederick. Draughtsman. *Annapolis Harbour* 1818 MS

Goederbergh, Gerrit van. Publisher of Amsterdam. Vlasbloem's *Leeskaart* 1664, *Zee-Spieghel* (1665)

Goedesbergh, Th. van. *Atlas Totius Orbis Tabulae* (1646)–1693

Goedkindt, Pieter. Engraver and publisher, after 1630. Published maps by Bolswert and Vorsterman

Goedsche, Bruno. Publisher. *Plan London* ca. 1838

Goeje, M. J. de. See **De Goeje**

Goekoop, Christiaan (1716–1767). Mayor and geographer of Goedereede

Goellnitz [Göllnitz], Abraham (fl. 1631–42). Traveller, geographer in service of Christian IV of Denmark, b. Danzig

Goepel, Paul. *Post Route Map Ohio & Ind.* 1870

Goere[e], Jan (1670–1731). Dutch designer and poet, b. Middelburg, d. Amsterdam, son of Willelm. Worked for de Fer's *Atlas Royal* 1699, for Aa 1713, *Japan* 1715, *America* 1729

Goeree, Willelm (1635–1711). Printer and publisher, b. Middelburg, d. Amsterdam. For Loon 1668, *Atlas* 1670, *Atlas ofte de geheele weerelt* 1677

Goerög (Gorög), Demeter (1760–1833). Cartographer and globemaker of Budapest. *Magyar Atlas* 1802, 1841

Goerringer [Görninger] *Hist. Atlas* 1840

Goës, B. *Hindu Kush,* Weller ∙1872

Goes, Damião de. Portolan maker and author. *Chronico da S. Principe D. Joao* 1567

Goeschen, Georg Joachim. Publisher of Leipzig. *Atlas von Amerika* 1830

Goessel, von. *Military route map Austrian army* 1866–71

Goethals. See Algoet

Goethe, Joh. Wolfg. von (1749–1832). German poet. *Die Höhen der alten und neuen Welt bildlich verglichen* (Engraved hypsometric map) pub. Weimar 1813

Goethem, G. van. *Polder van Oorderen* 1723

Goetz [Götz]. *Plan v. Moskau* 1812 (with Dietrich).

Goetz [Götz, Goetzio], Andreas (1698–1780). German cartographer and philologist. *Geog. antiqua,* Nuremberg 1729

Goetze [Götze], Ferdinand. *Naples,* Weimar 1801

Goetze [Götze], A. Frederick. Geographer in Weimar Geog. Instit. Maps for Gaspari 1804–11, revised Güssefeld's *N. America* 1812

Goetzio, A. See Goetz

Goffart, S. *Colonia Agrippina* 1753

Goffe, J. Surveyor in Jamaica. *Estates* 1688 MSS.

Gogeard. *Plan ville de Rouen* 1894

Goggins, Joseph. *Novel Carte of Europe* 1870

Gogh, J. van. *Vaarwater benoorden Makasser* 1849–(51)

Goghe, John. Irish cartographer. *Ireland* 1567 MS.

Gohlert, Vincent (1823–1899). Austrian geographer and statistician

Going, A. S. *Texada Is., Nanaimo Mining District* 1897

Going, Philip, Master, R.N. *Port Davey, Tasmania* 1850–(2)

Gold, Joyce. Printer, 103 Shoe Lane, Fleet St., London. *Voyage Anacharsis the Younger in Greece* 1806, *Otaheite* 1815, Rowe's *English Atlas* 1816

Goldbach, C. F. *Neuester Himmels-Atlas* 1803

Goldfrap, Ensign John George. *Plan Magdelain &c.* 1766, *Part Prov. Quebec* 1767, MSS.

Goldie, A. *New Guinea* 1878

Golding, John. Estate surveyor. *Finchingfield* 1820 MS.

Goldingham, John. *Survey Pulicat shoals* 1792–(4)

Goldmann, Charles Sidney. *Witwatersrand* 1895, *Philips' Atlas Witwatersrand* 1899

Goldschmidt, Martinus Martini. *Statt Lucern* 1597

Goldsmith, Rev. J. Pseudonym of Sir Richard Phillips, q.v.

Goldthwait, J. H. *Massachusetts* 1838, *Pacific States* 1865, *Railway map of New England* 1849

Goldtschmide, Matthias. Engraver, worked for Blaeu

Goldzweig, Jacob. *Biblical map Holy Land* (1893?)

Golescu, Iordache (1768–1848). Rumanian cartographer

Goliat[h] [Golijath], Cornelis (d. 1667/8). Surveyor and cartographer of Middelburg. *Siege Olinda de Pharnambuco,* Visscher 1648, *Plan Middelburg,* Meertens (1680)

Goll, Johann Jakob (1809–1860). Cartographer and engraver of Bern. *Suisse* 1850, *Thurgau* (1870)

Göllnitz, A. See Goellnitz

Golovnin, Vasili Mikhailovich. *Kurile Is.* 1818, *Duché de Schleswig* 1813

Gollowin, F. von. *Schleswig* 1806

Goltius, H. See Goltz[ius]

Goltz, Baron Colmar von der. *Umg. v. Constantinopel* 1897

Goltz [Golzius], Conrad. Engraver and publisher of Cologne, fl. late 16th century. *Great Brit.* pub. Overradt in Kaerius' *miniature maps of Eng.* (1617)

Goltz[ius] [Goltius] , Hendrik (1558–1616/7). Engraver, artist and glass painter of Haarlem, b. Mulbrecht

Goltz[ius] , Hubrecht (1526–1583). Painter, engraver, printer and antiquary, b. Würzburg, d. Bruges; active in Antwerp & Bruges

Goltz, Leonhard von der (1815–1901) *Pommern* 1851

Goltzius. See **Goltz**

Golzius, C. See **Goltz**

Gómara, L. de. See **López de Gómara**

Gombault, Louis. *Villehardouin* (1860?)

Gomboust, Jacques. Engraver and "Ingénieur du Roi". *Plan Paris* 1649–(52). *Rouen* 1655

Gomes, Estevão. Portuguese seaman and cartographer, early 16th century

Gomes, M. A. See **Gómez**

Gomez, Don. *Cuba,* Heather 1809

Gómez [Gomes] , Capitán Miguel Antonio. *Plano Manila* 1762, 1763 MSS, *Islas Babuyanes* 1781

Gómez de Arteche y Moro, José (b. 1821). *Atlas de la guerra de la Independ.* Madrid (1869–1901)

Gómez Holenda, Carmelo, Corps of Engineers, *Manila* 1864 MS.

Gómez y Parientos, Moritz Georg (1744–1810). Austrian general and cartographer

Gomme, Sir Bernard de (1620–1685). Engineer and surveyor, b. Lille, came to England with Pr. Rupert, Q.M.G. Royalist army 1642–6, "Engineer in chief King's Castles" 1661. *Plans fortifications Liverpool* 1644 MSS, *Portsmouth* 1665, *Survey Citty of Dublin* 1673 MS, *Plans Civil War battles, Survey Tilbury Fort*

Gönczy, Pál. *Magyar Korona* 1866, *Atlasza* (1897)

Gonichon, Sieur. *Fleuve Mississippi* 1731 MS.

Gonne. Engraver. For Morse 1792, *Ireland* 1795

Gonneim. 3 *planispheres* in 14th MS, *Imago Mundi* (attributed)

Gonsag, Père Ferdinand *Gulf of California* 1746

Gonsalves da Fonseca, J. *Governo de Pernanbuco* 1766

Gonzago, Curzio (ca. 1536–1599). Italian printseller in Rome and Venice

Gonzales, Alexandre. *Atlas maritimo Peru, Chile, costa Patagonia &c.* 1797 MS.

Gonzales, Andres. *Florida* 1609 MS.

Gonzales, F. *Nuevas Philipinas* 1705 MS.

Gonzales, T. Engraver. *Costas des Seno Mexicano &c.* 1808, *Isla San Martin* 1811

Gonzales Fernández, Ramón. *Plano Manila* 1875, *Arch. Filipino* 1875, 1877

González, A. y. See **Artero y González**

González, E. Garcia y. See **García y. Gonzalez**

González, G. See **González de las Peñas**

González, Joseph. Engraver. Venegas' *Mar del Sur* 1757

González, Juan de Dios. *Plan Prov. Yucatan* 1774

González, M. *Carta Topo. Uruguay,* Buenos Aires 1874

González, Nicolas. *Pen. Ibérica* (1871?) Madrid 1879

González de Agueros, Pedro, S. J. *Provincia de Chiloe* (1791)

González de Carvajal, Ciriaco. *Arch. Filipino* 1787

González de las Peñas, German. *Isla de Cuba,* Havana 1881

González de la Vega, Rafael. *Cuba* 1877

González de Mendoza, Juan. *Regno della China* 1589

Gonzz, J. See **González**, J.

Good, J. *12 miles round Berwick* (1806)

Goodenough, F. A. *Routes between India & China* 1869

Goodenough, Commodore James Graham (1830–1875). *Part of S. Pacific* 1876

Goodhue, J. H. *Concord, N.H.* 1868

Goodman, John. Engraver and printer of Frankfort, Kentucky. *Rapids of Ohio River* 1806

Goodman, R. *Norfolk* 1740 (with J. Goddard)

Goodman, Wm. Estate surveyor. *Parish St. Martin's Outwich, Essex* 1599, *Kelvedon & Little Coggeshall* 1605 MS.

Goodrich, A. T. & Co. Publishers, 124 Broadway, opposite the City Hotel, New York. Bought J. Melish's stock, New York. *Hudson between Sandy Hook & Sandy Hill* 1820

Goodrich, Rev. Charles Augustus (1790–1862). Congregational clergyman, geographer and historian of Worcester, Mass. *Atlas,* Boston (1826), *Outlines of Modern Geog.* 1827

Goodrich, Samuel Griswold [used pseudonym Peter Parley] (1793–1860). Publisher and geographer of Boston, brother of C.A. Goodrich. *U.S.* 1826, *Atlas* 1830, edited Bradford's *Atlases* 1841–2, *Nat. geog. for schools* 1845, *Shilling Atlas* (1859?), *Outline Atlas* (1871?)

Goodwill. Engraver. *Plan Kingston-upon-Hull* 1817 (with Consitt)

Goodwin, John. *Plans Limehouse* 1635 MSS.

Goodwin, M. P. Assisted A. Gray in maps *Guernsey &c.,* Faden 1816, *Guernsey* 1832

Goodwin, N. *City Hartford, Conn.* (1824?)

Goodworth, W. G. W. *Germany* 1899

Goor, D. Noothoven van. *Nieuwe School-Atlas* (1875)

Goos, Abraham (ca. 1590–1643). Engraver, mapseller, cartographer and publisher of Amsterdam. Engraved for Kaerius 1614, *Nieuw Nederlandtsch Caertboeck* 1616, 1625, engd. for C. J.

Visscher 1620–[34] & Speed 1626, *Geldria,* Hondius 1629, maps in 1630 Mercator-Hondius *Atlas; Ancient Sicily,* Jansson (1636)

Goos, Hendrik. Publisher, in Amsterdam, son of Pieter Goos. *Nieuwe Groote Zee-Spiegel* 1675, *Lightning Columne* 1689–92 (with Jacobsz.)

Goos, Pieter (ca. 1616–1675). Cartographer, engraver, publisher, printer and printseller of Amsterdam, son of Abraham Goos, "op't Water by de Nieuwe-brugh, inde Vergulde Zee-Spiegel" [Golden Sea Mirrour] . *Lichtende Colom* 1650, 1654, 1657, 1664, 1670 (Eng. edition 1669). *Nieuwe groote zee-spiegel* 1662, 1674, *Zee Custen van Europa* Portolan (1655). *Zee Atlas* 1666, 1668, 1674, 1675, *Atlas ofte Water-Weereld* 1666, *Le Grand et Nouveau Miroir de la Mer* 1667, 1671

——, Widow. Continued business at husband's death. Reissued *Nieuwe Groote Zee-Spiegel* 1676

Gorcum [Gorkum] , Jan Egbert van (1780–1862). Military cartographer of Arnhem, *Oud-Nederland* (1831?)

Gordon, A. Lithographer, 66 Paternoster Row, London. *Liverpool Canal &c.* 1831

Gordon, Charles. *Ferro Carriles de la Rep. Argentina* 1889

Gordon, Lieut. (later General) Charles George (1833–1885). Held important posts in China, Mauritus, S. Africa; Governor Equatorial Provs. Africa 1874–6, Gov. Gen. Soudan 1877 & 1884. *Road Constantinople to Adrianople* 1856, *Milit. plan country around Shanghai* 1865, *Suakin to Khartum* 1874, *Contour map Jerusalem* 1883

Gordon, Lieut. David McDowall. Admiralty surveyor. *Survey China* 1845–8, charts *Borneo* 1852, *Tam Sui Harbour* 1855

Gordon, Capt. Harry. Chief Engineer in N. America. *Road Stirling to Fort William* 1751, *Fort Edward to Crown Point* 1755, *River of Ohio* 1766 MS.

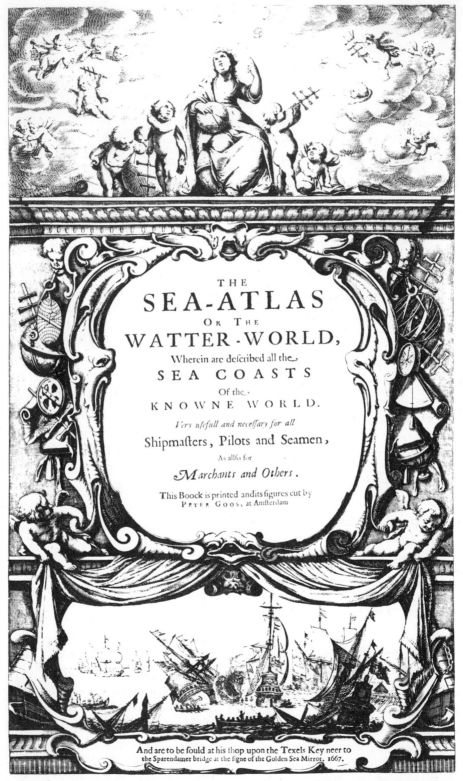

THE
SEA-ATLAS
Or The
WATTER-WORLD,
Wherein are described all the
SEA COASTS
Of the
KNOWNE WORLD.

Very usefull and necessary for all

Shipmasters, Pilots and Seamen,

As allso for

Marchants and Others.

This Boock is printed andits figures cut by
PETER GOOS, at Amsterdam

And are to be sould at his shop upon the Texels Key neer to
the Sparendamer bridge at the signe of the Golden Sea Mirror. 1667.

Title-page to the English edition of P. Goos' Sea Atlas (published Amsterdam, 1667)

Gordon, Rev. James (1615?–1686). Parson of Rothiemay, Banffshire, son of Robert Gordon, assisted father with revision of Pont's maps of Scotland. *Plan St. Andrews, Cupar* 1642, *Fife* 1645, *Survey Edinburgh* 1646–7, *Plan Aberdeen* 1661

Gordon, J. Bentley (1750–1819). English geographer and historian. *New System of Geog.*

Gordon, Patrick. *Geog. Anatomiz'd* 1702 &c.

Gordon, Peter. First Bailiff of Savannah 1732. *View of Savannah* 1734 (first plan of town)

Gordon, Peter. *Reception Bay, Tristan d'Acunha* 1814

Gordon, Robert. *Atlas Irrawaddy* 1879–80, *Course Sanpo* 1885, *Ruby Mines Burma* 1888

Gordon, Robert Jacob. Dutch soldier and explorer. *Cape Colony* 1777–95 MS.

Gordon, Thomas F. (1787–1860). *New Jersey* 1834, *Gaz. of State of New York,* Phila. 1836

Gordon, W. *Commercial map Scotland* 1785 (with J. Walter), *Trav. Directory through Scotland* 1792 (with N.R. Cheyne)

Gordon, Wm. Surveyor. *Huntingdonshire* 1730–1, pub. (1732?), *Beds.* 1736

Gordon, Comm. Wm. Everard Alphonso. *Bird Is. (S. Africa)* 1853

Gordon. See **Griffing, Gordon & Co.**

Gordon & Gotch. Publishers in Melbourne, Sydney & London. *Australia* 1877

Gore, Capt. Arthur, R.N. *Turon Harbour, Cochin China* 1764, pub. Dalrymple

Gore, Major Charles Wm. *Paths round Newcastle* 1893

Gore, Lieut. George Corbet. Maps *Afghanistan & Pakistan* 1880

Gore, J. Publisher of Liverpool. *Plan Liverpool* 1821

Gorges, Sir Ferdinando (1566?–1647). Governor of Plymouth, instigated Colony of New Plymouth 1628. *Fortifications Plymouth* 1596–7 MS.

Gorio di Stagio. See **Dati**

Gorjan, August (b. 1837). *România* 1881, *Atlas–géog. România* 1895 (with Luncan)

Gorkum, J. E. van. See **Gorcum**

Gorlinski, Joseph. Draughtsman to Gen. Land Office. *U.S.* 1867, *Railroad routes* 1869

Gormaz, F. V. *Süd Chile* 1880

Görög, D. See **Goerög**

Gorrell, T. D. *Pleasants Co., Va.* 1865

Gorries, Capt. Joannes. *Bremen,* Blaeu 1662

Görringer, M. See **Goerringer**

Gorton, John (d. 1835). Collaborated with S. Hall. *Topo. dict. of Gt. Britain & Ireland* (1831–3)

Gosling, Ralph (1693–1758). Topographer *Plan Sheffield* 1732 (earliest known), and 1736

Goslyng, John. *Plott building in Nevills Alley, Fetter Lane* 1670

Gosse, Pierre, Junior (1729–1765). Publisher at the Hague. *Atlas Méthodique Palairet* 1755. Then in partnership with D. Pinet; *Atlas Portatif* (W. Saxe) 1761, *military plans German towns & battles* 1763–6

Gosse, Wm. Christie. *Route Australian Govt. Central & W. Exploring Expedition* 1873

Gosselin, Col. *Plans Guernsey* 1795 MSS

Gosselin, B. *Plan Boulogne-s-Mer* 1863

Gosselin, Pascal François Joseph (1751–1830). Geographer of Paris, b. Lille. *Géog. des Anciens* (1797)–1813

Gosselmann, Capt. *Prov. of La Rioja* 1839

Gosselo, Edmund. *Scilly Is.* 1707–10 MS

Gosset, R. *Plan St. Helier, Jersey* (1850)

Gosset, Major Wm. R.E. *Tripoli* 1813, *Algiers* 1816

Gosset, Wm. Driscoll. *Ord. Survey environs Aldershot* 1856

Gosson, Henry. *The Carrier's Cosmographie* 1637

Gotch. See **Gordon & Gotch**

Gotendorf. *Rivière de Paris à Rouen* 1878

Gotham, Wm. *Manor W. Harting, Sussex* 1632 MS.

Gothus [Svart, Ornehufvud] , Olaus Joannis [Olof Hansson] (1600–1644). Swedish cartographer, Cosmographer to K. Gustavus. *Prussia, Brandenburg, Misnia,* used by Jansson 1641 & Blaeu. *Dantzig, Voightland, Marchia Nova & Ukraine, Ruppin & Prignitz, Elbing,* used by Blaeu 1662

Gotofred, J. L. See **Abelin,** J. P.

Gott, Capt. Reeve. *Chart Gulf of Finland* 1785

Gotteberg, E. de. Maps *Egypt* 1857–68, *Ostaegyptischen Wüste* 1859

Gottfried, J. L. See **Abelin,** J. P.

Gotthard. Maps for Meyer 1867

Gotthold, Aug. *Plan Kaiserslautern* (1883)

Gotthold [Gottholdt] , H. H. *Denmark* 1808, *Deutschland,* 55 sh. 1808–31, *S. Africa* 1810, *Cape of G. Hope* 1815

Gottlieb, August. *America,* Homann Heirs 1746

Gottorp Globe. 11 ft. in diameter, constructed 1654–64 for Duke Frederick of Holstein-Gottorp by A. Busch, under the supervision of Olearius

Gottschalck, Friedrich. *Plan v. Dresden* 1847

Gottschalk, Adolf. Engraver, for Meyer (1830–40)

Gottwald, Johann. *Österreichisch-Ung. Monarchie* (1876?)

Götz. See **Goetz**

Gotze, F. *Europa* 1815

Gotzmann, W. *Greiz u. Umgegend* 1882

Goubaut. *Plan Philipsbourg* (1750)

Goubet. Surveyor. *Plans fortifications in Ireland* 1690–5 MSS, *Plan Portsmouth* (1692?)

Goudriaan, B. A. [or B.H.] , *Hoofdrivieren op de Schaal* 1830–9

Gouge, J. Publisher and mapseller in Westminster Hall. Price's *30 miles round London* 1712

Gouge, James. Estate surveyor of Sittingbourne, Kent. *Boughton under Blean &c., Kent* 1803, *Glebe Cuddington, Bucks.* 1818, *Panfield, Essex* 1819, *Lake estates in Kent* 1832 MSS.

Gouget, Jerome T. Engraver, for Blanchard 1869

Gough, Capt. H. *Chart Andaman Is.* 1708, pub. Dalrymple

Gough, Richard (1735–1809). English collector, Director of the Soc. of Antiquaries 1771–97. *Brit. Topography* 1768, 1780, edited Camden's *Britannia* 1789

Gough [Bodleian] **Map.** *Map of Great Britain* ca. 1335, named after discoverer Richard Gough.

Goujon, Andriveau. See **Andriveau-Goujon**

Goujon, J. Marchand de Cartes Géographiques, Rue du Bac No. 6 Paris [1805– 20] , Rue du Bac No. 17[1845] , *Plan Vienna* (1805), Tardieu's *U.S.* 1812, for Brué 1815, Frémin 1820, *N. America* 1821, *Océanie* 1845. See also **Andriveau-Goujon,** J.

Goulart[io] , Jacques [Jacobo] (1580– 1622). Cartographer of Geneva. *Lake Leman,* used by Hondius 1606, Blaeu 1634, and Jansson 1638

Gould, Augustus Addison. (1805–1866). Physician and conchologist. *New Ipswich, N.H.* 1851

Gould, Charles. Govt. geologist. *Expedition W. Tasmania* 1860

Gould, Benjamin Apthorp (1824–1896).

Astronomer, founded Observatory at Córdova, Argentine (1870)

Gould, Francis. Ordnance draughstman. MS copies of various plans 1764—81

Gould, F. A. *Atlas Randolph, Wayne Counties, Indiana* 1847 (with Lake & Sandford)

Gould, G. See **Gauld**

Gould, Hueston T. *Atlas Franklin Co. & Columbus, Ohio* 1872 (with J.A. Caldwell), *Atlas Ross Co. & Chillicothe, Ohio* 1875

Gould, James. Estate surveyor. *Little Coggeshall, Essex* 1721 MS.

Gould, Jay (1836—1892). Financier. *Ulster Co., N.Y.,* 1854, *Delaware Co., N.Y.* 1856

Gountei [Ontagawa] Sadahide (1807—1873?) *Map Fuji* 1855

Gourdin. *Baie St. Nicholas* 1838, and other plans for Dumont d'Urville's *Voyage au Pôle Sud* 1842—8

Gourlay, Robert Fleming. *Edinburgh* 1855

Gourmet. *Atlas* (1770?)

Gourmont[ius] Jérome de [Hieronymus]. Mapseller of Paris. Catalogue 1536, *Germany* 1545, *Champaigne* 1546, *Britain* 1548, reissued Olaus' *Iceland,* 1548, *France* 1553, pub. maps of Finé. Postell's *Signorum coelestium* 1553

Gourmont [Gurmontius, Germontius], Jean de (fl. 1561—85). Engraver and publisher of Paris, succeeded father Jérome

Gournay, C. Publisher and engraver "à l'entré du Quay de l'orloge du côte du pont au change", Paris. Engraved for de Fer 1690, *Plan du siège de Namur* 1695, *Plan Liège* (1713)

Gourné, Pierre Mathias, Abbé de (1702—ca. 1770). Cartographer of Dieppe. *Géog. Méthodique* 1741, *Atlas abrégé,* Desnos 1763

Gouvion, Lt. Col. Jean Baptiste (1747—1792) French engineer. *Plan Yorktown campaign* 1781 MS.

Gouvion Saint-Cyr, Laurent, Marquis de (1764—1830). Marshal. *Atlas aux campagnes* 1828, *Atlas à l'histoire militaire* 1831

Gouwe[n], Gilliam [Gelliam, Wilhelm] van der. Dutch engraver, pupil of Picard. Worked for Allard, Visscher, Wit (1696), for Halma 1704, Braakman 1706, cartouche for Aa's *World* (1713)

Govantes, Felipe María de. *Filipinas* 1878

Gove, F. W. *Rico Pioneer Mining District, Colorado* (1880)

Gover, Edward. *Biblical Atlas* (1840), *Hand Atlas* 1850, *Atlas univ. hist. geog.* 1854, *Two shilling physical atlas* 1854

Gower, Richard Hall (1787—1833). Naval architect and inventor, b. Chelmsford, d. Ipswich. *Chart Cape Bank* 1791, pub. Dalrymple

Gowing, John Sewell. Publisher of Swaffham, Norfolk, *Swaffham* 1845

Gowland, Master (later Navig. Lieut.) John T. *Gulf of Patras* 1865—(6), *Survey E. coast Australia* 1866—71

Goyder, G. W. (fl. 1896) Surveyor General of S. Australia. *Mt. Denison Range* 1860 MS, *S. Australia,* Adelaide 1880

Graaf [Graaff], C. J. Van de. *Cape of Good Hope* 1790

Graaf, G. I. van de. *Caap de Goede Hoop* 1785—94

Graaff, H. W. van de. Capt. Ingenieur. *C. of Good Hope* 1786

Graaf [Graaff, Graff], Isaac de (1667—1743). Cartographer to the Dutch E. India Co. from 1705; produced 200 MS, charts. *Golfe du Bengale* 1709, *Détroit de Malacca* 1710, *La Sonde* 1711, *Mer de Java* 1718, *Atlantique* 1723, *Océan Indien* 1728

Graaff. See also **Graaf**

Graah, Capt. Wilhelm August (1793—1863), Danish Navy. *Vestlige kyst of Grönland*

1823–4, *Grönland* 1832, *Juliansehaabs District* 1844

Graap, H. Photo-lithographer. *Australien u. Neu-Seeland* (1876?)

Graberg, Jakob. *Mappamondo* 1802, *Imperio di Marocco* 1834, *Moghribul Acsa* 1834, *Kharesmia* 1840

Grabowski, Ambrozy. *Cracow* 1830

Graça, F. C. da. See **Calheiros da Graça**

Gracher, Johann Georg. *Saltzburg District* 1836 MS.

Gracht, Quentin van der. *Plan Bethune,* used by Braun & Hogenberg 1588

Gracie, P. See **Garcie**

Grack, Carl. Lithographic printer in Berlin. Fay's *Atlas* 1871

Gracroft, John. *Greenland, Norway, Nova Zembla &c.* 1740 MS.

Grad, Charles (1842–1890). Geographer and geologist, b. Turkheim. *Alsace* 1889

Graef [Gräf], Adolf. *Hannover &c.* 1857, *Atlas des Himmels und der Erde* (1863–6), *Eisenbahn Karte v. Deutschland* (1877?)

Graef [Gräf], Carl. Collaborated with Adolf Graef. *Sachsen* (1855?), maps for Kiepert 1871, *Schweiz* (1890?)

Graefé, C. *Preussen* (1855?), *Bayern* 1861

Graessl. Maps for Meyer 1852–7, *Texas* 1852, *Wisconsin* 1852

Gräf, A. & C. See **Graef**

Graf, C. Lithographer, Gt. Castle St., Oxford St., London. Friederich's *Plan London* 1850

Graf [Grave], Hans. *Statt Franckenfurt* 1552, used by Braun & Hogenberg

Graf [Graff], Urs, elder (ca. 1485–1527/8). Artist, engraver and goldsmith in Basle, b. Soleure

Graf & Soret. *Plan Parish St. Marylebone* 1833

Graff, F. *Plan v. Rostock* (1859?)

Graff, I. de. See **Graaf**

Graff, U. See **Graf**

Grafnetter, Josef. *Plan v. Prag* (1877?)

Graham, A. B. Photo-Lithographer in Washington for War Dept., 1899

Graham, Lieut. C. *Chart Sandusky Bay* 1826 (reduced version 1838)

Graham, C. A. E. *Military plans Dublin District* 1828 MS.

Graham, C. B. Lithographer, 4 John St., New York [with Graham, J. R., 1835], later Washington, D.C. *Detroit* 1835, *L. Superior* 1842, *New Mexico* 1847, *South Pass to Gt. Salt Lake* 1852, *surveys Michigan* 1857

Graham, Lieut. Cyril. *Palestine* [1864?]

Graham, G. Carleton's *Map of the District of Maine* (1790)

Graham, George. *Waikato District, N.Z.* 1864

Graham, H. Landscape painter. Corrected *Lakes Killarney* (1786)

Graham, James. Land surveyor. *Wicklands Level, Essex* 1858 MS.

Graham, James Duncan (1799–1865). Army Officer in U.S. Topog. Engineers. *Winchester & Potomac Railroad* 1831–2, *Extremity Cape Cod* 1833–5, *R. Sabine* 1840, *Boundary U.S./Canada* 1842–(3), *Ports on L. Michigan,* 12 sh., *Chicago* 1854–7, *L. Superior* 1863

Graham, J. R., Lithographer. See **Graham, C. B.**

Graham, John. *Franklin Co., Ohio* 1856

Graham, Dr. Patrick. *Stirlingshire* 1812, *Kinross & Clackmannan* 1814

Graham, W. R. M. *Maps Provinces E. India & Bengal* 1846–52

Grainger. Engraver. *Macao* (1785?)

Grainger, Thomas (1794–1852). Civil Engineer, b. near Edinburgh. *Railway surveys Scotland* 1825–32

Gram, Hans (1685–1748). Founder

Danish Royal Society, historian and mathematician to Christian VI

Grambo. See **Lippincott, Grambo & Co.**

Grammaye, J. B. (1579–1635). Belgian geographer and historian, b. Antwerp, d. Lübeck. *Africae Illust.* 1622

Gramolin, Alvise. Portolan maker of Venice. *Portolans* 1612, 1630

Granchain, Capt., French Marine. *Plan Baye de St. Lunaire* 1784–(5)

Grand, Le. See **Legrand**

Grand, Bey P. *Plan du Caire,* 4 sh. 1874

Grandchamps. *Dépt. des Alpes Maritimes* 1865

Grandi, Francesco, S. J. Italian globe-maker, mid 18th century

Grandidier, Alfred (1836–1921). French traveller and naturalist. *Prov. d'Imerina* 1880, *Madagascar* 1884, *Hist. de la géog.* 1885–92

Grandis, Alvise. *Veneta Laguna* 1799, 1820

Grandison, H. Script engraver for O.S. *Waterford,* Dublin 1842

Grandpré, L. de (1761–1846). *Plan du Cap (Africa)* 1793, *Dict. Univ. de Géog. marine* 1803

Grand Voinet. Geographer to the King. *Pyrenées* 1772

Granelli, Carlo, S. J. (1671–1739). Theologian, cartographer and mathematician in Vienna, b. Milan

Graner, Fredrik. *St. Clair Co., Ill.* (1854?)

Grange, Doctor. *Geological map Tasmania* 1847

Grange, M. C. *Poiters* 1882, *Chaumont* 1897

Granger. *Birmingham* (1860?)

Granges, G. des. See **Gagnières des Granges**

Granges, des. See **Desgranges**

Grangez, Ernest. *Navig. de la France* 1840, *Belgique & Hollande* (1877?)

Granier. *Montagnes de l'Arpette* (1760?)

Grant, A. A. *Standard indexed atlas* (1885)

Grant, Charles [Viscomte de Vaux]. *Uranographia* 1803 (star map)

Grant, Daniel. *Eccles. map England & Wales* 1851

Grant, E. *Plan Redlynch Park, Somerset* 1738

Grant, E. *Paris Monumental* (1855?)

Grant, E. S. & Co. Publishers of Philadelphia. Bradford's *Illustrated Atlas* (1838)

Grant, Lieut. J., R.N. *Australia* 1800–02

Grant, James. Surveyor and draughtsman. *Plan Piscataqua Harbour* 1774, *Plan Boston Harbour* 1775 (with J. Wheeler), *Plan Perth Amboy* ca. 1780 MS, *S.E. coast I. St. John* 1781 (in *Atlantic Neptune*), assisted Holland with *New Hampshire,* Faden 1784

Grant, Col. J. A. (18th century). See **Grante**

Grant, Capt. (later Lt.-Col.) James Augustus (1827–1892). Explorations Africa 1861–3, (with Speke), awarded Gold Medal of R.G.S. 1864

Grant, J. Murray. *Kaffraria* 1872, *Transkeian Territory* 1875 (with G.P. Colley)

Grant, Capt. P. W. *Martaban, Ye, Tavoy & Marguie (Burma)* 1870

Grant, Robert (1814–1892). Scottish astronomer. *Atlas of astronomy* 1869

Grant, Major Samuel Charles Norton. *Survey Cyprus,* 15 sh. 1885, *Anglo-Portuguese boundary E. Africa* 1893, *Communications in Natal,* 18 sh. 1897, *Atlas boundary Guiana/Venezuela* 1898 (with Ardagh)

Grant, T. M. *Gold Lakes Otago, Mount Cook:* both pub. in Wellington 1888

Grant, Ulysses Simpson (1822–1885). General, President of the U.S.A. *Field of Operations U.S.* (1865)

Grant, Vincent. *Deeping Fen, Lincs.,* ca. 1670 MS.

Grant-Wilson, J. S. *Chart Lochs Tay, Rannoch, Tummel & Earn* 1888

Grante [Grant], Col. James A. [Baron d'Iverque]. *France* (1745), *Routes of the Pretender Charles Edward in G.B.*, Jaillot (1747)

Grantham, Edward. Surveyor. *Mannour of Seasonscot, Glas.* 1704 MS, *Manor of Burston, Bucks.* 1720 MS.

Grantham, Capt. James. *Colony Natal* 1861, *Natal*, 4 sh. 1863

Grantham, John. *R. Shannon* 1830, *Plan Belfast Harbour* 1852

Grantham, Wm. *Mannor of Testwood, Hants.* 1755

Granton, G. R. *Midlothian* 1795

Grantzow, C. *Schlesien* 1855

Grapheus [Schryver, Scribonius], Cornelius (1482–1558). Flemish poet, b. Alost, fl. Antwerp. *Plan Antwerp* 1565 (with Virgile de Bologne)

Grapow, Capt. *Elbe River* 1868, *Schleswig Holstein*, 2 sh. 1869

Grass, J. See **Honter**, J.

Grasset, François (1722–1789). Printer and bookseller of Lausanne. *Switzerland* 1769

Grasset, F. *Plan Ports de Mombaze* 1851, *Côte de Syrie* 1855

Grasse, Comte de. *Mémoires du Comte de Grasse* 1782

Grasset de Saint-Sauveur, Jacques (1757–1810), b. Montreal. *Geog. works, Encycl. Voyages &c.*

Grassi, Raniero. Engraver of Pisa. *Città di Pisa* 1831

Grassmann, R. *School-Atlas*, Stettin 1853

Grassmüller, E. *Great Streams of World* 1834

Grassom, John. Surveyor, *Stirling*, 4 sh. 1817, *Town of Stirling* 1819

Grataroli, Guil. (1516–1568). Italian physician, b. Bergamo, d. Basle. *Itinerary*, Basle 1561

Graterix [Greatorex], Ralph. Surveyor, mathematician and instrument maker. Leake's *Plan London* 1667, *Ruines City of London* (1669?)

Gratia. Engraver. *Environs de Mascara* (1837)

Gratiani [Graziani], Paolo. Publisher in Rome; acquired some of the Lafreri-Duchetti plates. *Sicily* 1582, *Acquapendente* 1582

Gratiot, C. *Map Island of Michilimackinac* ca. 1843

Grattan, Edward (fl. 1824) Publisher, 51 Paternoster Row, London. *Guide through London* 1836, *Pedestrians' Companion* 1838

Gratton, J. Sterland. *Plan Ilkley* 1885

Graurock. *Klein-Asien u. Syrien* 1840

Grave, Hans. See **Graf**

Grave, Heinrich. *Plan v. Wien* 1871

Gravelot, Hubert François Bourguignon [d'Anville] (1699–1773). French engraver and designer of Paris, brother of J-B. B. d'Anville. In England from 1733, returned Paris 1754. *Chart St. Domingo* (1730?), drew charts for Pine's *Armada maps* 1739, designed cartouches for d'Anville 1751–63

Graves. Geologist. *Environs de Paris* 1865

Graves, R. W. *Flushing, Queen's County, N.Y.* (1871?)

Graves, Rev. Rosewell Hobart (1833–1912). S. Baptist Missionary in S. China, *Kwei Hong* 1866 MS.

Graves, Lieut. (later Capt.) Thomas, R.N. (d. 1856) Admiralty surveyor, entered Navy 1816. *Survey Mediterranean* 1832–50, *Plan of Troy* 1842, *Grecian Archipelago* 1849, *Cyprus* 1864

Graves, W. H. Assistant surveyor to D.L. Miller, *County atlases Mass. & N.Y.* 1895–6

Graves & Hardy. *Eureka pocket atlas Red River Valley* 1894

Gravesande, S. V. *Volkplantinge in Essequebo* 1749, *Rios Essequebe* 1750

Gravier, Giovanni. *Planta di Gibiltera* 1762

Gravier, Ivone [Yves] . Publisher and bookseller, 'sous la loge de Banchi, Genoa'. *Città di Genova* 1789, *Atlas Maritime* 1801

Gravius, N. T. *Zak-en Reisatlas* ca. 1760, *Kaart-Boekje* (1770), *Atlas van Duitsland* n.d.

Gravius, S. Dutch surveyor, latter 17th century. Maps for Schotanus à Sterringa's *Atlas Frisiae* 1664

Grawert, August Reinhold von (1746– 1821). Military cartographer of Königsberg

Gray, Alexander. Ensign, 40th Regt. *Fort Griswold (Conn.)* 1781 MS.

Gray, Andrew, Q.M.G. Dept. & Capt. Nova Scotia Fencibles. *Topo. map Guernsey, Sark, Herm & Jethou,* Faden 1816

Gray, Andrew B. U.S. surveyor and Civil Engineer. *R. Sabine* 1841, *L. Superior* 1845, drew Weller's *San Diego* 1849, *U.S./Mexican boundary* 1855

Gray, Capt. David. *Ice chart Arctic* 1881, *Arctic Ocean & Greenland* 1882

Gray, Hon. Frederick. *Pianta di Tripoli* 1842

Gray, George Carrington. *New Book of Roads* 1824

Gray, Henry. Surveyor. *Laurens District, South Carolina* 1820, for Mills 1825

Gray, John. *Part Chopwell Royalty* 1846 MS

Gray, John. *Marseilles, La Sable Co., Ill.,* 1868

Gray, John C. *Atlas of U.S. for use of Blind* (1837)

Gray, O. W. [& Son] (fl. 1882). Civil and topog. engineers, 10 N. Fifth St., Philadelphia; collaborated with H.F. Walling. *County maps Canada W. & N.Y.* 1859–64, *Atlas Windham & Tolland Co's., Conn.* 1869, *Atlas Mass.* 1871, *Penn., Ohio* 1872, *Atlas Md. & D.C.* 1873,

Atlas U.S. 1873, *Railroad map Mich.* 1874, *National Atlas* 1875, *Dutchess, Essex Co's., N.Y.* 1876

Gray, W. Surveyor in N. Carolina. *Boundary Survey N. & S. Carolina* 1735 MS

Gray, Warren. *Grant Co., Wisconsin,* 1868

Gray, William, Pennsylvania Regt. *MS maps* 1778

Gray & Son. Steel engravers, for Bell 1836

Gray. See also **Walling & Gray**

Grayson, John. Royal military surveyor and draughtsman. *Plan buildings Woolwich* ca. 1810 MS

Grayston. *Series of Slate Cloth Wall Maps* (1893)

Graziani, P. See **Gratiani**

Gream, Thomas. Land surveyor, Villiers St., Strand, London. *Environs Brighthelmstone* 1794, *Sussex,* 4 sh. 1795, (with Gardner), assisted Gardner & Cubitt's *Jersey* 1795, *Estate map Beckenham* 1809, *Brighton* 1817

Greathead. *Islands about Otaheite* 1797

Great Map of Muscovy. Made in the reign of Boris Godunov (1598–1605), later copies

Greatorex, Albert D. *Parish Sutton,* Surrey 1896

Greatorex, R. See **Graterix**

Greaves, C. Admiralty hydrographer. *Tor-Bay* 1836

Gredsted, F. *Kjøbenhavn med Forstaederne* 1875

Greebe, F. W. *Amsterdam* 1765

Greef, Goirt de. Dutch mapmaker 16th century; possibly author of *circular map Gooiland* 1524

Greeff, Prof. Dr. R. *Ilha de Sao Thome* 1884

Greeley, Aaron. District Surveyor. *Claims Michigan Territy* 3 sh. 1810, *Military ground at Detroit* 1809

Greeley, Carlson [& Co.]. *Atlas Hyde Park, Ill.* 1880, *Atlas town Lake, Ill.* 1883–92. *Atlas Chicago* 1884, 1891

Greeley, Horace & Co. Job Printers and Stereotypers, 29 Beekman St., New York. *Chicago Railway Route* (1854)

Green. See also **Longmans**

Green, Alex. Henry (1832–1896). *Geolog. map South Carr. Lincs.* (1896?)

Green, Francis (1731–1791). English cartographer

Green, F. A. *Route in Damaraland* 1858 MS

Green, James. Surveyor. *Nottingham Canal* 1791

Green, John [pseudonym for Braddock Mead] (d. 1757). Construction of maps and globes. *N. & S. America,* 6 sh., Jefferys 1753 (with *Remarks),* 2nd edition 1768

Green, Lowthian. *Tetrahedral map World* 1899

Green, Capt. M. *Course of the Shut ul Arab* 1858

Green, R. See **Greene**

Green, T. *Plan München* 1806

Green, Valentine (1739–1813). Draughtsman, engraver and publisher of London; Engraver to K. George III, b. Salford, Warwick; d. London. *Brampton Bryan Castle, Hereford.* 1778, *Plan Worcester* 1795

Green, (General Sir) William (1725–1811). Engineer in Newfoundland and Gibraltar. *Plans Bergen-op-Zoon* 1751, *Plan Louisbourg* 1755 MS.

Green, Wm. *Chart Cocos Is.* 1779, pub. Dalrymple

Green, Wm. (1760–1823). Surveyor, artist and publisher in Manchester, b. Ambleside. *Plan Manchester & Salford,* 9 sh. 1787–94

Green, Wm. *Picture of England illus.* 1804

Green, W. *City Melbourne* 1854

Green, Rev. Wm. Spotswood. *New Zealand* 1884, *Selkirk Range* 1888

Green & Wadham. *City of Adelaide* (1860?)

Green Globe. See **Quirini**

Greene. *Jeffersonville & Environs, Indiana* 1868

Greene, Francis Vinton (1850–1921). Soldier, historian and engineer. *Atlas Russian army in Turkey* 1877–8, pub. 1879

Greene, J. N. Civil Engineer. *Keweenaw Point* (1880)

Greene [Green], Robert. Publisher and mapseller, (a) "Near Ratcliffe Cross", London [1674], "at ye Rose & Crowne in ye middle of Budge Row" [1675]; collaborated with Morden. Hollar's *Plan London & Westminster* 1675, 1685, *Virginia & New Eng.* (1680?) (with Thornton, in the *Blathwayt Atlas), Wall map Canaan* 1682, *Scotland* 1686, *England with post roads* 1686, *Ireland,* Lea & Overton 1686, *Royal map England* ca. 1690

Greene. See **Sherrards, Brassington & Greene**

Greenhill, Henry (1646–1708). Governor Gold Coast; Commissioner of Navy 1691. *Plans forts W. coast Africa* 1680–2, *Chart coast Africa* 1682

Greenhow, Robert (1800–1854). Physician, scientist, linguist and historian; Translator to State Dept. *N.W. coast America* 1840, *W. coast N. America* 1844 MS

Greenleaf, Jeremiah (1791–1864). *New Universal Atlas,* Boston 1840 &c.

Greenleaf, Moses (1777–1834). American cartographer and lawyer. *Maine* 1815, 1829, *Atlas,* Portland (1829)

Greenough[e], George Bellas (1778–1855). F.R.S., 1st President Geological Soc. 1811, President Geog. Soc. 1839–40; d. Naples. *Geolog. map England,* 6 sh. 1819, 1839, *Wales* 1839, *Geolog. map British India* 1855

Greenshield. *Route on Niger* 1885–(97) (with Hamilton)

Greensted, Edward. *Estate plan Watering-bury* 1769 MS.

Greenwich Observatory. Founded 1675. John Flamsteed first Astronomer Royal

Greenwood Firm. Publishers in London with following imprints & addresses (map imprints & Directory entries differ here); (1) *Fowler, Greenwood & Co.* [no address, 1818–20], (2) (50 Leicester Sq. *[C. Greenwood & Co., 1819–20]*, (3) 70 Queen St., Cheapside [*C. Greenwood & & Co.; Greenwood, Pringle & Co.; G. Pringle Jr.,* 1820–7], (4) Piccadilly *[C. & J. Greenwood,* 1823–4], (5) 13 Regent St., Pall Mall [*Greenwood, Pringle & Co.; C. & J. Greenwood,* 1824–31], (6) Waterloo Place [*C. Greenwood and Co.* 1825–31], (7) *Sharp, Greenwood & Fowler* [no address, 1825–8], (8) 21 King St., Covent Garden [*C. Greenwood & Co.,* 1832], (9) Aldine Chambers, Paternoster Row [*C. Greenwood & Co.* 1833–4], (10) 3 Burleigh St., Strand [*C. Greenwood & Co.; Greenwood & Co.,* 1834]. Large-scale separate maps of the Eng. & Scottish counties 1817–31, *London* 1827, *Atlas counties of Eng.* 1834

Greenwood, Christopher (1786–1855). Surveyor of Wakefield, came to London 1818. Compiled and published large-scale maps of most of the counties of England 1817–31 [first one, *Yorks.,* surveyed 1815–17]; from 1821 with his brother John. See also **Greenwood Firm**

Greenwood, C. & H. *Monmouth* pub. Greenwood & Co. 1830

Greenwood, John (fl. 1821–40). Surveyor and publisher, became partner with brother Christopher in 1821

Greenwood, J. Engraver of Hull. *Lincoln-shire* 1836, *Holderness* 1840

Greenwood, Pringle & Co. See **Greenwood Firm**

Greer, M. William *Klondyke River* 1899

Greer, Wm. *Plan Surigao Bay* 1762, *pub. Dalrymple 1774*

Gregg, Josiah (1806–1850). Trader and writer. *Indian Territory, northern Texas & N. Mexico* 1844, for Morse & Breese

Grégoire, L. (b. 1818) French historian and geographer. *Planisphère Grégoire* 1875, *Atlas Univ., Paris* (1882)

Grégoire, R. P. *Plan Lyon,* Lyons 1740 (with Serancourt)

Gregoras, Nikephoros (ca. 1295–1360+). Byzantine scholar. Commentary on Ptolemy's *Geography:* perhaps author of *world map* in some MS versions of Ptolemy

Gregorii, Johann Gottfried (1685–1770). German pastor, historian and geographer, b. Toba, fl. 1713. Worked for Homann. *Geog. novissima* 1708–9

Gregory. Engraver of Liverpool. *Dublin* 1809

Gregory, Sir Augustus Charles (1819–1905). Australian explorer, b. Notts., England. Travels in W. Australia 1846–8, Queensland 1854–6, N.S.W. & S. Aust. 1857–8. Surveyor-General Q'land 1859–79, *N.W. Australia* 1856 MS, information used by Stanford 1860

Gregory, Charles C. Civil Engineer. *New Brunswick* (1867)

Gregory, Francis Thomas (1821–1888) Australian surveyor, b. Notts., England, d. Toowoomba. brother of A.C. Gregory, explored with him in W. Aust. 1846–8, & on own, Murchison R. 1857, N.W. Coast 1861. *N.W. Australia* 1862

Gregory, H. American Lithographer. *Niagara & Detroit Railway* 1858

Gregory, Henry, Junior. Publisher, 148 Leadenhall St., London. Reissued *Herbert's sea charts* 1777–80, reissued *New Directory of E. Indies* 1787

Gregory, I. Publisher of Leicester with W. Dawson of Prior's *Leics.* 1779

Gregory, Isaac. *Russian & Turkish armies* 1854, *War in Italy* 1859

Gregory, Rev. John. *Opuscula* 1649 (use of globes)

Gregory, John Bates. *N. Wales coalfield* 1879

Gregory, T. Surveyor. *Leasehold Estates Wellington St., Strand* 1826

Greipel, E. von. *Oesterreich* 1809

Greischer. Publisher. Juvigny's *Siege Budapest* 1686

Greischer [Gryscher] , Matthias (d. ca. 1712). Engraver in Vienna. Reiner's *Hungary* 1683

Greive, J. C., Junior. *Java* (1876?)

Gremion. Officer, Swiss Guards. *Bohemia and Moravia* 1779

Grenard, Joseph Fernand. *Asie Centrale* 1898, 1899

Grenet, l'Abbé (b. 1750). French geographer, Professor at Lisieux. *Atlas Portatif* (1779–82) (with maps by Bonne), Italian edition by Santini 1794. *Abrégé de Géog. ancienne et moderne* 1782

Grenfell, Rev. George. *MS maps Congo* 1880, *Surveys Congo River* 1884–9

Grenfell, John. Capt. under Clinton. *Hudson River Highlands* (1775) MS.

Grenier. Comissaire de la Rép. (Swiss). *Environs Geneva* on Martel's *Plan Geneva* 1743

Grenier, Jacques-Raymond, le Vicomte de Giron (1736–1803). French hydrographer. *E. coast Madagascar* 1768, *Foul Point* 1768, *Archipelago to north of Mauritius* 1776, *Chart Mahé Is.* 1776, *Chart currents S.W. Monsoon,* pub. Laurie & Whittle 1808

Grenier, L. Engraver, for Dépôt de la Marine (1861), for Dufour (1864)

Grent, Thomas. *World,* engraved 1625

Grenville [Greville, Greynvile] , Sir Richard (1541?–1591). Naval Commander, organized coastal defences in west of England 1566–8. *Dover with Pier* 1584 MS

Grepin, Lucien. *Douai* (1874?)

Gressien, Victor Amédée. Hydrographic Engineer. Charts 1822–3 in *Pilote de Brésil* 1826, charts for Dumont d'Urville 1833

Gressier, C. L. Hydrographical Engineer, *Baie de Todos os Santos* 1819–(23), *Atlant. Mérid. 1834, Partie Mérid. de la Mer du Nord* 1847

Greth, Julius. *Bodensee* (1862)

Gretter, P. *Boller Landtafel* 1602

Greubel, M. *Schulwandkarte Unterfranken* (1893?)

Greut[h]er, Matthäus (1556?–1638). Cartographer, astronomer, globemaker, artist, typecutter and engraver, b. Strassburg, worked in Lyons, Avignon, principally Rome, fl. 1615. *Roma Moderna* 1618, *Italy* (pre-1620), *Frascati* 1620, *Terrestrial & celestial globes,* Rome 1632, *Cel. globe,* Rome 1636

Greve, Wilhelm, Lithographer and printer of Berlin. *Plans Strassburg* 1874–8, *Plan Bremen* 1890

Greville, Sir R. See **Grenville**

Grew, Frederick. Engraver. *Warwickshire* (1870?)

Grey, Andre. *Guernsey,* Faden 1816

Grey, Capt. *Swan River to Shark Bay,* Arrowsmith 1840

Grey, B. Engraver. Makreth's *Plan Lancaster* 1778

Grey, Major R. *Limpopo R.* (1891)

Grey, Wm. *Survey Newcastle* 1649

Greydelin, G. *River at Lancaster* 1755 MS

Greynvile, Sir R. See **Grenville**

Grezel, E. Engraver, for Levasseur 1849

Gridley, Enoch G. Engraver, worked in Boston for Morse *American Gazetteer* (1797), in New York (1803–5), & Philadelphia (to 1818). Worked for Carey 1817, *Kentucky* 1818

Gridley, Lt.-Col. Richard (1710/11– 1796). Surveyor and artillery officer. *Plan*

Louisburg 1745, pub. in Boston 1746 & by Jefferys 1757

Griel. *Rade Nouvelle de Cherbourg* 1786

Grienfeld, Johann Disman Floriantschitsch de. Cartographer of Krain. *Carniola,* Laibach 1744

Grieninger, J. See **Grueninger**, G.

Grierson, Boulter. Publisher in Dublin, "Printer to the King", son of George Grierson, took over his business (1755). *English Pilot, Fourth Book* 1767

Grierson, George (d. 1753). Publisher, printer and bookseller "at the Kings Arms & 2 Bibles in Essex St., Dublin". "Kings Printer" (1727). Business passed to eldest son, who died 1755, & then to son Boulter. Moll's *World Described* 1733, Salmon's *Modern History* 1739, *English Pilot, Fourth Book* 1749, Marr's *Chart E. coast Scotland,* Moll's *Atlas Minor,* both n.d.

Grierson, George & John. Publishers in Dublin with Martin Keene. *New & Correct Irish atlas* (1820)

Grieve, Lieut. (later Comm.) Albany Moore, Indian Navy. *Survey India* 1845–58, *chart Gulf of Aden* 1847–8

Grieve, J. Zincographer, 33 Nicolas Lane, London. *Plans Embankment Fulham to Barking* 1841 MSS

Grieve, John. *Plan Pier Whithorn* 1792 MS

Griffen, Joseph. *Atlas to Elements of Mod. Geog.,* Glen's Falls, N.Y. 1833

Griffin. Publisher with J. Bumpus, 3 Skinner St., London. *London* 1831

Griffin, James. *Rotary planisphere* 1845

Griffin, P. Publisher, map & print seller "next to ye Globe Tavern Fleet Street", London. Roades' *map London,* 1748 edition, *County Middlesex,* engraved Oliver 1748

Griffin, W. Bookseller, in Catherine St., in the Strand, London. Speer's *West Indian Pilot* 1766

Griffin, Wm. M. *Primary geog. state New Jersey* 1884 (with C.E. Meleney)

Griffing, B. N. *County atlases, Illinois, Indiana, Kentucky, Ohio & Virginia* 1870–90

Griffing, Dixon & Co. *Atlas Davies Co., Ind.* 1888

Griffing, Gordon & Co. *Atlas Hancock & Jay Co's., Ind.* 1887

Griffith, Dennis. Surveyor. *Maryland* 1794, J. Vallance 1795 (large inset *plan of Washington)*

Griffith, E. & Son. *Plan Birkenhead* 1890, *District of Wallasey* 1890

Griffith, J. M. *Amator Co., Calif.* 1866

Griffith, Sir Richard John (1784–1878). Surveyor and geologist of Dublin. *Bog of Allen* 1810, *Leinster Coal district* 1814, *Roscommon* 1817 (with Edgeworth), *Geolog. map Ireland* 1838

Griffith, Wm. Junior. *Atlas Gallia Co., Ohio* 1874

Griffiths, Henry (d. pre-1849). Engraver and artist. *Scotland* 1846

Grigg, John. Publisher of Philadelphia. *New General Atlas* 1830

Grigg & Elliot. Publishers, No. 14 N. 4th St., Philadelphia, *New General Atlas* 1832, *Modern Geog.* 1848

Grigingerus. See **Criginger**

Grigny. Engraver for Malte-Brun 1812

Grigviger. See **Criginger**

Grijalvn, Juan de. Explorer. Early map *Brazil* MS

Grijp, D. See **Gryp**

Grilo, Miguel. *Atlas geog. de España* 1876

Grim, Franz. *Plan Wien* 1872

Grim, M. de, See **Grimm**

Grimaldi, Filippo, S. J. (1639–1712). Jesuit astronomer and cartographer in China. Succeeded Verbiest in Peking 1688. *Constellations* 1711

Grimaldi Casta, L. *Italia* 1893

Grimberghe, van. Belgian geographer, mid 17th century

Grĩmel [Grimmel] *Lake Ladoga, Ingria & Carelia* 1735, *Russian Karelia,* 7 sh. De l'Isle 1745

Grimes, Charles (1772–1858). Deputy Surveyor–Gen., N.S.W. 1791, Surveyor-Gen. 1802. Discovered Yarra River 1803. *Plan Settlements New South Wales,* Arrowsmith 1799

Grimm, J. *Siebenbürgen* 1855

Grimm, J. L. *Palaestina* 1830, *Atlas v. Asien* 1833–54, wall map of *World* 1854. Also worked with Berghaus

Grim[m], Maximilian von. *Plans Vienna* 1796–1814

Grimm, Simon. *Augsburg* 1679

Grimmel, I. See **Grĩmel**, J.

Grimminger, Georg Adolf (1802–1877). Lithographer in Stuttgart

Grimoard, Philippe Henri, Comte de (1753–1815). French General, b. Verdun. *Atlas to campaigns of Turenne* 1782

Grimstone, Edward. Translator of Acosta's *E. and W. Indies,* d'Avity's *World*

Grindlay, *Capt. Robert Melville.* India, Wyld 1840

Griner, S. See **Grynaeus**

Grinlinton, Lieut. *Plans Sevastopol* 1857, 1858

Gripenhjelm, Baron Carl (ca. 1655–1694). Poet and surveyor of Stockholm, Director General of Corps of Surveyors 1683. *Sweden* 1688 MS.

Grisco, M. See **Gruisco**, M.

Grisel[i]ini, Francesco (1717–1783). Cartographer and engraver. Re-drew mural maps in *Sala dello Studio* of the Ducal Palace, Venice (1762) *Città de Praga,* n.d.

Grist, C. *Winter estate Thanet* 1825 MS (with J. Grist)

Grist, George. Surveyor of Canterbury. *Gt. Shelford & Mudgrove Estates, Kent* 1824 (with J. Grist)

Plans Dunkirk, Kent 1811, *Gt. Barton etc., Canterbury* 1814, *Gt. Shelford & Mudgrove Estates* 1824 (with G. Grist), *Winter estate Thanet* 1825 MS (with C. Grist)

Grist, J. & Son. Surveyors of Canterbury. *Hardres Court, Kent* 1837

Grist, Jonathan. Land surveyor, Old Cavendish St., London. *Tillingham* 1799 MS, *Foulness Is. Levels* 1801

Grist, John. Surveyor of Canterbury.

Grive, Abbé de la. See **Delagrive**

Griwtonn, P. L. *St. Domingue,* Paris 1801

Groc, Alcide. *Dépt. de la Charente Inf.* 1876

Grodecki [Godreccius, Godrecio], Waclaw [Wenceslaus] (d. 1591). Polish cartographer and engraver. *Poland* ca. 1558 (lost), pub. Oporinus, Basle 1562, and used by Ortelius 1570

Grod[e]metz, J. D. Engineer. *Baye de Gibraltar* (1700?)

Groeger [Gröger], C. *Nord-friesischen Inseln* 1888

Groell [Gröll], Michal (ca. 1772–1806). Printer, publisher and bookseller in Warsaw

Groenewegen, Gerrit. *Atlas Zeehavens Batavia* 1805

Groenewegen, J. Publisher in the Strand, London. Mortier's *Nouveau Théâtre de la Grande Bretagne* 1728 (Vol. V, with Prévost)

Groenlund [Grönlund], Karl (1751–1815). Swedish land surveyor

Groenouw, D. A. Dutch surveyor, 17th century. *Central Holland* n.d.

Groesbeck, Gerard van (1508–1580). Bishop of Liège 1563; *Plan of Liège* used by Braun & Hogenberg 1572

Groffier, Valerien. *Missionary maps* 1883–6

Gröger, C. See **Groeger**

Grognard, Fr. *Mer Méditérranée* 1745

Grohmann, Paul. *Dolomit-Alpen* 1875

Grol, J. See **Groll**

Groll, Cornelius. *Noord Holland* (1853)

Groll [Grol], J. *Détroit de Sonde* 1840–1, pub. 1846

Gröll, M. See **Groell**

Grombchevsky, Col. *Alai Road* 1893

Gronden, A. van den. *Plan Edam* 1743, *Plan Monnikendam* 1743

Grondona, Nicolas. *Maps Provinces Argentina* (1865?), *Argentina* 1875

Gronen, Ed. Engraver and publisher. *Culmbach* (1853), *Schweiz* (1877?)

Grönlund, K. See **Groenlund**

Gronovius, Jacobus (1645–1716). Dutch classicist, b. Deventer, d. Leiden. *Geogr. Antiqua* 1697

Groom & Co., Foster. *Bohemian Campaign* 1866

Groom, S. Surveyor. *Barracks Brecon* 1825 MS.

Groot, Cornelius de. *Geolog. map Bilitong* 1887

Groot, J. de. *Handatlas* 1789, *Reisatlas door de Nederlanden* 1793 (with Warnars, maps by Tirion)

Groot. See **De Groot**

s'Grooten. See **Sgrooten**

Gropp, Capt. A *Transvaal* 1886

Gros, C. *Genealog., hist. & chron. atlas* 1807 (with Lavoisne), *Denmark* 1813

Grosdidier, F. E. *Europe Centrale* (1870)

Grose, Francis (1731? –1791). Antiquary and draughtsman. *Antiquities of Eng. & Wales* 1773–87 (52 maps), *Antiquities of Scotland* 1789–91

Grosier, l'Abbé Jean Baptiste Gabriel Alexandre (1743–1823), b. St. Omer, d. Paris. *Hist. gén. de la Chine* 1777–84

Grosjean, A. *Plan Melun* 1889

Grosley, Pierre-Jean [Jean Pierre] (1718–1785). b. Troyes. *Plan London* in his *Londres,* Lausanne 1770

Grosmann, Charles W. F. Donnelley's *Sectional Atlas Chicago* 1891

Gross. *Ost-Galizien* (1780?), *Plan Göppingen* 1783

Gross, Emanuel (1681–1742). Swiss mathematician and cartographer

Gross, I. M. *Kalamazoo Co., Michigan* 1861, *Lapeer Co., Mich.* 1863, *Allegan Co.* 1864

Gross, Rudolph (b. 1808). *Plan v. London* 1844, *Polytopischer Reise Atlas* (1845–52), *Geog. Schul-Atlas* 1847, *Neuer Schul-Atlas* 1862, *Neuester Atlas* (1868?), *Deutsches Reich* 1873

Grosselin-Delamarche. *Atlas de Géog.* 1869

Grossen, Johann Gottfried. *Orbis in Tabula* (1720?)

Grosset, C. S. C. *Watch Colony of Surinam* 1781

Grossmann. *Umgegend v. Ruppin* 1860

Grosvenor, H. C. *Silver Mines Arizona* (1860?)

Grosvenor, James. Pilot. *Mouth of the Thames,* Sayer & Bennett 1781, *Margate Road* 1794

Groth, J. L. See **Grotte**

Grothaus, Friedrich. *Plan Barmen* 1850

Grotte [Groth], Johan Larsson (d. 1686). Swedish surveyor

Grout, I. R. Surveyor. *Clinton River* 1849

Grove. Publisher with Canaan, Swinton & Ritchie. Wood's *Plan Lanark* 1825

Grove, Capt. C. F. *Charts of Norway* 1793–1817

Grove, Sir George (1820–1900). Civil engineer and musicologist. *Bible Atlas* 1868, *Atlas Ancient Geog.* 1872–4

Grove, John. *Plan Ipswich* 1761

Grover, Mrs. Bookseller in Pelican Court in Little Britain, London. *Carolina* 1682

Grubas, Giovanni. *Mare Mediterraneo* 1801, *Adriatic Sea* 1803

Grubb, Lieut. J. H. *Plan Mathurin Bay* 1818

Grube, A. *Plan Aschersleben* (1890?)

Grube, A. W. (1816–1884). German geographer

Grube, F. W. Publisher. *Düsseldorf* 1848

Gruchet, G. Publisher, Havre de Grâce Bougard's *Flambeau de la Mer* 1709

Gruchy & Co., H. G. Lithographers of Melbourne. Walch's *Tasmania* 1874

Gruendler [Gründler], August, Printer in Wroclaw. Helwig's *Silesia* 1627

Gruenewald [Grünewald], C. Engraver, for Schlieben 1828–30

Grueninger [Grüninger], Bonifacius. *Schwarzwaldes (Black Forest)* 1783, 1788

Grueninger [Grüninger, Grieninger], Johannes [Johann Reinhard] (1480–1528). Printer, publisher and wood engraver in Strassburg. Pub. wall maps by Waldseemüller: *Globus Mundi* 1509, *Europe* 1511, editions Ptolemy 1522, 1525, *Carta Marina* 1525

Gruenthal [Grünthal]. Swabian geographer, early 18th century

Gruenwald [Grünwald], Christoph. *Eisenbahn-Karte b. Deutschland* 1846

Gruisco, Matteo [Mateus] (fl. 1581). Portolan maker of Majorca

Grumbkow, von. *Mediterranean* 1799, *Black Sea* 1834, *Klein Asien* 1840

Grund, F. J. *Handbuch Wegweiser . . . Nord Amerika*, Stuttgart 1846

Grund, H. *Wilhelmshaven* 1872

Grundemann, Dr. Reinhold (1836–1924), *Missions-Weltkarte* (1865), *Allgemeiner Missions-Atlas* 1867–71, *Neuer Missions-Atlas* 1903

Gründler, A. See **Gruendler**

Grundt, Christoph Ludwig. Surveyor. (fl. ca. 1750–75)

Grundy, John, Senior (d. 1748). Mathematician and surveyor of Congeston, later Spalding, Lincs. *Atterton, Lordship* 1729, *Plan Spalding*, 1732, *R. Witham* 1743–4 (with John, Junior)

Grundy, John, Junior (fl. 1739). Surveyor and engineer of Spalding, Lincs. *R. Witham* 1743–4 (with John, Senior , & 1762, *East Fen* 1744, *Sutton sea-bank* 1787

Grunert. Engraver, for Stieler (1869)

Grünewald, C. See **Gruenewald**

Grüninger, B. & J. See **Grueninger**

Grüenthal. See **Gruenthal**

Grünwald, C. See **Gruenwald**

Gruss, Franz. "Ingénieur de la Cour de Vienne". *Plan Vienna*, 4 sh. 1770 (with Neussner)

Grynaeus [Griner, Gryner], Simon (1493–1541). Geographer and theologian, Professor at Basle, b. Veringen, d. Basle. Compiled with Huttich *Novus Orbis regionum*, Basle 1532 &c. (contains *World* map attributed to Münster (geog.) & Holbein (decoration)

Gryp [Grijp], Dirck. Engraver in Amsterdam. Speed's *France* 1626, for J. Hondius, Junior (pre-1629)

Gryscher, M. See **Greischer**

Guad, M. See **Quad**

Guadagnino. See **Vavassore**

Guadet, Joseph (1795–1881) *Atlas de l'histoire de France* 1833

Guarini, Cap. Giuseppe. *Terr. di Ravenna* 1770

Guarinoni, Luca. Publisher in Venice. *Augsburg* 1568

Guastaldo, J. See **Gastaldi**

Guattier, Capt., French Navy. *Chart Archipelago* 1825

Gubernatis, Enrico de. *Epiro* 1869–75, pub. 1879

Gucht, Michael van der (1660–1725). Engraver, b. Antwerp, worked and d.

London. Title-page to Moll's *System of Geog.* 1701

Gudenov, Peter Ivanovitsch. *N. Asia* (1668), *Siwerische Landchard* 1690

Gudme, A. C. (1779–1835) *Gegend v. Kiel* 1844

Gudmundi, Jonas. *N. Atlantic* 1570 MS

Gudmundsson [Gudmundus], Jon[as] (1574–1658). Icelandic chartmaker. *N. Regions* ca. 1650, pub. 1706

Guedeville, N. See **Gueudeville**

Guembel [Gümble], C. W. von. *Bayern* 1858, *geolog. map Bavaria* (1911)

Guenther [Günther], E. Lithographer. *Harz; Rügen; Thüringerwald,* all 1866

Guenther [Günther], Johann. *Hand-Atlas* 1822, *Volksatlas* 1842

Guenther [Günther], W. *Gera u. Umg.* (1895)

Guérand, J. See **Guérard**

Guérard, Augustin Frédéric Stanislas. *Charts Iceland* 1866–(7)

Guérard [Guérand], Jean. Chartmaker of Dieppe. *World maps* 1625, 1634, *Chart* 1631 MSS

Guérard, Nicolas (ca. 1648–1719). Artist and engraver of Paris. Titles & cartouches for de Fer 1690–1700, for Nolin, for De l'Isle 1700–09, for Placide (1714)

Guérard, Nicolas, Junior. Engraver. *Nouvel France* 1683, *Amér. Sept.* 1700, Frézier's *Plan Baye de Coquimbo* 1716, *Provincias de Quito* 1750

Guerber. *Services à vapeur dans l'Atlantique* 1867

Guerich, Dr. Georg. *Geolog. Skizze v. Afrika* 1887, *Schlesien* 1890

Guérin, Christophe (1758–1830/1). Artist and engraver of Strassburg

Guérin, Léon, *Atlas du Cosmos* 1867–8

Guérin, Victor. *Plan de Jérusalem* 1884

Guerin, Delamotte. *Environs de Lisbonne* 1821

Guérolt [Guervaldus], Guillaume. Born Caen. *Epitome de la Corographie d'Europe* 1553

Guerra, Giuseppe (fl. 1794). Engraver of Naples, for Rizzi-Zannoni 1785–92

Guerrera, Pedro. *Isle de Cuba* (1730)

Guerreros, Juan Antonio. *Rio de la Plata,* Buenos Ayres (1790) MS

Guersch [Gürsch], C. F. Engraver of Berlin. *Battle Plans.* 1782–1800, *Poland* 1791

Guervaldus, G. See **Guérolt**

Guesnet. *Mouillage de Scala-Nova* 1843–(4)

Guessefeld [Güssefeldt], Franz Ludwig (1744–1807). Worked for Homann Heirs; *Brandenbourg* 1773, *U.S.A.* 1784, *Moldau* 1785, *Amerika* 1796, *W. Indies,* Weimar 1800, for Gaspari 1804–11

Guether [Güther], F. *Haardt-Gebirge* (1889), Relief maps *Hohen Schwarzwald* (1891–3), *Umg. v. Karlsruhe* 1895

Guettard, Jean Etienne (1715–1786). French naturalist and geologist, b. Etampes, d. Paris. *Carte Minéralogique . . . des Terrains qui trav. la France et l'Angleterre,* Buache 1746 (first attempt to give geological observations on a large area). *Atlas et desc. Minéral. de la France 1778–80* (with Monnet; first geological atlas of France)

Gueudeville [Guedeville], Nicolas (ca. 1654– ca. 1721). Geographer and print-seller, b. Rouen, d. The Hague. *Atlas Historique* &c., Amsterdam, Chatelain 1705–20, Aa's *Nouveau théâtre du monde* 1713

Gueydon, Louis Henri de. *Plan Ensenada de Barragan* 1830–(3)

Gugliantini. *Città di Firenze* 1826

Guianotus, Fran. Engraver. *Town views* from *L'Ungheria Compendiata* 1686

Guicciardi, Raffaello. Dedicated map of *Acquapendente,* pub. Gratiani 1582

Guicciardini, Giovanni Baptista. Cartographer of Florence. *World map* (double-headed eagle shape), Antwerp 1549

Guicciardini, Luigi [Louis, Lodovico] (1523–1589). Cosmographer, b. Florence, d. Antwerp. *Descrittione de tutti i Paesi Bassi* 1567, editions to 1660

Guidalotto, Nicola di Mandario. Franciscan chartmaker. *Sea atlas* 1646 MS (4 maps)

Guido. Italian cartographer. *Geog. compendium* 1119 (with *TO map, Italy & circular world map*)

Guidotti, Giovanni Lorenzo. Engraver, b. Lucca, worked there & Genoa. *Genova* (1740?)

Guieysse, Capt. *African Is.* 1771, pub. Dalrymple

Guigoni, Maurizio. *Atlante geog. universale* 1875

Guijetus, L. See **Guyet**

Guilbaudière, Jouhan de la. French chartmaker. *Atlas of Pacific* ca. 1696 MS (35 maps)

Guilbert, P. E. Served under Dumont D'Urville. *Charts of Astrolabe, Torrent and Tasman Bays* 1833

Guillain. *Madagascar* 1843

Guillaume de Conches. (b. ca. 1080), b. Conches, near Evreux. Composed glosses on works of Boetius and Macrobius. *De philosophica* ca. 1130 (2 *world maps*)

Guillaume de Tripoli. *World,* 12th century

Guillaume-Maury. *Puy-de-Dôme* 1845. *Carte routière Puy-de-Dôme* 1868

Guillaumin, Amédée (b. 1826). French astronomer and geographer

Guillemard, Francis Henry Hill. *Philippine Is.* 1894

Guillemaro, Gilberto. *Entrada de la Bahia . . . de Panzacola* 1787

Guillemot, G. *Carte routière Puy-de-Dôme* 1868

Guillier, A. *Geolog. maps Sarthe* 1874–6

Guillo, H. Prescott. Surveyor. *Atlas Brookline, Mass.* 1893 (with Eliott &c.)

Guillon. *Mandchourie* 1894

Guillot, Joseph Frédéric. *Carte routière du Dépt. de Calvados* 1871–6

Guillotière [Guilloterius] , François de la, b. Berry. *France,* 9 sh., Le Clerc 1596– (1613), *Isle de France,* used by Ortelius 1598, Mercator 1606, Blaeu & Jansson

Guimpel, Friedrich (b. 1774, fl. 1830). Engraver and artist, b. Berlin. Worked for Sotzmann 1808

Guiter [Guittair] , Ch. A. (d. 1787). French engraver in Copenhagen. Worked for Wessel 1771, *Laaland* 1776, for Soc. des Sciences, Copenhagen from 1777, *Fyen* 1780, 1783

Guiterrez, D. See **Gutierrez**

Guitet, Mathurin (1664/5–1745). Dutch hydrographer and merchant. *Seekarte der Watt- und Aussenfahrt* (1708–10)

Guittair, Ch. A. See **Guiter**

Guittan, Philippe. *Bahia de tous les Saincts* 1647

Guldenstaedt, Jean Antoine (1745–1781). Russian naturalist and physician, b. Riga. *Map of the Caspian Sea* (1776), *Voyage en Russie et Caucase* 1787–91, *Black Sea* 1798

Guldenstein, Ant. Fr. von. *Plan Vienna* 1832

Guler von Weineck, Johann (1562–1637). Swiss geographer, historian and soldier, b. Davos, d. Coire. *Besch. von Rhaetia* 1616

Gulick, A. *Atlas des enfants* 1799

Gullan, E. Engraver. *Island of Capri* 1815

Gumbel, Karl W. V. (1823–1898). German geologist. Various works on geology of Bavaria

Gumilla, José, S. J. Spanish missionary.

Jesuit missions in New Granada (Orinoco) 1741

Gumoens, C. de. *Plan ville & baye de Cadiz* 1820

Gumpp, Johann Baptist (1651–1728). Engraver and engineer of Innsbruck, brother of J. M. Gumpp, engd. his *Tyrol* 1674

Gumpp, Johann Martin (1643–1729). Artist and architect of Innsbruck, brother of J. B. Gumpp. *Tyrol* 1674

Gumppenberg, Wilhelm. *Atlas Marianus* 1657–9

Gumprecht, Thaddäus Eduard (1801–1856). Geographer of Berlin

Gunby, Wm. *Chart harbour Wainfleet* 1809

Gundling, Jacob Paul, Fr. von (1673–1731). Historian and statesman, b. Nuremberg, d. Potsdam. *Brandenburgischer u. Pommerischer Atlas* 1714–24, *Desc. géog. de Magdebourg* 1730

Gunman [de Valle], Christopher (d. 1685). Master in Eng. Navy. *Algiers Bay* 1664 MS, *La Bouche de Valle, Guernsey* 1680

Gunn, Ortis B., of Wyandott, Kansas. *Kansas & Gold Mines* 1859

Gunnison, Lieut. (later Capt.) John Williams (1812–1853), U.S. Topog. Engineers. Surveyor and draughtsman, b. Goshen, Notts., killed Sevier River. *Surveys in Georgia* and *Northern Lake Regions* 1840–9, *Survey Utah* 1849–50, pub. 1862, *Surveys Great Salt Lake* 1852, *Mississippi-Pacific railway survey* 1855

Gunnlaugsson, Björn (1788–1876). Icelandic cartographer. *Iceland* 1844–(8), *Uppdráttr Islands* 1849

Gunst, P. *Plan Riga* 1879

Gunst, Pieter van. *Castrum Lipsiae (Leipzig)* (1720?)

Gunten, Nicholas. *Enfield Chase* (1658) MS

Gunter, Edmund. Wrote on Cross-staffe 1626

Günther. See **Guenther**

Guran, Alexander (1824–1888). General and cartographer of Vienna

Gurlach, P. See **Gerlach**

Gurmontius. See **Gourmont**

Gurney, R. Surveyor. *Buckingham property, Aylesbury* 1804 MS.

Gürsch, C. F. See **Guersch**

Güssefeld[t], F. L. See **Guessefeld[t]**

Gustawicz, Bronislaw (1852–1917). Cartographer

Gutbier, Ludwig von (fl. 1855). Lithographer and draughtsman of Glogau, for Flemming. *Gegend v. Hohnstein v. Schandau* 1857, maps in *Neuer Atlas* 1879

Gutch, George. *Plans Paddington* 1828–52

Gutenhag, Count of. See **Herberstein**

Guthe, Prof. Hermann. *Plan Hannover, Oldenburg &c.* 1868, *Palästina* 1890

Güther, F. See **Guether**

Guthrie, Wm. (1708–1770). Scottish geographer and historian, d. London. *New geog., hist. & comm. Grammar* (1770) &c., *Atlas to Guthrie's System of geog.* 1785, *Gen. atlas for Guthrie's geog.* 1820

Gutiérrez [Guiterrez, Gutiero], Diego, elder (1485–1554). Chart and instrument maker, b. Seville. Pilot Major in the Casa de la Contratación Seville 1547/9, Cosmographer to the King. *Atlantic* 1550 MS, *World* 1551 MS (based on Cabot), *America* (pre-1554)–1562, engraved Cock (first printed map to name California)

Gutiérrez, Diego, younger. Spanish cartographer, son and successor of above. *America* 1562 (accredited to his father, above) has been attributed to him.

Gutiérrez, Sancho. Cosmographer attached to Casa Seville 1553–73, son of Diego, elder. *America* 1562, accredited to Diego, elder, has also been attributed to Sancho

Gutierus. See **Gutierrez**

Gutschouen, Gérard à. *Louvain* in Beek's *composite atlas* ca. 1700

Guveto, Lucimo. Cartographer (fl. ca. 1625—50)

Guy. Master of Falmouth, *Cawely & Cagayanes Is.* 1764

Guy. Surveyor of Crewkerne. *Survey Longlond* 1824 MS.

Guy, Capt. J. M., R.N. Hydrographer. *Persian Gulf* 1820—30

Guy, M. S., 2nd Master, R.N. *Survey River Derwent, Tasmania* 1861—3

Guyen. Script engraver, for Dufour 1860

Guyet [Gajetius, Guijetus], Lécin [Licinio, Lézin] (1515—1580). Poet and geographer, b. Angers. *Anjou* 1573, another version engraved 1591 for Bouguereau but not included in atlas, used by Blaeu 1631 &c.

Guyot, A. *Zeekust der Banda Eilanden* 1871

Guyot, Arnaud. *Mural Atlas* 1856, *Wall Atlases* 1862—9, *Australia, New York* 1866, *Elementary Geog.* 1871

Guyot, Christoffle. Printer. Langenes' *Caert Thresoor,* Amsterdam 1602

Gwin, Roger. Land surveyor. *Gt. Henny & Bulmer, Essex* 1600 MS.

Gwyn, J. See **Gwynn**

Gwynn, George. *Ramsgate harbour* 1815

Gwynn [Gwyn, Gwynne], John (d. 1786). Architect. *Plan London after Great Fire* 1749 (based on Wren), *Proposed improvements Mansion House & London Bridge* 1766

Gwynn, Lieut. Walter. *Road Zanesville-Florence* (1828)

Gwynne, J. See **Gwynn**

Gy, André de. See **Chrysologue**

Gybertszoon, E. See **Gysbertszoon**

Gyer, Edward. *Estate map Erlington, Sussex* 1629

Gyger, H. C. See **Geiger**

Gÿger, H. J. See **Geiger**

Gylbert, H. See **Gilbert**

Gyldén, C. W. (1802—72). *Plan Helsingfors* 1838, *Finland* 1853

Gyoki Bosatsu (670—749). Japanese cartographer; type map of Japan named after him

Gyokuransai [Sadahide], Hashimoto (19th century) Japanese artist and mapmaker

Gysbertszoon [Gybertszoon], Evert. Chartmaker of Edam. *E. Indies* 1599, *North Sea* 1601 MSS.

H

H., A. *Waies from one Town to Another* 1619

H., C. *Plan der Stadt Neise* (1720)

H., C. *World map* (1514), pub. Jan Severszoon, Leiden

H., C. de. See **Hooghe**

H., E. *English Chron. (Table Roads)* 1618

H., E. D. *Stuttgart* (1740?)

H., E. V. D. *Plan Anvers* 1890

H., G. *Ichnog. Charleston,* engraved Toms 1739

H., H. *Stadt Pilsen* (1620?)

H., J. E. *Route to Gold Mines N. Platte* 1839

H., L. See **Homem, L.**

H., M. *Road map of Bamberg* (1545?)

H., Fr. M. A. *Italia Augustiniana* (1730?)

H., W. *Guide to Travellers* 1682

Haack. *Eisenbahn Trier–Mannheim* 1847

Haacma, S. A. (fl. 1678) collaborated with Schotanus on *wall map of Friesland*

Haan, D. B. de. See **Bierens de Haan**

Haan, Friedrich Gottlob (1771–1827). Cartographer and mechanic. *Erdkugel* 1821, *Planisphaeren* 1844

Haan, L. C. A. de. *Breda en Omstreken* 1853

Haan, Laurens Feykes. *Noord Ocean* Keulen *Davis Strait,* Keulen (1714)

Haardt, Vincent von (1843–1914). Austrian cartographer born Iglaa 1843. *Oesterreichish–Ungarischen Monarchie* 1879, *Nord-Polar-Karte* Vienne (1899), Alpine maps, *Süd Polen Karte* 1895, *Ethno. Asien* 6 sh. Vienna 1887, *Wandkarte Alpen* 1882

Haas, George. Engraver (1756–1818). *Partages de la Pologne* (1800?)

Haas, H. *Odenwald* 1808, *Rhine* 24 sh. 1815

Haas, Johann Baptist. *Schwarzenwald* 1783, *Niger River* 1788

Haas, Johann Heinrich (1758–1810). Military cartographer

Haas [Hase, Hasio, Haase], Johann Matthias (1684–1742). Prof. of Mathematics at Wittenberg, worked for Homann Heirs. *Africa* 1737, *Guinea* 1743, *Grund Staedten* 1745, *Atlas Historicus* 1750

Haas, Jonas (1720–1775). Engraver. *Türkisches Reich* 1747–8, *Iceland* 1772

Haas, Peter F. (1754–1804). Danish engraver. *Battle plans* 1782–1800

Haas, Wilhelm (1766–1838). Printer of Basel. *Sicily* 1777, *Basel u. Frickthal* (1795). Printed from movable types

Haase. Engraver for Stieler (1834)

Haast, Sir Johann Franz Julius von (1824–1887). German and British geologist. Surveyor Gen. Canterbury 1861–71. *Geolog. Sketch N.Z.* 1873, *Prov. of Canterbury & Westland, N.Z.* 1879

Haast, J. M. German scholar, b. Augsburg 1684, d. 1742. *Hist. Univ.* 1743, *Atlas Historicus* 1750

Habenicht, Hermann (1844–1917). German cartographer, compiled maps for Stieler and Perthes *Deutschland* 1866. *Africa* 1885, *Sea Atlas* 1894, *Wand Atlas* 1899

Habermel, Erasmus. *Armillary sphere* 1587

Habermel, Josua (16th century). Astronomer and instrument maker of Regensburg

Habersham, *Oregon* 1874

Hablitscheck. Engraver. *Pressburg* (1872?)

Hablitz, de. *Crimée* 3 sh. St. Petersburg 1790, *La Tauride* [1800]

Habrecht, Isaac (1544–1633). Mathematician and Doctor of Medicine of Strassburg. *Small world map on a clock face* 1589, *Pair globes* 1619, *Pair globes* 1625. Wrote pamphlet *Planiglobium*

Hachette, George. Editor, b. Paris 1828, d. 1892. *Tour du Monde, Géogr. de Reclus, Atlases, Dictionaries*

Hachette, Louis Christophe François (1800–1864). French publisher, b. Rethel in the Ardennes started publishing in Paris in 1826. *Ansart's Atlas Hist.* 1836, *Selves Geog.* 1843, *Cortambert's Géog. Mod.* 1847, *Meissas & Michelot, Petit Atlas Univ.* 1860

Hachette, L. (Libraire) et Cie, (1) Rue Pierre-Sarrazin No. 12 [1840], (2) Boulevard St. Germain 77 [1869]. Mapseller for Meissas & Michelot 1840 and 1869, for Charle 1868 etc.

Hack, J. B. *Three Brothers S. Australia* 1839, *Sources of the Para, S.A.* 1843

Hack[e], Capt. William (fl. 1680–1710). Seaman, pirate, and chartmaker. At the sign of Great Britain and Ireland near Wapping Stairs. Compiled various MS atlases of Western and Southern Seas, *MS atlas of* 183 *maps* (1686), *World Map* 1687, *Collection of Original Voyages* 1699. Pub. with Moll *Isthmus of Darien & Golden Islands* (1710)

Hacke, K. *Chemnitz – Wuerschnitzer Eisenbahn* (1859)

Hacke, W. See **Hack**

Hacker, Anton. *Railway & Telegraph Map Europe* 1857

Hackert, Jacob Philipp. *Map of the Sabine* 1780

Hackett, Thomas. Translated *Apian's Cosmographie* in 16th century and Thevet's *Antarcticke* 1568

Hackford, Solomon. *Chart Boston Deeps* (1847?)

Hackhausen, C. *Kreuznach und Umgebung* (1875)

Hackhausen, I. I. *Plans Cologne* 1837, *Mainz* 1837

Hacnel, de Cronenthal. *Guerre des Alliés* Berlin 1821

Hacq. J. M. (fl. 1820–50). Script Engraver, Rue de la Harpe, No. 35. *Plan Paris* 1825. Engraved for Duperrez 1827, Gouvion St. Cyr 1828, Dumont D'Urville 1833 and 1842–8, Ansart 1834, Dépôt de la Marine 1850

Hacquart, E. *Plan de Paris* 1827

Hacquet, Balthasar (1740–1815). French traveller, naturalist and cartographer. *Istria, Carniola* 1778

Hadaszezok, Johann. *Friedland* 1891

Hadden, G. E. Surveyor with J. O. Browne. *Chelmsford-Norwich Railway* 1846 MS.

Haddock, Ray. Contributed to *Michigan State Atlas* 1873

Haddon, R. Surveyor. *Property Princes Risborough, Bucks.* 1818 MS.

Haddon, W. T. *Chart Port Natal* 1835–8

Hadfield, J. Crown Surveyor, Georgetown. *British Guiana* 1838 and 1842

Hadji, Ahmas (16th century). Arab Cartographer. *Description of the Entire World* (1559–60)

Hadley, George (1685–1768). Scientist. *Theory of Trade Winds, Meteorological Diaries* 1729–30

Hadley, John (1682–1744). Math. and scientific mechanic. *Reflect Telescope* 1719–20, *Reflect Quadrant* 1734

Haeberlin, R. *Eisenbahnen Deutschlands* 1859

Haegel, J. J. *Wiener Welt-Industrie* (1873)

Haellstrom, Carl Peter (1774–1836). Finnish cartographer and geologist. *Maps* for Hermelin 1797–1807, *Kareliens* 1797, *Nyland* 1790, *Abo* 1799, *Finland* 1799, *Suède et Norwège* 1815

Haen, A. de. *Bergen op Zoom* 1739

Haenel, E. (1784–1843). German topographer

Haënke, Tadeo. *Santa Cruz de la Sierra (Bolivia)* ca. 1770 MS.

Haestens, Hendrick. Publisher of Leiden. *Plan Rotterdam* 1599, *Rhine & Heidelberg* 1608

Haesters, A. *Hand atlas* (1863)

Haeuw, Ernest. *Plan Dunkerque* 1862, 1878

Haevernick, H. *Geolog. Sketch Map S.E. Africa* Perthes 1884, *Zululand* 1885

Haeyen, Aelbert (fl. 1585–1613). Dutch cartographer. *Amstelredamsche Zee-caerten* 1585

Haffner, A. *Belfort et ses Environs* 1871

Haffner, J. Pr. W. (1836–1901). Chief, Swedish army topo. service. *Finmark Amt.* 1870

Hafner, Johann Christoph. Publisher. *Coblenz* (1720?), *Mannheim* (1730?)

Haftmann, K. F. V. *Würtemberg-Baden,* Stuttgart 1836

Hagar. *See* **Averill & Hagar**

Hagberg, Aug. *Filipstads Bergslag* 1873

Hagdorn, F. *Eifel-Gebietes* 1890

Hage, Holker. Civil Engineer. *City Harrisburg, Dauphin Co. Pennsylvania* (1862)

Hage, M. de la. See **Mengaud de la Hage**

Hagelgans, Johann Georg. *Atlas Historicus,* Franckfurt 1718 etc.

Hagelstam, Otto Julius (1784–1870). Military Cartographer Finland. *Sverige och Norrige* 1820–1

Hagen, Christian van der (1663–95). Engraver

Hagen, J. van. Engraver. *Bachiene Atlas* 1785

Hagen, Johannes Georg (b. 1847). German astronomer. Director observatory at Washington and at Vatican 1906

Hagenauer, Johann George (1746–1835). German architect and cartographer

Hagenmeyer, L. *Plan v. Heilbronn* 1877

Hagenow, Karl Friedrich von (1797–1865). Geologist and cartographer *Rügen* 1829, 1835, 1839. *Grund v. Greifswald* 1842

Hager, Johan Georg (1709–1777). Geographer and historian

Hagerup. *Charts coast Norway* 1835–48

Haggart, J. L. *Central America* 1856

Haggi, Ahmad. See **Hhâggy Ahmed**

Haghe. See **Day & Haghe**

Haghe, Louis. *Map Isère, High Alps* (1841)

Hagnauer, Robert. *Atlas Madison Co., Illinois* 1892 (with Dickson & Rinicken)

Hagner, Charles N. *Alleghany Co., Maryland* (1820?)

Hagrewe, J. L. *Plan Hanover* 1800

Hagstrom, Carl Peter (1743–1807) Swedish surveyor and cartographer

Hague, Arnold. American geologist. *White Pine Mining District* 1870–80, *Eureka District, Nevada* 1883, *Yellowstone North Park* 1904

Hahn, Georg Gottlieb (1756–1823). Military cartographer

Hahn, Gustav Leopold. *Environs de l'Escaut* 1785

Hahn, Jonas, Capt. Swedish Navy. Re-issued *Mansson Marine Atlas* 1748

Hahn, J. G. (1810–1869). Austrian traveller

Hahn, J. H. von. *Morava River* 1861

Hahn, Philip G. (1739–1790). German astronomer. *Terrestrial Globe* 1772–90

Hahn, Th. *Great Namaqualand & Damaraland* 4 sh., *Cape Town* 1879

Hahr, August (1802–85). *Maps of Sweden* 1845–80, *Sweden* 6 sh. 1875

Hahvich. *Trier* 1823

Haidinger, Karl (1756–1797). Austrian geologist and mineralogist

Haidinger, William Karl, Ritter von (1795–1871). Austrian, one of the founders of Geogr Soc. of Vienna. *Süd Amerika* 1854, *Oesterreich. Mon.* 1845–7

Haig, Capt. A. *Wabbs Harbour, V.D.L.* (1860)

Haigh. *Map of circuits of Wesleyan Methodists in England & Wales* 1824

Haigh, Capt. George. *Indian country west of Carolina* 1751 MS

Hailbrun, F. R. de. See **Renner de Hailbrun**

Hailer [Hailler] Martin (1640–78) Engraver of Frankfort. Cluver's *Geog.*

Hain, Jos. *Militär-Geographie.* Vienna 1848

Haines, D. American engraver and script writer. *Map seat of War* 1815, *Charles' U.S.* 1819, for Mitchell 1832–3

Haines, Comm. (later Capt.) S. B. Indian Navy. Admiralty Surveyor. *Arabian Sea* 1833–9, *Gulf Aden* 1853–4

Haines & Co. Publishers, 19 Rolls Buildings, Fetter Lane. *Plan London & Westminster* 1796, *England & Wales* 1797

Hains, P. C. *Country between Millikens Bend Louisiana & Jackson Mississippi* ca. 1863

Hainzel, Johann and Paul. Assisted Tycho Brahe in *Globe* construction 1568

Haipolt, John. Astronomer. *Celestial Globe* 1617

Haire, G. W. *Village of North Bend, Ohio* 1868

Haiward. See **Hayward**

Hajeck, I. F. (1836–50). Engraver. *Maps Saxony, Plan Dresden* 1839

Hajji, Ahmed of Tunis. See **Hhaggy Ahmed**

Hakewell, Jas. *Plan Buckingham Palace after John Nash* 1826

Hakluyt, Richard, of Oxford (ca. 1553–1616). British geographer. *Divers Voyages* 1582, *De Orbe novo P Martyr* Paris 1585, map by P. Galle; *Principall Navigations* 1589. Enlarged edition 1598–1600, maps. Assisted Molyneux with *Globe* 1592

Halaska, F. I. (1780–1847). Czech astronomer and mathematician

Halbfass, Wilhelm. *Arendsee* 1896, *Eifel Maare* 1879, *Pommersche Seen* 1901

Halcon, Capt. José M. *Ports Masingloc & Matabi (Luzon)* 1835–(73), *Mindoro* 1834–(98)

Haldane, Capt. John. *Coast India China* 1780, *S. Coast Hainan* 1776–7, *S. & B.* 1781, *Galloon Bay* 1781

Halde, J. B. See **Du Halde**

Halder, A. *Interlachen* (1867), (1896)

Halder, Albert A. *Zoutpanberg Goldfields* (1896)

Halder, Laurenz (1765–1821). Engraver

Haldingham. See **Richard of Haldingham**

Haldimand, Lieut. Frederick, Royal American Regt. *River Chaudière* 1761 MS, *Plan Magdelain &c.* 1766 MS

Hale, Edward E. *Map of Kansas,* Boston 1854

Hale, F. Chester surveyor, Assistant to D.L. Miller. *County Atlases* from 1895

Hale, George. *Lake Co., Illinois* 1861

Hale, Nathan. *New England States* 1826

Hale, Robert, with **Cole,** W. [Cole and Hale]. Estate Surveyors of Colchester 1813–1824, later 32 Oval Cottages, Hackney Road, London. *Copford* 1826 MS, *Hartfield Broad Oak* 1835 MS, *Rainham Levels* 1834–42 MSS

Halenbreck, L. *Elbe & Ems* 1883

Hales, John Groves (1785–1832). *Plan Boston* 1814, 1819, *Co. Essex, Mass.* 1825, *Northampton, Mass.,* 1831, *Town of Roxbury* 1832

Halfeld, Henrique Guilherme Fernando. *Atlas no de S. Francisco* 1852–4, pub. 1860, *Minas Geraes* 1860 MS

Halfpenny, H. E. *York Co., New Brunswick* 1878, *Atlas Oxford Co., Maine* 1880 (with J.W. Caldwell), *Atlas Somerset Co., Me.* 1883 with Colby

Halifax, John Holyrood of. See **Sacrobusco**

Hall. *Watering Places New England & Canada* 1869

Hall, Asaph (1829–1907). American astronomer

Hall, Comm. (later Capt.) Basil 1788–1844). *Linkin Is.* 1818, *Voyage to Corea & Loo Choo Is.* 1818

Hall, C. Engraver. *Plan Town & Harbour Boston* 1775. See **Costa,** J. de

Hall, Charles Francis. U.S. Navy. *Smith So. Kennedy & Robeson Channels* 1871–2, pub. 1875.

Hall, Daniel, Staff Comm. *Portland Harbour* 1873, *Shannon* 1877–9

Hall, David. Printer of Philadelphia. *Evans Map of Middle British Colonies* 1755

Hall, Elias. *Coalfield Lancs., York, Ches., & Derby* (1830)

Hall, F. *Plan Union Canal* 1818 (with H. Baird), *Shu benacadie Canal (Nova Scotia)* 1826

Hall, Henry. *S. Africa* Capetown 1857, *S.E. Africa* 1859, *E. Frontier Cape Colony* 1856

Hall, J. Engraver. Johnson's *Lanark* 1840

Hall, James. *Chart Northern Seas* 1606

Hall, Sir James (1761–1832). Scottish geologist. *USA* 1857, *Canada* 1864, *W. Mississippi River* 1870

Hall, James (1811–1898). State geologist of Albany. *Survey, State New York* 1826

Hall, John. *Survey Humber* 1821–2–(36)

Hall, Capt. John. *Straits of Sunda* 1788 (1794), *Malacca* (1798), *Sincapore* (1799), pub. by Laurie & Whittle

Hall, J. B. *Satiric Map World* Utrecht 1643

Hall, J. B. Publisher, 66 Cornhill, Boston. *N. America* 1849

Hall, J. H. American lithographer of Albany, N.Y. *Michigan Range 37–25,* 1846

Hall, Joseph L. (mid 19th century). Ordnance Surveyor

Hall, J. R. Lithographer of Albany, N.Y. *Lake Superior* 1846

Hall, Leventhorpe. Government Printer of Hobart. *W. Melbourne* 1875, *Tasmania* 1883, *Geolog. features Tasmania* 1888

Hall, L. W. *Lake Ontario to River Mississippi* 1800

Hall, Mary L. *Lessons in Geogr.* Boston 1872

Hall, Mathew, Junior. Estate surveyor. *Gt. Braxted* 1777 MS, *Purleigh* 1778 MS

Hall, Ralph. Engraver for English edition of Mercator, *Virginia* 1635

Hall, R. (19th century). Surveyor. *New College Estate Adstock, Bucks.*

Hall, Robert. *Plan Gloucester* 1780 (–2) with Pinnell, *Plan Boston, Lincs.* 1741–(2)

Hall, R. T. *S. African Republic* 1877

Hall, Sampson. *Rio Janeiro,* Laurie & Whittle 1794, *W. Coast Sumatra* ditto

Hall, Samuel Read (1795–1877). *School Geographies*

A NEW

GENERAL ATLAS,

WITH THE

DIVISIONS and BOUNDARIES

CAREFULLY COLOURED;

CONSTRUCTED ENTIRELY FROM NEW DRAWINGS,

and Engraved by

SIDNEY HALL.

London,

PRINTED FOR LONGMAN, REES, ORME, BROWN AND GREEN,

PATERNOSTER ROW,

1830.

Title-page to S. Hall's New General Atlas
(London, 1830)

Hall, Sidney. (fl. 1817–60). Engraver and publisher 14 Bury St. Bloomsbury, 18 Strand, London. Engraved for Arrowsmith 1817, Leake's *Egypt* 1818, *New Gen. Atlas* 1830, *New British Atlas* 1833, *Co. Atlas* 1842; for Black 1840 & *Atlas Australia* 1853; for Thomson 1831, *New Co. Atlas* 1847, *Trav. Co. Atlas* pub. Chapman & Hall

Hall, Thomas. Estate surveyor of Debenham. *Copford & Birch* 1739 MS

Hall, Thomas C. *Reese River in Nevada* 1864

Hall, William (d. 1614). Printer. Partner with John Beale. Speed 1611–12 and part 1614 edition

Hall, Lieut. William. *Upper Canada* 1799–(1800), *Building lots to landing on Niagara R.* 1798, *Harbour of York* (1800)

Hall, W. *Bullia Flats. Ganges* MS 1848

Hall, William & Co. Publisher of N.Y. *New Univ. Atlas* 1836

Hall & Elvan. *Meridian Hill* 1867

Hall & Mooney. Lithographers, Buffalo. *California & Oregon* 1849

Halle, S. Engraver. *Prussia* 1802

Hallett, H. S. *Parts Burma, Siam*. E. Weller 1886

Halleius, E. See **Halley**

Haller. *Allen Co., Ohio* 1871

Haller, Johann. *Map of Zürich* 1620

Haller, Johann Jacob. Son of J. Haller. Mathematician. *Terrestrial & Celestial Globes*

Haller, Leopold Franz. Pub. Brünn *Schlesien* 1810

Hallerstein, A. Jesuit Missionary in China. *Survey of China* 1770 used by Klaproth in his book

Hallerworden, Martin. Publisher *Alt & N. Preussen* 1684

Hallett, Holt S. *Burmah, Siam, & Shan States* 1886

Hallewell, Lt.-Col. *S.W. part Crimea* 1856

Halley [Halleius] Edmund (1656–1742). English Astronomer Royal and geographer. Succeeded Flamsteed as Astronomer Royal in 1720. Constructed first meteorological chart *Cat. Stellarum australian* 1679. Observed Comet 1680. Surveyed coasts & tides Bristol Channel 1682. *Meteorolog. Chart of Trade Winds* 1688. *Gen. Chart of Variations* 1699–71. *Astron. Cometicae Synopsis* 1705. First to use isogonic lines on map. First to give detailed description of Trade Winds. *Map of Surinam* 1733

Hallez D'Arros, Charles Henri Olivier. *Empire Français* 1870

Halliger, Johann I. *Grundriss von Putbus* 1834

Halligey, J. T. F. *Plan Freetown* 1881

Halliwell. Engraver. Title *Smith's general atlas* 1808

Halloran, Alfred Laurence, Master R.N., H.M.S. *Osprey, Port Moniganui, N.Z.* 1847, *Jeddo Bay, Niphon* 1849

Hallowel [late Carew], Sir Benjamin. *Description of Caiffe on the coast of Syria* 1799

Halloy, J. B. J. d'O d? See **Omalius D'Halloy**

Halls, Daniel. Estate surveyor. *Hatfield Peverel* (1723) MS

Hallström, C. P. See **Haellstrom**

Halma, François (1653–1722). Publisher and Bookseller of Amsterdam and Leeuwarden. *Univ. Tab. juxta Ptolemy* 1695, *Algemeene Wereld-Beschryving* 1705, *Geogr Sacra* 1703–4, *Maps Netherlands* 1718, *Tooneel der Vereen. Nederlanden* (1725)

Halma, H. Publisher. *Halley's World* 1700

Halma, Nicolas (1755–1828). Mathematician. French translation of *Ptolemaeum* 1828

Halpin, Patrick. Engraver of Dublin. Rocques *Plan Dublin* 1757, *map* for Harris' *Hist. Dublin* 1765, Richards & Scale's *Plan Waterford* 1764

Halpin, Wm. G. *United Jewish Cemetery (Cincinnati, Ohio)* 1862

Halse. (19th century). English globemaker

Halström, Jacob (1734–97). Swedish surveyor

Halsey, John. Surveyor. *Manor of Amersham* 17th century MS

Halsey, John. Surveyor. *Waterden, Norfolk* 1713–14 MS

Halstead, J. T. *Carver, Carver Co., Minnesota* 1857

Halsted, E. P. *Coast of Arracan. Akyab* 1842

Ham, Thomas **[& Co.]** Cartographer, engraver and publisher of Melbourne. *Map of Victoria,* 1851, 1864–5, *Queensland* 4 sh. Brisbane 1856, *Atlas of Queensland* 1865, *City of Brisbane* 1863

Hambden, John. Cartographer. *Map of Africa*

Hamdani, Abu Mohammed ul Husan (d. 945). Arab geographer. *Kitab Jazirat ul Arab (Geography of Arabian Peninsula)*

Hamdy, Mej. Achmet. *Carte Dabbe-Obeiyad* Cairo 1875, *Kordofan* 1876

Hamel, J. Corrected Hawkins' *City of Quebec* 1829

Hamel, P. W. Topographer and draughtsman. *Exped^n. S & S. E. Nevada* 1869

Hamer, Stefan. Publisher Nuremberg. *Siege Wolfenbüttel* 1542, *Plan of Wismar* 1547 used by Braun and Hogenberg, *Siege Brunswick* 1554

Hamersveldt, Everard [Evert] Symonsz van (1591–1653). Engraver Amsterdam. For Hondius and Jansson 1628–1658, for 1633 edition *Mercator-Hondius,* for Speed's *Prospect* 1627, for Blaeu 1631, for le Clerc 1619, for Jod. Hondius. II (ca. 1625)

Hamilton. *Route on Niger* 1885–(97) with Greenshield

Hamilton, A., Junior. Publisher *Country round Boston* New England 1776

Hamilton, Alex. Merchant East India Co., 1688–1723, d. 1732. *New account of E. Indies* 1727 (8 maps)

Hamilton, Archibald. Masters Mate R.N. *Nautical MS* 1750–66

Hamilton, C. C. *Leigh's New Pocket Road Book of Ireland* 1827, 1832, 1835

Hamilton [formerly Buchanan], Francis (1762–1829). Doctor and cartographer, worked in India

Hamilton, Sir. F. *Hinterindien* Perthes 1832, *British Army at Aladin* 1854, *Heights of Galata* 1856

Hamilton, Capt. H. G. *Sketch New England Beardy Plains N.S.W.* 1843

Hamilton, James. Surveyor. *Plan Chelsea* 1664

Hamilton, J. Successor to D. Lizars & succeeded by W. H. Lizars. *Scotland* 1831

Hamilton, J. W. *Middle Islands, N.Z.* 1849

Hamilton, N. E. S. A. *Nat. Gaz. of Great Britain* 1868

Hamilton, R. *Plan Edinburgh* 1827

Hamilton, S. Printer, Falcon Court, Fleet St. London *Atlas to Guthrie's System of Geog.* 1801

Hamilton, Theodore F. *People's pictorial atlas* 1873, *Hist. atlas* 1875 (with C. H. Jones)

Hamilton, W. I. *Routes Asia Minor* 1836

Hamilton, W. J. Geologist 1805–1867

Hamilton, Adams & Co. Publishers, Paternoster Row. *Canaan* 1828

Hamm, P. E. Engraver for Carey and Lea's *American Atlas* 1822, and for Butler 1831

Hammar, F. *Transvaal* Jeppe 1868, *Zululand* 6 sh. 1906

Hammel, L'. (19th century). German mathematician and cartographer

Hammer, Bernhardt Frantz (fl. 1746–50) Danish geographer

Hammer, C. F. Draughtsman. *Hohenlohe* 1806, *Franconie* 1813, *Mein* 1821, *Würtemberg* 1831, *Bayern* 1838

Hammer, Heinrich (fl. 1480—96). German cartographer

Hammer, R. *Grönland,* Gotha, Perthes 1883

Hammer, W. *Deutschland* 9 sh. Berlin 1881

Hammerdöfer, Carl. *Geogr. Hist.* Leipzig 1785—92

Hammershaimb, Wenceslaus Franciscus de. *Falstria* 1676 and 1682

Hammersveldt. See **Hamersveldt**

Hammett, Charles E. *Road map Rhode Island* 1849

Hammond, Capt. Sir Andrew Snape. *Chart Delaware Bay* 1779

Hammond, John. Engraver. Agas's *Cambridge* 1592

Hammond, Capt. John. *North Sea* 2 sh. (1720)

Hammond, J. T. American engraver. *Traveller's Map of Michigan* 1839

Hammond, John. *Birds eye view Cambridge* 1592, copied by Speed

Hammond Publishing Co. *Plat book Green & Jersey Counties, Illinois* 1893

Hamon, Pierre. Writing master to Charles IX of France. Executed in Blois 1569. *World* 1568, *France* 1568 MSS

Hampton, William. *Estate plan Manor of Allington* 1770 MS, *Essex plans* 1763 MS, *Plans Preston Andover* 1771 MS

Hampton, William. *Surveyor's Field Book, Co. Longford (Ireland)* 1813 MS

Hanchi, Fujii [O Daim] (late 17th century). Japanese surveyor. *Route maps* published at Edo 1690+

Hancock, Capt. Benoni. Marine surveyor. *Draught Cockle & St. Nicholas Gatts* 1751, pub. 1753

Hancock, E. J. Surveyor. *Thames to Collier Row Canal* 1818 MS

Hancock, Dr. John. *Chart interior British Guiana* 1811

Hancock, R. Engraver. Chantry's *Plan Bath* 1793, *Plan Worcester* (1780?)

Hancock, William. Surveyor. *St. Clair & Macomb* 1854, *Wayne Co. Michigan* 1854

Hancox, John. *Birmingham Canal Navigations* 1855

Hand, John. Publisher and mapseller 409 Oxford Street, near Soho Sq. Andrews' *Cities of Europe* 1772, *Costa's Plan Boston* 1775, Andrews & Drury *65 miles round London* 1777 and *30 miles round London* 1782

Handcock, W. Dept. Surv. Gen. of lands. *Estate Plan Clanwilliams, Tipperary* 1790 MS

Handley, B. *Mohammera (Persia)* 1853 MS

Handmann, Johan Jacob (1711—1786). Swiss goldsmith and globe maker

Handtke, Friedrich H. (1815—1879). German cartographer. *S. America* 1872, *European Turkey* (1876), *Africa* 1889, maps for Sohr (1844), *Crimea* 1854, 4 sh. 1855, *Australia* 1899, *Switzerland* 1895

Hane, Gerdt. *Plan of Ratzeburg* 1588 used by Braun & Hogenberg

Haneman, A. Engraver for Stieler 1857—85

Hanicle, Sr. Drew Guilbaudière's *Mer du Sud* ca. 1696

Hann, Julius (1839—1921). Austrian geographer and mineralogist. Berghaus *Physicalischer Atlas* 1887—92

Hanna, Andrew (d. 1812). *Plan Baltimore* 1801 (with Warner)

Hanna, Capt. James. *St. Patrick's Bay* 1786, Dalrymple 1787. *Plan Sea Otter Harbour* in Meare's *Voyages* 1790

Hannagen, Barth. Surveyor. *Estate Plan Queens Co.* 1792 MS

Hannak, Eman. 1841—1899. *Hist. Schul Atlas* Wien 1899

Hanno (fl. ca. 450 B.C.). Author of earliest surviving Periplus, *Gibraltar. W. Coast Africa*

Hannum, E. S. *Atlas Fairfield Co., Ohio* 1866

Hansard, Henry. Printer, House of Commons. *Arrowsmith N.Z.* 1860

Hansard, L. Engraver. *Section London Bridge* 1762, *Depth Thames* 1762

Hansard, Luke (1752–1828). English printer b. Norwich, worked compositor under John Hughes, printer to House of Commons, Partner 1774, sole owner 1800 in Great Turnstile, Lincolns Inn Fields

Hansard, Thomas Curson (1776–1833). Son of Luke Hansard, printed independently Parliamentary Debates

Hansard, James and Luke Graves (1777–1851). Younger sons continued printing Parliamentary Papers

Hansard, T. C. Printer, Peterborough Court, Fleet St. *Ostell's New Gen. Atlas* 1818

Hansen, C. F. V. *Plan Topo. de Formosa y del Chaco.* Buenos Aires 1889

Hansen, J. *Philippines* 1879–81, 1886; *Bassins du Haut-Nil* 11 sh. Paris 1893, *Course du Niger* 41 sh. 1898, *Djibouti* 1900

Hansen, Peter Andreas (1795–1874) Danish astronomer

Hanser, Anton. *Ober Bayern* Nürnberg (1860)

Hanser, G. *Post & Eisenbahn Reisekarte* Wien 1849, *Schul Atlas* Regensburg 1853

Hanson, Olaf (17th century). Swedish surveyor

Hanson, Thos. *Plan Birmingham* 1778 and 1781

Hanson, R. *China Tea Districts* (1875)

Hansteen, Christopher (1784–1873). Norwegian astronomer and explorer. Supervised trigonometrical & topograph. survey Norway 1837, *Magnetischer Atlas* Christiania 1819

Hansteen, Prof. Ridder M. M. *Kart over Christiania* Oslo 1844

Hanway, J. *Harwich* 1709 MS

Hanway, J. Junior. *Plans Stratford Place Property* 1732 MS, *Estate John Hope* 1732

Hanway, Jonas (1712–1786). English traveller and philanthropist. *Historical Account of British Trade over the Caspian Sea, with a Journal of Travels, etc.* 1753

Happel, Eberhard Werner (1647–1690). *World chart* in *Relationes Curiosae* 1675, early attempt to show ocean currents. Reissued in *Mundi Mirabilis tripatus* 1687

Haraeus, Franciscus, of Antwerp. *Globe* 1617, *Ancient World* for Horn, *Ortelius Paregon* used by Jansson (1653)

Harbaugh, John. *Canal around Harpers Ferry* (with N. King) (1803?)

Harboe, F. C. L. (1758–1811). Danish hydrographer. *Aalborhuus* 1791, *Aastrup* 1793, *Kort over Amter* 1793–1803 (4 sh.), *Schleswig Holstein* 1776–1806, *Atlas of Denmark* (with others) 1821

Harcourt, Robert. (1571–1631). Explorer in Guiana

Hard, C. de. Portuguese traveller. *Map of Brazil* for *Schöner* 1515

Hard, Wm. G. American surveyor. Assisted *Alleghany Co. Michigan* 1873, *Barry Co.* 1873

Hardcastle, Lieut. *Battles Mexico* 1847, *Gen. Worth's Operations Texas* 1850

Hardesty, Hiram H. *Hist. & geog. Encycl.* 1833

Hardesty, L. Q. *Illus. Hist. atlas Ottawa Co. Ohio* 1874

Hardie, James. *City of Philadelphia* 1794

Harding, C. L. Astronomer. *Atlas novus coelestis* 1822

Harding, F. G. Publisher. 24 Cornhill, London 1833

Harding, John. *Scotiae descrip. brevis* MS, 1450. *Chronicle of Scotland*

Harding, S. Publisher & Mapseller, Pavement in St. Martins Lane. Popple 1733, *Laws' Plan Cartagena* 1741 (with W. H. Toms)

Harding, W. D. *Land reclaimed Norfolk, Lincs., etc.* 1824

Harding Read, W. British Consul, St. Michael [Azores]. *Chart St Michael* 1806, *Heather* 1808

Harding, J. *Desert of Atacama (Chile)* 1877

Hardt, Lt. R. C. *British Front Prahsue (Ghana)* 1881

Hardy. *Celestial Globe* 1738

Hardy, Général. *Rec. Mil. Carte Rhin et Moselle Hundsruck* 6 sh. Paris (1798)

Hardy, Le Sieur Claude, Mareschal des logis du Roy. *Bretaigne,* Mercator-Hondius 1633

Hardy, G. Mapseller, St. Johns New Brunswick Wyld's *New Brunswick & Nova Scotia* 1845

Hardy, John. *New projection western hemisphere* 1776

Hardy, Lieut. Robert William Hale, R.N. *Sonora & Gulf of California* 1829

Hardy. See **Graves & Hardy**

Hare, Edward. Estate surveyor. *Warmington (Northants)* 1755 MS

Hare. See **Harrison, Sutton & Hare**

Hareieus, Franciscus. See **Haraeus**

Harenberg, Johann Christophe. Worked for Homann Heirs 1737–50. *Turkey* 1741, *Palestina* 1744

Harge, Jan [wife of Pieter Straat] *Diecheterlandt* 4 sh. 1735–36

Hargrave, I. L. *Land. z. Elbe u. Weser* Hanover 1812

Hargrave, J. *Chart Loch Eil* 1717

Hargreave, L. *Fly River, New Guinea* 1876 MS

Hargreaves, J. Surveyor of Burslem. *Stoke Trent* 1831 for Dawson, *Staffs. Pottery Burslem* 1832

Harigonio, Fra Bona, of Venice. *Figura totius orbis* 1509 MS

Haring, R. *Michigan* 1845

Hariot [Harriot] Thomas (1560–1621). British geographer and astronomer. *Virginia* De Bry 1590

Harkness, Olney. *Plan Philadelphia* and *Baltimore Railway* 1860

Harlee, Thos. Surveyor. *Horry District South Carolina* 1820 for Mills 1825, *Marion District* 1818, *Williamsburgh District* 1820

Harleian Mappemonde. Anonymous *World Map* 1536 (1542). Named after Edward Harley Earl of Oxford, the owner

Harley, D. S. & J. P. Surveyors. *Clinton Co., Mich.* 1864

Harliensis or Harlingensis, P. See **Feddes**

Harmer, T. Engraver. *Atlas to Cook* 1784, *Rennell's Hindooston* 1788, for Dalrymple 1803, *Script engraving for Admiralty* 1814

Harme, L. F. de. See **Deharme**

Harmon. *Hudson's Bay Countries* 1820

Harmsworth, A. *Franz Josef Land* 1897

Harnberger, Ch. *S.W. Ewe-Sprach-gebiet (Ghana)* Gotha 1867

Harners, Lieut. Henry D. Royal Engineers. *Maps for Railway Comm., Ireland* 1837

Harney, E. M. *Kent Co., Mich.* 1863

Harness, Lieut. Henry D. R. E. Compiled *statistical maps Railways Ireland Population, Flow of Traffic &c.* Dublin 1838, *Map of Ireland* 1837

Haro, Gonzalo López de. *Nootka Sound* 1789 MS, *W. Coast North America* 1790 MS

Harpe, Bernard de la. *Partie d'ouest de la Louisiane*

Harpe, J. F. de la. See **La Harpe**

Harper. *New Cheltenham Guide* (1827)

Harper, A. P. Surveyor. *Map New Zealand* 1893

Harper Bros. Publishers, 82 Cliff St., N.Y. *Texas* 1844

Harpff, Philip. *Saltzburg* 1630

Harradan, R. and Sons. *Plan Cambridge* 1810

Harraden, R. B. Designed cartouche for Baker's *Cambridge* 1821

Harrevelt, E. van. Publisher of Amsterdam. *Hist. Gén. des Voyages* Hague 1747–80, *Atlas Portatif* 1773

Harrewyn [Harrewijn]. Family of engravers of Brussels. Jacques (1660–1727) engraved for Fricx. Jacques-Gérard son of above. François (1700–64) engraved for Fricx. Jean Baptiste engraved for La Feuille 1685 and Aefferden 1709

Harrington, M. W. *Currents of Great Lakes,* Wash. 1894

Harriot, Thomas (1566–1621). Mathematician, and geographer. *Chart of the Virginia Coast* (1585)

Harris, Caleb. American surveyor. *State Rhode Is.* Phila. Carey 1795

Harris, Cyrus. *State of Kentucky,* Morse 1796

Harris, Harding. *State Rhode Is.* from surveys of Caleb Harris, 1795

Harris, Lieut. H.R., R.N. *Barbados,* Admiralty 1871

Harris, J. *Description & use of globe* 1751

Harris, John. Map & Printseller, No. 3 Sweeting Alley, Cornhill & No. 8 Broad St., London. *Plan London* 1779, Andrews' *30 miles round London* 1782, *Action Bay of Boukkier* 1798

Harris, John (fl. 1680–1740). Engraver and draughtsman, Bulls Head Court, Newgate St., London. Engraved for Collins 1693, Morden 1696, Gascoigne 1700–3, Prat 1705, Cole 1710, Senex 1719

Harris, John (fl. 1656–1746). English cartographer and engraver. *Plan London* 1700, *Navigation* 1705 (Moll maps), *Morden's Brit. Empire in America* (1695), *Cornwall* by Gascoyne 1700, Halley *W. & S. Oceans* 1701, *Holmes' Pennsylvania* 1687, *Mortier's Nouveau Théatre* 1715

Harris, John (ca. 1665–1719). Theologian, mathematician and surveyor

Harris, J. & Son. Publishers, St. Pauls Ch Y^d [London] . *America* 1822, *Africa,* J. Aspin 1823 and 1832

Harris, John. Publisher, Corner St. Paul's Churchyard, [successor to E. Newberry] . T. Smith's *Univ. Atlas* 1802

Harris, John & James. *County Maps* 1842

Harris, Joseph. Estate surveyor. *Wix* 1811 MS, *Little Bentley & Bromley* 1815 MS

Harris, J. S. *Plan Fort Jackson (USA)* 1852

Harris, I. *World,* John Garret (1676)

Harris, Moses. Surveyor. *Plan Chebucto Harbour* 1749 MS, *Plan Halifax, Ryland* 1749

Harris, T. (19th century) English globe maker

Harris, Walter. *Map of County Down Dublin* 1744

Harris, W. B. *N.W. Morocco* 1889–95

Harris, Capt. W. C. *Africa NE of Cape Colony* 1837

Harris, W. G. *S. Australia. London & Sydenham.* R. K. Burt 1862

Harrison, E. & Co. The West End Athletic Outfitters, 259 Oxford St., London. Published Lith. transfer of Carys' *Warwick* (1823)

Harrison, G. Engraver. *U.S.* 1816

Harrison, John. Printer, publisher, & engraver. No. 115 Newgate St., London. Rapin *Hist. Eng.* 1784–9, *Africa* 1787, *School Atlas* 1791, *America after D'Anville* 1791, *English Counties* 1791–2, *General & Co. Atlas* (1815)

Harrison, John (1693–1776). Clockmaker born Yorkshire, mathematician and instrument maker. Marine Chronometers 1765–71

Harrison, J. F. *Plan New York,* Dieppe 1867

Harrison, Joseph. *N. Boundary Rhode Island* 1750

Harrison, L. and **Leigh**, J. C. Printers 373 Strand. *Scripture Atlas* 1812

Harrison, P. *Plans Havana, Porto Bello & c.* 1740

Harrison, Lieut. R. *Communications Tsien-Tsin to Pekin* 1860 (with Wolseley)

Harrison, R. H. *Atlas Hamilton Co., Ohio* 1869, *Illus. Hist. Atlas Scotland* 1876

Harrison, Samuel. Engraver for Macpherson, *Atlas Ancient Geogr.* Phila. 1806, for Lucas 1812, Darby's *State of Louisiana* 1816

Harrison, Thomas (Crown Surveyor). *Isthmus of Panama* 1857, *Jamaica* compiled for George Henderson & Co. 1873, *Stamford* 1895

Harrison, W. Engraver of Philadelphia. *Map State Ohio* 1806, *US* 1819, *Plan Philadelphia* 1811

Harrison, W. Junior. Engraver for Stockdale 1789, for Carey's edition of Guthrie *Geogr.* 1795, *Travellers Direct.* 1802, Pinkerton's *Modern Geogr.* 1804

Harrison, Walter. *Plan London* 1777

Harrison, William (1534–1593) English topographer and antiquary. *Description of England* 1577

Harrison, William, Senior. Engraver, No. 12 Winchester St., Battle Bridge. Engraved for Dalrymple, for Rennell 1781–1788, for Purcell *East States* 1792, for Wilkinson 1794

Harrison, William, Junior. Engraver of Philadelphia for Carey 1795, for Wilkinson 1794, for Macpherson 1806

Harrison & Reid. Engravers. Shortland's *Track of the Alexander* 1788–(90?)

Harrison & Rushworth. Printers in Brooklyn. Darby's *U.S.* 1818

Harrison, Sutton & Hare. *Atlas Union Co., Ohio* 1877, *Atlas Marion Co., Ohio* 1878

Harrison & Warner. *County atlases Wisconsin, Iowa, Missouri* 1871–6

Harrison & Watkins. Publishers. of Corio Victoria, Fawkner's *Corio* 1841

Harrower, Henry Draper. *Pocket Atlas* (1887) with S. Mecutchen

Hart, Albert Bushnell (1854–). *Epoch maps illus. American hist.* 1891 etc.

Hart, Andrew (d. 1621). Publisher in Edinburgh. Pont's *Lothian* (1610) used by Hondius

Hart. See **Carey and Hart**

Hart, Charles. Lithographer, New York, 36 Vesey St. (flourished 1852–1855) *Maps Michigan* 1874–7

Hart, Henry. Architect and surveyor, 140 Pearl St., New York. *Kalamazoo* 1853

Hart, J. *Plan Bradford* (1870)

Hart, Joseph C. (d. 1885). American cartographer. *Modern Atlas* 1828 (5th edition)

Hart, Lt. R. C. *Survey Bussum Prah (Ghana)* 1881

Hart, Dr. W. *Sierra Leone* 1883

Harte, N. *Autriche*, 9 sh. Vienna (with W. Streit)

Hartert, E. *Route in Nigeria* 1886

Hartgers, Joost. Publisher, Amsterdam. *Oost Indische Voy.* 1648

Hartknoch. Christopher. *Preussen (Prussia)* 1684

Hartl, Heinrich (1840–1903?). Austrian military cartographer

Hartl, Martin. *North America* 1806. *Australien,* Wien 1815 (with Swoboda)

Hartl, Seb. Bookseller of Vienna, Singelstrasse. Doetsch's *map of Hungary*

Hartleben, Alois. *Atlas von Africa* 1886, *Volks atlas* 1888, *Hand atlas* 1889

Hartley. *Arizona from Official Documents* (1863)

Hartley, Ch. A. *Cartes du Danube* 1874

Hartley, David (1732–1813). British peace commissioner. *United States east of the Mississippi River* 1784

Hartlieb v. Wallthor, K., Freiherr (1786–1862). Austrian military cartographer

Hartman, J. W. Lithographer. *San Francisco* for Scholfield 1851

Hartmann, Carle Friederich Alexander of Weimar. *Californien* 1849

Hartmann, C. H. *Australien, Wolfenbüttel* 1824

Hartmann, George (1489–1569). Compass and globemaker. Early life in Italy. Settled in Nuremberg in 1518. Possibly author of the *Ambassador's Globe* ca. 1525. *Celestial Globe gores in 10 parts* 1535, earliest example of celestial engraved gores

Hartmann, Georgius (1489–1564). German instrument and globemaker of Nuremberg, first to establish magnetic variation of compass needle

Hartmann, Johannes Franz (1865–) German astronomer. Edited volume on astronomy in series "Die Kultur der Gegenwart" (1921), invented microphotometer (1899), and spectromparator (1904)

Hartmann, Johan Georg. Fried. (1796–1834). German military cartographer

Hartsinck, Jacob van. *Guiana* 1770

Hartt, Chr. F. American geologist b. 1840. Explored Amazon 1865–71

Hartwig, Eugen v. (1838–90). Cartographer from Glogau

Harvey, Aug. F. and Wm E. Civil engineers. *Nebraska gold mines* 1859

Harvey, Edward. *Vassam & Mayhem* 1777

Harvey, J. Publisher and bookseller, Sidmouth. *Country round Beer* 1837

Harvey, J. See **Darton & Harvey**

Harvey, J. S. Q M G 's office Madras. Drew Scott's *Peninsula India* 1855

Harvey, John. *List Cities & Towns in England* 1839

Harvey, Smuel. Estate surveyor. *Braxted* 1752 MS, *Birch* 1773 MS, *Mayland* 1775 MS

Harvey, T. Land surveyor of Ilford. *Noak Hill,* ca. 1850 MS

Harvie-Brown, J. A. *Naturalists' Map of Scotland* (1893)

Harwar, George. Cartographer. *St. Laurence,* Keulen (1682–6), *Draught of the River Canada,* Thornton 1689

Harwood, T. Etcher for Lewis, *Scotland* 1842

Harwood, W. *Plan Rome* 1865

Harzheim, Joseph. *Germany* Homann Heirs, 1760. *Map of Germany divided into Bishoprics* 1762

Has [Hase], J. M. See **Haas,** J. M.

Hase, F. Engraver for Stieler (1834)

Hase, J. M. See **Haas,** J. M.

Hase, W. Engraver for Stieler 1857

Hasebroek, J. *Paskaart . . . Oost Zee* Amsterdam 1740

Haselberg a Reichenau, Johann. *Map Turkish Campaigns* (1535)

Haselden, T. *Map of Known World* London, T. Page 1722

Hasell, Capt. John. *Mahe Is.* 1778

Hasemann, Henning. *Wolfenbüttel* 1628

Hasenfratz, K. See **Dasypodius**

Hashimoto. See **Gyokuransai**

Hasicoll, Wm. Printseller in Churchyard, Winchester. *Plan Winchester* 1756

Hasius. See **Haas,** J. M.

Haslop [Hasselop], Henry. Printer, Bookseller, London. Probably pub. *Mariners Mirrour* 1588

Hass, J. B. See **Haas,** Johann Baptist

Hassall, C. of Eastwood, Surveyor. Penn. *Road Milford–Gloucester,* Cary 1792 (with J. Williams)

Hassel, James. *Survey Carteret Grant* 1746

Hassel, S. G. H. (1770–1829) German statistician and geographer. Geographical manuals

Hassell, Capt. John. See **Hasell**

Hasselup, H. See **Haslop**

Hassenstein, Bruno (1839–1902). Cartographer of Berlin. *Ost Africa* 1864, *Deutsche Colonie Rio Grande do Sul* 1867, *Deutsches West Africa* 1884, *Polynesia* 1885, *Atlas von Japan* 1887, *Die Adamsbrück* 1891, *Buganda* 1891

Hassert, E. E. K. (1868–1947). *Polar Karte* 1891, *Geolog. Uebersichtkarte von Montenegro,* Gotha 1895, *Hydro. Karte von Montenegro* 1895

Hassler, Ferdinand Rudolph (1770–1843). Supt. U.S. Coastal Geodetic Survey

Hasted, Edward. *Hist. & topog. survey Kent* 1778–99 (maps of hundreds)

Hastings, E. J. *Stat. Atlas Comm. Geogr.* 1887

Haswell, Thomas. *Oyster rocks in Conway Bay* 1790

Hatch, F. H. *Geolog. Map S. Transvaal* Stamford 1897

Hatchard, J. Publisher, London. Green's *Picture of Eng.* 1804

Hatchard, T. *Diocese Ruperts Land,* London, T. Hatchard (1849)

Hatchett. Engraver for Walpole's *Eng. Traveller* 1794, for Faden 1798

Hatchett, J. Engraver. *Chart Indian Ocean,* Faden 1817

Hathon, A. E. *Detroit Plan* 1849

Hatsek, I. (1828–1902). Hungarian cartographer. *Oesterreich-Ungarn.* Gotha 1884, *Ethno. Karte Ungarischen Korona,* 1885

Hattinga, Anthony (1731–1788). Military engineer and cartographer. *Zélande* 5 sh. Amsterdam, Tirion 1753

Hattinga, David Willem (1730–1807). Engineer and cartographer

Hattinga, Willem Tiberius (1700–1764). Cartographer and physician. *Maps of Zeeland and Brabant,* issued later by Tirion

Hatton, T. Estate surveyor. *Fingrinhoe* ca. 1750 MS

Hatton, T. *British North Borneo* 1888

Hauber, Eberhard David (1695–1795) Worked for Homann Heirs, author book on cartography (Ulm 1724), *Württemberg* 1723

Haubold, G. Engraver for Kiepert (1869)

Haubold, Georg (1846–1881). Map lithographer

Haubold, O. Engraver for Meyer 1867

Haubold, O. Jr. Engraver for Kiepert (1869)

Hauchecorne, G. *Chemin de Fer d'Allemagne* Brussels, Van der Maelen 1862, *Chemins de Fer Europe* ditto, 1863

Hauducoeur, C. P. Engineer. *Head of Chesapeake Bay* 1799, *Plan Havre de Grace* 1799

Hauer, D. A. Engraver of Nuremberg for Homann Heirs, 1737–70 for Jäger 1789

Hauer, Franz. Ritter von (1822–1899) Austrian geologist *Geolog. Uebersichts-karte der Oesterreich-Ungarischen Monarchie* 12 sh. Vienna 1867–71. *Geolog. Karte Oest-Ungarn,* Vienna, Holder 1875

Hauer, Johann (1586–1660). Cartographer and engraver. *Silver globe* 1620

Hauke, Maurice. Lieut. Polish Artillery. *Plan Mantoue* 1800

Hault, Nickolaus David (17th century) Engraver and cartographer

Haultard, E. Engraver for Vaugondy 1735

Haupt, E. *Eulen-Gebirge Glogau* 1855

Hauser, S. T. Publisher N.Y. Lacy's *Montana* 1865

Hausermann, R. Engraver. *Possessions Anglaises et Françaises Golfe de Guinea* Paris 1883, *Tunisie et Algérie Orient* 1883

Hauslab, Fr. von (1798–1883). Austrian cartographer, one of the founders Military Geographical Institute in Vienna. Pioneer of uniform shading and contouring in maps.

Hauslab. Globemaker. *Wooden globe* 37 cm., 1514

Hausmann, J. F. L. (1782–1859). German geologist

Haussard, E. Engraver for Vaugondy 1749–57, for Bellin 1759, *Greece* 1752, *Luxemburg* 1753

Hausse, E. *Californie,* Paris 1850

Hausser, Wolfgang. *Steyr* 1584

Haussermann, (fl. 1877) Engraver

Haussknecht, Prof. C. *Routen im Orient . . . von H. Kiepert,* Berlin 1882

Haüz, Abbe René-Just. Assisted in founding metric system

Have, N. Ten, See **Ten-Have**

Havell, D. Engraver. Cartouche for Baker's *Cambridge* 1821

Haven. See **Delafield and Haven**
Haven, E. J. de. See **De Haven**

Haven, John. Publisher, No. 3 Broad St., N.Y. *Oregon, Texas & California* 1846

Haven, John. Publisher, 86 State St., Boston. Burr's *World* 1850

Havenga, W. J. *Atlas van Nederlandsch Oost Indië* 1885, *Eiland Sumatra* 1886, *Java en Madoera* 1888

Haver Droeze, F. J. *Bataklanden en eiland Nijas* door Hedemann en Voigt, Batavia 1890

Haviland, Dr. G. D. *Trusan River* MS 1885

Haviland, W. Engraver for Tanner 1833

Havilland, Th. de. Lieut 55th Regt. *Chusan* (1857) with Sargent

Haward, Nicholas. Surveyor. *Plan Estate of Searle in Dagenham* 1764 MS

Hawes, Comm. E. *Port Natal,* 1831

Hawes, J. H. Draughtsman to General Land Office 1865

Hawes, L. & Co. Bookseller, Camden's *Britannia* 1772

Hawkes, T. Surveyor. *Plan Glastonbury* 1844

Hawkes, W. Publisher, No. 59 Holborn Hill. Successor to T. Kitchin. *25 miles round N. York* 1776

Hawksworth, J. Surveyor. *Plan Road St. Mary Islington* 1735

Hawkesworth, John (ca. 1715–1773). *Voyages in South Seas* 1773

Hawkin, Lieut. *Parts of Vancouver* 1864 MS

Hawkins, Alfred. *Plan city of Quebec* 1845

Hawkins, Brigadier. Corrections to Mitchell's *N. America* 1776

Hawkins, James. Mapseller of Fenchurch St., London. Wimble's *Chart N. Caroline* 1738

Hawkins, Sir Richard (ca. 1562–1622). English seaman. *Observations in his Voiage into the South Sea* (1593)

Hawkins, Richard. *Estate survey of S. Weald* 1743 MS

Hawkins, T. B. of Brackley, Northants. Surveyor. *Yates Estate Turweston, Bucks.* 1842 MS

Hawkshaw, J. *Plan Holyhead New Harbour* 1873

Hawkworth, J. Engraver. *Plan Buckingham Palace by Hakewell after Noah* 1826

Hay, James. *Plan Musselburgh* 1824

Hay, Sir John Drummond. *N.W. Marocco* 1880

Hay, John Ogilvie. *Routes prop. for connecting China with India & Europe,* Stamford 1875

Hay, Thomas. Engraver. *Plan Amsterdam,* Andrews 1771

Hayashi Joho (d. 1646). *Route map Kawachi to Osaka,* engraved 1709. First map of Kawachi province

Hayashi Shihei (1738–1793) Japanese cartographer and engraver

Hayashi Yoshinaga of Kyoto. *Maps* 1618

Hayden, Ferdinand Vandeveer (1829–1887). American Geologist. *Geological Report of the Exploration of the Yellowstone and Missouri Rivers* 1859–60 (1869), 1877, *Geological and Geographical Atlas of Colorado*

Haydon, Wm. Engraver. Little Mayes Buildings. *Chart Delaware* 1776, Rennell's *Bengal* Dury, repub. 1794 Laurie & Whittle, Andrews' *30 miles round London* 1782

Haydon, W. Surveyor. *Plan San Miguel & Darien Harbour* 1853 MS

Haye, G. de la. See **Delahaye**

Hayes, E. L. Surveyor. Assisted *Lake Atlas Wrain Co. Ohio, Atlas upper Ohio River* 1877, *Sebastian Co. Arkansas* (1887), *County Atlases of Michigan* 1878–81

Hayes, Isaac Israel (1832–1881). American explorer. *Voyage North Pole* 1854–5, *Greenland* 1860–1, *Smith Sound & Kennedy Channel* 1865

Hayes, Capt. John. *Chart of Van Diemens Land* 1798

Hayes, L. des. See **Des Hayes**

Hayes, S. C. & Co. Publishers. *Cal., Texas, Mexico, Philadelphia* (1848)

Hayes, Walter (fl. 1651–92). English Instrument maker

Hayman, John, Lieut. 17th Infantry. *Plan Yorktown campaign* 1782 MS

Hayn, Ant. *Avinea S. Quentin e Peronne* 1537

Haynes, John. Surveyor of York. *Remains Roman Antiq. Yorkshire,* engraved Vertue 1744

Haynes, John. Surveyor and engraver at Michael Angelo's Head in Buckingham Court, Charing Cross, London. *Botanic Gardens Chelsea* 1753, *Park Lane to Half Moon St.* 1767

Haynes, M. B. *Atlas Renville Co., Minnesota* 1888

Haynes, Tilly. *New map of Boston* 1883

Hays, John C. Surveyor General, California. *Public surveys Calif.* 1855

Hays, Capt. John, See **Hayes**

Hays, Mr. Deputy Governor, African Company. *W. African Trading Posts,* London, E. Say 1745

Hayter, Capt. George. *Charts coast Burma* 1755–1778

Hayter, H. H. Geographer and statistician 1821–1895. Died Melbourne

Hayton, from Armenia. *Map* for *Relation of Tartar Empire* ca. 1520

Hayward, G. W. *Trans. Indus.* 1860 MS

Hayward, G. J. W. *E. Turkistan* 1870

Hayward [Haiward, E.] William. Titles all 1596 MS. *Draughts of Tower Liberties* 1597, reissued 1742, *Survey Fens* 1604 MS

Hayward, H. F. *Map of Canada West,* Toronto 1858

Hayward, John. *Charts in John Meares Voyages* 1790

Hayward, John. Publisher, Hartford, Conn. *Gazetteer of USA* 1853

Hayward, Lieut. Thomas. *Bay Selema (Ceram)* pub. 1800, *Timor to Ceram* 1801

Hayward, W. *Chart Kings Lynn Wisbech* 1591 MS

Hayward and **Howard.** *Atlas Brockton, Mass.* 1898

Hayward & Moore. Publishers, Paternoster Row. *Dower's New Gen. Atlas* 1838

Haywood, J. See **Heywood,** James

Haywood, James. Draughtsman. Some *maps* in *Harrison's School Atlas* 1787–91

Haywood, John. Draughtsman. *I. of White* 1781, *Warwick* 1788, *Maps* for Rapin and Tindal 1784–9, *Maps* for Harrison 1787–91

Hazama Shigetome (fl. 19th century). Japanese cartographer and astronomer

Hazen, William. *New Brunswick* 1791 MS

Hazen, W. B. *N. W. Alaska* by P. H. Ray 1854

Hazzan, Richard. *Northern Boundary Massachusetts* 1741

Hazzard, J. H. Engraver for Cowperthwait 1850

Hazzard, J. L. Cartographer and engraver. *Michigan* 1856, *Lake Superior* 1857, *Maps for Mitchell* 1859, *Maryland & Delaware* 1860

Head, H. N., Midshipman. *Maps for Parry's Journal* 1826

Headrick, J. *Map of Arran* 1807, *Angus* 1813

Heald, Henry. *Roads of Newcastle County Delaware* 1870

Heald, William. Estate surveyor. *Broomfield* 1834 MS

Heap, Capt. D. P. *Yellowstone* 1870 & 1871, *Montana Territory* 1872

Heap, George. *Plan City of Philadelphia* with N. Scull, engraved W. Faden 1777

Heap, Gwinn Harris, of Philadelphia. *Mississippi to California* 1854

Heaphy, Charles. *Auckland N.Z.* 1849, *North Island* 1861, *Conquered Territory North Island* 1864, *Military Settlements Waikato* 1868; all N.Z.

Heard, William. Land surveyor of Hitchin. *Netteswell (Essex)* 1848 MS

Hearding, W. H. Surveyor. *Michigan* 1854–67

Hearne, J. *Map of the Kingdom of Guatemala* 1824

Hearne, Samuel (1740–1792). English traveller and sailor. Explored N.W. America in service of Hudson Bay Co. Captured by La Pérouse 1782

Hearne, Thomas (1678–1735). Antiquary. Second Keeper Bodleian Library. Leland's *Itinerary* 1710–12, Leland's *Collectanea* 6 vols. 1715

Heart, Jonathan (1748–1791). First U.S. Regt. *MS maps Pennsylvania*

Heath, Lieut. G. P. Admiralty surveyor. *Tonga* 1852 *Vavu Group* 1855

Heath, J. *River Mersey* 1759, *Dover Castle-Sandgate* 1760

Heath, L. A. *Hong Kong* 1846–7

Heath, Robert. *A New and Correct*

Draught of the Islands of Scilly 1744, T. Hutchinson sculp. Published . . . by R. Manby and H. S. Cox, Booksellers on Ludgate Hill 1748/9

Heath, Thomas. Estate surveyor. *Woodford* 1757 MS

Heath, Thos. C. (1714–1765). Globe and instrument maker

Heath, (Mr.). "Next Fountain Tavern in the Strand." Price's *Bristol Channel* (1690)

Heath, W. Publisher. *Chart Is. St. Michael,* London, W. Heath 1808

Heathcote, J. Norman. Surveyor. *Is. St. Kilda* 1900

Heather, John. Estate surveyor. *Steeple* 1724 MS

Heather, William (fl. 1790–1812). Publisher at the sign of the Little Midshipman, later called Navigation Warehouse No. 157 Leadenhall Street. Joined by Williams 1768 [Heather & Williams] – up to 1800; succeeded by Norie 1812. Clements' *Thames* 1791, *China Seas* 1799, *Pilot American Ocean* 1795–1801, *Harbours Brit. Channel* 1801, *New Mediterranean Pilot* 1802, *Andaman & Nicobar Is.* 1803, *New North Sea Pilot* 1807, *Marine Atlas* 1808, *North Amer. Pilot* 1810, *Pilot London Spain* 1810, *Pilot Brazils* 1811, *World* 1812

Heather & Williams. See above

Heatherwick, Rev. A. *District east of Blantyre (Nyasaland)* 1877 MS

Heaviside, Capt. W. J. *Thal Chotiali Route Survey* Calcutta 1880

Heawood, Edward (1865–1949) with H. R. Mill. *Buttermere* 1893, *Bassenthwaite Lake* 1895, *Bathy Survey English Lakes* 1895

Heb[b], Andrew (fl. 1825–48). Bookseller and Printer at the Bell in St. Pauls Churchyard. Pub. Camden's *Britannia* 1637

Heber, Johan Jacob (d. 1725). Cartographer of Lindau

Heberlein, Adolf. *Geolog. survey Michigan* 1870

Heberstein. See **Herberstein**

Hebert, Edm. (1812–1890). French geologist. Works on geology, physical geography, and oceanography

Hebert, G. Geographer. *Berks, Cornwall, Lancs* for Ellis (1819)

Hebert, L. J. *British Guiana* 1887

Hebert, L. Worked for Q.M.G. Office. *Hyde Park* 1829, *Colony Cape of Good Hope* 1830, *Mauritius* 1830, *Antwerp* 1832, *Disputed Territories N. America* 1839

Hebert, L. Engraver [of Phila?]. Lewis Evans' *Middle Colonies* 1755

Hebert, L., Jr. Geographer. *S. Hemisphere* 1812. Drew for Pinkerton 1809–14, for Wallis's *British Atlas* (1814), *Cape of Good Hope* Arrowsmith 1834, *Gibraltar* 1825, *Ceylon* 1831

Hebert, L. J. Draughtsman. *Honduras* 1835, *S. Peru & Boliva* 1842

Hebner, John. 15 Great Maddox St. Hanover Sq. Phillips *Southern England* 1821

Hecataeus of Miletus (ca. 550–480 B.C.). Greek Historian, compiled first known geography B.C. 501, in the form of a Periplus or seaman's guide. Corrected Anaximander's map 518 B.C. *World map* engraved on copper 500 B.C.

Heck, G. *Atlas Geogr.* Paris 1842 chez Mallet, *Maps* for Stein 1852, Hinrich's *Atlas* 1853 and 1855

Heck, Johann Georg (1795–1828). Cartographer. *Atlas Géographique* 1830

Hecke, A. van der [Heckius] Engraver and cartographer. Worked at Frankenthal 1599–1608

Heckler, J. M. (18th century). Globemaker

Hector, J. *Province of Canterbury N.Z.* 1863, *Geolog. sketch map New Zealand*, Wellington 1869, also 1873, 1883

Hector, Dr. J. *Maps U.S. & Canada* 1857–8

Hector, T. *Ellipto-Polar Map World* 1872

Hedemann, E. W. *Bataklanden en eiland Nijas* 1890 (with F. J. Haven Droeze)

Hedemann, Friedrich v. (1797–1866). Military cartographer

Hedraeus, T. C. (17th century). Swedish surveyor

Heduus. See **Quintinus**

Heer, Daniel (18th century). Saxon engineer. *Stralsund* Homann 1715

Heer, Oswald (1809–1883) Swiss geographer

Heermans, Anna A. *Hieroglyphic geog. U.S.* 1875

Heger, F. J. with Jacob von Bors. *Postkarte Deutschland* 16 sh., Nuremberg, Homann Hered. 1764; English edition Faden 1785 on 1 sheet

Hegi, François. Aquatinter for Delkeskamp (1830)

Hegi, Hans Kaspar (1778–1856). Map lithographer

Heiden, C. See **Heyden**

Heidmann, C. *Europa geogr. vet.* 1658

Heijns, Zacharia. Mapseller. Mercator, *Magellan* 1613

Heiliger, I. F. W. *Länder zwischen Elbe u. Weser* 6 sh. Hanover 1812

Heilwig, M. See **Helweg**

Heim, Alb. and C. Schmidt. *Geolog. Karte der Schweiz* Berne 1894

Heimburger, A. *Maps* for Meyer 1830–40

Hein, Melchior Gottfried. Publisher *Köhlen Schlesische Chron.* 1710

Heine, P. B. Will. (b. 1827). American general, writer and geographer. *Central America, China, Japan &c.*

Heineken, C. A. *Territ. de Brême* 1798

Heinemann, E. *Plan town of Brunswick* 1584

Heinrich von Mainz. See **Henry of Mainz**

Heinrichschafen, Wilhelm. Publisher of Magdeburg. *Charte von Harz von I. H. Fritsch* 1833

Heinsius, P. See **Heyns**

Heintzelman, Major S. P. *Colorado River* 1851 MS

Heis, Ed. *Atlas coelestis novus* 1872, *Atlas coelestis eclipticus* 1878

Heitmann, Johan Hansson (1664–1740). Hydrographer. *Noord Zee.* Amstd. 1725

Heitor, de Coimbra (16th century), Cartographer. See **Coimbra**

Heitov, Antonio. *Planta de Macao,* Lisboa 1899

Hekel [Hekelius] , Johan Friedrich. Editor. Cluver's *Geogr.* 1686

Heksch, A. F. *Karte der Hohen Tatra* (1880)

Held, Carl. *Mappa di Brazil* 1878

Held, Leonz (1844–1925). Swiss topographer

Heldensfeld, Anton Meyer von. Quarter Master General. *West Gallizien* 12 sh. 1808

Heldring, Henry. Capt. 3rd Regt. of Waldeck, acting engineer Pensacola. *Plan Fort George Pensacola* MS, *Harbour of Pensacola* 1781, *Siege of Fort George* (1781)

Helfrecht, J. *Fichtelberg,* 1880

Heley, Richard. *Part Island of Axholme* 1596

Heliot, J. and **Boutigny**. *Rouen* 1817

Hell, Hommaine de. *Geological Section Crimea* 1855

Hell, M. *Pilote de L'isle de Corse* 1820–4, Paris, *Dépôt Gén de Marine* 1831

Hell, Maximilien. *Ungarn* 1790

Hell, Miksa (1720–1792). Cartographer and globe maker

Hell. *Charts of Majorca* 1823–9

Hellard, Am. Th. Norwegian geologist, b. Bergen 1846. *Structure du globe Terrestre* 1878

Helle, E. Engraver. *Ports maritimes de la France* 1871–98

Hellermann. *Frankfort on Oder* 2 sh. 1786–7 MS

Hellert, J. J. (19th century). French cartographer. *Atlas Ottoman Empire* 1843, *Nouvel Atlas physique* Paris 1843

Helloco, Le. *Cartes des Vents dans l'Océan Pacifique Mérid.* Paris, Dépôt de la Marine 1764

Hellström, Carl Peter (1774–1836). Finnish cartographer. *Finland* 1799

Hellwald, Fr. Heller v. (1842–1892). Austrian geographer

Helman, William. *Chart Is. & harbour Codgone* 1742

Helme, James. *Pautucket river . . . to Slocums Harbour* 1741 (with William Chandler)

Helmersen, Gr. von (1803–1885). Russian geologist, b. Livonia. One of founders of Imperial Russian Geogr. Society. *Empire of Russia,* St. Petersburg 1841

Helweg, [Helwig] Martin (1516–1574). German cartographer. *Maps of Silesia* 1561 and *Italy,* used by Ortelius & Blaeu

Hem, Laurens van der (1670–78). Map collector Amsterdam, 46 vols. supplementing *Blaeu Atlas,* now in National-Bibliothek Vienna also known as "Atlas of [Prince] Eugene" (q.v.)

Hemingway, Wm. Surveyor. *Georgestown District, South Carolina* 1820, for Mills 1825

Hemming, John. *Survey operations at Cape* 1851

Hemminga, Doco (d. 1555). Geographer. Professor at Louvain. *World map* (lost)

Hemmings, S. (18th century). British

cartographer. *Seat of War in America* 1789

Hen, Ioh. Engraver. *Prov. Nord in Sud Carolina*

Henchman, Daniel. Publisher of Philadelphia. *Louisbourg* 1758

Henderson, Capt. John. *Caithness, Sutherland, Wick & Thurso* 1812

Henderson, D. Lithographer. *Gold Fields Otago,* Otago 1864

Hendricks, Cornelius. *World* (1514) signed C. H. should possibly be attributed to him

Hendricks, Cornelius. *Map N.Y., N.J., & Penn.* 1616 MS

Hendrip, Hans. *Travels in Greenland* Godthab 1875

Hendry, James. Engineer. *Whitechapel to Thames Docks* 1842 MS

Hendschel, U. *Eisenbahn Atlas,* Frankfort 1846, *Eisenbahn Karte Cent. Europa,* Frankfort 1852, *Post Karte Deutschland,* Frankfort 1842

Henecy. Engraver for McCrea's *Monaghan* 1795, *Donegal* 1801

Heneman, J. C. See **Henneman**

Hengsen, G. Senior engraver for *Weiland Hand Atlas* 1848

Henion, J. W. Collaborated with C. M. Foote 1890–5 *Plat Books*

Henman. *Salcombe to Looe* 1791 (with Smith)

Henneberger [Hennebergen], Kaspar [Gaspar] of Elrich (1529–1600). German cartographer. *Livonia* 1555 (lost) reissued Blaeu 1613, *Prussia,* Ortelius 1576, used later by Hondius & Blaeu

Henneman, J. C. van. Cartographer and surveyor. *Plan Siege Brunsvic.* pub. Gosse & Pinet 1763, *Surinam* 36 sh. Hulst van Keulen & N. Vlier 1784

Hennepin, Fr. Louis de. Dutch missionary in N. America 1640–1701, *New France*

1683 in *Description de la Louisiane, N. America* 1698

Hennequin. Engraver "au Dépôt G.^{d} de la Guerre. Rue St. Landry No. 5 en la Cité". *St. Cloud* 1816. Engraved for Buchon 1825

Hennert. *Saxony & Bohemia* 20 sh. 1778

Hennessy, J. M. *Oro Bay (Pacific)* Brisbane 1892

Hennet, G. Surveyor. *Lancs.* 4 sh. Teesdale 1830

Hennicke, Johann Friedrich (1764–1848) Geographer of Gotha

Hennipin. See **Hennepin**

Henoy, Thomas Edward Campbell (1864–1948). Surveyor

Henrichs, Christoph. Artist, assisted *Praetorius Celestial globe* ca. 1576

Henrici, Alberti. Publisher of The Hague. Linschoten *Navigatio* 1599

Henricpetri, Sebastian. Publisher of Basle. Strabo 1571, Münster's *Cosmography* 1588

Henrion, Denis. (d. ca. 1640) French mathematician and cosmographer

Henry, Albert. Printer [Amstd?] *Thrésor de chartes* (1602)

Henry, Alexander, the elder (1739–1824). *Map of Lakes and Hudson's Bay* 1775 MS, *N.W. Parts of America* (1776)

Henry, Anson G. Surveyor General. *Public Surveys Washington Territory* 1863

Henry, B. Engraver for Dalrymple 1770–5. *Balambangan* 1770, *Bengal* 1772

Henry, John. *New & Accurate Map of Virginia.* Engraved Jefferys 1770

Henry of Mainz (d. 1153). *Oval World Map* ca. 1110

Henry the Navigator [Dom Henrique] (1394–1460)

Henry VIII encouraged navigation in England. Founded Trinity House 1514

Hensal, M. *E. Africa, Abyssinia* 1861–4

Hensel, Gottfried (18th century). Germany cartographer, worked for Homann Heirs

Henschall, J. Engraver and Printer. No. 1 Cloudesley Terrace, Islington. *Town Plans for S.D.U.K., Athens* 1832, *Copenhagen* 1837, *London* 1836, *Milan and Turin* 1835, *Birmingham* 1839

Henschall, W. Engraver for S.D.U.K. (1844–6)

Hensgen, C. Senior Engraver for Weiland (1848), for Kiepert (1869)

Henslow, John Stevens (1796–1861) English botanist and geologist. *Geology I. of Man* 1821

Hentschell, K. F. T. *Western Hemisphere* 1795

Henty, James. (d. 1882). Australian explorer

Hentzner, Paul. *Itin. Angliae* 1598, *Itin. Germaniae* 1612

Henwood, D. Engraver for Smith 1813

Hepburn, John. Surveyor. *Ellor District Aberdeen* 1848

Hepburn, L. Surveyor. *Lands of Bally-donagh Wicklow* 1762 MS

Hepe, Christoph. (17th century) Geographer and historian

Heracleitus (5th century B.C.). Greek philosopher

Herald Press. Lithographers. Birmingham. *Warwickshire* 1890

Herault. Engraver. Worked for Moithey and Philippe 1787, for Mentelle 1797–1801

Herawi. See **Al Heravi**

Herba, Giovanni de l'. *Itiner. delle poste . . . del mondo* Rome 1563

Herberstein, Sigismund, Freiherr von (1486–1566). Austrian statesman and writer. Imp. Ambassador to Moscow 1517. *Russia* engraved Hirschvogel 1546. *Plan of Moscow* 1547 used by Braun & Hogenberg *Rerum Moscoviticarum comment.* Vienna 1549 (includes map of Russia), Reissued Basle, Oporin 1556, Antwerp 1557, Basle 1567

Herbert, Charles E. *State of Sonora* 1885

Herbert, Edmond (1812–1890). French geologist

Herbert, I. *Australia* in Pinkerton's *Modern Atlas* 1813

Herbert, F. L. Lithographer of Saginaw, Michigan. *Bay County* 1869, *City Saginaw* 1870

Herbert, Col. John. *Plan fortification of Charlestown* 1721 MS, *Prov.S. Carolina* 1725 MS

Herbert, L. J. *British Guiana* 1842

Herbert, Louis. *Scotland* 1823

Herbert, S. Publisher and mapseller on London Bridge. Reissued Willdey's *30 miles round London* with Jefferys ca. 1755

Herbert, Sir Thomas (1606–1682). English explorer. *Travels Persia* 1634

Herbert, William (1718–1795) Surveyor, hydrographer, publisher and mapseller. At the Golden Globe on London Bridge and 27 Goulston Square near Whitechapel Bars. *Straits Malacca* 1752, *New Directory for East Indies* 1758, *W. Atlantic* 1757, Nicholson's *Ceylon* 1762, *Plan Canterbury* 1768, *Kent* 1769 (with Andrews & Drury)

Herbertson, A. J. (1865–1915). With J. G. Bartholomew *Atlas of Meteorology* 1899

Herbin, P. C. *France* (100 maps, with Chanlaire) 1802

Herbitz, Abbe. *Moldavia* 4 sh. Vienna 1811

Herbst, F. Draughtsman and engraver. *U.S. Mexican boundary* 1857

Herbst, Johann (1507–1568). Swiss publisher and bookseller

Herd, Capt. James, of Providence and Rosanna. *Chart N. Zealand* 1824 for Duperrey 1827, *Jokeehanger, N.Z.* Earliest chart of *Otago Harbour* 1826, *Wellington Harbour* 1826

Herdegren, Friedrich (1793–1843). German geographer

Herder, B. Lithographer and publisher of Freiburg. *France* 1833, Loweberg's *Hist. Geog. Atlas* 1839, *Kausler's Atlas des batailles* 1831–7, *S. Central Europe* 1844, Woerls *N. America* 1849

Herédia, M. G. de. See **Godinho de Herédia**

Hereford. *World Map* ca. 1280. By Richard of Haldingham

Hergesheimer, E. *Plan Fort Jackson* 1862, *Geolog. survey Yellowstone Lake* 1871, *Parts Idaho, Wyoming & Montana,* 1871

Hergt, C. *Palästina* Weimar 1869

Héricourt, R. de. *Voy. dans le pays d'Adel* (1840)

Hérigone, Pierre (17th century). French mathematician and geographer

Herisset. Engraver. *Plan Versailles,* Desnos, 1767

Herisson, Charles Claude François (1762–1840)

Hérisson. Géographe. *Océanie* 1837, *corrigée Frémin* 1854, *L'Europe* 4 sh. (1808), *France* 1818

Hérisson, Eustache. (b. 1759). Hydrographical engineer, geographer, pupil of M. Bonne. Maps for Grenet 1785, engraver for Bonne 1786. *Amérique,* Paris, Bassett 1806, revised 1807, ca. 1819 and 1823; *World* 1818, *Nouvel Atlas Portatif* 1811, *Atlas ou dictionnaire de géog. univ.* 1806, *Atlas Portatif* 1806, 1807, 1811

Herkenrath, A. *Bibel Atlas* 1881

Herklots, G. *Road thro. Bengal* 112 sh., Calcutta 1828

Herkner, J. (1802–1864). Polish geographer. *Atlas* Warsaw 1850, 1853, 1856, 1862, 1867

Herkt, Otto. *Samoa Inseln* **1899**

Herline & Hensel. Engravers of Michigan 1858–9

Herman[n], (Augustine) (1621–1685). Bohemian cartographer. *Virginia and Maryland* London, Aug. Herman & Thomas Withinbrook 1673; *Virginia* 1670

Herman, Enrique. *Pilot Manila to Acapulco* 1730 MS

Herman, Leonard David. Silesian cartographer, ca. 1812

Herman, Lichenstein. Printer. First printed edition of *Ptolemy* Vicenza 1475

Herman, A. See **Hermann**

Hermann, H. S. *Arktisches Amerika,* Berlin 1883

Hermannides [Hermannida], Rutger. *Britannia Magno* Amstd. Valckenier 1661 (31 town plans after Speed)

Hermelin, Samuel Gustav, Freiherr. (1744–1820). Swedish cartographer. *Geogr. Kartor ofver Swerige* Stockholm 1797–1812, 33 sh. (not completed) *Finland* 1797–8

Hermite, L'. See **L'Hermite**

Hermitte, d'. *Baie de Bombétok* 1732, *Baie d'Antongil (Madagascar)* 1733

Hermon, R. W. *Township Lount (Ottawa District).* Toronto 1878

Hermundt, Jakob (fl. 1697–1701). Engraver

Herndon, Lieut. W. L., U. S. N. *Maps rivers Brazil* Washington 1854

Hero of Alexander. Geometer

Herodotus of Halicarnassus. (486–406 B.C.), of Samos and Athens. Geographer and historian

Heroldt, Adam (fl. mid 17th century). Instrument maker. *Globe,* Rome 1649

Herport, Johan Anton (1702–1757). Swiss military geographer

Herrade, Abbess of Landsberg. *Zone map* ca. 1180

Herrenvelt, E. van. Publisher. Raynal's *Atlas portatif* 1773

Herrera, A. Alcedo y. See **Alcedo**

Herrera y Tordesillas, Antonio de. Historiographer of the Indies under Philip of Spain (1559–1625). *Hist. Gen.* 1601 (14 maps) and 1622

Herrera, Francisco Xavier de. *Plano de Manila* 1819

Herrera, Juan de. *Harbour of Cartagena, Zisipata Bay,* Jefferys 1768

Herrewyn. See **Harrewin**

Herrich, A. *Wandkarte des Weltverkehrs* 4 sh., Glogau 1894, *Schweiz* 1895, *Nordpol* 1896

Herrick, J. K. Publisher, Bell Allan & Co. *New Gen. Atlas* 1837

Herriset. Engraver for Desnos. *Atlas Gén.* 1767–90

Herrison, E. See **Herisson**

Herrle, Gustav (1843–1902). American hydrographer

Herrlein, Edward. Engraved on stone. Lawson's *Upper California* 1849

Herrliberger, David (1697–1777). Engraver *Topo. Eydgnossenschaft* 1754–8

Herrman [Bohemiensis] . See **Herman[n]**

Herschel, Caroline Lucretia (1750–1848). English astronomer, sister of Sir William Herschel

Herschel, Sir Frederick William (1738–1822). English astronomer, born in Hanover. *World* 2 sh. MS

Herschel, Sir John Frederick William Bart. (1792–1871). English astronomer. Son of Sir William Herschel. *Cape Observations* 1847, *Outlines of Astronomy* 1849, articles in Ency. Britannica

Herterich, C. H. Engraver for Berghaus. *Persian Gulf* 1832

Hertzberg. See **Herzberg**

Hertzel, C. L. *Halle* 1791

Hervagius, Jo. of Basle. Editor *Novus Orbis* of Grynaeus (1532)

Hervey. *New System of Geography* 1785

Herwarth, Johann Eberhard Ernst (1753–1838). German military cartographer

Herz, Ch. C., (d. 1879). French geographer

Herz, Johann Daniel (1693–1754).

Engineer and publisher of Augsburg. Müller's *Bohemia* 1720

Herzberg, Heinrich (1859–1931). Engraved for Meyer's *Univ. Atlas* 1830–40, for Kiepert's *Hand Atlas* 1878, and Stieler

Hesiod (8th century B.C.). Supported concept of circular world with outer ocean

Hesketh, James. Publisher, 13 Sweetings Alley, Royal Exchange. Andres' *30 miles round London* 1806

Hess. *Topographie* (1785)

Hess, Em. Lieut. in Royal American Regt. *Fort Littleton at Port Royal, South Carolina* (1758) MS

Hess, F. American surveyor. *Oakland Co. Mich.* 1872

Hess, Heinrich Freiherr v. (1788–1870). Field Marshal, military cartographer

Hess, Ludwig (fl. 1809–39). Engraver *Spain and Portugal* for Gaspari 1809, Stieler's *Nord-Amerika* 1815–24

Hess, Oskar (1863–1921). Engraver for Perthes of Gotha

Hesse, L. A. C. *Atlas Minimus Univ.* 1806

Hessel, Gerritsz. See **Gerritsz**

Hesselgren, Abraham (1671–1751). Swedish surveyor

Hesseln, Robert de. *France* 9 sh. Paris 1782

Hesselus Gerardus. See **Gerritsz,** Hessel

Heteln, Zacharias. Publisher. With Wiering, *Olearius' Voyages,* Hamburg 1696

Hettner, Dr. A. *Peru u. Bolivien* 1890, *Kordillere* 1892, Spamer's *Grosser Hand Atlas* 1896

Heubeldinck, Marten. *Oost-Ind. & West-Ind. Voyagien* 1617–19

Heuglin, T. von. *Ost Afrika* 1861, *Suakin & Berber* 1869, *Nilgebiete* 1869, *Ost Spitsbergen* 1871

Heumann, Georg Daniel (1691–1759). Nuremberg painter and engraver

Heunisch, A. J. V. (1786–1863). Military cartographer. *Baden* 1819, *Nassau* 1822,

Baden und Würtemberg 1835, *Baden* 1838, *Taschen Atlas* Carlsruhe 1843

Heurdt, A. van, *Meurs* published by de Wit

Hevelius [Hewelcke], Johannes (1611–1687). Polish astronomer. *Selenographia* 1647, *Prodromus Cometicus* 1665, *Cometographia* 1668

Hevenesi, Gabriel (1656–1715). Austrian cartographer. *Atlas of Hungary* 1689 (40 maps), first atlas of Hungary

Hevia, Deogracias. *Provincia de Salamanca* 1860

Hewelcke. See Hevelius

Hewes, Fletcher Willis. Scribner's *Statistical Atlas U.S.* (1883) with Henry Gannet. *Citizens Atlas of American Politics* 1789–1892

Hewett, Lieut. (later Capt.) William, R. N. Admiralty Charts *Lynn & Boston Deeps* 1828. *Cromer* 1828, *Flamborough Head* 1830, *North Sea* 1831–40

Hewett, William. *Chart Gariah Harbour* 1756

Hewitt, N. R. Engraver. Queen St. Bloomsbury (1812), 10 Broad Street Bloomsbury (1814), Grafton St. East Tottenham Court Road (1817), Buckingham Place, Fitzroy Sq. (1819). Macpherson's *Anc. Geogr.* 1806, Mathew & Leigh's *Scripture Atlas* 1812, Thomson's *Gen. Atlas* 1817, Johnson's *W. Lothian* 1820, Wyld's *Gen. Atlas* (1822), *Parish St. Giles* 1824 and 1828

Hewitt, Robert. Estate surveyor. *Beaumont Essex* 1688 MS

Hewitt. Engraver. 1 Buckingham Place, Fitzroy Square. For Thomson 1815

Hewson, Thos. R. *Townships Hodgins & Anderson (Canada),* Toronto 1878

Hexamer, Ernest, and Son. *Philadelphia* 1858–60 (with W. Locher), *Insurance maps Phila.* 1872–1914

Hexham, Henry (ca. 1585–ca. 1650). Soldier, scholar, surveryor to Earl of Arundel. Translator of English edition *Mercator* 1633–6

Hexham, John. Surveyor. *Fens* (1580), *Castle Rising Chase* 1588 MS

Heybrock, J. M. *Nautischer Hand Atlas* 1857

Heydanus, Carl. Cartographer. *Germany* Engraved by H. Cock ca. 1565. Also used by Ortelius 1570 sq.

Heyden [Heiden], Christian (1525–1576). Globe maker, scholar and Prof. of Mathematics, Nuremberg *Globe* 1560

Heyden, Gaspard, Van der (1496–1549) of Louvain. Flemish globemaker

Heyden, Gaspar, the younger (1530–80) of Malines. Globemaker

Heyden, Jacob (1573–1645). Printer, historian, artist and engraver of Strassburg and Brussels. Habracht's *Globes* 1619, Lazio's *Austria* 1620, Vopel's *Rhine* 1621, *Magdeburg* 1631

Heyden, Pieter (ca. 1530–1572). Engraver of Antwerp

Heydt, Johann Wolfgang (18th century). German surveyor and engraver. *Schauplatz von Africa u. Ost Indien* 1744

Heylin [Heylyn], Peter (1599–1662). *Geography* 1621, *Cosmography* Seile 1652 and 1657, *Chetwind* d. 1660 and 1670, *Passenger* 1682, *Microcosmos* 1621

Heyman, J. with **Roper,** J. *Plan Exeter,* Britton 1805

Heymann, Ignazio (1765–1815). Austrian geographer. *Italy* 1798, *Italie* 4 sh. 1806, *Postes d'Allemagne* 4 sh. 1813

Heyns [Heinsius], Pieter (1537–1598). Editor, publisher, geographer and engraver. Friend of Ortelius, assisted P. Galle with *miniature edition of atlas of Ortelius* 1577, 1579, 1583 &c.

Heyns, Zacharias (1566–1638). Printer, bookseller and engraver of Amsterdam "in de Hooft Deughden [enseigne des Trois Vertus"]. Son of Pieter. *Miroir du Monde* 1598 and 1599, *Nederlanden Landspieghel* 1599

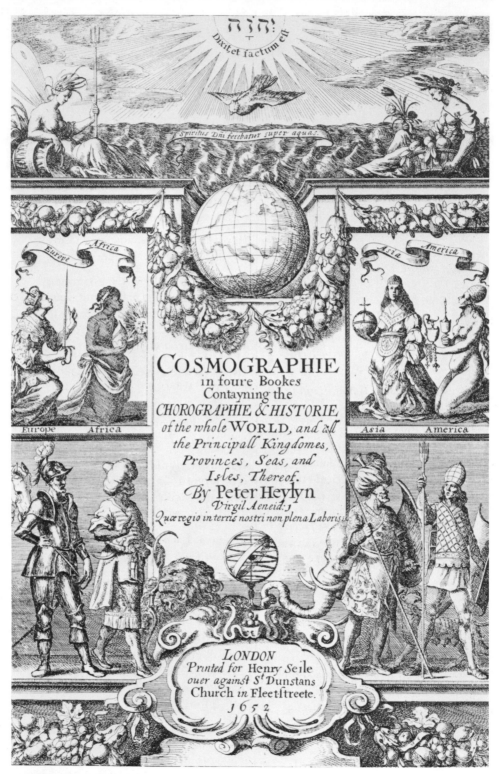

Engraved title to P. Heylin's Cosmographie (published London, 1652)

Heyse, D. (1818—after 1846). Lithographer and cartographer for the Dutch Topographical Service

Heyse, P. F. (1816—79). Map lithographer of The Hague

Heyteman, Capt. John, of Christiania. *North Sea,* Keulen (1740)

Heywood, Capt. R.N. *Chart Ceylon* 1822

Heywood, Abel and Son. Publishers, 56—58 Oldham St. Manchester. *Penny Guide Books* (1870)

Heywood, C. *Chart Merjee River* 1803

Heywood, [Haywood] James. No. 3 St. Martin's Churchyard. Draughtsman for Harrison's *County Maps* 1787—91

Heywood, John. Printer and Publisher, 141 and 143 Deansgate, Manchester and 3 Brazenose Street, Manchester. *Trav. Atlas England & Wales* (1860), *County Atlas Wales* 1879, *Co. Atlas* 1882

Heywood, Capt. Peter, Bombay Marine. H. M. S. *Dedaigneuse. Charts of Ceylon, India, & East Indies & China* 1798—1806

Hezeta, Bruno de. *W. coast of America* 1775

Hhâggy Ahmed of Tunis. *Arabic World Map,* cordiform woodcut based on O. Finé, engraved Venice 1559—60

Hibbart, W. Engraver. *Plan Bath* (1780), *Five miles round Bath* 1787

Hibbert, Lieut. *Prov. of La Rioja* 1839

Hibbert, Dr. Samuel. *Shetland* 1820

Hickey, Benjamin. Bookseller, Nicolas Street, Bristol. Rocque's *Plan of Bristol* 1743 and 1750

Hicklin, John. *Excursions N. Wales* 1847

Hickling, C. Publisher. Jenks' *Bible Atlas,* Boston 1847

Hickmann, Anton Leo (1834—1906). Czech statistician and geographer. *Geog. Stat. Taschen Atlas,* Vienna 1895, 1897

Hicks, F. Tapestry weaver. Five *Tapestry maps* (ca. 1570) now in York Museum and Bodleian

Hicks, Henry. Doctor and geologist, b. St. Davids, d. 1899

Hicks, Capt. John. *Bay of Bengal* 1794

Hicks, R. J. *Ngamiland* 1892 MS

Hieronymus [St. Jerome] (330—419). *Maps of Eastern Europe Asia & Palestine* in MSS of his works

Hiemcke, A. G. *Plan Paramaribo* 1850

Hietzinger, Karl Bernard Freiherr v. (1786—1835). Military cartographer

Higden [Hygden, Hugeden] Ranulph (ca. 1299—1364). Benedictine monk of Chester, Chronicler, 2 *World Maps* ca. 1350 used in Caxton's *Polychronicon* 1480

Higgie, George. Higgie's *map of the Isle of Bute* (1886)

Higgins. *Atlas Indiana* 1870—1 (with Asher and Adams)

Higgins, J. Printer, St. Michael's Alley. Cornhill. *Moorgate to London Bridge* Ca. 1831

Higgins, R. T. Assisted R. H. Hamson. *Atlas Hamilton Co. Ohio*

Higgins, S. W. Draughtsman, U.S. Depy. Surveyor. *Michigan &c.* 1846, *Official Map San Francisco* 1849

Higgins, Belden & Co. [Belden Higgins & Co.]. *Hist. Atlas Elkhart Co., Inc.* 1874, *St. Joseph Co.* 1875, *Kent Co., Mich.* 1876

Highmore, John. *Trav. Cities Europe* 1782

Hilacomilus, pseud. for **Waldseemüller** q.v.

Hilarides, Johannes (1649—1726). Painter, engraver and publisher. Designed title pages for Schotanus à Sterringa 1717, *Friesland* 1717

Hilary, St. Bishop Poictiers 353—68. Wrote of itineraries and cosmography

Hilbert, J. Engineer and printer of Hull. Scott's *Chart of the Humber* 1734

Hilbrants, G. (named Hopper). *Map of Groningen* MS 1578—95

Hildebrand, Henrik Robert Teodor Emil (1848–1919). *Hist. Atlas Germany and Sweden* (1883)

Hildebrandsson, H. H. (b. 1838). Swedish geographer and meteorologist

Hilder, Fr. Fred. Soldier, geographer, and ethnologist, b. Hastings 1836, d. Washington 1901

Hildt. See Macgowan and Hildt

Hilgard, Julius Erasmus (1825–1891). American hydrographer. Supt. Coast and Geodetic survey. *Agricultural Map Colville Region Wash.* 1883, *Yakima Region* 1883

Hilhouse, W. *Part of British Guyana* 1834

Hill. Engraver. Ham's *Rhode Is. & Conn.,* Boston (1796)

Hill, G. D. Surveyor General. *Dakota Territory* 1861

Hill, Lieut. I. 23rd Regt. Asst. Engineer. *Skirmish at Petersburg* 1781–84, *Plan Chesapeake Bay* 1781 MS

Hill, James S. H.M.S. *Britomart. Plan Harbours Akaroa and Waitemata N.Z.* (1840)

Hill, Jared. Estate surveyor. *Walthamstow* 1739 MS

Hill, Johann Jacob (1730–1801). Military cartographer of Darmstadt

Hill, John (18th century). English astronomer

Hill [Hills] , John. Mapseller, Exchange Alley, Cornhill. Seller's *New England* (1675), Oliver's *Plan London* (1680)

Hill, Joseph. *Plan Salford* 1740 MS

Hill, J. H. Publisher of Burlington Vermont. Thomson's *Vermont* (1840)

Hill, M. B. *Plan of Bristol* (ca. 1775)

Hill, Nathaniel. Engraver, Estate surveyor and globemaker (fl. 1742–1762). *Barking* 1742 MS, Vincent's *Scarborough* 1747, Morris' *Plans of Harbours* 1748, Warburton's *Kent* 1748, *Middlesex* 1749, Wing's *N. Level Fens* 1749, *Pair Miniature globes* 1754

Hill, P. Publisher, Edinburgh. *New Gen. Atlas* 1814

Hill, Peter **& Co**. *Trav. map of Scotland* (1820)

Hill, Robert T. *Texas,* Washington 1899

Hill, Samuel. American engraver of Boston. *Plan Washington* (with Ellicot) the first folio plan of the city. Engraved for Morse 1789–96, for Carey 1795–6 and for Malham's *Naval Atlas* 1804

Hill, S. W. American surveyor. *Cliff Mine* 1847, *Lake Superior* 1855, *Trap Range* 1863, *Isle Royale* 1871

Hillebrands, A. J. *Atlas van Noord-Amerika,* Groningen, J. Oomkens 1849–50

Hiller, Karl (1869–1943). Cartographer for Perthes

Hillestrom, Christopher. *Trollhatten Sweden* 1765 MS

Hillhouse, William. *Brit. Guiana* 1827 and 1834, *Massarooney River* 1834

Hilliard, **Gray & Co**. Publishers of Boston. Worcester's *Modern Atlas* 1821

Hillock. Contributor to Jefferys' *American Atlas* 1776

Hills, John. *Town of Haddonfield* 1778 MS

Hills, Lieut. John (fl. 1777–1817). 23 Regt. Assistant Engineer, surveyor and draughts-man. Later in Philadelphia *Revolutionary War Plans* MS drawn or copied by Hills 1777–82, *Plans* for *Faden's Atlas* 1784–5 *Plan Phila.* 1796, *North America* 1811 MS

Hills, J. See **Hill**, John

Hiltensperger, Johann Jost (1711–1792). Swiss engraver

Hilton, J. Engraver. *Canterbury* 1752

Hilton, William. (d. 1675). Sailor from Charlestown Mass. *Cape Hatteras to Cape Roman,* drawn by N. Shapley 1662 MS

Himburg, Christian Friedrich. Publisher of Berlin. *Kentniss des Himmels* 1777

Himly, L. Aug. (b. 1823) French geographer and historian

Himmerich, Johann (18th century). Military cartographer of Hamburg. *Elbe-strom,* Covens & Mortier (1730)

Hincke, Capt. P. A. W. von. *Magdeburg* 1809

Hind, H. Y. *Maps of Canada* 1858—64

Hind, John Russell. English astronomer, fl. 1850

Hinder, Thomas. *Chart coast N. America* 1732 MS

Hindermann, E. Lithographer of Basel. *Map of Malta* engraved J. Locherer

Hinderstein, G. F. D. van. See **Derfelden van Hinderstein**

Hingeston, M. Bookseller. Camden's *Britannia* 1772

Hinman. See **Mitchell & Hinman**

Hinman & Dutton. Publishers of Philadelphia. *Wisconsin* 1838

Hinrichs, J. C. Book, map seller and publisher of Leipzig. *Carte d'Allemagne* 1803, *Hamburg* 1810, *L'Empire Franç.* 48 sh. 1812, Stein's *Neuer Atlas* 1833, *Australia* 1847

Hinton, C. Publisher, Ivy Lane, Paternoster Row. For Wallis & Read 1820

Hinton, I. T. Publisher, London. *Maps of U.S.A.* 1832 (with Simpkin & Marshall)

Hinton, John (fl. 1745, d. 1781). Bookseller and Publisher at Kings Arms St. Pauls Churchyard (1745—52). Kings Arms Newgate Street (1752—65), Kings Arms, 34 Paternoster Row (1776—81) *Universal Magazine* 1747—81, Bowen & Kitchin *Large English Atlas* 1749—52

Hinton, John Howard. *Hist. of U.S.* 1830 and 1846

Hiouen-Thsang (629—46). Chinese traveller

Hippalus. Discoverer law of monsoon winds Indian Ocean

Hipparchus (180—125 B.C.). Greek astronomer and geographer of Rhodes. Proposed division of equator into 360°. Invented astrolabe. Constructed celestial globe

Hippodamus of Miletus (5th century B.C.) Greek engineer and architect. Plans for Alexandria

Hippokrates of Chios (fl. 480—450)

Hipschmann, Sigismund Gabriel (b. 1639). Engraver of Nuremberg

Hire, de la. See **De la Hire**

Hirsch, Adolph (1830—1901). Swiss astronomer and geodesian

Hirschfeld, G. (1847—1895). Geographer. Professor at U. of Königsberg

Hirschfogel, A. See **Hirschvogel**

Hirschgarter, Mathhias (1574—1653). Swiss mathematician and astronomer

Hirschhorn. *Business Map of London Suburbs* 1880

Hirschko, Carlos. *Mamore ò Madera* 1782

Hirschvogel [Hirsvogel, Hirssfogel], Augustin (1490—1553). Artist, engraver, mathematician and cartographer (b. Nuremberg, d. Vienna. *Turkish border* 1539 MS, *Austria* 1542, used by de Jode; *Carinthia* 1544, *Muscovy* for Herberstain 1546, *Plan Vienna* 1549, *Saxony* 1550, *Hungary* 12 sh. 1565

His, J. de. Engraver for Thevenot 1666

Hise, Charles Richard van. *Atlas Iron bearing district Michigan,* Washington 1896

Hishikawa, Moronobu (1648—1694). Japanese painter and woodcut engraver

Hislop, A. Publisher of Peebles. *County of Peebles* 1774

Hislop, Commandant. *Guyana* 1802

Hitch, Charles (d. 1764). Publisher and bookseller. Apprenticed to and succeeded A. Bettesworth at Red Lion Paternoster Row, then with Hawes 1754—63. Salmon's *Modern History* 1744—6, Badeslade's *Chorographia Britannia* 1743, Gilson's edition of Camden's *Britannia* 1753

Hitchcock, Charles Henry. *Atlas New Hampshire* (1877) with H. F. Whaling

Hitchcock, Chas. Henry (1836–1919). Son of Edward. *Vermont Geog. Survey*

Hitchcock, D. C. *Honduras & San Salvador*, N.Y. 1854

Hitchcock, Edw. (1793–1864). State geologist. *Massachusetts* 1830, *Survey of the state* 1837 and 1841

Hitchcock. See **Marvin & Hitchcock**

Hixon, W. W. & Co. *Racine & Kenosha Co., Wisconsin* 1899

Hjort, C. A. *Vastmanland Prov.* 1800

Ho, Ceng-t'ien (370–447) Chinese astronomer

Hoare, Edward and Reeves. Engravers, publishers and mapsellers, 13 Little Queen St., Lincolns Inn Fields and 45 Kirby St., Hatton Garden [1823], then 90 Hatton Garden [1823–1833], and later 14 Warwick Court Holborn. *Plans London* 1823–27 for Ebden 1828, for Murray 1830

Hobbs, John. Estate surveyor. *Mucking & E. Tilbury* 1687 MS

Hobbs, J. S. Hydrographer. *North Sea* 1845, *Charts* for Chas. Wilson successor to J. W. Norie, *English Channel* 1851, *Straits Malacca* 1852, *Atlantic* 1860, *North Sea* 1877, *Finland* 1880, *Caribbean* 1883, *Canary Is.* 1883

Hobday, Capt. J. R. *Mandate (Burma)* 1886, *Andaman Is.* 1888, *Upper Irrawaddy* 1892

Hobley, C. W. *River Tana* 1891, *Mount Masawa* 1897, *Mt. Eglon (Masawa)* 1899 MS

Hobson, Capt. Sam. *Plan Londonderry*, sold Rich. Chiswell (1690?)

Hobson, William Colling. Publisher, 9 Castle St., Holborn. *Durham* 1839 and 1840, *Yorks* 2 sh. 1843, *Fox Hunting Atlas* Walker 1850

Hochstetter, Ferdinand von (1829–1884). Austrian geographer and mineralogist. *Geolog-topog. atlas Neu-Seeland* 1863 (with A. H. Petermann) *Türkei* 1872

Hocker, Sir Joseph D. (1817–1911). Military cartographer

Hocquart, A. Engraver of Mountains for Andriveau-Goujon 1829

Hoctomanno, Conte. See **Freducci**

Hodges, J. Bookseller, joint publisher with T. Osborne of *Geogr. Magnae Brit.* 1748, Kitchin's *Geogr. Scotiae* 1749, 1750, Camden's *Britannia* 1753

Hodges, N. *Sikh Territory* (with Capt. Wade) 1846

Hodges & Smith. Publishers of Dublin. *Townland survey of Ireland* 1843, *Ireland* 6 sh. 1839, 1847, 1855

Hodgkin, R. S. *St. Clair Co., Illinois* 1863

Hodgkinson, J. See **Hodskinson**

Hodgson, J. A. *Hurriana District (India)* 1810–1811, *Bettiah Frontier* 1815

Hodgson, Orlando [& Co.]. Publisher, 10 Newgate St., 111 Fleet Street, Maiden Lane Wood St. [1820], Newgate St. [1825], Fleet St. [1838]. *Pocket Tourist & Eng. Atlas* 1820, *Plans London* 1823–51, Reid's *Panorama* (1825), Leigh's *Plan of London* 1843

Hodgson, Lieut. Robert. *Moskito Shore* 1760 MS

Hodgson, Thomas. Surveyor. *Westmoreland* 1828

Hodskinson [Hodgkinson], Joseph. Engraver and surveyor. Jefferys' *Beds.* 1765, Donald's *Cumberland* 1771 and 1774, *Wells* 1782, *Suffolk* 6 sh. 1783 and 1787

Hoeckner [Höckner], Carl. Engraver for Meyer's *Univ. Atlas* 1830–40, for Wieland 1842–6

Hoedl, Leopold Joseph. Publisher, Vienna. *Alte u. Neue Geog.* 1734

Hoefnagel [Hufnagel], Jakob (1575–1630). Revised many of Joris Hoefnagel's *plans* for Braun and Hogenberg. *Vienna* 6 sh. 1609, unique

Hoefnagel, Johann (17th century). Engraver, brother of Jakob Hoefnagel.

G. Hoefnagel (1542–1600); engraving by H. Hóndius

Hoefnagel [Hufnagel] , Joris [Georg] (1542–1600). Belgian painter, poet, miniaturist and topographer. Travelled widely and produced nearly 100 *views* for Braun and Hogenberg's *Civitates Orbis Terrarum* 1570–1618, *Plan of Cadiz* for Ortelius

Hoeg, Anders (1727–1796). Danish hydrographer. *Nord-soen* 1769

Hoeius, Franciscus. *World map* ca. 1600 republished later H. Allard ca. 1640; *Germany* 1632

Hoelzel [Hölzel] , Hugo Joseph (1852–95). Cartographer and publisher

Hoen & Co., A. Lithographers. *Surveys Michigan* 1855

Hoest, Georg. *Morocco* 1781

Hoet, G. Designed front. for Halma's *Geographica Sacra* Amsterdam 1704

Hoeye, François van den (1590–1636). Engraver and publisher of Amsterdam. Father of Rombout

Hoeye, Rombout van den (1622–1671) Printer, publisher, engraver and map-seller of Amsterdam "inde Kalverstrate" *Leo Belgicus* 1636

Hofel, Prof. B. *Wien,* Vienna, Sollinger (1850)

Hoff, Karl Ernest Adolf (1771–1837). German geologist

Hoffer, Andr. Engraved title for *Atlas Homannianus* 1762 after Preisler

Hoffgaard, Jens [Jans] . *Map of Iceland* 1723. The first Danish map of Iceland, 1724

Hoffman, Chas. F. *Survey Sierra Nevada* 1863–7

Hoffman, E. (1801–1871). Russian explorer and mineralogist. *Geology Urals*

Hoffmann, Erhard (fl. 1570). German astronomer

Hoffman, Friedrich (1797–1836). German geographer and geologist. *Geolog. maps of Germany,* Juttners, Prague (1820)

Hoffmann, Gottfred. *Scania* 1657

Hoffman, Johannes (1629–1698). Map publisher of Nuremberg. *Hungary* 1664, *Nürnberg* 1677, *Wien* 1683, *Rheinstrom* 1689, Dapper's *Africa* 1689

Hoffman, Karl Friedrich Vollrath (1796–1842). *Atlas für Schulen* 1835, *Himmels Atlas* 1835, *Würtemberg* 1836, *Orbis Terr. Antiq.* 1841

Hoffman, Tomas (1667–1750). Swedish surveyor of Gotland

Hoffman, Michael (17th century). Surveyor

Hoffmann. *Orbis antiq. Schul Atlas,* Leipzig 1853

Hoffmann, Lieut. *Prussia* 1831

Hoffmann, Wolffgang. Printer of Frankfort. *Besch. Schweden* Hulsius 1632

Hoffmeyer, N. H. C. (1836–1884) Danish meteorologist. *Meteorological maps*

Hofmann, Elias (d. 1591). *Environs Frankfort,* Cologne 1583

Hofmann, F. Anton. Engraver. *Battle Plans* (1836–50)

Hofmann, J. See **Hoffmann**

Hofmann, Wolfgang. Printer. Gottfried-Hulsius *Svecia* 1632

Hofmeister, Johannes. Publisher of Zürich, "an der Rosengass". Murer's *Zürich* 1566

Hogan, J. Sheridan's *Canada* 1855

Hogan, Capt. M. *Straits of Boeton* 1796 Laurie and Whittle, *Chart Straits westward New Guinea* 1796 Laurie and Whittle 1798

Hogan, William. *Tasmania* 1859

Hogben, H. Surveyor. *East Tilbury* ca. 1775–96 MS

Hogben, Thomas. (18th century) Estate surveyor

Högborn, A. G. *Geolog. K. Jamtlands Län* 1894

Hogeboom [Hogenboom] , A. Engraver for Jansson (1658). Visscher *Atlas Minor* (1684), De Wit 1690 and Ottens

Hogenberg, Abraham. Publisher of Cologne. *Strassburg* for Braun and Hogenberg 1572

Hogenberg, Frans [Franciscus] (1535–1590). Flemish artist and engraver b. Malines, d. Cologne. Engraved maps for Ortelius, Plantin &c. *Jerusalem* 1584, *America* 1589, joint publisher with Braun of *Civitates Orbis Terrarum* 1573–90

Hogenberg, Hans. Map publisher in Mechlin 1520

Hogenberg, Remy [Remigius] (1536–1587). Engraver and publisher b. Mechlin. *Munster* 1570, *Exeter* 1587. Refugee in England. Maps for Saxton 1579

Hogenboom, A. See **Hogeboom**

Hogg, Alexander (fl. 1778–1805). Publisher and bookseller, Kings Arms, No. 16 Paternoster Row. Moore's *Voyages & Travels* 1779, Miller's *System Geog.* (1782), *Plan London & West* (1784), Walpole's *British Traveller* 1784, and 1794, Anderson's *Voyages* 1795

Hogg. Hogg & Co. 1805–1818

Hogg, J. *Peninsula Mt. Sinai* 1849

Hogg, Thomas. Land surveyor, 34 Castle Street, Holborn. *Tolleshunt Major* ca. 1807 MS

Hoggar, Robert Syer. Engineer. *Map Oxford . . . showing . . . localities . . . Cholera* 1854

Hogius, C. de. See **Hooge**, C. de

Hogius, R. de. See **Hooghe**, R. de

Hogreve, J. See **Hogrewe**

Hogrewe, Johann Ludwig (d. 1814). Military cartographer. *Hannover* 1800, *Länder zwischen Elbe u. Weser* 6 sh. 1812

Hohagen, F. *Plan of Cusco,* Paris 1861

Hoheneck, G. A. E. von. See **Enenckel von Hoheneck**

Hohenkerk, L. S. *Chart mouth Waini River* 1898

Hohn [Höhn, Hoehn], August (1807–86). Lithographic printer and engraver of Baltimore

Höhnel, Ludw v. (b. 1857) Austrian hydrographer. Works on African lakes

Hoijaeus, F. See **Hoeius**

Hoirne [Horen], Jan van [Joannes à Horn, Jan de Beeldsnyder] . *Caerte van de Oosterscher Zee,* Antwerp 1526. The oldest known European printed sea chart. *Netherlands* 1526 (now lost)

Hoit, David. *Deerfield & Springfield, Mass.* 1794

Hoit, N. I. Engraver. Thomson's *Roxburgh* 1822

Hojeda, Alonso de (1471–1515). Spanish navigator. Second expedition Columbus·1499

Hokusai, Katsushika (1760–1849). Japanese artist, engraver and printer. *Panoramic map of the Kiso Highway* 1819, *Bird's eye views and map of China* (1840)

Hol [Holl, Holle] , Lienhart. Printer of Ulm. *Ptolemy's Geography* 1482

Holbein, Hans, the younger (1497–1543). Artist and engraver. Decor for *World Map* in Grynaeus *Novus Orbis,* Basle 1532. Decoration for reverse of maps in 1542 *Ptolemy*. Map in More's *Utopia* 1579 attributed to him

Holbrook, A. R. N. Hydrographer *Newfoundland* 1814–20

Holbrook, J. *Railway Survey Essex* 1835 MS

Holbrook, W. H. *Railway & Parliamentary map Ireland,* Dublin 1846

Holcroft, J. *Part State Michigan* 1858

Holder, Edw. S. (b. 1846) American geographer and astronomer. Founder Astronomical Soc. of the Pacific

Holdich, Col. Sir. Thomas Hungerford (1843–1929). Geographer and military cartographer. *Maps of India & Afghanistan* 1879–98

Holding, Rev. J. *Prov. of Tanibé (Madagascar)* (1870) MS

Holditch, George, of Lynn Regis. *Part E. Coast England* 1810

Holdredge, Sterling M. *Guide Book Pacific* 1865–6

Hole, William (fl. 1600–1646). Engraver. Camden's *Britannia* 1607, Smith's *Virginia* 1608, Drayton's *Polyolbion* 1612. Raleigh's *History* 1614

Holgarth, Ludwig, Graf von (fl. 1804–10). Cartographer of Vienna

Holinshed [Hollingshead], Raphael (d. 1580). *Chronicles of England* 1578

Holland, Frederick Whitmore (1837–1880). Canadian geographer

Holland, John. Surveyor. *River Colne & Tributary* 1842 MS

Holland, Capt. N. *Coasts Carolina,* Laurie & Whittle 1794, *Coast of Georgia,* Romans 1794

Holland, Philemon. *Translated* Camden's *Britannia* 1610

Holland, Capt. (later Major) Samuel (1728–1801). First Surveyor Gen. for British N. America. Taught Capt. Cook. *Plans* for Jefferys' *Topo. of N. America* 1768, *Island St Johns* 1765–75, *N.Y. & N.J.* 1775, *Cape Breton* 1779, *New Hampshire* 1784, *Navig. Halifax-Phila* 1798. Contributed to *Atlantic Neptune*

Hollar, Wenceslaus [Wenzel] (1607–1677). Bohemian artist and engraver, b. Prague, d. London. King's iconographer 1660. *Plan Dueren* 1634, *Plan Oxford* 1643, *England,* 6 sh. 1644, *Westminster & London* 1655, Leake's *Hungary* 1664, *America* 1666, Leake's *London* 1667, *London,* Blome 1673, *Tripoli* 1675, *Naples* 1676

Holle, Lienhart [Leonardo]. See **Hol**

Holle, Publisher of Wolfenbüttel. *USA* 1851, *Schul Atlas* 1854

Holliday, Capt. *Draught Bristol Channel* (1740)

Hollingworth, H. G. *Survey Po Yang Lake* 1868 MS

Hollingworth, J. Estate surveyor. *Chishall* 1769 MS, *Little Waltham* 1776 MS, *Pebmarsh* 1807 MS

Holloway, Capt. Charles. Royal Engineers. *Rivers Thames & Medway* ca. 1795 MS

Holm, G. F. *Christian IX land* 1886, *Gronland* 1888

Holm, John. Land surveyor, Colchester. *West Bergholt* 1810 MS

Holm, Saemundur Magnussen (1749–1821). Icelandic cartographer. *Rangaville county* 1777

Holman, A. J. & Co. *New Biblical Atlas* 1898

Holmberg, H. J. *Russisches America* 1854

Holme, Randle F. *Hamilton River Labrador* 1887 MS, *Peninsula of Labrador* 1888

Holme [Holmes], Thomas (1624–1695). Appointed surveyor general of Pennsylvania by William Penn in 1682. *Plan of Philadelphia* London, Andrew Sowle Shoreditch, 1683: first plan of the city. Reissued London, P. Lea 1687–90. Later issued Geo. Willdey

Holmes. Engraver. Krazeisen's *Plan of London* 1793

Holmes, John. Estate surveyor. *Layer Marney* 1698 MS, *Hadleigh* 1709 MS

Holmes, Philip. Engraver. Overton's *Africa* 1668

Holmes, Thomas. Estate surveyor. *Quendon* 1702 MS, *Paglesham* 1775 MS

Holmes, W. H. Engraver for Chamber's *Encycl. Philadelphia* 1869, *Mitchell's Mexico* 1859

Holst, Captn. Jacob. *N.W. coast Madagascar* 1738

Holstein, Lukas (1596–1661). Geographer, friend of Cluver

Holster Atlas. Sayer and Bennett's *Military Pocket Atlas* 1776

Holt, Adam. Estate surveyor. *Wanstead* ca. 1725 MS, *Epping* 1743 MS

Holt, Edward. *Ireland* 1838

Holt, George L. of Washington, D.C. *Wyoming* 1884

J.B. Homann (1663–1725)

Holt, Thomas. Surveyor. *Plan of Cork* 1832

Holt, Warren. Agent for J. H. Colton, 305 Montgomery Street, San Francisco; later 411 Kearney St. 1869 and 607 Clay Street, San Francisco 1873. Farley's *Tramontane River* 1861 for Ramson 1862, for De Groot 1863, *Nevada* 1866, *California* 1869—73

Holtrap, W. *Zak atlas* 1763—89, *Atlas du Rhin* 1798

Holtzwurm. See **Holzworm**

Holywood, John. See **Sacrobusco**

Hölzel, Eduard (1816—1885). Editor numerous geographic publications. *Atlas von Oesterreich-Ungarn* 1887

Hölzel, H. (19th century). Map publisher of Vienna

Holzmüller, Heinrich. *Nördlingen* for Münster 1550

Holzschuher, George. Supervised construction *Behaim's globe* 1492

Holzworm [Holtwurm], Abraham (17th century). Brother of Israel Holzworm. *Austria* 1628, used by Blaeu 1662

Holzworm, Israel (d. 1617). Austrian engineer. *Carinthia,* completed by Abraham 1636

Homann [Homanno], Johann Baptist (1663—1724). Engraver and mapseller, geographer to the Kaiser of the Holy Roman Empire 1715, member Prussian Royal Academy of Science. Born Kammlach, died Nuremberg. Engraved for Funck, Jacob von Sandrart and Scherer. Bear's *Gross Brit. (Scotland)* 1690, Cellarius' *Not. Orbis* 1698. Firm founded 1702. First *Atlas* 1704, *Neuer Atlas* 1707, *Pair globes* 1705, *Grosser Atlas* 1716, *Atlas Methodicus* 1719, *Grosser Atlas* 1731, 1737, *Kleiner Atlas,* Doppelmayers *Star Atlas* 1742, *Geogr. Maior* 1759, *Atlas Homannianus* 1762

Homann, Johann Christoph (1701—1730). Mapseller and cartographer, son of Johann Baptist. *Morocco* 1728, *Thuringia* 1729

Homann Heirs [Heredes, Héritiers, Homannischen Erben]. Publishers 1730—1813. 1730 Johann Michael Franz and J. G. Ebersperger, 1813. Collaborators, J. G. Gregory, J. Halmer Doppelmeyer, Lowitz, Hase, and Mayer. In 1813 taken over by Christoph Fembo. *Atlas Minor* 1732, *Atlas Novus* (1732), 1747, *Grosser Atlas* 1731, *Atlas Germanicus* 1735, *Atlas Silesiae* 1737, *Hand Atlas* 1754, *Atlas Geogr. Major* 1759—84, *Stadt Atlas* 1762, *Atlas Helvetiae* 1769, *Atlas Regni Bohemiae* 1776. Also collaborated with T. Bowles

Homby, Capt., R. N. *Chart Coast Guard Service* 1844

Home, Col. R. *Trans Caucasia* 1877

Homem, André (1570—1599). Portuguese cartographer. *World* 10 sh. Antwerp (1559) MS

Homem, Diogo (fl. 1530—1576). Portuguese portolan maker, worked in London, Lisbon and Venice. *Carta del Navigar,* Venice, Forlani 1569, first sea chart engraved on copper. *A Portuguese atlas* 1558—61 (attributed)

Homem, Lopo (fl. 1497—1572). *Atlas* ca. 1519, *World* 1554 MS

Homem, Lourenço. *Portugual* 1808

Homem de Melo Marcondes, Ignacio Francisco (1837—1918). Brazilian cartographer. *Imperio do Brazil* 1882

Homeria, Diego (17th century) Spanish hydrographer. *Sea chart* 1673 MS

Hommaire de Hell, I. X. Morand (1812—1848). French geologist and explorer

Hommeyer, H. G. *Zeichnung der Schweiz* 1804

Homolka, Josepf (1840—1907). Czech surveyor and cartographer

Hondius, Jodocus [Hondt, Josse de] (1563—1612). Geographer, engraver and publisher. Settled in London 1583. Engraved plates for *Mariners Mirrour* 1588, *World* 1589 (first world map engraved in England), *Molyneux Globes* 1592-3 (first globe gores

Grosser
ATLAS
Uber die
Gantze Welt.

Wie diese sowohl,
Nach Göttlicher Allweisen Schöpffung aus den heutigen Grund-Sätzen
der berühmtesten Astronomorum

NICOLAI COPERNICI und TYCHONIS de BRAHE,
In der Bewegung und unermeßlichen Weite

Des Himmels/
als auch
In dem Umfang unserer mit Wasser umgebenen allgemeinen

Erd=Kugel/
zu betrachten,
Samt einer kurtzen Einleitung
zur

GEOGRAPHIE,
Worinnen die Erde

1. Mathematice: Nemlich was sie mit der Himmlischen Sphæra für eine Correspondenz habe.
2. Physice: Wie sie in ihren natürlichen Stücken durch Wasser und Land unterschieden.
3. Historice: Wie sie in ihre darauf befindliche Monarchien, Königreiche, Staaten und Herrschafften, auch nach Aus-
breitung verschiedener Religionen eingetheilet,
deutlich beschrieben
durch

Herrn Joh. Gabriel Doppelmayr/der Kays. Leopoldinisch= und Carolinischen Academie Natur. Curios.
wie auch der Königl. Preussischen Societät der Wissenschafften Mit-Glied und Math. PP. Ordin. allhier,
und

Mit   auserlesenen/theils Astronomischen/ meistentheils aber Geographischen Charten (in welchen
alle bißher zu Wasser und Land geschehene Land-Entdeckungen aus denen berühmtesten Autoribus dieses
Seculi anbemercket worden,) in Kupffer gebracht und ausgefertiget
von
JOHANN BAPTIST HOMANN,
Der Röm. Kays. Majestät Geographo, und Mit-Glied der Königl. Preussischen Societät der Wissenschafften.

Nürnberg,
In Verlegung des Auctoris.   Gedruckt bey Johann Ernst Adelbulner.
MDCCXXXI.

Title-page to J.B. Homann's Atlas (published Nuremberg, 1731)

H. Hondius (1597–1651); engraved by F. Bouttats

engraved in England), *America* 1589. Returned to Holland 1593. *Europe* 1595, Acquired Mercator plates 1604, *Atlas* 1606, *World* 12 sh. 1608, for Speed 1611

Hondius, Jodocus. Business continued by his widow **Coletta van den Keere** in the Kalverstraat. Assisted by her sons Jodocus II and Henricus. *Ptolemy* 1618 edited by Bertius

Hondius, Jodocus II (1594–1629, Elder son of Jodocus senior. Cartographer, engraver, publisher and printer, "Calverstraat in den Wackeren Hondt". Married Anna Staffmaecker 1621. *Pair globes* 1613, Veen's *Scandinavia* 1613, *Pontanus* 1614, *Globe* 1618, *World map* 1618. On his death 34 of his maps sold to Blaeu

Woodcut printer's mark of J. Hondius ("De wackere Hondt" – the watchful dog)

Hondius, Henricus [Hendrick] (1597–1651). Younger son of Jodocus senior. Helped his mother and his brother (1619–29). Started on his own 1621 on the *Dam sub signo Atlantis.* Merula's *Cosmography* 1621. Reissued Blaeu's *World map* 1624. On death of his brother (1629) returned to parental home on the "dam in den Wackeren Hondt". In 1630 took into association his brother-in-law Jan Jansson. Last edition of *Mercator-Hondius* with his imprint, 1641

Hondius, Willem (ca. 1597–1660). Engraver for Le Vasseur in Dantzig. *Siege Smolensk* 16 sh. 1634, *Is. Anthony Vaaz, Visscher* 1640, for Le Vasseur de Beauplan 1648–50, for Zwicker 1650

Hondt. See **Hondius**

Hondt, P. de. Publisher of The Hague Horn's *Orbis Delin.* 1740, Bellin's *World* 1750

Hone, William. Publisher, *Plan St. Peter's Field, Manchester* 1819

Honius, Henricus. Engraver of Haarlem, worked in Italy 16th century for Ducheto. *Africa* 1579

Honkoop, A. Publisher *Atlas des Enfants* 1817

Honore, l'. François. Publisher of Paris. *Voyages Lahontan* 1703, *Chatelain's Atlas Historique* 1705–39

Honorius, Julius. Cosmographer. *World map* 5th century

Honorius Augustodunensis [of Autun] *Imago Mundi* ca. 1129 (maps)

Honter [Grass, Honterus] , Jan Coronensis (1498–1549). Cosmographer and engraver of Transsylvania. Published from his own press in Kronstadt. *Rudimentorum Cosmographia* 1530 (2 maps), 1542 (13 maps), many later editions. *Procli di Sphaera* 1561 (23 maps), *Globe* 1542, *World* 1546, *Swizzer Chronik* 1548

Hood, Edwin C. *Plat book Richland Co., Wis.* 1895, *Vernon Co., Wis.* 1896, *Counties in Wisconsin, Minnesota, Iowa* 1887–99 (with C. M. Foote)

Hood, John (1720–1783). English surveyor and inventor

Hood, R. V. Lithographer and publisher, Liverpool St. *Plan Hobart Town* 1854, *Unsettled part Tasmania* (1865), also Admiralty charts

Hood, Thomas (fl. 1577–98). English mathematician and geographer. *West Indies* 1592, *Use of celestial globes* 1590, *Chart N.E. Atlantic* 1592, *Chart Bay Biscay etc,* 1596 MS

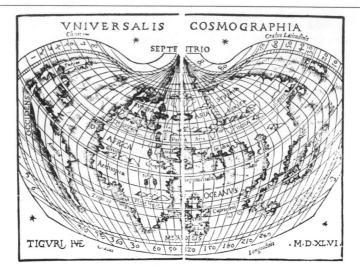

Woodcut world map by J. Honter (1546)

Hood, Lieut., Washington War Dept. *Pensacola Harbour* 1835, *Brandywine Shoal* 1836, *Lake Nicaragua* 1838, *Oregon* 1838. Drew for Bureau of Topographic Engineers 1834–39

Hood. See **Vernor, Hood & Sharpe**

Hooftman, Gilles Egidius (1521–1581). Shipowner, scholar, Antwerp merchant. Suggested *Atlas* to Ortelius

Hoog, de. Engraver. *Plan of Crete* (1670)

Hooge [Hooghe, Hogius], Cornelis de (1540–1583). Dutch engraver. Bastard son of Charles V; beheaded at The Hague. Worked for Saxton. *Norfolk* 1574, *Holland* 1565, based on Deventer, engraved for L. Guicciardini

Hooge, R. de. See **Hooghe**

Hooghe, C. See **Hooge**

Hooghe [Hooge, Hogius], Romain [Romanns, Romein] de (1645–1708). Engraver, born Amsterdam or The Hague ca. 1638, ennobled by John III of Poland in 1675. *English Ports* 1667, *Flanders* 1670, *World Map & Continents* ca. 1680, *Atlas Maritime,* Mortier 1693, *Zee Atlas* 1694, *Atlas François* pub. De Fer 1695

Hooijer, G. B. *Atlas van Nederlandsch Indië* (1895–7)

Hook, Andrew. *Province of Nova Scotia.* Philadelphia 1760

Hooke, Robert (1635–1703). Experimental philosopher, city surveyor and astronomer

Hooker. *Plan Exeter,* engraved R. Hogenberg 1587

Hooker. American engraver. Furlong's *America Coast Pilot* 1809

Hooker, William. Surveyor. *City New York* 1831, for Blunt 1817, for Derby 1818–19

Hooker, William. American engraver of New York. For Arrowsmith and Lewis 1804, for Furlong 1809, for Pinkerton (1804), for Phelps 1832

Hoole, J. London publisher. Collaborated with Henry Overton. *Dublin* (1730), *Pocket Map of London* 1731

Hooper, Samuel (fl. 1770–1793). Publisher, Bookseller, Stationer at New Church, Strand; 25 Ludgate Hill; 212 High Holborn. Sold Speer's *West India Pilot* 1766 and 1771, Collet's *Carolina* 1770, Byres' *St. Vincent & Tobago* 1776, *Attack Gibraltar* 1782, *Scotland* 1791. Succeeded by his widow Mary

Hoorenhout, Jacques. Dutch surveyor. *Hist. map Zeeland* 1540

Hoorn, Melchisedech van (16th century). Surveyor and engraver. *Plan of Utrecht* 1569, used by Braun & Hogenberg

Hoorn, N. S. van. See **Schouten**

Hoover, H. S. *Atlas Bremer Co. Iowa* 1875

Hopfgarten. *Atlas S. Preussen* 1799

Hopkins, B. B. Publisher, 66 South Fourth Street, Philadelphia 1815, *New Juvenile Atlas* 1815

Hopkins, E. Surveyor. *Nova Scotia Co. Halifax* 1889

Hopkins, Griffith Morgan and Co. Publishers in Philadelphia. Numerous town and country atlases of Eastern U.S.A. from 1870

Hopkins, Henry W. *Atlas Delaware Co., Penn.* 1870, *Atlas Germantown, Pennsylvania* 1871

Hopkinson, John. *Synopsis Paradisi,* Leyden 1598 (map)

Hoppach, C. *Plan Ostend* Fricx 1707

Hoppe. Engraver for Hartmann 1824

Hopper, Joachim. *Battles in Holland* used by Ortelius

Hoppner, Capt. H. P., R.N. Hydrographer. *N.W. Passage* 1824–25

Horatius, Juriaen Janssen. *Rivier van Siera Lione,* Sierra Leone 1666

Horatius, Andreas Antonius. Drew maps and plans for Santa Teresa's *History of Brazil,* Rome, Rossi 1700

Hore, Edward C. *Maps of Tanganyika* 1879–89

Horen. J. van. See **Hoirne**

Horenbault, Jacques. *Flandre Maritime* (ca. 1620)

Horman, Robert. *Mouth Thames* 1580 MS

Horn [Hornius], Georg (1620–1670). Professor of History and Geography at Leiden. *Orbis Antiquis* 1644, editions to 1741. English edition *Ancient Geography* 1700, *Acc. Orbis delineatio* 1660, *Description of Earth* 1700

Horn, Hosen B. of New York. *Guide to California & Oregon* 1852

Hornbeck, H. B. *St. Thomas (Virgin Is.),* Kobenhaven 1835–9 and 1846

Horne, J. à. See **Hoirne**

Horne, J. *Fiji Inseln* 1877–8

Horne, Robert. Publisher of London. *Description of Carolina (with map)* 1666

Horne, Tho. Co-publisher with others of Chiswell's *Map of London* 1707

Horner, Johann Kaspar (1774–1834). Zürich, mathematician and astronomer

Horner, W. B. *Gold Regions Kansas & Nebraska* 1859

Hornhovius, Cornelis Ant. *Utrecht,* Hondius 1599

Hornius. See **Horn**, G.

Horrebows, M. *Iceland* (1770)

Horrocks, Jeremiah (1619–1641). English astronomer

Horsburgh, Capt. James, R.N. (1762–1836). Publisher and hydrographer to East India Co. *Strait Macasser* 1800, *Atlas of E. Indies* 1806–21, with P. Steel. *E. India Pilot* 1817, *Atlas of India* 1827–33

Horsely, Capt. of Liverpool. *Lagos & its channels* 1789 for Norris

Horseley, John (1685–1732). *Northumberland* (completed after his death by R. Cay) 1753

Horsfield, R. Bookseller. Camden's *Brittania* 1772

Horsfield, Thomas. *Is. Banka* (1822), *Is. Java* (1852)

Horsley, Benedict. Surveyor, York. *Plan York* 1694, pub. Tempest 1697

Horsley, J. *British Antiq.* 1794

Horsley, Lt.-Col. W. H. *Madras Presidency* (scale 12 miles to inch) 1861. Chief Eng. dept. Public Works

Horstmann, Nicholas. *Rio Esquibo to Rio Negro* 1743

Horton, Surg. Maj. J. A. B. *Maps Dassay & Ashante Gold Coast* 1882

Horton, J. Estate surveyor. *Buckland Kent Lexden Manor* 1819 MS (with S. Horton)

Horton, Simon. Estate surveyor. *Lexdon* 1819 (with J. Horton), *Lexdin* 1821 MS

Horwood, Richard (ca. 1758–1803). Surveyor and publisher. *Plan London, 32 pl.* 1792–1799. *Plan Liverpool, 6 sh.* 1803

Hose, C. *Baram District, Borneo* 1893

Hosford, Oramel. Contributed to *Atlas Michigan* 1873

Hosken, Lieut. H., R.N. *Goro Is. (Fiji)* 1874

Hosking [Hoskins], William *Harwich railway* 1840–44 MS

Hoskins, W. See **Hosking**

Hoskold, H. D. *Mapa Topo. Rep. Argentina*, 10 sh. Philip 1893

Hoskyn, Commdr. R. F., R. N. *Ireland* 1853–62, *Banks Strait* 1888

Hospein, Michael (b. Strassburg 1565). *Maps of Hohenlohe* 1589–1607

Hospin, Michael. *Kirchberg an der Jagst* 1607

Hosser, D. S. C. E. *Mont des Géants* Vienne 1812

Host, S. de L'. See **Sorriot de L'Host**

Hosted, Dan. Surveyor. *Estate map Aunsby, Lincs.* 1744

Hotan. Japanese Buddhist priest. *World map* MS, *woodcut version* pub. by Bundaiken Uhei 1710, later edition 1744

Hotchkiss, J. *Augusta Co., VIrginia* 1870

Hottinger, Heinrich (1681–1750). Military engineer

Hottinger, J. H. Swiss engineer and cartographer, ca. 1783, compiled *maps of Drent, Groningen & Overyssel*

Houard, J. *Whitehaven* 1791

Houbraeken, Arnold (1660–1719). Amsterdam engraver

Houdan, d. See **D'Houdan**

Houel, Jean (1735–1813). Information used by Buache. *Martineco* Jefferys 1760, *Martinique, Voyage Pittoresque Sicile, Malte & Lipan* 1787

Hough, Benjamin. American surveyor. *Original Surveys of the Townships of Michigan* 1815, *Map of the state of Ohio* 1815

Hough, Charles C. Publisher. *Colford (Glos?)*, Atkinson's *Forest Dean* 1845

Houghton, J. *Min. Region Lake Superior* 1845

Houlen, P. Van. *Hambantotte Salt Pans* 1813 MS

Houlston & Sons. *Wilts.* 1869

Hourst, Emile Aug. L. (b. 1864). *Hydro. explor. of Niger* 1896

House, W. M. *Chart North Dakota & Richland country* 1897

House & Brown. Publishers. *Mexico, Texas, etc.* 1847, 1849

Housman, John. *Pocket Plan Manch. & Salford* 1800

Hout, J. van. See **Van Hout**, J.

Houten, G. Van. Designer title page Robyn's *Atlas* 1683

Houtman, Cornelius. *Descriptio Hydrog.* (Dutch route to E. Indies) 1597–7, (Claesz 1598)

Houtrijve, J. van. Publisher of Dordrecht. *Atlas van Noord Nederlanden* 1839

Houve, Paul de la. Publisher, Paris. Plancius' *Italiae, Illirici . . . descriptio* (1620), *Plan Florence*, Paris 1601, *Plan Messina*, Paris 1601

Houze, Antoine Philippe. French publisher. *Atlas Hist. de France* 1840, *Atlas Hist. de Espana* 1841, *Atlas Univ. Hist. et Géogr.* Paris 1848–9, *Atlas Univ.* 1854

Houzeau, J. Ch. (1820–1888). Astronomer and geographer. Director Observatory at Brussels

Hovart, J. A. *Reis Oost Indië* 1789–92 MS

Hovell, W. H. (1786–1876). Australian explorer. With Hume 1824

Hoven, George Christian von (1841–1924) *Lithograph maps*

Hövinghoff, J. (18th century) Danish engraver and globemaker

Howard, Henry, of Winchester. Surveyor. *Great & Little Pollicott* 1848, *Chandos Estate* 1848, *Stowe Estate* 1843 MS

Howard, Nicholas. Estate surveyor. *Dagenham* 1764 MS

Howard. See **Hayward & Howard**

Howden, J. A. *Atlas Warren Co., Penn.,* 1878 (with A. Odbert)

Howe & Spalding. Publishers, Newhaven. *Univ. Atlas* 1822

Howe, Henry. *Map of Great West* 1851

Howe, Samuel Gridley (1801–1876) *Atlas of U.S. for blind* 1837

Howe, Thomas. *Strait Singapore* 1759, *S. coast Madeira* 1762

Howell, D. J. with J. D. Hoffmann. *Maryland* 1895

Howell, Reading (d. 1827). Publisher and author. *Pennsylvania*, engraved J. Trenchard 1792

Howitt. Engraver, Buckingham Place. *N.S.W.* (1816), *Chart of New South Wales & Van Diemen's Land* 1828

Howland, C. W. *Atlas Abington & Rochland, Mass.* 1874 (with W. A. Shermay)

Howland, H. G. *Atlas Hardin Co., Ohio* 1879

Howlands, W. American engraver. *Western States* 1853

Howlett, Samuel Burt. Estate surveyor. *Stow Maries* 1818 MS

Hoxton, Walter. Cartographer. *Chesapeack Bay,* Mount & Page 1735

Hoyav, Germain. *Plan Paris* 1551 (with Truschet)

Hoyle, John. Engraver. *City of Norwich,* Chere (1720?)

Hsieh Chuang (421–466). Chinese cartographer

Hsu, Ching-Tsung (592–672)

Hsu, Lun. *Defense atlas China* 1538

Hu, Lin (1812–1861). Chinese cartographer

Huang, Peng-nien (1823–1891). *Atlas of Hopei* 1884

Huang Shang (12th century) Chinese astronomer

Huang, Tsung-Hsi (1609–1695) Politician and philosopher. *China* 1673

Huart, Johannes Marinus (1809–1855). Engraver and lithographer of Bergen op Zoom. Worked for Siebold 1840

Huart, M. L. *Kaart v. de Hoofdplaats Buitenzorg en Omstreken* 1880

Hubault, Gustave (b. 1825). *Atlas de Géographie* (1873)

Hubault, Jacques. Bookseller and Printer of Havre de Grâce. Bougard's *Petit Flambeau de la Mer* 1702

Hubbard, Bela. American geologist. *Lenawee, Calhoun & Jackson Counties* (1858)

Hubbard, Oliver P. (1809–1900). American geologist

Hubbard, William (1621–1704). *Narrative of the troubles with the Indians in New England,* Boston 1677 (woodcut map of New England by J. Foster)

Hubbard, W. of Dartford. Surveyor. *Kevington, St. Mary Cray, Kent* 1825, *Dartford & Crayford Ship Canal* 1835

Hübbe, H. See **Huebbe**

Hübber, Blasius. See **Hueber**

Huber, C. *Alpine Panoramas* 1869–72

Huber, E. Publisher of Philadelphia. *S. Lewis U.S.* 1816

Huber, Emil. *Selkirk Range B.C.* 1891

Huber, Gaspar. Engraver of Kilkenny. *Siege Kilkenny* 1645

Huber, J. *Sardinische Mon.* (1855). *Polen* 1858, *Schul u. Reise Karte* 1867

Huber, Johann Heinrich (1677–1712). Engraver with Schalck of Scheuchzer's *Switzerland* (1712)

Huber, Josef Daniel (fl. 1769—85). Military cartographer. *Scenographie Wien* 1769—72

Huber, Thomas. *Profil durch Deutschland* 1887

Hubert, R. Engraver. Cadomi Maps for Bochart, *Geographiae Sacrae,* 1651

Huberti, Adrian. Engraver. *Ville d'Amiens Arras, Cales, Groningen &c.* (1605)

Hubertis, L. A. See **Uberti**

Huberts, Wilhelmus Jacobus Arnoldus de Witt. (b. 1829). *Nieuwe Geogr. Atlas* (1870), *Hist. Geogr. Atlas* 1870

Hubinger, Alan. (19th century) Globe-maker

Hubley, Adam. (1744—1793) Lt. Col. Pennsylvania Regiment. *MS maps of army encampments*

Hubner, A. *Transvaal* 1877

Hübner, J. See **Huebner,** J.

Hübschmann, D. See **Huebschmann**

Huc, E.R. (1813—1860). French missionary, b. Toulouse. Visited China and Tibet

Huchet de Cintré, H. M. F. *Plan du Port de Teavarua* 1862

Huddart, Capt. Joseph (1741—1816). Hydrographer to the East India Company, Elder Brother of Trinity House 1791. *St Georges Channel* 1779, *New Pilot for E. Coast Scotland* 1792, *Charts Coast Ireland* 1794, *Coasting Pilot G.B. & Ireland,* 1791—1803, *New & Enlarged Baltic Pilot* 1809, *W. Coast Ireland* 1812—1813

Huddleston, John. *Parish of Alldingham Lancs.* 1848

Hudson, G. of Woolwich, Surveyor. *Dagnall Hill Farm, & Pedley Property* 19th century MS

Hudson, Henry (d. 1611). Navigator and cartographer. *Tabula Nautica* 1612. The first map to show his strait

Hudson, J. *Geogr. Vet. Script. Graeci Min.*

Hudson, John. *Guide to Lakes* 1843

Hudson, John. *Kyme Eau (Lincs.)* 1792

Hudsworth. See **Robinson, Son & Hudsworth**

Hue, Gustave. *Atlas de Géogr. Militaire,* Paris 1879

Huebbe [Hübbe], Heinrich (1803—1871). Maps for Stieler 1824—28, *N. Americanische Freistaaten* 1824, *Hand Atlas* 1834, 1840, 1843, 1848

Huebbe, J. *Mapa de Jucatan* with Perez 1878, *Halbinsel Yucatan* 1879

Huebbe [Hübbe], S.G. *S. Australian Stock Routes* 1896

Hueber [Hüber], Blasius (1735—1814). Compiled with Peter Anich an *Atlas Tyrolensis* 21 sh. 1774, *Prov. Arlbergica* 1783

Hueber, D. *Postkarte der Oesterreich. Monarchie* 1827

Huebinger, Melchior. *Atlas Scott Co., Iowa* 1894

Huebner [Hübner], A. *Natal & Orange Fluss Freistaat* Gotha 1871, *S. Africa* 1872

Huebner [Hübner(n)], Johann (1688—1731). Geographer of Hamburg. Worked with Homann and Palairet, Homann's *Method. Atlas* 1719, *Denmark, Germany, Lusatia, Switzerland* 1716, *Abrégé Géogr.* Otten's 1735, Homann's *Grosser Atlas* 1737

Huebschmann [Hübschmann], Donat (1540—1583). German engraver. Sambuco's *Hungary* 1566

Huebschmann, Gustav. *Mittel Europa* 1869

Huefer [Hüfer], H. *Nova Zemla* 1874

Huerne de Pommeuse. *Lac de Nicaragua* 1833

Huenerwadel [Hünerwadel], Gottlieb Heinrich (1769—1842). Military cartographer

Huener. *Oldenburg* 1804

Hues, Robert (1553—1632). Mathematician. *Tractatus de globis* 1592, the standard work

J. Hübner (1668–1731); mezzotint portrait by J. Kenckel

on globes in 16th and 17th centuries.
Dutch editions from 1597, French 1618

Huet, A. *Amsterdam* 1874

Huet, Luis. *Plano de la Villa de Pansacola* 1781

Huet, Pieter Daniel. *Terrestial Paradise,* Mortier (1700)

Huettenbacher [Hüttenbacher], Franz (1772–1826). Czech topographer

Hüffer, H. See **Huefer**

Hufnagel. See **Hoefnagel**

Hufty, S. American engraver of Philadelphia. For Fielding Lucas of Baltimore 1823, for Carey 1823

Hugeden. See **Higden**

Hugel, Baron Charles (1796–1870). *Map Punjaub* MS 1846, pub. Arrowsmith 1847

Huggins, Sir William (1824–1910). English astronomer

Hughes. *County adjacent to River Mississipi* (1713) MS

Hughes, Andrew, *S. Carolina* (1750), *Georgia* 1789

Hughes, E. *School Atlas phys., polit. & comm. geogr.* 1853, *Hand Atlas for Bible Readers* 1848 and 1856

Hughes, Mathew. *City Orange, N.J.* 1872, *Essex Co. N.J.* 1874

Hughes, Rev. Griffith. Rector St. Lucy's, Barbados. *Barbados* 1750

Hughes, Col. G. W. *Railroad of Panama* 1849, pub. Kiepert 1856

Hughes, Hugh. Publisher. 15 St Martins Le Grand. *Plan London* 1847

Hughes, James. *Town Crawford, Orange Co. NY* 1863

Hughes, John. *Map Baltic* (1854)

Hughes, J. H. *Mon. & Bucks. (Tasmania)* 1837

Hughes, John T. *New Map Mexico* 1848

Hughes, M. & J. Surveyors. *Part of New Jersey* 1867

Hughes, Luigi (b. 1836) Italian geographer. *Nuovo Atlante Geogr.* Roma 1889

Hughes, Michael. *Plumstead Township Penn.* 1859, *Town of Newburgh, Orange Co., N.Y.* 1864

Hughes, Philip. *Country round Fort Erie* 1814

Hughes, Price (d. 1715). Welsh adventurer. *Southeast (U.S.A.)* 1713 MS (lost)

Hughes, R. B. *Map of Rivers Parana & Paraguay,* W. Forshaw Lith., Liverpool 1842

Hughes, Samuel. Civil engineer. *Essex Railway Survey* 1835, M. Tyas' *Geology of England* 1841

Hughes, T. Publisher, Stationers Court. Lambert's *Plan London* 1806

Hughes, Thomas. *Union Township, N.J.* 1860

Hughes, William (1817–1876). Geographer and Engraver, Aldine Chambers, Paternoster Row, London. *Illust. Atlas of Script. Geogr.,* Knight 1840, *Atlas of Constructive Geogr.* 1841, Maps for A & C Black 1840–53 and Stamford, Mackenzie *Mod Geogr.* (1866), *National Gazetteer* 1868

Hughes. See **MacAuley & Hughes**

Hughs, John. (1703–1771). Government printer to House of Commons

Hughson, David. pseud. for Edward Pugh

Hugo, a Porta. Publisher. *Ptolemy* 1541

Hugunin, Robert. *Chart Lake Erie* 1843

Huilier, J. L'. See **L'Huilier**

Hulagu Khan. Constructed Observatory 1264

Hulbert, E. J. with J. C. Booth. *Geolog. Map Lake Superior* 1855–64

Hulet, John. Estate surveyor. *Parish Maps of Essex* 1654 MS

Hulett, J. Engraver for Le Bruyn 1744

Hulett, William. Engraver for Kaempfer 1727

Hulke, John Whitaker (1830–1895).
British surgeon and geologist

Hull, Abijah. Surveyor. *Plan Detroit* 1807

Hull, Edward. (b. 1829). *Geolog. Survey
Ireland* 1867–7, *Geolog. map Ireland*
1878

Hull, George L. *Morristown, Morris Co.
N.J.* 1874

Hull, Thomas A. Staff Comm. *Atlantic
Ocean Pilot,* London Admiralty 1868,
Current Charts Pacific 1872

Hull, W.H. with W.H. Johston. *Caloom-
byan Harbour* 1819, *Samangca Bay* 1819

Hullmandel, Charles (1789–1850)
Lithograph Printer. *City of Bunarus*
1822, *Vicinity St. Leonards* 1829

Hullmandel & Walton. Lithographers,
London. *California* 1849, *S. Essex
Estuary Plans* 1852

Hülmer, Johann. Member of Homann
Erben

Hulot, Baron (b. 1857). Secretary, French
Geographical Society. *Voyage de
l'Atlantique au Pacific* (1888)

Huls, Joa. *Plan Jerusalem* ca. 1480

Hulsbergh, Henry. Engraver. Mortier's
Cadiz 1703, De Fer's *War Savoy* 1703,
Senex' *World* 1711, *Plan Preston* 1715,
Plan Claremont (1720), *London after Fire*
1724, *Plan St. Pauls* 1726

Hulsius, Friedrich, Junior. Gottfrieds'
Svecia, Frankfort 1632

Hulsius [Hulst], Levinus [Lieven] (d.
1605). Publisher, geographer and instrument maker, *Danube* 4 sh. 1596, *Collection
of 26 Voyages to E. & W. Indies* 1590–1650,
Frankfort, *Totius Orbis Terrae* 1598. Translated *Epitome Ortelius* into German 1604

Hulst van Keulen. See under **Keulen**

Hulton, J. G. *Saudi Arabia & Yemen*
1836

Hulton, Robt. Mapseller at the Corner of
Pall Mall, St. James. Roades's *Plan London*
1731

Humann, Carl. *Asia Minor* (1872),
Bulgarien 1877, *Syrien* 1890

Humbert, Abraham (1689–1761).
Military cartographer

Humbert, C. J. von. *Belagerung von
Maynz* 1793, *Plan v. d. Insel Potsdam,
Berlin* 1800, *Stadt Hersfeld* (1830)

Humbert, Pierre. Publisher. *Frezier
Voyage,* Amstd. 1717

Humbert, Pierre, Junior. City Surveyor.
Plan Boston 1895

Humble, George (fl. 1599, d. 1640).
Publisher, Book and Print Seller, nephew
and partner of John Sudbury, with whom
he published Speed's *Atlas* 1611, 1614–
16. Published 1627–31 edition alone.
Addresses: White Horse, Pope's Head
Alley 1610–27 and Pope's Head Palace
1627

Humble, William (fl. 1640–59). Publisher
and Bookseller of Pope's Head Palace.
Published 1646–50 edition of Speed.
Son of G. Humble

Humboldt, Friedrich Heinrich Alexander
von, Baron (1769–1859). *Voyage de H. et
Bonpland* 1805–39, *Atlas de la Nouvelle
Espagne* 1811, *Atlas du Nouveau
Continent* 1814–34, *Atlas Cosmos* Paris
1867 (compiled by Traugott Bromme).
Invented system of isothermal lines

Humboldt, A von

Hume, Rev. Abraham. *Liverpool* 1858,
Religious Map of England 1860

Hume, Alexander. *Harbour Nangasaki*
pub. 1788

Hume, Hamilton (1797–1873). Australian
explorer

Hume, R. *British Honduras,* London,
Weller 1888

Humelius, Johannes (1518–1562).
German mathematician and geodesist.
Risse 1556–62 MS

Humfrey, John. Supposed author of
Sketch of Action of Bunkers Hill 1775,
engraved Faden & Jeffreys, Earliest
separate map of Battle of Bunkers Hill

Humfrey, Major John Hambly. Q. M. G.
Dept. *St. Sebastian & French Frontier*,
J. Wyld 1840

Hummel, Aug. (1839–1898). *Geographical
Manuals*

Hummel, Bernard Friedrich. *Handbuch
d. alten Erdbeschreibung*, Nürnberg
1785–93

Humphreys, Andrew Atkinson (1810–
1883). Topographer and hydrographer,
Chief of Engineers U.S. Army, b.
Philadelphia. *Surveys Mississippi Delta*
1850–1, *Mil. map Penin. of Florida*
1856, *Mil. Dept. of Oregon* 1858, *Army
of Potomac* 1869, *Geological Atlas* 1876–
81

Humphreys, Clement. County surveyor.
Northern Portion of San Francisco County
1852–3

Humphreys, Daniel. Publisher *Plan of
Attack on Fort Sullivan* 1776

Humphreys, Hugh. Publisher of Carnarvon.
Map of N. Wales as an old woman (1850)

Humphreys, William. Surveyor. *Caecil
County from survey of the roads in 1792*

Humphrys, F. Engraver for Tanner 1839

Hunaeus, Prof. Dr. *Hanover* 1864.

Hunckel, G. *Karte der Weser* (1846)

Hundeshagen, Bernard (1784–1849).
Mainz Ville 1815, *Maynz u. Umgebungen*
1843

Hunfalvy, Janos (1820–1888). Hungarian
geographer. President Geogr. Soc.
Budapest. *Magyar kezi atlasz* 1865

Hunnius, A. Cartographer. *Dept.
Missouri New Mexico* 1873

Hunsen, J. *Empire de Djambi* 1878

Hunt, C. C. Explorer *W. Australia*, Perth
1865

Hunt, F. W. Colton's *Hist. atlas* 1860

Hunt, James. *Mannor of Welles* 1668 MS

Hunt, John. *Draught of St George's Fort*
1607

Hunt, John P. *City of Philadelphia* 1875

Hunt, Nathaniel. *Miniature Terrestial
Globe* 1756

Hunt, Capt. Phineas. *Carnicobar Island*
1769

Hunt, Richard S. *Texas, N.Y.*, Colton
1839, with J. F. Randel *Guide to Texas*
1845

Hunt, Robert. *Mem. Geolog. Survey
Great Britain* 1855–82

Hunt, Thomas Carew. *Bay of Ponta*
1839, *Ponta Delgada* 1839

Hunt, Thomas Sterry (1826–1892).
American geologist and chemist

Hunt & Eaton. *Columbia Atlas* (1893)

Hunte, John. Pilot of Plymouth. *S. W.
Ireland*, pub. Gerritsz 1612

Hunter, Assistant Engineer. *Plan I. aux
Noix (Lake Champlain)*, 1780 MS

Hunter, A. Engraver. *Newcastle–Maryport Canal* 1795

Hunter, C. M. *Atlas Williamsport, Pa.*
1888

Hunter, E. *Joliet, Illinois* 1859

Hunter, George. Surveyor General.
Cherokee Nation 1730 MS

Hunter, Lieut. (later Captain) John, R.N.
(1738–1816). Governor of New South
Wales 1795–1800. *Delaware River* 1777,
N. York Harbour 1779, *Sydney Cove*
1788, *Port Jackson* 1789, *Stewarts Is.*
1791, *Port Jackson* 1803

Hunter, Joseph. *Stikine River* 1877

Hunter, M. *Plan & View of Asab* 1783,
Plan Gogo 1784, *Plan & View Moha* 1784

Hunter, R. L. *Wabaay Harbour [Ceram]*
1840

Hunter, Thomas. Printer of Philadelphia.
St. Clair Co. 1876

Hunter, Sir W. W. (1840–1900). *Gazetteer of India* 1894

Huntington, Eleazor, of Hartford Conn., Publisher. *World* 1829, *U.S.* 1830, *Map of Maryland* 1834

Huntington, F. J. Publisher of Hartford. *Western States* 1832

Huntington, Nathaniel Gilbert. *School Atlas,* Hartford 1833

Huntley, W. Land surveyor. *Lexdon Heath (Colchester)* 1830 MS

Huntte, John. Estate surveyor. *Sturmer Mere (Essex & Suffolk)* ca. 1575 MS

Huot, Jean Jacques Nicolas. Revised Malte-Brun's *Géog. Univ.* 1830 and 1837, *Karte der Krim* 1855, *Carte Géologique de la Crimée* 1853

Hurd, D. H. & Co. *Town & City Atlas of New Hampshire* 1892, ditto *Connecticut* 1893

Hurd, Capt. Thomas, R.N. (1757?– 1823). Hydrographer to the Navy 1808– 1823. *Brest* 1804, *Falmouth* 1806, *English Channel* 1811, *Aust. & Tasmania* 1814. Made first exact survey of Bermuda

Huré, Veuve [widow] Sebastian. Publisher, Paris, Rue St. Jacques "à l'image St. Jérome, près St. Séverin". Hennepin *Louisiane* 1683

Hureau de Senarmont, Henri. *Carte Géolog. Environs Paris* 1865

Hurez, J. F. J. Publisher Cambrai. *Plans attack Antwerp* 1817

Hurlbert, J. Beaufort, *Physical atlas of Canada* 1880

Hurley, R. C. *Tourists Guide to Hong Kong* 1896

Hurn, Jury van. *Plan of The Hague* 1616, used by Braun & Hogenberg

Hurrell. See **Sorrel & Hurrell**

Hurst, G. *Ramsgate* 1822 (with Collard)

Hurst, John. Publisher of Wakefield. Greenwood's *Yorks.* 1817

Hurst. See **Longmans**

Hurter [Hurtero] , Johann Christoph. *Alemannia sive Sueviae superioris Choro.* 1625, used by Blaeu 1634 &c.

Hus. Engraver. *Plan Hamburg* 1651 for Meyer

Husain, Sidi Ali Ibn (d. 1562). Turkish astronomer and hydrographer

Husen, Franciscus. Engraver. *Sicily* for Du Sauzet 1734

Huske, John. *N. America* 1755

Husman, Johann. *Samsoe* 1675, *Scania* 1677

Hussey. Surveyor. *Land near Horsenden* 1806 MS

Hussey, William. Surveyor. *Tilford House & Farms* 1855 MS

Husson, Pieter (1678–1733). Publisher and Bookseller at the corner of the Kapel brugh The Hague. *Les XVII Provinces* 1706, *Bataille de Ramelies* 1706, *Atlas Battlefields Hague* 1709, *Variae Tab. Geog.* (1710)

Hustler, John. *Canal Leeds to Liverpool* 1788

Hutawa, Edward. Lithograph Publisher, St. Louis (Missouri). *Oregon Territory* 1843, *Mexico & California* 1848

Hutawa, Julius. Lithographer. 2nd St. No. 45 St. Louis. Worked with Edward above. *Railroads U.S.* 1849, *U.S.* 1854, *Vicinity of Mexico* 1863

Hutcheon, T. S. Civil engineer. *Plan Elgin* 1855

Hutchings, W. F. Surveyor. *Staffs.* with J. Phillips 1832, *Cheshire* with W. Swire 1830

Hutchings & Rosenfield. Publishers San Francisco. *Overland mail routes* 1859, for De Groot 1860

Hutchins, J. N. *Almanack* for 1759 (with *Plan of Louisbourg* 1758)

Hutchins, Capt. Thomas. 60th Foot Regt. (1730–1789). First and only Geographer to the United States. *Ohio and Muskingum*

Rivers 1764, *Battle near Bushy Run* 1766, *W. Part Virginia* 1778, *Illinois Co.* 1778, *Seven ranges of townships* 1785

Hutchinson, T. Engraver for Heath's *Is. of Scilly* 1748–9

Hutchinson, Thomas. Engraver and draughtsman. *England & Wales* 1747, *Northumb,* for Osborne's *Mag. Brit.* 1748

Hutchinson, Lieut. W. C. *India District Hoshiarpoor* (2 miles to inch) 1853

Hutchison, John. *Port Jackson* 1859, *N. Australia* 1864, *Port Adelaide* 1869, *Ports Gulf St. Vincent* 1870

Huthwaite, Samuel. *Estate map Annsby Lincs.* 1718

Hutson, George, Junior. Estate surveyor. *Estate plans in Essex* 1771–92

Hutter, F. X. Cartographer. *Poland,* Augsburg 1796

Hutter, Otto. *Umgebung von Kempten* 1867

Hutter, P. H. Publisher Frankfort *Eveché de Paderborn*

Huttich, Johann. Alsatian scholar. Actually compiled text (attributed to Grynaeus) for *Novus Orbis regionum,* Basle 1532: world map attributed to Münster (geog.), Holbein (decoration). Material for Alsace used by Münster

Hutton. Engraver *De Witt's Plan Albany* 1794

Hutton, Charles. Publisher. *Plan Newcastle* 1770 and 1772

Hutton, Dr. Charles, F.R.S. of Royal Military Academy, Woolwich. (d. 1823). First to use contours in Britain. Proposed Mudge & Colby for Ordnance Survey

Hutton, James (1726–1797). Scottish geologist. *Theory of the Earth* 1785

Hutton & Corrie. *County of Dumfries Ordnance Survey* 1859

Huusmann, J. (d. 1711). Engraver of Copenhagen

Huveune, J. *Carte topo. Bruxelles* 1858

Huyberts, K. Engraver for De Fer 1699, *Meuse* Mortier (1700)

Huyett. See **Parker & Huyett**

Huygens, Christiaan (1629–1695). Dutch mathematician, astronomer and physicist. *Horologium* 1658, *Cosmotheros* 1698 (posthumous)

Huygens, Capt. Henry, R.E. *MS atlas* for Clive of India 1765–6

Huys, Pieter (ca. 1519–1581). Engraver for Plantin 1568

Huyser, C. V. de. *Plans of Arabia* 1774

Huysche, Capt. George Lightfoot. *Cape Coast Castle* 2 sh. 1873, *Routes into Ashantee* 1873

Huyssen, van Kattendyke W. J. C. *W. Coast Kiusiu (Japan)* (1860)

Hyacomilus, M. See **Waldseemüller**

Hyakuga (18th century). Japanese geographer. *Map of Yamashiro Province (Kyoto)* 1778

Hyde, G. W. *Sect. Map Oregon,* N.Y. 1856

Hyde, I. Mapseller "under ye N. Pizza of ye Royal Exchange". Price's *30 miles round London* 1712

Hyde, W. & Co. American publishers, Boston. Field's *American School Geog.* 1832

Hyde, Lord & Duren. Publishers, Portland, Maine. Warren's *Geog.* 1843

Hydrographical Office of the Admiralty. Founded 1795. A. Dalrymple first hydrographer

Hyett, William. Surveyor. Worked under Mudge 1815, Instructor in Surveying, Royal Military College 1824

Hygden, Ranulf. See **Higden**

Hyginus. Roman surveyor (81–96 A.D.)

Hyginus, Caius Julius. *Poeticon Astronomicon* Venice, Radtolt 1485. The first book to show the figures of the constellations

Hylacomylus, M. See **Waldseemüller**

Hylton, Edmond Scott. Engineer. *Plan St John's Harbour* 1751 (with James Bramham)

Hynmers, Richard. Translator Blaeu's *Sea Mirrour* 1625

Hyozo, Kabo (mid 18th century). Map publisher in Kyoto. *World map* 1744

Hyrne, Edward. *Plan Cape Fear River* 1749–(53)

Hyslop, J. M. *Vestiges of Assyria* 1855 (with F. Jones)

I

I. Often indicates letter **J.**, *q.v.*

I.B. See **Jenichen**, Balthasar

I., E. S. *Karte von Friaul* 1805

I, G. H. *Plan Besancon* 1788

I. Hsing (672–717). Chinese astronomer and cartographer

Iacubiska. *Plan Vienne* 4 sh.

Iaeck, Alberti. Engraver. *Mecklenburg & Ratzeburg* 1788–92. *World Map* Berlin, Schropp 1834

Iago, Rev. William. *Eccles Map of Cornwall* 1877

Iakinth, Nikita Yakovlevich Bichurin, Archimandrite. *MS Maps Tartary* 1823–5

Ibanez, Carlos (1825–1891). Spanish General, geodetician, founder Geog. Institute of Spain, Madrid 1874

Iberville, L. d'. *Environs de la Riv. Misisipi* 1688–9

Ibn Alwardi [Abu Hafs Zain al-Din] (1292–1349). Syrian writer. Compiled a *world map*

Ibn al' Arabi. *World Map* 1240

Ibn Al Faqib. (10th century) Arab geographer

Ibn Batouta [Abou Abdullah] (1302–1377). Arab traveller and author b. Tangier

Ibn Haukal, Aboul Quasim Mohammed (10th century). Persian geographer. *Kitab el Masilik (Book of Routes & Countries)* ca. 976 (with maps)

Ibn Husain. See **Husain**

Ibn Khaldoun, Abou Zeid (1332–1406) Arab historian and taveller, b. Tunis, d. Cairo. *Hist. Arabs & Berbers,* translated by Slane 1847–51

Ibn Said al Maghribi. *Somali Coast* 1250

Ibn Yunus. *World* 1008

Ibrahim, Efendi, Mutafarrikah (1674–1744). Hungarian, resident in Constantinople where he started the first printing press. *Gihannuma* 1732 (with maps)

Ibrahim, Ibn Said-as-Sahli. Astronomer. *Globe* 1080

Icano, E. See **Eitoku Kano**

Ichizaemon, Shimaya. Japanese explorer. *Portolan chart of the Bonin Islands* 1675

Ides, Evert Isbrand [Everardus, Ysbrants] (ca. 1660–1705). *Nova Tab. Imp. Russici* 1704, English edition 1706

Idiaquez, E. *Mapa Elam. de Bolivia* La Paz 1894, *Prov. de Caupolican* 1899

Idrisi [Al-Idrisi, Edrisi, Muhammad ibn Muhammad] (1099–1164). Arab geographer b. Ceuta. Compiled a geography for the court of Roger of Sicily. *World map* 70 sh. 1154. Abridged version in 1161

Ienefer, Capt. *Draught Golden Islands,* Darien (1700)

Ieremin. Engraver for Piadischeff 1823–6

Iglesias, Miguel. *Carta Hidro. Valle de Mexico* 1862

Ignaz, George, Freiherr v. Metzburg (1735–1798). b. Graz d. Vienna. Mathematician and cartographer. *Postkarte* 1782

Ihering, Dr. H. von. *Maps of Brazil* 1887

Ikeda, Tori. Japanese cartographer. *Shinano Kuni Kyoto* 1835

Ikku. Mapmaker of Osaka. *Character map of Japan* ca. 1823

Ilacomilus, M. See **Waldseemüller**

Iles, John Alexander Burke. *Island of Nevis* 1871

Iliffe & Son [later Iliffe, Sons, & Sturmey Ltd.). Map printers. *Warwick* 1893, 1899

Il'in, Aleksyei Aleksyeevich (1832–1889). Russian cartographer. *Russian Empire* 1871, *Kiev Dist.* 1875, *Asiatic Russia* 1880, *School Atlas* 1888, *Road map Vistula* 1889

Ilive, Jacob. *Aldersgate ward,* engraved R. W. Neale 1739

Illis, Thomas. *Walworth Manor* 1681

Illman, T. & Sons. Publishers. Lithog. for Eastman 1853

Illman [Ulm?] & Pilbrow. Engravers and publishers. *Oregon Territory* 1833, Walk's *North America* 1846

Illyn. See **Il'in**

Imbault, A. *Plan du Mans* 1862 (with Desgranges)

Imbert, Duca A. *Adriatisches Meer,* 4 sh. 1867–75, *Brindisi Harbour* 1875

Imbert, J. Leopold. *Possessions Angloises dans l'Amérique Sept.* Paris 1777

Imbert des Mottellettes, Charles. *Atlas syncronistique . . . Hist Mod. de l'Europe* 1834

Imfeld, Xavier (1853–1909). Swiss cartographer. *Alpen Pan.* 1878, *Mont Blanc* 1896, *Relief Karte* 1898

Imlay, Gilbert. *American Topography* 1795, *Plan Rapids of Ohio* 1797

Imle, Ober lieut. *Schul Atlas* 1862, *Neuester Atlas . . . der Erde* (1868)

Immanuel, Friedrich. *Alpine maps, Northern India* 1892–4

Imperial Academy of St. Petersburg. See **Akademia Nauk St. Petersburg**

Imray, James (d. 1870). Hydrographer, publisher and chartseller. Joined Blachford 1836 (Blachford & Imray), on own 1840

Imray, J. & Sons. Bayfield's *Gulf St. Lawrence* 1850 & 1853, *Newfoundland* 1855, *Virgin Islands* 1855

Imray, James F. (d. 1891). Son of above. Continued business, 1899 amalgamated with Norie & Wilson

Inberg, Isak Johan (1835–1893). Finnish cartographer. *Finland* 4 sh. 1876

Incelin. See **Inselin**

Inciarte, Dr. Ing. F. d'. *Orinoco* 1898

Indeisseff. *Plan de Moscow* 1852

Indicot, John. *Kennebek River* 1754

Inga, Athanasius. *West Indische Spieghel* 1624

Ingalls, Capt. Rufus. Q/M. U.S. *Salt Lake City to San Francisco* 1855, *Routes to Fort Vancouver* 1858 MS.

Inganni, G. *Citta di Piacenza* (1855)

Ingber, O. *Umgebung von Frauensee* 1899

Inger, W. D. S. Printer, Frankfort. *Pays Bas* 1784

Ingelheimensis. See **Münster**, Sebastian

Inghirami, Giovanni (1779–1851). Astronomer. *Carta Geomet. della Toscana* 1830

Inglefield, E. (1820–1894). Admiral, R.N. 1852–54. In search of Franklin, *Chart showing N.W. Passage* 1853, *Smith Sound* 1852

Inglis. See **Gall & Inglis**

Inglis, H. R. G. *Road Maps* 1894

Inglish, Robert (d. 1695) *Plan Chelsea Hospital*

Ingraham, Robert G. *New Bedford & Fairhaven, Mass.* 1857

Ingram, I. Engraver. *Planisphere* (1752)

Ingram, J. Engraver. *Carte Minéralogique de l'Election d'Estampes* 1757

Ingram, Capt. T. Lewis. *River Gambia* 1860 MS

Ingrey, C. Publisher and lithographer, 310 Strand, London. *Natal* 1835, *Laby. Londiniensis* 1830–37

Inks, William C. *Plan Franklin, Missouri* (1875)

Innes, C. E. **& Tinkham**. Civil engineers in Grand Rapids, Michigan. *Muskegon River* (1840)

Innes, James. *Ancient cities of London & West.* 1849 (with N. Taperell)

Innes, Robert S. *Kalamazoo Co. Michigan* (1870)

Innes, Lieut. Thomas. Plan for *Dock where no Tide Yarmouth Haven* 1759 MS, *Caledonian Harbour, San-Blas-Kays* ca. 1760 MSS

Innys & Knapston. *Coast Scotland* 1728

Innys, W. Publisher. *Chart Scotland* 1728, Bowen's *Complete Atlas* 1752

Innys, William. London bookseller, Princes Arms, St. Pauls Churchyard

Ino, Tadataka. See **Chukei**

Inselin, Charles. Parisian geographer, engraver for Froger 1699, De Fer 1701–5, Desnos, De Lisle 1707, Placide 1703, Jaillot 1715

Inskip, George H. *Vancouver Island from a Russian chart,* corrected Inskip 1856

Inskip, R. M. Hydrographer, Admiralty. *Puget Sound* 1846–9, *Port Simpson* 1856

Insulanus. See **L'Isle**, Guillaume de

Inu-Bo-Ye-Saki. *Maps of Japan* from 1882

Inverarity, Capt. David. Hydrographer. *Coast China* 1793, *Ceylon* 1800, *Madagascar* 1802–6, *Delagoa Bay* 1806, *Goa* 1812

Iona, Benjamin Ben. *World* 1160–1173

Ireland, T. Lithographer of Montreal. *Missouri to Walla Walla, Oregon* 1851

Iriarte, H. **y Cia.** Lithograph printers of Mexico. *Baja California* 1858

Irmer, Anton. *Plan Zerbst* (1880)

Irminger, C. L. C. (1802–1888). Danish admiral and hydrographer. *Works on ocean currents* 1854–5

Irrgang, Oswald (1880–1902). Carto-lithographers of Leipzig

Iruhae, Carol. *Orbis Terr. Ant.* 1827

Irvine, H. *Map Guilford Conn.* 1852, *Town Waterbury, (Conn.)* 1852

Irvine. *Survey Virginia & Maryland* 1728 (with Mayo)

Irving, Benjamin Atkinson. *Atlas Mod. Geog.* (1866)

Irving, R. D. *Geolog. Atlas Wisconsin* 1877–82

Irving, Washington (1783–1859). American writer and geographer

Irwin, George. *Ref. Map Street List Dublin* 1853

Irwin, Lieut. James R. *Ohio Boundary* 1835

Irwin, R. *Forrest Co., Penn.* (1868)

Irwin, S. M. *Venango Co., Penn.* 1860

Irwin, Samuel D. *Forrest Co., Penn.* (1868)

Irwin, W. *Chart River Kenmare* 1749–51

Isaacs, Abraham. *Harbour and city of Louisbourg* 1748

Isakov, J. A. (flourished 1862–1876)

Isbister, Alexander Kennedy. *Expedition to Peel River (Canada)* 1839–41

Isenmenger, S. *Libellus geographicus,* Tubingae 1562

Isennachcencis, Jodocus. See **Trutretter**, Jesse

Ise-No-Kimi (683 A.D.). First known Japanese surveyor. *Provincial surveys* (now lost)

Ishikawa, Rin Tan. Japanese cartographer. *Roads of Japan* 1730

Ishikawa, Toshiyuki [Ryusen] of Edo (fl. 1688–1713). *Maps of Japan and European type World map* 1688

Isidore of Seville (Bishop) (560–636). *Origines,* an encyclopedia with section on Geography. *TO map,* first printed edition Augsburg, Zainer 1472, The TO map it contains is the first printed map

Isingrin[ius], Michael. Printer, Basle. Collaborated with Heinrich Petri in publishing some of Munster's works

L'Isle, Claude (1644–1720). French geographer, father of Guillaume, Joseph Nicolas and Louis. Prepared *maps for Iberville's voyage* and an *Atlas Hist.* 1654

L'Isle [Insulanus], Guillaume de (1675–1726). French cartographer, pupil of Cassini, called Father of Modern Geography. Member Academy Sciences 1702. Premier Géographe du Roy 1718, Addresses Paris: Rue des Canettes près de St Sulpice (1700–1707), Quai de l'Horloge à la Couronne de Diamans (1707–8), Quai de l'Horloge à l'Aigle d'Or, 1708 onwards. Compiled *globe* 1699, *World map & continents* 1700. Produced over 100 maps and atlases

L'Isle, Joseph Nicolas (1688–1768). French cartographer, brother of Guillaume. Took service under Peter the Great and produced the first printed *Russian Atlas* 1745, Returned to France 1747

L'Isle de la Croyère, Louis (d. 1741). French astronomer, took service with his brother Joseph Nicolas at Russian Court, died while accompanying second expedition of Vitus Bering

Islenieff, Johanne Flu. *Irtisch* 1777, *Russia* 1779, *Tobolsk* 1780, *Asowensis* 1782

Isler, John B. *Map Caroline County, Maryland* 1875

Israel, A. B. *Treatise Use Globes* St. Louis 1875

Israel, Capt. S. *Damaraland & Namaqualand* 1885–6 MS

Issel, Arturo (b. 1842). Italian naturalist and geographer

Isselburg, Peter. *Plans of sieges in Germany* 1620–21

Issleib, Wilhelm. *Staaten Theile der Erde* 1869, *Deutschland* 1869, *Brandenburg* (1870?), *Central-Europa* (1870?) *Oesterreich-Ungarn* 1871, *Toutes les Parties de la Terre* 1873

Isslenief. *Map Russia* 31 sh. 1753–86

Istakhri, Abu Isak al Farisi. (10th century). Persian geographer. *Victab el Aqualin* (Book of Climates Islamic Atlas) 934 A.D.

I'Toya, Zuiemon. MS *Portolan of World* ca. 1590 (lost)

Ittar, Sebastiano. *Malta* 1791

Ivachinzov, Nicolas (1819–1871). Russian hydrographer and traveller. *Explor. Caspian* 1866–69

Ivanov, K. (d. 1666). Russian cartographer

Ivanov. *Buchara and Afghanistan* 1884

Ive [Ivye], Paul (fl. 1580–1602). Surveyor. Worked in Ireland, *Coasts England* 1600 MS, *Kinsale Harbour* 1601, *Corke* 1602, *Coasts W. Europe* (attrib.) ca. 1590

Ivens, Robert (1850–1898). Spanish explorer in Africa. *Corso do Rio Zaire* 1885, *Portugiesches Süd Afrika* 1887

Iverque, Baron d'. See **Grante**, J. A.

Ives, Edw. *Travels E. Indies and Persia* 1773

Ives, Joseph Christmas (1828–1869). Military cartographer and traveller. *Florida* 1856, *Rio Colorado* 1858

Ives, William. Cartographer 1844

Ivory, T. Engraver. *Scotland* in Chalmers' *Gazetteer* 1803

Ivoy, Col. d'. *Plan Bataille Hochstett* 1704, *Tirlemont* 1705

Ivry, Content d'. *Plan de la Madelaine* 1761

Ivye, Paul. See **Ive**

Iwanoff. Engraver for Piadischeff 1823–6

Izard. John Grafton. *Plan of the Parishes of St. Giles in the Fields and St. George, Bloomsbury* 1890

Izumiishi, Takami. *Takeguchi Shigesada* 1849, (Printed map translated by Takami)

J

J. J. *Danemark* (1866)

J., J. H. Ed. *Hübner's Museum Geog.* (1720)

Jablonowski, A. *Historical Atlas of Poland,* Warsaw-Wien 1889—1904

Jablonowski, Prince Joseph Alexander (1711—1777). Patron of Polish cartography. *Map of Poland* 1772 (with Rizzi-Zannoni)

Jachnick, Lieut. *Feste Koenigstein* (1792), *Belagerang Maynz* 1794

Jack, A. Lithographer. *Belag. bei Kehl* 1870, *Belag. bei Strasburg* 1870

Jack, Robert Logan. *Geolog. Map Queensland,* Brisbane 2 sh. 1886, ditto 6 sh. 1899; *Chillagoe* (1898)

Jack, C. See **Jaeck**

Jackson, Capt. *Plan of Lake George surveyed* 1756 (inset on Brasier's *Lake Champlain)*

Jackson. Engraver, No. 19 Abbey St. Dublin. Rocque's *Dublin* (1780)

Jackson, F. *Arctic Regions,* Phila. 1897, *Franz Josef Land* 1897

Jackson, H. *Triang. Prov. Wellington* 1872

Jackson, Capt. H. M. *Settlement Sierre Leone* 1884, *Sketch Plan Prahue* 1881, *Anglo Siamese Boundary Conm.* 1890

Jackson, John. *Philippine Is.* 1814

Jackson, Lieut. *Survey from Midnapur to Nagpur* 1818

Jackson, Lieut. R.N. *Honolulu Harbour* 1881

Jackson, Luke. Engraver for *Scull's Pennsylvania* 1778

Jackson, Major M. *Plan Siege Rangoon* 1852

Jackson, W. A. *Mining District of California* 1851

Jackson, W. Publisher in Oxford. Taylor's *Plan Oxford* 1751, *River Witham* (1750)

Jackson & Cowan, Publishers, Glasgow. Lothian's *Co. Atlas Scotland* 1826

Jacme, Mestre. See **Cresques**

Jacob, A. A. *Plan Nurbudda Mineral Districts Surat* 1854

Jacob, E. Surveyor. *City of Poughkeepsie* 1857

Jacob, Edward. *Town Faversham* (1770)

Jacob, F. G. *Blue Rapids, Kansas* 1870

Jacob, Capt, J. *Plan Battle Meanee* 1843

Jacob, J. Publisher of Vienna. *Nov. Silesie Theat.* (1750)

Jacob, Leon. *Colonie du Gabon* 3 sh. (1892)

Jacobi, C. Cartographer. *Bibel Atlas* 1891

Jacobs, J. S. *Theat. Bellorum a Cruce sig. gest.* 1842

Jacobs, Rudolf. *Oest. Erdhälfte* 1838, *Erdkarte* 8 sh. 1848

Jacobs, S. Engraver for Duflot de Mofras 1844, for Gide Paris 1847, *Geo. Map V.D.L.* 1847, *India* 1861, *Atlas Cosmos* 1867

Jacobs & Barthelemier. Engravers. *Europe Railroads* 1861

Jacobsen, Françoijs. *Chart of Tasmans Voyages* ca. 1666

Jacobsen, F. Surveyor. *Grand Plan Crysolite mine Iviktout (Greenland)* Godthab 1860

Jacobsz [Theunis, Theunits, Theunitz or Lootsman] Family of Amsterdam. Marine publishers and printers. Used the name Lootsman [Pilot] to distinguish themselves from others with the same name. Firm founded by:

—— Anthonie Jacobsz [Theunis] (ca. 1606–1650) "op het Water in de Lootsman tussen de Oude en Nieuwe Brugh." Pub. a *Zeespiegel* in 1643 and many later editions, French editions from 1666 and English editions from 1649. *Straets boeck* 1648 and many later editions, French editions from 1659 and English from 1678. Business continued by his widow until his sons old enough to take over.

—— Jacob and Caspar Jacobsz [Theunis, Lootsman] Sons of the above. Jacob died in 1679 and Caspar in 1711. Sometimes published separately, sometimes together. Jacob published *Water Werelt ofte Zee Atlas* in 1666, and the same Atlas jointly with his brother Caspar in 1676. English edition 1668. Caspar continued to publish *Water Werelt* after the death of his brother, with editions from 1681 onwards. French and English editions appeared in 1681 and later. A. Jacobsz's plates were sold to Pieter Goos and were copied by Doncker. His sons Jacob and Caspar collaborated with Goos and Doncker.

Jacobszoon, Anthony (fl. 1653–1663). Publisher in Amsterdam

Jacobus. Angelus of Scarparia. Translator. *Ptolemy* 1409

Jacobus, Belga. See **Bussius Belga** Jacobus

Jacobus de Mailo (16th century). Hydrographer

Jacobus, Russins of Messina, Ms. *Mediterranean* 1564, a second in 1588

Jacotin, Col. Pierre (1765–1827). French cartographer, followed Bonaparte to Egypt. *Basse Egypte* 1800, *Carte Gén. de l'Egypte* 47 sh. Paris 1807, *Egypte* 3 sh. 1818

Jacoubert, T. *Atlas Gén. de Paris* 1836

Jacovlev. *Plan v. Samarkand* Gotha 1865

Jacquemart, Alfred. *Atlas Colonial* Paris (popular edition) 1890

Jacquemart, C. *Plan Ville de Lyon* 1747

Jacques, Joao Candido. *Rio Grande do Sul* (1893)

Jacquot, E. *Carte Géolog. du Gers* 1869, *Carte agron . . .* Toul 1860

Jacubicska, Stephan (1742–1806). Austrian military cartographer. *Plan de Vienne* 4 sh.

Jaeck (Jäck), Carl (1763–1808). Engraver, publisher of Berlin. *Plan Insel Potsdam* 4 sh. 1800, *Teutschland* 1795, *Schlacht bey Pirmasens* 1797, *Ost Preussen* 1796–1809, *Westphalen* Berlin 1805, *Postkarte Deutschland* 1813, *Halberstadt* 1788–1794

Jaeck, W. Engraver 1860

Jaeger, Carl Friedrich Julius (1815–1857). Lithographer. *Map Java*

Jaeger [Jäger], E. Publisher of Stuttgart. *Schul Atlas* 1862

Jaeger, Franz Anton. *Rhoengebirg in Franken* 1802

Jaeger, F. W. (19th century). Map publisher. *Grosser Schul Atlas,* Hamburg 1838, *Kleiner Schul Atlas* 1837

Jaeger, J. *Zak Atlas* 1843, *Nederland* 1850, *Zak Atlas* 1853, *Schoolatlas* 1858, *Spanje* 1874

Jaeger, Johann Christian. Publisher in

Frankfurt. *Atlas* 1796, *Holland* 1784, *Kriegs-Schauplatz* 1788

Jaeger, J. G. A. *Grand Atlas d'Allemagne,* 81 leaves, Frankfurt 1789

Jaeger, Johann Wilhelm Abraham (1718– 1790). Artillery officer, cartographer, publisher and bookseller of Frankfurt am Main. *Théatre de la Guerre Russie-Turquie* 6 sh. 1770, *Allemagne* 1780, *Etats de Saxe* 1779, *Grand Atlas d'Allemagne* 1789

Jaegerschmid, A. *Q. de Ste Catherine,* Cuba 1834

Jaettnig, Ferdinand. Engraver. *Berlin* 1825

Jaettnig, Karl (1796–1835). Berlin engraver and publisher. *Pomerania* 1789, *Oostfries* 1804, *Westphalen* 1805, *Oesterreich* 1810

—— Karl Junior. Engraver (fl. 1830–1850)

Jaettnig, Wilhelm. Engraver of Berlin. For Perthes 1852, for Berghaus 1835–1850

Jaffray, Rev. John. *Sea coast about Peterhead* 1739

Jagellonicus Globe ca. 1510 in Univ. Library Cracow

Jagen, C. van (fl. 1706–1744). Engraver of *Amsterdam Map* together with his son

Jagen, Jan van (ca. 1710–1796). Amsterdam designer, mapmaker and engraver, son of Cornelius. *Atlas Bachiene* 1785

Jager, Robert. *Town & Port Sandwich* 1574

Jager, Dr. G. *Weltkarte,* Gotha 1865

Jagor, Fedor (1817–1900). *Luzon,* Berlin 1872, *Philippine Is.* 1875

Jahn, G. A. *Atlas novus coelestis* 1856

Jahn, H. B. *Nord Ostsee Kanal* Kiel 1886, *Karte v. Kiel* 1890

Jahncke. *Reise Atlas* 1869

Jaillot family. Alexis-Hubert **Jaillot** (1632–1712). Sculptor, engraver, publisher, and geographer was born in Franche Comté and set up in Paris using the im-

print "A Paris chez H. Jaillot joignant les grands Augustins aux deux globes". He died in Paris. He was the first to be created Géographe du Roi (Louis XIV) in 1678. He married first Jeanne, daughter of Nicolas Berey, publisher, map seller, and colourist to the Queen. By her he had seven children. She died in 1676 and Jaillot succeeded to the stock of his father-in-law. In 1676 he married again, his bride being Charlotte Orbane by whom he had eight children.

His works include a *map of Spain* 1688, a *map of Franche Comté* 1669. In the same year he reissued Blaeu's *wall map Africa.* Induced by the heirs of Sanson to join them, he redrew Sanson's maps on a larger scale with fresh embellishments, producing a large folio *Atlas Nouveau* 1674 (many editions including some reprinted in Amsterdam, first by Mortier then by Covens & Mortier). He also published a *Plan of Jerusalem* in 1678 (4 ft. by 6 ft.) In 1693 he issued his Marine charts in *Le Neptune François* and following year 1694 an *Atlas Français* in two volumes. In 1697 a a *map of Palestine* (4 ft. 8 in. by 2 ft 7 in.) and in 1708 *Liste générale des Postes de France*

—— Bernard Jean Hyacinthe **Jaillot** (1673–1739). Son of A.-H. Jaillot, Royal Geographer. *Postes de France* 1713, *Plans Paris* 1713, 411, *France Ecclésiastique* 4 ll., 1731 *Diocèse de Bayeaux* 1736

—— Bernard Antoine **Jaillot** (d. 1749), son of B. J. H. Jaillot. Royal geographer, worked with his brother-in-law Chauvigné-Jaillot. *Routes of the Pretender* (1747)

—— Jean Baptiste-Michel Renou de Chauvigné-**Jaillot** (1710–1780). Brother-in-law. Quay et à côté des Grands Augustins, *Prague* 1757, *Recherches topograph. Ville de Paris* 5 vols. 1775. After his death the stock of Jaillot was liquidated in 1781 (mostly being melted down but some plates bought by Buache and used later by Dezauche)

Jaisse, L. de la. See **Lemau de la Jaisse**

ALEXIUS HUBERTUS IAILLOT, Regis Christianissimi Geographus Ordinarius

A.H. Jaillot (1632–1712)

Jailly, A. *Grandes Routes Comm.* (1868)

Jakubicsk. *Grundriss Residenzstadt Wien* Artaria 1800

Jamar, A. Publisher of Brussels. *Atlas Gén. de Géogr. Moderne* Tarlier 1854

James, Sir Henry. *Escheated Counties in Ireland* 1861

James, Col. Sir H. (1803–1877). Director Topo. Dept., War Office, London. *County round Peking* 1859 (with A. C. Cooke), *Pensacola Bay* (1860)

James, Sir Henry (1803–1877). Director Gen., Ordnance Survey. Applied photo-zincography to ordnance maps 1859. Photozincography 1860. Comp. standards of lengths 1866. *Trig. Survey G.B.* 1858

James, Horton. *Ascension Harbour Adm. Chart* 1840

James, H. E. M. *Manchuria,* London 1887

James, H. F. Lithographic press Ridgefield Manchester. *Plans Market St., Manch,* 1821, 1822

James, J. A. and U. P. Publishers. *Map of Texas* 1836, *Cincinatti, Mexico, Calif., Oregon* 1847

James, J. O. N. Surveyor. *China Coast* 1860, *N. E. Frontier Bengal* 6 sh. 1865, *India* 1870

James, S. A. *Keokuk Co., Iowa* 1861

James, Thomas (ca. 1593–1635). British navigator. *Greenland & Hudsons Bay. Strange & Dangerous Voyage* 1633

James, Lt. Col. Thomas. *Attack Fort Sullivan* 1776

James, T. R. *Victoria Tele. Circuits* 1871

James, W. D. Surveyor. *Prov. Ogadyn* 1855, *Routes in Somaliland* 13 sh. MS 1885

Jameson, T. *Chart of Europe* 4 sh. (1793)

Jameson, Dr. W. *River Napo* MS 1861

Jamieson, Alexander. Astronomer.

Celestial Atlas (1822) etc., *Maps Heavens* 1824

Jamieson, Joan. *Roman Military Way in Muir of Lour* 1785

Jamieson, J. B. *City of Helena* 1859

Jan III, King of Poland. See **Sobieski**

Jan from Stobnicza (1470–1519). See **Joannes de Stobnicza**

Janellus. See **Gianelli**

Janesson, Nicolas (d. 1617). Cartographer and engraver of Amsterdam

Janicke & Co. Lithographers, 3rd St., St. Louis. *Pikes Peak Gold Region* 1859

Jankowsky, J. *Russo Turkish War Map* 1877

Janney, J. D. and E. Publishers. *Lucas Co., Ohio* 1861

Janocki, J. D. Catalogued Polish carto-graphic collection (1775) in Zaluski Library published later by Rastawiecki (Warsaw 1846)

Jansen, H. J. and **Perronneau.** Publishers and booksellers of Paris. Levaillant's *Carte de l'Afrique. Pays entre Mer Noire et Caspienne* 1795

Jansen, J. *Charte von Friedrichstadt* 1851

Jansen, J. F. *Kaart van Europa* (1874), *Wandatlas* 1876–7

Janson. Printer in Paris, for Escudero 1861

Janssen, J. See **Jansson**

Janssen, Leon. *Malacca,* Brussels 1881

[Janssen] Jansz. Father of Jan Jansson of Amsterdam

Janssen, P. J. C. (b. 1824). French astronomer and geographer

Jansson, Jan. Publisher Arnhem. Bertius' *Tab. Geog.* 1600 etc., *Ptolemy* 1617, *Magini* 1617, *Mercator-Hondius Atlas Minor* 1607–21 (joint pub. with H. Claesz). Repub. Heyn's *Nederlantsen Landtspiegel* 1615, *Wytfliet* 1615

Jansson[ius] [**Janssen Johnson**] (Jan)
(1588–1664). Born in Arnhem, married
Elizabeth Hondius, daughter of Jodocus
Hondius in 1612 and settled in Amsterdam.
On death of Jodocus II Jansson together
with Henry Hondius published a series of
Atlases. On death of Henry Hondius,
Jansson continued expanding the *Atlas*.
Maps of France & Italy 1616, *Grand
Atlas* in 11 vols. M. Elix. Hond started
business 1612, son-in-law J. Hondius,
on death H. Hondius 1638 (1651?)
succeeded to business. *Nouveau Phalott*
1637, *Theatrum Universae Galliae* 1631,
Theatrum Urbium 8 vols. 1657, *Atlantic
Major Appendix* 1630, *Theatrum Imp.
Germanicum* 1632. Pub. Blaeu's *Flambeau
de la Navig.* 1620 [piracy?] . Reissued
Braun & Hogenberg 1657. "Opt Water at
sign 'De Pascaerte'. Dwelling upon the
waterside by the Old Bridge at the sign of
the sea mappe" 1618

Jansson Heirs [**Waesburg, van Waesberghe**]
Janssonius haeredes Joannes Jansson van
W. d. 1681. Weyerstraet & Waesberger sons
in-law of J. Jansson. Waesberger acquired
much of J's property at auction 1676.
Sold up by heirs 1694. Pub. *Atlas
Contractus* 1666, J. van W. and Jan van
Someren (1672), *Atlas de Fabrica mundi*
1673

Janssonius, Jodocus Veuve, et Héritiers.
Hist. Vie Frédéric Henry 1656

Jansz, Harmen. "Chaertschryver tot Edam
inden Witten Os". *Map of Atlantic* 1604,
World 6 sh. 1606
—— Harmen and Marten. Chartmakers.
Edam Charts 1604, 1610 MS, *World* 6 sh.
ca. 1606, *World* 1610, *Europe* ca. 1618

Jansz [**Bilhamer**] Joost. Surveyor and
engraver. *Noort Hollandt* 1621

Janszoon, J. See **Jansson**

Janszoon, G. See **Blaeu**, G.

Janszoon, L. Schenk. See **Schenk**

Januensis, Nicolo de Canerio. See **Caneiro**

Janvier, Antide (1751–1835). Parisian
clockmaker

Janvier, Sieur Jean French geographer,
"Rue St. Jacques et l'Enseigne de la Place
des Victoires". *America* 1754 (with
Longchamps), *L'Amérique* 1772, *Paris*
Lattré 1783–1790, *France* 1751 (with
Longchamps), *Isles Brit.,* Bordeaux 1759

Janvrin, Daniel. *Guernsey* (1810)

Jappé, J. H. *Prov. Groningen* 4 sh. 1835

Jaquier. Draughtsman for Engel's
Memoires 1764–(5)

Jaraczewski, J. (1798–1867). Polish
military topographer and surveyor

Jardin, J. R. de M. *Prov. de Goyaz (Brasil)*
1875, *Rio Araguaya* 1879

Jardine, W. Johnston. *Plan Edin* 1850,
Coast of Japan (1859)

Jardot, A. *Chemins de Fer de l'Europe*
1842

Jarman, George. Engraver for Evans &
Sons, London 1857; for Stanford 1863,
for Reynolds 1862

Jarman, R. Collins' *Illust. Atlas of London*
1854, *Portsmouth* 1866

Jarrad, Lieut. F. W. R.N. *Ionian Is.* 1864,
Rangoon River 1877, *Bengal* 1879

Jarrett, John. Surveyor. *Abney Estate* MS
1767

Jarroll. *Plan of Norwich* (1887)

Jartoux, Pierre, S. J. (1699–1720).
Astronomer and missionary in China

Jarwood, Capt. *Plan Delagoa Bay* (inset on
Norie's *Cape of Good Hope* 1831)

Jasinski, Yakub. Polish surveyor. *Military
maps* ca. 1791

Jasolinus, Julius. *Ischia* 1590 used by
Blaeu 1647 and 1662

Jaspers, Jan Baptist (1620–1691).
Engraver, collaborated in Pitt's *English
Atlas*

Jastrzebowsky, Albert. *Carte clim.
Varsovie* 1846

Jättnig, C. See **Jaettnig**

Jaubert, Amédée (1779–1847). *Historical*

Geography. Revised Meyendorffs' *Bokhara* 1843

Jauncey, F. Admiralty surveyor. *Cun-Sing-Mun Canton River* 1840, *Port Ta-Outze* 1832

Jausz, G. (1842–1888). Hungarian cartographer of Jarek

Jay, G. M. Le. See **Le Jay**

Jean. Publisher Paris. Rue S^t. Jean de Beauvais No. 10 Brion de la Tour's *Afrique* 1814, Nolin's *America* 1818, *Africa* 1829, *Océanie* 1837. Collaborated with Mondhare

Jeanes, R. *Geo. Synop. Europa* (1840)

Jeanneret, Georges. *Ile St. Pierre* 1878

Jeanviliers. Engraver, for Nolin's *L'Amérique* 1720

Jeckyll, Thomas. Civil engineer. *U.S.A.* 1857

Jeekel, Lieut. C. A. *Dutch Possessions Gold Coast* 1873

Jeffers, Lt. W. N. J. *Honduras Eisenbahn* 1853, *Port Caballos* 1853, *San Salvador* 1858, *Mexico* (1863)

Jefferson, Peter. Surveyor and planter. With Fry compiled *Map of Virginia* 1671, Surveyed *Va. & N. Carolina boundary* 1749–51 (with J. Fry)

Jefferson, Thomas. Son of Peter Jefferson. *Virginia* 1786

Jefferson, Thomas (1743–1826). President U.S. *Map of the country between Albemarle Sound and Lake Erie* (1786)

Jefferson, T. H. Geographer and publisher, New York. *Emigrant road from Independence Missouri to San Francisco* 1849

Jeffery, James. *Admiralty Charts* 1861–69

Jefferys, Jn$^\circ$ Jn^r. *A New Map laid down from Surveys by ye Wheel of all the Great or Post Roads and principal cross roads throughout England and Wales . . .* by John Jefferys Junr. Teacher abroad of Writing, Arithmetick, Algebra and Geography 1750. Sold by the Proprietor

Jn$^\circ$. Jefferys Junr. in Chapel Street, Broadway, Westminster

Jefferys [Gefferys] Thomas (ca. 1710–1771) Engraver, geographer, and publisher. Red Lyon Street near St. John's Gate 1732–1750. In 1750 he married and moved to St. Martins Lane 5 Charing Cross with shop 487 Strand. One of the most prolific and important English map publishers of the 18th century. Appointed Geographer to Frederick Prince of Wales in 1748 and later to George III. His earliest known work a *Plan of London & Westminster* 1732. He collaborated with Kitchin *Small English Atlas* 1749, with Parson & Bowles *Map of Staffs.* 1747. He engraved maps for Salmon's *Geography* 1749 and for Cave's "Gentleman's Magazine". He issued the last edition of Saxton's *Atlas* in 1752 and *Maritime Ports of France* in 1761.

Between 1751 and 68 he produced important maps on America and the West Indies, Fry & Jefferson's *Virginia* 1751, *Nova Scotia* 1755, De Brahm's *Carolina* 1757 and *St. Laurence* by Capt. Cook (1760). A volume on the *Spanish Islands and West Indies* 1762, and *Topography of North America and West Indies* 1768.

He had surveyed and engraved large scale *county maps of Bedfordshire* 1765, *Hants* 1766, *Oxfordshire* 1766–7, *Durham & West,* 1768, *Bucks* 1770, and *Yorks* 1767–70. The expenses incurred in these ventures probably led to his bankruptcy in 1765, when Robert Sayer acquired a large part of his interests. Sayer in conjunction with Bennet published much of Jefferys' work posthumously, notably his *American Atlas, North American Pilot,* and *West Indian Atlas,* all in 1775.

Jefferys partly recovered from his bankruptcy being joined by William Faden who succeeded to the business and shop on Jefferys' death in 1771

Jefferys, T. Doharty's *Plan of City of Worcester* 1741, pub. 1742

Jefferys & Faden. See **Faden**

Jehenne, Capt. Louis Auguste. *Plans of Ports* 1841–65

Jehotte, L. Engraver. *Dépt. de l'Orte* 1801

Jehuda Ben Zara. See **Zara**

Jekelfalussy, J. de Jekel e Margitselva (1849–1901). Chief Statistical Service Hungary

Jekyll, Thomas. *Geolog. feature country W. Mississippi* (1870)

Jelle, I. *Cleve u. Umgegend* (1848)

Jelly, J. C. Royal Navy. *Port Denison Queensland* 1860 MS

Jelowiecki, Edward. Polish military surveyor 1820–31

Jemeray, Mr. de la. *Partie du Lac Superior*

Jemme, van Dokkum (1508–1555). See **Gemma Frisius**

Jemmett, William. Estate surveyor of Ashford Kent. *Roxwell (Essex)* 1773 MS

Jenderich, Christian. *Würtemberg* (1670)

Jengstrema, L. Publisher, St. Petersburg. *World* 1847

Jenichen, Balthasar (d. 1599). Engraver and publisher. *Palestine* (1570). Crininger's *Meissen & Thuringia* (1570), *Rottenham Franconia* 1571, *Cyprus* 1571, *Tunis* 1573

Jenifer, Capt. John. *Darien, Isle of Pines* MS 1686, *Golden Is.* (1699) *Draught of Golden & Adjacent Islands* (1720)

Jenings, Wm. Moedin & Bengal Artillery. *Plan Madura, S. India* 1755 MS

Jenkins, H. L. *Assam & Burma* 1827–69

Jenkins, John. Surveyor. *Rectory Wraysbury* 1829 MS, *Glebe Land Chapter Windsor* 1834

Jenkins, S. *Lizard to Land's End* 1810

Jenkinson, Anthony (1525–1611). Merchant and cartographer fl. 1545–77. Travelled in Russia and Persia 1557–60. *Russia* 1562, used by Ortelius, De Jode and others

Jenks, William (1778–1866). Cartog-

rapher and explorer. *Bible Atlas* Boston 1847

Jenkyns, Francis. *Plan for St. Ives Bay* 1800

Jenner, Thomas. Bookseller, engraver and publisher (fl. 1623–68). At the White Beare in Cornhill; south Entrance to Royal Exchange *Direction English Traveller* 1643, *New Book of maps, Europe* (1645), 1650; *Kingdome of Eng.* 1644, *Booke of names of all Parishes* (Langren maps) 1643, 1657, 1668; *Sea coast, etc.* 1653, *R. Thames* 1660–7

Jenner, Von Aubonne, F. von. *Canton Bern* (1820)

Jenner, Capt. *Battlefield Chattanooga* (1864)

Jennings, Cyrus. *Kentucky City* 1855

Jennings, D. *Use of Globes* 1747

Jennings, John. Collaborated with J. Leake in his *Plan London* 1667

Jennings, Joseph C. *City of Dubuque Iowa* 1852

Jennings, Robert. London publisher. *Anc. & Mod. Geog.* (1830)

Jennings, Capt. R. H. *W. Baluchistan, E. Persia &c* 6 sh. *Dehra Dun* 1886

Jennings, W. H. *Coalfields Ohio* (1874)

Jennings, William. Jun. Surveyor. *Hillfield & Scutts Farm (Dorset)* 1792 MS

Jennings & Branly, W. E. *Siwa Egypt* 1896–98

Jennings & Chaplin. Publishers, 62 Cheapside. Moule's *English Counties* 1830–34

Jenny, Ludwig V. (1719–1797). Austrian military cartographer

Jensen, Harald. *Frederiksborg* (1876)

Jensen, H. N. A. *Germany* 1847, *Sleswig* 1847

Jensen, J. L. *Skole Atlas* 1867

Jensen, Nicolas. Printer of Venice. *Solinus* 1473

Jenson, H. J. and **Perronneau**. See **Jansen**, H. J.

Jentzch, K. A. (b. 1850). German geologist. *O. & W. Preussen* 1891

Jenviliers. Engraver. Nolin's *Amérique* 1720, for Desnos 1767

Jep, I. Engraver. *Göttingen* 1641

Jeppe, C. F. W. *Transvaal* 1899

Jeppe, Frederick (1833–1898). Geographer. *Transvaal* 1868, 1877, and 1899 (4 sh.), *Kaap Goldfields, Weller* 1888, *Witwatersrand* 1888, *Südafrik Repub.* 1892, *Southern Goldfields* 1894

Jerez, F. M. Y. See **Moreno y Jerez**

Jerman [Germaine], Edward [J.]

Jérôme, J. *Environs Hanoi* (1885), *Tonkin* 1898

Jerome, Saint. *Palestine* 1150

Jerrard, Paul. Lithographer, 206 Fleet St. *Gold regions of California* 1849

Jersey, T. *Chart River Thames* 1750 MS

Jervis, F. P. *Berea District Basutoland* 1881, *Leribe District* 1881

Jervis, Capt. (later Lt. Col.) Thomas Best (1797–1857) Geodesist. *Operations Konkan Prov.* 1824, MS *Plan Pekin* 4 sh. 1843, *Turkey* 7 sh. 1854, *Circassia* 1855, *Khiva* 1857

Jervis, Chev. G. Son of T. B. Jervis. *New Cycloidal Projection* 1895

Jervis, William Paget. *Crimea* 1855–7 MS, *geological notes on Italian maps* 1860–1889

Jervois, Capt. (later Lt. Col.) Wm. F. Drummond, R.E. Ordnance Southampton. *Keiskan Hoek* MS, *Milit. Sketch British Kaffraria* (1850). Supplied information for Arrowsmith's *Eastern Front. C. of Good Hope* 1851

Jesse, Capt. William. *Plan Sevastapol* (1839)

Jessop[e], William. *Plan River Trent* 1782, *Ashby de la Zouche* 1792, *Harbour plans Bristol* 1803, *Plan River Avon & Frome* 1792, Faden 1793 (with W. White)

Jewell, R. J. *Gold Fields S.E. Africa* 1876

Jewett, C. F. *Atlas Col. Co. Penn.* 1876

Jewett, J. P. & Co. Publishers. *Whitman's Eastern Kansas* 1856

Jikoan, Mabuchi (early 18th century). Japanese mapmaker. *Japan* (with Okada Jiseiken)

Jilek, Aug. V. (1818–1898). Oceanographer

Jimbo, K. *Hokkaido* 1891

Jirecek, Josef (1825–1888). Czech historian and map editor

Jisbury, J. *Maps Burma* 1863–1871

Jiseiken, Okada (18th century). Japanese cartographer. *Complete map of Japan* (with Jikoan)

Joachimus, Albertus Lyttichins. *Palestine* Leipzig 1569

Joami, Kise (d. 1618) Japanese artist. *Map of Japan engraved on silver* ca. 1600

Joanne, Adolphe Laurent (1823–1881). French writer and geographer. *Atlas Chemins de Fer Français* 1859, *Atlas de France* 1870

Joanne, Paul B. (b. 1847). *Son of A. L. Joanne*, continued his father's guides

Joannes de Stobnicza. Professor at Cracow University. First map of planiglobes published in Poland. *Intro Ptol. Cosmographiam* 1512

Joannis, Georgius. See **Georjio**

Joannis, Gotho O. See **Gothus**

Joao, Bartolomeo. Engineer. *Map Madeira* 1655 MS

Joao, Pessoa. *Stadt Parayba* Amsterdam, Visscher 1635

Jobard, Ambroise. Lithographer and surveyor, worked in Amsterdam and Brussels 1827–30

Jobbins, John Richard. Engraver and publisher, 3 Warwick Ct., Holborn, London. *Environs of London* 1840, *Brighton* 1843, *Great Britain* 1843. *Alton Cal.* 1849

GERARDVS DE IODE. *Franciscus vanden Wyngaerde excudit.*

G. de Jode (1509–1591)

Jobin, André. *Ile de Montreal* 1834, *City Montreal* 1834

Jobson, Francis. *Cork* (ca. 1589)

Jobson, R. English traveller. *Discovery Gambia* 1623

Jocelyn, N. and S. S. (1796–1881). Engravers and publishers. *New Haven* for Morse 1823

Jode, Arnold de. Engraver. *Plans* 1713

Jode, Cornelis de. (1568–1600). Son of Gerard. Engraver and publisher, scholar. *World* 1589, *Gallia occidentalis* 1592, *4 continents* ca. 1595, *Speculum Ordis terrae* 1593, *Belgium* ca. 1598

Jode [Judaeis, Judaeus, Iuddeis] Gerard de (1509–1591). Engraver, printer, printseller, publisher, cartographer. Born Nijmegen, died Antwerp. Ortelius' *World* 1564, Gastaldi's *World* 1555, Musinus' *Europe* 1560, Seco's *Portugal* 1563, *Speculum Orbis Terrarum* 1578, reissued by Cornelis 1593

Jodot, Marc. *Dépt. du Nord* 1830–1834

Johanes, Utinensis. *World* (ca. 1350)

Johann, von Armsheim (end 15th century). Woodcut engraver

Johannes, de Villadestes (15th century). Catalan hydrographer

Johannes, Magnus (1488–1544). Bishop of Linkoping. Historian

Johannesen, Capt. E. H. *Karisches Meer* Gotha Perthes 1870, *Nowaya Semla* 1871, *W. Sibirisches Eismeer* 1879

John of Ephesus. *Nubia,* 6th century

John, Jesse. *Map San Francisco* (1857)

John of Holywood [Halifax]. See **Sacrobosco**

Johnes, E. Owen. *Port e Roque Terra Firma* 1827–8

Johns, D. J. *Tehuantepec* 1851

Johns, Lieut. E. Owen. Assisted *Admiralty Charts Africa* 1822–6

Johns, William Master. *Admiralty Charts Australia* 1823–1831

Johnson, Andrew. Engraver. Round Court London. Parker's *Plan London* 1720

Johnson, A. J. Publisher in New York. *County maps Repub. of N. America* 1860, *Family Atlas* 1862 (with J. H. Colton), *American Atlas* 1865, *Ontario* 1867

Johnson, Capt. *Survey Bhopal* 1821

Johnson, D. Griffing. Geographer, publisher and engraver of New York and Washington. Trinity Buildings 111 Broadway. *New World* 1853, *Railroad Map Europe* 1854, *Rail Map Illinois* 1858, *Middle States* 1860, *U.S.* 1863

Johnson, E. *Level of Axholme* 1791

Johnson, E. *Berwick to N. Sunderland* Heather 1812

Johnson, Edwin F. Civil Engineer. *Railroad to Pacific* 1853

Johnson, Comm. E. I., R.N. *Admiralty Charts Farm Is. to Berwick* 1831, *East Coast England* 1831–66

Johnson, E. V. *Prairie Region* 1879, *Railways Canada* 3 sh. 1891

Johnson, F. *Limerick* 1587 MS

Johnson, George of Portaferry, Pilot. *Strangford River* 1755

Johnson, George. *Apia Bay* 1843, *Cloudy Bay, N.Z.* 1840

Johnson, Colonel Guy. *Map of the Six Nations* 1771 MS

Johnson, Henry. *World* 9 sh. 1889

Johnson, Isaac. Surveyor and antiquary of Woodbridge, Suffolk. *Surveys of Essex & Suffolk estates* 1790–1850

Johnson, John. English name for Jan Jansson q.v.

Johnson, John. Land surveyor. *Wisbeck* 1597

Johnson, J. Publisher in St. Pauls Churchyard. Paterson's *S. Africa* 1789, *U.S.A.* 1791, *England Delineated* 1790, and 1809

SPE-
CVLVM
ORBIS
TERRÆ

ANTVERPIÆ.
Sumptibus Viduæ et Hæredū Gerardi de Iudæis

Engraved title to the second edition of de Jode's "Speculum"
(published Antwerp, 1593)

Johnson, J. *Antigua,* Smith Elder & Co. 1829

Johnson, J. Hugh, F.R.G.S. *Oceania,* A. Fullarton *Ireland* 1870

Johnson, J. M. & Sons. Printers and lithographers 56 Hatton Gdn., London. *P.O. Map Warwickshire* 1868, *Herts* 1871

Johnson, Joseph. Cartographer. *Warwick* 1788, etc.

Johnson, Rowland (fl. 1559–1587). Surveyor of Works, Berwick. *Plan Berwick* 1559, *Norham* 1576, *Portsmouth Harbour* n.d. MSS ca. 1584

Johnson, Samuel L. *Eastbourne* (1894)

Johnson, T. Engraver for Mortier, 1715–28

Johnson, T. Draughtsman for Gregory's. *Trincomalay* 1787, *Subec* 1787

Johnson, Thomas. Engraver of Boston. *Boston, New England* 1729

Johnson, Thomas. Publisher in Manchester. *Atlas of England* 1847

Johnson, William. *English form* for W. J. Blaeu

Johnson, William. Surveyor and publisher. Faulkner St. Manchester. *Plans Manchester* 1818–19, pub. 1870–1

Johnson, William. Surveyor. *W. Lothian* 1820, *Lanarkshire* 2 sh. engraved J. Hall 1822. Maps for Thomson's *Scotland* 1820–1832

Johnson, William P. *Morgan Co., Ohio* 1854

Johnson, W. & Son. *Canals Yorks, Lancs, Derby & Cheshire* 1825

Johnson, Rev. W. P. *Yao country* 1822, *Lake Nyassa* 1884

Johnson, W. W. *W. coast* 1864 MS

Johnson & Browning. Publishers. *U.S. & Mexico* 1859, *California Territory* 1861

Johnson & Browning. *Australia* (1860?)

Johnson & Ward. Publishers, N.Y. *N. America* 1863, *Nebraska* 1865

Johnson. See **Franks & Johnson**

Johnson. See **Gaston & Johnson**

Johnston. Firm of W. & A. K. Johnston founded by Sir William Johnston (1802–1888). Printer, 6 Hill Square, Edinburgh from 1825. In 1826 he moved to 160 High Street Edinburgh and in 1837 to St. Andrews Square, finally in 1879 to Edina Works, Easter Road. *Royal Atlas Modern Geography* 1861 and later editions, *Elementary Atlas* 1862, *Gen. Atlas Modern Geography* 1862

Johnston, Alexander Keith (1804–1871). Youngest brother of Sir William, joined the firm in 1826 and styled himself Geographer at Edinburgh in Ordinary to the Queen. *Physical Atlas* Edin. 1848, first British Atlas to give a synoptic view of physical geography. Based on work of German geographer Berghaus. *Handy Royal Atlas* 1868, later editions half crown, shilling, and sixpenny *Atlases* 1869, *New Cabinet Atlas* 1873, *People's Pictorial Atlas* 1873

Johnston, A. K. & E. Forbes. *Palaeontological Map. Brit. Isles* 1850

Johnston, Andrew. Cartographer and engraver. *Plan Edinburgh* after De Wit (1700), *N. Scotland* Camden 1722, *S. Scotland* Camden 1722

Johnston, Col. D. A. Published Ordnance Survey sheets 1881–8

Johnston, George. *Strangford River* 1755, *Ramsgate to Rye* (1757)

Johnston, J. See **Johnstone**

Johnston, J. Ruddiman of Murrayfield Edinburgh. *Oceania* (1880)

Johnston, Thomas. Engraver of Boston. Burgis' *Plan Boston* 1728, Blodget's *Battle fought near Lake George* 1755

Johnston, Thomas Bramby. Joined the firm at end of century and took over Bacon & Co. in 1941. *Geological map of Scotland* 1876, *S. Africa* 1880, *Handy Royal Atlas* 1887 and 1896

Johnston, T. B. *Map of South Africa,* Cape Town 1882

Johnston, W. Bookseller at the Golden Bull in St. Pauls Ch. Yd. for Badeslade's *Chorog. Brit.* (1747), *Geog. Magna Brit.* 1748

Johnston, William. *Geo. Hydro. Survey Madeira* 1788, Faden 1791

Johnstone, J. *Glasgow* 1842, *Nat. Atlas* 1843, *Road Map Scotland* 1848

Johnston[e], Master (later Captain) James. *Dover to Rye* 1805, *Dover to Dungeness* 1806, *Eng. to Canton* 1806 for Arrowsmith's *Pilot, Charts W. Coast America,* pub. Dalrymple 1787–9

Johnstone, John. Land surveyor. *Edinburgh in* Thomason's *Scotland* 1832

Johnstone, Q. *Township of Plummer Ontario* 1878

Johnstone, William. Land surveyor, attested maps in Thomson's *Scontland* 1832, *Edin. & Heddington & Buteshire Perthshire,* after Stobie, *Fife & Kinross Nairn & Elga Inverness*

Johnstrup, Fred (1818–1894). Prof. of Geology and Mineralogy at Copenhagen

Joho, Hayashi. See **Hayashi Joho**

Joliet, Louis. See **Jolliet**

Jolivet, Jean. Priest, cartographer, Geographer to Francis II (fl. 1545–69). *Maps of Bourges, Normandy, Picardy, General Map of France* 1560, reissued by Ortelius & de Jode. *Berry* engraved 1545, *Picardy* woodcut (1559), *France* woodcut 4 sh. 1560

Jolivet, Morise Lewes. Surveyor. *Park & Gardens West Wycombe* 1752 MS

Jollain, Claude. *Trésor des cartes géographiques,* Paris 1667

Jollain, Francis, the Elder. Mapseller and publisher, Rue St. Jacques à la Ville de Cologne. *Plan Siam* 1686

Jollain, Gérard. Publisher. Paris rue St. *Jacques à l'Enfant Jesus,* for Duval 1672, *French Am.* Du Val 1686, *Turin* 1713

Jolliet, Louis S. J. (1645–1700) of La Rochelle and Canada *Carte de la découverte du Sieur Jolliet* (1673)

Joly, Joseph Romain (1715–1805). Cartographer and publisher. *Atlas de l'ancien géogr.* 1801

Jomard, Ed. François (1777–1862). Cartographer in Paris. *Monuments de la Géographie* 1842

Jomard

Jombert, A. *Plan de la dernière guerre de Flandres* 1751

Jomini, Antoine Henri, Baron de (1779–1869). *Atlas portatif* 1840, *Atlas des guerres de la Rév.* (1811–16, 1840)

Jones, Lieut. *Plan Puerto Cavallo* 1741, Sayer & Bennet 1779

Jones, Lieut. Ordnance Surveyor. *Waterford* 1841

Jones, Abner P. *Illinois* 1838

Jones, Benj. Engraver. *Map part Rhode Is.* in Marshall's *Life Washington* 1807

Jones, C. *Craven Property Bayswater* 1779

Jones, Charles H. *People's pictorial atlas* 1873, *Historical atlas* 1875 (with T. F. Hamilton)

Jones, E. Engraver for Rees 1806, for Smith 1808, for Arrowsmith 1810, Espinosa 1812, Playfair 1814

Jones, E. & G. W. Newman. Lithographers, 128 Fulton St. *Seat of War in Mexico* for Bruff 1847

Jones, E. L. *Mapa de Patagonica,* Buenos Aires 1858 MS

Jones, Felix, Commr. (later Captain). *Vestiges of Assyria* 1855 (with J. M. Hyslop), *Enceinte of Bagdad* 1856 (with W. Collingwood), *Course Shut el Arab* 1857 ·

Jones, George. Publisher, London, Ave Maria Lane. *London* 1814, *Monmouth* 1817

Jones, H. Estate surveyor. *Buttsbury* 1774

Jones, Maj. Gen. Sir Harry David. *Defenses of Sevastapol* 1855, *War Crimea* 1855, *Aland Is.* 5 sh. 1856

Jones, H. A. *Springfield, Mass.* 1851

Jones, Henry. Surveyor. *Allotments Footscray, Victoria* 1858

Jones, J. Draughtsman. *Road & Distance Map N.S.W.* 1871

Jones, J. C. 2nd Master H.M.S. *Retribution. Sebastopol,* 1854 Arrowsmith

Jones, J. W. Surveyor. *Country N.E. of Eucla. Adelaide* 1880

Jones, James Alexander. *Bogs Lough Corrib, Galway* 4 sh. 1814, *Bogs Meath, Westmeath & Kings Co.* 2 sh. 1811

Jones, James Felix. *Gulf Cutch* 1834, *Assyria* 1855. *Ancient Babylon,* 6 sh. 1855

Jones, John (1812). Globe and instrument maker

Jones, Richard. Engineer. *Port Royal & Kingston Harbours* 1756

Jones, Lt.-Col. Rob. Owen, R.E. *Ordnance Survey Warwickshire* 1885, 1888

Jones, Samuel. Globe and instrument maker. *Globe* ca. 1800 (with W. Jones)

Jones, S. L. *Hillsdale Co., Mich.* 1857, *Branch Co., Michigan* 1858, *Kalamazoo Co.* 1861, *State Michigan* 1864

Jones, Stephen. *Hist. of Poland with map* London 1795

Jones, T. Engraver for H. Gregory 1787

Jones, Thomas. 62 Charing Cross London. Surveyor. *Cippenham Liberty* 1841 MS

Jones, Thomas. *Copper Globe in Relief* Chicago 1894

Jones, T. W. Cartographer. *Travellers Directory* (with S. S. Moore) Phila. 1802

Jones, Wm. Draughtsman for Telford & Douglas 1800

Jones, William. Globe and instrument maker (fl. 1820). *Globe* ca. 1800 (with S. Jones)

Jones. *Macomb & St. Clair Counties, Michigan* Phila. 1859 (with Geil), *Monro Co. N.Y.* 1859 (with Weil)

Jones, W. *Citie of Rochell* (1650)

Jones, Capt. W. A. *Campaign Map Nebraska & Wyoming,* Washington 1872–4

Jones, Wyndham C. *Oil Regions Butler County, Penn.* 1874

Jones & Co. Publishers. *Classical Atlas* 1830

Jones. See **Sherwood, Neely & Jones**

Jones & Smith. Engraver of Pentonville. Jones, Smith & Co., Beaufort Buildings, Strand. C. Smith's *New English Atlas* 1801, Arrowsmith's *West Indies* 1803, *London* 1803

Jongelinx, I. Baptista. Engraver. *Polder van Oorderen* 1723

Jongh, A. de. *Lake Shore Railway* 1870

Jongh, Jacob de. *Gibraltar* 1700

Jonghaus, Gustav (1807–1870). Printer, publisher and bookseller

Jonghaus & Venator. Publishers, Darmstadt. *Hand Atlas* 1861

Jonghe, Clement de (1624–1677). Mapseller and colourist in the Kalverstraet, Gekroonde Konstkaert, Amsterdam. *Brazil* 1664, *Tabula Atlantis* 1675, *Europae Novae Descriptio (Town views)* 1669, 1675

Jonker, H. J. W. *Java,* 1872, *Timor* 1873

Jonsson, B. See **Bjørn Jonsson**

Jönsson, J. Swedish geologist *Forsta och Gustafsberg* 1887

Jöntzen, W. Lithographer of Bremen. *Europe & North America* 1849

Joop, G. *Stockholm* 1876

Jordan, A. Cartographer. *Historical Atlas St. Petersburg,* many editions in 19th century; *Istoreciskij Atlas*

Jordan, Claude. *Voyages historiques de l'Europe* Amsterdam 1718

Joosten, Hend. *De Kleyne wonderlijke Werelt* 1651

Jordaens, L. Worked Amsterdam ca. 1650. 24 plates for *Theat. praecipuarum urbium Brabantiae* 1660, 36 plates for *Speculum Zeelandiae* pub. Visscher

Jordan, Carl. *Bad Harzburg* 1897

Jordan, James B. Geologist. Stanford's *Geolog. Map London* 1870, *Lib. Map London* 24 sh. 1878, *Geog. Sect. Brit. Isle* 1879

Jordan, Wilhelm (1842–1889). German geodesist. *Höhen Karte. Baden u. Würtemberg* 1871, *Expd. Libysche Cüste* 1875

Jorden [Jordanas], Mark (1521–1595) Danish cartographer and professor of Mathematics at Copenhagen. *Denmark* 1552 (now lost), *Denmark* used by Braun & Hogenberg 1585, *Holstein* 1559

Jörgensen, Gotfred E. *British Columbia* 1892, *N. coast British Columbia* 1893, *S.E. Vancouver Is.* 1895

Jorio, D. Andrea de. *Napoli e Contorni* 1819, *Pozzuoli* (1820), *Ville de Naples* 1826, *Pompei* 1829, *Naples* 1835

Jorius, P. See **Giovio**

Jose, Amaro. *Lands Cazembe* 1873

Josenhans, J. *Atlas der Evangel. Missions Gesell. zu Basel* 1859

Joseph, Charles. *Grand Trunk Road to Sutledge (India)* 1851, *Grand Trunk Railway Calcutta–Benares* 1855, *Part India Calcutta Lahore* 1857

Jouan, Henri (b. 1821) French sailor and geographer

Jouanny, P. V. *Plano de Lima* 1872

Joubert, Louis Martin Roch (1749–1786). Publisher and engraver of Lyons. *Plan Lyons* 1767 (with widow Daudet), *Lyon* 2 sh. 1784

Jouet, John. *Chart N. approaches Liverpool* 1851

Joumar, *Plan Ville de Mans* (1855)

Jourdan, Adolphe. Publisher. *Environs d'Alger* 1884

Jourdan, Justin. *Atlas Guide Pyrenées* (1875)

Jourdan, T. E. C. *Atlas Hist. Guerra Paraguay* 1871

Joutel, Henry (1640–1735). *Carte de la Louisiane* 1713, English edition 1714

Jouvency. Marine surveyor. *Plans for D'Entrecasteaux* 1807 and *Dépôt de la Marine* 1792–1807

Jouvenel, J. B. Engraver. *Bruxelles* 1810

Jouvet (with Fume). Publisher of Paris 1875–1800

Jouy, H. *Géog. Populaire* Paris 1834, *Atlas Universel* 1834

Jove, P. [Jovius, Paulus]. See **Giovio**

J[oyce], H[enry]. *Urbis Galviae tot. connatiae (Galway)* 17th century

Joyce, Patrick Weston. Revised Philip's *Handy Atlas Ireland* 1881

Jozsef, K. S. *Heves-Szolusk geolog. Budapest* 1868

Juan y Santacilia, J. (1717–1773). Spanish sailor. With Condamine and Bouguer *Voyage to S. America* 1748

Juan de la Cosa. See **Cosa**

Jubrien, Jean (1569–1641). French surveyor. *Théat. Géogr. France* 1621, *Retelois* 1624, *Nivernois* 1621, *Reims* 1623, used by Blaeu 1631, *Maps* for Tavernier 1634, Rethel 1635

Judaeus, C. de, See **Jode**, C. de.

Judd, James. *Shilling Gen. Atlas* 1854, *Gen. & Hist. Atlas* 1855, *Mod. Geog.* 1855

Judd, P. E. *Map Michigan* 1824

Judd & Glass. Phoenix Printing Works, London. *Overland Railway through British North America* 1868

Juegel [Jügel], Carl. *Post u. Reise Karte Deutschland* 1843, *Plan Frankfurt* 1844, *Eisenbahn Atlas* 1846

Juegel, F. *Der Kremlin* 1810

Juel, Rasmus. *Faroe Is.* 1709—10

Juettner [Jüttner] Joseph (1775—1848) Czech mathematician and military cartographer. *Haupstadt Prag* 1811—15

Jugge, Richard 16th century engraver

Juigne, R. Printer of Paris. For *Le Sage* (1818)

Jukai. (b. 1297). Buddhist priest, Japan. *Map of the Indies* 1364 MS

Jukes, Francis. Aquatint engraver. *Pelham's Plan Boston* 1777

Jukes, Joseph Bette. *Geolog. Map Canada,* Montreal 1865, *Geolog. Map Ireland* 1867 and 1878

Jukes-Brown, Alfred Joseph. *Geological Map of Barbados* (1890), *Map of neighbourhood of Cambridge* (1875)

Julien, Sieur Roth-Joseph (fl. 1750—80). Publisher & mapseller of Paris. At Hôtel de Soubise. Geographer to the King. *Atlas Géog. et Militaire de la France* 1751 &c. *De Boehme* 1758, *Théatre du Monde* 1768, *Guerre d'Allemagne* 1758, also collaborated with Homann, Rocque and Sayer & Bennett

Julius, Friedrich. *Karte v. Asien* 1813, *Harz Gebirge* 1817—1844

Jullien, Amédée. *La Nièvre* 1883, *La Nièvre et le Nivernais* 1884

Junboll, Th. G. J. (1802—1861). *Geographical Lexicon*

Jung, Ch. Emile (born 1836). *Lexikon der Handelsgeographie*

Jung. Family of cartographers from Rothenburg, Franconia. George Conrad (1612—1691) son of Johan George the Elder. *Franconia* 1638, *Würzburg* 1639, *Road map Germany* 1641, all with his father.

Jung, Johan George the elder (1583—1641). Painter, glass cutter and cartographer. *Franconia* 1638, *Würzburg* 1639, *Germany* 1641

Jung, Johann George the younger (1607—1648)

Jung, Johan Adam. Publisher of Mallet's *Hist.* 4 vols., Frankfurt 1719

Jung, G. Cartographer. *German Historical Atlas* Berlin 1859

Junghuhn, Dr. F. *Eiland Java* 4 sh. Breda 1855, *Kultur Karte Java* 1866, *Hohen Karten v. Java* 2 sh. 1845

Jungman, Carl. Engraver for Weiland 1848, Kiepert 1851, Stieler 1860

Jungmeister, A. Jn. Publisher of Atlases. *St. Petersburg* 1845

Junker, C. Engraver of Vienna. *Klagenfurter Kreis* 1790, *Unter Steyermark* 1789, *Inner Oesterreich* 1794, *Laybacher Kreiss* 1797, *Stadt Washington* 1796, Title for *Magyar Atlas* 1802

Junker, Dr. W. *N. & Cent. Afrika* 1880, *Central Afrika* 4 sh. Gotha, Perthes 1888, *Buganda* 1891

Junsai, Fujita (19th century). Japanese surveyor. *Map of Ezo* 1854

Jurand [Jurang], Wilhelm. Publisher of Lelewel's *Atlas* in Leipzig 1846

Jurien-Lagravière, Jean Pierre Edmond (1812—1892). *Baie de Palmas* 1845, *San Pietro* 1844, *Sardaigne* 1846—54

Jurrius, J. *Nederlands Indië* (1869)

Justinianus, A. See **Giustiniani**

Just-Moers. *Waldeck* 1575 and *Hildesheim*

Juta, J. C. & Co. Publishers, *Cape Town. S. Africa* 1866, *Excelsior Atlas* 1899, *Enlarged map S. Africa* 4 sh. 1891

Juttner, Joseph (1775—1848). Mathematician and military cartographer. *Plan Prague* (1820)

Juvenalis, Decimus Junias. *World*, 12th century

Juvigny, Charles de. *Siege Budapest,* Greischen 1686

Juzo, Kondo. Japanese explorer of Ezo and Sacchalin. *Map of Sachalin* known as the *sand map*

K

K., D. Kandel, David

K, P. See **Keere,** Pieter van den

Kabo Hyozo. See **Hyozo**

Kaczanowski, G. (b. 1769). Polish military topographer 1820—31

Kadar, D. von. See **Duka von Kadar**

Kado, M. (1764—1820). Prof. of Topo. Design at Wilno University

Kaempfer, Engelbert (1651—1716). *Reis-Kaarten van Japan* 7 sh. 1700, *Rivière de Meinam* 1729

Kaeppelin & Bincteau. Printers of Paris. *Atlas* 1844

Kaeppelin & Cie. Printers. *Lapie Atlas* 1842

Kaercher, Karl. Publisher. *Orbis Ant.* 1827, *Schul Atlas* Karlsruhe 1830

Kaerius. See **Keere,** Pieter van den

Kaesberg, D. Military cartographer. *Amt. Baden* 1786

Kaestner, Abraham G. (1719—1800). Mathematician and physician. *Globes* Nüremberg 1749

Kaestner, Sandor. *Palestine map in Hebrew* 4 sh. 1883

Kageyasu, Takahashi. Japanese astronomer and cartographer. *World map* 1810, printed from copper plates

Kahe, Daikiyoji, (mid 17th century). *Map of Japan*

Kaibara, Tokushin [Tok'sin] (b. 1629). Japanese cartographer. *Panoramas*

Kailhan, B. M. *Geog. Maps Norway* 1858—65

Kaiser, Dr. E. *Ost Afrik Expd.* 1880—82, *Tanganyika* Berlin 1884

Kaiser, George. *Baltimore Co., Maryland* 1863

Kaiser, Josef Franz. German geographer. *Globe* (1830), *Austria, Prussia, Russia, Poland* (1855), *Europe, N. Africa, Asia Minor* (1859)

Kal, P. van. See **Cal,** P. van

Kalbermatter, Johann (d. 1551). *Valesia* for Munster 1540

Kaler, Gab. von. *Wandkarte v. Tirol* 1872

Kaliwoda, Leopold Johannes. Publisher in Vienna. *Atlas Novus* 1736, *Philippines* (1748)

Kalleffel, J. L. *Geogr. Sueviae U. Desa,* 6 sh. Nuremberg, Homann Heirs (1755)

Kalm, Peter. *New Map part North America N.E., N.J.,* 1771

Kalt, Nicolao. Printer of Constance on Bodensee *Schwarzwald* by Joh. Georg., Schinbain [Tibianus] 1603, Seltzel's *Bodensee* 1603

Kaltbrunner, David (1829—1874). Swiss geographer

Kaltschmidt, Abraham (1707—1760). Engraver for Florian de Grienfeld's.

Carniola Laibach 1744, augmented (12 sh.) 1799

Kalussowski. Publisher of Lelewel's *Atlases* in Brussels 1837

Kamermaister, Sebastian. Associate publisher of *Nuremberg Chronicle* 1493

Kamkin, P. *Plan Moscow* 1856

Kammermeister. See **Camerarius**

Kampen, Albert von. *Historical Geography* 1878–93, *Atlas Antiq.* 1893

Kan, J. B. *Hist. Geog. Atlas* 1867 and 1869, *Volks Atlas* 1870, *Kleine Atlas* 1871, *Hist. Geog. Atlas* 1881

Kandel, David (1524–1596). Wood engraver and artist. Münster's *World* 1550, 1567

Kandlpaltung, Hans. *Amberg* 1583 woodcut

Kane, Dr. E. K. *Arctic Map* 1853–6

Kango, Takeda. Japanese geographer. *World map on Mercator projection* 1858

Kanitz, F. (1829–1899). *Bulgarien und des Balkans* 1870, 4sh. *Leipzig* 1870 *Donau Bulgarien* 1877

Kano, Eitoku. See **Eitoku**

Kanter, Cornelis de (18th century). Mapmaker. *Schouwen-Duiveland* 1747, pub. Tirion

Kanter, J. Jacob. Bookseller and publisher of Königsberg. *Poland and Lithuania* 16 sh. 1770

Kantselyast, E. J. V. *Towns Russian Empire* 1839

Kanuikow, Y. *Mer d'Aral* 1851

Kapfinger, Joseph. German cartographer, end 19th century

Kapp, E, (1808–1896). German geographer, died Texas

Kappel, F. Engraver 1826

Kappeler, Moritz Anton (1685–1769). Swiss surveyor

Karacs, Ferenc (1771–1838). Cartographer and engraver. *Maps* 1793–1806

Karacsay, Comte Fedor de. *Montenegro* (1838), *Albania* 1842

Karamzin, W. M. Russian cartographer. *Atlas of Russia* St. Petersburg 1845

Karbasinikov, N. P. Russian printer and publisher. *School Atlases* St. Petersburg (1888)

Kardt. Engraver. *Atlas Militaire* 1821, *France* 1845

Karjavine, Theodor. *Carte du Voyage de S.M.I. partie mérid. Russie* 1787–1788

Karge, Johann Friedrich (1691–1761). Military cartographer

Karow, W. *K. der Umgegend von Schwerin* 1899

Karpf, Alois. *Polar Regionen* 1878

Karpinski, Hilarion. *First Geographical Dictionary in Polish* (1766)

Karsten, K. J. B. (1768–1853). German geologist

Kartaro. See **Cartaro**

Karterus. See **Cartaro**

Karver, Felix. *Geolog. d. Insel Luzon* 1878

Kascones, G. *Epirus & Athens* 1897

Kashgari, Mahmud al. Turkish cartographer. *World map* (1076)

Kasson. See **Dixon & Kasson**

Kastell, Prosper von. See **Castell**

Kastner, A. G. See **Kaestner**

Katahiro, Tatebe. Japanese mathematician. *Map of Japan* ca. 1725

Katanchich, Matthias P. (1750–1825). Croatian geographer

Katib Celebi [Mustafa L. Abdullah, Hajji Khalifa] (1609–1657). Turkish historian, translated Mercator-Hondius *Atlas Minor* into Turkish

Katsushika, Hokusai (1760–1849) of Tokyo. Woodcut designer and publisher

Katte, Stephen. Surveyor. *Nairn Estate Burwash* 1772 MS

Katzenschläger, Michael. *Mittel-Europa* (1845), *Croatien und Slavonien* 1857, *Oesterreich Eisenbahnlinien* (1864)

Kauffer, Franz. *Mer de Marmara* 1784, *Carte de l'Egypte* 1799, *Plan Constantinople* Weimar 1807, *Canal du Bosphore* (1826)

Kauffer, Michael (1673–1756). Engraver and cartographer of Augsburg. Worked for Müller *Bohemia* 25 sh. 1720, 1726 for Köhler 1730, for Faber and Weigel.

Kaufmann, N. *Cosmographia Dantisci* 1651

Kaulbars, A. (b. 1842). Russian explorer and geographer

Kaupert, Johann August (1822–1899). Topographer and cartographer. *Atlas v. Athen* 1878, *Alt Athen* 4 sh. 1881, *Attika* 1881–94

Kausler, Franz G. F. von (1794–1848). Publisher. *Alte Welt* 1826, *Battle Atlas* (213 plans), Karlsruhe 1831–37

Kawerau, F. Th. *Wand Karte. O. u. W. Preussen* (1868)

Kawamura, Heieman. *World Map on Screen* late 16th century MS

Kawatsi ya Gisuke. Japanese cartographer 18th century. *Province Halima no Kuni* (1749)

Kayser, G. H. Cartographer. *Europa's Staaten* Augsburg 1816

Kayser, J. (1826–1895). Oceanographer

Kayserling, Alex. de. (1815–1891). Russian explorer, born Courland. *Reise in Petchoraland* 1846

Al Kazwini (1200–1283). Arab cosmographer

Kearney, James. *Geelong* 1855, *Melbourne* 1855, *Pensacola* 1858, *S. Australia* 1859

Kearney, James. Major in U.S. Topgraphical Engs. *Lower James R.* 1818 MS, *Pensacola Harbour* 1835

Kearsley, George. London publisher, Golden Lion, Ludgate St. In 1773 he moved to Fleet St. opposite Fetter Lane. *Strangers Guide London & Westminster* 1791, *Scandinavia* 1791, *Guide through Great Britain* 1801 and 1803

Keddie, Arthur W. *County Prince Edward U. Canada* 1863, *Owens River Mining County* 1864

Keeble, W. Draughtsman and engraver. *Waterford Harbour* 1835, *Africa* 1853, *S. America* 1853

Keef, H. W. *N. Wirral* 1800

Keefer, Thomas C. *Basin St. Laurence* 1853, *Prov. Canada* (1855)

Keeler, Wm. J. Civil Engineer. *Mississippi to Pacific* Wash. 1867

Keenan, H. Engraver for Q.M.G. Office, Dublin. *Ireland* 1822

Keenan (James). Irish cartographer and surveyor. *Kildare* in conjunction with J. Noble 1752

Keenan, W. Lithographer. *Maps in Ramsey's Annals* 1853, *Part U.S.* 1854

Keene, H. G. *Plan Agra* 1874

Keene. See **Grierson**

Keere, Coletta van den. See under **Hondius, Jodocus**

Keere [Keer, Kaerius, Coerius], 1571-1646. Cartographer, engraver and bookseller of Amsterdam. 'op't. Rockin' 1593–9, 'Kalverstraat in den onseeckeren tijd'' (Turning Tide), 1610 and 'a L'Enseign du Temps'. Worked in London 1584–1593 Brother-in-law Hondius. *Ireland* 1592, *Nordens Speculum* 1593, *Europe* 1595, *Kaert Thresoor* 1598, *English Counties* 1599, *Mercator-Hondius* 1606, *World* 1608, *America* 1614, *Germania Inferior* 1617, *Totius Rheni* 1632

Keershaw, John. Map publisher, 14 City Rd., London. Schneider's *Ceylon* 1826 (with A. Arrowsmith)

Keferstein, Chr. (1784–1866). Geographer and geologist

Kegel, C. See **Pyramius**

Kehr, J. P. *Coblenz u. Ehrenbreitstein* (1840)

Keil, Franz (1822–1876). Austrian cartographer. *Karte v. Salzburg* (1866)

Keil, Herman. *Rheinpfalz* 1896

Keil, W. German geographer (fl. 1875–1898)

Keilhau, B. M. *Geogn. Karte v. Norwegen* 1849

Keily, James. *Mercer Co., N.J.* 1849, *Middlesex Co., N.J.* 1850, *City Washington* 1851, *Union Co., Penn.* 1856. See also Thomas A. **Parchall**

Keir, Thomas & Co. Edinburgh publishers. Playfair's *New General Atlas* 1839

Keisai, Kunagata (1764–1824). Japanese artist and mapmaker. *Birds eye map of Japan* 1810

Keiss, William Campbell. *Caithness* 1822

Keith, A. *Denmark* 1800, *West Indies* 1800

Keith, Rev. R. *Firth of Forth* 1730, based on Adair

Keith, Thomas (1759–1824). London mathematician, geographer

Keith & Gibbs. *Aberdeen* 1862 and 1871

Keitz, Fr. v. (1810–1883). Hungarian geographer

Keizer, J. Dutch engraver for Tirion. *Arabia* 1731

Kelday, J. Junior Surveyor. 14 Water Lane, Blackfriars. *Tolleshunt Knights* ca. 1840 MS

Kelius, Barni. See **Barnikel**

Kellander, Simon (1724–1776). Swedish surveyor

Keller, Andreas (1656–1708). Würtemberg mathematician and geographer

Keller, C. (1638–1707). Geographer

Keller, Christoph. See **Cellarius**

Keller, E. *Sierra Leone* 1883

Keller, Ferdinand (1800–1881). Swiss geographer. *Kanton Zürich* 1863, *Ost Schweiz* 1874

Keller, François Ant. Edouard. *Debouq. de St. Domingue* 1844, *Mer des Antilles* 1850, *Grandes Antilles* 1861, *Entrée Dardanelles* 1855–62

Keller, F. C. *Oesterreich-Ungarn* 1833

Keller, Georg. *Kassaw* 1605, *Fori Julii* 1616, *Mannheim* 1618, *Thes. Philo. Politicus* 1625

Keller, Heinrich (1778–1862). Swiss geographer and publisher. (Firm of Keller & Fuesli in Zürich). *Reise Charte Schweiz* 1813–15, English edition by Cary 1817, 1825, 1840; *Atlas de la Suisse* 1829, *Panoramas* 1828–60, *Erdkarte* 2 sh. Weimar 1814, *Switzerland* 1813, revised 1819 (6 sh.)

Keller & Giesemann. Photolith. printers. *Spreewalde* 1867

Kellermann, I. L. *Charte von Kur-Hessen,* Nürnberg 1836

Kellet, Lieut. (later Capt.) Harris, R.N. *W. Coast Africa* 1833–5, *Pacific & N.W. Coast America* 1835–41 and 1845–51, *China* 1842–43

Kellett, Capt. Henry R.N. Hydrographer. *Admiralty Charts* 1836–1863, *River Guayquil* 1836, *Chapow Road* 1843, *Canton–Nanking* 1851, *W. Africa* 1860, *S. America* 1863

Kelley, Hall J[ackson]. *Oregon* 1830 and 1839 MS

Kelley, R. P. U.S. Deputy Surveyor. *Territ. Arizona* 1860

Kellner, Joseph. Engraver for Jaeger (1780), for Murr (Reisen 1785)

Kellog, B. C. & Co. Lithographers of Hartford Connecticut. *Mexico* 1847, for E. F. Johnson 1853

Kellog, E. H. *Denver, Colorado* 1871, *Colorado* 1872

Kelly, Capt. [later Col.]. *S. Countries India* Faden 1788, *Peninsula India* 1800

Kelly, Christopher. Author *Universal Geography* 1837

Kelly, Dionysio. *Plano contornos de Manila* 1775 MS

Kelly, Frederic. Post Office Directory Offices 19-20 Old Boswell Court, Temple Bar (1845—68), 51 Great Queen Street, Lincolns Inn Fields (1872—92), 182-4 High Holborn (1896). *Post Office Directory,* London 1st Edn. 1843, *Atlas England & Wales,* 1860, *Atlas Directory Birmingham* (1845)—1880, *Atlas Directory Warwick* 1863—1900

Kelly, J. *Plan Acre* 1850, *S.W. Arabia* 3 sh. War Office 1893

Kelly, J. A. *Tehuantepec survey* 1851, *Khiva* 1873, *Sierra Leone* 1888

Kelly, John G. *Desoto Mississippi* (1860)

Kelly, J. A. & W. J. Military draughtsmen and lithographers, Q.M.G.'s Office. *Chusan* 1857

Kelly, Capt. Marwood. *Man of War Bay, St. Thomas* 1822

Kelly, Santiago T. *Plano de Habana* 1837

Kelly, Thomas. Publisher, London, Paternoster Row, 1837. Maps for Barclay's *English Dict.* 1835—43, *Universal Geography* by Christopher Kelly (1837)

Kelly, W. J. *England & Wales* 1854

Kelsall, C. *Comp. Geog. Sicily* 1812

Kelsey, E. H. Surveyor, Bloomsbury Sq. *Ashford Lodge (Sussex)* 1850

Kelsey, Richard. Surveyor. *Plan Public Sewers London* 1841

Keltenhofer, Stephan. *Champaigne* (1544?) 1570

Kelton, John. *Map Ashdown Forrest* Engrd. Toms 1747

Keltzl, Abraham. *Swabia* 16th century

Kemal, Reis. See **Piri,** Reis

Kemble, W. of New York. *Drew & Engraved map of Texas* **1844**

Kemp, George. Land surveyor. *Ten Miles*

round Harrogate 1832, *England & Wales* 1839, *New Map England & Wales* 1845

Kemp, William. Surveyor. *Race Courses in England* 1825

Kempe, Abraham (d. 1638). Swedish surveyor

Kempen, Godefridus. Printer of Cologne. *Ptolemy* 1578 and 1584

Kempthorne, Capt. John Jn. (fl. 1669). *East India Co., Ballasor River India* 1680 MS, *Falmouth* 1686, *I. of Wight* 1688, *Tristan da Cunha* 1689 MS

Kendall. London print and mapseller of Bury Street. Took subscriptions for Hodskinson's *Suffolk* 1780, pub. 1783

Kendall, Lt. E. N. *Deception Is. Staten Is.* 1828, *New S. Shetland* 1829, *New Brunswick* 1832

Kendall, John. Estate surveyor. *Birch* 1702, *Aldhan* 1703, *Gt. Clacton* 1730, *Great Horkesley* 1735—7

Kendall, Joseph. Estate surveyor. *Copford, Marks Tey* 1717 MS, *White Colne* MS (with William Kendall) 1732

Kendall, William. Estate surveyor. *White Colne* 1724 (with Joseph Kendall), *Birch* 1725, *Ardleigh* 1726 MS

Kendrick, John. London publisher, 54 Leicester Square. Moule's *English Counties* 1834

Kennedy, A. *Massachusetts* 1838

Kennedy, E. B. Surveyor. *Australia* 1860—1863

Kennedy, J. P. *Bombay Baroda Railway* 1862

Kennedy, Major R. G. *Sherpore & Vicinity (Afghan)* 1880

Kennedy, R. J. *British New Guinea* 1894

Kennedy, W. *Coast Texas* 1844

Kennish, W. New York engineer. *Plan Canal Central America* 1834, *Kieport* 1858, *Darian* 1885

Kenny. See **Sands & Kenny**

Kensett, Thomas, of Cheshire,

Connecticut. *Upper & Lower Canada* 1812

Kenshin, Kitajima. Japanese geographer, published *Komo Tenchi Nizu Zeisetsu* (explanation of European maps) 1737

Kent, Henry. London printer and publisher, Finch Lane near Royal Exchange. *Directory City London* 1732 &c.

Kent, William. *Port St. Vincent, New Caledonia* 1805

Kent, William (1684–1748). Surveyor. *Plans House Lords* ca. 1735 MS

Kentish, N. L. *County of Southampton* 1826 (with C. & J. Greenwood)

Kentish, N. Surveyor. *Forret Farms Emu Bay V.D.L.* 1844

Kepler, J. See **Keppler**

Keppel, Karl. German cartographer. *Historical Atlas, School Atlases,* mid 19th century

Keppen, P. *Krimea,* St. Petersburg 1836, *Ethno. Map European Russia* St. Petersburg 1851

Keppler, Johann (1571–1630). German mathematician and astronomer. *Rudolphine Tables.* Nürnberg 1630, with *World Map* by Eckebrecht. First map to show part of Australia

Kergariou, A. de. *Yu-lin-Kau* 1817, *Hainan* 1819, *Baie de Gaalong* (1820)

Kerguelen, Tremarec, Yves J. de (1734–1797). *Voyage Pacific* 1772, *Kerguelen Land*

Kermovan, Gilles Jean (1740–1817). Military engineer, served in America and prepared plans for coastal defense

Kern, Edward M. Drew for Simpson *Surveys New Mexico* 1849

Kern, F. Engraver for Kiepert 1857 and 1869, for Meyer 1867

Kern, Richard H. Surveyor and draughtsman. *Rio Pecos* 1850 MS, *New Mexico* 1851 (with Parke), *Zuni & Colorado Rivers* 1852. Assisted Gunnison 1855

Kerner, R. *Grundriss Hamburg* 1839

Kerner-Marilaun, A. von. *Phys. Hist. Pocket Atlas of Austro. Hungary* Wien 1887

Kernot, J. H. Engraver. Illustrations for Tallis' *Africa* (1851)

Kerr, James. Staff Comm. R.N. *Admiralty Charts* 1867–1877, *Newfoundland* 1867, *Ireland* 1875, *Solway Firth* 1877

Kersey, John. Estate surveyor. *Theydon Common* 1652 MS, *Stapleford Abbots* 1654 MS

Kershaw, Arthur. Revised John Walker's *Universal Gazetteer* 1807

Kershaw, John. Map publisher, 14 City Road, London. *Schneider's Ceylon* 1826 (with A. Arrowsmith)

Keschedt, Petrus. Printer of Cologne. *Ptolemy* 1596–1597

Kessler, Franz (ca. 1580–1650). Painter and cartographer of Cologne, born in Wetzlar, died in Dantzig

Kettle, C. H. *Chart Harbour Otago* 1846

Kettle, Asst. Surveyor to New Zealand Co. *South part North Island, N.Z.* 1843, used by Johnston

Kettle, D. & W. Publishers, successors to Findley of Laurie & Whittle business

Kettler, J. L. *Deutsche Kolonien* 1888, *Ostafrika* 12 sh. 1892, *Baden* 1892, *Der Ettersberg* 1893, *Belvedere* 1895

Keuchenius, S. J. Hydrographer. *Eems* 1853, *Friesche Zeegat* 1859

Keulen, [Ceulen] Johannes Jansz. van (1634–1689). Mapseller and colourist

Keulen, Van. Family of chart and instrument makers, publishers nautical, text books Sea Law, shipbuilding, almanacs &c. Existed as a firm for over 200 years. Founder of the firm was

—— [Ceulen], Johannes van (1654–1715). Publisher "over de Nieubrug in de gekroonde Lootsman". *Zee Atlas* 1680, with text and 38 maps by Claes Jansz Vooght. Immediately popular, 9 editions

DE
NIEUWE GROOTE
LICHTENDE
ZEE·FACKEL,

Behelſende

'tEerſte, 't Tweede, 't Darde, 't Vierde, Vijfde of
't Laetſte Deel.

Alwaer klaer en volkomen in vertoont wort , alle bekende Zee-Kuſten
van de geheele Noord Oceaan,

En deſzelfs inboeſemen

Met een partement

BESCHRYVINGE,

Van alle bekende Haavens, Bayen, Reden, Drooghten,
Strekkingen van Kourſſen, en op-doeninge van Landen
alles op haare Waare Pools-hooghte geleyt.

Uyt ondervindinge van veele ervaaren Stuurlieden, Lootſen, en
Liefhebbers der Zeevaert.

Vergaedert met groote koſten , en op het Nieuw in beter order geſtelt als voor deſe , en verrijckt met het
Vijfde, of laetſte Deel, dat noyt voor deſe in 't licht is geweeſt. Te ſamen gebracht,

DOOR

JAN VAN LOON, en CLAES JANSZ. VOOGHT. *Geometra, Leermeeſter*
der Wis-Konſt.

'AMSTERDAM,

Gedruckt by JOHANNES VAN KEULEN, Boeck en Zee-Kaart-Verkooper , aen de
Ooſt-zyde van de Nieuwe-Brugh, in de Gekroonde Lootsman, 168 .
Met Privilegie voor 15. Jaer.

Title to J. van Keulen's sea-atlas (published Amsterdam, 1681)

published within 5 years, including French and English editions. *Zee Fakkel,* 5 parts appeared by 1683. In 1693 he acquired the stock of one of his principal competitors Hendrik Doncker the Elder. Both the *Zee Atlas* and the *Zee Fakkel* continued to be issued in expanding and revised editions. In 1704 the elder Keulen retired, and the business was taken over by his son, Gerard.

—— Gerard van (1678–1727). More skilled than his father, he was an engraver and mathematician as well as publisher; he took over the scientific side formerly done by Vooght. He was appointed Hydrographer to the Dutch East India Company in 1714. Before his early death at the age of 49, he revised and reissued his father's atlases and himself added new maps, some of exceptionally large size, culminating about 1710 with an edition of over 180 charts. Later editions appeared with smaller numbers of selected charts, and some copies were made up to customers' requirements. On his death the business was continued by his widow Ludwina Konst

—— Johannes [II] van. Took over control in 1726. He revised the *Zee Fakkel* and added a 6th volume in 1753 with important large *charts of Africa, Asia, and Australia.* He also increased the resources of the firm by the purchase of the marine works of Jan Loots. On his death the business was continued under the title Johannes van Keulen en zoonen [sons] 1757–1779. He had two sons, G. H. and C. B. van Keulen.

—— Gerard Hulst van, and Cornelius Buys van, the latter dying in 1788. Gerard Hulst van Keulen took control from 1779 to 1801 when he died, the firm passing to his widow who continued direction till 1810. She married Joannes van de Velde. Between 1810–1823 the firm was known as Johannes Gerard Hulst van Keulen van de Velde. He was the last of the Keulens, the business being sold to Staats Boonen in 1823, then to Jacob Swart; the stock

was finally dispersed at public auction in 1885

Keur, Jacob and Hendrik. *Paradisus Canaan* (Dutch Bible) 1648

Keux, J. Le. See **Le Keux**

Keyl, Christian Karl Maximilian. Engraver. *Sachsen* (1765), *Sachsen* 1809

Keyl, Julius. *Grund. Dresden* 1845, *Umgegend v. Dresden* 1873

Keyl, K. J. (1805–1870). Engraver of Dresden

Keyl, M. Polish engraver for Michael Groell, publisher in Warsaw (1779)

Keyley, James. *Jefferson Co., Ohio* 1856

Keymer, W. Printer, publisher, and map-seller of Colchester. Chapman & André *Essex* 1785, Sparrow's *Plan Colchester* 1767

Keys. See **Moore, Wilstach, Keys & Co.**

Keyser, Jacob. Engraver and Draughtsman. *Petrograd* 1703, *Terre Neuve* 1715, *Hungary* 1717, Tirion *N. Pole* 1735, *Breda* 1739 for Lat 1747

Keyserling, Count Alexander von. *Russia in Europe & Ural Mts.* 1845, *Petschora* 1846

Keyzere, Pierre. Printer, Ghent 1538

Khakim, Aros. *Côtes Dalmates* 1694

Khanikoff, Nicolas de. *Aderbeijan* 1851– 55, *Aral Sea* Fullarton (1860)

Kharikov, Jacob. *Aral Sea & Khiva* 1851

Khatov, A. *Georgie* 9 sh. 1826

Khegel, C. See **Pyramius**

Khistler, P. F. von. *Mittel Franken* Munich 1840, *Nieder Bayern* 1841

Khonsa. Egyptian deity, the Planmaker

Khrypffs, N. See **Cusa**

Khurdabah (825–912). Arab cartographer. *Atlases of Islam* 844–885

Khwarizmi. See **Alkharizimi**

Kichibe, Nakabayashi (17th century). Japanese mapmaker. *"Sedan chair" map of Japan* 1666 (used for travelling)

Title to the "Zee-Fackel" (Sea Torch) by J. van Loon and C.J. Vooght, published by J. van Keulen, Amsterdam 1687

Kiddle, Staff. Comm. W. W. *St. Domingo* 1873

Kieboom, J. van den. Publisher of geographical works, maps etc., The Hague (1736)

Kielisinski, K. W. (1810–1849). Polish engraver, designed *ex libris* for Pawliskowski cartographic collection in Medyka

Kienle, M. von. *Bavaria* 1826

Kiepert, Heinrich C. (1818–1899). *Atlas v. Hellas* 1846, *Mexico Texas* 1847, *Australien* 1849, *large map of Poland* 1849, *Neuer Hand Atlas* Berlin 1855, *Central America* 4 sh. 1858, *Grosser Hand Atlas* 1871, *Neuer Handatlas* 1878, *Grosser Handatlas* 1893, many maps pub. by Industrie Comptoir, Weimar, to accompany voyages

Kiepert, Richard (1846–1915). Geographer, son of J. S. Heinrich Kiepert. *Luzon* 1872, *Schulwandkarte Prov. Posen* Berlin 1873, *Nyasa Expd.* 4 sh. 1895, *Adamana* 1896, *Deutsch Ostafrika* 1899

Kies, Eberhard. *Erbach* Frankfurt 1620

Kiesdorff, F. M. J. (1777–1855) Cartolithographer

Kieser, Andreas (ca. 1620–1688). Military cartographer of Stuttgart. *Forests of Alt-Würtemberg* 1680–7 MS

Kiesling, Alexius. *Plan Berlin* 1875, *Gruenewald* 1875

Kiesling, Aug. *Sachsen* 1875, *Dresden* 1878

Kijenski, Ignacy (1797–1835). Polish topographer

Kikkert, J. E. Solicitor and surveyor. *Map Island of Texel* 1846

Kilbourne, J. *Map Ohio* 1820

Kile, James C. *Monticello Kansas* 1857

Kilian, George Christophe (1709–1780). Engraver of Augsburg. Supplement to *Atlas Curieux* 1738, *Kleiner Atlas* (1757), *Kriegs Atlas* 1758, *Théat. Guerre Allemagne* 1760, *America Septentrionalis* (1760)

Kilian, Wolfgang (1581–1662). Engraver. *Augsburg* 1626, *celestial charts for Schiller* (1627), *Chiemsee* 1640

Kilian, Wolfgang Philip (1654–1732). Engraver

Killaly, John. *Canal maps* 1808–1815

Killian [Kilian], John. Surveyor General. *Estate Plan St. Mary Antigua* 1787 MS, *Plan Town St. John* 2 sh. 1788

Kilroe, J. R. *Geolog. Map Ireland* 1882

Kimber & Sharpless. Publishers of Philadelphia. Howell's *Pennsylvania* 1817

Kinahan, G. Henry. *Maps of Ireland* 1863–1891

Kinbe, Yamazaki. Map publisher. *World.* pub. Osaka ca. 1785

Kincaid, Alexander. *Edinburgh* 2 sh. 1784, *Behring Straits* 1790, *Scotland* 1802

Kincaid, T. *Southern Hemisphere* (1790), *New Holland* (1790)

Kinchoff, A. German cartographer. *School Atlases* Leipzig, 19th century

Kinckelbach, Q. von. See **Quad.**

Kind, G. B. American engraver. *U.S.* 1820

Kinderley, Nathaniel. *Palestine* 1727, *Asia Minor* 1729, *Great Level Fens* 1751

Kindermann, Joseph Karl (1744–1801). Hungarian cartographer. *Klagenfurter Kreiss* 1790, *Inner-Oesterreich* 11 sh. 1790–1800, *Boehmen* 1803, *Atlas Oesterreich Kaiserthums* 40 sh. 1805

King, B. *Harbour Archangel* 1836, *Old Calabar River* 4 sh. 1842 MS

King, Charles. *Proposed Jungfrau Railway* 1897

King, Charles R. B. Surveyor. *Church Plans,* London and Totnes

King, Clarence. *Yosemite Valley* 1865, *U.S. Geolog. Expd.* 1871, *Topo. Atlas 40th Parallel* 1876

King, Daniel. Engraver. *Smiths Vale Royal Cheshire* 1656, *Ground Plot Chester* 1656, *Map Seat of Wars in*

Germany (1670), *Conquests of French* (1680)

King, D. O. *Siam & Cambodia* 1857–8 MS

King, Eusebius Francis. *California* 1794

King, F. Estate surveyor. *Chadwell etc.* 1802 MS (with W. Cole)

King, Frederick W. *Yorks* 1888, *Contour Map Yorks* 1892

King, George. *Plan Tortola*, Wilkinson 1798, *Virgin Is.* 1802

King, Gregory. London bookseller and engraver, East Corner Piazza House of James Street, Covent Garden. For Archer's *Alsace*, W. Berry, London 1676; for Adams *England*, Lea (1692)

King, G. B. Engraver for Seamans. *U.S.* 1820

King, H. *Map Town Petersburg* 1857

King, James. Surveyor. *Lowndes Estate Hanslope* 1774 MS, *Brixham* 1781 MS

King, James K. *Hancock Co., Indiana* 1875

King, John. Mapseller and Publisher at ye Globe in the Poultry London. Seller's *Kent* 1710, Price's *London* 1712. Joint publisher with others of later editions of *Moll's large atlas* 1725 onwards

King, John. Surveyor. *Grenville Estate* 1745, *Parish Ashton Clinton* 1814, *Bierton Parish* 1821, *Tithe Buckingham* 1752 MS

King, John & Son. Estate surveyors of Saffron Waldon, Essex. *Chesterford* ca. 1804, *Saffron Walden* ca. 1823, *Sturmer & Kedington* 1849 MS

King, Joseph William. *New Road Bristol to Clifton* 1874

King, Nicholas. Copied *maps of Lewis & Clarke's expdn.* 1805–6, *Canal Harpers Ferry* (1803)

King, Lieut. (later Admiral) Peter Parker, F.R.S., R.N. (1793–1856). Hydrographer to the Admiralty. Surveyed coasts of Australia, correcting and filling in Flinders gaps in 1817–1822. *Coasts of*

Magellan 1825–1835, *W. Coast S. America* 1835

King, Samuel, Senior. Publisher and mapseller near Charing Cross Norwich. *Plan Norwich* 2 sh. 1766

King, Samuel, Junior. Publisher and Mapseller, corner of Grange Ct., Clements Lane, Clare Market, London. *Plan Norwich* 1766

King, Samuel D. *State Indiana* 1853

King, Thomas. Estate surveyor. *Stanway* 1807–8 MS (with W. Cole), *Mayland* 1812 MS

King, Thomas. *City Pekin Illinois* 1872

King, T. K. *England & Wales* 1854

King, William (fl. 1598–1610). Engraver for Norden, Camden's *Britannia* 1610

King, William. Surveyor. *Ardwell Nately, Basing & Mappledurewell* 1787 MSS, *New Forest* (with Richardson & Driver) 1789, *Belvois District Leicestsh.* 1806

King, William (1833–1900). English geographer. Director Geological Survey of India

King, W. Falmouth 1824

King, W. H. Estate surveyor. *Barling (Essex)* 1804 MS

Kinghorne, Alex. Engineer and surveyor. Attested maps for Thomson's *Atlas Scotland, Roxborough & Selkirk* 1832

Kinghorne, James. Land surveyor. Attested maps for Thomson

Kingman, A. L. *Vanwert County* 1872, *Wells Co. Indiana* 1873, *Delaware Co.* 1873

Kingman, N. H. *New London* 1877

Kingsbury, John. Land surveyor of Melford. *Bulmer* 1794 MS

Kingsbury Parbury & Allen. *Madras* 1824, *India* 1825

Kingsford, William. *Railway systems E. Canada* 1876

Kingston, Sir G. S. *S. Australia (Rainfall) S.G.O.* 1874

Kinneir, John Macdonald. *Countries between Euphrates & Indus.* Arrowsmith 1813, *Asia Minor* 1818

Kinnersley, E. Publisher of Bungay, Suffolk. Collaborated with C. Brightly. *W. Indies* (1807)

Kino [Chino], Eusebio Franciso (1644–1711). Jesuit MSS *California* 1683 (1702), *Mapa del paso por tierra del California* 1698

Kinsbergen, Capt. Jan Hendrik van. *Crimea* 4 sh. 1776, ditto 1 sh. 1787

Kinsui, Shotei (19th century). Japanese mapmaker. *Panoramic map of Tokaido Highway* ca. 1845, illustrated by Kuwazata Shoi

Kip, John (1653–1722). Engraver. *Amsterdam* 1685, *Bridgetown* 1695, *Britannia Illust.* 1707, *Dodington* 1708, *London* 1710, *Views Scotland* used as side borders by Moll 1714, *Althorp* 1720, *Windsor* 1720

Kip, William (fl. 1598–1635). Engraver. *Hartfordshire* 1598, *World* for Hakluyt 1599, *British Isles* for Woutneel 1603, Camden's *Britannia* 1607, *Anglia Descript.* 1635

Kipferling, Karl Joseph. Austrian cartographer. *Salzburg* 1803, *Oest. Mon.* 1803, *Empire Autriche,* 1805, *Reise Atlas,* Wien 1804, *School atlas for ancient history,* Wien 1806. Worked with Kindermann

Kiprianov, Vasily (d. 1723). Russian publisher, engraver and cartographer of Moscow. Librarian to Peter the Great. *Maps of Russia* 1706–1718, *Globes* 1707, *World* 2 sh. 1707

Kips, Joseph H. *Utrecht* 1850, *France* 9 sh. 1863, *Liège, Namur &c.* 1890

Kips, T. Engraver. *Luxemburg* 1870

Kirby, Capt. B. *Gold Coast* London, Weller 1884

Kirby, James P. *Fiskill* 1872, *Property map Cornwall, Orange Co., N.Y.* 1873

Kirby, John. Surveyor. *Suffolk* 4 sh. 1736, engraved Basire

Kirby, Joshua and William. *Suffolk* engraved Ryland 1776

Kirby, S. R. *E. Saginaw* 1870

Kirchebner, Anton (1750–1831). Tirolese cartographer

Kirchenstein. *Map of Ukraine* 1782

Kircher, Athanasius (1602–1680). Jesuit geographer, mathematician and engraver. *Mundus Subterraneus* Amsterdam 1665. Contains earliest map to depict ocean currents. Later editions

Kircher, I. V. *Travellers Guide Road Map England* (1720)

Kirchmair, F. *Regensburg* 1589

Kirchmayr, C. *Pianta di Trieste* 1871

Kirchmayr, Francesco. *Prov. di Venezia* 32 sh. 1876

Kirchner, E. *Florenz* 1876

Kirchner, F. *Lombardy & Venice* 24 sh. (1830) MS.

Kirchner, J. D. N. Map publisher. *Schul Atlas* Berlin 1833

Kirchner, Moritz *Elsass im Jahre 1789,* Kirchner 1880 *Lotharinga* 1882

Kirchoff, Alfred (b. 1838). German geographer. *Schul Atlas* 1893, *Wandkarte des Weltverkehrs* 4 sh. 1894, *Formosa* 1895

Kirilov, I. See **Kyrilov**

Kirk, Sir. John. *Shupanga (Zambezi)* 1858

Kirk, R. Surveyor. *Tajoora to Ankoter (Abyssinia)* Bombay 1841

Kirk, and **Mercein.** Publishers, 22 Wall St., N.Y. Darby's *U.S.* 1818

Kirk. See **Everts & Kirk**

Kirkall, Elisha. Engraver. Cartouche title for Moll's *World* 1719, for Beighton's *Warwick* 1728

Kirkham, Major R. N. *Plan Boston* ca. 1807 MS.

Kirkor. Publisher, Vilna. *Atlas* 1859

Kirkpatrick, Capt. William. *India* 1800, *Nepaul* 1811

Kirkwood, James. Engraver. Lendrick's *Antrim* 1780, Forrest's *East Lothian* 1802, for Thomson's *Atlas* 1814, *Edinburgh District* 1817, Siborne's *History* 1844

Kirkwood, James & Sons. Engravers, and publishers of Edinburgh. Housman's *Plan Manchester & Salford* 1800, Crawford's *Dumfries* 1804, Wood's *Plan Ayr* 1818, *Kilmarnock* 1819, Cleland's *Glasgow* (1822), *Maps of Scotland* 1808–1829

Kirkwood, John. *Dublin* 1839, 1849 and 1860

Kirkwood, Robert (d. 1818). Read's *St. Helena* 1815, *Edinburgh* 2 sh. 1817, *Environs Edinburgh* 2 sh. 1817

Kirmani, Gafur Ibn. Uman. (14th century) Persian globe maker

Kirmani, Mahammed Ibn Gafer. Globe maker, 15th century

Kirsbaum, V. Publisher of *Atlases* in St. Petersburg, end 19th century

Kirton, William. *Pulo Pisang* 1780, *W. Sumatra* 1782, *Borneo River* 1787

Kitajima Kenshin. See **Kenshin**

Kise, Joami. See **Joami**

Kitchen [Kitchin] Thomas (1718–1784). Engraver, publisher, hydrographer to the King, worked from following addresses: Clerkenwell Green 1755, Opposite Ely Gate Holborn 1756, Charing Cross 1758, Holborn Hill, 1768. Prolific output, worked for various publishers: Bowen, Dalrymple, Dunn, Elphinstone, Jefferys, London Magazine, Luttrell, Mackenzie & Willdey. Elphinstone's *Scotland* 1745, Willdey's *Highlands* 1746, *English Atlas* 1749, *London Mag.* 1760–1765, Mitchell's *America* 1755, *England Illust.* 1764, *Pocket Atlas* 1769

Kitchen [Kitchin] Thomas Junior. Hydrographer to His Majesty. Engraved for Mackenzie 1745

Kitchen was succeeded by W. Hawkes

Kitchener, H. H. Lieut. (later Field Marshall). *Palestine Survey* 1872–77 O.S. *Southampton* 26 sh. 1880, *W. Palestine* (6 sh.) 1881, *Cyprus Trig. Survey* 15 sh. 1882, *Nile* 1884

Kitchin. See **Kitchen**

Kitching, William. Surveyor. *Plan Sapton House* 1865

Kittensteyn, C. van (1600–1638). *Plan Siege of Haarlem* 1626

Kittle, Robert. *City of Fremont, Nebraska* 1870

Kittoe, George. Purser H.M.S. *Termagent Is. Scilly* 1788

Kizer, T. *Clarke Co., Ohio* 1859

Kjellander, Jonas A. (1614–1666). Swedish surveyor

Kjeliman, Per. (1719–1795). Swedish surveyor

Kjellström, C. J. O. (1855–1913) *Grönlands Ostkust* 1844, *Sverige-Finland* 2 sh. 1888, *Beeren Is.* 1899

Kjerluff, Tr. (1825–1888). Norwegian geographer. *Geolog. Kart Söndenfjeldske Norge* 1866

Klahr, J. H. *Sachsen* 1863

Klaproth, Heinrich Julius von (1783–1835). Traveller, orientalist, cartographer. Born Berlin. *Lac de Baikel* 1806, *Asia Polyglotta* 1823

Klasing. See **Velhagen & Klasing**

Klaudian, Nikulas [Nicolaus Claudianus]. Pub. *first map of Bohemia* 1518

Klauser, Andrew Bernard (1656–1721). Czech surveyor

Klausner, Jacob Joseph (1744–1791). Swiss engraver, and geometrician

Klein, Anton. *Germany* 25 sh. 1822, *Bayern* 1842, *Deutschland* 1844

Klein, Herman. *Star Atlas* 1888

Klein, J. B. Bookseller of Leipzig. *France* 1792

Klein, Lt. Col. John Baptist (d. 1789). Chief Engineering Corps Poland, founded 1775

Kleiner, Solomon (1703–1761). Architect and engraver in Vienna. *Würtzburg* 1725, *Frankfurt on Main* 1738, *Views Vienna* 1724–1737

Kleinhans, Caroline. *Relief maps Mt. Blanc, Dept. Nord Savoie* 1874–5

Kleinig, W. Engraver. *Umgegend Dresden* 1854

Kleinknecht, L. V. Engraver for Weiland 1848. Published *Atlas der merkwürdigsten Staedte* 1844 (only 1 part pub.), *World Map* 1844

Kleinschmidt, Samuel (1814–1886). Greenland missionary and cartographer. *Maps of Greenland,* Godthab 1859–60

Kleinstraettl, George William. *Bregenz* 1647

Klemann, C. Publisher. Humboldt's *Central Asia* 1844

Klemensowski, Marcin (1791–1869). Prof. of Topography in Poland

Klencke. *Atlas* 1660. Put together by Amsterdam merchants headed by Jan Klencke; presented to Charles II. *37 Wall maps* now in British Museum, 6 ft in height

Klengel, Wolf Kaspar (died 1691). Geographer and military cartographer of Dresden

Klerc. See **Clerck**

Kleshnin, Y. Russian surveyor. *Map of Karelia and Keksholm* (with A. Zhikhamnov) pub. 1724

Kliewer, d'A. Fried. Wilh. (1829–1879) Cartolithographer of Berlin. *Reise Karte Europa* 1845–1867

Kliewer, d'A. Heinrich (1793–1840). Engraver. Berlin. *Ost Preussen* 1796–1809, for Gottholdt 1808

Klinckowström, Axel Leonhard (1775–1837). Swedish military cartographer. *Atlas Stockholm* 1824

Kling, C. *Hessen* (1870)

Klinger, Johann Georg (1764–1806). Painter, engraver, globe maker. *Pair globes* 1790–2, *World map,* Erlangen 1812

Klingsey, H. C. *Jylland Slesvig* 1836

Klingsey, Poul (fl. 1827–1860). Military cartographer of Copenhagen

Klingsey, Peter E. (1817–1887). Military cartographer and lithographer of Copenhagen

Klinkenberg, Dirk (1709–1799). Mathematician, astronomer, surveyor

Klinkens, B. Publisher of Haarlem *Atlas 23 Prov.* 1788

Klint, Erik Gustaf af (1801–1846). Hydrographer. *Pater noster Skasen* 1795

Klint, Gustaf af (1771–1840). Hydrographer of Stockholm. *Sveriges Sjö Atlas* 1795–1840. *Ostersjön* 1801, *Baltic* 1827, *Finsk Wiken* 1855

Klipstein, Aug. (1798–1894). Austrian geologist. *Vogelsgebirge* 1826, *Odenwald* 1827, *Hessen* (1850)

Klisin, Leontij. (17th century). Russian cartographer

Klobucaric, Ivan (1550–1605). Cartographer and engraver

Klockhoff, H. Engraver of Amsterdam. For Keulen *Environs Berlin* 1780, *Suriname* 1784

Kloeden [Klöden], G. Ad. V. (1814–1855). Geographer of Berlin. *Geberges u. Gewässer Karte v. Europa* 1814, *Altmark* 1816, *Wegen Karte Europa* 1830

Kloeden [Klöden], K. F. *Celestial Chart,* Weimar 1849

Kloenne, Julius C. *Logansport, Cass Co., Indiana* 1872

Kloot, Isaac van der. Publisher in The Hague, early 18th century

Klose, J. G. B. *Maps for Sohr* 1842–4

Kluepfel, Carl. *Heilbrun* 1889

Kluk, Father Christopher. *Diocese Luck* 1792

Klukowski, Ignacy. Designer of *Polish Town Views* in Paris 1835–1838

Klun, V. F. (1823–1875). Austrian geographer and publisher. *Atlas zur Industrie* 1866, *Hand u. Schul Atlas* 1869

Klyn, C. W. M. Surveyor. Revised Edition of Dou's *map of Holland* 16 sh. 1825

Knaplock, R. Printer. Chiswell's *London,* 1707

Knapp, Jan Danl. *Platte grond . . . Rio de Berbice,* H. de Leth. Amsterdam (1720).

Knapp. See **Sarony**

Knapton, Charles. *Nasipore River* 1751

Knapton, James (fl. 1687–1738). Publisher. At Queens Head, St. Pauls Churchyard; At the Crown in St. Pauls Churchyard, London. Dampier's *Voyages* 1705–1717, *Atlas Maritimus* 1728

Knapton, John (d. 1770). Publisher with James Knapton of Camden's *Britannia* 1730; with Paul Knapton *Geogr. Classica* 1747, *N. America* 1752; with Knaplock, Moll's *Atlas Manuale* 1723

Knapton, Paul (d. 1755). Worked with John Knapton. Camden's *Britannia* 1753

Kneass. American engraver, worked for Macpherson 1806, for Carey (with G. Delleker) 1817, for Carey & Lea 1822, for Lucas Fielding Jr. (1823)

Knight, Charles & Co. Publishers, London, 22 Ludgate St., and later 90 Fleet Street. *Plan Geneva* 1840, *Spain* 1845, *India* 1846, *Atlas for Society Diffusion Useful Knowledge* 1846, 1849 and 1852

Knight, Lieut. R.N. *Chart Delaware River* for Des Barres 1779 and *New York Harbour* 1779

Knight, J. American engraver for Tanner 1828–43

Knight, Capt. (later Rear Admiral) Sir John (1748?–1831). Hydrographer. *Road Bastia* 1793, *Ajaccio* 1795, *Mediterranean* 1795, *Needles,* 1797, *Portsmouth* 1799, *Brest* 1802, *British Channel* 3 sh. 1804, Steel's *Charts Ireland* 1828

Knight, P. *Bellmullet* 1834–36

Knight, Val. *Scheme for rebuilding London after Fire* 1666

Knight, William (fl. 1710–40). Printer and mapseller at Queens Head on Snow Hill, London. Morden's *W. Indies,* De Lisle's *Artois,* Sutton Nicholl's *Scotland* (1710)

Knight, Rev. W. Church Missionary. *Atlas* 1857

Knight, William Henry. *Washoe Silver Region Nebraska* 1861 MS, *Pacific States* for Bancroft 1863, *Rocky Mt. States* 1866

Knipe, J. A. Geologist. *Stamford Lincs.* 1834, *Gt. Brit.* 1840, *Brit. Isles & part France* 1844, *Scotland* 1858

Kniphausen. *Oldenburg* 1804, *Ostfriesland* (1815)

Knipping, Irwin. Stanford's *Lib. Map of Japan* 1879 Justus Perthes, *See Atlas* 1894

Knittel, Franz Anton (1671–1744). Map maker of Linz.

Knittel, Franz Jakob (ca. 1700–1770). Map editor

Knittel, Johann Ernst (1805–1831). Engraver. *Postkarte Deutschland, Polen Preussen* pub. Nuremberg 1809, 1812, 1814

Knobel, J. Surveyor of Uitenhagen. *District of Albany (Cape)* Faden 1820, *Graham's Town* 1821

Knobelsdorf, B. von. *Frankreich, relief map* (1862)

Knoblauch, Conrad. *Plan of Leipzig* 1595, used Braun & Hogenberg

Knoblauch, Hugo (fl. 1878–9). German geographer

Knoll, Fr. *Plan Stadt Braunschweig* (1881)

Knonau, G. M. von. See **Meyer von Knonau,** Gerald

Knopf, Thomas Hans Henrich. Norwegian Army surveyor. *Map of Iceland* 1731–34

Knoor, Andreas Wittib. Printer of *maps and Town views* in Nuremberg for Christ. Riegel (1685–1690)

Knorr, Capt. E. *Pelew Is., Admiralty Chart* 1877

Knorz. Printer of Nuremberg for Christ. Riegel. *Paffendorf* 1697

Knothe, C. *Zoutpansberg (S. Africa)* Perthes 1890

Knowles, Capt. C. *Plans Port Antonio Jamaica* 1732–3 MSS, *Plan Gibraltar Bay* 1744

Knox, George. *Tyrone* 1813 (with McCrea)

Knox, Henry N. *Admiralty Charts, Nanaimo Harbour* 1856, *Ports Q. Charlotte Is.* 1856

Knox, Maj.-Gen. Henry (1750–1806). MS *Plan America Camp at Morristown* 1777

Knox, James. Surveyor. *Commercial Map Scotland* 1788, *Midlothian* 4 sh. 1812, 1816, 1821, *Basin Forth* 1828, *Edinburgh & Environs* 1829

Knox, John. *Collection of Voyages* 1767

Knox, Robert (1740?–1720). *Hist. Acc. Island of Ceylon* 1681

Knox, Robert. Surveyor. *15 miles round Scarborough* 1821 and 1849

Knyff, Leonard (1650–1721). Topographer. *Plan St. James Park* 1706, *Burlington House* 1708, *St. James Palace* 1710

Kobayashi, Heihachi. Japanese engraver, end 19th century

Kobayashi, Kanshun. Japanese cartographer, end 19th century

Koberger, Anton (1455–1513). Printer and publisher. *Nuremberg Chronicle* 1493

Koch, Augustus. *Wairo & Poverty Bay District* 1868, *Middle Island New Zealand* 1874, *New Zealand* 4 sh. 1876

Koch, Comm. A. *Afgan.* Paris 1886, *Cochinchine* 1889, *Dahomey* 1890, *Maroc* 1891

Koch, Christoph Wilhelm (1737–1813). *Maps & Tables of Chron. & General* 1831

Koch, E. *Altenburg & Ronneburg* 21 sh. 1813

Koch, Felix A. (b. 1816). French cartographer. *Maps on Algeria, Cochin Chine, Morocco*

Koch, J. *Elbe* 1802, *Cuxhaven* 1817

Koch, Dr. Karl. *Schwarzes Meer* 1843, *Klein Asien* 1844, *Kaukasia* 1850

Koch, Wilhelm. German geographer, end 19th century

Kocherthal, Josue Wm. *Landschafft Carolina* 1709 with map *Virginia, N. & S. Carolina*

Koczizk, A. *Plan Krakau* Olmütz 1847

Koeber, G. *Sans Souci* 1836, *Charlottenhof* 1839, *Thiergarten* 1840

Koehler [Köhler] A. H. Military engineer in Saxony. *Asia* 1839, *Neuer Atlas,* Leipzig 1846, *Australien* 1847, *S. America* 1851, *World Map* 1848

Koehler [Köhler, Koeler], Johann David (1684–1755). Professor in Altdorf and Göttingen. Collaborated with Weigel in publication of School Atlases. *Schul & Reisen Atlas* 1718, *Orbis Antiqui* (1720), *Atlas Manualis* (1724) &c.

Koeke [Köke] (1852–1897). Viennese cartolithographer

Koelle, Rev. S. W. *Tropical Regions Africa* 1853

Koenig, Charles Dietrich Eberhard (1774–1851). Mineralogist

Koenig, Erhard Georg. *Erfurt* 1740

Koenig, H. A. Engraver. *Gotha* 1745

Koenig, Theophil. German cartographer. *Weltkarte* Berlin 1851, *Post & Eisenbahn Karte v. Mittel Europe* 1877, *Hist. Geogr. Hand-Atlas* (25 maps) 1850–55

Koeppen, Pierre (1793–1864). Russian geographer, *Geographical, statistical and ethnographic charts Russia*

Koerner [Körner] Hans Heinrich (1755–1822). Geographer of Zürich

Koetsveld, C. E. van. Preacher of The Hague. *Biblical Maps* 1854

Koffman, O. *Patagonien* Perthes 1882. Maps for Stieler 1885–1889

Kogge. See **Siebert**

Kohl, J. G. (1808–1878). Historical geographer

Köhler. See **Koehler**

Kohn, Alexis (1820–1900). Translator Russian geographical works

Koho, Kobayashi. Woodcut *world map* mid 19th century

Kokan, Shiba (18th century). Japanese cartographer. *World map* 1796, the first Japanese map printed from copperplates

Koksharov, Lieut. *Russia in Europe* 1845 with Murchison, Verneuil & Keyserling

Kolb, J. G. *Kenia* Perthes 1896

Kolb, Peter (1675–1726) Astronomer and meteorologist. *Description of the Cape of Good Hope* (1727) (six maps engraved by B. Lakeman), German edition 1719, French ed. 1741, 1742, 1743; English ed. 1731

Kolbe, Karl (1792–1849). Engraver, Berlin

Kolberg, C. H. J. See **Colberg**

Kolberg, W. von. *Railroad Maps, Warsaw– Cracow–Wien*, pub. Leipzig 1859

Koldewey, Ch. born 1837 in Hanover. Polar explorer. *Greenland, Spitsbergen* 1868–74

Koldewey, R. and H. Kiepert. *Lesbos* 1890, *N.W. Syria* 1891

Kolecki, Theodor. American cartographer. *N.W. Boundary U.S.,* Washington 1866

Koler, Johann. Printer of Nüremberg (fl. 1563–1578). Published German edition of *Ortelius*

Kolitz, Ed. Engraver on stone. For Kiepert's *Central America* 1858

Kollataj, Hugo (1750–1812). Clergyman, philosopher, politician and reformer. Author of works on reconstruction of Poland, initiator and contributor to plan of Cracow known later as the *Kollataj plan of Cracow* 1785

Kolleffel, J. L. V. Austrian cartographer. *Schwaben* Homann Heirs (1755)

Kolman, G. *Epirus.,* Athens 1897

Kolner, J. Engraver in Dantzig, end 16th century

Kombst, Dr. Gustaf. *Ethnographic Map Europe* 1844

Komensky, Jan. See **Comenius**

Kondet. See **Condet**

Kondratenko, E. *Cis-Caucasus region Tiflis* (1896)

Koner, W. (1817–1887). German geographer

Konig [König] . See **Koenig**

Komarzewski, Jan. General. Pub. *map hydrograph. of Poland,* Paris 1809

Konomura, Shosuke (19th century). Japanese globemaker

Konungess, H. M. *Skandinavien* Stockholm 1815–16

Konrad, Als. *Turquie d'Europe* 1816

Koops, Matthias. *Map Rhine, Meuse & Scheldt* 1796–7

Kondo Juzo. See **Juzo**

Kopernicki, Walery. *Hydro. map of Old Slav Lands* 1882

Kopernicus. See **Copernicus**

Koppe, Johann. Dantzig merchant. Printed Wied's *Map of Moscovy* 1542

Koppin, Ludwig. *Maps of Vistula and ports. Baltic, Elbing* 1811, *Berlin* 1850

Koppmayer, Jakob. Printer and publisher of atlases in Augsburg 1686–1710

Korabinsky, I. M. (1740–1811). Historian and geographer. *Hungarien,* Hague 1801, *Hungary,* London 1810, *Reg. Hungariae* 1817

Korb, Joannes Georgias. *Itiner. Moscovae,* Vienna 1698

Kordt, Ven. Pet. Collected material for Russian cartography. *Kiev* 1899

Korff. See **Mayer & Korff**

Korenski, Pusch. J. B. *Geognostische Besch. von Polen* and *atlas* 1833–36

Koriot, J. (1785—1855). Military cartographer. *Plan Warsaw* 1819

Korista, K. Fred. Ed (1825—1906). Born Moravia. Austrian geographer. *Mähren u. Schlesien* (1870)

Koristka, C. *Maps of Tatra Mountains,* Gotha 1862

Korkunov, N. *Maps of Russo Polish wars of 16th century and plans of fortresses,* Petersburg 1837

Korn, with Ernst. Publishers Berlin 1855—1889

Korn, W. B. *Oldest Polish School Atlas* pub. Breslau 1806, enlarged 1818

Kornatzki, Ed. V. (1813—1880). Carto-lithographer of Breslau

Koro, Inagaki. Japanese cartographer. *World map* 1708

Korten, Jonas. *Reise, Maps near East* 1751

Kortmann, E. Publisher of *School atlases* in Berlin 1837

Korzeniowska. *Hist. Atlas of Poland,* Warsaw 1831

Kosack, Capt. William. *Civil War Map* 1865

Kosciuszko, Thaddeus (1746—1817). Polish patriot, Colonel in American War of Independence. *MS plans of West Point, Halifax Battle of Saratoga &c.*

Kosmas, Indicopleustes. See **Cosmas**

Ko-Shun-King. Mongolian astronomer. *Armillary Sphere & globe* 1274

Kosinski, W. *Geolog. maps of Poland,* Warsaw 1873

Kosminski, Jan. Polish cartographer. *General Atlases,* Warsaw before 1850

Kotomin, A. M. Publisher of *Atlases,* St. Petersburg 1882

Kotzebue, Otto von (1787—1846). Russian Admiral and oceanographer. *Three Voyages round the World. Atlas,* St. Petersburg 1823

Kovalski, Egor. (1811—1868). Russian traveller and geologist, b. Karkov

Kovalski, M. H. (1821—1884). Russian astronomer and naturalist

Kowalski, Marian. Polish cartographer with Russian Geogr. Soc., 19th century

Kowerski, Edward (1837—1916). Polish military cartographer. *Great map of Russian Empire* 1894

Koyabashi, Chu. *World on screen* late 16th century

Kozaburo, Kikuya (mid 19th century). Map publisher of Edo. *Panoramic map of Shinano Province by H. Settei,* woodcut ca. 1850

Kozenn, Blasius (1821—1871). *Tyrol* (1872), *Schul Atlas* 1894 and later editions

Kozhin, A. Russian surveyor. *Gulf of Finland* 2 sh. 1713 (with Travin, Myaskow and Capt. Lein)

Kraatz, Leopold. Lithographer of Berlin. For Lange 1854, for Kiepert 1860—1888

Kraay, E. *Guiana (River Marowine)* used by Voogt 1682

Krachenikov, E. P. (1713—1755). *Kamtchatka* 1723—40

Kraeuter [Kräuter], Nic. (1711—1793). Firework and map maker

Krafft, G. W. German architect in Russian service. *Plan of St. Petersburg* 2 sh. 1737, later edition Seutter 1745

Krafft, Johann. See **Crato**

Kraft, [Craft, Hermagoras] (1527—1548). Cartographer

Krais. Publisher *Atlases* in Stuttgart 1850—1862

Kramer, Geogr. *Handboeck Rotterdam* Zuidema (1887—9)

Kramer, Ferdinand. *Road Map Biela—Lemberg* 1819 MS

Kramm, G. Engraver for Tanner 1836

Krase, J. W. Engraver, designer. *Atlas von Lievland,* Riga 1798

Krasilnikov, And. Dmitrie (1705—1773). Astronomer and geodesist

Krasilnikov, Vasilij (1740—1780). Russian geodesist

Krasinski, Josef W. (1783–1845). *Guide du Voyageur en Pologne,* Warsaw 1820 (French text), 1821 (Polish text), *map* and *10 town views* by Dietrich

Kraskowski, Julian. Polish surveyor and author. *Map of Pinsk region* 18th century

Krasopolski, Karol. *Road atlas of Polish Kingdom,* 148 maps, Warsaw 1894

Kratz, C. Engraver of Weimar. *Australien* 1849

Kratz, E. (1861–1869). Engraver

Kratz, W. Engraver for Wieland 1848, *Australien* (with C. Poppey) 1870

Kratzer, Nikolaus (1487–1550). Mathematician, cosmographer and cartographer

Kraus, Johann Ulrich (1655–1719). Printer and engraver for Seutter, *London, Copenhagen, Strassburg, Augsburg &c.* and for Valck

Kraus, Ph. (18th century). Surveyor

Kraus, Johann Thomas. Engraver. *Augsburg* 1745, *Strasburg* 1745

Kraus, C. *Brazil* 1865

Krause, Dr. A. *Passen zum Yukon* 1882–92

Krause, Frank. *Freedom Township* 1874

Krauseneck, Wilhelm Johann (1775–1850). Prussian General and cartographer

Krayenhoff, Cornelius Rudolphus Theodorus, Baron van (1758–1840). Minister of War under Lous Napoleon, military cartographer. *Triangulation of Holland* 1802–11, *London,* Arrowsmith 1840

Krazeisen, Lewis. Publisher, 41 Leicester Sq. *London,* 1793

Krebs, A. C. *Kiobenhavn* 1806

Krebs, Nikolaus of Kues. See **Cusa**

Kreibich [Kreybich] Franz Jacob Heinrich (1759–1834). Astronomer and cartographer, priest of Ziterice. *Bavaria* 1807, *Karlsbad* 1828, *Bohemia* 1833, *Töplitz* 1834

Krekwitz, Georg. *Hungariae . . . descrip.,* Frankfurt 1685

Kremer, Gerard. See **Mercator**, Gerard

Kreummer, H. *Wandkarte v. Afrika* Breslau (1850), *Asia* 1870

Krevelt, A. van. Engraver of Amsterdam, for Raynal 1773, for Elwe & Langeveld 1786

Kribber, Cornelius. *Belgii pars Sept.* Utrecht 1751

Kriegner. See **Criginger**

Krieger, S. *Reg. Hungariae,* Vienna (1780)

Krille, K. F. Engraver (1850)

Kringerius. See **Criginger**

Krockow von Wickerode, Carl, Graf *Ost Sudan* (1866)

Kromer, Josef. Prof. Cracow University, surveyor and cartographer. *Plan Cracow* 1783, *Map Poland* 1787

Kromer, Martin. Publisher. *Mag. Duc. Lith., Liv., et Moscoviae* 1589, *Polonia-Libri duo* (1555 unauthorized ed. Frankfurt), 1577 (authorized edition in Basel and Köln), Spanish translation published Madrid 1588, German trans. Leipzig and Dantzig 1741, Polish trans. Wilno 1853

Kron, H. Wood engraver. Helwig's *Silesia* 1561

Kronevelt, Henri de. Publisher and printer at Delft. *Maps and town views of Levant* 1700

Kropatscheck, H. Worked with Kirchoff, q.v.

Kropotkine, Alex. (1841–1886). Russian astronomer and physician

Kropotkine, Pierre (b. 1842). *Geogr. Studies Siberia*

Kruger, Capt. S. *New Chart of False Bay,* inset on Heather's *Cape* 1796

Kruger [Kruginger] . See **Criginger**

Kruikius. See **Cruquius**

Krukowski, Jan. *Topo. Atlas of Polish Mining Region,* Warsaw-St. Petersburg 1860–63

Krumbholz, Carl. Map publisher. *Schul Atlas,* Dresden 1853

Krümmer, H. German mapmaker mid 19th century *18 wall maps* for school use

Kruse, Karsten Christian (1753–1827). Cartographer. *Atlas . . . Europäschen Länder* 1822

Krusenstern [Kruzenstern] , Baron Ivan (Adam) Fedorovic von (1770–1846). Russian Admiral, b. Esthonia. Marine cartographer. *Atlas de l'Océan Pacifique,* St. Petersburg 1813 and 1824, *Beyträge zur Hydrographie* Leipzig 1819

Krzewski, Jan. B. Surveyor. *Map of Posnan district, Plan Posnan* 18th century

Kuang-Yu-Tu. *China* 1555

Kubary, Jan Stanislas (1846–1896). Polish cartographer and explorer. *Fiji, Marshall Is., Samoa etc.*

Kubn [Kühnovio] , Friedrich. Silesian cartographer, d. 1675. *Silesia* Blaeu 1662, *Schweidnitz* 1662

Kuchel, C. C. Lithographer of San Francisco for Holt 1864. See also **Nagel Fishbourne & Kuchel**

Kuchel & Dresel. Lithographers, 176 Clay St., San Francisco. *Overland route* 1858

Kuehn, E. Engraver for Meyer 1867

Kuehn, George-Wilhelm. Publisher of Ulm. *Vorstellung der gantzen Welt* 1692, *Kleiner Atlas* 170(2)

Kue Kenthal, Dr. W. *Ost Spitsbergen,* Gotha 1890

Kuemmerly [Kümmerly] , Gottfried (1822–1884). Swiss cartolithographer

Kuemmerly, Herman (1857–1905). Swiss painter, publisher and relief cartographer

Kuendig [Kündig] , A. *Canton Basel* 1830

Kuesel, J. G. Engraver. *Henneberg* 1743

Kuesel, Melchior (1626–1683). Publisher and engraver. Zeiler's *Brandenburg & Pomerania* 1652, Visscher's *Austria* 12 sh. 1669

Kuestner [Küstner] Heinrich **& Co.**

Publishers of Leipzig. *Eisenbahn Karte . . . Nord Amerikas* (1865)

Kuffner, Paul (1713–1786). Engraver in Nuremberg

Kuhl, C. *Delta du Danube,* Leipzig 1887

Kuhl, F. Printer for Watson 1850

Kuhlemann, Lieut. *Meissen,* Dresden 1811, *Plan Dresden* 1813

Kühnovio, F. See **Kubn**

Kuiper, J. *Atlas v. Nederland* Leeuwarden 1872

Kukiel, Adam. Royal Geographer to the King of Poland, 18th century

Kulczewski, Antonie (fl. 1825–31). Polish military topographer

Kulczycki, Adam (1809–1882). Polish cartographer in French Colonial service. *Maps of Tahiti and Moorea* 1850–59

Kulhanek, M. See **Claudianus**

Kult, Nic. Publisher of Constance. *Georgius Schwartzwald* 1603

Kummel, C. A. Publisher of Halle. *Preussischen Staate* 24 sh., 1820, *N. Deutschland* 2 sh., 1826, *Europa* (1850)

Kummerer, V., **Kummersberg**, C. Publishers of *admin. maps* Wien 1855

Kunagata Keisa. See **Keisai**

Kündig, A. See **Kuendig**

Kunike, W. Publisher. *Greifswald* 1842

Kunsch, H. Lithographer. *World Map,* Leipzig 1853

Kunst, Christian Ludwig. Publisher of *Atlas,* Berlin ca. 1790

Kuo-Shen (1279–1368). Chinese astronomer

Kuo, Shou-ching (1231–1316). Chinese astronomer and geographer

Kupffer, Ad. Th. (1799–1865). Russian mineralogist and meteorologist. Founder and Director Meteor. Observatory, *Carte de l'Oural* 1833

Kurnatowski, Viktor Adam (fl. 1810–1846). Lithographer, author and publisher. *Map of Great Poland* (1843–1861), *Plan of Poznan*

Kurowski, Teodor S. *Map of Cracow District* 1797

Kurtz, Joh. Heinrich. *Hist. Atlas Germany,* Berlin 1859

Kurzboeck, Joseph. Publisher and printer. *Regni Bohemiae Vienna* 1760

Kurzewski, Erazm (fl. 1827–31). Polish military topographer

Kusa, N. See **Cusa**

Kusell, Melchior. Engraver for Zeiller 1652

Kushelev, Admiral Graf. *Frontier between Russia & Turkey,* St. Petersburg 1800

Küstner, W. See **Kuestner**

Kutscheit, J. V. *Hist. Pocket atlas* 1843, *Maps of Poland, Russia,* Berlin 1872

Kuyper, J. (1821–1908). Dutch map

editor. *Wereld Atlas* 1857, 1874, and 1880; *Nederlandsche Reiskaarten, Gemeente Atlas* 1865–69

Kuznetsov. *Tsarstvo Pol'skoe* 1861

Kvilcinski, L. I. *Maps of Vistula near Warsaw* 1896

Kwiatkowsku, Martin. Translated into Polish *descrip. of Livonia, A. Henneberg's map* pub. Königsberg 1564

Kypseler, Gottlieb, of Munster. *Délices de la Suisse* (maps and views) Leide 1714

Kyrilov [Kivilloric] , Ivan (1689–1737). Russian cartographer. Director of first surveys in Russia from 1717. *First Atlas of Russia,* St. Petersburg 1737

Kyrott, Joh. David. Printer of *maps and atlases,* Nuremberg 18th century

Kyser, J. Engraver. Moll's *Carolina* 1721, Keulen's *Nouvel France* (1740)

L

L., A. *Plan de Paris* 1829

L[upton], D[onald]. *Map World in 4 plain maps* 1670

L, E. C. *Missions of d'Indo-Chine* 1879

L., J. E. *St. Domingo* 1792

L, J. G. *Grundriss Dresden* 1801—09

L, L. *Citta di Gotha in Sassonia* (1565?)

L, M. L. *Memingen* 1622

L, S. *Cölln am Rhein* 1667

L, T. F. *County of Savannah* 1749

Laan, Adrian van der. Engraver of Amsterdam (1720). Title page for Schenk's *Flambeau de de la guerre* 1735

Laan, Anne van der. *Friesland* 1874

Labanna [Labaña, Lavaña, Lavanha], Joäo Baptista (1582—1624). Portuguese Royal cartographer, d. Madrid. *Maps of Spanish Provinces Aragon* 1610—14 pub. 1616, 1620, 6 sh. used by Jansson & Blaeu; *Regimento Nautico* 1596

La Barre, Fébure de. See **Barre**

La Barre Du Parc, Nicolas Edouard de (b. 1819). Various geographical publications

Labarthe, Charles. *Plan Hanoi* 1883

Labarthe, Pierre (1760—1824). French geographer

Labat, J. B. (1663—1738). Dominican traveller. *Iles d'Amérique* 1722, *Afrique Occ.* 1728, *Ethiopie* 1732

Labatt, J. E. *Galveston, Texas* 1869

La Baume, Charles de. Civil engineer, surveyor. *Geolog. Map Boise River Basin, Idaho* 1865, *Alturas County* 1866

Labberton, Robert Henlopen (1812—1898). *Historical atlases* 1872 to 1887

Labelye, Charles. *The Downs* 1736 and 1737, *Lands between Sandwich & Shore* 1736, *Westminister Bridge* 1739

La Bella. See **Domenico Machaneus**

Laberge, A. M. de C. *Arpentine Railway* 1863

Labhart, F. *Lombardy & Venice,* 24 sh. (1830) MS

Labiche, *Maps* for Freycinet 1826

Labillardière, Jacques Julien Houton de. French naturalist. *Carte Mer des Indes* 1800, *Atlas Voy. recherche la Pérouse,* Paris, F. Schoel 1811

La Blottiere (18th century). *Pyrénées.* English edition Arrowsmith 1809

La Bodega y Cuadra, J. F. de. Maps and charts coast of California 1791

La Boissière, Gilles de. *Geographical Playing cards* 1669—71

Laborde, Alex. Louis Joseph, Comte de (1773—1842). *Magellan* 1790, *Atlas de l'Itiner. de l'Espagne* Paris, Nicolle 1809

La Borde, Jean Benjamin de. *Charts South Seas* in *Hist. Mer du Sud* 1791, *Toscane* 2 ll. 1786, *Umbria* 2 ll. 1786, *Détroit de Malacca* 1791

Laborde, Léon Emmanuel Simon Joseph, Marquis de (1807–1869), Historian. *Comment. Géogr. sur l'Exode* 1841 (19 maps), *Ville Petra* 1829, *Arabie Petrée* 1834, *Sinai Halb Insel* 1859

Laborde, M. J. *Excursion Colombie* 1872 MS

Labouche, Hor. *Ville de Toulouse* (1875)

Labre, Col. A. R. P. Expedition *Bolivia* 1889 MS

La Bretonnière; See **La Couldre La Bretonnière**

Labrosse, Paul. *Plan de la Ville de Montreal* 1766 MS

La Bruyère, Jean Anthème Fayard de. *Atlas Universel* 1877–96

Lac, P. du. See **Perrin du Lac**

La Caille, Jean de (d. 1720). *Ville de Paris* 1714

La Caille, Nicolas Louis, Abbé de (1713–1762). Mathematician and astronomer. With Cassini de Thury revised arc meridian of Paris 1739–40. *Cape of Good Hope* 1752, *Isle de France* 1753

Lacam, Benjamin. *N. part Bay Bengal* 1785, *Hughley River* 1779, *Tracks ships Monsoons* 1784, *Pilot Bay of Bengal* 1795, later edition 9 sh. 1803

La Camp, J. de. See **Camp**, J. de la

Lacarole, Charles. *Plan de Montpellier* 1877

La Carrière-Latour, Arsène. *Atlas War in W. Florida & Louisiana* 1816

La Caze-Duthiers (d. 1901). French geologist

La Cavada y Mendez de Vigo. See **Cavada y Mendez de Vigo**

Lacey, Edward. London publisher. *Map London* (1834)

La Chaudière. *Plan Kennebek & Sagadahok* 1775

Lachmann, W. *Braunschweig & Harz-Gebirge* 1852

Lackington, **Allen & Co**. Publishers &

booksellers. Temple of the Muses, Finsbury Square, London. *New Pocket Atlas England & Wales* 1806, *Walkers Univ. Atlas* 1820

Lackner, Ig. K. von. Engraver for Schraembl. *Vienna* (1789) *Polynesia*

La Combe, Q. de. See **Querenet De La Combe**

La Concha Miera, F. de. See **Concha Miera**

La Condamine, Charles Marie de (1701–1774). Geographer, b. Paris. With Godin & Bouguer measured the arc of the meridian (1736–9). *Figure de la Terre* 1749, *Journal du Voyage . . . à équateur . . . mesure . . . degrés de Méridien* 1751

La Cosa, J. de. See **Cosa**

Lacoste, Charles. *Plan Toulouse* 1874, *Dijon* 1875, *Maps* for Migeon *Australia* 1874, *Océanie* (1880)

Lacouchy. See **Lahanier**

La Couldre La Bretonnière, Lt. *Charts coast of France* 1776–1808, *Rade de Dunkerque* 1792, *Côtes de France* 1797 &c.

La Crenne, V. de. See **Verdun de la Crenne**

La Croix, A. P. de. See **De la Croix**

Lacroix, E. *Charente Inf.* 1864, *Arrond. Cognac* 1865

Lacroix, Fred. *Malta* 1848, *Afrique* 1864

La Croix, F. de. See **Croix**

Lacroix, J. *Lot et Garonne* 1867–8

La Croix, L. A. N. de, Abbé. See **Delacroix**

Lacroix, Silvestre François (1765–1843). Mathematician and military cartographer of Paris

Ladomin, F. *Vestung Bonn* (1689)

La Croyère, D. de. See **Lisle de la Croyère**

La Cruz, J. de. See **Cruz**, J. de la

Lacy, John. *Covent Garden with part St. Martin in Fields* 1673 MS

Lacy, W. *Solar System* (1798)

Lacy, W. W. de. See **De Lacy**

Lade, Capt. Robert. *Voyages* Paris 1744

Ladebour, *Atlas zu Ladebour's Reisen* Berlin, Reimer 1830

Ladyjenski, Col. *Plan Peking & Environs,* St. Petersburg, Mil. Top. Dept. 1848

Laemmert, E. & H. *Rio de Janeiro* 1876

Laet, Herrerd de. *Terre Ferme, Perou, Brésil, Amazons* (with Acuna & Rodriguez) Paris 1703

Laet, Joannes de (1583–1649). Geographer and naturalist. *Nieuwe Wereld,* Leyden, Elzevier 1625 (10 maps). Later editions: Dutch 1630, Latin 1633, (14 maps), French 1640. The maps by Hessel Gerritsz

La Favelure. *Portsmouth Haven* 1600

Lafayette, Marie Paul Roche Yves Gilbert de Motier, Marquis de (1757–1834). Commander French Forces in American War of Independence

La Faille, P. de. See **La Feuille**

La Ferté, Papillon de. Astronomer. *Système de Copernic* 1783

La Feuille, Daniel de (1640–1709). Goldsmith, clockmaker, engraver and publisher of Amsterdam. *Atlas Portatif* 1701 and 1706, *Véronois Padouan* (1705), *Military Tablets* 1707

La Feuille, Jacob de [Jacques] (1668–1719). Mapmaker, engraver and publisher. Married widow of De Ram. *Maps of London & Paris* 1690, *Malta* 1696, *Atlas Amstd.* ca. 1710, *Danube* 1717

Le Feuille, Paul de (d. 1727). Succeeded his father Daniel. Published military pocket atlases. *Dunamonde en Livonie* (1710), *Tablets Guerrières* 1706 & 1717, *Geographisch-Toneel,* Amsterdam Ratelband

La Feuille, Jeanne, sister of Paul, continued business

La Fleche, J. de L. de. See **Le Loyer de la Fleche**

Laffont, J. *Tehuantepec* 1851

Lafitte, Sollon. *Bearn* 1642

Lafon, Barthéleme. Geogr. engineer of New Orleans. *Portion du Territ. du Mississippi* 1805, *Plan New Orleans* 1806

Lafon, Gabriel. *Costa Rica* 1851

Lafon, Guillaume. *Narbonne* 1704 & 1739

La Fontaine. *Island Diego Garcia* 1784

La Fontaine, Sr. *Plan de Candie* 1668

La Fontaine, Eugène. *Seraing* (1879)

La Fora, Nicolas de. *Mexico–U.S. boundary* 1771 MS

La Fosse, Chatry de. *Orlenois* 1761, *Lorraine et Bar* 1762, *Cherbourg* 1785–6

La Fosse, J. B. de. See **Fossé**

La France, Joseph. *Map N. America* in *Dobbs Hudson's Bay* 1742

Lafreri [Lafrèrie, Lafrery], Antonio (1512–1577). Cartographer, publisher, map and printseller born Besançon in Burgundy. Antoine du Perac Lafrèrie emigrated to Rome about 1540 and set up in business in the Via de Perione in 1544. Partner with Salamanca 1553–63, and continued alone until his death in 1577. Succeeded by Claude Duchetti. *Catalogue* of his publications in 1572. One of the first to issue collections of maps in atlas form, variable in contents; to the later examples he added a title page. *Geografia: Tavole moderni &c., World* n.d., *Europe* 1560, *Northern Regions* 1572, *Asia* 1561, *Cyprus* 1570, *Genoa* 1573, *Lombardy* 1564, *Malta* 1565, *Rome* 1577, *Milan* 1573 &c.

La Gallissonnière, R. M. Barrin, Marquis de (1693–1756). Sailor and director of Dépôt de Cartes et Plans. Contributed to the voyages of Chabert, Bory & La Caille

La Garde, J. de. See **De la Garde**

La Gironière, Paul de. *Lac du Bay* 1853

Lagnia, Giacomo A., of Trapani. *Portulan* 1539

Lagniet [L'Agniet], Jacques. Publisher "sur le Quai de la Mégisserie au pont l'Evesque" Paris. *Dunkerque* 1646, *l'Isle Cayenne* (1652), Du Val's *America* 1661, *Environs de Paris* 1665

World map published in Rome by A. Lafreri

Lagny, Thomas Fantet de (1660–1734).
Mathematician and oceanographer

Lago, A. F. P. *Cidade do Desterro (Brasil)*
1876

Lagrange, Guiseppe Luigi di (1736–1813)
of Turin. Mathematician and astronomer

La Grive, Abbé Jean de. See **Delagrive**,
Abbé Jean

Laguille. *Projections du Globe Terrest.*
9 sh. 1848

La Guilbaudière, Johan de. French chartmaker. *Atlas Pacific coast* ca. 1696 MS
(35 maps)

Laguillermie. Engraver of Paris, Rue de
Noyen 56, and Rue St. Jacques No. 82.
For Monin's *Atlas Classique* 1844–5,
for Levasseur 1854–61

La Guillotière, F. de. See **Guillotière**,
F. de la

La Hage, M. d. See **Mengaud de la Hage**

Lahainaluna, School. *Map Hawaiian
Islands* 1838

Lahanier, Lacouchy et Cie. Engravers
for Dufour 1864

La Harpe, Jean François de (1739–1803).
Histoire générale des Voyages 1780, 1820,
1821, 1825. An abridgement of *Prévost's
Hist. Gén. des Voyages* 1746

La Haye, G. de and J. de. See **Delahaye**

La Haye, le jeune [the younger]. *Paris*
1790, *Avignon* 1791, *Louvre* 1792

La Herty, F. D. and H. S. **Tobin**. *Route
Edmonton to Yuka River* 3 sh. (1898)

La Hire, Gabriel Phillipe de (1640–1719).
Planisphère céleste 2 sh. 1705, *Globe
Terrestre* 4 sh. 1708

La Hoeye, F. van den. See **Hoeye**

Lahont, J. L. M. Lavenère. *Cantons
de Moulins Est et Quest* 1842

Lahontan, Louis Armand, Baron de
(1666–1715). *Nouveaux Voyages Amérique
Sept.* 1703, and English edition 1703;
Carte Gén. de Canada (1705) and later
editions

La Houve, Paul de. See **Houve**, Paul de la

Lahovari, G. L. Rumanian geographer, b.
Bucharest 1838

Lahr, Leopold. *Eisenbahnkarte Elsass-
Lothringen* 1871

Laicksteen, Peter (fl. 1556–70). Cartographer and astronomer. *Palestine* (with
Christian Sgroth) 1556, *Palestine* 9 sh.
Hieronymus Cock 1570. Used by Ortelius
1584. *Plan Jerusalem* used by De Jode

Laidman, J. London publisher, 119
Chancery Lane. *Davies' map of London*
1848

Laillet, E. *Madagascar* by Laillet & L.
Superbie, Paris, Challamel 1889 & 94,
Madagascar 3 sh. 1895

Laing, Major Alexander Gordon (1794–
1826). African explorer. *Expd. Soudan*
1826, *W. coast of Africa* Wyld 1830

Laing, James. Surveyor Australia, mid
19th century

Laing, John. *Hastings & St. Leonards*
1859

Laing, Joseph. Lithographer, 107 Fulton
St. N.Y. For Bancroft 1870

Laing. See **Lang & Laing**

Lainpacher, Matthaeus Petter. Engraver
for Lotter ca. 1745

Laird, James Stewart (1842–1928).
Canadian surveyor.

Lairesse, G. de. Engraver for Visscher
1680 & 1710, for Schenk 1705, for
Braakman 1706

La Jaisse, Pierre Lemau de. See **Lemau
de la Jaisse**

La Jonchère, Antoine Simon Lecuyer
de. See **Lecuyer de la Jonchère**

Lajouvane, Felix. Publisher. Calle del Pera
51 & 53, Buenos Aires. *Paz Soldan's Atlas
Repub. Argentina* 1888

Lake, D. J. American civil engineer.
County Atlases of Ohio 1870–6: *Guernsey
Co., Jackson Co., Vinton Co. &c.*

Lake, D. J., **and Co**. *County Atlases*

1874–1884, *Indiana, Michigan, Kentucky*

Lake, Edward. *Church Missionary Atlas* 1873

Lake, H. *Johore Territ.* 1893

Lake, J. W. *Guide to London* 1827

Lake, W. Lithographer, 170 Fleet St., London

Lakeman, Balthazar. Publisher of Amsterdam. *Kaap de Goede Hoop* (1690?), *Besch. Kaap Goede Hoop* 1727

Laker, Mathew. *Dover Harbour* (with Pett) 1583

Lalande, [La Lande] Joseph Jérome le Français de (1732–1807). Astronomer. *Figure du Passage de Venus* 1760, *World* 1770, *Celestial Globe* 1775, *Voyage en Italie* (atlas, 35 Town Plans &c.), *Bibliographie astronom.* (1803)

Lale. Letter engraver for Freycinet (1807–16), Giraldon and Vuillemin (1847)

Laler, Richard P. *Kildare Hunting District* 1893, *Meath Ward* 1895

Lalleau de Bailliencourt, Alphonse Marie Florimond de. *Géogr. Hist. de France, Atlas* (1859)

Lallemand. Writing engraver, Rue St. Jacques No. 66 Paris. For Frémin 1820 for Perrot 1823, for Lapie 1823–33. *Plan Panorama de Paris*

Lallemant, G. A. *Prov. S. Luis (La Plata)* 1822, Expd. *Minere Argentina B.A.* 1886

Lallemant, H. *Map of St. Lawrence* 1669 MS

Lallemant, Nicolas (late 17th century) *Atlas forests in Alsace* MS

Lallier, Justin. *Amérique Centrale* 1850, *Paris* 1867

Lamairesse. *Etudes Hydro. Mont. Jura* 1874

Lamal, Prosper. *Bruxelles* 1875

La Malle, A. J. C. A. D. de. See **Dureau de la Malle**

La Marche, Charles François de *Africa* 1790, *Lorraine* 1792, *Théatre de la Guerre d'Allemagne* 2 sh. 1793, *Empire Turc* 1802, *Carte Gén. France* 1818

Lamarche. See **De la Marche**

Lamare, F. *Plan de Genève* 1860

La Mare, Nicolas de. *Lutèce ou Paris* 9 sh. 1705   1738

Lamarre, Joaquin Raimondo de. Corrected charts of *Bahia do Rio de Janeiro* 1847

La Martinière, A. A. B. de. See **Bruzen de la Martinière**

Lamb, Daniel W. *De Kalb Co., Illinois* 1864

Lamb, David. *Plans Castle St. Johns* (Central America) (1780) MS

Lamb, Francis (fl. 1665–1700). Engraver and publisher at Pewter Plate Alley, Gracechurch St. and after 1690 at Little Montague St. in Little Britain. For Blome, *Barbary* 1667, *America* 1669, Ogilby 1676 and Seller 1671, for Speed 1676, Ogilby & Morgan 1678, for Greene & Pitt 1688, Petty's *Ireland* 1685, Morden & Berry 1690

Lamb, Lieut. H. *Island of Perin* 1856, London. 1858

Lamb, James. Land surveyor, attested Renfrew for Thomson's *Scotland* 1832

Lambach, H. *Union Pacific Railroad* 1869

Lambarde, William (1536–1601). Historian. Perambulation of Kent, (the first county history) *Map of Kent* ca. 1570, *Schyre of Kent* London, Ralph Newberie 1576, *Carde of the Beacons* 1576

Lambert, Ainé [the elder]. Engraver for Freycinet 1808

Lambert, B. *Plan London before Fire* 1806

Lambert, G. C. *Badische Eisenbahn* 1860

Lambert, John. American draughtsman.

Cape Diamond 1802, *Part Quebec* 1810, *Sorrel Lower Canada* 1810

Lambert, Johann Heinrich (1728–1777). German astronomer and mathematician, invented various map projections

Lambert, M. W. Engraver of Newcastle. Fryer's *Northumberland,* Bell's *N. & D. Coalfield* 1843–61, *Great Northern Coalfield*

Lambert, M. J. Gustave Ad. (1824–1871). *Map Behring Strait*

Lambert, T. See **Gemini**

Lambert of Saint Omer (ca. 1050–1125). Chronicler and mapmaker. *Liber Floridus* ca. 1120 (with maps)

Lambert, William. Surveyor. *Wood Green Estate Middx.* 1853

Lamberti, Archangelo. *Totius Colchidis* 1652

Lambien, Antoine (ca. 1635–83). Military cartographer

Lambilla, Guillaume de (1649–1699). Jesuit from St. Malo, mathematician and cartographer

Lambilly, G. de. *Esveché de Nantes* 1706, *Loire River* 1774

Lambrechts, G. Engraver. *Bruce & Mengden's S. Russia* 1699

Lambreschtsen, T. A. *Atlas Prov. Zeeland* 1877

Lambton, Lt.-Col. W. Surveyor *India* 8 sh. 1802–14, *Horsburgh* 1827, *Nizam's Dominions* (with Capt. George Everest) 2 sh. 1827

Lameau, P. J. *Turquie d'Europe* Paris, Piquet 1827

Lamelina, Albert. Draughtsman and lithographer. *Grundriss Wien* 1858

Lamothe, A. de. *La France des Bourbons* (1873)

La Motte, C. *Plan de Versailles* 1783

Lamotte, Guérin de. *Environs de Lisbonne* 1821

Lampen, M. Surveyor. *St. Joseph Co., Michigan, U.S.A.* 1858

Lampert, Ignacius. *Kirchenkarte von Unterfranken* 1863

Lampert, P. *Walcheren* 1852

Lamplugh, George William. *Geological Survey England & Wales* 1898

Lamprecht, Johann. *Bisthumes Linz* 1841, *Land ob der Ens* (1872)

Lampton, John. *Hardtville, Kansas* (1857)

Lamrinck, Jan. *Frisia Occid.* 1622

Lamson, A. C. American surveyor. *Chart Mouth Detroit River* 1874

Lamson, G. W. & C. A. *Tawas Harbour* 1857, *Lake St. Clair* 1870

Lamsvelt, J. Engraver for Van der Aa. *Galérie agréable du Monde* 1729

Lanagan, J. *Jefferson & Oldham Co. Kentucky* 1879 (with D. G. Beers)

Lancaster, A. B. Belgian geographer and meteorologist. *Carte Pluviométrique de la Belgique* Brussels 1895

Lancaster, Sir James (ca. 1550–1618). *Maps of Carolina & Albemarle River* in *Blathwayt Atlas* 1679 MS

Lancefield, Alfred. Civil engineer. *Plan Edinburgh,* W. & A. K. Johnston 1851

Lanchester, Henry Jones. Architect and surveyor. *Stanford Estate House* 1865

Landale, C. *Part of Angus showing proposed Railway Dundee–Strathmore*

Landale, D. *Geology East of Fife Coalfield* 1837

Landavus. See **Ziegler**

Landbom, Karl. (18th century) Swedish surveyor

Lande, J. J. de F. la. See **Lalande**

Landen, J. *Karte . . . Aachen* 1833

Lander, O. *Sverige, Norge och Danmark* 6 sh. Stockholm 1880

Lander, Richard. *Route Clapperton* 1828, *Course Quorra* 1851

Lander, William. *Bristol* 1826, *Electoral District Map Bristol* 1840

Landmann, Capt. George R. E. *Plan Cadiz Harbour* 1809 MS, *Cadiz-Marabella* London, Arrowsmith 1811, *Greenwich & Gravesend Railway* 1835, *N. Kent Railway* (1845)

Landmann, G. *Universal Gazette* 1835

Lands, Heber. Estate surveyor. *Burstead* 1720 MS

Landsberg, Leopold Franz. *Queensland* 1860

Landsborough, W. (d. 1886). Australian explorer. *Australia* in *Proeschels Atlas* 1863 MS, *Surveys* 1859−62

Landteck, Zacharias. *Geographical Clock* 1705, In Homann's *Atlas*

Landts, C. *Post Roads London to Edinburgh* 1668

Lane, J. C. Engineer. *Prov. of Choco Colombia* 1854

Lane, Lieut. Michael, R.N. Marine surveyor trained by Cook, assisted him in his Newfoundland Surveys. *North American Pilot* 1775, Second edition 1779, French edition 1778, *Fogo Is.* 1785, *Steel's N. America* 1807, *Pocket Globe* 1818

Lane, N. *Terrestrial Globe* 1776, *Pair Miniature Globes* London 1825

Lane, Nicholas. Estate surveyor. *Branshelye (Kent)* 1632, *Romford* 1633, *Horton* 1746 MS

Lane, W. Capt. *Parry's Discoveries Polar Regions* (1821)

Lanée. Editor of Paris, Rue de la Paix No. 8. Successor to Longlet. *Dufour's Algeria* 1867

Lanessan, G. M. A. de (b. 1844). Governor Indo China. *Geogr. publications Indo China*

La Neuville. See **Buache de la Neuville**

Lanfranconi, E. *Wasserstrassen Mittel Europa* 6 sh. 1880

Lang, A. *St. Croix (W. Indies)* (1800)

Lang, Carl. *Heilbron am Neckar* 1705

Lang, F. C. *Prospectors map Kootenay* 1897

Lang, George. Engraver and publisher. *Altenberg* (1550), *Augsburg* (1595), *Erd Karte* 4 sh. 1592, *Elsass Lothringen* 2 sh. 1596

Lang, Hartlieb. *Europa* 1859, *Bayern* (1867), *Deutschland für Schulen* 1867

Lang, I. Doctor. *Neighbourhood Torquay & Teignmouth* 1842

Lang, I. Caspar. *Schaffhausen* 1641

Lang, Johann Anton (1765−1811). Mathematician. *Helgoland* 1787, *Elbe & Weser* 1795

Lang, Major. *W. coast Africa* 1843 (with O'Beirne)

Lang, Moritz (fl. 1649−64). Engraver of Augsburg. *Stier's Hungary* 1664

Lang & Laing. Lithographers 117 Fulton St., N.Y. (1850). For Colton 1860

Langara, Juan de. *Costa del Pera* 1798, *Canal Viego de Bahama, Costas de Chile, and Antilles* 1799

Langau, Piotr. Mid 17th century surveyor and cartographer, b. Dantzig. *Map Pomerania* 1659

Langdale, Thomas. *Topo. Dict. Yorkshire* 1822

Langdon, C. H. C., Lieut., R.N. Admiralty surveyor. *Calabar River* 1870, *Sherbro River* 1870, *Dar es Salaam* 1874

Langdon, Rev. L. *New Hampshire* 2 sh. 1761 (with Col. Blanchard)

Langdon, Samuel. *Map of New Hampshire* 1761 (with Joseph Blanchard)

Langdon, Thomas. Surveyor. Estate maps. *Langley (Bucks)* 1596, *Gamlingay (Camb)* 1601, *Estates in Leicester* 1607, *Estates of Corpus Christi College, Oxford* 1606−1616

Lange, D. A. *Plan Suez Canal* 1869

Lange, F. E. Geographer and surveyor. *France.* Leipzig 1792, *Industrie-karte*

Preussischen Staaten 1796, *Sachsen* (1800)

Lange, F. R. A. *Westpostverens* 1876

Lange, Gunardo. *Prov. de Catamaca* 1893

Lange, G. H. *Minahassa (Indonesia)* 1878

Lange, Henry (1821–1893). German cartographer, one of the founders Geogr. Soc. Leipzig. *Atlas v. Nord Amerika* Braunschweig 1854, *Reise Atlas* 1857, *Bible Atlas* 1860, *Atlas von Sachsen* 1860, *Neuer Atlas* 1863, *Hand Atlas* 1865, *Atlas zur Industrie* 1866, *Atlas de Géographie* 1875, *Süd Brasilien* 1879

Lange, J. E. German engraver. *Leipzig* 1788, *Königreich Dänemark* 1791. *Preussischen Staaten* 1796

Lange, J. H. de. Publisher *World* ca. 1810 (with Mortier & Oomkens)

Lange, S. L. *Liselund paa Moen* 1805

Langen, Capt. A. L. *Key Islands* 1886, London R.G.S. 1888

Langenbucher, Jacob (1649–1712). Globemaker of Augsburg

Langenes, Barent [Bernardt]. Dutch cartographer and publisher. *Caert. Thressor* Middelburg 1598, set the fashion for miniature atlases. Reissued, copied, pub. for many years. French translation 1602. Burger records an edition of 1597

Langenschwarz, Frederick. *Plans of Halifax, Nova Scotia* 1779 MSS

Langeren. See **Langren**

Langeveld, D. M. Book and art dealer. Collaborated with J. B. Elwe, *Zak-atlas zeventien Nederlandsche Provincien* 1786

Langevin. Engraver of Paris, 6 Rue du Foin St. Jacques. For Charle *Océanie*, for Vuillemin *Afrique* 1857, for Andriveau-Goujon 1874

Langeweg, D. I. *Bergen op Zoom* 1747, S' Gravenhage 1747, *Plan de la Haye* 1773 and 1776

Langford, George. Australian surveyor. *Township of Ararat (Victoria)* 1857, *Township Stawed* 1858

Langfort, Teodor. Mid 18th century military cartographer. *Hydrographic map of Poland*

Langguth, Eduard. *Thüringen und Harz* 1857

Langhans, Paul (b. 1867). German cartographer. *Kleiner Handatlas* 1895, *Staatsburger Atlas* 1896, *Colonial Atlas* 1897, *Deutscher Marine Atlas* 1898

Langlands, George & Sons. *Taymouth District* 1786 MS, *Kintyre* 1793, *Argyll* 1801 with *plans of Campbelltown & Inverary*

Langle, F. de. See **Fleuriot de Langle**

Langley, E. Publisher. 173 High St., Borough. *Metropolis Displayed* (1820)

Langley, Edward, & **Belch**, William. Publishers and engravers, 173 High St., Borough. *Maps of London* 1802, 1812, 1816, 1818, 1820; *Bedford* 1817, *County Atlas* 1817–18. *Essex* in collaboration with Phelps. Belch publishing on his own in 1820 from 1 Staverton Row, Newington Butts

Langley, J. *Tokar sub district [Sudan]* 1893

Langley, James. Master, R.N. Admiralty surveys. *Bay & town Kingston, St. Vincent* 1817 MS (with Elliot) 1820, *Gros Isle Bay, St. Lucia* 1820, *Great Courland Bay, Tobago* 1820

Langley, Thomas. Engraver. *Plan Windsor Castle* 1743 and 1750

Langlois, A. D. *Algérie* 2 sh. 1884

Langlois, Achille. *Saône et Loire* (1865)

Langlois, Hyacinthe. Cartographer and publisher. *Rhaetie* 1812, *Roy. France* 1828, *Grand Atlas Français Départmental* 1856 (30 sh.)

Langlois Fils, Hyacinthe, Rue de Savoie, Paris. Geographer and editor. *Terrestrial Globe* 1815

Langlois [l'Anglois], Nicolas. Bookseller and publisher of Paris, Rue Jacques à la Victoire. *Hierusalem* (1640), *Du Val's Géogr. Univ.* 1682, *Morée* 1687

Langlois, Victor. Engraver for Placide de Saint Hélène 1714

Langlumé et Pelletier. Publishers in Paris, Rue du Foin St. Jacques No. 11 *Océanie* 1834

Langner. *Süd Preussen* 1803

Langren [Langeren]. Family of Dutch cartographers and globemakers. Arnold Florent van (1580–1644). Son of Jacob, settled in Antwerp. Globemaker to Albert & Isabella of Spain. Maps for Linschoten 1596, *Delineatio totius Australis partis Americae* 1596, *Globes* 1609, 1620, 1622 and 1644

—— Floris van. Amsterdam engraver, son of Jacob

—— Hendrick Florent van (fl. 1574–1604). Grandson of Jacob. Engraved for Linschoten with Arnold Florent. *Planisphere* (1600)

—— Jacob Floris van (fl.1570). Founder of the dynasty. Born in Utrecht, settled in Amsterdam as engraver. *Terrestrial and Celestial globes* 1589, maps for Linschoten 1595

—— Jacob Floris van, II. Grandson of Jacob. Engraver. *Thumbnail map for English Traveller* 1635, 1643

—— Michael Floris van (1612–1675). Mathematician and astronomer to King of Spain. Lived in Brussels. Maps for Blaeu 1635: *Brabant Louvain, Antwerp, Mechlin*

Langres, Sieur de. *Plan de Malte* 1680

Langsdorff, H. *City Cleveland, Ohio* 1856

Lanier, Lucien (b. 1848). French geographer

Lannoy, Ferdinand de, Comte de la Roche (1542–1579). *Burgundy* 1563, *Franche Comté* 1565 used by Ortelius in 1579

Lannoy de Bissy, Reginald de. Military engineer b. 1844. Large *map Africa* 37 sh. 1881–88, *Madagascar* 1885

La Noe, Gaston Ovide de (1836–1902). Military cartographer

La Nouerre, De Fer de. See **De Fer de la Nouerre**

Lanoye, Ferd. Ch. Aug. (1806–1870). Geographical writer. *Sudan* 1858

Lans, O. C. *Zee Kust Banda* 1871

Lansraux. Engraver for Dufour (1864)

La Pagerie, G. Mariette de. *Diocèse de Coutances* 4 sh. 1689

La Pallière Christy, Seigneur de. Capt. of *Joseph Royal. Journal of voyage to South Seas* 1719–20 (48 charts) MS

La Passe, Joseph Marie Henri d. *Iles Chincha* 1857

La Paz, Principe de. See **Paz**

La Penne, J. A. Barras de. Plan *Détroit de Gibraltar* 1756

La Pérouse, Jean François Galoup, Comte de (1741–1788). French explorer and hydrographer *Voyage autour du monde . . . Atlas* 1797

Laperuta, Leopold. *Itin. Milit. Bologna-Napoli* 1809

La Pezuela, J. de. See **Pezuela**

Lapham, Increase A. *Topographical Description of Wisconsin* 1844, *Milwaukee* 1856, *Geolog. Map Wisconsin* 1869

Lapi, Domenico de. Printer of Bologna Edition of *Ptolemy's Geography* 1477, first edition with maps

Lapide, E. Surveyor. *Aveley to Childeraitch Canal* 1833 MS

Lapie, Alexandre Emile, Colonel de l'Etat Major and premier Géographe du Roi. *Océanie* 1809, 1812, *Atlas Universel* 1829–37–51 &c., *Orbis Romanus* 6 sh. 1834, *Atlas Univ. Lehuby* 1841, *Atlas Militaire* 1848–50

Lapie, M. fils. Capitaine de l'Etat Major, Géographe de S.A.R. le Dauphin

Lapie, Pierre (1779–1851). Geographer to the King and publisher of Paris, Rue de

Bussy No. 33. *Paultre's Syrie* 1803, *Atlas Classique et Univ.* 1812, *Océanie* 1816, *Atlas Univ.* 1829

La Pierre, Alfred. *Plan Bordeaux* 1898

La Pierre, M. Officer of Marine. Hydrographer with Jeanneret. *Port Jackson* 1828, used *Bougainville* 1837

La Place, Capt. Cyrille Pierre Théodore (1793–1875). Voyage *Favorite Atlas Hydrographique* (1830–2), Paris 1833–9, *Circumnavigation Artemise* (1838–40), Paris 1844–8

La Place, P. Simon, Marquis de (1749–1827). French astronomer and physician. *Système du Monde, Mécanique Céleste*

La Plaes, A. B. de. See **Plaes**

Lapointe, D. Engraver for Duval 1664, *Royaume du Nord* 1670

Lapointe, F. de. See **Delapointe**

La Popellinière, Lancelot du Voisin Seigneur de. French cartographer and writer. *World map* in *Trois Mondes* Paris, 1582, French translation of *Mercator's Atlas* 1609

Laporte, Etienne. *Carte Astron. de l'univers* (1875)

Laporte, Joseph de (1713–1779). *Atlas Mod. Portatif* 1780,–81,–86

Lapparent, Alb. Aug. C. French geologist b. 1839

La Quadra, Juan Francisco de. *Inlet of Bucarelli* 1789

Larcom, Lieut. (later Sir) Thomas Aiskew (1801–1879). R.E. 1820 posted Ordnance Survey, 1824 under Colby. Took over the Survey in Ireland, retired 1846 on completion of 6 in. survey. Irish Privy Councillor

Lardner, Dionysius (1793–1859). Encyclopaedist. *Handbook Philosophy and Astronomy*

Lardner, John. Australian surveyor. *Plans in Melbourne* 1871–6

Laree, V. *Carta hydro. da Bahia* 1837

Larenaudière, Ph. Fr. de (1781–1845). French geographer, Director Soc. Géogr.

Largent, Auguste. *Senegal* 1868

Larionoff, Lieut. *Khanat of Kuldja (China)* 1871

Larken, J. Engraver. Somerset Street, London. Diston's *Thames* 1767, Williamson's *St. George's Channel* 1767, *W. Coast Newfoundland* 1768, engraved for Capt. Cook 1766–7

Larkin, Thomas O. *Valley Sacramento* 1848

Larkin, William. Irish cartographer. *Post Roads Ireland* 1805, *West Heath* 1808, *Meath* 6 sh. 1812 and 1817, *Sligo* 6 sh. 1816; *Waterford* 6 sh. *Leitrim* 6 sh., *Galway* 16 sh. 1819

Larkins, John Paskall, Capt. *Coasts Is. between St. John's and Ladrone* 1786, *Shitoe Bay China* 1786, *Hastings Track* 1788–9, *China Sea* 1794

La Rocha of Figueroa. See **Rocha y Figueroa**

La Roche, C. F. de. See **Delaroche**

La Roche, H. von. Geographer. *Charte Rheinlaender* 1825

La Rochemaillet, Gabriel Michel de (1561–1642). Geographer

La Roche Poncie, Ferdinand Antoine Jules de.*Baie de Madelaine au Spitsberg* 1841, *Reikiavik* 1842, *Miquelon* 1843, *Cours de Gironde* 1856, *Sicile* 1856, *St. Pierre* 1863

Larochette, C. *Cherbourg* 1858, *France* 1877

La Rochette, L. S. d'Arcy de. See **Delarochette**

La Roncière, Le Noury (1813–1881). French Admiral. Promoter geog. works

La Roquette, Dezos de (1784–1868). French geographer. Historical Geographical and Biography. *Essai Hist. Pays Bas* (1831)

Larrazabal, Juan. *Prov. de Caracas* 1855

Larrit, E. Australian surveyor. *Stratfieldsaye* 1858

Larrit, P. Assistant surveyor *Northcote (Victoria)* 1855

Larrit, R. W. District surveyor *Axedale* 1858, *Sandhurst* 1856

La Rozière, Marquis de. See **Carlet**, Louis François

Larré, J. R. Y. See **Rajal y Larré**

Larsen, Nicolas Aug. (1839–1893). Norwegian Geogr. writer. *Voyage Autour du Monde* 1871

Larsson, D. J. M. Swedish geographer. *Atlas öfver Sverige* 1870–74, *Upland* 1872

Larsson, Werner. *Charts Sweden* 1888

Lartet, Ed. A. Is. H. (1801–1871). French geologist. *Maps Pyrenees and Palestine*

Lartet, Louis (1840–1899). French geologist. Expd[n]. to Dead Sea

Lartigue, Joseph (1791–1876). French hydrographer, charts and nautical instructions. *Routes de Galice* 1808, *Arica* 1824, *Mollendo* 1824, *Pérou* 1824

Lartigue, Pierre (1747–1827). Hydrographical engineer. *Charts and maps in relief*

La Rue, Philippe de (1683–1761). French cartographer. Maps in *Le Clerc's Atlas Antique* (1705), *Armenia, Assyria, Canaan, Palestine* &c.

La Rue, P. de. See **De la Rue**

La Ruelle, Claude de. *Nancy* 1611

La Ruelle, J. Lithographers. *Aachen* 1614, (1875)

La Salle, Fr. L. de. *Diocèse de Sées* 1718, *Nouvelle Thébaide* 1700

La Salle, Cavalier Robert de (1643–1687). Born Rouen, died Texas. *Louisiana* 1685 MS

Lasaulx, Const. P. Fr. Arn. de (1839–1886). French geologist and mineralogist

Lasaux, P. de *Diocese Canterbury* 1782

Las Casas, Bart. (1474–1566). Spanish missionary and historian

Las Cases, Emmanuel Marie Joseph Auguste Dieudonné Comte de (1766–

1842). *Atlanté storico* 1813, *Atlas historico* 1826, *Atlas Historique* (1803)

Lascelles & Co. *Warwick* 1850

Lash, Joseph. *Draught of Senegal* (1780)

Lasius, George Sigismund Otto (1752–1833). *Harz Mountains* 1789 and 1798

Laskoffsky, Ju. Russian cartographer. *Maps of Russia* for Imp. Academy of Sciences, St. Petersburg 1866

Lasne, Jean Etienne. *Plan de La Rochelle* 1899

Lasne, Michel. *Plan of Fontainbleau* 1614 used by Braun & Hogenberg

Lasius, G. S. O. (1752–1833). German military cartographer

Lasor a Varea, Alphonsus. *Univ. Terr. Orbis* 1713

Las Peñas, G. G. de las. See **González de la Peñas**

Lassailly, Charles. *Troop stations* 1888–95, *Carte spéciale Maroc* 1892

Lassay, Sr. de. See **Calamaeus**

Lassen, Christian. *Alt Indien* 1853

Lasso, Bartolemeo (fl. 1564–90). Portuguese cartographer. *MS portolan atlas* ca. 1590. Believed to have furnished source material for Linschoten's *map East Indies*

Last, J. T. *Tanganyika (Nguru)* 1881, *Eastern Africa* 1890, *Madagascar* 1890–5

Lat, Jan de. (fl. 1734–50). Publisher from Deventer. Associated with Jacob Keizer, pub. miniature atlases. *Nieuw Kaart Boekje* 1734, *Atlas Portatif* Deventer 1742, *Pocket Atlas* 1734, *Nederl. Provincien* 1741 (23 maps), *Pais Bas Autrichiens* 1746 (20 maps)

Lateranus. Pseud. for **Ziegler?**

Latham, F. P. Engraver. *Longman's London & Environs* 1818

Latham, G. R. *Sedashegar (India)*1862 MS

Latham, Capt. Roger. *Cameroons River* in Laurie & Whittle's *African Pilot* 1797

La Torre, F. M. de. See **Martinez de la Torre**

Latomo, Sigismund. *Belagerung Hydelberg* (1622)

La Tour, Jose Maria de. *Cuba* 1851, 1862, *Spain* 1864, *World* 1861

La Torre, M. de. See **Torre**

La Touanne, M. E. B. de. See **Touanne**

Latour. *Plan of New Orleans* 1720

Latour, Arsène le Carrière (d. 1839). American cartographer. *Atlas to war in W. Florida & Louisiana* Phila. 1816

La Tour, Le Blond. See **Le Blond de la Tour**

Latour, H. *Plan de Pau* 1865 & 1874

La Tour, M. de. *Plan New Orleans* 1760 and 1777

La Tourette, John. *Alabama & W. Florida* 1838, *Mississippi* 1847, *Louisiana* 1853

Lathrop, J. M. *American County Atlases: Frederick Co. Va.* 1885 (with A. W. Dayton), *Highland Co. Ohio* 1887 &c.

Latrie, M. L. de Mas. *Chypre* 1862

Latrobe, B. H. (1806–1878). *Projected Railroad City Washington* (1840), *Phila. & Baltimore Railroad* (1845)

Latrobe, B. H. Surveyor. *River Blackwater* ca. 1794 MS

Latski, Ivan. *Muscovy* 1542

Lattré, Jean. Engraver and publisher. Graveur Ord. du Roi 1776–82, "Rue St. Jacques au coin de celle de la Parcheminerie à la ville de Bordeaux". Engraved for Vaugondy 1743, Janvier 1760 and Bonne.

Lattré

Malte 1752, *Chambry* 1754, *Bordeaux* 1759, *Dijon* 1762, *Paris* 1765, *Atlas Moderne* 1771, 1783, 1793, *Etats Unis* 1784

Lattré. Madame. Engraver for Madame le Paute Dagelet 1764

Latz[en], W. See **Lazius**

Latzina, Francisco. *Argentina* 1883, *Dict. geogr. Argentino* 1891

Laube, Gustave Ch. (b. 1839). Geologist. *Voyage Pole* 1869–70

Laubrie. Designed *Plan Vienna* (1805)

Laubscher, Heinrich. *Bienne, Switzerland* 1654

Lauchen, J. von. See **Rhaeticus**

Lauder, John. Land surveyor. Attested maps for Thomson's *Scotland* 1832

Laudien, G. *Königsberg* 1876

Laudonnièrre, René Goulaine de. (fl. 1562–82) French colonizer in Florida

Lauen Zonen, Van. *Terrestrial Globe* 1745

Laug. *Map Heligoland* 1809 (with Rheinke)

Laughton, G. A. *Isle of Bombay,* 8 sh. 1882

Laumont, Fr. Nicolas Gillet de (1747–1834). French mineralogist

Launay, Adrien. *Atlas des Missions* Lille 1890

Launay, L. de. French geographer. *Lemnos* 1894, *Geolog. map France*

Launay. See **Belin de Launay**

Launders, J. B. *Big Bend Mines* 1866, *British Columbia* 1871

Launsky. (19th century) Cartographer

Laure, P. Jesuit Missionary *Carte du domaine du Roy en Canada* 1731 MS, printed 1732

Lauremberg, Johannes Wilhelm [or Wilms] (1590–1658). From Rostock. Author, mathematician, historian. *Mecklenburg,* used by Blaeu 1630 and Hondius 1633, *Macedonia* 1647, *Sea Atlas Greece and*

Aegean 1650, *Cyclades* 1662, *Atlas Graecia Antiqua* 1656

Laurence, Edward. Young surveyors Guide 1717

Laurent, Alexandre. *Planisphère* 1899

Laurent, Arthur. *Belgique* 1878, *Buenos Aires* 1892

Laurent, Charles. Engraver and geographer. New Road, St. Georges Fields, London. *Plan Manchester and Salford* 1793

Laurent, J. *Fleuves de l'Europe* for La Harpe 1753, *Berry & Bourges* 1770, *Greenland* 1770, maps for La Harpe 1780, Coutances 1790

Laurent. Printer, Rue St. Jacques 71, Paris. For Charles 1865–8, for Meissas & Michelot 1869

Laurent de Lionne. *Picardy* 1781

Laurents, Lorentz [Lorenzen], Henry. Amsterdam publisher. Frankfurt edition of Langenes' *Caert Thresoor* 1605–13 (with titles in German also)

Laurie, A. H. *Chart N. Coast Scotland* 1835, *Railways and Inland Navigation* 1836

Laurie, E. P. *Botanic Gardens Adelaide* 1874

Laurie, Frank. *School Atlas* 1870

Laurie, J. S. *Atlas Physical Maps* 1877

Laurie, John. Surveyor. *Lothians* 1745, First British map to show altitudes. *Mid Lothian* 4 sh. engraved Baillie 1763, *Edin. District* 1766, *Forth and Clyde Canals* 1785

Laurie, Richard Holmes (fl. 1822–1858), son of J. Robert. Chartseller to Admiralty and publisher. Joined firm Laurie & Whittle in 1814 and took over control in 1818. Died in 1858. *Ireland* 1817, *Norfolk & Suffolk* 1827, *Chart W. Part Pacific Ocean* 1822, *Australia* 1841, *London* 1854

Laurie, Robert (1755–1836). Publisher and engraver. Various atlases in conjunction with James Whittle. Successor

to Sayer & Bennett in 1794. *East India Pilot* 1798, *Imp. Sheet Atlas* 1808

Laurie & Whittle. Publishers and engravers. 53 Fleet Street, London, 1795 moved to Great Eastern Street and in 1803 amalgamated with Imray, Norie & Wilson. Acquired Sayer's Geographical business in 1794. Robert Laurie retired in 1812, his place being taken by Richard Holmes Laurie 1812–18. Whittle retired 1818. R. H. Laurie died 1858. Business passed to Findaly 1873 and D. & W. Kettle 1875. *American Atlas* 1794, *New & Elegant Atlas* 1796, *Oriental Pilot* 1797, *Universal Atlas* 1798, *Country Trade* 1799, *East India Pilot* 1800, *African Pilot* 1801, *General Atlas* 1804, *New Travellers Companion* 1807, *Juvenile Atlas* 1814

Lauro, Jacomo. Roman author and engraver. *Aquila* 1600, *Nice* 1625, *Roma Antiqua* 1630, *Valetta* 1635, *Roma* 1642

Laussedat, Aimé (1819–1907). French surveyor and military cartographer

Lautensack, Hans Sebald. *Nürnberg* 3 sh. 1552

Lauter, Capt. *Danube* 1789

Lauterbach, Johan Balthasar (ca. 1654–90). German astronomer and architect

Lauterbach, Johan Christoph (1674–1744). Cartographer. *Territ. Ulm* (1715)

Lauterer, George L. von (1738–84). Military cartographer

Lautreren, Fernando. *Estado de Sinaloa* 1862

Lauzan. Assisted Gardner & Cubitt *Jersey* 1795

Laval, Ant. de, Sieur de Belair (1550–1631). Royal Geographer

Lavallée, Théophile (1804–1866). French historian and geographer. *Géog. physique de la France, Géogr. Militaire. Atlas de Géogr. Milit.* 1852, and 1859

La Valliere. *Atlas Maritime* 1546 MS

Lavaña, J. B. [Lavanha]. See **Labanna**

Lavars, J. *Country round Bristol* (1854), *9 miles round Bristol* 1858, *Milford Haven* 1860, *Raglan* 1862

Lavasseur, Victor. *Italie* 1854

Lavaud, C. F. *Bancs de Terre Neuve* 1663

Lavaux, A. de. See **De Lavaux**

Lavelli, Arcangelo. *Citta di Milano* 1788

La Verendrye, P. G. La Varenne de. (1685–1749) b. Trois Rivières. Explorer in W. Canada and United States

La Vega, R. G. de. See **González de la Vega**

Lavertine, R. A. *Stella Landes* 1885

Lavezzi. *Isle de Corse* (1720)

La Vigne, Nicolas M. de. *Buda* (1700)

Lavis, Henry James Johnston. *Geolog. map Vesuvius* 1891

Lavizzari, D. L. *Profondità del Ceresio* 1859

Lavoisne, C. V. *Genealogical Hist. and Chron. Atlas* 1807 (no maps), 1820 first American edition (with maps), later editions

Law, B. Bookseller. *Geog. Magnae Brit.* 1748, *Bowen's Univ. Hist.* 1766

Law, E. F. *Plan Northampton* 1847

Law, James Thomas. *Diocesan maps England & Wales* 1864, *Lectures* 1868

Law, W. W. Publisher. *Plan Northampton* by J. Wood and E. F. Law

Law & Whittaker. Booksellers London. *Ewing's Atlas* 1817

Lawes, Rev. *Eastern New Guinea* 1879

Lawrance, Lieut. George Bell, R.N. Admiralty Surveyor West Indies. *Gorda Sound* 1850, *Tortola* 1850, *Campeche* 1852, *St. Thomas* 1853, *Virgin Is.* 1856

Lawrance, Jeremiah. *Entrance Arrakan River* 1784

Lawrence, F. A. von. Lieutenant. *Plan Hamburg* engraved Pingeling

Lawrence, Lieut. G. B., H.M.S. *Centaur. Scandinavian charts* 1807–8 MSS, *Port of Spain, Trinidad* 1849, *Admiralty* 1869

Lawrence, Capt. H. M. Surveyor *District of Allahabad* 1838, *Furruckabad* 1837–9

Lawrence, H. L. American lithographer 88 John Street, N.Y. For Marcy 1853. *Survey Lakes* 1857, *Military Map U.S.A.* 1857

Lawrence, Sir Henry. *Military Plan Lucknow* 1857

Lawrence, I. *Orbis imperantis* 1685

Lawrence, Thomas. *Borough of White-church* 1730

Lawrie. Publisher. *New Plan London* 1841

Laws, Capt. John M. Admiralty Surveyor. *Charts Bengal* 1835–1875

Laws, Capt. William. *Plan Harbour, Town, & Forts, Carthagena* 1741. Pub. Harding & Toms, Wills Coffee House

Lawson, A. Engraver for Pinkerton (1804)

Lawson, I. *Chart Solway Firth* 1825

Lawson, John. Surveyor General of North Carolina 1706. *Map Carolina* 1709, 1711, 1714

Lawson, J. T. *Mines Upper California* 1849

Lawson, W. *School Maps counties of England* 1876

Lawson, W. T. G. *Lagos & adj. Territories* 1881, *Plan Town Lagos* 2 sh. (1887)

Lawton, W. Surveyor. *Map Brikenheed Estate* 1823, pub. 1893

Lay, Amos (b. 1765). Geographer and Map Publisher. *N. part State New York* 1812, *Seat War Lower Canada* Phila. 1814 (with J. Webster), *N.Y.* 1817, *U.S.* 1832

Layard, Sir Henry Austen (1817–1894). English diplomat and explorer. *Hamadan to Julfa* 1840 MS, *Mesopotamia* 1853

Lazare. *Plan Bois de Boulogne* 1866

W. Lazius (1514–1565)

Lazaro, Luis. Portuguese cartographer, 16th century

Lazius, George Sigismund Otto *Petrographische. Carte des Harz Gebirges* 1789

Lazius, [Lazio, Latzen] Wolfgang (1514–65). Hungarian cartographer, Prof. Medicine Vienna, and Secretary to Thomas Bakocz, Archbishop of Esztergom. Oldest known *map of Hungary* known as *Lazar's map* (78.3 × 54.8 cm.) published by Tanstetter in 1528 includes *Slovakia*. His maps of Austria, Hungary and Bavaria were used by Mercator, Ortelius &c., *Austria* 1545, 1556 (4 sh.), *Hungary* 1556 (10 sh.), *Bavaria* 1545, *Greece* 1558 (4 sh.), *Peloponese* 1558, *Tyrol* 1561, *Atlas of Austrian Provinces* 1561

Lea, Ann. Continued business of Philip Lea from Atlas & Hercules in Cheapside in partnership with Robert Morden 1700–1703, with William Berry 1708 and Richard Glynne 1725. *Maps of London and Westminister* (with Richard Glynne) 1716–25, Morden's *West Indies*

Lea, Isaac (1792–1886). American naturalist. Contributions to geology 1833, see **Carey & Lea**. *Complete Hist. Chron. & Genealogical American Atlas* 1822 and 1827

Lea, Philip (fl. 1666–1700, d. 1700). Cartographer, globe and instrument maker, publisher and mapseller. At Atlas & Hercules in Cheapside, near Friday St. and at his shop in Westminster Hall, near the Court of Common Pleas. Collaborated with other publishers e.g. Robert Morden, John Overton, John Seller, Ogilby & Morgan and Bassett & Chiswell. Issued *Shires of England & Wales* (reissues of Saxton's plates 1689–93), John Adams' *England and Wales* 1690. *Atlas World* 1690, *Hydro. Univ.* 1700, with Seller *Barbados* (1685), with Moll *Ireland,* with Overton *Asia, Penn. & W. Jersey* 1698

Lea, Thos. *Atlas* (23 maps) 1700

Lea & Blanchard. Publishers of Philadelphia. *Stuart's California* 1849

Leach, E. P. Col. *Bazar Valley (Pakistan)* Calcutta 1879, *Chart Nile* 1889

Leach, H. W. *City Osh Kosh, Wisconsin* 1895

Leach, Capt. John. *Gambia River* 1732

Leach, Lieut. Employed on Ordnance Survey. *Ireland* 1846

Leake, John. Surveyor (fl. 1650–86). *City of London* engraved Hollar 1667

Leake, J. Publisher and bookseller of Bath. Rocque's *Plan of Bristol* 1743

Leake, Lt.-Col. W. M., Royal Artillery (1777–1860). Explorer, topographer and archaeologist. *Egypt* 2 sh. Arrowsmith 1818, *Antiq. Athens* 1823, *Asia Minor* 1840

Leal, Manuel Fern. *Repub. Mexicana* 4 sh. 1899

Lear, Tobias. *Plan city Washington Observations on the River Potomack* 1793

Leard, John. Marine Surveyor and publisher. Master of *Centurion. Charts for Jamaica & Windward Passages,* Mount & Davidson 1793

Leardo, Giovanni of Venice. Cartographer. *World maps* 1442, *Verona* 1447, (lost) 1448, *Vicenza* (1452–3).

Leavenworth, E. American geographer. *Maps Minnesota* (1857), *Delaware* 1866

Le Baron, J. F. *Repub. de Honduras* 1886, 1888

Lebas, Philippe. *Atlas Hist. Etats Europe* 1836

Le Beau, Philippe (17th century). Royal Engineer and mathematician of Lyons

Lebedeff. Russian cartographer with Smirnoff & Strauch. *Newest atlas of the World* St. Petersburg 1878

Lebiano, Anders. Maps for G. Blaeu's *Novus Atlas*

Leblond, *M. Guyane Française* 1814

Le Blond de la Tour. *Nouvelle Orléans* 1722

Le Boucher, Odet Julien (1744–1826). *Atlas . . . de la guerre de l'indép. des Etats-Unis* 1830

Lebour, G. A. *Geolog. Map Northumberland* 1879

Le. Bourguinon-Duperré. Hydrographer. *La Plata* 1833, *Charts Coasts France, Italy, Spain* 1848–64

Le Breton, Auguste. *Canal Paris–Dieppe* 1864

Le Brun, G. Raymond (1825–1887). Swiss Sect. Soc. Geogr. Berne

Le Brun, M. *Moselle* 3 sh. 1772

Le Brun, Piotr. (1802–1879). Military cartographer. Contributed to *Topographical map of Poland* after 1815 Congress of Vienna

Le Bruyn [Bruin, Brun], Cornelis. *Travels Levant* 1700, *Turkish Empire* (1710), *Travels Muscovy &c.* 1737

Le Camus de Mezières, Nicolas. *Plan des Halles* 1763

Le Camus de Moffet, H. *Carte commerciale et douanière de France* 1876

Leccensis. See **Ferrari,** A.

Lechard, A. *Plan ville & environs de Vannes* 1898

Le Charpentier, Gené Jacques (1733–1779). Engraver and publisher for Brion de la Tour 1766, for Desnos 1768

Le Chevalier, Armand. Publisher and engraver, Rue Richelieu 60, Paris. Engraved for Dufour 1856–63, *Océanie* 1860, partner with Paulin to 1860

Le Chevalier, J. B. *Atlas du Voyage de la Troade* 1802, *Plan Constantinople* Weimar 1807, 1812, 1836. Engraved for Dufour *Océanie* 1857

Lechmere, Capt. R.N. *Cape of Good Hope* 1822

Lechner, Ernst *Karte der Bernina Gruppe* 1865

Le Clair (1731–1787). Military cartographer

Leclanger, V. *Plan de Hanoi* 1890

Leclerc. *Atlas of Russia with plans of St. Petersburg* Paris 1794

Le Clerc, Jean (1560–1621). French geographer, publisher, and engraver. *Four continents* engraved by Hondius 1602. Acquired plates of Bougereau and re-issued them as *Théatre Géogr. du Royaume de France* 1619, 1621 and other editions to 1631, *England* 1605. Business continued by his widow Veuve Le Clerc

Le Clerc, Jean. *Empire François* 9 ll. 1640

Le Clerc [Clericus], Jean (1657–1736). Born Geneva, d. Amsterdam. *Atlas Antiq.* Mortier (1705), *New France* 1691–2

Le Clerc, Nicolas. *Princip. Liège* (1780)

Le Clerc, Sebastian. *Plan Siège Bude* (1686), *Plan Vienne* [1740]

Leclercq. C. *Plan Paris* 1850

Leclercq. Engraver for Andriveau-Goujon 1874

Le Clerq, J. H. W. *Nederlanden* 9 ll. 1841

Leclerq, Mécanique & Co. Lithographers, Rue Martel 6, Paris. *Oregon* 1846

L'Ecluse, C. de. See **Clusius**

Le Cointe, G. *Charts Antarctic* 1899

Le Comte, C. *Plan de Spa* 1780

Le Conte, Jos. (1823–1901). American explorer and geologist. *Portion Sierra Nevada* 1893

Le Coq, Major Karl Ludwig von (1754–1829). *Topo. Karte v. Westphalen* 22 ll. Berlin 1805, *N.W. Deutschland* 1815

Le Coq, Henri. *Atlas géolog. Dépt. Puy-de-Dome* 1861

Lecuyer de la Jonchère, Antoine Simon. *Ville et Environs Clairmont-Ferrand* 1739

Leczycki, Pawel. Rev., Translated into Polish *Boter's Gen. Geography* 1613, 1659

Ledenfeld, R. V. Born 1858 Gratz. Geologist. Explorations New Zealand

Lederer, John. *Map of whole Territories Traversed by John Lederer (Carolina & Virginia)* 1672

Lediard, Thomas. Engraver. *Plan Horse-ferry Road to Whitehall* 1739, *College St to Whitehall* 1740 MSS, *Plan Westminister* 1740

Ledoux, Etienne. Bookseller. Rue Guenegaud No. 9, Paris 1823

Ledyard, John. Journal 1783 with *Chart of Cooks Voyage*

Lee, A. N. Lieut. American surveyor, U.S. Engineers, *Lake Erie* 1852, *St. Clair Ship Canal* 1874

Lee, E. F. *Map of Texas* 1836

Lee, John. Estate surveyor. *Pittlewell* 1724, *Hockley* 1745, *Great Baddow* 1754 MSS

Lee, Lieut. R. P. *Morocco City* 1891

Lee, Sir Richard (1513–75). Military engineer and surveyor of the Kings work at Calais. *Orwell Harbour* 1533 MS, earliest dated chart of a harbour in the British Isles. Later chief surveyor of the Queens [Elizabeth] works at Berwick. *Card of Berwick* ca. 1560

Lee, Robert E. *Harbour of St. Louis, Mississippi* 1837, *Des Moines Rapids* 1837

Lee, Lieut. T. J. *Boundary U.S. & Brit. Provinces* 1843, *River Sabine* 1840

Leech, Major R. Surveyor for Walker's *N.W. Frontier* 1841

Leech, William. Estate surveyor. *Great Waltham* ca. 1800 MS

Leeghwater, J. A. *Haarlemmer Meer* (1651)

Leek, F. H. Mapseller of Bath. *Hibbart's Plan of Bath* 1780

Leenen, Juan. *Bengal* 1700

Leer, G. Antonovich, Lieut.-Gen. *Strategical maps, tactics in field* 1886–7

Leeuw, Arent Martenez. *New Guinea* 1623

Lefèbure, Simon (1712–1770). Mapmaker

Lefebure de la Barre. See **Barre**

Le Febure. Script engraver. *Ville de Candie* (1760), *Leo Belgicus* (1672)

Leferre d'Etaples. Teacher of Cosmography, Univ. of Paris 15th century

Lefevre (d. 1839). French geologist. *Basin Nile*

Le Fèvre, I. A. *Atlas Manhattan & Staten Is.* 1895–6

Lefranc, Em. *Géographie Mod.* (1845), *Géogr. Ancienne* (1845)

Lefrancq, B. *Nieuwen Atlas der Jeugd.* Brussels 1780, *Carte Gén. de la Guerre* 1792

Lefroy, Sir J. H. (1817–1890). English General. *Magnetic Observations* St. Helena, Canada, Tasmania, Bermuda

Lefroy, V. S. *Perak* 2 sh. Stanford 1892

Legagneur. Engraver, Rue de la Harpe No. 35, Paris. Perrot's *London* (1827)

Legatt, John. Publisher and printer. Speed's *Prospect* 1650

Legendre, A. M. Worked with Cassini on triangulation. *Reims et Environs* 1769

Le Gendre-Decluy. *Vignobles de la France* Paris 1852

Le Gentil de la Gallissonnière. See **La Gallissonnière**

Le Gentil, G. I. H. (1725–1792). French astronomer and traveller. *Voy. l'Inde* 1780–1782

Le Gentil, Labarinus. *Amérique Sept.* 1713

Legnani, G. *Prov. of Cremona* 1820

Legrand, Augustin. French geographer. *Exposition Géographique* 1839

Legrand, Louis. Engraver for Bonne & Lattré *Atlas Moderne* 1771

Legrand, P. *Atlas général toutes les Russies* 1795 (with Ancelin)

Legrand, P. *Terrestrial Globe* 1720

Le Gras, A. *Atlas de Vents et Courants* 1870

Leguat, Fr. (1637–1735). French traveller. *Voyages* 1708

Lehman, H. Lithographer. *Goedshe's plan London* (1838)

Lehman, Johann Gotlieb (1719–67). Mineralogist

Lehmann, J. G. (1765–1811). Cartographer. *Plan Dresden* 1813 and 1821

Lehmann, Jan Krystian (end 18th century). Polish military cartographer

Lehnert, Josef R. von (1841–1896). Hydrographer in Vienna

Lehuby, P. C. Publisher of Paris, Rue de Seine, No. 48, later 53. Lapie's *Océanie* 1842

Le Huen. *Palestine* 3 sh. 1488 MS

Leichardt, Dr. Ludwig F. W. Born in Prussia 1813. Explorations in Australia, *Moreton Bay to Port Essington* 1844–5, pub. 1846. Leichardt disappeared in 1848

Leidig, Johann Christoph (18th century) Bohemian engraver

Leif Erikson. See **Lyef Erikson**

Leigh, J. C. See **Harrison & Leigh**

Leigh, M. A. Bookseller, publisher and stationer, 421 Strand, London. *New Atlas of England and Wales* 1833, *New Plan London* 1834, *Metro. Distance Map* 1834. Continued as Clarke, Leigh & Son 1835

Leigh, Samuel. Publisher 18 Strand. *New Pocket Atlas England & Wales* editions 1818–43, *Pocket Road Book England & Wales* 1825

Leigh, Valentine. Surveying 1577, 1578, 1596

Leisten. Land surveyor. *Zuyd Cust van Africa* 1779

Leith & Smith. Engravers. Ray's *Elga* 1838

Leitner, Johann Sebastian (1715–1795). Engraver of Nuremberg, worked for Homann Heirs

Leiviska, Livari Gabriel (19th century). Finnish geographer

Leivong, J. (18th century). Swedish surveyor

Le Jay, Guido Michael. *Religious maps of Palestine &c.* 1630

Lejean, G. *Adoua* (1867), *Lake Tana* 1867, *Nord Thessalische Grenzlandschaft* 1879

Lejean, M. G. *Liva de Trnova (Bulgaria)* 1857

Lejeaux, A. French geographer. Worked with E. Levasseur 1879

Lejeune, T. Publisher of The Hague. *Malte Brun's Atlas Complet* 1837

Le Keux, J. Engraver. *Findlay's Plan Cambridge* 1842

Leland, Alonzo. *Mining regions Oregon, Wash. Territory* 1863

Leland, John (1506–1552). Antiquarian to Henry VIII. *Itinerary Oxford* 9 v. 1710, *Collectanea* 6 v. 1715

Leleu, A. *Prov. Antioquia (Colombia)* 1819 MS

Lelewel, Joachim (1786–1861). Polish historian, Professor Wilno University, Librarian Warsaw. Minister Polish Provisional Government 1830–1. Political exile in Paris and Brussels. *Géographie du Moyen Age* 1848–52. (with Atlas 35 maps). Designer of many maps and atlases. Pioneer writer on cartography. Collector

Le Lievre, C. *Channel Islands* (1850)

Le Loupe, Auguste. b *Dept. de l'Aube* 1864

Le Loupe, Eugéne. *Chemins de Fer Belgique* 1859

Le Loupe, Lucien. *Itin. Mexique* 1882

Le Loyer, Jacques (1619–1704). Cartographer

Le Loyer de la Fleche Jean (d. 1688). Cartographer *Anjou* 1652

Le Maire, F. Missionary from Paris. *Louisiane et pays circonvoisins* 1716

Le Maire, [Jacob] , Jacques (d. 1616). Dutch navigator. Sailed with Schouten 1615. *Spieghel der Australische Navigatie* Amstd. Colijn 1622, three maps including map of passage of Straits of Magellan

Lemaitre, Ger. Engraver in Paris, Rue des Fosses St. Victor No. 32 for Frémin 1820, for Lapie 1821

Lemare. *Maps* for Julien 1768

Lemau de la Jaisse. French cartographer. *Plans des principales places de guerre et villes marit. de France* Didot 1736

Lemercier. Printer, Rue de Seine 51 Paris, for Andriveau Goujon (1875), for Furne et Cie. *Océanie* (1842), for Levasseur *Océanie* (1838), Duflot du Mofras *Californie* (1849)

Le Mire, Aubert (1573–1640). Church historian and geographer

Le Mire, C. *Nouvelle Calédonie* 1878

Le Mire, N. Engraver of cartouches for D'Anville 1754–63

Lemonnier. Publisher of Paris. For Fournier 1837

Le Monnier, Franz Ritter von. *Oesterreich*-Ungarn 1887

Lemos, Pedro de (16th century). Portuguese cartographer. *Portolan* ca. 1590

Le Moyne, F. See **Franciscus Monarchus**

Le Moyne de Morgues, Jacques. Huguenot artist of Dieppe, d. 1587. *Map Florida* 1564 MS bought by De Bry and published in 1591 *Floridae Americae Prov . . . desc.*

Lempriere, C. Designed cartouche for Popple 1733, *Highlands of Scotland* 1731 MS

Lempriere, Capt. Clement. Engineer. *Portsea Is.* 1716, *Plan Tower London*

1726, *Cartagena* 1741, *Minorca* 1753, *Bermudas* Sayer 1775, *Jersey* 1786. Collaborated with W. H. Toms and H. Gravelot

Lempriere, T. *Island of Jersey* 1734

Lemstrom, K. S. (b. 1838). Finnish physician. Works on magnetism and meteorology

Lenc, Russian cartographer. *Atlas physical geography St. Petersburg* 1851, 1865

Lenczearski, Antoni. Worked with John Witt. Colonel in the Crown Artillery on a team of Polish surveyors who mapped boundary line between *Grand Duchy of Lithuania and Russia* 1766–7

Lendenfeld, Dr. R. V. *Maps routes Australia* 1887

Lender, C. *Carte statist. de France* 1851

Lendrick, John (fl. 1780–1810). *County Antrim* 1780, pub. 1782

Leneus, Joannes. Drew *Holy Land* for Swedish Bible 1618

Le Neve, Peter. Estate surveyor of Ardleigh. *High Rookine* 1786 MS, *Barline* 1794 MS, *Ardleigh* ca. 1800 MS

L'Enfant, Pierre Charles (1754–1825). American military engineer, town planner. *Plan Washington* 1791 MS

L'Enfant. Printer, 12 Rathbone Pl., London. *Travellers Europe* 1852

Lenglet-Dufresnoy, Abbé Nicolas (1674–1755). Diplomat, geographer. *Methode pour étudier géographie* 1716, *Kinder Geogr.* 1764, *Geogr. Ant. et Nova* 1768

Lenny, Isaac. Surveyor. *Lavendon Mills Bucks.* 1856 MS

Lenny, J. G. Surveyor of Beccles, Suffolk. MS *estate plans of Boston Lincs.* 1778–1806, *10 miles round Bury,* engraved Hall, 1823

Lenny, J. *Bedford Level* 6 sh. 1842

Leno, Francesco di (fl. 1561–7). Printer, Venice. Reissued Bordone's *Isolario* ca. 1567

Lenormant, François, [i.e. Charles François], (1837–1883). *Atlas d'Hist. Anc. de l'Orient* (1868)

Lenox. *Globe* ca. 1510. In N.Y. Public Library

Lens [Lense], B. Artist. Designed cartouche for Moll's *large map of N. America & S. America*

Lens, Ed. Estate surveyor. *Hornchurch* ca. 1750 MS

Lense, B. See **Lens**

Lenthall, John. Map Seller and Stationer next the Mitre Tavern, against St. Dunstan's Church Fleet Street 1709 and at the Talbot against St. Dunstans Church in Fleet Street 1728. *N. America* 1718 (with P. Overton and T. Taylor), Geographical Playing Cards (1724)

Lentulus, Robert Scipio v. (1714–86). Military cartographer

Lentz, P. C. (18th century). Surveyor and cartographer

Lenz, Herm-Oscar (b. 1848). Geologist and explorer. *Voyages in Africa* 1874–85

Lenz, Ionath. *Archipel du Nord* Stuttgart 1774

Leo, Africanus [Alhassan ibn Mohammed Alvazzan] (1483–1552). Arab geographer *Description of Africa*

Leon, B. F. y P. de. See **Francia y Ponce de Leon**

Leon, M. P. de. *Atlas de Colombia* 1865, *Carta choro. Bolivar* 1864

Leon y Pizarro, R. G. de. See **García de Leon y Pizarro**

Leonhard, C.C. von. *Vulkaner Atlas* c 1850

Leonhard, H. Engraver for C. Glaser Mainz 1840

Leonard, Dr. R. *Insel Cerigo* 1891

Leonard, W. *Plans estate Oxfordshire* 1799

Leonardi, Antonio. Cartographer. *Circular world maps* 1457 and 1466, both now lost

Leonardo da Vinci (1452–1519). Painter and inventor. Drew many *plans for military fortifications* and a *sketch map of the world in gores*

Leonhardt, K. C. V. (1779–1862). German geologist *Geologie der Erde, Lehrb. d. Geologie*

Leonov (18th century) Russian engraver *Rossisskoi Atlas* 1792

Leontius, Mechenicus, I. (8th century). Byzantine globemaker, and writer on globes

Leopold, Johann Christoph (1699–1755). Engraver and publisher of Augsburg

Leopold, Joseph Friedrich (1668–1726). Publisher and engraver of Augsburg. *Grundriss Pultana* 1709, *Russia* 1711

Leopold, Nikolaus (16th century) Globemaker

Leowitz, Cyprian (16th century). Astronomer

Le Page du Pratz. *Histoire de la Louisiane*, 3 vols., Paris 1758 (maps)

Le Parmentier. Engraver Rocque's *Plan Montpellier*

Le Paute Dagelet (Madame). *Passage de l'Ombre de la Lune au travers de l'Europe . . . Ier Avril A Paris chez Lattré Graveur.* (Title cartouche *Gravé par Mad. Tardieu)*

Le Pautre, Pierre. Engraver. *Ville et Rade de Carthagene* 1698, *Plan Versailles* Amsterdam A. & H. de Leth (1750)

Lepechin, I. Russian naturalist. *Map of the Dwina Estuary*

Le Pere et Avualez. Printsellers of Rouen. Rue St. Jacques près la f^e. S. Sévérin. *Brion Africa* 1775

Lepine, Louis de. *America* Paris 1595, (1695 Wheat?)

Le Predour. *Côte d'Or* 1828, Dépôt de Marine

Lepsius, R. *Maps of Egypt* 1845–59

Lerch, Johann Martin (1659–84). Engraver of Vienna

Le Rouge, George Louis (fl. 1740–80). Publisher, Ingénieur Géographe du Roi. Rue des Grands Augustins vis a vis le Panier fleuri. *Atlas Général* 1741–62, *Recueil des Cartes Nouvelles* 1742, *Guerre en Europe* 1743, *Atlas Portatif* 1748, *Villes d'Angleterre* 1750, *Intro. de Géographie* 1756, *Recueil des Plans de l'Amérique Sept.* Paris 1755, *Pilote Américain* 1778

Le Rouge (Sr.). *Recueil des Cartes Nouvelles Paris* 1742, *Large scale map Munster* 1762

Leroux. Engraver for Lorrain 1836

Leroy, Charles. Surveyor. *Atlas Univ. et Classique* 1867 etc. (with C. J. Drioux)

Le Roy, Jacques, Baron (1633–1719). Belgian geographer and historian, b Antwerp, d. Lierre. *Antwerp* 1768, *Topo. Gallo. Brabant* 1692, *Castella Brabant* 1697, *Brabantia Illustrata* 1705, *Atlas Brabant* 1730

Leroy. *Plan Namur* (1840) with Debussy

Le Roy, Henri. Engraver. Briet's *Palestine* 1641, Lochom's *British Isles* 1639

Le Roy, Sr. Hydro. Engineer. *Lagos* 1737

Le Roy. Engraver for Desnos. *Western Ocean* 1761

Le Roy-Beaulieu, H. J. B. Anatole (b. 1842). French writer and geographer

Lerpenière, D. Etcher. Employed Dépôt de Marine 1770, Dalrymple 1770–5

Lery, Cha[u]ssegros de. See **Chaussegros de Léry**

Le Sage, Eman. Aug. Dieudonné. See **Las Cases**

Lesaulnier de Vauhello. *Chart Bay St. George Nept. Fran.* 1822, *Harbour Alexandria* 1840

Lescan, J. Fr. (1749–1829). French hydrographer and mathematician

Lescarbot, Marc (1590–1630). *Hist. de la Nouvelle France* Paris 1609, three maps engraved by Millot

Le Scellier, M. *Diocèse de Beauvais par Delisle* 1709

Leschenault de la Tour, J. Bl. Cl. Th. (1733–1826). With Baudin on the *Géographe* 1800–3

L'Escluse, Charles de. See **Clusius**

Le Seney, Sebastian (fl. 1537–1547). Clock and instrument maker. *Planisphere* for the King

Leseur. Surveyor, with M. Perron. *Coupang (Timor)* 1803

Leski, Jozef (1760–1825). Mathematician and topographer at Military College, later Cracow. Translator of Hogrewe's textbook

Leslaeus, J. See **Leslie**

Lesley, Capt. Charles. *Coringa Bay (Golconda)* 1774, pub. Laurie & Whittle 1798

Leslie [Leslaeus], John. Bishop of Ross 1527–1596 . *Map Scotland* Rome 1578 [7 × 11 in.], engraved Bonifacius. *Scotland* Rouen 1578 (1586), (15 × 20 in.)

Leslie, John. *Topo. Map Loch Rannoch* MS 1756

Leslie, William. *Nairn & Moray* 1813

Lespeyres, H. *Saar-Rhein Gebiete* 1868

Lespinasse de Villiers. *Coasts of Brittany* Paris 1760

Lessacher, Peter. *St. Florian* 1649, *State St. Veit* 1649, *Volckelmarkt* 1649

Lessar, P. M. Merv. 1884

Lessel, Caspar (1779–1834). Geographer

Lesseps, Ferdinand de (1805–1894). French engineer and diplomat. *Percement de l'Isthme de Suez Atlas* 1856. President Soc. Géog. Paris

Lesueur, Ch. Alex. (1778–1857). Traveller and naturalist. *Voyage de la Coquille* 1800–1804. Maps for Freycinet 1807–16

Leszxyn [Leczynski], Franciszek (b. 1797). Military cartographer, member of team which produced large map of *Polish Kingdom* after Congress of Vienna

L'Etanduère. *Plan Ile Royale, Nept. Franc.* 1780

Le Tellier. Engraver for Vaugondy 1737

Le Testu, Guillaume (1509—1572). Pilot and hydrographer of Le Hâvre. *World map* 1555—6 MS, *Cosmographie Universelle* (56 charts) 1556 MS

Leth, Andries de (1662—1731). Engraver and map maker in de Beurssluys in de Visser, Amsterdam. Father of Hendrik (younger), *Mer du Sud* (1730), *Gibralter* 1727. Took over business of N. Visscher

Leth. Hendrik (elder) (1692—1759). Engraver and publisher of Amsterdam

Leth, Hendrik (younger) (1703—1766). Mapseller on the Beurssluys in Amsterdam. Engraver and Publisher à l'enseigne du pêcheur. *Wall maps of the continents* and *two miniature atlases: Nederlandsche Zak Atlas* ca. 1740 and *Nieuwe Geogr. en Hist. Atlas* (1740). *Pocket map London* 1739, *Plan Rotterdam* 1733 and many topographical works

Leth, Henry de, II. Continued business of father Hendrik (younger). *Atlas* 1788

Letienne, L. Publisher in Paris. *Dépt. de Moselle* 3 sh. 1838

Letoschek, Emil. *Geogr. repet. u. Zeichenatlas* 1888

Letronne, J. Ant. (1787—1848). Works on historical geography. *Atlas de Géogr. Ancienne* 1827

Letronne, Ludwig (mid 19th century). Lithographer in Warsaw including map publications

Lettany, Franz (1792—1863). Military cartographer

Letterman. See **Scatcherd & Letterman**

Letts & Son. Publishers, 8 Cornhill, London. *Davies Plan London* 1840, Letts Son & Steer, 8 Royal Exchange. *Plan London* 1848, *Thomas Letts London* (1858), Letts Son & Co., Publishers, 8 Royal Exchange. *Popular County Atlas,* London Bridge E.C., 1884 (1885). Agents for Ordnance Maps

Leucho, J. P. de. See **Pentius de Leucho**

Leusekan, Manlinius. *Maps settlements Ceylon* 1719 MS

Leutemann, H. Engraver of Leipzig. Worked for J. C. Hinrichssche. Buchandlung *Australien* 1830, *Ganze Erde* Leipzig 1847, *Illustrierter Atlas* 1863

Leuzinger, Rudolf (1826—1869). Swiss geographer. *Schweiz* 1893, Cantons 1865—1890

Levaillant, F. *Carte partie Mérid. de l'Afrique* 1795

Levanto, Francesco Maria of Genoa. *Specchio del Mare.* Genoa 1664

Le Vasseur, A. Nephew and successor to A. Pilon. Publisher, Rue de Fleuris 33, Paris, for Dufour (1870)

Levasseur, Emile (1828—1911). French geographer and economist, Geogr. didactic economic, demographic. Revised *Byrne's Atlas* (1880), *Atlas Univ.* (1876), *Grand Atlas Géogr. Physic.* 1890

Levasseur, E.

Levasseur, Guillaume (d. 1643). Norman hydrographer and pilot. *Chart Atlantic* 1601 MS

Levasseur, R. *France* 12 sh. Paris, Delagrive 1876

Levasseur, Victor. French cartographer 19th century. *Atlas National . . . de la France* 1845 and later editions

Le Vasseur de Beauplan, Guillaume (1595—1685). Norman engineer in service King of Poland; son of Guillaume Le Vasseur. *Ukraine* 1648 and 1650 (8 sh.), *Normandy* (12 sh.), 1667, *Brittany* 1666 MS, and Navigation guides

Levassor, M. *St. Dominique* 1803

Lévêque, Pierre (1746–1814). Mathematician and hydrographer of Nantes

Leverett, William. Surveyor. *Newport Pagnell* 1690, *Church End & Brook End Shenley* 1693–1698 MSS

Leverger, Augusto. *Rio Paraguay* 1857

Leverton, T. *Estate plan Regents Park* 1812

Levilapis. See **Lichtenstein**

Levrault, F. G. Publisher, Strasbourg *Outline map of America* 1833, *Europe* 1833

Levy, Michel (b. 1844). French geologist

Lewes, T. Surveyor. *Estate map of Hunnington (Lincs.)* 1753 MS

Lewicki, Jan Nepomucen (1802–1871). Polish painter, lithographer and cartographer. Worked for Dépôt de la Guerre in Paris, and for Portuguese in Lisbon. Adapted with Bojarski Polish edition of *Putzger's historical atlas*

Lewis, A., Master H.M.S. *Phaeton.* Charts maps, plans 1814–17. American campaign 1814, *St. Helena, Ascension* 1816 in logbook of Admiral Sir Pulteney Malcolm

Lewis, Lieut. (later Capt.) George. *Coast India, Mt. Dilly–Pondicherry* 1782–3, pub. Laurie & Whittle 1798; *Palk's Straits (Negapetam)* 1783, pub. 1794

Lewis, G. W. Lithographer and publisher, corner Beekman & Nassau St. N.Y., later 50 Lemond St. 1858. For Folson 1842, for Lewis 1858

Lewis, I. T. Surveyor of Winchester. *Alton* 1829

Lewis, John, of Nettlebed (Oxon.). Surveyor. *Assendon Cross Farm Fawley Bucks* 1730 MS

Lewis, J. Engraver for Willetts 1814

Lewis, J. W. (d. 1881). Australian explorer. With Warburton 1873–4

Lewis, Capt. Meriwether (1774–1809). American explorer and surveyor with W. Clark 1803–6

Lewis, Samuel. Estate surveyor E. Bergholt, Suffolk. *Langham* 1769 MS, *Navestock* 1770 MS, *Stanway & Copford* 1774 MS, *Lexden* 1781 MS

Lewis, Samuel & Co., 87 Aldersgate St. London and 13 Finsbury Place South. English cartographers. *Topogr. Dict. of England,* editions 1831–1848, *Plan of London & Environs* (1850)

Lewis, Samuel. American geographer, draughtsman to Plantation Office. *W. Florida &c.* 1774 MSS, *N. Carolina* 1795, *U.S.* 1795, *Atlas* in conjunction with Mathew Carey 1795 and with Aaron Arrowsmith 1804, *N. Y. Island* 1807

Lewis, T. *St. Johns, Bay of Fundy* 1756

Lewis, William (fl. 1819–36). Publisher, 21 Finch Lane, London. *New Travellers Guide* 1819 and 1836

Lewis & Brown. Lithographers of 272 Pearl Street, New York

Lex, Joseph L. (1791–1866). Polish painter, designer and lithographer. Worked on team which produced *large map of Poland* after Congress of Vienna 1815

Lexlin, Susanne. Printer, *Salzburg,* Woodcut 1565

Leybourne, William (1626–1716). Printer and surveyor. *Surveigh of London* (with Leake, Jennings, Marr and others) 2 sh. 1667, *Nine Geometrical Exercises* 1669, *Panorganon* 1672, *Goldsmith's Company Lands East Acton* 1683, *Stepney* 1684, *The Compleat Surveyor* 1653, 1674, 1679 &c. Wrote *Planometria* under name of Oliver Wallinby 1650

Leyderdorp, A. (1789–?). *Map Purmerend* ca. 1838

Leymarie (1809–1844). Engraver and lithographer. *Topo. Maps and views of Lyon and other French Towns*

Leymerie, Alex. P. G. Achille (1801–1871). French Geologist. *Eléments de Géologie* 1861

Leynslager (Capt. H.). *Gibraltar,* Amstd., Keulen (1695)

Leypold, J. (1806—1874). Painter and engraver. *Views of Dresden and other towns*

Leyritz, M. *St. Domingue* Paris 1803 (with Levassor & Bourjolly)

Leysten. Surveyor. *S. Coast of Africa* 1779

L'Herba, G. de. See **Herba**

L'Hermite. Engineer, hydrographer. Worked for Dépôt de Marine *Baie des Chaleurs* 1780, *Plan Ile Royale* 1780

L'Honoré, F. Publisher of Amsterdam. *Atlas Historique* with Chatelain 1705—29 and 1732—9, 7 vols.

L'Host, S. de. See **Sorriot de l'Host**

L'Huilier, Jan. [Joannes]. Engraver of Amsterdam. Sanson's *Arabia* 1655, Du Val's *America* 1655, De Wit's *East Indies* 1662, Mariette's *Transilvania* 1664, *Russia* 1682

L'Huillier, Pierre. Publisher of Paris. La Popellinière's *Trois Mondes* 1582

Lhuyd [Fluddes, Fludus], Humphrey. Welsh cartographer and physician b. 1527 in Denbigh, d. 1568. *Map of Wales & England* 1568, used by Ortelius 1573

Li Chao Lo (1769—1841). *China* in 66 sh. 1842

Li, Chi-fu (758—814). Chinese geographer. *Map Chinese Empire*

Li, Hui (15th century). Chinese cartographer

Li, Ju-ch'un. *Atlas China* 1645

Li, Ling (100 B.C.). Chinese cartographer

Li Po (7th century). Chinese cartographer, Taoist

Li, Shun feng (7th century). Chinese mathematician and geographer

Li, Te Yu (787—849). Chinese cartographer

Li Tse-Min (14th century). Chinese geographer

Liagre, J. B. J. (1815—1891). Belgian surveyor

Liais, Eman (b. 1826). *Hydro. du Haut San Francisco (Brazil)* Paris and Rio de Janeiro 1865

Libert, E. (1820—1908). Engraver and painter. *Town Plans and Views of German and Danish Towns*

Libri, Francesco. Veronese globemaker (fl. 1530)

Liburnau, J. R. Lor de (b. 1825). Works on physical geography

Lichenstein, H. *Kaart van de Kaap de Goede Hoop* engrd. J. C. Bendorp Reizen in Zuid Afrika 1815. English edition engraved Thomson, pub. Colburn 1815

Lichfield, Dean of. See **Nowell** Laurence

Lichtenstein [Levilapis], Herman [us]. Printer of Cologne. *Ptolemy* 1475

Licinius [Licinio], Fabio (1521—1565). Venetian engraver and publisher. Worked for Gastaldi 1556—65. *Europe* 1559, *Asia* 1559, *Greece* 1560, *Italy* 1561, *Jerusalem* 1559

Liddel, Colin. *Map of Jamaica* 2 sh. Stamford 1888 and 1897, *Jamaica* 1 sh. 1895

Liddon, Lt. M., R.N. Hydrographer. *Baffins Bay* 1819—20

Lidington, Richard. Surveyor. *Biddleston Park Estate* 1755 MS

Lidl, Johann Jacob (1696—1771). Maps for P. Probst's *Reisebesch. Missionaries.* Wien 1748, *Land Karte Ungarn* 5 ft. 3 in. × 2 ft. 5 in., *Silesia* (1760), *Views Vienna*

Liebano y Trincado, Evaristo. *Islas de Luzon* 1882

Liebaux, Jean Baptist. Engraver for Thévenot 1681 and for De Fer. *Globe Céleste* 1697, *Cartagene* 1698, *De Lisle* 1700—15, Maginis *Mantou* 1702, for Esquemeling 1744

Liebaux, Le Fils. *De L'Isle* Scandinavia 1788, *Russia* 1780, *Hungary* 1780

Liebe, Christian Gottlieb (1696—1753). Engraver, of Halle. *Grundriss v. London* (ca. 1780)

Liebe, Gottlob August (1746–1819). Engraver of Halle

Lieber, K. W. (1791–1861). Mapmaker for Bartuch's Geogr. Inst. Erfurt.

Liebenow, W. (1822–1897). German cartographer and topographer. *Central Europa Eisenbahn* 1886, *Rhein Prov.* 1890, *Oesterreich-Ungarn* 1896

Liebhart, Matthias (1850–1927). Austrian cartographer

Liechtenstern, Joseph Marx, Freiherr von (1765–1828). Geographer. *Maps of Austria & Hungary* 1793–1809. *Hand Atlas Erde* 1807

Liechtenstern, Maximilian Joseph Freiherr von (1792–1850). Military cartographer in Vienna. *Europe* 36 sh. Vienne 1807–09, *Venice* 4 sh. 1805, *Austria* 12 sh. 1810

Liechtenstern, Theodore Freiherr von (1800–57). *Central Europe* 64 sh. (1800), *Britischen Inseln.* Magdeburg 1829, *Schul-Atlas* 1853

Liefrinck, Hans of Augsburg. Engraver and publisher in Antwerp. *Maps, Plans and Views Antwerp* 1589

Liefrinck, Jean [Jan van] (1518–1573). Cartographer, engraver, mapseller. Boechel's *Ditmaers* 1559, Van Noort's *Plan Leiden* 1574, used by Braun & Hogenberg

Liefrinck, Mynken (16th century). Flemish engraver and colourist.

Lienard, J. B. (1750–1806?). Maps to St. Non's *Voy. de Naples,* and Choiseul's *Voy. en Grèce*

Lienhardt, J. Lithographer, of Mainz. *Nord Amerika* 1849

Liens, L. Des. See **Desliens**

Liesch, J. B. *Luxembourg,* Brussels 1862

Liesganig, Jos. (1719–1799). Ex Jesuit, supervised in 1772 a political cartographic survey for military mapping carried out by Austrian authorities on territories annexed from Poland: *map of Galicia* (and Lodomeria 49 sh.)

Liesvelt, Jacob van (1489–1545). Publisher in Antwerp. *Champagne* 1549 (with S. Keltenhofer)

Lieude de Sepmanville, M. de. See **Sepmanville**

Lieussou, J. P. H. A. (1815–1855). French hydrographer. *Etudes ports Algériens* 1849, *Variations Marche pendules* 1854

Lievre, C. Le. See **Le Lievre**

Lieuwendal, F. Engraver and publisher of Helsinki. *Europe* 1855

Ligar, C. W. *Map Auckland N.Z.* 1848

Light, Col. W. Surveyor-General of South Australia. *District Adelaide* 1839, *South Australia,* Arrowsmith 1839, *Plan Port Adelaide* 1859

Ligon, Richard. *Yland of Barbados* 1657

Ligonier, John Louis, Earl of (1680–1770). Military cartographer, Field Marshal. Huguenot origin

Ligorio, Pirro (1496–1583). Architect, artist, cartographer, b. Naples, d. Ferrara. *Maps of Rome* 1552, *France* 1558, *Belgium* 1558, *Hungary & Spain* 1559, *Greece* 1561, *Friuli* 1563, *Rome* 1570

Ligter, H. J. Dutch Seaman, 17th century. Worked for Van Keulen family. *Sea Chart S. Ireland* ca. 1750

Liguera, Juan de. *Cayos de la Florida* 1742

Ligustri, Tarquinio. Italian cartographer, end 16th century

Lilienstern, R. von. *Sachsen* 1810, *Deutschland* 1826

Lilius, A. See **Lily, G.**

Lilius [Lilio], Zacharius. Cosmographer, Scientist and scholar, of Florence. *De Origine et laud. scient.* 1496, *Orbis brevarium* 1493 (with T-O maps)

Lillie, H. *Plan Township of Laurier,* Toronto 1879

Lilly, Charles (fl. 1715–20). Military engineer. Surveyed S. Coast harbours 1715–19, *Dartmouth, Falmouth, Plymouth, Portland, Scilly Is.*

Lilly, Christian. *Plan de la ville de Kingston* 1764

Lilly, C. *Plan Fort Charles, Jamaica* 1699 MS

Lily, George. [G(eorge) L(ilius) A(nglus)] (fl. 1528–1559). *Britanniae Insulae,* Roma 1546, later editions 1556, 1558

Lily, Henry. Surveyor. *Manor Ellesborough* 1629, *Wendover Borough* 1620 MS

Lima, J. J. L. de. *Carta Hydro da Guiné Portugueza* 1844, *Golfo de Guiné* 1845

Lima Lopes, de. *Territ. das Velhas e Novas Conquistas* 1860

Limburg, Remade van, Canon of Liege. *Plan of Hoei* (1574) for Braun & Hogenberg

Limiers, H. P. Collaborated with Gueudeville. Wrote text Vol. VII of *Atlas Historique*

Linant de Bellefonds, Louis Maurice Adolphe (1800–1883). Engineer, editor, Inspector General of roads and bridges. Chief Engineer of Suez Canal. Topo. studies. *Isthmus de Suez* 1844 MS, Earliest plan of the Canal *Alexandria Harbour Egypt* 1869, *Basse Egypte* 4 sh. 1882

Linck, Norbert Wencel von. *March Flusz* 1719

Lind, W. van (d. before 1725). *8-sheet map of Crimpenerwaard* Dutch polder Map pub. Covens & Mortier

Linde, Luca di (17th century). Geographer

Lindeman, O. Engraver. R. and J. Otten's *Dalmacie* (1730)

Linden, Gerard onder de. Publisher, of Amsterdam. For Valentyn 1726

Linden, Lieut. W. van der. *Special Karte . . . Ostfries u. Harrlingerlande* Berlin 1804

Lindenau, Baron Bernard August von (1779–1854). Statesman and astronomer

Lindenkohl, Adolph (1833–1904). Oceanographer and cartographer *N. W. America* 1867, *Küstendistrikt v. Süd Ost Alaska* 1894, *Temperaturen . . . im Golfstrom* 1896

Linder, Rudolf (fl. 1863–1879). Cartographer, lithographer. *Battle Plans, Nachod, Skalitz, Trautenau &c.* 1866

Lindesay, A. See **Lindsay**

Lindhistrom, Petr. (b. 1632). Fortifications engineer. *Maps of Swedish Colonies in Delaware,* copperplate, 1654–5, *Plan Fort Christina* 1654, *Nova Suecia* 1654–5

Lindley, Joseph (1756–1808). Surveyor, merchant &c., 10 Surrey Place, Kent Road. *Surrey* 1789–1790, 2 sh. (with W. Crossley) 1793, second edition 1814

Lindmeier, D. (Halberstadt ca. 1600). Engraver. *Maps & Views of Goslar, Brunswick &c.*

Lindner, Cornelius (1694–1740). Mathematician

Lindner, Dr. F. L. *Karte von Neu Holland* (1830)

Lindner, F. P. Engraver and print dealer. *Map of Regensburg* 1714

Lindsay [Lyndsay, Lindesay], Alexander. Pilot to James V of Scotland. *Chart Scottish Coast* (1540) MS

Lindsay, David. *Central Australian Ruby Field* Adelaide 1889, *Explor. Northern Territ.* 1885–6, London 1889, *Explor. S. Australia* 1891–2, *Adelaide* 1892

Lindsay, Rev. J. *Chart Cork* 1759, for Collins 1760

Lindsay, R. S. *Chart Hartlepool* 1846

Lindsey [Lindsay], Capt. J. S. *Part Straits Malacca* 1798, *Singapore St.* 1799

Lindstram, Axel and Leon. *Geolog. maps Sweden* 1871

Lines, D. *Guerre dans l'Inde* 1781

Lines, W. *Sketch Shat el Arab* 1857

Lingard, John. Laurie's *New Plan London* 1841

Linghangen, D. G. (b. 1819). Swedish astronomer

Linnard, Capt. T. B. *Plan Battle Buena Vista* 1847 (with Lieuts. Pope and Franklin)

Linnerhjelm, Gustaf Frederic (1757–1819). Swedish Royal Geographer. *Lands Wagarne genom Sodra* 1792

Linnemann, B. *Hinterland v. Hatz-feldthafen* 1895

Linnig, J. T. (1815–1891). Engraver, of Antwerp. *Topo. engrs. Antwerp &c.*

Linschoten, Jan Huygen van (1563–1610). Dutch traveller and historian. In Goa 1583–88. Returned Holland 1592. *Itinerario* 1596 (Dutch text), 1599 (Latin); *Hist. de la navig. aux Indes Orientales* 1619. Maps used in Linschoten possibly based on Bartolomeo Lasso

Linsley, James H. Surveyor. *Plan of Stratford, Connecticut (U.S.)* 1824

Linth, A. E. von der. See **Escher von der Linth**

Linton, S. Benton. American engraver and lithographer, 148 S. 4th St., Philadelphia. For Holt 1869

Lintot, H. Bookseller & Publisher at ye Crossed Keys agst St. Dunstans Church in Fleet Street. *Budgens Sussex* 1724, *Camden's Brittania* 1753

Lipinski, Ignacy (fl. 1802–1836). Polish military cartographer

Lippens, Philip. *Lettering* for Vandemaelen 1827

Lippert, Dr. P. Text for Kiepert's *Grosser Hand Atlas* Berlin 1893

Lippincott, Grambo & Co. Publishers in Philadelphia, No. 14 N. 4th St. 1853, No. 20 in 1855, W. Williams *U.S.A.* 1852, For Eastman 1853

Lipszky, Johannes de (1736–1826). Slovak military cartographer. *Karkof* 1789 MS, *Hungary, Budapest* 1806, 12 sh., reduction to 1 sh. 1827

Livelli, Salvador. *World* 1780 and 1786

Lisboa, Joao de. MS *Atlas* 20 maps), 1565–75

Lisiansky, Capt. Urey (1773–1837). *Voyage round World* 1803–6, London 1814, with *charts of N.W. America*

L'Isle, Claude de (1644–1720). Cartographer, father of Guillaume

L'Isle, Guillaume de [Insulanus] (1675–1726). Premier Géographe du Roi 1718. Pupil of Cassini. Member of Academy 1702. He worked from the following addresses: Rue des Canettes près de St. Sulphice 1700–07, Quai de l'Horloge à la Couronne de Diamans 1707–08, Quai de l'Horloge à l'Aigle d'or 1708 onwards, Quai de l'Horloge only, and Quai de l'Horloge au Palais. His first works were a *globe, map of the world and the 4 continents* 1700; *Atlas de Géographie* 1700–12; *Mississippi* 1701 MS. Posthumous *Atlas Nouveau* 1730 and later editions, Italian edition 1740–50. Business was continued by his widow, who took into partnership Philippe Buache

L'Isle, Joseph Nicolas de [1688–1768] Elder brother of Guillaume. Educated College Mazarin, pupil of Cassini. Astronomer and cartographer. Took service in Russia and jointly with Kirilov produced the first Russias atlas, *Atlas Russicus* 1745, *N. Pacific* 1731 MS, printed 1750; *Georgie* 1766, In Russia he founded the Academy of Sciences of St. Petersburg. He returned to Paris in 1747 with an important map collection

L'Isle de la Croyère, Louis de. Astronomer, brother of Joseph Nicolas. Took service in Russia with his brother in 1726 and worked with him until his death in 1741 on expedition with Chirikov

Lislet-Geoffroy, Jean Baptiste (1755–1836). Hydrographical engineer. Captain of Corps of Military Draughtsmen. Worked for Dépôt des cartes et plans. Cartographer for *Mauritius Isle de France et Reunion* 1798, revised 1802, 1807, 1814, *Chart of Madagascar* 1819

G. de L'Isle (1675–1726)

List, Joseph. Publisher and engraver of Vienna. *Schlesien, Brunn* 1810, *Carte Hydro. de l'Europe* 1816, 1818

Lister, John. *Plan Stratford Place* 1779 MS

Lister, M. Ingenious Proposal for a new sort of Maps of Countrys [Geological] Philo. Trans. of Royal Soc. 1684

Litchfield. See **Auld & Litchfield**

Lithov, Isak (18th century). Finnish surveyor

Litke, Fedor Petrovic, Graf (1797–1882). Hydrographer and cartographer, St. Petersburg

Littledale, St. G. *Map of Hoang Ho* 1894, *Route across Tibet* 3 sh. 1896

Littrow, Henri V. (1820–1895). Astronomer and oceanographer

Littrow, J. J. v. (1781–1840). Austrian astronomer and mathematician. Astron. works applied to geography. *Atlas des gestirnten Himmels* (1830), *Chorographie* (1833)

Liu, Ching Yang (11th century). Chinese geographer

Liu, Hsun. *Atlas Honan Province* 1870

Livesey, C. Engraver, of Leeds. For Teal's *Intended Navigable Canal from Calder W. Riding* 1792

Livingstone, Rev. Dr. David (1813–73). Explorer and missionary. Contributed various maps to Royal Geographical Society 1853–5. *Route across Africa* Arrowsmith 1857, *Africa* 1866–73, pub. 1880 &c.

Lizars, Daniel (fl. 1776–1812). Engraver and publisher, Parliament Stairs Edinburgh. Apprenticed to Andrew Bell, started engraving 1776. Engraved for Armstrong 1777, for Brown's *Scotland* 1806, *Plan Leith* 1806, Kerr's *Collection of Voyages* 1811–24. Succeeded by his son William Home Lizars

Lizars, William Home (1788–1859). Scots painter and engraver, son of Daniel

Lizars above, apprenticed to his father and produced with him *Cape of Good Hope* 1818. Left the business but returned to take charge on death of his father and worked from 61 Princes St., 5 David St., and 3 St. James Square. He engraved for Ewing 1828, Blackwood 1830, Thomson's *Scotland* 1838, *Gen. Atlas* ca. 1850, *W. Lothian Reg.* 1824, *Dunbarton* 1841, *County Map Scotland* 1845, *Caledonian Canal* 1852

Ljatkoj, Ivan (16th century). Russian soldier and cartographer

Ljunggren, Erik Gustaf (1817–88). *Atlas ofver Sveriges* 1853–61

Llasso, Bart. de. *Charts* sold by Plancius 1591

Lloyd, A. See **Lhuyd**

Lloyd, Capt. *Straits of Banca,* Laurie & Whittle 1796

Lloyd, H. H., & Co., 21 John St. N.Y. *Georgia* 1864, *Pacific States* 1865

Lloyd, J. A. *Maps Panama & Columbia* 1829–1881

Lloyd, Lieut. Colonel J. A. English surveyor. *Trigonometrical Surveys Panama* 1830. Surveyor Gen. Mauritius, *Madagascar* 1850

Lloyd, J. T. Publisher of London. *U.S.A.* 1863

Lloyd, Comm. (later Capt.) Richard. Indian Navy Surveyor. *India* 1833–41, *River Hoogly* 1836, *Gulf of Bengal* 1840

Lo, Hung-hsien (1504–1567) Chinese astronomer and geographer

Loader, George. *Plan Chichester* 1812

Loan, Jeremiah. Estate surveyor. *Tolleshunt d'Arcy* 1728 MS

Lobeck, Tobias. Engraver and publisher. *Atlas Geographicus* (with Lotter's maps) 1762, *Kurzgefasste Geographie* Augsburg 1762

Lobo, Jeronimo (1595–1678). Portuguese priest. *Hist. de Ethiopia* 1659, *Haute Ethiopie* Van der Aa (1710)

Locher. Missionary. *Goldküste W. Africa* (with Plessing) (1855)

Locher, William. *Philadelphia* 1859—60 (with E. Hexamer)

Lochner, C. F. (1761—1805?). Engraver of Nuremberg. Maps for Homann Heirs, Schneider & Weigel

Lochner, Zacharias. *Tract. Geomet. zu dem Feldmesser* 1583

Lochom, Michael van (1601—47). Engraver and Royal Printer, Rue St. Jacques à la Rose blanche Couronné *Paris* 1630, *British Isles* 1639, *Asia* 1640

Lock, Major. *S.W. Crimea* (with Barnston & Hammersley) 1856

Lock, W. G. *E. Coast Iceland* 1881

Locke, Joseph. Engineer. *Railway surveys* 1843—6, *Grand Junction Railway* 1836

Lockett, Col. *Carte Gén. d'Afrique* 1877

Lockman, J. *Travels of the Jesuits . . . Spanish settlements in America* 1762

Lockwood, Anthony, Master, R.N. (d. 1855). *Falmouth* 1806, *Nova Scotia* 1818

Lockwood, Lieut. J. B. U.S. War Dept. *Maps of Arctic, Coast of N. Greenland* 1884

Loczy, L. (b. 1849). Hungarian geographer, traveller. *Asia* 1877—1880

Lode, G. von (18th century). Engraver of Copenhagen. *Map Stralsund*

Lodesano, Francesco. Italian Portolan maker 16th century

Lodewick van Bercheyck, Laurens. *River Demerary* (1770)

Lodewijcksz [Lodewycksz], G. M. A. Willem. *Dutch Voyages to East Indies* 1595—97 (1598)

Lodge, Benjamin (d. 1801). *Pennsylvania Regiment Military Maps* 1779—80

Lodge, John. Surveyor 17th century. *Village of Westgerardston* 1656

Lodge, John (fl. 1754—96). Geographer and engraver, 45 Shoe Lane, London. *S.*

Carolina 1771 Parish *Stephen* 1773, Adair's *Amer. Indians* 1775, *Polit. Mag.* 1780—90

Lodge, J., Junior. Engraver, of Islington. For Banke's *System of Geography,* for Bew, Laurie & Whittle, Rees & Arrowsmith ca. 1780—1810

Lodge, William (of Leeds 1649—1689). *Book of Divers Prospects York* 1678

Loeff, Steph. van der. *Sea chart Hoek van Schouwen to sluys.,* Keulen en Zoonen 1799

Loehr, F. *Californische Halbinsel,* Gotha 1868

Loelling, H. (d. 1894). Topographer

Loew, Conrad. *Meer oder Seehaven Buch* Cologne 1598

Loew, Dr. C. *N.W. Texas* Gotha 1873

Loewenberg, Julius (1800—1893). *Hist. Geogr. Atlas* 1839, *Geogr. Länder Bibel* Berlin 1846

Löwenorn, P. See **Lovenorn**

Löffler, J. B. *Königsee,* Munich, *Eyb see, Wann See,* from Riedl's *Strom Atlas von Bayern* 1812

Loftus, W. K. (1821—1858). English geologist and explorer. *Maps of Irak & Persia* 1849—55

Logan, James Richardson (d. 1869). Scientist. Geological papers to Asiatic Society 1846. *Geological map of Singapore* 1851 MS

Logan, Sir W. E. (1798—1875). Canadian geologist. *Map of Canada* 8 sh. Washington 1866

Logerot, Aug. Publisher, Quai des Augustins 55 Paris. For Fremin's *Océanie* (1840) and Tardieu *Africa* (1840), *Carte Postale de la Répub. Française* 1848, *Europe* 1878

Loggan, David (1635—1700?). Draughtsman and engraver. *Plan Oxford* 1675, *Cambridge* 1688

Loggie, T. G. *New Brunswick* 6 sh. St. John 1885

Loghlan, Maj. Gen. N. M. *Part Somaliland* 1863 MS

Logie, Andrew. *Aberdeenshire* 1746 MS, *City Aberdeen* 1746 MS

Lohr, Jacob (18th century). Surveyor of Wertheim

Lohrmann, W. G. (19th century). German astronomer, surveyor and cartographer from Leipzig

Loisel, L. Engraver, of Paris (17th century). Plates and maps to *Les glorieuses conquêtes de Louis le Grand* 1643

Lok, Michael. Traveller, Governor Cathay Co. *Hakluyt's Divers Voyages,* London 1582 (with *map N. America*), *Hist. West Indies* 1613

Lollianus. Roman geographer

Loman, J. C. Publisher of Amsterdam 1876

Lomba, R. L. Uruguay. *Montevideo* 1884

Lombard, H. C. (1802–1895). Geographer and astronomer. *Traité de climatologie* 1877–1880

Lombard, R. P. *Fleuve Me-nam (Thailand)* Paris 1879

Lomnicki, Marian (1845–1915). Co-author *Polish geological atlas of Galicia*

Lomonosov, Michael (1711–1765). Russian geographer. Director St. Petersburg Acad. of Sciences 1757–65. Corrected J. N. De Lisle *Atlas of Russia, globe gores* 1763

Londerseel, J. (d. 1625). Dutch engraver. *View Hague* 1614

London Magazine (1732–1885). Contains many maps by Kitchin, Baldwin, &c.

Long, A. *Persia* S.D.U.K. 1831

Long, Capt. *Wrangles Land (Arctic)* 1867 MS

Long, George. *Atlas Classical Geogr.* 1854

Long, Major S. II. (19th century). U.S. Topo. Engineer, cartographer. *Arkansas* &c. for Carey & Lea 1827, *Washington to Rockies* 1821 MS

Longchamps, Sieur S. G. French geographer and publisher, Rue St. Jacques à l'Enseigne la Place des Victoires. *France* (with Janvier) 1751, *Canada & Louisiane* 1756, Brion's *British Isles* 1756, for Bonne 1779

Longchamps, Jacques François des (1747–1843). French royal Geographer. *Gibraltar* 1779, *L'Europe* 1817

Longhi, Guiseppe. Publisher of Bologna (fl. 1670–80)

Longle, J. A. *Mozambique,* Lisbon (1886)

Longlol, Paulo Daniel. Edited M. F. Cnopf's *Princip. Brandenburg* Homann (1735)

Longmans. House founded in 1724 survives today as Longmans Green & Co. Ltd. 1724 Thomas (1699–1755). 1725 J. Osborn & T. Longman. 1734 T. Longman. 1745 T. Longman & T. Shewell. 1747 T. Longman. 1753 T. & T. Longman (nephew Thomas Longman [1730–1797] joins firm). 1755 nephew Thomas succeeds to control of business. 1797 T. N. Norton (Thomas Norton Longman [1771–1842] son above, succeeded). 1799 T. N. Longman & O. Rees (Owen Rees taken into partnership). 1804 Longman, Hurst, Rees & Orme. 1811 Longman, Hurst, Rees, Orme & Brown. 1823 Longman, Hurst, Rees, Orme, Brown & Green. 1825 Longman, Rees, Orme, Brown & Green. 1832 Longman, Rees, Orme, Brown, Green and Longman (Longman Rees & Co.). 1838 Longman, Orme, Brown, Green and Longman. 1839 William Longman (1813–1877) became partner. 1840 Longman Orme & Co. 1841 Longman Brown & Co., Longman Brown Green & Longman. 1856 Longman, Brown, Green, Longman & Roberts. 1859 Longman, Green, Longman & Roberts. 1862 Longman, Green, Longman, Roberts & Green. 1865 Longmans, Green, Reader & Dyer. 1889 Longmans Green & Co. Longmans issued among others Salmon's *Modern History* 1744–6, *Camden* 1753, Arrowsmith's *Pacific* 1802, Pinkerton's *Modern Geogr.*

1802, Humboldt's *Atlas* 1810–11, Pinkerton's *Modern Atlas* 1812, Walker's *Universal Atlas* 1812, *General Atlas* 1817, Hall's *Atlas* 1818–60, *British Atlas* 1837–79, Butler's *Mod. Geogr.* 1834 and various *maps of London*

Longmate, B. Engraver, No. 11, Noel Street, Soho, London. *Mineral Map W. Counties of England* 1797

Lognon, Auguste Honoré (1844–1911). *Atlas Hist. France* 1885–89

Longobardi, Niccolo (1566–1654). Italian Jesuit and geographer. *Chinese globe* 1623 (with Dias)

Longomontanus, Severin. Pupil of Tycho Brahe (16th century)

Longtree, S. D. Lithographer. *Burr's N.W. coast America* 1840

Longuemar, A. Le Touzé de (1803–1881). Geographer and archaeologist. *Regional Geography Poictou*

Longus, Josephus. Publisher. Corrected edition of de Wit's *large World map* ca. 1670

Lons, D. E. (1599–1631?). Dutch engraver. Topo. Engrs.

Loomis, Elias. American professor. Produced *synoptic weather maps* 1843 in Trans. of Amer. Philo. Soc.

Loon, Gillis van. Publisher, of Amsterdam. J. van Loon's *Zee Atlas* of 1661

Loon, Herman van (fl. 1667–74). Engraver and publisher, of Amsterdam and after 1686 in Paris. Engraved maps for De Fer, *Quebec* 1694, *Forces de l'Europe* 1695, *Asia* 1700. For De Lisle 1700–1 and for Nolin 1689

Loon, Jan [Johannes van] (ca. 1611–1686). Mathematician, astronomer, chartmaker and engraver. Worked for Jacobsz' *Sea Mirrour* 1649, *Pilot Books* for Jansson 1650–54 and produced his own atlases: *Zee Atlas* and *Klaer Lichtende Noordster* 1661 (with Gilles van Loon), reissued Jansson 1666. *Flambeau de la Mer* 1682–8.

Later he collaborated with Keulen and Robijn

Loose, Arnold de. Engraver. *Plan of Jerusalem* 1584

Loots, Johannes (1665–1726). Amsterdam publisher of marine books and charts at the "Nieuwebrugsteeg in de Jonge Lootsman". Studied under Doncker. *Boek van de Noord en Oost Zee* 1697, *General Atlas* ca. 1710, *Great Sea Mirror* 1717

Lootsman [Sea Pilot]. Name used by the Jacobsz family. See **Jacobsz** (Theunis)

Lopes [Lopez], Sebastiao [Bastian]. Portuguese cartographer. *Charts* 1558, (1570), 1583 MS

Lopez, Don Juan. Geographer to the King of Portugal 1795. *Atlas* 1780, *Terra Firme* 1785, *Nueva Granada* 1795, *N. America* 1801, *Plano de Madrid* 1812

Lopez, D. T. *Spain* 1782

López de Gómara, Francisco (1510–60). *Tóde la Tiérra de las Indias,* Woodcut *Map World* 1552

Lopez de Vargas Machuca, Don Tomas (1730–1802). Spanish geographer. *Atlas de Espana* Madrid 1757, *Atlas de la Amer. Sept.* 1758, *Atlas Geogr.* Madrid 1758, *La Louisiane* 1762, *Pacific Constanso* 1771, *Quito* 1786, *Atlas Elemental* 1792, *Asia* 1794, *Terre Firme* 1802

Lopez de Velasco, Juan. *Map of the Division of the Indies* 1575 MS

Loputzkij, Stanislav (fl. 1663–68). Russian cartographer and painter. *Map of Lithuania* 1664

Lorain, A. *Carte Itin. de la France* 1837

Lörcher, Rev. T. J. *Prov. of Canton* 1879

Lord, Lieut. W., R.N. *Liverpool* 1839–40, *Approaches to Liverpool* 4 sh. 1846

Lord, W. B. *Routes to Diamond & Gold Fields (S. Africa)* London, J. B. Day 1870

Loredano, Pietro, of Venice. *Portolan* 1444

Lorenzen [Lorentz], H. See **Laurents**

Lorichs, Melchior (1527–90). Painter and engraver. *Elbe* 1568 (MS 41 ft.) and *View Constantinople* (11 meters)

Lorin. *Map for Nouvel atlas portatif* 1811

Loring, Hector, R.N. *Admiralty Chart of Trincomalee* 1832 (with Lieut. J. Cannon)

Loritz [Loritus, Loriti]. See **Glareanus**

Lorrain, A. *La France* 1836

Lorrain, N. Père. Worked for Dépôt de la Guerre. *Amérique Sept.* 1845

Lory, Ch. (1823–1889). French geologist. *Text and maps,* particularly on Dauphiny

Lory, Gabriel (1763–1840). Swiss painter, engraver and publisher. *Views and Prints Swiss Towns*

Losada, Juan C. *Tradados de esferica y geog.* Madrid 1844 (6 maps)

Loscher, Valentinus Ernest (18th century). German cartographer

Lose, F. (18th century). Painter and engraver. *Maps N. Italy, Views Milan, Genoa, Turin &c.*

Losenau, L. C. L. de. *Galizien u. Lodomerien* (1780)

Losi [Lose], Carlo. Printer, of Rome, *Pianta da Roma* 1784

Los Rios Coronel, H. de. See **Rios Coronel**

Loss, Johann Kaspar von der (1664–1711). *Saxony in form of a stag*

Los Santos, B. de and D. J. de. See **Santos**

Lossef. *District of Irkoutsk* 1811

Lossieux, Gaetano. *Pianta . . . di Palermo* 1818

Loth, Johann. (19th century). Bohemian cartographer

Loth, W. L. *Kaart van Guiana* 1888, *Surinam* 1899

Lothian, John (fl. 1825–46). Geographer and publisher, 41 St. Andrew Square, Edinburgh. *Plan Edinburgh* 1825, *Atlas Scotland* 1826–30, *Pocket Bible Atlas* 1832, *County Atlas Scotland* 1835 *Revised People's Atlas,* Glasgow 1846

Lotter, Andreas. Engraver for Lotter's *Atlantis Minoris*

Lotter, Gabriel (1750–1800) *Charta ofwer Skane*

Lotter, Georg Friedrich. Engraver for T.C. Lotter's *Portugal* 1762, *Sardinia* 1764, *N. America* 1784, *World* 1787

Lotter, Gustav Conrad. Engraver for T. C. Lotter 1769–74

Lotter, Mathias Albrecht (1741–1810). Engraver. Worked for his father Tobias Conrad and succeeded to his business. *Palestine* 1759, *Philadelphia* 1777, *Sauthier's New York* 1777, *Mappemonde* 1782 and *N. America* 4 sh. 1784

Lotter, Tobias Conrad (1717–1777). Engraver and publisher. Apprenticed to Funk, married daughter of M. Seutter 1740 and succeeded him in 1756 jointly with Albrecht Seutter's son and G. B. Probst, the son-in-law. *Terra Sancta* 1759, *Sardinia & Corsica* 1764, *Russia, Turkey &c.* 1769, *World* 1775–78 and–82, *New England* 4 sh. 1779, *Pacific Ocean* 1781, *Russia* 1788, *Atlas Minor* n.d. *Atlas Geogr. Portatilis*

Lottin, V. Ch. (1795–1858). Navigator, accompanied Dumont Durville. *Maps Pacific* 1827–33, *Baie Shouraki* 1827, *L'ile Ika Na Mawi* 1827

Louis, H. *Maps parts of Thailand* 1890–4

Lounder, Dan. *Draught of Cockle & St. Nicholas Gatts* 1751, pub. 1753

Loup, Samuel (d. 1728). *Occid. de l'Oberland* Dury 1766

Lour De-Rocheblave, L. *Pyrenées* 1874

Loureiro, F. de. See **Ferreira de Loureiro**

Lous, Prof. Christian Charles (1724–1804). Born and died in Copenhagen. *Kattegat* 1776, *Sound* 1777, *North Sea* 1784, *Baltic States* 1801

Louvemont, Franc. de (1648–1890). Engraver Couplet's *China* 1686

Love, Benjamin. *Plan Manchester* 1839

Love, John. *Geodaesia or Art of Surveying* 1686, 1715, 1720, 1731, 1768

Love, John. *Carolina,* used by Crisp (1711)

Loveday, R. Kelsey. *Lyndenburg Gold Fields* 2 sh. Pretoria 1883

Lovell, J. Publisher, of Montreal. *Navigat. of St. Lawrence,* Toronto 1856, Boxer's *Montreal* 1859

Lovenorn, Paul de (1751–1826). Danish Officer of Marine, Founder Danish Hydrographical Office 1784. *Sailing Directions Kattegat* 1800, *Chart Baltic Straits* 1807, *Kaart over Noord Soen* 1815, *Norske Kyst* 1816

Loveringh, Jacobus. Publisher, Amsterdam. *Zak Atlas* 1764

Lovering, Joseph (1813–1892). Physician and astronomer

Lovett, Capt. Beresford. *Part Baluchistan* India Office (1872), *Central-Persien* Gotha Perthes 1874, *Alburz Mts.* 1883

Lovisa, D. Italian engraver. *Album Venice* 1720

Low, A. P. *Geological Surveys in Canada* 1895, *Labrador Peninsula* 4 sh. Ottawa 1896

Lowe, Alexander. *Berwickshire* 1794, *Plan Coldingham* 1828

Lowell. *Plan Seckford Estate, Clerkenwell* 1764

Lowell, Daniel W., & Co. *Washington Territory* 1862

Löwenberg, Julius (1800–1893). German cartographer. *Historisch. Geogr. Atlas* Freiburg 1839

Löwenorn, P. See **Lovenorn**

Lowitz [Lowizio], George Moritz. Prof. of Mathematics (1722–1774). Member of Homann Firm. *Philippines* (1750). *World Map in Homann's Atlas of America* 1752–1759

Lowndes, Sam. Publisher "over against Exeter-house in the Strand". Joint publisher with John Overton of *Map of London* 1675

Lownes, Caleb. American engraver, of Philadelphia. *Plan Boston Harbour* 1775

Lownes, Mathew. London bookseller. Drayton's *Polyolbion* 1612, *Tab. Hist. Geogr.* ca. 1617

Lowry. Engraver for Arrowsmith's *Pacific* 1802

Lowry, John. American surveyor *Chesterfield District S. Carolina* 1819, for Mills 1825

Lowry, Joseph Wilson (1803–1879). Draughtsman and engraver, F.R.G.S. For Sharpe's *Corresponding Atlas* 1847–(9), *Table Atlas* Chapman & Hall 1852, *Universal Atlas* 1853, *Atlas of Physical & Hist. Geography* London, J. W. Parker & Son (1858), *Weekly Dispatch Atlas* 1858–9

Lowry, Lieut. R. *Exmouth Bar* 1829

Lowther, Lieut. H. C. *Somaliland* War Office 1894

Loyer, J. le. See **LeLoyer**, Jacques

Lozier. See **Bouvet de Lozier**

Luar, Dato Bintara *Johore (Malaya)* 4 sh. Adelaide 1887

Lübberas, Baron Ludwig von. Surveyor employed by the Russian government for charting the *coasts of Esthonia and Ingria* 12 sh. 1726

Lubbin, Augustin (1624–1695). Geographer, painter and engraver. *Orbis August* Paris 1659, *Canariensis* Paris 1659

Lubbin [Lübbin, Lubinus], Eilhard, [Eilhardus, Eilert]. Cartographer and Professor of Mathematics of Rostock. *Rügen* Mercator 1609, Blaeu 1631, *Pomerania* 12 sh. 1618 used Hondius, Blaeu &c.

Lubbock, Sir John William (1803–1865). Astronomer and mathematician. *Six Star maps for S.D.U.K.* 1830, *Eclipses & Occultation* 1835, *Perturbation Planets* 1833–61, *Treatise on Tides* 1839, *Gnomic Projection* 1851

Lübeck Chronicle. See **Rudimentum Novitiorum**

Lubieniecki, Stanislaw. Polish cartographer. *Theatrum Cometicum* Amsterdam 1666–68 *apud Fran. Cuperum,* 58 maps and 2 views

Lubienski, Wladyslaw Aleksander (1703–67). *The World geogr., chron., & historically described* (in Polish), 2 vols., 13 maps Wroclaw, (Breslau) 1740

Lubin. See **Lubbin,** A.

Lubini, E. See **Lubbin,** E.

Lubinus, E. See **Lubbin,** E.

Lumbrecht, Charles. *Pict. & Comprehensive Atlas* 1855

Luca, F. de (ca. 1793–1869). Geographer and mathematician

Lucas, A. (1803–1863). German painter and engraver. Introduced zincography in map making

Lucas, Claude. Engraver for Bretez' *Plan of Paris* 1739

Lucas, C. P. *Empire Atlas,* Edinburgh Johnston 1897

Lucas, D. (1802–1881). Pub. Engrs. *English Landscape* 1846

Lucas, Fielding, Jr. (1781–1854). American publisher of 138 Market St., Baltimore, Md. *New & Elegant General Atlas* 1816–23, *Atlas W. Indian Islands* 1824

Lucas, George Oakley. *Plan Paddington* 1842, *Marylebone* 1846

Lucas, J. *Hydro. Survey London,* Stanford 1878

Lucas, Jean François. *Plan d'Anvers* (ca. 1710)

Lucce, Guia (b. 1819). Italian geographer

Luccock, John. *Tableland of Brasil* 1820

Luchini [Lucini], Vincenzo [Vicentio]. Publisher, of Rome. *Spain* 1559, *Ancona* 1564

Luchtenburgh [Lugtenburg], André van. Mathematician. *Circular world map* on 4 sheets, *World map,* Rotterdam 1706. Worked for Dankerts & Le Clerc (1705–1710)

Luchtenburgh, Joannes van. Engraver and designer for Jaillot 1700, for Covens & Mortier &c.

Luchtmans, S. & J. *Reise Atlas* 1804

Lucini [Luchini], Antonio Francesco (b. 1610). Engraver and printer. *Malta,* Bologna 1631. All the plates to *Arcano del Mare,* Florence 1646 and 1661

Luckas, John. *Birdseye-view Dover* (1612)

Luckombe, O. *Traveller's Companion* 1789

Luckombe, Philip. *Ten miles round Reading* 1790 (with T. Pride), *Englands Gazetteer* 1790

Lud, Gaultier (1448–1527). Scholar, canon and printer. Founded printing press in St. Dié. *World* 1512

Lud, (17th century). Globemaker in Bologna

Luddecke, R. (1859–1898). Cartographer and editor. Revisions for Stieler's *Hand Atlas* 1890–1, Petermann's *New Zealand* 1889, *Aust.* 1890, *Afrika* 10 sh. 1892

Ludewig. Publisher, of Gratz. *Nord Amerika* 1849

Ludolf[us]. Jobi (1624–1704). Scholar and historian of Magdeburg. *Abysinnia* MS, edited by his son Christoph [Christianus]; Frankfort 1683

Ludovicus, Georgius. See **Barbuda**

Ludvigsson, Rasmus. Swedish surveyor (16th century)

Ludwig, Karl. *Bohemia* 1799 and in 4 sh. Vienna, Artaria 1808

Ludwig, Rudolph. *Hessen* 1867

Ludwiger. German engraver (fl. 1830–47)

Lueben, E. See **Lubbin,** E.

Luffman, I. Geographer, engraver and publisher, No. 2 Lam Buildings, Nassau St., New York. *Rhode Is.* 1813, *U.S.A.* 1819

Luffman, John (fl. 1776–1820). London engraver, publisher, goldsmith. Worked from 98 Newgate St. 1776, 85 London Wall 1780, Finsbury Square 1789, Inner Sweeting Alley, Royal Exchange 1799, Little Bell Alley, Coleman St. 1800 and finally 377 Strand from 1807. Taylor & Skinner's *Roads* 1776, Prior's *Leics.* 1799, Armstrong's *Rutland* 1781, *Antigua* 4 sh. 1788, *N.Z.* 1800, *Select Plans Cities* 1801–3, *New Pocket Atlas* 1803 and 1806, *Geogr. & Topo. Atlas* 1809, *Rhode Is.* 1813, *Univ. Atlas* 1815, *Topo. Atlas* 1815–16

Lucini, Vincenzo. See **Luchini**

Lugtenburg, A. van and J. van. See **Luchtenburgh**

Luiken, J. & C. See **Luyken**

Luis, Francisco (1591–1603). Portuguese instrument maker and hydrographer

Luis [Luiz] de Barbuda, Jorge. See **Barbuda**

Luis [Luiz], Lazaro. Portuguese cartographer. *Atlas,* Lisbon 1563 MS (13 maps)

Lukacs, Denis (1816–68). Globemaker

Luke, Rev. J. *Nigeria* 1893 MS

Luksch, Josef (1836–1901). Geographer and hydrographer

Lull, F. (ca. 1780). Surveyor for Count of Fünfkirchen in Bohemia

Lulow, Otto (19th century). Cartographer, of Hamburg

Lumsden, Ernest. Born Edinburgh 1883. Engraver. *Australian, Japanese & Indian Towns*

Lumsden & Son, James. Publishers, of Glasgow. Collaborated with W. & K. Johnston mid 19th century. *Map Glasgow* 1842, Duncan's *Itinerary* 1820, *Steamboat Companion Western Isles* 1839, Pub. for Philips (1870)

Lumsden, Maj. Gen. Sir P. *Part of Afghanistan* 1885

Luncan, I. *Atlas géog. Romania* 1895 (with A. Gorjan)

Lund, Johan Michael (1735–1824). *Telemark* 1784

Lundgren, E. S. (1815–1875). Swedish

painter and engraver. *Battle plans India* 1857

Lupo, Thomas. Portolan maker ca. 1600

Lusignano, Stefano (1537–ca. 1590). Vicar Gen. of Dominicans. *Cyprus,* pub. Bertelli 1576

Lupton, Donald. *Map World* (1677?)

Lupton, F. *Bharel Ghazel (Sudan)* 1844

Lupton, T. G. (1791–1873). Engraver. *Ports England* ca. 1840

Lutke, Count Feodor Petrovich (1797–1882). Russian Admiral and explorer. *Voy. autour du Monde* 1817–18, 1826–28, *Atlas voy.* 1826–9, pub. 1835

Lutke, L. E. (1801–1850). Lithographer. *Views Berlin, Hamburg, Rügen, Pomerania &c.*

Lutolf (1824–1879). Swiss geographer and historian

Lutsch von Luchsenstein, Col. Stefan (1710–92). *Transilvanie* 1735, 1777, *Walachia* 1738, *Siebenbürgen* 1751, *Transit-Moldavian Boundary Maps* 1751–75

Luttrell, Hon. John. *Chart of Honduras* (with Dalrymple) 1779

Lützenkirchen, Wilhelm (1586–1613). Publisher and bookseller, of Cologne. Vopel's *Europe* 1597

Lux, Anton. (1847–1908). Austrian cartographer and geographer

Lux, Ignatius (ca. 1650–93). Painter and engraver of Amsterdam

Luyken [Luiken], Jan (1649–1712). Engraver, painter and publisher, of Amsterdam. Title for Keulen's *Nouvel Atlas* 1682, Cartouches for *Caertboeck Oost & West Voorne* 1701, for Van der Aa

Luymes, Johan Lambert Hendric (b. 1869). Hydrographer

Luynes, Duc de. *Course du Jordain* 1865 (with M. Vignes), *Wady Arabah* 1865

Luytenburg, J. van. See **Luchtenburg**

Luyts, Jan of Utrecht (1655–1720). Cartographer. *Intro. ad geographiam* 1692

Lyaskoronsky, Basil I. Russian historian. Researches early Russian cartography, S. Russia, Kiev 1898

Lyatskoi, Ivan. Russian general. *Russia* engraved 1555, used by Munster & Herberstain

Lyborn, W. *Plan Parish St. Paul Covent Garden* 1686

Lycosthenes, Conrad. Astronomer

Lyef [Lief] Erikson. Icelandic seaman, son of Erik the Red. Discovered *Newfoundland, Nova Scotia* and *N. America (Helluland, Markland and Vinland)* ca. 1000 A.D.

Lyell, Charles (1797–1875). Geologist. *Principles of Geology* (1838), *Travels in N. America* (1845)

Lynch, F. *Plan Siac (Sumatra)* MS (1815)

Lynch, Henry Blosse. *Maps Persia* 1890

Lyndsay, A. See **Lindsay**

Lyne, James. *City of New York* 1828

Lyne, Richard. Engraver. *Map of Cambridge* 1574, *Angliae Heptarchia* in Lambord's *Kent* 1576. Employed by Parker as map engraver

Lyne, S. Print and Map Seller at the Globe in Newgate St. *Pocket Map London* 1742

Lynker. *Fortress of Graudenz* 1807

Lynslager, Hendrik. Navigator. *Map* for *Voogt* 1682, *Gibraltar* 1726, *Entrance to Mediterranean,* Keulen 1738

Lyon, Rev. C. J. *St. Andrews* 1843

Lyon, Comm. (later Capt.) G. F. *Hudsons Bay* 1821–23, *Arctic* 1824

Lyons, H. G. *Murrat Wells (Sudan)* 1895

Lyon, Lucius. American surveyor. *Island Bois Blanc* 1827, *Surveys in Michigan* 1849–50

Lyskirchen, Constantin von. Hanse merchant. Supplied *views of African and Asian towns* to Braun & Hogenberg

Lysons, D. (1762–1834). Author and engraver. *Environs London* 1792, *Magna Brit.* 1806

Lyter. See **Andreas, Lyter & Co.**

Lythe, Robert (fl. 1556–1574). Military engineer, surveyor, cartographer. *Calais* 1556, surveyed *S. half of Ireland* and part of *E. Ulster* 1567–70, *Dublin & Carrickfergus* 1567, *Plat Munster* 1571, *Isle of Sheppey* 1574

Lysulin, Capt. Corps of Military Cartographers, St. Petersburg. *Maps Asiatic Russia* 1867, 1873, 1878

M

M., C. C., Engraver *Port Natal* 1843

M., G. F. *Posnan* Jansson (1658)

M., H. See **Mallet-Prevost**, H.

M., I. See **Millard**, John

M., L. B. D. See **Montchel**, **Barentin De**

M., P. *Plan of Preston* (1715)

M., P. I. *Plan of Belgrade,* "P.I.M. fecit"

Maag, J. N. (1724—1809). German engraver. *Military maps of Munich* (1786)

M[aas], Abraham. *Caspian Sea* 1735

Maas, Abraham. *River Dessekebe* 1706, *Gulf of Finland* 1700

Maasburg, Ludwig Freiherr von, and G. Mack. *Railway map Austria—Hungary* Vienna 1881

Maasch, Otto. *Plan Hamburg* Leipzig, L. Voss 1882

Maaskamp, Evert (1769—1834) *Atlas Portatif pour la Hollande* 1822, *Gen. Kaart Nederlanden* 1816, *Kaart van Palestina* 1828

Maass, Albert. *Railway map E. Europe* Vienna, Pohl & Widinsky 1895

Mabre-Cramoisy, Sebastian. Printer to the King, Rue S. Jacques, aux Cicognes. *Relation de Voyages* 1666

Mabyre, Maxime. French geographer, end 19th century. *Carte de France* Paris, Mabyre. 1895 & 1897

MacAleer, Joseph. *Delaware Trust Lands* 1860, *Route Gold Mines Missouri* (1860)

MacAlpine, William J. *State of New York* 1854

MacAlvin, J. H. *Jackson Co., Indiana* 1856

McArthur, John. *Loch Tay* 1769 MS, *Plans Breadlebane estates* 1769, *Plan Glasgow* 1778

McArthur, J. J. & J. **Doup**. *N. Saskatchewan* 1885, *Boundary Yukon* 1900

Maccari, Giovanni. (17th century) Italian instrument maker

MacCarthey, O. *Plan Town Algiers* (1878)

MacCarthy, D. *Towanda, Pennsylvania* 1854

McCarthy, J. C. & S. *China* 1877—9, *Siam* 1888

MacCarthy, Jacq. French geographer b. Cork, d. 1835. *Trait. élém. de Géogr.*

MacCarthy, Oscar. French geographer. *Géogr. phys. de l'Algérie* 1858, *Plan d'Alger* 1888

McCarthy & Davis. Publishers of Philadelphia. *U.S.A.* 1824

Macartney, Lieut. John. 54th Regt. Bengal Native Cavalry. *Kingdom of Caubul* 1809, pub. 1815

McCarty, Capt. Richard. *Lakes & Hudsons Bay* 1775 MS

Macauley, J. J. *Ornaments for Ordnance Survey Sheet Waterford,* Dublin 1842

Macaulay, Zachary. *Railway maps* 1851– 1897

MacAuley & Hughes. Publishers, No. 10 George's Quay, Dublin. *Charts coast of Ireland* 1790–4

McBride, Admiral. *Chart of Smith's Knoll,* Laurie & Whittle 1796

MacCabe. Lithographer, Parliament St., Westminster. H. T. Smith's *New California* 1849

MacCabe, P. *Plan Madison, Wisconsin* 1855

McCallum, Duncan. Estate surveyor. *Romford Market* 1832 MSS

MacCallum, P. *City of St. Louis, Missouri* 1874

Mach, Gottfried. With Ludwig von Maasburg. *Railway map Austria & Hungary* Vienna 1881

MacChesney, U. B. *Fulton Co., N.Y.* 1856

MacClean. Engineer. *Ulverston & Lancaster Railway* 1850

Maclear, Sir Thomas (1794–1879). Royal Astronomer at Cape of Good Hope. *Remeasured and extended Lacaille's arc* 1837–47

MacClelan, A. R. *New Brunswick* 1870

MacClellan, Capt. George Brinton (1826– 1885). *Battle Plans Mexico* 1847, *Siege Vera Cruz* 1847

MacClellan, C. A. O. *American County Maps* 1863–4

McClelland, D. Engraver and publisher. *City Washington* 1846, *Post Map Michigan & Wisconsin* 1871

MacClellan, D. Lithographer, 26 Spence St., N.Y. *State of Florida* 1846, *U.S.A.* 1853, for Colton 1854

McClellan, Lt. John. 1st Regt. U.S. Artillery. *U.S.* 1830 MS

McClelland. See **Smith & McClelland**

MacCleverty, Henry. Admiralty Hydrographer H.M. Ship *Superb. Sproe Is.* 1809, *St. Johns Harbour* 1813

MacClintock, Sir Francis Leopold. Hydrographer to Admiralty. *Chart N. Polar Sea* 1848–9, *Discoveries N. Coast America* 1859, *Labrador* 1860

MacClure, Rev. Edmund. *Relief Maps* end 19th century. *Historical Church Atlas* 1897

MacClure, John. Lieut., R.N. Hydrographer for Admiralty Charts 1785–1826, *Arabia, Persia, India, Pelew Is. etc. Entrance Jaffrabut* 1834, *Survey of Diu* 1834

MacClure, Sir Robert J. le Mesurier. Comm., R.N. (1807–73). Navigator and explorer Arctic. *Survey N. of Behring Straits* 1849–52

Macomb, Alexander. Major Gen. (1782– 1841). *Michigan Territory* (1819) MSS

McComb, John (1763–1853). *Plan New York* 1789

Macomb, J. N. *Delta St. Clair* 1842, *Surveys New Mexico* 1860

MacConary, C. A. *American Co. Maps* 1874–6

MacCormack, Walter S. *Atlas of Philadelphia* 1874, *World* 1878

McCoy, Isaac. U.S. surveyor. *Surveys West of Arkansas & Missouri,* used by Phelps 1832

McCoy, John C. of Fayette M.D. Drew for Isaac above. 1832–8 MSS

McCracken, S. B. *Michigan,* Lansing Mich. 1876

McCrea [McRea], Irish surveyor. *Dartrey* 1795, *Donegal* 4 sh. 1801, *Monaghan* 4 sh. 1795, *Tyrone* 4 sh. 1813 (with G. Knox)

MacCrohon, Jose. Spanish hydrographer. *Coasts of Africa, G.B.* and *Peru* 1858–9

MacCullagh, J. R. Lt. Col. *Triangulation Madras District* 1893

McCulloch, John (1773–1835). Doctor

and geologist to the Trigonometrical *Survey 1814. Descrip. Western Isles* 1819, *Geolog. Map of Scotland* 1843 (posthumous)

McCulloh. *Land in N. Carolina* 1736 MS

Macculloh, William J. *Louisiana* 1855, 1860

Maccullogh, N. *Fort Madison, Iowa* 1859, *Lee Co., Iowa* 1861

Macdermott, Philip. *Ancient Ireland* 1846

McDonald, A. B. *Road Map County & City of Glasgow* 1894

MacDonald. Italian military cartographer. *Maps N. Italy* (with the French) 1803

McDonald, A. L. *N.W. Alaska,* San Francisco 1898

Macdonald, Major C. *River Benue (Nigeria)* 1889 MS, *Sketch Map Niger* 1891

Macdonald, Donald. *Huron District* 2 sh. 1846

Macdonald, H. W. American surveyor. *Ohio* 1867

Macdonald, Sir James Ronald Leslie. *Mombasa–Victoria Railway* 7 sh. 1893, *Uganda* 1894 and 1895

Macdonald, John. Capt., R.N. *Charts in E. Indies* 1788–9

Macdonald, Stephen. Surveyor for Admiralty. *Macasser* 3 sh. 1795

Macdonald-Kinneir, J. *Carte de l'Asie* (1817)

Macdonell, A. H. *Atlas* 1881

McDonnell, Lieut. Thomas, R.N. *New Zealand* 1834 and 1837, *Harbour Kiapora* 1841

Macdougall, P. L. *Guernsey* 4 sh. 1848

MacDougall, G. F. Hydro. surveys for Admiralty. *Ireland, India, Ceylon* 1861–76

Macdougall, Stephen. Land surveyor. *Map Islay* 1749–50 and 51

Macé. *Rade de Moka* 1798

Maceachen. A. *Brass River, Nigeria* 1873 MS.

Macelfatrick, S. *Fort Wayne, Indiana* 1860

Maceroni, A. *Pianta della Citta di Roma* 1837 and 1843

Macevoy, James. *Geological maps. Canada* 1889–1898

Macfarland, Robert. *Lee Co., Iowa* 1850

Macfarlane, Rev. S. *S.E. New Guinea* 1881, *Fly River* 1875 MS, *E. New Guinea* 1879

Macfarlane, W. H. Printer of Edinburgh. Printed Bartholomew's maps. For Black's *School Atlas* 1860

Macfarlane. See **Schenck & Macfarlane**

Macgeachy, Edward. Surveyor. *Port Morant Jamaica* (1833)

Macgee, John. *Maps of Ohio Districts & Virginia* 1864

Macgee, W. J. *Geologic Map of New York* 6 sh, 1894

Maggill, Alexander. *Plan River Spey* (1760)

McGlashan, Alexander. Estate surveyor of Colchester. *West Bergholt* 1841 MSS.

Macgonigale & Bergmann. Surveyors. *Kentucky* 1860–67

Macgowan, D. and G. H. **Hildt** of Missouri. *U.S. west of Mississippi* 1859

Macgowin, R. E. *Pittsburgh* 1852

Mcgregor, Samuel. *Baptist Meeting House Lot,* Newcastle Co. Maryland 1788

Mcgregor, Sir W. Explorer. *E. New Guinea* 1898, *W. New Guinea* 1899

Mcguigan. See **Wagner & Mcguigan**

Machado, G. F. (18th century). Portuguese engraver. *Maps* for Feliz Independante

Machajeff, M. I. (1718–1770). Russian engraver. *Master mapmaker* 1757

Machau, M. Engraver. *68 battle plans* (with Dosseau)

Machenry, Morris. *Denison, Iowa* 1858

Macherl, Peter. *Bishoprick Seckau,* 4 sh. 1886

Machin, John. Prof. astronomy, Gresham College, 18th century

Machin [Macham], Robert. Alleged English discoverer of Madeira (fl. 1344)

Machin, Thomas (1744–1816). *Map of Hudsons River* 1778

Macht. Engraver for Meyer's *Hand Atlas* 1867

Maciej [Miechowa, Miechowita] (Matthias of Miechow) (1457–1523). Author of a classical work on geography of Eastern Europe published first in Latin (1517). *Tract. de duabus Sarmatiis,* second edition (1521), under changed title *Descriptio Sarmatiarum Asianae et Europianae* in 1535, translated and published in Polish later, also in German, Italian and Dutch. 18 editions in the 16th century.

Macintosh, H. Surveyor. *Plan St. Mary Lambeth* (1892)

Mcintyre, A. Scottish engraver, 3 East Rose Street, Edinburgh. *Scotland* (1770), *Africa* (1790), *Scotland* 1793, *Edinburgh* 1793, *Map for Brown's Gen. Atlas* 1801

Macintyre, Henry. Surveyor, No. 17 Doune St. Boston and N.E. Corner of Perry & Pine Sts., Philadelphia. *Plans of Cities of Norwich* 1850, *Newburyport* 1851, *Boston* 1852, *Danvers* 1852, *Lynn* 1852

Mackau, Ange René Armand de. *Arica* 1824, *Côte du Pérou* 1824

Mackay, Alexander (1815–1895). English geographer. *Manual of Modern Geogr.*

Mackay, Arthur. Deputy to Surveyor Gen. *N. Carolina* 1763, *Survey Coast about Cape Lookout, N.C.* (1756)

Mackay, A. M. *Caravan Route Tanzania* 1880

Mackay, Rev. D. J. *Diocesan Map England* 1880

Mackagy, H. D. *Plan Louisburg* 1758 MS

Mckay, J. Surveyor. *Nigeria* 1894–5 MSS

Mackay, John. Capt. U.S. Army. *Seat War Florida* 1839

Mackay, John, the elder. Surveyor, St. George, Hanover Sq., 1725

Mackay, John, the younger. Surveyor. *Roman camps* 1726, *River Dee* 732

Mackay, William. *Nova Scotia* 1834

Mackeand C. Lithographer. *Follet, Victoria, Australia* 1866

Mackeller, Patrick. *Plan Quebec* enlarged from Bellin, *Lower Falls Saratoga* (1756)

Mackellar, Capt. William, R.N. Surveyor for Admiralty 1804–21. *St. Domingo* 1821

Mackena, B. V. (b. 1831). Chilian geographer

Mackenzie, Alexander (d. 1820). Explorer in N. Canada (Mackenzie River) and publisher, 38 Norfolk St. *Voyages* 1801, *America* 1801, *Plan of Croft Town* 1770 MS

Mackenzie, Capt. Colin. Engineers *Dominions of Nizam Aly Khan* 1798

Mackenzie, E. A. *Relief Map of Transvaal* 1899

Mackenzie, F. Draughtsman. *England & Wales* 8 sh. Arrowsmith 1852, *European Railways* (1857)

Mackenzie, Frederick. *Position of Concord* 1775 MS.

Mackenzie, Sir G. S. *S.W. coast Iceland* 1810–11, 2 maps in *Travels in Iceland* 1812

Mackenzie, Murdoch, the elder (d. 1797). Hydro-surveyor to the Admiralty. *Orcades* 8 charts 1747–50, Fourth edition 1791, *Treatise on Marine Surveying* 1774, *Charts W. Coast Scotland* 1775–6, *Ireland* 59 charts 1776 and 1800

Mackenzie, Murdoch, the younger (1743–1829). Hydrographer to Admiralty, succeeded his uncle (above). *Bristol Channel* 1771 and other *charts south coast England* to 1810

Mackenzie, William. Publisher, 69 Ludgate Hill, London, with branches in Edinburgh and Dublin. Hughes' *New Comprehensive Atlas* 1866, *Comprehensive Gazetteer of England & Wales* 1895

Mackenzie, W. H. *Montreal* (1860)

McKerlie, John. Maps for *Communications between England and Ireland* 1809

McKerrow, James. *Maps New Zealand* (1860) 1879

Mackey, Donald J. Diocesan maps *England, Ireland, Scotland, India* 1878–1885

Mackie, A. Publisher for Walker's *Univ. Atlas* 1820

Mackinlay, A. *Nova Scotia, Halifax* (1869), *Maritime Provinces of Canada* 4 sh. 1886

McKinnon, L. *Plan Town Colinton, E. Florida* 1821

Maclaren, I. *County South of Adelaide* 1840

Maclauchlan, Henry. *Map of Watling Street* 1850–51, 6 sh. 1852; *Map Northcumberland* 1866

Maclean, A. Lithographer and publisher *Kentucky* 1862

Maclean, A. G. Surveyor General. *Portion New South Wales* 1861

Maclean, L. A. *Kansas Territory* 1857

Maclean, Robert. Editor. *New Atlas of Australia* 5 vols., Sydney, Sands 1886

Mclean, Thomas. Publisher. *English Topography* with James Goodwin 1830

Maclear, T. *Africa S. Coast* 1857

Macklecan, Capt. Newfoundland Regt. Mil. Surveyor. *Entrance to St. Johns Harbour* 1784

Mackou, Baron de. *Côte du Pérou* 1824. Dépôt de Marine. See also under **Mackau**

Mackoun, J. Estate surveyor. *Barrington Hall* 1766 MS

Maclear & Co. Lithographers, Toronto. Denne's *N.W. Canada* 1857

Maclear, Capt. J. S. *Levuka Harbour, Fiji* 1880, *Philippines* 1886

Maclehose, J. *Sydney* 1839

Macleland, John. Description for Blaeu's *Galloway* 1654

McLeod, L. *Maps E. Africa* 1857–60

Macleod, N. *Corea* 1879

Macleod, Walter. Geographer. *Wall Maps* 1863, *Middle Class Atlas* 1863, *Scripture Geogr.* 1853, *Pupils Atlas* 1869

Macleod, W. C. *British Burmah* 1870

Maclot, Jean Charles (1728–1805). Geographer, cosmographer. *Atlas Général*, Desnos 1770, 1786

Maclure, A. *Crimea* 1855, *Cronstadt* 1854, *Delhi* 1857, *War Italy* 1859, *Sicily* 1860, *Grampian Mountains* 1875

Maclure, H. H. *Glasgow* (1859)

Maclure, W. (1763–1840). Explorer and geologist

MacMahon, J. Ponte Couland (d. 1837). *Draught Fort Howard* 1819, *Fox River* 1819 MS

Macmillan & Co. Publishers. *Short Geography of British Isles* 1879

Macminn, J. M. *Williamsport, Penn.* 1857

McMurdo, Col. E. *Gold Mines French Guiana* 1888

McMurray, W. Late Asst. Geogr. to U.S. *The U.S. According to Treaty Sept. 3, 1783, Plan of the State* 1780

McNair, Major. *Map of Perak & Selangor* 1875

Macnair, Thomas, S.N. Civil Engineer. *Hazleton, Penn.* 1870

Macnair, W. W. (d. 1889). Cartographer, worked on trigonometrical survey India. *Afghanistan* 1880

Macnally, Rand & Co. See **Rand, McNally & Co.**

Macneil, Henry. *Watertown, Illinois* (1860)

Macneill, Sir. John. *Railway maps* 1837–45

Macniff, Patrick. *Plan Lake Erie* 1791, *Settlements at Detroit* 1796

Macphaill, Miles. *America* 1844

Macpherson, Alexander. Engraver for Salt 1814 and for Smith's *Class. Atlas* 1835

Macpherson, Alexander. Geographer. *Atlas of Anc. Geog.* Phila. 1806, Humboldt's *New Spain* London 1810

Macpherson, D. *Historical Map Scotland* 1769

Macpherson, D. Cartographer. *Atlas of ancient geogr.* Phila. 1806

Macpherson, W. H. Engraver. *Italy* 1806

Macpherson, W. W. *Maps for Playfair* 1814

McPhillips, A. *Plan Winnipeg* 1881

McPhun, W. R. Publisher. *Plan Glasgow* (1860)

McQueen, James. *Niger* 1839, *Africa* Arrowsmith 1840, *Aethiopia*, 1843, *Central Africa* 1859

Marquet, Ph. Engraver. For Bonne 1786, for Grenet 1790, for Mentelle 1797

Macquisten. *Glasgow to Carlisle* 1813 (with Abercrombie)

Mcrae, Lieut. Archibald, U.S.N. *Passes across Cordilleras* 1854–5

Mcrae, John & James. Publishers. *New plan of London* (1847)

McRea, W. See **McCrea**

Macredie, Alexander. Publisher of Edinburgh. Playfair's *Atlas* (1814), Murphy's *Pocket Atlas Scotland* (1832)

Macrobius, Ambrosius Aurelius Theodosius (399–423 A.D.). Roman grammarian, philosopher and geographer. Numerous geographical MSS of different periods extant from 1200 to 1500 A.D. *In somnium Scipionis,* standard work in schools in middle ages

Madaba. Mosaic. *Plan Jerusalem,* 6th century A.D.

Maddock, G. *Arracan* 1858

Mad[d]ocks, Hannah. Mapseller at the Red Lion in New Row, Thomas St., Dublin. Brodkin's *Plan Dublin* (1728), *Plan Dublin* (1830). With Overton & Hoole

Madel. See **Maedel**

Madeley. Engraver. *Estate Plans Lord Southampton* 1840

Madenie. *Rade de Cherbourg* 1787

Madiis, Hannibal de [Annibale de Maggi]. *Padua* 1449 MS

Madison, James, D. D. *Map of Virginia* 6 sh. 1707

Madison, John. Estate surveyor. *Ingrave* 1596, *Woodham Ferrers* 1622 MS

Madocks, H. See **Maddocks**

Madou, J. B. (1796–1877). Lithographer. *Maps for Belgian Army,* later with Jobard, Brussels

Madox, Owen. Discovered Atlantic coast of America in 12th century, according to Welsh tradition.

Madoz, F. *Islas Canaries* 1849, *Isla de Cuba* Madrid 1853

Madox, R. Chaplain. *Journal of Fenton's Voyage* 1582 MS

Maedel, C. G. Engraver. *Weimar* 1857

Maedel, J. (Junior). Engraver for Weiland, *N. America* 1829, 1843, 1848

Maedel, [Mädel], K. Engraver for Kiepert 1851

Maedel, Karl Joseph (1823–1859). For Gaspari 1804, for Weiland's *Hand Atlas* 1848

Maedel II, L. Engraver for Stieler 1850, 1855 and 1861

Maedler, J. H. (1794–1874). Astronomer and physician. *Math. & Phys. Geogr.* 1843, *Mappa Selenographica* 1834

Maelen, Philippe van der. Belgian cartographer. *Atlas Universel de Géogr.*

Physique politique 6 vols. 1827, *Plan Bruxelles* (1836), *Atlas Admin. et statistique Belgique* 1844, *Carte Topo. de la Belgique* 1857

Maehler, Matthias (1782–1828). Austrian Military cartographer

Maehli. See **Mähli**

Maehrlin [Mährlin], Johannes (1778–1828). Surveyor of Ulm.

Maerschalk, F. *Plan New York* 1755

Maes, J. *Grande Carte de Flandre* Antwerp 1882

Maestre, M. Rivera. Cartographer. *Atlas Guatemaltico* 1832

Maffei [Maffeius] Peter [Petrus]. Jesuit traveller and writer, Sect. to the Republic of Genoa. *Historiarum* Venice 1589 and Cologne 1593, contains map *Indiarum orientalium* (after Ortelius)

Magallon, Francisco. *Mapa de Aragan* (1893)

Magdeburg, Friedrich Freiherr von (1783–1870). Military cartographer in Vienna

Magdeburg[us] Hiob (1518–1595). *Misnia & Thuringia* 1562, *Saxonia & Thuringia* 1566

Magellan, F. (1479–1521). Portuguese navigator, the first circumnavigator

Mager, H. *Atlas Colonial* (1860)

Maggi, A. de. See **Madiis**

Maggi, C. Editor, Turin. *Oceania* 1849, *Amer. Sept.* 1849

Maggi, Giovani (1566–1618). *Fiorenze* 1597, *Rome Churches* (1610), Engraver for Mauro's *Carta della Sabina* (1617)

Maggi, Gia. Batt. Publisher of Turin. Stucchi's *Oceanica & Africa* 1830, *America* 1849

Maggini, G. Swiss cartographer, end 19th century

Maggiolse, A. Dutch cartographer. *Map of Zeeland* (1860)

Maggiolo [Maiolo, Maillo]. Family Genoese chartmakers.

—— Vesconte (fl. 1504–49). Cartographer to Genoese Republic MS *atlas of* 10 *maps* 1511, *Atlas of charts* 1519, *Europe* 1524, *World* 1527, *Med.* 1546–1648

—— Giovanni Antonio (fl. 1525–1605). *Carta nauicatoria* (with Baldassare) 1605

—— Giacomo [Jacopo] (fl. 1551–1567). 8 charts including *Adriatic* 1561

—— Baldassare (fl. 1583–1605). MS *chart* 1583, *Navig. chart* 1586, *Carta nauicatoria* (with Giovanni Antonio) 1605

Magilton, A. L. *St Mary's River (Canada)* 1855

Magin, M. Marine engineer. *Golfe Gascogne* 1756, *Rivière de Loire* 1757, *Riv. de Bordeaux* 1785

Magini, Giovanni Antonio (1555–1617). Italian mathematician, cartographer, Prof. of astronomy in Bologna. Edited edition of *Ptolemy* Venice 1596, *Italia* (61 maps) published by his son Fabio in 1620. His maps used by Blaeu in 1640

Magini, Fabio. Son of G. A. Magini. *Nova descritt. d'Italia* 6 sh. (1650)

Magnan, Dominique (1731–1796). *City of Rome* 1779, *Geogr. Dict.*

Magnani, F. *Plan d'Alexandrie* 1877

Magnelli, Francesco. *Pianta di Firenze* 6 sh. (1783)

Magnus, Albertus. Geogr. writer. *Opera* 1494, *De natura locorum totius orbis* 1515

Magnus, Albertus. Bookbinder of Amsterdam, offered Blaeu *Atlases* for sale 1689

Magnus, Charles. Publisher and map seller, 22 N. William St., New York. *U.S.A.* 1850, 1851, *Historical War Map* 1865

Magnus, Johannes (brother of Olaus). *Historia Gothorum etc.* Rome 1554

Magnus, Olaus (1490–1558). Swedish archbishop and cartographer, b. Linkoping,

G.A. Magini (1555–1617)

d. Rome. *Map of Northern Regions* 9 sh.
Venice 1539, reissued Rome 1572; *Hist.
de rebus Septent.* Rome 1555

Magowan, J. Publisher. *Chart Atlantic*
1780

Magoths, W. *Voyage of the Delight to
Magella* Hakluyt 1600

Magra, Perkins. Lieut. 15th Regt.
Michulimackinac 1766 MS

Magson, John. Watchmaker. *Road London
to Tunbridge Wells* 1711

Maguire, C. E. Engraver for Allen's *Wicklow* 1834

Maguire, Jas. Neale. *Middle Brit. Empire
(N. America)* ca. 1756 MS

Mähli, J. F. (1805–1848). *Plan of Basle*
1847

Mahlmann, Heinrich (fl. 1830–73)
Europa 1841, *Amerika* 1847, *Süd Ceylon*
1850, *Asien* 1865, *Deutschland* 1880

Mahmoud, Pacha el Falaki (1810–1885).
Egyptian geographer and astronomer

Mahon, Charles. *Alluv. Region Mississippi*
1861

Ma, Huan (15th century). Chinese
hydrographer

Maida, J. Engraver for Vaugondy 1750

Maidment, C. *Geological Map Zoutspanberg.*
Pretoria 1889, *Witwatersrand* 1890

Maier, T. Mathematician. *Cercle d'Autriche*
Hered Homann 1747, *Ducatus Silesia* 1745

Maier, T. See **Mayer**, Tobias

Maierski. *Topo. Karte Gegend Berlin*
(1844)

Maillard, Jean. Cosmographer, Mathe-
matician, and Poet Royal to Francis I of
France, later to Henry VIII of England.
Premièr Livre de Cosmographie MS with a
World Map 1543

Maillard, L. *Carte de Réunion* 1878 and
1884

Maillart, J. C. Engraver. *Théatre de la
Guerre* 1793–5

Maillart, Ph. (1764–1857). Engraver for
Bouge's *Pays Bas* 1789, *Bruxelles* (1799)

Maillart, J. C. Sister of P. H. Maillart. Maps
signed 'Ph. J. Maillart et soeur' [sister]

Maillebois, Marquis de. See **Desmarets**

Maillet, Benoit de. *Carte d'Egypte* 1740

Maimonides, Moses (1135–1204). Jewish
traveller, philosopher and geographer

Maingy, A. *Postal map of Ontario* 1883

Mainz, H. of. See **Henry of Mainz**

Maiolo. See **Maggiolo**

Mainvilette. See **Après de Mannevillete**

Mainwaring, Major H. G. *Routes in
Somaliland* 1894–5

Maioc, Presbiter H. *Palermo,* Duchet 1580

Maior, Johann. Vienna bookseller. *Map of
Hungary* 16th century

Mair, A. *Colony of Natal* 4 sh. Stanford
1875

Mair, Alexander. Engraver. *Hungary* 1594,
Augsburg 1602, *Uranometria* 1603 (51 pl)

Mair, G. J. J. Surveyor. *St. Giles & St.
George's Bloomsbury* (1867)

Mair, J. See **Marr**, John of Dundee

Mairan, J. J. Dortus de (1678–1771).
Mathematician, astronomer. Studies in
geology

Maire, *Géographie Océanie* after Arrowsmith, Weylor and Faden; Paris 1834

Maire. *Louisiane,* used by De Lisle 1718

Maire. *Environs de Paris* 1827

Maire, Christopher (1697–1767). *A New
Map of the County Palatine of Durham by
Christopher Maire* (1711). Ded. to John
Montague Dean of Durham . . . by Christop.
Maire* (aged 14 years)

Maire, F. I. Publisher in Vienna. *Galicia
and Lodomeria* 12 sh. 1780, *Carte
Hydro. des Etats Maison d'Autriche*
1786, *Vienna* 1801, *Theatre War Russia–
Turkey* 1807

Maire, J. Le. See **Le Maire**

Maire, M. N. Geographer. *Paris* 1808, *Atlas de Poche* Paris, Deuesle n.d., *Carte Itin. d'Europe* 1816, *Carte Itin. France* 1820

Maire, N. *Map London* 1818

Maire, Padre, C. *Stato Eccles.* 3 sh. 1745 & 1752, *Carta Geogr. Stato della Chiesa* 1770

Maire, R. P. *Plan de Liège* n.d.

Maisch, C. Publisher in Vienna. *Schlesien* 1816

Maisey, F. C. *City Delhi* (1857)

Maison Neuve. Engraver of titlepage for Vaugondy's *Atlas Portatif* 1748

Maisonneuve. *Mauritius* 1854

Maissiat, Michel (1770–1822). Military cartographer

Maistre, G. *Port of Madagascar* 1895

Maitland, A. Gibb *Geolog. Map Northampton (Australia). Observations B. New Guinea* 1892, 2 sh. Perth 1898; *Collie Coalfield* 1898

Maitland, Capt. R.N. *Streights of Dryon* 1765, pub. Herbert

Maius (Monk of San Salvator de Tavara). Illuminator, ca. 926 A.D. Beatus' *World Map*

Majer, Johann (1608–1674). *Württemberg* 1710, used by Homann

Majer, Johann (1641–1712). Publisher

Majer, Tob. *Stato della Chiesa. Toscanna–Corsica*

Majerski, Stanislaus (1852–1928). Author for *great map of Galicia and Ludomeria* (1894)

Major, Daniel. *Indian Territory Kansas* 1866

Major, Capt. Joyn. Marine surveyor. *Draught of Cockle & St. Nicolas Gatts* 1751, pub. 1753, *Pakefield Roads* 1751, *Waxham to Pakefield* 1754

Major, Joan. Editor, *Map Tyrol* 1612

Major, J. & D. Lithographers, 177 Broadway, N.Y. for Parke 1851

Major, Richard Henry (1818–1891). Geographer, Keeper Maps B.M., Ed. Hakluyt Soc., Bibliogr. Columbus 1st Letter (1872)

Major, Sieur. Engraver. Cartouche for d'Anville 1746

Major, Thomas (1714–1799). Engraver

Makita, Okabe. *Map of the Kurile Islands* from native sources

Makoto, R. *Corea* Tokyo 1880

Makowski, Maciej (1796–18–). Military topographer. Collaborated on *Topo. Maps of Polish Kingdom*

Makowski, Tomasz (1575–1630). Polish cartographer and engraver. *Lithuania* engraved Gerritsz 1613 falsely attributed to Prince Radziwill, used by Blaeu 1613. First detailed and accurate map of Lithuania

Makreth, Stephen. *Plan Lancaster* 1778

Makriski, Ahmed al (ca. 1360–1442). Arab writer on geography and history of Egypt

Malaise, C. H. J. L. (b. 1834). Belgian naturalist and geologist

Malard, G. *Cahiers Géographiques* 1867

Malaspina, Comm. Alessandro. Commanded Spanish expedition in Descubierta and Altrevida (1789–94). *Plan of Islands Maluynas (W. Falkland)* 1789 MS, *Filipinas* 1792–3, pub. 1808; *Bay Manila* 1798 MS

Malassis. Printer of Brest, Libraire de la Marine. Joint publisher with Demonville of *Neptune Oriental* 1775

Malaver, A. E. *Plano . . . Buenos Aires* 1867, *Propriedades rurales B.A.* 1864

Malaver, Domingo. *Garrido de Puerto de San Luis [Mariannas]* 1738

Malavolti, Orlando (1515–1596). Historian of Siena. *Siena* 1599

Malby. *Celestial Globe* 1846, *London Colossus Globe* Wyld 1852

Malby & Sons. Lithographers. *Celestial Globe* 1856, *Terrest. Globe* 1857, *Charts for Admiralty* 1872, *for Stationary Office* 1884

Malby, Thos. H. *Durham* 1859

Malcho, S. Danish engraver. *Panoramic View of Lake of Geneva* (1780)

Malcolm, Capt. *Reconn. Map Canton* 1861 MS

Malcolm, Sir John. *Central India* 1822

Malcolm, J. S. (1767–1815). *Views within 12 inches,* London (1800)

Maldonado, Pedro Vicente (1704–1748). Spanish geographer

Malecot. *Hesse Cassel* 4 sh. Paris, Denis 1760

Malfalti, Bart. (1828–1892). Italian geographer

Malfeson, Ignace Balthaser. *Plan Ghent* 1756

Malglaire, Marie Charles Louis Joseph de. *Indo-Chine* 4 sh. 1893

Malham, Rev. John. *Isle Wight* 1795, *Naval Gazetteer* 1795

Maline[s], Fr. de. See **Franciscus Monarchus**

Mall, J. *Diocese of Berchtolsgaden* 4 sh. 1628, woodcut

Mallat de Bassilan, Jean (1806–1863). *Philippines* Paris 1846

Malle, A. J. C. A D. de la. See **Dureau de la Malle**

Mallet, Alain Manesson (1603–1706). Engineer, took service Portuguese Army. Returned France and served Louis XIV. *Description de l'Univers,* 5 vols., Paris 1683 (maps, plans and views). German edition 1686

Mallet, Capt. F. *Trinidad,* Faden 1802

Mallet, H. *Switzerland* 1802

Mallet-Prevost, Henri (1727–1811). Swiss cartographer, engineer and surveyor. *Environs de Genève* 1776, *Suisse Romande* 4 sh. 1781, *Suisse* 1798

Malombra, Gio. Revised 1574 edition of *Ptolemy*

Mals Frères. Engravers, Rue des Grues No. 10 près la Sorbonne, Gauthier's *Amérique* 1820

Malte-Brun, [Brunn], Conrad (1775–1826). Danish geographer and publisher. Settled in Paris 1800, *Géogr. Mathématique* (with Mentelle) Paris 1804, *Atlas Complet* 1812, *Géogr. Univ.* 1816

Malte-Brun, Victor Adolphe (1816–1889). Son of above. *Géogr. Univ.* 1861, *France Illust. &c., Yucatan* 1864, *Siam* 1869

Malthus, Thomas. Printer. *Ground Plot of Hamburg* 1686 (with H. Moll)

Malton, J. (d. 1803). Engraver. *Topo. Maps & Views* Dublin 1792

Maltzan, A. von. *Atlas of Witwatersrand* 1899

Maly, T. & Co. *Terrest. Globe* 1845

Miamiya, Rinzo (1780–1844). Japanese cartographer and traveller

Mamora, F. della. See **Ferrero della Mamora**

Man, E. H. *Andaman Is.* 1880

Man, John. Chief Surveyor. *Jamaica* 1662 MS

Man, J. Alex. *Manchuria* 1879

Man, Thos. Printer. Reed's *City & Liberties of Philadelphia* (1774)

Man, T. J. & Son, 4 New Bridge St., London. *Colchester–Harwich Railway* 1842 MS

Manasser, Daniel, of Augsburg. *Bavaria* 1623

Manazzale, Andrea. *Roma* 1805

Manby, R. Publisher. Heath's *Scilly Is.* (1749)

Manceli, Antonio. *Regnos d'Espana* 1642, *Cataluna* 1643

Manche, E. (1819–1861). *Album d'Ostend* 1840

Mandelslo, Jean Albert de (fl. 1616–44). *Voy. Célèbres* 1727

Manderson, Lieut. James. *Falmouth Harbour* Dalrymple 1805

Mandeville, Sir John de (ca. 1300–1372). *Boke of John Mandeville . . . of ways to Jerusalem,* MS. Printed edition by Pynson ca. 1496

Mandrot, Lt. Col. A. de. *Terre Sainte* 1867

Mandrot, Louis Alphonse de (1814–82). Swiss military cartographer. *Neuchâtel* 1870

Manetti, Gaspers. *Gran ducato di Toscana* 1846

Manfredus, Hieronymus. Astronomer, edited maps in 1477. Bologna edition of *Ptolemy's Geography* (with Bonus)

Manganari (1801–1887). Russian sailor and hydrographer. *Sea Azof & Marmora.* Yalta 1836

Mangin, Joseph. *Plan Shore of Long Island* 1813 MS.

Manier, J. *Departmental maps of France* 1867

Mankell, Julius. *Stockholm* 1874

Mann, A. T. *Charts of the North Sea* 1789

Mann, Gother (1747–1830). R.E. Inspector General of fortifications. Employed *survey N.E. coast England* 1781, *Plans for fortifying Canada, Fort St. John* 1790, *Lake Champlain* 1790, *Fort Ene* 1803

Mann, J. *Map of White Cloth Hall at Gomersall . . . the different manufactures circuit of 8 miles*

Mann, Robt. J. (1817–1886). English meteorologist

Mann, Thomas. *Survey Port Mahon* 1811

Mann, William (1817–1873). Astronomer, assistant Royal observatory Cape of Good Hope

Mannasser, David (17th century). Engraver together with his brother Daniel M. *Bird eye View Graz* (Austria)

Manneberg, Karl (1729–1860). Surveyor in Finland

Mannert, Conrad (1756–1834). Historian and geographer. *Great Britain* 1795, *America* 1796 and 1812, *Hungary* 1803, *World* 1806. Also engraved for Schneider & Weigel

Mannevillette. See Après de Mannevillette

Manning, H. Estate surveyor. *Quendon* 1645 MS

Manning, Joseph. *Plan of Norwich* 1830

Manning, J. (19th century). Globemaker

Manning, William C. *Newmarket Training Grounds* 1882

Mano, J. C. (1831–1886). Geologist and archaeologist in service govts. Colombia & Guatemala

Mansa, Jacob Henrick (1797–1885). Danish military cartographer. *Bornholm* 1851, *Kjobenhavns* 1857, *Denmark* 1875, *Atlas Denmark* 1878

Mansard, G. *Fürstenthum Waldeck* (1846)

Mansel, Jean. Miniaturist. *TO map* ca. 1455

Mansell, Lieut. Arthur Lucas (fl. 1834–66). *Admiralty Charts Greek Archipelago* 1849, *maps of Egypt & Mediterranean* 1854–64, *Black Sea* 1854, *Coast Egypt* 1864

Mansell, F. Engraver. *Rideau Canal* 1844

Mansell, Lieut. G. H. *Beaver Harbour, Admiralty Chart* 1851

Mansfeld, Johann Ernst (1739–1796). Engraver. *Tyrol,* Vienne 1774

Mansfeld, S. *Wien* 1800

Mansfield, John F. *Map State Ohio* 1806

Mansi, Gregorio. Geographer of Todi. *Territ. di Todi* 1612

Mansing, A. Dutch cartographer, 19th century. *Map of Alblasserwaard* 1840

Manson, Lieut. John. Worked on *survey Scotland* 1749–51

Manson, Peter. Land surveyor Caithness. Thomson's *Scotland* 1832

Mansson, Johan (d. 1658). Swedish hydrographer. *First printed sea chart of Gulf of Finland* (1644)

Mansveldt, J. I. van (1761–1802). Engraver and engineer. *Military surveys of Utrecht* (ca. 1790)

Manteau, Petri. *Thinehausen* pub. Visscher 1633 (with Culenburgh)

Mantell, Gideon A (1790–1852). English geologist. *Rape of Bramber (Sussex)* 1830, *Wonders of geology* 1838, *Geolog. Survey* 1840

Manuel, Hans Rudolph [Deutsch]. *Berne Helvetiae* 1549

Manuel, J. *Sources du Nil* Paris 1870

Manwaring, Sir H. *Sea Mans Dictionary* printed 1786

Manzano, Jesus P. *Plan Mexico City* 1867

Manzieri, Luigi. *Bassa Romagna* 1750

Manzini, Guiseppe. *Citta di Bergamo* 1816

Manzoni, Desiderio. *Carto Topo. Prov. di Como* (1860)

Manzoni, R. *Sanah* 1878, *Yemen* 1885

Manzotti, Luigi. *Reggio di Lombardia* 1817

Mao Kun (1511–1601). Chinese scholar. *On the necessity of defending coasts* 1562 (with maps)

Mao, Yuan-i (17th century). Military cartographer

Maplet, John. (d. 1592). *Dial of Destiny . . . situation of Countreys* 1582

Maqueda, Juan Diaz de. 1st Pilot. *Carta Esferica* 1793

Maquisset, J C. (ca. 1800). Engraver. *Views & Plans of Cassel*

Mar [John?]. Mariner of Dundee. *Chart entrance to Tay for G. Collins* 1693

Mar, William. Surveyor. *Map of Norwood* 1678 MS

Marak, J. E. (1832–1899). Engraver. *Maps of Austria*

Maraldi, Giovanni Domenico [Jean Dominique] (1709–1788). French scientist, astronomer. *Carte de France Levée par Messrs. Maraldo et Cassini de Thury* 1744

Maraldi, Jacques Philippe. Astronomer, worked with Cassini & La Hire on triangulation

Marchais [Mareechais], Chev. R. de. *Plan Cayenne,* Jefferys 1760

Marchand, Etienne (1755–1793). *Voy. autour du monde* 1790–2, pub. 1798–1800. *Isle de la Révolution* 1791 MSS

Marchand, J. C. (1680–1711). Mathematician and engraver of Dresden, *maps and illust.*

Marchant, J. Writing Master and surveyor. *Plan Brighthelmstowe* 1808 and later editions

Marchetti, D. *Polesine de Rovigo* 1786

Marchetti, Pietro Maria. Edited edition of *Ortelius* 1598

Marchbank, R. Printer of Dublin. *Drury's Surveys of Irish Harbours* 1789

Marche, A. *Luçon et Palawan* 1887, *Philippines* 1898

Marche, De la. See **Delamarche**

Marchesi, E. *Pianta Cagliari* (1832) (with Cominetti)

Marchisio, Domenico. *Carta Geogr. Postale d'Italia* 1878

Marcianus, Heraclensis (5th century). Geographer. *Description coasts Europe*

Mareinkiewiez, A. *Carte physique et routière de Belgique* 1840

Marcolini [Marcollino], Francesco (16th century). Venetian publisher

Marcou, Jules (1824–1898). French explorer and geologist. *Geological Map of U.S.* 1853, *Geological Map of World* 1861

Marcumost. *Gulf Finland* 1777

Marcy, Capt. R. B. 5th U.S. Inf., Ft. Smith. *Arkansas to Santa Fe* 1850, *Arkansas to New Mexico* 1853, *Bravos & Wichita Rivers* 1855

Mardelet. Engraver for Vivien 1827

Mare, Carl (1804–1863). Berlin engraver. Schmidt's *Australia* 1820 and 1828

Mare, Carl. Berlin publisher. *Ost Preussen* 1796–1809, *Danzig* 1813, *Lanzarotte* 1820, *Bosphorus* 1825

Mare, Nicholas. *Portolan Chart* 1487

Mareau. *Maps* for Freycinet 1826

Mareechais, R. de. See **Marchais**

Marelius, Nils (1707–1791). Swedish cartographer. *Various charts of Sweden* 1773–1779. English edition *Chart of Baltic* 1794

Marelli [Marrelli], Michelangelo (f. 1570– 80). Engraver from Ancona. *28 circular maps* for Ferretti, *Dialoghi* 1579

Marenti, Philip. *Geog. antiqua et nova* 1742 (Cellarius maps)

Marés. *Coblentz* 1796

Marescaux, Lieut. G. C. A., R.N. *Delta R. Niger* 1895

Mareschal, Jean Philip (1689–1770). Surveyor

Mareschal, Mathias. *Théàtr. Géogr. de France* 1626

Margarita Philosophica. See **Reisch**, G.

Margraff, Georg. German geographer. *Brasilia* 1643, 1664

Mariani, L. *Plan Callao* 1855

Marie, P. and A. Bernard. Publisher of Paris, rue des Grands Augustins No. 1. *Plan Paris* 1850

Marie-Davy, Fr (°1816–1893). French astronomer

Marieni, J. (1783–1767). Military cartographer

Marieschi, M. (1696–1748) Engraver. *Urbis Venetiarum prospectus* (1741)

Mariette, Pierre the elder (1603–1657).

Bookseller, printseller, and publisher. Rue S. Jacques à l'Espérance, Paris. *Palestine* 1646, Sanson's *Atlas* 1650, *Maps* for Briot

Mariette, Pierre the younger (1634– 1716). Engraver and publisher. *Madagascar* 1667, *Maps* for Sanson's *Atlas* 1669, 1690

Marigny, E. Taitbout de. Portolan maker. *Mer Noire* Odessa 1830

Marillier, Clemont Pierre (1740–1808). Designer and engraver. Titlepage for Bonne-Lattré's *Atlas Moderne* 1771. Also landscapes and illustrations

Marin, Franzisco Anthonie. *Messico* 1758 MS.

Marin. Engraver for De Lisle 1723–25

Marinari, Horatius. *Stato Bologna* (1630)

Marinari, Michelangelo. Engraver *Territ. di Roma* 1674

Marinari, Onorio (1627–1715). Engraver, astronomer, *18 star maps. Fab. dell Annulo Astron.* 1674

Marine. Dépôt des Cartes et Plans de la. See **Dépôt**

Marineo, Lucio (1460–1533). Sicilian writer. *Chronica de Espana* 1539

Marini, Hieronymus. *World* 1513, probably forgery

Marino, G. Publisher of Milan. *Levanto's Specchio del Mare* 1664

Marino, Girolamo. *Planisphere* 1511

Marinoni, Giovanni Giacopo di (1670– 1755). Mathematician, astronomer and cartographer. *Plan Vienna* in conjunction with L. Anguissola 1704 and 1706

Marinos [Marinus] of Tyre. Phoenician geographer and cartographer. First to introduce map projections (ca. 100 A.D.) criticised by Ptolemy

Marion-Dufresne, Mark Joseph (1726– 1772). Navigator, companion of Bougainville, later leader of expedition to New Zealand. *Charts New Zealand* 1772, pub. Crozet

Marion. See **Foster & Marion**

Mariotti, M. L. *Dépt. de l'Aisne* (1865)

Mariotti, Vincenzo. Engraver for Rossi's *Mercurio Geografico* 1688

Mariz Carneiro, A. See **Carneiro**

Markgraf, H. (1574–1581). Surveyor

Markham, Capt. A. H. *Melanesia* 1872 MS, *Nova Zemla* 5 sh. 1879, *Hudsons Bay* 1888

Markham, C. R. *Maps, parts of Brasil & Peru*, 1853–97

Markham [Rev. G.] and **Whittle**. Late edition of Jefferys *Yorks* 1800

Markwell, Joseph. Estate surveyor of Billericay. *Wickford* 1808 MSS, *Rawrel* 1824 (with Clayton)

Marlay, W. Lt.-Col. Military cartographer. *Invest. Cairo* 1803, *Egypt Postes Milit.* 1838, *Walcheren* 1810

Marlianus, J. B. (16th century). Italian antiquary. *Urbis Romae Topographia* 1534

Marlow, Michael. Publisher at King Edward Staires in Wapping. *Brest* 1680 (with Morden), title to *Claris Commercials* 1674

Marlow, William (1740–1813). *Views & Plans, Naples, London and Somerset*

Marmocchi, C. P. (d. 1858). Italian geographer

Marmol, L. de (16th century). Historian and geographer. *Descrip. Gén. de l'Afrique* 1573

Marmora, Gen. *Isola da Sardegna* Turin 1845

Marnef, Hieronymus de. Mapseller. *World map* 1534

Marny, N. B. Barbot de (1837–1877). Russian explorer and geologist

Marot the elder, Daniel (1662–1752). Publisher and map engraver

Marques, A. *Isles Samoa* 1889

Marquette, Jacques (1637–1675). *Carte de la découverte faite l'an 1673 dans l'Amérique Sept.*

Marquez, Feliciano. *Map plans Philippines* 1767–9 MSS

Marr, D. R. Publisher of Karlsruhe. *Baden* 1838

Marr [Mair], John of Dundee. *Chart E. coast Scotland* Doncker (1670) and Grierson (1740)

Marr, William. Collaborated with J. Leake in his *Plan of London* 1667

Marr, John. *MS Plans of Canadian Buildings* 1761–1772, *installations Nova Scotia & Quebec* 1761–72

Marra, John. *Southern Hemisphere* 1775

Marras, V. *Railway Map Italy* 1878

Marrat, William. *Map of Lincoln* 1817, corrected 1848

Marre, Jan de (1696–1763). Member of Dutch East India Co. and examiner of pilots. Maps for Keulen's *Sea Atlas of E. Indies* 1753

Marret (P. Veuve de) [widow] . Publisher *Paraguay* 1703

Marriot, John. Publisher. *Drayton's Poly Olbion* 1612

Marsden, William. *Map of Sumatra* 1811

Marsh, E. D. & H. G. Publishers in Philadelphia. *Alabama* 1875

Marsh, Thomas. Estate surveyor. *Plan road Paddington to Tottenham Ct. Rd.* 1771–1755, *Various plans* (with J. Crow) Fobbing & Cunningham 1781 MS, *Marylebone* 1789

Marshall, I. & Co. Publishers and mapsellers. Aldermary Churchyard. Succeeded Dicey and reissued some of his plates. *Plan of London* 1782, 1789. See also **Marshall**, R.

Marshall, J. Publisher at the Bible in Grace-Church-Street London. Seller's *History of England* 1703

Marshall, J. Publisher. *Plan Bath* 1835

Marshall, John. *Rutter for the Sea About Scotland* ca. 1545, *Times of full seas, low water . . . havens about the coasts of France, Flanders, Britain &c.* MS ca 1545

Marshall, John (1755–1835). American historian and atlas publisher. *Atlas to Life of Washington* 1807 and 1832, reduced French edition 1807

Marshall, Capt. John. *Chart Track of Scarborough* 1789

Marshall, Richard. Publisher and bookseller. Partner with Dicey in Aldermary Churchyard *Catalogue* 1764–5, *N. Britain* 1778, Smith's *Trav. Guide England & Wales* 1786

Marshall, Thomas. *Survey Hertford Castle* 1608 MS

Marshall, Thomas. Engraver for Pinkerton 1804, B. Edwards 1818

Marshall, William. Engraver for Fuller's *Palestine* 1640, *Gad* 1648, *Ruben* 1650

Marshall. See **Pratt & Marshall**

Marshall. See **Simpkin & Marshall**

Marsigli, Conte Luigi Ferdinando (1658–1730). Italian military cartographer and scientist. *Surveyed Hungary and the Danube 1696–9, 37 maps 1726* (with J. C. Müller) and 1741, *Golf du Lion* 1725, *Hist. Physique de la Mer* 1725

Marstaller, J. T. (second half 18th century). Captain in service of Polish king, of Dutch origin, engraver. *Reduced edition of great plan of Warsaw* (orig. by Tirregailbe) 1762

Marstaller, G. J. Engraver. *Map Poland* 1755

Martel, Peter (1701–1761). *Fribourg* 1740, *Plan Geneva* 1743, *Fort Louis* 1744, *Turin, Luxemburg, Hamburg* 1746, *Panoramas of Swiss Alps*

Martel de Gayangos, Antonio. *Mindanao* 1850 MS

Martelleur, François. *Plan Marseilles* (1605) MS

Martelli, L. Italian engraver. *Military maps* 1835

Martellus, Henricus [Germanus] (fl. 1480–96). German cartographer, worked in Florence, perhaps with

Roselli. MS of *Ptolemy's Geography* (with 13 modern maps), *Insularum* 3 MSS, *World Map* (1489)

Marten, Benjamin. *Nat. Hist. of England* (with county maps) 1761

Marten, James Forrest. *Map India* 1886

Martenet, Simon J., of Bath. *County maps of Maryland* 1858–65, *Atlas of Maryland* 1873, *City of Baltimore* 1874

Martenet & Bond. *Montgomery Co., Maryland* (1865)

Martens, F. French engraver. *Paris* 1832, *Cologne, Frankfurt*

Martens, J. Ḥ. Military engineer. *Map of Cassel* (1803)

Martens, W. Swiss engraver. *Plans & Views Lausanne* (1860)

Martier. French engraver, 19th century

Martin. Engraver for Wessel 1771, *Siaeland* 1768–71

Martin. *Pitcairn* 1773, *Geog. Mag.* 1782

Martin. French engraver. *Maps Lorraine* ca. 1760

Martin, Alexander. Surveyor of Cupar. *Corrected map of Fife* for Thomson 1832

Martin, Amédée. *Plan Paris* 1814, –15 and –17

Martin, Benjamin. *Trading Countries Europe,* Bowen 1758; *Solar System* 1757, *New principles of Geogr. and Navigation* 1758

Martin, C. Engraver for Schlieben 1828

Martin, Camille. *Panorama of Venice* ca. 1880

Martin, D. Engraver for Winterbotham. *Pennsylvania* 1796, *Engagement White Plains* 1797

Martin, George. Glasgow publisher *Johnston* 1842

Martin, Henry. 4 Swinton St., Gray's Inn Road, Kings Cross. Allen's *Plan*

of London 1830, Drew and
engraved *Warwickshire* for Atlas
Newspaper 1833, and Reynolds'
London 1847

Martin, J. *Milford Haven* (1780)

Martin, John. *Plans for Aqueduct,
Sewage & Railway* 1828–46

Martin, J. A. *War in East* 1877

Martin, J. B. (18th century). Engraver.
Surprise de Cremorne ca. 1750

Martin, Michael. Estate surveyor.
Roydon 1700 MS.

Martin, P. Publisher London, 198 Oxford
St. *Maps* for Wallis (1819)

Martin, Peter Francis. *River Demerary*
1781

Martin, Philip. Printer of Augsburg.
Stocklein's *Trav. Miss. Soc. Jesu* 1728

Martin, P. F. *St. Eustatia* 1781 MS

Martin, P. L. *Geolog. map part W. Sussex*
1828

Martin, R. Bookseller and publisher
London, 47 Great Queen Street, Lincolns
Inns Fields. Bowen's *English Atlas* 1794

Martin, R. Publisher London, 124 High
Holborn and 51 Carey St. *English Atlas*
(1794), for Dawson 1832

Martin, Reinhard Jacob. *MS maps on War
Independence in America* 1799–80

Martin, R. Montgomery. Hist. of Brit.
Colonies. Tallis' *Illustrated Atlas* 1851,
Index Gazeteer of World

Martin, S. D. Surveyor of Leeds. *Leeds &
Selby Railway* 1829, Thorps' *Leeds* 1831

Martin, Thomas. *Draught Religious
Houses Thetford* MSS, pub. 1779

Martin, William. Geographer. *Palestine*
(1842)

Martin de Bohemia. See Behaim

Martin de Lopez, Pedro (fl. 1840). Spanish
globemaker. *Globos Geograficos de
Bolsillo* Madrid 1840

Martin & Fox. *Railway surveys* 1846 MSS

Martineau du Plessis, D. *Nouvelle Géogr.*
3 vols., 1730

Martinelli, A. Italian engraver. *Bridges
over Nera & Tevere* 1676

Martineau, Harriet. *Tourist Atlas of the
Lake District* (1875)

Martineau, M. A. *Nouvel Atlas Illustré,*
1891

Martines [Martinez] Joan [Giovanni]
(1556–1590). Miniaturist & chartmaker
of Messina. *World* 1562, *MS Atlas* 1567,
Atlas 1570 (18 maps), *Atlas Atlantic*
1572 MS, *Black Sea* 1579. In all 8 charts
and 18 atlases survive

Martinet, A. Designer and engraver.
Carte de Picardie 1766

Martinet. Bookseller, Rue du Cocq St.
Honoré. *Plan Vienna* 1805

Martinet, G. *Piedmont* 1799

Martinet, Simon J. Publisher of Baltimore.
Scott's *City of Baltimore* 1858

Martinez, Fernando. *Florida* 1765 MS

Martinez, Juan (16th century). Spanish
cartographer. *Portolan atlas, 5 charts*
1582, *Atlas Messina* 1587

Martinez, J. Garcia. See **Garcia Martinez**

Martinez de Castro, Marino. *Mapa Official
del Estado do Sinalon* Mexico 1891

Martinez de la Torre, Fausto. *Plan Madrid*
1760

Martinez de Zuniga, Ioaquin. *Islas
Philippines* 1814

Martinez Fortun, Jacques. Pilot. *Islas
Barbados & Ladrones* 1565

Martinez, Martin. Translator. *Breydenbach*
1498

Martinez Vigil, Ramon. *Geog. Isles
Filipines* 1895

Martini, Aegidius [Gilles]. Cartographer,
lawyer and mathematician. *Limburg*
1603 for Mercator-Hondius, also used by
Keere 1616, 1617, 1622

Martini, E. *Railways Argentina* 1891

Martini, Fr. Martino (1614—1661). Italian Jesuit, Superior of Hangchow. Maps for Blaeu's *Atlas Sinensis* 1655

Martini, J. G. (b. 1785). *Thuringian Towns* 1815

Martini, Martin (1566—1610). Swiss engraver. *Map of Lucerne* 1596, *Fribourg* 1606

Martinière, A. A. B. de la. See **Bruzen de la Martinière**

Martinon, Giovanni. *Sicilia* 1812

Martinot, Henri (1646—1725). Parisian clock and globe maker

Martins, A. A. *Plan Lisboa* 1887

Martins, Ch. Fred. (1806—1889). Explorer and geologist. *Voy. Laponie* 1838—9, *Asia Minor* 1856

Martire, P. See **Martyr**

Martius, Carl Friedrich. Ph. von. *Schwerin* 1819, *Amérique Mérid.* 1825, *Atlas zur Reise in Brasilien* Spix & Martius, Munich 1834, *Amazonen Strome* 1832, 1837

Martszen, Jan (1609—1647). Painter and engraver. *Plan Gennep* 1641

Martyn, Benjamin. *The South East (America)* 1733

Martyn, Thomas. Surveyor. *Map Cornwall* 9 sh. 1748, *Index to Martyns Map* 1816

Martyn, W. F. *Geographical Mag.* 1782

Martyr [Martire] Peter d'Angieri [Anglerius] (1459—1526). Italian geographer of Arona. *World map* in his *Decades* Seville 1511. First printed Spanish map of America and first to show Bermuda. *De rebus oceanicis et de orbe nova* Paris 1536, *Les Isles nouvellement trouvées* 1532

Marvin & Hitchcock. Publishers of San Francisco, for Scholfield 1851

Marx, A. R. (b. 1815). *Danube—Main Canal* 1852

Marx, D. R. Publisher. *Herzogt. Nassau* 1852

Marx, H. G. Author of *maps of Polish-Prussian boundaries* (end 18th century)

Marx, J. Freiherr von Lichtenstern. *Venedig* 1805, *Wien* 1809

Marx, J. *Comm. Papua, Admiralty* 1852

Marx, J. A. (end 18th century). Military topographer, author of *hydrographic maps and plans*

Marzo, Isidoro de Antillon y. See **Antillon**

Marzolla, Benedetto. Geographer of Naples. *Sicile* 1836, *Atlante Chorografico* 1837, *Atlante Geografico* 1844—58, *Atlas Naples* 1848—54

Mas, Sinibaldo. *Islas Filipines* 1842

Masatsuma, Kuchiki. Japanese geographer. *Geography of European countries* pub. 1789 (with maps and town plans)

Mascall, Lieut. Joseph. *Crotchey Bay* 1744 pub. Dalrymple, *Eagles Track* 1774—84, *Plan Harbour & Road Suez* 1777—82

Mascardi, Vitale. Printer of Rome. *Territ. di Roma* 1674

Mascaro, Manuel Agustin. *Nueva Espana* 1777, *N. America* 1782 MS

Maschek, Alex. *Nord America* 1849

Maschek, Rudolf (1843—1887). Viennese cartographer and engraver. Worked for Artaria & Scheda & Steinhauser. *Salzburg* 1881

Maschop [Maskop, Mascopius], Gottfried (fl. 1560—77). Geographer and surveyor. *Munsterland* 9 sh. *Osnaburg* 9 sh. 1568. *Munster* used by de Jode 1578

Masi e Compagni, G. T. Publisher. *Atlante dell' America* Livorne 1777

Masius, Arnoldus. *Plan of Namur* 1575 for Braun & Hogenberg

Maskelyne, Nevil (1732—1811). Astronomer Royal 1765. *Established Nautical Almanac* 1766

Mas Latrie, M. L. de. See **Latrie**

Mason, Capt. *Map of Gen. Worth's Operations Texas* 1850

Mason, Col. A. M. *Maps of Egypt & E. Africa* 1875—80

Mason, Charles. Worked with Jeremiah Dixon on measuring boundary between Maryland and Pennsylvania 1768

Mason, Charlotte M. *London Geographical Series* Stanford 1881

Mason, D. *Chart approaches to Liverpool* 1846

Mason, I. Engraver for Anson's *Voyage. St. Julians River* 1748

Mason, P. Steel's *chart Cape Verde Is.* 1809

Mason, John (1586–1635) of Kings Lynn Norfolk, sailor, later Governor of Newfoundland. The first separate *map of the island of New Foundland.* Reissued in Vaughan's *Golden Fleece* 1626

Mason, L. D. V. *Map of Town of Morisania N.Y.* 1871

Mason, Michael John. Estate surveyor, Mountnessing. *W. Hanningfield* 1811, 1842 MSS

Mason, P. *Plan of Belfast* 1820

Mason, William G. American engraver, active 1829–45 in Philadelphia. Designed cartouche for Mitchell's *U.S.* 1835

Mason & Payne. Publishers. 41 Cornhill London, Proprietors, Letts' *Popular County Atlas* 1887

Masquelier, Louis Joseph (1741–1811). Engraver

Massa [Massart, Massaert], Isaac (1586–1643). Dutch cartographer and traveller to Moscow. *Plans of Moscow* 1610, 1618; *N. Russia* 1612 and *South Russia,* used by Blaeu & Jansson

Massaio [Massajo], Pietro da. Florentine painter and miniaturist, composed extra maps for a *Ptolemy* codex MS, ca. 1458

Massaloup, I. V. *Fürstenthum Rumänien* 1876

Massart, I. See Massa

Masse, Claude (1650–1737). French topographer and engineer. *Fortifications plans France & Spain* MS

Masser, Joseph F. Lithographer, 25 Boar Lane, Leeds. *Map of the Colony of Port Natal* 1862

Massey, C. *Bay Coche Guzzerat Portolan* (1670)

Masslowsky, P. (b. 1783) Russian engraver, illustrated for military textbooks.

Masson, Alexandre Frédéric Jacques, Marquis de Pezay. *Cartes géographiques (Campaign in Egypt)* fol 1775

Masson, Ch. P. Ph. (1761–1807). *Cours de Géog. &c.*

Massoudy. See Mas'udi

Massuet. Engraver for Brué 1833

Mast, Crowell & Kilpatrick. American publishers. *Atlas of World* 1892

Masters, C. Harcoart and Day, W. Surveyors. *Somerset* 1782 & 1800, *Ilchester to Langport* ca. 1794

Mas'udi, al (885–957), Ali Aboul Hassan. Geographer, traveller, historian. *Meadows of Gold, a description of the lands of Islam,* A.D. 912–30

Mat, P. J. de. Publisher. *Brussels* 1827

Matal[lus], Jean. See Metellus

Matham, Adriaen (1599–1660). Amsterdam cartographer, engraver and publisher. *Maps and plates* for *Begin ende Voortgang* (1646)

Matham, Jacob (1571–1631). Engraver, of Haarlem. Plates for S Ampzing, *Desc. of Haarlem* (1621)

Matham, Theodor [Dirk] (1606–1676). Engraver, of Amsterdam. *Siege of Heusden* (1625)

Mathé, J. M. Ensign in Spanish Maine. *Port de Castro Urdiales* 1849

Mather, Benjamin. *Plan Town of Lowell, Mass.* 1832

Mather, Cotton. *Magnolia Christi Am.* 1702 *(map of New England and New York)*

Mather, Henry T. Surveyor of Surbition. *Slough House Farm* 1900

Mather, John. Surveyor. *Town of Wigan* 1827

Mathes. See **Mathysz**

Mathew, Felton. *Plan city Wellington, N.Z.* 1841, *Plan Auckland* 1842, *Bay of Islands* 1842

Mathew, Francis. Drew *River Thames,* Engraved by Jenner 1667

Mathew, Paris (ca. 1195–1259). Benedictine writer and cartographer, historian monk of St. Albans, succeeded Roger of Wendover as chronicler to the Abbey. *4 maps England* MSS. *Chronicon Angli Hist. Major* London, R. Wolf, 1571

Mathew, William. *Town Plan Newcastle,* used by Speed

Mathewes, Augustine. Printer. *Drayton Polyolbion Part II* 1622

Mathews, Francis. *Town lands Templeoge (Co. Dublin)* 1783 MS

Mathews, J. R.N. *Twenty-one plans actions West Indies* 1784

Mathews, John. Publisher. *Book V English Pilot* 1701–2, with Jermiah Seller & J. Price

Mathews, Maurice. Surveyor Gen. of Carolina 1677–1683. *Plan of the Province of Carolina* 1680

Mathews, Thomas. *Map Flints Croft* 1773, *Plan Bristol* 1794

Mathews & Leigh. Publishers, 18, the Strand. *North America* 1808, *Atlas* London 1812 *(20 col. maps of Tribes)*

Mathieu. Engraver. Headpiece for *Après de Mannevillette* 1775

Mathieu, Ae. Lieut. *Charts of Majorca* 1823–4–5

Mathieu, A. Engraver. "Les Armes Triomphantes du Duc d'Espernon" (1656)

Mathieu, C. *Reunion* 1884, *Senegal* 1884

Mathijsz. See **Mathysz.**

Mathison, R. *District of Midnapoor* 1849

Mathonière, Alin de. Publisher of Paris. Fines' *France* 1557, Jolivet's *France* 1565

Mathýs, Jan (fl. 1657–85) of Amsterdam. *Plans and views; Gouda, Utrecht*

Mathysz [Mathiez, Mathes] Jan. Engraver (Plaatsnyder) of Amsterdam. *4 large wall maps of the 4 continents* ca. 1650. Reissued Plancius's *World*

Matius. See **Metio**

Matkovic, Peter (1830–1898). Serbian geographer

Maton, William George (1774–1835). *Mineralogical map W. Counties of England* 1797

Matos, Don Antonio de. Hydrographer. *Portole in Gulf of Mexico* 1740 MS, *Costas de Tierra Firma* 1749

Matosaburo, Matsudaira (18th century). Japanese surveyor. *Chart of Ryuku Archipelago* 3 sh. 1756

Matriti, Pinilla. *Obispado de Caxtaxena (Murcia)* 2 sh. 1724 (with P. Vidal)

Matson, Henry I. *Lobito Bay, W. Africa* 1844

Mattei, Innocenzo (1626–1679). Papal geographer, b. Rome, d. Faenza. *Antico Latio* (1650), *Campagna di Roma* 1666, *Territo Distretto di Roma* 1674

Mattern, J. *Plan Breslau* (1867)

Matthaeus. *Westmonasteriensis Flores Historiarum* London, Jugge 1567

Matthews, John. Sailor. *21 plans of actions in W. Indies,* Chester 1784, *Voy. to Sierra Leone* 1788

Matthews, Mary. Publisher. Morden's *Britannia* 1772

Matthews, Northrap Co. Publishers of Buffalo, N.Y. *Travel Atlas of U.S.* 1893, *Phillipine Is.* 1898

Matthias. *France* 1815 (with Schmidt and Klöden)

Matthias, Wilhelm Heinrich. *Süd Preussen* 1803, *Frankreich* (1815), *Travel Map Europe* 1830

Mattioli, Pietro Andrea (1500–1577). Physician of Siena. Translated *Ptolemy* into Italian for edition of 1548

Matusowski and Paulinov. *Trav. W. Mongolia* 1873

Matuta, Temple. *Map of Sicily made for it* 174 B.C.

Maty, C. *Dict geogr. Univ.* Amstd. 1701

Matz, Major O. H. M. *Military Maps, Field of Shiloh* (1862), *Siege of Vicksburg* (1863)

Matzinger, L. *Map State Iowa* 1850

Mauborgne. *France* 2 sh. 1818

Mauch, C. *Berlin* 1833, *S. Africa* 1872, *Transvaal* 1877, *Natal & Orange Free State* 1871, *Mozambique* 1874

Maucroix, d'E de. See **Estrémont de Maucroix**

Maudsley, A. P. *Traverse Brit. Honduras* 1886 MS, *Ruins Copan* 1886

Maud, W. S. *District round Shanghai* 1864

Maugein, Charles. *Maps for Philippe de Prétôt* 1768

Maughan, Philip. *Maps of China* 1808– 40

Maule, William. Surveyor general. *Is. Roan-Oak,* N. Carolina 1718

Maund, E. A. *Map of Matabililand and Mashonaland* 1891

Maunder, S. Publisher, 10 Newgate St. London. *Atlas of English Counties* 1828

Maunder. See **Pinnock & Maunder**

Maunsell, F. R. *Fort of Attock* 1849–52, *Kurdistan* 1893

Maupertuis, P. L. Moreau de (1698–1759). Cosmographer, born St. Malo. *Figure de la Terre* 1738, *surveyed Meridian arc in Lapland*

Maupin, J. *Railway Maps France* 1885–99

Maupin, Simon (17th century). Cartographer and architect. *Ville de Lyon* 1659

Maurelle, Francisco Antonio. *Costas y mares septent. de las Californias* 1766

Maurer. Engraver. *Plan Geneva* 1743

Maurer, Christoph, Hans and Joost. See **Murer**

Maurer, H. *Cantons de Mulhouse* 1862

Maurer, Heinrich. *Goldau* 1806, *Zürich* (1830)

Maurer, Jonas. *Plans town of Zürich* 1566 and 1576

Maurer von Maurersthal, Joseph Freiherr (1787–1857). Viennese general and cartographer

Maurice, James. *S. shore Lake Erie* 1830 (with T. Forster)

Mauritius. See **Myritius**

Mauro, Fra. (d. 1549). Cosmographer. *World map* Murano, Venice 1457–9. Diameter 6 ft. Now in Bibl. Marciana, Venice

Maurolico [Maurolyan], Francesco (1494–1575). Benedictine mathematician of Sicilian origin. *Cosmographica* Venice 1543

Maury, Guillaume. *Atlas Géometrique et Topo., Puy de Dôme* 1844

Maury, Matt. Fontaine (1806–1873). American hydrographer U.S.N. *Track charts & wind charts North Atlantic* 15 sh., *N. Pacific* 10 sh.

Maury, W. L. Lieut., R.N. *Charts of Japan* 1859–62

Maverick, Peter (1780–1831). American publisher, engraver, lithographer. *Gen. Atlas New York* 1816 (with Durand), Lay's *New York* 1812, *Hudson & Mohawk Rivers* 1810

Maverick, Peter Rushton (1755–1811) Engraver, 65 Liberty Street, N.Y. *McComb Purchase* 1791, *N.E. part Town of Mexico* 1796

Maverick, Samuel. Engraver. *Map Present Day (U.S.)* 1826

Mawman, Joseph. Bookseller of York, later publisher London. Succeeded C. Dilly between 1799 and 1801. Joint publisher with Guthrie, G. G. & J., *Robinson's System of Geography* 1808, Walker's *Univ. Atlas* 1820, *Travels of Anacharsis* 1805, *Map of S.W. and Middle U.S.* 1805

Maxfield, Lieut., R.N. Hydrographer. *Red Sea* 1804–6

Maxfield, Capt. W. *Survey False Point Palmiras* 3 sh. 1818, *Chart entrance to Hooghy River* 1820, *Survey Is. Thwart the Way* 1820, *Track of Nearchus* 1822

Maximilian. Transilvanus. (d. 1538). Teacher and secretary to Maximilian I, *Globe*

Maximilian, Prince of Wied-Neuwied. *Reise nach Brasilien* 1815–17, pub. 1820–1 etc.

Maximinus à Guchen, Father (d. 1655) *Atlas of Capucin order* 1649

Maximowitch. *Plan St. Petersburg* 1821

Maxon, John H. *Delaware City, Nebraska* 1856, *Fort Kearney* (1860)

Maxwell, George. *Survey of River Congo Africa Pilot*, Laurie & Whittle 1797, *English Road Isle of Ascension* 1797, *River Zaire* 1817

Maxwell, John. Publisher. Partner with Senex 1708–11, in Salisbury Court near Fleet Street. Geographer to Queen Anne, also worked with Chas. Price. *Atlas* 1708–12, *World* 1711, *Ireland* 1712

Maxwell, W. Lieut. *Maps of Districts in India* 1846–62

Maxwell, Staff. Comm. W. F. *Admiralty surveys coast of Japan* 1869–72, *Labrador* 1875–79

May, C. *Umgegend von Halle* 1880

May, Jan. Cornelisz. Capt. under Schouten and Le Maire. *Molucques* in Spilbergen's *Journal* 1619

May, D. J. Master, R.N. *Charts Africa, Nupe & Yoruba* 1858 MSS, *River Rovuma Admiralty Chart*

May, H. von. *Umgebung von Aachen* (1890)

May, Peter. Surveyor of Aberdan. *River Findhorn* 1758, *River Spey* 1760 (1761)

May, Patrick. Publisher in Bristol, Massachusetts. *Sacketts Harbour* (1815)

Mayer, Alexander. *Maps of Austrian Kingdom* 1860–76, *W. Galicia* Vienna 1808

Mayer, Andreas. *Pomerania & Rügen* (1763)

Mayer, Christian S. S. (1719–1783). Astronomer of Mannheim. Pupil of Cassini. *Basle & Mountains* (1762) MS, *Chart Palatinatus . . . Bavariae* (1750) 2 sh. (1780), *Schwetzingen and Umgebung* 1773

Mayer, Ernst. *World Maps* 1879–1893

Mayer, Ferd. & Co. Engravers and lithographers of New York, 96 Fulton St. For Preston 1856. *Railway map Calif.* 1868, *U.S.* 1870

Mayer, Joan. *Rhetiae Alpestris*, Ortelius 1573

Mayer, Dr. J. *Chemins de Fer de France* 1866

Mayer, J. R. American surveyor. *Keweenaw Point, Lake Superior* 1865

Mayer, Johann Tobias, elder (1723–1762). Mathematician and astronomer. Prof. of Math. at Göttingen. *Gegend um Esslingen* 1743, *Switzerland* 1750. Collaborated with Homann Erben 1746–51, *G.B.* 1749 and for Kilian 1752. *Silesia* 1759

Mayer, Johann Tobias, younger (1752–1830). Mathematician and cartographer. *Poland* 1778, *Indes Orientales* ca. 1780

Mayer, M. Johann. *Ducatus Württemberg* 1710

Mayer & Korff. Lithographers, New York, 7 Spruce St., for Creutzbauer 1849

Mayer, George. *Maps of Tyrol* 1838, *of Bavaria* 1841–53, *Salzburg* 1841, *Palestine* 1842

Mayerne-Turquet, Louis de. *Nouv. manière de représenter le Globe Terrestre* 1648

Mayerne-Turquet, Théodore de. *Somm. Descrip. France . . . Guide des Chemins.* Geneva 1561

Mayer von Heldensfeld, Anton. *West Gallizien* 1808

Maynard, J. *Island of Portsea* 1859, 1860

Mayers, W. S. F. *Sketch Map Kwantung* Canton 1873

Mayne, A. de. *West Indies & Gulf of Mexico Admiralty* 1824

Mayne, Capt. Richard Ch. (1835–1892). English sailor, hydrographer. Four years in B.C. & Vancouver 1862. *Coquinbo Bay* 1868

Mayne, Capt. Wm. *Township mouth of Detroit River* 1796

Maynier, Ludovicus de. *I'Isle St. Honoré de Lerin* 1635

Mayo, James. *Plan Delagoa Bay* 1787–9

Mayo, Robert (1784–1864). American cartographer. *Atlas Philadelphia* 1813

Mayo, Major William (b. 1684). Land surveyor. *Barbados* 1722, 1756 and 1794, *Virginia–Carolina line* 1728 MS. *Estate N. Carolina* 1733 MS.

Mayol. Bookseller, Calle de Fernando 7, Barcelona. *Alabern's Amer. Merid.* 1845

Mayr, C. Swiss cartographer. *Karte der Alpen. Atlas der Alpenländer* 1871–2

Mayr, Johann George (1800–1864). Cartographer and topographer. *Atlas der Alpenlaender* 1858–65. *Deutschland & O. Frankreich* (1870)

Mayr, Josef. *Prag* 1869, *Bohemia* 1871

Mazo, German. *Plano de Malaga* 1863, *Valencia* 1863

Mazza [Maza], Giovanni Battista. Venetian engraver and publisher. Rasciotti's *America* (1590), *Padua* 1590, Rosaccio's *World* 1597

Mazzini, Carlo Massimiliano. *La Toscana Agric.* 1882

Mea, Paul. *France, Stations Balneares* (1879)

Mead, Bradock (18th century). Geographer and cartographer. *Globes,* used pseud. John Green. *Chart N. & S. Amer.* 1755, *Seat War N.E.* 1776, *most inhibited part of N.E.* 1755

Mead. See Dodd, Mead & Co.

Meakin, H. W. *Villiers Co. Victoria (Aust.)* 1867

Mears, John (1746–1801). *Voyages* 1790, *Chart Northern Pacific Ocean* 1790

Mears, W. D. Statist. *Chart New Zealand* 1889

M[ease], E. *Draught River Mississipi,* 3 sh. (1860)

Mechain, P. F. A. (1744–1805). Astronomer, hydrographer to Marine Dépôt. *Charts coast of France* 1776, pub. 1797. Worked with Cassini and Delambre in establishing precise length of a meter

Mechel, Christian von (1737–1817). Publisher, engraver and art dealer of Basle. *Canton Basle* 1766, Ryhiner's *plan of Basel* 1786, *Carte gén. Suisse* 1799, *Hauteur princip. du monde* 1806

Mechlin, J. R. P. *Tehuantepec Survey* 1851, *Basins of Mississipi* 1861

Mechliniensis, F. M. See Franciscus Monarchus Mechliniensis

Mechow, Major Alex. von. *Karte der Kuango Expedition (Angola)* Berlin 1884

Mecutchen, S. *Pocket Atlas* (1877) (with H.D. Hanover)

Medary, Samuel A. Engraver. *Pacific Wagon Roads* 1858, *Fort Ridgely & S. Pass Road* 1858

Medau, C. W. Publisher. *Reise Karte Sächsische Schweiz* [1875]

Medan, C. W. Czech publisher. *Teplitz von Kreybich* 1834

Medcalfe. *Map Burgoyne's campaign* Faden 1780

Medebach, C. See Vopel

Medici Atlas. Dated 1351 on calendar but possibly later. *World maps showing S. Africa*

Medina, Fray Baltasar de. *Prov. de San Diego* (16th century), *Plano Geografico de Mexico* 1618

Medina, Cristiano. *Guatemala* Paris 1890 (with F. Bianconi)

Medina, Pedro de. Cosmographer and mathematician. Examiner of Pilots for the Indies 1493–ca. 1567. *Arte del Navegar* Valladolid 1545 (with *map of New World*) and later editions. *Regimento de Navegacion* 1552, *Globe* Lyon 1569,

Medina y Cabrera, Nicolas Joseph *Sonora* 1768

Medland, T. Engraver for Stockdale. *Port Jackson* 1789

Medlicott, H. M. *Geolog. K. von Vorder Indien* 1879

Medrano, S. F. de. See **Fernandez de Medrano**

Medtman, M. van. *Map* for Husson (1709), *Louvain district* 1706

Mée, L. du. Engineer. *Draught of Vigo* 1702

Meecham, Comdr. *Sketch Volume Sandwich Is.* 1858 MS.

Meek, James Mackain. *Hist. & Descriptive Atlas of British Colonies* 1861

Meek, Michael. *Roads & Villages within 5 miles of Northallerton* (1810)

Meerans, J. Publisher of Amsterdam. Goliat's *Plan of Middelburg* (1680), *Zealand* (1700)

Mees, G. *Hist. Atlas Noord-Nederland* 1851–61

Megasthenes. Greek historian. Described *Punjaub* ca. B.C. 300

Meggen, Jodocus à *Peregrinatio Hierosolymitana* 1580

Meheux, James N. *Prospect of Colchester* 1697

Meighen, J. G. *Carte Topo. Mairie de Stolberg* 1811

Meigs, M. C. Assisted Robert & Lee's *Map Harbour St. Louis* 1837, *Potomac River* 1858

Meijer, A. *Map* for Allard (1698)

Meijer, B. *Abriss. Territ. Ulm* (1720)

Meijer, Peter. Publisher of Amsterdam. *Kleine Atlas,* 1768 (maps, copied from Eman. Bowen)

Meikle, See **Stewart, Meikle & Murdoch**

Meikleham, Edward. *10 miles round Glasgow* (1858)

Meine, Davidt de. Publisher of Amsterdam. Geilkercken's *World* 1610

Meinecke, K. Ed. (1803–1876). Geographer, b. Brandenburg

Meinhold, C. C. [and **Meinhold und Sohne**]. Publishers. *Plans of Dresden & District* 1850–96

Meirtonitz, Major von. *Oesterreich Italien z. Etsch u. Piave* 4 sh. 1804 MS

Meisel, Bros. Lithographers of Boston. For Tappen 1861–66

Meisner, Daniel (1585–1625). Bohemian artist and topographer. *Thesauras Philo. Politicus . . . Schatzkästlein* (52 plates of cities) 1623, 1624–6 (8 parts, 416 views). Title changed to *Sciographia* 1642 and plates increased to 800. Reissued Nuremberg 1678

Meissas, Achille Pr. de (1799–1874). Geographer, worked with A. Michelot as bookseller, Paris, Rue du Condé 14. *Océanie* 1840, *America* 1865, *N. América* 1859

Meiss, Gaston. *Europe, France, Palestine &c.* (1875–6)

Meissel, K. See **Celtes Protucius**

Meissner, G. B. *Finland* 1808

Meister, C. See **Stimmer**

Meister, G. *Odenwald* 1808, *Choro. K. von Hessen* 1816

Meister, P. *Schwabenkriegen* 1499, 1505

Mejer [Meyer, Meiro], Johannes (1606–1674) of Husum. Royal Mathematician, surveyor and cartographer. *Denmark* 1650, *40 Maps* in Danckwerth's *Atlas of Schleswig Holstein* 1652, used by Blaeu 1662

Mel, J. de. Publisher, Bruges. *France* 1647

Mela, Pomponius (1st century A.D.). Roman geographer. *Cosmographia* first printed edition 1471 many later editions. *De Orbis situ* 1540 (world by Oronce Finé)

Melander, S. (d. 1706). Swedish surveyor

Melchert, F. L. *Carto. topo. de la Pampa* 1875–6

Melchinger, J. W. *Geogr. Stat. Topo. Lexicon von Baiern* 4 vols. 1796–1802

Melchiori. *Carta topo. Dalmazia* 2 sh. 1787

Melchor, Alfaro de. *Santa Cruz Yucatan* 1579

Meldemann, Nikolaus. *Siege Vienna* 1529, *Budapest* 5 sh. 1541

Meleney, C. E. *Primary geog. state N.J.* 1884 (with W.M. Griffin)

Melhuish, Thomas. Axford surveyor. *Parish of Wood Ditton* 1823

Melin. *Atlas Melin. Hist. et Geogr.* Paris 1895 and later

Melish, John (1771–1822). Scots traveller, cartographer and publisher. In Philadelphia from 1811, succeeded A. T. Goodrich. *Geog. Descrip. U.S.* 1816, 1818; *Mil. & Topo. Atlas U.S.* 1813, 1815, *N. U.S.A.* 1816

Melleville, Max. (1807–1872). French geologist

Mellin, G. *Atlas World* 1894

Mellin, Ludwig August, Graf. *Atlas von Liefland,* Riga 1798, *Atlas of Livonia* 17 sh. 1793–1804

Mellinger, Johann (1540–1603). Theologian, physician and cartographer of Halle. *Thuringia* 1568, *Mansfeld* 1571, *Limburg* 1593, used later by Hondius & Blaeu. *Plans of Halle & Hildesheim* for Braun & Hogenberg 1598

Mello, Barao Homen de. *Atlas do Imperio do Brasil* 1882

Mellor, T. K. *Handy map of the Moon* (1886)

Mellvill, B. D. *Plan Johannesburg* 1897, *Pretoria* 4 sh. 1899

Melvill, P. *Mozambique* (1816)

Melvill van Cambee, Pierre, Baron. *Atlas Nederlandsch Indië* 1853, *Kaart van Java* 1855

Melvin, A. Contributor Blaeu's *Atlas Scotland* 1654

Memhard, Johan Gregor. *Oranienburg* 1652

Memije, Vicente de. *Aspecto Geogr. del Mundo Hispanico* 1761

Menard. See Barbier de Menard

Menard & Deseune [Desennes]. Publishers. Rue Git le Coeur No. 8 Paris, Rue Haute Feuille No. 10, Place Sorbonne No. 3 Viviens' *Atlas Universel* 1823, 1828, 1843

Menassier, C. *Côte d'Or* (1868)

Menchero, Juan Miguel. *New Mexico* ca. 1745 MS

Mendana de Neyra, Al (1541–1595). *Voys. Indian Ocean* 1568–9. Credited discovery Solomon Is.

Mende, Gen. Maj. *Topo. boundary map Tver. Govt.* 1848–9 (147 sh.) Moscow, pub. 1853

Mende, T. *Nordfriesischen Inseln* 1888

Mendel, Ed. Lithographer of Chicago, *Maps of Lakes* 1856–7, *Iowa Railroad* 1858, *Mineral regions Lake Superior* 1864

Mendelsohn, Herman. Publisher of Leipzig. For Mollhausen 1858

Mendenhall, Edward. Publisher of Cincinnati, Ohio. *Kansas* 1859, *Trav. Map of W. States* 1866, *Shipping Atlas of W. States* 1871, *Map Cincinnati* 1874

Mendes, Alfonso, *Carte d'Ethiopie* 1672

Mendes de Almeida, Candido. *Atlas Imp. do Brazil* 1868

Mendez, Jose. *Luzon* 1883

Mendez de Vigo, A. de la C. y. See Cavada y Mendez de Vigo

Mendenhall, E. Publisher of Cincinnati. *Cinn.* 1865, *Kansas* 1857, *Ohio* 1866

Mendoza, Diego Hurtado de (1503–1575). *Spain* for Gastaldo 1544, *Peru* for Ortelius 1575

Mendoza, Felis. *Chart part coast China.* (1780)

Mendoza [Mendosa], Juan Gonzalez

(1550—1620). Castilian chronicler. *De regno China historia* 1585

Mendoza y Rios (ca. 1763—1816). Sailor and astronomer. Treatise on navigation, tables etc.

Mendozza, Antonio di. *Lettera del discoprimento della Terra Firma* 1539, Ramusio *Viaggi* 1556

Mengaud de la Hage, M. Lieut., French Navy. Surveyed *Coast Madagascar* 1775—6 *(Neptune Orientale)*

Mengden, George von (1628—1702). Livonian cartographer. *S. Russia,* printed Holland 1699 by order Peter the Great

Mengel, C. G. *Faroe Island* 1789

Menges, J. *Maps of Ethiopia, Somaliland, and Sudan* 1884—8

Menges, R. *Hannover* (1875)

Mengold, G. W. *Kanton Graubünden* 1864

Mengs, D. Rafael. Capt., Engineers. Designed title for *Torfino de San Miguel* 1788

Menke, Th. (1819—1892). Historian and geographer. *Historical & Classical Atlases* 1865—80, *Bibel Atlas* 1868
Menlos, Peder Mansson (17th century). Swedish surveyor

Menteath, James Stuart. *Geolog. Map of Dumfrieshire* 1844

Mentelle, Edme (1730—1815). Geographer and historian, member Royal Inst. of Sciences. Prof. Central Schools Dept. of Seine, Cour du Louvre No. 7. *Dannemark* 1782, *Atlas Pruss.* 1788, *Atlas National* 1790—1811 (with P. G. Chanlaire), *Atlas Elément.* 1798, *Atlas Univ.* 1797—1800 (with Grégoire) 1808, *Atlas des tableaux et cartes* 1804

E. Mentelle

Mentelle, Fr. Simon (1731—1799). Brother of Edme. Collaborated with Cassini on *map of France*

Mentelle, Mad^me Ve. Publisher, Rue des Petits Augustins No. 18; widow Simon M. Published Mentelle and Chanlaire's *World* Paris 1819, *Mappemonde in Atlas d'Etude* pub. Delaval 1820

Mentz, C. F. (1765—1832). Swiss lawyer, topographer. *Oldenburg* 1804

Mentzer, Th. A. von. *Atlas till Sveriges Hist.* 1856, *Atlas öfver Sveriges Län* 1869, *Svensk Historisk Atlas* 1871

Menzel. Lithographer of Cincinnati, end 19th century

Menzies, John (b. 1780). Engraver of Edinburgh. *Scotland* (for Kincaid's Grammar) 1797, *Scotland* (for Fairburn's Guide) 1798. In 1819 became firm of J. & G. Menzies

Menzies, J. & G. Engravers of Edinburgh, 201 High Street. Engraved Forrest's *Lanarkshire* 1816 8 sh. Thomson's *Atlas* 1816

Meon. Designed headpiece for *Après de Mannevillette* 1775

Mercator, Arnold (1537—1587). Surveyor, son of Gerhard elder. *Plan of Cologne* 1571 for Braun & Hogenberg

Mercator, Barthelemy (fl. 1540—68). Geographer, son of Gerard the elder

Mercator [Kremer, Cramer], Gerhard. Geographer, cartographer and mathematician, b. 1512 Rupelmonde, d. 1594 Duisburg. *Globes, large scale maps, Atlas* 1595 (posthumous). Studied under Gemma Frisius at Louvain. Moved to Duisburg 1552. *World* 1538, 1569 (1st with Mercator's Projection). *Globes* 1541—51. Assisted Frisius with globe, 1537. *Palestine* 1537, *Europe* 1554, *G.B.* 1564, *Flanders* 1540, *Edn. Ptolemy Geog.* 1578

Mercator, Gerard, younger (ca. 1565—1656). Grandson of G. elder. *Africa, Asia* 1595, *Westphalia* 1598

Mercator, Joannes (ca. 1562—1595), fl. 1575. Eldest son of Arnold, assisted grandfather. *Mörs* 1591, in 1606 *Mercator-Hondius Atlas*

G. Mercator (1512–1594); engraving by F. Hogenberg

World map by G. Mercator (published 1538)

MEXICANA

AFRICA.

GERARDI MERCATORIS

EVROPA

ASIA.

PRVANA.

MAGALA NICA.

ATLAS
SIVE
COSMOGRAPHICÆ
MEDITATIONES
DE
FABRICA MVNDI ET
FABRICATI FIGVRA.

Iam tandem ad finem perductus, quamplurimis æneis ta:
bulis Hispaniæ, Africæ, Asiæ & Americæ auctus ac
illustratus à Iudoco Hondio. Quibus etiam additæ
(præter Mercatoris) dilucidæ & accuratæ omnium tabu:
larum descriptiones novæ, studio et opera Pet. Montani.

Excusum in ædibus Iudoci Hondij Amsterodami. 1606.

Title to G. Mercator's Atlas, in the edition published 1606 by J. Hondius.
G. Mercator coined the name "Atlas" for a collection of maps

Mercator, Michael (ca. 1567–1600). Son of Arnold. *America* for *Atlas* 1595

Mercator, Rumold (ca. 1545–1599). Son of Gerhard, elder. *Cart.* Directed public[n] of *Atlas World* 1587, *Continents, Germany* 12 sh. 1590, *Europe* 1595

Mercein. See **Kirk & Mercein**

Mercer, J. M. *Tehuantepec* 1851

Mercer, J. M. Surveyor, *Yazoo Co., Missouri* 1874

Merchant, Ahaz. *Map Cleveland* Ohio 1835 and 1850

Merchant, G. W. American engraver of Albany. *Map U.S.* 1825

Mer *des Hystoires* Paris 1488. See **Rudimentum Novitiorum**

Merensky, Rev. A. Superintendent Berlin missions Transvaal. *Map Transvaal* 1875, *S. Africa* 4 sh. 1884 and 1887, *Der Shire-Fluss van Matope bis Nyassa* 1894

Merian, Caspar. Publisher of Frankfurt. *Topo. Galliae* 1655, *Frankfurt am Mein* 1660

Merian, Matthäus (1593–1650). Topographical engraver and publisher. Studied in Zürich. Worked in Frankfurt. Married daughter of Th. de Bry. *Theatrum Europaeum* 1629–1718, *Topographie* 31 parts 1642–1688, *Gotofred Newe Archontologia Cosmica* 1638 and 1649

Merian, Matthäus

Merian, Matthäus (1621–1687). Son of above, continued business. Worked with Sandrart & Van Dyck

Merica G. & P. à. See **Myrica**

Merit, C. Surveyor of Kings Lynn. *Chart of the Wash* 1693, *Chart jurisdiction Kings Lynn*

Meriton, G. *Geogr. Description of World* 1674

Merkel, G. *Grosvenor Atlas Eisenbahn Mittel Europa* 1892

Merklas, Vaclav Michal (1809–1866). Czech cartographer, engraver and globe maker

Merle, Paul von. See **Merula**

Merlini, C. *Pianta della Citta di Firenze* 1818

Merlo, Giovanni. *Citta di Venetia* 1656

Merlugo, Giov. *Territ. Vicenza* 1775

Merovingian [Albi]. *Mappa Mundi,* 8th century rectangular world map preserved at Albi

Merriman, J. K. *Map of Basutoland* 1871 (with W. H. Surmon)

Merrington, Wm. *Plan Cuff estate Whitechapel* 1818

Merrett, H. S. *Plan Cheltenham* 1814

Merry, F. C. *Westchester Co., N.Y.* 1858

Merry, John. Stationer, N°. 122 next the London Tavern Bishopsgate Street. Adam's *Seat War Turkey* 1774

Merryweather, F. *Plan Old Sarum* (1761)

Merryweather, Henry. American surveyor. *Township* 51 *N. Michigan* 1864, *Geolog. Section Eagle River* 1864

Mersich, Andreas. *Siebenbürgen,* Vienna 1854

Merula [Merle], Paulus G. (1588–1607). *Cosmographia* (49 maps), Amstd. 1605. Later editions to 1636

Merveilleux, David François de (1652–1712). Geographer and cartographer. *Neuchâtel* (1710), reissued Seutter 1760

Merz, Johann Ludwig (1772–1850). Military cartographer

Messager, Jean. French engraver and publisher (fl. 1615–36), d. Paris 1649. For Tessin 1636

Messhala, Arabus (9th century), Jewish astrologer. *De elementis et orbibus celestibus.* Issued Nuremberg 1549

Messeder, Isaac. *Runwell* 1774 MS

Messerschmidt, Count Daniel G. (1685–1735). Scientific explorer of Siberia 1719–26. Together with Strahlenberg Ms. *maps of Siberia*

Messier, Charles (1730–1817). French astronomer, map and globe maker, b. Lorraine. Various *celestial globes* 1780–1800

Messina, Joan Martinez de. See **Martines**

Messinger, John. *New Sectional Map Illinois* 1835

Messner, Franz. Engraver. *Grundriss . . . Stadt Wien* (1770)

Mestdagh, P. G. De Vey. *Stad Vlissingen* 1875

Metcalf, R. Engraver, City Road. *Plans City of London* 1800–06

Metcalfe. Engraver. *Bedford Estate Russell Sq.* 1802

Metelka, Jindrich. *Moravie Nova . . . delin.* 1892

Metchnikov, L. J. (1838–1888). Prof. of geography and publisher, St. Petersburg

Metellus [Matal] Natalius Sequanus (Germanus) (1520–1597). Geographer of Louvain, later Cologne. *Speculum Orbis terrae* 1600 with maps, *Insularum* Cologne 1601, Map America *America sive novus orbis* 1600 (20 maps after Wytfiet)

Meteren, Emmanuel van (1535–1612). *Nederlanden* 1633

Metio, [Metius, Matius], Adriano (1571–1635). Mathematician and astronomer. *W. Frisia* 1630, used by Hondius and Blaeu

Metz, C. See **Cadet de Metz**

Metz, G. de. See **Gautier de Metz**

Metzburg, George Ignaz Freiherr v. (1735–1798). Doctor of Philosophy and Theology in Vienna, mathematician and cartographer. *Map Galicia* 1771, *Road Map* 1782

Metzeroth, Karl (1824–1875). Engraver for Perthes and for Stieler 1856

Metzeroth, T. and Gustav. Maps for Meyer 1830–40

Metzger, Joseph. Goldsmith. *Stadt Görlitz* 1566

Metzger, K. *Stuttgart und Umgebung* (1889)

Meurer, J. *Touristenkarten Oesterr. Alpen,* Vienna 1890

Meurerus, Noe, *Wasser vecht . . . Rheni fluminis et aliorum fluminum . . . Frankfort* 1570

Meurs [Meursius], Jacob van (1620–1680). Amsterdam map publisher, engraver and bookseller. Pub. for Montanus & Dapper 1670–1680. *Africa* 1670, *America* 1671, *Vrankrijk* 1666, *Wandering Israel* 1677. Business continued by his widow, see infra

Meurs, Weduwe [widow] Jacob van. *Brasil zee en lant-reise* 1682

Meusnier, Jean Baptist Marie. Worked on triangulation with Monge, Cassini etc.

Meybom, P. I. M. *Lijst. gedruckte Kaarten . . . in het Archief der Genie* 1857

Meyendorf, George von, Baron (1790–1863). Russian officer. *Excursions Siberia & Caucasus* 1820, *Boukhara* 1843

Meyer, Adalbert. *Military plans Germany* 1864–89

Meyer, A. B. Publisher. Dresden *Neu Guinea Reise* 1873

Meyer, Daniel (ca. 1671–1710). Swiss surveyor and cartographer. *Gemarkungs plan von Hessenthal* 1701

Meyer, Elias. *Plan Fort Augusta* 1756

Meyer, Ernest L. *Maps New Jersey* 1862–4.

Meyer, F. Andreas. *Maps Hamburg & Elbe district* 1865–75

Meyer, Georg Friedrich (1645–1693). Mathematician, engineer, surveyor and engraver of Basle. Son of Jacob below. *County of Pfirt Alsace* 1667, *Alsace Superior* 1677

Meyer, Dr. H. *Kilimanjaro (Tanzania)* 1887–90, *Kenya* 1891

Meyer, H. Ad. (1822–1889). Oceanographer

Meyer, J. German sea captain in Russian service. *Maps of the Caspian Sea* published 1725

Meyer, Jacob (1614–1678). Surveyor of Basle. *Alsace* 1667

Meyer, Johann. Royal mathematician. Surveyed the duchies of Schleswig & Holstein for Christian IV of Denmark 1638–48, pub. 1652

Meyer, Johann Conrad. *Nova descrip. Tiguriane (Zürich)* 1685

Meyer, Johannes. Mathematician *Schonen & Bleeking,* 1659

Meyer, Joseph. *Maps of Holland & Germany* 1866–95, *Hand Atlas* 1866 (100 maps)

Meyer, Joseph (1796–1856). Publisher. *Neu Univ. Atlas* 1830–40, *Grosser Schul Atlas* 1830–8, *Grosser Hand Atlas* 1846, *Neuester Zeitungs Atlas Hildburghausen* (1849), *Auswanderungs Atlas* (1856)

Meyer, J. H. *Leventina* 1784

Meyer, J. T. See **Mayer**

Meyer, J. R. (1739–1813). Swiss cartographer. *Carte de la Suisse* 1788–96, *Atlas* 1786–1802 (with J.H. Weiss)

Meyer, Pieter. Dutch publisher. *Platte Grond,* London 1760

Meyer & Sons. Publishers, New York. *U.S.* (1870)

Meyère, L. *France et pays voisins,* 4 sh. 1889

Meyerpeck, Wolf (d. 1598). Engraver of Leipzig, Vienna and finally Prague. *Misnia & Thuringia*

Meyer von Kronau, Gerold. *Hist. Geogr. Atlas der Schweiz* 1846–55

Meyler, W. Publisher and mapseller. *Plan Bath* 1793 (with W. Taylor), editions to

1805; Hibbart's *5 miles round Bath* 1787

Meyn, Ludwig. *Geological maps Schleswig Holstein & Insel Sylt* 1876–81

Meza, L. G. *Bolivia,* La Paz 1898

Mezentsof, A. I. Russian cartographer, 17th century. *MS map* by order of the Czar ca. 1627

Mezières, N. le C. de. See **Le Camus de Mezières**

Mezzetta, Girolamo di Tomeo. *Estate Plans Lucca* 1640 MS

Mialaret, Charles. *Dépt. des Ardennes* 1876

Miani, Giovanni. *Bassin du Nil* (1862) orig. *del Nilo* 1864

Michal, Jacques. *Swabia* 9 sh. 1725, *Alsace III Tab.* 5 ft. 9 in. × 2 ft., Seutter 1735

Michaelis, Ernst Heinrich (1796–1873). Cartographer. *Maps Canton of Aargau* 1843–60, *Schwaben & Würtemberg* (1793–1830), *Canton Tessin* 1859

Michaelis, F. *Paris mit Umgebung* 1870

Michaelis, Johann Wilhelm (18th century). Engraver. *Ground plan Frankfort on Oder* 1706

Michaelis, Julius. *Railway maps Central Europe* 1859–73

Michaelis, Laurentias (d. 1584). *Friesland* ca. 1540, *Frisia Orientalis* 1579, used by De Jode 1593, *Oldenburg* used Ortelius 1584

Michaelis & Braun. Publishers, Long Street, Cape Town. Merensky's *South Africa* 1884

Michaelson, George David (1680–1765). Military cartographer

Michailescu, Nicolae. *Romanie* 4 sh. (1896)

Michalet [Michallet], Estienne. Publisher of Paris. For Thévenot 1681. For Jaillot 1689, *Egypt* 1692

Michaud. See **Rassau & Michaud**

Michault, R. Engraver for Duval 1666, for Modiford's *Jamaica* 1674, Sanson's *Géogr. Univ.* 1675, *Costes et Riv. de Virginie* 1674

Michel, Sr. *L'Indicateur Fidèle ou Guide de Voyages* 1767, 1722, 1780

Michel. l'Hôtel de Soubise, Paris *Possessions Angloises en Amérique* 1778, *Routes de France* 1767 and 1780 (with Desnos)

Michel, Chr. *Alpine Maps Bavaria & Tyrol* 1868–96

Michel, F. P. Engraver. *Carte Topo. Militaire des Alpes* 12 sh. 1820, *Environs Coblentz* (1825), *Nederlanden* 1816

Michel. Engraver for Dépôt de la Marine first half 19th century

Michelena y Codazzi. *Parte de Venezuela y Colombia* 1884

Micheli du Crest, J. B. (1690–1766). Geographer. First to execute panorama of glaciers in relief

Michelin, F. Lithographer, 111 Nassau St., New York. *Oregon &c.* 1846

Michell. See **Mitchell**

Michell, Lieut. A. *Lisbon Harbour* 1735 MS

Michell, Major C. C. *Cape of Good Hope* 1836

Michell, C. H. *Possess. Angloises et Portugaises Afrique Australe* 1843

Michell, Robert. *Khiva & Turcoman County* 1873

Michelot, Auguste (1792–1866). *Europe* 1876, *Atlas Universel* (1877), *Petit Atlas Universel* (1878), all with A. Meissas

Michelot, Henri. *Plans of Towns, Barcelone, Cette, Ferraro, Gibraltar, Livorno* 1727, *Recueil de plusiers Plans Ports et Rades de la Méditerranée* (with Brémond) 1727–30

Micheoro, Mattheo di. Polish author, student, Cracow. *Descrittione delle due Sarmatia.* Ramusio *Viaggi* 1559

Michielson, W. J. M. *Oost Kust van Sumatra* 1891

Michler, N. *American Civil War Maps* 1867–9

Michie, R. S. Engraver for Morison's *Clackmannen* 1848

Miechowa, M. V. (1457–1523). Polish geographer, physician and astrologer. *De Sarmatia Asiana et Europa Novus Orbis* 1518

Michu, N. Engraver for Du Val 1676

Michurin, I. Russian surveyor. *Plan of Moscow* (1741)

Middendorff, A. von. *Norden & Osten Sibiriens* 13 sh. St. Petersburg 1859

Middleton. *Complete system of Geography* (1779)

Middleton, Abraham. Surveyor, Whitehaven. *Admiralty Chart* 1847 (with H. M. Denham)

Middleton, Capt. Christopher. Publisher and cartographer in service of Hudson's Bay Co. Exped. to N.W. Passage 1741–2. *Hudson's Bay* 1743

Middleton, Empson Edward. *Maps World* 1876–1886

Middleton, Francis. *Plan Forest of Holt* 1790

Middleton, Hendrik. Engraver for Van der Aa (1720)

Middleton, Sir Henry. *Voyages for E. India Co.*

Middleton, John. *View of Agriculture of Middlesex* 1798. Land use map of county in 3 colours to distinguish arable, pasture and nursery grounds

Middleton, John. Estate surveyor of Lambeth. *South Weald* 1788–9 MS

Middleton, John. *Celestial Atlas* 1842

Middleton, R. D. *Sea of Marmora Admiralty Chart* 1839

Middleton, R. W. E. *Port Victoria, Mahé (Mauritius)* 1878

Middleton, Strobridge & Co. Printers for Ehrenberg 1858

Middleton, Wallace & Co. Lithographers of Cincinnati. Boynton's *Kansas* 1855; for B. Gray 1856

Middnagten, Christoffel of Harlingen (1659–1723). Hydrographer. *Paskaart vant Noorder Deel van Europa* (1690), composed charts for Johannes Loots 1706–1717

Midwinter, D. Publisher at the Rose & Crown St. Pauls Churchyard and later at Three Crowns in St. Pauls Churchyard. Partner with T. Leigh at first address. Moll's *Scotland* 1714, *N. America* (1720), *Africa &c.*

Miera y Pacheco, Bernardo de. *New Mexico* 1773–9 MSS

Mierisch, Bruno. *Geolog. maps Nicaragua* 1893–5

Miet, Leon Pierre. *Dyre Fiord (Iceland)* 1856

Mieullet, Capt. H. *Chain of Mont Blanc* 1871, *Galilée* 1870, *Massif de Mont Blanc* 1875

Migeon, J. Publisher: (1) Rue du Chemin des Plantes 34, Paris, Monte Rouge, (2) 11 Rue du Moulin Vert, Paris. *Géogr. Univ.* (1864) and 1874, *Nouvel Atlas Illust. Géogr. Univ.* 1891

Migneret, Madame. Engraver for Depping 1828, for Perrot 1823 *(county maps G.B.)*

Migue. *Environs de Nancy* 1778

Miguel, Vincente Torfino de San. See **Torfino de San Miguel**

Mikhailov. Russian engraver 18th century *Rossiiskoi Atlas* 1792, *Mappa von Servien* 1788 MS

Mikoviny, Samuel (1700–1750). Military cartographer. *Comit. Posnaniensis* (1735)

Miles, William, H. *Plan Johannesburg* 1890

Mikulas. See **Claudianus**

Milau Y Marval, Francesco (1727–1805). Spanish hydrographer

Milbert, J. Publisher of Paris. Freycinet's *Atlas* 1807–16

Milcent, F. D. *Plan City of Lisbon* 1785

Milet de Mureau. Possibly the first to use contour lines on land, about 1749

Milhauser. See **Milheuser**

Milheuser, Julius. *Liège* Blaeu (1650), *Madrid* (1660), *Venetia* (1670)

Mill, Henry. *Hydro. map River Colne* 1766

Mill, Henry (18th century). Surveyor

Mill, Hugh Robert. *Atlas Comm. Geogr.* 1889, *Map of English Lake District* 1895

Millan, B. C. *Estados Unidos de Venezuela* (1871)

Millan, J. Publisher next Scotland Yard, Whitehall. *Plan Aberdeen* 1746, *Battle Plans Culloden & Falkirk* 1745–6, *Dunkirk Road* 1746

Millan of Villanueva, Camillo. *Prov. de Ilocos Norte (Philippines)* 1891

Millar, Andrew (1707–1768). British publisher, Katherine St. in the Strand. Elphinstone's *N. Britain* 1745, *Geogr. Magna Brit.* 1748, 1750, Camden's *Brit.* 1753, Mitchell's *Brit. Colon. in N. Amer.* 1755, *Universal History* 1766. Retired in 1767 and business passed to Thomas Cadell, his partner from 1765

Millar, George H. *New & Univ. system of Geography* pub. by A. Hogg (1782)

Millar, William. *Florence Douglas Co., Nebraska* (1860)

M[illard], I. *Famous city of Bristol & its suburbs.* Sold by John Overton at the White Horse in London (1673)

Millard, John. Master, R.N. *Admiralty Charts Ionian Is.* 1866, *Ben Ghazi* 1863, *Dar es Salaam* 1874, *Suez Canal* 1875–7

Millard, Thomas. Print and mapseller at ye Dial and Three Crowns next ye Globe Tavern in Fleet St. *Plan London* 1741,

Oliver's *Middlesex* 1742 De Wit, Edinburgh. *Roads of England* 1742

Millard, W. S. *Plan City Norwich* 1830 (with J. Manning)

Miller, Allister, M. *Swazieland* 1896

Miller, G. W. *Minneapolis* 1873

Miller, J. Estate surveyor. *Faversham, Kent* 1774 MS

Miller, James. Publisher, 554 Broadway, New York City (1857)

Miller, John. Estate surveyor. *Layer de la Haye (Essex)* 1735 MS, *Birch* 1750

Miller, John. Publisher, London. *American Atlas,* Phila., Carey & Lea; London, John Miller

Miller, J. *Terrestrial Globe* 1793

Miller, Robert. Publisher, 24 Old Fish Street, London *New Min. Atlas* 1810, 1820, 1825

Miller, R. J. *Gold Fields S.E. Africa* 1876

Miller, T. Engraver. *Maps India, Madras, Fort St. George* 1778

Miller, Wm. Publisher, Albemarle St., London. Outhett's *Port Mornington* 1809, Salt's *Alexandria* 1809

Miller, W. *Lincolnshire* 1805

Miller, W. L. *Isthmus of Tehuantepec* 1851

Miller. Script engraver of Paris, Rue Jacques No. 52. For Poirson 1799, *Carte Londres* 1801

Miller & Boyle. Lithographers, 102 Broadway, New York. *N. America* 1849

Miller & Co. Lithographers. Booth's *Liverpool & Manchester Railway* (1830)

Miller & Hutchens. American publishers of Providence, R.I. Drury's *Geography for Schools* 1845

Miller, Orton & Mulligan. Publishers of New York and Aubern. *Great West* 1856

Millerd, I. See **Millard**, I.

Milles, T. *Dover Harbour* (1580)

Milleson, M. *Sierra Mining District* (1865)

Millingen, Major Frederick. *L'Arménie Orientale et Kurdistan Sept.* (1868)

Millo, Antonio (fl. 1557–90). Venetian cartographer. *Isolarii* 1557–90, *Carta Marina* 1582, *Arte del navegar* 1590, MS *Sea Atlases*

Millot, Jean. Engraver for Lescarbot 1609. *Figure de la Terre* 1609

Mills. Surveyor. Assisted Ogilby & Morgan, London, 1677

Mills, C. *Plan dockyard Sheerness* 1830 MS

Mills, George. Engraver. *Plan Leman Estate Goodmans Field* 1815, *Ship Canal Bridgewater to Seaton* 1815

Mills, James. *Mail Rd. between Edinburgh & Morpeth* (1820)

Mills, J. B. Civil engineer. Drew Moury's *Sonora & Arizona* 1863

Mills, John Edward, midshipman (d.1814). *Lynhaven Bay & Hampton Roads Virginia* 1812 MS

Mills, Robert (1781–1855). Survey *Atlas of S. Carolina,* Lucas 1825. The first state atlas of U.S. *Country between Atlantic and Pacific* 1848

Miln, Robert. Engraver. *Map Shetland* 1739

Milne, Alexander. *Plan City Aberdeen* 3 sh. 1790

Milne, J. Master, R.N. *Port Anna Maria [Marquesas]* 1814 MS (with W.J. Prowse)

Milne [Mylne], Thomas. Surveyor, engraver and publisher. 7 New St., Knightsbridge. *Hampshire* 6 sh. Faden 1788–90, *Bldgs., St. George's Fields* 1794, *Norfolk* (with T. Donald) 1790–94 pub. 1797

Milne, William C. *Country round Pekin* 1859

Milner, Rev. Thomas (d. 1882). Astronomer. *Descriptive Atlas of Astronomy* (1850),

Atlas Physical Geography 1850, *Atlas Political Geogr.* 1851

Milton, Thomas. Dockyard plans: *Woolwich Deptford* 1753, *Portsmouth* 1754, *Sheerness* 1755, *Plymouth* 1756, *Views Seats Ireland* 1783, *Views Egypt* (1801)

Mimpress, Robert. *Biblical Maps* 1834–54

Minchin, Juan B. *Repub. Boliviana* 1879, *Paria y Carangas* (1880), *Map Bolivia* 9 sh. MS

Minde, J. See **Mynde**

Minet. Engineer to La Salle, credited with map *Louisiana* (1685) MS

Mingay, Capt. *Selsey Bill* 1827

Mingucci, Francesco. *Territ. Urbino* 1636

Minguet y Irol, P. Spanish engraver and publisher. *Madrid* ca. 1750

Minikin, George. Bookseller and publisher, London, at the Kings Head in S. Martin's. Morden's *52 Counties of England & Wales* 1676

Ministère des Travaux Publics. *Ports maritimes de la France* (1871), *Atlas des canaux de la France* 1879, *Routes Nationales* 1884

Minor, Hans Erik. Surveyed parts of Iceland 1776–7

Minorita, P. See **Paolino**

Minter, K. F. (1780–1847). Painter and lithographer of German descent settled in Warsaw where the great *topographic map of the Polish Kingdom* was executed in his lithographic shop

Minutelli, Frederico (1846–1906). Geographer

Miot [Miotti], Vincenzo (1712–1787). Astronomer and globemaker. *Celestial Globe*

Miranda, Joseph Acosta. See **Costa e Miranda**

Mirandulanus, A. C. *Armillary Spheres* 1676

Mirail. *Plan Géomet. Bordeaux* 1755

Mirbeck, C. I. B. Publisher, London. *Hamburg* 1803

Miricio [Miritius], G. or J. See **Myritius**

Mirkovich, M. F. *Ethno. Map Slavonic Peoples* St. Petersburg 1875

Miscomini, A di B. *Pub. Orbis Brev.* 1493 with T. O, Map

Missy, J. R. de. See **Rousset de Missy**

Mitchell, A. *Canada East* 1846

Mitchell, Sir Arthur (1826–1909). Historian and geographer

Mitchell, C. *Newspaper map of South Africa* 1895, ditto *Australasia* (1895)

Mitchell, C. C. *Admiralty Charts S. Africa* 1839–52

Mitchell, D. T. *Maps Kansas* 1857–60

Mitchell, E. Engraver for Thomson's *Atlas* 1817

Mitchell, Ephraim. American surveyor. *Counties of Mecklenburg & Tryon* 1772 MS

Mitchell, George. *Is. Canso & Sable* (1750), *Massachusetts Boundary* 1741

Mitchell, Capt. John. *Entrance Thames* 1733, *Survey Thames York Stairs to West Horse Ferry* 1737–8

Mitchell, John (d. 1768). Cartographer, physician and botanist. Emigrated Virginia in 1720. Published his *map of America* 1755, French edition 1756, German edn. 1775

Mitchell, J. *Triangulation Prov. Wellington, N.Z.* 1872

Mitchell, Robert. *Sea coast Spurn Head to Wells* 1749, *Draught River Humber* 1761 and 1778

Mitchell, Samuel Augustus (1792–1868). Philadelphia publisher, N.E. corner Market and 74th Street. Succeeded Tanner. Various atlases 1839–1866. *Ancient Atlas* 1844, *Universal Atlas* 1847 and later editions, *Virginia & Maryland* 1832, *School & Family Atlas* 1832

Mitchell, Major (later Col.) Sir Thomas L. (1792–1855). Surveyor Gen. New South Wales and of Cape Colony. *Colony of*

N.S.W. 1834, *S.E. Australia* pub. Arrowsmith 1837, *Expd. E. Australia* 1838, *E. Frontier Cape of Good Hope* Arrowsmith 1848, Proeschel's *Atlas Australasia* 1863

Mitchell, Thomas. Road surveyor. *Selkirk* in Thomson's *Scotland* 1832

Mitchell, W. *Cape May New Jersey* 1856, *Nantucket* 1838

Mitchell, W. F. *Geological Survey Ireland* 1879—91

Mitchell & Hinman. Publishers of Philadelphia, No. 6 North 5th Street. Mitchell's *Reference Map of U.S.* 1834, 1836; *Tourist Map of Michigan* 1835

Mitchelson, S. P. *Mineral Map Matabeleland* 1898

Mitelli, Agostino. *Pianta Citta di Bologna* 4 sh. 1692

Mitra, Rajondralala (d. 1891). Indian historian and geographer

Mittelbach, R. *Geological Maps* 1885—1900

Mittnach, Freiherr von (fl. 1848—53). Cartographer of Stuttgart

Mitzopulos, C. Mittel. *Grieschisches Erdbeben* 1893—4

Miyasaki, Riuto *[Map of Japan]* 1876

Mizaldus [Mizauld], Antonius (ca. 1520—1578). French astrologer and cosmographer. *De Mundi Sphera Sive Cosmographia* 1549, *Miroir du Temps* 1547

Mocetto, Girolamo. *4 maps and plans Nola* in Leo's *Nola* 1514

Mockler Ferryman, Capt. A. F. *Maps of Nigeria, Niger & Benin* 1889, *Kebbi River* 1891, *Donga River* 1899

Modiford [Modyford], Sir Thomas, Bart. *Isle de la Jamaique* 1674

Moehl, Heinrich. *Maps of German States* 1867—81

Moellendorf. *Plan Berlin* 1843

Moeller, A. W. *Karte des Heiligen Landes* 1825

Moerentor. See **Moretus**

Moering, A. *Donau Strom* (1873)

Moermans, J. (1602—1653). Engraver and publisher of Antwerp. *Town views and maps*

Moers, Justus (1572—1617). German surveyor and cartographer. *Waldeck Marburg* 1575

Moesch, Casimir. *Geological Maps Switzerland* 1861—66

Moetjens, Adrian. Publisher at The Hague. Blaeu's *Savoy & Piedmont* 1700

Moeyaert, Claes (1597—1655). Painter and engraver. *Views Amsterdam,* engraved Nolpe

Moffat, J. Engraver of Edinburgh, (fl. 1795—1820) for Laurie & Whittle 1800. *Ken Travels* 1811, *Dumfries* 1812, *Caracas* 1814, *N. Atlantic* 1815, Thomson's *Atlas* 1817, *Kirkcudbright* 1821, *Shetland Is.* 1827

Moffat, Robert. *Part Orange River* 1857 MS, *Little Namaqualand* 1856

Moffat, William. Comm. of *Phoenix. Chart of Red Sea* 1801, *Indonesia* 1815

Moffat, William. *Sixpenny Atlas* 1876, *Penny Atlas* 1882

Moffet, H. le C. de. See **Le Camus de Moffet**

Mofras. See **Duflot de Mafras**

Mogami, Tokunai (1754—1836). Japanese cartographer and geographer

Mogg, Edward. Publisher, engraver and mapseller. Addresses: 1804, 61 Margaret St. Cavendish Sq. 1805, 14 Little Newport Street, Leicester Sq. 1814, 51 Charing Cross. 1829, 14 Great Russell St. *Plans London* 1803—28, *Twenty four miles round London* 1805—36, *London in Min.* 1806—44, *Guide London* 1806, *Plan Mexico* 1811, *Forty five miles round London,* 1821, *Roads England & Wales* 1822, *England & Wales* 1825, *New London Guide* 1848

Mogg, William. 537 New Oxford St., Bloomsbury. Publisher. 24 *miles round London* 1859

Mogiol, Francesco (16th century) Engraver *Nice earthquake* 1564

Möglich, L. P. Steel engraver. *Town views Austria*

Mohl, H. Publisher, London. *Asia* (1700)

Mohr, C. A. *Geologische Schul-Wandkarte von Deutschland* 1889

Mohr, Eduard. *Natal, und Orange-Fluss Freistaat* 1871

Mohr, Edwin. *Süd Africa* 1872, *Gold Fields Africa* 1876, *Transvaal* 1877

Mohr, Ernst Rudolf (1821–1885). Swiss cartographer

Mohr, H. (b. 1835). Norwegian meteorologist and oceanographer. *Storm Atlas* 1870, *Jan Mayen Is.* 1878

Mohun, E. *British Columbia, Victoria, Comm. of Land & Works* 1884

Moir, Wm. Publisher. Jaffray's *Peterhead* 1739

Moises, Franz. *Deutsch Ost Afrika* (1889)

Moissenet, Leon. *Etudes sur les Filons du Cornwall* 1874

Moisy, Alexandre. Parisian engraver. Place St. Michel No. 129. Worked for Meissas & Michelot and Ansart. *Plan de Londres* (1810), *Océanie* 1840, *N. America* 1869

Moithey, Maurille Antoine (1732–1810). Rue de la Harpe vis à vis la Sorbonne. Ingénieur Géogr. du Roi 1780, *Poland* 1769, *English Poss. America* 1777, *America* 1786, *Dict. Hydro. de la France* 1787. Collaborated with Crépy 1769–77 and with Philippe de Pretot 1787. *Atlas Nat. Portatif de France* 1792, *Globe Terrestre* 1793

Moithey, P. J. Engraver. *Description de l'Egypte* 1818

Mojsisovics & Mojsvar. Dr. E. von. *Geolog. Übersichtskarte des Tirolisch-Venetian-ischen Hochlandes zwischen Etsch und*

Pierre 6 sh: Vienna, A. Holder 1878; *Bosnien-Herzegovina* 1880

Mol, P. Engraver for Janvier *(Histoire Universelle)*

Moland, Thomas. Estate surveyor in Ireland early 19th century

Molanus, J. See **Molijns**

Mole, G. E. de. *South Australia Lacipede Bay* 1859 (with P.A. Nation), *Rivoli Bay* 1858

Mole, U. (18th century). *Map of Madeira* 1760

Moleen. Script engraver with Wizzell for Arrowsmith's *World* 1794

Molenda, C. G. See **Gomez Molenda**

Molengraaff, G. A. F. *Stromkarte von West Borneo* 1895

Moler, J. Douglas. *Clarke Co., Ohio* 1859

Moletius [Moleti, Moleto], Josephus [Gioseppe] (1531–1558). Italian mathematician, b. Messina, d. Padua. *Tab. Gregorianus ed.* edition of *Ptolemy* 1566, *Ephemerides* 1563 and 1580, *Discorso Universale* 1561

Molich, C. G. *Spogkart over Sonderjylland* 1889

Molijns [Molinus, Mollijns, Molanus, Molyns] Jan (1565–1599) Antwerp publisher. *British Isles* 1549, *Siege of Malta* 1565

Molinax, E. See **Molyneux**

Molinero, J. A. Spanish military surveyor. *Teatro de la guerra de Oriente* 1876

Molineux, Emery, See **Molyneux**

Molineux, T. *Knowledge of Globes* 1834

Molini, Giuseppe. *Italia* 1802

Molinus. See **Molijns**

Molitor, C. A. *Erfurth Haupstadt, Thüringen* 1745

Molitor, Albert. American surveyor. *S, Shore Keweenaw Pt. Lake Superior* 1865

Molitor, Edward. American lithographer. *N.W. Lakes* 1866, *Lake Superior* 1868,

Mich & Wis. 1871, *Mouth Detroit* River *1874*

Moll, B. P. Map collector, 18th century. A 50 vol. collection of maps brought together by him preserved in Czechoslovakia

Moll, Herman (d. 1732). Engraver, geographer and bookseller of Dutch origin. Came to England 1678. Worked as engraver for Moses Pitt, Greenville Collins, John Adair, and Seller & Price. Started selling maps in Vanleys Court Blackfriars but later settled in Devereux Court Strand. *America & Europe* for Moore's *Geogr.* 1681, Six *charts* for Collins 1689–9, for Adair 1688, *A System of Geography* 1701, *Globe* 1703, *Ireland* for Philip Lea Large *atlas* ca. 1710, *Atlas Minor* 1727 and 1729, *World Described* 1727

Moll, José. *Isla de Menorca* 1887

Moller [**Mollero**], Christian. Cartographer. *Fluvii Albis (Elbe) nov. delin.* 1628, used by Hondius & Blaeu

Moller, G. *Austrian monarchy* (with F. Pilsak) Vienna 1822

Moller, H. P. C. *Kart over den Deel af Godthaabs District (Greenland)* Copenhagen 1840

Moller, J. J. Engraver. *Town views* pub. Schenk ca. 1710

Møller, Lars. Lithographer. *Maps of Greenland* printed in Godthab 1858–75

Mollero, C. See **Moller**, C.

Möllhausen, Baldwin. *Mississippi* 1858

Mollineux, Globe. See **Molyneux**

Mollison, James. Cartographer and engraver. *Perforated Planisphere* 1852

Mollo, Edward (1797–1842). Engraver and publisher of Vienna

Mollo, Tranquillo (fl. 1800–1830). Viennese engraver, printer and bookseller. *Teutschland* 1800, *Turkey* 1808, *Bohemia & Galicia* 1809, *Austria* 1816, *Vienna* 1821 and 1830. Also published for Sotzmann 1812 and Dirwald *N. America* 1823

Mollijns, J. See **Molijns**

Mollis, H. von. See **Glareanus**

Mols-Marchal, L. *Chemins de Fer de l'Europe* Brussels 1863, *Carte Admin. Belgique* 1868, *Wall map Belgium* 1869, *Brussels* 1875

Moltke, Adolf, Freiherr von (1822–1884). Austrian military cartographer

Moltke, Herman Carl Bernard, Count von (1800–1891). German general and cartographer of Berlin. *Constantinople* 1842, *Klein Asien* 1844, *Bosphorus* 1849, *Rome & Environs* 1852

Molyneux [**Mollyneux, Molinax, Molineux, Mullineus**] Emery (fl. 1587–1605). Mathematician and instrument maker. *Terrest. & celest. globes*, engraved Hondius. First globes engraved in England, now in Library Middle Temple

Molyns, J. See **Molijns**

Momo, Giuseppe. Engineer geographer to Dépôt des Cartes. *Stati . . . di Sardegna* 4 sh. Turin 1819 and 1840

Mommsen, T. *Galliae Cisalpinae* 1877

Mon, J. F. *Portolan chart* 1629 MS

Monachus, F. See **Franciscus Monarchus**

Monachus, Stephanus. *Florentini Domini . . . descriptio* 1595

Monaldini, Venanzio. Publisher of Rome. Nolli's *Pianta di Roma* 1816 and 1823

Monarchus, F. See **Franciscus Monarchus**

Monath, Johann Kaspar (1763–1810). Publisher and bookseller, Nuremberg. Married Barbara Ebersperger and became part of firm of Homann Heirs

Monath, Peter Conrad (1695–1748) of Nuremberg. *Grundriss Belgrad* 1717

Moncel, T. A. L. (1821–1884). *Views Athens* 1849

Moncornet, B. (1600–1668). Publisher and engraver, town plans and views

Moncrief, James. *Maps of Florida* 1764–5

Mondhare & Jean. Publishers of Paris, Rue St. Tea de Beaurais près celle des Noyers. Sanson's *Sçandinavia* 1788, *Carte d'Amérique* Nolin 1789–92 and 1818. Clouet's *Géogr. Moderne* 1767 and 1793

Mondhare, L. Publisher of Paris, Rue St. Jacques près la Fontaine St. Sévérin, also Rue St. Jacques à la ville de Caen, and in Cadiz, Casa de Monhare. *Eng. Poss. N. America* 1777, *Antilles* 1782, *Africa & Asia* 1788, *Europe* 1789

Monecke, C. Printer of Berlin. Kiepert's *Central America* 1858, *Railway Map Europe* (1864)

Monegal, G. *Uruguay,* Montevideo 1882

Moneypenny, George. *Plan of Derby* 1791

Monfiroli, G. *Pianta di Roma* 1853

Mongenet [Mongenetto], François de. (d. 1592). Globemaker, doctor, cosmographer *Globes* 1544, *Globe* 1552 (1560), *Globes Duchetti* ca. 1580

Mongez, l'Abbé André. *Mappe Monde Physique* Paris, Mentelle 1779

Mongonetto. See **Mongenet**

Moni, Claude. *Autun* 1834

Monin, Charles V. (fl. 1830–80). French geographer of Caen and Paris. *Petit Atlas National* 1835, *Atlas Classique* 1844–5, *France* (1868), *Mappemonde* 1875

Monk, Jacob. Publisher of Baltimore Md., later Philadelphia 1862. *North America* 1852–3, *New American map* 1857, *Central America* 1859, Ebert's *Colorado* 1862, 5

Monk & Scherer. Publishers, College Building, Cincinnati *U.S.* 1847

Monnet & Guettard. Publishers, Paris. *Atlas et Description Minéral de la France* 1778–80, the first geological atlas of France

Monnier, H. *Cul de Sac Marin* 1828, *Marée dans la Manche* 1839, *Martinique* 1828 and –31

Monnier, Louis Gabriel (1739–1804). Cartographer

Monnier, P. (d. 1843). *Atlas de la Martinique* 1827–31, *Atlas des côtes Mérid. de la France*(1850)

Monno, Giovanni Francesco, of Monaco. *Chart of Medit.* 1613, ditto 1622, *Collection of Charts* 1633

Monreal, B. *El Globo Atlas Classico Universal* Madrid 1870

Monrócq. Printer of Paris for Andriveau Goujon 1876

Monserrate, Juan. *Boundary between Guiana & Venezuela* 1890

Monhousen, Nicolaus. *Trier* 1865

Monson, Frederick John, Baron. Map Dept. *Isère & High Alps.* (1841)

Monson, Sir William. *Discourse concerning N.W. Passage* MS, *Lecture of Navigation* MS (1612)

Monster, R. van (18th century). Dutch surveyor and poet. *Map of Sevenburgen*

Mont, Pierre du. See **Montanus,** Petrus

Montague, Major Edward. *Plan of Seringapatam* 1792

Montalboddo, Fracan [L. Fracan[zano]], Antonio. *Itinerarium Portugallesium* 1508. Includes the first separate printed map of the whole of Africa

Montano, J. *Mindanao* 1880–1

Montano-Hansen. *Mindano* 1891

Montanus, B. Arias. See **Arias Montanus**

Montanus, Petrus [Pieter van den Berg]. Amsterdam cartographer and publisher, brother-in-law of Jodocus Hondius. Wrote text for Mercator-Hondius *Atlas* 1606, and for Keere's *Germania Inferior* 1617, *Orbis Tabula* 1571, *Pays Bas* 1613

Montargues, Peter v. (1690–1730). Military cartographer

Monaru, Louis Mathurin François. *Plan Baie Barracouta (Tartary)* 1859

Montbazin, Maps for Freycinet 1826

Montchel, L. Barentin de. *Geographie Anciénne* 1807

Monte, Urbano. See **Monti,** U.

Montecalerio, Joannis à. *Choro. Descrip Prov. Capucinorum* Milan 1712

Monteculloli, Graf Raimund v. (1609–80). Military writer, cartographer, and collector

Monteiro, Alberto. *Mappa de Portugal* (1896)

Monteiro, Pinto da Fonseca Vaz, Jao. *Mappa do Rio Zambese* 1881

Monteiro de Salazar. *Map of Brazil* 1777

Monteith, James. *Intermed. Geogr.* 1866, *Manual Geography* 1869 and 1870

Monteith, Lt.-Col. W. M. Surveyor. *Plan Canal thro. reefs at Pamban (India)* 1839 MS, *Armenia, Georgia & Caucasas range* 1832 MS, *Parts of Georgia & Armenia . . . from Trig. Survey* 4 sh. London, Arrowsmith 1833

Montemont, Albert. *Colonies Françaises* (1860)

Montensi, J. Surhonis. See **Surhone**

Montero, Capt. Claudio. Spanish navy. *Archipelagos de Calamianes* 1856, *Estrecho de Iloilo* 1858, *Isla de Luzon* 1860, *Arch. Filipino* 1875, *English editions* published by Admiralty 1866–9

Montero, D. *Plano Topo . . . Ciudad de Valencia* 1864

Montero, Ignacio. See **Moreira**

Montero, Juan Nepomuceno. *Atlas Geogr. do Ciudad de Moyobamba* 1865

Montero y Vidal, José. *Islas Marianas* 1886

Monteros, A. E. de los. See **Espinosa de los Monteros**

Montesson. See **Dupain de Montesson**

Monteverde, Tomas. *Atlas de Geogr. Militar de Europa* (1888), *Division Militar de Espana y Portugal* 4 sh.

Monteverde y Sedano, Federico de. *Military plans of the Philippines* 1897

Montgomerie, Thomas G. (1830–1878). Military surveyor, India. *Route Nepal to*

Lhasa 1868, *Upper Basins of Indus* 1868, *Route accross Pamir Steppe* 1871

Montgomerie, W. *Eastern Portion of British Burmah* 1862 MS

Monthucon, H. de. Engraver of Altona. *Middle States of America* 1794

Monti [Monte] Urbano (1544–1613). Cartographer of Milan. *World Map* 1589 MS

Monticelli, Gaetano. *Reyno Lombardo, Veneto* 1827, *Prov. di Como* (1860)

Montigny, G. F. F. de. See **Févre de Montigny**

Montmartin, Sr. de. See **Avity**

Montojo, Fabian. *Arch. de Jolo* 1876 (8)

Montojo, Patrico. *Ciudad de Cardenas y su Porto* 1881

Montojo y Salcedo, Capt. Hydrographer, Spanish navy. 19th century. English editions of his *charts,* 1875–8

Montpereux, F. D. de. See **Dubois de Montpereux**

Montresor, James Gabriel (1702–1776) Col. R.E. Surveyed *Lake Champlain & Environs* 1756, Constructed Fort George 1759

Montresor, Capt. (later Major) John (1736–1788?). English military cartographer, son of above. *Nova Scotia* 4 sh. 1768, *Prov. New York* 4 sh. 1775, *Action Bunker Hill* 1775, *Boston & Environs* 1777, *City of New York* (1776)

Montri, Phya Surasak di. *Map of Siam* 1897

Montriou, Lieut. Charles William. Indian Navy. *Survey India* 1844–46, *Calicut Roads* 1856, *Rajapoor River* 1851

Montulay. Engraver *Freiburg* 1708

Monty. Map seller of Geneva. *Environs de Genève* 1776

Mooburger, G. L. *Plan Battle Oudenarde* 1708

Moodie, G. P. *S. African Republic (Transvaal)* London, Weller 1880

Moody, Charles. *Plan Estates parish St. Ives* 7 sh. 1824, *Plan of the Town* 5 sh. 1824

Moody, R. Lithographer, Brinklow, Warwick 1838

Moody, R. B. & Co. Lithographers and printers, 12 Cannon St., Birmingham. Lascelles & Co.'s *Warwickshire* 1850

Moon, J. *New Geographical Table* 1794

Moon, Richard. Surveyor of Wye. *Land in Newchurch Romney Marsh* 1751 MS, *Mannors of Butlers in the Parishes of Boughton under Blean and Selling* 1786 MS

Mooney, John. Surveyor of Greashill Co. Down (fl. 1740–50). *Kings County (Down)* n.d.

Mooney. See **Hall & Mooney**

Moor, Lieut. H. H.M.S. *Hobart. Chart of Moluccas* pub. Laurie & Whittle 1794, *Augusta & Pigeon Is.* 1798, *Booro Bay* 1801, *Moluccas & Eastern Islands* 1801, *Bantry Bay Castletown Harbour* 1803

Moor, Jonas. See **Moore**, Sir Jonas

Moor, J. *Map of London* ca. 1640

Moorcroft. *Surveys used in N. Punjab* Walker 1846

Moore, Lieut., R.N. *Fiji, Admiralty Chart* 1822

Moore, C. Engraver. For Kaempfér's *Japan* 1727

Moore, Francis. *Draught of the River Gambia* 1738

Moore, George. *Passe Caballo* 1841–56

Moore, Gordon. Surveyor. *York District, South Carolina* 1820, for Mills 1825

Moore, H. B. *Plans of Parishes in Victoria (Australia)* 1858–9

Moore, J. Surveyor of Hoddesdon, Herts. *Part of the River Gade* 1786 MS, *Epping* 1800 MS

Moore, John. *British Channel* 1786

Moore, John Hamilton (d. 1801). Chartseller to H.R.H. the Duke of Clarence and H.S.R. the Empress of Russia. Tower Hill, London. Founded the firm 1763, which later became Imray. Succeeded by his son-in-law Robert Blachford. *Complete Collection of Voyages,* A. Hogg (1779). *S. Foreland to Humber* 1789, *Practical Navigator Chart N. America* 1784, *Chart Downs* 1793, *Navig. from Philadelphia* 1793

Moore, Sir Jonas (1611?–1679). Mathematician and surveyor. Tutor to Duke of York. Surveyed Fen Drainage System 1649–63. *Fens on ½ inch scale* 1654, *on 2 inch scale* 16 sheets 1684. *Mapp River Thames* 1662, *City of Tangier* 3 sh. 1664, *Prospect & Map of London* 1662

Moore, R. Publisher, Store St., Bedford Square. Davies' *London* 1843

Moore, Samuel. *Schema corporis Solaris* (1676)

Moore, S. S. *Travellers Directory,* Phila. 1802 (with T.W. Jones)

Moore, Thomas. Lithographer of Boston. *Plan of the Harbour* 1837, *Village of Augusta* 1838

Moore, Thomas. *Houat,* Dalrymple 1800

Moore, Tyrell. *Travajos Geodesicos . . . Antioqua Nueva Granada* (1855)

Moore, W. *Town lands of Salisbury, North Carolina* 1823

Moore, William. Estate surveyor. *Bobbingworth* 1761 MS, *High Ongar* 1774–5 MS, *Norton Mandeville* 1776 MS

Moore, Lieut. William U. *Fiji Islands Admiralty Chart* 1878–9

Moore, **Wilstach, Keys & Co.** Publishers of Cincinnati. Boynton's *Kansas* 1855

Moore. See **Hayward & Moore**

Mooring, James. *River Avon* on *Naish's Plan Salisbury* 1716

Moorsom, Capt. Robert. *Track of the Ariel* 1805, *Andaman Islands* 1835

Moorsom, W. S. *Southampton & Dorset Railway* 1844

Moosmair, Adolph. *Eisenbahn Karte Württenberg* (1880)

Mooy, Hendrick. Publisher, Amsterdam. Reissued Middagten's *Schagerrack* ca. 1730

Moraes Sarmento, A. de. *Delta do Zambeze* 1880 and 1891

Moraes E Souza, L. de. *Carta de Angola* 1885

Moraleda y Montero, J. de. *Costa Occidental Patagonica* 1794 MS, *Puerto de Valdivia* (1795)

Morales, Andres de. *Carta de la Isla Espanola* 1509

Morales, Jose Pilar. *Plano de Madrid* 1866, *Turquia* 1876

Morando, Giovanni Visconti (d. 1777). Austrian military cartographer

Morant, Alfred W. *Geological Chart* (1854)

Morant [Morand], Conrad (fl. 1561–73), b. Basle, d. Strassburg. *Plan of Strassburg* 1548

Morant, Philip. *Geogr. Antiqua et nova* London, Knapton 1742

Morant, R. L. *Siam & Her Neighbours* Bangkok (1895)

Morata, Antonio. Pilot in Philippine Marine. *Peso* 1835, *Filipinas* 1852

Morata, J. Engraver for Antillon. *Antilles* 1804, *Buenos Aires* 1812

Moravius, Jacobus. *Plan of Breslau* (1587) for Braun & Hogenberg

Morden, Robert (d. 1703). Geographer and publisher at the Atlas in Cornhill. Collaborated at times with Thos. Cockerill, William Berry, Philip Lea, Christopher Browne, J. Overton, Paske, R. Green, and R. Walton. *Plan of London* engraved Hollar 1675, *Geographical Playing Cards* 1676, *Geography Rectified* 1680, *Globe* 1683, *Atlas Terrestris* 1695, Camden's *Britannia* 1695, *Sea Atlas* 1699, *London* 1700, *New Description England* 1701 and separate *large scale maps* of various parts of the World

More, John. *Mintem & Hartley, Dorset* 1616 MS, *Stoke Park* 1613 MS

More, John. *Canaan* begun J. More, continued by J. Speed 1611. *Canaan* in Speed's *Genealogies* 1651. *Canaan* revised and published R. Green 1682

More, Roger. *Plan Quebec* 1759, *Fort du Quesne and adjacent country* 1759

Moreau, Edme. *Charleville* (1655), *Reims* (1675)

Moreau, Fr. Jacobus. *Peregrinatio Israelitarum* 1621

Moreau, Jean Victor (1763–1813). Military cartographer. *Souabe* 1801

Moreau de Saint-Méry, M. L. E. (1750–1819). *St. Domingue* 1796

Moreira [Monteiro, Morera, Montero], Ignacio (1724–1812). Portuguese cartographer in Japan. *Japan* ca. 1592 MS

Morel, E. *Planisphère Hydrographique* (1864)

Moreland, J. S. Surveyor. *Kingston upon Hull* 1834

Morell, George Webb. *Presqu'Isle Bay* 1835

Morell, Johann & Daniel. *Buch der Badischen Markgrafschaften* 1668

Moren, F. P. *Maps Patagonia* 1869–99

Moreno, Licenciado Antonio. *Magellan* 1618

Moreno, M. Engraver for Aguirre 1775. *Terra Firme-Antilles* 1805, *Oceano Atlantico* 1813

Moreno y Jerez, Federico. *Archipelago Filipino* 1875

Morera, I. See **Moreira**

Moresby, John. *Discoveries & surveys New Guinea* 1876

Moresby, Lt. (later Capt.) Robert, Indian Navy. Hydrographer. *Straits Durian* 1823, *Red Sea* 1830–33, *Maldives* 1838–39, *Chagos Arch.* 1839, *Malacca* 1840

Moreschi, S. *Citta di Bologna* 1822

Moretti, Dionisio. *Venezia* 1843

Moretus, Balthasar. Son of Jan (d. 1641). Editor and publisher of Antwerp. Edited 1612 Edition of *Ortelius* for Plantin and the 1624 edition

Moretus [Mogrentorf], Jan [Johannes], son-in-law, and succeeded Plantin in 1589 (with Raphaelengius). *Theatrum* 1601, 1612 bought plates from Vrints, 3 editions 1612

Moretus, Jan. Widow and son. Dutch edition of *Ortelius*. Continued as firm till 1704

Morey, W. *Military Map Cuba* 1897, *Havana Province* 1898, *Puerto Rico* 1898

Morgan, Augustus de (1806–1871). Mathematician and astronomer

Morgan, B. Assisted in *Plan part of Province of Pennsylvania* 1778

Morgan, Charles Carrol. *Criton & Fitch's Modern School Geography* 1870

Morgan, Charles S. *Plan of Richmond, Manchester & Springhill, Virginia* 1848

Morgan, David, of Llandaff. (15th century) *Geographia*

Morgan, Delmar. *Central Asia* 1884

Morgan, E. *Sketch of Kororareka* 1845

Morgan, Griffith. *Map of Bible History* (1865)

Morgan, Henry. *Mapp Lordship of Eburie. Parish St. Martins in Fields, London* 1675 MS

Morgan, Jacques de. *Mer Caspienne* 2 sh. 1895, *Perse* 1895

Morgan, Silvanus. Estate surveyor. *Chadwell & Little Thunock* 1646 MS

Morgan, William. Surveyor next the Blew Boar in Ludgate Street. Succeeded Ogilby as cosmographer Royal. *Map of City of London* 20 sh. 1676 (with John Ogilby), *London* 12 sh. 1681; *Middlesex* 1677 and *Essex* 1678 (both in conjunction with Ogilby)

Morgan, W. H. *Borough of Towanda, Bradford Co., Pa.* 1869

Morgan & Co. Engravers. Union Street,

Bristol 1800. *Juan Fernandez* in Selkirk's *Voyages*

Morghen, Filippo. Engraver and mapseller. Naples 1766, *Plan Naples* 1770, *Views & Plans Naples* 1772, *Magno di Napoli* (1810)

Morgues, Le Moyne de. See **Le Moyne de Morgues**

Moriarty, Henry A, *Santa Maria Mole, Admiralty* 1845

Morin, F. H. Publisher. *Stettin. Die Insel Rügen* 1835

Morin, Jean Baptiste (1583–1656). French scientist and astrologer. With Gassendi, fixed upon Ferro as prime Meridian for Louis XIII. *Science des Longitudes* ca. 1634, *Nova Mundi sublimaris anatomia* 1619

Morison, S. N. Surveyor. *Clackmannan* 1848

Morisot, Claude Bartholemy (b. 1592 Dijon, d. 1661). *Orbis Maritimi* Dijon 1643

Morley, D. H. *Townships Ohio* (1864)

Morley, Col. Francis. *World* 1884–5

Morley, William. *Plan Borough Haslemere* 1735

Morley. *New Mexico* 1873

Morlot, A. von. *Geolog. Karte zur Reise von Wien . . . bis München* Wien (1847)

Mornas. See **Buy de Mornas**

Moro, G. de A. y. See **Gómez de Arteche y Moro**

Moroncelli, Silvester Amantius (1652–1719). Benedictine abbot, globemaker of Fabiana. Cosmographer to Queen of Sweden and Roman Academy. MS *Globes* 1672?, 1677, 1681, 1713, 1716

Morozzo, G. *Il Patrimonio di S. Pietro* 4 sh. Rome 1791

Morphew, J. Bookseller, publisher, near Stationers Hall, London. Morden's *Magna Britannia* 1720

Morren, Edouard. Univ of Liège 1882–3

Morris, Charles. Chief surveyor, Nova Scotia. *Bay of Fundy* 1748, *Northern*

English Colonies 1749 MS, *Penin. of Nova Scotia* 1755, *Harbour Halifax* 1759, *Passamaquoddy* 1874

Morris, Billingsley. *Map Borough of Dudley in Worcester* 1836

Morris, C. A. F. *City of Bloomington, Ill.* 1855. *Map of Minnesota* 1861, *Route to Idaho* 1864

Morris, E. R. *Sydney & Environs* 1892

Morris, Lewis (b. 1701 Anglesea). *Plan of Harbours &c. St. Georges Channel* 1748, revised 1801. *Coast Wales* 1737–44, *Cambrian Coasting Pilot* 1737–42, *Estate Surveys* 1724, *Lead Mines N. Cardiganshire* (1750)

Morris, William. Son of Lewis Morris. *Revised Plans of Harbours and Charts* and added some of his own to 1801 edition *St. George's Channel* 1800

Morris, William E. *Bucks Co., Penn.* (1850), *Montgomery Co., Penn.* 1849, *Map Penn.* (1850)

Morris, W. R. *Plan St. Alphage Greenwich* 1834, *Southend* 1829

Morrison, Lieut. C. C. *New Mexico, Fort Leavenworth* 1875. *Wheeler Survey* 1878

Morrison, F. M. *City of Boston* 1896

Morrison, G. S. *Route maps of China* 1878 MS

Morrison, Hector. Surveyor. Corrected *Map Co. of Sutherland* by Burnett & Scott

Morrison, J. Publisher of Glasgow. *New Edinburgh General Atlas* (1844)

Morrison, Samuel. *Five N.W. States* Cincinnatti 1859

Morrison, William. *Plan Part Estate Lochiel* 1772

Morrison & Sons. Publishers. *Scotland* 1795

Morrison & Williams. *Hamilton Co., Ohio* 1835

Morse, Charles Walker. *Illinois* 1854, *Indiana* 1855, *Iowa* 1856, *Gen. Atlas* 1856

Morse, H. Engraver for Edwards 1832, for Baldwin 1837

Morse, H. B. *Prov. of Kwangsi* 1897

Morse, Rev. Jedediah (1761–1826). American geographer. *American Geography* 1792 (2 maps), 1795 (25 maps). *Amer. Univ. Geography,* Charlestown 1819. *Modern Atlas* 1822

Morse, Sidney Edwards (1794–1871). Son of Jedidiah. Publisher of New York. Divised technique of wax engraving. *Atlas of U.S.* New Haven 1823, *Universal Atlas* 1825, *Geogr. Atlas of U.S.* 1842, *Specimens of American Cerographic Maps* 1843. Collaborated with Samuel Breese *Michigan* 1844

Morse & Gaston. Publishers 115 and 117 Nassau St., N.Y. *Kansas, Nebraska* 1856

Mortier, Corneille [Cornelis] . See **Covens, Jan**

Mortier, David. Book, Map, and Print-seller at ye Sign of Erasmus's Head near the Fountain Tavern in ye Strand. *West Indies* 1692, *Plans of Vigo* 1702, *Cadiz* 1703, *Toulon* 1707, *Lille* 1708, *Nouveau Théâtre de la Grande Bretagne* London 1715–28, *Paraguay* 1720

Mortier, Pierre (1661–1711). Publisher and mapseller in the Vygendam, Amsterdam. Pub. Sanson & Jaillot's maps from 1690, *Neptune François* 1693, also pub. for N. de Fer 1696 and Fred. de Wit 1703. Acquired De Wit's stock in 1706 and Jansson plates from Schenk and Blaeu's plates of *Towns*

Mortillaro, Carlo and Vicenzo. *Atlante Gen . . . di Sicilia* (1859)

Mosberger, William A. *Sect. Map Arkansas* 1872

Mortimer, C. *Chart Virgin Islands* 1739

Mortlock, Henry. Bookseller of London at the Phoenix. *Playing card maps* (1676)

Morton, John. *Northumberland* 1712

Morton, Robert H. American surveyor. *Alabama* 1889 and 1895

Moryson, Sir Richard. *Survey of Castles, Forts . . . Thames . . . South Coast* MS 1623

Mosburger, G. L. *Battle plans Belgium* 1684–1709

Moseberg, J. H. *Colonie Suriname* (1780) and 1801

Mosely, Edward (d. 1749). Surveyor General North Carolina. *Estate Survey* 1706, *Va.–N. Carolina Line* 1709–28 MS, *N. Carolina* 1733, German edition 1736

Moses of Khorene (5th century). Armenian historian and geographer

Mosley, Charles. Engraver. *Citadel Plymouth* 1737, *Ukraine* (1740), Rutherford's *Highlands Scotland* 1745

Mosquera, Tomas Cipriano de. *Curso del Rio Magdalena* 1849, *Republica Granada* 1852, *Estados Unidos de Colombia* 1864

Moss, C. D. *Oberlin,* Ohio 1870

Moss, William. Surveyor. *Plans in Monk Sherborn and St. Lawrence Wootton* 1790

Mosse, R. *Harwich–Lexden Railway* · 1844 MS

Mossner, A. G. Engraver for Schlieben, 1828

Mossner, Jean Michel. Engraver. Worked for Gaspari 1804–11, for Schlieben 1828

Mosting, Hermann (17th century). Engraver of Lüneburg

Mosto, A. C. de. See **Cadamosto**

Mot, Ch. de. *Prov. de Buenos Aires* 4 sh. 1880

Mott, Frederick T. *Charnwood Forest & Environs* (1861)

Mottu, Fried. August (1786–1828). Painter and cartographer

Mouchez, A. Ernest B. (1824–1892). Hydrographer. *Tartary* 1854, *Plans coasts of Brasil* 1862–70

Mouchin [Mukhin], General major.

Military Map Crimea 10 sh. 1817, London edition 4 sh. 1854

Mougenot, F. *Carte géologique de France* 1878

Mould, Thomas R. *Defensible Works Auckland, N.Z.* 1860

Moule, Thomas (1784–1851). Bookseller and publisher. *English counties delineated* 1837

Moullart-Sanson, P. See **Sanson**

Mount & Page. Publishers of marine charts, Postern Row on Tower Hill. R. Mount & T. Page 1698–1712. R. & W. Mount & T. Page 1713–22. R. Mount I (d. 1722). T. Page & W. & Fisher Mount 1723–28. F. Mount (d. 1728). T. Page & W. Mount (1729–33). T. Page I (d. 1733). W. Mount & T. Page 1733–48. W. & J. Mount & T. Page 1748–55. W. & J. Mount & T. Page & Son 1755–62. T. Page II (d. 1762). W. & J. Mount & T. Page (again) 1762–3. W. Mount III ret. 1763? (d. 1769). J. Mount & T. Page 1764–73. J. Mount, T. Page, W. Mount, T. Page 1775. J. Mount, T. Page, W. Mount 1778. J. Mount, T. Page, W.M. T.P. (again) 1780. Messrs. Mount & Page 1781–4. T. Page III (d. 1781). Mount & Davidson 1789–1800 James Davidson 1800–01. Davidson, Smith & Venner 1801, Smith & Venner 1802. J. Mount (d. 1786). T. Page IV, ret. T. Page IV (d. 1797). W. Mount IV (d. 1800). S. Smith [Leard's *Jamaica*] (1805). The firm of Mount & Page published succeeding editions of the *English Pilot* and Collin's *Coasting Pilot* throughout the 18th century

Mount & Page, 1770. *First chart of Gulf Stream* pub. order Benjamin Franklin, 1770

Mount, Richard (d. 1722). Publisher and bookseller. Apprenticed to W. Fisher 1669. Married into the family and took over part of the business on Fisher's death in 1691. With Fisher published Seller's *Atlas Maritimus* 1685. Sold Collins *Coasting Pilot* 1693, *English Pilot* 1698

with Thornton. Collaborated with Jeremiah Seller 1699. *Sea Coasts, France* 1701, *World* in *Harris' Voyages* 1705

Mount, William. Son of Richard. Brought into partnership Mount & Page 1712. *English Channel* 1730

Mountain, C. *Plan Kingston upon Hull* 1817

Mountaine, William. *Hydro. Chart of World* based on Halley 1745 and 1758

Mouqueron, P. Arsène. *Atlas du Pérou* 1865

Mouravieff, Capt. *Map of Country of Chiva and the Turkomans* Moscow 1822, *Tiflis to Chiva* 1822

Moureaux, Alphonse. Dépôt de la Guerre *Globes* 1873

Mourelle, Francisco Antonio. *Gulf of California* 1777 MS, *Inlet of Bucareli* 1789

Mourik, B. *Graafschap Holland* 1761

Mourilyan, Thomas L. Navig. Lieut. H.M.S. *Basilisk. Port Moresby & Fairfax Harbour*, Admiralty 1873, *S.E. End New Guinea* 1873

Moussaint, H. J. *Versailles and environs* 4 sh. Paris 1788

Moussy, V. Martin de. *Description Géogr. de la Confédération Argentine 1873*

Moutoux, Jules. *Stadt Mannheim* 1847, *Baden* (1850)

Mouzon [Mouzen], Henry, Jr. (1741–1807). *North & South Carolina* 1775, French edition 1778, *Southern British Colonies* (with Capt. Collet) 1776, *Parish St. Stephen Craven Co.* 1773

Mowat, J. L. G. *Properties in Oxfordshire* 1888

Mowry, Sylvester. *Map of Town of Fairhaven* 1855, *Sonora & Arizona* 1863, *Union Co., Ohio* 1870

Moxon, James the Elder and Younger (fl. 1647–96). Engravers and publishers on Ludgate Hill at the Signe of the Atlas. Engraved for Lea 1657, *Carolina* 1681,

for Adair 1688, for Ogilby *(Carolina)* 1671, for Collins *Coasting Pilot* 1693

Moxon, Joseph, of Wakefield (1627–1700). Instrument maker, map publisher, and engraver. At the Sign of the Atlas in Cornhill. Tutor in astronomy and geography. Hydrographer to the King 1670. *World* 1655 for Wright's *Certain Errors. Book of Sea Plats* 1657 (the first English attempt to break the Dutch monopoly in chart publication), *Treatise on Globes* 1659, *Americae Sept. Pars* 1644 (first map to show Manhattan Island after British Occupation and first on which name *New York* appears), Adair's *Strathern* (1690), *Garden of Eden* 1690. Engraver for Wells' *New Set of Maps* 1701, *Sacred Geography* 6 plates 1671

Moya, Carlos (with E. de P. Beltram). *Plano Ciudad de Merida* 1864–5

Moyne, F. Le. See **Franciscus Monarchus**

Moyne de Morgues, Le. See **Le Moyne de Morgues**

Mubeck, C. I. B. *Germany* Hamburg 1803

Muchnitsky, M. I. *Plan St. Petersburg* 4 sh. St. Petersburg 1868

Mudge, R. L. *Portion New Brunswick* (1845)

Mudge, Lt. Col. William (1762–1820). Director of Ordnance Survey 1798. *Acc. Trig. Survey* 1799–1811, *Kent* 1801, *Essex* 1804–5, *Isle of Wight* 1810, *Sussex* 1813, *Cornwall* 1809–13, *Map Co. of York* 9 sh. Wakefield 1817

Mudge, Lieut. William, R.N., later Comm. Admir. surv. Assisted Admiralty Charts. *Africa* 1822–6. *Ireland* 1827–37

Mudie, Robert (1777–1842). London publisher

Muela, Andres. Engraver. *Nueva Andalucia* 1778

Muffing, General Friedrich (1775–1851). Introduced a form of contouring

Muelich [Mülich], Hans. *Plan of Ingolstadt* 1546

Muflin, Joannes. *Plan of Bilbao* 1544, used by Braun & Hogenberg

Mugeraner, Franz, Pfarrer. *Diocesan Land und Postkarte Oberwienerwald* Linz 1844

Muhry, A. J. W. E. Adolphe (1810—188). Geographer and meteorologist

Mujia, J. M. *Mapa Repub. Bolivia* 1859

Mukhin, Gen. Maj. See **Mouchin**

Mulder. Engraver for Van der Aa. *Galérie Agréable du Monde* 1724

Mulheuser, J. Julius (1611—1680). Engraver

Mülich, H. See **Muelich**

Mullan, Capt. John. *N. Pacific States* 1865

Mullens, Rev. Joseph. *Central Africa* 1879, *Madagascar* 3 sh. 1878

Müller, Carl Friedrich (1776—1821), Engraver of Leipzig. *France* 1792. Worked for Gaspari 1804

Muller, Caspar. *Koningryk der Nederlanden* The Hague 1816

Muller, Charles. *Ancient Geography* 1874, maps for *Ptolemy* edition (1901)

Muller, Christian Frederick (1776—1821) Publisher of Karlsruhe. *Bayern* 1814, *Rheinbayern* 1817

Muller, Christian Helfrich (1621—1691). Military cartographer

Muller, Curt (1865—1928). Map lithographer

Muller, C. T. *Map of Pompeii* (1819)

Muller, F. Engraver. *Land ob der Enns* 12 sh. 1787

Muller, F. *Reg. Hungariae, Croatia, Sclavonia . . . Transylvaniae* 2 sh. 1792

Muller, Fr. Th. *Greece* 12 sh. Vienna 1800

Muller, Gaspard. *Map Netherlands* 3 sh. Amsterdam 1816

Müller, Georg Wilhelm (1785—1843). Military cartographer of Hanover

Müller, Gerard Friedrich, (b. 1705, d. 1783 Russia). *Nouvelle Carte des Descouvertes faites par des Vaissaux*

Russes 1754, reissued 1754 and extensively copied

Müller, Gottfied (d. 1803). Military cartographer

Müller, H. *Eisenbahn Karte von Mittel Europa* Glogau 1861

Muller, (I.). *Mappa Geogr. Hungariae* 12 sh. 1769

Müller, I. C. Engraver for Gaspari 1804—11

Müller, Joachim Eugen (1752—1833). Panoramas and cartography in relief

Muller, J. F. Valerien. *Plan Koenigsberg* 4 sh. 1815

Muller, J. J. A. *Map of environs of Maastricht* (1871)

Müller, Johann (Joh. de Monteregio). See **Regiomontanus**

Müller, Johann Christoph (1673—1721). Austrian engineer and cartographer. Helped Marsigli survey of *Danube* 1696 and *Hungary* 1699 and 1709 (4 sheets), *Schweizerischer Atlas* (1712), *Bohemia* 24 sh. Augsburg 1720, *Moravia* 1720

Müller, Johann Jakob (1743—1795). Engraver, publisher and cartographer in Vienna

Müller, Johann Ulrich (1633—1715). Geographer and cartographer. *Geogr. totius orbis* Ulm 1692, *Strassenkarte v. Deutschland* 1692, *Kriegskarte d. Ulmer Gebietes* 1693

Muller, John B. *Civil War Map* 1865

Muller, S. *Maps Indonesia* 1841—45

Muller, Thos. *Lithog. plan New Orleans* ca. 1840

Muller, T. C. *Terrest. Globe* ca. 1800

Muller, Wilhelm. Painter and lithographer in Warsaw (early 19th century)

Muller, W. *Atlas Ethnographique de l'Europe* 1842

Muller, W. *Königreich Hannover* 20 sh. Hannover 1818, *Postkarte Hannover* 12 sh. 1821

Müllersche Buchhandlung. Erfurt *Atlas Königreichs Preussen* 1831

Mullet, John. Surveyor. *Plan Détroit* 1830

Mullhaupt, Fritz (1846–1917). Geographer and cartographer

Mullhaupt, H. H. (1820–1894). Swiss cartographer and colourist. *New map of Switzerland* 1860. *Chemins de Fer, Postes et Télégraphes de la Suisse,* Berne 1872; *Front Franco–Allemand* 1890

Mulligan. See **Miller, Orton & Mulligan**

Mulliner, Thos. Surveyor, Old Broad St. *Ware Park* 1850

Mullineus, E. See **Molyneux**

Mullon. *Plan de défense de New Port dans l'Isle Rhode* 1780

Mumbrute, William B. *Map of Wisconsin* (1860)

Mumford, Daniel. Estate surveyor, 10 Greville St., Hatton Gdn. *Thunderley* 1780 MS, *Felsted* 1796 MS

Munch, Peder Andreas (1810–1863). Norwegian geographer. *Hist. Geogr. Beok. over Norge* 1849, *Nordlige Norge* 1852, *Orkneys* 1862

Munday, A. Enlarged Stow's *Survey London* 1618

Mundel, R. Estate surveyor. *Harlow* 1782 MS

Munich, Geog. Anstalt. See **Geographische Anstalt**, Munich

Munk, F. de. See **Franciscus Monarchus**

Munnich, Maréchal. *Maps and plans of Lake Ladoga* 22 sh. St. Petersburg 1764

Munnickhausen, Johann van (1641–1729). Engraver

Munnicks, F. See **Franciscus Monarchus**

Munoz, Ignacio. Fr. missionary. *Descrip. Géogr. Florida* 1765, *Manila* 1671 MS

Münozguren, J. M. Mexican engraver (fl. 1863)

Munro, A. Printer, Queens Head Yard, Great Queen Street, Lincolns Inn Fields. *Chidley New Gen. Atlas* (1825)

Munsell, Luke (1790–1854). American surveyor and publisher, Kentucky. *Frankfort* 1818

Munson. See **Doolittle & Munson**

Munthe, Gerard (1795–1876). Norwegian military cartographer and historian. *Map Scandinavia* 1826

Münster, Sebastian. Cosmographer and scholar, b. 1489 Ingelheim, d. 1552 Basle. *Geographia* 1540, *Cosmography* 1544, MS maps from ca. 1514. Married widow of Adam Petri 1530, edn. Etzlaub's map 1525, Commentaries of *Solinus* and *Pomponius Mela* 1538

Münster, Seb.

Münster, Sebastian. *Statt Basel mit um Ligender Landschafft.* 79 × 38 cms. in *Baszler Chronick Henricipetri* 1580

Munthe, Gerhard (1795–1876). *Scandinavia* 1842

Muntzer [Münzer] (Hieronymus) (1437–1508). Physician in Nuremberg, compiled modern *map of Germany* after Cusanus for *Nuremberg Chronicle* 1493

Muos, Heinrich Ludwig (1657–1721). Swiss topographer and engraver

Murchison, Sir Roderick Impey (1792–1871). Geologist. *Silurian system* 1838, *Geolog. Map England & Wales* 1843–4, *Geolog. Russia & Ural Mts.* 1845, *Geolog. Atlas Europe* 1856, *New Geolog. map Scotland* 1861

Murdoch, J. See **Stewart, Meikle & Murdoch**

Murdoch, Patrick (d. 1774). Mathematician, pub. geographical works

Murer [Maurer], Christoph (1558–1614). Swiss painter and engraver. Son of J. Murer. *Switzerland* 1582

Murer [Maurer], Hans. Son of Josias. Pastor of Rickenbach. *Thurgau* 1620

Murer [Maurer], Joost [Joseph, Josias, Josen] (1530–1580). Swiss writer, glass

S. Münster (1489–1552); mezzotint portrait by J. Haid

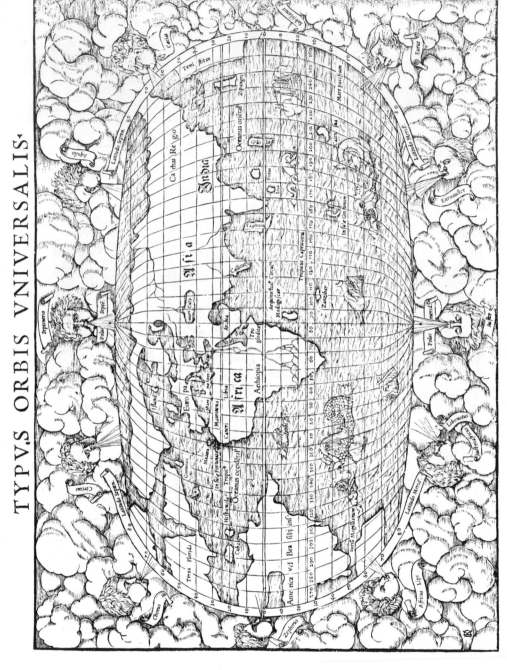

TYPVS ORBIS VNIVERSALIS.

Woodcut world map by Seb. Münster (published Basel 1545)

painter, surveyor and woodcut engraver. *Map Zürich* 6 sh. 1566, *Plan Zürich* 1576 used by Braun & Hogenberg, *Black Sea* Geneva 1577

Muri, Ibrahim (15th century). Arab hydrographer

Muriel, Ed. A. y. See **Almonte y Muriel**

Murillo, Velarde Pedro de (1696–1753). Spanish cartographer. *Map Philippines* 1754

Murillo. Maps for Laurie & Whittle's *Pilot* 1800

Murphy, Dan. Architect. *Plan Cook* 1789

Murphy, E. J. *Map of Germantown, Pa.* (1851), *Greene County (Ohio)*, (1855), *Oneida County, N.Y.* (1852)

Murphy, W. (b. 1800). Engraver and printer, 209 High St. Edinburgh. Wood's *plans of Annan* 1826 and *Greenock* 1825, *Pocket Atlas of Scotland* 1832, *Biblical Atlas* 24 maps drawn and engraved by Murphy 1835, *Scotland for Scottish Tourist* 1850?

Murr, Christoph Gottlieb v. (1733–1811). *Amazonum Fluminis* 1780

Murray, Alexander. *Diocesan map C. of E. in Newfoundland & Labrador* (1877)

Murray, H. *Sketch Action Bladensburg* 1814

Murray, John (b. Edinburgh 1745, d. 1793 London), founded firm of John Murray first at 32 Fleet Street London 1789 and at 50 Albemarle St. from 1820. *Political Mag.* 1789, Parry's *Journal* 1821–4 Lyon's *Journal of Fury & Hecla* 1824, Franklin's *Journal* 1828, *Journal R.G.S.* 1832, *Handbook Shropshire & Cheshire* 1879, *Playing Cards* 1892

Murray, Lieut. J. Admiralty surveyor. *Charts of S. Coast Winchelsea to Shoreham* 1804–07, *Australia* 1802–3

Murray, T. L. British publisher. *Atlas of English Counties* 1831 (44 maps) and later editions

Musculus, Johann Conrad (d. 1651). Cartographer

Muse, W. T. *Hydro. Chart James River, Va.* (1859)

Musgrove, J. *Ancient Chester* (20 plts.) (1880)

Musi, Giulio (fl. 1535–58). Venetian engraver

Musil, A. *Arabia* 4 sh. 1906

Musinus, B[artholomew?]. Chartmaker. *Europe* 6 sh. 1560, pub. de Jode

Musis, Agostino de [Agostino Veneziano]. Venetian. *Plan Tunis* 1535

Muth, Brothers (fl. 1720). *Celestial globe* 1721

Muth, Heinrich Ludwig (1673–1754). Instrument maker in Cassel

Mutlow, H. Engraver. Hall & Pinnell's *Plan Gloucester* 1780–2

Mutlow, I. Engraver, 6 James St., Covent Gdns. For Morse 1794, Hebert's *Lakes* 1816

Muzinger, W. *Pays au nord de l'Abyssinie* 1857 *Ost Afrika* 1861

Myer, Albert J. (1828–1880). American geologist

Myers, J. F. *Parish Halifax, W.R.* 1834–5

Myers, Thomas. British geographer. *Modern Geography* 1822 (49 maps)

Myers. See **Tofswill & Myers**

Mylius, Printer and editor. Translated Ptolemy's *Geographia* 1584, *Map Lithuania* 1589

Mylne, R. W. *Geolog. Map London & Environs* 1871

Mylne, Robert, F.R.S. *Sicily* 1757, Laurie & Whittle 1794 and 1799

Mylne, T. See **Milne**

Mynde, James. Engraver. Wimble's *N. Carolina* 1738, *Par Harbour* (1742?), *Ramsgate* 1755, *Draught of Humber* 1761, *Spurn Head to Wells* 1749, *Havana* (1760), *Middlesex & Essex Turnpikes* 1740, Hammond's *North Sea* (1720) for G. Collins, *Choro. Chart & Kent* 1743

Myrbach v. Rheinfeld, Karl (1784—1844).
Military cartographer

**Myrica [Merica, Ameringius, Amyrilius,
Van der Heyden]**, Pierre? Gaspard (1530—
1572). Engraver of Louvain/Antwerp.
Collaborated with Gemma Frisius *Globes*
1530—37, *World and Holy Land* 1573 in

Polyglot Bible

Myritius [Miricio, Mauritius], [Joannes]
Opusculum Ingoldstadt 1590 *World map*

Myslakowski, Ignacy (b. 1795). Military
topographer and lithographer of maps in
Warsaw.

N

N. B. Natalis Bonifacius. See **Bonifacius**

N., I. de [Anne Joseph de Neuville] *Great River Maranon or Amazons,* pub. in Quito 1707; London, Senex 1712

N, St. N. Stopius

Nabert, H. *Die Deutschen in Europa* 8 sh. (1891)

Nabholz, J. C. Engraver. *Charte des Causkasis Gebirges* (1760), *Plan St. Petersburg* 1786

Nachtigal, Dr. Gustave (1834–1885) Sahara and the Sudan. *Africa* Gotha, Perthes 1874, *Lake Chad* 1876

Naftel, C. O. *West Indies* 1898

Nagaev, [Nagahieff] Capt. Aleksyei Ivanovic (1704–81). *Atlas of the Baltic,* Russian Admiralty 1757, 1789, 1792. *Tab. Hydro. Sinus Finnici.* 1777

Nagakubo, Sekisui (1717–1801). Japanese geographer

Nagel, A. A. M. *Grundriss von Hamburg* 1845

Nagel, Heinrich. Cartographer. *Salzburg* 1590, *Westphalia* 1590, *Scotland* 1595. Maps for Quad (1600)

Nagel, Joseph Anton (1717–ca. 1800). Viennese cartographer

Nagel, Louis. Lithographer of San Francisco, 529 Clay St. For Dixson & Kasson 1859, for Ranson 1862

Nagel, Fishbourne & Kuchel. Lithographers, Clay St., San Francisco. *Fraser River Route* 1862

Naghtegaal, Aernout. Composed title for Keulen 1682

Nagy, Karoly (1797–1876) Hungarian globe maker. *Terrestial* 1840, *Celestial* 1840

Nairn[e], Capt. Thomas (d. 1715). In Carolina 1702. Additions to Crisp's *Carolina* 1711

Naish, Thomas. *Inset River Avon* on W. Naish's *Plan Salisbury* 1716

Naish, William. Surveyor. *City of Salisbury* 1716 and 1751

Nakamura, Lieut. K.U. *World map* 1873, *Japanese Islands* London, Admiralty 1878

Nakashima, K. *Geological Survey Japan* 1884

Nakoshin, Capt. *District Velige* 16 sh. 1785 MS

Nakovalnik, Sergej Fedorovic (1732–1790). *South Russia* 1686 (with Solomein)

Nalkdwski, Waclaw (1851–1911). Geographer, author of many works on Polish and world geography, text books, reviews, and chronicles, 1881–1900. Published with Andrew Swietochowski 1844–1928. *Great Atlas of World*

Nancarrow, John. *Chart of Mounts Bay* 1751 and 1793

Nancrede, Publisher of Boston. Malham's *Naval Atlas* 1804

Nancy Globe (1530) Silver ball 16 cm. In Lorraine, Mus. Nancy.

Nansen, Dr. Fridtjof b. 1861, *Ost Grönlandischen Kûste* 1888–1892, *Arctic Regions* 1892, *Süd Grönland* 1892, *N. Polar Map* 1893

Nansen, Joh. (1598–1667) Traveller in Russia. *Compendium Cosmographicum* Copenhagen 1633

Nansouty, Ch. M. E. (1815–1895) founded observatory at Pic du Midi

Nantevit. Engraver for Blaeu 1668

Nantiat, Jasper. *Russian Dominions in Europe* 2 sh. 1808, *Spain and Portugal* 4 sh. 1810

Napier, Lieut. Col. Edward D. H. E. (1808–1870). *Plan Troad* 1839, *Road across Mt. Lebanon* 1840 MS, *Plan Alexandria* 1842 MS

Napier, Capt. Hon. George E. *N. Front Khorasan* (16 miles to inch) 1876

Napier, Lt. the Hon. H.D. *Iran* W.O. 1894

Napier, John. *Island of Engano, Admiralty Chart* 1822

Napier, Richard H. *Admiralty Charts Coasts of China and Malaya, Fiji* 1867–78

Napier, Robert Cornelis Lord Napier of Magdala. *Line march Annesley Bay to Magdalla* 1869, 5 sh. Ord. Survey 1869

Napoli, Zuan de. Portolan chartmaker 15th century

Narborough, Admiral Sir John (1640–1688) *Chart South Seas* 1673, engraved Thornton. *Magellan Straights,* Thornton 1694

ᴎarciss, Theodor August (1721–1773). Military cartographer

Nares, Capt. Sir George Strong, R.N. *Admiralty Charts Sicily* 1864–70, *Gulf Suez* 1873, *Fiji* 1875

Naronowicz-Naronski, Josef (d. 1678). Mathematician and cartographer, worked for the Radziwill family of Nieswiez; after 1660 worked for geographical survey of Prussia. Author first Polish textbook

on cartography (did not appear in print)

Narstin, John L. Philadelphia. Printed and coloured Robinson's *Mexico* etc. 1819

Narvaes, [Narvaez] José Maria. Pilot. *California* 1823 MS, *Dept. de Jalisco* 1840

Narvaes, Juan. *Great River Maranon* Quito 1707

Narwoysz, Franciszek. Jesuit. One of the surveyors of the western regions of Lithuania. His observations used later by Perthes in 1770 for maps of Poland and Lithuania

Nash, John. Numerous Plans of parts of London: *Regents Park* 1812, *Marylebone Park* 1812, *Regents Park Estate* 1826, *St. James'* 1827

Nash, John (17th century). Instrument maker

Nash, Joseph D. *Maps of Michigan & Indiana* 1861–4

Nash, Lieut. R.N. *Admiralty Charts E. Africa, Seychelles, Mauritius* 1825–7

Nasir Al-Din Tusi (1210–1273). Founded Observatory in Malaga 1279. *Islam Atlas* 1261

Natalis, Bonifacio (16th century). Italian engraver

Natalius, J. See **Metellus**

Nathorst, A. G. (b. 1850). Swedish geographer. *Geolog. Karta King Karls Land* Stockholm 1899

Nation, P. A. Surveyor. *Rivoli Bay* 1858, *Lacipede Bay, S. Australia* 1859

National Society London. *Occupations of People B.I.* 1851, *British Isles* 4 sh. 1877, 2 sh. 1879; *Europe* 1878

Natoroff, W. & Co. Berlin publishers. *Atlas von Europa,* 2 vols. 1834–7

Nav, Anthony. Drew for Pike's *Louisiana* 1810, *Mississippi* (1806) MS

Nav De Champlouis. *Carte de l'Afrique sous la domination des Romains* 1864

Naudin. Family of French engineers attached to Dépôt de la Guerre,

Naudin, the elder (fl. 1688–1746). Engraver-in-chief. *Meuse* 1688, *Anvers* 1703, *Théâtre de la guerre en Allemagne* 1726 MSS

—— C., the younger. *Lombardie* 1702 MS.

Naumann, Bernard. *Die Küste der deutschen Nord See* 1894

Naumann, Edmund. *Geolog. Survey Japan* 1884–94

Naumann, Karl Friedrich (1797–1873). Geologist and mineralogist. *Maps of Saxony* 1845 and 1871

Nautonier, Guillaume de, Sieur de Castelfranc. *Orbis Terrae* 1603

Nautz, F. J. *Kaart der Stad Haarlem* (1822) and 1836

Navarrette, M. Fern. de (1765–1844). Spanish sailor and geographer. *Collection Voyages and Discoveries* 1825–1837

Navarro, Francesco. *Teatro de la Guerra entre Espana in Francia* 1794

Navas, Francisco de. *Plano de la villa de Panzacola* 1781

Nax, Jan Ferdynand (1736–1810). Polish economist and author of hydrographic maps

Naylor, John. *Astron. and Geogr. Clock* (1680)

Naymiller, F. *Maps of Austria and Italy* 1857–59

Neal, Daniel. *Map New England* 1720, *Harbour of Boston* 1747

Nearchus. Greek navigator. Sailed from the Indus to Arabian Gulf, 330 B.C.

Nebolssin, Paul. (1817–1893). Russian geographer

Nebot, Pasqual. *Seno Mexicano* 1781

Neck, Jacobus Cornelius van. *Tabula Itineria* 1600 (voyage to East Indies)

Neckam [Neckham], Alexander. English monk of St. Albans, b. 1157. Wrote on the magnetic needle in Paris Univ. ca. 1180. *De Utensilibus*

Necton, William. *Platte of Tottenham Courte* 1591

Neef, Martin. *Road map Thuringia* 1666 MS

Neele, George. Engraver. 11 Judd Pl., West St. Pancras. *Allotments New South Wales* 1814 for Pawley's *Gen. Atlas* 1819 (with Samuel). Later at 352 Strand (1826)

Neele, J. Publisher, 3 Burleigh St., Strand, London. Sutherland's *Ajerbayjaun* (1810)

Neele, James. Engraver. *Plan of Cheltenham* 1834, *Topo. Survey Thebes* 6 sh. 1830, *Africa* 4 sh. 1823

Neele, Josiah. Publisher & engraver, 352 Strand, London. For Poirson 1826, Greenwood 1827–9, Johnsons *Antigua* 1829, Seatons *Palestine* 1835, Fowler's *Yorks.* 1836, for Colton 1845. Also worked with George and James Neele

Neele, Samuel John (1758–1824). Engraver 352 Strand, London. Antiquarian views and maps. Worked for Dugdale, Faden, Laurie, Leard, Playfair, Stackhouse. Thomson &c. *Scotland* 1782, *Saffron Waldron* 1787, Cole's *Fens* 1789, Albin's *Isle of Wight* 1795, Beaufoy's *Meath* 1797, *Ordnance Survey* 1804, *Lisbon* 1814

Neele & Son. Engravers, 352 Strand. For Greenwood 1818–24 &c.

Neely. See **Sherwood, Neely & Jones**

Neeper, Alex. M. *Election Districts cities Pittsburgh & Allegheny* 1886

Neer, Cornelius van der. Pilot of Amsterdam. *Chart Coast of Holland* 1781, 1794 and 2 sh. 1801

Neff. See **Sidney**

Nègre, Louis. *France Protestante* 1878

Negri, Christoforo [1809–1896]. Italian, historian and geographer

Negri, Domenico Mario (15th century). Venetian geographer

Nehrlich, W. Engraver for Kuypers *Atlas* 1855–7

Neiperg, Count of. See **Herberstein**

Neiss, P. *Hoffstatt Arrakan* (1680)

Nell, A. M. (b. 1824). Work on projections and astronomical observations

Nell, Johann Peter. *Post Roads Germany and Low Countries* 1709–11. Worked for Homann *Atlas German.* 1753

Nell, Louis. Civil engineer. *Colorado* 1880

Nell. See **Frey & Nell**

Nelli, Nicolo. Venetian engraver. *Malta* 1565, *Africa* 1565 3 part, *Asia* 1565 after Gastaldi, pub. Bertelli. *Gotha* 1567, *Cyprus, Morea, Poland* 1570

Nellson, Robert. *Admiralty Charts Adriatic* 1803

Nelson, E. W. *Northern Alaska* 1882

Nelson, Horatio, Lord. As Lieut. drew chart. *Chart of Atlantic* 1805

Nelson, Newell. *Town of Cumberland Rhode Island* Boston 1838

Nelson, H. S. *Galway Bay* 1818

Nelson, Thomas. *Wall Maps* 1859–64

Nelson, T. *Atlas* 1859 (with T. Davies)

Nepveu, Charles (1791–1871). Military cartographer

Nerenburger, Ad. G. (1804–1869). Dutch geographer, director Dépôt de la Guerre. Works on construction of maps

Neron, Mathieu. Cosmographer. *California* Florence 1604, (Buache 1754)

Ness, W. W. van. *Economic Geolog. Map Matabeleland,* Stanford 1896

Nessi, Ernesto. *Alta Lombardia* (1888)

Netherelift, J. Lithographer. *Parish of Lambeth* 1824, Dawson's *Plan Manchester* 1832, *Prov. Minas Geraes* 1835, *Anglers Map Thames* (1841)

Netherlitt, J. Lithographer. 23 King William St., West Strand. *Map* in *Strangers Guide to London* 1835

Nettebroth, Henry. *Louisville* 1869, *Kentucky* 1873

Neu, Andreas, Freiherr von (1731–1803). General, cartographer in Vienna

Neual, Isaac. Engraver for Cook (1773)

Neugebauer, J. G. *Fluss & Hoehenskizze von Deutschland* 1838

Neugebauer, Salomon. German geographer, 17th century

Neumann, C. J. H. (1823–1880) German geographer and Hellenist. *Kurland* 1846, *Livonia and Kurland* 1867

Neumann, Wenzel August. *Maps Italy and Adriatic* 1859

Neumann, Wilhelm. *Cassel* 1877–8

Neumayer, George Balthasar von (1826–1909). Founded and directed observatory at Melbourne. *Gen. Map of Australia* 1863, *Orographische Karte Victoria* 1871, *Süd Georgien* 1886

Neumayr, Melch. (1845–1890). German geologist and palaeontologist

Neussel, G. J. W. *Teil Prov. Ciedad Real* (Spain) 1884

Neussel, Otto. Artist. *Atlas Geogr. Univ.* 1877

Neussner, Joseph. Engineer to Court of Vienna. *Plan Vienna* 4 sh. 1770 (with F. Gruss)

Neuvel, S. van den. See **Novellanus**

Neuville, Buache de la. See **Buache**

Neve, P. le. See **Le Neve**

Nevil [Nevill, Neville], Arthur Richard. Irish cartographer. *Wicklow* 1790, *Wexford* 1798

Nevil [Nevill], Jacob. Irish cartographer. *Wicklow* 2 sh. 1760

Nevil, James. Publisher. Scull's *Pennsylvania* 1770

Nevot, Pasqual. *E. coast Florida &c.* Madrid 1784

New, G. R. Surveyor. *The Parish High Wycombe* 1840 MS

New, W. Publisher, 11 Strand. Froggett's *Survey 30 miles round London* 1831

Newberry, J. Str. (1822–1892). U.S. Geological Survey. *Geolog. Map Ohio* 1870

Newbery [Newberry], Francis John (1743–1818). Publisher & bookseller, 65 St. Paul's Churchyard. Son of John Newbery. Partner with stepbrother Thomas Carnan from 1767. *W. Hemisphere* 1776, *Travelling Dict.* 1773, *Counties of England & Wales* 1779

Newbery [Newberry], John (1713–1767). Bookseller and publisher. Left printing business by J. Carnan and married his widow 1737. Worked from Bible & Crown, Devereux Court and later Bible & Sun, St. Pauls Churchyard. *Atlas Minimus* by Gibson 1758, *Counties England & Wales* 1759

Newbery [Newberry], Nathaniel. Printer and map seller of Pope's Head Alley, London. *Bergen op Zoom* 1622

Newbery [Newberie], Ralph [Rafe]. English publisher. Hakluyts *Voy.* 1582 Stow's *Annals* 1580, Lambarde's *Kent* 1576

Newbold, Lieut. J. J. *Malacca Territory* 1837

Newcomen, Robert. *Plott of Droughedagh* 1657

Newcourt, Richard (d. 1679). Topographical draughtsman. Born Somerton in Somerset. *Plan London & Westminster,* engraved William Faithorne 1658. Only complete copy Paris, Bib. Nat. *Map Sedgemoor* 1662

Newcourt, Richard (d. 1716). *Dioceses London* 1708

Newell, D. P. East India Co. Service 1815–16. *Survey Canton River*

Newland, Capt. Charles. *Coast of Ava* 1784

Newlands, James. *Borough of Liverpool* 1849

Newman, A. K. & Co. Publisher, London. *New Edin. Gen. Atlas* (1844)

Newman, C. S. Surveyor. *Tithing of West Overton* 1862

Newman, G. Engraver. *Holy Land* in Swedish Bible 1618

Newman, G. W. See **Jones, E. & Newman**

Newman, H. Bookseller at the Grasshopper in the Poultry, London. Seller's *History of England* 1697

Newman, J. T. *Parish of West Ham* 1883

Newman, O. *First Steps Geography* (1887)

Newman & Co. Booksellers. Walker's *Univ. Atlas* 1820

Newsam & Son. *Estate Map Leeds area* 1853

Newson, Arthur. *Township of Bolinbroke* (1855), *Village of Oaklands, Victoria* (1858)

Newton, Sir Charles Thomas. *British & Roman Yorkshire* 1847

Newton, George. Globemaker, 66 Chancery Lane. *Pocket globe* 1817, *Pair globes* 1832

Newton, Henry. *Atlas Black Hills of Dakota* 1879

Newton, Sir Isaac (1642–1727). Mathematician and physician, a great number of his works concerned directly with physical geography

Newton, J. Estate surveyor of Chancery Lane, London. *Haverhill* 1737 MS

Newton, J. Estate surveyor. *Dunton etc.* 1805 MS, *W. Ham* 1816 (with W. below)

Newton, J. and W. *Hampstead* in Park's *History* 1814

Newton, J. & Sons. Globemakers. 66 Chancery Lane, London. Firm later changed to Newton Son & W. Berry. *Celestial and Terrestial Globes* 1816, 1836, 1849

Newton, John. *Mathematical Elements* 1660 (11 plates of globes)

Newton, W. Surveyor. *Land belonging Foundling Hospital* 1763

Newton, William. Estate surveyor. *E. Ham* 1816 MS, *Felsted* 1822 MS

Newton, William. *London in reign Henry VIII* 1855

Newton. See **Palmer & Newton**

Newton. See **Woodrow & Newton**

Newton & Sons. See **Newton, J. & Sons**

Newton Son & Berry, 66 Chancery Lane, London. *Miniature Globe,* n.d.

Neyder, I. L. *Preszburg* 1820

Nezon, De. *Maps* for Julien 1768

Nibby, A. *Roma Antica* 1818 and 1824, *Environs of Rome* 1861 (with Sir W. Gee)

Nicephorus, Blemmida (13th century). *Geographical compendium*

Nichol, John Pringle. *Physical Atlas* 1848

Nichol, W. & Co. Lithographers, Edinburgh. *Nova Scotia* 1847

Nicholay, N. de. See **Nicolay**

Nicholls, C. G. R.N. *Calay Strait* 1807

Nicholls [Nichols], Jeremiah. Estate surveyor. *Bookine* 1730 MS, *Rivenhull Parsonage* 1732, *Coptford* 1735 MS

Nicholls, Sutton (fl. 1680–1740). Draughtsman and Engraver at London Wall near the Weavers Arms against the Postern, and later at ye Golden Ball in St. Pauls Churchyard and at the Royal Jelly House in ye Pell Mell near St. James. Engraved for Lea's reissue of Saxton 1687, for Camden's *Britannia* 1695, Petty 1699, Wells 1700, Overton 1706–39, *Siege Namur* (1708) for Price 1719

Nicholls, William. *Plan of the Calcutta's Track* 1784

Nichols, Beach. *Maps of Counties in Penn., Ohio and New York* 1868–1876

Nichols, B. & Son. *Essex* 1831

Nichols, Francis. American publisher. *New Atlas Philadelphia* 1811

Nichols, G. *I. of Wight* 1844

Nichols, John (1745–1826) Printer, Antiquary and Author, of Red Lion Passage, Fleet St. *Gent's Magazine* 1792–1826, Camden's *Britannia* 1806

Nichols, John Bowyer (1779–1863). Son of John Nichols, continued the business. Published most of the large county histories: Lipscomb's *Bucks,* Ommerod's *Cheshire,* Surtees' *Durham,* Clutterbuck's *Herts* &c. *Family Topographer* 1835 &c.

Nichols, Richard. Publisher of Wakefield. *Walker's Inland Navig. G.B.* 1830

Nichols, Robert C. *Alpine Map Switzerland* 1874

Nichols, William. *Turnpike Road, Kensington* (1814)

Nichols. See **Dewhurst & Nichols**

Nicholson, I. Engraver. *Map England and Wales* 1823

Nicholson, James M. Australian surveyor. *Township . . . Allotments at Framlington (Victoria)* 1855, *Wild Duck Creeks* 1854, *Township Pyalong* 1857

Nicholson, John. (fl. 1686–1715). Bookseller and Publisher at the Kings Arms, Little Britain. *Hist. Europe* 1698, Gage's *W. Indies* 1699, *Index Villaris* 1700, Morden's *New Descript. England* 1704, *Maps Europe* 1707, *Maps G.B.* 1708, *Brit. Empire in America* 1708

Nicholson, Robert. Printer and publisher. Morden's *Survey,* engraved Whitwell 1594

Nicholson, Thomas. Estate surveyor. *Debden* 1777 MS

Nicholson, T. E. Engraver for Soc. Diff. Useful Knowledge. *Warsaw* 1831, *Pompeii* 1835

Nicholson, William. See **Nicolson**, Bishop William

Nicholson [Nichelsen], William. Master HMS *Elizabeth, Mathurin Bay Diego Reys* 1761, *Trincomalay* 1762, *Island of Timoan* 1763, *Madagascar* 1767, *Bombay Harbour* 1794

Nicholson, William. *London Postal District Office Map* (1857)

Nicholson, W. L. Topographer of P. O. Dept. *Mountain Region N. Carolina* 1863, *Delaware. and Maryland* 4 sh. 1869, *Post Route map Michigan and Wisconsin* 1871

Nicklin, Philip H. Publisher of Philadelphia. *General Atlas* ca. 1812 (with F. Lucas)

Nickolaus Germanus. See **Nicolaus Germanus**

Nichols, J. See **Nicholls**

Nicol, G. Map seller of Pall Mall, London. *Plans of Capital Cities* 1772, Taylor & Skinner's *Roads Ireland* 1777, Rennell's *India* 1788, *Embassy to China* 1796. Nicol, G. & W., publishers of Flinders' *Charts* 1814

Nicol, James (1860–1879). *Geological Map County of Roxburgh* 1847, *Geological Map Europe* 1856

Nicol, John M. *N.E. Nicaragua* 1898

Nicolai, Arnold (fl. 1550–65). Engraver and publisher of Antwerp. Anthonisz's *Oostland* 9 sh. 1543

Nicolai, Cornelius. *Nova Tab. Insul. Javae* (1595)

Nicolai, Gulielmus Belga. *Globe* Lyons 1603

Nicolai, Nicolaus. See **Nicolay**

Nicolas, C. See **Claesz**

Nicolas of Cusa [Nicolaus Cusanus]. See **Cusanus**, N. de

Nicolaus [Nikolaus], Germanus, Donis [Donnus] (ca. 1470–1490), Benedictine humanist, cartographer, printer. Editor Ulm editions of Ptolemy 1482 and 1486. Added *5 modern maps* to Ptolemy's *Geography*

Nicolay, Charles Grenfell. *Africa* 1857, *Eton College Atlas* 1858

Nicolay, Corneille. See **Claesz**

Nicolay [Nicolai, Nicholay], [Del Dolfinatto, Dolfinado, Daulphinois] Nicholas de. Sieur d'Arfeuville, (1517–ca. 1583). Géographe du Roy and cartog-

rapher to King of Spain, *Map New World* 1554 (in Medina's *Arte de Naviguar*), *Berry* 1565–7 MS, *Bourbon* 1569, *Calais* Ortelius 1570, *Navigation du Roy d'Ecosse* Paris 1583

Nicolini, F. Publisher of Venice. *Boschini Arcipelago* 1658

Nicollet [Nicholet], H. Maps for Andriveau Goujon 1841. *Mouvemens apparens du Soleil* 1842, *Atlas de physique et météorologie agricoles* Paris 1855

Nicollet, J. N. *Iowa and Wisconsin* (1845), *Upper Mississipi* 1843

Nicolo, Nicolaus de. Portolan maker, flourished 1470

Nicolosi [Nicolosius], Gio. Batt. Cartographer to Propaganda Fide Rome 1610–1670. *Dell' Hercole Roma* Muscardi 1660; 2nd edn. 1671

Nicolson, Bishop William (1655–1727). *Descriptions of Germany* for Pitts' *Atlas* 1681–3

Nicols, Nicolas Norton. *Isla de Mindanno* 1757 MS

Nieberding, H. *Memphis and suburbs (Tenn.)* 1830

Niebuhr, Carsten (1733–1815). Danish engineer and geographer. *Scientific voyages in Arabia* 1761–67, *Yemen* Hafniae 1771. Worked for Homann Heirs & Schraembl

Niederlein, G. *Paraguay B.A.* 1893

Niedzwiedzki, J. *Geolog. Karte Bukovina* 1876

Niehausen, C. F. *Lippe–Detmold* 1806

Niel, Adolphe, General. *Siege of Sevastopol Atlas* 1858

Niemann, A. *Atlas über den Feldzug* 1870–1

Niemeyer, Conrad Jacob de. *Carta do Brazil* 1857, *Imp. do Brazil* 1864, *Carta do Imp. do Brazil* 1873

Niemeyer, G. W. *Unter Elbe* 1837, *Hamburg* (1840)

Niemyski, Wozeiech. Supervisor of topographical survey of mining district of Kielce in 1827—35

Nienborg, Samuel Augustus. *Saxony* 1704 MS

Nieprecki, P. Jean (1719—54). Jesuit, teacher in Warsaw College. *Lithuania* Nuremberg, Homann Heirs 1749, an adaptation of Radziwill-Makowski map of 1613

Nierop, Dirck Rembrandtsz van. Nautical writer and editor. Revised text and composed almanac for Doncker's *Zeespiegel* 1664 and later editions to 1685. Edited Vlasbloem's *Lees Kaert* 1670 and a *map of Europe* by Goos

Nieuhof, Jan (1618—1672). Dutch traveller. Went on Embassy to China for Dutch East India Co. 1656. *Reiskaerte van de Ambassade naer China,* used by Van de Aa. *Cabo de Bona Esperanca* (166-)

Nieukerken, J. J. van. *Kaart von Walcheren* 1875

Niezonanski, Ferdynand (b. 1796). Military topographer, collaborated in the mapping of the Kingdom of Poland after 1815

Niger, Claudius. See **Clavus**

Nightingale, Joseph (fl. 1806—20). English topographer. *County maps in English Topography* 1816

Nigrin[us], Jonas. Czech cartographer. *Tesin, Silesia* 1724

Niles, J. M. and L. T. Pease. *Map of Mexico & Republic of Texas,* engraved Twitchet 1838

Nimmo, Alexander. Admiralty charts. *Iveragh Co. Kerry* 1811, *Bogsonhiver Cashen* 1814, *Carlingford Loch* 1821, *Sligo* 1821, *Valentia* 1832, *Youghall* 1835

Nimwegen, Dirk Wolter van. Dutch Admiral. Edited *plan of St. Jago* for Keulen (1700)

Niness, W. *Goldfields Mozambique* 1894

Niox, G. Leon (b. 1840). French General and geographer. Military geographer 1876—1900. *France* 1880, *Allemagne* 1882, *Atlas Gén.* 1891, *Afrique Central* 1894

Nipanicz, Jan (mid 19th century). Military topographer. *Mappa Krolestwa Polskiego Warzawa* 1863 and 1870

Nipher, Francis E. *Topo. Map Missouri* 1880, *Magnetic Declination Missouri* 1881

Niquet. Engraver for Desnos 1767

Nisbet, James & Co. 21 Bemen Street, Miss. *Map of World* (1865) by John Gilbert

Nishikawa, Seikyu (1693—1756). Japanese surveyor

Nispen, M. van. *De Stadt Dordrecht* 1718

Nixon, C. *S.E. coast Arabia* 1846—50

Nixon, John Browne. *Plan Bouchain* 1812 MS

Nixon, Robert. *Map Submarine Telegraph* 1858

Nixon, Thomas. *Moorngay* 2 sh. 1864

Njuren, Nils Eriksson. Swedish surveyor 1653—77

Nobili [Nobilibus], Pietro de [Petrus de]. Acquired some of Lafreri-Duchetti & Tramazini plates end 16th century, *Asia, City Augusta, Frisland, Elbe, Golfo di Venezia, Lombardy, Rhodes* Undated, 1560—70

Noble, A. W. Publisher of Louisville Ky. *Mexico* 1847 (with Bauer)

Noble, Charles. Surveyor Gen., Public Surveys in Michigan. *Detroit* 1852

Noble, E. Engraver for Cary's maps in Camden's *Britannia* 1789 and Stockdale's *New British Atlas* 1805

Noble, G. R. Surveyor of Woodford. *Grants in the Manor of Wanstead* 1857 MS

Noble, John. Surveyor. *Barking* ca 1735 MS, *Plan Northampton* (with Butlin) Jefferys 1746, *Manor of Aldersbrook* 1740 MS, *Kildare 2 sh. 1752 (with J. Keenan)*

Noble, Joseph. Gazetteer. *Lincolnshire* 1833

Noble & Butlin. Surveyors. *Plan Northampton* 1746

Noblet, Ca. *Carte géolog. de l'Europe* (1855)

Nockells, Christopher W. *Port Morant (Jamaica)* (1833)

Nockolds, Arthur. Estate surveyor of Stansted Mountfitchet. *Berden* 1841 MS

Nockolds, Martin. Estate surveyor. *Essex estate plans* 1790 MS

Nockolds, Martin, Junior. Estate surveyor. *Harlow* 1795 MS, *Saffron Waldron* ca. 1800 MS, *Lady Manners Estate* 1824

Nodel, Bartolomae Garcia and Goncalo de). Fl. early 17th century Spanish navigators. *Relacion Del Viaje* 1621, *Map of Tierra del Fuego*

Noe, Henri (1835–1896). Writer and geographer. *Geolog. Uebersicht der Alpen* 1890

Noel, Geographer. *Carte routière de la France* 1824, *Empire Ottoman* 1827

Noel, Joseph. *Map City of Carondelet, Missouri* 1853

Noeldeke, Theod. (b. 1836). Translated Arab geographical works

Noellat, J. B. *Dépt. de la Côte d'Or* 1820, *Dépt. du Rhône* 1827, *Carte de France* 1834, *Ville de Lyon* 1842

Noetling, Dr. Fritz. *Maps Oil Fields Burma* 1889–92

Noettelin, Jorg (d. 1567). Surveyor, Nuremberg

Noetzli, Johann Caspar (1689–1753). Surveyor and mathematician

Noguera, C. *Costas Orient. de la America Sept.,* Madrid 1828

Noha, P. de. See **Pirrus de Noha**

Noia, Duca di. See **Carafa**

Nolan, Joseph. *Notes on Geological Survey of Ireland* 1870–90

Noli, Agostino (16th century). Genoese cartographer

Nolin, Jean Baptiste (1657–1725). French geographer, publisher and engraver of Paris. Quay de l'Horloge du Palais vers le Pont Neuf à l'Enseigne de la Place des Victoires. Royal geographer 1693. *Siam* 1687, *Canada* 1688–9, *Amér. Sept.* 1789, *Celestial Globe* 1693, Cassini's *World* 1693, *Languedoc Canal* 1697, *Paris* 1699, *World* 1700 (pirated from De L'Isle), *Bavaria* 1704, *France* 1705. Convicted of plagiarizing maps of De Lisle 1705

Nolin, Jean Baptiste, Junior (1686–1762), son of above. Géographe du Roi. First at father's address, later Rue St. Jacques. *Amérique* 1720 and 1740, *Canada* 1756

Nolin, Widow. *Theatre of War in Italy* 1718

Noll, E. P. & Co. Map publishers of Philadelphia. *Atlas of Kewaunee Co.* 1895

Nollet, Jean Antoine (1700–1770). French physicist and cartographer. *Globes* 1728–30, *Celestial* 1730

Nollet, Jean Baptiste. *Globes* with Desnos. *Terrestrial* 1728, 1757 and 1772

Nolli, C. Engraver. Chiesa and Gambarini's *Tiber* 1744

Nolli, Giovanni Battista (ca. 1692–1756). Architect and engraver of Rome

Nolloth, Mathew S. Admiralty charts. *Greytown Harbour* 1850, *River Salween* 1845

Nolte, Ernst. Publisher, Buenos Aires. *Paraguay* 1886

Nonius, P. See **Nunes**

Noone, J. Lithographer, Melbourne. *Dandenong State Forest* 1872

Noordhoff, P. Dutch publisher. With M. Smit, *School Atlases* 1877–8

Noort, Juan de. Engraver. *Nueva Espana* 1649

Noort, Lambert. *Hungary* 1546, *Plan Antwerp* 1569

Noort, Oliver Van. *Voyage round World* 1602

Noorthouck, John. *Plan London & West.* 1772

Noothoven van Goor, D. Dutch publisher of Leiden, for Kuyper 1863, for Frijlink 1872, for Schmidt 1874, *Nieuwe Atlas* 1865–71

Nops, I. G. *Wangaruru Harbour* 1845

Norberg, J. E. *Plan of Revel* 1816

Nordberg, Lieut. John. (fl. 1758–1775). *Fort Michilimackinac* 1769 MS

Norden, John, the elder (1548–1625?). Surveyor, religious writer, topographer. The first to plan a series of County histories and first to insert roads on English maps. *Speculum Brit. (Middlesex)* 1593 (2nd edn. 1723); *Northants.* 1591 MS, printed 1720; *Surrey*, engraved Whitwell 1794; *Essex* 1594 MS, printed 1840; *Sussex*, engrd. Shwyt 1595; *Hants.* 1595, *Herts.* 1598 (2nd edn. 1723); *London* 5 sh. 1600; *Cornwall* 1605. Some maps in Camden's *Britannia* 1607; *Honour of Windsor* 1607; *Surveyors Dialogue* 1607, and numerous estate plans MSS in *Hants., Herts., Kent, Middlesex, Northants, Suffolk, Surrey and Sussex.* He was Surveyor of crown woods in 1600 and surveyor to Duchy of Cornwall in 1605

Norden, John, the younger. MS plans Manors in Berkshire: *Blewberie* 1617, *Shippon* 1617, *Laleham* 1623

Nordenankar, Jan. de (1722–1804). Admiral. *Maps of Baltic* 1788–90, *Finland* 1789, *Sweden* 1789, *Sjö Atlas* 1795, Used by Keulen in 1799 and 1800

Nordmann, A. P. H. *Roy. de Pologne* 9 sh. Vienna, Artaria 1813; *Allemagne* 4 sh. 1821

Norgate, J. W. T. Hydro. surveyor. Coasts Australia: *Port Campbell* 1879, *Port Philip* 1882, *West Channel Port Philip*1883

Norie, John William (1722–1843). Hydrographer, publisher, teacher of navigation at the Navigation Warehouse and Naval Academy, No. 157 Leadenhall Street, London. *Chart of the Downs and*

Margate Roads 1797, *Planisphaerum Coeleste* 1801, *Nautical Tables* pub. Heather 1803. Bought Heather's business in 1812 and J. Steel's in 1819. He retired in 1839; the business passed to J. Wilson, and continued as Norie & Wilson. *East India Pilot* 1816, *Chart Pacific* 1826, *British & Irish Coasting Pilot* 1830, *Chart English Channel* 1839, *Chart for Ships Voyage from England to East or West Indies* 1839

Norman, B. M. *Plan New Orleans & Environs* 1845

Norman, I. [John]. Printer and publisher. Collaborated with Osgood Carleton, prod. maps of *Maine* 1790, *American Pilot* 1792, *Nantucket* 1794, *Massachusetts* (1795)

Norman, J. Engraver *Jerusalem Bible* 1791

Norman, Joh. Alb. *Charts of Sweden* 1868

Norman, John & William (1748–1817). Publishers and mapsellers, *American Pilot,* Boston 1792 and 1794, Reprinted W. Norman 1798 and 1803, Carleton's *Maine and Mass.,* Both 1795

Norman, Robert (fl. 1560–96). Surveyor of Ratcliff and compass maker. *Estuary of Thames* 1580 MS, *Safeguard of Saylen*

Norman, William. Publisher and mapseller of Boston, 75 Newberg Street. *Pilot for West Indies* 1795, *Tobago* 1791, See also John Norman

Normand, Charles. Designed and engraved cartouche for Coutan's *Environs de Paris* 1800

Norris, Jacob, of Wisbech. *Lynn-Deeps* n.d. in 1722 Camden

Norris, Sir John (Foulweather Jack) (1660?–1749). English hydrographer and admiral. *A compleat sett of New Charts North Sea & Baltic* 1723

Norris, Robert. African trader. *Windward and Gold Coasts* Sayer 1785; *Africa Cape Verga–Cape Formoso* 3 sh. 1789; *African Pilot* Laurie & Whittle 1794–1804

Norroy, Sieur de. See **Du Pinet**

Norsworthy, George. Master, H.M.S. *Pylades*. *River Peitti* 1840, *Admiralty* 1841, *Coast China* 1841

Norton, John (fl. 1587–1612). Bookseller and printer, Master Stationers Company. English edition of *Min. Ortelius* 1602, English edition of *Folio Ortelius* 1606 with John Bill, Camden's *Britannia* 1607

Norton, Robert (d. 1635). Engineer and gunner. *Various MS plans of Algiers* 1620, *Gunners Dialogue* 1643

Norton, William. *Platte of Tottenham Court* 1591 MS

Norwood, Andrew. Surveyor General. *Map of St. Christopher* Thornton (1690) and later edition

Norwood, J. G. *Geolog. Map Minnesota and Wisconsin* 1852

Norwood, M. *Chart Milford Haven* 1689

Norwood, Richard (1590–1675). Mathematician and surveyor. Surveyed *Bermuda* 1616, used by Speed

Nott, William. Stationer and Map Seller. Supplied maps to Plantation Office 1670's–80's, joined in publication of Moses Pitt's *English Atlas* and with Will Berry's Sale of Morgan's maps in 1687

Noual, Isaac. Engraver. Bontein's *Jamaica* 1753, Hawkins' *Maidenland* 1775

Nouerre, De Fer de la. See **De Fer De La Nouerre**

Nouet, Sieur. Contributed to *Map of Egypt* 53 sh. 1807–15

Nourse, J. Publisher with P. Vaillant of Palairet's *Atlas Méthodique* 1755

Nova, P. de. See **De Nova**

Novak, Joseph J. *Johnson Co., Iowa* 1889

Novellanus [Neuvel], Simon. Engraver and artist of Mechlin. Worked for Braun & Hogenberg's *Civitates* from 1571

Novigmagus, Joannes [Johann Bronchost]. Translator. *Ptolemy Geogr.* Cologne 1840

Nowack. Atlases and maps of the western parts of Poland (Bydgoszy and Posnan regions). *Town and City Plans* published 1834, 1839, 1840, 1843, and 1861

Nowell [Nowel], Laurence, B.A., 1542, d. 1576. Dean of Lichfield 1560. Antiquary. Proposed map all the English counties. Left 2 MSS of regional maps: one in B.M., one Lord Lansdowne, *British Isles* 19 sh. 1563

Nowicki, Felitis. Military cartographer in the service of Poland. Author of detailed and accurate *map of the Palatinate of Podolia* 1776–87. Probably at one time in the map depot of St. Petersburg but now lost

Noyan, Chev. de. *Pensacole* 1769

Noyer, G. V. du. See **Du Noyer**

Nugent, Lieut. *Plan Alexandria etc.* 1840

Numajiri, Bokhusen (1774–1856). Japanese globemaker

Nunes [Nunez, Nonius], Pedro, Portuguese cosmographer. First to use rhumb lines 1534. *Tratado da Esphera* 1537

Nunez, Pedro (1492–1577). Mathematician and cosmographer of Coimbra

Nunn, J. Bookseller. For Walker's *Univ. Atlas* 1820

Nunn, Capt. W. *Draught of Downs* 1723, in Collins' *Pilot*

Nuremberg Chronicle, Koberger 1493. Contains *map of World and map of Central Europe*

Nuscheler, I. Caspar. *Zürich* 1654

Nutt, E. R. Printer in the Savoy. Morden's *Magna Britannia* 1724

Nutt, M. Publisher, Bookseller, Exeter Exchange in Strand. *Magna Britannia* 2 vols. 1700, 1720

Nyberg, Claes Henrik (1799–1883). Surveyor of Helsingfors

Nyon the Elder. Publisher of Paris. Rue du Jardinet. Philippe de Prétot's *Recueil de Cartes* 1787, *Atlas Universel* 1787

Nyon. Engraver for Dépôt de Cartes, 19th century

O

O[akden], C[harles]. *New map country round Leicester* 1830

Oakley [Ockley, Okley], Edward. Publisher and architect. John Street, Golden Square. *Thionville* 1753, *Rochefort* 1757, *Brest* 1757, *Quebec* 1759, *Dunkirk* n.d. *Metz* 1754, produced in conjunction with Rocque's widow

Oakley, F. F. Lithographer of Boston. *5th Ward Grand Rapids, Mich.* 1856

Oakley & Thomson. Lithographers 46 Water Street, Boston. For Westcoatt 1864

Oakley, T. *Kenilworth Castle Plan* 1885

O'Beirne, Brian. Surgeon. *West Coast of Africa*, Wyld 1830

O'Beirne, P. *Map of Richland County, Ohio* 1856

Obel, P. B. *Aarhus Haven* (1876)

Oberlander, Richard (1832–1891). Writer and traveller. Geographical publications Australia

Oberle, H. *Repub. del Paraguay y . . . Brasil* 1893

Oberlercher, P. *Hochalpen Spitze* (1893)

Obermeier, W. *Die Deutschen Bundesstaaten* 1843

Ober-Muller, W. *La France et La Belgique en Relief* 1844

Obernetter, Johann Baptist. *Post u. Reise Karte von Bayern* (1866)

Oberreit, General Jacob Andreas

Hermann (d. 1856). *Topo. Atlas Sachsen* 1836–50. Begun in 1819 by Gen. Oberreit

Obert, Franz. *Schul Wand Karte von Siebenbürgen* 1861

O'Brien, John and Joseph. *Twenty-six MS Estate Maps of County Cavan, Ireland* 1758–80

O'Brien, J. J. *Union Pacific Railroad Map of part of Colorado Territory* New York 1864

O'Brien, R. L. *Piano de la Ciudad de Cordova (La Plata)* (1879)

O'Bryn, Friedrich August, Baron. *Maps Brazil* 1859–60–65

Obruchev, V. *Mountain System Nan Shan (China)* (1895)

O'Bryne. See O'Beirne

Ockerson, John A. and C. W. Stewart *Mississippi River* 1892

Ockerse, P. M. *Resid. Soerakarta* 6 sh. 1873, *Topo. Kaart/Resid. Japara* 1894, *Resid. Rembang* 4 sh. 1897

Ochikochi, Doin. Japanese cartographer, 17th century

Ockley. See Oakley

O'Connor, R. *Van Dimens Land* 1834

Ockley, Sr. *Plan Metz* 1754

O'Conor, Charles. *Ireland* (1792)

O'Corrain, T. See Carve

Odbert, A. *Atlas Warren County, Penn.* 1878 (with J.A. Howden)

Oddi, Angelo degli, of Padua. *Viaggio* 1584, *Codice Cartaceo* 1587, *Città, fortezze di Candia* 1603 MS

Oddi, Jacopo. Cartographer and mathematician, b. Narni 1600. Studied in Naples. *Carta del Ducat di Bracciano* 1637. Used by Blaeu 1662. *Patri. di San Pietro* 1646

Oddy, S. A. Publisher of No. 20 Warwick Lane. *General Atlas* 1811, Wallis' *New British Atlas* 1811–(23), *Intended Stamford Junction Navigation* 1810

Ode, H. Lithographer. *Atlas Universel* by Philippe Vandermaelen, Brussels 1827; *Haut Canada* 1825

Odebrecht. *Kolonie Grao Para . . . Santa Christina Eisenbahn* Gotha Perthes 1888

Odeleben, Lieut. Col. Ernst Otto Innocenz von. *Gegend von Bautzen* 1817, *Dresden* (1840), *Sächsisch Böhmischen Schweiz* [1845]

Oder, George (d. 1581). Surveyor of Saxony. *Amt Schwarzenberg* 1531 MS

Oder, Mathies (fl. 1586–1607). Son of George Oder. *MSS maps*

Odlanicki, Poczobut, Marcin (1728–1810). Mathematician and astronomer, director Astron. Observatory in Wilno. Member Royal Soc. London and Acad. Roy. des Sciences Paris. Royal astronomer 1767. Helped project of *Modern Map of Poland* for Stanislaus Augustus

Odoric of Pordenone (1286–1331). Italian missionary to China

O'Donovan, John. *Irish Topographical Dictionary* 1830

Odorici, Frederico. *Mappemonde* 1367

O'Dowd, J. J. *Ceylon E. Coast* 1880

Oeder, Matthias (d. 1607). *Sachsen* 1586–1607

Oelschlager, A. See **Olearius**

Oernehufud, O. H. See **Gothus**

Oernskoeld. See **Ornskold**

Oertel. See **Ortelius,** A.

Oeser, H. Lithographer. *Türk. Kriegsschauplatze* (1877)

Oesfeld, Carl Ludwig von (1781–1843). *Ober Barnimsche Creis* 1786, *Storckowsche Creis* 1788, *Halberstadt* 1794, *Deutschland* 1821, *Harz Reisende* 1834

Oesterreicher, T. von. *Adriatisches Meer* 1867–73

Oetjes, F. W. Engraver for Greenville. Collins 1723. *England* (1685), *Sea Coasts England France* (1690)

Oettinger, Johann Friedrich (fl. 1745–65). *War on the Rhine,* Seutter (1730), *Campagne du Haut Rhin* (1735), *Theatrum Belli* (1738)

Oetzel, Franz August von (1783–1850). Berlin cartographer, b. Bremen. *Hydrographical Atlas; Völkerkarte von Europa* 1821, *Atlas von Asien* 1833–54, *Inner Asiens* 4 sh. 1841, *Hand Atlas von Africa* 1831

Oexmelin, A. O. See **Esquemeling**

Oeynhausen, C. von (1795–1865). *Geogr. Karte v. Ober Schlesien* 1821, *Rheinlander* 1825, *Laacher See* 1847

Offor, George. *Tower Liberty* 1830

O'Flaherty, Edmond. *City Kansas* 1860

Ogden, A. B. *Hudson, Macon Co. Missouri* 1860, *Montgomery City* (1858)

Ogden, Peter Skene. *Snake River Expedition* 1828–9 MS

Ogee, Jean (1728–1789). Geographical works on Brittany. *Atlas Itin. de Bretagne* 1769, *Brittany* 4 sh. 1771

Oger, Felix. French historian and geographer, 19th century. *Atlas de Géogr. Gén.* 1874, *School Atlases* 1874–78

Oger, [Ogier, Ogierius], Mathieu, French Priest. *Maine* 1539. Used by Ortelius 1579 and by Blaeu 1634 and 1647

Ogerolles, Ian d'. Printer of Lyons. Pinet's *Plantz* 1564

Ogg. *Plan of Huntly* (1845) (with Farquhar)

Ogierius, M. See **Oger**

Ogilby, John (1600–1676). Born Edinburgh, d. London. Geographer, historian, publisher, Royal Cosmographer (1671). Master of Revels in Ireland. Compiled the *Britannia* 1675, the first book of road maps, the first to use the standard mile of 1760 yards. *Ipswich* 9 sh. 1674, *London* 1671 (with Morgan), *Essex* 1678 (with Morgan). Translator of Montanus. *Africa, America* 1670, *China, Japan, Persia, Asia* 1673 (with maps and views)

Ogilvie, William. *Peau River (Canada)* Ottawa 1884

Ogilvy, David, Junior. *Itinerary of England and Wales* 1804

Ogilvie, W. *Sea Coast U.S. from Cape Cod–Saughkonnet Point* 1857

Ogle, George A. **and Co**. *County atlases U.S.* from 1892

Ogle, Edward. *River Thames* 1796

Ogle. See also **Alden, Ogle & Co.**

Ogleby, J. *Plans of Tower and St. Catherins Stow* 1754, *Tower Liberty* 1754, *Mary Whitechapel* 1755

Ogleditsch. *Bulgarien* 1877

Ogles, Duncan & Co. Publisher for Walker's *Universal Atlas* 1820

Oglethorpe, James. *Plan of the Town, Castle and Harbour of St. Augustine,* London 1742

O'Grady, G. *Westliches Russland,* 4 sh. Cassel 1884

O'Hagen, James. *Plan of Belfast* (1855)

O'Hea, Richard A. *Town of Greenville, Wash. Co. Miss.* 1867

Ohlin, Axel. *Antarctic Regions. Ymer* 1890, *Terra del Fuego* 1896

Ohlin, Ole. *Map of Reykjavik* 1801 (with Ole. M. Aanusm), *Sydlige Kyst af Iisland* 1823

Ohmann, C. L. (1817–1868). Engraver and lithographer. *Environs Berlin* 10 sh. 1816–19, *Wand Karte des Preussischen Staates* 9 sh. 1857, *Candia* 1866, *N.*

America mit West Indien 1869, *Palestina* 1868, *Deutschland* 9 sh. 1881. Engraved for Kiepert

Ohmaun, E. Engraver for Kiepert 1869

Ohsen, F. W. *Postcharte Chur Braunschweigischen Hanover* 1774, 1797 and 1805

Ojea, Ferdinand. *Reyno de Galizia (Ortelius)* 1598. Used Blaeu 1634 and Jansson 1658

Ojeda, Alonzo de (1465–1510). Spanish navigator

Okell. Publisher of London. S. Parker's *World* 1740 with J. Cluer

O'Kelly, J. *Maps Geolog. Survey Ireland* 1865–6

Okem, J. G. *Carte d'Egypte &c.* (1880)

Okley. See **Oakley**

Olano, P. R. de. See **Ruiz de Olano**

Olascoaga, Manuel J. *Territ. de la Pampa* (1882)

Olaus, Magnus Gothus. See **Magnus**, Olaus

Olavide, Martin José de. Spanish sailor. *Costa de Luzon* 1794

Olavius. *Nye Carte over Island* 1780

Olbers. *Special Karte von . . . Ost Fries* 1804

Olbers, E. W. *Geolog. Karta Inlands Sodre* 1859

Olbers, Nicolas Heinrich (1708–1794). Military cartographer of Hamburg

Olbrich, G. *Special Karte des Kreises Waldenburg* 1892

Olbrich, Th. *Floetz-Karte Saar-bruecker Stein hohlen District* 1865

Oldfield, H. J. *Yezo E. Coast Akishi Bay* (with Comm. St. John). *Admiralty Chart* 1872

Oldham, Hugh. *Plan Manchester and Salford* 1780 (lost)

Oldham, Richard. *Platt public roads in S. Milford, Cecil County* 1796

Oldham, R. D. *Geolog. Map India*, Calcutta 1895

Olearius [Oelschlager], Adam (1599– 1671), German diplomat. *Russia* 1636, *Persia* 1646, *Volga,* Jansson 1658; Blaeu 1662. Supervised Gottorp *Globe* 1654–64

Oleatus, G. See **Oligiato**, G.

Olesnicki, Zbigniew. Polish Cardinal, statesman 1449. Patron Jan Dlugocz *(Chorographia Poloniae* 1466)

Oleszcynski, Seweryn (1794–1876). Engraver and lithographer at Bank of Poland, lithographic shop. *Map of the Kingdom of Poland . . . and free city of Cracow* 8 sh. Warsaw 1810 (with Julius Colberg)

Olewinski, Wincenty. Collaborator at Inst. Geografico Militar in Buenos Aires mid 19th century

Oligiato [Oleatus], Girolamo [Hieronymus] Engraver in Venice 16th century. *Brabant and Holland* 1567, *Padua* 1568, Gastaldi's *Asia* 1570

Oliphant & Son, William. Publishers of Edinburgh. *Biblical Atlas* 1835

Olivo. See **Oliva**

Oliva [Olives]. Family of Portolan makers in Majorca, Messina, then Marseilles
—— Bartolomeo (fl. 1532–88) in Majorca. *Atlas of 11 maps* 1532, *Atlas* ca. 1561 MS, *West Europe and Mediterranean* 1575 MS
—— Domingo. Son of Giovanni. *Portolans* 1561–8
—— Francesco (1587–1601). *Atlas* (6 maps) 1602 MS, *Atlas* 1659, *Mediterranean* 1609–13 (made in Messina)
—— Giovanni [Juan Johannes]. *Oval World Map* 1580 MS, *Charts* 1587, *S.E.* 1596, *Mediterranean* 1599, *Europe* 1601, *Mediterranean* 1601, *Mediterranean* (1634?). Published in Messina
—— Giovanni Battista Caloiro e Oliva. *World Map* 1673
—— Jaume, of Majorca. *Chart* 1550, *Europe and Atlantic* 1557, *Mediterranean* 1559

—— Joanne, of Livorno. *Marine Atlas World* 1638
—— Placido Caloiro e Oliva. *Mediterranean* 1615, 1622, 1626 Messina, *Spain to Syria* 1629
—— Rienzo Juan de. *World* 1591, *America* 1596
—— Salvatore, of Messina. *Atlas* (7 maps) 1620, *Atlas* (3 maps) 1631

Olivan, Rebolledo J. *Mexico* 1717 MS

Olive, Richard. English surveyor in Russian service. *Plan St. Petersburg* 1727–8 MS

Oliveira, Antonio August d'. *Caminhos de Ferro . . . Loanda* 1884, *Africa Austral* 1885

Oliveira, B. J. *City New Orleans* 1897

Oliveira, E. J. da Costa. *Guine Portugueza* Lisboa 1890, *Curso do Rio Zaire* 1891

Oliveira, Fernando. *World Map* 1570

Oliver, Claudio. *Atlas Geogr.* 1865

Oliver, I. Map and Printseller at his shop at the Corner of the Old Bailly (sic) on Ludgate Hill, London. *Siege of Buda* 1686, *County Essex* 1696

Oliver, John. Surveyor, Engraver, Glass Painter at the Eagle and Child on Ludgate Hill. *Draught Island of Buss* by J. Seller 1671, *Oxford* (with Seller) 1675, *Middlesex* 1679–80. *Kent, and Herts* (with Seller 1680), proposed *Atlas Anglicanus* with Seller and Palmer ca. 1681, *Essex* 1696, *Hungary* 1699. Engraved Overton's *Oxford* 1715, *London* 1726

Oliver, J. Engraver. Griffin's *Middlesex* 1748

Olover, J. W. Lithographer, 43 Amen St., New York. *Gold Regions W. Kansas* 1859

Oliver, S. Pasfield. *Route from Tamative* 1862, *Madagascar* 1886

Oliver, Thomas. Architect. *Town and County of Newcastle* 1831, *Newcastle and Borough of Gateshead* 1833, *Plan Borough Newcastle* 1844

Oliver, Thomas. Revenue surveyor. *District of Goorgaon (India)* 1848, *Panneput* 1848

Oliver & Boyd. Publishers, Edinburgh, for Ewing 1821 and 1828, Reid 1837. Joint publishers with Philips of *Commercial Atlas of World* 1856, *School Atlas* 1876, *Junior Atlas* 1676

Olives. See **Oliva**

Olivetano, F. M. *Stato di Milano* 1790

Oliviari, C. L. *Routes de Poste de l'Europe* 1836

Olivier, Lieut. *Carte de la Bouche du Niger* 2 sh. 1898

Olivier, Arsène. *Dépt. de Seine* (1877)

Olivier, C. B. *Nieuwe Merwede* 2 sh. 1864

Olivier, E. *Koningrijk der Nederlanden* 6 sh. 1864

Oliver, F. *Carte de la Mer Méditerranée* Toulon, Boery 1746

Olivier, G. A. *Atlas Empire Othoman* 1807

Olivier, H. *Carte Postal Repub. Argentina* 1865

Olivier, Jean. *Ports et Rades de la Mer Méditerranée* (1796)

Olivier, Victor. *Plan de Saigon* (1800)

Oliviera, Antonio Augost d. *Carta da Africa Merid. Port* 1886, *Possessies Portuguezas da Africa Merid.* 1890, *Angola* 1890, *Goa* 1891, *Macao* 1891

Olivieri, Bernardino. *Marca di Ancona* 1803, *Perugia* 1803, *Camp. di Roma* 1802, *Umbria* 1803, *Urbin* 1803

Olivieri, Charles I. *Tab. Routes d'Europe* 1836

Ollefen, L. van. *Dutch Towns & Villages* 8 vols. 1793–1801 (with Bakker)

Ollero [Olleros], Anselmo. Colonel, Philippine Army. *Luzon* 1882

Ollivier, Jules Marie. *Mer Méditerranée Tetouan* 1853–1857

Olmadella, Juan de la Cruz Cano y. See **Cruz Cano y Olmadella**

Olmo, Joseph Vicente. *Del Nueva desct. del orbe* Valencia 1681

Olmo, Rocco dall' (fl. 1542). Italian portolan maker. *Chart Europe* 1542

Olmsted, Frederick Law. *Central Park N.Y.* (1863)

Olney, Jesse (1798–1872). American geographer. *School Atlas* 1829 and 1844

Olsen, Emil. *War Maps France* 1870–77

Olsen, Olaf Nikolas. *Maps Denmark & Iceland* 1841–50, *Climat de l'Italie* 6 sh. Copenhagen 1839

Olsen, O. T. *Piscatorial Atlas North Sea* 1883

Olsen, and **Bredsdorff**. *Europe* 1830

Olshausen, Theodor. *Staat Iowa* 1855, *Staat Missouri* 1854

Olsvig, Viljam W. *Norway* (1889)

Oltmanns, J. (1783–1833). Astronomer and geographer, collaborated with Humboldt

Omalius d'Halloy, Jean Baptiste Julien d'. (1783–1875). Belgian geologist. *Carte Ethno. du Globe* 1853

Omerat, M. *Yu-Lin-Kan Bay* 1760, pub. Dalrymple 1774

Ommanney, Capt. R. *Discov. in the Arctic Seas betwn. Ban Bay and Melville Is.* 1851

Ondarza, Juan. *Repub. Bolivia* 1842–59, *N.Y.* 1859

Onder de Linden. See **Linden**, G.

O'Neill, H. E. *Coast of Mozambique* 3 sh. 1882 MS, *Port Kisima* 1883, *E. Africa, Zambesi–Rovuma rivers* 1885

Onofri, Francesco. Printer of Florence. Dudley's *Arcano del Mare* 1647

Oomkens, J. Publishers and booksellers of Groningen, passing from father to son, all called Jan. From the 18th to 19th centuries mostly *School Atlases*. Published for Jaeger 1843, *U.S.* 1849, for Wijk Roelandszn 1851. *Gemeente Atlas* 1862, *Atlas der Aarde* 1858

Oort, Henricus. *Bible Atlas* 1884

Oostwoudt, Govert. *Dregterland* 1723

Opell, Peter. *Ratisbon* 1590

Opitz, C. *Repub. de Chile* 1891, *Nord Ostsee Kanal* 1895, *Eisenbahn Karte N. Deutsch.* 1897, *Fichtelgebirge* 1898

Oporino [Oporinus], Giovanni. Publisher of Basle. Sophiano's *Greece* ca. 1544

Oppel, Alwin. *Erdkentnis vom Mittelalter* (1893), *Neuguinea* Bremen 1893

Oppenheimber, Theodor Hermans. *Hohle nächst Kirchberg am Weichsel* 1868

Opperman, Gen. Maj. *Detailed map Russian Empire* 100 sheets, St. Petersburg Imperial Map Depot (1801–4). (With General Quartermaster von Sukhtelen) German edition *Russisches Reich* 9 sh. 1812

Opperman, Herman. *Neuester Plan Hannover* (1864)

Oppolzer, Th. von (1841–1886). President, Austrian Commission International Geodesy

Orbigny, Alcide de (1802–1857). Naturalist and geologist, traveller. *Carte de l'Amérique Mérid.* 1838, *Partie Répub. Argentine* 1835, *Bolivia* 2 sh. 1839

Orbigny, Charles (1806–1876). *Geological Manuals*

Ord, Edward Otto Cresap, Lieut. *Gold and Quicksilver District California* 1848

Ord, Harry St. George. *Plan of Kertch* (with Major Crease) 1855

Ord, T. *Portsmouth Harbour* 1726

Orell, Füssli & Co. *Atlas of Switzerland* 19 sh. 1816

Orgebozo, J. de. *El Istmo de Tehuantepec* 1825 MS

Ordnance Survey. Founded 1791 by Duke of Richmond, with headquarters in the Tower. Scale one inch to mile. First county *Kent* published 1801 (4 sheets). One inch completed 1853. Six inch *Ireland* 1825–47. *England* six inch 1846–96; twenty five inch 1846–93

O'Reilly, Henry (1806–1886). American publisher

O'Reilly, M. of Adelaide. *South Australia Plan Co. of Hindmarsh* 1860

O'Reilly, Montague. *Town and Harbour Balaklava* 1854

Orescu, G. A. *Planul Orasului, Bucuresci* 1892

Oresme, Nicolas d' (d. 1382). Theologican, economist, astronomer and cosmographer, b. Normandy, Bishop of Lisieux. *Globe* 1377

Orfelin, J. (1750–1804). Polish engraver. Maps and illustrations in Rajic's *History*

Orfelin, Z. (1726–1785). Polish engraver. *Map of Russia* with armorial decoration

Orgiazza, A. Publisher. *Repub. Nueva Granada*, Paris 1847

Orgiazzi, C. C. Engraver. *Londres* 1801 (with Thuiller)

Orgiazzi, J. Alexis. *Pyrenées* 1793–5, *Mineralogique de l'Italie* 1816, *Mod. Map Italy* 1825

O'Riley, Capt. 13th Regt. *Pigeon Island St. Lucia* MS ca. 1750

Orio, A. (1737–1825). Engraver. Worked for Remondini

Orlandi, Giovanni. Engraver and publisher of Rome (fl. 1600–1604). Inherited some of the stock of Luchini, Lafreri, and Duchetti. *Ancona* 1600, *Calais* 1602, *Germany* 1602, *Sicily* 1602, *Malta* 1602, *Puteoli* 1603

Orlebar, John. Admiralty Charts. *Beaver Harbour* 1859, *Louisburg Harbour* 1859, *Newfoundland* 1861, *Catalina Harbour* 1862, *Bay of Bulls* 1864, *E. Coast Newfoundland* 1868 and 1870

Orlich, Leopold Ludwig (1804–1860). Military cartographer

Orliens, David d'. *Map of Ostend* 1802 MS

Orlitsek, F. *Topo. Karte von Wien* (1840)

Orlov-Chesmensky, Count Alexei. *Map of Russian Turkey* (1771)

Orlovius. *Umgegend von Posen* (1847)

Orme, Capt. William *Six plans of the English Army (America)*

Orme. See **Longmans,** Rees Orme Brown and Green

Ormerod, R. M. *Sketch Map Korokoro (Kenya)* 1895

Ormes, C. F. des. See **Fournier Des Ormes**

Ormsby, G. O. *Manukau Harbour New Zealand* 1845, *Admiralty* 1846

Ornehufud, O. H. See **Gothus**

Ornskold, P. A. *Charta ofver Angermanland, Medelpad och Jamtland* 1771–97

Orojiazzi, J. A. *Carte Stat. Polit., et Minéral. de l'Italie* 2 sh. Paris 1816

Orontius, See **Finé,** O.

Orosius, Paulus. Cosmographer, 8th century. *Cosmography* MS

Orozco y Berra, Manuel (1816–1881). *Islas Filipines* 1659–(63), *Repub. Mexicana* 1860

Orpen, F. H. S. *Griqualand West* 1872

Orr, J. W. Engraver. *Chicago Railroad* 1859

Orr, N. and Co. Engravers for Deforrest 1859

Orr, William S. **& Sons.** Publishers of Glasgow. *New Edinburgh General Atlas* 1844, *Hydro. Map Brit. Isles* 1849, *Palestine* 1849

Orr and Company. Publishers, Amen Corner, Paternoster Row. Davies' *London* 1847, Teesdale's *Travelling Atlas* 1852, *Australia* (1860)

Orr and Smith. Publishers, Amen Corner, Paternoster Row. *East India Isles,* engraved Dower (1840)

Orta, Domingo Martins d'. (16th century) Portuguese cartographer

Orta, Don Bernardo de. Captain, Spanish Navy. *Vera Cruz and Mexico* 1798, English edition 1805

Ortel[1]. See **Ortelius,** A.

Ortelianus. See **Colius**

Ortelius [Ortel, Ortell, Oertel, Wortels], Abraham (1527–1598).Cartographer and publisher, born and died in Antwerp. Set up as map colourist in 1547. Maps of *World* 1564, *Egypt* 1565, *Asia* 1567, *Spain* 1570. Published first modern uniform Atlas, *Theatrum Orbis Terrarum* 1570 [first issue May 22], editions continued to 1612. *Roman Empire* 1571. His map of *Morocco* used by Blaeu 1662

Ortlepp, A. A. *Witwatersrand Gold Fields* 1898, *Route Map Rhodesia* 1899

Orto, A. Dell. See **Dalorto**

Ortoli, A. *Telephone map* (1898)

Orton, Edward (1829–1899). American geologist

Orton. See **Miller, Orton and Mulligan**

Ortt, J. R. T. Engineer. *Atlas Dikes and Polders Zeeland* 1861

Os, Pieter van. Publisher in The Hague. Brouckner's *Nieuwe Atlas* 1759

Osanu, Oka. *Yochi Kokai-zenu (Map for Navigators),* Tokyo 1872

Osborn, Henry S. *Map Palestine* (1855) and 1868, *Landscape Map Egypt & Palestine* 1874

Osborn, O. S. *City Calais, Washington County Maine* 1856

Osborn, Sherard (1822–1875). *Voyages in the Arctic, Tientsin* 1858 MS

Osborne, Henry. Surveyor. *Olney Park* 1608 MS

Osborne, Joseph. *Chart of Hull* 1668 MS

Osborne, J. & T. Publishers. *Geogr. Magna Brit.* 1748 and 1750 (county maps). Salmon's *Modern History* 1744–6. T. Osborne joined with Hondius to publish an *Ancient Geography* in 1741

Osborne, J. W. *Victoria Census Districts* 1857

Osborne, T. See **Osborne,** J. & T.

Oschmann, R. H. *Maps of Brandenburg and Cologne* (1900), *Dortmund* (1897)

A. Ortelius (1527–1598); engraved by Ph. Galle

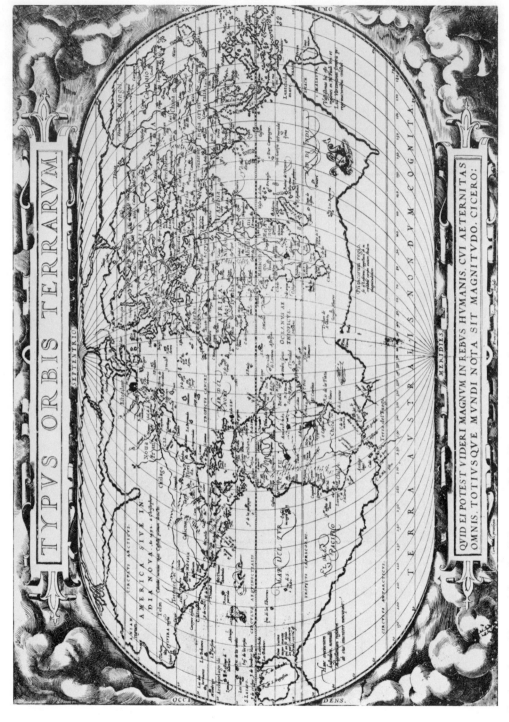

*Engraved world map from the "Theatrum Orbis Terrarum" by A. Ortelius
(published Antwerp, 1570)*

Oseguera, Andres. *Nouvelle Carte du Mexique* Vuillemin 1862

Osler, G. *Teatro de la guerra de Oriente* 1876

Oss, S. F. van. *American Railroad Maps* 1892

Ossenbrug, J. V. *Oesterr. Herzog. Venedig,* Vienna 1804

Ossorio, *Manuscript Plans of Portuguese Fortresses,* 17th century

Ossowski, Gotfryd (1835–1897). Polish archeologist and geologist. *Carte Archeologique de la Prusse occid . . . et Posen* (1875–8), 1880

Ostell. Publisher in London. *New General Atlas* 1810, 1818 & 1819

Osten, Carl Heinrich (d. 1691). Engineer and military cartographer

Ostendorfer, Michael. Engraver. *Ratisbon* 1632

Ostening, H. van d'. See **d'Ostening**

Osterberg, E. Engraver. *Atlas Juvenilis* 1789

Osterhausen, Ch. *Malta* Augsburg 1650, *Plan Valetta*

Ostertag, Heinrich Jonas. Engraver *Augsburg* 1729, *Plans of Aschaffenburg, Würzburg, Augsburg.* Worked with J. Probst

Ostervald, Jean Frédéric d. *Neuchâtel* levé1801–06, *Carte topo. et routière de la Suisse* (1820)

Osterwald , G. (1603–1884). *Panorama of Rome*

Osthoff, H. L. *West Kust Sumatra* 1839, *Gasper Straaten* 1858

Otera, Philipe. *Puerto de Bolinao (Philippines)* (1812)

Oterschaden. See **Otterschaden**

Otley, J. W. *Burlington Co., N.J.* 1849, *Columbia Co. N.Y.* 1851, *Genesee Co. N.Y.* 1854, *City Philadelphia* 1852

Otley, Jona. *District of the Lakes* 1813, 1818, 1849

Otley, Bookseller. See **Fox and Otley**

Otridge, W. Bookseller for Walker's *Universal Atlas* 1820

Ottagio. *Cartes géogr. . . . plans des marches . . . en Italie* 1745–6, 1775

Ottens Family. Dutch publishers of Amsterdam

—— Joachim (1663–1719). Founder of the firm. Engraver and publisher. *Austria* 1720

—— Widow of J. Ottens and Sons 1719–25). "Op de Nieuwen Dyk inde Wereldkaart"

—— Reinier and Josua (1725–1750). Publishers, map and booksellers. Op de Nieuwen Dyk in de Wereld Kaart later Kalverstraet au carte du Monde. *Mer Caspia* 1723, *Copenhagen* 1728, *Pocket Atlas* 1723, Renard's *Atlas de Navigation* 1739 and 1745 *Atlas Major.* Atlases of Dutch maps assembled to order were issued by Ottens varying from 4 to 15 volumes. Reinier died 1750. Josua died in 1765. The business was continued by his widow

—— Frederick (18th century). Plates and maps for Valentyn *Oost Indien* and Marsigli *Danube*

—— Josua & Reinier Ottens II (1750–1765). *N. America* 1755, *Guerre en Amérique* ca. 1760

—— Widow of J. Ottens and Son (1765–93). *St. Eustatius* 1775

Otter, Lieut. (later Comm.) H. C. R.N. Admiralty Surveyor. *Scotland* 1842–3, *West Coast Scotland* 1844–53, *Baltic* 1754–6

Otterloo, A. van. Dutch editor. Frijlinks' *Hand Atlas* 1872

Otterschaden, Johann. German globe-maker. *Globes* between 1580 and 1613

Ottersky, F. *Eisenbahn u. Post Karte Mittel Europa* 1874

Ottinger, Johannes Fr. *Servia,* Homann Heirs (1760)

Engraved title-page to the last (and most extended) edition of
A. Ortelius' famous Theatrum Orbis Terrarum

Letter by A. Ortelius, in which he advised his correspondent to take uncoloured rather than coloured maps. Should he prefer coloured copies, Ortelius counsels him to take copies coloured by his (Ortelius') daughter.

Ottmer, C. T. *Map Williamsburg* (1836)

Otto, J. C. Engraver. *Pyrmont* 1738

Ottomon. See **Freducci**

Ottonius, C. P. *Sweden* 1868

Ouctomsky, A. *Map Russian Asia* 1798

Oudemans, I. Abr. Dutch geographer and explorer, 19th century

Oudney, Dr. *Discoveries in Central Africa* (1825)

Ouerduyn, P.A. *Map of Ameland* 1809

Oughtred, William. Writer on mathematical instruments, cosmography 1832–6

Ourand, C. H. *Maps Cuba, Panama, Philippines* 1897

Outghersz, Jan. Dutch pilot. *chart Magellan Strait* (woodcut), Amsterdam 1600

Outhett, John. Surveyor. *Dorset* 1826, *Warwick* 1828. Drew Salt's *E. Coast Africa* 1814, Laurie's *Plan London* 1837

Outhier, L'Abbé Reginald [or Renaud] (1694–1774). Born Lamare Jousserand, d. Bayeux. Astronomer. *Globe* 1731, *Map Diocese Bayeux* 2 sh. 1736, *Archbishopric Sens* 2 sh 1741

Outram, B. Engineer. *Canal Huddersfield–Ashton under Lyne* (1793)

Ovalle, Alfonso d'. S. J. (1601–1651). *Regni Chile* 1646

Overbeke, Pauwels van. *Plan Antwerp* 1568

Overman, Capt. L. C. *Route Texas to Fort Yuma* 1870

Overradt, Peter, Publisher and Printseller in Cologne, late 16th–early 17th century. *Miniature map of British Isles* in Kaerius' *Min. Atlas* 1617, *Gloria Germanicae Typus* 1630

Overton, Henry (fl. 1706–1764). Publisher "at the White Hors *(sic)* without Newgate near the Fountain Tavern." Succeeded his father John Overton in 1707. Reissued the series of anonymous *County Maps* 1708, reprinted Speed's *County Maps* (with additions) 1713, reissued 1743; *Plan London* 1739; *Trading Part W. Indies* 1745, Bowen's *Royal Atlas* 1763

Overton, John (1640–1713). Married daughter of William Garrett. Acquired stock and presses of Peter Stent in 1665. Addresses: "White Horse without Newgate at the corner of Little Old Bailey near St. Sepulchre's Church" (1665–1666). "At the White Horse in Little Britain next door to Little St. Bartholomew's Gate" (1666–1669), "White Horse without Newgate near the Foutain Tavern" (1669–1707). In 1707 sold his stock to his second son Henry. Employed W. Hollar to copy maps. Made up sets of County Maps. In 1700 acquired John Speed's plates from Christopher Browne. Hollar's *Hungary* 1664. *England and Principality of Wales* taken out of J.S. and sold by John Overton 1673. Mapseller for Ford's *Barbados* 1681

Overton, Philip (d. 1751). Printer and Map Seller "Ye sign of ye Golden Buck by ye Mitre Inn Fleet Street" (1720). Later 53 Fleet Street over against St. Dunstans Church. *Oxford* 1715, *Essex, Middx., Herts.* 1726; *Yorks.* (with T. Bowles) 1728, *Sussex* 1740. Took in Robert Sayer as partner 1745

Overton, T. Engraver. *New Shoreham* 1815

Overton and Hoole. *New plan garrison Gibraltar* 1726

Overweg, Adolph. Explorer. *Libya* 1850–2, *Central Africa* 1854

Oviedo y Valdez, G. F. de (1478–1557). *Hist. Nat. et Gén. des Indes* 1535

Owen, A. K. *Mexico* 4 sh. 1884

Owen, David Dale. Surveyor. *Geolog. Map. Wisconsin, Iowa & Minnesota* 1851, *Tennessee* 1866

Owen, Francis. Master, R.N. *St. Johns Harbour* 1799, *Trinity Harbour Newfoundland* 1801, *Chart Spithead,.* engraved Baker 1801, *Island Navassa* 1814

Owen, George, of Henllys (1552–1613). Historian and cartographer. *Description of Pembroke* 1603, *Map of Pembroke* in Camden's *Britannia* 1607

Owen, H. *Chart Plymouth* 1841

Owen, John. *Britannia Depicta* 1720 (with E. Bowen) and later editions

Owen, Lt. (later Comm.), R.N. Hydrographer. *Delagoa Bay* 1822, *Africa* 1823–7, *W. Indies* 1829–37

Owen, Richard. *Great Delagoa Bay* 1827, *Belize Harbour* 1831, *The Colorados* 1838, *Bahama Bank* 1839, *Honduras Gulf* 1844, *Florida Strait* 1848

Owen, Stanley. *E. Coast Australia* 1855

Owen, W. *Victoria* 1862

Owen, William. *Wales* 1788

Owen, William (d. 1793). Bookseller and publisher, at Homers Head near Temple Bar Fleet Street. *Fairs in England and Wales* 1756, *Gen. Magazine of Arts & Sciences* 1755–1765. *Map of New England* 1755

Owen, Capt. W. F. W., R.N. *Port of Lake Ontario,* 1815, *Survey River Detroit* 1815

Owen, William Fitzwilliam (1774–1857) Vice admiral. *Charts of Africa for Admiralty* 1821–6, *East Indies* 1803–8, *Canada* 1815–16, *Bay Fundy* 1842–7

Owen, W. O. *Albany Co., Wyoming Territory* 2 sh. 1886

Owens, G. *Mining Regions Idaho and Montana* 1866

Oxenham, Edward Lavington (1843–1896). *Hist. Atlas Chinese Empire* Shanghai 1888 and 1898

Oxholm, Peter Lotharius. *Danske St. Croix* 1799 and 1800, English edition 1809 and 1848; *Danske St. Jan* 1800

Oxley, John (1781–1828). Surveyor General New South Wales 1812. Explorations 1817 to 1823. *Chart part Interior of N.S.W.* 1822, Revised 1825, pub. Arrowsmith 1829

Oya, Gaiko (1839–1901). Japanese globemaker

Oyarvido, Lieut. Andres. *Chart Coast Montevideo* 1819, *Buenos Aires* 1800–03, published 1812

Oyarzun, Juan. *Plano del Estero Coman* 1866

Oyley, W. d'. See **D'Oyley**

Ozanne, Edward. Charles. *Statist. Atlas Bombay Presidency* 1889

Ozanne, J. F. (1735–1795). *Views Paris*

Ozanne, Nicholas Marie (1728–1811). *Ports de France* (1783–91)

Ozersky, A. *Russian geolog. map* (1849)

P

P.D. [David Powell]. *Brief Rules of Geography* 1573

P.E. *Nova Daniae Regni Tab.* 1763

P.J. *Aachen (& Environs)* 1875

P.L. *Mirandola* (1565), *Napoli* (1565), *Agria* 1568, *Hierusalem* (1570), *Tripoli* (1575)

P.L. *Ciudad Buenos Aires* 1878

P.M.A. *Partie des Alpes voisines Mont Blanc* 1731

P.P. See **Pfinzing**

P.R. *London as it lyeth in ruins* 1666

P.T. *England & Wales* 1668

Paasche, W. (b. 1865) German cartographer and publisher

Paauw, Gerrit van der. *Omstreeken der Stadt Haarlem* 1805

Pacar, Petro. *Maragnon Fluminis* 1785

Pacheco, B. de M. y. See **Miera y Pacheco**

Pacheco, Carlos. *General Bosquejo de una Carta Geolog. Mexicana* 1889, *Carta Gen. Mexicana* 1890

Pacheco, Joao. (16th century). Portuguese pilot

Pacheco Pereira, Duarte (16th century) Portuguese navigator

Pacher, Ober Leut. *Plan Trieste* 1843 (with Frantzl)

Pacho, J. R. *Marmarique et Cyrénaique* 1827

Pacholowiecki [Pacholowic], Stefan . Scribe to Chancellor of Stefan Batory, King of Poland. *Plans of Polock,* (fortresses and castles conquered in 1579), *Ducatus Polocensis* Rome 1580

Pachoux, J. J. Script engraver for Tardieu's *U.S.* 1806 and 1812

Packe, Charles. *Les Monts Maudits* Weller 1866, *Guide to Pyrenees* 1867

Packe, Christopher (1686–1749). Physician and cartographer. The first to attempt a physiographic map. *Philosophico Chorographical Chart of East Kent* 4 sh. 1736–43 (scale 1½ in. to mile). First English map to use spot heights and to distinguish different types of soil

Packer, Thomas. *Panoramic maps of Russia and Turkey* 1855–77

Packx, H. (1602–1658). *Plans of s'Hertogenbosch*

Padley, James Sandby. Surveyor. *Plan Fosdyke Navigation from Lincoln to the Trent* 1826, *Plan City Lincoln* 1842, new edition 1868

Padley, P. *Plan Orange Hill Estate . . . Is. St. Vincent* 1817

Padoani, Dom. Mapseller, Ponte di Rialto Venice, for Coronelli maps 1689–92

Padtbrugge, H. (d. 1687). Maps and plates for *Suecia antiqua et hodierna*

Padua, Seminario Vescovile. *Tipografia. Tab. Geogr.* 1699

Paep. See **Pape**, Jehan de

Paes Clemente, Adelino (d. 1891). Mathematician and surveyor of Lisbon

Paez, D. E. *Bahia de Sta Isabel, Fernando Poo* 1860

Pagan, Blaise François, Comte de. *Mag. Amazoni Fluvii nova delin* 1655

Pagan, Matteo. See **Pagano**

Pagano, Matteo (fl. 1538–62). Publisher, wood engraver, and mapseller in Venice. "In Frezaria al Segno della Fede." *Crete* 1538, *Cyprus* 1538, *Piemonte* (1539), *Tuscany* n.d., *Ungheria* 1564, *Cairo* 1549 (used by Braun & Hogenberg, *Portolan* 1553.

Page. See **Page**, Thomas, also **Mount & Page**

Page. (1814–1879). English geologist

Page. Bookseller of Ipswich. Took subscriptions for Hodskinson's *Suffolk* 1780, pub. 1783

Page, F. R. *Newfoundland* 1859

Page, J. O. *Fulton Co., N.Y.* 1856

Page, Lieut. Thomas Hyde (1746–1821). Engineer. *Action at Bunkers Hill* 1775, *Plan Town Boston,* Faden 1777, *Boston its Environs and Harbour* 1777 and 1778

Page, Thomas. Publishers in partnership with R. Mount. See **Mount & Page**. Succeeded by 3 further generations, all Thomas Page

Page, Thomas, of Hilgay. *Estate plans Hilgay Thetford &c.*

Page, T. H. *Chart Yarmouth to Rye* 1783

Page, Thomas J. U.S. Navy. *La Plata* 1858, *Paraguay* 1858, *Survey Rivers Parano (Uruguay and Paraguay)* 17 sh. 1856

Page du Praté. See **Le Page du Pratz**

Pagerie, G. Mariette de la. See **La Pagerie**

Pages, A. *Atlas Universel* 1873

Pagès, Pierre Marie François, Vicomte de (1748–1793). French naval officer and explorer

Pagitt, Ephraim. *Christianographie . . . sundray sorts of Christians (with maps)* 1586, 1636, 1640 and 1674

Pagnano, Carlo. *Decret. superflumin. Abduae* Milan 1520, woodcut map.

Pagnau, E. *Dépt. de la Gironde* 1867

Pagteel, P., of Prague. *Island of Leyte* 1788 MS, used by Dalrymple

Paguenard, E. Improved maps for Lavosine's *Atlas* 1820, for Carey & Lea 1822

Pahnke, E. *Topo Karte der Kreise des Reg.-Bezirks Münster* 1889–92

Paillard, H. P. (1844–1912). *Views Paris & Algeria*

Pain, Capt. H. *Route betwn. Pra & coast (Ghana),* W.O. 1881

Paine, J. Surveyor, draughtsman, architect. *Estate plans* 1734, *Plans Mansion House Doncaster* 1751

Paine, J. A. *World's Great Empires* N.Y. 1883

Paine, J. D. *Parish St. Mary Lambeth* 1841

Painter, J. W. *Settled Districts S. Australia,* (1857) (with G. Higginson)

Pais, Peter S. J. *Sources of the Nile* 1618, used by Kircher 1678

Pajares, J. Lithographer of Madrid. *Mapa Geolog. Europa Occidental Chromolithographia* (1875), *Liebano y Trincalo* 1882

Palacios, Geron. Martin. *MS Atlas 33 charts* 1603

Palacios, J. A. *Plano Topo. Rio Madera* (1846)

Palacky, F. *Histo. map Czechoslovakia* (1874)

Palairet, Capt. *Taway River* 1753, used by Dalrymple as *Admiralty Chart* 1774

Palairet, Jean [John] (1697–1774). English cartographer, b. Montauban, France. *Atlas Méthodique* 1755–(63), English edition 1775, Bowles *Universal Atlas* (1775–80)

Palazzi, Giovanni. *Isola Schut Ungaria* 1684

Palazzo Vecchio, Florence. *Vaval maps* by Egnazio Danti 1563–75 and Buonsignori 1576–89

Palbitzke, Frederick. Corrected Lubino's *Map of Pomerania* for Hondius 1633

Palestrina, S. de. See **Pilestrina**

Palfrey, John Gorham (1796–1881). American writer and cartophile. *New England in 1620–1644 . . .* engraved for P.'s *History of New England,* J.G.P. del 1858; *New England in 1689,* engraved for P.'s *History of New England,* J.G.P. del (1858)

Palgrave, W. G. *Map pf Arabia,* illust. Palgrave's *Journey* 1865

Palin, T. W. *Sketch map North Island of New Zealand* Wellington 1869

Pallas, Peter Simon (1741–1811). *Entdeckungen zwischen Sibirien und America* 1781

Pallavicino, Leone (1590–1616) and Lucio (1590–1610). Engravers of Brescia. *Territ. Bresciano* 6 sh. 1597, Monte's *World* 1604

Palleau, Sr. *Carte de Poitou* 1785

Palliser, Sir Hugh (1723–1796). Admiral, Governor of Newfoundland. Authorized Cook's surveys

Palliser, J. Capt. (1807–87). *Gen. Map of Routes in British North America* 1857–60, London, Stanford 1865

Pallota, Felipe. *View of Almansa* (ca. 1710)

Palm, B. Magdeburg. Woodcuts for Bunting's *Chronicle of Brunswick* (ca. 1580)

Palm, J. Publisher of Munich. *Plan München* 1841, *Reise Karte Bayerischen Hochlandes* 2 sh. 1841

Palma, Gaetano. *Turquie d'Europe,* 2 sh. 1811, *Provinces Illyriennes* 1812, *Turquie d'Europe* Trieste 1814

Palmatary, J. J. *Birds-eye view of St. Louis, Missouri* (1857), *View of Springfield* 1860

Palmer. *Plan Liverpool* in Aitkin's *Manchester* 1795

Palmer, David. Estate surveyor in Cornwall 1789

Palmer, E. H. *Route map of Negeb* 1871

Palmer, Capt. Edmund. *Military Sketch Is. St. Helena* 1859

Palmer, Capt. George. *Chart of Boddams Track to Island Bouro* 1799, *Tonghou Cove* 1799

Palmer, H. R. Civil engineer. *Plan Burnham Salt Marshes* Norfolk 1822, *London Ipswich Railway* 1825 MS, *Limehouse* 1832 MS

Palmer, Lieut. (later Captain) Henry Spencer. *British Colombia* 1861, *Gold Regions Fraser River* 1861, *Sinai Penin.* 1869, *Ordnance Survey* 1873

Palmer, I. Engraver. *Vermont* 1798, Green's *Sussex* 1799. For Dalrymple 1805

Palmer, James. *Ireland* 1821, *World* 1825

Palmer, James. Engraver. *St. Pancras* 1804, Arrowsmith's *Upper and Lower Egypt* 1807

Palmer, Joseph. Surveyor. *Olney Pasture* MS, n.d.

Palmer, Loomis. *Alabama,* Chicago, Fairbanks & Palmer 1889

Palmer, R. E. *Plan City Vancouver* 1891

Palmer, Richard. Engraver, fl. 1680–1700. Worked for Blome (Speed's maps epitomised). 1681 proposed *Atlas Anglicanus* (with Seller and Oliver); *Herts.* 1680, *Surrey* 1680, *France* 1890, *Hungary* 1700

Palmer, Richard. Geographer. *Arabia* 1844, *Palestine* 1845

Palmer, Sir Roger. Surveyor. *Ashendon . . . Estate Plans* 1624–28

Palmer, Sir Thomas. *Coast of Sussex* 1587 (with W. Coverte), pub. 1870. *How to make our travailes into foraine Countries more profitable* 1606

Palmer, Capt. W. R. U.S. Topo. Eng. *Central America* 1856

Palmer, W. Surveyor. *Survey River Dunn to Sheffield* 1722, *River Ouse* 1728, *Swale and Ouse* 1735

Palmer, William. Engraver, No. 128 Chancery Lane. Worked for Ellis *English Atlas* 1766, Dalrymple 1774, Faden 1784–1803, Wilkinson 1808, Cook's *New Caledonia* 1777, Sayer's *World* 1792, Faden's *Atlas Minimus* 1798, Laurie & Whittle's *World* 1800

Palmer and Newton. *Terrestrial Globe* 1783

Palmeus [De Palmaus], A. F. Gervais de. Engineer, Geographer to the King, mapseller, Rue Tictonne. *Malta* 1752, *Valetta* 1757. Chambray, *Gozo* 1754, *Malta and Gozo* 1799

Palmier, L. A. (19th century). Belgian engraver

Palmieri, Luigi (1807–1896). Physician and meteorologist

Palmquist, Eric (d. 1676). Swedish military cartographer. *Descript. Moscoviae* 1674

Palmstruch, J. von. (1770–1811). *Maps of Sweden* (1802)

Palombo, Pietro Paolo. *Isola di Malta* 1565. Succeeded by Albertus

Palomino, Diego. *Peru* 1549

Palomino, Ja. Engraver *Islas Philipinas* 1727

Palsson, Sveinn (1762–1840). Icelandic cartographer

Palumbi. See **Palombo**

Pampani, Francesco. Engineer, geographer. *Citta di Ferrara* 1836

P'An, Wang. *China, Korea & Japan* (1549)

Panades, Banet (16th century). Portolan maker of Majorca. *Mediterranean* (1557)

Panagathus. See **Algoet**

Panckouche, C. L. F. Publisher. Rue et Hôtel Serpente, Paris. Lapie's *Océanie* 1816, *Isle of Staffa* 1831

Panfili, P. *Pianta Bologna* (1700)

Panhuijs, Jan Ernst. *Provincie Friesland* 1874

Pannartz, Arnold (1464–1490). German engraver and publisher (with Sweynheim)

Panser, Symon. Astronomer. *Hemel Spiegel* 1738, *De Loop van Mercurius om de Zonne* 1743

Pantoja, Juan. *Various maps and plans. W. coast of America* ca. 1780. *Plan San Diego* 1782, used by Espinosa y Tello

Panouse, Jacqueline. Engraver. De Fer *Milan* 1705

Paolino [Paulinus]. Minorite Friar, d. 1345. *Chronicke* ca. 1320 *(maps of World & Palestine)*

Paon, W. See **Pawne**

Papa, R. *Maps Austria and Germany* 1870–99

Papasogolu, Dimitrie (1811–1893). *Maps Roumania, . . . Bessarabia* 1865–9

Papaut, M. *Etabliss. français de la Guinée* (1885)

Pape [Paep], Jehan de [Jan] (fl. 1515–31). Miniature painter, of The Hague. Surveyor and cartographer. *Holland* 1513–31 MS

Papen, Augustus (d. 1858). Hanoverian military cartographer. *Hannover u. Braunschweig* 78 sheets Hannover 1832–47, *Central Europe* 9 sh. 1858–9, *Oldenburg* 1865

Papendieck, G. E. (1788–1835). *Maps & views of Bremen, Hannover etc.*

Pappasoglu, Dimitri. See **Papsogolu**

Papworth, J. B. *Plan Hygeia* in *Bullock's Travels in N. America* 1827, *Travels in*

N. America 1827, *Plan Streets occupying Valley of River Fleet* (1823)

Paracelsus [Huhenheim] (1493–1541). Physician, alchemist, philosopher. *Astronomia Magna* Frankfort 1571

Paradin, W. *Anglicae Descript. Compendium* 1545

Paradis, Sr. *Plan Madras* 1750

Parant, L. *Plan de la Ville de Bourg* 1869

Paravey, Ch. Hipp. de (1787–1871). Astronomer and geographer

Parboni, Achille and Pietro. *Roma* 1829

Parbury. See **Black, Kingbury, Parbury, & Allen**

Parbury, Allen & Co., Leadenhall Street, London. Publishers. *Chart England to Canton* 1829, *China* 1833

Parbury, George. *Countries between England and India* 1841

Parcar, Petrus. *Amazon* 1780

Parchappe. *Partie de la Répub. Argentine* 1835

Parchwitz, P. *Scenographia urbium Silesiae* (1737–9)

Pardo, Carlos. *Repub. Argentina Chile y Boliva* 1896

Pardy, J. R. *Beira Harbour* 1891

Parea, Carlo. *Carta Topo. Prov. di Milano e Pavia* Milan (1820)

Parenti[us]. See **Parenzio**

Parenzio, Gellio. Military engineer. *Tuscany* 1572, *Spoleto* 1597

Pareto, Bartholomew, of Genoa. *MS Planisphere* 1455

Pareto, M. *Carte géolog. d'Italie* 1846

Parides, Ignatius and Gaston (17th century). Astronomers

Parientos, M. G. G. de. See **Gómez y Parientos**

Parijs, Sylvester. See **Paris,** Sylvester

Parijs, William van. Antwerp engraver, end 16th cnetury

Paris. Gilt Globe [Globe Doré]. In Bibliothèque Nationale, Paris. Engraved on copper and gilded 1530–40

Paris, E. Ensign on the *Favourite* (La Place voyage). *Tonga-Tabou* 1827, *Papous* 1828, *Anamba Island* 1831. *China Seas Natuna Islands,* Admiralty 1831

Paris, Louis Philippe Albert d'Orléans, Comte de (1838–1894). *Guerre civile en Amérique* 1883

Paris, Mathew (d. 1259). (Benedictine) monk and chronicler, Abbey of St. Albans, carried on the *Chronica Majorca* to 1259. Compiled *four maps of England* ca. 1250, *Palestine, route map Italy to Apulia,* and *World*

Paris, Sylvester (ca. 1500–1576). One of the oldest engravers of Antwerp. Worked with Camotius in Venice and Plantin in Antwerp

Paris, H. W. *Map* for Staunton 1798

Parish, Sir Woodbine. Consul General, Buenos Aires. *Argentine* Arrowsmith 1834–39, –42. *Falkland Is.* 1830 MS

Parisi, G. *Itin. Militari da Bologna a Napoli* Milan 1809

Parisio [Parisius], Prospero. *Calabria* 1589 (used by Ortelius 1595), *Regno di Napoli* 1591

Parisot, L. *Carte d'Algérie* 1876, *Environs de Belfort* 2 sh. 1878

Park, J. D. *W. Part City Covington Kentucky* (1857)

Park, Moses. Draughtsman, *Connecticut* 1761, *Plans Central America* (1780) MS

Park, Mungo (1771–1806). Africa Explorer. *The Niger*

Parke, J. G. Lieut. U.S. Topo. Engineers. *New Mexico, Santa Fe* 1851, *Reconn. Zuni & Colorado Rivers* 1852

Parke, R. Tanslated Mendoza's *History of China* 1588

Parker, Charles H. (d. 1819). Engraver. *Maps of Pennsylvania* 1812

Parker, George (1697–1764) Second Earl of Macclesfield, astronomer

Parker, Henry. Bookseller and publisher. With E. Bakewell opposite Birchin Lane in Cornhill. Maps for Bowen's and Kitchen's *Royal English Atlas* (1765). *Plan of London* 1758 and 1760, *Kent* (1672), Cook's *South Carolina* 1773

Parker, Hyde. *Luabo River* 1851

Parker, James. Surveyor of Thetford. *Nazeing* 1767, *Estate of Barningham (Suffolk)* 1771 MS

Parker, John Henry. *Rome Ancient & Modern* (1882)

Parker, Joseph. Surveyor. *Canal Basingstoke to River Wey* (1787)

Parker, Joseph. Publisher, Oxford. Cranmer's *Ancient Greece* 1827

Parker, L. M. *Town of Shrewsbury, Worcester County, Mass.* 1859

Parker, Mathew, Archbishop (1504–1575). Collector and patron of arts, published *Bishops Bible with map of Holy Land* by Humphrey Cole (the earliest map engraved in England). Parker also employed Remigius Hogenberg and Richard Lyne *(1st plan of Cambridge)*

Parker, Nathan H. *Geolog. Map Iowa* 1856, *Missouri* 1865

Parker, P. Master, R.N. *Additions to chart Maldonado Bay* 1820

Parker, Robert. Estate surveyor. *Writhe* 1724 MS, *Roxwell* 1728 MS

Parker. Engraver. Hayward's *Descript. Fenns* 1604, engraved S. Parker 1606; Warburton's *Yorks* (1720)

Parker, Samuel. Engraver and draughtsman. *Kent* 1719, *Romney Marsh* 1719, Warburton's *Yorks*. Warburton's *Essex* (1726), Gaudy's *Newfoundland* 1728. *N. Celestial Hemisphere* (1728). Engraved for Senex, *English Pilot* 1751

Parker, Rev. Samuel. *Oregon Territory* 1838

Parker, William. *Voyage to Margarita, Jamaica &c.* Hakluyt 1600

Parker, William Hyde. *North Bank Congo, Admiralty Chart* 1866

Parker and Huyett. *Pike's Peak Gold Regions* 1859

Parker & Nourse. Publishers for Patterson 1766

Parkes, W. *Plan Newhaven Harbour* 1846

Parkin, G. R. *Map British Empire* 1893, *World* 1894

Parkinson. Surveyor and engraver. *Topo. map of Humboldt Co.* 1866

Parkinson, F. B. *Somaliland* 1897–8 MS

Parks, Murray.T., Lieut. *N. Approaches to Liverpool* 1857, *River Mersey* 1857 and 9 sh. 1864, *Liverpool Bay* 1859

Parkyns, G. Lieut. *Admiralty Charts S.E. Coast China* 1825, *Killon Harbour* 1835

Parkyns, Mansfield. *Part Ost Afrika* 1861, *N. Abesinien* 1864, *Abysinnia and Nubia* 1867

Parley, Peter, pseud. for S. G. Goodrich, q.v.

Parmenter, L. Lithographer. *Plan building Lands Ash Surrey* (1850)

Parmenter, Stephen C. *Map of Newburgh, Orange Co. N.Y.* 1850

Parmentier [Permentier], Jean (1495–1530). Mathematician, chartmaker, cosmographer of Dieppe

Parmentier, M. *Carte d'Amérique* 1750

Parmentier, Théodore. *Théatre Guerre Russo-Turque* 1854

Parmentier, Le. See **Le Parmentier**

Parr, Nathaniel. Engraver. *Universal Traveller* 1752–53

Parr, Richard (fl. 1723–51). Engraver. Dickinsons. *S. Part Co. York* 1740, *Plan Dunkirk* 1743, *Survey London & Westminster* 1746, 1748, 1751; Rocque's *Shrewbury* 1746, De Lavaux *Chester* (1745)

Parreus Map. See **Barre**

Parilla, Col. Diego Ortez. *Parte d'el Seno Mexicana* 1766

Parrocec, I. J. (1667–1722). *Views of Battley & Town, Views for Prince Eugene of Savoy*

Parrot, George B. Draughtsman. *Boston & Maine Railway* 1849, *Winship Estate* 1856

Parrott, T. S. *Colony of Queensland* (with Teage), *Melbourne* 1869, *Illawarra Coaling District* 1894

Parrott, William P. *Map Boston and Maine Railway* 1849, *Survey Boston Harbour* 1850, *Boston Harbour* 1851

Parry, Lieut. (later Admiral) Sir William Edward, R.N. (1790–1855). Hydrographer to the Navy 1823–29. *Baffins Bay* 1817–20, *Hudsons Bay* 1821–23, *N.W. Passage* 1824–5, *Arctic* 1827

Parsones, S. See **Parsons**, S.

Parsons, Frederick James. *Map Hastings and St. Leonards* (1894)

Parsons, Horatio A. *Map of Niagara Falls* (1830)

Parsons, John. R.N. Admiralty Charts. *Survey W. Indies* 1854–57, *Grenadines Bequia* 1863, *Barbados* 1869, *Dover Bay* 1873, *Folkestone Harbour* 1874, *Harwich* 1879

Parsons, Philip. *Coast of Aracan* 1743–4, 1785

Parsons, Capt. R. M., R.E. *Maps British Columbia* 1861–3

Parsons [Parsones], Samuel. Estate surveyor. *Little Coggeshall* 1639 MS

Parsons, Samuel. Publisher Jeffery's *Staffs.* with J. Bowles 1747

Parsons. *Virgin Islands* 1848–52 (with Lawrence & Tuson), French edition 1874

Parthey, G. F. K. (1798–1872). German collector, bookseller, and bibliographer of W. Hollar

Partridge, S. Estate surveyor. *West Thurrock* 1645 MS

Partsch, Joseph. *Africae Veteris* (1874), *Korfu* 1887, *Kolonial Atlas* 1893

Partsch, Paul. *Geogn. Karte Wien u. der Gebirge* 1843

Parvin, Theodore S. *Iowa* 1859

Pasaueus, S. See **Pass**

Paschall, Thomas H. *Union Co., Pennsylvania* 1856

Pascoe, James. Horton's *Addition to San Diego (California)* 1868

Pascoe, John. 2nd Master H.M.S. *Blenheim. Kintang and Blackwell Channels (China)*, 1840, *Admiralty* 1841

Pascoe, T. *Chittagong River* 1877

Paseus, Simon. See **Pass**, Simon

Pask, Joseph. Stationer and publisher. (1) At ye Stationers Arms and Ink Bottle on the North side of the Royal Exchange. (2) At ye Three Inkbottles on Castle Alley at ye West End of the Royal Exchange. *Pocket Book of Counties* (1680), Ford's *Barbados* (1681), *Essex* with R. Morden (ca. 1700)

Paske, Robert. Stationer. At the Stationers Arms and Inkbottle in Lombard Street. Bookseller for Ogilby's *English Atlas* 1679

Pasley, Sir Charles William. *Part of the City of Worcester* 1844

Pasolini, Andrea. *Topo. della Romagnola* 18th century

Pasquali, Giovanni Battista. Engraver for Zatta's *Atlante Noviss.* 1775–85 (with G. Pitteri) and for Rossi 1786

Pasqualin, Nicolo de. *Archipel. Coast Asia Minor* 1489 MS

Pasqualini, G., of Venice. *Portolan* 1408

Pasquier, J. J. Engraver. *Plan Paris* 1758 (with L. Denis), Palmeus' *Malta* 1752, *Forêt de St. Germain* (1770)

Pasquin, Manuel. *Plan Puertos de Naos y Arrecife* 1860, *Canary Is.* 1853, *English edition Admiralty* 1881

Pass, J. Engraver for *Encyclopedia Londinensis* 1827–8

Pass [Passi, Passeus], Simon van de (1595–1647). Map for Smith's *New*

England, Printed J. Reeve 1614–(16)

Passard, F. L. *Carte d. rasses Europ.* (1873)

Passaroti, Aurel. Surveyor. *Plan of Lwow (Lemberg),* early 17th century

Passe, Engravers. See also **Pass**

—— Crispyn I (1564–1637). Engraver. Worked for Plantin

—— Willem van (1589–1637). Engraver

—— Crispin II (1598–1670). Engraver. Nautical Instruments, etc.

Passenger, T. At the Three Bibles on London Bridge. B. Tooke at the Ship in St. Pauls Churchyard. Sawyer at the Three Flower de Luce in Little Britain. Joint publishers of Heylin's *Cosmography* 1682

Passini, J. N. (1798–1874). Austrian lithographer. *Maps and plates.* Austrian topography

Passmore, Walter N. *Railway Map Natal* 1878

Passy, Ant. *Carte Géolog. des Environs de Paris* 1865

Passy, Christoph von (1763–1837). *Maehren und Oesterreich Schlesien* 1810

Pasteur des Amadis. See **Boileau de Bouillon,** Gilles

Pastoret. French pilot. *MS Atlas of 78 maps* 1587

Patavia. Seminary. Published Sanson's maps of *Ancient Geography* 1694–6

Patch, Thomas (1720–1782). Engraver. *Views Florence*

Paterson, Daniel (1739–1825), Lieut. Col., Governor of Quebec 1812. Cartographer. *Plan Island of Grenada* [as Lieut. Paterson] pub. Faden 1780, reissued 1796 and 1825. *Direct and Principal Cross Roads in England & Wales* 1786, *British Itinerary* 2 vols. 1786, *Travelling Dict.* 1787, Bowles *New 4 sheet map England & Wales* 1796, *HM Forces in N. America* 1766 MSS (Asst. Q.M.G.)

Paterson, G. & W. *Survey of Old & New Aberdeen* 1746

Paterson, Lt. Col., of New South Wales Corps. *N.S.W.* Arrowsmith 1799

Paterson, J. W. *North Formosa* 1882

Paterson, Lt. William. *Southern Extremity Africa* 1789

Patersson, B. Swedish engraver. *Maps St. Petersburg* (1799)

Paton, A. *Map Sydney and Environs* 1892

Paton, R. (1717–1791). Engraver of *English Ports*

Patriceo, Joze. Pilot. *Chart N. Coast Brasil,* Faden 1809

Patrickson, Major Samuel. *Battle Waterloo,* scale 4 inch 1815 MS

Patrini, G. (1711–1786). *Map of Parma* 1770

Patrokles (300 B.C.). Greek chartmaker. *Periplus of Caspian Sea*

Patten, John. "Traders Maps": *W. of Alleghany Mts., and S. of Lake Erie* 1753 MS

Patten, Nathaniel. Printer *Seat of War near New York* 1776, *Investment of Yorktown* 1783

Patten, William. *Exped. into Scotland* 1548 (3 maps), *Calendar of Scripture* 1575

Pattenden, John. Surveyor of Brenchly, Kent. *Payne Estate* 1639 MS, *Beltringe Farm* 1656, *Cats Place* 1656 MS

Pattergill, G. Delevan. *Grant Co. Wisconsin* 1868

Patterson, Dan. *Disposition British Forces in America* 1766

Patterson, Mathew. *Durham* 1595

Patterson, R. *Map* for Bowles 1775

Patteson, Rev. Edward. *Gen. & Classical Atlas* 1804, *England & Wales* 1804, *Egypt* 1806

Pattison, Juan. *South Central Railroad (U.S.)* 1866

Pattison, Thomas. Civil engineer. *Dearborn Co. Indiana* 1860

Patton, F. *Plan Town & Harbour of Mahon*

Patton, Lt. *Map Siege Vicksburg* 1863

Paul, C. M. *Geolog. Karte Bukovina* 1876

Paul, George. *West Hickory Creek, Venango Co. Pennsylvania* 1866

Paul, Hosea. *Summit Co. Ohio* 1856, *Atlas of Wabash Co. Indiana* 1875

Paulet, A. J. de S. *Prov. von Ciara (Brasil)* 1831

Paulet, Giovanni. Revised *miniature edition of Ortelius* 1612

Pauli, A. (1600–1639). Engraver. Plates for *Triumphal Entry of the Queen Mother in Holland* (1632)

Pauli, Christopher, Lieut. 60th Regt. (fl. 1761–75). *Harbour Cape St. Nicole, Hispaniola* 1775

Pauli, Seb. *Malta* 1733–7

Paulin. Publishers, Rue Richelieu 60, Paris (with A. Le Chevalier to 1860). Dufour's *Atlas Univ.* 1856–60

Paulin, N. *Carte Géolog. de la Moselle* (1876)

Paulinc, Jacob, Joseph. *Maps of Croatia, Germany, &c.* (1855–98)

Paulini, Giacomo. *Carta di Montenegro* (1860)

Paulinski, P. *Plan St. Petersburg et Environs* 1860

Pauling, Jacob Joseph (1827–1899). Topographer. Worked in Egypt and the Balkans. *Special Karte des Salzkammergutes* 1860, *Deutschland* (1866), *Croatia* 1876, *Dalmatia* (1879), *Schneeberg* 1898

Paulinus. See **Paolino**

Paulis, Johann Anton de. *Venetia* (1620), *Ritratto del antich. cella de Tivoli* (with M. A. Gozzadino) 1622

Paulitschke, Philip (1854–1899). Austrian geographer. *Expd. Ghika, Pays des Somalis* 1898

Paulleau, Sr. *Carte du Poictou* 1785

Paullus, Simon (1603–1680). Danish geographer

Paulmier, L. A. (19th century). Brussels engraver

Paulsen, E. (1749–1790). Danish engraver

Paulson, Paulus (18th century). Military cartographer

Paulson. See **Paulusz**

Paultre, Charles. Artilley officer. *Syrie* 1803

Paulus, Ageminius (Paul the engraver). *World map "du coffret"* engraved on copper 1511

Paulus, H. E. G. German edition of *D'Anville,* Nuremberg 1785–93

Paulus, von. Topographer. *König. Würtemberg* 1841–50

Paulusz [Paulson], Paulus. Born Karlskrona Sweden. Worked for Dutch East India Co. 1723–50 in Batavia. *South Coast of Java* 6 sh. 1739 MS

Pauly, Jean Pierre (mid-18th century). Military geographer

Paur, Hans (15th century) 2 woodcut maps of *Nuremberg and Bavaria*

Pausanias. Greek geographer, 2nd century

Paute, Mme. Le. See **Le Paute**

Pauter, M. *Serbia* 1788 MS

Pautz, Th. *Plan von Stettin* 1869

Pavia, J. B. F. de. See **Funes de Pavia**

Pavia, Jose F. Capt. of Frigate. *Costa de Mindaneo* 1832

Pavie, Auguste. *Mission Pavie Indo Chine* 1892

Pavin, Robert. *Plan of Kemar Road* 1801

Paw, Corneille de (1739–1799). Dutch geographer

Pawley, G. *General Atlas* engraved S. & G. Neele 1819 and 1822, *Twenty five miles round Oxford* 1805

Pawley, George. Surveyor. *Boundary North–South Carolina* 1764 (with others)

Pawlikowski, Gwalbert (1792–1852). Polish collector of atlases and maps

Pawlowski, J. N. *Diozese von Culm* 1848, *Alten Preussen* (1855), *Kreis Culm* (1890)

Pawlowsky von Rosenfeld, Wenzel (d. 1778). Military cartographer

Pawne [Paon], William. Survey engineer under the Tudors, fl. 1485–1523 for Henry VII and Henry VIII

Paxton, J. A. *Plan Philadelphia* 1811

Payart, R. *Plan de la Ville de Ham* 1867

Payen, J. F. *Valle de Montjoie et Bains de St. Gervais* 1859

Payer, Julius. *Ortleralpen* 1868–9, *Tiroler Fjord* 1873, *Ost Grönland* 1874

Payer, Richard. *Rio Napa and Rio Curaray* 1894

Payne, Albert Henry. *Eisenbahn Atlas Mittel Europa* 1860, *Illust. Plan London* 1846, *Orbis Pictus* (1851), *Panorama* (1859)

Payne, Charles Wynn. *Ceylon* 8 sh. 1850

Payne, G. H. *Imp. Plan London* 1852

Payne, J. Publisher in London. Paternoster Row. *Plan Fort du Quesne* 1755

Payne, John. *Univ. Geogr.* 1792, *Atlas N.Y.* 1798–1800

Payne, John Willet (1752–1803). *Harbours Port Royal and Kingston, Jamaica* 1772 MS

Payne, M. Mapseller at the White Hart, Paternoster Row. Kitchen & Jefferys small *English Atlas* 1749, *Fire in Exchange Alley London* 1748

Payne, Richard. Surveyor. *Plan Borough Bury St. Edmunds* 1834

Payne, Robert. *Brief description of Ireland* 1589

Payne, T. Publisher. *Upper & Lower Egypt* 1809

Payne, T. & Son. *Britannica* 1789

Payne, W. H. Letts *Birds eye View Soudan* 1884

Payne. See **Mason & Payne**

Paynell, Thos. Edited Benese's *Book of Measuring Land* 1537

Pays [Pais], Pero. *Carte d'Ethiopia* 1672

Payte, John. Surveyor of Mayfield, Sussex. *Estate Plans* 1850 MS

Payter, J. W. *Victoria Telegraph Circuits* 1871

Payton, T., R.N. *Carlisle Bay Barbados, Admiralty* 1821

Payton, Walter. *Voyage to the East India* 1612–14

Paz, Manuel Maria. *Estados Unidos de Colombia* 1864, *Atlas Geogr. et Hist. de Colombia* 1889

Paz, Principe de la. *Maps* for Torfino di San Miguel 1787, *Islas Antillas* 1802

Paz Soldan, Mariano Felipe (1821–1886). Peruvian geographer and politician. *Mapa del Peru* 1862, *Atlas Geogr. del Peru* 1865, *Diccionario geogr. del Peru* 1877, *Repub. Argentine* Buenos Aires 1877

Pazzi, Giuseppe. Script engraver. *Maps for Gazettiere Americano*, Livorno 1763

Peabody, M. M. Engraver of Utica, N.Y. Parker's *Oregon Territory* 1838

Peach, Benjamin Neeve. *Geolog. Survey Perthsh. & Clackmannansh.* 1878

Peach, Joseph. Lieut. *Plans Canada & St. Lawrence* 1760 MS

Peacham, Henry. *Graphice or Art of Drawing and Limming (including maps)* 1612

Peacock, George. *San Juan de Nicaragua, Admiralty Chart* 1833

Peacock, J. *Plan Thames Legal Quays* 1796 & 1803

Peacocke, William. *Southern Turkestan* 1884, *Country between Rud & Murghab Rivers* 2 sh. 1885

Peak, A. Surveyor (19th century)

Peake, Robert. Surveyor of Waltham Abbey. *Surveys of Nasing* (1824) MS

Peake, Robert Edward. *Mouths of the Amazon* 1896

Pearce, Sir Edward Lovett. Surveyor Gen. H.M. Works Ireland *Various plans* 1728–9

Pearce, Lieut. Robert. *Passages into Sandwich Bay* 1821, *Harbour Grace, Newfoundland* 1825

Pearce, Stewart. *Luzerne Co., Penn.* 1860

Pearcey, W. R. Engraver. Laurie & Whittle's *Coast of Norway* 1812

Pearman, Fountain. Publisher. *Nebraska Gold Mines* 1859

Pearson, Charles H. *Historical Maps of England* 1869–70

Pearson, Isaac. *Bay of Matanzas* 1729, *Port Royal Rattan* 1743

Pearson, J. W. Printers and publishers of Melbourne. *Map of Victoria* 1869, *Australasian School Atlas* (1870)

Pearson, Robert. *Bay of Matanzas, Cuba* 1729 (with Isaac)

Pearson, R. M. *Map S.E. portion of New South Wales* 1860

Pearson, W. Surveyor. *Coast of Chittagong* 1869

Peary, Comm. R. E. USN (b. 1854). Arctic explorer

Pease, Andrew. Estate surveyor. *Mountnessing* 1622 MS

Pease, Joseph. *Travelling Map of G.B.* (1825) (marking Friends meeting places)

Pease & Cole, of Chicago. *W. Kansas & Nebraska* 1859

Pease, L. T. See **Niles**, I. M.

Peasley, A. H. American engraver and mapmaker

Peat, A. & Co. *Map Roads Scotland* 1822

Peat, Thomas. *Plan Nottingham* 1744

Pecci, Matthieu Neron. Cosmographer. *Wall map America* 1604, copied by Tatton

Pecciolen, Mathew Neron. Cosmographer. *La Californie* 1604

Pechar, Johann. *Kohlen Revier* 1864 & 1873, *Baumwolle* (1874)

Pecheur, Frederic. *Bains de Mer Guérande* (1868)

Pechmann v. Masser, Ed. (1811–1885). Austrian surveyor and topographer

Pechold, Gustav. *Military Railway map Austria, Hungary* (1883)

Pechon, Major. *St. Domingo,* Wilkinson 1799

Pecht, J. A. *Umgegend von Konstanz* 1898

Pechy, Imre (1832–1898). Map editor and publisher

Peck, J. M. *Illinois & part of Wisconsin* 1835, *New Sect. Map of Illinois* 1837

Peck, Lieut. W. G. Topographical engineer. *Territ. of New Mexico Wash.* 1845–7 (with Albert), *Military Recon. map Arkansas* 1847

Peck, William. *Constellations & how to find them* (1884), *Observers Atlas of the Heavens* 1898

Peckham, Edward. *Ightham Moat* 1803

Peckham, William & Edward. Surveyors part of Dorset. *Estate near Maidstone* 1797 MS

Peddie, Lieut. William. *Maps & Plans* 1808–14, *War in Spanish penin.* (1840)

Pedischeff. See **Piadischeff**

Pedro, Don de. *Portugal Libro del Infante Caragoca* 1538

Pedro de Medina. See **Medina**, Pedro

Pedrone, Carlo. *Cyrenaica* Gotha, Perthes 1881

Peek, Cuthbert E. *Maps Arctic Regions* 1822–4

Peers, Richard (1645–1690). *Low Countries* 1682, for Pitt; *English Atlas* Vol. IV 1683

Peeters, Jacques [Jacobus] (1637–1695).
Engraver and publisher of Antwerp.
L'Atlas en abrégé 1692, 1696

Peeters, Ian. *Asia* 1709, for Medrano

Pegg, S. M. *Anders Co., S. Carolina* 1877

Pegues, F. *Dépt. Bouches du Rhône* 1895

P'ei H'siu (224–271). Chinese cartographer

Peip, Chr. *Taschen-Atlas von Berlin* 1893
& 1896, *Taschen Atlas von Wien* 1896

Peirce, John. *County of Worcester
(Massachusetts)* 1793 (with C. Baker)

Peirse, Samuel. *Estate Digswell* 1599 MS

Pejer, Capitaine. *Schafhouse,* Homann
Heirs 1753

Pelerin. ["Viator"] (1445–1525).
French geometrist. *De Artiiavali Perspectiva* (1505)

Pelegan e Miraro, Antonio. *Chart coast
Dalmatia* 1459

Pelet, Jean Jacques Germaine, Baron
(1779–1858). French military cartographer from Toulouse. *Etats de 'Europe*
1832, *Versailles* 1837. *Campaign
Napoleon* Dépôt. Gén de la Guerre, Paris,
Piquet 1844, *Bataille de Laon* 1849,
Dresden (1850), *War Spanish Succession*
1836–62

Pelet, Paul. *Nouv. Atlas colon. Françaises*
1891

Pelham, Edward. *Map Greenland in Gods
Power and Providence 8 Englishmen left
by mischance in Greenland* 1630, 1631

Pelham, Henry. Surveyor. *Plan of Boston
in New England* 1777, *Clare (Ireland)*
12 sh. Faden 1787

Pelham, P. *Louisbourg* 1745

Pelham, William. Surveyor General, New
Mexico. *Public Surveys in New Mexico*
1855

Pelicier, T. Script engraver for Delamarche
1819–20. *Koningryk Nederlanden* 1816

Pellegrino, Danti de Rinaldi. See **Danti**

Pelletier, David. Engraver for Champlain's *Map of Canada* 1612

Pellicer de Touar, J. See **Touar**

Pellicier. Letter engraver for Lapie 1809

Pellizer, Joseph Emanuel. *New System
of World* 1798

Pellizzaro, J. *Plan de Gand* 1851

Peltzer, Jean Adam. *Stolberg* 1811

Pemberton, Henry L. *Cantonments of
Moulmein Calcutta* 1880, *Town Moulmein*
1880

Pemberton, J. Despard. *N. America* 1860

Pemberton, James Jeremiah. *District of
Bhaugulpoor* 1852, *District Rajshahee*
1855, *Revenue Survey India* 1861

Pemberton, John. Bookseller. Camden's
Britannia 1730

Pemberton, R. Boileau. *Islands Prov. o
Arracas* 1832, *Eastern Frontier British
India* 1838, *Muneepoor* (1862)

Pemble, Wm. *Brief Intro. to Geography*
1630

Pena, Guillermo. *Plano del Rio Lebu*
1862, *Puerto de Yanez* 1866

Pena, Is. *Mapa de la California* 1757

Pena, Jose Antonio de la. *Puerto Santa
Cruz* 1782 MS, *Carta Hidrografico del
Valle de Mexico* 1862

Pena, Petrus. *Map of Europe & Part
Asia* 1570

Penafiel, Antonio. *Lago de Chalco* 1884

Penaguiao, J. E. *Prov. do Alemtejo,
Algarve* 1851

Penalosa, Comte de. Governor of New
Mexico. *Map New Mexico* 1665 MS

Penaranda, Jose Maria. *Prov. de Tayabas
(Philippines)* 1821 MS

Peñas, G. G. de las. See **Gonzalez de la
Peñas**

Penate. Pilot to Pizarro and Almagro.
Coast Panama MS

Penck, Albrecht. *Oesterreichischen
Alpenseen* 1895–6

Pencz, Georg (1500–1550). Engraver
and woodcutter. Worked for Martin
Behaim *seven planets*

Pender, Staff Comm. Daniel. Admiralty hydrographer. *N. America W. Coast* 1866, *Anchorages Port Simpson* 1868, *Fitzhugh & Smiths Sounds* 1872

Pendleton, John B. Lithographer. *Maps and plans Boston area* 1820–36

Pene, Charles. Assisted production of *Neptune François* 1693

Penez, Alexander. *Hannoniae Comit. Delin.* (1735)

Pengelly, William (1812–1894). English geographer

Penha, Lauriano Jose Martins. *Rio Paraguay* 10 sh. 1857, *Imperio do Brazil* 1833, *Rep. do Brazil* 2 sh. 1892

Penhallow, Samuel. *Part of ye Spanish & Musketor Shore* (1735)

Penn, Mrs. Vine Street, Bristol. *Bristol Channel,* by Charles Price

Penn, William. *Map of the improved part of Pennsilvania begun by William Penn in 1681*

Pennay, Agostino. *Views Rome* 1840

Pennant, Thomas (1726–1798). Antiquary, traveller and naturalist. *Map of Scotland, Tour of Scotland* 1777

Pennefather, C. *Batavia River (Australia),* Brisbane 1880, *Parker & Bayley Points* 1880

Pennesi, Giuseppe (1854–1909). *Atlante Scholastico* 1894, 1897, 1898

Pennethorne, James. *Street plans for London* 1840–53

Penney, Stephen. *Navigating Directions for Thames* 1892

Pennier. *Plan de Bataille de Nerwinde* 1693, *Album du Mareschal de Montmorency*

Penniman, J. *Susquehannah River* 1840)

Penning, William Henry. *S. E. Kalahari* 1891 MS, Stanford's *Map of the Transvaal Goldfields* 1893

Penningen, Daniel. Dutch engraver and mapmaker, 1696

Pennington, Joseph. Land surveyor. *Ipswich* 1778, *Hainault Forrest (Essex)* 1791–3 MS

Pennington, P. Estate surveyor. *West Ham* 1787 MS

Penny, R. Engraver. *U.S.* 1812, *New South Wales* pub. Whittaker 1820

Penny, William. *Cumberland Isle* 1839, *Admiralty* 1840, *Arctic America* 1851, *Chart N.W. Passage* 1853

Pennythorn, Jas. *Plan Record Repository, Chancery Lane* 1850 and in conjunction with T. Chawner. *Plan New Street from Oxford St. to Holborn* 1840, *From Long Acre to Stiles* 1840, *Battersea New Park* 1852

Penrose Map. *An MS chart of the East Indies,* Portuguese ca. 1545

Penrose, Cooper. *Maps Zulu War* 1881

Penson, Thomas. *Turnpike Roads No. Wales* 1838

Penstone, J. J. Steel engravings, 1835–44

Penther, Johann Frederick (1693–1749). Mathematician and geographer

Pentius de Leucho, Jacobs. Printer of Venice. Edition *Ptolemy* 1511

Pentland, J. B. *Laguna di Titicaea, Admiralty* 1840 & 1848

Penuelas y Vazquez, Manuel. *Isla de Cuba* 2 sh. (1890)

Pentz, Inspector. *Hannover* (1770)

Penzler, Johannes. *Geogr. Stat. Lexicon* 1895

Pepingue. Publisher of Paris. Du Val's *Géogr. Univ.* 1662

Pepys, Samuel (1633–1703). Collected maps and plans of London, and early charts for his projected history of the English Navy. His collection in Magdalen College, Cambridge

Peguin. *Dépt. de la Vendée* 1862

Pérac, Etienne [Stefano] de. See **Du Pérac**

Peralta, Manuel Maria de. *Atlas Hist.*

Geogr. de Costa Rica 1890, *Mapa Costa Rica of Veragua* 1892

Percolasco, F. *Lima* 1687

Percy, Algernon Heber. *Tide Charts English & Bristol Channels* (1891)

Percy, George. *Southern Colony in Virginia* 1606 MS

Percy, Butcher & Co. *Series of Directory Maps* (1873)

Perczel, Laszlo (1827–1897). Hungarian globemaker

Perdiguier, M. de. *Maps Hanau & Strassburg* (1726)

Perdoux. *Plan d'Orléans* 1773

Pere. See **Le Pere et Avualez**

Peregoy. See **Pollard & Peregoy**

Peregrinus, Petrus *De Magnete* 1558

Pereira, Antonio. Portuguese mapmaker. *America* 1545 MS

Pereira, Duarte Pacheco (1450–1533). *Esmeraldo de Situ Orbis* ca. 1505 (sailing directory for E. coast Africa), published 1892

Pereira, F. See **Francisco dos Prazeres Maranhão**

Pereira, Jose Joachim. *Entrée du Para* 1822

Pereira Da Silva, Francisco Maria. *Enseada de Perniche* 1855), *Leiria e seus arredores* 1859, *Porto de Lisboa* 1878, *Porto da Figueira* 1881

Pereira Da Silva, M. J. *Prov. Sao Joze (Brasil)* 1828

Pereira dos Reiss, Andre. *Indian Ocean Macao* ca. 1654 MS

Perell, William. Surveyor. *MS Lists of Crown Lands* 1610–29

Perelle, Adam. *Plans and maps of Towns in French Provinces* 1667–1685, *Maisons de France* 1685

Perelle, Gabriel. Engraver for Jollain, ca. 1640, Corbie 1640

Perelle, Nicholas. *Plan d'Armentières* (1680), *Lens* (1680), *St. Venant* (1680)

Peres, Alonso. *Atlantic* 1648

Perestreto, Manuel de Mesquita (16th century) Portuguese cartographer

Perez, Andres Aznar. *Yucatan* 1878–9 (with J. Hubbe)

Perez, Felipe. *Atlas de la Repub. de Colombia* 1889

Perez, J. E. *Farms in Calhoun Co., Michigan* 1869

Perez, P. J. B. de. *Hydro Kaart Zeegat van Tjilatjap* 1858

Perfert, Johann. Publisher of Breslau. *Helweg's Silesia* 1627

Perham, B. F. *Plan Harbour of Boston (Mass.)* 1837

Periccioli, Cesare. Map maker in Siena, 17th century

Periccioli, Francesco. Writing Master. *North, Northwest, and South Coasts of Europe* (1650)

Pericciulo, Borghese A. *Malta* 1830

Periegetes, D. See **Dionysius Periegetes**

Perier. Engraver. De Beaurain's *hesse Cassel* 1760. See **Perrier**

Perignon, N. (1726–1782). Plates for *Zurlauten Suisse*

Perigot, Charles. French geographer. Prof. Lycée St. Louis. *Plan Paris* 1867, *Germany* 1870, *Europe* (1873), *World* (1873), *Terrest. Globe* (1873)

Pering, Thomas. Navigator. *Expeditions* 1814–15

Perini, L. (1685–1731). Engineer and topographer at Verona

Perint, Charles. *Forêt de Compiègne* (1876)

Perisse, L. Publisher. *Atlas Moderne Portatif* 1799

Perisse Frères. Parisian Publishers, 8 Rue du Port de Fer, Saint Sulpice. Monin's *Atlas Classique* 1844–5

Perissini, Marco. *Venezia* 1841, 1849, 1866

Perizot. Map for Homann Heirs' *Stadt Atlas* 1762

Perkin, Henri. *Atlas de l'Europe* 165 sh. 1833

Perkins, A. *Pewar Kotal (Afghanistan)* 1879

Perkins, Charles Carroll. *Boston Water Works* 1851, *Boston Proper* 1895

Perkins, Frederick B. *Vote map of the United States* 1880

Perkins, George R. *Railroads State N.Y.* 1858

Perkins, H. I. *British Guyana (gold industry)* 1895

Perkins, Jos., American engraver for lettering, in Philadelphia. Worked for Fielding Lucas Jr. 1823

Perks, Joseph. Surveyor. *Watts Estate Hanslope* 1779 MS

Perks, W. *Sweden, Norway, Denmark* 1793

Perlaska, D. Engraver. *Map Budapest* 1838

Perley, George Hayward. *New Brunswick St. John N.B.* 1853

Perley, Henry F. *Eastern Portion British N. America* 1853

Permentier, J. R. See **Parmentier**

Peroglio, Celestino. *Nuovo Atlante Cosmografico* 1878

Peron, M. F. *Maps* for Freycinet 1807–16

Perot [Perrot] d'Ablaincourt, Nicolas (1606–64) of Chalons sur Mer. *Zee Atlas* Mortier 1700

Perouse, J. F. de la. See **La Perouse**

Perret, A. *Carte des Pyrénées* 1850

Perret, I. I. Engraver. Rocque's *Plan Camp Thur.* 1755, Rocque's *Dublin* 1757

Perret, Petrus. *Plan of Escorial* 1587, used by Ortelius and Braun & Hogenberg; plates to *Esfera del Universo* by G. de Rocamora (1599)

Perrier. Engraver and mapseller of Paris, Rue des Fosses S. Germain l'Auxerrois. *Hesse Cassel* 2 sh. 1760, *Théatre de la Guerre* (1779), *Plan de Lyon* 1813, *Nouveau Plan Paris* 1824. Worked for Bonne and Clermont

Perrier et Verrier. Publishers of Paris. Pupils and successors to M. Julien, Royal Geographer Hôtel de Soubise. Faden's *New Jersey* 1777

Perrin, Joseph. *Carte de la Savoie* 1869

Perrin, Maurice. *Carte des Forts et Camps du Nord Est* 1888

Perrin du Lac, François. *Missouri* 1802 in his *Voyage dans les deux Louisianes* 1805

Perrine, C. O. *War Map Southern [United] States* 1863

Perring, J. S. *Pyramids of Ghizeh* 1837

Perris, William. *Map of the Seat of War in Virginia* (1862)

Perron, A. du. *Cours du Gange* 1784

Perron, C. E. (1837–1909). Swiss mapmaker

Perron, J. du. See **Duperron**

Perrot, Aristide Michel (1793–1879). French geographer. *County maps of Great Britain* 1823, *Atlas des Départments de la France* 1825, *Atlas des Routes de France* 1826, *Atlas Général de France* 1837

Perrot D'Ablancourt. See **Perot Ablaincourt**

Perry, C. N.   American draughtsman. *Map Co. Los Angeles* 1898

Perry, E. W. of New York. *Nevada Silver Mines* 1865

Perry, George. *Plan Liverpool* 1769, *Environs Liverpool* 1773, *Lancs.* 1786

Perry, John. *Improvements Roads S.W. side London* 1838

Perry, Capt. John. Assisted Moll's *Muscovy* 1719, *Stopping Daggenham Breach* 1721 (with map). *N. Level Marshes* 1725, *Dublin* 1728 (with Burgh)

Perry, Stephen Jos. (1833—1889). English astronomer, Jesuit

Perry, W. *Map Margate Ramsgate &c.* 1861, *Visitors' Map Margate* 1871

Person, Nicolaus (ca. 1660—1710) of Mainz. Architect, cartographer and engraver. *Erfurt* 1675, *Alsatiae choro.* (1685), *Bell-Isle* (1690), *Atlas Archbishopric of Mainz* 1694, *Palatinatus* 1697

Personne, Edu. Carl. *Johans Stad* 1812, *Belagenheten of Kongsbacka* 1816

Personne, L. E. Swedish engraver. Riksgran's *Sverige och Ryssland* 1809

Persoy, Pieter. Publisher and engraver over t'Hamburger Post Comptoir op den Dam op de Hoek van de Beurs Straat, Amsterdam. Revised *Normandy, Aquitane,* 1695, *Brittany,* plates *Amsterdam* (1694), *Tempel van Jerusalem*

Persyn [Persijn], Reiner van (1614—1668). Engraver of Amsterdam. Worked for Hendrik Doncker. *Map of Vroone*

Pertees, Charles. Publisher Bakalowitz's *Plan Cracow* 1772

Perthées, Bernard (1821—1857). Publisher and bookseller of Gotha

Perthées, Charles de (1739—1815). Royal geographer to King of Poland (Stanislas Augustus), cartographer. *Maps of Polish Palatinates. Two large maps Poland and map Cracow* (now lost). *Carte Gén. et Itin. de Pologne* Warsaw, Groll 1773, *Poland* 1780, *Description Poland* 12 vols. (1790—6) with *maps Warsaw* Paris, Tardieu 1794, *Pologne* 1809

Perthes, Johann George Justus (1749—1816). Publisher and bookseller

Perthes, Justus. Publisher and geographer of Gotha. *Chinesisches Reich* 1833, *Atlas Asia* 1835, *Deutschland* 1846, *Africa* 1875, *Staatsburger Atlas* 1896, *Geschichts Atlas* 1898

Pescheck, C. J. L. Engraver (1803—47). *Views Berlin* (1836), *Teplitz Oybin* etc.

Peschel, Oskar (1825—1875). German geographer b. Dresden. *Physik-Statistischer Atlas* 1878, *Geschichte der Erdkunde*

Peschka, J. *Plan Badestadt Teplitz-Schonau* (1875).

Pesina, Benedetto, of Venice. *MS chart of Archipelago & coast of Asia Minor* 1489

Pesnel, E. *Plan Ville d'Argentan* 1876

Pesson, L. & B. Simon. American lithographer (19th century)

Pestalozzi, Heinrich (1790—1857). Swiss engineer and road inspector

Pestel. *Arrondissement de Dunkerque* (1866), *Seine et Oise* 1868

Pet, Arthur. Discovery made by Arthur Pet and Chas. Jackman 1580, pub. in Hakluyt 1589

Petavius, P. *Globe terrestre* (1350)

Petch, William. Admiralty surveyor. *Greytown Harbour* 1856, *Hillsborough Bay* 1857

Peteghem, Louis van. *Plan d'Anvers* 1892, *l'Agglomération Bruxelloise* 1894

Peter (1482—1525). Map painter of Munich

Peter, J. A. *Post und Strassen Charte Oesterreich* 1819

Peter, J. F. *W. End Lake Erie & Detroit River* 1849

Petermann, Adolf. *Das Gen. Gouvern. Elsass* 1870

Petermann, Augustus Herman (1822—1898). Physical geographer to the Queen. Worked in London 1847—54 experimenting in thematic mapping. *Population Map British Isles* 1849, pub. Orr & Co. *Atlas physical Geogr.* 1850, *Industrial Map distribution of Trades in Great Britain* 1852, *Medical Map Cholera Map of British Isles* (1852). Worked for Fullarton. Returned to Germany and worked for Perthes. Founded Geogr. Mittheilungen in Gotha. *Sleswig* 1864, *Arctic & Antarctic* 1865, *Atlas of New Zealand* 1864, *United States* 1874, *Australia* 9 sh. 1875 &c.

Petermann, J. C. Mapmaker of Leipzig (fl. 1788—1809)

Peters, C. A. F. (1806–1880). Astronomer, Director Observatory of Kiel

Peters, Chr. H. Fr. (1813–1880). American astronomer, b. in Schleswig

Peters, Jan. See **Peeters**, Jacques

Peters, Geo. Midshipman. *Bombay Harbour* (with Lt. Cogan) 1833

Peters, Samuel. *City of New London (Conn.)* 1856

Peters or **Peterson**. Engravers, goldsmiths, and publishers of Husum in Jutland. *Steinborch on Elbe* (1650), *Hamburg* 1651, *Maps* for Mejer's, *Schleswig Holstein* Blaeu 1662.

—— Claus.

—— Matthias (d. 1676)

—— Nikolaus (1620–1705). With Matthias, engraved *Carte Ambte Kiel und Bords Holm* 1665

Petersen. See **Peters**

Petersen, S. H. (1788–1860). Danish mapmaker. *Danske Atlas* 1832–60

Peterson, Carl S. (1826–1890). Norwegian geologist

Peterson, J. C. *St. Thomas (Virgin. Is.)* Kobenhaven 1846

Peterson, J. van. *Kaert van Amstelland* (1720)

Peterson, Matthias and Nicolas. See **Peters**

Petherick, J. *Nile* 1865

Petiet, Jules. *Chemin de Fer, Paris à Meaux* 1839

Petit. Surveyor. *Carte de l'Escaut* (1864)

Petit (fl. 1775–99). Engraver for the Dépôt de la Marine *Atlantic* 1775, for *Neptune François* (1775–80), *Mers du Nord* 1776, *Havre du Casco* 1779, *Jamaique* 1799, *St. Pierre* 1782 &c.

Petit, Jean de. *Carte l'Isle Cadiz et Detroit de Gibraltar* (1740), *Antilles* (1740)

Petit [Bourbon], Pierre (1598–1667). Mathematician and Geographer to Louis XIII. *Gouvern. de la Capelle* (1630)

in Blaeu 1631, Hondius 1633, Jansson 1658

Petit-Thouars, A. A. See **Du Petit-Thouars**

Petite, Jean (1619–1694). *Diocèse de Bayeux* 1675

Petitot, E. *Bassin du Mackenzie* (1873), *Athabasca* 1883

Petitpas. *Terre Neuve* 1874

Petitville, E. (1815–1868). Lithographer. *Alsatia, Rhine, France*

Petley, W. H. Navigation Sub. Lieut. *Balabac Strait* 1868–9, pub. 1870

Petowe, Henry. *Descript. of Surrey . . . a geographical account* 1611 MS

Petrarch [Petrarca], Francesco (1304–1374). Italian poet and humanist, composed map of Italy

Petrell, Frederik Magnus (1748–1777). Swedish surveyor in Abo.

Petri, Edmund. *Ichnog. of Charleston (S. Carolina)* 1790

Petri, Eduard Juleric (1854–1899). School geographer. St. Petersburg

Petri, Heinrich [Henri] (1508–1579). Printer and map publisher of Basle, son in law of Sebastian Munster. Published editions of Munster's *Cosmography* 1540–52

Petri, Isaak Jakob v. (1701–1776). Military cartographer. *Saxony* 38 sh. 1763; 12 sh. 1765; *Zwickau bis Würzburg* 8 sh. 1759

Petri, Otto. Publisher of Rotterdam. *School Atlases* 1848–70

Petri, Sebastian. Publisher of Basle. Editions of Munster's *Cosmography*

Petri de Nobilibus. See **Nobili**

Petrich, Andras (1768–1842). Hungarian military cartographer. *Budapest, Belgrade, Orsova* etc

Petrie, Edmund. See **Petri**

Petrie, Capt. Martin. *Operations River Pei Ho* (1860), *Defences of Besançon* (1870)

Petrini, Nicolo. *Pozzuoli* 1750

Petrini, Paolo. Publisher and map seller of Naples "a S. Biaggio de Librari". Baundran's, *Greece* (1670), *N. America* (1700), *Naples* 1718

Petrino, O. von. *Geolog. K. Bukovina* 1876

Petrle, Michael. *Plan Prague* 1562, used by Braun & Hogenberg

Petroff, Ivan. See **Petrov**

Petroschi, Giovanni [Joannes]. *Suizzeri e Savoja* 1703, *Paraguay P.S.J.* 1732, Pauli's *Malta* 1733–7, *Toscana* 1745, *Rome* 1754

Petroschi, I. *Province del Chaco* (1745)

Petrossi, F. *Austria* 1865

Petrov, Ivan. *Alaska* 1880 & 1882

Petrovits, L. E. *Wien* 1873

Petrucci, Guilio di Cesare, of Siena. Portolan maker. *Mediterranean* ca. 1550, *Chart* 1571

Petrus ab Aggere [Henri Montanus]. Flemish historian and cartographer b. 1556 in Arnheim

Petrus & Mercia. See **Heyden**, Pieter van de

Petrus De Alliaco. See **Ailly**, Pierre d'

Petrus de Medina. See **Medina**, Pedro de

Pett, P. *Dover* (with Laker) 1583

Petten, Pieter Cornelis van. *Cayenne* 1598, *Gulf of Paris* 1598

Pettenkofer, Max von. *Cholera Map East Indies* 1871

Petters, Hugo. *Tirol* 1880, *Alpen vom Bodensee* 1894, *Coburg u. Umgegend* 1895

Petters, K. L. (1827–1881). Austrian geographer and geologist

Petterson, Carl [Karl] Anton. *Lappland* 1871, *Kong Karl Land* 1889, *Tromsö* 1890

Petterson, N. P. *Map Sweden* 1874–7

Pettersson, Wilhelm. *Orsa* 1892

Pettibone, D. A. *City of Ann Arbor, Michigan* 1854

Pettiplace, Wm. *Relation of Virginia* 1609 MS

Pettit & Co., T. Lithographers of London. Maps of Livingstone's *Discoveries* 1874

Petty, Sir William (1623–1687). Cartographer, physician. A founder of Royal Society. Knighted 1662. *Hiberniae Delineatio* 1685, the first atlas of Ireland. This survey was finished in 1657, engraved in Amsterdam in 1673, and pub. 1685. Min. set, *Geograph. Description Ireland* 1685. *Maps of 214 Baronies* (now lost)

Petyt, Jacques. Lithographer. Belgian cities, *Franc de Bruges* 1852

Petyt, John. *Chart of the White Sea* 1755

Petzold, W. (1848–1897). German geographer. Various manuals

Petzoldt, G. Lithographer (1810–1878). *Salzburg and Tirol maps and plates*

Pencer, Gaspar. *De Sphaera* Wittenberg 1551, *De dimensione Terrae* 1554

Peurbach, George de (1423–1461). Astronomer, pupil of Nicolas a Cusa and teacher of Regiomontanus. *Theoricae novae planetarum* 1495, *Tabulae Eclipsium* 1514

Peutinger, Konrad (1465–1547). Classical scholar, collector and humanist, was born in and became Town Clerk of Nuremberg. Completed Cardinal Cusa's *Central Europe*. See Peutinger Table below

Peutinger Table. Road map of Western Roman Empire about 250 A.D. bought by Conrad Peutinger and published in Venice by Aldus in 1591 (2 sections only), in Antwerp 1598 (12 sections), in 1753 entire map printed for the first time. The only record of Roman map making that has survived to the present day

Peutzoff, M. V. *Chinese Turkestan* 1889–90

Peyer, Heinrich (1621–1690). *Repub. lib. Helvetiae* Nürnberg, Homann Heirs 1753; *Schaffhausen Gebiet* 4 sh. 1747

Peyer, Joh. Ludwig (1780–1842). State archivist *Stadt Schaffhausen* 1820

Peyre, de. *Ville de Carenage St. Lucie* 1784

Peyrounin, A. Engraver for Sanson & Mariette. *Bern* 1646, *Denmark* 1646, *Amérique Mérid.* 1650, *Natolia* 1657 &c.

Peyrouse. See **La Perouse**

Peyton, J. Engraver. *Hill shading on Turkestan* 1879

Peyton, Richard. Estate surveyor. *Navestock* 1835 MS

Peytret, Jacques. *Amphithéatre d'Arles* 1686

Pezuela, Jacobo de la. *Islas Filipines* 1871

Pezze, Giuseppe. *Lombardo. Veneto* 1831, *Contorni di Milan* 1825, *Citta di Milan* 1827

Pezzi, A. *Plano Campo de Melilla* (1893)

Pfaeffin, A. *Map of Tell City (Indiana)* 1858

Pfaff, Friedr. (1825–1886). German geographer. *Allgemeine Geologie.* Wrote against Darwin

Pfann, Wilhelm. Engraver. *Geogr. Playing Cards* (1678), *Transilvania* 1688, *Area between Rhine & Mosel* (1700)

Pfau, Conrad. Engraver. *Mittelfranken, Unterfranken*

Pfau, J. H. *Plan Siege Dillenburg* 1760. *Gen. Karte von Polen* 1793

Pfau, Theodr. Philipp v. (1725–1794). Military cartographer

Pfaundler, Johann Anton. *Prov. Arlbergica* 1783

Pfautz, Johann Gottfried (1687–1760). Engraver and publisher

Pfeffel, Johann Andreas (1674–1750). Studied in Vienna and later settled in Augsburg as publisher, engraver, and art dealer. *Vienna* 1706, *Austrian Provinces* 1709, Scheuchzer's *Physica Sacra* 1731–5 and *Switzerland, Frankfort am Main* 1738

Pfeffel, Johann Andreas, the younger (1715–68). Engraver and publisher, son, preceding. *Large map of Swabia*

Pfeiffer, A. *Plan von Hildesheim nebst Umgebung* 1881

Pfeiffer, F. *Citta di Trieste* 1867

Peiffer, G. *Mapa de Espana y Portugal* 1872

Pfeiffer, Johann Baptist. *Bayerischen Alpen* (1841), *Bayern* (1845), *Bayern* (1865), *München* 1869, *Eisenbahnkarte Bayern* 1875, *Regensburg* 1878

Pfenning, Johann Christoph. (1724–1804). Preacher and geographer

Pfetten, Baron. Lithographer. *Topo. maps Austrian army* 1817–1820

Pfintzing. See **Pfinzing**

Pfinzing, Paul (1554–1599). Cartographer of Nuremberg, mathematician and surveyor

Pfister, Ferdinand von (1800–86). Military writer and cartographer

Pfister, Francis. Ensign, 1st Bn. Royal American Regt. *New York, New England, & New France* 1758 MS, *Canada Creek* (1760), *Niagara* 1771

Pfister, Johan. *Canton St. Gallen* 1840

Pfister, Karl (1724–1800). Military cartographer

Pflummern, Karl, Freiherr von (d. 1850). Military cartographer of Mannheim

Pfnor, Rodolphe. *Palais de Fontainebleau* 1863–85

Pfyffer, Franz Ludwig (1716–1802) of Lucerne. *Reliefs der Zentralschweiz* 1762

Pfyffer von Altischofen. *Atlas de la Suisse* 1829

Pfyffer von Wyher, Ludwig (1783– · 1845). Topographer and surveyor of Lucerne

Phantamour, Emile (1815–1882). Swiss astronomer, meteorologist

Pharoah & Co. *Gazetteer of S. India* 1855

Phelipeau[x], Sieur René. Geographer and mathematician. *Boulonis* 1748. *Colonies Angloises Amér. Sept.* 1778, Supplement 1783, *St. Domingo* 1785–9

Phelps, Humphrey. Publisher, No. 144 Fulton St., New York. *U.S.* 1832, *Traveller's Map of Michigan &c.* 1838, *Ornamental Map U.S.* 1847

Phelps, Joseph. Publishers, London. Paternoster Row. Langley's *New County Atlas* (1820)

Phelps, Capt. J. *Chart Havana,* Mount & Page (1720)

Phelps & Ensign. Publishers. *New York* 1839

Phelps, Ensigns & Thayer. Printers of N.Y. *U.S. & Mexico* 1846

Philesius, M. See **Ringmann**

Philip, A. *New South Wales* 2 sh. 1791

Philip, George (b. 1799). Geographical publisher and globe maker. Started business in Liverpool 1834, joined by his son 1848. Moved to London 1856, 32 Fleet St. *New Gen. Atlas* 1856, *Imperial Atlas* 1864, *Atlas Counties of England* 1865, *Select Atlas* 1870, *Handy Atlas* 1874, *Wall Maps &c.*

Philip, William. Translator. Houtman's *Voyage E. Indies* 1597, Linschoten's *E. & W. Indies* (9 maps by Rogers) 1598, *Voyage Schouten round World* 1619 *(map Tierra del Fuego)*

Philippe De Pretot, Etienne André (1708–1787). Royal Censor, Prof. Hist. Roy. Acad. Sciences, Paris. *Cosmograph. Essai de Géographie* 1744, *Univ.* 1768, *Atlas Univ.* 1787, *Recueil de Cartes* 1787 (124 maps)

Philippi, B. E. *Valdiva* 1846, *Atacama* 1854

Philippson, Afred. *Ethno. Karte Peloponnes* 1809

Philips, Caspar Jacobsz (1732–1789).

Engraver and surveyor. Many maps and town views: *Amsterdam, London* etc.

Philips, H. *Geometrical Seaman or Art of Navigation,* 17th century

Philips, Jan Caspar (1700–1773). Engraver *Hedendaagsche Historie* and Kolbe *Cape Good Hope*

Phillips, G. G. *Survey Aberporth Bay* (1860)

Phillips, H. *Plan Worthing* 1814

Phillips, J. Engraver for Sayer & Bennett. *Gen. Atlas* 1757–94, for Rennell 1788

Phillips, J. Geological surveyor. *Geolog. Survey Victoria Ballarat* (1860)

Phillips, J. (1800–1874). Nephew of Wm. Smith. *Geolog. Map E. Yorks.* 1829, Teesdale's *Staffs.* 1832, *Principal features geolog. Yorks.* 1853, *British Isles* 1862

Phillips, James. Mapseller. George Yard, Lombard Street. Howell's *Penn.* 1792

Phillips, John. Surveyor. *England & Wales* 1799, *Staffs.* (with Hutchings) 1822, *Plan Improv. Sewerage, Marylebone* 1847

Phillips, J. C. Surveyor. *Catlettsburg Kentucky* [1868]

Phillips, J. H. *Beverley Burlington Co., N.J.* 1874

Phillips, M. Civil engineer and surveyor. *Grand Southern Tour England* 1821

Phillips, N. C. Surveyor. *Tutukaka Harbour New Zealand* 1837, pub. 1840

Phillips, Richard (later Sir) (1767–1840). Publisher, writer, and bookseller, Bridge St., Blackfriars (used pseudonym Rev. J.Goldsmith) and 71 St. Paul's Churchyard 1808. *London* 1804, 1807; *School Atlas* 1803 & 1813, *Scotland* 1820, *Grammar Gen. Georgr.* 1821, *Geogr. & Astron. Atlas* 1823. Later R. Phillips & Co.

Phillips, R. C. *Mackinaw City* 1857, *Cincinnati & Vicinity* 1865, *Hamilton Co., Ohio* 1865

Phillips, Capt. Thomas (d. 1693).
Engineer to Charles II. *Survey Baronie
Enish Owen* London, P. Lea, engraved
Sutton Nichols; *Dublin* 1685, *Athlone*
1685, *Charts of Channel Is. & Ireland*
1680 MSS

Phillips, Thomas. *Chart Humber* 1720

Phillips, W. (1775–1828). Printer,
bookseller, geologist. *Geolog. Map England
& Wales* 1816, *E. Yorks* 1829, *Geolog.
of Yorks* 1853

Phillips, W. *Richmond Co., Georgia* 1869,
Fulton Co., Georgia 1872

Phillips, Sampson & Co. Publishers, New
York. *Kansas & Nebraska* 1854

Phillpotts, Arthur Thomas. *Water commun.
Lake Erie to Montreal* 1842

Philp, J. B. Lithographer. *Map of Melbourne*
1853

Phinn, Thomas. Engraver, Edinburgh.
British & French Settlements N. America
(1750), *Scots Map* (1750), *Clyde* 1759,
Scotland 1760, *River Spey* 1761

Phipps, Constantine John, Baron
Mulgrave (1744–1792). English navigator
and explorer, b. Ireland, d. Liege. *Voyage
towards the North Pole (with charts)*
1774, *World* 1787

Phipps. See **Spurrier & Phipps**

Phrijsen, L. See **Fries,** L.

Phrisius [Phrysius], L. See **Fries,** L.

Phryes, L. See **Fries,** L.

Piale, Lewis. *Pianto Topo. citta di
Roma* 1851

Piacenza, Francesco (d. 1688). Neapolitan
jurist and geographer. *L'Egeo Redivivo*
1688

Piadischeff [Pyadischeff] . *Atlas Géogr.
de Russie* St. Petersburg 1828

Pian Del Carpine. See **Carpini**

Piat, A. Engraver. *Routes de Postes
d'Europe* 1835

Piattoli, Cesare. *Strade Ferrate Italiane*
4 sh. 1886

Piatti, G. *Pianta di Firenze* (1800)

Piazzola [Paciola] . *Panfilo Po delta* 1563

Picard, Hughes (1587–1662). Map
engraver

Picard, Jean (1620–1682). French
geodesist. Worked on Triangulation
of France with De la Hire. Worked on
arc of the meridian 1669–70

Picard, Pieter (1670–1737). Engraver
of Amsterdam. Entered Russian
service 1702

Picardus Bernardus, of Vincenza. Co-
editor with Vadius of 1475 edition
Ptolemy

Picart, Bernard. Designer and engraver.
Frontispieces for Chatelain 1718–20

Picart, Hughes. Engraver of Paris. Le
Clerc's *Lacus Lemani* 1619, *Théatre
Géogr. de France* 1621, Hondius's
Europe (1630), *Terre Sainte* 1637

Picart, Jean. See **Picard,** Jean

Picart, Jean Michael. *Afrique* 1671

Picart, R. Engraver for Chatelain 1710
and Davity 1637

Picaud, Michael. *Révolution de l'Univers.*
Paris 1763, 1775 (30 maps)

Picault, Lazare. *Islands discovered* 1744,
Dalrymple 1784

Piccoli, G. Italian engraver. *Territ. Veronese
e sua diocese* 2 sh. 1720

Piccolomini, Alessandro (1508–1578).
De la sfera del mondo Venetia 1540, *De
le Stelle Fisse* (47 star maps) 1570, *Della
Terra et dell acqua* Venetia 1558

Pichardo, E. *Dept. occid. de Cuba* 4 sh.
(1853)

Pichon. *Plan Routier de Paris* 4 sh. 1792

Picinino, Felix. *Map Gulf of Naples* 1767

Pickart, P. Engraver for J. Van Keulen
(1700)

Picke, C. J. *Atlas Provincie Zeeland*
(1877)

Pickel, C. See Celtes **Protucius**

Pickernell, J. *Cork* 1764, *Whitby* 1791, *Coast Yorks.* 1791, *Belfast Loch* 1794

Pickersgill. *Grönland* 1832

Pickett, A. J. Engraver Johnson's *Plan Manchester* 1820 (with J. Pickett)

Pickett, J. Engraver with A.J. Pickett at Bridgewater Sq., London. Arrowsmith's *Ceylon* 1805, Smith's *France & Germany* 1817, for Rees 1820, Smith's *Northumberland* 1822, *Gloc.* 1822

Pickins, John. Surveyor. *Boundary S. Carolina & Cherokee Indian Territ.* 1766 MS

Pico, D. *Montevideo* 1846

Picquet, Charles (1771–1827). Geographer to the King and the Duc d'Orleans. Engraver and publisher. Quai Malaquais (1803), Quai de la Monnaie Paris (1812), Quai de Conti No. 17 près de Pont des Arts (1820). Seul chargé de la vente des cartes du Dépôt de la Guerre. Revised Coutans *Environs de Paris* 16 sh. 1800, Paultre's *Syrie* 1803, *Plan Routier de Paris* 1814, *Empire Français* 1815, *Moncenis* 1821

Picquet, Charles. Geographer, publisher; Successor to Brué (1835). *Atlas Classique* Dufour, corrected and amended by Piquet; *Asie* 1839, *Mappemonde* 1840, Lapie's *France* augmented Picquet 1840, *Africa & America* 1863, *Océanie* 1865, *Atlas Universel* 1869

Picquet, F, *Accroissements puissance de France* 1804

Picquet, U. *Plan de Tournai* 1838

Pictet, E. *Lac de Genève* 1877

Pieck, L. van der Voordt. *Noord Brabant* 1841

Piedra, J. de la. *Puerto e Bahia Sn. Joseph (Argentine)* 1779 MS

Pielyman, Wm. *Northern Borneo* 1879

Pieman, Samuel. *Stadt Groningen* 1652

Piemontesi, Antonio. *Citta e Porto di Livorno*

Piepgras, H. *Schleswig-Holstein* (1867)

Pierce. *Plan Portsmouth MS* (Elizabethan)

Pierce, John. Surveyor General *Colorado Territory* 1863

Pierce, Mark. *Manor & Lordship of Laxton* 1635 MS

Pierce, Richard. *Straits of Malacca*, Admiralty 1779

Pierce, W. A. *Napa City (California)* 1869

Pierotti, Ermete. *Jerusalem* 1860), *Paris* 1871 and 1875

Pierre, Pierre Joseph Gustave. *Baie du Cap Normand* 1855, *Hâvres de St. Julien* 1856, *Baie de Saint Lunarie* 1859, *Tête de Vache* 1860, *Marquesas Is.* 1875, pub. Admiralty 1883

Pierre & Jeanneret. *Port Jackson* 1828

Pierrepoint, Major. *Maps & Plans War* 1808–12 *in Spanish Penin.* (·1840)

Pierriers, A. Painter and engraver. *Plan Spa* (1559)

Pierron. *Paraguay* (1820), *Cuba* 1825, *Possessions Russes* 1825

Piers, William. *Fortifications Portsmouth* 1585 MS

Pierse, Mark. Estate surveyor. *Ham* 1625 MS

Piesse, L. (1814–1900)

Pietersen, Arnoldus. Engraver. *Plan of Hamburg with 26 coats of arms* (1644, reissued 1688)

Pietersz, Dirck. Publisher of Amsterdam. Haeyen's *Zee Caerten* 1613

Pieterszoon, Claes, the elder (1550–1602). Dutch astronomer and mathematician of Amsterdam

Pieterszoon, Claes, the younger. Dutch cartographer. *Chart Atlantic* 1607 MS

Pietesch, Chr. D. Engraver. *Plans & Views Elbing, Braunsberg, Thorn &c.* (18th century)

Pietra Santa, Gasparo. Engraver. Cantelli's *Bavaria* 1688, *Morea* (1688), *Marea Anconitana*, Rossi 1711

Pietro, Marco di. Engraver for Rossi (1820)

Pietroi, J. *Siege of Brimstone Hill* 1782

Pietschmann, R. *Lucayischen Inseln* (1882)

Pietzsch. *Insel Alsen* 1864

Pigafetta, Filippo [Philip] (1533–1603). Italian historian and traveller. Chamberlain to Pope Sixtus V. *Rel. del Reame di Congo* Rome 1591 (map of Africa). English edition 1597 (4 maps by Rogers). Translated *Ortelius* text into Italian

Pigafetta, Francesco Antonio (1491–1534). Sailed with Magellan. *Filipine e Molucche* 1521, *Strait of Magellan* and other charts

Pigafetta, Marco Antonio. *Itin. de Vienna a Constantinople* 1581

Pigeon, Jean (1665–1750) Mathematician and globemaker

Pigeon, R. H. *Atlas City of New York* 1855 (with E. Robinson)

Pigeonneau, Henri. (1834–1892). French geographer. *Four continents, France &c.* 1878

Piggot, F. F. *City Peoria (Illinois)* (1857)

Pigot, James. *Plan Manchester and Salford* 1804, 1809, 1825

Pigot, James & Co. Draughtsmen, engravers and publishers, London. Fountain St. (1811), 18 Fountain Street, Manchester (1821), 24 Basing Lane, London (1824–9). *British Atlas* 1828 (editions to 1842) *London & Prov. New Direct.* 1827, *Pocket Topographer* 1835, *Comm. Director Scotland* 1825–6. Continued as Pigot & Son; succeeded by Slater 1829

Pigot & Slater. Engravers, printers and publishers of Manchester. *Pocket topo. gazetteer of England* 1842

Pigou, Peter. *Island Johanna (Comoro Is.)* 1762, pub. Dalrymple 1774

Piguenit, W. C. Draughtman. *Seat War New Zealand* 1863, *Fingal Gold Fields Tasmania* 1867

Piil, C. *Leipzig* 1846, *Copenhagen, Helsingfors Rosskilde (1865)*

Pijnacker [Pynacker], Cornelius (1570–1645). Dutch cartographer. *Drent* 1634, used by Hondius & Blaeu

Pijnappel, Jan. *Atlases East Indies* 1855–6 & 1883

Pike, Thomas, W. R. *Chart Vourla Bay* 1829 MS

Pike, Lieut. (later Major) Zebulon Montgomery (1779–1813). *Expedition up Mississippi* 1805–6, *Missouri Rivers & Rio Grande* 1807 MS, *Louisiana* 1810

Pike. See White & Pike

Pilaja, Paolo. Engraver. Worked for Petroschi. *Maps and views* 18th century

Pilbrow. See Illman & Pilbrow

Pilestrina, Salvatore de (fl. 1503–11). Portolan maker of Majorca. *Atlantic* 1503 MS, *Mediterranean* 1511 MS, *Ancient World* 1511

Pillinski, Adam (1810–1887). Designer, lithographer, engraver. *Maps and Views of Polish Towns*

Pillans, J. & J. Printers for Thomson, Baldwin & Craddock 1828

Pillans, P. J. *Sabine Pass & Mouth River Sabine (Texas)* 1840 (with Lt. Lee)

Piller, Piotr (early 19th century). Lithographer of Lwow. *Country views,* particularly *Galicia*

Pillet, L. *Carte Géolog. Savoie* 1869

Pillet, Louis Ernest. *Maps Newfoundland* 1864

Pilliet, V. Publisher. Lelewel's *Moyen Age* 1850

Pilliet, W. H. *Wakatipu New Zealand* (1860)

Pillod. *Environs de Bordeaux* 1870

Pillot. Publisher of Paris. Reissued Nolin's *America* 1789

Pillwein, Benedict (1779–1847). Topographer and cartographer

Pilon, Abel & Cie. Booksellers. *Océanie* (1870)

Pilot. See **Pillot**

Piloty, Ferdinand (1786–1844). Lithographer of Munich

Pim, Bedford. *Map Central America* 1860 & 1866

Pimentel, Luiz Serrao (1613–1679) of Lisbon. Royal Cosmographer

Pimentel, Manoel (1650–1719). Portuguese Royal Hydrographer. *Arte de Navegar* Lisbon 1762 (18 charts), *Brazil Pilot* 1809 (14 charts)

Pinadello [Pinadellus], Giovanni [Joannes]. *Treviso* 1595, used by Ortelius

Pinargenti, Simon. Italian engraver. *Palmosa* 1573, *America* 1574

Pinart, Alphonse L. *Recueil de Cartes Plans et Vues relatifs aux Etats Unis* 1893

Pinchard, A. *Plan de la Ville de Caen* 1875

Pinchetti, C. *Regno Lombardo-Veneto* 1831

Pinchetti, G. *Carta Milit. di Genova* (1780), *Mantoue et Environs* (1796), *Citta di Mantova* 1800, *Milano* 1801

Pinchin, R. *Geolog. Sketch Cape Colony* (1875), *Geolog. Sketch S. Africa* (1876)

Pinching, H. N. *Survey Harbours Red Sea* 1830–4 (with Comm. Haines)

Pinder, W. *Relation of Ormuz* 1621–2 MS, printed 1625

Pine, John (1690–1756). Engraver, publisher, mapseller, Bluemantle pursuivant at arms. Old Bond St. Prospectus for *Britannia Depicta* (1719), *Tapestry Hangings House of Lords* (with maps Armada) 1739, *Plan Bristol* 4 sh. 1743, *London & Westminster* 1746 (with Tinney), *Nottingham* 1751

Pineda, Lieut. Carlos. *Philippine Is.* 1869–(98)

Pinel, M. *Grenada* 1763, *New Plan Granada* 1780

Pinet, Antoine du. See **Du Pinet**

Pinet, Daniel. Publisher of The Hague, partner with P. Gosse. *Atlas Portatif* (W. Saxe) 1761, *Military plans, German Towns & battles* 1763–6

Pinetti, G. A. F. *Regno d'Italia* (1805), *Regno Lombardo-Veneto* (1815), *Regno d'Italia* 9 sh. (1815), *Stradale d'Italia* (1830)

Ping, J. *Environs of London* 1847

Pingeling, F. Engraver for Wangensteen's *Norway* 1761

Pingeling, G. C. (1688–1769). Engraver. *Maps of German Provinces: Hamburg* 1739, *Cuxhaven* 1751

Pingeling, T. A. Engraver of Hamburg. *Charte von Nord America* 1776, *Texel* 1781

Pingre, Alexander Gui., L'Abbé de. *Peking* 1765, *Mer du Sud* 1769, revised plates for *Neptune Oriental* for Après de Mainvillette 1775, *Cap Verd* 1744, *Antilles* 1775, *Mer du Nord* 1776

Pinheiro, Fernandez (1825–1876). Brazilian geographer. *Costa Occid. Africa* Paris 1825

Pinheiro, Furtado. *Congo & Angola* 1790

Pinistri, S. *Plan New Orleans* (1841)

Pink, George. Surveyor. *Is. of Jersey* 4 sh. 1795

Pinkar, Julio. *Lineas Telegraficas de Bolivia* 1896

Pinkerton, John (1758–1826). Geographer and publisher of Edinburgh. *History Scotland* (1797), *Voyages and Travels* 16 vols. 1807–14, *Modern Geography* 3 vols. 1802, 1807; *Collection of Voyages* 1808–14, *World on Mercators Projection* 2 sh. 1812, *Modern Atlas* published in parts 1809–14 and published as a whole in 1815

Pinkham, Capt. Paul. *Nantucket Shoals* 1791, Map for Norman's *American Pilot* 1792

Pinnell, T. Surveyor. *Plan City Gloucester* (with R. Hall in 1780) 1782

Pinney, Charles. *Map Town of Windsor* 1857

Pinnock & Maunder. Publishers, 267 Strand, London. Pawley's *General Atlas* 1820. See also **Maunder**, S.

Pinot & Sagaire. Lithographers. *Guerre en Allemagne* 1866, *en Italie* 1866, *Collection des Cartes* (1868)

Pinson, Felix Joseph. *Cadastral Plans Loire Inférieure* 1850–58, *Carte Géolog. Loire Inf.* 1851, *Bains de Mer Guérande* (1868), *Plan de Nantes* 1868 and 1875

Pinto, Antonio Corveia. *Amazon River* (1675)

Pinto, Chevalier. *Columbia Prima or South America* extracted from MSS of Chev. Pinto. London, Faden, 8 sh. 1807

Pinto, Jaco Soares. *Rio Amazones* 1865

Pinto, Serpa. *Africa Oriental, Moçambique* 1884–6, *Tropical S. Africa* 1880

Pintz, J. G. (1697–1767). Engraver. *Views Augsburg and Vienna*

Piper, Lt. Col. *Alogoa Bay,* Wyld 1847

Piper, Frederick. *Japan Kagosima Harbour* (with W. H. Parker) Admiralty 1863

Piper, S; Bookseller of Ipswich. Revised Kirby's *Suffolk* 1766

Piquet. Engraver for Ruelle's *Planisphère* (1785)

Piquet, R. C. (19th century). Publisher of Paris

Piranesi, Francesco. *Via Adriano* 1781, *Piazzi di Padoa* 1786, *Icnografia del Ciro di Caracalla* (1790), *Citta di Pompeii* 1792

Piranesi, G. B. Engraver (d. 1778). *Rome* 1748, *Plan Rome* 1778

Pirckheimer, Willibald [Bilibaldus] (1470–1530). Humanist and geographer of Nuremberg. Edited 1525 edition of *Ptolemy, Bodensee* 1505

Piré, L. *Atlas Classique . . . aug. de notions de Zoologie et de Botanique* 1876

Pires, Caldeira. *Plan of Lisbon* 1898

Piri, Reis [Kemel Reis] . Turkish Admiral and cartographer. *World* 1518, *Kitabe Bahrige* (guide to Eastern Mediterranean) 1521

Piringer, B. (1780–1826). Engraver. *Ecole de Paysages* (1823)

Pirrus de Noha. *World* ca. 1438 in MS of Pomponius Mela's *Cosmography*

Pisani, Octavio (b. Naples 1575). Cosmographer, astrologer, and mathematician. Worked in Antwerp between 1613–1637. *World* in 12 sheets (1613), 1637. Ded. to Albert Archduke Burgundy, *Astrologia* 1613

Pisani, V. *Bahia de Ancon* 1884

Pisanti, F. (d. 1889). Engraver. *Views of Naples* for Bourcard (1868)

Pisato, Giovanni. *Lombardy* 1440 MS

Piscator, N. J. Latinized form of Visscher. See **Visscher**, N.J.

Pischon, F. A. *Hist. Geogr. Hand Atlas* 1853

Piskart, P. Engraver for Sanson 1695

Pissis, Aimé (1812–1889). Chilian geographer. *Geogr. fisica de la Repub. de Chile* 1875, *Prov. de Santiago* 1857

Pissot, fils. Parisian bookseller. Vaugondy's *Atlas Portatif* 1748

Pistoja, Francesco and Giocondo. *Relief models Italian Mountains* 1876–8

Pitcarne, J. *Town St. Jago de la Vega (Jamaica)* 1785

Pithoea, C. L. V. Petrus. *Isle de France*, Jansson 1636 (with Guilloterius)

Piton, E. *Carte d'Afrique* (1868). *Deux Amériques* (1868)

Pitschner, W. *Relief Maps* 1862

Pitt, Moses (d. 1696). Publisher at the Angel in St. Pauls Churchyard. Map publisher. Planned a 12 vol. atlas, only

4 vols. issued, then imprisoned for debt in the Fleet 1689–91. *English Atlas* Oxford and Amsterdam 1680–3, Vol. I. *World & Northern Regions,* Vol. II, and III. Germany, Vol. IV. *Netherlands*

Pitt, M.

Pitt, William Drew. *Nebraska Railroads* 1857

Pitteri, G. Engraver. Worked (with G. V. Pasquali) for Zatta's *Atlante Novissimo,* Venice 1775–85

Pittman, Daniel. *Fulton Co., Georgia* 1872

Pittman, Lieut. Philip. Asst. Engineer *Plan Canada & St. Lawrence* MS 1760, *Entrance to Apalatay* (with Gould) (1767) MS (with Gould); *Plan Point Ibbérville* Cascasgias 1760, *Fort Rosalie* (1770)

Pittoni, G. B. (1520–1553). Engraver. *Views Rome* 1561

Pitts, M. *Sheppey & Grain* 1784

Pizarro, R. G. de Ly. See **Garcia de Leon y Pizarro**

Pizzigano, Francesco (fl. 1367–1373). Venetian cartographer. *MS Planisphere* 1367 (with brother Marco) and 1377. *Atlas* 5 maps 1373

Pizzigano, J. *Planisphere* 1597

Pizzigano, Marco (14th century). Portolan maker

Pizzigano, Z. *World* 1424 MS

Pizzighelli, R. *Kaap Gold Fields* 1887

Pjadysev, Vasili Petrovic (1768–1835). Cartographer and engraver

Plaat, E. G. *Straat Sunda tot Batavia* (1850)

Plaats, François van der. Bookseller, inde Capersteeg, Amsterdam. *Plan Turin* 1706

Place, Francis (1647–1728). Engraver, pupil of Wenzel Hollar. *Views Castles & London*

Placide, de Sainte Hélène, Le Père (1649–1734). Augustine monk, geographer to Louis XIV (1705), pupil and brother-in-law of Pierre Du Val, Géographe Ordinaire du Roi. Published an Atlas with revised versions of Du Val's maps. *Flandre* 1685, *Siam* 1686, *Germany* 1690, *Cours du Po* 5 sh. 1702–4, *Cours du Danube* 3 sh. 1703, *Combat du Parme* 1734

Placiola, P. See **Piazzola**

Plaes [**Plaets, Plaetz**], A.B. de la. Engraver for Sanson 1640–58

Plaetsen, Abraham de la. *Ant. Italie de Illyrici . . . descrip.* 1640

Plaisted, Bartholomew. *St. Augustine's Bay* (1775), *Bay of Bengal* (with Ritchie) Dalrymple 1772, *Coast Chittagong* 1784, *Chittagong River* 1785, *Pilot N. part of Bay Bengal* 1803

Plamann, Johann Ernst (1771–1834).

Plancius [**Plat[t]**], Petrus (1552–1622). Flemish cartographer and theologian. Map maker to the Dutch East India Company. *World* 1590, 1592, 1596, 1604, 1607. Maps for Linschoten 1596. *Nova Francia* 1592, *Anglia* 1592, *Maps for Bible* 1609, *Gores* 1614–15. Plancius issued no atlas but had a prolific output of over 100 separate maps

Planck, Leon. (17th century) Bohemian engraver

Plant, Johann Traugott. *Polynesien* Leipzig 1793

Plantade, François de (1670–1741). Astronomer and cartographer

Plantaganet, Beauchamp. *Descrip. Prov. New Albion* 1648

Plantin [**Plantijn, Platevoet**], Christopher [Christoffel] (1514–1589). Publisher and printer. Worked in Lyons, Caen, Paris. Settled in Antwerp 1549. Published for Ortelius 1570 (Theatrum), and later editions for Waghenaer (Spieghel) (Thresoor) 1592. Succeeded by his son in law F. Raphelengius, and J. Moretus

P. Plancius [or Platevoet] (1552–1622)

Planudes, Maximos (1230–1310).
Greek monk, possibly drew 26 maps
for Ptolemy's *Geography* ca. 1300

Plas, David van der (1647–1704).
Engraver, worked for P. Mortier

Plat. See **Du Plat**

Plate, H. F. (1824–1895). Lithographer.
Two plans of Hamburg 1846–7

Plate, William Henry F. *Arabia* 1847,
Arabia & Syria 1849

Platen, C. G. *Chatham Co., Georgia*
(1878)

Plater, Stanislas, Comte (1784–1851).
Polish Battle Plans 1825, *Atlas Hist.
de la Pologne* 1827

Platevoet. See **Plantin**

Plath, C. Engraver for Jüttner's *Plan of
Prague* 1815

Platigny, Jean. Engraver. *Plan des Villes,
Fortresses . . . que le Turc a prix . . .
dans le Royaume de Candie* 1647

Platt, Albert (1794–1862). Cartographer
and publisher of Magdeburg. *Braunschweig*
1836, *Erfurt* 1840, *Hydrog-Atlas of
Europe* 16 sh. (1840), *Magdeburg*
1843, *Asia* 1846, *Sachsen* 1848, *Nord
America* 1848, *Süd America* 1856

Plast, Sir Hugh. *Certain Inventions*
1593, *Fire of Coleballes* 1603

Platt, J. B. *City of Augusta, Georgia*
1869

Platt, W. *Stromkarte der Elbe und
Moldau* 1889

Plattes, Gabriel. *Discovery Infinite
Treasures Subterranean Treasure* 1639,
Profitable Intelligences 1644

Platus, Cardus. Metal globe maker.
Celestial Globe Rome 1578, 1598

Platzer, L. and T. *Grundriss Karlsbad*
(1850)

Playfair, Dr. James (1738–1819).
Historiographer to the Prince of Wales.
Rector of St. Andrews. *General Atlas*
1808 and 1814, *System of Geogr.*
1810–18, *New General Atlas* 1814

Playfair, Sir Robert Lambert. *Morocco*
1892

Playfair, William. *Commercial & Polit.
Atlas* (with Corry) 1786, *Geogr. Hist.
& Polit. Empire of Germany* 1800

Playford, J. See **Godbid**, A.

Playse, John. *Jnl. of Henry Hudsons
Voyage,* Purchas 1625

Playter, C. G. (d. 1809). Printer.
Walker's *Coast Guyana* 1799

Pleasanton, Capt. A. 2nd Dragoons.
Cascade & Rocky Mts. 1850 MS

Plechawski, Emil. *Railway map Middle
Europe* 1885

Pleep [Plep, Plepp], Joseph (1595–
1642). Swiss painter, surveyor, and
architect. *Berne and Environs* 1638
later used by Merian

Plees, William. *Map Is. of Jersey* 1817

Pleitner, J. Belgian military engineer
in Polish service. *Plan of Smolensk*
(early 17th century)

Plessis, D. M. du. See **Martineau du
Plessis**

Pleydell, J. C. *Sketch Map part S.
Ireland* 1774

Pleydenwurff, William (d. 1494).
Painter, part illustrator of the
Nuremberg Chronicle 1493, together
with his stepfather Wohlgemut. Teacher
of A. Dürer

Pli, Pierre (d. 1565). French pilot
in Spanish service. Hanged for mutiny
in Philippines. *Isles Filipinas* 1565 MS

Pliny [Plinius] the Elder. Caius
Secundus (23–79 A.D.) *Natural
History,* English edition 1601

Ploix, Alexander Edmond. *River Peiho*
1859, *Point Yulinkan* 1860, *Côte
Cochin Chine* 1862

Ploix, Charles Martin. *Maps of the
Balkans* 1848–65, *of West Africa*
1856–7; *Isles Vièrges* 1874

Plonnies, Johann Dietrich, Capt.
(d. 1745). Surveyor

Plot, Robert (1640–1696). First keeper of the Ashmolean Mus., Oxford. *Nat. Hist. of Oxfordshire* 1677 and *Staffordshire* 1680, both with a large map of the county

Plotho, A. T. Freiherr von (fl. 1820–44). Cartographer

Plovich, Vedastus du. Mathematician, cartographer, and engraver. *Salae et Castell Iprensis* 1647. *Ambach* (Blaeu) 1649, *Brouchburch* (1690)

Plumb, H. S. *Oil district Venango* 1865

Plumbe, John Junior. *Iowa Territory* 1839

Plume, E. Estate surveyor of Lexdon. *Lexdon & Stanway* 1797 MS

Plumley, John. *Plan Bristol & Suburb* 1813, completed by G. C. Ashmead in 1828, pub. 1829

Pobéguin, Henri. *Côtes du Congo* 1893, *Côte d'Ivoire* 4 sh. 1895

Pobuda, Wenzel (1797–1847). Bohemian engraver. Maps and steel engravings together with J. Rees in Stuttgart (among others, maps of *Ulm*

Pock, R. *Slavonia Inf.* 1871

Pocklington, Joseph. *Vicars Island, Derwentwater* 1783

Pocock, Ebenezer. *Patent Paper Globe* Bristol (1835)

Pocock, Nathaniel. *Appearances Coast of Carolina* 1770, pub. Le Rouge 1777, repub. 1794

Pocock[e], Richard. *Map Egypt* 8 sh. London, Overton 1743

Podolsky of Podoli, Simon (1561–1617). Bohemian surveyor

Podoski, Franciszek. Military surveyor and author of *maps of different districts of Poland* 1791

Poelitz, K. H. L. *Hist. Geogr. Atlas* 1839–42

Poerbus P. See **Pourbus**

Pogonowski, Piotr (1799–1847) Col-

laborator. *Survey of Poland* 1831

Pograbius. See **Pograbski**

Pograbski, [Pograbka, Pograbius] Andrzej [Andreas Pilsnensis] (d. 1602). Physician and cartographer. *Map of Poland* Venice 1570. Used by Ortelius 1595 onwards

Pohl, Anton the Elder (1825–1855). Cartographer of Glogau

Pohl, J. M. *Maps Brasil* 1820

Pohl, Johann. *Railway Map E. Europe* (1881) and 1889

Pohl, L. *Telegraph map Austria Hungary* 1877

Pohl, O. *Lombandy & Venice* 24 sh. (1830) MS

Pohl, Richard. *Umgegend von Baden Baden* 1893

Poinsart, Jean. *Environs de la Ferre* 1650, *Environs de Lestan* 1650

Pointe, F. de la. See **Delapointe**

Poinsignon, M. *Atlas de Géogr . . . de la Marne* 1877

Pointe, François de la (fl. 1666–90) in Paris. *Map Alsatia*

Pointis, Sr. de. See **Desjeans,** J. B. L.

Poiree, Ernest. Text to *La France et ses Colonies,* Vuillemin 1851

Poirson, Jean Baptiste (1760–1831). French geographer and engineer. *Atlas France* 1790, *Cours du Rhin* 1793, *Carolina* 1799, *Mississipi* 1803, *N. Amérique* 1808, *Mexique* 1827, *Atlas to Malte-Brun* 1830

Poker, Mathew. *Romney Marsh* 1617 engraved James Cole (1737)

Pokorny, A. (1826–1886). Austrian geographer and naturalist

Pokorny, W. *Die Kon. Militär-Grenze* 1847

Pokotilof, D. *China, Corea & Japan* 1895

Polack, Abraham Isaac (fl. 1757–80). Dutch mapmaker

Pol, Wicenty (1807—1872). Poet and Prof. Geography, Cracow University. *Maps & books geography of Poland* 1850—7, *Holy Land* 1863, *Historical Geogr. Poland* 1869

Polak, J. E. *Umgebung von Teheran* (1876)

Polakiewicz, Tomasz (1787—1837). Military topographer and lithographer

Polakowsky, H. *Repub. de Chile* 1891

Polani, Gabriele. Venetian cartographer. *World* 1513 MS

Polanzani, Felice. Engraver. *Golfo di Venezia* (1750). Engraved for Vallemont 1748, for Tentiro (1750)

Pole, Sir William. *Collections towards a description County Devon* 1635, printed 1791

Poley y Poley, Antonio. *Prov. di Sevilla* 1890

Policardi, Domenico. *Corsica* 1769

Political Magazine 1780—91. Includes maps by Bew, Lodge etc.

Polk, Thomas (1732—1794). Surveyor. *North & South Carolina* 1772

Pollack, John. *Road map of Lanark* 1830

Pollard. *Relations of the Bermudas* MS 1618. Part used by Smith

Pollard & Peregoy. Lithographers. *California* 1851

Polley, William (fl. 1805—1840). Surveyor of Southlands. *Sandon Woodham Ferrers* 1860 MS

Polo, Nicolo and Maffeo. Merchant travellers to Asia 1260—1271

Polo, Marco (1254—1324). Accompanied above on their second journey to China 1271—95

Polter, Richard (fl. 1578—1605). Chartmaker and map seller. *Thames Estuary* 1584 MS, Master of Trinity House 1599, *Pathway to Perfect Sayling* 1605

Poly, Crétien de. *Baie de la pointe à Pitre* 1839

Polyakov. Engraver. *St. Petersburg* 1853

Pomar, Luis. *Maps of Chile* 1866—79

Pomarede, Daniel (fl. 1742—1765). Huguenot silversmith and map engraver. Noble & Keenan's *Kildare* 1752

Pomba, Cesar (1830—1898). Cartographer of Turin. *Relief maps*

Pombart, L. *Madagascar* 1895, *Sketch map S. Africa* 1899

Pomel, A. *Carte géologique d'Alger* 1881

Pomeroy, A. Publisher of Philadelphia. Lake & Beers' *Philadelphia* 1860

Pommeuse, H. de. See **Huerne de Pommeuse**

Pomponius Mela. Roman geographer fl. ca 43 A.D. *Cosmographia de Situ Orbis:* First edition Milan 1471 (no map), first edition with map, Venice 1482; first to contain printed Spanish map of World, Salamanca 1498; English version by Golding 1584

Pompper, Hermann. *Hand Atlas* 1846

Ponce, J. *Carte Routière du Pas de Calais* 1867

Ponce, Nicolas (1746—1831). Publisher and engraver to the Count of Artois. Rue St. Hyacinth 19, Paris. *St. Dominique* 1795

Ponce de Leon, B. F. y. See **Francia y Ponce de Leon**

Ponce de Leon, Manuel. *Los Estados Unidos de Colombia* 1864

Poncet, A. & J. *Nile* 1868

Pond, Capt. Peter (1740—1807). Fur trader and explorer of Milford, Conn. *N. America* 1784 MS

Pond, P. *Hudson Bay* 1785 MS

Pondman, J. B. *Noordzee Kanaal* 4 sh. 1893

Pongratz, Johann. *Postkarte Europ. Staaten* 1801, *Karte von Deutschland* 6 sh. 1800

Ponheimer, Kilian. Engraver for publishing house of Reilly, Vienna. *America* 1795, *Switzerland* 1796

Poniatowski, Michal Jerzy (1736–1794). Brother of king, archbishop, and patron of cartographic projects

Ponickau, F. L. von. *Gegend um Erfurt* 1844

Ponsonby, W. *Report Sir Anthony Sherley's Journey to Persia* 1600

Pont, Timothy (fl. 1579–1610). Clergyman born ca. 1560. Compiled maps for Blaeu's *Scotland* 1595–1608, revised by Gordon of Straloch, used by Blaeu 1654

Pontano [Pontanus], Giovanni Giorgio (1421–1503). Historian and poet of Naples. *Naples,* one of first maps engraved in Italy.

Pontanus, H. See **Petrus ab Aggere [Terbruggen]**

Pontanus, Johannes Isaksen (1571–1639) from Helsinki. *Tabula Geograph.* Amstd. 1611 (maps), Hues' *Use Globe* 1617

Pontault, Seb. de. See **Beaulieu**

Ponte Ribeiro, Duarte da. *Imperio do Brazil* 1873

Pontoppidan, Christian Jochum (1739–1807). Danish cartographer. *Mappa Daniae, Norveg. et Sueciae* 1781, *Sydlige Norge* 1785, *Nordilige Norge* 1795

Pontoppidan, Erik (1698–1764). Danish theologian, naturalist and geographer. *Norges naturlige Historie* 1752–3, English edition *Nat. Hist. Norway* 1755, *Plans of Copenhagen* 1760, *Danske Atlas* 1763–9, *Supplement* 1774–81

Ponzi, Giuseppe. (1805–1885). Prof. Geolog. Univ. Rome. *Carta Geolog. Campagna Romana* 1880

Ponzoni, Fabio A. *Plano del Puerto Vera Cruz* 1816

Pool, Matthys (1670–1732). Engraver.

Views of villages in the Amstel (ca. 1700). Also worked for I. Tirion

Poole. *Chart Rye* 1743

Poole, Benjamin (19th century). American surveyor

Poole, Charles R. *Map City of Roxbury (Mass.)* 1856, *Route Map Michigan– Wisconsin* 1871

Poole, Henry W. *Topo. Map Mine Hill Railway (Penn.)* 1855

Poole, Jonas. *Voyages to Cherie Island,* Purchas 1626, *Voyage Discovery Greenland* 1611

Poole, P. G. *Klerksdorp Gold Fields* Johannesburg 1890

Poole, Reginald Lane. *Hist. Atlas of Modern Europe* 1896–1902

Poole, W. Engraver. J. Man's *Borough of Reading* 1802

Poole, W. G. *Parish of St. Giles Camberwell Surrey* (1834)

Poopard, James. See **Poupard**

Poor, Henry V. *Map of all the Railways in U.S. and Canada* 1857

Poorbus. See **Pourbus**

Pope, Lieut. & **Franklin.** *Plan Battle Buena Vista* 1847

Pope, John. Topographical engineer. *Territ. of Minnesota* (1850). *Survey Pacific Railway* 1855

Pope, Thomas. Land surveyor. *Ramsden Bellhouse* 1615 MS, *Finchingfield* 1618 MS

Popelinière, L. V. la. See **La Popelinière**

Popham, Lieut. Henry. *Great Fish Bay (W. Africa), Admiralty Chart* 1847

Popham, Capt. H. R., R.N. *Prince of Wales Island,* pub. Laurie & Whittle 1798

Popham, Sir Home Riggs. *Chaves Bay* 1789, *Red Sea* 1804, *British Possessions in East Indies* 1804

Popinjay, Richard (fl. 1560–87). Military

engineer and surveyor. *Chart
Jersey* 1563, and various *charts of
Portsmouth* 1576–87

Popp, Philippe Christian (1805–79).
Painter, cartographer, publisher. *Atlas
Cadastral Flandre Occid.* (1860)

Poppel, Johann (1807–1882). Engraver.
Ansichten v. Nürnberg (1841), *Galerie
Europ. Städte* (1845), *Preussen* (1842),
Salzburg (1845)

Poppele, E. *Post-Reise und Zoll-Karte
v. Deutschland* 1835

Poppey, Carl (1829–1875). Engraver
for Wieland 1848, for Stulpnagel
Australia 1858, for Perthes *Aust. u.
Polynesien* 1860

Poppey, Carl, the younger (1840–80)

Popple, Henry (d. 1743). Geographer,
Clerk to the Board of Trade, and
Cashier to Queen Anne, 1713. *Map of
British Empire in America,* 20 sheets and
index map 1733

Poppleton, Thomas H. City surveyor.
Plan city of New York 1817 pub. Prior
& Dunning, *Plan City of Baltimore*1823

Por, Stanislav. See **Porebski**

Porcacchi, Tomaso (1530–1585), of
Castiglione Aretino. *L'Isole del Mondo,*
Maps engraved by Girolamo Porro 1572.
Editions up to 1686

Porebski [Porinski], Stanislaw. Poet and
cartographer, compiled *first map of
Oswicum and Zator* Venice 1563 [Sta
Por], later used by Ortelius

Poro, G. See **Porro**

Porro, Girolamo, of Padua. Publisher
and engraver. Worked in Venice. *Map of
Europe and N. Africa* 1567, maps for
Porcacchi's *Isole* 1572. Editions of
Ptolemy 1596, Mercator *Atlas Minor*
1596, published for Veer 1599

Portal, A. du. Engineer. *Plans of Strasburg
and Breisach* (ca. 1700)

Portantius [Portant], Joannes [Jean].
Astrologer, mathematician, and geog-

rapher of Antwerp, *Livonia,* Ortelius
1573, also used by De Jode 1578

Porte, Abbé Joseph de la (1713–1779).
Edited atlases

Porter, Augustus. Surveyor. *Connecticut
Reserve* 1796 MS

Porter, Benjamin. *Shropshire* 1734

Porter, David (1780–1834). *Massachusetts
Bay, Nuka Hiva* 1815

Porter, Judge. *Genesse Co.* 1790

Porter, Peter. *Lake Huron* 1820, *Lake
St. Clair* 1828, *Detroit* 1820

Porter, Sir Robert Ker. *Borders of Black
Sea*1822, *Caracas* 1825

Porter, Thomas. *Map of London & Westminster,*
Printed & Sold By Robt. Walton 1654,
New Booke of Mapps 1655, *England &
Wales,* Walton 1668 (one of the first
maps of England to show Roads)

Porter, William E. *Parish of Carrara-
garmungee* 1875, *Govt. House
Melbourne* 1876

Portinari, John (fl. 1525–60). Survey
engineer of Italian origin, worked for
Henry VIII

Portlock, Capt. Joseph Ellison. *Report
on Geology of Co. of Londonderry* 1843

Portlock, Nathaniel. *Voyage* 1789, *Charts
W. Coast America*

Portman, Lodewyck (1772–1813).
Engraver. *Leiden* (1807), Topogr.
engravings in Holland

Portolan [Portulan, Portolano]. Sailing
directions developed out of the compass
in conjunction with sandglass. Oldest
surviving example 1294 (copy of a lost
example of 1254) now in Cagliari
Univ. Oldest dated example surviving
1311 (P. Vesconte) in Venice. First
printed example Venice 1490

Pory, John. Translator. Leo's *Geogr.
Hist. Africa* 1600. Reprinted in Purchas.

Poschacher von Poschach, Ferdinand
(1819–66). General, military cartographer

Poseidonius, (135—51 B.C.). Greek philosopher and geographer of Rhodes

Possart, Paul Anton. *La Suisse* 1850

Posselt, Heinrich. Geographical engineer. *Munich* (1770)

Post, Frans. *Brasil* 9 sheets, pub. Amsterdam by Allard 1689

Post, L. W. *Map Jersey City* 1870

Postans, Capt. T. *Maps Pakistan* 1841—4

Postel [Postellus], Guillaume [Guilelmus] (1510—1585). French astronomer and cartographer. *De Etruria* Florence 1552. *Cosmographia,* Oporin 1561. *Terre Sainte* 1562 (lost). *World* 1587, 1596 and 1621, *Gallia,* Bouguereau 1594, *Théatre Géogr. de France* 1626

Posthumus, N. W. *Atlas der geheele Aarde* 1872, *Atlas van Nederland* 1878

Postlethweyt, Malachy. *Dict. Trade & Commerce* (maps by Bolton) 2 vols. 1766, another edition 1774

Potain, N. M. *Partie de la ville de St. Germain en Laye* (1740)

Potanin, N. W. *Mongolie* 1881

Potel, A. *Uruguay* 1887

Potel, Felice. Printer for Weingartner 1834, *Atlante Univ.* Naples 1850

Potenti, G. *Railway Map Central Europe* 1846

Potgieter, Bernard. *Freti Magellanici* (1640)

Pothignon. Engraver. *Commune de Lyon* 1873

Potier, Nicolas (18th century). French surveyor

Potinger, Capt. E. *Country Nmai Kha (Burma)* 1897

Potiquet, Alfred. *Carte Hydro. du Dépt. de la Seine* 1858, *Vincennes* 1867 and 1878

Potkanski, Adolf (1801—1835). Military topographer, collaborated in *Survey of Poland, Warsaw area*

Potocki, Jan (1761—1815). Russian

cartographer. *Study on Charts of Middle Ages* 1796, *Atlas arch. de la Russie* St. Petersburg 1823

Pots, Richard. *English Colony in Virginia* 1606—12

Potter, Elisha Reynolds (1811—1882). American jurist and map editor

Potter, J. D. 31 Poultry and 11 King St. Tower Hill. Agent for Admiralty Charts 1850—73

Potter, Jan Janszoon (d. ca. 1590) Dutch surveyor of *Delft Land onder Poeldick* 1526 MS, *Platte gronden van Rotterdam* 1567

Potter, John E. Potter Bradley *Atlas of the World* 1894

Potter, Peter. Surveyor. *Plan Grosvenor Estate from Chelsea Hospital to Penitentiary* 1815 MS, *Turnpike Rd., Kensington* 1814, *Parish St. Marylebone* 1820

Potter, T. *Dunlieth NW Terminus Illinois Central Railway* 1854

Potter, Paracette & Co. Publishers of Poughkeepsie (19th century). *Atlas* 1820

Pottier, Eugene Jerome. *Maps Iceland* 1857—8

Pouchot, M. de. Capt., Regt. Bearn. *Plan Fort Niagara. Attack on Fort Levis* 1760, *Frontiers English French from Montreal to Fort Du Quesne, Mémoires dernière guerre* 1781

Poulinaire, L. *Madagascar* (1890), *Siam* 1890

Pound, Thomas. *New Mapp of New England* 1691, London, P. Lea (1692)

Pourdoux, Sieur. *Plan Ville d'Orléans* 1773

Pouqueville, F. C. H. L. *Grèce Moderne* 1821, *Grecia Antica* 1828

Poupard [Poopard], James. Engraver of Franklin's *Chart of the Gulf Stream* (1770), for *Pennsyl. Magazine* 1775—6 (plans 7 maps), *Baltimore* 1792

Pourbus, [Poerbus, Pourbusch, Poorbus, Poerbuasse, Pourbasse, Poerbuus] Pierre (1510–1584). Painter and cartographer. *Watervliet* 1549–51 (lost), *Franc de Bruges* 1551–71, *other maps Bruges area* 1578

Poussin, Guillaume Tell (1794–1876). *Map, Plan & Profiles Canal to Mississippi with Lake Pontchartrain* 1827, *Etats Unis* 1834

Povedano, Diego Lope. Spanish soldier. *Negroes Is. E. Indies* 1572 MS

Powell, Capt. *Map of New England* 1641

P[owell], D[avid]. *Brief rules of Geography for the understanding of maps and charts* 1573, *History of Cambria* 1584

Powell, E. J. Draughtsman for Admiralty. *Harbours of Refuge* (1585), *Chesapeake Bay* 1859, *N.E. China* 1860, *Jerusalem* 1864, *Seat War Waikato N.Z.* 1864

Powell, Lieut. Frederick Thomas, Indian Navy. *Maldeeve Is.* 1838–9, *Paumben Pass* 1838, *Chagos Archipelago* 1839. *Palks Straight* 1852

Powell, George. *Chart S. Shetland* 1822

Powell, John. Draughtsman, mate of the *Osterley. Surveys Kerry Estates* 1764, *Chart Islands between Borneo and Banca* 1758–9, *W. Herbert* 1767

Powell, J. W. *Atlas Geology Ulinta Mountains (U.S.)* 8 sh. N.Y. 1876

Powell, T. K. Engraver. Maps for Turner's *Views of Earth,* London 1787; *World, England, Asia & Africa*

Powell, W. Angelo. *Official Map State of Virginia* 1862

Powell, Wilfred. *N.E. Portion New Britain* 1881

Power, William. "Printer in Fletestrete at the sygne of the George next to Saynt Dunstones church." *Chronicles of Yeres* 1530–1 (Early Road Book)

Pownall, Thomas (1722–1805). Governor of New Jersey and Mass. *N. America* 1776, 1783, 1794; *New York & N.J.* 1776, *Chart Gulf Stream* 1787

Poyda, H. von. *Charte von Sachsen, Schlesien* 1836

Prado, Casiano de. *Provincia de Madrid* 1864

Pradzynski, Ignacy (1792–1850). Polish military geographer

Praet, Stephan de (ca. 1650 at Danzig). Engraver. *Map of Danzig* pub. by Hondius

Praetorius, Johannes. (1537–1616) of Nuremberg. Born Joachimstal. Globe maker, inventor of surveyors plane table 1590. *Various globes* 1566–68

Prahl, Arnold Friedrich (1709–1758). Map publisher. *Stift Ellwangen* (1746)

Prahl, G. C. C. W. *Australia* (1840)

Prasch, Christian (18th century). Map maker of Baden

Prasser, Johann Baptist. *Wien* (1745)

Prat, Wyllyam. *Description of the country of Afrique* 1554

Prato, Rinier da. Italian publisher. *Ancona* 1569

Pratt, F. W. *Real Estate Directory City of Washington* 1874 (with E.F.M. Faentz)

Pratt, Henry. *Survey of Marylebone & Barrow Hills* 1708. Improved Moll's *Ireland* 1714

Pratt, John J. Civil engineer. *Kansas Gold Mines* 1859, *Central City, Colorado* 1862

Pratt, W. *Arithmeticall Jewel . . . use of the small table* 1617

Pratt & Marshall. Booksellers in Milsom St., *Plan of Bath,* engraved Ashby (1780)

Pratz, Le P. du. See **Le Page du Pratz**

Prault, le fils. Quay de Conti, Paris. Joint publisher with Le Rouge of *Intro. to Geography* 1748 and *Atlas Nouveau Portatif*

Pré, G. du. See **Du Pré**

Predetich, *Kadastral Karte von Dalmatien* 14 sh. 1842

Predour, Le. See **Le Predour**

Préfontaine, Bruletout de. *Cayenne* 1762

Preisler, I. Just. Engraved title for *Atlas Homannianus* 1762

Prejevalsky, N. M. *Central Asia* 1877, *Tibet* 1887, *Chinese Turkestan* 1890

Prempart, John. *Siege of Busse (Bois le Duc) maps & plans* 1630

Prescott, Cyril Jackson. *Maps of India* (1860–67)

Pressler, Charles W. *Map Texas* 1858

Preston, Capt. Amias. *Voyage to West Indies with Capt. George Sommers,* Hakluyt 1600

Preston, J. H. Publisher. *Oregon & Washington* 1856

Preston, Capt. Thomas. *Shetland Is.* 1743–4

Prestwich, Sir Joseph (1812–1896). *Geolog. Map Thames Basin* 1870, *Temp. Sea . . . Geology* 1886

Pretot, E. A. P. de. See **Philippe de Pretot**

Pretty, Francis. *Voyage of Thos. Caundish,* Hakluyt 1589

Preuschen, August Gotlieb (1734–1803). Geologist and geographer

Preuschen, Ernst Ludwig (1783–1842). Surveyor of Darmstadt

Preuss, Charles (1803–1854). Assisted Fremont in his map of *Road from Missouri to Oregon* Baltimore 1846, *Surveys in Utah* (1852), and drew for Williamson 1855

Preussen Sea Atlas, Berlin 1841

Prévost, Ant. F. (1697–1763). *Histoire Gén. des Voyages* 25 vols, Paris 1747–80 (maps by Bellin)

Prévost, Augustine (1725?–1786) *Part of S. Carolina* 1779

Prévost, Jacques. Engraver, fl. 1537

Prévost, Louis Const. (1787–1856). French geologist

Prévost, N. Publisher, Strand, London. Vol. V. Mortier's *Nouveau Théatre de la Grande Bretagne* 1721 (with Groenwegen)

Prévost. Publisher of Paris. Cook's *Mer Pacifique* (1780)

Price, B. *Plan Dorchester* (1800)

Price, Charles (fl. 1680–1720). Publisher, draughtsman, surveyor and globemaker. Worked from following addresses: Hermitage in Wapping (with Seller 1699–1703), Next Fleece tavern in Cornhill (with Senex 1690), Whites Alley in Coleman St. (with Senex), Archimedes & Globe in Ludgate Street (with Brandeth 1711). In partnership with Jeremiah Seller in 1700, with Senex and Maxwell 1708–12, and with Willdey 1711–13. *Bristol Channel* 1690, *English Pilot* with Seller 1701–3, Seller's *System of Geography* 1703, *Thirty miles round London* 1712, *World* 1714 (with Willdey), *Globes* 1715

Price, Edward. *Gen. Map of Australia* 1859

Price, Franklin. *Bloomington, Illinois* 1855

Price, H. *Herefordshire* 2 sh. 1817

Price, Capt. J. *Charts Guinea, St. Thomas & Annabona* 1796

Price, J. Engraver. Well's *Port Philip* 1840

Price, Jacob. *Delaware* 1850

Price, Jessie. *North Wales Coalfield* 1879

Price, John. Chief mate of the *Kent* 1786. *West Coast of Sumatra,* Laurie & Whittle 1794

Price, Jonathan. *Cape Fear River* (1800)

Price, Owen. *England Displayed* 1769

Price, Rev. R. E. *Africa* (1876)

Price, T. T. *Atlas of New Jersey Coast* 1878

Price, William. *Plan Boston* 1739 and later editions

Price, William. Master H.M.S. *Theseus* (later Captain). *I. of White,* Laurie & Whittle 1798; *Skagerack,* revised 1794, *Coasts of Holland & Friesland* 1801

Price & Strother. *N. Carolina* 1808

Prichard, James Cowles (1786–1848). *Ethnographic maps* 1843

Pricke, Robert (fl. 1666–98). Engraver, print and mapseller, pub. in Whitecross Street Cripplegate, London. *Exact map of London* (1666)

Prickett, John. Land surveyor of Highgate, later Castle St. Holborn. *Various estate plans* 1747–8 MS, *Kelvedon* 1792 MS

Pride, J. *Travellers companion* 1789 (with Luckombe)

Pride, Thomas. Surveyor. *Woodham Ferrers* 1771 MS, *Manor Wenlock Barn Shoreditch* 1799, *Ten Miles round Reading* (with Luckombe) 1790

Priestley, J. *Grand Canal Liverpool-Leeds* (1780), *Inland Navigation* 1830

Prime, Frederick (1846–1904). Geologist

Prinald. Engraver for Speer's *West India Pilot* 1766–71

Prince, Thomas. *Louisbourg* 1747

Principe de la Paz. See Paz

Pringle, G. *Aberporth Bay* (1860)

Pringle, George & Co. Publishers 1816

Pring, Martin. *Journal two Voyages E. India,* Purchas 1625

Pringle, George Senior & Junior. Partners with Greenwood 1824–31 (Greenwood Pringle & Co.), 13 Regent St. Pall Mall, London

Pringle, Capt. *Penin. of India,* Faden 1800

Pringle, George Jr. Publishers, 70 Queen St. Cheapside. Published many of Greenwood's *large scale county surveys* 1821–3

Pringle, J. W. *Mombassa–Victoria Lake Railway* 1893

Pringle, Robert. *Plan Toulon* 1766 MS

Prins, Joh. Hubertus (1757–1790). Engraver. *Dutch town views and plans*

Prins, P. Bloys van Treslong *Straat Lagoendi* 1855

Prinsen, P. J. *School Atlases* 1832, 1840

Prinsep, James. Surveyor. *City of Bunaras* 1822

Prinsep, Thomas. *Calcutta* 1830

Prior, Rev. John. Schoolmaster and surveyor, of Ashby de la Zouche. *Leicestershire* 1779 (with J. Whyman), 2nd Edition 1804

Prior & Dunning. Publishers of New York. Poppleton's *Plan New York* 1817

Priorato, Graf Galaezzo Geraldo. *Vincennes* 1608–78, *Teatro del Belgico* 1673

Prisius, J. *Hist. Brit. Defensis* 1573

Pritty, John. Estate surveyor. *Woodham Mortimer* 1759 MS

Prixner, G. Engraver. *Galicia* 1800

Pro, Joachim. Engraver for Torfino de San Miguel 1786

Probst, George Balthasar (1673–1748). Engraver and publisher of Augsburg

Probst, George Balthasar. Married Seutter's daughter in 1745, joined firm of Matthew Seutter 1756. Later became independent publisher. *Gross Brittannischen Städte* 1782

Probst, Johann Friederich. Publisher, of Augsburg. Successor to Jeremiah Wolf

Probst, Johann Michael (d. 1809). Publisher and engraver of Augsburg. Worked with Seutter & Lotter. *General Karte der Schweiz* (late 18th century), *Gent* 1780

Probst, J. S. Engraver. *Map of Wörlitz* (1788)

Probst, J. Draughtsman for Bruff 1847, *Seat of War in Mexico* 1847

Probst, Petro. *Reise Beschreibung Miss. Soc. Jesu* 1748

Probsthayn, J. F. G. (1719–1779). *Plans of Meissen* engraved C. G. Werner 1767

Prochaska, Carl. *Austria Hungary* 1877, *Teschen* 1897

Prockter, J. (fl. 1750–76). Engraver. Worked on globe for Doyle's *Brit. Domin.* 1770, Taylor & Skinner's *Roads of Scotland*

Prockter, I. *W. Indies,* Sayer 1762

Proclus. *De Sphaera* 1499, 1539, 1547, 1561

Procter, John. *Town of Danvers* 1832

Proctor, Richard Anthony (1837–1888). Astronomer and cartographer. *Star Atlases* 1866–95

Proeschel, F. *Map Melbourne* 1852, *Road to Mines* 1853, *Gold Fields* 1859, *Atlas of Australia* London 1863

Progenie, Francesco. Engraver of Naples. *Calabria* 9 sh. (1784)

Prohaska, Joseph Edler von (1758–1835). Austrian military cartographer

Prokesch-Osten, Anton, Graf von (1795–1876). Military cartographer of Vienna

Prompt, M. *Vallée du Nil* 1898

Pronck, Cornelis (1691–1759). Dutch artist. Many *topographical and town plans* for *Het Verheerlÿkt Nederland* engraved by Spilman, and *Atlas van Zeeland* (1760)

Pronner, C. M. Maps for Homann Heirs. *Kleiner Atlas* 1803

Pronostel, Pierre. *Diocèse d'Alby* (1650

Pronti, D. Engraver, Mannazzale's *Rome* 1805

Prony, Gaspard Clair François Marie Riche, Baron de (1755–1839). Mathematician, engineer of Paris. *Atlas des Marins* 1823, *Marais Pontins* 1823

Prooyen, Adriaan Gerrit. van (1796–

1854) of Middelburg. Engraver and map draughtsman

Propert, T. *Pencaer Pembroke* (1798)

Propper, George N. *Township map of Kansas* 1857

Pross, Jacobus. *Hungary* 1594

Prosser, W. *Gold Farms Transvaal Republic* 1883

Protucius, Celtes. See **Celtes Protucius**

Proude, Richard. *The New Rutter of the sea for North Part* 1541

Prouhet, J. A. *Rade de Mogador* 1842

Prowd, John. *Account of Navigation (to Batavia)* 1631 MS

Prowig, Lieut. *Plan Dresden* 1813

Prowse, Lieut. W. Jones. *Port Anna Maria (Marquesas)* 1814 MS (with J. Milne)

Prudent, F. French military geographer end 19th century. With F. Schrader, *Atlas de Géogr. Mod.* Paris, Hachette 1889

Prudhoe, Lord. *Prov. of Seunar* 1835

Prugner, Nicolas. Translator of Frisius's *Hydrographia* 1530

Pruneau, Reinato. *Costa de Ylocos y Luzon* 1791 MS

Prunes, Family. Portolan makers of Catalonia, fl. 1532–1560

Prunes, Battista de (1532–60). Catalan hydrographer

Prunes, Joan B. *Mediterranean* 1649

Prunes, Matteo. *Portolan charts* 1560–99

Prunes, Pietro Giovanni. *Portolan chart* 1651

Prussia. See **Academiae Regiae Scient. Boruss.**

Pruthenus. See **Somer[s]**, Jean

Pruyssenaere, E. E. J. M. de (1826–1864). Belgian geographer and traveller

Pryce, Benjamin. Estate surveyor of

Dorchester. *Southminster (Essex)* 1755 MS

Pryer, W. B. *British N. Borneo* 1888 and 1894

Ptolemy [Ptolemaios], Claudius [Klaudios] (87–150), of Alexandria. Earliest surviving MS. 12th or 13th century. MS. brought to Italy and translated into the Latin tongue by Jacopo Angelo in 1406. First printed edition 1475 (no maps). First edition with maps, Bologna 1477. Important first German edition, woodcuts with 5 new maps added, Ulm 1482. Martin Waldseemüller's edition with 20 new maps added, 1513. Gastaldo's miniature edition (maps increased to 60), 1548. Mercator's edition, of classical maps only, 1578. Porro's edition with new engraving of the maps, 1596. Ptolemy has been called the father of geography. While the instructions or text are his, doubt has been expressed as to the authorship of the maps, which have been stated to be either by Agathodaemon, Planides or an editor as late as 1450. Ptolemy's *Almagest* was a popular version by Sacrobusco

Puche, C. Engraver. *Gibraltar* 1727

Puchova, Zigmund (16th century). Czech map publisher

Pufendorf, Samuel, Baron von (1632–1694). Swedish historian, professor University of Heidelberg. *Sueciae Regis* Nürnberg 1696 (maps and prospects), *Hist. Gen. et Polit. de l'Univers* Amsterdam 1743

Puget, Rear Adm. Peter (1765–1822). *Port Stewart Alaksa* 1795 MS

Puillon-Boblaye, Emile. *Carte Trigono-métrique de la Morée* 1832, *Ile de Tine* 1833

Puis, C. du. See **Du Puis**

Pugh, Charles A. *Township Map county Campbell, Virginia* 1870

Pugh, Edward. David Hughson pseud

Puiségur, Vcte. de. See **Chastenet-Puiségur**

Puissant, Louis (1769–1843). Parisian mathematician and geographer

Pujol, J. J. *Dépt. de l'Allier* 1845

Puke, J. Engraver for Wilkinson 1794, for Arrowsmith's *N. America* 1795

Pulham, J. *Harwich* 1778–80, *River Orwell* 1780

Pulitz, von. *Gegend um Potsdam* 1811

Pullen, J. *Map St. Mary Rotherhithe* 1755

Pullen, J. W. *Mouth Murray* 1840, *Lake Alexander* 1841

Pullen, T. F. *Jamaica* 1877

Pullen, William of Chelmsford. Estate surveyor. *Terling* 1734 MS, *Springfield* 1735 MS

Pulman, R. *Chart Gallions Point to Hookness* 1718–19

Punnett, J. M. American cartographer. *Bay Counties California* 1893

Punnett Bros. American publishers, 625 Mission St., San Francisco, Cal. *Napa Co., Cal.* 1895

Punt, Jan. Dutch engraver (1711–1779). *Map of Rotterdam, Views of Schoon-hoven* (1762)

Püntener, F. L. (d. 1720). Engineer and mapmaker at Altdorf

Purcell, Joseph Southern. *Indian District of N. Amer.* (1735) MS, *Va. N. & S. Carolina & Georgia* 1792, Moore's *Geography,* pub. Stockdale 1792

Purcell, Mathew A. *Transvaal* 1872

Purchas, Samuel (ca. 1575–1626). Theologian and travel writer. *Purchas his Pilgrimes* 1625

Purdy, Isaac. *Charts* 1844–65

Purdy, John (1773–1843). Hydrographer of Norwich. *N. Sea* Laurie & Whittle 1806, *Africa, Asia* 1809, *Canada Cabotia* 1814, *W. Indies* 1823, *World* 1828

Purey-Cust, Lt. Comm. H. E. *Surveys Tasmania* 1893–4

Puydt, Lucien de. *Surveys Darien* 18th century

Purgstall, Wenzel Karl, Graf von (b. 1681). Czech cartographer

Pursell, Henry D. Engraver of Philadelphia. Filson's *Kentucky* 1784, *U.S.* for Bailey's *Pocket Almanac* 1785

Pursglove, Wm. *Voyage to Petchora* 1611–12 MS

Pusch-Korenski, Jerzy Bogumil (1791–1846). *Geological Atlas of Poland*

Puschner, J. G. (ca. 1700–1750). Engraver at Nuremberg. *Map of Anspach* (1743)

Puschner, Johann George the elder (1680–1749). German cartographer and engraver, worked for Homann Heirs. *Plan Belgrade* (1717), Dopplemayer's *Globe* 1728, *Celestial Globe* 1730

Puschner, Johann George, the younger (1706–1754). Nuremberg globe maker and engraver

Putbaux. Script engraver, Jaillot's *plan Paris* 1713

Puteanus, B. See **Putte**

Putnam, Israel. *Stations from Albany to Lake George* (1756)

Putnam, Rufus (1738–1824). American military engineer. *State Ohio* 1804

Putnam. See **Wiley & Putnam**

Putsch [Bucius], Johann (1516–1542) of Salzburg and Innsbruck. *Map Europe in shape of woman* 1537

Putte [Puteanus], Bernard van der (1528–1580). Engraver, printer and

publisher of Antwerp. *France* 12 sh. 1557, *Antwerp & Brussels* 1565, *Holland* 1553, *Friesland* 1559, Vopel's *World* 1570, *Europe* 1572

Putter, A. de *Malta* 1729

Putzgel, F. W. *Historischer Schul Atlas* 1891

Puya, y Ruiz, Adolph. *Phillipines* 1805

Puységur, Vte de. See **Chastenet-Puiségur**

Pye, J. London publisher, 18th century. *Earthquake in Shropshire* 1773

Pyle, Stephen. Engraver, Angel Court, Snow Hill, London. *Ayrshire* 6 sh. 1775, for Armstrong's *Lincs.* 1778, Lendrick's *Antrim* 1780, for Herbert's *Atlas* 1708, for Taylor & Skinner's *Roads of Scotland* 1775, for Dalrymple 1774

Pynacker, C. See **Pijnacker**

Pynchon, W.L. *Kansas* (1857)

Pyne, James B. (1800–1870). English topographer. *Windsor* (1828), *Lake District* (1853), *Lake Scenery* (1870)

Pynnar, Nicholas. *Survey Co. Cavan, Fermanagh and Derry* 1618–19 MS

Pyper, William. *New General Atlas* 1839

Pyramius [Kegel, Khegell], Christoph. Secretary to Charles V. *Map Germany* Brussels 1547

Pyrrhus, Ligorius. See **Ligorio**, Pirro

Pythagoras. (ca. 582–490 B.C.). Philosopher and mathematician of Samos

Pytheas. Greek traveller. Visited Britain ca. 333 B.C.

Q

Qazwini, Hamdallah Ibn Abi Bakr Ahmad (ca. 1281–1349). Persian historian and geographer

Qazwini, Zakariya Ibn Muhammad al' (1203–1283). Arabian cosmographer

Quackenbush, Eugene. *Topo. Atlas city New York* 1874

Quacq. Postmaster, Rotterdam. *Maas Rotterdam & Hook* Rotterdam 1665, *Plan Rotterdam* 1665

Quad [Quadt, Quaden] von Kinckelbach, Matthias (1557–1613). Geographer, humanist and engraver of Cologne, b. Deventer. *Braunswick Ducatus* 1593, *Europae* Cologne 1592, 1594, 1600; *Geogr. Hand Buch* 1600, *Fasciculus Geographicus* 1600, 1608; *Deliciae Galliae* 1603

Quadra, J. F. de la B. y. See **Bodega y Cuadra**

Quartermaster's Map. Garrett (1688)

Q.M.G. Office, Lithog. Press. *Wyld's Settlements New South Wales* 1820, *Zululand* 1879

Quast, Mathys. *Japan* 1639

Quatrefages de Breau, J. L. A. (1810–1892). Geographer and anthropologist

Quayle & Lusker. *Island of Oparo (Pacific) Admiralty Chart* 1868

Queboren, Crispiaen van den (1604–1652). Dutch mapmaker and publisher of The Hague

Quefada, A. X. de. See **Ximenes de Quefada**

Queiros, Pedro Fernandes (ca. 1560–1604). *Pacific Ocean* 1598

Quenet, Engraver. Title for Dezauche's *Afrique* 1827

Quenstedt, P. R. von (1809–1889). Geologist

Quentell, H. Publisher. *Sacrobosco Opus* 1500

Quentin, François Alexandre. *Admiralty Chart South Pacific* 1869

Querenet [Querneal] **de la Combe.** MS map *Yorktown campaign* (1781)

Querneal de la Combe. See **Querenet de la Combe**

Querret, Jean. *Carte Comté de Bourgogne* 1748 (5 ft. 10 in. by 3 ft. 7 in.)

Querret, J. J. (1783–1839) Mathematician and hydrographer

Quesada, A. F. See **Faneau Quesda**

Quesada, José Maria de. Hydrographer. *Banco de Bahama* 1858, *Bermuda* 1858, *Costas Merid. de Francia* 1858, *Seno de California*

Quesada y Carvajal, José. *Distance map Spain* 1869

Map from M. Quad's "Geographisch Handt Buch" engraved by J. Bussemacher

Quesado. See **Coello y Quesado**

Quesnay, A. M. *Plan Bataille de Jemappest* 1792

Quesnel, G. *Nouv. Atlas Classique* 1883

Quétin, Louis. *Nouvel Itin. de G.B.* 1825, 1828, 1837; *Tab. Itin. et Postal France* (1847)

Quilley, J. P. Engraver. *Hampstead from various surveys* 1814

Quin, Edward (1794–1828). *Historical Atlas* 1828, 1830; *Atlas Univ. Hist.* (1859)

Quin, Richard. *Leavenworth Co., Kansas* 1857

Quin, W. H. *Admiralty Chart W. Africa* 1835

Quinby, T. *Plan Lands Saco Water Power Co.* 1848

Quintinus, Haeduus Joannes. *Malta* 1536

Quirini, Vincenzo. *Chart or Atlas* 1654

Quirini. *Green Globe* ca 1514, in Bibl. Nat., Paris

Quist, H. Engraver for Fester

R

R., A. M. *Imp. Citta d'Augusta* (1600)

R., F. W. *War Map* 1877

R., G. *British Isles* (1570)

R., I. C. *Figure de la Terre* (1761)

R., I. E. *Vestung Dansburg* (1730), *Vestung Trankenbar* (1730)

R., J. *Etats Romains* (1868)

R., P. du G. D. *Geog. game* De Fer 1670

R., T. *Draught Castle San Lorenzo*, Gents Mag. 1740

Raab, C. J. C. *Eisenbahnen Mittel Europas*, A. Logan 1856, edition 6 sh. 1889

Raab, G. F. *Eisenbahn Karte v. Russland* 1867–1900

Raabe, H. *Geog. Special Karte von Schaumberg* 1867 (with A. Franke)

Raaz, C. *Photolithograph relief maps* (1867–73), *School Atlas* (1868)

Rabaut. *Partie des Illinois Renaut* 1733

Raben, Peter. *Charte over Island* 1721

Rabot, Charles (1856–1944). French geographer and traveller

Rabuin. Publisher for Desfontaines 1849

Rabus, Jac. *Terrestrial Globe* 1546

Rabusson, A. *Golfe Arabique* (1850), *Mer Egée* 1846

Rac, Emil (1853–1900). Czech cartographer. *School Atlases*

Rachel, L. *Maps of Tübingen & Würtemberg* 1868–1890

Rad, Christopher the elder (d. 1710). German goldsmith and globemaker

Radakoff, B. *Hand Atlas . . . Europ. Russland* 1876

Radcliffe, E. H. *Township and Business map of Montgomery Co., Penn.* 1873

Radclyffe, Thomas. Engraver. *Plan Cheltenham* 1825, *map County of Gloucester* 1826, *Topo. works* 1830–34

Radclyffe & (Son), T. Publishers 100 New St., Birmingham, *Warwick* 1842, *Ham's Railway map Midland Counties* 1847

Raddatz, H. *Transvaal & Swaziland Goldfields* 1886

Radefeld, Carl Christian Franz (1788–1874). Cartographer from Hildburghausen. *Atlas der Erdbeschreibung* 1841, contributed to Meyer's *Neuster Universal Atlas Grosse Ocean* 1849, *Bayern* 4 sh. 1864

Rademacher, H. *Karte v. Grünewald* 1867

Rademacher, Jacob Cornelis Matheus (1741–1783). Dutch geographer of The Hague

Rademacher, Johann (1538–1617) of Aachen. Historian and merchant in Antwerp

Rademaker, Abraham. *Stadt Alk-maar* 1725

Radics, P. von Arch. *Karte v. Krain* 1864

Radiguel, Adolphe. *Canal Maps France* 1862

Radimsky, Wenzel (1834–1895). Austrian geologist

Radkiewicz, Stanislaw (1801–1875). Polish cartographer and military surveyor

Rados, Louis. *Plan de Cadiz* (1820)

Radziwill, Prince Mikolaj Krysztof (1549–1616). Grand Marshall of Lithuania. Collector and protector of cartography. *Lithuania* (attributed to Makowski) engraved Hessel Gerritsz and first pub. by Blaeu 1613

Rae, John (1813–1893) Arctic explorer. *N.E. Coast America* 1847, *N.W. Passage* 1853

Raefe (2nd half 16th century). French word engraver

Raemdonck, Jean Hubert van (1817–99) Doctor, archaeologist and map historian

Raffel, H. Draughtsman. *Central Ceylon* 1860)

Raffelsperger, Franz (1793–1861). Hungarian cartographer. *Austria* 1844, *Europe* 1844, *Street, Post & Railway map of Hungary Croatia &c.* 1849

Raffles, Sir Thomas Stanford (1781–1826). Lieut. Gov., Penang. *Map Java* 1817

Rafinesque, C.S. Surveyor *Ancient Works* in Squier's *Ancient Monuments* 1848

Rafn, Carl Christian (1795–1864). *Discoveries Scandinavians* 1837, *Kort over Grönlands* 1845

Raglovich, Lieut. von. *Military map S. Germany* [1820]

Ragozin, Victor Ivanovich (1833–1901). *Volga* (1881)

Ragusinus. See **Gazulić**

Rahden, Wilhelm von (1793–1851). Military cartographer of Breslau

Rahnke. Lithographer. *Siege Warsaw* (1840)

Raiband. Lithographer. *Plan Territ. Marseille* (1860)

Raidel, H. F. Engraver. *Reichs statt Memmingen* 1622

Raignauld. *Isle de Cipre* (1590)

Raignauld, Henry. *Malta* (early 17th century)

Raimondi, A. D. (1826–1890). Peruvian geographer. *Mapa del Peru*

Raimondo, Annibale (b. 1505). Mathematician of Verona. *Flusso e Reflusso del Mare* Venice 1559

Raimurez, Jos. Ant. de Alzate y. *Plan de la Nueva Espana* 1769–70

Rainaldi, Carlo. *Piemont & Monferrat* 1615

Rainaud. Engraver for Dufour 1864

Raine, John. Surveyor. *Godolphin Estates Farnham Royal* 1801–2 MS

Raine, Richard. Land surveyor. *Ilford* 1816 MS, *Vicarage Oakley (Bucks.)* 1819 MS

Rainer, Gemma Frisius. See **Gemma Frisius**

Rainerus. See **Gemma Frisius**

Rainold, Carl E. *Lexicon Böhmen* 1836

Rajal y Larré, Jonquin. *Isla de Mindanao* 1881

Raleigh, Sir Walter (1552–1618). *Discovery of Empyre of Guiana* map 1595, *History of World* 1614 (with maps)

Ralph, B. *Portugal e Algarve*, Jefferys 1790

Ralph, W. *Campaign on the Meuse* 1792, N. York 1797

Ram, Johannes de (1648–1693). Engraver, publisher, globe maker and art dealer of Amsterdam. His widow married J. De la Feuille in 1696, who reissued his plates. Ram issued *atlases with maps by various geographers* as well as issuing some maps himself including a *world map*

Ramage, Capt. W. *Balta Sound* 1815, *N. Norway* (1850)

Ramberg, F. *Plan von Elbing* 1804

Ramberger, J. J. *Vestung Gibraltar* 1782

Ramble, Reuben. *Travels thro. counties of England,* 40 maps 1845

Rameau, P. A. Engraver. Bellin's *Isle de France* 1763. Script engraving for Lapie 1817

Ramelleti, Giovani Domenico. *Pianta . . . di Torino* 1756

Ramhofsky, Jiri (end 16th century). Czech globemaker

Ramière, René Augustin Constantin de. *Manila Bay* 1725

Ramirez. See **Alzate y Ramirez**

Ramirez y C. Lithographer for Millan y Villanueva 1891

Ramirez, Capt. Diego. Spanish sailor. *Bermuda* 1603 MS

Ramiro, Carlos. *Plano Topo ciudad de Merida (Yucatan)* 1864–65

Ramis, Giovanni. *Stato di Milano* 1777

Ramm, L. A. *Panoramas* (1850–70)

Ramm, N. A. *Agerhuus Amt.* 1827, *Grevskabernes Amt.* 1832, *Hedemarkens Amt.* 1829

Ramminger, Jacob (1535–1596). Mathematician and surveyor of Stuttgart

Ramon, Juan. *Prov. del Rio de la Plata* 1683

Ramon, Manuel, S. J. (18th century). Spanish mapmaker

Ramon del Moral, Tomas. *Maps Mexico* 1851–2, *Estado de Mexico* 1852

Ramoni, Cesare. *Ferrovie Italiane nel* 1890

Ramos, Bernardo. *Amazones* 1899

Ramos, Manuel. *Japan* 1635

Ramsay, Sir Andrew Crombie (1814–1891). Geologist. *Orograph maps N. & S. America* 1876, *Africa* 1877, *Asia* 1878, *Ordnance Survey Antrim* 1874, *Cumb. & Dumbarton* 1878 &c.

Ramsden, Jesse (1735–1800). Yorkshire mechanic, optician, and engraver. Astronomical instruments

Ramsey, N. A. *Chatham Co., No. Carolina* 1870

Ramsey & Carter. *City of Stillwater, Minn.* (1856)

Ramshaw. Printer, Fetter Lane. *Binns' Plan Lancs.* 1825

Ramus, Joachim Frederick (1686–1769). Danish mathematician and cartographer of Copenhagen. *Delin. Norwegiae noviss.* 1719, *Regni Norvegici delin.* 1753

Ramus, Melchior (1646–1693). Theologian and cartographer of Drontheim

Ramusio, Giovanni Battista (1485–1557). Venetian historian and geographer, Secretary to the Council of Ten in Venice, translator and editor. *Raccolta di Navigationi et Viaggi* Venice 1550–59, editions to 1623, with woodcut maps incl. *first plan of Montreal*

Rancher, Joseph Rosalinde. *Ville et Campagne de Nice* 1825

Rand, McNally & Co. Publishers. Atlases and maps from 1862 to present day, inc. *Business Atlas* 1878, *Cuba* 1882, *Florida* 1885

Randall, A. Draughtsman in Tasmania. *Triabunna* 1863, *Buckland* (1868)

Randall, G. A. *Kent Co., Michigan* 1863, *State Wisconsin* 1865

Rande, Thomas. Engraver and mapseller. No. 326 Oxford St. Koop's *Rhine, Meuse etc.* 1796–7

Randegger, Johannes (1830–1900). Swiss engraver and map lithographer. *Schweiz* (1880), *Alpenland* (1885), *Reisekarte* 1898

Randel, Jesse F. *Texas,* Colton 1839 (with R.S. Hunt)

Randel, John. *Delaware Railroad* 1836

Randolf, J. A. *City Pekin, Illinois* 1872 (with T. King)

Randolph, Bernard. *Greece* (1665), *Morea* (1686)

Randolph, L. B. *Jacksonville, Missouri* (1860)

Randon. Engraver. For Michelot (ca. 1720), *Marseilles* 1730 and 1743

Randon, Carlo. Architect. *Citta et Territ. Torino* (1800)

Rands, W. H. *Geological Maps goldfields* 1896–9

Rango [Rangonis] . *Hellas seu Graecia* (1680)

Rankin, Charles. *Counties Grey & Bruce* 1855

Ransom, Leander. *California* 1862, 1863

Ransonnette, N. Engraver. *Arabian Desert* 1717, *Grand Cayman* 1727

Ranty, G. I. de. *L'Introducteur à la cosmographie,* Paris 1656–7

Rantzen [Ranzovius, Rantzau] , Heinrich, (1526–1598). *Schleswig Holstein.* Drew and supplied plans for Braun & Hog. 1583–1598; *Denmark* 1585

Raoul, A. M. A. Hydrographer for Dépôt de Marine. *Cadiz Bay* 1807–11, *Bassin d'Arcachon* 1817, *Côtes de France* 1818

Raphelengius, Francis. Printer of Leiden. See also **Plantin.** Printed Waghenaer's *Spieghel* 1586, 1588, and *Thresoor* 1592 on his own

Rapilly. See **Esnauts & Rapilly**

Rapin-Thoyras, Paul de (1661–1725). Maps to Tindal's *History of England* 1730–51

Rapin, Conrad **& Co.** Pubs. *Washington*

City, Arrowsmith & Lea's *General Atlas* 1804

Rapkin, John. Drew and engraved maps for Tallis 1845–51

Rappard, F. von. *Topo. Maps and plans of Germany* (1851–77), *Mainz* 1851, *Environs Berlin* 1868, *Düsseldorf* 1877

Rappenhoener. *Stadt Neuss* 1873

Rasch, Jacobus. *Gronlandiae Antiq.* 1706–15

Rasche, Emil. *Historischer Atlas* 1874

Rascher, Charles. Publisher in Chicago. *Atlases parts of Illinois* 1873–90, *Michigan* 1891, *Kansas* 1891–3, *Chicago suburbs* 1885

Rasciotti [Rascicotti] , Donato of Brescia. Publisher in Rome. *Stampata in Borgo di Roma Parigi* (1575), *N. & S. America* (1590), *Brescia* 1599, *Teatro . . . citta del Mondo* (1600)

Raspe, Gabriel Nikolaus (1712–1785). Publisher and bookseller of Nuremberg. *Schauplatz der gegenswaertigen Kriege* 1757–64, *Geschichte der Kriege* 1776

Rassau & Michaud. St. Louis. *Rocky Mountains* 1840

Rastowiecki, Edward (1805–1874). Historian and author of annotated list of maps of Poland entitled *Mappografia* pub. 1846

Rastell, John (d. 1536). Printer and lawyer. Master of the Revels under Tudors. Cosmological pageants *New Interlude of the Nature of the Four Elements* (ca. 1519), the first English work on contem. geography

Rastrick, John U. *Manchester-Cheshire Railway* 1836

Rastrick, William. *Plans Kings Lynn* 1725

Ratelband [Ratelbant] , Johannes (1715–1793). Amsterdam theologian and publisher, "op den hoek van de Kalverstraat aan den Dam." *Geographisch-toneel* 1732, *Kleine en*

Beknopte Atlas 1735, *Geolog. Tooneel* 1735

Ratdolt, Erhard (d. ca. 1528). Publisher of Augsburg and in Venice. *Pomponius Mela's situ orbis* 1482

Rath, Gerh. de (1830–1888). Geologist. Studies on Volcanoes

Rathborne, Aaron. *The Surveyor* 1616

Ratzel, Friedrich (b. 1844). Geographer and anthropologist. *Anthropogeographie* 1882–91, *Verein. Staat N. Amerika* 1878–80

Ratzer [**Ratzen**], Bernard. Lieut. 60th Royal American Regt. *Draught Passmaquody Bay* 1756, *Plan part Oneida Lake* 1760 MS, *Plan Fort Schuyler* 1760 MS, *Plan Fort Pitt* 1761, *Plan Indian Countrys* 1765 MS, *Prov. New York*, Faden 1776, *City New York* 1776, *Province of New Jersey* 1777

Rau, C. von. *Preussischen Staate* 4 sh. 1828

Rau, Johann Jakob (1715–1782). Theologian and astronomer of Ulm

Rau, Karl Ferdinand von (1783–1833). Major Director, Topographic Bureau Berlin

Rauben, Johannes Andreas. *Lindau* Blaeu 1662

Rauch[en], Johannes Andreas (fl. 1608–35). German cartographer and painter. *Wangen & Lindau* 1611–28, engraved 1643–47, *Plan Reichenbach* 1628

Rauchen, Mary Magdalen. Publisher in Munich for Scherer 1702–3

Raulin, Victor (b. 1815). *Crete* 1845 (pub. 1868), *Carte géolog. Environs Paris* 1865, *Dépt. Yonne* (1867)

Raumer, Carl von (1783–1865). *Part Silesia &c.* 1811, *Palaestina* 1844 (with Stulpnagel)

Raupach, Johann Friedrich (1775–1819). Mathematician and physician

Raus, Muhammed. Portolan maker. *Archipelago* 1590 MS

Rausch, I. Engraver of Nuremberg, for Güssefeld-Homann. *N. America* 1797, *America* 1796

Rausche. *Maps of East Prussia* (1845)

Rauw [**Rauis**], Johannes (d. 1600). *Cosmographia* Frankfurt 1597 (with Quad's maps), editions to 1672

Ravell, Anthony. *St. Christophers*, Sayer 1770, 1775, L. & W. 1794

Raven, J. Estate surveyor of Terling. *Terling & Hatfield Peveral* 1774 MS, *St. Laurence* 1778 MS

Raven, Thomas. *Plat Lands at Mounemore Londonderry* 1622 MS, *River Lough Foyle & City Londonderry* 1625 MS

Ravenel, Henry. Surveyor. *Charleston & Beaufort District, S. Carolina* 1820 with Vignoles, for Mills 1825

Ravenna. Geographer. Anonymous geographer of 7th century. *World and Egypt*

Ravenshaw, W. *Plan Madras* 1824

Ravenstein, E. G. *S.W. USA* 1869

Ravenstein, Ernest George (1834–1913). *Abyssinia* 5 sh. 1867, *Africa* 3 sh. 1863, *Training College Atlas* 1880, *Phillips Systematic Atlas* 1894, *Phillips Handy Vol. Atlas* 1895

Ravenstein, Friedrich August. (1809–1881). Cartographer and publisher of Frankfort. *Umgegend Frankfurt* a. M. (1845), *Frankfurter Gebiet* 1851 and 1853

Ravenstein, Hans. *Schweizer Alpen* 1897

Ravenstein, Ludwig. *Deutschland* 4 sh. 1866, *Railway map Germany* 1869, *Taringen* 1869, Meyer's *Hand Atlas* 1866

Ravenstein, Simon. *Plan Frankfort* 1867

Raville, Huyn de. Capt. Eng. *Albanie* (1820) MS

Ravis, J. See **Rauw**

Ravn, Niels Frederik. *Popul. map Danish Monarchy* 1845

Rawdon, R. Engraver. *N.W. Territories, U.S.* 1821

Rawson, A. L. *Palestine* 1869

Ray, Lieut. F. J. *Balabac Strait* 1868–9, pub.–70; *Canton River, Ladrone Islands* 1868

Raymond, John. *Mercurio Italico* 1648

Raymond, J. B. S. *Mont-Blanc* (1800), *Carte Topo. Milit. des Alpes* 1820, *Lombardie Vénétie* (1865)

Raymond, Rossiter W. *Geolog. Map U.S.* 1873

Raynal, Guillaume Thomas François, S. J. (1713–1796). Philosopher and author. *Atlas Portatif* 1773 became *Atlas de toutes les parties connues du Globe* (1780); *Atlas des Deux Indes* (1780)

Raynaud. For Dufour (1864)

Raynolds, W. F. *Yellowstone & Missouri Rivers* 1860

Razaud. Royal Engineer. *Plan Marseilles*, pub. J. B. Benoit, engraved Randon 1743

Re, Marc Antonio dal. Publisher in Milan. *Chafrions Liguria* 1685, *Regno di Corsica* 1731

Re, Sebastiano di [Sebastinus de Regibus Clodiensis] (16th century). Engraver. *Galliae Belgicae* (1558), *Crete* (1559), *Galliae descrip.* (1558). *Britann. Insulae* (1558), *Nova descrip. Hungariae* 1559

Rea, Roger, the elder and younger (fl. 1660–1667). Booksellers and publishers at (1) Golden Cross in Cornhill; (2) Gilded Cross in Westminster Street near Greshams College. Published edition of Speed's *Atlas* in 1662 and 1665, and miniature edition 1668

Rea, Samuel M. *Newcastle Co., Delaware* 1849, *Alleghany Co.,* 1851, *State Delaware* 1850, *Genesee Co., N.Y.* 1853

Read, Mr. Surveyor. *Land Stowe Kiln* MS (19th century)

Read. John. Surveyor. *Ham Green* 1846 MS

Read, Mary. *Map India* 1857, *North West Provinces* 1857

Read, R. P. *St. Helena* 1815

Read, Thomas. Publisher, Dogwell Court, White Fryars Fleet St. *English Traveller* 1746

Read, T. Surveyor. *Herefordshire* 1753, *Manor Boarstell* 1817 MS

Read, William. *Chart Boston Deeps* (1847) (with S. Hacking)

Read, W. H. See **Harding Read**

Read. See **Rice & Read**

Reade, Thomas. Surveyor. *Glebe Hillesden* 1847 MS, *Margins Buckinghams Seat at Wotton* 1789 MS

Reader. See **Longmans**

Ream, Robert L. *Nebraska Territory* 1857, *Kansas* 1867

Rease, William H. Lithographer printer, N.E. corner 4th & Chestnut St., Phila. *Harkness Railway Map* 1860, *Pratt & Buells Central City* 1862, *Oil Territories Gaspé Bay* 1864, *Anthracite Coal Fields* 1864, *Plymouth Rock, Penn.* 1862

Rebein, J. P. *Regno di Corsica* 1731

Reben, Johann Baptist von. Hungarian engineer. *Bosphorous,* Homann Heirs 1764

Reber, Rudolf. *S. Tyrol* 1862

Rebolledo, Juan de Olivan. See **Olivan, Rebolledo**

Rebuelta, Joaquin Da. MS map *Chulumani Boliva* 1810

Recco, Nicolosa da (14th century). Italian hydrographer

Reckwill, Wolfgang (1574–1582). Goldsmith and cartographer. *Buchavia & Fulda* for Ortelius 1574

Reclam, C. H. *Railway map Germany* 1871

Reclus, Jean Jacques Elisée (1830–1905). French geographer. *Nouvelle*

Géographie Universelle, 19 vols., Paris, Hachette 1876—94

Reclus, Onésime (1837—1916). Brother of above, geographer

Record, Robert. Cosmographer and mathematician. *Pathway of Knowledge* 1551, *Castle of Knowledge* 1556

Rector, Henry M. *Survey Arkansas* 1856

Rector, William. Surveyor of U.S. for Missouri and Illinois. Supervised Roberdean's *Western U.S.A.* 1818 MS

Recupero, Giuseppe. *Carta di Mongibello* 1815

Recutti, A. Engraver for Seminario Vescovile, Padua 1699. *Empire Charlemagne* 1697

Redding, T. B. and Watson. *Racince Co., Wisconsin* 1858

Reden, F. W. von. *Railway map Germany* 1847

Redfield, W. C. (1789—1857). American geographer

Redford, John. *Straits Malacca.* Used by Laurie & Whittle 1802

Reding, H. *School Atlas* 1846, *Plan de la Haye* (1872)

Redknap, J. *Casco Bay Marblehead Fort* 1705, *Salem Port* 1705

Redman, E. *Worthing & Vicinity* (1830)

Redmayne, Wm. Mapseller. *Playing cards* (1676), (1677)

Reed, Abner (1771—1866). American engraver. *Connecticut* 1812, Kilbourne's *Ohio* 1820 and 1831

Reed, A. L. & Co. Publisher and mapseller N.Y. and Chicago *California & Nevada* 1863

Reed, Hugh T. *Pocket Atlas* 1889

Reed, John. *City & Liberties of Philadelphia* Philadelphia 1774

Reed, J. G. *Ship Canal. Lakes Erie & Ontario* 1835, *Canal round Niagara* 1835

Reed, J. W. Staff Commander. *Singapore Roads* 1865, *Rhio Strait* 1869, *Singapore Main Strait* 1876

Reed, Capt. Mackey. *Bay of Mayumba in Norris's Africa Pilot* (1804)

Reed, Theron. *Silver Mt. Mining Districts (Cal.)* 1864

Reed, Thomas H. *Putnam Co., N.Y.* 1876

Reed, William. *New England* (1685) MS

Reed & Barber. Publishers, Hartford, Conn. *U.S.A.* 1849, 1850

Reekes, Thomas. *Town & County Poole* 1751

Reeland, Adrian (1676—1718) Theologian and orientalist in Utrecht

Rees, Abraham (1743—1825). Welsh mathematician, encyclopedist, and map publisher. *Cyclopaedia,* Phila. (1806), London 1819

Rees, Capt. *Northumberland Straits* 1783

Rees, James H. *Chicago* 1849, *Co. Cook & Dupage* (1860)

Rees, John. Admiralty Surveyor. *Cunsingmun (Canton River) Port Ta-Outze* 1832, pub. 1840

Rees, Thomas. *River Min (China)* 1840, *Port Taoutze* 1840

Rees. See **Longmans**

Reese, George W. *City Buffalo* (1872)

Reeuwich, Erhard (15th century). See Reuwich

Reeve, Benjamin. Estate surveyor 1751

Reeve, James. Printer, London. Smith's *New England* 1614

Reeves, J. Drew *plan Lambeth Palace* 1750

Reeves, Edward Ayearst. Sutton's *Geogr. Cards* 1892

Reeves, J. Engraver. *Hodgson's 15 miles round London* 1823

Reeves, James. *Makiloto Bay, Finland* 1810

Reeves [& Hoare]. See **Hoare & Reeves**

Reeves. Engraver for Murray. *Eng. Atlas* 1831

Regazzoni, Giocondo. *Nuovo Planisfero* 1860

Regelmann, C. (1842–1920). *Ludwigsburg* (1891), *Würtemberg* 1891, 1893

Régemortes. Family of engineers and surveyors: Antoine de (1723–45); Jean Baptiste de (d. 1725); Noel de (ca. 1710–1790); the Elder Louis de (18th century); the Younger Louis de (1715–1776)

Reger, Johann (end 15th century). Printer of Ulm. *Ptolemy* 1486

Regibus, Sebastianus a. See **Re, Sebastiano di**

Regiomontanus, Johann Müller [Joh. de Monteregio] (1436–1476). Astronomer and mathematician of Nuremberg. *Tabula directionum* 1475

Regis, Jean Baptiste, S. J. (1663–1738). Geographer and missionary to China. *Map China* 1718

Regler, Ludwig Wilhelm von (1726–1792). Military cartographer

Régley, Abbé. *Atlas Chorographique* 1763

Regner [Regnier] . See **Gemma Frisius**

Rego, Luis do (16th century). Portuguese cosmographer and cartographer

Rego, Mateus do. (16th century). Portuguese cartographer

Regowill [Regewill] . See **Reckwill**

Regewill, Wolfgang. See **Reckwill**

Rehlin the Elder, Philip (ca. 1550–1598). Cartographer

Rehlin the Younger, Philip (1569–1605). Painter and cartographer of Ulm. Son of the preceding

Reich [Reych] , Erhard [Erhart] of Tyrol. *Map of the Palatinate Bavaria* pub. Zell, Nürnberg 1540. Used by Ortelius 1570 and De Jode 1578

Reich, G. See **Reisch**, G.

Reichard, Christian Gottlieb Theophil (1758–1837). Cartographer. *Atlas des ganzen Erdkreis* Weimar 1803, *Neuer Hand Atlas* 1832, *N. America* 1802, 1813, 1826; *Orbis tem. anti.* 1824, *Orbis terr. veter.* 1830

Reichard, Friedrich von. *Russia* 1821, *Turkey* 1823

Reichard, Heinrich August Otto (1752–1828) of Gotha. Writer and atlas publisher, brother of preceding. *Atlas Portatif* 1818–1821

Reichardt, Lieut. *Environs Berlin* 10 sh. 1816–19

Reiche, Ludwig von (1775–1854). Military cartographer in Berlin

Reichel, L. T. (1812–1878). *Chart coast Labrador*, 1862, *Missions Atlas* 1860

Reichelt, Julius (1637–1719). Strassburg mathematician. *Germany* 1658, *Holy Roman Empire* (1690), *Germany* 1703

Reichenbach, Ludwig (1793–1879). Geologist and botanist

Reicherstorffer [Reichersdorff] , George von. Historian and geographer. *Crimea* Cologne 1595

Reichert, Lieut. *Strassburg & neighbourhood* (1873)

Reid, Albert. *Walsh's plan Hobart Town* 1870, *Torquay (Tasmania)* 1863

Reid, Alex (1802–1860). *Mod. Geog.* Edinburgh 1837, *School Atlas of Mod. Geog.* 1848

Reid, Andrew. *York-Berwick Railway* (1848), *Newcastle & Gateshead* 1879, *Plan Newcastle* 1896, *River Wear* 1898

Reid, John. *Otago River, New Zealand* (1860)

Reid, John. *Mapp of Rariton River* 1683. The earliest known copperplate map engraved in America

Reid, John. Publisher. *American Atlas* New York 1796. The first general atlas engraved and published in USA

Reid, J. Engraver for Stockdale. *Hunter's Plan Port Jackson* 1789

Reid, W. H. and J. Wallis. Publishers. *Panorama or Trav. Instructive Guide* (53 maps) 1820

Reidig, M. German cartographer. Maps in Stein's *Neuer Atlas,* Leipzig 1834

Reiffenstuell, Ignaz, S. J. (1664—1720). Austrian theologian and map editor

Reigart, J. Franklin. *Plan Lancaster City, Penn.* 1848

Reilly, Franz Johann Joseph von (1766—1820). Viennese art dealer, map maker and publisher. *Schauplatz der Welt Atlas* 1789—91, *Deutscher Atlas* 1796, *Atlas Univ.* 1799

Reimer, Dietrich (1818—1899). German publisher, Berlin. Numerous geographical publications. Kiepert's *Hand Atlas* 1857 and 1878

Reimer & Olcott. *Insurance Map city Orange N.J.* 1875

Reim, Melchior. Engraver of Augsburg. *Planisphaerium Coeleste* (1730), *Germany* 1730, *Augsburg* 1745, *Louisiana* 1745

Reinagle, H. *Fort Ticonderoga* 1818

Reinaud, Joseph Toussaint (1795—1867). French Orientalist and writer on maps. Translated *Geógraphie* of Aboulfeda

Reinecke, Johann Matthias Christoph (1768—1818). Cartographer of Halberstadt and Coburg. Cartographer to Geog. Inst. in Weimar. *Iceland* 1800, *Australien* 1801, *Africa* 1802, *Nieder Guinea* 1804, *Turkey* 1807

Reineke, M. F. *Lapland* 1855

Reinel. Family of Portuguese cartographers

Reinel, Jorge (ca. 1518—1572). *Iceland* 1519, *Atlantic* (1540). Son of Pedro R.

Reinel, Pedro (fl. 1485—1535). Royal cartographer. *W. African Coast* (1485), *N. Atlantic* (1516), *World* 1522. With

Jorge above and Lopo Homen prob. joint author of the Miller *Atlas* (1520)

Reiner, Ignatius. *Strait Gibraltar* 1824

Reiner, Johan Alexander (17th century). Hungarian engineer. *Hungary* 12 sh. 1683

Reiner, V. L. See **Reinner**

Reinerus. See **Gemma Frisius**

Reinhard, F. *Plan Munich* 1829

Reinhard, Johann (1480—1528). German bookseller and publisher

Reinhard, Karl von (1798—1857). Cartographer. *Kamtschatka* 1838

Reinhard, Magister of Prague. Astronomer (early 15th century)

Reinhardt, Andreas. Engraver of Frankfurt. *Plan Cartagene* (1741), *Attack Weissenburg* (1744), *Siege Tournay* (1745), *New Scotland* 1750

Reinhardt, Carl. *Helgoland* (1871)

Reinhardt, I. von. *Plan Cologne* 1752

Reinhart, T. A. *Ysla de Menorca* (1783)

Reinhart, Johann Ludwig (1684—1747). German mathematician, surveyor and military cartographer

Reinhold, C. L. *Plan Osnabrück* 1767

Reinhold, Johannes (fl. 1584—92). Instrument maker of Augsburg. *Celestial & Terrestrial Globes* 1586—88

Reinhold the Elder, Erasmus (1511—1533). Astronomer

Reinhold. See **Roberts, Reinhold & Co.**

Reiner. See **Gemma Frisius**

Reinke, J. T. *Rivers Elbe & Weser* 1795, *Zeekaart Helgoland* 1787

Reinner [Reiner] , Vadar Laurence of Prague. Designed cartouche for Müller's *Bohemia* 1720

Reinold, John. Editor *Pomponius Mela* 1739

Reinoso, José. *Prov. Toledo* 1877, *Teatro de Guerra de Oriente* 1876

Reinsberger, Johann Christoph (1711–1777). Painter and engraver in Vienna for Homann's *Silesian Atlas* 1752

Reis, A. M. dos. *Porto de Lisboa* 1878, *Plano hydrografico Porto do Rio Guadiana* 1881, *Oceano Atlantico Norte* 1887

Reis, A. P. dos. See **Pereira dos Reis**

Reis, Piri. See **Piri Reis**

Reis, Seydi Ali (d. 1561). Turkish astronomer and navigator

Reisch, Gregorius (ca. 1470–1525). Mathematician and cartographer, Abbot of Carthusian Monastery, Freiburg. *World maps* in various editions in *Margarita Philosophica* 1503, 1504, 1513

Reisch, Johannes. See **Ruysch, J.**

Reisner, Adam. *Plan Jerusalem* 1559

Reiss, R. *Wandkarte Rhein-Provinz* 1875, *Wandkarte Deutschland* (1877)

Reiss, William (b. 1838). Geographer and traveller. *Tenerife* 1867, *Süd Colombia* 1899

Reisser, Franz (1771–1836). Viennese engraver. *Post map Germany* 1801, *Moravia* 1802

Reitmayer, I. *Plan Regensburg* (1830)

Reitzenstein, Carl von. *Plan citadel Antwerp* (1833)

Reitzner, V. von. Cartographer. *Greece* 1880

Reland [Relandus], Adrien [Hadriano], (1676–1718). Cartographer and Orientalist. *Maps of Ceylon, Palestine, Persia, Java & Japan* 1705–35

Rembielinski, Eugeniusz & Juliusz (ca. 1814–1860). Polish lithographers. French departmental maps: *Canton Giron* 1845, *Ville de Lyon, Environs de Constantine* (1850), 1847

Rembielinski, Jan (18th century). Military surveyor involved in work on Polish-Russian demarcation line

Remezov, Semjon Ulanovic (ca. 1662–1715). Historian and cartographer. Redrew Witsen's *Map of Siberia* in 1672, *Atlas Siberia* 1698–1701 MS

Remezov, Ulijas. Father of S.U. Remezov. *Siberia* (with Godunov) 1667 MSS

Remington, George Jr. Engineer. *Colchester-Harwich Railway* 1840 MS, *Dagenham Dock* 1854

Remington, T. J. L. *Co. Winnebago, Illinois* 1859

Remmey, John. *Disposition of the English and French Fleets Aug. 1, 1798*

Remond, V. *War in Germany & Italy* 1866, *Central Europe* 1870

Remondini, Antonio (18th century). Venetian publisher and cartographer

Remondini, Giuseppe Antonio (1747–1811). Venetian cartographer and publisher. Reissued Santini's *Atlas Univ.* (1784), *Atlas Geogr.* 1801

Renard, Charles. Plat U.S. Surveys. *Saint Louis (Missouri)* 1859

Renard, Louis. Publisher and bookseller of Amsterdam. Married daughter of D. de la Feuille. *Atlas de Navigation et du Commerce* 1715, Reissued R. & J. Ottens 1739 and 1745

Renat, Johan Gustaf (1682–1744). Swedish military cartographer. *Dzungaria* 1738

Renaud, H. Belgian globemaker. *Terrestrial Globe* 1835

Renaudus. *Gennep* pub. Visscher, 1641

Renault, Jn. F. *Plan Yorktown* 1825

Rendel, James Meadows. *Charts Torbay* 1836, *Plan Docks Birkenhead* 1844

Rendziny, Stanislaus. *Carte von Polen* 1810

Rennell, Major James (1742–1830). Surveyor General to East India Company, Bengal 1764–1777. Drew maps for Clive of India. *MS Atlas* 1765–6, *Bengal Atlas* 1779, *Hindostan* 2 sh. 1782, *Bengal* 1776, *Delhi Agra Oudh* 1794, *Geogr. System Herodotus* 1800, *W. Asia* 1809–11

Renner, F. *Karte von Africa* 1845

Renner, Lieut. L. *Atlas Hildburghausen* 1841 (with Radefeld)

Renner de Hailbrun. Franciscan publisher of Venice. *P. Mela's De situ orbis* 1478

Rennie, George (1791–1866). Civil engineer, partner with his brother John. *Map country between Liverpool and Manchester* 1826

Rennie, James. *Chittagong River* 1842, *W. Coast India* 1866

Rennie, John (1761–1821). Civil engineer. *Prop. Navigation from Bishop's Stortford* 2 sh. 1790, *Proposed Lancaster Canal* 1792, *Proposed Rochdale Canal* 1794, *Plan Clyde* 1807, *River Nene & North Level* 1813, *Part Dublin Bay* 1820

Rennie, Sir John (1794–1874). *Plan Belfast Harbour* 1852, *Dagenham Dock* 1860

Renou. *Régence de Tripoli* (1850)

Renouard, Antoine Augustin. Publisher, Rue St. André des Arts No. 55, Paris. Gauthier's *Amérique* (1820)

Renouard, Vve Jules. Rue de Tournon 6, Paris. *Atlas sphéroidal & Univ.* 1860

Renshaw, James. *Fingringhoe* 1815 MS, *Railway survey Essex* 1835 MS

Renshaw, John. *Survey British Channel* 1743

Rensz, Alois. *Railway map Austria Hungary* 1895

Rentone, James, of Charing Cross. *Plan Porto Bello* 1740, *Plan City & Harbour Havana* 1760, Homann *Portus Pulchri Panama* (1762)

Repton, Humphrey (1752–1818). Landscape gardener and surveyor, Hare Street near Romford Essex. *Stansted Hall* 1791 MS, *Calford Suffolk* 1791–2 MS, *Barton Seagrave Northants.* 1793–4 MS, *Woodford* 1801 MS, *Design for Brighton Pavillion* 1805

Requena, Don Francisco. *Amazon* MS maps 1788–9

Rerick Bros. Publishers. *County maps of Ohio & Indiana* 1893–96

Resen, Hans Poulsen. *Map Atlantic* 1605, *Greenland* 1605

Resen, Peder. (1625–1688). *Atlas Danicus* Iceland 1666, *Bornholm* 1677, *Faroe Is.* 1684

Resniczek, Franz. *Stadt Salzburg* (1869)

Restorff, C. von (1783–1859). *Mecklenburg-Schwerin* 1858

Restrepo, José Manuel (1780–1864). *Historia de Colombia* 1827

Retter, E. *Bayern* 5 sh. 1892

Reuschle, K. G. *Neuer Schul Atlas* 1862, *Volks Atlas* 1875

Reuss, F. H. & J. L. Browne. Surveyors and Architects, 134 Pitt Street, Sydney. *N.S.W.* (1862), *Subdivisions in and about Sydney* (1856)

Reusse, H. *Kurhessen* 13 sh. 1839, *Kurhessen* 1853 (with A. Schwarzenberg)

Reuter, Christian Gottlieb (1717–1777). Moravian surveyor, b. Steinbach, Germany. Royal surveyor to Prince of Hanover, emigrated to America 1756, d. Salem. *Wachau N. Carolina* (1760) MS, *Wachovia or Dolbs Parish* 1766 (MS)

Reuter, C. *Helsingsfors Stad* 1869, *Elsas-Lothringen* 1896

Reuvens, Casparus Jacobus Christianus. *Nederland* 1845

Reuvens, L. A. *Polder District Guelderland* 1871

Reuwich [Reych], Erhard of Mainz. Publisher and illustrator of Breydenbach's *Peregrinations* 1486, *Arabia, Egypt, Palestine* 1484

Revell, (Anthony) St. Christopher. *London,* Jefferys (1768), 1775, and 1794

Revell, William. *North America . . . boundaries Ontario* 1878

Reverley, W. *Plans for improvement to Thames* 1796–1803

Revillas, D. *Marsovum Dioecesim* 1735

Rewich, E. See **Reuwich**

Rey, E. G. *Nord de la Syrie* 1885

Rey, Marc. Michael. Publisher of Amsterdam. Muller's *Russian Discoveries (N.W. Passage)* 1766

Rcych, E. See **Reuwich**

Reyes, José Maria. Col. of Engineers *Uruguay* 1846 and 1893

Reyher, E. Engraver for Kiepert 1860

Reyher, Ernst (fl. 1825–49). Engraver

Reyher, Ferdinand Julius. Engraver. *Hohnstein u. Schandau* 1830

Reymann, D.G. *Russisches Reich* Berlin 1799–1802, *Hesse* 1804, *Estland* 1812, *Gegend um Berlin* 1818, *Grundriss v. Berlin* 1835

Reymann, Gottlob Daniel (1759–1837). Military cartographer

Reynard, Joseph. *Dépt. Puy-de-Dôme* 1881

Reynold, Nicholas (fl. 1570–80). Engraver

Reynolds, G. S. *Harbour of Kittie (Ascension)* 1839 and 1845

Reynolds, James. Publisher 174 Strand, London. Booth's *Plan London* 1845, *Travelling Atlas England* 1848, *Suffolk* 1850, *Astronomical Diagrams* 1851, *Geological Atlas G. B.* 1860, *Portable Atlas England & Wales*. Business continued as James Reynolds & Sons

Reynolds, John. *Map of Georgia* (1756)

Reynolds, Nicholas. English engraver and draughtsman. Saxton's *Herts.* 1579. *Portolan of Mediterranean* 1612

Reynolds, R. Engraver for Palairet 1760, *Denmark* 1760

Reynolds, T. *Roman Britain* 1798

Reynolds, W. H. *Chart of Thames Estuary* 2 sh. 1887

Reynolds Moreton, Henry John, Earl of Ducie. *Plan Malta and Gizo* 1856

Rhaeticus [Lauchen] , Joachim Georg (1514–1574). German astronomer, mathematician and cartographer. Pupil of Copernicus. Professor at Wittenberg and Leipzig. *Map of Prussia* (now lost) ca. 1521, *Guide to Surveying* 1540

Rhaetus, P. William C. *Danube*, Wagner 1685

Rhea, Matthew. *Tennessee* 1832

Rheim, M. Engraver for Lotter's *Atlas* 1778

Rhein, Heinrich von (fl. 1810–20). Cartographer

Rheinhard, Hermann. *Classical maps* (1868–78)

Rhenanus, Beatus. Supplied material for Münster's *Alsace Brisgoia* 1540

Rhigas (1753–1798). Historian and cartographer

Rhijn, A. van. *Marine maps & charts Holland* 1853–70

Rhind, William. Philips *Commercial Atlas of the World* 1856, *Treatise Physical Geog., Elements of Geology*

Rhinewald, J. S. *Theatre war on Rhine &c.* Mannheim 1794

Rhins, J. L. D. de R. See **Dutreuil de Rhins**

Rhode, C. E. *Hist. Schul Atlas* 1861 and later editions

Rhode, Johann Christoph (1713–1786). Cartographer, Berlin. *Mappa Mundi* (1745), *N. & S. Hemispheres* 1753, *Theatrum Belli in America Sept.* (1755)

Rhode, Jorge J. *Republica Argentina* 1889

Rhode, L. I. *St. Croix* (1800), *Harbour St. Thomas (Virgin Is.)*, Copenhagen 1815

Rhoden, F. C. von. *Berlin* 1783

Rhodes, Joshua. Surveyor. *Tobago* 1760, *Parish Kensington* 1766, *Mickleham* 1770 MS, *St. James Westminster* 1770

Rhodes, S. Land surveyor of Wellington St., Strand. *Tunbridge Wells* 1828 and 1838, *Islington Parish* 1828

Rhodes, William. *Parish St. James Westminster* 1770

Rhodes. See **Dobson & Rhoades**

Rhys, E. L. Supplied information on *Brit. Honduras* used by Abbs 1867

Rhys, Morgan John. *Plan Beula Penn.* (1797)

Ribary, Ferenc (1827–1880). *School maps*

Ribaut [Ribault], Jean (ca. 1520–1565). Navigator of Dieppe. Killed in Florida. *French Colony in Florida* 1562

Ribbius, J. *Geogr. Hist. Wereld Besch.* 1683

Ribero [Robeiro], Diego [Diogo] (d. 1533). Portuguese Royal Cosmographer in Spanish Service; instrument maker and pilot. Voyages to India as pilot. *Carta Universel* 1529 MS, *Africa* 1529, *America* ca. 1532

Ricart, Giralt, José (b. 1847). Spanish hydrographer

Ricart, Robert. *Plan Bristol* (oldest known) 1479

Ricarte, Hipolito. Letter engraver of Madrid 1770–5

Ricaud de Tirregaille, Piotr. *Plan Varsovie* 4 sh. 1762

Ricaudy, T. de. *Golfe de Mexique* 1862

Ricci, Matteo, S. J. (1552–1610). Missionary and cartographer. Apostle of China. *Map China* printed in Shao-King 1584, Nanking 1599, Pekin 1602 (woodblock 12 X 14 feet), *Mappemonde* 1602

Riccioli, Giovanni Battista (1598–1671). Italian astronomer and geographer in Bologna. *Systema Mundi Tychonicum* (1640)

Riccioli, R. P. *Italy* 1742

Rice, E. *Bennington Co., Vermont* 1856

Rice, G. Jay. *Dakota* 1872, 1875;

Minnesota 1874, *Minneapolis* 1884, *Map St. Paul* 1887

Rice, G. Jay & **Read**. Lithographers of St. Paul Corinne. *Utah to Ft. Shaw Mount.* 1870, *Dakota Territ.* 1872

Rice, William McPherson. Admiralty Charts: *Algoa Bay* 1801, *Mossel Bay* 1801, 1806; *Plettenberg Bay* 1801

Rich, Barnby. *New Descrip. Ireland* 1610, *News from Virginia* 1610

Rich, John. Engraver. *Grand Roads of England* 1679

Richard, C. G. *Dépt. de l'Eure* 1867

Richard, Joannes (fl. 1542–56). Antwerp publisher

Richard, Joseph. *Tunbridge Wells* (1856)

Richard de Haldingham [Richard of Belleau, Bellau]. *World map* ca. 1285, owned by Hereford Cathedral

Richards, Bernard. *Plan Waterford* 1764

Richards, Eugen Lamb (1838–1912). American globe maker

Richards, F. D. Liverpool. *Utah* 1855, *Fairview Mining Camp* 1897

Richards, Rear Admiral Sir George Henry (1820–1896). Hydrographer to Admiralty 1863–74. *Survey New Zealand* (with Stokes), *Haro & Roseno Straits* 1858–9

Richards, Capt. J. H., R.N. *Survey China* 1853–58, *Lake Huron* 1865

Richards, John. Surveyor. *Cotley Wood* 1745

Richards, L. J. & Co. *Atlases of Towns & Counties in Massachusetts* 1893–98

Richards, R. *Chart of Chusan* 1840

Richards, Lieut. R.N. *Fiji Islands Admiralty* 1822

Richards, Thomas. Government Printer, Sydney. *Sketch Map New South Wales* 1876

Richards, William (d. 1766). Collaborated with Scalé *City & Suburbs Waterford* 1764, *Ireland* 1762, *Wexford Harbour* 1764, *Chart coast Wicklow Head* 1765

Richards, William Hamilton (1833–1895). Military topographer. *Borough of Dudley* 1865

Richards, W. A. *Murchison Gold Fields* 1889, *Cape Town* (1893)

Richards, W. A. **& Sons**. Printers, Cape Town. Tompkins' *Plan Johannesburg* 1889

Richards. See **Everts & Richards**

Richardson, A. R. *10 miles round Newcastle* 1838

Richardson, Bartholomew. *Burlington* 1831

Richardson, C. S. *Texas* 1860, *W. Virginia* (1862) and 1864

Richardson, D. *Central British Burmah* (1870)

Richardson, F. G. *Hundred of Kooringa (S. Australia)* 1860

Richardson, Gabriel. *State of Europe* 1627

Richardson, Capt. G. G., R.N. *Mocha Harbour* 1795, *Manilla Bay* 1798, *Malacca Strait* 1802

Richardson, George. *Plan Marylebone Park Farm* 1794

Richardson, J. Bookseller. For Walker's *Univ. Atlas* 1820

Richardson, James. Publisher. *Map of Passamaquaddy* 1808

Richardson, John. *Manor of Lye (Surrey)* 1627

Richardson, John, Senior. Surveyor. *Hamlet Seer Green* 1753 MS

Richardson, John. Surveyor. *Estates Manor Cookham* 1762, *Manor W. Wycombe* 1767

Richardson, Sir John. *Chart N.W. Passage* 1853

Richardson, Ralph. *Forests and Mines of Corsica* 1894

Richardson, Richard. Surveyor of *Enfield Chase* 1777, *Manor of Athington* 1798 MS

Richardson, Thomas. Land surveyor. *Chelsea* 1769, *Manor of Richmond* 1771, *Barking* 1781, *New Forest* 10 sh. 1789, *Trav. Map Scotland* 1803

Richardson, William. *Porto Bello,* Overton 1740

Richardson, W. Lithographer. Smith's *Syria* 1840

Richardson, Son & Corfield, Lincolns Inn Field. *Leyton (Essex)* 1807 MS

Richardson, Corfield & Wharton. *Ilford* 1811 MS

Richelieu, A. de., Comm. Siamese Navy. *Admiralty Charts Siam* 1876–77

Richer, L. *Perspective View Arras* Paris, Jollain (1640) *Isle des Faisans* 1659

Riches & Gray. *Map Bexhill* 1895

Richie, John. *Plan New Harbour* 1779

Richie, W. W. American surveyor and publisher 1870–74. *Indianapolis* 1874

Richmond, Mathew. *Map Caribee & Virgin Is.* 1777, *New Map Europe* 1805

Richmond, Van Rensselaer. *New York Canals* 1858–68

Richomme, Joseph Théodore (1785–1849). Script engraver for Tardieu 1819–20

Richter. Draughtsman. *Plan Mainz* 1852

Richter, Eduard (1847–1905). Austrian geographer. *Austrian Alps* 1895–6, *Wörther See* 1897

Richter, Franz Xavier (1759–1840). Military cartographer in Vienna

Richter, Gustav. *Wandkarte Pommern* 6 sh. *Schlesien* 1895, *Würtemberg* 1896

Richter, Johannes (1537–1616). Czech mathematician and instrument maker

Richter, J. C. Engraver. *Berlin* 1811, 1826

Richter, Leopold. *City Peoria* 1857, *Sect. Map State Illinois* 1861

Richter, Wolf. Printer of Frankfurt. De Bry's *Voyages* 1600

Richthofen, Ferdinand, Baron von. Geologist and explorer. *Atlas China* 1885

Rickauer, von. *Umgebungen von München* 1812

Rickeseler, Capt. Eng. *Plan (siege) Belgrade* (1736)

Ricketson, I. *Meetings of Friends (New York)* 1821

Rickey, Joseph M. *Jefferson Co., Ohio* 1856

Ricour. *Dépt. de la Sarthe* 1875

Ricquier. *Partie du Ganges* 1726

Riddell, Jo. Engraver. *Route Moscow to Pekin* 1725

Rider, W. *Atlas World* 1764

Ridge, J. Engraver in Ireland. *County Cork* 1750 for Gillies, *Down* 1755, for Johnson's *Strangford River* 1755, Charlevoix's *Brit. Dominions in N. America* Dublin 1766

Ridgway, J. Publisher in Piccadilly. *U.S.* 1818

Ridley, Dr. Mark. *Magnetical Bodies & motions* 1613

Ridout, Thomas. *Prov. Upper Canada* (1827)

Ridwan, Ibn Muhammud (13th century). Arab astronomer

Riecke, Friedrich. *Kleiner Schul Atlas* 1855

Riecke, Gustaf Frid. Engraver. *Inundation of Rhine Ettlingen-Kethsch* 1734—5. Engraved for Kilian 1738

Riecker, Paul. American cartographer. *County Santa Barbara, Cal.* 1888

Riedel. See **Riedl**

Riedig, Christian Gottlieb (1768—1855). Globemaker and map seller

Riediger [Reidiger], Johann Adam (1680—1756). Swiss surveyor and military cartographer. *Brandenburg* (1710), *Freyen Aemter* 1727, *Globes* 1733

Riedl, Adrian von (1746—1809). Topographer and cartographer in Munich. *Strom Atlas v. Bayern* 4 sh. 1806

Riedl, C. (1701—1783). Mathematician and cartographer

Riedl, J. *Moldau* 1811, *Servia Bosnia* 4 sh. 1810, 1812

Riedl von Levenstern (fl. 1811—40). Mathematician and cartographer

Riegel, Christoph. Map publisher of Nürnberg

Rieger, Albert. *Vienna* (1870)

Rieger, G. *Costa Occid. dell Istria* Trieste 1845, *Isole di Dalmazia* 1863

Riehl, W. (1823—1897). German geographer

Riemer, C. *Ost Afrika* 12 sh. (1892), *Inner Afrika* (1892), *Deutsche Colonien* (1888)

Rienzi, L. G. Demeuy de (1789—1843). *Univers pittoresque. Dict. scient de Géogr.* 1840

Ries, Piri. See **Piri Reis**

Riese, G. H. *Principauté de Halberstadt* 1750

Riesel, Carl. *Grünewald* 1867

Riess, Richard von (1823—1890). *Bibel Atlas* 1887

Rietveld, S. A. G. *Straat Sunda Java* 1855

Rigaud, Isaac. *Maps of Paris & Algiers* 1864—68

Rij, Danckerts de. See **Danckerts**

Rijk, J. C. *Texel stroom* 1816, *Goeree en Maas* 1823, *Schelde* 1825

Rijkens, R. R. *School Atlases of Netherlands* 1872—79

Riley, C. R. *Plan Dublin*, Faden 1797

Riley, I. Publisher. *Map of U.S.* 1810

Rimmington, Capt. Ordnance surveyor. *Waterford* 1841

Rinaldi, Pellegrino Danti de. See **Danti**, E.

Rinewald. See **Rhinewald**

Ringelberg, Joachim. Astronomer of Antwerp. *Instit. Astronomicae* 1535

Ringgold, Cadwalader (1802–1867). *Surveys of the Farallones* 1851, *State California* 1852

Ringgold. George H. Drew Greenhow's *N. America* 1844

Ringle, M. I. I. *Liesthal* 1654

Ringmann [Philesius], Matthias. (1482–1522). Scholar, b. Vosges, studied Heidelberg and Paris. Text for Waldsee-müller's *Europe* 1511, edited Ptolemy's *Geog.* 1513

Ringrose, John. *Plans Rivers India* 1784–1811

Riniker, H. *Atlas Madison Co., Ill.* 1892 (with Dickson & Hagnauer)

Rink, Hinrich (1819–1893). *Map Disko-Fjord Greenland* 1859

Rinteln. *Special Karte Lippe* 1824

Rinzo, Mamiya. Japanese explorer. *Map of Sachalin* (after 1809)

Rio [Cosa J. del Rio], José del. Capt. Spanish Navy. *Havana* 1798 (English edn. 1805), *Part Cuba* 1805 and 1821, *S. Cuba* 1824

Rios, Andrea. *Catalan Portolan* 1612, *MS Atlas* 1607

Rios Coronel, Hermando de los. *Luzon y Hermosa* 1597 MS

Riou, Edward (1756–1801). Sailed with Cook's 3rd voyage. *Harbour of St. Peter & St. Paul* 1779 MS, *Southern Promontory of Africa* (1792)

Ripking, B. Improved Moll's *map of Brunswick Lüneburg*

Rischwin, Simon. Information for Munster's *Eifel* 1550

Risdon, Orange. American surveyor. *Albany* 1825, *Michigan* 1825, *Sagenaw* 1830 MS

Risdon, Tristran. *Choro. Descrip. Devon* 1630 MS

Risdon & Cogger. American engravers. Cornell's *Western States* 1855

Riss, H. See **Rüst**

Ristori, C. *Livorno* 1828

Ritchie, J. Publisher High St., Edinburgh. Collaborated with Canaan & Swinton. Wood's *town plans Scotland* 1823–5

Ritchie, Capt. John. Marine Surveyor to East India Co. *Nicobar Is.* 1771, *Bay Bengal* (with Plaisted) Dalrymple 1772, *Aracan* 1784, *Coast Choromandel* 1784, *Andaman* 1785, *Andaman Is.* 1799

Ritchie & Dunnavant. Lithographers. *Richmond Va.* for Freyhold 1858

Ritsuzo, Tezuka. Japanese geographer. *Japanese version of Colton's Atlas* 1862

Ritter, Carl. See **Ritter**, Karl

Ritter, Ed. *Hamburg Altona Ottensee & Wandsbeck* 1872

Ritter, Francis. *Astrolabium* 1620, *Dials* 1640 and *World Map*

Ritter, F. C. Publisher of Hamburg. *Nord America* (1776), *Broadsheet of Seat of War in America* 1776

Ritter, Karl (1779–1859). Berlin geographer. *Atlas von Asien* 1833–54, *Inner Asiens* 4 sh. 1841, *Palaestina* 1842, *Jerusalem* 1858

Ritterhouse, David. *Plan river Schuylkill* 1773

Rittershus[ius], Nicolas (1597–1670). *Newenmarck* Blaeu 1662, *Franconia* Jansson 1642, *Palatinate* Jansson

Rittmann, André (18th century). French geographer

Rittner, Heinrich. *Dresden* 1800

Rivas, Benito. *Bataan, Luzon* 1848

Rivaux, Auguste. *Plan Tours* (1850)

Rivelli, Francesco Giacomo. *Isola di Corfu* 1850

River, Charles. Copied various early American maps, mid 18th century

Rivera, Julian. *Guatemala* 1842

Rivera, Vicente Talledo. *Prov. de Cartagena* 1820

Rivera Cambus, Manuel. *Atlas Repub. Mexicana* 1874

Riverdi, John Joseph Ulrich (d. 1808). American engineer

Rivière. Engraver. Mariette's *World* 1651

Rivington, Charles. Bookseller at the Bible & Crown in St. Paul's Churchyard. Camden's *Britannia* (1730), Moll's *New Descr. of Eng. & Wales* 1724, 1733

Rivington, F. C. Bookseller and publisher, St. Pauls Churchyard. Walker's *Universal Atlas* 1821, Salt's *E. Africa* 1814

Rivington, J. Publisher, St. Pauls Churchyard. Partner with J. Hinton 1745, partner with J. Rivington 1746. Salmon's *Modern History* 1744–6, Camden's *Britannia* 1753

Rivington, J. Bookseller. Walker's *Universal Atlas* 1820. Salt's *E. Africa* 1814

Rivington, J. & J. See above

Rivius, Gerardus. Printer, Louvain. Wytfliet *Supp. to Ptolemy* 1598

Rizzi-Zannoni, Giovanni Antonio (ca. 1736–1814). Geographer to the Republic of Venice. Born in Padua, studied astronomy and surveying. First to carry out triangulation in Poland. Assisted in survey French-English boundary in N. America 1757. First modern map *Naples* 1769, Chief Hydrographer Dépôt de Marine 1772. *Atlas Kingdom of Naples* 32 sh. 1812, *Atlas Géogr.* 1762, *Atlante Marittimo Sicile* 1792

Roades, William. Engraver. *Pocket Map of Cities of London & Westminster* 1731, Seymour's *History of London* 1733, Foster's *Norfolk* 1739, for Goddard & Goodman's *Norfolk* 1740, *Africa* 1754

Roake & Varty. Publishers. Later **Varty** q.v.

Robaert, Augustin (17th century). Cartographer to Dutch East India Co. *Charts* 1600 MS

Robaut, Félix. *Plan de Douai* 1851

Robb, James. *Geological map of Canada* 1865 (with others)

Robbe, Jacques (1643–1721). Geographer. *Méthode pour apprendre géographie* 1678, *Philippines &c.* for De Fer 1703

Robe, A. W. *Plan Lundy Island* 1820

Robe, Frederick H. *Plan Jaffa* 1841, *Syria* (1846)

Robelin, L. *Carte du Haut Niger* 4 sh. (1890), *Columbia* (1899)

Roberdeau. U.S. Top. Eng's. *Western U.S.A.* 1818 MS, *Sketch Crown Point* 1818 MS (with J. Anderson)

Robert, Merzoff. *Plan de Constantinople* 1829

Robert, Sieur. See **Robert de Vaugondy**

Robert de Vaugondy, Didier (1723–1786). Royal Geographer 1760 and Censor. Quai de l'Horloge du Palais près le Pont Neuf. *Mexico* 1749, *Maps in Atlas Univ.* 1750–57, *Nouvel Atlas portatif* 1784, *Amér. Sept.* 1761. His atlases reissued by Delamarche

Robert, D.

Robert de Vaugondy, Gilles (1686–1766). Succeeded Pierre Moulart Sanson 1730. Géogr. Ord. du Roy, Quai de l'Horloge du Palais proche la rue de Harlai. Sanson's *Guerre Rhein* 1735, *Atlas Portatif* 1748, *Petit Atlas* 1748, *Atlas Universel* (assisted by son) 1757, *Amérique* 1767

Robert de Vaugondy, G.

Roberts. Engraver. *Vermont* 1796

Roberts, Alfred James. *Parish Hammersmith* 1853

Roberts, B. Publisher. *Ichnog. Charles-Town* 1739 (with Toms)

Roberts, Charles. *Chart Gulf Florida,* Faden 1794; *Windward Passage,* Faden 1795

Roberts, David. *Plan Is. Porto Santo* 1802

Roberts, Lieut. Henry. *N.W. Coast America & N.E. Coast Asia* 1784, Drew charts to illustrate Cook's 3rd Voyage

Roberts, I. L. *Coast Labrador* 1822

Roberts, James. *Poland* (1790)

Roberts, John. Engraver and geographer. Taylor & Skinner's *Roads of Scotland* 1776, *Prussia* 1794, for Laurie & Whittle 1796

Roberts, John of Amboina. *Plan of Malloodoo (Borneo)* 1779

Roberts, Lewes. *Merchants map of Commerce* 1638, 2nd Edn. enlarged 1671; *Map Glamorgan* 1620

Roberts, Solomon. *Route N. Penn. Railway* 1857

Roberts, Thomas, Sautelle. *Entrance Waterford Harbour* 1796

Roberts, W. F. *Forest Dean* (1840)

Roberts, W. F. *Anthracite Regions Penn.* 1849, *Plan Schuylkill Valley Coal Co.'s Property* 1863

Roberts, W. J. *Whitman Co., Washington* 1895 (with J.F. Fuller)

Roberts. See **Longmans**

Roberts, Reinhold & Co. Publishers, Montreal. Laurie's map *N.W. Territories* 1870

Robertson. Lithog. with Ballantine. *Railway surveys Scotland* 1825

Robertson, A. *Maps Victoria* 1854–61

Robertson, A. *Road London-Bath-Bristol* 1792

Robertson, George. Hydrographer. *Bay*

Honduras 1764, *MS Tahiti* 1767, *MS Charts China Navigation* 1788, *Chart China Sea* 1791

Robertson, George. Agriculturalist. *Kincardin Shire* 1813

Robertson, H. *Varna & Silistria Railway* (1870)

Robertson, James. *Poland* 1790, *Counties of Cornwall, Middlesex, Surrey in Jamaica* 1804

Robertson, James. Master H.M.S. *Jason. Port Savanilla* 1872

Robertson, James. Surveyor. *Aberdeen* 6 sh. 1822

Robertson, William. *Mexico, Golf van Mexico, Zuid Zee* 1778

Robertson, Wm. Surveyor. *Abbeville District, S.C.* 1820, for Mills 1825

Robijn, Jacobus (d. ca. 1710). Amsterdam publisher, hydrographer and colourist. Worked for Keulen. Published his own *Zee Atlas* 1683, 1688, 1689. Roggeveens *New Enlarge'd Lightning Sea Columne* 168(9); *Atlas de la Mer* 1696

Robin, E. *Dépt. de l'Indre* 1862

Robin, Jacques. *Partie du Terre neuve* 1717

Robin, Paschal. *Descrip. France* 1576

Robinot, La Veuve. Publisher. Quay des Augustins, Paris. Joint publisher with Le Rouge of *Intro. à la Géogr.* 1748 and *Atlas Nouveau Portatif.* Roussel's *Plan Paris* [pre 1748]

Robins, James & Co. (d. 1836). Publisher. Albion Press, Ivy Lane, P'noster Row. *New Brit. Trav. Atlas England* 1819

Robins, J. *Atlas of England & Wales* 1819

Robinson, Arthur. *Chart Tor Bay* (1710)

Robinson, Lieut. (later Capt.) C. G. Admiralty surveyor. Surveys in *Wales* 1833–38, *Scotland* 1838–53

Robinson, Charles Gepp. *Approaches to Liverpool* 1834, *Coasts N. Wales* 1837, *St. George's Channel* 1850

Robinson, D. F. & Co. American publishers. Maps in Olney's *School Atlas* 1829, and later editions

Robinson, Edward. *Plan Jerusalem* 1845, *Relief map Jerusalem* 1858

Robinson, Elisha. *Atlas of city of New York* 1885 (with Pigeon), *Borough of Richmond city N.Y.* 1898, *Atlas of Chicago* 1886, *Jefferson Co.* 1888, *Kings Co.* 1890

Robinson, Eugene. American city surveyor. *Plans of Detroit* 1865, 1876

Robinson, E. L. *New Map of Victoria* 1858

Robinson, Lt. George. Royal Marines. *Monte Video* 1807

Robinson, G. G. [George]. Publishers, Paternoster Row. *U.S.* 1783 (with Dilly), Guthrie's *System of Geog.* 1795, 1799, 1801 (with J. & J. Mawman), Moore's *Chart N. America* 1784

Robinson, G. G. & J. Publishers. Paternoster Row. *Grand Ocean* 1798, *Saldanha, Table & False Bays* 1798

Robinson, I. Bookseller. *Geog. Magna Brit.* 1748, 1750

Robinson, J. J. *Banca* 1819

Robinson, John (fl. 1819). Cartographer. *Saffron Walden district* 1787

Robinson, Brig. Gen. John H. *Mexico, La. etc.* 1819

Robinson, Richard. *South Level of the Fens* 3 sh. 1758

Robinson, Capt. William, R.N. *Gulph Persia* 1781, *Falkland Is.* 2 sh. 1785, *Eagle Is.* 1810, *Falkland Is.* 1841, *S, America* 1851

Robinson, W. Draughtsman. *Ireland* (1700)

Robinson, W. Engraver. *British Possessions in N. America* 1814, *Chili* 1814, *Parish Edmonton* 1819

Robinson, William. Surveyor of Reigate. *Chilmeads* 1778

Robinson. *Victoria, Melbourne* 1862

Robinson, Son & Hudsworth. Pub. of Leeds. Greenwood's *Yorks* 1817

Robiquet, Aimé. Hydrographer. Paris, Rue Pavée St. André 102. *Atlas Hydrographique* (1844–51). *Carte de l'Archipel* 1854, *Mer Méditerr.* 1859, *Australie* 1879

Robiquet, F. *Corse* 1835

Roblet, Desiré. See **Desiré Roblet**

Robotham, Francis. *Chart of the White Sea* 2 sh. 1755

Robsahm, C. A. Swedish cartographer 1797

Robson, Capt. of the *Kent. Plan of Delagoa Bay*, inset on W. Herbert's *Chart of S.E. Coast of Africa* 1767

Robson, M. *Berkeley Sound*, used by Norie, 1822

Robustel, Jean François. Printer and publisher, rue de la Calendre près le Palais, à l'image Saint Jean, *Neptune Oriental*, D'Après de Mannevillette 1745

Robyn. See **Robijn**

Rocayrol, A. *English Plantations in America* (1720)

Rocco dell Olmo. Portolan maker 1542

Rocha, Felix de. *Map of Asia & Europe* (Chien Lung Atlas) 1775

Rocha, J. Joaquim. Cartographer. *Portuguese Brasil* 1809; *St. Catarina, Brazil* used by Faden 1809

Rocha, J. J. da. *Colombia Prima* 1828

Rocha y Figueroa, Geronimo de. *Frontera de Sonora* 1750, *Expd. contre los Apaches* 1784

Rochambeau, Jean Baptiste, Comte de (1725–1807). *Plan de la défense de New Port dans l'Isle Rhode* 1780, *Amérique Campagne* 1782

Roche, C. F. de la. See **Delaroche**

Roche, J. Engraver in Gibson's *Atlas Minimus* 1798

Roche, J. E. F. H. *Map Pamirs* 1893

Rochefort, Albert Jouvin de. *Plan Paris* 1675, 1694, 1697; *Plan de Tolose* (1700)

Rochefort, P. de. Engraver. For De Fer (1746–50)

Rochefoucauld Liancourt, François Alex. Fred. Duc de la. *U.S.* 1799

Roche-Poncié, F. A. J. la. See **La Roche-Poncié**

Roches. See **Brochet des Roches**

Rochette, L. S. d'A de la. See **Delarochette**

Rochlitz (19th century). German cartographer

Rochon, L'Abbé. *Mahe Is.* Sayer & Bennett 1778

Rock & Co. *Illust. Map Brighton* 1851, *London* 1840

Rock, Miles. *Lehigh Valley Railroad* 1868

Rockwell, Capt. Cleveland. American surveyor, end 19th century. *Part thousand Islands of St. Lawrence River* 1875

Rockwell, H. E. *Connecticut for Schools* 1848

Rockwood, C. H. *Atlas of Cheshire Co. N.H.* 1877

Rocque, John [Jean]. Huguenot surveyor, engraver and publisher, fl. 1734–62. Topographer to Prince of Wales. Most notable works: *London* 24 sh. 1746, *Environs* 16 sh. 1746, *Bristol* 1750, *Shropshire* 1752, *Middx.* 1754, *Dublin* 1756, *Armagh* 1760, *Berks.* 1761, *Surrey* 1765, *N. America* 1761

Rocque, Mary Ann. Wife of above continued business. *Environs of London* 1764, *A set of plans & forts in America* 1765, *Surrey* 1770 and 1775

Rodbertus, J. T. *Chart Hawaii Arch.* 1840

Rodd, Jane. Publisher. Reissued Rennell's *W. Asia* 1831

Rodgers, H. E. *Railroad Boston to Providence* 1831

Rodowicz, Theodor. *Helgoland* 1849 and 1856

Rodrigues, Eugenio. *Navig. Coste Settent. ed Oriental America del Sud* 1857

Rodrigues, Francisco (d. 1537). Portuguese hydrographer. *Portolans* (1515–30)

Rodrigues, Joao (16th century). Portuguese cosmographer and cartographer

Rodriguez, Antonio. *Puerto de Apra (Ladrones)* 1796

Rodriguez, Esteban Pilot. *Islas Barbados y Ladrones* 1565 MS

Rodriguez, Eugenio. *Atlante gen. . . . America del Sud* 1857

Rodriguez, Manuel. *Mapa de la America Venegas* 1757

Rodriguez, Nicolas. Engineer. *Plan Gatan Panama* 1745

Rodriguez, Ventura. *Plano de Madrid* 1757

Rodriguez Campomanes y Sorriba, Pedro. *Noticia Geografica del Reyno, y Caminos de Portugal* 1762

Rodtgiesser, Andres. *Trittow, Reinbeeck (Holstein)* 1649

Rodwell, M. M. *Geogr. of Brit. Isles* 1834

Roe, F. B. (1845–1905). Surveyor. *Sunbury* 1869, *Altona* 1870, *Aroostook County, Maine* 1877 (with G. N. Colby), *City Wilkes-Barre Pa.* 1882 (with Alva D. Roe), *Atlas Rockford* 1892

Roe, John Septimus, Lt. R.N. Surveyor General, W. Australia. *Survey Port Jackson* 1822–26, used by J. Arrowsmith 1832. *Cockburn Sound* 1840, *Exped. Perth to Russell Range* 1848–9, published Arrowsmith 1852

Roe, Nathaniel. *Tab. Logarithmicae* 1633

Roe, Sir Thomas. Supplied information for Baffin's *Map of Moghul Empire* 1619

Roe Bros. *Maritime Prov. Canada* 1878

Roeder, A. *Berlin* 1825

Roeder [Röder] , Friedrich Erhardt von (1768–1834). Military cartographer of Berlin

Roeder, T. C. *XIII Prov. der Vereenigde Staaten N. America* 1782

Roel, Lieut. *Environs Berlin* 10 sh. 1816–19

Roelafsz, Roelofs. See **Deventer**

Roelandts, Jacob Thomas (fl. 1721–39). Royal cartographer of Antwerp

Roelofs, Jacob. See **Deventer**

Roemer, Ferdinand (1818–1894). Geolog. explor. *Texas* 1845, 1847, *Oberschlesien* 1867

Roemer, Friedrich Adolph. *Harz Gebirge* 1859, *Geolog Karte Rheinprovinz* 1855–65

Roemer, H. *Königreich Hannover* 1852

Roemer, Ole. *Seeland* 1697

Roesch, Carl. *Freiburg in Breisgau* 1838

Roesch, Jakob Friedrich, Ritter von (1743–1841). *Collection of battle plans*, Frankfurt 1796

Roesch [Rösch] , Johann Georg (1779–1845). Surveyor

Roese, W. Engraver for Scheda & Steinhauser. *Hand Atlas* 1879

Roesel [Rösel] , Stephan (ca. 1475–1533). Mathematician and astronomer in Cracow and Vienna

Roeser, C. Draughtsman. *Maps in Geographical and Political Atlas* Washington 1876: *California, Florida, Kansas, Louisiana, Michigan, Oregon, Washington &c.*

Roesler, Gotzhein. *Plan der Stadt Posen* (1866)

Roessel, Th. *Vogtland* 1875

Roessler, Michael. Engraver for Homann 1707. *Carlsruhe* 1739

Roetsch, A. *Dresden* 1846

Rofa, Martinus (with Valegio). *Citta di tutto il mondo* (1600)

Rogel, Hans. *Reichstatt Augspurg* 1563, *Kempten* 1569. Used by Braun & Hogenberg

Roger, A. *Estado Oriental del Uruguay* 1841

Roger, the Elder. *Atlas Portatif* 1823

Roger, Pierre. See **Rogier**

Roger, Salomon. See **Rogiers**

Roger, S. Engraver. Rossi's *Danimarca* Rome (1677)

Rogers, Daniel. *Description of Ireland in Verse*. Sent *map of Ireland* to Ortelius 1572

Rogers, Darius. *Osage City, Kansas* (1857)

Rogers, David. Capt. of Artillery. *York Harbour* 1760

Rogers, Henry Darwin. Geographer of Boston. *Atlas of U.S.* 1857 (with A. K. Johnston), *Geolog. Map Penn.* 1858, *People's Pict. Atlas* 1873

Rogers, John (fl. 1522–76). Engineer surveyor. Worked in Hull 1537, Berwick and Boulogne, Le Havre 1547, Cork 1551

Rogers, J. Engraver. For Tallis' *Atlas* 1851

Rogers, Paul. Pilot U.S. Coast Survey. *Chart Long Island Sound* 1857

Rogers, Thadeus M. Surveyor. *Marion Co., Missouri* 1875

Rogers, William (fl. 1589–1604). Engraver Speeds' *Cheshire* (Early version pre 1605), *Map* for English Edition of Pigafetta 1597, for Linschoten 1598, for Camden's *Britannia* 1600

Rogers, William Barton (1804–1882). Physician and geologist of Philadelphia

Rogers, W. E. *Harbour of Bahrein* 1828, *Town of El Khatiff* 1858

Rogers, Woodes (d. 1732). Voyage round world. *Map of World with tracks, ships Duke & Duchess* 1712, *The Pacific or S. Sea* 1718

Rogerson, A. E. *Borough of Allentown*

(Penn.). 1850, *Vicinity Baltimore* (1854), *Township Cheltenham, Penn.* 1867

Rogg, Gottfried. Engraver for Seutter's *Africa, America, Palestine,* 1725; *Netherlands* 1741, *Louisiana,* Augsburg 1745, *Turkey* 1778

Roggeveen, Arent (d. 1679). Dutch mathematician, navigator and hydrographer. *Het brandende Veen* Amsterdam 1675, Spanish edn. 1680, *Nieuwe Groote Zee Spiegel* 1685, English Edition 1687

Roggeveen, Arnold. Hydrographer (fl. 1675–1700). *Monte de la Turba ardiente (America)* Amsterdam, Goos 1680

Rogier, J. de. See **De Rogier**

Rogier [Roger] Pierre. *Poictou* for 1579 edition of Ortelius, also used by *Bouguereau; Théat. Géogr. de France* 1626

Rogiers [Roger] , Salomon. Engraver of Amsterdam for Bertius 1600, for Blaeu 1630, for Jansson 1658. *France* 1621, *Bourdelois* 1626, *Braunswyck* (1633), *Venezuela* 1642, *Paix de Caux* (1650)

Rogozinski-Szolc, Stefan. Polish explorer in Cameroon 1882–3 and 1887–90

Roguski, M. *Plan Hamburgh* 1868

Rohlfs, A. *Wandkarte von Europa* 1859–60

Rohman, Assar (d. 1703). Swedish surveyor

Rohr, Charles. *Bay Seven Islands* (1757), *Harbour Gaspee* (1757)

Roiz, Pascual, of Lisbon. Portolan maker. *Atlantic* 1633

Roksandik, Daniel (1845–1899). Military cartographer

Rokwashi (1633–1738). *World map* 1710

Roldan, Manuel. Spanish navigator. *Port Palompoy* Philippines 1862

Rolf, John. *State of Virginia* 1616 MS

Rolffsen, F. N. Engraver. *Minorca* 1756

Roll, Georg (d. 1592). German clock and instrument maker of Augsburg. *Celestial Globe* 1586 (with J. Reinhold)

Rollin, Charles. *Rome ancienne* 1738, *Ancient Geography* 1757, *Gaules Cisalpines* 1874

Rollinson, William (1762–1842). American engraver. *Universal Geography* 1798–1800, *Post Road between N.Y. & Albany,* Goodrich 1820

Rollinson. Engraver for Malham 1804, Rayne 1798–1800

Rollos, G. Engraver, mapseller and geographer (fl. 1754–89). *Madagascar* 1754, for Herbert 1757, Rider's *Atlas of the World* (1764), *England Displayed* 1769, *Univ. Traveller* 1779, *Seat War India* 1785

Rolph, J. T. Engraver of Toronto. *Chart Lake Superior* 1870

Romagnoli, P. *Pianta topo. della Citta di Bologna* 1822

Romain [Van Roomen] , Adrien. born Louvain 1561, d. Mayence 1615. *Speculum Astronomicum* 1606

Roman, A. & Co., 417 & 419 Montgomery St., San Francisco. *Railroad map California* 1868

Roman, J. Publisher of Amsterdam. Schenck's *Hecatomopolis* 1752

Romano, G. F. A. *Lazio,* Rossi 1693

Romans, Bernard. Dept. Surv. Gen. of S. Districts of N. America under De Brahm. *Pensacola Harbour, Mobile Bay* (1771), *W. Florida* 1773 MS, engraved 1774; *Southern British Colonies in America* 1776, *Windward Pilot* 1794

Romans, Matteo Gregorio de. *Nouveau plan de Rome* (1750)

Romanus, Adrianus (1561–1615). *Parvum theatrum urbium* 1595

Romat & Baillet. Lithographers, Rue St. Etienne 2, Paris, & Rue St. André des Arts 59. *California* 1849

Rome, John. Surveyor. *Estate Plans in Jamaica* 1761–67

Romero, Cleto. *Paraguay* 1898

Romero, Admiral Francisco Diaz. *Islas Philippinas, Manila* 1727 (with Chardin)

Rommel, Ernst August. *Plan von Leipzig* 16 sh. (1873) and 1878

Romstet, Christopher. *Terrestrial globe* 1692

Ronand, Manuel. *Atlas du Pérou* 1865

Roner, Col. W. W. *Hudsons or North River* 1700, *Draft Castle Island* 1705

Ronquillo, Pedro (d. 1691). *Costa de Florida* 1687

Ronsard, Marine Surveyor. *Charts of Australia* for Freycinet's *Atlas* 1811

Ronzelen, J. J. van. *Bremerhaven* 1849

Rook, T. Printer, Phila. Filson's *Kentucky* 1784

Rooker, Edward. Engraver. Gwynn's *Plan London* 1749, *View Quebec* (1760)

Roomen. A. van. See **Romain**

Roon, Albr. van (1803–1879). Military topographer and geographer

Rooney, W. T. **& Schleis.** *Kewaunee Co. Wis.* 1895

Roos. *Plan von Coeln* 1877

Roosen, Carl B. *Foeroerne* 1806, *Norge* 1820

Roosing, Pieter (1794–1839). Engraver and publisher of Amsterdam

Roost, Johann Baptist. *Maps of Bavaria* 1864–68, *Deutsche Bundesstaaten* 6 sh. 1842

Root, George M. *Allegany River* 1864, *Edgewater Staten Is.* 1866

Roper, J. Engraver for Wilkinson 1800–1805, for *Beauties of England & Wales* 1807; *Plan Norwich &c. British Atlas* (with G. Cole) 1810

Ropes, John Codman. *Atlas Campaign of Waterloo* 1893

Roppelt, Johann Baptist Georg (1744–1814). Topographer and physician. *Bamberg* 1800, *Franconie* 1806

Roquette, M. B. M. A. D. de la (1784–1868). Hydrographer

Roquevaire, R. de F. de. See **Flotte de Roquevaire**

Rominger, C. American geologist. *Survey Michigan* 1876

Rosa, Francesco de la. *City Puebla de los Angles* (1796)

Rosa, I. de. *Kaart van Batavia* 1860

Rosa, Martino. *Modon* 1574

Rosa, R. *Empire City* 1864, *Amer. Continent* 1865, *Venezuela* 1866

Rosa, Vincenzo (18th century). Italian globemaker. *Globes* 1793

Rosa. Publisher. Gran Patio del Palacio Real, Rue Calle de Montpensier No. 5, Paris. *Méjico* 1822, 1837

Rosaccio, Aloisio. *Geogr. della Tuscana* (1608)

Rosaccio, Giuseppe (ca. 1530–1620). Italian cosmographer. *Teatro del Cielo e Terra* Florence 1594, *World* 1597. Edited 1598 edition of Ptolemy *Monde element.* 1604, *Teatro del Cielo* 1615

Rosamel, Joseph de. *Plan de la Rade de Panama* 1843

Rosapina, B. Engraver. *Citta di Firenze* 1826

Rosatio, G. See **Rosaccio**

Rosazza, Calixto. *Ferro-Carriles Argentina* (1892)

Roscio, Francisco Joao. *Porto e Entrada do Rio de Janeiro* 1778

Rose, Edward. *Survey mouth Galien River* 1835

R[ose], Frederick W. *Serio comic war map (Europe)* 1877, *Angling in Troubled Waters* 1899

Rose, G. F. Publisher, Ann St., Birmingham. Radclyffes' *map Warwicks* (1851)

Rose, Hugo von. *Umgegend von Dresden* (1863)

Rose, J. See **Rotz**

Rose, T. F. Historian and biographer. *Atlas New Jersey Coast* 1878

Rose, W. *East coast of Sumatra* 1824

Roseli [Roselli] , Francesco. See **Rosselli**, Francesco

Roselli [Roseli] , Piero [Pietro, Petrus] . Majorcan hydrographer. *Charts* 1447– 89, *Chart Medit.* 1489

Rosen, Per Gustaf. *Spitsbergen* 1891

Rosenberg, W. *Umgegend von Ems* (1876)

Rosenfält, B., Vice Admiral. Swedish Naval Commander. *Atlas of Baltic* pub. St. Petersburg 1714 (12 charts), New editions 1720 (15 charts) and 1723 (16 charts)

Rosenfeldt, Werner von (1639–1710). Hydrographer

Rosenfield. See **Hutchings & Rosenfield**

Rosetti, Emilio. *Mapa de la Cordillera* (1866)

Rosiers, Des. See **Desrosiers**

Rosili, Vice Admiral. *Baie de Manille, Mer de Chine* 1798; *Golfe de Suez* 1799

Rosili-Mesros, François Etienne. *Charts of China, India & Philippines* 1798

Rosini, Pietro. Olivetan monk of Lendinara. *Globe* 1762

Roskeiwicz, Johann (1831–1902). Cartographer

Roslan, M. *Seychelles*

Rosmaesler, J. A. *Planiglobium Boreale, Australe* 6 sh.

Rosny, L. Leon de. *Formose* 1856, *Cochin Chine* 1860, *Romains d'Orient* 1884

Ross, Alexander. *Columbia River Basin* 1821 MS

Ross, Al Irvine. Land surveyor. Attested *Aberdeen* for Thomson's *Scotland* 1832

Ross, Charles. Surveyor. *Lands Cessnock* (1770), *Map Lanark* 4 sh. 1773, *Renfrew*

1754, *Dumbarton* 2 sh. 1777, *Stirling* 1780

Ross, Capt. Charles and John Klinefelter. Revised Cumings' *Western Pilot* 1848

Ross, Capt. Daniel. Bombay Marine, hydrographer. *Arroa Islands (Malacca),* Admiralty 1820; *Singapore surveyed* 1827, pub. 1840; *Sunda Strait* 1819, pub. 1840; *Hainan (Cochin China)* 1817; *Singapore Harbour* 1830, pub. Admiralty 1840; *Canton River* 1815, pub. 1840

Ross, David *Courland Bay* inset on Jeffery's *Tobago* 1775

Ross, E. H. *Sect. Map Arkansas* 1872

Ross, Capt. James C., R.N. *Antarctic exploration* 1839–43

Ross, Jean. See **Rotz**

Ross, John. Lt., 34th Regt. Surveyor. *Map Mississipi* 1765, (1772), and 1775

Ross, (J.). Engraver and publisher. *Map settled part of Van Diemen's Land* 1830

Ross, John L. *N.Z. Gold Fields* 1861, *Gipps Land (Victoria)* 1864, *Ballarat* 1868

Ross, Sir John (1777–1856). *Heel of Dantzig* 1810, *Countries round N. Pole, Baffins Bay* 1819, *Grönland* 1832, *Discoveries Arctic Sea* 1851

Ross, Joseph. *Chart of the Downs* 1779

Ross, Melville G. H. W. *Bass Strait* 1857, *Corio Harbour* 1860, *Corner Inlet* 1862

Ross, Robert (1800–1868). Canadian surveyor

Ross, William. Geographical writer. *De terrae motu circul.* 1634, *New Planet no Planet* 1646

Rossari, Carlo. Publisher. *Nuovo Atlante Milano* 1822

Rossel, E. P. E. Chev. de (1765–1829). Librarian, Dépôt de cartes et plans de la marine

Rosselli, Alessandro (fl. 1500–27). Book printer and cartographer of Florence

Rosselli, Francesco (1447–ca. 1513). Cartographer, engraver and mapseller. Shop in Florence, the earliest known map shop. *World* ca. 1490; engraved Contarini's *World map* 1506, *World map* in 2nd edition of Bordone 1532, also engraved for Dalli Sonnetti's *Isolario*

Rossetti, Domenico. Engraver and publisher. *Castelnuovo* 1684, *Gulf of Kotor* 1685

Rossi, Dario Giuseppe. *Carta Geogr. Milit. della Lombardia* 1859

Rossi [Rubeis], Domenico de (1647–1719). Publisher in Rome and Templo S. Marine de Pace. *Mercurio Geografico* (with Giacomo Giovanni Rossi) 1692–4, Greuter's *Globes* 1695, *Italy* 1695, *France* 1697

Rossi [**Rubeis**], Giacomo Giovanni. Publisher of Rome, Stamperia all Pace. *Naples* 1649, *Venice* (1650), *World* 1674, *Rome* 1676, *N. America* 1677, *Russia* 1678, *India* 1683, *Buda* 1683, *Mercurio Geografico* (1690). Succeeded by Domenico

Rossi, Giovanni Battista de (1576–1656). Rome. Piazza Navona urbis Romae (1650), Basilicae Vaticanae 1660. Reissued Greuter's *Globes* 1638

Rossi [Rubeis Mediolanensis], Giuseppe di (17th century). Globemaker. Reissued Hondius' *Globes* 1615, *Sabina Italy* 1716

Rossi (Luigi) (1764–1824). *Nuovo Atlante Univ.* Milano 1820

Rossi, Matteo Gregorio de. Cartographer. *Pianta di Roma* 1665, *Fabrica della Basilica* 1682

Rossi, Tommaso (16th century). Publisher and bookseller of Venice Olaus Magnus' *Scandinavia* 1539

Rossi, Veremondo. Maps in *Gazzettiere Americano* Livorno 1763, *America Merid., Barbados, Prov. Cartagena, Jamaica, Quebec &c.*

Rossi. Publisher, Milan. *Pole* 1820, *Impero* 1820

Rossieyskoey. *Atlas of Russian Empire* 1800

Rossley, W. R. *Madras* (1860)

Rosskopf, J. M. *Schoolkaart van Java* 1875

Rosso, Pietro (15th century). Portolan maker

Rostan [Rostand], Alex. P. Jesuit teacher in Warsaw and Grodno, worked with K. Perthes (18th century)

Rostovchev, Aleksej Ivanovic. Engraver and globemaker fl. 1723–44. *Ingria* 1727, used by Homann 1734, *Terrestrial Globe* 1723

Rota, Martino [Martinus] of Sebenico. Engraver. *Zara & Sebenico, Modon, Spalato* 1570; *Famose citta di tutto il Mondo* 1595 (with Valegio)·

Rotenhan, Sebastian von (1478–1534). German cartographer. *Franconia* Apian 1520, used Ortelius 1570

Roth, C. F. *Grundriss v. Stuttgart* 1811

Roth, Christopher Melchior (d. 1790). Engraver of Nuremberg. *Moldavia* 1771, *St. Petersburg* 1776

Roth, Just. Ludw. Ad. (1818–1892). Geologist

Roth, Magnus (1828–1895). *Scan.* (1875), *Atlas Sverige* 1876, *World Atlas* 1871, *Geografisk Atlas* (1884)

Roth, Matthias (18th century). Engraver of Nuremberg. *Atlas Novus* (Seutter's maps) 1730

Rothenberger, Heinrich. *Lothringen* 1875

Rothenburg, Friedrich Rudolph von (1796–1851). Engraver of Berlin. For Meyer 1830–40, *Prov. Pommern* 1851

Rothenburg, R. von. Engraver for Meyer's *Univ. Atlas* 1830–40

Rothgeister [Rodtgeister], Christian Lorensen. Engraver of Husum. *Maps of*

Schleswig-Holstein for Mejer (1652), pub. Blaeu 1662; *Olearius Hivermia* 1640

Rothwell, H. G. *Michigan* 1860

Rothwell, John. Printer. At the signe of the Sun and Fountain, in St. Pauls Churchyard. *Woodhouse's Guide to Strangers Ireland* 1647

Rothwell, R. P. *Lehigh Coal Co's property near Mauch Chunk, Pa.* 1869

Rotie, Denis de. *Portolan of the Atlantic* 1674

Rotteck, Carl von. *Hist. Geogr. Atlas* 1839–42

Rotterdam, Pieter. Bookseller sur le Vijgendam, Amsterdam. *Siege Ostende* 1706

Rotz [Rose, Roze, Ross], Jean [John] (1534–1560), born in Dieppe. Hydrographer to Henry VIII 1542. *Boke of Idrography* 1542, *World Map* MS

Rotzamer. *Plan de Metz* 1781

Rouand y Paz Soldan, Manuel. *Atlas del Peru* 1865

Rouge, G. L. le. See **Le Rouge**

Rouge, J. B. de. *Dépt. du Nord* 1793

Rougemont, P. H. de. *Railway map Chili* 1883

Rouille, Guillaume. Printer of Tours. Symeone's *Auvergne* 1560

Roulet, Régis de. *Cours de la Rivière aux Perles* 1732

Roulhac de Rochebrune, Laurent. *N.E. coast Iceland* 1856

Roulin, F. *Environs de Hondamines de Santana* 1825

Rous, Francis. *Description Athenian Territory* 1637

Rousseau, F. F. *Plan de Lille* (1822), *Dépt. du Nord* 1827

Rousseau, J. *Pyrenées* 1793–95

Rousseau, L. B. T. Engraver for Nolin and for Desnos. *Théâtre de la guerre sur le Haut Rhein* (1745)

Rousseau, N. L. *Plan d'Allah-Abad* 1821

Rousseau, T. Engraver. *Castille* 1762, *Galicia* 1762

Rousseau d'Happencourt Leopold, Count (1796–1878). Military cartographer

Roussel, C. Engraver. Placide's *Siam* 1686, De Granges' *G.B.* 1693, Baudrands' *Brit. Isles* 1701

Roussel, E. *Nancy* 1879

Roussel, Capt. Engineer. *Nouvel France* 1683, *Pyrenées* 1730 (English edition Arrowsmith 1809); *Paris* Rocque 1748, *Aragon* 1765, *Cataluna* 1776

Rousselet, Louis. *Nouv. Dict. de Géogr. Univ.* 7 vols. 1879–95

Rousset, J. *Plan Hague* Beek (1705)

Rousset, P. Letter engraver. *Atlas général* 1835, *Chemins de Fer d'Europe* 1850. For Andriveau-Goujon 1856, for Meissas & Michelot 1840–69

Rousset de Missy, Jean (1686–1762). *Nieuwe astron. geog. atlas* (1742), *Nouvel Atlas géog.* Amstd. (1742)

Roussin. Map maker of Toulon. *MS maps France, Spain, &c.* 1646

Roussin, Albion René Baron de, Contre amiral Admiralty Surveyor. *Pilote du Brésil* 1826, *W. Coast Africa* 1846, *Rio Grande* 1852

Roussin, Augustin, of Marseilles. *Gulf of Mexico* MS

Roussin, J. Fr. *Atlas 5 maps Atlantic, Aegean, Medit. Europe & Africa* 1663

Routledge, George & Co. Publishers in London, Soho Square; later Faringdon St. *California* 1849, *Kansas Territory* 1857, *British Columbia & Vancouver Island* 1862

Rouve, T. Richard de. *Dipartimento dell Adige* 1812

Rouvier, Abr. *Suisse,* revised Rocque 1760, *Congo Français* 1887

Roux, Joseph. *Carte de la Méditerranée* Marseilles 1764, *Plans et Rades de la*

Méditerranée, Marseilles 1764 (enlarged 1804); *Costes de Provence* 1770, *Golfe de Lyon* 1771

Rovelascus, S. *West Africa* (with L. Teixeira) engd. B. Doetecum ca. 1600

Rovere, G. Della (1441–1513). Cartographer

Rovere, G. F. *Terrestrial sphere* 1575

Rovere, Paolo. *Trivigiano* 1591

Roverea the Elder, Isaac Gamaliel de (1695–1766). Surveyor of Geneva

Row, R. *Sussex* 1821

Rowan, Mathew. Surveyor with others of the *Carteret Grant* 1746

Rowe, Edwin. *Owaarte Harbour* 1853

Rowe, I. Engraver for A. Arrowsmith's *Germany* 1812

Rowe, John E. *City of Newark, New Jersey* 1873

Rowe, Robert (1775–1843). Engraver, geographer, publisher. No. 11 Crane Court Fleet St. and 19 Bedord St., Bedord Row 1811–15. *Plan London* 1804, *Europe* 1807. *Island Walcheren* 1809, *English Atlas* 1811–(16), taken over by Teesdale 1829

Rowe, R. R. *Cambridge* 1872

Rower, Geo. C. American surveyor *Ventura Co., Cal.* 1897

Rowland, Johnson (fl. 1559–87). Survey engineer. *Plan Berwick*

Rowland, J. W. *Lagos* (1892)

Rowland, Richard. *The Post of the World* London 1576

Rowley, Alexander S. *City Hudson, Columbia Co. N.Y.* 1871

Rowley, John (d. 1728). Globe and instrument maker

Roxas, A. F. de. See **Fernandez de Roxas**

Roy, B. du. See **du Roy**

Roy, J. T. *County of Banff* 1812

Roy, Le. See **Le Roy**

Roy, William, Major General (1726–

1790). Military cartographer, one of the main participants in the Ordnance Survey. He began the trigonometrical survey of England after the 1745 rebellion in the Highlands. *Survey Highlands* 1747–55, *Whitstable to N. Foreland* 1758, *Battle Tonhausen* (1760), *S. Ireland* 1765. Surveyed base line on Hounslow Heath 1784

Royal Geographical Society. Founded 1830 by John Barrow

Royce. Engraver. *Canal plans surveyed* 1769–75

Roycroft, Tho. Printer of London. Blome's *Britannia* 1673

Royer, Augustine (17th century). French astronomer

Royer, E. *Haute Marne* 1859

Rozas, Joaquin Perez de. *Ciudad de Malaga* 1863, *Almeria* 1864, *España y Portugal* 1866

Roze, J. See **Rotz**

Rozet, Cl. Ant. (1798–1858) Traveller and geologist

Rozière, Marquis de la. See **Carlet**

Rozzell, B. *Solar System* (1856)

Rubeis, J. de. *Terrestrial Globe* 1615

Rubeis. See **Rossi,** G. B. and **Rossi,** G. G.

Rubens, J. F. *Weinkarte von Europe* (1875)

Rubeus, Pietro (15th century). Hydrographer of Messina

Rubie, G. *British Celestial Atlas* 1830

Rucker. *Chicago* 1849

Rudbeck, Johannes (1581–1646). Bishop of Uppsala. Orientalist

Rudbeck, Olof. *Atland eller Manheim* 1679 *(map of Scandinavia)*

Rudbridge, Frederick Preston (1806–1898). Canadian surveyor

Ruddock, Samuel A. Surveyor. *Colleton District, S.C.* 1820, for Mills 1825

Rudemare. *Ville d'Angers* 1813, *Lyon* 1818, *Lille* 1820

Rudimentum Novitiorum. Lübeck Chronicle, Lucas Brandis 1475. World history famous for its maps *World* and *Palestine.* One of a few non Ptolemaic printed maps of 15th century. French edition *Mer des Histoires* Paris 1488—9

Rudolph, A. *Stadtplan von Zittau* 1879

Rue, J. *Middlesex Co.* 1781 MS (with Dunham), *Penn., East and West New Jersey* 1778

Rue, P. De la. See **De la Rue**

Ruecker, G. C. *Livland* 6 sh. 1839, 1857, 1867; 4 sh. 1890

Rueda, Manuel de. *Atlas Americano* Havana 1766

Ruehl, O. *Plan Stadt Worms* (1880)

Ruehle von Lilienstern, Jean Jacques Othon Auguste (1780—1847). *Sachsen* 2 sh. 1810, *Allgemeiner Schulatlas* 1825, *Böhmen* 1833

Ruelle, Alexandre. *Planisphère Astronomic-Géog.* Dezauche (1875), *Nouvelle Uranographie* 1786

Ruelle, Claude de la. *Plan of Nancy* 1611, used by Braun & Hogenberg

Rueppel, Eduard. *Aethiopien* 1859 and 1867, *Sinai Halbinsel* 1859

Ruesta, Sebastian de. *Canada* (1654), *N. Coast S. America* 1654

Ruff, E. & Co. Publishers. London, No. 2 Hind Court, Fleet St. *Kent* 4 sh. 1837, *Greenswoods Warwick* 4 sh. 1832, *London* 1839

Ruffell, William. Land surveyor, Colchester. *Elmstead* ca. 1843 MS

Ruffoni [Ruffonio, Ruphon] (16th century). Cartographer of Padua. *Padua* 1509

Ruga, Pietro. Engraver. *Campagne de Roma* 1816, *Citta di Roma* 1823

Rugell, Chr. *Map of the Kalmius River in Russia* (1699), *Map Sea of Azow* (pub. 1701)

Ruger, Edward. *Chicago* 1872

Rugg, Rowland. *Matabele Gold Fields* 1890, *W. Nyassaland Gold Fields* 1892, *Klondyke Gold Fields* 1898

Ruggieri, Ferdinando. *Citta di Firenze* 1731

Ruggles, E. *City Marietta confluence Ohio and Muskingam* (1780)

Rughesi, Fausto (1597—1605). Architect, cartographer of Montepulciano. *Maps of World, Europe, Asia, Africa, America* 1597

Ruhedorff, F. Jos. *Walachia* 1788

Ruil, Jan Luities. *Ooster en Wester Eemze* 1797

Ruiz, Anton. *Golfo de Mexico* 1799

Ruiz, A. P. y. See **Puya y Ruiz**

Ruiz de Olano, Pedro. *Plano Fortin de San Francisco de Pupo* Florida 1738 MS

Rule, Robert. *Map of Norham (Durham)* 1824

Rulhière, Louis Amedée de. *Plan de la Baie de l'Oues* 1845

Rulison, Duane. 33 S. 3rd St. Phila. *U.S.A.* 1860

Rumball, H. *Land Surv. St. Albans Eastwood (Essex)* 1816 MS

Rumpf, Hieronymus. *Post u. Dampfschiffahrt Reise Carte der Schweiz* 1840

Rumsey, E. Engraver for Gill's *Waxford* 1811

Rumsey, Elisha Walker (d. 1827). *Ann Arbour* 1825 MS

Rundall, F. H. *River System S. India* 1886

Rundall, J. Wharton. *N. Polar Discoveries* 1849, *Empire of Japan* 1850

Rundell, W. W. Surveyor. *Plymouth, Devonport & Stonehouse* (1840)

Runge, Christopher. Publisher, Berlin. Helwig's *Silesia* 1738 edn.

Runge, F. R. Engraver for Meyer's *Hand Atlas* 1867

Runge, W. *Geolog. Karte Nieder-schlesischen Gebirge* 1862–8

Runstrom, Johann (18th century). Swedish cartographer

Ruphon. Engraver for Seminario Vescovile, Padua 1699

Ruphon. See **Ruffoni**

Ruprecht, Mathias. *Plan Memmingen du Chaffat* 1737

Ruremundanus, Joannes. Printer of Cologne. *Ptolemy* 1540

Ruscelli, Girolamo (ca. 1504–1566) Editor in Venice. Ptolemy's *Geogr.* 1561

Rush, Hayward. Land Measurer, writing master and teacher of Navigation & Maths. *Wivenhoe* 1734 MS, *Asheldham* 1742 MS

Rushworth. See **Harrison & Rushworth**

Russe, J. *Nord Amerikanske Fristater* 1830

Russegger, Joseph. *Aegypten* 1842, *Libanon* 1842, *Nubien* 1843, *Ost Sudan* 1843, 1846

Russell. Printer, Penge. For S.D.U.K. 1833–6

Russell, B. B. Publisher. *Boston* 1868

Russell, F. Royal Surveyor. *Enfield Chase* 1776–7

Russell, George. Estate surveyor, Rochester. *W. Thurrock & Stifford* 1699 MS

Russell, J. *Selenographie* 1797

Russell [Russel], John. Engraver and draughtsman, 9 Constitution Row, Grays Inn Road, and 32 Fetter Lane. For Dalrymple 1744–5, for Cook 1733–79, for Guthrie's *System of Geography* 1785; *American Atlas* 1795, for Burney's *Voyages* 1813

Russell, J. C., Junior. Worked with John above. Guthrie's *System of Geogr.* Ostell's *New General Atlas* 1810,

Robinson's *Archaelogia Graeca* 1827; *Portland, Maine* 1869

Russell, Capt. Lockhart. *Harbour St. Mary Madagascar* 1782

Russell, Michael. Surveyor. *North of Icknield Way* 1810 MS

Russell, P. and Owen Prince. *England Displayed* 1769

Russell, S. I. *Pedestrians Companion* 1837

Russell. See **Fisk & Russell**

Russells, John Scott. *Physical Atlas* (with A. K. Johnston) 1848

Russinus, Augustinus. *Portolan* 1590

Russo [Russus], Giacomo [Jacobus] (fl. 1520–70). Miniaturist and chart maker of Messina. *Atlantic & Sea of Azof* 1520 MS, *Mediterranean* 1549, 1546, 1568; *Black Sea* 1565

Rüst [Rist, Riss], Hans (d. 1484). Wood engraver in Augsburg and Nuremberg. *World map* (1480)

Rüst [Risk, Rio], Hars (d. ca. 1484). Wood engraver, Augsburg. *World* 1460

Rüstel, F. *Lombardy & Venice* 24 sh. (1830) MS

Rutherford, A. *Roads through Highlands of Scotland* 1745

Rutherford, John (fl. 1767–72). *S. & N. Carolina* 1772

Rutherford, J. H. *Corporation of Ithaca N. Y.* 1872

Ruthner, Anton de (1817–1897). Austrian geographer

Ruthven, John. *Map English Lakes* 1855

Ruthvin, C. R. Lithographer. *Plan Jamaica Place, Commercial Road* n.d.

Rutowski, Antoni. Military surveyor, Worked on *great topographical map of Poland* 1820

Rutlinger, Jan. (d. 1609). Flemish engraver working in London at the Mint. Engraved map for Saxton, *Galicia* in English edition of Waghenaer 1588

Rutt, William. Surveyor. *Eye Draught Aylesbury* 1809 MS, *Little Brickhill Parish* 1810 MS

Ruuth, Hans Persson (d. 1677). Swedish surveyor

Ruysch, Johannes (d. 1533) Astronomer and cosmographer. *Universalior cogniti Orbis Tabula Ptolemy* 1508

Ruyter, B. Engraver for Covens & Mortier, De Lisle, and Visscher

Ruyter, Gerard de. Maps in Laurie & Whittle's *E. India Pilot* 1800, *S. coast Africa,* Sayer 1790

Ruyter, Michiel Adriaenszoon de (1607–1676). Dutch Admiral. *Gold Coast* 1666

Ruzek. Engraver. *Schlacht bei Pr. Eylau* 1808

Ryall, J. Publisher with others *Royal English Atlas* (1762)

Rycaut, Lt. Gen. *Plan attack Fort Louis, Guadeloupe* 1760, *Plan attack Basseterre* 1760

Ryckov, Petr. Ivanovic (1712–1777). Russian geographer. *Atlas of Orenburg Province* 1762

Ryder, Capt. (later Admiral) Alfred Phillipps. *Plan Bay Eupatoria* 1853

Ryder, Charles (b. 1858). Danish astronomer

Ryder, John (1777–1832). Canadian surveyor

Ryder, Robert (1670–81) *Long Is., N.Y. & Conn.* MS in Blaythwait *Atlas* 1675

Rye, Egidius van der. *Town plans* for Braun & Hogenberg 1617, *Cracow* 1657

Ryhiner, S. *Grundriss Stadt Basel* Basel 1786

Ryland, E. Publisher. Harris' *Plan Halifax* 1749

Ryland, John. Engraver. *Kirby's Suffolk* 1766, *Turnpike Roads* 1767, *Cook's Strait N.Z.* 1773 and 1820

Ryland & Bryer. Engravers and publishers at Kings Arms Cornhill. *Plan London* (1765)

Rylke, Stanislaw (10th century). Military cartographer

Rymers, Petrus Josephus. *Atlas Kleynen . . . der wereld.* Antwerp 1743

Ryne, I. van. Bombay 1794

Ryo, Wakabayashi (19th century). Japanese surveyor. Detailed *maps of Treaty Ports of Kobe and Hyogo* 1860

Ryther, Augustine (fl. 1576–1595). English engraver and map seller, born Leeds. *Maps of England, Gloucester, Yorkshire, Durham, Westmoreland and Cumberland* for Saxton's *Atlas,* 3 maps for Waghenaer, *London* 1587, *Oxford* 1588, *Spanish Armada* 1590, *Cambridge* 1591, Hood's *Chart Atlantic* 1592, *London* 1604

Ryusen. See **Ishikawa**, Toshiyoki

Ryves, Capt. George Frederick. *MS printed chart Magdalen Is. (Sardinia)* 1802, *Admiralty Chart* 1804

Ryves, P. Thomas. *Plan Battle Maida* 1807

Rzepecki, Jan (18th century). Royal Geographer, author *plan Poznan & environs*

Rzetkowski, Mikolaj. Military surveyor. Worked on Polish-Prussian demarcation line 1818–1822

S

S., J. E. *Topo. Karte Friaul* 1805, *Mähren* 1802

S., J. H. *Siège Dardanelles* ca. 1680

S., L. *Plan Boston Harbour* 1775 MS

S., R. A. Engraver of Homann 1728. "R.A. sculpsit"

S., T. *Lympsfield & Environs* 1838

S., V. *Nonantola* ca. 1650

S., W. D. *Maps of Holland* 1784 pub. Jaeger

Sá e Faria, José Costodio de. *MS maps of Brazil* 1773—5

Saad, Lamec. *Plan Smyrne* 1876

Saaybe, H. E. G. *Copenhagen* 1871

Sabbadino, Christoforo (1496—1560). Venetian cartographer. *Laguna Veneta* (ca. 1552)

Sabine, Edward (1788—1883). Physician and Arctic explorer

Sabir, C. de. *Bassin d'Amour* 1861

Sabucus, J. See **Sambucus**

Sacenti, Camillo. *Territorio Bolognese* 1651—87

Sachs, Fr. *Map Schwarz Wald* 1896

Sachs, Hans. *Stadt Altenburg* (ca. 1550)

Sack, Albert, Baron von. *Reise nach Surinam* 1821

Sackett, W. J. *Ely* (1845)

Sacrobusco [i.e. John of Holywood or Halifax] (fl. 1220—56). Taught astronomy in Paris. *Tractus de Sphaera.* Original MS written ca. 1233; several MS and 65 printed editions; first 1495

Sadebeck, Prof. *Plan Breslau* 1868

Sadeler, Egidius (1570—1629). Dutch engraver, b. Antwerp, d. Prague. *Bohemia* ca. 1605 & 1620, *Prague* 1606 *Valtelina* ca. 1610

Sadeler, Tobias. Viennese engraver. *Battaglia di Lutzen* ca. 1670

Sadler, Justus. *Italy* Magini 1662

Sadler, Thomas. Engraver. *Hatfield House* 1707

Saenredam, Jan Pieterszoon (1565—1607). Dutch engraver

Saevry, S. Engraved maps for Hornius' *Ancient Geogr.* 1660, 1684 & 1700

Safford, J. M. *Geolog. Maps Tennessee* 1866—8

Safft, J. C. W. (1778—1849). Dutch map seller

Sagansan, L. *Railway maps Europe* 1855—83

Sage, A. Le. See **Las Cases**

Sage, Rufus B. *Oregon etc.* 1846

Sage & Sons. American lithographers of Buffalo, N.Y. *Maps of Lakes* 1858, Reed's *Calif.* 1863

Sagra, Ramon de la'. *Hist. Cuba* (13 maps) 1842

Sahili, Ibrahim (fl. 1067—85). Arab geographer in Valencia. *Globes*

Saidi Ali Ben Hosein (d. 1562). Turkish astronomer and navigator. *Mohit (guide to Indian Ocean)* 1554

Saile, F. X. *Maps of Lorraine &c.* 1876—77

Sailmaker, J. Engraver for Mortier 1715—28

Saint Claire Deville, Ch. Jos. *Voyages géologiques aux Antilles* 1856—64

Saint-Cyr, de Gouvian. See **Gouvion Saint-Cyr**

Saint-Fond, B. F. de. See **Faujas de Saint-Fond**

St. Genevieve. *Map. World* ca. 1370, now in B.N., Paris

Saint-Germain A.M.P. de. See **Ducros de Saint-Germain**

Sainte Hélène, P. de. See **Placide**

Saint Hilaire, Aug. Fr. César (1779—1853). Traveller and naturalist notably in Brazil

Saintin, C. *Nouvel atlas des enfans* Paris 1811

St. Jerome. See **Hieronymus**

St. Lys. *Plan of St. Louis (Missouri)* 1796

Saint-Martin, D. F. de. See **Fournier de Saint-Martin**

Saint-Martin, L. V. de. See **Vivien de Saint-Martin**

Saint-Omer, L. See **Lambert de Saint-Omer**

St. P., R. L. *Terrestrial Globe* 1725

St. Petersburg. See **Akademiia Nauk St. Petersburg**

Saint-Sand, J. M. K. A. (b. 1853). Work in Pyrenees triangulation & itineraries

Saint-Sauveur, J. G. de. See **Grasset de Saint-Sauveur**

Saint Simon, E. J. A. D. de. See **Dexmier de Saint Simon**

Saint-Victor. *Plan Paris* 1555

Sala, Salvator. Engraver *Arch. Filipino* 1884

Salamanca, Antonio (ca. 1500—1562). Born Milan, d. Rome. Bookseller and engraver, publisher. Collab. with Lafreri 1553+. *Hungary* 1553, *Palestine* 1548, *Switzerland* 1555, *Toscana* 1560

Salamanca, Francesco. Son of Antonio. Publisher of Rome. Sofiano's *Greece* ca. 1560

Salam Din. *Maps of Hindostan* 1887—8

Salaverria, Santiago. *Islas Philipinas* 1789 MS, *Isla da Panay* 1797 MS

Sala y Gomez. *Easter Is.* 1876

Salazar, Ramon. *Plano de Valparaiso* 1848

Salberg, C. L. *Etats Prussiens* 4 sh. 1813

Sale, John. *Estate Plans Foxgrove Farm & Manor* 1777, *Plan Rochester* 1817

Salinas Bellver, Salvador. Spanish geographer 19th century

Salinas Vega, Luis. *Cartes Bolivie* 1887

Salingen [Salinhen], Simon van. Dutch cartographer. *Coast Lapland* 1567 MS, *N. Russia* 1601 MS

Salisbury, Capt. I. *Scilly Is.* 1788

Salisbury, Rollin D. *Geological maps New Jersey* 1893

Salkield, R. *Chart Cork Harbour* 1807

Salle, R. de la. See **La Salle**

Sallieth, Mathias de. *Atlas Zeehavens Bataafsche Republieck* 1805

Salmon, G. Surveyor. *John Knapp's Estate* 1771 MS

Salmon, George. Surveyor. *Greenville District, S.C.* 1820, for Mills 1825

Salmon, M. A. *Atlas* 1725, Italian Edn. 1740—54

Salmon, Thomas. *Modern History* (maps by Moll) 1725—38, *Geog. & Hist. Grammar* 1754, *Mod. Univ. Gazetteer* 1781

Salomon, C. E. Surveyor, 19th century

Salomon, F. T. *Plan Père Lachaise* 1855

Salomon, J. F. C. *Water Work District Columbia* 1851

Salt, G. B. *Entrance Great Belt* 1808

Salt, Henry. *Plan Alexandria* 1809, *Abyssinia* 1811, *Chart E. Coast Africa* 1814

Salter, Albert P. *Crown Land Survey Canada N. Shore Lake Huron* 1860

Saltonstall, C. *The Navigator* 1613, pub. 1636

Saltonstall, Wye. Translator, English edition of *Mercator* 1635

Saltzenberg, J. F. Engraver of Hanover. *English Channel* 1801

Salway, Joseph. Surveyor for Kensington Turnpike Trust. *Hyde Park Corner to Counter's Bridge* 1811

Salway, T. *Plan Estate John Gregory Camden Town (London)* 1806

Salzmann, Fried. Zacharias. *Plan de Sans Souci* 1779

Samball, P. *River Truro* 1838

Sambucus [Sabucus, Zsambocky, Zsamboki] , Johannes [Janos] . (1531–1584). Geographer. Born Hungary, d. Vienna. Worked with Ortelius. *Transilvania* 1566, *Hungary* 1566, *Siebenburg* 1570, *Poland* 1573, *Forum Julii* 1573

Samper, Juan Antonio (1799–1840). Columbian geographer. *Nueva Granada* n.d.

Sampson, George. Surveyor. *Maner of Lordeshippe (Norfolk)* with R. Agas 1575

Sampson, George Vaughan. Surveyor. *Londonderry* 4 sh. 1813

Sampson. See **Phillips, Sampson & Co.**

Samson. See **Sanson**

Samson, John. *S. America* 1894

Samuel, Peirse. *Plot Manor House Norton (Kent)* 1599 MS

Sanbiasi, Father Francesco, S.J. (1582–1649). *World Map* on rice paper ca. 1648

Sanches, Alejandro. *Prov. maps Philippines* 1834–(45)

Sanches [Sances] , Antonio [Tonio] (fl. 1523–1641). Portuguese cartographer. *World* 1623 MS, *Pacific* 1641

Sanches [Sanchez] , Cipriano. Portuguese cartographer. *Atlantic* 1596 MS, *Ceylon* used by Hondius 1606

Sanches, Domingos. Portuguese cartographer. *Atlantic* 1618 MS

Sanchez, Alexandro. Publisher of Manila. *Mapa general de las almas* Manila 1845

Sanchez, C. See **Sanches**

Sanctis, Gabrielle de. *Atlante corografico del regno delle due Sicilie* 1843

Sandberger, F. *Geolog. Karte von Baden* 1856–70, *Würtemberg* 1870

Sandby, Paul. Artist and draughtsman to Ordnance ca. 1742. Drew *Survey Scotland* 1747–53

Sandelyn, Peter. *Maldives* pub. Dalrymple 1774

Sander, Anton. See **Sanderus**

Sander, W. Engraver. Sotzmann's *New York* Hamburg, Bohy 1799

Sanders, Lieut. F. W. Hydrographer. Worked on Admiralty Charts with Cheyne

Sanders, I. *Draught Torrington Harbour (Chibucto)* 1747

Sanders, John. Surveyor. *Roxwell* 1788 MS

Sanders, John Park. Comm. Indian Navy. *Survey India* 1844–8, *Arabia* 1850–4

Sanders, S. Surveyor. *New Shoreham Harbour* 1844

Sanders, William. Geologist. *Bristol Coal Fields* 19 sh. 1864

Sanderson, E. J. *Easter Is.* 1869

Sanderson, G. Surveyor. *Derbyshire* 1836, *Twenty miles round Mansfield* 1835, *Notts.* 1836

Sanderson, William. (ca. 1541–1631). Merchant of London, supported Wright & Molineaux' work 1599–1600, also Norden's.

Sanderus, Antoine (1586–1654). Flemish chorographer. *Flandria illustrata* Cologne 1641–4, *Ypres* 1641, *Oudenard* 1650, *Alost* 1650, *Courtrai* 1667, *Sacra Brabant* 1659. Maps used by Blaeu

Sandi, Antonius. Engraver. *Polesine de Rovigo* 10 sh. 1786

Sandow, George. American draughtsman, end 19th century

Sandoz, Ernest. *Wall maps* 1863–7

Sandrart, Jacob von de (1630–1708). Born Frankfurt, d. Nuremberg. Painter and engraver. *Danube* 1644, *Bohemia* 1666, *Austria* (1670), *Moscovia* (1670), *Hungary* (1670), *France* (1675), *Franconia* (1680), *Hibernia* (1690)

Sands, John. Publisher, 374 George St. Sydney. *Atlas Australia* 5 vols. Sydney 1886

Sands, William. Surveyor. *Manor Henry Magna Essex* (1660) MS

Sands & Kenny. Publishers, Melbourne and Sydney. *Victoria* 1859

Sandtner, Johann. Lithographer. *Carlsbad & neighbourhood* (1862)

Sandvoort, Abraham (1636–64). Amsterdam publisher

Sandys, George. *Voyage Holy Land* (map) 1621

San Filippo. See **Amat di San Filippo**

Sanford, E. F. Surveyor and publisher. *City of Concord (New Hampshire)* 1868

Sanford, George P. Surveyor and publisher. *Warren Co., Ohio* 1867; *Ulster Co., N.Y.* 1870; *Randolph & Wayne Co., Indiana* 1874

Sanford Beers & Co. *County Atlases U.S.* 1869

Sanford Evert & Co. *County Atlases U.S.* 1871

Sanford Gould & Lake. *County Atlases U.S.* 1874

S'Angelo. See **Angelo**

Sangallo, Antonio de. *Plates of Rome* 1540–8

Sangster, George [& Son]. Estate surveyors. *Estate plans in Essex* 1755–85

Sanis, Jean Léon. *Maps France* 1859–70

Sanitor, H. Engineer. *Plan Rostock* (1859)

San Miguel, Torfino. See **Torfino de San Miguel**

Sanson, Adrian (d. 1708). Son of Nicolas. Continued business with his brother Guillaume.

—— Gullaume (d. 1703). Son of Nicolas. Géographe Ordinaire du Roy. With his brother Adrian continued Sanson business. Took Hubert Jaillot into the business ca. 1670

—— Nicolas (1600–1667). Born Abbeville, moved to Paris: Rue de l' Arbe Secq. près de St. Germain l'Auxerrois (1652); Dans le cloitre de St. Germain de l'Auxerrois près et joignant la grande Porte du Cloistre (1657–62); Rue St. Jacques à l'Espérance (1670), Aux Galleries du Louvre (1676). Géographe Ordin. du Roi (1630–1665). Called founder of French School of Geography. Tutor to Louis XIV. *Ancient Gaul* 1618, pub. 1629; *Cartes générales de toutes les parties du Monde* 1658, 64–66, 67, post. 1670, 76; *L'Asie* 1652–3, *L'Afrique* 1656, *Amérique* 1657. Business continued by his sons Nicolas, Adrian, and Guillaume.

—— Nicolas Fils (d. 1648). Eldest son of Nicolas, assisted his father. Compiled *Estats du Czar* engraved by Peyrounin.

Sanson Heirs. Moullart-Sanson, Pierre. Grandson of Nicolas the elder. Géographe du Roi. From 1695, A Paris aux

Galléries du Louvre vis à vis St. Nicolas. From 1710, Rue Frément près le Cloître de St. Nicolas du Louvre. *Mappemonde* 1695, *Hémi. Sept.* 1696, *Géogr. Sacrée* 1716, *Plan Montréal* 1723, *Intro. à la Géographie* 1728. Succeeded by his nephew G. Robert de Vaugondy

Santa, Gasparo Pietra. See **Pietra Santa**

Santacilla, Juan y. *Compendio de Navegacion* Cadiz 1757

Santa Cruz, Alonso de (ca. 1500–1572). Spanish Cosmographer Royal to Charles V. Sailed with Cabot 1526–30. *Globe* 1542, *Totius Orbis descriptio* 1542 MS, *Atlas* in Madrid (109 maps) 1545 MS, *Isolario* 1560

Santarelli, Francesco. *Vedute del Regno di Napoli* 1829

Santarem, M. F. de Barrios Sousa, de M. de M. L., Visconte de (1791–1865). Portuguese cartographer. *Essai sur l'histoire de la cosmog.* 1849–52

Santa Teresa, Joao José de [Giovanni Giuseppe]. *Istoria delle Guerre del Regno del Brasilia* Roma, Corbelletti Heirs 1698, Rossi 1700 (with maps)

Santiago, J. Script engraver. Figueroa's *Philippines* Madrid 1816

Santiago, Fr. Pedro de. *Ms map Indios Chiriguanos.*

Santini, François. Publisher in Paris, Rue St. Justine près la dite Eglise Venice. Re-issued *maps* of Vaugondy, Homann Heirs, De Lisle, d'Anville

Santini, Marco. *Mappa Citta di Venezia* (1847)

Santini, P. *Atlas Universel* Venice 1776 & 1783, *Atlas Portatif* 1782. Collaborated with Bonne

Santo, A. di S. See **Stefano**

Santos, Bernardino de los. *Plano de Manila* 1877

Santos, Domingo Juan de los. *Manila Bay* 1802

Santucci, Antonio. Italian globemaker, 16th century

Sanudo, Marino. Credited with *portolan World map* of 1306, but probably by Vesconte

Sanuto, Giulio (1540–1580). Venetian engraver. Engraved plates for 1564 & 1574 editions of *Ptolemy's Geography* & for his brother Livio's *Geographia* 1588

Sanuto, Livio (1520–1576). Venetian geographer and mathematician. Constructed a *globe* with his brother Giulio 1561–74. Published *Geographia* Venice 1588 (12 maps), the first printed atlas of Africa; engraved by his brother Giulio

Sapper, Karl. *Maps Guatemala & San Salvador* from 1893

Saraxin. Printer of Paris, Rue Git le Coeur. Garnier's *Atlas Sphéroidal* 1860

Sargeant, George. Surveyor. *Plan Ashenden* 1641 MS, *Grenville Manor* 1649, *Audley End* 1666, *Saffron Walden* 1666 MS

Sargent, Charles Sprague. *Maps on Forestry N. America* 1884–6

Sargent, E. W. Lieut. 18 Royal Irish Regt. *Chusan* 1842–3–(57) with Havilland

Sarin, John (1580–1643). *Voyage to Japan* 1611

Sarjent, F. J. Draughtsman and engraver. *Hull* 1810, *St. Michael* 1811

Sarony & Co. [Sarony, Major & Knapp]. Lithographers of New York, 19th century

Sarret, G. B. C. Engraver for Lapie 19th century

Sarson, T. *Bounty Bay* 1765

Sartin, M. *Coast of France* 1769

Sartori, Franz (1782–1832). Austrian geographer

Sartorius von Gmünden, Johannes (1383–1442). Theologian, mathematician, astronomer and instrument maker of Vienna

Sarychev, G. A. (1763–1831). Russian hydrographer.

Sasso, G. Engravers for L. Rossi's *Nuovo Atlante* 1821, and for Bonatti

Sastron, Manuel. *Prov. de Batangas* 1895

Satchell, John. Revised Lavoisnes's *Atlas* 1840

Sattler, Christ. Fried. (1705–1785). Topographer

Sauerbrey. Engraver. For Prussian Acad. of Science *Maris Pacifici* (1755)

Sauermann, P. J. See **Saurmann**

Saulcy. See **Caignart de Saulcy**

Saulnier de Vauhello, L. See **Le Saulnier de Vauhello**

Sauly, A. *Carte Puy de Dôme* 1845

Saumarez, Henry de. Invented Marine Surveyor (Sea Log) ca. 1714. *Channel Is.* (1727)

Saunders, Sir Charles, Admiral. *River St. Laurence* 1760, *Attack on Quebec* 1759

Saunders, G. *Plans of Westminster* 1835

Saunders, George Henry. Land surveyor, 12 North St. Westminster. *Ulverston & Lancaster Railway* 1850, *Ilford* 1854 MS

Saunders, James. *Map Parish Kingham (Oxon)* 1840

Saunders, Trelawney. Geogr. assistant India Office. *Central America* 1853, *Australian Goldfields* 1853, *Ancient Babylon* 6 sh. 1885, *Map Civil Divisions India* 1885, *Meteorological Map India* 1885, *Military Map India* 1885 *Mountains of India* 1885

Saunders & Otley. Publishers of London. Gell's *Rome & Environs* 1834

Saunderson, Charles. *Ireland* (1863)

Saunderson, William. Publisher *Molyneux' Globes* 1592

Saunier. French engraver. Leynadier's *Paris* 1855

Sauracher, Adelbert. *Rhetiae et Helvetiae Tab.* 1584 (original lost)

Saurij, Salamon. *Topo. Madrid* 1656

Saurman[n], P. J. S. Publisher of Bremen. Hennepin's *Chart America* 1698–9

Saussure, Hor. Ben de (1740–1799) Geologist and Alpinist

Sauter, C. *Post & Reise Karte von Deutschland* 1843 & later

Sauthier, Claude Joseph. Surveyor and topographer. *Plan Hillsborough, N.C.* 1768 MS, Many *MS plans of Towns in North Carolina* 1769–70, *Hudson River* 1776, *Prov. of N.Y.* 1776 & 1777 (Lotter), 1779 large scale, *Operations Kings Army in N.Y. & East N.J.* 1777, *Canada* 1777, *Part N.Y. Island* 1777

Sauzet. See **Du Sauzet**

Savage, Henry. *Map of Baroche Purgunnals* 1782

Savage, John. Engraver. *Bath* 1697 for J. Gilmore, Later ed. 4 sh. 1717

Savanarola, Raffaelo, 18th century Italian cartographer. Pseudonym of Lasor à Varea. *Universus Terrarum Orbis* 1713

Säve, P. A. *Gotland och Wisby* (1875)

Savery, Samuel. Surveyor. *Boundary Georgia with Creek Indians* 1769 MS

Savery, Solomon (1594–1670). Dutch map publisher and bookseller

Savi, Paolo. *Carte géolog. d'Italie* 1846

Saville, J. Estate surveyor. *Little Hedingham* 1834 MS, *Finchingfield & Stambourne* 1855 MS

Savinkov. Russian engraver 18th century. *Rossiiskoi Atlas* 1792

Savory, E. W. Publisher Steam Press, Cirencester. Eclipse series of *County Maps* 1895

Savry, Jacob. *Four maps Palestine and Near East* and *Plan of Jerusalem* dated 1647, *High Dutch Bible* pub. Amsterdam, Dircksz van Baardt 1648

Savry, Salomon. Engraver of Amsterdam. Portrait of Speed for 1631 *Theatre, Bloody*

Fight Rocroy 1643, *Stadt Groningen* 1652

Sawbridge, T. Printer and Bookseller at the Three Flower de Luces in Little Britain. *Heylin's Cosmography* 1682 (jointly with T. Passinger & B. Tooke)

Sawkins, James Gay. *Geolog. Map Trinidad* 1860, *Geolog. Map Jamaica* 1869, *British Guiana* 1875

Sawyer, Frederick Ernest. *Map Brighton* (1884), *Domesday Sussex* (1886)

Sax, Karl. *Turkey* 1878

Saxton, Christopher (ca. 1542–1606). First *English County Surveys* 1574–9, *Atlas* 1579. Large *map England & Wales* 1583; repub. 1687, reissued by Lea 1693, Willdey 1749, Jefferys 1770. Assisted Lythe *Survey of N. Ireland. Carvickfergas Bay* 1569 MS, Numerous *estate plans. County Atlas* reissued by Webb 1642, edn. by Lea 1690 (town plans added)

Sayer, Henry. *Parish Kingsbury* 1819

Sayer, Robert (1725–1794). Publisher, map and print seller. St. Dunstans Church Fleet Street opposite Fetter Lane; at the Golden Buck opposite Fetter Lane Fleet Street; No. 53 Fleet Street. Sayer was taken into partnership with Philip Overton in 1745; Overton d. 1751 and Sayer started on his own. Sayer pub. Rocque's small *British Atlas* 1753, *map of the Atlantic* 1757, and *Large English Atlas* 1760. He collaborated with Herbert and reissued works of Senex. Sayer acquired some of Thomas Jefferys' assets on the latter's bankruptcy in 1768, and pub. with Jefferys *Gen. Topo. map N. America & West Indies* 1768 and *Middle Brit. Colonies in America* 1768 & 1775. In 1770 Sayer was joined by John Bennett. Jefferys died in 1771, his business passing to William Faden, but Sayer & Bennett acquired part of Jefferys' stock particularly his charts & issued *Gen. Atlas* 1773, *North American Atlas* 1775, *North American*

Pilot 1775–6, *American Military Pocket Atlas (Holster Atlas)* 1776, *West India Atlas* 1775, *Complete Channel Pilot* 1781. Bennett retired and d. 1787. Sayer continued on his own and retired ca. 1792, selling his business & shop to Robert Laurie & James Whittle.

Sayler, Nelson. *Geolog. maps Ind., Kentucky, Ohio* 1865; *Tenn.* 1866

Scacciati, Andrea. Engraver for some of the maps in *Il Gazzettiere Americano* pub. Livorno 1763

Scaihi [Scaiki, Scaichi], Gottfried de. Of Utrecht. Publisher in Rome. *Blaeu's wall maps of 4 continents*

Scala, E. *Vienna in Ungheria Compendiata* Modena 1686

Scalé, Bernard (fl. 1760–87). Surveyor and topographer of Lower Abbey St. Dublin; brother-in-law of J. Rocque. *Plan Trinity College* 1761, *Ireland*, 1762, *Waterford* (with W. Richards) 1764, *Bay Dublin* 1765, *City & Suburbs Dublin* 1773, *Hibernian Atlas* 1776

Scale, R. W. Engraver. For Univ. Mag. ca. 1780

Scaltaglia, [Scattaglia], Pietro (fl. 1780–84). Engraver for Bonne. *Globes* 1784

Scarlett & Scarlett. *Five Maps of U.S. Counties* 1889–91

Scatcherd & Letterman. Booksellers. For Walker's *Univ. Atlas* 1820

Scato, Peter. *Siege plan Gennep* 1642 MS

Scatter, Francis. Engraver for Saxton. *Cheshire & Staffs.* 1579

Scattergood. *16 miles round Philadelphia* (1876)

Scaupae, L. *Plan von Warschau* 1831

Scavenius [Laurids Clausen], (1562–1626). Danish geographer, Bishop of Stavanger. *Stavanger* Blaeu 1640 etc.

Scellier, Le. See **Le Scellier**

Scepsius, T. See **Schoepf**

Schabuecher, Hans. German cartographer, 17th century

Schach, C. *Alpine maps Bavaria,* Salzburg 1866

Schade, Theodor. *Maps Germany* 1860–75

Schaefer, H. W. (1835–1892). German geographer

Schaeffer, Major de. *Operations Swedish Army Pomerania* 1769

Schaeffer, Joseph & Peter. *Freystadt Presburg in Ungarn* 1787

Schaffnit, F. K. *Atlas König. Hannover* 1847

Schagen, Gerrit Lucaszoon van (1642–1690). Amsterdam engraver and art dealer at the Haerlemmerdyck inde Stuurman. *Africa* (1660), *America* (1670), *Asia* (1690), *Denmark* (1700), *Europe* (1720)

Schalbacher, P. J. Publisher in Vienna. *Grand Atlas Universel* (1786), Schraembl's *Allgemeiner Grosser Atlas* 1800

Schalbeter, Johann. *Valesia (Swiss)* used by *Münster* 1545, the first Swiss Canton map. Used later by de Jode

Schalch, Emanuel. Engraver for Scheuchzer's *Switzerland* 1712

Schalck, Agnes. Daughter of H.H. Coentgen. Married Peter Schalck, gilder in Frankfurt (1785)

Schalekamp, Matthijs. Publisher of Amsterdam. Bachiene's *Atlas* 1785

Schall von Bell, Johann Adam (1591–1666). Jesuit Missionary in China, succeeded Ricci at Peking Observatory. *World in Silk* 1628, *Globe* 1636

Schaller, G. *Street & Railway map Germany* 1864

Schallern, Ludwig Edler von (1793–1843). German military cartographer

Schallrooth, Olof A. Swedish surveyor, 17th century

Schamboky, J. See **Sambucus**

Schanternell, Christoph. Globemaker of Augsburg, 18th century

Schantz, C. A. Cartographer. *Austria* 1669

Schapuzet, Charles. *German State maps* Nuremberg 1748

Schardinger, J. *Karte der Braun-Kohlen Bergreviere von Elbogen-Karlsbad* 6 sh. 1889

Scharenberg, W. *Schlesischer Baden Atlas* 1846

Scharfenberg, Georg. *Stadt Görlitz* 1566

Schatz. Engraver for Homann Heirs, 18th century

Schatzen, Johann Jacob. *Schul Atlas* Nürnberg 1745 (after Homann)

Schaudt, Phil. Gottfried. Instrument maker, 18th century

Schaup, C. *Kriege Russlands am Kaukasus* 1846

Scheck, Caspar. *Coelum Stellatum Christianum* 1627

Scheda, Joseph (1815–1888). Austrian cartographer. *Karte von Europa* 27 sh. Vienna 1845–7, *Austria* 20 sh. 1856, *Balkan Länder* 1880, *Oesterreich-Ungarn* 4 sh. 1891

Schedel, D. Hartman (1440–1514). Settled Nuremberg 1484, pub. *Nuremberg Chronicle* 1493. Illust. by Wolgemut & Pleydenwurff. *First modern map of Germany* after Cusanas (Cardinal Cusa) by Hieron. Muntzer. *World map & town plans.* Schedel's Library sold in 1552 to Hans Jacob Fugger.

Schedius, Lajos. (1768–1847). Hungarian geographer

Schedler & Liebler. Engravers, 129 William Street New York. For Colton 1854

Schedler, J. *Seat of War Middle & Southern States America* 1861, pub. (1863)

Scheel, H. I. von. *Islandske Kysts* 1818–24

Scheel, Heinrich Otto von (1745—1808). Danish officer, geographer and topographer

Scheepers, Alois. *Plan d'Anvers* 1868, *Nouveau Plan* 1875

Schefer, Ch. H. Aug. (1820—1898). Orientalist, writer on history and geography

Scheffel, P. J. (1789—1868). Cartographer

Schefler, Joachim Ernst. *Urbis Lipsiae delin.* 1749

Scheibers, J. C. See **Schreiber**, J. C.

Scheidnagel, Manuel. *Distrito de Benguet* 1876, *Islas Filipinas* 1879

Scheidt, von. *Environs of Berlin* 20 sh. 1816—19

Scheiger, Jos. von. (1802—1887). Austrian topographer

Scheiner, Christoph (1575—1650). German mathematician and cartographer

Schenck, Fr. Printer of Edinburgh. Colour printing of maps for Black's *Atlas* (1860)

Schenck & MacFarlane. Colour printers for Black (1857)

Schenck, Herman. *Plan der Stadt Halle* 1866

Schenck, J. H. *Map of Long Branch N.J.* 1868

Schenck, P. *Map village Fulton, Oswego N.Y.* 1855

Schenk, A. J. *Swansea Harbour* 1880

Schenk Jansz, Leonard (1732—1800). Engraver of Amsterdam. *N. America* 1755, Bawr's *Moldavia* 1771, Meijer's *Africa* 1785

Schenk, Pieter (1645—1715). Publisher and engraver of Amsterdam sur le Vygendam. Bought plates from Blaeu 1694, Janson 1694, Visscher 1712, *England* 1690, *Hungary* 1700, *Hecatompolis* (100 pl.) 1702, *Atlas Contractus* 1705, *Théatre de Mars* 1706

Schenk, Pieter Junior (ca. 1698—1775).

Took over business at father's death ca. 1718. *Persia* 1722, *Theatre War Poland* 1733, *Flambeau de Guerre* 1735, *Atlas Saxony* 1752. Worked with G. Valck.

Schenkel, Conrad. *Maps of Moravia* 1845—51

Schepero, Sebastian. *Sphaera automatica* 1711

Scherer, Heinrich (1628—1704). Jesuit, Prof. of Maths in Munich, geographer. *Géogr. Naturelle et Physique,* and *Géogr. Hiérarchique* 1703; *Atlas Marianus* 1702, *Atlas Novus* 8 vols. 1702—1710

Scherer, Leonhard. Augsburg. *Raetiae Veteris* 1616

Scherer, Wilhelm. *Plan von Elberfeld* (1876)

Scherer. See **Monk & Scherer**

Scherm, Lauwerens. Draughtsman and engraver. Title page to De Wit's *Atlas* (ca. 1700), *Plans Towns in Flanders* (1700)

Scheuchzer, Johann Gaspar. *Japan, Miaco, Jedo &c* for Kaempfer

Scheuchzer, Johann Jakob (1672—1733). Swiss mathematician, physician and geographer of Zürich. *Switzerland* 4 sh. 1712, *Nova Helvetiae Tab.* Amsterdam Schenk 1716, Covens 1735

Scheurleer, Henri. Publisher in The Hague. *Plan Hague* 1755, D'Anville's *Atlas de la Chine* 1737

Scheurmann, Johann Jacob. Engraver. Weiss' *Atlas Suisse* 1786—1802, *Helvet. Repub.* 1799, *Plans de Zürich* (1814)

Scheurer, Christoph. Maps for Gluck's *Nuremberg* 1733

Scheus, Herm. Publisher, Rome. *Strada* 1640 & 1653

Scheyb, Franz Christoph von. Engraver of *Roman road map (Peutinger Tab.)* 1753

IOANNES IACOBUS SCHEUCHZER, HELVETIO TIGURINUS,
MED. D. MATH. IN LYCEO PATRIO PROF. ACADEMIÆ CÆSAREÆ
LEOPOLDINO-CAROLINÆ ADIUNCTUS DICTUS ACARNAN, NEC
NON SOCIETATUM REGIARUM ANGLICÆ ET PRUSSICÆ MEMBRUM.
ÆTAT. ANN. LIX.

J.J. Scheuchzer (1672–1733)

Schiaparelli, Giov. v. (b. 1835).
Astronomer

Schickhar[d]t, Heinrich (1558–1634).
Architect. *Würtemberg,* Monteliard 1616

Schickhardt, Wilhelm (1592–1635).
Prof. Maths, astronomer and mapmaker.
Nephew of Heinrich Schickhardt.
Würtemberg 13 sh. 1634, now lost.
Land Tafeln Tübingen 1669

Schieble, Erhard. Engraver of Paris,
Rue Bonaparte. *Mont Blanc* 1865, *Nice*
1866, *N. Pole* 1867, *Paris* 1867 &c.
Engraved for Dufour & Andriveau-Goujon

Schiedt, Jacob E. *Atlas* Philadelphia
1892

Schiegg, P. Ulrich (1752–1810).
Surveyor

Schiepp, Christoff. Engraver of Augsburg. *Universi orbis descrip. Copper
globe* 1530

Schiffler, Johann. (fl. 1528). Cartographer
of Nuremberg

Schilde, J. See **Schille**, J.

Schilder, Cornelis. *S. Holland* 1537 MS

Schilder, Matias. Swedish surveyor, 18th
century

Schilker, George. Bookseller of Vienna.
Fabricius' *Moravia* 1575

Schill, J. *Geolog. Karte v. Baden* 1856–
70

Schille, B. von. See **Scultetus**

Schille [Scillius, Schilde, Schiller] , Jan
van (1533–1586). Cartographer of
Antwerp. Military engineer, maps for
Ortelius & De Jode. *Liège & Trèves*
1578, *Lorraine & Luxemburg*

Schillemann. *Trinité Dahomey* 1893

Schiller, Julius. Star maps by Kilian.
Coélum Stellatum Christ. Strassburg
1627

Schiller, K. *Railway map Austria-
Hungary* 1877

Schilling, Friedrich. *Railway map
Europe* 1843

Schimek, Max. (1748–1798). Map for
Schraembl (1786–1800), *Atlas War
Austria-Russia-Turkey* 1788

Schimmelfinnig, A. Drew for Whipple
1855

Schimmelpenning. German military
cartographer, 18th century

Schinbain [Tibianus] , Johann Georg.
(1541–1611). *Bodensee* 1578, 1603,
Schwartzwald 1603

Schindel, Johannes (1370–1442).
Austrian astronomer

Schindelmeyer, Karl Robert. Austrian
engraver, 19th century

Schindler, Albert Houten. *Maps Persia*
1879–86

Schindler, Otto. *Stadt Carlsbad* 1876

Schippan, Heinrich Adolph *Neighbourhood Freiburg* 1823, *Topo. plan
Freyberg* 1837

Schirach, Gottlob Benedikt von. *Sumatra*
(1780)

Schissler, Christoph (1530–1608).
Instrument maker of Vienna. *Sundials
with maps* 1558–68, *Terr. Globe* 1597,
Augsburg 1602

Schjelkrup, H. F. (1827–1887) Danish
astronomer

Schlachter, H. *Sverige* 1870

Schlachter, J. *Central Europe* 1875;
Greece 1886

Schlagintweit, Adolph von. *Atlas Physic.
Geogr. u. Geolog. der Alpen* 1854

Schleenstein, Johann Georg. (1660–
1732). German military cartographer

Schleich, C. S. Junior. Engraver for
Schlieben 1828–30

Schleich, Johan Karl. German engraver,
19th century

Schleuben, J. I. (1740–1774). Engraver
of Berlin. *Preussische Länder* Berlin,
for Petri (1760)

Schley, Jacob van der. Engraver. *Battle
Plans Germany* pub. by Goose & Pinet.

1763–66 maps for Prévost's *Hist. des Voyages* 1747, Bawr's *Moldavie* (1770)

Schlicht, Rudolph. Publisher of Mannheim. *Poland* 1831

Schlieben, Wilhelm Ernst August. von (1781–1839). Cartographer of Leipzig. *Norwegen u. Schweden* 1825, *Atlas von Europa* 1829, *Atlas von America* 1830

Schlomach, Melchior of Dresden. *Saxony* 1679 MS

Schlozer, August Ludwig. *Amerika,* Göttingen u. Leipzig 1777

Schmalkalder, Samson. Military cartographer, 17th century

Schmelling, Hugo von. *Maps of Brandenburg, Pommern, Posen, Preussen &c.* (1865)

Schmeltzer, Lieut. Engineer. *Insel Rügen* 1841, *Münster* 1883

Schmettau, Comte de. *Naples* 1719–21

Schmettau, Fried, Wilhelm Carl, Baron von. Maj. Gen. military cartographer. *Berlin* 1749, *Bohemia* 1789

Schmettau, Samuel Graf von (1684–1751). *Topo. Karte Sizilian* 25 sh. 1719–21, *Piemont* 1745

Schmid, Christian Andreas. *Goslaria* 1732

Schmid, Ernst E. *Topo. Geogr. K. Umgebungen von Jena* 1859

Schmid, Franz. *Swiss panoramas* 1825–68

Schmid, J. *Petit Atlas Suisse* 1840

Schmid, Ludwig. *Oesterreich* 1800

Schmid, Sebastian. *Zürich* (with Maurer), *Cart. manual* 1566

Schmid de Grueneck, Col. *Pays des Grisons* (with P. Cluver) for Ottens (1715), for Cluver 1724

Schmidburg, G. R. von. Worked for Gaspari 1804–11

Schmidio, Jacobo. See **Schmidt, J.**

Schmidt, A. R. *Geogr. Karte van Vorarlberg* 1839–41

Schmidt, E. C. & F. E. *Dresden u. Umgebung* (1871)

Schmidt, Edward. *Maps Bohemia* 1869

Schmidt, F. W. *Dioc. Aggerhus* 1783, *Plan Conde* 1780

Schmidt, Georg. Fried. (1712–1775). Berlin engraver

Schmidt, Georg. Gottlieb (1768–1837). Mathematician and astronomer

Schmidt, H. *Mosquito-Staater* 1845, *Kort over Esbierg* (1898)

Schmidt, I. I. C. *Plan Village d'Oelper (Brunsvic)* (1762)

Schmidt, I. L. *Deutsches Reich* 1876

Schmidt, I. M. *Postkarte Deutschland* 1786

Schmidt, Jacob F. *Govern. Wiburg* 1722, *Novogorod* 1772, *Tartary* 1774, *Plescow* 1773, *Walachia* 1774, *Gulf Finland* 1779

Schmidt, J. F. Julius. *World Maps* 1878

Schmidt, James F. *Maps Baltic & Russia* 1770–1787, *Reval* 1770, *Moldavia* 1774, *Sinus Finnici* 1777

Schmidt, J. H. *Esthland* 2 sh. 1844 & *Liv-Esth-Kurland* 1867

Schmidt, Johann Marias Friedrich (1776–1849). Cartographer, Prof. in Berlin. *Deutschland* 1816, *Australia* 1820, *Africa* 1827, *America* 1830, *Europe* 4 sh. 1829

Schmidt, Johann Michael. Engraver for Homann Heirs 1784

Schmidt, Ludwig. *Maps Austria* 1798–1814

Schmidt (O.F.) *Post Reise Karte Deutschland* Berlin 1824, Schropp 1831

Schmidt, Paulus. *Berlin* 1795, *Preussische Staate* 1859, *Hemispheres* 1837

Schmidt, R. of Leipzig. Engraver for Weiland 1848. Drew *Ober Californien* 1849

S. Schmettau (1684–1751)

Schmidt, Wilhelm. Theophil (1756–1819). Mathematician and cartographer

Schmidtfeldt, J. *Taschen Atlas* Vienna, T. Mollo (ca. 1820)

Schmidtmeyer, Peter. *Travels into Chile* 1824

Schmirmund, H. *Plan Mainz* 1878 (with Happenberger)

Schmitfeldt, Georg. *Bohemia* 1831

Schmitz, J. P. *Special Karte der Schweiz* (1860)

Schmitz, Leonhard. Worked for publisher Collins. *Class. Geogr.* 1873, *Internat. Atlas* 1873, *Library Atlas* 1875

Schmölder, Capt. B. *N. America* 1848

Schmoll, Friedrich. *Boehmen* 1809

Schmollinger, W. Engraver. *Plan London* 1833, Moule's *English Counties* 1836, *Inland Waterways* 1837

Schmück, Michael (1535–1606). *Thuringia* 1593
Schmude. *Umgegend von Posen* 4 sh. 1844

Schmuzer [Schmuter], Johann Adam Engraver. *Regnum Sclavoniae &c.* Vienna (1718)

Schnapp, R. *Hessen* 1870

Schnebbeli, R. B. *Plan Blackwall Hall, Guildhall* 1819 MS, *Plan Old Guildhall Chappel* 1819 MS, *Ludgate Prison* 1819

Schneider, Adam Gottlieb (1745–1815). Carto. Publisher of Nuremberg. Worked with Weigel. *W. Hemisphere* 1797, *Netherlands* 9 sh. 1787

Schneider & Weigel. Founded 1746 Nuremberg publishing house

Schneider, Capt. A. Engineer, Surveyor General of Ceylon. *Ceylon* 1816 & 1822

Schneider, F. L. *Schlesien u. Glatz* 1845

Schneider, Friedrich. Translated *Sailing directions for Kattegat* 1800

Schneider, J. F. *Plan of Berlin* (1798), *Topo. carte Berlin area* 1811

Schneider, J. H. *Atlas des enfans* Amsterdam 1773

Schneider, Nicolas. Printer of Legnica. Helwig's *Silesia* 1605

Schneider, R. A. Engraver for Homann. *Hildesheim* 1727, *Hanau* 1728, *Cadiz & Gibraltar* (1730)

Schnell, Fidel. Surveyor, 18th century

Schnell, Paul. *Maps Morocco* 1892–8

Schniepp, Christoff. *Terrest. Globe* 16th century

Schnitzer, Johann. Engraver of Arnsheim. *World Map* in 1482 edition of Ptolemy, earliest signed engraved map

Schnitzler, Johann Heinrich (1802–1871). *Atlas Hist. et Pittoresque* 1860–85

Schnoedt, Carl Ludwig. *Anhalt* 1757

Schnyder von Wartensee (1750–1784). Topographer, Luzern

Schoch, Conrad. *Suisse* 1818

Schoda, Josef Ritter von. *Oest.-Ungar. Monarchie* 20 sh. 1870

Schoel, Hendrik van. Printer. *Nettuno* 1557, *Rome* 1601, *Algieri* 1601, *Augusta* 1602, *Zighet* 1602

Schoel, Hiero. *Stockholm* Jansson 1657

Schoell, F. Bookseller of Paris. Humboldt's *Mexique* 1811

Schoell & Co. Publishers. Weiss' *Schweiz* 2nd edition 1803

Schoen, Field Marshall Anton, Freiherr von (1783–1853). Austrian cartographer

Schoen [Schoett], Erhard. *Plan Buda* 1541 for Braun & Hogenberg

Schoenberg [Schönberg] & Co. Publishers of New York. *World* 1861, *Mexico* 1866, *Standard Atlas* 1868

Schoenborn, A. *Klein Asien* 1844

Schoene, Gustav. *Wein Wasser und Alkohol Karte (Rheinlandes)* (1868)

Schoenemann, Friedrich. Engraver and publisher of Hamburg. *Panorama Lisbon* 1756

Schoener [Schöner, Sconer], Johann (1477–1547). Geographer and astronomer of Nüremberg, Prof. Maths. *Globes* 1515, 1520, 1523, 1533

Schoenerer, M. *Railway map Austria* 1835–42

Schoenfelder [Schonfader] von Feuersfeld, Franz. *Post & Road map Bohemia* 1852

Schoennagel, A. *Würtemberg Railway* 1870

Schoenning [Schönning], Prog. G. *Iceland* 1771–2, *Norwegia* 1778

Schoenwetter [Schönwetter], Godfried. Publisher of Mainz. *Strada* 1651

Schoenwetter, J. B. Publisher in Vienna *Germ. Aust. Topo.* 1701

Schoenwetter, Johann Martin. Publisher of Frankfort am Main. *Strada* 1699

Schoepf [Scepsius], Thomas (1527–1577). *Berne* Strassburg 1578

Schol, Hieronymus. *Stockholm & Bergen* in Braun & Hogenberg 1588

Scholey, R. Bookseller. For Walker's *Univ. Atlas* 1820

Scholfield, Nathan. Civil engineer. S. *Oregon & N. California* 1851

Scholl, Louis. Draughtsman. For Ingalls 1855

Scholz, J. C. F. *Schlesien* (1850)

Schomberg, A. W. *Harbour of Cefalonia* 1823

Schomburgk, Sir Robert Hermann. *Anegada* 1832, *Brit. Guiana* 1840, *Barbados* 1847

Schonaich, Georg. *Lower Silesia,* Blaeu 1662

Schonberg, L. Lithographer. *Map Tasmania* 1837

Schönberg & Co. See **Schoenberg**

Schöner, J. See **Schoener**

Schön[n]ing, Prof. See **Schoenning**

Schönwetter. See **Schoenwetter**

Schoonebeck, Adrian (1661–1705). Engraver

Schoonebeek [Damianus], Pieter Damiaan (fl. 1688–1705). Engraver and mapseller of Amsterdam. In Russian service 1699–1705. *Armenian Atlas* 1695 MS, Cassini's *Planisphere,* Nolin 1696

Schooten, W. C. See **Schouten**

Schöpf. See **Schoepf**

Schooley, David. *Co. of Luzerne, Penn.* 1864

Schorman, P. Engraver of Amsterdam. Adrichom's *Palestine* 1665

Schort, Joh. Jac. Engineer. *Schenckenschans* 1636 pub. Visscher

Schotanus van Sterringa, Bernhardus. *Frisia* for Visscher, used by Jansson ca. 1658; *Atlas Friesland* 1664 (35 maps) used later by de Wit 1698

Schoten, Francis van. Prof. Mathematics at Leiden, cartographer. *Bergen op Zoom,* Blaeu 1630 etc. and Jansson (1636)

Schott, E. See **Schoen**, E.

Schött, Joh. Publisher and printer Reisch' *Margarita philosophica* 1503, *Ptolemy* Strassburg 1513, 1570

Schouten [Schooten, Schotenio], Willem Cornelisz [Gul. Cornel] (1567–1625) of Hoorn. *Voyages* 1615–17, *Map of Schouten & Le Maire's passage thro. Straits of Magellan* 1618, English Edn. 1619. *New Guinea* 1619

Schouw, Joakim Frederik (1789–1852). *Pflanzen geog. Atlas* 1823, *Physikal Atlas* (1850)

Scottus, Carolus. Engraver of Bologna 1680

Schraag, Fred. *Monterey. Tenn. Corinth Miss.* (1862)

Schrader, C. *Hatzfeld Haven* 1887

Schrader, F. (1844–1924). *Atlas de Géogr. Mod.* Paris, Hachette 1889, *Géogr. Hist.* 1896

Schrader, Theodore. Publisher of St. Louis. Kelley's *Arizona* 1860

Schraembl, Franz Anton (1751–1803). Austrian cartographer. *Allgemeiner deutscher Atlas aller Länder* 138 II. Wien 1786–94, *Schweiz* 2 sh. 1789, De la Rochette's *Cape of Good Hope* 1789, *Allgemeiner Grosser Atlas* 1800

Schreiber, Ch. *Atlas Militaire* 1848–50

Schreiber, Charles. Surveyor. *Whitehead, Kansas* (1857)

Schreiber, Johann Christoph. Engraver and cartographer of Leipzig, 17th century

Schreiber[n], Johann Georg (1676–1750). Son of above. Cartographer and publisher. *America* (1720), *Lusatiae Sup.* 1732, *Atlas Selectus* Leipzig 1749

Schreiner, Luiz. Engineer surveyor. *Rio de Janeiro* 1879

Schrenck, Albert Philibert, Baron von (1800–1877). *Karte Aldenburg* 1856, 1865; *Grossherzogthum Oldenburg* 1872

Schreyer, Sebald. Associate publisher of *Nuremberg Chronicle* 1493. See also under **Schedel**, H.

Schreyvogel. Publisher. *Dalmatia etc.* 1810 (with Riedel)

Schroeder, Gillius. Swedish surveyor, 18th century

Schroeder, John Frederick. *Diocese of New York* 1851

Schroeter, A. Engraver and geographer for Disturnel 1853, *U.S.–Mexican Boundary* 1853

Schroeter, E. *Landkarte v. Mansfelder See Kreis* (1890)

Schroetter, Freiherr Leopold von (1754–1816). *Preussen* 25 sh. Berlin 1803

Schroeter, G. *Cal., Oregon, Wash., Utah* 1853

Schroeter, Johann H. (1745–1816). Astronomer

Schroll, Z. *Danmark* 1856

Schropp, G. Publisher of Berlin. Schmidt's *World* 1820

Schropp Simon & Co. Map sellers and publishers of Berlin. *Schweiz* (1798), Schmidt's *America* 1820, Bonne's *N. America* 1826, *Australia* 1828

Schrot [Schrott, Schrotenus], Christian. See **Sgrooten**, C.

Schryver, C. See **Grapheus**

Schubach, D. A. *Stadt Bergedorf* 1875

Schubach, E. W. & Fr. Eugen. *Unter Elbe* 1837

Schubach, Fr. Eugen. *Flensburg* 1843, *Plan Hamburg* 1839, *Hamburg u. Altona* 1856

Schubarth, Matthaeas (fl. 1723–58). Lieut. Engs. Successor to Weiland, revised his *survey Silesia* 1735–50

Schubert, G. L. *Russian maps* 1832–56

Schubert, Theo. Fried. von (1789–1865). Military geographer

Schubert, Gen. Major. *Kriegsstrassen Karte Russland* 16 sh. 1837

Schuberth, Julius. *Neuester Atlas* 1847, *Hand Atlas* 1850

Schuchart, Johann Tobias. *Anhalt* 1710, 1746

Schuchman, William. Lithographer of Pittsburgh, for Gunn 1859

Schuckburgh, J. Publisher, Fleet St. Salmon's *Modern History* 1744–6

Schudi, G. See **Tschudi**

Schueltz, Balt. Fried. (1664–1734). Military cartographer

Schuetz, Carl. Engraver and designer. *Land ob der Enns* 1786–7, *Kriegstheater Österreichs &c.* 1788, *Wien* 2798

Schultz. See **Sculteto**

Schultz, Lieut. Contributed to *map Environs Berlin* 10 sh. 1816–19

Schultz, Bart. (1540–1614). Mathematician and cartographer

Schultz, F. *Canal Map Sweden (Stockholm-Gotheburg)* 1837

Schultz, Fried. *Orograph. Carte von Europe* 1803

Schultz, J. *Gegend um Berlin* 1870

Schultz, William P. *Whitfield Co., Georgia* 1879

Schultz, Waldemar. *Deutsche Colonien . . . Prov. Sao Pedro do Rio Grande do Sul* 1865

Schultzen, Caspar. *Bremen* 1664

Schulz, C. *Plan Station* 1869

Schulz, E. G. *Plan v. Jerusalem* 1845

Schulz, G. *Provincia de Oviedo* 1878

Schulz, R. A. *Africa* 1842, *Post u. Strassenkarte Illyrien* 1848, *Post u. Strassenk. Deutschland* 1859, *Öesterreich* 1866

Schumacher, E. *Umgegend v. Strassburg* 1883–4

Schuman, Edouard. *Carte de la Télégraphie Electrique Europe Centrale* 1857

Schurig, Kurt. *Plan v. Plauen* 1874, *Stadtplan Lobenstein* 1885

Schurtz, Con. Nico. Engraver. Beer's *Ireland* (1690)

Schusser, Vincenz. *Reise Karte Oesterreich* 1862, *Salzburg* 1874

Schut, A. *Map* for Danckerts ca. 1710

Schütz, C. See **Schuetz**

Schwaerzle, J. *Maps of Egypt & N. Africa* 1837–56

Schwan, Wilhelm. *Jerusalem* 1629 (after Adrichom)

Schwantz, Fridericus. *Walachia* 1717

Schwartzkopf. *Environs Berlin* 10 sh. 1816–19

Schwarz, Joseph of Heidelberg. Surveyor, 18th century

Schwarzenberg, Adolph. *Kurhessen* 1853 (with Reusse)

Schwarzenfeld, E. von. *Seestadt Libau* 2 sh. 1887

Schwarzer, Ernst von. *Indust. k. Boehmen* 1842

Schwarzmann, Joseph. *Maps Brasil* 1823–31

Schweickhardt, Franz Xaver (1794–1858). Viennese cartographer

Schweizer [Suicerus], Joh. Heinrich (1553–1612). *Nova Helvet. Descrip.* (1605), *Chronolog. Helvetica* 1607

Schwenck, J. (ca. 1699–1735). Cartographer of Ulm

Schwengeln, Georg von (ca. 1590–1668). Swedish military Eng. *Siege Riga* 1621, *Estonia & Livonia* 1627 MS, *Arensburg* 1645 MS

Schwensen, Carl. Norse lithographer in Oslo. Godtkjob's *Atlas* 1859

Schwilgue, Jean Baptiste. *Celest. Globe* 1842

Schwinck, G. *Mappa Coelestis* 1843

Schwytzer [Schwyter], Christopher. Engraver. Norden's *Sussex* 1595 and for Speed

Schynvoet. Engraver for Chatelain. *Atlas Historique* (1720)

Scillius, J. See **Schille**

Scobel, A. *Africa* (with R. Andrée) 4 sh. 1884

Scobie, Hugh. Publisher and lithographer of Toronto. *Lake Huron* 1850

Scofield, Horace G. *Atlas City Bridgeport Conn.* 1876

Scolari, Stefano (fl. 1598–1650) Venetian engraver and publisher "all' insegna delle Tre Virtu a San Guliano." *Maps* of Gastaldi & Bertelli end 16th century, *Territ. Veronese* 1612

Scholes. Engraver of New York. Worked for Reid 1795, Winterbotham 1796, Payne 1800

Scoles, John. Engraver for Payne's *Univ. Geog.* 1800

Score, Edward. Bookseller over against the Guildhall, Exeter. *Plan Exeter* 1709

Scoresby, William (1783–1857). English mariner and explorer. *Polar Regions*

Scot, R. Engraver. McMurray's *Map of U.S.* 1783–4, Thomson's *Atlas* 1817

Scotin, Gérard Jean Baptiste. Jaillot's *Paris* 1713, *Plan de Rheims* 1722

Scott, A. *Porto Rico* 1846

Scott, B. Engraver for Bell 1833–6

Scott, Capt. C. Rochfort. *Plan Castle St. Juan de Ulloa* MS, *Sketch course River Panuco Tampico* MS (ca. 1830), *Map Palestine & Syria* 1846–58

Scott, Major Francis Henry. Depy. Q.M.G. Madras Army. *Penin. India* Madras 1855

Scott, George. Publisher. Adair's *Clyde* 1731

Scott, Capt. James. *Strait Pinang* 1786

Scott, James D. American cartographer. *Hampden Co., Mass.* 1854, *Philadelphia* 1855, *Baltimore* 1856, *Burl. Co.* 1876

Scott, John. Customs Officer Hull. *River Humber* 1734, first printed chart Humber

Scott, Sir John (1585–1670). Scot statesman, prepared Blaeu's *Scotland*

Scott, Joseph. Engraver and cartographer. *United States Gazetteer* Phila. 1795, *Atlas U.S.* 1796. Worked for American publisher Carey.

Scott, R. *New Geogr. (County maps)* 1681

Scott, Robert (1777–1841). Engraver. *U.S.* 1783, *Philadelphia* 1794, *Lanark.* 1798, *Glasgow* 1801, *Ross* 1813. Engraved for Thomson (1817) and Fullarton (1834)

Scott, Capt. Robert. *Charts S. America* 1784–88

Scott, Thomas. Assistant Surveyor General of Lands in Tasmania. *Chart of Van Diemen's Land* 1824

Scott, William. *Plan Harwich* 1819 MS

Scott, William. Surveyor. *Sutherland* 1833, in conjunction with G. Burnett

Scott, W. H. *Central Asia* 1866

Scotti [Scottus], Carlo [Carolus]. Engraver. *World* ca. 1680

Scotti, Vinc. *Flags of World* 1804

Scotto, Benedetto. *Globe Maritime* 48 sh. Paris 1619 & 1622

Scotto, Giacomo [Jacobo] Levanto. *Portolan of Mediterranean* 1589 MS, *Atlas* 1593

Scowden, T. R. *Cincinnati Water Works* (1865)

Scratchley, Peter Henry. *Plan Sevastopol* 1857 (with Capt. Cooke)

Scribling. Surveyor. *Pendleton District, S. C.* 1820, for Mills 1825

Scribner, C. & Sons. Publisher, New York. *Statist. Atlas U.S.* (1883)

Scribonius, C. See **Graphaeus**

Scull, James. *Lehigh Co., Penn.* 1816 (with I.A. Chapman)

Scull, Nicholas. Surveyor Gen. of Penn. 1748–61. *Penn.* 6 sh. 1759, *Phila.* (with G. Heap) pub. Faden 1777, another edition pub. Lotter

Scull, William. Son of N. Scull. *Map Penn.* 1770 and 1775, French edition 1778

Scully, William. *Brazil* 1866

Scultetus [Sculteto, v. Schille, Schultz], Bartholomeus (1540–1614). Cartographer of Görlitz. *Meissen & Lausitz* 1568. Used by Ortelius 1573, Blaeu &c. *Ober & Nieder Lausitz*, de Jode 1578

Scultetus, Jonas (1603–1664). Silesian cartographer. *Glatz* 1626 used by Jansson 1636 & Blaeu 1640, *Silesia* used by Jansson & Blaeu, *Breslau* Jansson 1649

Scylax of Caryanda (4th century B.C.). Greek geographer. *Periplus of Erythrean Sea*

Scymnos (2nd century B.C.). Greek geographer. *Description of world*. Fragments only

S. D. U. K. See **Society for Diffusion of Useful Knowledge**

Seale, Richard William (1732–1785). London engraver and draughtsman. Engraved for Popple 1733, Tindal &

Rapin 1744–7, Pine & Tinney 1749,
Bolton *N. America* 1750, Stow 1756,
Universal Mag. 1747–63, *War N. America*
1757

Seall, Robert. *Voyage of Thos.
Stuckley . . . Terra Florida* 1563

Seally, John. *Geogr. Dict.* 1787

Seaman, James V. *New General Atlas
N.Y.* 1820

Searcy, I. G. *Florida* 1829

Searl, A. D. *Eastern Kansas* 1856 (with
Whatman)

Searle, S. W. *New Haven, Conn.* 1859

Searle, Daniel. *Manor of Ockham* 1706

Searles, L. Estate surveyor. *Wanstead*
1779 MS

Sears, See **Fenner, Sears & Co.**

Seaton, John. *Plan Vizagapatnam* 1783

Seaton, Robert. Hydrographer to the
King. *England & Wales* (1830),
Palestine, Neele (1835)

Sebastiano di Aragona. Painter of
Brescia. *Brescia* 1571

Sebastianus di Regibus Clodiensis [da
Re di Chioggia]. Italian engraver. For
Tramezini, *Ligorio &c., Crete* 1554,
Belgium, France, G.B. Naples 1558,
Hungary, Spain 1559, *Greece,
Portugal* Rome 1561

Sécalart. Pilot and cosmographer.
Treatise on Cosmography 1545

Secco [Seccus], F. A. See **Alvares Seco**

Seckford, Thomas. Patron of Saxton.
Commissioned his *survey of England &
Wales* 1579

Secsnagel [Secznagel, Zecsnagel,
Setznagel], Marcus [Mark]. Cartographer
b. Salzburg. *Map & plan of Salzburg*
1551, used by Ortelius, revised 1595,
used by de Jode 1578 & Braun &
Hogenberg

Sedano, F. de M. y. See **Monteverde
y Sedano**

Sedille, J. *Carte du Théatre de la Guerre*
1870

Seefried, Friedrich (1549–1609).
Surveyor, engraver

Seehusen, P. J. Engraver for Bredschorff
& Olsen. *Esq. orographique de l'Europe*
1830, *Maps Denmark* 1841–6

Seelstrang, Arthur de (1838–1896)
Argentine geographer. *Repub. Argentina*
1875, *Atlas* 1886–92

Seely, Capt. John B. *India* 1826

Seener, P. *Plan Regensburg* (1890)

Seetzen, Ulrich Jasper. *Imp. Map Palestine*
(1864)

Segato, Girolamo. *Maps Egypt, N. Africa
& Tuscany* 1823–44

Seger (d. 1583). Painter and woodcarver.
Rivers Aarach & Main (ca. 1574) MS

Segner, Johann Andreas von (1704–
1777). Mathematician and astronomer

Segresser, J. P. *Stadt Luzern* (1848)

Seguin, Jean. Engraver and draughtsman.
France (Cassini) 1744–60, *Bresse* 4 sh.
1766, *Montagnes de Vosges* 1769,
Bourgogne 15 sh. 1771

Ségur, Louis Philippe, Comte de (1753–
1830). *Atlas Univ.* 1822, *Atlas Géogr.
Ancienne* 1827

Segurola y Linares, Juan. *Travel Maps
Spain* 1874–5

Seibel, J. Bombadier. *Plans* for Therbu
1789–91

Seidel, A. *Railway map Germany* 1889

Seidel, L. W. *Austria Hungary* 1883

Seidel, R. Engraver for Meyer's *Hand
Atlas* 1867

Seiffert, Moritz. German cartographer.
Dresden 1873–4, *Loessnitz* 1876,
Chemnitz 1877(–80)

Seignelay, Jean Bapt. Colbert, Marquis
de (1651–1690). Hydrographer

Seile, Henry. Publisher "over against St.
Dunstans Church in Fleet-streete." Heylin's
Cosmographie 1652, 1657

Seile, Anna. Widow of H.Seile. Heylin's *Cosmographie* 1669, *Asia* 1671, *Africa* (ca. 1673)

Seiller, Johann Georg. (1663–1740). Engraver of Schaffhausen. *Schaffhausen* 1681, *Zürich* 1698, *Switzerland* 1718

Seitz, C. Engraver for Kiepert's *Hand Atlas* 1860, *München* 1871

Seitz, Johann Bapt. (1786–1850) Topographer & Engraver

Sekell, A. C. Draughtsman. *Plan City Grand Rapids Mich.* 1872

Selander, N. *Atlas öfver Sverige* 1880

Selander, N. J. T. *Karta öfver Sverige* 14 sh. (1884)

Selby, Lieut. (later Commander) William Beaumont. Indian Navy Hydrographer. *Survey India* 1841, 1846–50. *W. Coast India* 1855. *Bombay Harbour* 1858, *Part of Mesopot.* (1868)

Seligmann, Joh. Michael. German Edition of Catesby with *map Carolina, Florida & Bahamas* 1775

Seljan, D. (fl. 1810–48) Cartographer

Seljana, Dr. Agutina. Geographer. *Austria &c.* 1847

Seller, Jeremiah. Hydrographer to the Queen, Publisher at the Hermitage Stairs Wapping. Succeeded to John Seller 1698? Bankrupt by 1705. *Pract. Navig.* 1699 (with R. Mount) *English Pilot 5th Book* (with C. Price) 1701. *System of Geogr.* 1703 *English Pilot 4th Book* 1703

Seller, John (fl. 1664, d. 1697). Hydrographer to Charles II & James II. Publisher, surveyor & compass maker, seller of Nautical Instruments. (1) Exchange Alley in Cornhill (also called shop "at the West side of the Royal Exchange") later Popes Head Alley (2) Sign of the Mariners Compass at the Hermitage stairs in Wapping. *Map New Jersey* (1664) & 1677 (with William Fisher), *English Pilot Parts 1 & 2* 1671–2, *Atlas Maritimus* 1675, *Oxford* 2 sh. 1675

(with J. Oliver), *Atlas Minimus* 1679, *Middx. Kent & Surrey* 1680; Proposed *Atlas Anglicanus* (with Oliver & Palmer) 1681, *Atlas Terrestris* 1680, *Hydro. Universalis* (1690), *Atlas Contractus* 1695

Seller, John Junior. Collaborated with father 1685

Selma, Fernando. Engraver fo Aguirre 1775, Torfino 1788, Bauza 1788, Antillon 1804

Selous, Frederick Courtney. *Mashonaland* (1887), *Matabeleland* (1890)

Selss, Ed. *Terrestrial Globe* 1841

Selter, J. C. *Grundriss v. Berlin* (1800) & 1845

Seltzlin, David. Engraver of Ulm. *Circul. Sveviae* 1573, *Franconia* 1576, *Schwäbischer Kreis* 1579

Selves, Henri. *Atlas Géogr.* 1822–29

Selwyn, Alfred Richard Cecil. Government geologist. *Maps Canada* 1861–84

Selwyn, Charles Henry. *Corea* 1894

Semedo, Alverez. *China* 1655

Semen. Engraver for Langlois 1811

Semen, younger. Engraver for Delamarche 1827

Semenoff, Petr. Petrovich (b. 1827, fl. 1850–75). Russian geographer. *Dict. Russian Empire*

Semper, Carl. *Philippines* 1869

Sems, Jan (1572–1656). Dutch surveyor of Friesland

Senefelder, Johann N.A. (1771–1845). Lithographer and publisher. Founder of lithography

Senex, John. Surveyor, engraver, publisher. Geographer to Queen Anne. Addresses: Next to the Fleece Tavern in Cornhill (with Price); Against St. Clements Church in the Strand 1703; Salisbury Court in Fleet Street (with Maxwell); Whites Alley in Coleman St. (with Price); At the Globe against St. Dunstans Church

Title-page to J. Seller's "Atlas Terrestris"

Fleet St. *Stars N. Hemi.* (1690). Pub.
Seller's *System of Geography* (with
Jeremiah Seller & C. Price), *Gen. Atlas*
fol. 1708–12, Lawson's *Carolina* 1709,
English Atlas 1714, Ogilby's *Roads* 1719,
New Gen. Atlas 1721; Engraved Mayo's
Barbados 1722, Norden's *Herts.* 1723,
Budgen's *Sussex* 1724, Elected to Royal
Society 1728. *Surrey* 4 sh. 1729, *First
Settlements America* 1735, *Surrey,
Sussex, Kent, Hants. & Berks.* 1746

Senex, Mary. Widow of John. Continued
business at Globe over against St. Dunstans
Church in Fleet St. *Catalogue of Globes,
maps &c.*

Senger, C. I. *Plan Stadt Liegnitz* 1868

Sengre [Senghre, Sengher, Senger] , Henri.
Cartographer, Secretary to Marquis de
Vauban. *Environs Strassburg* 1680, *Rhein
& Mein* (1685), *Lorraine &c.,* Jaillot
1692; *Basle-Bonn* 8 sh. 1705

Sengteller, L. A. (1846–1889). Polish
cartographer and engraver. Worked for
Migeon 1873–82

Sengteller, M. A. (1813–1883). Polish
engraver. *Géogr. Univ.* 1874

Senior, William. "Practioner and
Professor Arithmetique, Geometry,
Astronomie, Navigation and Diallinge."
Estate plans for Earl of Devon 1609–28

Senn, Johann. Engraver. *Plan Zürich*
1804

Sepmanville, M. de Lieudé de. *La Gonave,*
Dépôt de Marine 1788

Sepp, Christ. Cartographer, bookseller,
and publisher of Amsterdam. *Vereenigde
Nederl. Prov.* 1773

Sepp, Jan. Christian (1739–1811).
Worked with his father. *Nieuwe geogr.
Nederl. Reis & Zak Atlas* 1773

Septala [Settala] , Joannes Georgius.
Cartographer, b. Milan, worked in Spain.
Mediolanensis Ducatus 2 sh. 1560. Used
by Ortelius 1570. *Map Spain* (now lost)

Sequanus, J. M. See **Metellus**

Serane, J. R. *Terrestrial Globe* 1804

Serapion. Greek geographer, B.C.

Seraucourt, Claude. *Plan de Lyon* 6 sh.
1735, *Plan de Lyon* 1746

Serela, Carlo. *Prague* 1648

Serigny, M. de. *Côtes de la Louisiane*
1719–20

Serlin, W. *Hungary* Frankfort (1680)

Serne, S. H. *Java* 1866, *Nederl. Indië*
1869

Serra, Padre Junipero. *California Ant.
y Nueva* 1787

Serrano, José. *Mexico* 1867

Serrano, P. Andres. *Nuevas Philipinas*
(1707)

Serres, Domingo. *Pensacola Santa Rosa*
(Jeffreys 1763), *St. Lucie* 1780

Serres, John Thomas (1759–1825).
Marine painter and draughtsman to
Admiralty. Additions to Bougard's
Little Sea Torch 1801

Serres, P. Marcel Toussaint de (1783–
1862. Geologist

Serres, Unal. *Carte Vinicole Dépt.
Gironde* 1856

Sersteuens, T'. See **T'Sersteuens**

Serth, Ernst. *Südbaiern* 1868, *Handels
Geogr.* 1872, *Deutsches Reich* 1875 &c.

Servet[us] [Villannovus] , Michael
(b. 1509). Editor editions *Ptolemy*
1535 & 1541 (woodcut maps)

Sessa. Heirs of Melchior. Printers.
Venice edition of *Ptolemy* 1598–9

Sesti, Giovanni Battista. *Pianta della
citta Milano* 1707, 1711

Seton, Christopher. Engraver in
Ordinary to His Majesty at the Golden
Head. *Plan Battle Preston Pans* 1745

Settala. See **Septala**

Setznagel, M. See **Secsnagel**

Seutter Family. Artists, painters, goldsmiths and engravers.

Seutter, Albrecht Karl (1722–1762).
Map maker and engraver. Son of

George Matthäus. Married daughter
of John Balt. Probst. *Oettingen* (1741),
Partie Orient. Nouvelle France (1750),
Neuchâtel (1760). Business then passed
to T. C. Lotter & G. B. Probst

Seutter, George Matthäus the elder
(1678–1757). Cartographer and
publisher of Vienna and Augsburg.
Apprenticed to Homann 1697, engraved
for Wolff. *Globes* 1710, *Atlas Geogr.*
1725, *Atlas Novus* 1728, 1730, 1736;
Atlas Minor (1744), *New Ebenezer*
1747. Business passed to son Albrecht
Karl

Seutter, Georg. Matthäus the younger
(1729–1760). Engraved maps for *Atlas
Novus*

Sever. *Haut-Sénégal* 6 sh. 1882

Severin, O. C. *Kjobenhavn* 1840, *Island*
1858

Severszoon [Severinus, Zepherinus].
Publisher of Leyden. Jan Veen's *World*
1514, *De Kaert vander Zee* 1532

Severt, Jacques. *N. Hemisphère De
Orbis* Catoprici, Paris 1598

Seville, Isidore of. See **Isidore** of Seville

Sevin-Talive, L. de. *Atlas Lot et
Garonne* 1873

Sevon, J. F. *Helsingfors Stadtplan* 1893

Sewall, J. F. Surveyed portion of
Minnesota 1857

Seward, William Wenman. *Hibernian
Gazetteer* 1789, *Topo. Hibernia*
Dublin 1795

Seydlitz-Kurzbach, E. V. (1784–1849).
School Geography

Seyfferth, J. A. *Biblischen Laender*
(1876)

Seyfrid, J. H. *Playing Cards* engraved
W. Pfann (1678)

Seyler, Johann Christian (1651–
1711). Surveyor

Seymour, J. H. Engraver of Philadelphia.
For Carey's *American Atlas* 1801, for
Marshall 1807, for Lewis *(New York*

Is.) 1807, *Cincinnati Plan* 1815, for
Edwards 1818

Seymour, Stephen. Mate of the *Centurion*.
Surveyed with Luard various harbours
in Jamaica: *Montego Bay* 1790, *Lucea
Harbour* 1790, *Port Antonio* 1792, *St.
Anns Bay* 1792

Seyra y Ferrer, Juan. *Regni Aragoniae
des.* 1715

Sgrooten [Schrotenus, Schrott], Christian
(ca. 1532–1608). Royal Geographer to
Philip II of Spain 1557. Lived Calcar in
Germany. *Germany* 9 sh. pub. H. Cock
1565, *Guelders* 1558, *Antwerp* 1560,
MS *Atlas* 1573, *Palestine* 9 sh., *Jerusalem*
2 sh. Many maps used by Ortelius

Shadwell, Charles F. A. Mate HMS
Princess Charlotte. Surveyor for
Admiralty 1838

Shaefer, P. W. *Maps on Coalfieds &
Railways of Pennsylvania* 1849–70

Shall, D. F. *Railway & Township map
Arkansas* 1860

Shallard. See **Gibbs, Shallard & Co.**

Shallus, Francis. Engraver of Philadelphia.
For Marshall *N. Part New Jersey* 1776,
Delaware 1801, Maps for *Life of Washing-
ton* 1807, for Carey's *Atlas* 1817

Shapley, Nicholas. Surveyor. Hilton's
Cape Hatteras to Cape Roman 1662 MS

Sharks, C. B. Surveyor. *Town of Molyneux
(New Zealand)* 1862

Sharbau, H. Hydrographer. *Chart of
Tsis Anchorage*, Admiralty Chart 1872;
Burma-Siam China Railway (1888)

Sharfi. Family of Arab cartographers. *Sea
Atlas* 1551

Sharman, J. *Scotland* (1800), *Switzerland*
(1800)

Sharp, G. W. Engraved for Thomas
Richards of Sydney. *Port Jackson &
Sydney* 1867, *N.S.W.* 1876

Sharp, John. Publisher of Warwick.
Yates' *Warwick* 1793

Sharp, Thomas. *Coventry* 1807

Matthaeus Seutter (1678–1757)

Sharp, W. C. Lithographer. 251 Washington St., Boston. *Kanzas & Nebraska* 1854

Sharp, Greenwood & Fowler. Publishers, 19th century

Sharpe, Capt. Bartholomew. Sailed with Dampier. MS *Atlas S. Seas* (1682)

Sharpe, John (1777–1860). Cartographer. *Corresponding Atlas* 1849, *Students Atlas* 1850

Sharpe, Samuel. *Ancient Egypt* 1848

Sharpe. See **Vernor, Hood & Sharpe**

Sharpless. See **Kimber & Sharpless**

Shatzen, Johann. Jacob. *Atlas Homann illustratus-Hübner's Methode* 1747

Shaw, Benjamin F. *Comprehensive Geography* 1866

Shaw, H. *Plan Part Barnsbury Manor Estate Holloway* 1843

Shaw, John. Estate surveyor. *W. Ham* 1819 MS

Shaw, Dr. Norton. *Royal illus. atlas,* Fullerton (1860)

Shaw, W. *Philippine Is.* 1886

Shawe, G. Surveyor. Cockmead's *Denham* 1602 MS

Shawe, William. *Maps Africa &c.* end 19th century

Sheahan, James Washington (1824–1883). *Univ. Hist. Atlas* 1873

Sheardown, William. *Country road Doncaster* 1805

Shearer, W. O. *Carroll Co., Md.* Phila. 1863, *York Co., Penn.* 1860

Sheasby, T. Engineer. *Intended Navigable Canal from Swansea to Nen Noyadd* 1793

Sheerbarth, Hugo O. *Alton Madison Co., Ill.* 1851

Sheffield. Surveyor. *Jamaica* 1730–49 (with others), used by Browne 1755

Sheldon, William, of Weston Warwick. *Tapestry Maps* ca. 1588–1624

Sheldrake, W. *Neighbourhood of Aldershot* 1871

Shelley, Mortimer M. *Washington Markets* 1874–5

Shelvocke, George. *Voyage* London 1726 *(World map)*

Shene, Robert. Estate Surveyor. *Little Canfield* 1590 MS

Shepperd & Sutton. Publishers. *London in Olden Time* 1844

Sherard, Thomas. Surveyor. *Estate map Ireland* 1777–1810

Sheriff, Charles. *Scotland* 1766 MS

Sheringham, Lieut. (later Capt.) W. L., R.N. Hydrographer. Admiralty Charts *coasts of England & Wales* 1830–1855

Sherman, Lt. A. *Atlas Alington & Rockland, Mass.* 1874 (with C.W. Howland)

Sherman, C. & Son. Publishers in Philadelphia. *Wilkes' Atlas to U.S. Explor. Expd.* 1858

Sherman, Edwin A. *Silver Mines Nevada & California* 1865

Sherman, George E. Publisher and engraver (with J.C. Smith) (1843–50) 122 Broadway N.Y. Engraved for Goodrich 1841, Smith's *U.S.A.* Bouchette's *Canada* 1846, Pub. with Stiles, Sherman & Smith for Colton in 1839

Sherman, J. M. Surveyor. *Washington Co.* for Barker 1856

Sherman, W. A. *Norfolk Co., Mass.* 1876, *Penobscot Co.* 1875

Sherman, Will. Tecumseh, General (1820–1891). *Maps Civil War in America* (1865)

Sherrard, Thomas. Surveyor. *Estate maps Ireland* 1797–1800

Sherrards, Brassington & Greene. *Estate survey Kildare* 1816 MS

Sherren, C. K. *Flaunchford* 1727 MS

Sherrick, J. B. R. *Map of Decatur, Illinois* (1860)

Sherriff, James. Surveyor and publisher, Crescent Birmingham. *25 miles round Birmingham* 2 sh. 1798, *Environs Liverpool* (1800) & 1823

Sherriff, Thomas. Surveyor. *Estate Map Dudley* 1744

Sherwill, Walter Stanhope. *Maps of Districts in India* 1849–66

Sherwood, Michael. Estate surveyor. *High Easter* 1731 MS

Sherwood, W. S. *Plan Liverpool* 1821

Sherwood Neely & Sons. Publishers, Paternoster Row London. *Beauties of England & Wales* 1810, *Plan London* 1814, Walkers *Univ. Atlas* 1820, Myers' *Mod. Geogr.* 1822, Goodwin's *Guide Van Diemen's Land* 1823

Sherwood [Jones] **& Co.** Gray's *Roads* 1824

Sherburne, Jonathan W. U.S. Navy. *Annapolis Harbour* 1818 MS

Shevelev, A. Russian geographer 1872–90

Shewell, T. Publisher, Paternoster Row London. Partner with Longman 1745. Salmon's *Modern History* 1746

Shiba, K. (1738–1813). Japanese engraver and cartographer

Shier, John. *County Ontario* 1860, *County Durham, Upper Canada* 1861

Shillibeer, Lieut. John, Royal Marines. *Chart Port Anna Maria (Marquesas)* 1814 MS

Shober, Charles. Lithographer, 109 Lake St. Chicago. For Pease & Cole 1859

Sholl, Charles. Civil and topo. engineer Milit. *Topo. Map E. Virginia* 1864

Shortgrave, Richard. Collaborated with J. Leake in his *plan of London* 1667

Shortland, Comm. Peter F., R.N. Admiralty Hydrographer. *Charts of Nova Scotia* 1857–1867, *Sicily* 1867 and 1872

Shortland, Thomas George. *Track of the Alexander (Port Jackson-Java)* 1788–90, *N. End of Ceylon* 1805

Shuckburgh, Sir George, Bart. *Chart Mountains border Geneva* (1777)

Shufeldt, Rear Admiral Robert Wilson, USN. *Yellow Sea* 1861, *Surveys for ship canal Atlantic to Pacific* (20 maps) 1872

Shumard, B. F. *Geolog. Map Franklin Co., Miss.* (1860)

Shury, John. Printer and engraver. For S.D.U.K. 1829–49

Shuttleworth, J. Lithographer. *Plan Marshalsea Prison* 1843

Shwytzer. See **Schwytzer**

Sibbald, Alexander. *River of Persaim* 1775, *Coast of Ava* 1784

Sibbald, Sir Robert, M.D. (1641–1722) Royal Geographer of Scotland, assistant to Adair. *Orkney Islands* pre 1694, reissued 1711, 1750; *Roman Wall* 1739, *Shetland* 1739

Siborne, William (1797–1849). Historian. *Atlas to war of 1815* (1844)

Sibrandus [Sÿbrands], Leo. *W. Frisia*, pre 1545, used by Deventer & Ortelius 1579

Sibree, James. Drew *Warwicks.* 1855

Sicard, Claude. *Route des Hébreux* 1727, *Egypt* 1753

Sick, Paul. *Königreich Würtemberg* 1850

Sickinger, Gregorius (1558–1631). *Statt Freyburg* 1589

Sickler, Friedrich Carl Ludwig. *Campagne de Rome* 1811 and 1816, 1828 &c.

Sidell, William Henry. *Isthmus of Tehuantepec* 1870

Sideri, Giorgio. See **Callapoda**

Siderocratis, Samuel. *Geographia Tübingen* 1562

Sides, William. *Baltimore* 1852, 1853, 1860; *City Richmond* 1859

Sidney, Lieut. (later Comm.) Frederick William, R.N. Hydrographer. *Charts coast of Australia* 1864–70, *Channel Is.* 1869, *Buenos Ayres* 1857

Sidney, J. C. of Philadelphia. *10 miles*

round Phila. 1847, *Wilmington (Del.)* 1850

Sidney, J. E. *Abington, Montgomery Co., Penna.* 1849

Sidney & Neff. Engineers and surveyors. *Map Westchester Co., N.Y.* 1851

Siebert, Albrecht. *Würtemberg* 1843, *Westphalia* 1845, *Hanover, Bavaria* 1850

Siebert, Henry & **Bros.** Lithographers, 93 Fulton St. N.Y. *S. Boise Gold Mines* 1865

Siebert, N. *Plan Tacna* 1861

Siebert, Selmar. Engraver and printer, Wash. D.C. Engraved for Perkes 1832, Stieler's *Hand Atlas* (1540), *Miss.- Pacific Railway* 1855, *U.S.A.* 1857

Siebold, Philip Franz von (1796—1866). *Nagasaki* 1828, *Japanisches Reich* 1840, *Nippon* 1852

Siedamagrotzky. *Aachen* (1875)

Siedentopf, F. R. *Terrestrial Globe* 1825

Siegen, P. M. *Carte Géolog. Luxembourg* 1877

Siegried, Hermann (1819—1879). Swiss topographer and cartographer. *Topo. Atlas Schweiz* 578 sh. 1870—1908

Sieglin, Wilhelm. *Atlas Antiquus* 1893

Siegman. *Environs of Berlin* 10 sh. 1816—19

Siethoff, J. J. Ten. *Atlas der Nederl. Oost Ind.* 1883—5

Signot, Jacques. *Descrip. des Passages entre Gaules et Italie* 1515 (map N. Italy)

Siguenza, Antonio. *Plano Prov. Camarines Luzon* 1823

Sijpe [Sypes], Nicola van. *Le Heroike Enterprise faict par Le Signeur Draeck* 1589—1641

Sikkena, Jan. *Carte marine toutes les côtes de l'Amérique* Amstd., Keulen (1700)

Silbereysen [Silbereisen, Silberstein], Andrea. Engraver for Seutter's *Scotland, Tyrol, Rhaetia* Probst (1765)

Silberman, Joseph. 19th century Globe-maker

Silberstein, A. See **Silbereysen**

Silfverling, Jonas. Swedish engraver. *Map Pennsylvania* for Biörck, publ. Upsala 1731

Silva, D. Franciscus Cassianus de. Engraver for Buleifons *Naples,* Naples 1692

Silvan, N. H. Swedish surveyor, 18th century

Silva, E. *Plan Tangier* 1888

Silva[no] [Sylvanus], Bernardo. Portuguese of Eboli. Worked in Venice, edited 1511 edition of Ptolemy. *World map,* first to be printed in two colours

Silva Paulet, A. José de. *Provinz von Ciara* (Spix & Martius Brasilien) 1823—31

Silvares, José R. F. *Republica do Brazil* 2 sh. 1892

Silver Map of World 1581, to celebrate Drake's circumnavigation.

Silver & Co. 9 & 10 Corhill London and 4 & 5 St. George's Crescent South Liverpool. Pub. Cruchley's *London* 1834 on silk

Silvestre, Israel (1621—1691). *Recueil des vues de villes* 1751

Silvestre, Pierre François. Translator. Keulen's *Flambeau* from Flemish into French 1699

Sim, W. New York. Engraver for M. Dermot & Arden 1814

Simao, Fernando (16th century) Portolan maker

Simencourt, Edouard de. *Carte Routière de France* 1828, 1831, 1841

Simeones [Symones], Gabriel. *Limoges,* Blaeu 1635

Simerlus [Simmlers], Josias. *De Repub. Helvet. et Descrip. Vallesiae* 1576, *De* 1574

Simes, William. *Plan City of Wells* 1735

Simitière, Pierre Eugène du (ca. 1736–
1784). Born Geneva, d. Philadelphia.
Maritime parts Virginia. Penn. Mag.
1775, Aitken's *Virginia* 1776

Simmes, Lieut. *N. coast Mexico* 1844–64

Simmonds, John Henry. *Dist. of Goorgaon*
1849, *Dist. of Hissar* 1858, *Panipat*
1858

Simmons, Capt. David. *Plan Bushier
(Persian Gulf)* pub. Dalrymple 1774

Simmons, F. *Chart Scilly Isles* ca. 1700

Simmons, I. L. A. *England & Wales*
1852

Simmons, Mary & Samuel. Printers.
Speed's *Prospect,* Roger Rea 1662

Simms, Robert. Estate surveyor. *Birch*
1750 MS

Simms, Fred. Walter. *Proposed S.E.
Rly.* (1836), *Plan of Calcutta* 1858

Simon, A. Draughtsman and engraver.
Egypt (1882), *Madagascar* 1899

Simon, C. *Geogr. et Hist. Puerto Rico*
(1875)

Simon, Charles. Engraved for Andriveau-
Goujon 1840–60

Simon, Fernando. See **Fernando**

Simon, L. *Plan d'Angers* 4 sh. 1736

Simonet de Maisonneuve, L. A. A. *Rade
de Mazaghan* 1851

Simoneo, Gabriel. See **Symeone**, G.

Simonneau, Charles. Designed and
engraved *cartouches for De Lisle's
maps* 1700–5, also *globe* 1701

Simonneau, Charles. Paris, Rue de la
Paix, No. 6 vis à vis le Timbre. Book-
seller and joint publisher with Brué of
Atlas Universel 1822

Simonov, Feodor. (1682–1780).
Russian sailor. *Caspian Sea* 1731,
Baltic 1738

Simons, Fred. H. A. *Maps of Colombia*
1887–94

Simons, Mathew (d. 1654). Bookseller

and printer. *Direction for English
Traveller* 1635

Simonson, P. See **Symondson**, P.

Simony, Friedrich (1813–1896).
Geologist and geographer *Atlas Austrian
Alps* 1862

Simpkin & Marshall [Simpkin Marshall
& Co.] Publishers and Stationers. Moule's
English Co. 1830–34, *Atlas U.S.* 1832,
Dower's *New Gen. Atlas* 1838, Reynold's
Trav. Atlas 1848, 1856. Simpkin,
Marshall, Hamilton, Kent & Co. in 1893

Simpkins. Engraver for Rapin 1784–9

Simpson. Engraver. *Lash's River Senegal*
(1700)

Simpson, Emilius. *Fraser River* 1827

Simpson, George B. *Route to Pacific*
1856

Simpson, James. Estate surveyor. *Great
Clacton* 1742 MS

Simpson, James. *Jamaica* 1763 (with T.
Craskell) *Co. Middx., Co. Cornwall, Co.
Surrey* 1763

Simpson, Lieut. (Later Lt. Col.) James
H. Corps U.S. Topo. Engs. *Portland
Harbour (Lake Erie)* 1838, *Milit. Map
Kentucky & Tennessee* 1863

Simpson, R. *Twofold Bay New South
Wales* (with Flinders) 1798, pub. by
Arrowsmith 1801

Simpson, S. *Agreeable Historian or
Complete English Traveller* 1746 (46
maps)

Simpson, William. *Weymouth &
Portland* 1626 MS

Simpson, William. Surveyor. *Plan
Stoney Point* 1784

Simson, Richard. *Plan Port of Sali* 1637

Sinapius, Daniel. Silesian cartographer,
early 17th century

Sinck, Lucas Jansen. Surveyor and
cartographer. *Purmer* pub. Visscher
1622, used by Hondius 1633

Sinclair, Andrew. Government surveyor.
New Zealand 1859

Sinclair, J. *Collectorate of Trichinopoly* 1835

Sinclair, Sir John. *Scotland* 1814

Sinclair, Thomas. *Plans Ports Queen Charlotte Is.* 1856

Sinclair, Thomas. Lithographer, 79 S. 3rd St. Phila. *Gillian's Travels in Mexico* 1846, for Ebert 1862

Sinclair, W. T. *Map Kansas* (1870)

Sineck. *Plans Berlin* 1856–67

Singer, Joseph. *Cardiganshire* 4 sh. 1803

Sinner, Carl von. *Grundriss v. Bern* 1790

Sinner, Rudolph Sam. Gott. von (fl. 1779–99). Swiss military cartographer

Sinnett, F. *Colony S. Australia* 1862

Sintzerrich, Heinrich (1752–1812). Engraver

Siradot. *Dépt. de Jura* 1864

Sirotica. *See* **Radziwill**

Sirven, B. *Europe Centrale* (1875)

Sismondi, A. *Carte géolog. d'Italie* 1846

Sismondi, J. C. L. (1773–1842). Swiss historian and editor

Sitgreaves, Capt. L. U.S. Topo. Engs. *Zuni & Colorado Rivers* 1852, *Kentucky & Tenn.* 1863

Sitwell, Honorius Sisson. *Montenegro* 1860

Sitzka, J. *Livonia, Kurland* 1896

Sizlin, Johann. *Würtemberg, Tübingen* 1559 (attributed)

Skakov, K. A. Russian geographer, 19th century

Skanke, H. (1766–1807). Danish Cartographer. *Moen* 1776, *Siaelland* 1777, *Fynen* 1783

Skead, Fred. *Admiralty charts coast Africa* 1858–78

Skead, A. H. Master of H.M.S. *Modeste. Hula-Shan Bay* 1840–1

Skene, A. J. *Maps of Victoria (Aust.)* 1872–6

Skinner, Andrew (with Taylor) *Post Roads London–Bristol* 1776, *North Brit. & Scotland* 1776, *Roads Ireland* 1778, *Co. Louth* 1778

Skinner, Capt. *Plan Funchal* (inset on Johnstone's *Madeira)* 1791

Skinner, George. Surveyor. *Estates Essex* 1681

Skinner, T. See **Skynner**

Skinner, James. *London* 1861, *Pocket Map London* (1866)

Skinner, R. J. *Allen Co., Indiana* 1860, *Wabash Co.* 1861

Skinner, W. *Moray Firth* 1752

Skottowe, Nicholas. *Ballambouang Bay* 1779

Skribanek, Josef von (1788–1853). Austrian military cartographer

Skrzeszewski, Adolf von. Military cartographer in Austrian service. *Maps Austria Hungary* 1861–78

Skylax. Greek geographer and mathematician, B.C.

Skynner [Skinner], Timothy. Estate surveyor. Numerous *estate plans Essex* 1713–68

Skynner, Lieut. William Augustus. *Broach River* 1773, pub. Dalrymple 1775; *Gulf Cambay* 1778, *Tannah River* 1778

Skyring, Charles Francis. *Jaffa & St. John Jean D'Acre* 1843, *S. America* 1863

Skyring, Comm. W. G. Admiralty Surveyor. *W. coast Africa* 1833–4, *Patagonia* 1861

Slade, James. *Back Bay Boston* 1861

Slade. See **Stirling & Slade**

Slaney, Edwd. *Jamaica* 1678, pub. London, W. Berry (1680)

Slater, Isaac. Engraver and publisher of Manchester, successor to Pigot. In Manchester & London 1844, Portland St. Manchester & London (1862). *Manchester* 1843, *New British Atlas* 1844, editions to 1862; *England & Wales* 1848, *New Plan*

Manchester & Salford (1879), *Royal Nat. & Comm. Directory* 1876—7. See also **Pigot & Slater**

Slater, Lieut. (later Comm.) Michael A. Hydrographer. *Admiralty Charts Mediterranean* 1824—6, *G.B.* 1828, *St. Abbs Head* 1830, *Blyth to Eyemouth* 1832, *Tees Bay* 1838, *Sunderland Harbour* 1840, *Firth Inverness* 1846, *Port & Vicinity of Wick* 1857

Slater, Thomas A.B. *Firth of Pentland* (1854)

Slater & Co. *Queensland* 1874

Slater. Publisher, High St. Oxford *25 miles round Oxford* 1831

Slaterus, J. J. *Nederlanden* 1876

Slator, John & Thomas. *5 miles round New York* 1855

Slaughtor, A. *Lexington, Kansas* (1860)

Sleater, Rev. Matthew. *Civil & Eccles. Topo . . . , Table Cross Roads* Dublin 1806

Sleath, J. *Brighton* 1820, 1830 and 1853

Sleight, C. Estate surveyor. *N. Ockenden* ca. 1775

Slella, Paulus. Publisher. Milan, *Switzerland* 1622

Slezer, John. *Theatrum Scotiae* 1693

Slier, C. Engraver for Stieler's *Schul Atlas* 1855

Slight, James. Engraver in Melbourne, end 19th century. *Map Victoria* (1883)

Slight, William. Engraver, Melbourne. *Victoria* 1872 & 1879, *Triangulation Victoria* (1888)

Sloane, Charles Jr. Estate surveyor, Wardour St. *Gt. Bardfield* 1761 MS

Sloane, Sir Hans (1660—1753). Physician. Map collection acquired by British Museum

Sloane, Oliver. *Queens Co.* 2 sh. (1789)

Sloniewsky, J. *Constantinople* 1887

Slotboom, Richard. *Speculum Nauticum* 1591, *Spieghel der Zee* 1596, *Mirror des Voyages* 1605

Sloten, P. K. P. J. van. *Arnhem* 1874

Slotie, J. Additions to Knight's *Bear Banty Harbour* 1806

Slowacznski, Andrzei. Polish cartographer, 19th century

Sluyter, P. Engraver (1692—1711). Front for Chatelain's *Atlas Hist.* Vol. IV after Schynvoet 1737, *Polder map Delft*

Smagalski, John von. *Statist. Charte v. Galicien* 1817

Small & Son. *Surrey* Dover (1844)

Smallfield, George. *Rivers of World* 1829

Smart, John. *Tables of Time* (1720)

Smeaton, John. *River Calder* (1758), *Plan of Bristol* 1765, *Ramsgate Harbour* 1790

Smedley, Capt. Henry. *Chart N. Coast Java,* Laurie & Whittle 1794; *Road & City of Batavia* ditto

Smedley, Samuel L. *Atlas of Philadelphia* 1863

Smellie. *Salt Is.* 1779 MS

Smet. See **De Smet**

Smids, Ludolph. Preface to Allard's *Orbis Habitabilis Oppida* (1680)

Smiley, Thomas T., of Philadelphia. *Improved Atlas* 1824, *New Atlas* 1830 etc., *Atlas for Use Schools* 1838

Smirke, Robert. Architect. *Plan Royal Mint* 1842, *Plans & Views British Museum* 1780—1858

Smirke, Sydney. *W. Cliff Folkestone* 1849

Smit, Jan. *River Maps Guelders* 1741

Smith. Engraver. For Andriveau Goujon's *Atlas Class. et Univ.* 1875

Smith. Engraver. *Gloucester & Mon.* 1801, *Cambrai* 1815

Smith, (Mr.) Engraver, pupil of Mr. Whitehurch. *Islington* for Dalrymple 1774

Smith. *Salcombe to Looe* 1791 (with Henman)

Smith, Anthony. Pilot of St. Marys. *Chart Chesapeake Bay* 1776 (A revision of Hoxton's map of 1735) & 1794, Reissued Norman's *Amer. Pilot* 1803

Smith, A. Surveyor of Aberdeen. *Aberdeen Railway* 1845

Smith, A. A. American engineer. *Modoc County, Cal.* 1887

Smith, Rev. A. C. of Minnesota. *Winona County, Minn.* 1867, *100 m. round Avebury* 1884

Smith, Alfred. *York*, 1822, *Sheffield* 1822, *W. Riding* 1822, *Kingston upon Hull* 1823

Smith, Augustus Frederick. *Mil. sketch Hobart Town* (1849)

Smith, B. Engraver of Walthamstow. Laurie & Whittle's *America* 1805. For Wilkinson 1806–1818, for C. Smith 1808 for Playfair 1814

Smith, Benjamin E. *Century Atlas* 1897

Smith, Capt. Carmichael. Royal Engineers. Information for Arrowsmith's *Cape of Good Hope* 1805

Smith, Charles. Map publisher, New York. *Quebec* 1796

Smith, Charles, of Waterford. *County of Cork* 1750, *City of Cork* 1750, *County of Kerry* 1756, *Waterford* 1745, *County of Waterford* 1746

Smith, Charles [Smith & Son 1845] Publisher, map & globe seller. Engraver and Map Seller Extraordinary to H.R.H. the Price of Wales, No. 172 Corner of Surrey Street Strand (1803–1862); from 1864 at 63 Charing Cross. *Plan London* 1803, *New English Atlas* 1804, editions up to 1864; *England & Wales* 1806, *General Atlas* 1808, *Canals & Rivers England* 1815, *Roads England & Wales* 1826, *Plan Bristol* 1829. *Terrestial Globe* 1834, *Yorks.* 1841, *London* 1856 and 1862

Smith, Ch. Engraver for Barbié du Bocage. *L'Océanie* 1843, and for Andriveau Goujon *Océanie* 1854

Smith, Charles C. Surveyor. *Isthmus Tehuantepec* 1851, *Village Mina* 1851, *Titusville* 1864

Smith, Charles Roach. Illusts. for *Roman London* 1859

Smith, C. M. Engraver. 42 John St. N.Y. Devines *Canada* 1873

Smith, D. Engraver of Birmingham. *Map of the City* (1860)

Smith, General David. *Tennessee* for Guthrie's *Geogr.* Phila. 1794–5

Smith, D. *Plan Glasgow* 1839 (with Collie)

Smith, Capt. Daniel. *Clinch River* 1774 MS

Smith, David. Surveyor. *Plan of Glasgow* (1822), *Crinan Canal* 1824, *Estate Knoch Castle* 1850

Smith, E. D. Engraver for Edward Wells 1700–1738

Smith, Capt. Edward, R.N. *Coast of Syria* 1840

Smith, Eli. *Plan von Jerusalem* 1845, *Jerusalem & Environs* (1856), *New Map Palestine* 1862, *Wandkarte Biblischen Geschichte* (1869)

Smith, E. A. *Plan Stirling* (1815)

Smith, Capt. Francis. Voyage for discovery N.W. passage 1748

Smith, G. Engraved *views* on Rocque's *Environs of Dublin* 1757

Smith, G. Astronomer of York. *Geogr. of Solar Eclipse* 1835, 1848

Smith, G. Surveyor. 14 Finch Lane, London. *Cavendish Estate Latimer* 1850 MS

Smith, G. Campbell. *Glengarry* 1840, *Lands Inverlochy Castle* 1840

Smith, George John. Spalding's *map of Cambridge* 1888

Smith, Capt. G. M. Royal Artillery, Surveyor Gen. N. Z. *Plan Port Nicholson* (1845)

Smith, G. R. Publisher of Reading. *Street Map Reading* 1889

Smith, H. M. *Maps of India* 1833–57

Smith, H. T. *Gold districts New Calif.* 1849

Smith, J. Cartographer. *Switzerland* 1818, *France Post Roads & Relays* 1828

Smith, J. Print & Map seller next ye Fountain Tavern in ye Strand. *New & Exact Plan London* 1724, Oliver's *Middx.* 1724

Smith, J. Engraver for Thomson (1820)

Smith, J. Surveyor. *Plan Ashby de la Zouche Canal.* R. Whitworth W. Jessop Engineers. Downes sculp n.d.

Smith, Capt. J. *A New Chart of St. Augustines Bay, inset on Heather's Cape* 1796

Smith, James, Master, R.N. *Tiberon Road* ca. 1768, *Windward Passages Bahamas* ca. 1768, *Pensacola Bay* ca. 1768

Smith, J. A. Publisher of Philadelphia. Owen's *Mexico* 1884

Smith, James Alexander. *Gen. Gazetteer* 1869

Smith, J. L. *Ridgeley, Maryland* 1870

Smith, J. Piggott. Surveyor. *Map Birmingham* 1826, 1828; *Street Map* 1862

Smith, J. Welber. *Interior of Hedjaz* (1850)

Smith, Capt. John (1580–1631). *New England* 1616: the foundation of New England cartography; nine different states exist. *Map Virginia* 1612, 1625

Smith, John. Surveyor. *Plan Aberdeen* 1810

Smith, John Calvin [& Son]. Surveyor. Publisher & Engraver 122 Broadway N.Y. *Guide Ohio,* Mich. 1839, N.Y. 1840, Colton's *map U.S. & Texas* 1843–54, *N. Amer.* pub. Disturnell N.Y. 1850, *Naval & Milit. Map U.S.* 1862

Smith, J. Mott. *Geogr. Hawaiian Is.* (1889)

Smith, John S. *Admiralty Chart coast Arabia* 1783–4, *Worthing* (1877)

Smith, J. T. *Plan Parish St. Margaret* 1806, *Plan Parish St. John (Milbank), Improvements Great George St.* 1806, *Plan Deans Yard (West)* 1807

Smith, J. W. *Survey Derwent River* 1806

Smith, Joseph. Draughtsman, Publisher, Engraver & Mapseller at the sign of Exeter Exchange in Strand. Johnston's *Plan Edinburgh* (1710), *British Isles* 1720, Jansson's *Atlas Anglois* 1724, *Twenty miles round London* 1724

Smith, Joseph Wood. *Maps of Australian coasts* 1854–64

Smith, L. Engraver for Migeon. *Australia* 1874 & 1891

Smith, Lewis. *Barony of Kenry (Limerick)* 1657 MS, *Barony of Poblebriane* 1657

Smith, Marcus Dundas. *Canada West* 1851, *Hartford Conn.* 1850, *Rochester N.Y.* 1851

Smith, N. *Dedham Norfolk* 1851

Smith, Payler. See Smyth, Payler

Smith, Philip. Engraver for S. D. U. K. 1838–49

Smith, R. Engraver for Laurie & Whittle. *America* 1813

Smith, Rae. Lithographer, 120 Nassau St. N.Y. *Montana* 1865

Smith, Ralph. Publisher at the Bible under the Piazza of the Royal Exchange Cornhill. Morden's *State of England* 1701, *Fifty six new & accurate maps G.B.* 1708

Smith, Richard. *River Medway* 1633 MS

Smith, Richard Baird. *Canals Lombardy & Piedmont* 1852

Smith, Robert Pearsall. *Henrico Co. Virginia* 1853

Smith, Roswell C. *Gen. Atlases* 1835–68

Smith, Roswell Pearsall. *Atlas for Schools* 1835, *10 miles round Phila.* 1847

Smith, S. Publisher, Tower Hill. . Successor to firm Mount Page & Davidson. Leard's *Pilot for Jamaica* (1805)

Smith, Sam. Printer to the Royal Society, St. Pauls Churchyard. Narborough's *Straits of Magellan* (1694) (collab. with B. Wallford)

Smith, Sam. Surveyor. *Wm. Hart Estate Wing* 1817 MS

Smith, Samuel **& Co.** *Plan Calcutta* 1854

Smith, Seth. *Discov. Arctic* 1873

Smith, T. Engraver. For Elphinstone's *Lothians* 1744, *Plan Norwich* 1783

Smith, T. *Travellers Guide England & Wales* pub. J. Bowles (1760)

Smith, T. *Turnpike Roads Shoreditch–Enfield* 1798

Smith, Thomas. Surveyor of Hartford. *Glebe Lands Puckler Hedge Field* 1831 MS

Smith, Rev. Thomas. *Univ. Atlas* 1802, pub. Harris & Cooke

Smith, Thomas, of Philadelphia. *Dauphin & Lebanon Counties, Penn.* 1818

Smith, Thomas. Lieut. (later Comm.) R.N. Admiralty Hydrographer. *Newfoundland* 1821–28, *River Exe* 1833, *W. Indies* 1837–41, *S. Coast England* 1853

Smith, W. Surveyor of Grimsby. *Plan of the Town* 1812

Smith, William (ca. 1550–1618). Pursuivant Rouge Dragon. *Norwich* (after Cunningham) 1559, *Bristol* 1568, *Cambridge* (after Lyne) 1574, *Vale Royal* 1585, *Canterbury* 1588, *Cheshire* for Speed (1608) engrd. Rogers; *Particular Descrip. England* 1588 MS, *Vale Royal (Cheshire)* 1585 MS

printed 1656. *Description Norenberg (maps)* 1594 MS

Smith, William, of Birmingham. *Hist. Co. Warwick* 1830

Smith, William. *Ancient atlas* 1872–4 (with Sir A. Grove)

Smith, Capt. Wm. M. *Plan Port Nicholson (N.Z.)* 1840, *Port Cooper & Point Levy N.Z.* used by Johnston (1844)

Smith, William. Publisher, 18th century. *Suisse* (with R. & J. Wetstein)

Smith, William (1769–1839). Geologist and mineral surveyor. *Strata of England & Wales* 1801 MS, pub. 1815; the first geological atlas. *Prop. Geolog. Atlas England & Wales* 1819–24, *Geological Atlas* with Cary's maps 1819–24

Smith, William. Estate surveyor. *Shortgrove* 1786 MS, *Granta River (Camb.)* 1788 MS

Smith, William, Capt. R.N. *Coast Arracan* 1759, 1785

Smith, William. Surveyor to Royal African Company. *Coast of Africa* 1744, Thirty different *draughts of Guinea* (1730)

Smith, W. S. Civil engineer. *Chart Niagara River* 1838

Smith, William **& Son.** Publishers. *Crimea* 1855, *Overland Routes India* 1859, *Commer. Map England & Wales* 1863, *New Series Co. Maps* 1864, *Railway Map* 1865, *Seat War N. Italy* 1866

Smith, W. H. **& Son.** Publishers. *Egypt & Suez Canal* 1882, *Reduced Ordnance Maps* 1894

Smith, William Henry. *Canada West* 1856

Smith & Damoreau. Engravers. 131 Washington St. Boston. *Great Western Railway* 1860

Smith, Elder & Co. Publishers 66 Cornhill, London. (19th century)

Smith & McClelland. Engravers of Washington. *Rocky Mts.* 1845

Smith & Venner. See **Mount & Page**

Smith. See **Faulkner & Smith**

Smith. See **Hodges & Smith**

Smith. See **Jones & Smith, & Co.**

Smith. See **Orr & Smith**

Smith. See **Sherman & Smith**

Smither, James. Engraver of Philadelphia. *Boundary line Md.—Penn.* 1768. Engraved for Carey 1795—6, for Jones 1802

Smits, Gerard (b. 1549). Printer of De Jode's *Speculum* Antwerp 1578

Smollinger. *Durham* (1840)

Smulders, J. *Maps of Netherlands* 1864—1881

Smyth, David William. First surveyor Gen. of Upper Canada 1793. *Upper Canada* 1799, 1800, 1813, pub. Faden 1818, Wyld 1835 & 1838

Smyth, Henry Lloyd *U.S. Geolog. Survey Michigan* 1896

Smyth, Capt. Hervey. *Quebec.* 1760 & 1766

Smyth, Payler. Surveyor. *York* 1720 (with Bland & Warburton), Copy of Hayward's *Map of Fenns* 1727, *Essex, Middx., & Herts.* (with Warburton & Bland) 1726 & 1734

Smyth, Capt. R. *Cuttack* 1849, *Allipoor & Baraset Divisions (India)* 1857

Smyth, R. Brough. *Aust.* 1859, *Geolog. Map Australia* 1875

Smyth, Sir Warrington Wilkinson. *Mines Wicklow* (1855)

Smyth, Capt. (later Admiral) William Henry (b. 1788). Hydrographer and astronomer. *Surveys Sicily & Malta* 1814—16, 1823; *Medit.* 1821—24

Smythe, H. W. *Plan Launceston (Tasmania)* 1855

Smythe, John. *Upper Canada* 1837

Snape, John. Surveyor. *Plan Lichfield* 1781, *Birmingham* 1884, *Dudley Canal* 1792, *Estates Wootton Wawen* 1800

Snare, John. *Ten Miles round Reading* 1846

Snedecor, W. Gayle. *Greene Co. Alabama* 1856

Snell, Robert. *Map of the County of Monmouth* 4 sh. 1785

Snell, William. Estate surveyor. *Chingford* 1782 MS

Snellius [Snel van Royen], Rudolph (ca. 1546—1613). Astronomer and cosmographer

Snellius [Snel van Royen], Willibrord (1580—1626). Mathematician and geographer. Measured arc of the meridian 1615. Son of preceding.

Snodham, Thomas (fl. 1603—1625). Printer. Editions of Speed

Snook, Samuel. *W. coast Ceylon* 1797, *Gulf Manaar* 1798

Snow, Dr. John. *Map show distribution Cholera* 1854

Snowden, George & William, of Derby. Improved version of Burdett's *Derby*

Snyder, G. Lithographer, 138 William St. N.Y. For Lawson 1849

Snyder, Van Vechten & Co. *Atlas Wisconsin* 1878

Sobreviela, Fr. Manuel. *Huallaga y Ucayali* 1791, *Colombia Prima* 1828, *Sacramento corrected* 1830

Society for Diffusion of Useful Knowledge. Series of *maps of World* originally issued in parts by Baldwin & Craddock 1829—32, Chapman & Hall 1844, C. Knight 1844—52, G. Cox 1852—3, Stanford 1857—70; later editions revised.

Society for Promoting Christian Knowledge. 77 Great Queen St., Lincolns Inn Fields. *Cottage Maps* (ca. 1850)

Society for the Propagation of the Gospel in Foreign Parts. Colonial church after 1842

Society of Anti Gallicans. *English Empire in N. America* 1755

Socin, A. *Jerusalem mit Umgebung* (1881)

Soeckler, J. M. Engraver, 18th century

Soedeberg, A. Engraver. *Atlas Juvenalis* 1789

Söderling, C. F. Swedish surveyor, 19th century

Södermark, Olof J. (1790–1848). Cartographer

Sørenson, Jens (1646–1723). Danish hydrographer. *Danske Søkort* 1646–1723

Sofiano, N. See **Sophianos**

Sohier, H. *Dépt. de l'Indre* 1862

Sohm, Isaak (1705–1767). Surveyor

Sohr, Carl (1844–1901). *Hand Atlas* Glogau 1842–4, Berghaus' *Handatlas* 1872, *Afghanistan* 1855

Soimonov, S. I. *Caspian* 1720, used by De L'Isle

Sojmonov, Fedor Ivan (1682–1780). Russian hydrographer

Solaric, Paul (1779–1821). Yugoslav writer and atlas publisher

Soldan, M. F. P. See **Paz Soldan**

Soldner, J. G. von (1776–1833). Mathematician and astronomer

Solé, P. *Pompeii* (1847)

Solem, Johannes (b. 1841). Norwegian officer and geographer

Soler[i] [Solerus, Soleri] , Guilielmus. Portolan maker of Majorca (fl. 1380–85). *Medit. & Black Sea* (ca. 1380), *Chart Atlantic to Azof* 1385 MS

Soli, Achille. *Pisa* (1625)

Soligo, Cristoforo. *Guinea Portugalese* ca. 1490 MS

Soligo, Zuan. *Chart Italy & Ionian Is.* 1489 MS

Solinus, Caius Julius. *De situ orbis terrarum* Venice, Jenson 1473. Joannis Camertis Minortani, Vienna 1520; *Polyhistor Rerum totu Orbe* Basel

1538 (with 2 woodcut maps, one the first to show *N.W. coast America)*

Solis, Hermando de. *America* 1598, used by Botero 1599

Solme, Edward. Estate surveyor. *Tolleshunt Major* 1709 MS

Solomein. *S. Russia* 1686 (with Nakoval'nik)

Solverini, Ilario. *Citta di Parma* 2 sh. 1716

Somer, Jan van. *Coste de Barbarie* 1655, *Le Canada* 1656

Somera, José. *Islas Palaos* 1710–11

Somers, Sir George. *Somer (Summer) Isles,* Bermuda, named after him. *Bermuda* (1609) MS

Somer[s] [Sommer] Pruthenus, Jean. Engraver, worked for Sanson & Mariette 1653–7

Somer, Math. van (1648–67). Amsterdam engraver

Somervell & Conrad. Publishers. Petersburg U.S.A. Arrowsmith & Lewis' *General Atlas* 1804

Sommer, Adolph. *Boehmen* 1862

Sommer, C. (1823–65). Cartolithographer

Sommer, I. Engraver. De la Rue's *Asia Minor* 1652

Sommer, J. See **Somers Pruthenus**

Sommerson, Robert. *Road of Jasques (Persia)* 1795 MS

Sonetti. See **Bartolomeo Dalli Sonetti**

Sonis, J. *St. Domingue* for Ponce 1795

Sonklar (Edler von Innstaedten), Carl (1816–1885). Geographer and Alpine cartographer

Sonneman, J. T. G. *Die Statt Ofen* (1700)

Sonnenston, Maxmilian von. *Guatemala* 1859, *Nicarague* 1859, *America Central* 1860, *Nicaragua* 1863

Sonnet, L. *Brest* (1875), *Algérie* (1881), *Réunion* 1883

Sonnius, Claude. Publisher in Paris, Rue Saint Jacques, à l'Escu de Basle & au Compas d'or. Davity's *Afrique* 1637

Soon, J. Engraver. *Karte von Nord America,* Schenk 1755

Sophianos [Sofiano] Nikolaos (fl. 1520–52). Scholar and cartographer of Corfu, lived in Venice from 1533. *Maps of Greece* engraved 1536 (lost), 1549, used by Ortelius

Sopwith, Thomas. *Coal & Iron Mine Forest Dean* 1835

Sora, Miguel. *Islas de Menorca* 1887

Soret. See **Graf & Soret**

Sornique, Dominique (1708–1756). French engraver

Sornique, J. Engraver of Paris. Labat's *St. Christophle* 1722

Sorrel, Cen. Ingénieur. *St. Domingue* (1803)

Sorrell, William. *Element. Geogr.* 1849

Sorrel & Hurrell. Estate surveyors. *Little Waltham* 1798 MS

Sorriba, P. R. C. y. See **Rodriguez Campomanes y Sorriba**

Sorriot de l'Host, Andreas von (1767–1831). Austrian General and military cartographer. *Europe* 1816 & 1818

Sorte, Christoforo (ca. 1510–1595). Venetian engineer and cartographer. *Bresciano* engd. 1560, *Tyrol* (now lost), *Friuli* MS, *Padua & Trevisa* 1594

Soto, Hernando de (ca. 1500–42). Discoverer of the Mississippi

Sotzman, Daniel Friedrich (1754–1840). Cartographer of Berlin and Hamburg. Published in Berlin, Leipziger Strasse 36. *Maps Prussia* 1788–1808, *Europe* 16 sh. 1792, *Stadt & Vestung Toulon* Strassburg 1793, *N.H., Conn., Vt.* 1796; *Penn.* 1797, *Deutschland* 1803, *France* 16 sh. 1831

Sotzmann & Franz. *Terrestrial Globes* 1808, 1821

Sotzman, (J.D.P.) *Suède et Norvège* Vienna 1812

Soulard, Anthoine (1766–1825). First Surveyor Gen. of Upper Louisiana. *Maps Missippi & Missouri*

Soulasie, J. L. G. (1752–1834). Engineer and geographer

Soulier, E. *Mouvements de la Lune* 1839, *du Soleil* (1842), *Atlas Elémentaire* (1868), *Etats Allemagne* (1870), *Monarchie Prussienne* (1870)

Souter, J. Publisher in Islington. Map for Evans *Van Diemens Land* 1822

Southack [**Southicke**], Capt. Cyprian (1662–1745). Born London, d. Boston. *Boston Harbour* 1694, *Brit. Empire in N. America* Boston 1717 (the first chart engraved on copper in N. America), *Annapolis Royal* 1719 MS, *Plans Quebec & St. Laurence Casco Bay* 1720–28

Southwood, Henry. *Coast Newfoundland* 1675, *English Pilot* 1728 & 1737

Southwood, Sam R. Civil engineer Eton, surveyor. *Tangier Mill Eton* 1856 MS

Souvent, A. J. B. (1794–1864). Austrian military cartographer

Souza, Emile J. d'. *Malay Peninsula* 6 sh. 1879

Souzy. *Emb. du Congo* 1864 MS

Sower, Barnes & Co., of Philadelphia. *Gold region Pikes Peak* 1860

Sowers, Thomas. Engineer, accompanied Braddock's Expd. to N. America 1754. *Plan Montreal* 1756 MS

Spaengler, Andreas (1589–1669). Austrian engraver

Spaeth, I. D. *Billiapatam Rivêr* Admiralty 1792

Spaeth, Johann Leonhard (1759–1842). *Nürnberg* (1810)

Spafarev [**Spafarieff**], Leonti Vasilevich (1765–1845). Military cartographer. *Atlas Gulf of Finland* 1823

Spaight, William Fitz Henry. *Egypt* 1884–7

Spalart, S. R. *Plan battle Dettingen* 1743

Spalding, Charles C. *Map of Kansas* 1854–7

Spalding, W.P. *Plans of Cambridge* 1881–8 & later

Spamer, Otto. *Hand Atlas* (1896)

Spanier, E. *Plans of The Hague* 1855–65

Spano, Antonio, of Tropea (d. 1615). *Globe* 1593

Spark, Alex. Surveyor. *Saddocks Farm Eton* 1778 MS

Spark, Thomas. *Entrance Poole Harbour* Admiralty 1829

Sparke, John. *Div. & Districts Inland Letter Carriers* (1797)

Sparke, Michael. Publisher, London, "are to be sold in green Arboure." Mercator's *Hist. Mundi* 1635, 37, 39. Partner with Cartwright 1632

Sparrman, Anders. Prof. Physics, Stockholm. *Chart Cape of Good Hope* 1779, 1785

Sparrow, Joseph. Estate surveyor. *Kelvedon Hatch* 1776 MS

Sparrow, S. Engraver. *Palace of St. Cloud* 1794

Sparrow, Thos. Land surveyor in Hammersmith. *Haggerston Estate* 1757, *Survey Town & Borough Colchester* pub. Keymer 1767

Sparry, Humphrey. *Plan Manor Edgbaston* 1718

Sparwenfeld, Johann Gabriel (1655–1727). Historian and linguist, educated Upsala. *Maps of Moscovy & Siberia.* For Blaeu & Cantelli

Spatural, Nic. M. (1636–1708) of Moldavia. Cartographer

Spear. Engraver, Star Alley Fenchurch St. For Horwood 1792

Specht, Caspar. Printer, bookseller, and engraver of Utrecht. *Urbis Traiecti ad Rhenum Delin.* 1695, *Maps* for Hennepin 1698, *Nederl. Prov.* (1710), *Landen en Staaten in Europa* (1740)

Specht, F. W. *America,* Braunschweig 1821

Speck, H. *Kiel u. Umgegend* 1866

Speckle [Speckel, Speckin], Daniel (1536–1589). Architect and engineer of Strassburg. *Alsace* 1576, *Strassburg* used by Ortelius 1582

Speed, John (1552–1629). Historian and cartographer. *County maps of England & Wales* begun ca. 1603, published as *Theatre of Empire Great Britain* 1611, 1614, 1616, 1627, 1632, 1646, 1650–62, 1676. *Prospect of World* 1627,–46–62–76. *Palestine* 1617

Speeding, William C. *Maps of Lagos & Yoruba Areas* 1886–7

Speer, Capt. Joseph Smith. *West India Pilot* 1766 (13 pl.) and 1771 (26 pl.), *Chart W. Indies* 8 sh. 1771, another edition 11 sh. 1774; *Gen. Chart W. Indies* 1796

Spehr, F. W. (1799–1833). German mathematician and topographer

Speilbergen, J. van. See **Spilbergen**

Speke, John Hanning. *Equat. Nile* 1861–3 MS, *Route Capt. Speke,* Stanford 1863

Spekel, Daniel. See **Speckle**

Spelman, Sir Henry. *Description & map of Norfolk* MS n.d. (used by John Speed)

Speltz, Alex. *Brazil sul.* 1885

Spence, Graeme. Admiralty. Surveyor assistant to M. Mackenzie Jr. *Charts S. & S.E. coasts* 1788–1804, Revised Mackenzie's charts 1807–8, *Scilly Is.* 1791–2 MS, printed 2 sh. 1792; *Entrance Thames* 1812, *Orkneys* 1812

Spence, J. K. *Postal Map Bombay Presid.* 5 sh. 1801

Spence, W. Ensign 2nd West India Regt. *Plan Sierra Leone,* Wyld 1825

Spencer, Lt. G. Queens Rangers. American War of Independence. *Landing at Burrells* 1781, *Action at Osburns* 1781, *Skirmish at Petersburg* 1781

J. Speed (1552–1629); engraved by C. Lacey

Spencer, Thomas. Estate surveyor. Wickhambrook. *Haverhill (Suffolk)* 1767 MS

Spencer, R. Engraver. *Plan Holborn & Finsbury Sewers* 6 sh. 1846

Spengler. Lithographer. Berney's *Plan Lausanne* 1838

Spengler, Capt. W.A. v. Marine surveyor. *Arabia* 1825

Spergs, Joseph de. *Tyrolis Merid.* 1762

Sperlinga, Jeron (1695–1777). Engraver of Augsburg. Cartouche for reduced version of Müller's *Bohemia* 1726

Spiegel, J. Baptist, of Augsburg. *Terrestrial Globe* 1722

Spiegel the younger (1776–1823). Engraver for Artaria of Vienna

Spiering, Lt. Col. J. H. C. *Colony Surinam* (1750)

Spies, Fried. *Baden Baden* (1893)

Spiess, B. *Reise Karte der Rhon* (1868)

Spiess, F. *Topo. v. Thüringen* (1874)

Spiesshaimer [Spiesshamer]. See **Cuspinianus**

Spilbergen [Speilbergen, Spilberghen], Joris van. Dutch admiral and circumnavigator. *Speculum Orient. Indiae Navigationum* 1614–18, pub. Leiden 1619; *Strait Le Maire, Valparaiso* 1619

Spilbergh, J. *Map* for D. Mortier 1715–28

Spilberghen, Joris van. See **Spilbergen**

Spilsbury, J. Engraver and printseller, Russell Court. For Nicholson (1762), for Herbert 1767. *Bombay* 1703, *Mathewren Bay Diego Rayes* 4 sh. (1761), *Trincomalay* 1762, *Warsaw* 1763

Spinder, Adrian. *Haarlem* 1742

Spindler, Ant. *Krakow* 1827

Spinelli, Gio. Batt. *Citta di Napoli* 35 sh. 1775

Spinetti, Gaetano. *Prov. di Roma* 1863

Spirinx, Nicolas (fl. 1606–43) of Lyons. *Pair globes* 1610

Spitz, Karl. Engraver, 19th century

Spix, Johann Baptist von (1781–1826). *Gen. Charte Süd. America* (with Martius) Munich 1825

Spleiss, Thomas (1705–1775). Swiss astronomer and physician

Spofforth, Robert. Engraver nr. Broad St. Blackfriars, London. Engraved for Wells 1700, Morden 1701, *Lincs.* 1704

Spole, Anders (1630–1699). Swedish astronomer, assisted Dahlberg

Sponholz, O. *Saal Eisenbahn* (1877)

Spooner, William. Publisher, 379 Strand, London. *Travellers of Europe* 1852

Sporer, E. Engraver for Meyer's *Atlas* 1830–40

Sporer, F. Engraver for Kiepert's *Hand Atlas* 1860

Sporer, Hans (fl. 1472–1500). Wood engraver, printer and print colourer of Nuremberg, Bamberg & Erfurt. Additions to Rust's *map of Nuremberg* ca. 1500

Spotswood, Alexander. Lt. Gov. of Va. Copied Hughes' *Country adjacent to R. Misisipi* (1720) MS

Sprange, J. Publisher and bookseller of Tunbridge Wells. *14 miles round Tunbridge Wells* ca. 1770, Revised Budgeon's *Sussex* 1779

Sprague, Rob. W. *Glengarriff Harbour* (1866)

Spratt, Comm. (later Capt.) Thomas Abel Bremage. Admiralty surveyor. *Teignmouth* 1836, *Medit.* 1851–55, *Tripoli* 1863, *Crete* 1866

Sprecher von Berneck [Sprechero, Sprechio]. Fortunato. *Foed. Rhaetia* (1600) used by Hondius (1625), Visscher 1630, Blaeu 1639; *Tyrol* 1638

Sprecht, C. *Germany* Utrecht 1706, *Military Fortifications* Utrecht 1703

Sprent, James. Master of HMS *Wellesley*. *Toong-Koo* 1838, Admiralty Chart 1839

Sprent, James. Surveyor Gen. *Entrecasteaux Channel* 1851, *Derwent River* (with Calder) 1851, *Tasmania* 4 sh. 1859

Sprint, John. Bookseller and publisher 1698–1727, At the Bell in Little Britain. Morden's *fifty six new & accurate maps of Gt. Brit.* 1708

Sprotta. See **Scultetus**

Sproule, G. 59th Regt. Foot. Worked under Des Barres. *Environs of Charlestown* 1780, *S.W. New Brunswick* 1786, *New Brunswick* 1791, Assisted Sam. Holland *map New Hampshire* 1784

Spruner, Karl von. *Hand Atlas* 1846, *Atlas Antiq.* 1850, *Hand Atlas* Gotha, Perthes (1855); *Schul Atlas* 1866, *Hand Atlas* 1876, *Hand Atlas*, Perthes 1880

Spruner von Mertz, K. See **Mertz**, J. L.

Spruytenburgh, Jan. Engraver in Amsterdam 1734

Spry, William. *Canada & St. Laurence* 1760 MS

Spurrier [Spurrie?] **& Phipps**. Estate surveyors. *Beauchamp Roothing* 1794 MS, *St. John's Wood State* 1794

Spyser, William. *Fortifications Portsmouth* (ca. 1584)

Squier, E. G. *Nicaragua* 1851, *San Salvador* 1853, pub. Kiepert 1858; *Honduras &c.* 1854, *Central Amer.* 1856

Stabius, Johann (fl. 1497–1522). Court Astronomer to Maximilan, wood engraver. Cuspinianus' *Austria* 1506, *World Map* 1515 (decor. Dürer)

Stable, Robert Scott. *Maps of Wanstead* 1862–81

Stace, J. F. *Hoogly . . . . Ganges &c.* 1847

Stackhouse, Thomas (1706–1784). Publisher, London. *Universal Atlas* 1782, 1798; *Zodiacal Chart* 1823

Stadelmann, J. W. Engraver in Nuremberg for Homann 1711

Staeckel. *Brandenburg* (with Struebing) 1855

Stafford, Robert. *Geogr. Description empires of Globe* 1607 & 1618

Stafford, Thomas. *Pacata Hibernia* 1633, *Plans of Cork, Youghall, Battle Plans &c.*

Stage, G. H. *Atlas für die Jugend* Augsburg 1791

Stagnon, Ant. Maria. *Citta e Territ. Torino* (1800)

Stagnon, Jacopo. *Carta . . . di Sardegna* 1772

Stahl, Johann Ludwig (1759–1835). Surveyor and engraver

Stainforth, George. *Chart coast China* 1779

Stairs, W. H. *Davies Co. Indiana* 1870

Stalker, C. Publisher Stationers Ct. for Hervey's *Geography, North America* 1787, *Cape of Good Hope* 1788

Stalker, E. Engraver. *Views on Somerset* 1822, *Berks.* 1824, *Cumb.* 1842

Stallybrass, James S. *Maps illust. Caesars Gallic War* 1879

Stalpaert [Stalpart], Daniel (1615–1676) of Amsterdam. Architect. *Amstelodami vet. et noviss.* (1676)

Staltonstale, W. See **Saltonstall**

Stamfer, Hans Jacob (ca. 1500–79). Zürich goldsmith and engraver

Stampioen, Jan Janszoon. Surveyor Rotterdam. *Heemraedtschap van Schielandt* 9 sh. 1718, *Sphere* 1711

Standidge & Co. Lithographers London, Bethnal Green 1838. In Old Jewry London 1857

Stane[s], William. Estate surveyor. *Ingrave & W. Horndon* 1689 MS, *S. Ockendon* 1691 MS, *Romford* 1696 MS

Stanford, Edward (1827–1904). Publisher, engraver, and founder of Geographical Establishment, 6 Charing Cross. Took over the Society for Diffusion of Useful Knowledge in 1857 and issued the *Family Atlas* 1857–75, *Library Map Australia* 1859,

England & Wales 1881, *London Atlas* 1882, *Handy Atlas* 1892 &c. Still publishing as firm

—— Edward, younger (1856–1917) Continued business

Stanghi, V. *Div. milit. di Alessandria* 2 sh. (1860)

Stanier, R. Cartographer. *Thames Estuary* 1790

Stanley, Staff Comm. G. *Roseau Roads* 1873, *West Indies* 1876, *Admiralty Charts*

Stanley, Henry J. Staff Comm. R.N. *Maps of Australia* 1870–81, *Victoria* 1870, *Banks Strait* 1877, *Entrance Port Dalrymple* 1881

Stanley, Sir Henry Morton. Explorer. *Route to Tanganyika* 1871 MS, *Lake Region E. Africa* 1874, *Bassin du Congo* 1886, *Equatorial Africa* 1874–77, *Route Emin Pascha* 1890

Stanley, Capt. Owen, R.N. Admiralty surveyor *Harbour of Waitemata* 1840, *S.W. Pacific* 1847–50

Stanley, T. Engraver. *Liverpool,* S.D.U.K. 1836

Stannard & Dixon. *Ordnance map India* (1857)

Stansbie, Alex C. *Counties of Salem & Gloucester, N.J.* 1849

Stansbury, Capt. M. Howard. Corps Topo. Engs. *Oregon* 1838, *Survey Utah* 1849–50, *Great Salt Lake* 1852

Stansbury, J. F. *Plan Bristol* 1830

Stanton, Lt. Col. E. A. *Adarama (Sudan)* 1897

Stapleaux, C. Lithographer. *Carte d'Angleterre* (1835)

Stapleton, G. H. *Maps of Cawnpore* 1849

Starback, C. George. *Scandinavia* 1870–4

Starckman, P. Engraver for De Fer, De Lisle, Desnos. *Alsace* 1705, *Rhein* 4 sh.

1717–20, *Poland* 1716, *Paris* 1717, *Pais Bas* 1716, 1737; *Poictou* 1751

Staring, W. C. H. *Geolog. Map Netherland* 1858–67

Starkenburg, Ludolf Tjarda van. *Groningen* (1700), worked for Ottens (1750)

Starkey, J. B. *Sketch map of Zambesi* 1888 MS

Starkweather, George A. *Sunday School Geogr.* 1872

Starling, Edmund A. *Wallamette Valley* 1851 MS

Starling, Thomas. Draughtsman and engraver, 13 Norfolk St. Lower St. Islington. *Rape of Chichester* 1815, *Norwich* 1818, *Family Cabinet Atlas* 1830, *Geogr. Annual for Lewis* 1831–3, *Gage's Plan Liverpool* 3 sh. 1836, *Scotland* 1839, *Lewis Topo. Dict.* 1842, *Royal Cabinet Atlas* 1850

Starovolski, Simon. *Poland* 1720, 1730

Starr, J. W. *Atlas Knox Co., Ohio* 1871, *Champaign Co., Ohio* 1874

Staszio. *Geolog. map Poland* 1806

Stateler. See **Curtice & Stateler**

Stand, Jo. *Stralsund* in Zeiller's *Topo.* 1652

Staunton, Sir George Leonard (1737–1801). *Embassy to China . . . . Atlas* 1798

Staveley, Christopher. *Maps on Inland Navigation* 1790–2

Stavely, Edward. *Nottingham* 1831, *Delaware Railraod* (1836), *Canada* 1846 & 1851

Staveley, Col. William. *War in Penin. Spain* (1840)

Stavenhagen, W. *Hydro. Karte Europ. Russlands* 1842

Staszio, Stanislau (1755–1826). Geographer

Stead, H. W. N. *Lincs. Midland Junction Railway* 1845

Stead, J. W. *Jersey* 1799

Steains, W. J. *Map of Rio Doce* 1888

Stealey, George. Civil engineer. *Mexico* 1847

Stebnicki, Hieron. (1832–1897). Polish cartographer

Stechi, Fabrico. *Piemonte e Monferrato* (1640)

Stedman, John Gabriel. *Map Surinam* 1791, *Map Guiana* 1793

Steedman, John. Land surveyor, Edin. *Situation Bell Rock Lighthouse* 1820, Thomson's *Scotland* 1832

Steel & Co. Hydrographic Publishers At the Navigation Warehouse Union Row Little Tower Hill. (1782–1835). Founded by R. Steel 1782, succeeded by David Steel 1799, Penelope Steel 1803. J. Steel became head of firm 1819; business bought by Norie but firm continued issuing charts.

Steel, David. *White Sea* 1796, *River Thames* 1799, *W. Coast Ireland* 1800, *E. coast England* 1803, *Medit.* 1804, *W. Indies* 1804, *S. Part N. Sea* 1805

Steel, Penelope. Publisher At the Navigation Warehouse Tower Hill. *Borda's Canary Is.* 1803, *Dessiou Tor Bay* 1814, *E. India Pilot* (with Horsburgh) 1817, *Bristol Channel* 1820, *St. George's Channel* 1823, *Ireland* 1828

Steel, T. Dyre. *Survey River Usk* 1870

Steele, Charles B. *Goldfield Queensland* 1899

Steele, J. Dutton. *Transport Coal to New York* 1856

Steele, O. A. Publisher of Buffalo *Michigan* (1834)

Steen, Helge. Norwegian engineer and cartographer, 19th century

Steen, Lieut. (later Major) Enoch, U.S. Dragoons. *W. Missouri to Rockies* 1836

Steenke, George J. *Elbing. Oberlaendisches Canal* 1862

Steenstrup, K. J. V. *Kusten des Waigatte* 1874, *West Kuste Gronlands* 1883

Steenwyck, Hendrick. *Plan Aachen* 1576, used by Braun & Hogenberg

Stefani, D. *Florence* 1594 & 1660

Stefano, Alberto di Stefano o Santo, of Genoa. *Atlas* (14 maps) 1645

Stefano, Hiero. de San. *Viaggio nelle Indie* 1499

Stefansson, Sigurd. See Stephanius, S.

Steff, L. *Dépt. Moselle* 1870, *Environs Metz* (1871), *Carte Géolog. Dépt. Moselle* (1876)

Steffen, Hans. *Süd Chile* 1894

Steffens, G. *Carta Topo. contorni di Roma* 1859

Steger, Fried. *Hist. Schul Atlas* 1845

Steger, H. *Jefferson Co., Wisconsin* 1862

Steiger, E. *Karte von Californien* 1867

Steiger, W. T. *U. S. A.* 1854

Steigner, L. de. *Ohio City, Kansas* 1857

Stein, Carl. Engraver. *Tyrol* 1823, *Umgebungen Wien* 1823, *Grundriss von Berlin* 1824, for Artaria 1837

Stein, Christian & Gottfried Daniel (1771–1830). *Geogr. Manual* 1816, *Neuer Atlas* Leipzig 1830, 1846 &c.

Stein, Fred. *Township Map Indiana* 1865

Stein, Gottfried. *Frisinga* (1700)

Stein, H. *Geogr. Schul Karten* 1865

Stein, Col. Lieut. von. *Neu Alt Preussen . . . Warschau* 13 sh. 1807, *Tyrol* 2 sh. 1831

Steinaecker, Franz von. *Herero Land* 1889

Steinau, Ernst. *Pommern* 1899

Steinberger, Johann Christoph (1680–1727). *France* for Lotter

Steinberger [Steynbher], Manasse. Engraver. *Plan Leipzig* 1595

Steinbronn, J. *König. Württemberg* 1891 & 1893

Steinegger, H. Engraver for Britton 1857

Steiner, Joh. Caspar, of Zürich. *Helvetiae Tab. Geogr.* 1679, 1685

Steiner, P. *Gold Coast* 1885

Steinhausen, J. *Atlas der Rheinprovinz* 1894

Steinhauser, Anton (1802–1890). Viennese mathematician and cartographer. *Maps of Austria* 1865–82, *Balkan Halb Insel* 1880

Steinhoeffer, Carl. *Maps Austria* 1869–74

Steinmann, George (1824–1885). Swiss cartographer

Steinmann, I. T. *Environs de Basle* 1798

Steinmann, J. G. *Schweiz* 4 sh. 1867–73

Steinmetz, J. C. R. *Sumatras West Kust* 1852

Steinmetz, W. *Plan von Breslau* (1660)

Steinwehr, A. W. A. F. von (1822–1877). *Atlas* 1869

Stel, Simon van der. *False Bay S. Africa* 1687 MS

Stella, Cherubinus de. *Poste diverse parti del mondo* Lyons 1572

Stella [Stolz, Stoltz, Sigenis], Tielemano (1527–1589), d. Wittenberg. *Totius Germaniae* 1546, *Maps for Bible* 1552–7, *Jerusalem* 1557, Revised Münster's *Germany* 1560. His maps used by Ortelius, de Jode, Hondius

Stelle, James Parish. *Indiana Gold Regions* (1867)

Stelliola, N. A. See **Stigliola**

Stellwagen, A. W. *Nederland. Oostindische Bezittingen* 1879

Stelzner, H. *Bayreuth* 1870

Stemfoort, J. W. & **Siethoff**. *Atlas Oost Indië* 1883–85

Stemmen the Elder, Jan (1695–1734) of Amsterdam. Engraver for Covens & Mortier. *Voorne* 1701, *Mexico* 1722

Stempel[ius], Gerrit, of Cologne (fl. 1587–98). *Territ. of Kerpen & Lommerschem* 1587, *Oppidi Bonnae* 1588

Stendel. *Plan Magdeburg* 8 sh. 1882

Stengel, J. Rudolf (1824–57). Swiss cartographer

Stenger, Johann (1767–1802). Engraver for Reilly & Schraembl 1787–8

Stenklyft, Jakob Christofersson (ca. 1620–1686). Swedish surveyor

Stent, Peter (fl. 1595–1662). Publisher and engraver at the Crown in Giltspur St. nr. Newgate; later at White Horse in Giltspur Street betweixt Newgate & Pye Corner. Norden's *Hants.* engraved Stent 1595, *Series of Anon County maps* (1650), Employed by Hollar (1653–62), *Symonson's Kent* 1659, Catalogue of his Books, Prints & Maps 1662

Stephani, A. A. F. von. *Oro-Hydro. Karte von Europa* 1840

Stephanius [Stefansson], Sigurdus. Died 1594 or 5. Icelandic cartographer. *Terrarum Hyperborearum* 1570

Stephanus. See **Stephanius**

Stephanus Florentinus. See **Buonsignori**

Stephen, H. V. *Maps parts India* 1846–65

Stephens, E. F. & R. J. *Map of Leicester* 1877

Stephens, Capt. E. L., R.N. *Southampton Rly.* 1831

Stephens, J. *City Appleton, Wisconsin* 1872

Stephenson, George [& Son]. *Stockton-Darlington Railway* 1822, *London & Birmingham Rly.* 1833 & 1835 (First Trunk Line in World). *York & Midland Rly.* 1835 (with F. Swannick)

Stephenson, Sir Henry Frederick. *Arctic Sea* 1875–6

Stephenson, J. Engraver for J. W. Norie. *Coasts & Harbours G. B.* 1781–1802, *Channel Pilot* (1791), *Ireland* 1807, *Entrance Thames* 1817, *Chart Indian & Pacific Oceans* (1831), Norie's *Coasting Pilot* 1830–6

Stephenson, John, Master R.N. *Channel Pilot* 1786, *Coasting Pilot* 1794 (both with George Burn), *British Channel* 1800

Stephenson, Magnus (1762—1833). Cartographer

Stephenson, Robert. Engineer. *London & Birmingham Railway* 1835, *Rly. Essex—Cambridge* 1839 MS, *Chester—Holyhead* 1843, *Trent Valley* 1845

Stephenson, Sir Rowland Macdonald. *Railways Turkey* 4 sh. 1859, *Railways Persia* 2 sh. 1878

Sterling & Slade. Publishers in Edinburgh. For *Walker's Univ. Atlas* 1820

Sterm, Severin. Danish cartographer. *Kjobenhavn* 1843

Stern, Heinrich. *Jerusalem* (1630)

Sterne, T. Publisher. *Description East India* 1618

Sterne, Thomas (fl. 1619—31). Globe-maker

Sterneck, Johann K. von. *Oesterreich* 1874—1876

Sterringa, S. à. See **Schotanus**

Stessels, A. *Admiralty Charts Belgium* 1863—71

Stetter, Johann Jacob. *Nassau* used by de Wit (1710), *Wetteravia* (1710)

Steub, Ludwig (1812—1888). German geographer

Steudel, Albert (1832—1890). Geologist. Text & *maps on Alps*

Steudner, Johann Philip. *Totius Rheni* (1685)

Steutner von Sternfeld, Heinrich (1676—1720). Military cartographer

Stevens. Engraver for Gosset (1850)

Stevens, Capt. B. J. *Jamaica* Kingston 1898

Stevens, C. A. *Endermo Harbour Admiralty Chart* 1859

Stevens, George. Publisher *Directory Maps English Counties* 1881—91

Stevens, George. Surveyor. *N. Dansville, N.Y.* 1857, *Norwalk, Conn.* 1858

Stevens, Lt. Col. G. S. *Vicinity of Aden* 1880

Stevens, James. *Maps Masonic Lodges* 1880—1

Stevens, R. T. *Table Distances World* (1886)

Stevens, R. White. *Chart Plymouth Sound* 1843

Stevens, T. *County Monteagle, New South Wales* 1870

Stevens, William. Surveyor. *Plans at Mursley* 1750 MS

Stevens, William H. *Geolog. Map Trap Range, Lake Superior* 1863

Stevenson, Alan. *Chart coast Scotland* 1832, *Skerryrae Rocks* 1836

Stevenson, C. L. *Waterways Lynn* 1864

Stevenson, James. *Nyassa Tanganyika* 1888

Stevenson, Robert. *Chart of North Sea* 1820

Stevenson, V. K. *New York City* 1872

Stevinkhof, I. *Map* for Voogt (1700)

Steward. Engraver. Brook St. Ipswich. Ogilby's *Plan Ipswich* 1674

Stewart, Col. C. E. *Afghan Persian Border* 1886, *Khorasan* 1881

Stewart, Charles W. *Mississippi River* 1892

Stewart, D. J. *County Atlases U.S. Seneca Co., Ohio* 1874, *Logan Co., Ohio* 1875, *Cumberland Co., N.J.* 1876

Stewart, H. *Country round Delhi* 3 sh. 1861

Stewart, J. *Cincinnati & Vicinity* 1861

Stewart, J. *Lake Nyassa* 1883

Stewart, Meikle & Murdoch, J. Publishers, Glasgow. *Scotland* (1800)

Stewart, William. *Chautauqua Co. New York* 1867

Stewart and Page. *Atlas Erie Co., Ohio* 1874

Stewart. See **Everts, Baskin & Stewart**

Steyaert, A. Flemish cartographer, 17th century

Steynbher, Manasse. See **Steinberger**, M.

Stichart, H. Engraver for Stieler 1857–69

Stickney, J. W. *Geolog. Map of Plymouth Windsor County, Vermont* 1864

Stickney, Robt. *Course rivers Baine & Waring* 1792 (with S. Dickinson), *River Witham* 1792

Stieler, Adolf (1775–1836) of Gotha. Geographer and cartographer. *Schwarzes Meer* 1798, *England* 1804, *Hand Atlas* issued in parts 1816, went into many editions with corrections

Stieler, August. *Holland* 1801, *Allgemeiner Hand Atlas* 1804. Worked with Adolf Stieler

Stieler, H. F. A. *Charte von China* 1804

Stieltjes, Thomas Joannes. *Eiland Java* (1865), *Overijssel* 1872

Stier, C. Engraver for Stieler 1850–69

Stier, Martin (1630–1669). Austrian military engineer. *Hungary* 1664, revised 1687

Stiernstrom, M. R. *Westernorrlands Lan* 1865

Stiessberger, Jos. *Plan von Salzburg* (1864)

Stiffe, Lieut. Arthur William. *Abu-Shahr Penin.* 1875, *Coast near Maskrat* 1878 MS

Stigliola [Stelliola], Nicolo Antonio (1547–1623). Astronomer and engraver of Naples. *Atlases Kingdom of Naples* (with Cartaro) (1576)

Stileman. *Ulverston & Lancaster Railway* 1850

Stiles & Co., S. Publishers and engravers of New York. *Colton's Atlas* 1833, *Taylor's Canada* 1834, Engraved for Goodrich 1841

Stiles, Sherman & Smith. Engravers, New York 1839–44

Stiller, Curt. *Argentina* 2 sh. 1876 and 1883

Stimmer [Meister], Christoph. Engraver. *Map Greece* for Sophianos (1554)

Stinck, C. W. Bookseller and publisher. Bolt's *Kjφbenhavn* 1858

Stirling, Capt. (later Admiral) Sir James (1791–1865). First Gov. W. Australia (1829–39). *Chart Swan River* 1827

Stirling & Slade. Publishers. *Edin. School Atlas* (1840?)

Stirrup, Thomas. Surveyor. *Wabridge Park* 1672, *Kimbolton* 1673

Stitz, F. Austrian engraver. For Scheda-Steinhauser 1879

Stivers, Aaron. *Meigs Co., Ohio* 1874

Stizza, Antonio. *Plan of Catania* 1592, used by Braun & Hogenberg

Stoane, C. Estate surveyor. *E. Tilbury* 1735 MS

Stober, Jas. Engraver. For von Reilly of Vienna 1796

Stobie, James. Surveyor. *Perth & Clackmannan* 9 sh. 1783, 1787 & 1805

Stobie, Mathew. Land surveyor. *Roxburgshire. Tiviotdale* 4 sh. 1770

Stobnicza [Stobnicensis], Johannes de. *Introductio in Ptolomei Cosmographiam* Cracow 1512. *Woodcut World map*, first map printed in Poland

Stobo, Robert, *Plan Fort le Quesne (Pittsburg)* 1754 (attributed)

Stockdale, John (1739–1814). Bookseller and publisher, Piccadilly. *Andrews Plans* 1792, *American Geogr.* 1794, *London* 4 sh. 1797, *Chauchard Germany* 1800, *British Atlas* 1805, *England* 32 sh. 1806–9, *Spain* 1812

Stockdale, W. Publisher of Piccadilly. *London* 1797

Stockhausen, Ferdinand de. *Plan Ville de Béthune* (1700)

Stöcklein, J. See **Stoecklein**

Stockley, S. Engraver. *N. Sea* 1809, *Chart Sleeve* 1810, *Blackford's America* 1818, *Wyld's World* 1840, *Wyld's Atlas* 1851

Stoeberer, Johann (ca. 1460–1522). Mathematician and cosmographer

Stoecklein, Hans. *Plan von Erlangen* (1890)

Stoecklein [Stöcklein], Joseph. *Reis. Besch . . . Miss. der Gesellschafft Jesu* 1727–48

Stoeffler [Stöffler], Johannes (1452–1531) Prof. in Tübingen. Taught Sebast. Münster, Schöner, Rheticus. Re-issued *Ptolemy's World & Germany. Calendar* 1518, *Terr. Globe* 1493, *Elucidatio . . . Astrolabii* Oppenhein 1513 & 1524, Mainz 1534, *Cosmographicae descrip.* Marburg 1537

Stoeltzlin, Johann (1594–1680). Painter and engraver.

Stoer, Johann Wilhelm (1727–55). Engraver of Nuremberg

Stoessner, Edward. *Elemente der Geographie* 1876–8

Stoifier, E. *Philadelphia* 1869 & 1875

Stok, A. J. van der. *Noordzeekanaal* 1875, *Plan Amsterdam* (1877)

Stoke, J. Publisher. Andrews & Dury's *65 miles round London* 1777

Stokes, Charles. *Plan Estate Lord Stafford Pimlico* 1725

Stokes, Gabriel. Surveyor. *Dublin* 1750

Stokes, Comm. (later Capt.) John L. Admiralty surveyor, joined navy 1826. *Australia* 1841–3, *Bass Strait* 1843 & 1846, *New Zealand* 1847–51

Stokes, N. W. *St. Joseph Co., Indiana* 1863

Stokes, William. *Guide River Shannon* 1842, *Mnemonical Globe* 1868

Stoliczka, Ferd. (1838–1874). Geologist

Stoll, Lieut. *Berlin* 1796, *Siege Breslau* 1757, *Camp bei Düsseldorf* 1760

Stolle, Edward. *Maps on Sugar Beet* 1850–3

Stolnicul, C. C. (1645–1716). Cartographer

Stolpe, C. *Stadt Constantinople* 1863, *Bosphorus* (1864), *Constantinople* 1867

Stoltz. *Plan Residenzstadt Oldenburg* (1875)

Stol[t]z. See **Stella**, T.

Stolz, J. H. *Topo . . . . Carte von Tegernsee* (1820)

Stone, Abraham. Surveyor. *Hundridge Chesham* 1741 MS

Stone, C. K. Publisher, Phila. *Lake & Beer's Philadelphia* 1860 (with Pomeroy), *Oswego County* 1867

Stone, O. C. *Country round Port Moresby* 1876 MS, *Baxter River* 1875

Stone, Thomas. *Caldecot (Hunts.)* 1791

Stone, W. J. Engraver for War Dept. Washington, 19th century. *Oregon* 1838, *Texas* 1844, *Cape Cod* 1836

Stone & Co. Publishers, Denver. *Nell's Colorado* (1880)

Stone & Stewart. *County Atlases* 1866, *Erie, New York* 1866

Stonhame, John. *Plote of Romny Marshe . . . . and Rye Harbour* 1599

Stonestrell, John. Surveyor. *Withen Farm Burwash* 1726 MS

Stooke, Edwin. *Map River Wye* 1892

Stoop, Dirk. *River Tagus* 1877

Stoopendael, Daniel. *Amsterdam* 1700, engraved for Van der Aa's *Galérie Agréable du Monde* 1729

Stopius, Nicolaus. Cartographer, 16th century

Stoppani, Antonio. *Carte Géolog. Suisse* 1867

Storace, Jorge. *Costa del Sur* 1769 MS

Storck, William. *Barony of Rathdown (Dublin)* in conjunction with William Farrand 1656–7

Storer, G. Engraver for Cole & Roper 1805–08

Storer, John. Estate surveyor of Halstead. *Waltham* 1765 MS

Storey, Thomas. Engineer. *Plan North England Railway* 1836

Stork, W. *Plan St. Augustine* (1769)

Storrs, C. Surveyor. *Humboldt, Kansas* 1857

Stout, Jas. D. American engraver for *Smyth's Upper Canada* 1813, for Willet 1814, for *Darby's U.S.* 1818

Stout, Peter F. *Maps of West Virginia* 1864–5

Stow, John. *Survey of London* 1598–9, 1603, 1618, 1633

Straalen, Hendrik van. *Canal & River maps Holland* 1772–80

Straat, Pieter. *Drechterlandt* 4 *sh.* Amsterdam 1735–6

Strabo (ca. 50 B.C.–A.D. 25) Greek geographer and historian compiled *Geographia* Treviso 1480, Basle 1539, & later editions

Strachey, John. Surveyor. *Somerset* 2 sh. (1736)

Strachey, J. *Acc. of Strata in Coal-Mines &c.* 1725

Strachey, Sir Richard. *Battle Plan Aliwal* 1846, *Sabraon* 1846, *Allahabad* 1862

Strachey, William. *Travel into Virginia . . . Cosmography & Commodities of the country* 1615 MS

Strachowsky. Engraver. *Battle plans Bresslau* 1758, *Görlitz,* 1757, *Leuthen* 1758

Strack, Anton Wilhelm. Draughtsman and engraver. *Schaumburg* (1798), *Tecklenburg* 1803

Strada, Famiano, S. J. *De Bello Belgico* 1632, with Leo Belgicus *(map Belgium in form of Lion);* many later editions

Stradling, Sir John. Translated *Direction for Travellers* 1592

Strahan, Col. Charles. *Distrib. Religions India* 1886, *Index N.W. Provinces* 1899

Strahan, Capt. G. *Simla & Jutog* 1874

Strahan, William. Publisher, New St. Shoe Lane London. Camden's *Britannia* 1753, *Cook's Voyages* (with Cadell) 1773–7, *Kitchen's Mexico* 1777

Strahlenberg, Philip Johann Wm. of Stralsund (fl. 1715–30). Military cartographer. *Siberia* 1715, 1718, 1723 MSS lost. Description and large map of Siberia published in Sweden 1750

Straker, F. *Seat War Europe* 1870

Straker, S. Lithographer. *Nizell Estate Kent* 1837, *Swaffham Norfolk* 1845

Strantz, C. von. *Karten zur Topo. Athen* 1868

Strass, F. *Chronological Maps & Tableaux* 1809–60

Stratford, Ferdinando. Engineer. *Estate plans Castle Meads Gloc.* 1765, *in Monmouth* 1757–8

Stratford, J. Publisher. *Plan London* 1806

Stratford, Richard of Amersham. Surveyor. *Estates at Amersham* 1800– 1814 MSS

Straton, James. 2nd Lieut. Royal Engineers. (American War of Independence). *Plan of Portsmouth* 1781 MS, *Post at Great Bridge* 1788 MS, *St. Johns New Brunswick* (1800)

Straub, H. Lithographer. *Plan von Constanz* 1844

Straube, Julius. Berlin cartographer and publisher. *Eisenbahn K. Mittel Europa* (1867), *Umgegend Berlin* (1871), 1872 and later; *Glogau* 1871, *Reval* 1881, *Brandenburg* 1899

Strauch. *Coalfields Pa.* 1872 (with Cochran)

Strauch, Lorenz. *Nürnberg* 1599, *Innspruck* 1614

Strauss, Felix A. *Boston & Vicinity* 1874

Street, J. G. *Plan Hull* 1836

Street, Samuel M. Surveyor. *Whitfield Co., Georgia* 1879

Street, Thomas. Collaborated with J. Leake in his *plan of London* 1667

Streffler, V. R. von (1808–1870). Austrian military cartographer

Streit, Friedrich Wilhelm. Born Vienna, settled Weimar, d. 1839. Engineer Prussian Artillery, mathematician, and military cartographer. *Kroazin* 1808, *Teutschland* 1810, *Africa, America, Aust.* 1817; *Atlas von Europe* 1834–7, *Atlas der Erde*

Strel'bitskiy, Col. *European Russia* 1876 (6 sh.) and 1887

Stretch, R. H. *San Miguel. Prop. Mexico* 1868, *U.S. Geolog. Explor.* (1871)

Streulinus, I. H. *Zürich. Gebiet* 1698

Stribley, Capt. J. *Chart of Baltic & Gulf of Finland* (improved W. Heather) 1797

Strickland, G. *River Jordan and Red Sea* (1849)

Strickland, William. *Plan Phila.* 1811, *Phila. & Baltimore Railway* (1845), *Cairo* (1860)

Stridbeck, Johann, elder. (1640–1716). Draughtsman and publisher Edn. of Apian's *Bavaria* 1684, *Plan London* (1700), *Hildegung in Frankfurt* (1706), *Atlas curieux* (maps engraved by his son)

Stridbeck, Johann, younger (1707–1772). *Mantova* (1710), *Tyrol* (1710), *Fretum Oresund* (1720), *Poland* (1730), *Barcelona* Berlin (1740). Business passed to Bodenehr

Strik, L. *Univ. Tab. juxta Ptol.* 1695 (with F. Halma)

Strippleman. See **Davenport & Strippleman**

Strobridge. See **Middleton, Wallace & Co.**

Strode, Edward. Master HMS *Centaur. Chart Copenhagen* 1807 MS

Stroem, C. F. *Sverige-Norige-Danmark* 1856

Stroff, Lieut. Collo de. *Plans fortifications I. of Wight* c. 1660

Strömerona, Nils. *Charts of Sweden* used by Claret de Fleurieu 1787

Strombeck, A. von. *Braunschweig* 1856

Strong, J. G. *Queen Anne's Co., Maryland* 1865

Strong, Moses. *Atlas of Geolog. Survey of Wisconsin* 75 sh. 1877–82

Strousberg, Dr. *Plan Docks & Quays Antwerp* 1871

Strout, Richard. *Terre Haute, Vigo Co., Indiana* 1872

Struben, Frederick Pine Theo. *Geolog. map Africa S. of Zambesi* 4 sh. 1896, *Geolog. map Transvaal* 1899

Strubicz, Maciej (ca. 1520–1589). Polish historian and Royal Geographer. *Poland eastern Frontiers* MS (now lost), *Mag. Ducatus Lithuaniae* 1581

Strubius, Johannes. *Descrip. Vetus Univ.* 1627, *Hierosolyma* 1629, *Orbis Terr. Veter.* 1664

Struckmann, C. *Umgegend Hannover* (1874)

Struebing. *Wandkarte Brandenburg* 1855

Struve, Fr. G. Guill de (1793–1864). Geographer and astronomer

Struve, H. Post Director Berlin. *Railway map middle Europe* 2 sh. 1888, *Rly. map Germany* (1877)

Struve, Henri. *Carte pétrographique du St. Gothard* 1791

Struyck, N. *Inleiding tot de algemeene Geographie* Amstd. 1740

Struys, Jean. *New Card of Caspian Sea* 1668, *Voyage en Moscovie Tartarie Perse* 1681

Stryienski, A. *Canton de Fribourg* 1855, *Rassemb. de Troupes Fribourg* 1873

Strype, John. *New Plan London* 1720

Strzelecki, E. P. de. *Australian Alps and Gipps Land* 1840, *Map New South Wales* 1845

Stuart, A. K. *Map of Klondike* 1897

Stuart, Andrew. *Chart River Tyne* (1800)

Stuart, Carl Magnus (1645–1705). Swedish military engineer

Stuart, Charles. Brig. Gen. *Seat of War Spain & Portugal* 1828

Stuart, Lieut. D. D. V., U.S.N. *North Shore Long Island* 1886

Stuart, Fred. D. *Is. Zante* (1825), *California* 1849

Stuart, J. *Plan Liverpool* 1795

Stuart, James. Engraver. *County Chester* 1794 (reduced from Burdett)

Stuart, James. *New Chart Delagoa Bay*, inset on *Heather's Cape* 1796

Stuart, John. Surveyor. Superintendant Indian Affairs, Southern District 1762. Born Scotland, emigrated to Charlestown 1748. Directed surveys of Roman Gauld &c. to 1776. *Map Cherokee Country* MS (1761), *S. India District* 1764, *Brahm's S. Carolina* 1780

Stuart, J. H. & Co. *Atlas Maine* (1890)

Stuart, John MacDouall (1818–1866). Explorer, d. Adelaide. *Stanford's new map of Australia* 1860, *Route of J. MacDouall* 3 sh. 1861

Stuart, Oliver J. *Map City Rochester, New York* (1860)

Stubba, A. *Wandkarte Prov. Sachsen* 1858

Stubbin, T. Estate surveyor. *Gt. Oakley* 1825 MS

Stuber, Josef Anton (1684–1741). Surveyor

Stuber, J. D. (1718–1787). Surveyor

Stucchi, Adone. *Italia* 1834, *Carta Fisica-dell Italia* 4 sh. 1859, *Sardegna* 1856

Stucchi, Stanislao, of Milan. Draughtsman and engraver. *Italia* 1812, *Prov. de Lodi* 1818, *Grande Atlante Univ.* Milano 1826, *Milano* 1830, *Mappa Mondo* 1861, *Africa, N. America, Oceana* 1862

Stucchi, S. *Pianta citta Milano* n.d., *Stato di Milano* 1764

Stuck, J. Cooper. *Map Douglas Co., Kansas* 1857

Studer, B. and A. Escher. *Carte Géolog. de Suisse* Winterthur 1853, *Carte Géolog. de la Suisse* 26 sh. 1859–87

Studer, Gottlieb (1804–1890). *Karte Süd. Wallisthäler* 1853, *Jahrbuch Schweizer Alpenclub*

Studt, C. *Plans Breslau* 1853, 1858 & 1867

Stuebel, Alphons. *Hochland von Bolivien* 1892, *Süd Colombia* 1899

Stueck, H. *Plans Hamburg* 1868 and 1873–6

Stuelpnagel [**Stülpnagel**], Fr. von (1781–1865). Draughtsman and engraver. *Deutschland* 1838, *Scotland* for Stieler 1837, *Deutschland* (with J. C. Bar) 1846, *Rly. Atlas Germany &c.* 1848, *Australien* 1853, *U.S.* Gotha 1861, *Europa* 1871, *Ost Indien Insel.* 1875

Stuer, D. *Geolog. . . Alpen v. Oesterreich* 1855, *Umgebungen Wien* 1860

Stuers, François V. H. R. de (1792–1881). French explorer. *Atlas Guerre de l'Ile de Java* 1833

Stuhlmann, Franz. *Maps German East Africa* 1890–6

Stukeley, Dr. *Plan Roman London* 1722, *Caesar's Camp St. Pancras* 1750

Stukeley, William (1687–1765) Antiquarian. *Ground plot fathers dwelling Holbech* 1703, *Cambridge* 1704, *England* 1708 MS, *Grahams Dike* 1720, *St. Albans* 1721, *Roman Amphitheatre Dorchester* 1723, *Levels in Lincs.* 1723

Stülpnagel, F. van. See **Stuelpnagel**

Stumpe, J. H. *Orbis Veter. Notus* 1832

Stumpf, Johann (1500–1576). Swiss cartographer, d. Zürich. *Upper & Lower Vallais* (1544), *Schwyzer Chronick* 1548, *(town plans* used later by Braun &

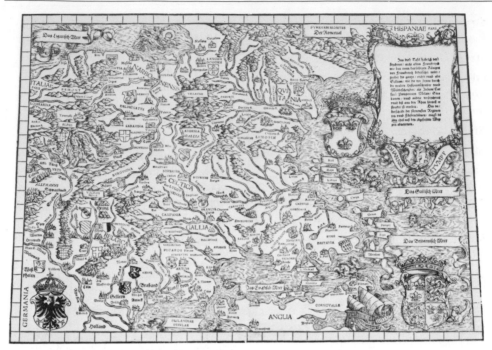

Woodcut map of France by J. Stumpff, published in Zürich in 1547

Hogenberg), Reprinted in Zürich as *Landtafflen* 1552 (the first atlas of Switzerland)

Stunghi, V. *Contorni di Firenze* 1870

Stur, D. *Geological map Stelermarck* 4 sh. 1865

Sturdevant, W. H. *Luzerne Co., Penn.* 1860, *City Wilkes-Barre, Penn.* 1894

Sturgen, Lt. Col. Henry. *Maps & Plans Brit. Army in Spanish Peninsula* (1840)

Sturm, L. C. *Stadt Franckfurt an der Oder* 1706

Sturmhoefel, A. *La Suisse* (1830)

Sturnegh, Fred. *Sect. Map Dakota Territory* 1872

Sturmey. See **Iliffe, Sons & Sturmey Ltd.**

Sturt, Capt. Charles. Surveyor Gen. of S. Australia 1839. *Explorer Route Adelaide into central Australia* Arrowsmith 1849, *Country north Mt. Remarkable* 1857, Stanford's *New Map Australia* 1860

Sturt, F. *Cycling map 20 miles round Farnham* (1887)

Sturt, John (1658–1730). Engraver at the Golden Lion Court in Aldersgate Street. *Various plans* 1689–94, *Map for Narborough* 1694, *Two maps in Camden's Britannia* 1695

Stuurlieden. Engraver and cartographer. *Map for Voogt* (1700)

Stuurman, Jan Pieterse. *Chart N. Europe,* Keulen 1730

Style, F. Engraver. *Plan Cuff Estate Whitechapel* 1818

Suberbie, L. *Madagascar* 1889, and 3 sh. 1895 (with E. Laillet)

Subias, Joaquin. *Carta . . . . del. S. Fé de Bogota* 1760 MS

Suchet, Louis Gabriel, Duc d'Albufera (1770–1826). French Marshall. *Atlas to Campagne en Espagne* 1808–14, pub. 1828–9

Sucholdez. *Map for Reilly* 1796

Suchtelen. *Russisches Reich in Europa* 9 sh. 1812, another edition 12 sh. (1820)

Sudbury, John (fl. 1599–1619). Printseller at White Horse (later Pope's Head) Alley London against the Exchange. *Boazio's Ireland* (1599). In partnership 1603 with his nephew George Humble with whom he produced Speed's *Theatre* 1610–11

Sudlow, E. Engraver for *Rapin's History* 1784–9 and Harrison's *Counties* 1787–91

Sueyoshi, Magozaemon. *Portolan of World* (1615?)

Suitor, J. Publisher of London. Evans' *Van Diemens Land* 1822

Sukhtelen, Gen. vön. *Russian Empire* 100 sh. St. Petersburg 1810–4

Sullivan, Lieut. (later Capt.) B. J., R.N. Admiralty surveyor. *Falkland Is.* 1838–45 2 sh. pub. (1885); *Rio de la Plata* 1844, *Parana* 1847, *Baltic* 1855–6

Sullivan, J. W. Mapseller. News Agent, San Francisco. *Frazers River* 1858

Sullivan. See Whittaker & Sullivan

Sulman, T. *Pictorial Plan London* 1893–4

Sulton, John. *Forest of Bere* 1800 MS

Sulzeberger, Johann Jakob (1802–1855). Military cartographer

Sulzer, J. Engraver for Kiepert's *Hand Atlas* 1860

Sumiya, Schichirobe. *Portolan chart World* early 17th century MS

Sumner. *Survey of Dee* 1755 (with Eyes)

Summerfield, R. K. Surveyor. *Estate Smarden Park Farm* 1809

Suntheim, Ladislaus von (ca. 1440–1513). Historian and geographer

Supan, Alexander. Geographer, worked for Petermann 1886–99

Sureda, Juan. *Atlas Geogr. del Uruguay* 1891

Surbey, Thomas. Surveyor of London. *Plan York to mouth of Humber* 1699 MS

Surhon[ius] [Surhonio, Surchon-] , Jean and Jacques. Cartographers, goldsmiths and engravers, b. Mons. *Hainault* 1548

(used by Ortelius 1579, Hondius 1633); *Luxembourg* 1551, *Namur* 1553, *Artois* 1554, *Picardy* 1557 (used by Ortelius 1559), *Vermandois* 1577 (used by Blaeu 1631)

Surmon, W. H. *Basutoland* 1871

Surperbie, L. *Carte de Madagascar* 1889

Surplice, William Henry. Surveyor. *Maps of Lots & Townships in Victoria* 1857–9

Surville, Luis de. *Nueva Andalucia* 1779

Susemihl, Johann Conrad (1767–1816). Engraver

Sutherland, George C. *Historical Atlas Clark County Missouri* 1896

Sutherland, James. *Azerbaeejaun, Armenia, Georgia* 1833

Sutherland, John. Publisher, 12 Colton Street, Edinburgh. *Lothians Scotland* 1838

Sutherland, P. C. Surveyor General. *Natal* 1864 & 1875

Sutton, William. Estate surveyor. *Manor Bathwick Somerset* 1727 (1735)

Sutton. See Shepperd & Sutton

Sutton. See Harrison, Sutton & Hare

Suzuki Hanbei (1815–56). Japanese globemaker

Suzuki, T. *Geolog. Survey Japan* 1884, *Topo. Map Japanese Empire* 1899

Suzuki, Yokuen. *Roads in Japan & Kurile Is.* 1871

Svabo, Jens Christien. *Faroe Islands* 1784

Svart, Claudius. See **Clav[i]us**

Svart, Olof Hansson. *Brandenburg* 1644, *Prussia* 1644

Swaan, P. *Nieuw Guinea* 1879

Swab, Anton. *Charts Sweden* 1796–7

Swaef, S. de (1597–1662). Engraver and publisher

Swaine, Francis. *Plan Port Royal S. Carolina* (1729) MS, copied from Gascoigne

Swaine, J. Engraver. *World on Mercator's Projection* 1827

Swale, Able (1665–1699). Bookseller and map publisher. Partner with Child at the sign of the Unicorn at the West end of St. Pauls Churchyward. 1678 edition of *Ogilby's Roads*. Collaborated with Churchill for Morden's *Britannia* 1695 and Moll's *Thesaurus Geogr.* 1695

Swan, Charles H. *Map Frazer's Gold Regions* 1858

Swan, John. *Speculum Mundi* 1635, 1643

Swan, John. *Survey . . . Carteret Grant* 1746

Swan, John H. *Paris St. Giles in Fields* (1867)

Swan, Owen. Estate surveyor. *Braintree* 1771 MS

Swan, Engraver. *Mr. Phinn's Plan Glasgow* (1840)

Swan, R. M. W. *Matabele, Mashona and Manicaland* 1892

Swann, Charles E. *Military Map Kentucky & Tenn.* 1863

Swannick, F. See **Stephenson**, George

Swanston, Charles. Surveyor. *Roads of Mauritius* 2 v. (1811) MS

Swanston, George H. Engraver and daughtsman. *Black's Plan Edinburgh* (1848), *Fullarton's Australia* (1851), *Africa* 1858, *Atlas to Gazetteer of World* (1860)

Swanwick, Fred. See **Swannick**

Swart, C. See **Clav[i]us**

Swart, Dirk, *Amsterdam* 1623

Swart, Dirck Cornelisz. Engraver for Du Sauzet 1734–8

Swart, Jacob Junior. Continuation Van Keulen's business 1863–85. *N. Atlantic Ocean* 1876, *Java* 5 sh. 1887

Swart, Nic. Clausen. See **Clav[i]us**

Swart, Stephen. *Eugeniana Canal* 1627, one of publishers of *Pitt's English Atlas* 1680

Swedish Land Survey. Founded 1628. Organized systematic mapping of Sweden. Headed by Andre Bure 1628–33

Sweerts. Cartographer. *Maps for Wit's Atlas* (1690)

Sweet, Homer D. L. *Onondaga Co., N.Y.* 1860 & 1874

Sweet, S. H. *Erie Canal* 1862, *Railway N.Y.* 1864, *N.Y. Canals* 1862 & 1868

Sweitzer, H. Publisher of Philadelphia. *Carey's American pocket atlas* 1801

Swelinck, Jan. *Terre Neuve* 1609 and *Port Royal* 1609 in Lescarbot's *Histoire de la Nouvelle France*

Sweny, M. A. *Liverpool Bay* 1892–3

Swidde, Willem (1660–1697). Engraver

Swieten, J. van, Lieut. Gen. *Kaart van Boni* 1860

Swieten, W. van. *Kaart Oud en Nieuw Roosenburg & Blankenburg* 1727

Swift, W. H. *Florida* 4 sh. 1829, *Woodson Illinois* (1857)

Swigters, Sara (d. 1743). Widow of Johannes Loots, continued his business 1726 with help of her brother Isaac.

Swinburne, T. R. *Road Accra to Mansue* 1881

Swinden, Henry. Surveyor. *Plan Great Yarmouth* 1779

Swingley, C. D. *Crawford Co., Ohio* 1894

Swinton, T. *Plan crown Property Tower Hill* 1815 MS, *Postern Gate* 1818

Swinton, William. Publisher. Partner with Canaan from 1823 at 60 Princes St. Edinburgh 1823, at 63 Princes St. 1825. *Scotland* 1820. Collaborated with Canaan and Ritchie *Woods'. Plans of Scottish Towns* 1823–5

Swire, Comm. H.S. *Tidal Charts Is. of Wight* 1893

Swire, William. Survey. *Liverpool* 1823–4, *Manchester* 1824, *Cheshire* (with Hutchings), *Teesdale* 1830

Swiss, A. H. *Hunting Maps* 1887–92

Swithin, James. *Chart Celebes* (1765)

Switzer, Christopher (fl. 1593–1611). Swiss engraver

Swoboda, Franz, of Vienna. *Australien* 1815 (with Martin Hartl)

Syberg, A. *Phila. & Baltimore Railway* 1856

Sydenham. *Environs of Bournemouth* (1890)

Sydow, Emil von (1812–1873). Geographer of Gotha, worked for Justus Perthes. *Gradnetz Atlas* 1874, *Hydro. Atlas* 1847, *Orograph. Atlas* 1855, *Schul Atlas* 1856

Syers, Llewellyn. *Cab Fare Plans Liverpool & Manchester* 1868–9

Sylva, Rodrigo Mendez. *Hispania* 1730

Sylvanus, B. See **Silvano**

Sylverio. Engraver. Villaseño's *N. America* 1746

Sylvester, Bernardus (12th century). *Cosmographia* MS

Sylvius, Aeneas. *Cosmography Asia & Europe,* pub. Cologne 1477, Paris 1509

Symans, P. See **Symondson**

Symeon[e], Gabriel (1509–1575). *Limagna d'Overnia* Lyons 1560, *Auvergne* Tours 1560. His maps used by Ortelius 1570 and Bouguereau 1635

Symes, M. Publisher of T. Wood's *River Irrawaddy* 1800

Symes, Richard Glascott. *Explan. Memoir Geolog. Survey Ireland* 1871–88

Symon, Edward. Publisher in Cornhill London 1720 (jointly with others)

Symonds, Rev. Alfred Radford. *India* (1857)

Symonds, Capt., Royal Engineers. *Chart Boulez Bay Jersey* 1839, *Jerusalem* 1849

Symonds, H. D. Publisher in London, No. 20 Paternoster Row. *Russell's N. America* 1794, and *Kentucky* 1794

Symonds, James. Estate surveyor. *Havering* 1578 MS

Symonds, J. F. A. *Jerusalem* 1841, *St. Jean d'Acre* 1843, *Syria* (1846)

Symondson, Philip (fl. 1577–98). Surveyor, cartographer, Mayor of Rochester 1597–8. *Dover Harbour* 1577 MS, *Rye Harbour* 1594 MS, *New Description Kent* 1596, reissued by Stent (1650)

Symons [Symonsz], Evert van Hamersveldt. Engraver. Hondius' *Bourdelois* (1630), *Bearn* (1633); *Rhaetia,* Blaeu 1649

Symons, Lieut. T. W. *Dept. of Columbia* 1881

Sÿmonsz. See **Symons**

Sympson, S. Engraver and mapseller, Catherine St. Strand. Hollar's *London* (1666)

Sype, Laurentius van de. Engraver. *Gratz* 1634

Sypes, N. van. See **Sijpe**

Sypesteyn, J. v. *Surinam* 4 sh. 1849

Szajnocha, Wladyslaw. *Galicia* 1895–6

Szatkowski, S. Austrian, engraver for Scheda-Steinhauser 1879

Szén, Engraver for Kiepert's *Atlas Alten Zeit* 1857–69

Szepeshazy, Carl von. *König. Ungarn* 1825

Szultz, Lt. Col. *Ville de Kaire* 1846

T

T. Av. *Gegend bey Danzig* (1850)

T. G. See **Tavernier**, Gabriel

T., G. M. See **Terreni**, G. M.

T., J. See **Trumvall**, J., also **T[app]**

T., J. Engraver. For Horwood 1794–9

T., M. B. C. *Plan de la ville du Cap de Bon Espérance* 1770

T., R. *Use of the Globe* 1616 and 1620

T., R. T. *Central & Southern Manchuria* 1891 MS

Tabourot, Stefano (1547–90) Advocate of Dijon, cartographer. Drew 2 maps, *France* and *Burgundy* MS

Tabula Rogeriana. See **Idrisi**

Tachard, Rev. Père Guy (d. 1714). French Jesuit. *Voyage de Siam* 1689, *Carte du Cap* 1686, *Chart Suratt to Malacca* Paris, Nolin 1701

Taché, Eugène. Assistant Commissioner *Carte du Province de Québec* 1870

Tache, Jules. *Dominon of Canada* 1883, *Province de Québec* 1884, *Carte régionale de Québec* 1885

Täck. See **Taeck**

Tackabury. *Atlas Dominion of Canada* 1875

Tackabury, Mead and Moffett. *Summit County, Ohio* 1874

Tadataka, Ino. See **Chukei, Ino**

Taeck [Täck]. Engraver. Probst's *Crimea* 1784

Taintor Bros. and Merrill. *American household and commercial atlas* 1874

Tait, Capt. R.N. *Harbour St. Martin (Columbia)* 1816

Taitbout de Marigny, E. *Portolan de la Mer Noire* 1830, *Atlas de la Mer Noire* 1850

Taitt, David. *Rivers in West Florida* 1772 MS, *East Florida* 1773

Takahashi, Kageyasu (1785–1829) Japanese astronomer and geodesist. *World map*

Takashiba, Eizaturo. *(Provinces of Japan)* Edo 1849

Takebara, Sjun. Japanese cartographer. *Province Idsumi* 1769

Takebe, Katahiro (1664–1739). Japanese cartographer

Takehara, Shunchosai (d. 1800). Geodesist and engraver of Osaka

Takeshiro, Matsuura. Japanese explorer and mapmaker, Meiji period. *Map of Ezo* (1859)

Talbe-Bougionan. *Dominions of the King of Fez* 1711

Talbot, Capt. M. G. *Approaches to Herat* 1885, *Central Sudan* 1890

Talbot, Col. (later Lt. Gen.) Sherrington. *English Antigua* 1750 MS

Talcott, Andrew. *Delta of Mississippi* 1837, *Ohio Boundary* 1835

Talcott, Lt. Col. George. *Mineral Lands adjacent to Lake Superior* (1845)

Tallis, John. Publisher 15 St. John's Lane, Smithfield (1835). *London Street Views* (1839), in 1846 imprint London and Glasgow. *Illustrated plan London* 1851

Tallis, John & Co. Publishers: London 97–100 St. John Street and Bluecoats Buildings New Gate St., and New York, 40 John Street. *Illustrated Atlas (of world)* 1851

Tallis, L. Publishers, London, 3 Jewin St. City (1835), Warwick Sq. Paternoster Row (1860)

Tallon, C. *La France divisée par provinces Ecclésiastiques* (1869)

Tammar Luxoro Atlas. ca. 1300

Tamon, D. F. V. See **Valdes Tamon**

Tanesse, I. *Plan New Orleans* pub. Del Vecchio & Maspero, N.Y. 1817

Tanner, Benjamin (1775–1848) of New York. Engraver. For Carey 1795, Winterbottom 1796, Marshall 1807, *Plan Yorktown Va.* 1825

Tanner, H. C. B. *City of Nagpoor surveyed* 1865, pub. 1870; *Sketch Sikkim* 1886

Tanner, Henry Schenck (1786–1858). Draughtsman, engraver and publisher: 177 Chestnut Street Philadelphia, 156 Fulton Street N.Y. (1849), 201 B'way New York (1850). *New American Atlas* 1818–23, *New Universal Atlas* 1828 (editions to 1846, then issued by S.A. Mitchell), *Charleston district* for Mills 1825, *Oceana* 1843, *California* 1849, *United States* 1829, 1830, 1832

Tanner, R. American engraver of New York. *China* for Carey's American Edition of *Guthrie's Geog.* Phila. 1796

Tanner, Robert. *Mirror for Mathematics . . . how to make an Astrolabe* 1587, *Treatise usage Sphere* 1592

Tanner, Thomas R. *Maps* for Tanner, Vallance & Kearny's *American Atlas* 1818 & *Michigan* 1830

Tanner, Vallance, Kearny & Co. Publishers of Philadelphia. *New American Atlas* 1818–23, *Pocket Atlas of U.S.* 1828, *New General Atlas* 1828, *Universal Atlas* 1833–4, *Atlas Classica* 1840

Tannour & Compan. Lithographers in Paris, R. Rousselet 15. For Dufour (1850)

Tannstetter. See **Collimitius**

Taperell, Nicholas. *Ancient cities of London & Westminster* 1849 (with Innes)

Tapner, John. Surveyor. *Hartwell Estate* 1842 MS

T[app], J[ohn]. *Seamans Kalendar* 1601

Tappan, L. N. Publisher. *Colorado City* 1861

Tappan & Bradfords. Lithographers of Boston, Mass. For Dearborn 1849

Taramelli, Torquarto. *Carta geologicica della Lombardia* 1890

Tarassevich, O. A. *Ukrainian map* Cracow 1687

Tarde, Jean (1561–1636). Canon of Sarlat, mathematician and map editor. *Théâtre géographique du Royaume de France* 1626, *Diocèse de Sarlat et Haut Périgord* 1624. *Sarlat* used by Hondius 1633 and Blaeu 1635

Tardieu, Ambroise (1788–1846). Cartographer and engraver. Engraved for Dépôt de la Marine, for Langlois' *Nouvel Atlas* 1811, for La Harpe 1821 *(Hist. Gén. des Voyages),* for Dumont D'Urville 1833, *Atlas Universel* 1842, *Carte routière d'Italie* 1846

Tardieu, Antoine François (1757–1822). Geographer and engraver

Tardieu, Pierre Antoine. Son of preceding (1784–1869). Cartographer, inventor of new processes for engraving maps

Tardieu, Eug. Amédée (b. 1822). Physician and geographer

Tardieu, Jean Baptiste. Engraver. *Carte de Puerto Rico* 1810, *Carte du Brésil* 1815, *Carte Itinéraire d'Europe* 1817, *Carte d'Alger* 1840

Tardieu, Jean Baptiste Pierre (l'ainé) (1746–1816). Engraver and draughtsman of Paris, Rue Poupée No. 9 and Rue de Sorbonne No. 385. For Poirson's *Afrique* 1803 and *Mer des Indes. Statistique . . . . de la France et de ses colonies* 1804, *Environs de Versailles* 1808 (12 sheets with Bouclet), *Atlas de Géographie Universelle*, Malte-Brun 1816

Tardieu, Madame. Engraver for Le Paultre's *Passage de l'ombre de la lune* 1764

Tardieu, Pierre Antoine. Engraver for the imperial posts. *Carte des Routes de Postes de l'Empire Français* 1812, *Altenburg and Ronneburg* 21 sh. 1813, *Louisiana and Mexico* 1820, *Paris* 1821, *Atlas for Comte de Ségur* 1824, *Carte des Routes de Postes de France* 1831, *Atlas Géographique* (1845), *Demidoff Voyage* 1853

Tardieu, Pierre François. Used sometimes by Pierre François (1711–71) and also by Antoine François and Pierre Antoine

Tardieu, Pierre François. Engraver and publisher of Paris, Place de l'Estrapade No. 1 and No. 18 (1752–98). For Diderot's *Encyclopédie* (1779), *Golfe de Finlande* 1785, *Plan de Kronstadt* (1790), *Atlas de la Monarchie Prussienne* 1788, *U.S.* 1802 and 1812 (2 sh.), *Mentelle's Atlas d'Etude* 1820, *Bald's Mayo* 1830, for Lapie 1837–42

Tardieu, Deuesle. Publisher and bookseller of Paris, Quai des Grands Augustins No. 37. *Maire's Atlas de Poche* (ca. 1820)

Tardo, Joannes. See **Tarde**

Tardy de Montravel, Louis Marie François. *Maps of Brazil* 1840–80

Tarleton, Sir Banastre (1754–1833). *Westchester County* MS

Tarobe, Yoshida (17th century). Published first *Japanese Atlas* (1666)

Tarrant, Charles. Engineer and surveyor. *Island of Goree* 1758 MS (with F. Mulcaster)

Tarros, Raimondo. *California* in Clavigero Storia 1788–9

Tartaglia, Niccolo (1546) Savant of Brescia. Writer on Mariners Compass

Tasin. See **Tassin**

Tasman, Abel (1602–1659). Dutch navigator and discoverer of Tasmania & New Zealand. Tasman's *Kaart van zijn Australische ontdekkingen* 1644

Tassin, Jean Baptiste. Cartographer and lithographer. *Simla* 1831, *City & Environs Calcutta* 1832, *Frontier British India* 1838, *Singapore* 1837, *Upper Assam* 1839, *Eastern Asia* 1840, *New Bengal Atlas* 1841

Tassin, Leonard (17th century). Physician and geographer

Tassin, Nicolas (d. 1660). Royal cartographer, b. Dijon. *Plan des Villes de France* 1634, *Cartes générales d'Allemagne* 1634, *Provinces de France* 1634, *Costes de France* 1634, *Cantons de Suisses* 1635, *Cartes générales de la Geographie Royalle* 1655

Tastu, J. Publisher. *Voyage Dumont D'Urville* 1833

Tasuisan, Tetsu. Japanese globemaker. *Globe* (1670)

Tate, George. *Geological map of the Roman Wall* 1867

Tate, James. First *classical maps* 1845–7

Tate, J. C. and F. C. *Atlas of Des Moines (Iowa)* 1869

Tate, T. B. *Map City of Ogdensburg* 1869

Tate, Capt. W. A., Bombay Engineers. *West Coast of India* 1827, *S. Maria Harbour, S. America* 1833

Tatham, T. J. Estate surveyor at Lambs Conduit St., later Bedord Place. *Mayland* 1813 MS, *Burnham* 1823 MS

Tatikian, B. *Plan of Smyrna & Aidan Railway* (1859)

Tatiscev, Vasilij Nikitic (1686–1750). Russian statesman, geographer and surveyor. *Historical map of Tartar Russia* (ca. 1740)

Tattersall, George. Drew and engraved *plan of Ascot Heath* 1840

Tatton, Gabriel. English cartographer "at the Signe of the Goulden Gunn at the westende of Ratcliff." *Portolan Chart Mediterranean* 1596, *Guiana* 1599 MS, *Maris Pacifici* 1600, *California and New Mexico* 1600. Engraved Benjamin Wright, also 1616; *Atlantic &c.* 1602, *Nova Francia* 1610

Taufer. Revised *Lauter's Danube* 1785

Taulois, P. L. *Provincia de Santa Catharina (Brazil)* 1867

Taunt, Henry William. *Map of River Thames* 3 sh. (1875)

Tavernier, Daniel. 17th century traveller. Son of Gabriel

Tavernier, Gabriel, Baron d'Aubonne (1566–1610) of Antwerp, engraver and mapseller in Paris, brother of Melchior. Engraver for *Bouguereau's Théâtre François* 1594

Tavernier, Jean Baptiste (1605–1689). Traveller, son of Gabriel. Paris "sur le quai de l'Horloge aux trois Estoilles." *Treize Cantons de Suisse* 1668, *Les Six Voyages* Paris 1678, first English edition 1680

Tavernier the elder, Melchior (1564–1641) of Antwerp. Engraver and printer to the King in Paris. Map dealer and publisher "Sur L'Isle du Palais sur le quay du grand cours d'eau à la sphere Royalle." Sold maps for Schouten 1618, *World* 1628, *Siege of Rochelle* 1628, Cooperated with Jansson (1633–6) Hondius and Danckerts *Théâtre*

Géographique du Royaume de France for Tassin 1634, Bertius' *Asia* 1640, *Théâtre . . . de tout le Monde* 1642

Tavernier the younger, Melchior (1594–1665). Cartographer and engraver in Paris. Son of Gabriel

Taylor, A. *Topographical Map Barbados* by W. Mayo improved A. Taylor 1859

Taylor, A. Publisher. *Chantry's Bath* 1796 & 1801 (with Meyler)

Taylor, Lieut. Alexander, 81st Regt. (later Royal Engineers). Cartographer. *County of Kildare* 1783, *New Map of Ireland* 1793, *Environs of Dublin* (1800), *Dublin surveyed* 1811, Completed *Vallancey's Survey of Ireland* pub. A. Arrowsmith 1811

Taylor, Alexr. *Sketch of water of Druie. Part lands of Kinkardine* 1776

Taylor, Lieut. (later Commander) A. Dundas, Indian Navy. *Surveys India* 1853–88, *Arabian Sea* 1853

Taylor, C. Engraver. *Brooks London* (1812)

Taylor, David. *Travelling & Commercial map Canada* 1834

Taylor, E. *Navigable Rivers & Canals in vicinity Wormersley Railroad* 1821

Taylor, F. *Siacca river* 1805

Taylor, G. *Holyhead* 1807

Taylor, George. Surveyor. Joined with John Skinner to publish *Road books* 1775–7: *General map Roads of Scotland* 1775, *Roads North Britain and Scotland* 1776, *Great Post Roads London-Bristol* 1776, *Roads of Ireland* 1777, pub. 1778; *County of Louth* 1778

Taylor, H. E. *New Zealand* 1891

Taylor, H. E. B. *Barnes Driving map of Philadelphia and surroundings* 1867

Taylor, Isaac (1739–1829). Engraver. *Plan of Hogsden (Hoxton)* ca. 1770, *Plan Town and Harbour of Kingston upon Hull* 1792, *S.E. view Kingston upon Hull* 1798, *Road from Woodbridge to Debenham* 1800

Taylor, Isaac (1730–1807). Surveyor. Many *MS estate plans in Dorset* 1765–77, *Dorsetshire* 6 sh. 1765 and 1795, *County of Gloucester* 6 sh. 1777 and 1786 and 1800; *Map of Hampshire* 6 sh. 1759, *County of Hereford* 4 sh. 1754 and 1786, *Oxford* 1750, *Worcester* 4 sh. 1772 and 1800

Taylor, J. Publisher at ye Golden Lion in Fleet Street. Joint publisher with T. Brandreth, T. Bowles & J. Gouge of Price's *30 miles round London* dated 1712

Taylor, J. Publisher. 30 Upper Gower St. *Borough of St. Marylebone* 1837

Taylor, J. *County of Northumberland* 1864

Taylor, J. N. *New Railroad map of the United States and Canada* 1864

Taylor, James. *Turnpike roads from Manchester to Liverpool* (1790)

Taylor, James. Estate surveyor (fl. 1752–76). *Dengie* 1776 MS

Taylor, Janet. *A Planisphere of the Fixed Stars* 1846

Taylor, John. American land surveyor. *Map town of Petersburg, Menard Co., Illinois* (1857)

Taylor, John. Cartographer and engraver of Dublin. *Wicklow* 1790, Engraved *Nevill's Wexford* 1798, *Ten-fourteen miles round Dublin* 1816, *New and correct Irish Atlas* 1825

Taylor, John. Junior. Publisher Essex Standard Office Colchester. *Plan Colchester* ca. 1845

Taylor, John Hamlet. *Suburban Allotments west of Ballarat* 1856

Taylor, Josiah. Estate surveyor. *Fyfield Willingale* 1749 MS

Taylor, J. H. *Plans of Parish of St. Margaret* London 1824–8

Taylor, Norman. *Stawell Gold Field* 1878

Taylor, R. *Religious Houses in Norfolk* Norwich 1834

Taylor, Rich Cowling (1789–1851). Topographer and geologist. Surveyor *maps of U.S. & Cuba*

Taylor, Robert. *City and County of Baltimore, Maryland* 1857

Taylor, S. *Ancient work Rock River Wisconsin* 1848

Taylor, S. *Chart Harwich* 1667 MS

Taylor, Capt. T. I. *Post Office Stations & Routes India* 1832

Taylor, Thomas. Land surveyor. *Barony of Newcastle and Upper Cross* 1655 MS

Taylor, Thomas. Maritime surveyor. *Negrais Harbour* 1753, published Dalrymple 1774

Taylor, Thomas. Publisher of Liverpool. *Map of Liverpool* 1837

Taylor, Thomas (fl. 1670–1730). Bookseller, map and print seller at ye Golden Lyon in Fleet Street. Blome's *England exactly described* (1715), *North part of Great Britain called Scotland* (1715), *North America* 1718, *Plan London* 1720, 1723; *Gentlemans Pocket Companion* 1722, *Travellers Companion* 1724

Taylor, Thomas Glanvile (1804–1848). Astronomer

Taylor, William. Surveyor. *Proposed route from Boston to Framingham* 1807

Taylor, W. Publisher. *Chantry's Plan of Bath* 1793 (with Meyler and A. Taylor), *Hibbert's Plan of Bath Five miles round Bath* 1787 (with Meyler)

Taylor, William. Land surveyor. *Merydams Farm Sussex* 1748

Taylor, William. Book- and Mapseller at ye Ship in Paternoster Row. *Sacred Geography* 1716, *Naish's Plan of Salisbury* 1716 (with Senex), *Camden's Britannia* 1722

Taylor, Y. *Loudoun County, Virginia* 1853

Tayspill. See **Gilbert & Tayspill**

Tchihatchef[f], P. See **Chikhachev**

Tchitchagov, Peter. *Course of Irtish, Oby and Jenisée* rivers (1730) MS

Tchoubinski, P. (1839–1884). Russian geographer

Tchoulkof. *Coast of Russia on Baltic* 1808 MS

Teage. *Queensland* 1869 (with Parrott)

Teal, J. Surveyor. *Intended Canal W. Riding from River Calder* 1792

Teare & Son. Lithographers. George Street Tower Hill, *Tower Liberty* 1830

Teasdale, James. *Ground Plan Arundel Castle* 1789

Tebb. See **Davies & Tebb**

Tebbett. See **Didier & Tebbett**

Tebyen'kob, Michail Dimitrievic (d. 1872). Russian hydrographer. Governor of Russian America 1848–50. *Atlas of N.W. America* St. Petersburg 1852

Techo, Nicolas. *Carte du Paraguay*, De Lisle 1703 (with Ovalle) and later editions

Tedesco, Angelo. *Lombardia*, Venezia 1859

Teesdale, Henry [later **and Co.**] Publishers, 302 High Holborn (address deleted between 1830–40); 2 Brunswick Row Queen Square (1842). *York* 1826, 1828 (9 sh.). Published *Hennet's Lancashire* 4 sh. 1830, *New British Atlas* 1829 (editions to 1842) (taken over by Rowe, then Collins:) *New Travelling Atlas* 1830 (editions to 1860), *General Atlas* 1831

Tegg, T. (1776–1845). Publisher and bookseller, Cheapside. *New Plan of London* 1812, 1823, 1830

Tegg, T. T. & J. Publishers, Cheapside. *Travelling map of England and Wales* 1833

Tegg, William [Peter Parley] (1816–1895). Publisher, bookseller and author.

Tegg, W. & Co. *Findlay's Classical Atlas* 1847

Teiler, M. *Reysbeschreibung durch Engelland und Schottland* 1634

Teisera, L. See **Teixera**

Teive e Argollo, Miguel de. *Mappa do Estado da Bahia* 1892

Teixeira, Domingos (fl. 1565–95). Portuguese cartographer (brother of Luis.) *World Map* 1573, *Chart Atlantic* MSS

Teixeira, Albernas Joao, elder (1602–1666). Cosmographer to the King of Portugal. Patent as master of making of seacharts. Map in *Thévenot's Relations* Paris 1649 and 1664 Numerous *MS maps* (6 of *Brazil*, 5 of *India*, 4 of *Portugal*, 4 of *World*); *America* (1630)

Teixeira, Albernas Joao, younger (1627–1666). Lisbon cartographer

Teixeira [Teisera, Tesseira, Tiexiera], Luis [Ludovico, Luiz] (1564–1604). Portuguese Jesuit. Mathematician and cartographer to Spanish crown. *Voyages Brasil* 1573–8, *Atlas Azores* (6 maps) 1587 MS, *Japan* 1592, used by Ortelius 1595; *Magnum Orbis Terrarum Univ.* engraved Van den Ende 1612

Teixeira, Albernas Pedro (1575–1640). Son Luis. Portuguese cartographer. *Estrechos de Magallanes* 1621, *Survey of Portugal* 1622–30, engraved 1662; *Topographia de Madrid* 1656

Tejada, J. M. Engraver. *Costas de Chile* 1799

Tejera, Miguel. *Estados Unidos de Venezuela* 1875, *Alto Orinoco* 1886

Telford, Thomas (1757–1834). Engineer. *Plan Port of London* 1799–1800, *Intended Caledonian Canal* (1800), *Road map of Scotland* 1803, *Communication between England and Ireland* 1809, *Stamford Junction Canal* 1810, *London and Edinburgh Mail Roads* 1827

Tellier, Le. See **Le Tellier**

Tello, J. E. y. See **Espinosa y Tello**

Tempel, Ernest W. L. (1821–1889). Astronomer

Tempel Guglielmo. *Pianta di Venezia* 1854

Tempest, Pierce. Publisher. *Horsley's Plan York* 1697

Tempesta, Antonio. *Roma,* Merian sculpsit 2 sh. (1593), *Recens . . . urbis Romae . . . delin.* 1593, another edition 1606

Temple, Sir Grenville. *Shores and Islands of Mediterranean* (1850)

Temple, Lieut. G. T. *Russian Lapland* 1880

Temple, Sir R. *Sikkim* 1881, *Mahratta Country* 1882

Temple, Lieut. R. C. *Andaman Is.* 1880

Temple, V. du. See **Du Temple**

Templeman, Thomas. *Atlas Minor* (Moll), *New Survey of the Globe* 1729

Templeux, Damien de, Sieur de Frestoy (d. ca 1620). Cartographer. *Maps in Théâtre géographique du Royaume de France* 1621, *Isle de France* Blaeu 1635, Jansson 1641, *Beaune* (1650), *Valois* (1650)

Temporius [Temporal, du Temps, Temporarius], Joannes [Jean du]. *Pais Blaisois* 1591, published Bouguereau 1594, used by Ortelius 1598, Blaeu 1631, Jansson 1645; *La Perche* (Blaeu 1635)

Temps, J. du. See **Temporius**

Ten Brink, C.F. See **Brink**

Ten-Have, Nicholas. *Overyssel,* Jansson 1658, Blaeu 1662

Tenner, Karl Ivanovic (1783–1859). Astronomer and military cartographer

Tenney, Lucius E. *American Local Atlases* 1895 onwards: *Northampton* 1895, *Utica N.Y.* 1896, *Yonkers* 1896

Tentivo, Gasparo. Venetian chart maker and pilot. *Golfo di Venezia* (1750)

Terasson, H. Draughtsman and engraver in London. *Willdey's North America* and *World* 1714, *Bowen's Turkey in Europe* 1715, *Jefferys' N. America* (1760)

Terborch, [**Borch**] Gerard (ca. 1616–1681) painter of Deventer

Terbruggen. See **Petrus ab Aggere**

Terentiev, Kozina. Engraver of New Archangel. *Maps for Tebyenkob's Atlas N.W. America* 1845–50

Terreni, G. M. Engraver. *Pianta del Porto d'Acapulco* (1763), *Porto di Boston* (1763), contributed to *Il Gazzettiere Americano* 1763, *Istmo di Darien* 1763

Terry, Garnet. Engraver and Jeweller, 54 Paternoster Row (1778). Worked for Taylor and Skinner 1777–8, for Armstrong 1776, for Mackenzie 1775, *Map Louth* 1778, for Rapin 1784–9, for *Harrison's Atlas* 1787, for Turner 1794

Terry, H. *La Haute Savoie* 1866

Terwoort [Tervoort, Terwood], Lenaert [Leonard]. Flemish engraver. Worked in England for Saxton, *Warwick and Leicester* 1576, *Cornwall, Somerset and Southants.*

Terzi, Andrea. Engineer. *Provincia di Lodi e Crema.* Engraved Stucchi 1818 (2 sh.)

Tesi, Charles. *Ports et Rades Méditerranée* 1858

Tessan, A. Dortet de. See **Dortet de Tessan**

Tessan, U. de. *Atlas Hydrographie* 1845, *Valparaiso* 1882

Tesseira. See **Teixeira**

Tesseyman, W. Bookseller of York. *Rocque's Plan of York* 1766

Tessing, [Thielsing, Tising], Jan (d. 1701). Printer and publisher in Amsterdam. *Atlas Volga–Don Canal* 1703–4, Bruce & Mengden's *South Russia* 1699

Test. William. *Survey of Lake Champlain* (with William Brasier) 1776, *Plan of Pondicherry* 1778 MS

Testard, J. Alph. *Guide de Paris* (1867)

Testevuide. *Carte Topo. de l'Ile de Corse* 1824

Testone, Giulio. Engraver of Rome. *Plan Rome* 1773

Testu, G. Le. See **Le Testu**

Tettau, Freiherr von (1816–48). Cartographer of Berlin

Teullere, Citoyen. *Entrée Rivière de Bordeaux* 1801

Teunisse. See **Antonisz**

Texada, Moreno. Engraver. For Torfino de San Miguel 1786

Texeira. See **Teixeira**

Textor, Lieut. Johann Christoph. (d. 1812). Mathematician and military cartographer. *Ost Preussen* 15 sh. Berlin 1808

Teylingen, Laurens van (d. 1637). Dutch engraver

Thackara, James. Engraver. *Pennsylvania & Maryland* 1797 (for Proud's History), *Ellicott's plan of City of Washington* Philadelphia 1792 (with Vallance), the first official engraved plan of the city; *World* for Guthrie 1795

Thackara, James & Son. *Headwaters of James River* Philadelphia 1814

Thackara & Vallance. American engravers. *City of Washington* 1792, *World* in Carey's American edition of *Guthrie's Geography,* Phila. 1796, 1804, and later

Thacker, W. *Survey of India* 1891

Thadden, W. von. *Plan der Umgegend von Glogau* (1862)

Thalbitzer, Lieut. *Waldeck und Pyrmont* (1860)

Thales of Miletus (ca. 624–546 B.C.). Philosopher and mathematician

Thalheimer, Marie Elsie. *Eclectic Historical Atlas* (1874)

Tharp, John Allen. Surveyor. *Fairfield District S.C.* 1820, for Mills 1825

Thatcher, Edwin. *City of Louisville, Jefferson Co., Kentucky* 1868

Thayer, H. L. *Map of Denver, Colorado* 1874, *Map of Colorado* 1875

Thayer & Co. Lithographers of Boston. *Oregon County* 1845

Thayer, Bridgman & Fanning. Publishers and mapsellers, 156 William Street N.Y. For D. G. Johnson 1853

Thayer & Colton. Publishers, No. 18 Beekman St. New York. *Colton's U.S.A.* 1859

Thayer. See **Ensign Thayer & Co.** and **Phelps Ensign & Thayer**

Theinet, A. *Rhein Provinz* (1850), *Provinz Schlesien* (1850), *Plan von Magdeburg* 1870

Theinert, S. *Map for Sohr* 1842–4

Thelot, J. *Guiana* Frankfurt 1669

Thenot. *Plan itinéraire de Paris* (1850)

Theobold, G. *Carte géologique de la Suisse* (1867)

Theodorus, Petrus (17th century). *Celestial globe* (attributed)

Theodulfus. (ca. 760–821). Bishop of Orleans. Reputed to have drawn *World map* on the wall of his house

Therbu, L. Engineer Lieutenant. *Plans de la guerre de sept ans* 33 sh. Frankfurt 1793, *Attaque des forts de Chouagen* 1792

Therin, Nicolas. Pilot. *Carte de la Méditerranée* 1691–1700, *Golphe de Lion* Amsterdam, Mortier (1700) (with Michelot)

Theunisz, Jan. Printer of Leiden and from 1604 Amsterdam. Probably printed early editions of Mercator-Hondius *Atlas* to 1608

Theunits [Theunitsz], Jacobsz. See **Jacobsz**

Thévenot, Melchisédech, (1620–1692). *Relations de divers Voyages Curieux* Paris 1663–72, *Recueil de Voyages* 1681, 3 maps included. *Carte de la découverte faite l'an 1673,* the first printed map of Mississipi, and first to use word Michigan

Thevet, André (1502–1590). Franciscan monk and cosmographer to the King. Born Angoulème, d. Paris. *France Antartique* 1558, *Cosmographie Universelle* 1575, *Maps of the four continents, France, World in shape of a lily, Nouveau Monde* 1581

Thew, R. *Plan Kingston-upon-Hull* 1784

Thiel, Jan van (fl. 1715–31). Engraver of Amsterdam

Thiel [Theylen], Johann (1595–1603). Dutch surveyor

Thiel, L. Engraver for Stieler's *Hand Atlas. Australien* 1835, *Neu-Süd-Wales* (1834)

Thielsing, J. See **Tessing**

Thierry. Engraver of Paris, Rue des Mathurins St. Jacques No. 1. For Perrot 1823, *Atlas géogr. et topo. de France* (1825), for Malte-Brun *Mélanésie* 1835, *Océanie* 1837). Also worked with Berthelmier and J. Blumenthal *Océanie* 1842

Thierry Frères. Lithographers of Paris. *Cité Bergère I. Alta California* 1849, *U.S.* 1851

Thierry, Sieur. *Plan de la Nouvelle Orléans* 1758

Thiers, Louis Adolphe (1797–1877). Statesman. *Atlas de l'Histoire du Consulat et de l'Empire* 1859, *Campagnes de la Révolution* 1846

Thinot, J. *Carte du Théâtre de la Guerre* (1870)

Thiollet, Auguste. *Exposition Universelle* 1867, *Plan de Rome* (1878)

Thirion, A. *Plan de St. Denis près Paris* 1878

Thoaldus, J. See **Toaldus**

Tholing, Theodoro. *Persia* 1634

Thom, Ambrosius. Prussian Ambassador to Sweden. *S. Sweden* 1564 MS

Thom, A. Engineer, draughtsman, and engraver for Fullarton (1860)

Thom, George. *Rio Bravo del Norte* (1854)

Thoma, G. Engraver. *Lands between Black and Caspian Seas* 1825

Thomann, Jos. Lithographer. *Umgebung von Landshut (Bavaria)* 1871

Thomas, Caleb. *Underground survey of mines, Sandhurst (Victoria)* 9 sh. 1881

Thomas, Christian Ludwig (1757–1817). Geographer and cartographer. *Crimea* 1788, *Plan Belgrade* 1788

Thomas, David. *Seychelles* 1771, *Vincennes District Auburn* 1819

Thomas, Edward J. B. *Oil Territory, Venango Co., Penn.* 1864

Thomas, Edwin. Assistant Engineer. *Plan of London . . . intended railway to Limehouse* 1860

Thomas, Lieut. F. W. L., R.N. Admiralty surveyor. *Orkneys* 1845–8, *Scotland* 1846–9

Thomas, Gabriel. *New Jersey Coast and Delaware Bay* 1698

Thomas, G. F. *Appleton's Railway map of U.S. & Canada* 1865, German edition 1871

Thomas, George (d. 1846). Admiralty Hydrographer, Maritime Surveyor for Home Waters (1810), *Scheldt River* 1809–10, *Fowey Harbour* 1811, *Shetlands* 1810–36, *Downs* 1819, *Coast Suffolk* 1824, *Orkneys* 1837–46

Thomas, George H. *Field of operations under Grant & Sherman* (1865)

Thomas, H. W. *Advertising map of Coventry* 1891

Thomas, Issiah. *Chart World* Boston, Thomas & Andrews 1798; *Jerusalem* 1791

Thomas, I. V. Publisher of New York. *Alexandria in District of Columbia* 1789

Thomas, John, the elder. Engineer. *Latchingdon* 1723 MS, *Fortifications St. Simon's Island* 1738, *Map islands St. Simon and Jekyll* 1740 MS

Thomas, John, the younger. Engineer. *Map islands St. Simon and Jekyll* 1740–3

Thomas, John. Hydrographer. *Mounts Bay* 1751 & 1793, *Collection of charts of Harbours G.B. and Ireland* 1781–1802

Thomas, (John). Solider. *Siege of Enniskillen* 1594 MS, *Map Femanagh* 1594 MS

Thomas, Louis. *Schul Atlas* (1868), *Kartennetze* (1873)

Thomas, Owen. *Tract in Frederick Co., Virginia* 1751, *Survey part Aberporth Bay* (1860) with Fitzwilliams

Thomas, Lieut. R. *Falmouth* 1820

Thomas, Richard. *Geological map mining district of Cornwall* 2 sh. 1819

Thomas, Lieut. Samuel. *Chart part coast France* 1801

Thomas, W. Engraver. *Green's plan of Manchester and Salford* 1794

Thomas, W. *Dartmouth* 1841

Thomas and Andrews. Publishers of Boston. *Carleton's Maine* 1795, *Maps* in *Morse's American Universal Geography* 1796, Arrowsmith & Lewis 1805–19

Thomas, Cowperthwait & Co. Publishers. *Maps* for Augustus Mitchell 1839–59. See also **Cowperthwait, Desilver & Butler**

Thomas & Whipple. Publishers. *World* and *North America* for *Parish's New System of Geography* 1810

Thomassen, E. S. *New Guinea Maclay Coast* (1882)

Thomassen, Thomas. *Stadt Haerlem* (1578)

Thomassin. *Carte du Bosphore* 1828

Thomassin, Philippe. Draughtsman and engraver. *Spoletum Umbriae Caput.* (1600)

Thomassino, Filippo. Publisher of *printed map of Diogo Homem's Medit.* 1606

Thomasz, Thomas. See **Thomassen**

Thomaszoon. See **Thomassen**

Thomlinson, William. *Slocan Lake District, British Columbia* 1897

Thompson. Draughtsman. *London & Environs* 1823

Thompson. *Region between Gettysburg–Appomattox* 1869 (with Camp & Weyss)

Thompson, Alexander. Land surveyor. *MS surveys of farms and lots in Columbia* 1790–1801

Thompson, A. K. *The Holy Land* 1841

Thompson, Charles. *Province of Pennsylvania* 1759

Thompson, Charles. Engraver. *Basin of the Firth of Forth* 1828

Thompson, Daniel. *Part city of Pekin, Illinois* (1854)

Thompson, David. Astronomer and surveyor. *North West Territory* 1814

Thompson, Capt. Edward. *Coast of Guyana* 1783 and 1820 (Delarochette)

Thompson, (F. P.) *Map of Chelsea* 1836

Thompson, G. Publisher, 43 Long Lane West Smithfield. *America* 1799, *New Map of Europe,* 1810, *New Map of World by Kitchin* 1817

Thompson, G. Master, R.N. (fl. 1790 94) Hydrographer. *Surveys in Scotland & Ireland* 1791–4

Thompson, G. F. *Manchester Ship Canal* 1883

Thompson, G. M. *Russia, Turkey, the Baltic and Black Sea* 1854

Thompson, George. Surveyor. *Plan street between Pall Mall and Portland Place* 1813, *New Map London* 1824, *Parish St. James Westminster* 4 sh. 1825

Thompson, George Alexander. *Geogr. and Hist. Dictionary of America and West Indies* (English edition Alcedo) 1812–15, *Atlas to Thompson's Alcedo* 1819

Thompson, Henry. Second Master, HMS *Saracen. Bilbao* Admiralty Chart 1836

Thompson, H. M. *Blanchard's Township map of Illinois* 1864

Thompson, Harcourt. *Plan Manchester Ship Canal* 1883

Thompson, Isaac. *Plan Newcastle upon Tyne* 1746 MS

Thompson, J. Engraver. *Armstrong's Lincolnshire* 1781, *Plan Norwich* 1779, *Parish St. Pancras* 1804

Thompson, James. Mariner. *North Sea* 4 sh. Sayer and Bennett 1777

Thompson, James. Engraver, for *White's York and Ainsty* 1785

Thompson, J. C. *State of Rhode Island and Providence Plantations* 1892

Thompson, James J. *Dept. de Puno (Peru)* 1863, *Titicaca-See und seine Umgebungen* 1893—5

Thompson, P. Publisher. *Royal Bazaar, St. Andrews Norwich* 1834, *Map Religious Houses in Norfolk*

Thompson, Palliser. *Chart of England & Scotland (chart River Tees)* 1812—25, *Bay & Entrance River Tees* 1815

Thompson, R. Surveyor. *Union District South Carolina* 1820, for 1825

Thompson, S. Engraver, Dame Street, Dublin. *O'Connor's Ireland* (1775)

Thompson, Samuel. Surveyor. *Middlesex Canal* 1795 MS, *Charleston* 1794 MS, *Town of Wilmington* 1794 MS

Thompson, Thomas H. *Sonoma County (Calif.)* 1877, *Tulare County (Calif.)* 1892

Thompson, Thomas S. *Coast Pilot for the Upper Lakes* 1861, *Marquette Bay* 1861, *Eagle Harbour* 1861, *Ontonagon Harbour* 1861

Thompson, W. *Studland Bay and Poole Harbour* (1770)

Thompson, W. T. *Maps of Canada* 1878—84

Thompson, Zadock. *Vermont* (1840)

Thompson and Everts. *County Atlases of American States* 1870—2

Thompson and West. *Historical Atlas Santa Clara County, California* 1876, *Alameda Co., (Calif.)* 1878

Thompson. See **Oakley & Thompson**

Thoms, A. See **Thom**, A.

Thoms, G. *Bank of Soundings St. Helena* 1817

Thoms, W. H. See **Toms**

Thomsen, Claud. *Ogeechee River Georgia* (1756?)

Thomsen, J. *Skole Kaart over Europa* 1863

Thomson, A. E. *Johnston's map County of Ayr* (1840)

Thomson, Charles. Engraver for *Scott's Van Diemens Land* 1824

Thomson, J. Engraver. *Plan Town of Belfast* (1820)

Thomson, J. A. *Plan Town port Lincoln South Australia* (1845)

Thomson, J. P. *Kadavu Fiji Islands* 1889

Thomson, James. *Plan Hobart Town* (1850)

Thomson, John, & Co. (fl. 1813—69). Publishers of Edinburgh. *New general atlas* 1814, 1817, 1819 and 1828, *Classical and Historical Atlas* 1829, *Atlas of Scotland* 1831

Thomson, John Turnbull. Government surveyor. *Province of Otago* (1860), *Plan of Singapore* 1844, *Straits of Singapore* 1855

Thomson, Joseph. *Maps Central Africa* 1879—95

Thomson, Draughtsman and engraver, 14 Bury St. Bloomsbury. Drew and engrave for *Walker's Universal Atlas* 1816, *Macartney's Caubul* 1815

Thomson, Manuel T. *Rio Biobio* (1866)

Thomson, T. *Bluff (New Zealand)* 1875

Thomson, William (1727–1796). One of the certifiers of *Cook's boundary of Carolinas* 1772

Thomson & Hall. Engravers, 14 Bury St. London. Langsdorff's *World* pub. Colburn 1814

Thonning, P. Danish cartographer. *Danish Guinea* 1802

Thonus, Labert. *Ville et faubourgs de Liège* 4 sh., n.d.

Thoré. *Dépt. de la Sarthe* 1875–6

Thoreau, Henry David. *Walden Pond* 1846 MS

Thoring, Gabriel (d. 1706). Swedish surveyor

Thoring, Johan Persson (d. 1666). Swedish surveyor of Upsala

Thorlacius, G. See **Thorlaksson**

Thorlaksson [Thorlacius], Gudbrandur (1542–1627). Bishop of Holar in Iceland. *Celestial globe* 1575, *Iceland* used by Ortelius 1595, *Norse discoveries America* 1606 MS

Thorlaksson [Thorlacius], Thordur (1637–1697). Bishop of Skalholt, grandson of preceding. Copied *Norse map* and drew *stars* 1668. Used by Torfaeus in *Gronlandia antiqua* 1706

Thorndike. Surveyor. *Estate map Aunsby (Lincs.)* 1732 MS

Thorn, W. *Island of Java* 1812

Thorn[e], (Robert). Merchant *World map,* woodcut 1527, in *Hakluyt's Voyages* 1582

Thorne, James. *Handbook Environs London* 1876

Thornhill, Sir James (1675–1734). Painter, drew plates for *Flamsteed's Star Atlas* 1729

Thornton, Edward. *Gazetteer of the Countries adjacent to India* 1844

Thornton, G. Engineer. *Lyttelton Harbour New Zealand* 1873

Thornton, John. Engraver. *Green's Plan of Manchester and Salford* 9 sh. 1794

Thornton, John. Surveyor. *District of Seebpoor* 1849

Thornton, John. Publisher, chartseller, engraver. Hydrographer to the Hudson Bay and East India Company. At the signe of England Scotland and Ireland [or the Platt in the Minories]. Collaborated with Fisher, Seller and Mount. *Chart Humber to Goodwins* 1652 MS, *Chart East Coast* 1667, engraved *Narborough's South Seas* 1673, *World* 1683, Supplied *maps* for *Seller's Atlas Maritimus* 1675 (editions to 1710), and to the *English Pilot* 1689 (and later editions). *Mapp of Virginia, Maryland, New Jersey, New York and New England* 1680; *North America* 1699 MS, *Newfoundland* 1700, *Chart Antegua* 1689 and 1704, *Charts of the East Indies* 1699–1701

Thornton, Joseph. Brother of John above, apprenticed 1665

Thornton, Richard. Surveyor. *Manchester & Salford* 1807, pub. 1810; *Revised edition by Adshead* on 24 sh. 1851

Thornton, Samuel (fl. 1703–39). Hydrographer at the signe of England Scotland and Ireland in the Minories. Took over his brother John's business in 1706. Reissued *English Pilot* 1706 (plates taken over later by Mount & Page, *Coast of New Holland* (1740), *Bombay & Sallset* (1750)

Thoroddsen, Thorvaldur (1855–1921). Icelandic cartographer. *Geologische Karte Inneren Island* 1892, *Südostliches Island* 1895

Thorowgood, E. Engraver No. 25 Cheapside. *General Chart of St. George's Channel* 1767

Thorowgood, E. *Map of Lincoln* 1848

Thorp, Joshua. *Ten miles round Leeds* 1823, *Yorks.* 1822

Thorpe, Thomas. Land surveyor and publisher. *Survey City of Bath and five miles round* 1742, 1771; *Improved edition* 1772, editions to 1800. *Parish Barrington (Cambridge)* 1800

Thouars, Abel du Petit. See **Du Petit-Thouars**

Thoulet, Marie Julien Olivier (1843–1936). Oceanographer

Thourneyser, Johann Jakob. *Berna Metropolis* (1720)

Throop, D. S. Engraver of New York. *Lay's U.S.* 1832

Throop, J. V. N. Engraver for Dashiell (1800), for Fielding Lucas Junior 1823

Throop, O. H. Engraver in New York, later Washington. *Lay's New York* 1832, *Oregon* 1846

Thoyon, Alfred Jean Pascal. *Portendick* 1851

Thoyras, Paul Rapin de. See **Rapin de Thoyras**, Paul

Thrale, Willis. Publisher of Hartford. *U.S.* 1834

Thropp, James. *Plan City of Lincoln* 1883

Throsby, John. *Part Roman Route London to Lincoln* 1791, *Leicester* 1777

Thuillier, Col. (later Gen.) Sir Henry Edward Landor. Surveyor General of India (active 1858–90). *Plan City of Juggurnauth* 2 sh. 1849, *Great Trunk Road Calcutta to Benares* 1857, *Bengal and N.W. India* 1860, *Punjaub* 1870, *East Bengal* 1871, *Patna* 1889 *Railways India* 1890

Thuillier, L. *Arabie Pétrée* 1882, *Jerusalem* 1882, *Palestine et Liban* (1885), *Pyrénées* 1887. Engraved for Vuíllemin 1890

Thuilier. Engraver for De Wit (1690)

Thummel. *Altenbourg and Ronneburg* 22 sh. 1813

Thunet, E. **and Co.** Publishers. *United States* 1842

Thurneysser, Leonard (1530–1596). Physician and astrologer. *Map of Brandenburg*

Thurston, Thomas. Land surveyor of Ashford. *Hainault Forest* 1851, *Proposed Maidstone–Ashford Railway* 1863, *Gilham Farm Smarden* 1820

Thury, Cassini de. See **Cassini de Thury**

Thwaites, Christopher. Jarrold's *Map of Norwich* (1875)

Thym, Moses (fl. 1609–17). Wood engraver. *Thuringia* 1609

Thys, Albert. *Le Kassai et la Louloua* 1888

Tibell, Gustaf Wilhelm af Freiherr (1772–1832). Swedish military cartographer

Tibianus, Johannes Georgius. See **Georgicus**, Johannes

Ticho Brahé. See **Brahé**, Tycho

Ticknor, William D. Printer of Boston. *Bradford's Atlas* 1835

Tiddeman, Mark. *Draught of New York* 1737 (not credited) and 1753 (credited), *English Pilot Draught of Virginia* 1737 and 1753, *West coast of Scotland* 1730 MS

Tideman, Philip. *Turkish Empire* (1696) in De Wit's *Atlas*

Tiebout, Cornelius. American engraver, New York. *Colles' Roads of United States* 1789 (73 maps). Engraved for *Carey* 1794 and *Payne* 1798–1800, *Italy & Sardinia* 1794, *Plan New York* 1789, *Seven United Provinces of Holland* 1795, *State New Hampshire* 1810

Tiedemann, D. F. *O'Fallon Station St. Clair Co., Illinois* 1854

Tiedermann, H. *Plan City Panama* 1850

Tielenburg, Gerrit. *L'Escosse* (1730)

Tiexiera. See **Teixeira**

Tietz, R. *Umgebung von Wien* 1866

Tietze, Emil. *Bosnien* 1880, *Umgebung von Lemberg* 1882, *Montenegro* (1844)

Tiffin, R. Land surveyor. *Braintree & Black Notley* 1809 MS

Tilbrook, T. Surveyor. *Survey Sijdling St. Nicholas* 1824 MS

Tilden, S. D. *New map Connecticut* Hartford 1878

Tilden. See **Baker & Tilden**

Tillaeus, Petter (1679–1754). City Engineer Stockholm, cartographer. *General charta ofver Stockholm* 12 sh. 1733

Tillberg, Johann (1756–1826). Swedish surveyor from Abo

Tilleman, S. See **Stella**, T.

Tillemon[t]. See **Du Trallage**

Tilliard, (J. B.). Engraver. Frontispiece for Après de Mainvillette 1775

Tillinghast, David H. Surveyor General South Carolina (1809). *Plat of Charlestown* 1809 (copy of Bull's plan)

Tillo, Aleksej Andreevic v. (1839–1899). Russian geographer and military cartographer

Tillotson, Miles D. *Pocket atlas and guide of Chicago* 1893

Tillson, O. J. *Ulster County New York* 1853

Tilman, Giorgio. Italian engraver in Venice. Gastaldi's *Lombardia* 1570

Tilton, James. Surveyor General. *Territory of Washington* 1855, 1858 & 1860

Timberlake, Lieut. Henry (ca. 1765). *Draught of Cherokee Country* 1762–(5) in *Memoirs*

Timmons, E. *City & Suburbs of Lafayette Indiana* 1854

Timosthenes (180–200 B.C.). Geographer of Rhodes. *Compas card*

Tindal, G. A. *Kaart van Sumatra* 1872

Tindal, Nicholas (1687–1774). *Maps* for Rapin's *History. Battle of Hochstet* (1751), *Towns, forts, & Harbours in Ireland* (1751), *Battle Ramillies* (1751), Tindal and Rapin's *History of England* (1745)

Tindell, Robert. *Chesapeake Bay* 1607–8 MS, *Draught of Virginia* 1608

Tingle, James. Engraver. *Improvement West End London* (1830)

Tinkham, John F. *Business portion Grand Rapids* 1867, *County of Grand Rapids* 1872

Tinkker, T., Junior. *Plan of Manchester and Salford* 1772

Tinkler, W. Engraver. *Plan of River Humber* 1841

Tinney, John (d. 1761) Bookseller, Engraver and Publisher at the Golden Lion Fleet Street. *New Map of the Ukrain* (1740), Rocque's *London* 1746, *Blair in Atholl* (1750), Bowen and Kitchen's *Large English Atlas* 1753 (1760), *Fontainebleau* 2 sh. 1794

Tippelkirch, Ernst Ludwig v. (1774–1840). Military cartographer of Elbing. *Elbing and its surroundings.* 1804

Tipton, John. *Fort Wayne Indian Agency* 1824

Tirion, Isaak (d. 1769). "Boekverkooper voor aan in de Kalverstraat Amsterdam." *Historie . . . Vereenigde Nederlanden* 1742, *Nieuwe en Beknopte Hand Atlas* 1744–69, *Atlas van Zeeland* 1760. His stock sold 1769

Tiroff, H. F. Engraver. *Stadt Ronneberg* (1795)

Tirone, Enrico. *Carta dell' Alta Italia* 1859

Tisch, G. Engraver. *Kort ofver Sundet* 1821

Tischbein, Georg Heinrich (1755–1848). Engraver

Tisdale, T. W. M. *Plan Town Shrewsbury* 1875 & 1890

Tisenhausen, Baron. *Maps and plans of war between Russia and Turkey* 54 sh. MS

Tising, J. See **Tessing**

Tisley, S. London globemaker, 19th century

Tisserand, Felix (1845–1896). Astronomer

Tissot. Geographer. *Franche-Comté*, Jaillot 1669

Tissot, Nicolas Auguste (1824–1895). Mathematician

Tiswell, R. Surveyor, 16th century

Titeux. *Plan d'Alger* 1870

Titi, Filippo. *Legatione del Ducato di Ferrara* 1697, *Romagna* 1699

Titon du Tillet, Evrard (1677–1762). Globemaker of Paris

Titus, C. O. *Atlas of Hamilton Co. Ohio* 1869

Tjarde. *Maps* for Ottens heirs (1740))

Tjader, J. J. *Karta ofver Fahlu* 1845

Toaldus [Thoaldus] , Joseph. Engraver. For Sanson 1695–6

Tobar y Tamariz, José. *Road del Principe* 1786, *Plan of Monterey* 1789

Tobler, John. *Plan of Quebec* 1760

Tobzigliano, G. *Armillary Sphere*, 17th century

Tod, Oliver. Surveyor. *Harbour of Bridgeport, Conn.* (1835)

Todd, Elliot d'Arcy. *Route Ahhar–Kazveen* 1837

Todd, John. *Plan Kendal* 1787

Todd, R. *Pictorial Map Luton* 1853

Todescho, Nicolo. Printer of Florence. Edition of *Ptolemy* 1482

Todleben, Eduard Ivanovich, Count (1818–1884). *Defense Sevastopol* 1863

Toeppen, Max Pollux (1822–1893). *Atlas von Preussen* 1858

Toeput, Lodovico. *Plans of Acquapendente, Treviso* for Braun & Hog. 1598

Tofino de San Miguel. See **Torfino de San Miguel**

Tofswill & Myers. Engraver for Catlin 1841

Tognallo. Engraver. *Plan d'Hanoi* 1883

Tokei, Aou. Japanese geographer and publisher, 19th century. *General Atlas of the Provinces of Japan 2 vols.* (1828)

Tokunai. Japanese cartographer, 19th century

Toler, George, Junior. Draughtsman to Royal Engineer Dept. *Province of Nova Scotia* (1805) MS

Tolly, William. *Poolo Bay near Bencoolen* pub. Dalrymple 1774

Tollenare, Charles de. *Atlas de Plans Dépt. Loire-Inférieur* 1850–8

Tolman, B. *Maps Zeleznic Cech Moravy* (1876)

Tolmatcheff, Pilot. *White Sea* 1757, pub. 1780

Tolomeo, Claudio. See **Ptolemy**

Tomaschek, William (1841–1901). Geographer and cartographer in Vienna

Tomaszewicz. Cyprian. *Plan Kamienic-Podolsk* ca. 1675

Tombleson. *Panoramic plan Thames & Medway* 1846; *Plan of Rhine* (ca. 1850)

Tomka-Szaszky, Jan (1700–1762). Historian and geographer of Slovakia

Tomkins. Engraver. For Mathew and Leigh's *Scripture Atlas* 1812

Tomkins, D. D. (1774–1825). *Land Holdings Staten Is.* 1816 MS

Tomkins, George. Surveyor. *Plan of Johannesburg and suburban Townships* 4 sh. 1889

Tomkyns, I. Capt. Royal Dragoons. *Mapa da Cidade de Lisboa* Drawn & published 1814 by I. Tomkyns

Tomlinson, John. Surveyor. *Pigott Estate (Warwick)* 1767 MS

Tompkins, A. P. *Plan of Johannesburg & suburbs* 1890

Tompkins, C. *Seat of War in Florida* 1839

Thompson, J. *Map Parish St. Pancras* 1804

Toms [Thoms] , William Henry (fl. 1723–58). Engraver, printseller and publisher. Engraved and sold for Popple 1733, *Havana* 1733–4. Engraved and published Badeslade's *Chorographia Britannica* 1741–2 (editions to 1747). *Law's Plan Cartagena* 1741 (with Harding). Engraved and sold *Barnsley's Rattan* 1742, *Geographia Antiqua* 1785

Tondu. Astronomer. *Bosphorus and Canal of Constantinople* 1785, 1804; *Constantinople and Dardanelles* 1807

Tone, John F. *Railways on Tyne, Wear and Tees* 1864

Tonegutti, G. B. *Carto Topo. Laguna Veneta* (1866)

Tonge, Winckworth. *Draught of Fort Cumberland* 1757, *Isthmus of Chignectou* 1755

Tonsberg, Chr. *Norge* 1855, *Oscarshall* 1860

Tonson, J. *Gardner's English Traveller* 1719

Tooke, B. At the ship in St. Pauls Churchyard. Published *Heylin's Cosmography* 1682 (with T. Passinger and T. Sawbridge)

Tooke, Benjamin. Publisher and Bookseller at the Middle Temple Gate. *Geographia Classica* 1712 (with C. Browne)

Tooker, Arthur. Mapseller over against Northumberland House in ye Strand. *Travelling map of England* (1680)

Tootell & Sons. *Street Plan Maidstone* (1894)

Topolnicki, Jan. *Maps Poland* 1864–88

Topping, Michael. *Coringa Bay* 1789

Tordesillas, Herrera y. See **Herrera y Tordesillas**

Torelli, Luigi. *Carta Idrografica del Mar Rosso* 1865

Toreno, Nuño Garcia de. Spanish pilot and chartmaker. *East Indies* 1522 MS, *Carte de Turin* 1525

Toresano, Federico. Publisher of Venice. *Bordone's Isolario* 1547

Torfaeus, Thormodus. *Gronlandiae Antiquae* 1706

Torfino [Tefino] **de San Miguel**, Don Vincente (ca. 1732–1795). Mathematician, astronomer, cartographer. *Estrecho de Gibraltar* 1786, *Atlas Maritima de España* 1787–9, *Derotero de las Costas de España* 1787, *Puerto do Cadiz* 1789, *Islas Baleares* 1807

Torin, Robert. *Charts S. Africa and East Indies* 1785–97

Toriti, Ikeda (1788–1857). Author and calligrapher of Kyoto. *Town plan of Kyoto* 1835

Torlacius. See **Thorlaksson**

Tornauw, Nikolas Nikolavic. Russian geographer, 19th century

Törnsten, Johan. *Charta ofwer West Norrlands Lähn* 1771

Torquato, Antonio (1480–1540). Astrologer and physician

Torquemada, Fr. Juan de. *Yndiae Occidentales Madrid* 1723

Torre, José Maria de la. *Mapa del Mundo* 1864, *Isla de Cuba* 1841, *España y Portugal* 1864

Torre, F. M. de la. See **Martinez de la Torre**

Torre Campos, Rafael (1853–1904). Spanish geographer

Torre, P. de la. See **Turre**

Torreno, Nuño Garcia de (1519–27). Spanish hydrographer, pilot and examiner of navigation. *32 maps of Magellan's Voyage, Carta Universal* 1527

Torricelli, Giuseppe. Globemaker of Florence, 18th century

Torry. Engraver. *Praya Bay, St. Jago* 1782

Tory, Thomas. *Fortifications plan Pellican Point, Nevis* ca. 1665 MS

Toscanelli, Paolo du Pozzo (1397– 1482). Florentine astronomer and cosmographer. *World showing route to Cathay* (lost). Possibly author of Genoese *World Map* of 1457

Toshiyuki Ishikawa. See **Ryusen**

Tosinus, Evangelista. Brescia bookseller. Published 1507 & 1508 *editions of Ptolemy*

Töth, Agoston (1811–1889). Military cartographer

Totleben. See **Todleben**

Touanne, M. E. B. de la. Lieut. de Vaisseau. *Plan du Port Jackson* 1828, engraved by Tardieu, Dépôt Gén. de la Marine

Touar, Don Joseph Pellicer de. *Cataluna & Rosellon* Madrid 1643

Touche, M. de la. *Plan Fort Magebigueduce (Penobscot)* 1780

Toula, Franz. *Hand Atlas Oesterreich-Ungarn* 1887

Toulmin, Harry (1766–1823) *Kentucky* London (1792)

Toulorge, G. *Carte de Madagascar* 1895

Tour, C. de la. See **Cagniard de la Tour**

Tour, Louis. See **Brion de la Tour**

Tour, Le Blond de la. See **Le Blond de la Tour**

Tourmente, A. *La Republica Argentina* 1875–6

Tourné, E. *Ville d'Elbeuf* 1895

Tournetsen, E. *Suisse* 1764

Tourrier, J. *New Plan Paris* 1855

Toussaint, A. Geographer. Revised J. B. Poirson's *Carte de l'Océanie* 1836, *Carte d'Afrique* 1837, *Amérique* 1837, *Palestine* 1837

Toussaint, G. Alvar. *Nouveau Plan Paris* (1850)

Toutain. Engraver. *Le Vasseur de Beauplan's Normandie* 12 sh. 1667

Tovey, Abraham. *Is. Scilly* 1750, 1751, MSS; 1753, 1779. *Plan Cheltenham* 1826

Tovey, J. *Map of the County of Gloucester* 1826

Towne. See **Whitton, Towne & Co.**

Townly, E. W. *Lands at Grove Hill* (1800– 7) MS

Townsend, George. *Town of Portland (Victoria)* 1855

Townsend, L. K. O. *Tehuantepec Survey* 1851

Townsend, Thomas Y. Assistant Surveyor. *Melbourne & River Glenelg* 1840, *Murrambidgee Squatting District* (1870)

Townshend, John. *Islands of Tinian and Saypan* 1765

Townson, R. *Hongaryen* The Hague 1801 (with Kovalinsky)

Tozer, C. *Torquay Harbour* 1796

Tpyekotoml, I. *Russia* 1769

Traas, Oscar. *Plan Géologique Jerusalem* 1869

Tracy, William. *Bescod parke* (1650)

Trail, John. Engineer. *Map fortified part of Barbadoes* 1746, *Country between Dublin and Shannon* 1771

Traitteur, W. von. (1788–1839). Engineer. *Plan Mannheim* engraved Wolff 1813

Tral[l]age. See **Du Tral[l]age**

Trama, Salvatore. Piloto di Vascello. *Chart* for Rizzi Zannoni 1785

Tramezini, Francesco (fl. 1528). Publisher in Rome, partner with his younger brother Michaele. *Ancient Rome* 12 sh. 1561

Tramezini [Tramissin], Michaelo (fl. 1539–62). Venetian printer and publisher with two workshops, one in Rome with his brother Francesco and one in Venice "all Insegna della Sibille." *World* 1554, *Northern Regions* 1558, *Belgium* 1558, *Brabant* 1556, *France* 1558, *Friesland* 1558, *Friuli* 1563, *Guelders* 1566, *Hungary* 1559, *Portugal* 1561, *Rome* 1552, *Spain* 1559, *Ancient Rome* 12 sh. 1561

Trampler, R. (1845–1902). *Kartennetz-Atlas der Oesterreich-Ungarischen Monarchie* 1874

Tranchot (1752–1815). Military cartographer

Transilvanus, M. See **Transsylvanus**

Transsylvanus [Transilvanus], Maximilian [Max of Transylvania] (d. ca. 1526). *Globe* (lost), *De Moluccis Insulis* 1523

Trap, I. P. *Kongeriget Danmark* 1874

Trapezuntios, Georgius. See **Amiratzes**, G.

Trapp, Werner Freiherr von (1773–1842). Austrian military cartographer

Trask, J. B. *California* 1853

Trastour, P. E. *Boca-Barra* 1850, *Isthmus Tehuantepec* 1850

Traur, Max. *Europe* Vienna 1827

Trautman, V. S. *Holy Land* (in Swedish Bible) 1618

Trautwein, Th. *Kaisergebirge in Tirol* 1880, *Plan von München* (1875)

Trautwine, John C. *Philadelphia & Baltimore Rail Road* (1845), *Interoceanic Canal* 1852, *Port Cortez* (1860)

Traux, Maximilian de (1766–1817). *Dalmatie* 1810, *Europe* 1818, *Germany and Italy* 6 sh. 1821, *General und Post Karte Deutschlands* 1827, *Bouche de Cattaro et du Monténégro* 1830

Trechsel, Friedrich (1776–1849). Mathematician and astronomer of Bern

Tredwell, W. *Gulph of Lepanto* 1739

Treibe, W. C. *Livland* (1760)

Tremaine, George C. *County Brant*

Canada West 1859, *County Durham* 1861, *County Prince Edward* 1863

Tremaine, George R. *County of Peel* 1859, *Tremaine's Map Upper Canada* 8 sh. 1862, *Co. Elgin* 1864

Tremble, A. V. *Chemung County New York* 1853

Trench, Richard. *Geological Survey of England* 1875

Trenchard, J. Engraver for *Howell's Pennsylvania* 1792

Trenckman, Johann Paul *Comit. Schoenburgensis delineatus* 1760

Trendelenburg, Theodor (1800–1846). Engraver of Dresden

Trenkner, W. *Umgegend von Osnabrück* 1881

Trenks, Johann Friedrich. Bombardier. *Plan St. George d'Elmina* 1774, *Coenraadsboerg* 1774, *Fortres tot Taccorary* 1774

Trépied, Charles (1845–1907) Astronomer

Treschel, Gaspar and Melchior. Printers of Lyon. *Ptolemy's Geography* 1535 & 1541 (Gaspar only)

Treskot [Trescotti, Tresskott, Trouskot, Truscott]. *Map Northern Russia* 1772, *Siberia* 1775, *Government Irkoutsk* 1776, *Poland and Moldavia* 1769, *Russia* 1780, *Plan St. Petersburg* 9 sh. 1753, *Astrachan* 1770, *Mare Baikal* 1770, *Oesel* 1770, *Orenburgensis* 1772, *Imperii Rusici* 1776, *Irkutensis* 1776, *Cuban* 1783

Treskow, A. von. *Gegend bei Königsberg* (1850)

Tress, Richard. *Ward Farringdon Within* (1851), *Parish St. Michael le Querne* 1853

Tress, W. *Railway survey Essex* 1835 MS

Treswell [Trussell], Ralph. Surveyor Estate plans: *Brown Candover* 1538

MS, *Plans Woodmancott* n.d., *Elkington* 1587, *Bury Green Little Hadham* 1588

Tretow, O. *Güstrow und Umgegend* (1895)

Treuer, A. F. *Fürstenthum Halberstadt &c.* Berlin 1788

Trevethen, William. Engraver. *Maps of the 4 continents for Heylin's Cosmographie* 1652, *Roberts Merchants map of Commerce* 1677, *Mandelslo's East Indies* 1662

Trew, Abdias (1597–1669). Mathematician and astronomer

Trew, John F. *Map of Bristol* (1876)

Trewendt, Eduard. *Karte von den Sudeten* 1846

Treyer, J. *Karte von Ungarn* 1878

Trezier, Amedée François. *Voyage de la Mer du Sud* 1716

Tribout, Alexandre. *Carte routière physique et administrative de France* 2 sh. 1840

Triere, Ph. Engraver, for La Pérouse 1797

Triesnecker, Franz von Paula, S. J. (1745–1817). Astronomer and cartographer of Vienna

Triger, J. *Géologique du Dépt. de la Sarthe* 1874

Trimble, J. R. *Battle of Gettysburg* (1863)

Trimen, Andrew. Surveyor. *Railway plans* 1845 MSS

Trincado, E. L. y. See **Liebano y Trincado**

Trinity House. Established 1514 by Henry VIII at Deptford

Tripoli, G. de. See **Guillaume de Tripoli**

Tripp, William B. *Southern part South America* 1889

Triscott, S. P. R. *City of Worcester, Mass.* 1878

Trock, J. D. *Pas Caart van Texel* 1781

Troeltsch, Eugen Freiherr von (1828–1901) of Würtemberg. *Military maps* 1875–91

Trollope, A. *Map of Colony of New South Wales* Stanford (1873)

Trollope, Comm. H. *Survey Behring Strait* 1853–56

Tromelin. *Rivière d'Abord* 1798

Trommer, J. F. K. *Königreich Sachsen* 1856 and 1859 (9 sh.)

Troncoso, Diego. *California Mexico* 1787

Troschel, Hans (d. 1612). Compass maker of Nuremberg

Troschel, Eugen. *Umgebungen von Danzig* 1859

Trost, Andreas. Engraver. *Residenz Stadt Graz* 1703, *Styriae Ducatus* 1678

Trotter, George Master R.N. of *Swallow* 1776. *Harbour of Bareedy,* Dalrymple 1802, *Suez* 1777, *Tor Harbour* 1777, *Negapatam Road* 1782

Trotter, Capt. H. *Eastern Turkestan* 1875

Trotter, Henry Dundas. *River Niger. Arrowsmith's London Atlas* 1842(48)

Trotter, Lt. Col. James Keith. *Sierra Leone* 1893, *Nile Valley* 1894, *Madagascar* 1895

Troughton, Thomas. *Plan Liverpool* 1807, *Country surrounding Liverpool* 1807

Troye, G. A. Proprietor of *Jeppe's Transvaal* Johannesburg 1889, *Postal Map Transvaal* 1890, *Witwatersrandt Gold Fields* 1890

Trozzo, A. Dal. See **Dal Trozzo**

Truchet, O. See **Truschet**

Truebenbach, R. *Plan von Chemnitz* 1862

Truguet, Lieut. *Détroit de Constantinople* 1804, *Constantinople and Dardanelles* 1807, *Isle of Tenedos* 1785 MS

T[rumball], J[ohn] (1756–1841). Col. Connecticut Regt. *New London Harbour* 1776 MS, *Ticonderoga* 1777, *Boston and surrounding country* 1841

Trumbull, Benjamin. *Map of Connecticut* 1797

Trumper, J. Land surveyor of Harefield, Middlesex. *Colne Engaine* 1811 MS

Trumper, William. Surveyor. *Parish Farnham Royal* 1846 MS

Trumpet, James. Surveyor. *Coplands Estate* 1820 MS

Truschet [Truchet], Olivier. Publisher in Paris, Rue Mont Orgueil, au Bon Pasteur. Reproduced *early plans of Paris* (with Hoyau) 1877

Trusscott, Johann. See **Treskot**

Truscott & Son, James. Lithographers, Suffolk Lane Cannon Street, City. *Port Maldon Tasmania* 1858

Trusler, John. *Portable Atlas (of World)* (1798)

Trusler, Dr. L. *Map of Africa* 1793

Trussell, R. See **Treswell**

Trutch, J. W. *County map Oregon and Washington* 1856

Trutvetter, Jesse [Isennachcensis Jodocus]. *World map* 1524

Truxton, Thomas. *Chart of Globe* 1794

Tryer, (T.) Engraver of Islington. *Evans' chart of Van Diemens Land* 1822

Tschernoi, Theodor. Russian cartographer. *Maps for a proposed atlas of Russia* 1770–83, *Map of Kasan* 1779

Tschierschky, A. (fl. 1844–58). *Map for Sohr* 1842–4

Tschirikow, Capt. Discovered *N. W. coast America* 1741, *map World* 1741

Tschischwitz, W. von. *Umgegend von Neisse* 1860

Tschudi [Schudi], Aegidicus [Egidius] von (1505–1572). Swiss cartographer, Ambassador to Augsburg 1559. *Switzerland* 1538 (used by Ortelius 1570), *Helvetia and Rhaetia* pub. by Münster 1538. A collection of his MS maps is in the Library of St. Gallen

Tschudi, Iwan von (1816–1887). *Schweizer Karte fur Reisende* (1859)

T'Sersteuens. *Phillippines* 1726 MS

Tsurumine Hiko-Ichiro. *Province of Awa* 1849

Tua, Ant. *Plan Milan* 1817

Tubb, William & Son. Surveyors of Fisherton, Salisbury. *Combe Bissett* 1806 MS

Tuck, *Railways of Great Britain* 1847–8

Tucker, A. *Environs Weymouth* (1880)

Tucker, Arthur. Publisher at the Globe over against Salisbury House in the Strand. Joint pub. with Seller & Morden. *Map of Poland* 1672

Tucker, Burton. *General plan Witwatersrand Gold Fields* 3 sh. 1892

Tucker, Capt. Ordnance surveyor. *Waterford* 6 inch 1841

Tucker, F. C. *Town of Natick, Middlesex County Massachusetts* 1874

Tucker, H. *Windsor and Park* (1855)

Tucker, William. *County and Town of Poole* (with T. Reekes) 1751

Tucker, William Henry. *Maps of India* 1858–61

Tuckey, Capt. *Chart of River Zaire* 1817

Tuckey, J. H. Surveyor. *Port Philip in Bass's Strait* 1804 MS

Tudor, E. C. B. and A. B. Colin. *Plan Town & Port of Goole* (1844)

Tuerst [Türst], Konrad (1450–1503). Mathematician, artist and physician of Zürich. *De situ confoederatorum descriptio* 1495–7 with map; the oldest surviving map of *Switzerland*

Tueur, J. W. Designer. Cartouche for *Theatre War N. Provinces Turkey* 1790

Tueur, T. W. Designed cartouche for *Dury & Bell's Black Sea* 1769

Tufts, Peter, Junior. *Plan Town Medford* 1794

Tuke, John. *County of York* 4 sh. 1787, and 1794, *Index* to same 1792, *Country round Chesterfield* 1798

Tulla, Johann Gottfried (1770–1828). Military engineer of Baden. *Grand Duchy of Baden* 1812

Tulloch & Brown. Australian engravers. *Victoria* 1859

Tung, Fang-Lo. Chinese cartographer, 19th century

Tunnicliff, William. *Topo. surveys of counties of England* 1789, 1787–91

Tunstall, John. Estate surveyor. *Tolleshunt d'Arcy* 1692, *Latchingdon* 1696 MS

Tuomey. *Geological map of Tennessee* 1866

Tuppen. Engraver. *Chart of the Moluccas and Eastern Islands* 1801

Tupper, J. *Cuckmere Haven* 1809

Tuppin, J. Engraver. *Intended Improvements Bedford Estate after J. Burton* 1806

Turelles [Turreau], Petrus. French astrologer and philosopher. *France* 1575 MS

Turgis, M. de. Vve [widow]. Publisher in Paris, Rue St. Jacques No. 16 (later Rue Serpente 10); also in Toulouse, Rue St. Rome No. 36. *Océanie* (1830), *Océanie* in Dufour's *Atlas* 1851 (Rue Serpente)

Turgot the younger, Michel Etienne (1690–1751). *Plan Paris* 1739

Turin Papyrus (B.C. 1400). *Topographical map of goldmines in Upper Egypt*

Türke, Juan. *Atlas de Chile* 1895

Turmair, Johannes (1477–1534). Cartographer

Turmor, (John). Surveyor. *Plot of severall farms Co. Dublin* 1656 MS

Turnbull, Major. *Battles of Mexico Operations of U.S. Army* 1847, *Siege of Vera Cruz* (1847)

Turnbull, Charles. *Use of Celestial Globes* 1585 and 1597

Turnbull, J. Estate surveyor. *Little Warley* 1770 MS, *Dagenham* 1790 MS

Turnbull, William. Bookseller of Glasgow. *Ewing's Atlas* 1817, *Potomac River* 1832

Turner, Rev. *View of the Earth* 1771

Turner, A. R. Surveyor. *Estate Cockhill (Kent)* 1792 MS

Turner, George. *The World in a Pocket or a little roome* 1644 MS

Turner, James. Engraver and Printer near the Town House Boston. *Harbour Louisburg* 1745, *Cape Hatteras to Boston Harbour* 1747, *Evans' Middle British Colonies* 1755 and 1756, *Scull's Pennsylvania* 1759, *Seat of War in the Middle British Colonies* 1776

Turner, Jason. Engraver. *Chart Delaware Bay by Joshua Fisher* 1756

Turner, John, Junior. Publisher of Coventry. *Aston's Warwickshire* (1830)

Turner, Joseph Malloral William. *Hindostan* 1845

Turner, Richard, Junior (ca. 1724–1794). Geographer. *Introduction to Geography* 1794

Turner, Robert. Bookseller of London. *Playing card maps* (1676)

Turner, T. Lithographer. *Tress & Chamber Plan Farringdon Within* 1851, *Turner & Son's Local map of Deptford* 1892

Turner, William John. *Tropical South Africa* 1876, *British Colonies S. Africa* 1879, *Persia, Afghanistan & Beluchistan* 1891

Turner & Co. Engravers of Edinburgh. *Wood's Plan Lyme Regis* 1841

Turpin, H. Publisher and Bookseller, 104 St. Johns Street West Smithfield. *Morden's Brief description of England and Wales* 1770

Turpin, Jakob (1772–1798). Engraver of The Hague

Turre [Torre], Petrus de [Pietro de la]. Printer of Rome. *Edition Ptolemy* 1490

Turreau, Pierre. Philosopher and astronomer, 16th century

Turrell, Edmund. Engraver. *Plan Town St. Alban* 1815, *Plan City of Bristol* 1829 and 1833, *Edinburgh* SDUK 1834, *Florence* SDUK 1835, *Dublin* SDUK 1836

Turrettini, Théodore. *Ville de Genève* 1883–9

Turrini. Publisher of Venice. *Miniature edition of Ortelius* in Italian (1655)

Türst, K. See **Tuerst**

Turtle, C. Printer. *Port Phillip* 1840

Tusi, Nasir ad-din. Persian savant, 13th century

Tuttle, Stephen. *Reconnoissance of Mississippi & Ohio Rivers* 1821

Tuxen, N. H. *Hydro. Kaart Over Europa* 1833

Tuyn, Petrus Nicolai (1812–1867). Cartographer and lithographer of Amsterdam

Twardowski, Kasimir (b. 1797). Polish topographer and lithographer

Tween, Joseph. Estate surveyor. *Wethersfield* 1815 MS

Twiss, Lt. Col. William. Royal Engineers. *Woolwich Warren* 1797 MS

Twyford, T. W. Surveyor of Dagenham. *Barking* 1827, 1836 MSS

Twynam, T. H. *Chart S. coast Ceylon* 1833

Twyne, Thomas. Publisher. *Survey of the World* 1572, *Breviary of Britain* 1573

Tyas, Robert. *England & Wales* 1841, *Geology of England & Wales* 1841

Tycho-Brahé. See **Brahé**

Tyers, C. J. Surveyor. *Melbourne & the River Glenelg* 1840, *Township of Portland (Victoria)* 1855

Tyler. Engraver. *Regents Canal* 1819

Tyler, H. D. *Grants from Dutch West India Company* 1897, *Plan city New York* (1897)

Tymme, Thomas. *Briefe description of Hierusalem with a beautiful and lively map* 1595

Tumer, George. *Travellers Guide thro. Ireland* Dublin 1794 (with map)

Tyrer, James. Engraver, Chapel St. Pentonville. *Plan Parish St. James Clerkenwell* 1805, *Salt's Alexandria* 1809, *Coast of Kent* 1812, *N. Sea* 1813. Engraved for Wilkinson 1819

Tyrer, T. Engraver. *Evans' Van Diemens Land* 1821

Tyroff, Martin (1705–1758). Engraver and publisher

Tyson, J. Washington. *Atlas of Ancient & Modern History* Phila. 1845

Tytler, William Fraser. *Country N. & W. of Candahar* 1838–42, published 1875

U

Ubaldini, Petruccio. *Scotio Descriptio* 1576 MS, *Relazione d'Inhgilterra* 1552 MS, *Regno di Scozia* 1588, *Adams Discourse Spanish Armada* 1588

Ubbone. See **Emmius**

Ubelin, Georgius. Co-editor with Jacobus Eszler of 1513 edition of *Ptolemy*

Uberti [Hubertis], Lus Antonio degli. Venetian publisher, 16th century. *Lombardy* Venice 1525

Ubicini Brothers. *Terrestial globe* Milan 1826

Uchida, Itsumi (1805–1882). Japanese astronomer

Uffenbach, Peter. Translator. *Atlas Minor* 1631

Ugarte of Llano, Tomas de. *Seno Mexicano . . . y de Bahamas* 1794

Ughi, Ludovico. *Iconografia della citta di Venezia* 8 sh. 1729, *Nuova Pianta di Venezia* 1787

Ugolini, U. Engineer. *Part of Ethiopia* 1887

Uhei, Bundaiken. Kyoto map publisher, 18th century. *World map* (woodcut) 1710

Uhlenhuth, E. *Europe and North America* Bremen 1849

Uhlmann, C. Th. *City St. Louis, Missouri* 1865

Uittewael, Paulus van (1570–99). Painter and engraver of Utrecht

Ujfalvy, E. D. *Région du haut Oxus* 1879

Ujinaga, Hojo (17th century) Soldier and scholar. *Map of Japan* 1655, *Provincial and route maps*

Ukert, Friedrich August. *Atlas der alten Welt* 1823

Ulhart, Johann Anton (1571–1610). Publisher and papermaker

Ulloa, A. de (1716–1795). Spanish navigator. *Relacion historique del viaje a la America merid.* 1748, *Observations astronomiques Bahia y ciudad Portobello* (1792)

Ulloa. See also **Varela y Ulloa**

Ulpius [Vulpius, Volpaja], Euphrosynus [Eufrasino] (d. 1552). Painter, globe-maker and cartographer of Rome. *Globe* 1542, *Roman Campagna* 1547

Ulrich, C. F. *Grundriss Frankfurt am Mayn* 1811, *Post Karte Deutschland* 1826

Ulve. *Karta ofver Spetsbergen* 1874

Umbreit, Emil (1860–1904). German cartographer of Leipzig

Umlauft, Friedrich (1841–1899). Geographer of Vienna. *Schul Atlas* 1885, *Railway & Telegraph map Asia* (1888), *Universal Handatlas* (1892), *Kleiner Volks Atlas* 1896

Underhill, Arthur. *Yachtsman's map River Dart* 1887

Underwood, J. Surveyor. *Burford and Signet Fields* 1823

Underwood, J. A. *Caldwell's Atlas of Holmes Co., Ohio* 1875

Underwood, T. Publisher of London, Fleet St. *Dr. Playfair's Geography* 1814

Unger, August Wilhelm. Publisher of *Bobrik Atlas* 1838

Unger, Franz. *Neighbourhood of Gratz* (1845)

Unger, John F. Engineer. *Skinners Eddy and Little Meadows Rail Road and connecting lines* 1868

Ungler, Florian. Publisher of Cracow. For Stobnicza 1512, for Wapowski 1528

U.S. Geological Survey. Established 1878

U.S. Navy Department. Issued first charts 1837

Universal Magazine. See under **Hinton** John

Unschuld von Melasfeld, Wenzel (1814– 1896). Austrian military cartographer

Unterberger, Leopold Freiherr von (1734–1818). Military cartographer of Vienna

Unvertzagt, (C. I.) Engraver. *Plan St. Petersburg* 1737. For J. N. Delisle 1745

Unwin, C. *Position of Army on the Sutlej* 1846

Unwin, William Jordan. *Homerton's College Atlas of World* 1861

Unwin. Publisher, London. *Gardner's Plan Kingston* 1889

Upjohn, A. *Calcutta & Environs* 4 sh. 1794, *Post Roads Bengal &c.* 1795

Upjohn, William. *Plan of Shaftesbury* 1799

Upton, William. *Star Atlas* 1896

Urban, B. d'. See **D'Urban**

Urdi ad Dimasqi. Persian astronomer, 13th century

Ure, Frederick John (b. 1863). Canadian surveyor

Uricoechea, E. *Map to Teco Colombiaha* 1860

Uring, Nathaniel. *History of Voyages* 1726 *(maps of West Indies)*

Urintio, G. B. See **Vrients**

Urlsperger. *Georgia with part South Carolina* (1775)

Urooman, Rev. D. *City & Suburbs Canton* 1865 (MS)

Urquhart, (W. S.) Dept. Surv. Genl. *Gold Region Mt. Alexander, Victoria*, J. Arrowsmith 10 Soho Sq. n.d.

Urrutia, José de. *Frontier Mexico-U.S.* 1771 MS

Ursinus, Johann Heinrich (17th century)

Urville, J. G. D. d'. See **Dumont D'Urville**

Usborne, A. B. Admiralty Surveyor. *Survey of New Zealand* 1843

Ushakor, Col. *Plan Fortress of Kars* 1855

Ushakov. Russian engraver. *Rossilis-koi Atlas* 1792

Usher, Alfred. *Map British Honduras* 1888 & 1891

Usodimare, Antonio. *Itinerarium* 1455

Usselinx, W. *West Indische Spieghel* Amstd. Broer Jansz .... Wachler 1624, Contains *t'Noorder deel van West Indien* engraved Goos, the first map to mark Hudson River

Ussher, James. Archbishop of Armagh. *Geographical and Historical disquisition touching Asia* 1641

Usteri, Heinrich (1752–1802). Art and map dealer

Usteri, Paul (1746–1814). Relief cartographer of Zürich

Utinensis, J. See Johanes Utinensis

Utting, James. Surveyor. *Layer de la Haye (Essex)* 1818 MS, *Plan of Kings Lynn* 1848

Uz [Uzt], Georgia Frederico (1742–1796). Military cartographer and globemaker. *Mappa Geografica Regni Poloniae* 4 sh. Nürnberg 1773, *General Karte von Polen* 1793

V

V., D. *Volterrae nova descriptio* (1565)

V., E. See **Vico**, E.

V., E. G. *War in Spain* 1874

V., F. *Carte routière Environs Paris* 1830

V., P. *Environs de Dunkerque* (1714)

V., R. D. *Fransche Merkurius* 1666

Vaca. See **Cabeza de Vaca**

Vacani, Camillo. *Military maps Italian campaign in Spain* 1808–13, published 1820–23

Vaccari, Andrea. Publisher of Rome (fl. 1600–20)

Vaccaria, Lorenzo della. Publisher of Rome. *Ancona* 1582, *Castello S. Angelo* 1600

Vaclik, Jan. *Turecko Europske* 1875

Vadagnino, G. A. di'. See **Vavassore**

Vadianus, G. A. di. See **Vavassore**

Vadianus [Gallus Pugnans von Watte], Joachim (1484–1551). Swiss geographer, mathematician, man of letters. *Epitome trium terre partium* Zürich 1534 (with *World map*), *Typus cosmographicus Universalis* 1546

Vadius, Angelus. Editor of *Ptolemy* Vicenza 1475

Val, Pierre du. See **Du Val**

Valck. See **Valk**

Valckenborch. See **Valckenburg**

Valckenburg, Lucas van. *Gmunden* 1594, *Lintz* 1544 (Braun & Hogenberg)

Valckenier, A. Publisher of Amsterdam. *Hermannides' Britannia Magna* 1662

Valdes, D. Dionisio Galiano y D. Cayetano. *Costa N.O. de America* 1795

Valdes, Juan. *Gulf of Paria* 1776

Valdes Tamon, Fernando. *Ciudad de Manila* 1717, *Yslas Filipinas* engraved Bagay 1734

Valduggia, G. B. F. da. See **Falda**, G. B.

Vale, R. W. *Map City of Sydney* 1892

Valegio, Francesco. *Territorio di Cremona* (1600), *Vicenza* 1611, *Bolognese* (1620), *Segna* 1616, *Piemonte* (1640)

Valegio, Nicolo. Publisher of Venice, "al segno del Pozzo." *World* 1576, *Ducatus Carniolae* (1614), Took over Zaltieri's plates, end 16th century

Valentine, R. Surveyor. *Lord Orentone's Estate* 1851 MS

Valentyn, François. *Oud en Nieuw Oost Indien,* 8 vols. Dordrecht and Amstd. 1724–6

Valerianos, J. de la Fuca. See **Fuca**

Valesio, Francesco. See **Valegio**

Valet, Jean. Engraver for Mentelle 1797–1801

Valet, P. J. *Carte de la Florida et Georgia* 1806

Valgrisi, Vincenzo. Printer of Venice. *Ptolemy* 1561 and 1562, *Universale descrittione di tutta conosciuta* 1576

Valk, Gerard (ca. 1650–1726). Publisher and engraver of Amsterdam sur le Dam. In partnership with Schenk he acquired some of Blaeu's copperplates in 1683 and some of Jansson's plates in 1694. He was joined by his son Leonard. *Globes* 1700–15, *Atlas* 1702, *Caspian Sea* 1721 (wtih son Leonard), Cellarius' *Harmonia Macrocosmica* 1720, *Large maps of World and 4 continents* (1680)

Valk, Leonard (1675–1755). Son of Gerard above. Worked with his father in compiling an atlas and continued the business on the death of his father in 1726. *Orbis Terrarum* (G. & L.) 1680, *Repub. Veneta* (1690). *Circulus Saxoniae Superioris* (1710)

Valkenburg, Lucas van. See **Valckenburg**

Vallance, J. (1770–1823). Engraver and publisher, 145 Spruce Street, Philadelphia. *Ellicott's Plan City Washington* 1792 (the first official engraved plan), *Griffith's Maryland* 1795. Worked for Carey 1795, *Mayo* 1813 and *United States* 1814 (with Mellish and Tanner), *China, Macao* 1795, *N. Carolina* 1795, *World* 1795 (for Guthrie), *Boston* 1806

Vallancey, Lt. Col. Charles. Director of Engineers in Ireland. *Military Survey and Itinerary of Ireland* 1776–85; the survey completed by A. Taylor and published by Arrowsmith 1811. *South Ireland Cork* 1778, *Sketch of Hobkirks Hills* 1781, *Royal map of Ireland* 1785, *Military Survey S. Ireland* 1795 MSS

Vallard, Nicolas. Chartmaker of Dieppe. 16th century. *Sea Atlas* 1547 MS

Vallard, Ant. Publisher, Via Sta. Margharita 9, Milano. *Oceania* 1881

Vallardi, Pietro & Giuseppe. Publishers, Ca. S. Margharita No. 1101, Milano. *North America* 1810

Vallarroel, Domenica de. Spanish hydrographer, worked in Venice. *Sea Atlas* ca.

1580 MSS, *Atlas of 7 maps* 1590 MS

Valle, C. G. See **Gunman [de Valle]**

Valle, G. *Il Polesine di Rovigo* Venezia 1793, *Ferrara* Venice 1793

Vallegio, F. See **Valegio**, F.

Vallemont, Abbé Pierre le Lorrain (1649–1721). *Atlante Portatile* Venice (1748)

Vallesecha [Vallsecha, Valseca], Gabriele de. Of Majorca. *Portolan chart* 1439, *Chart of Mediterranean* 1447 MS

Vallet, J. E. T. Engraver for Desnos & Moithey 1766–7, for Philippe 1787

Vallet, Abbé P. J. *Carte Géologique Dépt. de Savoie* 1869

Vallière, La. See **La Vallière**

Vallnight, F. *Plan Harbour of Port Louis* 1748

Vallsecha, G. de. See **Vallesecha**

Vall-Travers, Rod. de. *Pacific* 1752 MS

Valluet, J. B. *Bruxelles et Environs* 1838, *Canton de Genève* 1857

Valseca, Gabriel de. See **Vallesecha**

Valvassore, Giovanni Andrea. See **Vavassore**

Valvasor, Johann Weikhard, Baron von. *Hertzogthums Crain* 1689, *Ducatus Carnioliae* 1679

Valverde, Emilio. *Plana del Rio Biobio* 1863

Valverde y Alvarez, D. Emilo. *Atlas Geogr. Descriptivo de la Penin. Iberica* 1885

Van Adrichem. See **Adrichem**

Van Aelst. See **Aelst**

Van Alphen. See **Alphen**

Van Anse. See **Anse**

Van. See under terminal name for many other names commencing with Van or Vander

Vance, D. H. *Massachusetts, Connecticut & Rhode Is.* 1825, *Maps in Hart's Modern Atlas* 1828

Vancouver, Capt. George (1758–
1798). English navigator and explorer,
served under Capt. Cook. *Chart of N.W.
Coast of America* 1798, *Voyage of
Discovery North Pacific Ocean and
Round the World* 1798 (with *folio
atlas of 16 charts)*

Vandeleur, Lieut. C. F. S. *Map of Routes
in Somaliland* 1894, *Nandi Country*
1896, *Nupe Country* 1897

Vandergucht, D. Engraver for Willdey.
Great Britain and Ireland 1715

Vandermaelen, Philippe Marie
Guillaume (1795–1869). Belgian
cartographer and publisher. *Atlas
Universelle* 400 sh. 1827, the first
atlas on a uniform scale. *Atlas de l'Europe*
165 sh. 1833, *Belgique* 1840, *Pays Bas*
1841, *Carte Topo. de la Belgique*
250 sh. 1854, *Atlas de l'Europe* 1864

Van der. See under terminal name for
· other names commencing with Van der

Van de Velde, C. W. M. *Eiland Java*
1845, *Terre Sainte* 1865, *Holy Land*
1865

Van Dyke, Sir Anthony. Drew *view of
Rye on Symonson's map of Kent* 1596

Van Esents. See **Fabricius**, D.

Vanhavermaet, Henry. *Bruxelles* 1875

Van Keulen. See **Keulen**

Van Ness, Sherman. Surveyor. *Map of
the city of Hudson* New York 1871

Van Ness, William W. *Matabeliland*
1896–8

Van Vechten. *Senatorial map of
Illinois* Chicago 1854

Vanni, Violante. *Maps in Gazzettiere
Americano del Nuovo Mondo* 1763–
77. *Havana, Pensacola, Porto Bello,
Sant Agostino, Vera Cruz*

Vanrenen, A. D. *Lucknow* 1870

Vanrenen, D. C. *Maps of Districts of
India* 1850–70

Vapovsky, B. See **Wapowski**

Varanderie, M. de la. See **Veranderie**

Varea, L. à See **Savanarola**

Varela y Ulloa, Josef (18th century)
Spanish cartographer. *Plano del Rio
Uruguay* (1784), *maps for Torfino de
San Miguel's Atlas Maritimo de España*
1787

Varen, Bernard. Dutch physician and
geographer

Varenius, Heinrich. *Gegend von Wismar*
(1715)

Varennes, Olivier de. *Voyage de France*
Paris 1639

Varentrap, François. *Plan Carthagène*
Frankfort 1741

Varese, Giovanni Antonio de (fl. 1562–
96). Painted *mural maps in Logia
Cosmografica* 1562–4

Vargas, Federico de. *Seno de Gunayangan
Luzon* 1830 MS

Varle, Charles. *Map of City and Environs
of Baltimore* 1799, 1801; *Frederick and
Washington Counties, Maryland* 1808

Varle, P. C. (fl. 1790–1820). Geographer
and engineer. *Partie françoise de St.
Domingue* 1814

Varrentrapp, François. See **Varentrap**

Varley, John. Surveyor. *Plan Intended
navigable canal from Chesterfield to
River Trent* 1769

Varty, Thomas. *Educational Series of Maps
Ireland* pub. Samuel Arrowsmith 1878,
British Isles 1878, *Asia,* Stanford 1850;
Europe (1850), *Africa,* Samuel Arrowsmith
and *America,* Stanford 1879

Vasconcellos, C. B. de. *Plano Hydro. Porto
de Lisboa* 1878

Vasconcellos, Ernesto de. *Carta de Angola*
1885, *Cabo Verde* 1887

Vasconcellos e Noronha, Carlos de.
Oceano Atlantico Norte 1886

Vasi, Guiseppe. *Pianta di Roma* 1781

Vaslin. *Carte de la Mer Rouge* ca. 1780
MS

Vasquez, Antonio. Engraver. *Puerto de Sn. Carlas* 1790

Vasquez, Graf. Carl. *Hand atlas . . . Wien* (ca. 1825)

Vasquez, Josef. Engraver. *Isla Juan Fernandez de Tiena* 1788, *Rios Huallaga y Vcayali* 1791

Vasquez, Joseph. *Islas Philipinas* 1773, *Costa Nueva España* 1773

Vasquez, Joseph. *Terreno entre Lima y el Callao* Lima 1794

Vassalieu, Nicolai. *Plan Paris* 1609

Vasseur de Beauplan. See Le Vasseur de Beauplan

Vat, L. *Nouvel Atlas* (1864), *Atlas of France* (1864)

Vatace, See Battazi

Vauban, Séb. le Prestre, Seigneur de (1633–1707). Franch military engineer and Marshall of France. *Maps for De Fer's Atlas Royal* 1699–1702, *Fortification Neu Brisach* 1698 MS

Vaudeclaye, Jacques de. *N. E. Coast S. America* 1579

Vaughan, David. *Chorographical map province New York* 1849, *Rail Roads New York* 1856, *Improvement Allegany River* 1864

Vaughan, Robert (1600–1666). Engraver and antiquary of Hengwrt. *Smith's Virginia* 1624, *Drake's World Encompassed* 1628, Reengraved *Brecknock* for Camden's Britannia 1637, for *Fuller's Palestine* 1650, for Heylin *Europe* 1652, *Warwick* 1656, *America* 1657, *Africa* 1666

Vaughan, Samuel. *Cowley's Coasts of Florida, Carolina* (1740), *Mount Vernon* 1777

Vaughan, William. *Newfoundland* 1625 in *Cambrensium Caroleia*

Vaugondy, Robert de. See Robert de Vaugondy

Vauhello. Le Saulnier de. See Lesaulnier de Vauhello

Vaultier, Sr. *Maps for Jaillot's Atlas Nouveau* (1690)

Vaulx, Jacques de (16th century). Norman seaman and cartographer of Hâvre de Grace. Pilot to the King. *Première oeuvres* 1533, *America* 1583 MS, *Manual of Navigation* 1583 MS

Vaulx, Pierre de. Younger brother of Jacques. *Atlantic Ocean* 1613 MS

Vause, Richard. Town Surveyor. *Pietermaritzburg. S.E. Africa routes to Victoria Gold Fields* 1868

Vaux, Viscount de. See Grant, Charles

Vaux, Calvert. *Map of Central Park* 1871–2

Vaux, John. *Patent Globes* (1746)

Vaux, P. See Vaulx

Vavassore [Valvassore, Vadianus, Valdagnino, Guadagnino], Giovanni Andrea [sometimes shown as Zoan Andrea or Zav]. Venetian wood engraver, printer, and artist (fl. 1510–72). *Rhodes* 1522, *Nova descrip. Hispaniae* 1532, *Galliae descrip.* 1536, *Britanniae Insul. descrip.* 1566, *Universalis Orbis descrip.*, pub. Vopel 1558

Vaz, Bartolomeo. Engraver, for Torfino de San Miguel 1787–8

Vaz Dourado [Durado], Fernão (ca. 1520–1580). Portuguese cartographer. *World* 1568, *Atlas Goa* 1568 (14 maps), *Atlas Goa* 1571 (15 maps), *Sea Atlases* 1568–80

Vaz Figueira, Andre. *Carta topo. Ciudade de S. Sebastiao do Rio de Janeiro* 1750

Vazie, Robt. Surveyor, George & Vulture Tavern, Cornhill London. *Proposed Archways through Highgate Hill* 1809, *Thames to Collier Row Canal* 1820 MS, *Thames to Romford Canal* 1824 MS

Vazquez, Bartolomeo. Engraver *Golfo de Sn. Jorge* 1794–5

Vazquez, Francesco. *Atlas Elementar* Madrid 1786, 1795

Vechner, Georg. *Ducatus Breslau* (with Scultetus) Jansson 1641, Blaeu 1662

Vedel, A. S. See **Velleius**

Vedova, G. dalla. See **Dalla Vedova**

Vedorelli-Breguzzo, Carlo. *Carta della Sierra Nevada (Colombia)* (1892)

Veelwaard, Daniel. Engraver. *Kust van Africa (Cape)* (1795), *Baij Fals* 1797, *Platte Grond Rotterdam* 1839

Veen [Veno, Venone], Adriaen [Adrianus Aurelius]. Dutch cartographer. *Scandinavia,* published Hondius 1613, *Globes* 1613

Veen, Pieter. *Maps for Atlas Royal* 1706—10 in Dresden Library

Veer, Gerard de. *Voyages undertaken by Dutch 1594—6* Amsterdam 1598, Italian edition 1599, English edition 1609

Veer [Vera], Gerrit. [Gerard] de, of Amsterdam. *Nova Zembla* 1598, in *Diaruim Nauticum,* Claesz 1598

Veersteeg, W. F. *Nederlandsch Oost Indië* Arnhem 1876—7

Vega, R. G. de la. See **Gonzalez de la Vega**

Vegetius. *De re Militari* (1645), maps engraved Dalen

Veitch, John. Merchant of Selkirk, sold *Ainslie's Selkirk* 1772

Vélain, Charles. *Carte géologique l'Ile Amsterdam* 1878, *Ile de la Réunion* 1878

Velarde, D. Francisco. *Rebuelta Chiquitos, Mojos &c.* 18th century MS

Velarde, Fr. Pedro de Murillo, S. J. (1696—1753). Spanish Jesuit. *Yslas Philipines* 1734 & 1744

Velasco, Juan Lopes de. *Maps of Mexico and South America* 1601: *Guatimala, Panama, Chili, Nueva Espana*

Velde, C. W. M. van de (1817—1898). Dutch cartographer and painter. *Java* 1842, *Environs of Jerusalem* 1858, *Heilige Land* 1867

Velde, Jan van den. Artist. Engraved *maps for Hugo Allardt* 1665

Velde, Johannes Gerrard Hulst van Keulen van de. See **Keulen,** Gerard Hulst van

Veldeck, Heinrich. *Plan von Göttingen* (1824)

Velden, W. Vonden. See **Vondenvelden**

Velez de Escalante. *Nuevo Mexico* 1788

Velhagen & Klasing. Publishers, 19th century. Founded by August Velhagen & August Klasing 1835

Velho, Bartolomeo. Portuguese chartmaker working in France. *World* 4 sh. 1561—$\dot{4}$, *Cosmography* 1568 MS, *Corpos Celestes* 1568

Velius, Theodorus. *Attica etc.* for Jansson (1638)

Velleius [Vedel], Andreas [Anders Sörensen Vedel] (1542—1616). Friend of Tycho Brahe. *Iceland* Ortelius 1585

Velserus, Marcus. *Theatrum Geographiae Veteris* 1618—19, *Opera* 1682 (with *Peutinger Map*)

Vendrasco, A. *Carta topo. Citta di Venezia* 6 sh. 1889

Venecia, G. de Bo da. See **Bo da Venecia**

Venegas, Padre Miguel. *Noticia de la California* Madrid 1757 (4 maps), first history of California. First English edition 1759, *Mar del Sur, Seno de California*

Venerable Bede. See **Bede**

Venetiano, D. See **Zenoi,** D.

Venetus de Vitalibus, B. See **Vitalibus,** B.

Veneziano, A. See **Musis,** Agostino de

Venner. See **Mount & Page**

Veno. See **Veen**

Venturi. *Reggio di Lombardia* 1817

Venyukov, M. I. *Boundaries between Russia and China* 2 sh. 1877, *Ethnographic maps Asiatic Russia* 1876, *Russisches Reich* 1877

Vera, G. de. See **Veer**

Veran. *Plan ville d'Arles* 1867

Veranderie, M. de la. *Nouvelles découvertes dans l'ouest du Canada* 1750

Verbeek, R. D. M. *Geolog. beschrÿving van Java Madoera* 1896, *Vulcan Karte von Java* 1898

Verbeke, Eugène. *Carte de la Belgique* 1832

Verbiest [Verbist], Ferdinand, S. J. (1623–1688) of Courtrai. Astronomer, cartographer, missionary to China, d. Pekin. Reconstructed Pekin Observatory 1672. *Celestial Globe* 1673, *World map in Chinese* Pekin (1674)

Verbiest, Isack. *World* published P. Verbiest 1636, *Totius geogr. Regni Galliae descriptio* 1653

Verbiest [Verbist, Ver Bistus], Pieter [Petrus] (1605–1693) of Antwerp. Cartographer, engraver, publisher. *Tab. Geographicarum Belgicae Liber* 1636 (19 maps), 1644 & 1652. Also *maps of World* 1636, *Great Britain* 1629, *Spain* 1639, *Germany* 1634, *Italy* 1634, *France* 1636, *Artois* 1639, *Lant van Waes* 1656

Verborcht, P. See **Borcht,** van der

Verburg, Dirck. *Cadiz.* Keulen (1715)

Vercelli. *Fragment World map* (13th century)

Vercruyse [Verkruys], Teodoro. Engraver. *Citta di Parma* 1716, *Etruria* 1724

Verdaguer, C. Lithographer of Barcelona. For Sala 1884

Verden, Carl [Karl] van. Russian sailor. *Mer Caspienne* 2 sh. 1722, first accurate map of Caspian, later pub. by Homann and Moll

Verdun de la Crenne, Jean René Antoine, Marquis de (1741–1805). *Voyage Flore* 1771–2 (with Borda & Pingre), *Iceland* 1774, *Antilles* 1775, *Beschreibung von Island* 1786

Vere, de. See **De Vere**

Vereker, Commdr. F. C. P. *Sulu* 1883

Verelst, Harman. Accountant to Trustees for establishment of Province of Georgia. *Country of Carolina* 1739 MS

Vergerius, Ludwig. *Carniola and Istria* used by Münster 1550

Verges, de. *Partie de la Louisiane, Nouvelle Orléans* Broutin Deverges 1740

Verhaer, Franciscus. *Lumen Historiarum per Occidentem* (1630)

Verhees, Hendrik. *Bataafsch Braband* 1794

Verhoef [Verhoeven], Pieter Willemsz [Pierre Guillaume]. *Manilles of Philippines Eylanden* 1609

Veris, Jacob. *Nieuwe Caerte . . . Haarlemner . . . Meer* 1641

Verjus, Bishop. *St. Joseph District British New Guinea* 1890

Verkruys. See **Vercruyse**

Verlinden, Rev. Father. *Ortos country (Mongolia)* 1876

Vermaas, J. C. *Amsterdam Haarlem &c.* 4 sh. (1883)

Vermale, Sieur. Ex Cornette de Dragons. *Louisiana* 1717 MS

Vermeersch, Capt. *Haut Dahomey* 1898

Vermes, Capt. A. B. *Tschagos Archipelag (Indonesia)* 1837

Vermeyen, Jan. *Plan Tunis* 1536 (used by Braun & Hogenberg)

Vermyden [Vermuiden], Sir Cornelis. *Map of the Fens* 1642

Vernaci, Juan. *Archipel de Filipinas* 1807

Vermandus, [Fernando] Secus. See **Alvares Seco**

Vernon, Edward. *American Railroad Manual* 1873

Vernor, Hood & Sharp (19th century) Publishers of London. *Cole and Roper's*

British Atlas 1810. Succeeded in 1819 by George Currie & Co., 31 Poultry

Vernieux, L. P. *British Sikkim* 1853

Verniquet, Edme. *Plan Ville de Paris* 1791

Vernon, August Joseph. *Plan du Hâvre de Grone Fiord (Iceland)* 1858

Verplank. Surveyor. *Plan Attack forts Clinto & Montgomery* (with Holland & Metcalfe), Faden 1784

Verrazano, Giovanni (ca. 1485–1530). Italian navigator sent by Francis I to explore New World. *World Map* 1530

Verrazano, Hieronymus (fl. 1522–29). Italian cosmographer. *MS World map* ca. 1529. Used information supplied by his brother Giovanni

Verreyke[n]. *Piedmont* published by Hieronymus Cock 1552

Verricy, Gio. *Rigo. Napoli* (1615)

Versi, Pietro de. Portolan maker of Venice 1444

Versteeg, W. F. *Nederlandsch Indië* 1853–62, *Nieuwe Atlas Nederlandsch Oost Indië* 1876–7

Vertue, George (1684–1756). Engraver for *Haynes Roman Remains. Yorkshire* 1744, *Surveigh Ruins of London after John Leake in 1666,* Vertue 1737 (8 sh.)

Vertue. See **Virtue**

Vesconte de Maiolo. *Portolan chart* 1512

Vesconte, Petrus, of Genoa. *Portolan Charts* (1311–27), *Portolan Carta Nautica* of 1311, oldest surviving example. *Atlas of maps* 1318, *Maps for Sanudo* (ca. 1320) attributed.

Vespucci, Amerigo (1451–1512). Florentine navigator and explorer. Pilot Major, Casa de la Contratacion 1508–12. Voyage to America. *Venezuela* 1499, *Brasil* 1501. America named after him by Waldseemüller

Vespucci, Juan Giovanni. Florentine pilot, nephew of Amerigo, also Pilot of the Casa de la Contratacion (1512).

Totius Orbis Descriptio 1523–4, *Mappemonde* 1526 MS

Vetch, Capt. J. Royal Engineers. *Projection of Globe on the Cylinder of a Meridian* 1820

Veth, D. D. *Maps of Indonesia* 1878–80

Vetzel, F. A. *Inner Asien* 1840

Veuve de Marret. See **Marret**

Vezou, Louis Claude de. Geographer, publisher and mapseller, "Rue St. Martin près la Rue aux Ours." *Détroit de Gibraltar* 1756

Viana, A. D. *Bisdom Utrecht* 1598

Vianen, Jan van. Engraver (fl. 1695–1729). *City Utrecht* 1695, *Plans of Hochstett* 1704, *Blenheim* (1705), for 1707 edition of *Ptolemy's geography,* for Van der Aa's *Galérie Agréable du Monde* 1729

Viani, Matheo. Publisher in Venice, Campo S. Bartholomeo. *Globes* 1784

Viazoffsky, Lieut. *Khanat of Kuldja (China)* 1871

Vibe, A. *Norske Kyst Jomfrueland-Graendsen med suerrig* 1860

Vico. (fl. 1804–24). French engraver. For *Atlas* of La Pérouse and for Dépôt de le Marine. *Gibraltar* 1804, *Bouche du Rhône* 1817, *Hainan* 1819, *Paris* 1820, *Neptune Français* 1822, *Côtes du Mexique* 1823, *Baie de la Table* 1824

Vico, Enea. Italian cartographer of Parma. *Budapest, Perpignan* 1542, *Nice* 1543, *Celestial Globe* 1544, *Germany* 1552

Vicuna Mackenna, Benj. (b. 1831). Chilean geographer

Vidal, Lieut. (later Capt.) Alexander Thomas Emeric. Hydrogrpher. *Porto Grande off Cape Verd* 1820, *Charts Africa* 1821–6, *Cape of Good Hope* 1828, *West coast Africa* 1835–8, *Azores* 1841–5, *Menai Strait* 1845, *Madeira* 1855

Vidal, Alfredo. *Estado do Rio Grande do Sul* (1893)

Vidal, E. E. *Detroit frontier to the head of Lake St. Clair* 1816, *Detroit River to River Thames . . . . up Big Bear River* 1815

Vidal, Frederico Perry. *Cidade do Porto* 1865

Vidal, J. M. y. See **Montero y Vidal**

Vidal, Philippus. *Obispado de Cartaxena Reino de Murcia* 2 sh. 1724

Vidal-Gormaz, Francisco and Ramon. *Puerto de Quintero* 1866

Viegas, Gaspar Luis (fl. 1534–7). Portuguese cartographer. *Chart Atlantic* 1534 MS, *Sea Atlases* ca. 1537, *Caribbean Sea* 1537

Viel, Charles François. *Plans of the Hôtel Dieu* 1781

Viele, General Egbert L. *Military Campaign Charts* 1861, *Topo. Atlas City of New York* 5 sh. 1874

Vielle, Vin. Engraver *Flags of World* 1804

Vien, Charles. Bookseller, Paris. *Virgin Islands* (1790)

Vienot, P. *Plan d'Amiens* (1867) and 1874

Vierenklee, Jean Ehrenfried. *Carte du Consistoire de Wittenberg* 1749

Vierge, M. Engraver. *Teatro de la Guerra* [in Spain] 1874

Viero, Teod. Mapseller in Merceria, Venice. *Plan Gibraltar* 1781

Vieth, Gerhard Ulrich Anton (1763–1836). *Atlas der alten Welt* Weimar 1819 (with Funke)

Vieth, Otto. *Herzogthum Braunschweig und Umgebung* (1897)

Vietor; Hieronymus. Printer of *Cracow edition of Ptolemy* 1519

Vigil, R. M. See **Martinez Vigil**

Vigliani, Paul Honoré. *Frontière Anglo-Portugaise dans la région de Manica* (1897)

Vigliarolo, Domenico (fl. 1530–80). Portolan maker of Stilo, Calabria. *Atlas of Mediterranean*

Vigne, G. T. *Northern Part of Punjab and Kashmir* 1846

Vignola, Cantelli da. See **Cantelli da Vignola**

Vignoles, Charles. Surveyor. *Charleston District and Beaufort District South Carolina* 1820 (with Ravenel) for Mills 1825

Vignoles, Charles. *Map of Ireland* 1835, *Plans Railways Ireland* 1837

Vigo, A. de la C. y M. de. See **Cavada y Mendez de Vigo**

Vila, Vicente. *San Diego* 1769 MS

Vila Franca, Balthasar de. *Mappemonde* ca. 1315

Viladestes, Mecia de. Catalan cartographer. *Chart* 1413 MS

Vilavicencio, Cipriano Sanches. *Atlantic* 1596 MS

Villa, Ignacio. *Planisferio Villa* (1879)

Villa, V. *Globes* 1815

Villa Abrille, Faustino. *Laguna de Bulman (Mindanao)* 1878 MS

Villadestes, Johannes de. Catalan portolan maker. *Chart* 1428

Villalpando, Juan Bautista (1552–1608). Spanish architect and cartographer, b. Cordova, worked in Rome. *Urbis ac Templi Hierosolymitani* 3 vols. 1596–1606, large plan of Jersusalem copied many times by later publishers, e.g. Van der Aa in 18th century

Villamena, Francesco. *Ager Puteolanus* 1602, *Urbis Romae Sciographia* 8 sh. 1574

Villanovanus, Michael. See **Servetus**

Villanueva, C. M. y. See **Millan y Villanueva**

Villar, James. Surveyor. *Plan Cheltenham* 1892

Villard, J. du See **Du Villard**

Villard, R. A. de. *Map Yangtse-Kiang* 13 sh. 1895

Villardo, Antonio. Publisher of Milan, Via S. Margherita 9. *America Settent.* 1850

Villaret. *Diocèse de Cambray* 4 sh. 1769, *Haut Dauphiné* 1758, *Cours du Var* 1760

Villaroel, Domingo de (fl. 1589) Portolan maker

Villaseno y Sanchez, Joseph Antonio. *North America* 1746 and 1754, engraved in Rome by Petroschi

Villavicencio, Manuel. *Repub. del Ecuador* 1858, *Luzon Is. Port Cavite* 1876

Ville, Ludovic. *Maps Sahara* 1872

Villefranche, Major. *Hudson River* 1780, *Plan of West Point* 1780

Villeneuve, Sr. de. *Carte de la comté de St. Laurens en la Nouvelle France* 1689

Villeray. Engraver. *Dictionnaire complet géographique* 1844

Villiers, A. J. M. Brochant de. See **Brochant de Villiers**

Villot, J. N. *Strassburg* 1870

Vilquin. Mapseller. Cour du Palais Royal No. 20, Paris. For Frémin 1820

Vinc Boons, J. See **Vingboons**

Vince, Samuel. Astronomical introduction to *Pinkerton's Modern Geography* 1807

Vincendon-Dumoulin, Clément Adrian (1811–1858). Hydrographer, French Marine Capt. *Solomon Is.* 1836, *Isles Gambier, Samoa* 1838, *Atlas Hydrographique* 1847

Vincent, H. Engraver. *Maps Brasil* for *Santa Teresa's Istoria delle Guerre del Regno del Brasile* Rome, Rossi 1700

Vincent, J. *Atlas Ancient Geography* 1828

Vincent, William. Publisher. *Plan Scarborough* 1747

Vincent, William. *Commerce and Navigation of Ancients in Indian Ocean* 1807

Vinchio, Pietro Maria da. Franciscan monk of Casale. *Celestial & terrestial globes* 1739–51

Vinci, Leonardo da (1452–1519). Celebrated Florentine painter, known also for his contribution to cosmography. Sketch for a *World map in gores* (ca. 1512), *Fortification plans*

Vincke, Baron von Klein. *Danube* 3 sh. 1840, *Asien* 1844 & 1854

Vinckeboons, J. See **Vingboons**

Vindelicor, A. Map for Homann Heirs *Atlas Géographique*

Vine, Stephen. *Survey of Durridge in Rotherfield Sussex* 1777

Vingboons [Vinc Boons, Vinckeboons], Jan [Johannes]. Dutch cartographer and engraver. *MS atlas* ca. 1639 made for Dutch West India Company, *Brasil* 1665, *Angola* (1665), *Nieuw Nederlandt* (1665), *Stadt Mexico* (1665)

Vinten, C. *Map of Petroleum District of West Virginia* 1868

Vion, Blaize. *Carte de la Rivière de Canada* 1699

Viquesnel, Aug. *Carte de la Thrace* 1854

Virga, Albertinus de. Venetian cartographer. *World map* (1414)

Virgile de Bologne. *Plan Antwerp* 1565 (with Grapheus)

Virtue, George. Publisher of London, 26 Ivy Lane Paternoster Row. *Moule's English Counties* 1830–7, *Atlas* 1836–9, *Barclay's Universal Dictionary* 1842–8

Virtue & Co. *Hughes' National Geogr.* 1868, editions to (1886)

Virtue, James S. Publishers of London. *Complete Universal Dictionary* (1850), (1852)

Visconte, Pietro. See **Vesconte**, Petrus

Visconti. Austrian military engineer. *Transylvania* engraved 1699

Visconti, Pietro Ercole. *Iconografia di Roma antica* (1835)

Visentini, Antonio. *Urbis Venetiarum Prospectus* 1742

Vismara, Pietro. *Strade Ferrate dell alta Italia* 1880

Vischer, George Matthäus (1628–1696). Tyrolese priest, cartographer and engraver. Geographer to Emperor Leopold of Austria. *Austria* 12 sh. 1669, *Lower Austria* 1670, *Styria* 1678, *Hungary* 1685

Visscher, Cornelis. *Plan of Kiel* 1585 (attributed)

Visscher, Claes Jansz[oon] (1587–1652). Engraver and publisher of Amsterdam, "In de Visscher on the Kalverstraat." Founder of the family business. Also known as Piscator. Published a *Panorama of London* by Hollar (1600), *Germania Inferior* 1634 and later editions, and a *Tabularium Geographicarum* 1649

Visscher [Piscator] , Nicolas I (1618–1679). Son of Claes, continued the business from the same address. *America* 1646, *Atlas Contractus* 1657, several later editions to 1677, with contents varying according to the customer's request. *England and Wales* 1686. Also issued an *Atlas minor*

Visscher [Piscator] , Nicolaes Jansz. II (1649–1702). Son of Nicolas I, whom he succeeded in 1679. *Atlas Minor* (1682) and later editions; *Germania Inferior* 1684 and editions to 1698; *Angliae Regnum* (1695), *Spain & Portugal* (after 1700), *Nova Sueciae tab.* 1702. The business was continued by his wife Elizabeth until 1726. Several editions of the *Atlas Minor* up until 1716, an *Atlas Major* 1702, and various

Battle Plans. Most of business passed to Schenk in 1717

Visscher, Nikolaes III (1639–1709)

Vitalibus, Bernardinus Venetus de. Printer in Rome. *Editions of Ptolemy* 1507 and 1508

Vitzthum, Gen. Major. *Plan de St. Petersburg* 1821

Vium, Poul. *Politisk Farvekort over Danmark* (1876)

Vivares, Francis, Junr. Engraver. For Mackenzie 1775, for Kitchen 1799

Vivien de Saint Martin, Louis (1802–1897). French geographer. *Dictionaries, Manuals, Carte Electorale* Paris 1823, *Atlas Universel* Paris 1825 & later, *Histoire de la Géographie* 1873–4

Vivien, F. *Environs de Paris* 2 sh. 1706

Vivien, L. Geographer. *L'Amérique Méridionale* 1825

Vlam, B. Publisher and bookseller. Kalverstraat, Amsterdam. *Nouvel Atlas des enfans* 1772 and 1776

Vlasbloem [Vlasblom, Vlasboem] , Louis (fl. 1646–70). *North sea* 1656, *Nieuwe Lees-kaart* Amstd. 1664, *Star chart* for Keulen (1680)

Vlier, Nicolas. Publisher. Collaborated with J. Hulst van Keulen *Henneman's Surinam* 1784

Vocke, Carl. *Reisekarte von Thüringer Waldgebirge* 1859

Voegeli, A. *Plan der Stadt Zürich* (1869)

Voegelin, J. Conrad. *Atlas der Schweiz* 14 sh. 1846–55

Voeikoff, A. (b. 1840). Russian meteorologist

Voelter [Völter] , Daniel (1814–1865). *Schul-Atlas* 1840, *Wand Karte von Deutschland* 1864, *Historischer Atlas* 1864, *Königreich Württemberg* (1867)

Vogel, Carl (1828–1887). *Thüringer Wald* 1867, *Hand Atlas* 1871–5, *Stieler's Kleiner Atlas* 1872, *Frankreich* 4 sh. 1889, *Balkan Halb-Insel* 1895

Vogel, F. *Hand Atlas Austria-Hungary* 1891

Vogel, G. *Ost und West Preussen* (1891)

Vogel, Julius. Editor *Official Handbook New Zealand* 1875

Vogel von Falckenstein, Capt. *Gegend Berlin* 1831–41, *Düsseldorf* 1842, *Schlachtfeld von Schleswig* 1863

Vogeler, F. W. *Schul Atlas* 1869

Voghera, Luigi. *Citta di Cremona* (1830)

Vogler, E. A. *Salem and Winston, North Carolina* 1876

Vogler, T. H. See **Aucuparius**

Vogt. *Grundriss der Stadt Köln* 1815

Vogt, Charles (1817–1895). Swiss geologist

Vogt, Capt. I. *Corsica* 1735 and 1740

Vogt, Johann George (b. 1669). Cistercian monk, adopted name of Maurice. *Nova totius regni Bohemiae tabula* 1712

Vogtherr, Conrad. Mapseller "à la grande Morskoy dans la maison de Mons. Kunthson, St. Petersburg." *Moldavie* 1771

Voight, B. F. Publisher and lithographer of Weimar. *California* 1849

Voigt, F. *Schul Atlas der alten Geographie* 1871

Voisin, Abbé. *Carte Archéologique Dépt. de la Sarthe* 1853

Voisin, Lancelot du. See **La Popellinière**

Volbeding, Hermann. *Plan von Leipzig* 1869

Volcius [Volcie, Voltius] Vincentius Demetrius (fl. 1563–1607). Portolan maker of Ragusa (Dalmatia). *Charts* 1592–1601 MS

Volck, Johann Melchior. *Stadt Minichen* 1616

Volckmann, Erwin. *Umgegend von Rostock* 1895

Volckner, Tobias. *München* (1893)

Volkaert, Jean F. *Brandenburg* (1740)

Volkamer, Johann Christoph (1644–1720). Botanist of Nuremberg. *Europe* 1697 MS, *World* 1702

Volkert, August. *Post Map Central Europe* Munich 1848, *Königreich Bayern* 1852

Volney, C. F. *Etats Unis* 1803, *Continent de l'Amérique* 1803, *Eastern Hemisphere* 1796

Volpaia, Eufrosino della. See **Ulpius**

Voltelen, J. *Zakatlas der geheele Aarde* (1871), *Kleine School Atlas* 1875, *Atlas Nederland* 1876

Völter, D. See **Voelter**

Volz, Eduard. *Eisenbahnen Deutschlands* 1864

Volzio. See **Volcius**

Von Aigner. See **Aigner**

Van Aitzing. See **Aitzing**

Von Altischofen, P. See **Pfyffer von Altischofen**

Von Amman, I. A. See **Amman**

Von Baggesen. See **Baggesen**

Von Bell, J. A. See **Schall von Bell**

Von. For futher names commencing with von see Terminal name

Vondenvelden, Wilhelm. *New Topo. map Province Lower Canada* (with L. Charland)

Vooght [Voogt] Claes Jansz (d. 1696). *Nieuwe groote lichtende Zeefakkel.* G. van Keulen 1687, *Flambeau de la Mer* 1691, *Antorcha de la Mar* (1695)

Vopel [Vopell, Vopellius, Medebach] Caspar (1511–1564) died Cologne. *Globes* 1532, 1536, 1541, 1542, 1543, 1544, 1545. *World* 12 sh. (lost) 1545, 1549, 1552, 1558 (woodcut version Vavassore), *Rhine* 5 sh. 1555, *Europe* 10 sh. 1555, *Europe Van de Putte* 1572, *Europe* (12 sh.) 1597, *Rhine* 1555, 1560

DE NIEUWE GROOTE
LICHTENDE ZEE-FACKEL,
't TWEEDE DEEL.

Vertoonende de Zee-Kusten van het Zuyderste gedeelte van de Noord-zee, 't Canaal, 't Westersche
gedeelte van Engelandt en Schotlandt, Yrlandt, Vranckrijck, Spanjen, Marocco, Gualata,
Genehoa en Gambia, met de onder-behoorende Eylanden, mitsgaders de Vlaamsche,
Canarische en Soute Eylanden.

ALS MEDE

De Beschrijvingh van alle Havenen, Bayen, Reeden, Drooghten, Diepten, Streckingen en Opdoeningen van Landen.
Op de ware Pools hooghte geleyd. Uyt ondervindinge van veel ervarene Stuurlieden, Lootsen en Liefhebbers der Zeevaert, vergadert.

Door JAN van LOON, en CLAAS JANSZ. VOOGHT,
Geometra Leermeester der Wis-konst.

t'AMSTELDAM, Gedruckt voor JOHANNES van KEULEN, Boeck- en Zee-Kaert-verkoper,
aen de Nieuw-brugh, in de gekroonde Lootsman. 1687.
Met Privilegie voor 15 Jaren.

*Title to the "Zee-Fackel" (Sea Torch) by J. van Loon and C.J. Vooght,
published by J. van Keulen, Amsterdam 1687*

Vorsterman, Lucas. *Castella et praetoria nobilium Brabantiae* 1696

Vorzet, E. *Chemins de Fer et Lignes Maritimes France* 1869

Vosburg, W. M. *Padonia Brown County Kansas* (1857)

Voss, Wilh. *Karta öfver Halmstad* 1847

Vossius, Isaac. *De Nili. . . . origine* (with map) The Hague 1666

Vou, J. de. See **De Vou**

Vouillemont, Estienne. Engraver to the King. "En l'Isle du Palais au Coin de la Rue de Harlet A la Fontaine de Iouvence." *Isle de Matthe* 1672, *St. Christophe et Cayenne* 1667, *Le Jeu de France,* Fer 1671

Voyeard, E. Engraver for Esnaut and Rapilly. *Phelippeaux's Possessions Angloises dans l'Amérique Sept.* 1777

Vredia [Vredius], Oliver [Olivarius]. Publisher of Bruges. *France* 1647

Vrients [Vrints, Vrientius, de Vriendt, Urintio] Jan [Johannes] Baptista (1552–1612). Engraver, publisher and mapseller. *Plancius' World map* of 1592, *Mappemonde for Linschoten* 1596. About 1600 he acquired stock and plates of Ortelius and De Jode. Issued an *Atlas of Belgium* 1602–3. Published *editions of Ortelius* in 1603, 1606 and 1609

contributed some fresh maps, including *England* and *Ireland* after Boazio. These maps were slightly larger in format than the rest of the atlas, and were folded in.

Vries, Claes de. *Zee Atlas* Amstd. 1698

Vries, David Pietersz de (fl. 1593–1655). *Voyages to America* 1631–44

Vries, Martin. Gerritsen. *Bay of Good Hope* 1643, *Japan* 1643, *Yezo Island* 1643

Vries, Nikolas de. See **De Vries**

Vriese, L. J. de. *Kaart van Java* 1842

Vrints, J. B. See **Vrients**

Vuillaume, R. *Maps of French canals* 1886–92

Vuillemin, Alexandre A. (b. 1812). Parisian cartographer. *Atlas Universel* 1847, Revised *Brue's Antilles* 1860, *Atlas Géographique* (1868), Revised and corrected *Atlas Migeon* 1874

Vulparia. See **Ulpius**

Vulpius, E. See **Ulpius**

Vusin, Caspar (1664–1747). Publisher and engraver of Prague, son of Daniel. Reissued *Aretin's Bohemia*

Vusin [Wussim] Daniel (1626–1691). Publisher of Prague. Reissued *Aretin's Bohemia* 1665

W

W. *Carte des Environs de Gènes* (1747)

W[hitwell], C[harles]. *Geographical Description of France* (1655)

W., C. P. *Italia Benedictina* (1740)

W. F. *Europe Centrale* 1852

Waal, J. de. *Kaart der Nederlanden* (1884)

Wachendorf, Adam. *Plan of Dantzig* for Braun and Hogenberg 1575

Wachsmuth, M. B. Swiss engraver. Tournetsen's *Suisse* 1764

Wachter, Alfred Oscar. *La guerre de 1870–71,* 1873

Wächtler, F. F. (16th century) German silversmith, *Globes*

Wackenreuder, Vitus. *Mines of Treasure Hill (Nevada)* 1869

Waddington. *Karte des Nil-Thals* 1859

Waddington, Major C. *Plan of Battle of Meanoe* (1847)

Waddington, John. Surveyor. *Plan Lands Rob. Langford in Hornsey Lane* (1750)

Waddington, Robert. *Mercator's Chart of the West Indies* (1740), *Chart coast Guayana* (Mount & Page 1750?)

Wade, Capt. C. *N. Punjab,* Walker 1846; *Sikh Territory* 1846

Wade, George. Field Marshall. *Roads Highlands Scotland* 1746

Wade, William. *Panorama of the Hudson River* 1845

Wadham. See Green & Wadham

Wadstrom, Carl Bernhard. *Nautical Map . . . . Sierra Leone and island of Bulama* 1795, *Plan Island of Bulama* 1795

Wadsworth. *California gold region* San Francisco 1862

Wadsworth, Alexander (19th century). American surveyor of Boston. *Narraganset Bay* 1832

Waerck, Fridries. Engraver. *Lon's Copenhagen Sound* 1771

Waesberger [Waesburg], Johann (d. 1681). Son in law of J. Jansson. With Weyerstraet, as Jansson's heirs, published *Atlas Contractus* 1666

Waesburg. See Waesberger

Wagener, H. T. *Touristen Karte von Potsdam* 1876

Waghenaer [Wagenaer, Aurigarius], Lucas Jansz[oon] (1533–1606). Dutch hydrographer and pilot of Enkhuizen. *Sphieghel der Zeevaert* 1584–5, the first printed marine atlas, Dutch text. Translated into Latin as *Speculum Nauticum* 1586, English edition *Mariners Mirrour* 1588, German text 1589, French text *Miroir de la Navigation* 1590. *Thresoor der Zeevaert* 1592, French edition 1601 and later editions. *Enchuyser Zeecaertboeck* 1598 and 1601

Waghorn, Thomas. *Overland Routes for Indian Passengers* 1846

Wagner, C. T. *Plan Stadt Glauchau* 1882

Wagner, E. *Plan Darmstadt* 1836

Wagner, Edward. *Bayerische Rheinpfalz* 1867, *Odenwald* 1869

Wagner, Emil. *Touristen Karte von Konstanz* (1894), *Atlas de poche de la Suisse* (1898)

Wagner, Fr. *Schul u. Reisekarte von Stuttgart-Cannstatt* (1896)

Wagner, Fridolin. *Orbis Terrarum Antiquus* (1864), *Post u. Eisenbahn-Karte Sachsen* 1876

Wagner, H. Publisher of Leipzig. *Neuer Handatlas* 1895 (with E. Debes)

Wagner, Heinrich. Cartographer of Darmstadt. *Weinheim* (1871)

Wagner, Hermann. *Deutsches Reich* 1874, *Schul Atlas* 1893

Wagner, Jan Edward (1835–1904). Czech cartographer. *Maps of Czecho-Slovakia* 1862–96

Wagner, Johann Christoph (17th century). Instrument maker and cartographer

Wagner, J. H. Surveyor. *Florence, Nebraska* (1860)

Wagner, Jul. *Joachimsthal* 1873

Wagner, K. Th. *Atlas der Erde* 1873

Wagner, Matthias (1648–1694). Publisher, bookseller and cartographer of Ulm. *Danube* 1685 (4 ft. by 1 ft. 8 in.)

Wagner, M. C. *Counties of Clinton & Gratiot, Michigan* 1864

Wagner, O. *Dresden und Umgebung* (1871)

Wagner, R. *Elsass* 1874, *Strassburg* 1874

Wagner, Thomas S. Lithographer, 38 Hudson Street, Philadelphia. *Pacific Wagon Roads* 1858

Wagner, Wilhelm. *Karte der Priessnitz-Wälder* (1872)

Wahl, Richard. *Atlas physique* 1829, *Map Italy* 1836

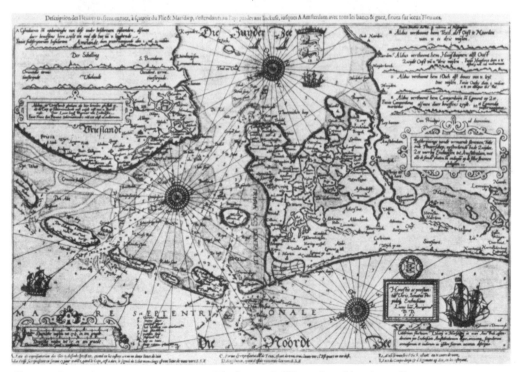

Engraved sea chart from Lucas Jansz Waghenaer's "Speculum Nauticum"

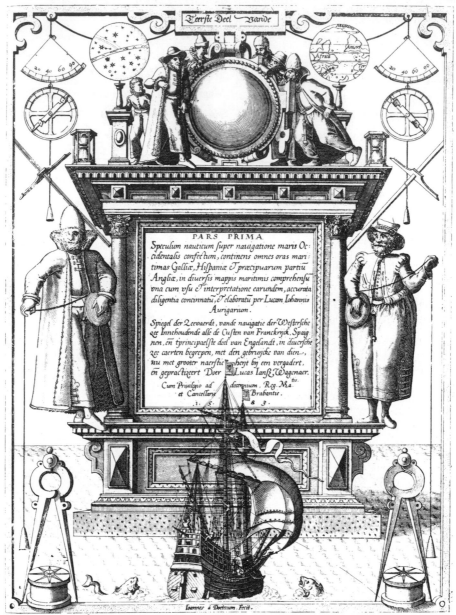

Engraved title-page to the first Latin edition of the Speculum Nauticum by Lucas Jansz Waghenaer.

Wahlfeldt, Charles Gustavus. *They Bay. Eugano Island* 1771, published Dalrymple 1774; *Cawood* 1771, *Bencoonat Bay* 1774, *Saubat (Sumatra)* published Dalrymple 1774

Wailly, Charles de. *Plan du Port Vendres* 1779

Wait, B. *Van Diemens Land* 1843

Waite, J. F. Publishers. *New International Office and family atlas* (1896)

Wakabayashi, Tokusaburo. *map of Korea* Osaka 1882

Walbeck, Joannes. *Terrae del Fuego*, Hulsius (1620)

Walch, Georg. *Reise der Kinder Israel aus Egypten* (1640)

Walch, Hans Philip. Engraver of Nuremberg. *Noua orbis terrarum* by Eckebrecht 1630

Walch, Jakob (ca. 1445–1516). German painter and engraver in Venice

Walch, Jean. *Charte de l'Afrique* Augsburg, Martin Will (1790); *La France* 1794

Walch, Johann. Cartographer and map publisher (fl. 1757–1824). *Augsburg* 1784, *World* 1803, *Australia* 1802, *America* 1807, *Augsburg* 1820, *America* 1824

Walch, J. & Sons. *War in New Zealand* 1863, *Hobart Town* 1870, *Map Tasmania* (1874)

Walch, Johann Philipp (fl. 1617–31). Engraver of Nuremberg

Walckenaer, Charles Athanase, Baron (1771-1852). *Recherches sur la Géographie Ancienne* 1822 and 1839. *Gallia* 1844, *Carte Physique de la France* 1847

Waldenström, E. R. *Karta öfver Lulea Stad* 1878

Waldersee, Gustaf Graf von (1826–61). Cartographer of Potsdam

Waldkirch, Konrad (1549–1612). Printer and publisher of Basle

Waldseemüller [Hylacomilus, Hilacomilus, Ilacomilus] , Martin (1470–1521). Geographer at St. Dié in Lorraine. Suggested the name America for the New World. *Cosmographiae Introductio* 1507, *Globe* 1507, *Europe* 1511, *Maps* to 1513 edition of *Ptolemy, Carta Marina* 1516

Wale, (Pieter) de. Printer of Antwerp. *Tipus Orbis Universalis* 1530, first World map printed in Antwerp. Copied from Apian & possibly printed by Hier. Cock.

Wale, Samuel. Engraver. *Titlepage* for *Geogr. Magnae Brit.* 1748

Wales, John. *Limbe Strait* 1798, *Aluer Straits* 1799, *Amahoy Bay* 1801

Wales & Co., William. Publisher Castle Street, Liverpool. *Swire's plan Manchester* 1824, *Liverpool* 1824

Walger, Heinrich. *Relief maps* 1867–72, *Koeniggratz* 1868, *Paris* 1871

Waligorski, Aleksander (1794–1873). Military cartographer in Norway and Frarrce. *Veikart over Norge* (1855)

Walis, George. *Plan N.E. portion of Metropolis* 1850

Walker. *New Geographical Game,* Darton 1810; *Geographical Pastime* 1834

Walker. Surveyor. *Project for Map of the Roads of the State of S. Carolina* 1787 (with Abernethie)

Walker, Lieut. R.N. *Graham's Island* 1835

Walker, Anthony. Engraver. *Map of England & Wales* (1760)

Walker, Arthur F. *Underground survey of Mines Sandhurst. Hustlers Line of Reef* 3 sh. 1877

Walker, Francis A. *Statistical Atlas of U.S.* 1874, *Atlas of Massachusetts* 1890

Walker, G. H. & Co. American publishers. *Town and County Atlas of Massachusetts* 1879–99

Walker, H. Estate surveyor of Windsor. *Barking* 1831 MS

Walker, Henry. *Map of the Curragh of Kildare* 1807, *Map of District Corps of Ireland* 1798

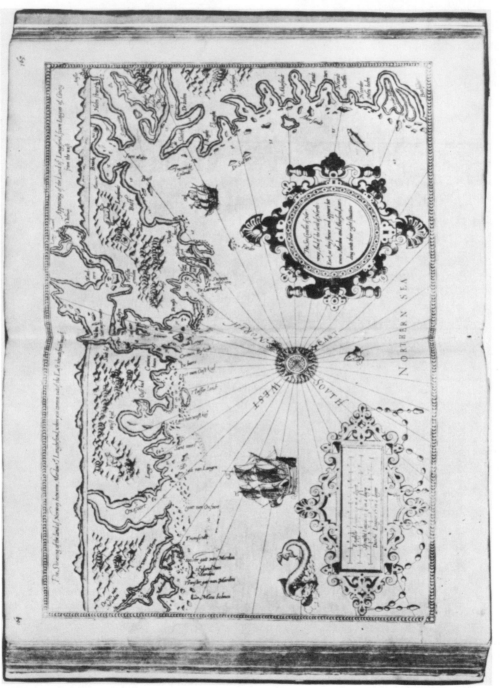

L.J. Waghenaer's "Mariner's Mirrour" (published London, 1588)

Walker, Henry B. Publisher. *Standard Atlas Dominion of Canada* 1878

Walker, J. and A. Publishers, 47 Bernard St. Russell Sq. and 33 Pool Lane Liverpool. *Map of U.S.* 1827, *Stephenson's Railway Liverpool-Manchester* 1824, *Canada* 1827, *Liverpool-Manchester Railway* 1829, *Bancks' plan Manchester* 1832, *Chart from Liverpool to Holyhead* 1841

Walker, J. & C. (fl. 1820–95). Engravers, draughtsmen and publishers, 47 Bernard Street, Russell Sq. (1830), 3 Burleigh St. Strand (1837), 9 Castle Street Holborn (1841–4), 37 Castle Street Holborn (1875). *For Admiralty* 1823–64. *For East India Company* 1839–59. *For W. H. Allen* 1842–8. *For Greenwood* 1831. *British Atlas* 1837, editions to 1879. *Royal Atlas* 1837–73. *Fox Hunting Atlas* 1845–95

Walker, James. Engraver. For Rennell 1799. *Route of Mungo Park* 1799, *Sumatra* 1811, *Mouth River St. John* 1818, *Guatemala* 1832

Walker, James. Land surveyor of London, E. India Road. *Improvements to Port of London* 1805, *East India Docks* 1807–8, *Little Ilford Level* 1818 MS

Walker, James Thomas (1826–1896). Major General, Surveyor General India, and Superintendant of Trigonometrical Survey of Dehra Dun. *North Trans-Indus Frontier* 4 sh. 1852, *Turkestan* (1868), *Afghanistan* 1885

Walker, Johann Gabriel (d. ca. 1850). Cartographer of Solothurn

Walker, John. Surveyor. *Salisbury Estate Berchampton* 1771 MS

Walker, John (1759–1830). Engraver to the Admiralty. *Universal Gazetteer* 1798, *Atlas to Walker's Geography* 1802, *Wood's Irrawaddy* 1800, *Plan Madras* 1815, *Universal Atlas* 1816, *World on Globular Projection* 4 sh. 1818, *Track of Fury & Hecla* 1824

Walker, John (1786–1873). Geographer of Cockermouth

Walker, John, elder (fl. 1584–1626). Surveyor and architect of West Hanningfield (Chelmsford) *Boxted* 1586, *Chelmsford* 1591 MSS; *West and East Hornedon* 1598, *Househam Haule* 1609

Walker, John, younger. Assisted his father from 1600. *Ingatstone* 1600–1, *White Roothing* 1609 MS, *E. Hanningfield Tye* 1615 MS

Walker, John, Junior. Geographer to East India Company and later (1839) to Secretary of State to India. [*Map of India* Horsburgh 1825], *North West Frontier* 1841, *Sikh Territory* 1846, *Gulf of Aden* 1854, *Bengal, Behar* 1858; *Map of India* 6 sh. 1864

Walker, John. Surveyor of Wakefield. *Inland Navigation Canals & Railways* 1830, *Navigations Humber and Mersey* 1826

Walker, Josef. *Canton Solothurn* 1832, *Schweiz* 1841

Walker, Juan Pedro. *Texas-New Mexico* 1805–17 MS

Walker, M. Engraver at Hydrographical Office 1813

Walker, Oscar W. *Atlas of Gardner Town, Massachusetts* 1886; *Atlas of Massachusetts* 1891

Walker, Obadiah. One of the compilers of Moses Pitt's *English Atlas* 1680

Walker, Ralph. Engineer. *West India Wet Docks* 1801, *Plan East India Docks* 1804, *Romford Canal* 1809–12 MS

Walker, Robert. Printer of Fleet St. *Simpson's Agreeable Historian (with County maps)* 1746

Walker, R. *River Lea to London Bridge* 1796

Walker, R. B. N. *Portion Gold Coast Colony* 1884

Walker, Samuel (fl. 1620–30). Estate surveyor of West Hanningfield. *Good Easter* 1623 MS, *Little Dunmow* 1631 MS

Walker, Samuel. Publisher of Boston. *North America* 1846

Walker, [Thomas, Lt. Col. Light Infantry]. *Plan Fort George & Mount Charity in Barbados* 1780 MS

Walker, Capt. Thomas. *Chart Coast Guyana* 4 sh. 1798

Walker, Thomas. Cartographer. *Chart Cape Blanco to River Sierra Leone* (1760)

Walker, Thomas. Civil engineer. *Upper Canada* 1865

Walker, Thomas. Engraver. *Estates & Farms belonging to Greenwich Hospital* 1805–25 (S. & T. Walker)

Walker, W. Engraver. *Cartouches for Jeffery's Yorks.* 1771–2

Walker, W. *Chart S.E. coast Africa,* Horsburgh 1818

Walker, William. *Map of British Guiana* 1875

Walkup & Co., W. B. American map publishers, San Francisco, late 19th century

Wall, G. P. *Geological Maps of Trinidad* 1860

Wall, H. C. T. van de. *Plattegrond Rotterdam* 1860

Wall, Mann & Hall. *Atlas of Noble County, Ohio* 1879

Wallace, Cosmo. Estate surveyor. *Covington (Hunts.)* 1764, *Kimbolton Parish* 1764, *Parish Swinshead* 1765

Wallace, (J.) Cartographer. *Orkneys* 1693

Wallace, James. *Charleston Harbour, South Carolina* 1777

Wallace, J. B. and T. Shillington. *City Lot book atlas of Manhattan Island* 1873

Wallace. See **Middleton, Wallace & Co.**

Waller, William. *Lead Mines of St. Carfery Prise* 1693

Wallford, Benjamin. Publisher and printer to the Royal Society at the Prince's Arms in St. Pauls Churchyard.

Narborough's Streights of Magellan (1694)

Wallinby, Oliver [i.e. W. Leybourne] *Planometria* (1650)

Walling, Henry F. (1825–1888). Civil engineer and surveyor. *American state and city maps: City Providence* 1849, *Attleborough, Mass.* 1850, *Augusta, Maine* 1851, *Maine* 1862, *Illinois* 1870, *Maryland* 1873, *City Salem* 1874, *Canada* 1875

Walling & Gray. *Atlas Massachusetts* 1871

Wallis, Edward. *Travelling map England and Wales* 1815, *New Railway Game* (1830), *Guide for Strangers through London* 1824 to 1841

Wallis, George Olivier, Comte de. *Carte militaire et itinéraire de Servie* 1788

Wallis, James. Engraver and publisher, 77 Berwick St. Soho. *Pocket Edition of English Counties* 1810, *New British Atlas* 1812, *Panorama* (1820), *Plan of London* 1827

Wallis, John. *Twenty-two miles round London* 1783, *Picturesque Round Game* 1795, *Post Roads England & Wales* 1796–8, *Seat War Germany* 1799, *Plan London & Westminster* 1795, *Guide London & Westminster* 1813

Wallis, John. Publisher. Map Warehouse, Ludgate Street London, *Roads and Inland Navigation Pennsylvania* 1791, *Map of the U.S.* 1783

Wallis, John. Vicar of Bodmin. *Cornwall* 1848

Wallis, Samuel. *New Universal Atlas* 1799

Wallis, Samuel, Capt. of H.M.S. *Dolphin* (1728–1795). Marine cartographer. Discovered Tahiti or Sandwich Islands. *Earl of Egmonts Island* 1767 MS, pub. Sayer & Bennett 1775

Wallman, Clas. *Charts ôfver Wästterbottn* 1796

Walpo[o]le, George Augustus. British topographer. *New British Traveller* London, Alex. Hogg *1784 & 1794*

Walseck, Georg. *Eisenbahn Karte von Deutschland* 4 sh. 1876, and 1891

Walser [Walserum] , Gabriel (1695–1776). Swiss cartographer and chronicler. *Atlas Novus Helvetiae,* Homann Heirs 1763–9, *Schweitzer Géogr. Zürich* 1770

Walsh (19th century) British cartographer. *World* 1802

Walsh, Jacob. Translator Giraldus's *Topography of Ireland* 1581 MS

Walsh, John. *Tonqueen Bar* pub. Dalrymple 18th century

Walsh, Thomas W. *Map of Talbot District* 1846

Walsperger, Andreas. Benedictine monk of Salzburg. *Circular World map in Constance* 1448 MS

Waltenberger, A. *Maps of the Alps* 1872–5

Walter von Pfeilsberg, Constantin Johann (1720–1781). Austrian military cartographer

Walter, B. *Geological map Bukovina* 1876

Walter, Henry. Surveyor. *Windsor Forrest* 1823, *Templewood Estate Stoke Poges* 1834 MS, *Railway Survey Essex* 1835 MS

Walter, J. Publisher. *Chart of Northern Pacific Ocean,* Meare's Voyages 1790

Walter, J. *Commercial map Scotland* 1785 (with W. Gordon)

Walter, Robert. *Plan Lunenberg Harbour* 1753

Walther, Bernard (ca. 1430–1504). Astronomer of Nuremberg

Walther, Johann Georg (d. 1697). Engraver and publisher. *Trèves* (1680)

Walthern, J. F. *Plan Berlin,* Homann 1737

Walthoe, J. One of the joint publishers of *Well's set of Maps* (1700)

Walton, Robert (fl. 1647–87). Publisher and mapseller. At the Rose and Crown

at the West End of St. Pauls (1666), The Dyall in Little Britain, The Globe and compasses in St. Paul's Churchyard (1673). First to put roads on printed *map of England (Porters Map)* (1668). *Maps of the continents* after Kaerius, Visscher and Blaeu

Wamser, Albert. *Darmstadt* 1894, 1896; *Friedberg* 1895, *Hessen* 9 sh. 1893, *Mainz* 1895

Wandelaer, J. Designer and engraver. *Palestine for Leland* 1714

Wang Chih-Yuan (13th century) Chinese cartographer

Wang, Chung-ssu (8th century) Chinese geographer.

Wang, Ming-yuan (7th century) Chinese cartographer

Wangesteen, Ore Andreas (d. 1763). Norwegian military cartographer. *Aggershuus Stift* 1763, *Regni Norvegici delin.* 1753, *Kongeriget Norge* 1761, English edition 1796

Wangersheim, William. *Railroad and Township map of Indiana* 1887

Wang P'an. See **P'an**

Waniek, W. W. *Plan Wiener Neustadt* 1800

Wanka von Lenzenheim, Joseph Freiherr von (1828–1907). Austrian military cartographer

Wapowski [Vapovsky] , Bernard (1475–1535). Royal Polish cartographer and historian. Educated Cracow, friend of Copernicus. In 1506 in Rome; worked with Beneventano in revising Cusa's map of Central Europe for 1507 edition of *Ptolemy's Geography. Sarmatia* published by Ungler in Cracow (now lost). *Map of Poland* Cracow 1525. His work used by Mercator and Munster

Warberg, O. *Kort over Haureballeg* 1789, *Kort over Anholt* 1792, *Maps of Schleswig Holstein* 1776–1806

Warburton, John. Somerset Herald and cartographer. *County of Northumber-*

land 4 sh. 1716, *Essex, Middlesex & Herts.* 1720 (with Bland & Smyth), *Yorkshire* 4 sh. 1720 engraved Parker, *London & Middlesex* 1749

Warburton, Thomas. *Estates of John Earle of Sandwich* 1757

Ward, Aaron, Bookseller. *Camden's Britannia* 1730

Ward, B. S. *Collectorate of Trichinopoly* 1835

Ward, Caesar. Bookseller with R. Chandler at the Ship Without Temple Bar, shops also at Scarborough Spaw and Coney St. York. *Morden's Magna Britannia* 1738

Ward, Capt. Charles. *Plan defences Algiers* 1816 MS

Ward, Comm. C. Y. Indian Navy. *Survey Malacca Straits* 1852–59

Ward, D. Surveyor. *Plan of Lucknow* 1858

Ward, E. T. *Improvements West End of London* (1830)

Ward, Ebenezer. *S.E. District of South Australia* 1869

Ward, H. E. *Map of Lagoon in Albert Park Melbourne* 1871

Ward, John. Surveyor. *London Estates Goldsmith's Company* 1691–3, *Manor of Wadden (Surrey)* 1692

Ward, J. *Camden's Britannia* 1753

Ward, J. Clifton. *Geological survey England & Wales* 1878–9

Ward, John Ferris. *Tremaine's Map of the County of Prince Edward Upper Canada* 1863

Ward, John Petty. *Rajmuhal Hills . . . . Damin i Koh* 1866

Ward, Marcus. *Universal Atlas* 1895

Ward, Samuel. *New Plan of London & Westminster* 1820

Ward, T. Humphrey. *Oxford illustrated* 1889

Ward. See **Johnson & Ward**

Ward & Reeves. Lithographers, New Zealand. *Various maps* 1853–64

Wardle, Thomas. Publisher, Philadelphia. *Atlas United States,* 1832

Ware, Joseph E. Engraver. No. 31 Locust St., St. Louis. *Route to California* 1849

Ware, R. Bookseller. *Camden's Britannia* 1753

Ware, Lord Thomas de la. *Voyage to Chesapeake Bay and state of the colony* 1610, MS

Wareham, J. *Map of the Waikato* 1866, *City of Auckland* 1866–7, *Bay of Plenty* (1873)

Warin. Script engraver. For Gouvion Saint-Cyr 1828, for Dufour 1836

Waring, (Capt. James). *Virgin Islands improved from Jefferys* London 1797

Warnars, G. *Reisatlas* 1793 (maps by Tirion, with J. de Groot)

Warne, Charles. *Dorsetshire* 1866

Warne, Frederick **& Co.** *Junior Atlas* 1857

Warner, B. Publisher, Philadelphia. *U.S.A.* 1820

Warner, C. S. *Auglaize County Ohio* (1864), *Hamilton County Indiana* 1866

Warner, Dn. Estate surveyor, Bishops Stortford. *Stansted Mountfichet (Essex)* 1783 MS

Warner, George E. *Plat Book of Fayette County Iowa* 1879, *Houston County, Minnesota* 1878, *Clayton County, Iowa* 1886, *Howard County, Iowa* 1886. Also collaborated with F.W. Beers and C.M. Foote

Warner, Henry. *Roads & footpaths in parish St. Mary Islington* 1810

Warner, I. Engraver. *Steel's Chart North Sea* (1803)

Warner, John. Estate surveyor, London. *Hornchurch* 1717 MS

Warner, J. Hydrographer. *Chart of River La Plata* 1817

Warner, John. Engraver. *Gardner's Survey Guernsey* 1787, *Vancouver's chart Sandwich Islands* 1798, *Ordnance Ground Woolwich* 1810

Warner, (John). *Course of Rivers Rappahanok and Potowmack in Virginia* 1736–7

Warner, L. Engraver. *Borda's Canary Islands* 1803, *Azores,* Steel 1802

Warner, Samuel. Estate surveyor, Kirby Street, Hatton Garden. *Arkesden* 1733 MS

Warner, Thomas C. Deputy Surveyor. *Edisto River, Carolina* (1753) MS

Warner, William. *Albion's England or an Historical Map of the same Island* 1586

Warner, Lieut. W. H. *St. Clair Delta* 1842 (with Lieut. Macomb), *Santa Fe to San Diego* 1847

Warner. *Plan Baltimore* 1801 (with Hanna)

Warner. See **Harrison & Warner**

Warnicke, John C. *District of Maine* 1814

Warr, J. & W. N. Engravers for *Tanner's New Universal Atlas* 1828, 1833

Warren. *Plan of City of Winchester* 1890

Warren, D. M. *Common School Geography* 1869

Warren, Francis. *Saffron Walden* 1752 MS

Warren, Lieut. G. K. Topographical Engineers. *Routes Pacific Railroad* 1855, *Dakota Country* 1856, *Nebraska & Dakota* 1858, *Territory of U.S.* 1868

Warren, H. *Illustrations for Tallis's Atlas* 1851

Warren, James. *Map of the County of Bruce* 1896

Warren, Michael & William Thomas. Surveyor, *Parish of Twickenham* 1846, *Parish of Isleworth* 1850

Warren, Moses. Surveyor. *Connecticut* 1813 (with Gillet)

Warren, Thomas. Estate surveyor of Bury. *Plan Bury* 1747, *Little Chesterford* 1774 MS, *Audley End* 1783 MS, *Ampton Suffolk* 1792 MS

Warren, Thomas, Gregory. *Palmiras to Calcutta* 1748, *Chart sands shoals and anchorages . . . River Hughley* (1770)

Warren, William (1806–1879). American geographer. *System of geography* Portland 1843

Warthabeth [Wuscan Theodorus Werthabeth von Herouanni]. Armenian bishop, established printing works in Amsterdam. *World map in Armenian* 1695

Warton, Capt. R. G. *Bechuanaland* 1884

Washbourne, Henry. Publisher, London. *Netherlands* (1844)

Washington, Capt. (later Admiral), R.N. Hydrographer of the Navy 1855–63. *Dover Rye* 1844, *Filey Bay* 1857, *Scilly Is.* 1863, *East coast England* 1841–47

Wasme, P. de. See **De Wasme**

Wasse, George. Estate surveyor. *Writtle* 1715 MS

Wassenberge, Jan. (16th century) Belgian cartographer

Watanabe, K. (19th century) Japanese surveyor

Water, Abraham. E. *Malling, Kent* 1699 MS

Waterhouse, Edward. *State of the Colony of Virginia* 1622

Waters, Edward. Surveyor. *Plan Parish St. Martins* London 1799–1800

Waters, H. C. *Boston and Lowell Railroad* 1836

Waters, Lieut. *Survey Harbour of Louisbourg* 1758

Waters, T. G. *Kildare Hunt and National Steeple chase Course* 1866

Watford, Alex. Surveyor. *Estates in Essex* 1825–27

Watkins. See **Harrison & Watkins**

Watson, Capt. C. Surveyor. *Plan Nattal (Sumatra)* 1762, Corrected *Nicholson's Plan Bombay Harbour* 1794

Watson, Col. David (d. 1762). Deputy Quarter Master General in North Britain. Director *survey of Highlands* 1747—54

Watson, Fred. *Geographical Dictionary* London 1773

Watson, Gaylord. Publisher, 16 Beekman Street N.Y. *Railroad map of U.S.* 1867, *New Commercial Atlas of U.S.* (1875), *New and Complete Illustrated Atlas* 1885

Watson, Captain George. *Milford Haven* 1758, *Cattegat* 2 sh. 1778, *Baltic Sea* 4 sh. 1779, *South & North Jutland* 1788. Worked for Laurie & Whittle's *Channel Pilot* 1794—1803

Watson, H. *Hughley River* 1777, *Plan Docks Calcutta* (1780)

Watson, J. Military Surveyor. *Harwich Harbour* 1752, *Rye* 1756

Watson, James. *Plan Edinburgh* 1793

Watson, John. Bookseller on Merchants Key near the Old Bridge by J. Gowan in Back Lane, Dublin. *Almanack* 1733—1810, contains *Principal Roads Ireland*

Watson, Capt. John. Admiralty Surveyor. *Plan Croce (Sumatra)* 1744, *Island Mayotta* 1774, *Charts of Sumatra* 1762, published Dalrymple 1774; *Straits of Salayr* 1781

Watson, Joseph. Estate surveyor. *Newton Hall* 1807 MS, *Boreham & Springfield* 1809 MS

Watson, J. F. Lithographer and publisher, Corner 4th and Walnut St. Philadelphia. *California* 1850

Watson, Justly. Director of Engineers. *Survey of the coast from Fort William . . . to Mosquito River* 1743, *Plans on West Coast of Africa* 1756

Watson, Samuel. Bookseller, Bury Street E. *Downing's plan of Bury St. Edmunds* 1740

Watson, S. *Stiffkey to Cromer*, Heather 1793

Watson, T. *Town of Sharon, Litchfield Co. Conn.* 1853, *Vincennes, Indiana* (1850)

Watson, William. *State of New Jersey* 1812

Watt, John. Surveyor. *River Clyde* 1734, 1759

Watt[e], J. von. See **Vadianus**

Watte, John. British cartographer. *Map of Sands vested in Bedford Level Corporation* 1777, *Kings Lynn* 1791

Wattenwyl, Alexander von (1735—1813). Swiss cartographer and engineer

Watts, John. Surveyor. *Pen Court (Kent)* 1697

Watts, L. J. *Portion of Colne Fishery Essex* 1896

Waugh, Lt. Col. A. Scott. *Survey Himalaya Mts.* Calcutta 1855

Wauters, Alphonse Jules. *Maps of Congo* 1888—1900

Wauthier. Pupil of Abbé Gaulthier. *America* 1797, 1799; *Battle map England* 1807

Wautier, A. de *Bruxelles et environs* 1810

Way, A. N. Vignette engraver for Tallis 1851

Way, John M. *Moosehead Lake and Headwaters of Penobscot* 1874

Waybred, Robert. Surveyor of Prittlewell (fl. 1564)

Waymouth, Capt. George. *Voyage of Discovery to the North West* 1602 MS, *Town and Castle of Gulicke* 1610

Wayne, Anthony (1745—1796) American General. *Action of 20th August 1794*

Wayne, C. P. Publisher of Philadelphia. *Life of Washington* (with maps) (1807)

Weatherhead, Matthew. *Jervis Bay* 1794

Weatherhead, Capt. Thomas. *Survey*

Amphila Bay 1814, *Chart Annerley Bay* 1814

Weatherley, Nicholas. Surveyor. *Map Manor of Scremerston (Durham)* 1824, *Spindleston and Outchester* 1824

Webb, Francis. Surveyor of Stow in the Wold. *Estates in Wilthshire* (with Thomas Webb) 1779

Webb, Capt. Mathew. *England Dungeness to Thames*, Admiralty 1872

Webb, Thomas. Surveyor. See **Webb**, Francis.

Webb, Thomas William. *New Star Atlas* 1872

Webb, William. Surveyor. *Parish Tattenhoe* 1613 MS

Web[b], William. Bookseller at the Globe in Cornhill London. Reissued *Saxton's Atlas* 1645–6

Webber, Capt. of Artillery. *Carta topgraphica di Roma e . . . . contorni* 1852

Webbers, L. Drew and coloured title pages for Ottens (1756) and Elwe 1792

Weber. Engraver. *Umgegend von Glogau* 1862

Weber, Professor. *La Suisse* 1871

Weber, C. A. *Eisenbahn Karte von Baden* 1870

Weber, Gottfried. *Hellas seu Graecia Universa* (1680)

Weber, H. L. and C. D. Swingley. *Atlas of Crawford County, Ohio* 1894

Weber, J. *Dal Lago Maggiore per Lago di Lugano al Lago di Como* (1890)

Weber, Jean Henri. *Plan de la Ville d'Aire* 1711

Weber, K. *North Eastern China* (1894)

Weber, W. Hydrographer. *Arabischer Meerbusen* 1888

Weber & Co. Lithographers of Baltimore, Maryland. *Missouri to Rockies* 1843

Webster. Topographical *Dictionary of Scotland* 1804

Webster, C. Land surveyor. *Skirsgill Estates (Cumberland)* 1866

Webster, J. Publisher, Philadelphia. *War in Lower Canada* 1814 (with A. Lay)

Webster, Thomas. Professor of Geology. *Geological map of Isle of Wight* 1815, Prepared and engraved *Greenough's Geological Map of England & Wales* 1819

Webster, Thomas. *Plan Docks & Warehouses Birkenhead* 1850

Wechel [Wechelius], Christian (d. ca. 1553). Bookprinter and seller

Wechter, Hans. *Reichs Statt Nürnberg* 1599

Wedell, Capt. J. *Berkeley Sound,* used by Norie 1822

Wedell, Rudolph von. *Hand Atlas* Glogau 1843

Wedgbrough, John. Bombay Marine. *Charts* for Meare's Voyage *Nootka Sound* 1790, *Charts of India & Ceylon for Dalrymple* 1795–6, *Ceylon* with Heywood 1822

Wedgbrough, S. *Track of the Snow Experiment* 1789

Wedge, Charles. Surveyor. *Chelmer and Blackwater navigation* 1792–4

Weekly Dispatch. 139 Fleet St. Published a *World Atlas* 1858, 1863

Weeks, H. R. *Jennings County, Indiana* 1859

Wees, H. J. van. *Kaart der Kerk-Provincie van Nederland* 1855

Wehrt, A. *Eisenbahnen Deutschlands* 1866

Weibl, E. Engraver of Weimar. *Australien* 1849

Weidenbach, Hofrath. *Herzogthum Nassau* (1871)

Weidler, Johann Friedrich (1691–1755). Mathematician and astronomer

Weidner, I. G. A. Worked for Gaspern 1804–11. *Gross Britannien und Irland* Weimar 1801

Weigel, Erhard (1625–1699). Globe maker and astrologer. *Celestial Globe* 1699

Weigel[ius], Christoph the Elder (1654–1725). Settled in Nuremberg 1698. Goldsmith, engraver, and publisher with Kohler. *Atlas Scholasticus* 1712, *Bequemer Schul und Reisen Atlas Kohler* (140 maps) 1718–9, *Orbis Antiquus* 1720, *Reise Atlas* 1724

Weigel, Johann Christoph (1654–1726). Brother of Christoph. Engraver, publisher, and map dealer of Nuremberg

Weigel, Johann Christoph younger (d. 1746). Continued business. Published three *atlases: Germanicus, Astronomicus,* and *Universalis* (1720). Business continued by Schneider-Weigel. *Australia* 1792, *America* 1796, 1812

Weigel, Widow. *Atlas* (ca. 1760)

Weigel and Schneider. *Special-Karte Vereinigten Niederlanden* 6 sh. 1787

Weiger, H. C. *Helvetia* (1695)

Weiland, Carl Ferdinand (1782–1847) of Weimar. *Grossherzogthum Frankfurt* 1812 *N. America* 1821, *Atlas von Amerika,* Weimar Geogr. Inst. 1824–8, *Hand Atlas* 1828–9, *Senegambia* 1831, *Europa in 4 Blättern* 1844 (pub. 1849)

Weiland, G.H. *Lombardo-Veneziansche Königreich* 1866, *Umgebung Wiens* (1867)

Weimar Chart. *Anonymous Spanish World Chart* (1527)

Weimar Geographisches Institut. *Atlas Minimus* 1805, *Plan St. Petersburg* 1807, *N. & S. Hemispheres* 1812, *Allgemeiner Hand Atlas* 1813, *Atlas von Amerika* 1825, *Mexico, Texas &c.* 1851

Weindel, Jos. *Plan der Stadt Baden* 1873

Weineck, J. A. von. See **Guler von Weineck**

Weiner, Peter (d. 1583). Publisher and engraver. Revised version of *Apian's Bavaria* 1579 (copper engraving)

Weingartner, (Ant.) *North America* 1834

Weingartner, Joannes. Engineer. Drew

Schwartz's Walachia Vienna 1738 MS

Weinhart, Ignatius. *Tyrolis . . . delineata* 20 sh. 1774

Weinher, Peter. *Chorographia Bavariae* 1579

Weinhold, Carl Wilhelm. *Plan von Freiberg* (1858)

Weinrib, T. *Plan de Paris* 1886

Weir, C. A. Surveyor and civil engineer of Greenwich. *Southend Pier* 1828 MS

Weir, Duncan. *Amboyna* 1796 MS, *Chart of Banda Harbour* 1796 MS, *Strait of Sincapore* 1805

Weir, Robert. Publisher of Glasgow. For Johnston (1844)

Weis, Frans Joseph. *Veteris Alemanniae descriptio* (1720)

Weis, Jean Martin. *La Sicile et partie du Royaume de Naples* 1783

Weisauer, W. *Nord-Luzon* 1878

Weise, Friedrich. *Helmstad* 1726

Weiss, A. *Umgegend von Münster* 1870, *Topo. Karte der Kreise des Regierungs Bezirks Münster* 1888

Weiss, Ed. *Zwei Stern Karten* 1874

Weiss, F. Lithographer for Bugarsky 1845

Weiss, Franz von (1791–1858). Austrian general and military cartographer. *Europaeische Türkey* 1829

Weiss, J. H. (1759–1826). Engraver and geographer of Strassburg, in French army. *Alsace* 1781, *Plan Strassburg* 1791, *Atlas Suisse* 1786–1802. 16 sh. *Carte de la Suisse* 1796, *Helvetien* Vienna 1806, *Chaine des Hautes Alpes* 1815

Weiss, Victor (1803–1870). Military cartographer of Zug. *Canton de Berne* 1830

Weissandt, Edouard. *Carte des Excursions dans les Vosges* (1868)

Weissenburch [Weissenburg], Wolfgang (1496–1575), Mathematician, geographer and cartographer of Basle. *Maps in*

Ziegler's Terra Sancta 1536, *Plan Jerusalem* 1542

Weitersheim, Alfred. *Pressburg* (1872)

Welch, Andreas. *English Channel* 1630 MS, *River Medway* 1667

Welch, B. T. American engraver. *Colden's New York* 1825, *Fielding Lucas', Baltimore* 1823

Welch, G. W. Publisher in Nevada. *Elliot's Central California* 1860

Welch, Henry. Surveyor. *London-Edinburgh Mail Roads East Retford to Morpeth* 1827

Welch, Joseph. Draughtsman. *Military Reconnaissance of the Arkansas, Rio del Norte and Rio Gila* 1847. Engrd. on stone by E. Weber & Co. Baltimore

Welcker, A. (d. 1888). Cartographer. *Ost Europa* 1867

Welden, Franz Ludwig, Freiherr von (1782-1853). Austrian military cartographer

Weldon [Welden], Daniel. Appointed commissioner (with Charlton) for South Carolina to extend *South Carolina-Virginia line* 1749

Welladvice, Thomas. *Sketch of Sapy Bay* 1783, *Track of the Glatton* 1786

Welland, George. *British & French Empires* 1858, *Railways Great Britain* 1857, *New Plan London* 1862

Welland, (J.) Engraver. For *Godfrey's Jersey* 1849

Welland, Ralph. *Emmanuel Parish Holloway* 1894

Weller, Edward (d. 1884). Engraver, publisher, and cartographer in London: 34 Red Lion Square (1861), Duke St. Bloomsbury (1872). *Dispatch Atlas* 1858, *Admiralty Charts* 1868, *Atlas Scripture Geography* 1853, *Elementary Atlas* 1854, *Crown Atlas* 1871, *Pocket Atlas* 1872, *Students Atlas* 1873

Weller, Francis Sidney. Lithographer, Red Lion Square. *Jeppe's Transvaal*

1889, *Cassell's Popular Atlas* 1890, *Longmans' School Atlas* 1890

Weller, John B. U.S. Commissioner. *Port San Diego* 1849

Wells, A. Provincial land surveyor. *Eastern Townships of Lower Canada*, 2 sheets J. Arrowsmith 1839

Wells, B. *Elevation of the Grand Aqueduct near Lisbon* 1792

Wells, Edward (1667–1727). Mathematician, geographer and Divine. *New Set of Maps both of Ancient and present Geography* 1700; several editions up to 1738; printed Oxford or London

Wells, G. *Thirteen Miles round Chipping Norton* (1835)

Wells, H. See under **Wells**, S.

Wells, J. S. Master, R.N. *Survey of Lizard with Capt. Williams* 1851, *Falmouth Harbour* 1853

Wells, James W. *Physical Map of Brazil* 1886

Wells, John. *Surveyor of Pennsylvania. Map of Somerset County, Penna.* 1819

Wells, L. *Canterbury* 1782

Wells, L. J. *Greenwood Cemetery [Brooklyn]* 1872

Wells, Lionel B. *Canals and Navigable Rivers England & Wales* (1894)

Wells, Philip. Surveyor. *Survey part Harlem River* 1683 MS

Wells, Samuel. *Map of Bedford Level* 1829 (Reissued by H. Wells 1878), *Great Level of Fens* 1829

Wells, Thomas. *Parochial Map Diocese of Canterbury* (1755)

Wells, Walter. *Lessons in Physical Geography* (1865)

Wells, William Henry. *County of Cumberland, New South Wales* 1840, *Port Philip* 1840, *Geographical Dictionary or Gazetter of Australian Colonies* Sydney 1848

Wellsted, J. R. *Trigonometrical survey of Socotra* (with S. B. Haines 1835)

Welser, Marcus (1558–1614). Fleming Prepared *Tabula Peutingeriana* for Ortelius 1598

Welsh, A. *Sea Coast of Bell-Isle from Pointe de Pierre to the Guard House de St. Foy* 1761

Welsh, George P. *American Arctic Expedition in search of Franklin* 1851

Welsh, Marine surveyor. *Sullivan Cove Tasmania* 1825, *Charts of Australia* 1825–31

Welsh, John. *River Tamar in Van Diemens Land* 1831

Welsh, M. *Plan Beachamstead and Dillington, Hunts.* 1796

Welzbacher, C. *Provinz Oberhessen* 4 sh. (1892), *Odenwald* (1885)

Wen, Wang. Chinese cartographer. *Map China* 1125

Wendelken & Co. *Atlas of Towns in Suffolk County, New York* 1888

Wenham, Richard. Estate surveyor. *Frating* 1772 MS

Wenker, G. *Sprach Atlas von N. u.M. Deutschland* 1881, 1895

Wennersten, C. S. Engraver. *Elfsborgs Lan . . . Skala* 1856

Wenng, C. F. *Kaisereich Fez und Marokko* 1845

Wenng, C. Gustaf. *München* (1848), *Mittel Europa* (1867), *Eisenbahnkarten von Deutschland* (1869), *Augsburg* (1874)

Wenng, G. Ludwig. *Königreich Bayern* (1890), *Ober Bayern* (1898)

Wenng, Karl Heinrich (1787–1837). Lithographer of Munich

Wentworth, Horace. Publisher, 86 Washington Street, Boston. *U.S.A.* 1852 and with Dayton 1854

Wentzel, Carel David. Surveyor. *Map of S. & S.E. coast Africa* 1752 MS, *Route of Governor Tulbagh*

Wenz, Gustav. *Mittel Franken* (1872), *Schwaben* (1873), *München* 4 sh. (1890)

Wenzely, A. von (fl. 1789–1806). Viennese cartographer. *Siebenbürgen* 1789, *Ungarn* 1790, *Schraembl's Oestreichische Niederlanden* 1790

Werden, Jacob van. *Isla Tercera* Hollar 1645

Werden, Karl von. Dutch captain in Russian Navy. *Chart of Caspian* (published 1720); Issued by De Lisle 1721 and Ottens 1723

Werdenhagen, Johann Angelius (1581–1652). *Scandinavia* Frankfurt 1641

Wergeland, N. *Veikart over Norge* (1855)

Werming, N. G. Engraved for Akerlund and E. Personne. *Helsingfors och Sveaborg* 1808, *Carl Johans Stad* 1812

Werner, (A). *Gingerah (Malabar Coast)* pub. Dalrymple 1774

Werner, Frederick Bernard (1690–1778). Engraver. *Antwerp* 1729, *Augsburg, Basle, Berlin, Mainz, Padua* 1730; *Pressburg* 1732, *Pisa* 1740, *Scenog. Urbium Silesiae* 11 sh. Homann Hered. 1752

Werner, Johannes (1466–1528). Mathematician, astronomer, and cartographer. Translated *Ptolemy Book I* into Latin 1574; *Map projections*

Werner, J. W. Lieut. Hessian Artillery. *Battle of Brandywine* 1777, published Faden 1778

Werner, Oscar. *Katholischer Kirchen atlas* 1884, French edition 1886

Werner, Wilhelm. *Harz-Gebirge* 1843, *Plans Düsseldorf* 1848–74

Wersebe, August von (1751–1831). *South India,* Faden 1788

Wersebe, Hermann Martin Christian von. Engineer 14th Regt. Hannoverians. *Plan Fort and Environs of Poligautcheri* 1783

Werthabeth, W. T. See **Warthabeth**

Werveke, L. van. *Deutsch-Lothringen Geologische Uebersichtskarte* 1886

Wery, J. *Carte topographique de la Belgique* (1857)

Wescoatt, N. Civil engineer. *Comstock Range, Nevada* 1864

Wessel, C. (1745–1818). Danish cartographer and mathematician. *Kiøbenhavns Amt* 1766, *Fierdedeel af Siaelland* 1770, *Siaelland og Moen* 1777

Wessenaer, N. *World in hemispheres* 1661

West, C. R. Sackville. *Plan Battle Ferozshah* 1845, (1849)

West, H. Engraver for *Arrowsmith's America* 1804

West, Joseph A. *Utah* 1885

West, Lane. *Terrestial globe* (1820)

West, William. *Marshall & Adjacent Fens* 1799 MS

West, W. Engraver. *Alpine country South Europe* 1804, *Coast Norway* 1812, *Laurie & Whittle's Roads* 1815

West. See **Ford & West**

Westbrook, J. B. Marion. *County maps in Illinois* 1870–74

Weste, W. Surveyor. *Map of Cumberland Gap (Tennessee)* 1863

Westenburg, (Joanne). Doctor Medicine & Mathematics, cartographer, *Bentheim* Blaeu 1635, 1649 & 1662; *Bentheim,* Hondius 1633

Westermann, George. Publisher of Braunschweig. *Lange's Atlas von N. Amerika* 1854

Westermäyr, C. Worked for Gaspari 1804–11

Westley, F. Publisher 10 Stationers Court. *S. Africa* 1822

Westley, William. Surveyor. *Plan Birmingham* 1731

Weston, James A. *Map city Manchester, New Hampshire* 1870

Weston, Samuel. *Survey City of Chester* 2 sh. 1789

Weston, William. *Plan proposed London and Western Canal* 2 sh. (1790)

Westphal, Julius Georg (1824–59). Astronomer

Westphalen, H. *Post und Reise Carte der Wege durch Franckreich* (1720)

Wetstein, H. Publisher Amsterdam. *Alting's Notitia Germaniae, Inferioris* (1697)

Wetter, George. *Anspach, Eichstätt* 18th century

Wetterstrom, G. C. *Lake George* 1756

Wey, William. *Itineraries in Palestine and Spain* 1458 MS

Weyerman, Jacob Christoph. Designed titles for Seutter's *Atlas Minor,* and Lotter's *Atlas Novus* (1780)

Weygand, F. Publisher of Amsterdam and The Hague. Map seller to the King of the Netherlands. Published maps for Lapie. *Océanie* 1827

Weyprecht, Karl Georg Ludwig Wilhelm (1858–81). Marine cartographer

Weys, Franz, Sappelführer. *Karte von Salzburg* (1840) MS, *Wildbad Gastein* (1810) MS

Weyss. *Region between Gettysburg and Appomattox* 1869 (with Camp and Thompson)

Weyssenburger, W. See **Wissenburg**

Wharton, S. *Survey Field of Waterloo* 2 sh. 1815

Wharton, Admiral Sir William J. L. Hydrographer to the Navy 1884–1904. *Isthmus Panama* 1885

Wheatley, Francis. *Plan Ceuta* Admiralty Chart 1813

Wheatley, (S.) Engraver. *Southern Part South America* 1748, *Dublin* by Stokes 1750

Wheeler, George. *Achaia Vetus et Nova* (1700)

Wheeler, Lieut. (later Capt.) George Montague. U.S. Corps of Engineers *Topographical and geological maps of America* 1869–81, *Topographical Atlas* 14 sh. 1876–9

Wheeler, H. F. Engraver. *Territory of Oregon* 1844

Wheeler, J. Assistant to S. Holland. *Chart Rhode Island* (1770), *Chart Boston Harbour* 1775 (with J. Grant)

Wheeler, James Talboys. *Afghan-Turkestan* 4 sh 1869, *Geography Old & New Testament* 1860

Wheeler, Thomas. *Plan of Narrows East River (N.Y.)* 1758 MS, *New England Coast (Atlantic Neptune)* 1780

Wheelock, Harrison. Publisher, San Francisco. For Milleson 1864

Whidbey, Joseph. Master of *Europa*. *Port Royal* 1788, published Leard 1792 (with Vancouver), *Torbay* 1799

Whipple, Lieut. A. W. U.S. Corps of Topographical Engineers. *Rio Gila* 1849 MS, *Pacific Railroad* 1853–5

Whipple, Lieut. W. D. *Navajo County* 1858 MS

Whish, Lieut. (later Commander). R. W. Indian Navy. *Bombay Harbour* 1863

Whishaw, Francis. Civil engineer and surveyor. *Hendon* 1828, *City London* (1740)

Whiston, J. Bookseller and publisher. *Camden's Britannia* 1753

Whiston, William. *Solar System* 1708–25, *Transit of Moon over Europe* 1724

Whitaker, Thomas. Engineer. *Proposed Exeter-Exmouth Railway* 7 sh. (1850)

Whitbread, Josiah. *Baltic Sea* 1854, *Crystal Palace Map of London* (1854, editions to 1864); *Parish St. Marylebone* 1855, *New Map London* (1881)

Whitchurch, William. Engraver, Bartholomew Lane, Royal Exchange, Pleasant Row, Islington. Engraved for Cook, and script engraver for Dalrymple. *New South Wales* 1770, *Bay Bengal* 1772, *Borneo* 1775, *Easter Island* 1777, Cook's second Voyage 1777

Whitcomb. *Carmarthen Estate, Tottenham Court Road* 1781

White, Maritime Surveyor. *Chart N.W. Part Madagascar* 1784

White, Benjamin Jr. Estate surveyor. *Maldon* 1819 MS

White, Charles A. *Washington Territory* 1870

White, Edward. Military draughtsman, England. *Military Brigade Districts* 1855, *Artillery defence East Suffolk* 1855

White, Edward. Surveyor of Ipswich. *Map of Ipswich* 1867

White, Francis. *Plan city of York and Ainsty* 1785, *Jeffery's Yorks.* (with Markham) 1800, *Warwickshire* 1850, *Nottinghamshire* 1854

White [With], John. English cartographer (fl. 1585–93). *Map of Virginia* 1585–7, in De Bry's *Voyages* 1590 (First separate map Virginia)

White, John. *Elementary atlas of modern geography* (1852)

White, John. Surveyor. *Duke of Portland's Estate* 6 sh. 1797–99 MS, *Plan Parish of St. Marylebone* 2 sh. 1809, White & Son 1813

White, Capt. (later Commander) Martin, R.N. Hydrographer. Admiralty Charts: *S.W. Coast England and Wales* 1817–29, *English Channel* 1812–28, *Channel Island & Guernsey* 1822, *Alderney* 1824, *Salcombe* 1825, *Dartmouth* 1829, *Coast France* 1831

White, R. Engraved *title for Speed* (Basset & Chiswell) and Overton editions; *Plan of Tangier* (1670)

White, R. N. Engraver. *English Prairie Illinois Overland route to California* 1852

White, Robert. *Red Sea* 1795

White, W. Surveyor. *Plan part Rivers Avon & Frome* 1792, pub. Faden 1793, *Harbour of Bristol,* Faden 1803 (with Jessop)

White, William, Gent. *Isle of White,* used by Speed 1611

White, William. Publisher of Sheffield. *New map of Lincoln* 1872

White, Gallaher & White. Publishers of New York. *Mejico* 1828

White & Pike. Publishers of Birmingham. *Warwickshire* [Quaker] 1873, 1886

Whitehead, A. G. Engraver. *Map County of Fife* 1800

Whitehead & Co. E. Engravers and lithographers, 67 Collins St. East Melbourne. *Atlas of Australia* 1870, *Melbourne & Suburbs* 1877, *New Map Victoria* (1868)

Whitelaw, Rev. J. *Grand and Royal Canals of Ireland* 1812

Whitelaw, James. *State of Vermont* 1793 and 1796 (3 sh.)

Whitelock, Lieut. H. H., Indian Navy. Admiralty surveyor. *Surveys of India* 1833–36, *Dio Harbour* 1834

Whiteway, Capt. John. *Maps of parts of East Indies* 1750, published by Dalrymple

Whiting, Henry L. *Atlas of Massachusetts* 1890

Whiting, R. E. K. *Plan of Isthmus of Tehuantepec* 1851

Whiting, W. B. *Wind and Current Charts N. Atlantic* 1848

Whitley, T. W. *Plan Parish Churst St. Nicholas and Ruins Abbey of St. Mary Kenilworth* (1890)

Whitney, J. D. State Geologist California. *Geological map Lake Superior Land District* 1847, *Geological Survey Wisconsin* 1860, *Topographical and Railway Map Central part of California* 1865, *Geological Survey California* (1868)

Whitney, Thomas R. Engraver. *Chart of the Falkland Islands* 1829

Whitpain, Robert. *Survey Lordship Applesham (Sussex)* 1677 MS

Whitson, J. Surveyor. *Jamaica Estates* 1671–4

Whittaker. See **Law & Whittaker**

Whittaker, G. & W. B. Publishers of London, 13 Ave Maria Lane. *Walker's Universal Atlas* 1820, *Travellers Pocket Atlas* 1821–3, *Pawley's General Atlas* 1822, *Capper's Topographical Dictionary* 1824–(5)

Whittaker & Sullivan. Lithographers and publishers, 22 Chancery Lane. *Daines County Courts District* (1846)

Whittle, James (d. 1818). Publisher in conjunction with **Laurie**

Whittlesey. Engraver. *Agas's Plan Oxford* 1578–88, 1728

Whittlesey, Chas. *Geological, Raiload and Township map of State of Ohio* 1856, *Physical Geology Lake Superior* 1875

Whittlesey, Robert. Surveyor. *Cow Mead & Swinfell Farm (Berks.)* 1726

Whitton, J. Surveyor for Mills. *Spartanburgh District South Carolina* 1820–25

Whitton, Towne & Co. Printers. 125 Clay St. San Francisco. *Miner's Map Frozen River* 1860

Whitwell, Charles. Engraver, 16th century. *Norden's Surrey* 1594, *Symonson's Kent* 1596, *Tymme's Jerusalem* 1595

Whitworth, Robert. *Plan Canal Mary le Bone to Moor Field* 1773, *Trent Severn Canal* 1775

Whitworth, Robert P. *Baillière's South Australian Gazetter* 1866, *New South Wales Gazetteer* 1870, *Baillière's Victorian Gazetteer* 1865

Whyman, Joseph. Surveyor. *Leicester.* 1775–77, (1779)

Whyman, Thomas. *Estates Knighton (Leicester)* 1771 MS

Whyte, W. *Map of soil of Aberdeenshire* 1811

Wibe, N. A. Surveyor. *Charts Coast Norway* 1812

Wiber. *Christiansund og til Stadt Land* 1793

Wibergs, Carl Fredrik. *Atlas till Sveriges Historia* 1856, *Atlas över Allmänna Historia* 1875

Wiborgs, A. Publisher of Stockholm. *America* 1818, *Sweden* 1819, *Russia* 1821

Wicheringe, Bartholdus. Cartographer. *Groninga Dominium* 1630, used by Blaeu and Hondius

Wichman, (J.) Engraver. *Map Polar Regions* Hamburg 1675

Wichmann, Ernest Heinrich (1823–1896). Austrian geographer. *Hamburger Gebiet* 1866

Wicker, Christian. *Plan von Braunschweig* (1841)

Wickham, Comm. J. C. R.N. Admiralty surveyor. *Surveys in Australia* 1837–41, *Entrance Moreton Bay* 1846

Wickings, W. *Plan Old Street Turnpike* 1809, *City Road* 1809

Wicksteed, John. Estate surveyor. *Little Ilford* 1775 MS

Wicsink, J. J. *Pascaarte van de Zuid West cust van Africa,* J. Hondius, Amstd. 1652

Widdows, Daniel. Translator. *Description of the World* 1621 and 1631

Widdows, T. *Additions to Hudson's Voyage in search of a North West Passage* 1611 MS

Widimsky, Bohuslav. *Eisenbahn Karte des oestlichen Europa* 1881, *Russische Eisenbahnen* 1899

Widman, Georgio. Engraver for *Rossi's Mercurio* (1685), *Geografico Scandinavia* Roma 1678, *America Settentrionale* 1677, *Terra Sancta* 1679

Widney, John. *Map of Presque Isle or Erie* (1813) MS

Wiebeking, Karl Friedrich R. von (1762–1842). *Holland u. Utrecht* 8 sh. Darmstadt 1796, *Topo. Carte Herzogthum Berg* 1789–92

Wieckmann, F. *Plan der Stad Riga* 1867

Wied, Anton (ca. 1500–1558). Artist and cartographer of Danzig. *Russia* compiled from material supplied by Lyatskoi 1555

Wiedenmann, P. *Glockner Gruppe* 1871, *Dolomit Alpen* 1874

Wiedmann. *Provincia del Sinu* 1892

Wiegrebe, Ernest Heinrich (1793–1872). Military topographer

Wieland, (C. F.) *North America* Weimar 1843

Wieland, Johann Wolfgang (d. 1757). Lieut. Austrian Engineers, successor to J. C. Müller. *Survey Silesia* 1723–32, published Homann Heirs 1752; Revised *Müller's Bohemia* 1726, *Principatus Silesiae Ligniciensis* 1739

Wieldy, A. See Willdey

Wieticx, (Antoine). Engraver. *Title & plates to De Jode's atlas* 1593

Wierix, Jan. Engraver. For *Ortelius' Parergon*

Wiesbaden Map. Fragments of a 13th century *map of World*

Wies, N. *Carte géologique de Luxembourg* 1877

Wigand, C. F. *Plan von Pressburg* (1864)

Wigate, John. Clerk to Middleton 1741–2. *Chart of the Seas thro' which H.M. Sloop Furnace pass'd for discovering a Passage from Hudson's Bay to the South Sea* 2 sh. 1746 (Dedicated to Arthur Dobbs)

Wigg, E. S. and Son: *Map of Australia* Adelaide (1870)

Wiggers, H. *Plan der Stadt Emden* 1880

Wiggins, John Junior. Estate surveyor. *Great Coggeshall* 1801 MS, *Great Britain* 1805 MS

Wight, Edward. Surveyor. *Plan Fort Edward at Pesaquid* (1758)

Wightman, H. M. *Plan Boston, Massachusetts* 1864, *Boston and its Vicinity* (1870)

Wightman, Thomas. Engraver for Harris. *Tour into Territory of North West* Boston 1805, *State Ohio, Alleghany*

River, and for Dickinson Geography, Boston 1813

Wigram, John. American surveyor (fl. 1790–97). Various MSS *estates in New York*

Wigram, Rev. Joseph Cotton. *Map of Palestine* (1853)

Wigzell [Wizzell] . Script engraver for *Arrowsmith's World* 1794 (with Mozeen), for Sayer and Bennett. *British Channel* 1788, *Channel Pilot* 1791, for Laurie & Whittle 1794

Wijk [Wyk] , Jan Roelandszoon, van (1781–1847). *Arctic* 1823, *Australia* 1825, *Nieuw Nederland* 1827, *Atlas* with R. G. Benet 1827, *Atlas van Europa* (1847)

Wijkberg, Maurus. *Maps of Finland* 1871–77

Wilamowicz, I. *Cairo and Fulton Railroad* (1853)

Wilbar, A. P. Surveyor. *Public Surveys New Mexico* 1880

Wilbraham, R. Surveyor. *Syria* (1846)

Wilbrecht, Alexander. Russian cartographer. *Environs de St. Petersburg* 1792, *Discoveries in the Pacific* 1787, *Rossiiskoi Atlas,* St. Petersburg 1792

Wilckens, C. *Carte von dem Niederstift Münster* 1796, *Fürstenthum Hildesheim* (1804)

Wild, Henry. *New System of Geography* 1821

Wild, James. Publisher and cartographer. *Tasmania* (1846)

Wild, Josef. *Military map Bavaria* (1868)

Wild, Joseph (fl. 1697–1701). Bookseller at the Elephant at Charing Cross. *Camden's Britannia Abridged* 1701 (Seller Maps)

Wilde, Charles. *Chart of N.W. Coast Madagascar* 1784, *Fort St. George Coromandel* 1779

Wildeboaı, Tobias. Surveyor. *Lynby* 1692

Wildey. See **Willdey**

Wiley & Putnam. London publishers. *Greenhow's North America* 1846

Wilford, J. Publisher behind the Chapter House. *Moll's New description of England and Wales* 1733

Wilford, S. *Map of Sydney* 1873

Wilkes, Charles (1798–1877) of New York and Washington. Marine cartographer. Commander United States. Exploring Expedition 1838–42, *Atlas Philadelphia* 1845, *Chart World* 1842

Wilkes, J. Publisher of London (with P. Barfort 1791) *Plan London* 1791, *British Colonies North America* 1797, *Berkshire* 1801, *County Atlas* 1801–3, *Encyclopedia Londinensis* 1810–28 (40 Maps)

Wilkie. *North West Passage* (Cook) 1785

Wilkins, Henry Musgrave. *Junior Classic Atlas* (1871)

Wilkins, W. H. *Gorakhpur and Benares Divisions* 3 sh. 1891

Wilkinson, Berdoc Amherst. *Map City of Dublin and its Environs* (1873)

Wilkinson, Charles Smith. *New South Wales* 1876, *Geological Sketch map of New South Wales* 8 sh. 1880, *Geological Map New South Wales* 1893

Wilkinson, Francis. Surveyor. *Map of the Level lying upon the River Axkholme* (with I. Fotherby) 1662

Wilkinson, George R. *Seat of the War in the Krimea* 1855, *Sevastopol* 1855

Wilkinson, Henry Fazakerly. *Dungeness to the Thames* Admiralty Chart 1872

Wilkinson, J. Surveyor. *New Brunswick* 1858–59

Wilkinson, James. *Affair of Bladenburg August 24th* 1814, *Philadelphia* 1816, *Affair of Princetown* 1777, *Battle Bridgewater View Diagrams and Plans* 1816

Wilkinson, John. Surveyor. *Agression of Maine and Massachusetts upon New Brunswick* 6 sh. 1839, *Part of the River*

St. John 1840, *British Provinces New Brunswick Nova Scotia* 1859

Wilkinson, J.

Wilkinson, Sir John Gardner. *Topographical Survey of Thebes* 1830

Wilkinson, Miss. *Atlas for use of Schools* 1816

Wilkinson, Robert (fl. 1785–1825). Publisher, No. 58 Cornhill (1794–1816) and 125 Fenchurch Street 1817–Jan. 1823. *Bowen and Kitchin's Large English Atlas* 1785. Collaborated with Bowles and Carver, successor to J. Bowles 1779. Reissued *Speer's West Indies* 1796, *Atlas Classica* 1797, *General Atlas of World* 1794, 1802 and 1809; *New Holland* 1820, *North America* 1823

Wilkinson, W. C. Lieut. 62 Regiment. *Maps for Faden's North American Atlas* 1780, *Braemus Heights on Hudson River* 1780, *Encampment General Burgoyne Stillwater* 1780

Wilks, Charles. See **Wilkes**, Charles

Will., Joh. Martin. Engraver of Augsburg. *Position Prussian army in Bohemia* (1778), *Festung Duynkirchen* (1780), *Belagerung Giberaltar* 1782, *Walch's Afrique* (1790)

Willard, A. Publisher of Hartford. *Connecticut* 1842

Willard, Emma Hart (1787–1870). American geographer. *Atlas Hartford* 1826, *Atlas to accompany a System of Universal History* 1836

Willdey, George. Publisher and Bookseller with T. Brandreth at the Archimedes and Globe in Ludgate Street (1712). *Muscovy* 1711, *Bowen's Asia* 1714. At ye Great Toy shop next the Dog Tavern in Ludgate Street. *Map Great Britain & Ireland* 1715, *Atlas of World* 1717, *Thirty miles round London* (1720), at the great Toy, Spectacle, Chinaware and Print Shop the corner of Ludgate Street next St. Pauls Church London

Willdey, Thomas. Publisher. Ludgate Street. *Popple's British Empire in America* 1733. At the Great Toy Shop in St. Pauls Churchyard. *Kings Roads in the Highlands of Scotland* (with S. Austin) 1740–46

Willemsen, Jan (1782–1858). Amsterdam cartographer and publisher

Willemsz [Willemsen], Giovaert. *Le Livre de Mer* MS, Origin of the *Caerte vande Oost ende West Zee* 1587–88–90, and –94

Willer, Jacobus. Dane, from Copenhagen, in English Service. *Fortification plan* 1665 made in Constantinople

Willer, P. Architect. *Danzig in plano* 1687

Willerman, Capt. W. *Maps & Plans British Army in Spanish Peninsula* (1840)

Willetts, Jacob. *Atlas of World* Poughkeepsie 1814, 1820

Williams, Alexander. *Map of Boston (Massachusetts)* (1850)

Williams, Anthony. *Draught of Port Royal* on *Lea's Jamaica* (1685)

William, B. Engraver. *Map Westphalia* in *Royal Magazine* 1760

Williams, C. P. *Geological Map of Trap Range Lake Superior* 1863 (with W. H. Stevens)

Williams, C. S. Publisher corner of Market Street and 7th Streets Philadelphia. *Map of Texas, New General Atlas, New Haven* 1832, *Map of the United States* New Haven 1833

Williams, Dionysius. *Chart Mounts Bay* 1751, 1793

Williams, Col. Edward. Director of Ordnance Survey 1791–98. *State Citadel Montreal* 1774 MS

Williams, E. C. S. *Province of Pegu* 1856

Williams, E. & T. Publishers and map-sellers London, No. 11 Strand. *W. Wilson's Plan Dublin* 1798

Williams, G. *Plan Oxford engraved Toms* 1733

Williams, Lieut. (later Capt.) G. Admiralty Surveyor. *St. Ives Bay* 1848, *Lizard* 1851, *Falmouth Harbour* 1853, *Scilly Islands* 1857–63

Williams, George Clinton. *North half Township of Woodstock* (with Child and others) 1772

Williams, G. M. *Plan District of Adelaide,* Hansard 1841

Williams, H. *Map of Sierra Leone Estuary* 1825

Williams, Hiram. *Sarawak River (Borneo)* Admiralty 1847, *Santubong Mountain Borneo* (1860)

Williams, J. *Province of Pennsylvania* 1778

Williams, J. of Glamorgan. *Road Milford–Gloucester,* Cary 1792

Williams, J. J. (1818–1904). Civil engineer. *Plan Isthmus of Tehuantepec* 1851, *Isthmus of Tehuantepec* 1852 (16 plates, 9 maps)

Williams, James. County Auditor Chester. *Map Cheshire* 1888

Williams, James. *Map of ye Strand of ye River Anna Liffe* (1812)

Williams, Jesse. *Surveyed part of Iowa* New York, Colton 1840

Williams, John. Publisher at ye Crowne in St. Pauls Churchyard. *Fuller's Palestine* 1650

Williams, Capt. John. *Track of the Hector* 1759, *Chart Bristol Channel* 1760

Williams, Lieut. John. Worked on *survey Scotland* 1749–51.

Williams, J. David. American publisher, 46 Beekman St. *People's Pictorial Atlas* 1873

Williams, John Francon. *Philips Standard Atlas* (1884), *Philips Elementary Atlas*

1882, *Jubilee Atlas British Empire* 1887, *Philips Handy Volume Atlas of Australasia* 1888, *Popular Atlas British Empire* (1898)

Williams, John L. *Bar Pensacola Bay* 1827, *Western Part Florida* Tanner, Philadelphia 1827

Williams, John. Sub. Engineer. *Plan Fort Stanwix Onieda Station* 1758

Williams, Jonathon Junior. *Atlantic Ocean,* American Philosophical Society Transactions 1793

Williams, P. *Plan Dram Road Browelty, Monmouth* 1799

Williams, Richard. Estate surveyor. *White Rooking* 1772 MS, *Plan Boston* 1776

Williams, Capt. R. *Chart Eastern Straits (Java),* Laurie & Whittle 1798; *Straits Westward New Guinea* 1796–7, pub. Laurie & Whittle 1798

Williams, R. D. American draughtsman. *Central America* 1856

Williams, R. Price. *Railway Map England and Wales* 1872, *North Wales Watersheds–London Aqueduct from Bala* 1893

Williams, Samuel. Estate surveyor. *Barline* 1766 MS, *Dagenham & Barking* 1772 MS

Williams, Samuel (1786–1859). *Michigan Territory and Great Lakes* (1819) MS, *State of Vermont* 1793

Williams, T. See **Williams,** E. & T.

Williams, W.[illiam] . *Denbigh & Flint* ca. 1720

Williams, William. Publisher of Utica, New York. *Tourist map N.Y.* 1830, *Cities of New York & Brooklyn* 1847, *North West States* 1849

Williams, W. Engraver, geographer and draughtsman of Philadelphia, 33–35 Street. For Mitchell 1839–53. *U.S.A.* 1852, 1855

Williams, W. A. *Sketch of Charleston Harbour* (1862), *Pensacola Navy Yard* (1862)

Williams, W. B. *Plan Isthmus Tehuantepec* 1851

Williams, Capt. W. G., U.S. Topo. Engineers. *Survey Ship Canal around Falls of Niagara* 1835, *Dunkirk New York Harbour* 1838, *St. Clair Delta* 1842

Williams, Sir William Fenwick. *Turco—Persian Frontier* 1855

Williams. See **Heather & Williams**

Williamson, Charles. *Map of the Middle States* 1798

Williamson, James. Surveyor. *Environs of Belfast* 1791, *Map Antrim* 1808, *Down* 4sh. 1810

Williamson, Robert. *Directions for a general chart of St. George's Channel* 6 sh. 1767

Williamson, Lieut. R. S. Topographical Engineers. *Sacramento Valley and Sierra Nevada* 1850, *Oregon & California* 1851 MS, *California and Railroad Survey* 1855

Willie, H. *Norske Kyst, Christiansand til Lindesnaes* 1857

Willington, Francis Howard. Surveyor. *Foxley Parish and Norton Farm* 1760

Willis, James A. C. *Port Jackson* 1867, *County of Harden N.S.W.* 1864, *County Bathurst* 1869, *Map New South Wales* 1881

Willis, John. Surveyor. *Ten miles round Newbury* 1768, *Estate at Highgate* 1772

Willis, W. *Pictorial Plan of Chester* (1860)

Willius, P. *Danube* 3 sh. 1686, *Circulus Suevicus* 1689

Willmann, Edward (1820–1877). Engraver of Karlsruhe. *Der Rhein Mannheim-Cöln*

Willmott, A. Surveyor. *Plan Government House &c. Melbourne* 1890

Willock, I. Surveyor. *[Twickenham Park]* engraved W. Darling

Wills, Roger. *Map of Manor of Benwell* 1637 MS

Willson, T. Estate surveyor. *Burnham* ca 1800 MS

Willson, Thomas Benjamin. *Handy map Norway south of Trondhjem* 1891

Willson & Hooper. Publishers 815 Montgomery Street, San Francisco, California (end 19th century)

Willyams, Cooper. *Maps Island of Martinique* 1796, *Plan Fort Bourbon* 1796

Wils, Pieter. *Haerlem dimersa a P.W.* 1646 (1670)

Wilse, I. N. Worked for Reilly 1796, *Norway* 1796

Wilson. *Gazetteer (of Scotland)* 1854–7

Wilson. *Travelling map roads of Ireland* 1803

Wilson, Alexander. *Suggested Railway connections East end Glasgow* 1883

Wilson, Charles. Chart publisher at the Navigation Warehouse and Naval Academy, No. 157 Leadenhall St. Successor to Norie & Wilson 1840. *Complete British and Irish Coasting Pilot* 1845, *Hobbs' World* 1860

Wilson, C. L. *California Central Railroad* 1860

Wilson, Col. Sir Charles William (1836–1905). Surveyor of Liverpool. *Jerusalem* 1865, *Ordnance Sheet Peninsula Sinai* 1869, *Palestine* 21 sh. 1890, *Ordnance Sheets* 1890–2

Wilson, George W. *Maps of Ionia County, Michigan* 1861

Wilson, H. Engraver, copperplate and lithographic printer, 197 Trongate, Glasgow. *Eight Miles Round Glasgow* 1829

Wilson, Henry. *A Mercator's Chart of the German Ocean and North Baltick* 1720, *Europe,* Senex (1719)

Wilson, H. Hydrographer. *Laurie & Whittle's New Chart coasts of Greece* 1812

Wilson, James. Engraver. *Correct map state of Vermont* 3 sh. 1810

Wilson, James. Surveyor and civil engineer. *Grant County, Wisconsin* 1857

Wilson, James, Capt. Ship Duff. *Maps South Sea Islands Fiji, Marquesas and Gambier Islands* published 1799

Wilson, James [later **Sons & Co.**] (1763–1855) of Bradford, Vermont; worked in Albany. First American globe maker. *First Globe* 1810

Wilson, John. Hydrographer. *Charts for Heather's Mediterranean Pilot* 1802, *Sea Coast County Down* 1818

Wilson, John. Engineer. *Plan Siege Savannah* 1779, *Map of South Carolina* (used by Carey & Lea 1822)

Wilson, John. Surveyor. *Foxholies State (Worcs.)* 1809 MS

Wilson, John. Land surveyor. *Map of Iroquois and Kankakee Counties (Illinois)* 1860

Wilson, Joseph S. Commission. General Land Office. *United States* 1870

Wilson, Lestock. *Chart Strait east of Banka* 1787–8, *Carnaticks Track* 1789

Wilson, Richard. *Dover* 1747

Wilson, T. Surveyor. *Map County Lancaster* (1840)

Wilson, T. Publisher of Kendal. *Map of Lakes Cumberland* (1870)

Wilson, William. *Otaheite* 1799

Wilson, William. Bookseller, 6 Dane Street, Corner of Palace Street, Dublin. *Taylor and Skinner* 1778, *Post Chaise Companion Dublin* 1784 to 1813, *Plan to Dublin* 1798

Wilson, William (1755–1828). American surveyor

Wilson, W. A. Surveyor. *Map District of Hoogly* 1852

Wilson, W. C. *Official map Albany County, Wyoming Territory* 2 sh. 1886

Wilson, W. C. B. *New frontier of Modern Greece* 3 sh. 1834, French edition 1836

Wilson, W. H. D. *Jefferson County Arkansas* 1872, *Map of Pine Bluff* (1865)

Wilson & Sons. Publishers of York. *Walker's Universal Atlas* 1820

Wilstach. See **Moore, Wilstach, Keys & Co.**

Wiltsch, Johann Elieser Theodor. *Kirchenhistorischer Atlas*. Gotha 1843

Wiltse. *Map Omaha City Nebraska* 1857 (with Byers), *St. Vrain (Colorado)* (1860)

Wilster, F. *Kort over Bornholm* 1805

Wiltshaw, Capt. Thomas, Commissioner of the Navy. *Ports of South West Coast England* (17 plans) 1698 MS

Wilutsky, Adolph. *Plan Residentzstadt Königsberg* (1855)

Wily, Lieut. William Henry. *Part Colony Cape of Good Hope* 1818

Wilzleben, Arthur de. Compiled *Kelley's Arizona* 1860

Wimberger, Karl. *Passau* 1826 (1830)

Wimble, Capt. James. Chartseller, cartographer, and mariner of Boston, Massachusetts. *North Carolina* 1738, sold by author, Mount and Page, and J. Hawkins

Wimmer, Friedrich. *Special Karte von Kurhessen* 1848, *Der Thüringer Wald* 1846

Winch, John. Surveyor. *College Eyght (Eyot)* 18th century MS

Winchel, Alexander. *State of Michigan* 1873

Winchell, Alex. American geologist. *Geolog. map Michigan* 1865, *Map Grand Traverse* 1865, *Township map Michigan* 1865

Winckelmann, Edward. Draughtsman and lithographer. *Wandkarte Deutschland* 1857, *Palestina* (1864), *Würtemberg Baden und Hohenzollern* 1897

Winckler, J. F. *Plan von Dresden* (1878)

Winckler, Johann Christoph. See **Winkler, J. C.**

Windels, D. *Atlas Royal . . . de la Belgique* (1876)

Winder, Rider H. *District of Columbia* 1816

Windsor, G. A. Draughtsman. *Victoria* 4 sh. Melbourne 1872

Windt, J. A. (1759–1844). Surveyor of Lüneburg

Wing, John. *Survey of the North Level, . . . of the Fens call'd Bedford Level* (1749)

Wing, Tycho. Surveyor. *Manor of Tedford county Lincoln* 1732, *Edenham (Lincolnshire)* 1745

Wingaerde, A. van den. See **Wyngaerde**

Wingate, Francis Reginald. *Map of Nile Basin* 1894

Winkler, E. *Eisenbahn Routen Karte Deutschlands* 1874

Winkler, G. *Situations-plan . . . Thiergartens* 1823

Winkler, Johann Christoph. *Temeswarer Banat* (1760)

Winkler, J. G. *Lou's Chart of Copenhagen Sound* 1771

Winkles [Winckles] , H. Set up an engravers workshop in 1824. Drew and engraved vignettes for Tallis 1851

Winnecke, C. *Exploration N.E. of Alice Springs (S. Australia)* 1878

Winniett, Sir W. *Die Dänische Goldküste auf Guinea* (1865)

Winstanley, Henry. *Audley End* 1688, Supplement to *Nouveau Théâtre de la Grande Bretagne* 1728

Winston, Ed. *Map of Indus River* 4 sh. 1835

Winter. *Topographische Karte von Osnabrück* 1880

Winter, Antony de. Engraver for Sanson 1683–1715, for Loots 1696–1707, Vries' *Zee Atlas* 1698

Winter, Capt. James Marine surveyor. *Draught Coakle & St. Nichols Gatt* 1751, pub. 1753

Winter, Peter (17th century) Globemaker of Augsburg

Winterberg, R. *Maltischen Inseln* 1879

Winterbotham, William (1763–1829). *Historical, Geographical, Commercial and Philosophical view of the United States* 1796, *American Atlas* 1796

Winterfeld, Ludwig von. *Cape Gracias Adios und Hafen* 1845, *Karte des Mosquito Staates* 1845

Wintersmith, Lieut. Charles. Brunswick Troop, Assistant Engineer. *Plan Ticonderoga* 1777

Winzelberger, David, of Dresden. *Postal map Germany* 1577, 1597

Wirminghaus, F. W. *Plan Pennsylvania Mine* 1865

Wirsing, G. Engraver. *Karta öfver Nyköpings Lans . . . Skala* 1866

Wise, Henry (1653–1738). Royal Gardener to William III, Queen Ann and George I. *Plans of Plantations Hampton Court, Kensington Gardens, St. James Park,* and *Windsor*

Wise, J. S. *Atlas of Rivers and Seas* 1883

Wiselius, Jacob Adolf Bruno. *Manila* 1876

Wislizenus, A. *Country between Atlantic & Pacific Oceans* 1848, *New Mexico and California* 1848

Wisner de Morgenstern, François. *République du Paraguay* 1873

Wisse. *República del Ecuador* 1858

Wissenburg [Wyssenburger] , Wolfgang (1496–1575). Cartographer, Professor of Theology at Basle. *Map Palestine* 1538

Wit, A. F. *Atlas* 1730

Wit, Frederick de (1610–1698). Cartographer and publisher, "In de Calverstraet bij den Dam inde Witte Paskaert." Founded his business 1648, bought some Blaeu plates 1674. *Atlases* from 1670, *Atlas Minor* 1670, *Zee Atlas* (1675), *Atlas Belgium* 1666–7, *Atlas Major* (1690)

Wit, Frederick de, Junior (d. 1706). Son of above. Bought plates of *Jansson's Town Plans,* Issued 1694

Witchell, George. *Map of passage of ye Moon's shadow over Britain* (1763)

With, J. See **White**

Withalm, Andreas (1777–1835). Engraver. For Reilly 1789–91, *Oesterreich* 1823

Witham, Thomas. Printer and mapseller at the Golden ·Ball in Long Lane nr West Smithfield, London. *Siege of Colchester* 1648

Withinbrook, Thomas. Publisher, London. Hermann's *Virginia and Maryland* 1673

Witkamp, P. H. *Atlases of Netherlands and the Netherland Indies* 1866–1880, *Noord Celebes* 1898, *Oost Java* 1892

Witsen, Nicolaas (1641–1717). Bourge-maistre of Amsterdam. *Noorder en Ooster deel van Asia* 4 sh. 1687, *Siberia* 1687, 1698; *Charts of Dutch Coast* MS, *Texel* 1712, *Tartary,* Halma 1705

Witt, Simeon de (1756–1834). American surveyor

Witt. See **De Witt**

Witteroos, Thomas (fl. 1569–75). Surveyor

Witteveen, J. W. *Provincie Friesland* 1844

Wittich, A. von. *Umgegend von Mainz* 1858

Wittich, G. W. *Bird's eye plan of Circlesville Ohio* 1836

Witzleben, Arthur de. *Kelley's map Territory of Arizona* 1860

Witzleben, Benno von (1808–1872). Military cartographer. *Reise Karte Deutschland* 1834, *West Deutschland* 1859, *Deutsch Alsace* (1871)

Wizzell, George Sr. Engraver. *Laurie's World* 1829

Woeiriot de Bouzey, Pierre (1532–1596+). Engraver

Wocher, Adolph. American lithographer. *Drake's Systematic Treatise* 1850

Woensel, J. van. *Paskaart van de Zwarte Zee* 1775

Woerl, Dr. J. E. (1803–1865). *Atlas* *von Central Europe* 1838, *Umgegend von Mainz* 1842, *Schweitz* (1850), *Tyrol und Vorarlberg* 12 sh. (1865), *Freyberg* 1891

Wogan, Patrick. Publisher, 23 Old Bridge, Dublin. *O'Connor's Ireland* 1773

Wohlers, Capt. Cornelius Martin (1720–1813). Engraver of Hamburg. *Charte vom Elbe-Strom* 1774, *Carte de la Mer Baltique* 1791

Wohlgemuth, George Elder von (1791–1859). Viennese cartographer

Wohlgemuth [Wolgemut] , Michael (1434–1519). Painter, master of Dürer. Part illustrator of *woodcuts in Nuremberg Chronicle* 1493

Woinovits, Illia. *Plan der Umgebung von Peterwardein* 1872

Woldermann, G. *Plastischer Schul Atlas* (1876), *Neuer Schul Atlas* (1877), *Königreich Sachsen* (1890), *Umgebung von Ems* (1894)

Wolf, Carl. *Atlas Antiquus* 1884, *Historischer Schul Atlas* 1889

Wolf, Gustav. *Radfahrer Karte von Deutschland* (1890)

Wolf, Jeremias (18th century) Mapmaker

Wolf, Julius Rudolf (1816–1893). Mathematician and astronomer

Wolf, Theo. *Mapa del Ecuador* 1882, *Plano de Guayaquil* 1887, *Carta geografica del Ecuador* 6 sh. 1892

Wolfe, E. B. Mercantile Marine. *Coast Isle of Panay* 1851

Wolfe, James. *Plan River St. Laurence* 1760, *Attack Quebec* 1772, *Quebec and Environs* 1781

Wolfe, John. Publisher and Printer to ye Honorable Citie of London. *Linschoten's Voyages* 1598

Wolfe, Lieut. (Later Commander) J., R.N. Admiralty Surveyor. *Surveys Mediterranean* 1831–32, *Lough Erne* 1835–38, *Shannon* 1839–42, *Ireland* 1843–49

Wolfe, Reyner (d. 1573). Printer and publisher of London

Wolff, Aug. J. *Plan af Kjöbenhavn* 1874

Wolff, C. R. *Salzburg und Bertholdsgaden* 1836, *Preussische Staate* 1846, *Umgegend von Coblenz* 1854

Wolff, Carl. *Mitteleuropaeische Staaten* 1872, *Königreich Polen* 1872, *Historischer Atlas* 1875–7

Wolff, F. Engraver of Mannheim. *Plan Mannheim* 1813, *Gegend von Mannheim* 1814

Wolff, Fr. W. *Der Stadt Kreis Magdeburg* (1845)

Wolff, Godefroy Guillaume. Engraver. *Mecklenburg* 1780

Wolff, H. *Diepenau* 1840

Wolff, J. E. *Vicinity of Hibernia*, N.J. (1893)

Wolff, J. L. *Carte géologique Dépt. de l'Ourte* 1801

Wolff, Jeremias (1663–1724). Clock-maker, printseller, publisher and engraver of Augsburg. *Views in Nürnberg, France and Italy* 1682–1700. *Teutschland Post Stationes* (1705), *Fuch's Rhine* 1707, *De L'Isle's North America*

Wolfgang, Abraham. Engraver and publisher of Amsterdam. "opt Rokin aen het opgaen van de Beurs." *Blaeu's Brazil* (1660), *Atlas Minor* (1680)

Wolfgang the Elder, George Andreas (1631–1716). Engraver of Augsburg

Wolfgang, M. *Territorii Ulmensis. . . . descriptio* 1653

Wolger, H. *Relief map of Helgoland* (1890)

Woll, John. *Draft Bay of Fundy* 1714

Wollaston, Francis. *A moveable Planisphere* 1823

Wollheim, H. J. *Karte vom Herzogthum Lauenburg* 1852

Wollnecker, J. A. *Eisenbahn Karte Mittel Europa* 1875 and 4 sh. (1877)

Wolmuth, Bonifaz. *Grundriss der Stadt Wien vom Jahre 1547,* 1858

Wolseley, Lt. Garnet Joseph. *Communications Tien-Tsin to Pekin* 1860 (with R. Harrison), *Stanford's map Battle of Tchernaya* 1855

Woltens, Jan. Publisher and bookseller of Amsterdam "op't Water." *Cluver's Introduction Universal Geography* 1697

Woltman. *Charte der Unter Elbe* 1837

Wonham, W. C. *Tremaine's Map of Oxford County Canada West* 1857

Wood. *Plan of Nottingham* (with Staveley) 1831

Wood, Capt. *Burlingham Bay* (ca. 1680)

Wood, Basil. *Shropshire* (1710)

Wood, David. Surveyor of Chelmsford. *Wirttle & Widford* 1815 MS

Wood, Francis H. *Map of Palestine* (1888)

Wood, John. Land surveyor of Edinburgh. *Plans of Scottish Towns* engraved 1818–28, *Town Atlas* 1828 (47 maps), *Plan of Lyme-Regis* 1841

Wood, John. Estate surveyor & Excise Officer. *Hornchurch* 1729 MS, *Stisted* ca. 1750 MS

Wood, John. Engraver. For *Gardner's Dunwich* 1754, *Anson's Voyage* 1748

Wood, John. Land surveyor of Grantham. *Estate Plans Uppingham* 1839 &c.

Wood, Lieut. *North West frontier* 1841

Wood, Lieut. (later Commander), R.N. Admiralty Surveyor. *Pacific* 1845–51, *Scotland* 1854–60

Wood, T. Estate surveyor in Sussex, 1709 MS

Wood, Thomas. Ensign of Engineers, Bengal Establishment. *River Irrawaddy* 1800

Wood, William (ca. 1580–1639). *New Englands Prospect. Map of the North East* 1634–5, *South Part of New England* 1635

Wood, William. Pilot. *Chart Palmiras to Calcutta* 1748

Wood, William Bryan. Land surveyor of Chippenham. *Hainault Forest* 1858 MS, *Barking Dagenham* 1863 MS, *Forest & Parlieus of Whichwood, County of Oxford* 1854

Wood, William H. *Historical Map of the United States* 1886

Wood, William R. *Map State of Minnesota* 1861, *Region round Lake Superior* 1857

Woodard, C. S. *City of Ypsilanti Michigan* 1859

Woodard, D. *Charte von der Insel Celebes* 1805

Woodbridge, William Channing (1794–1854). Publisher and geographer of Hartford, Conn. *School Atlas to accompany Rudiments of Geography* 1821 (with *World map*, the first to show Antarctic discoveries of Nathaniel Palmer of Stonington). *Woodbridge's larger Atlas* 1822, *Modern Atlas* 1831, *Political Map of United States* 1845

Woodcock, John. Surveyor. *Inner Ward Mead Eton* 1795 MS

Woodcock, William, of Clements Inn. Surveyor. *Estate in Iver* 1810–20 MS

Woodford, E. M. *Town plans in America: Goshen Conn.* 1852, *Saybrook* 1853, *Vernon* 1853, *Randolph Norfolk County Massachusetts* 1854, *West Boylston* 1855

Woodhouse, John. Surveyor. *Guide Strangers in Ireland* 1647 and 1653

Woodman, George. *Mining Sections Idaho & Oregon* 1864

Woodman & Mutlow. Engraver, Russell Court. *Maps for Meare's Voyage* 1790

Woodnough, J. *Blackport-Duddin* 1818

Woodriff, Lieut. D., R.N. *Chart Windward Passage from Jamaica,* used Homann 1794

Woodroofe, Thomas. *Chart of Caspian Sea* 1753

Woodrow & Newton. Estate surveyors of Norwich. *Little Coggeshall* 1832 MS

Woods, Alfr. Th. (1841–1892). Surveyor General S. Australia

Woods, John B. *City of Louisiana, Missouri* (1860)

Woods, John Charles. *Map Town of Singapore* 1881

Woods, Richard. Estate surveyor. *Hatfield Peverel* 1765 MS, *Padworth, Berks.* 1767 MS, *Bardfield Saling* 1781 MS

Woodthorpe, R. G. *Chitra, Hunza* 1888, *Chindwin River* 1889

Woodthorpe, V. Engraver, 29 Fetter Lane. For Laurie & Whittle 1804, *Plan Port of London* 1800

Woodville, William. Charts for Sayer, *Coast of Africa* 1797, for Laurie & Whittle 1800, for Norris 1794–1804

Woodward. Publisher with Bale, of Globes. *New Celestial Globe* 1845

Woodward, E. F. Engraver of Philadelphia. *State of Illinois* 1836, for Greenhow 1844, for Mitchell 1834, 1849

Woodward, Capt. E. M. *Malta* 1895 and 1897

Woodward, Harry Page. *Geological Map Para Wirra* (1888), *Geological Sketch map Western Australia* 1894

Woodward, John. *Portman Estate Marylebone* 1741

Woodward, John. Surveyor. *Manor Drayton Beauchamp* 1736 MS

Woodward, S. P. Drew *geological maps* for Murchison 1843

Woodward, William. Land surveyor. *Milton Abbas* 1769 MS, *Runwell (Essex)* 1769 MS, *Plan Piddletrenthide* 1771–2 MS, *Manors Bareton, St. Cross & Marston* 1776 MS

Woodcombe, B. Mate H.M.S. *Alligator. Lowang Channel* 1840 (with Drury)

Woolcott & Clarke. *Sydney & environs* Sydney 1854

Woolaston, M. W. *Plan of Calcutta* 1825

Woolley, A. Sedgwick. *Diamond Mines in Griqualand West* 1896

Woolley, W. W. *Map Civil War in Poland* 1863

Woolman, Harry C. *Map City of Burlington, N.J.* 1875

Woolnoth, William. Engraver. *County maps Suffolk* 1825, *Bedford* 1826, *Devon* 1827, *Principality Wales* 1828

Woolsey, Robert. *Celestial companion* 1802

Woore, Lieut. R.N. *Admiralty Chart Penang* surveyed 1832, pub. 1840

Worcester, George P. *Map of the City of Norfolk with Portsmouth and Gosport, Virginia* 1850

Worcester, Joseph Emerson (1784–1865). American cartographer of Boston. *Outline maps* 1829, *Historical Atlas* 1863

Worcester, Samuel (1793–1844). American school geographer

Worde, W. de. See **Wynkyn de Worde**

Workman, Benjamin. *Elements of Geography* 1790

Worms, Anton Woensam von. *Cologne* 9 sh. 1531

Worms, H. Engraver. *Album des Rheins* (1850), *Heidelberg* (1860)

Worpitzky, F. *Plan Ostseebades Heringsdorf* 1876

Worrall, John. *Map of Oldham* (1875)

Worsley, Benjamin. Surveyor General, Ireland. His *survey* replaced by Petty's (i.e. pre-1657)

Worsley, E. *School Atlas Classical Geography* 1867

Worsley, Lt. J. W. *Malta Drawn at the Royal Engineers Office,* Valetta 1824

Wortels, Abraham. See **Ortelius**

Worthen, A. H. *Campbell's State of Illinois* 1870

Woulfe, Vincent. *Dover Harbour* ca. 1540 MS

Woutneel, Hans (fl. 1580–1614). Flemish engraver, worked in London. A collection of English County Maps engraved and printed 1602–3, sometimes known as the *Anonymous Series of Maps* (6 in British Museum, 7 in Royal Geographical Society and 4 manuscript in British Museum)

Woycke, Robert. *Eisenbahnen Elsass-Lothringen* (1874)

Wragg, James Gray. *Chart West Coast of Andaman* 1784

Wragge, Clement L. *Meteorology of Australasia* 1891–2, *Standard Weather Charts of Australasia* 19 sh. 1898

Wren, Sir Christopher (1632–1723). *Plans London* 1744–49, after the Great Fire and rebuilding

Wren, Mathew. *Plan Newry* 1761, *Louth* 4 sh. 1766, *Hartford* 1766, *St. Albans* 1766 (with Andrews), *Canterbury* 1768 (with Andrews). Assisted Rocque

Wrigglesworth, John. *Ballarat East* 4 sh. 1877, *Maryborough* 2 sh. 1891

Wright, A. Drew and engraved *maps for Reid's Atlas* 1837

Wright, Benjamin (1575–1613). Engraver for *Blagrave's World, Astrolabium* 1596, for Langenes 1599, for *Tatton's Pacific* 1600, for *Waghenaer's Thresoor* 1600, for De la Haye 1602, and *Waghenaer's Sphieghel* 1603

Wright, D. Thew. *Plat Town of Morris, Indiana* (1861)

Wright, D. Engraver. Richard St. Islington. For *Wicking's Plans* 1809, for Faden 1813–16

Wright, Edward (1558–1615). Mathematician and hydrographer, Caius College Cambridge. Born Garveston Norfolk. *Certaine errors in Navigation* 1599 (with *map of Earl of Cumberland's Voyage to Azores).* Helped with *Molyneux Globe* 1592, *True Hydrographical description of the World* 1599, *World for Hakluyt* 1600, *Towne of Fayall* 1599, *Platte of Fens* 1604

Wright, E. T. American surveyor. *County of Los Angeles* 1898

Wright, G. (1740–1783). London instrument and globe maker. *Terrestrial Globe* 1783

Wright, George. Surveyor General of Prince Edward Island. *Map of Prince Edward Island* 1852

Wright, George Newnham. *New and complete Gazetteer* 1834–8

Wright, H. C. Surveyor, No. 10, Charles Street, St. James London. *Plan Stubbington, Portsea Island* 1811 MS

Wright, John. Surveyor. *Manor of Symeons (Berks.)* 1789 MS

Wright, J. W. *Streets of Windsor* (1887)

Wright, Paul. *Map of the Holy Land* 1795

Wright, P. H. Navigation Sub Lieut. *Balabac Strait* 1868–9, published 1870

Wright, R. Land surveyor. *Romford* 1828 MS

Wright, T. *Plan Market Street Manchester* 1821

Wright, Thomas. Instrument maker. *Eclipses of the Moon* 1735, *Synopsis of Universe* 4 sh. 1741, *New Improved Terrestrial Globe* (1790)

Wright, Thomas. Surveyor General of Island of St. John. *Georgia and Florida* 1783 MS, *Topographical Map of New Hampshire* 1784, *Chart Gulf and River St. Lawrence* 2 sh. (1785) and 1790

Wright, William. Engineer of Frimley Surrey. *Navigable Canals Basingstoke-River Wey* 1790, *Birmingham-Autherly* 1791

Wright, William. Surveyor in Ireland. *Barony of Coolock* 1655, *Barony of Knocktopher* 1657

Wrightson, T. Engraver for Tallis 1850

Wrigley, Henry E. *Oil Creek Township Venango County Penn.* 1868, *Alleghany River Oil Region* 1870, *Pennsylvania Oil Region* 1874

Wrotnowski, Feliks (1803–1871). Polish writer. With Dufour produced and published *maps of Poland: Atlas de l'ancienne Pologne* 1850, *Maps Rzeckypospolitej Polskief* 1864

Wroughton, Capt. Robert. Revenue Surveyor. *District of Agra* 1846, *District of Allygurh* 1837, *District of Budavon* 1848, *District of Mympoora* 1862

Wuerbel, Franz. *Plan Wien* 1845

Wurm, Hans. *Nürnberg* 1559 (woodcut), *Frankfurt* 1583

Wurstemberger, Johann Ludwig von (1783–1862). Military cartographer of Bern

Wurster, Johann Ulrich (1814–1880). Topographer and cartographer

Wussim [Wussin] . See **Vusin**

Wyatt, A. *Map District Tirhoot* 1851

Wyatt, S. *Plans of the Docks* 1794–1803

Wyburd, James. Estate surveyor, Symonds Inn London. *Geometrical Survey of Alice Holt Forest Hants.* 1787–90

Wyeth, William. Estate surveyor. *Barking* 1735–7 MSS

Wyk. See **Wijk**

Wyeth, William. *Manor of Bathwick* 1735, *Manor of How* 1737

Wyld, James, the Elder (1790–1836). Geographer to His Majesty and H.R.H. the Duke of York. Founder member of the Royal Geographical Society 1830. Worked in the Quarter Master General's Office. Introduced Lithography into map printing in *Plans of the Peninsula Campaign* 1812. Acquired William Faden's business in 1823. Addresses up to 1823: Corner of St. Martins Lane, 5 Charing Cross East. 1823 took over Faden's Shop and address 457 West Strand Duke of York, Geography to King only. He drew the maps for *Mathews and Leigh's Scripture Atlas* 1812. *General Atlas* 1819, *Settlements in New South Wales* 1820, *St. Giles in the Field* 1824, *General Atlas* (1825), *Alderney* 1833, *North America* 1836. He died from overwork.

A NEW

𝕲𝖊𝖓𝖊𝖗𝖆𝖑 𝕬𝖙𝖑𝖆𝖘

OF

MODERN GEOGRAPHY,

CONSISTING OF

A COMPLETE COLLECTION

OF

MAPS

OF

THE FOUR QUARTERS OF THE GLOBE;

DELINEATING THEIR PHYSICAL FEATURES

AND

𝕮𝖔𝖑𝖔𝖚𝖗𝖊𝖉 𝖙𝖔 𝖘𝖍𝖔𝖜 𝖙𝖍𝖊 𝕷𝖎𝖒𝖎𝖙𝖘 𝖔𝖋 𝖙𝖍𝖊𝖎𝖗 𝖗𝖊𝖘𝖕𝖊𝖈𝖙𝖎𝖛𝖊 𝕾𝖙𝖆𝖙𝖊𝖘;

INCLUDING ALSO THE LATEST

GEOGRAPHICAL AND NAUTICAL DISCOVERIES.

𝕷𝖔𝖓𝖉𝖔𝖓:

PUBLISHED BY JAMES WYLD, CHARING CROSS EAST,

FOUR DOORS FROM TRAFALGAR SQUARE,

GEOGRAPHER TO THE QUEEN,

AND TO HIS ROYAL HIGHNESS PRINCE ALBERT.

Title-page to James Wyld's New General Atlas

Wyld, James, the Younger (1812–1887).
Geographer to the Queen and H.R.H.
Prince Albert, Master of the Clothworkers
Company. M.P. for Bodmin 1847–52
and 1857–68. Joined his father's business
in 1830. Queen Victoria succeeded 1837,
and his imprint reads Geographer to the
Queen only. Victoria married Albert
1840 and imprint changed to Geographer
to the Queen and H.R.H. Prince Albert.
Addresses: Charing Cross East next door
to the Post Office and 457 Strand (1835–
60), Second shop at Regent Street 210
(1839), Royal Exchange 2 (1846–89),
Model of the Earth. Leicester Square
added to imprint (1853–61), 1861
Albert dies and name removed from
imprint; Charing Cross East Nos. 11 &
12 and Royal Exchange (1863–89).

New General Atlas 1838, *Atlas to
show stations of Protestant Missionaries*
1839, *Lenny's Southwold* engraved
Wyld 1839, *China* 1840, *Spanish
Peninsula Battles* 1840–1, *Atlas* 1840,
Atlas Modern Geography 1842, *North
Coast Africa* 1844, *Emigrants Atlas* 1848,
Popular Atlas of World 1849, *Lowestoft*
1853, *Globe* (6 feet high, stood in
Leicester Square 1853–61), *Baltic* 1854,
India Post Roads 1854, *Surrey* 1874

Wyld, John Cooper Wyld [III] (1845–
1907). Only son of J. Wyld younger,
took over his father's business, Sold the
family business 1892 to G. W. Bacon.
*Wyld's Military Staff Map of the Nile
District* 1888

Wylensis. See **Fickler**

Wylie, A. *Geological map Cape Colony*
(1875)

Wyly, Samuel. *Catawba Indians Land*
1764 MS

Wyman, Robert H. *Wind Current Charts*
1859

Wyngaerde, Anthonis van den (d. 1572).
Flemish draughtsman. *Dordrecht* (1545),
Genoa 5 sh. 1553, *London in 1543*

Wynkyn de Worde (d. 1534). Printer.
*Almanack with coasting charts of
England* 1520·

Wysocki, Czeskaw Jordan (1839–1883).
Polish military topographer. After Polish
uprising of 1863, as exile director
Topographical Institute in Buenos Aires

Wyss, Johann Rudolf (1781–1830).
Hand atlas Bern 1816

Wyssenburger, W. See **Wissenburg**

Wytfliet [Wytfleet] , Cornelis (d. 1597).
Geography, Sect. To council of Brabant
mid-16th century. *Descript. Ptol.
Augmentum* Louvain 1597, the first
printed atlas relating exclusively to
America. *Hist. univ. des Indes Orientales
et Occid.* 1605

X

Xantus, Jancs. *Kalifornia* 1858

Xenodochos, Giovanni of Corfu. *Atlas of 3 maps: W. Europ., Medit., Archipelago* 1520

Ximenes de Cisneros, Francisco

(1437–1517) Cardinal. *Polyglot Bible* 1514–?

Ximenes de Quefada, Gonzalo. *Portolan Colombia and Central America* 1620 MS

Ximenez, Hipolito. *Topographie de la ciudad de Manila* 1717

Y

Yaagov, Abraham Bar. Jewish engraver working in Amsterdam. *Holy Land* 1696 (first map with Hebrew lettering)

Yamamoto, Kirchiemon. *World on a screen* (late 16th century) MS

Yamashita, Shigemasa. Japanese cartographer. *Omi* 1742, *Harima* 1749, and other *provincial maps of Japan*

Yamazaki Kinbe. See Kinbe, Y.

Yanagi, N. Y. *Notske Anchorage* 1872, *Harbours & Anchorages East Coast of Japan* 1874, *Admiralty Charts*

Yanagida, Takeshi, Tokyo 1889 *[map of Japan]*

Yang, Chia (12th century) Chinese printer

Yang Ma-No. See Dias, Manuel

Yaqub, Ahmad (d. 897). Arabian geographer. *Atlas of Islam* 891

Yarranton, Andrew. *Map of Inland navigation* in *England's Improvement* 1677, *Chart Christchurch* 1677, *Town and Citadel of Dunkirk* 1681

Yates, George. Surveyor. *Glamorgan* 4 sh. 1799

Yates, Thomas. Surveyor. *Oak end Gerrards Cross* 1680 MS

Yates, Thos. Mapseller Stowe, Staffs. *William Yates' Staffordshire* 1775

Yates, (T.) Surveyor. *Cardiganshire* 1803

Yates, William. Customs Officer and surveyor. *County Palatine of Lancaster* (1786), 1800 and 1816; *Environs of*

Liverpool 1768 (with G. Perry), *County of Stafford* 6 sh. 1775, 1798 and 1799

Yates and Sons, W. *Warwickshire surveys 1787–9,* published 4 sh. 1793

Yates & Co. *Snig Lane Colliery* 1828

Ydrontius, Aloysius Cesanis. See **Cesani**

Yeager, J. American engraver for Lavoisne 1820–1, *Atlas to Marshall's Life of Washington* 1822, for Carey & Lea 1822, for Fielding Lucas 1823, for Mitchell 1834

Yeakill, Thomas, the elder. Chief Draughtsman of Ordnance 1782–7. Engraver and surveyor with Gardner. *Certain estates in the county of Sussex* surveyed 1776 MS, *Goodwood and Halnaker Parks* ca. 1770, engraved 1776; *Topographical Survey County of Sussex* 4 sh. 1778–83 and on 1 sheet 1795, *Brighthelmstone* 1779

Yeakill, Thomas, the younger. Ordnance draughtsman 1787, retired 1833. *Environs of Brighthelmstone* (1785), *Accurate survey of the Island of Jersey* 4 sh. 1795, *Plan shewing Ordnance Ground* 1810

Yeates, Nicholas. *Chart of the Port of Leith for G. Collins* 1693

Yeates, William. *Scarborough Port Tobago,* Admiralty Chart 1847

Yen, Te-chih (19th century) cartographer

Yen, Yung (18th century) Chinese cartographer

Yeoman, Thomas. *Plan of the River Chelmer* 1762, *Plan of Ground belonging to Lord Foley in Marylebone* 1764

Ygel, [Ygl], Warmund (1564–1611). Cartographer of Prague. *Tyrol* 1604, reduced 1621

Yoda, Yuho (1864–1909) Japanese cartographer

Yolland, Capt. William, R. E. *Hainault Forest* 1844

Yonge, Henry. Surveyor General *Georgia* (with De Brahm) 1760–3, *Rivers of Savannah* 1751 MS, *Plan Island of Sappola* 1760 MS

Yorke, Edmunde (fl. 1588–90). *Plan Yarmouth* 1588, *charts Yarmouth to Cley* 1588 MS

Yoshimura (19th century) Japanese globemaker

Yoshinaga, Hayashi. See **Hayashi Yoshinaga**

Young, (Francis) *N. America Engrd. Becker* published S.O. Beeton (ca. 1850), *British Columbia and Vancouver Island* 1862

Young, George. Surveyor. *Earthquake Coalbrooke Co. Salop.* 1773, *Plan city and suburbs of Worcester* 1779 and 1795

Young, G. H. Engraver of Philadelphia, worked with G. Delleker. For Macpherson 1806, Carey & Lea 1822, Fielding Lucas Junior 1823

Young, J. *St. Georges Grenada,* Admiralty Chart 1821

Young, James H. American engraver with Delleker. *Varle's United States* 1817, *Finley's North America* 1826, *Indiana* 1834, *Carolinas* 1835, *Virginia* 1837, *Mitchell's National Map of American Republic* 1846, *Kentucky* 1850

Young, John. Master H.M.S. *Garland. Bay of St. Lucia Madagascar,* Laurie & Whittle 1801

Young, John. Surveyor and draughtsman. *Pacific railroad survey* 1855

Young, Joseph. Surveyor. *Estates Pitchcott* 1694 MS

Young, R. E. & **Brownlee,** J. H. *Winnipeg District map* 1888

Yousse, P. See **Hughes,** Price

Ysanti Antonio. *Part of Nova Hispania* 1682

Ysbrants-Ides. See **Ides**

Yüan, Chen (799–831). Chinese cartographer

Yule, Lieut. C. B., R.N. Admiralty Surveyor. *Survey Australia* 1842–50, *North Entrance Moreton Bay* 1846

Yunus, Ibn-al-Hussain (12th century) Arabian astronomer

Yves. Publisher of Genoa. *Roux's Plans et Rades de la Mer Méditerranée*

Yvounet, Paul. *Grand et Nouveau Miroir ou Flambeau de la Mer* 1667, 1671 and 1680

Z

Z. I. A. G. [i.e. Jost Amman Gradiere von Zürich] . See **Amman**

Zabreski, Alex. Lithographer, 100 Marchant St. Bolton Barron's Building. For Ehrenberg 1854

Zach, Anton, Freiherr von (1747–1826). Austrian military cartographer. *Carta Topografica d'Italia* 1805, *Ducato de Venezia* (1810)

Zach, Franz Xavier, Freiherr von (1754–1832). German astronomer. *Neuester Himmels-Atlas* 1803

Zacuto, Abraão (1450–1520). Jewish mathematician and astronomer

Zahn, Georg. *Wandkarte von Bayern* (1868), *Deutschland* (1869), *Wandkarte von Europa* (1875)

Zahn, Johannes. *Celestial charts* 1696

Zakariya ibn Muhammad. See **Qazwini**, al'

Zakrzewski, Aleksander (1799–1866). Polish military topographer. Exile in Madagascar, Tahiti and United States of America. *Maps and city plans San Francisco &c. Arizona* 1857, *Atlas Militaire* 1848–50

Zalesk, Falkenhagen, Piotr (1809–1893). *Atlas of Poland,* Wyld 1837 (with J. M. Bausemer)

Zaltieri [Zalterius] , Bolognini (fl. 1550–80). Venetian publisher, collaborated with Forlani. *Venice* 1565, *Bohemia* 2 sh. 1566, *Totius Galliae* 1566, *Hibernia* 1566, *Citta di Tunisi* 1566, *Discoperto*

della Nova Franza 1566 (first to show Straits of Anian), *Brabantiae descriptio* 1567, *Hollandiae descriptio* 1567, *Palestine* 1569, *Carnolia* 1569, *Citta et fortezze del Mondo* 1569, *Gastaldi's Natolia* 1570. His plates passed to N. Valeggio

Zaluski, Jozef Andrzej (1702–1774). Bishop of Kiev, founder public library Warsaw, map collector, and promotor of cartography

Zambelli . Engraver. For Clavigero 1789

Zambelli, J. Engraver for *Tarros's California* 1788

Zamberti, Bartolomew. See **Bartolomeo dalli Sonetti**

Zamorano, (Rodrigo). *World Map* in *Arte de Navegar* Seville 1581,–82, &–88; English Edition 1610

Zandt, H. E. van. Surveyor. *Map of Fair Haven, New Haven Co. Conn.* 1859

Zane, Domenico de. *chart Mediterranean* 1489 MS, *Greece & Asia Minor* 1489

Zannoni, D. G. A. R. See **Rizzo-Zannoni**

Zanotto, Francesco. *Mantova* 1852

Zara, Jehuda Ben. Portolan maker of Alexandria. *Portolan chart* 1480 and 1497

Zaremba, Adam (mid 17th century) Royal geographer. *Map of Smolensk Dukedom* (1621)

Zatta, Antonio (fl. 1757–97). Venetian publisher. *Atlante novissimo* 4 vols. Venice 1775–85, *Europe* 1777, *Raynal's*

Amérique Septentrionale 1778, *Nuovo Atlante* 1799

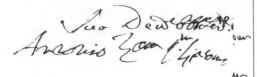

A. *Zatta*

Zavala, Miguel. *Atlas Géographique du Pérou* 1865

Zavoreo. *Provincia di Dalmazia* 1787

Zebrawski, Teofil. *Maps of Poland* 1836–62

Zeccus. See **Alvares Seco**

Zechetnev, Josef (1729–1812). Austrian military cartographer

Zecsnagel, M. See **Secsnagel**

Zedlitz-Neukirch, Leopold, Baron von. *Oesterreich* 1830, *Stadt Antwerpen* 1835, *Preussischer Staat* 1840, *Eisenbahnen in Central Europa* 1844

Zeegers, W. *Atlas of Netherlands* 1872

Zeelst, Adner Louvain. *Astrolabe* 1569

Zehender, Abraham. *Platte Grond der Stadt Middelburgh* 1739

Zeichner, Friedrich. *Oesterreich* 4 sh. 1884

Zeiler [Zeiller], Martin (1589–1661). Geographer and surveyor of Ulm. *Zeiller-Merianschen Topographieen* 1642–1736. *Sueuia* 1643, *Alsatiae* 1663, *Mainz* 1646, *Westphalia* 1647, *Franconia* 1648, *Austria* 1649, *Saxony* 1653, *Galliae* 1655, *Helvetia* 1654, *Italy* 1640

Zejszner, Ludwik (1805–1871). Polish geographer and geologist, professor at Cracow and Warsaw Universities. Author *Carte géologique de la chaine de Tatra* (1844)

Zeleny, Vaclav (1825–1875). Czech cartographer. *School atlases*

Zell, [Cel, Cella], Christoph. Publisher and wood engraver of Nuremberg. *Germany* 4 sh. 1560, Reissued by de Jode 1568

Zell, Heinrich (d. 1564). Born in Cologne,

worked in Dantzig and Königsberg, *Europe* Antwerp 1535, published by Christoph and Urlanus; *Bavaria* 1540, *East Prussia* 1542, used by De Jode 1578

Zell, T. Ellwood. *Hand Atlas of World* 1873

Zeno, Antonio & Nicolo (14th century). Venetian Navigators, *mythical islands Frisland*

Zenoi [Cenoi, Zenoni, Zeno], Domenico (fl. 1552–70). Publisher and engraver in Venice. *El Pignon* (1540), *Tripoli* (1551), *Flanders* 1559, *Spain* 1560, *France* 1561, *Greece* 1564, *Malta* 1565, *Tokay* 1566, *Gyula* 1566, *Vienna* 1566, *Italy* 1567, *Hungary* 1567, *Europe* 1568, *Constantinople* 1570

Zepherinus, J. See **Severszoon**

Zeroldis, G. See **Giroldi**

Zerolo, Elias. *Atlas Geografico Universal* 1891

Zertahelly, L. (1815–46). Cartographer of Munich

Zetter, Paul de. *Halberstadt,* Jansson (1659)

Zeune, Johann August (1778–1853). Geographer of Wittenberg

Zidek, Vinzenz. *Plan Linz* (1872)

Ziegler, J. M. *St. Gallen* 1853, *Madeira* (1855), *Schweitz* (1856), *Allgemeiner Atlas* 1857, *Italia superiore* (1859), *Geological Map of World* 8 sh. 1861, *Neuer Atlas* 1863, *Kanton Zürich* (1884), *Ticino* (1894)

Ziegler [Ciglerus, Landavus], Jacob (1470–1549). Born in Landau, Professor in Vienna. *Palestine* 1532 (Maps of Palestine and one of *Northern Regions,* the first printed map to show magnetic variation)

Zijl, B. van. *Plan de Béthune* (1710)

Ziletti, Giordano. Printer of Venice. *Ptolemy* 1564 & 1574

Zimmerman, Carl. *West Persien und Mesopotamien* 4 sh. (1840–43), *Inner*

Asien 4 sh. 1841, *Indusländer* 4 sh. 1851, *Süd Iran* 1850, *Jerusalem mit Umgebung* (1881)

Zimmerman, Eberhard August Wilhelm von (1743–1815). Geographer

Zimmermann, G. *Sachsen* (1877)

Zimmermann, G, P. H. *Kaart van de Rivier de Suriname* 1877

Zimmerman, Wilhelm Peter (d. 1630). Engraver of Augsburg. *Bohemia* 1619

Zinck, M., **Zinthio**, M. See **Zyndt**

Zintly, F. G. *(N. Europe)* 1869, *Plankarta öfver Stockholm och Omgifningar* (1875)

Zipter, J. Engraver for Meyer 1830–40

Ziraldis, G. See **Giroldis**

Zirbeck, C. *Plan von Berlin* 1837, *Sicilien* 1839, *Uckermark* 1865

Ziredis, Ziroldis. See **Giroldis**

Zittart, Dr. *Ducatus Westphaliae* 1757

Zittlow, A. *Mappa do Imperio do Brazil* 1878

Zittwitz, J. F. von. *Gegend um Erfurt* 4 sh. 1839

Zocchi, Cosimo. *Planta della Citta di Firenze* 1783

Zoebl, G. *Hafenplan von Barcelona* 1873

Zollmann, Friedrich (1690–1762). Dutch cartographer, worked for Homann. *Erfurt* 1717, *Hanau* 1728, *Saxony* Homann Heirs

Zollmann, Johann Wilhelm. *Vestung Metz* 1739, *Thuringiae Orientalis* 2 sh. 1747

Zollmann, Philipp Heinrich. *Stockholm* 1725, *Ducatus Saxoniae* 1731

Zonen, van Lauen. See **Lauen**

Zoppitz, Karl (1838–1885). Mathematician and geographer

Zoppino, Nicolo. Printer of Venice. *Bordone's Isole del mondo* 1528, 1534

Zorzi, Alessandro. Possibly author of *charts* ascribed to Bartolomeo Columbus

Zsambocky, J. See **Sambucus**

Zuan de Napoli. See **Napoli**

Zubov, Alexi (1682–1743). Russian engraver, pupil of Schoonebeck. *Russian maps* 1705–36

Zuccagni-Orlandini, Attilio. *Atlante Geografico degli stati Italiani delineatio* 4 vols. 1844–5

Zuccheri, Edmund von. *Royaume de Hongrie* 1811, *Deutschland* 1828

Zuelow, von. *Environs of Berlin* 10 sh. 1816–19

Zuerner [Zürner], Adam Friedrich (1680–1742). Geographer and publisher. *Africa* 1700, *America* 1709, *Saxony* 1712, *Carlsbad* 1715, *Europe* (1735). Produced nearly 1000 maps used by Schenk and Valk, Weigel &c.

Zuerner, G. A. *Empire Ottoman* 1805

Zuerner, H. C. *Inner Oesterreich* 1805

Zuev, Nikita Ivanovic (1823–1890). Russian cartographer

Zugler. See **Ziegler**

Zuichen, Viglius de, of Louvain. Collector of maps catalogued 1575

Zuidema, E. R. *Schoolkaart Groningen* (1872), *Handbook Rotterdam* (1887–9)

Zulauf, G. H. *Plan der Stadt Mainz* (1877)

Zuliani, Felice. Publisher. *Heymann's Italy* (1798)

Zuliani, G. Engraver for *Zatta's New Zealand* 1778, *Ulster* 1778, *Scotland* 1779, *Greece* 1782

Zuliani, Z. Engraver. For *Zatta's Africa* 1776

Zündt. See **Zyndt**

Zuñiga, J. M. de. See **Martinez de Zuñiga**

Zurara, G. E. de. See **Azurara**

Zurcher, (J. C.) Engraver for H. Frÿlink 1858

Zürner. See **Zuerner**

Zwan, Kazimierz (b. 1792). Military topographer, supervised engraving of *great map of Poland* after the 1815 Congress of Vienna

Zweibach, Ernst (19th century) Sino-Cartesian demographer

Zweidler, Petrus (1570–1613). Chorographer and surveyor. *Plan Bamberg* engraved 1602

Zwick, Johann (1496–1542). *Lake Constance* (with Blauret) used by Münster (1540)

Zwicker, Daniel (1612–1668). Doctor of Dantzig. *Pinsk marshes* 1650, *Polesie Regio* 1650

Zwikopf, Christoph. *Ingolstadt* 1546, used by Braun & Hogenberg

Zylinski, Josef (1834–1921). Polish military topographer in Russian service

Zyndt [Zündt, Cynthius, Zinthio, Zinck], Mathis. Engraver and printseller of Nuremberg. *Cyprus* 1550, *Hungary* 1566 (used by De Jode), *Belgium & Holland* 1568, *Gotha* 1568, *Grodno* 1568 used by Braun & Hogenberg